Springer Proceedings in Physics

Volume 275

Indexed by Scopus

The series Springer Proceedings in Physics, founded in 1984, is devoted to timely reports of state-of-the-art developments in physics and related sciences. Typically based on material presented at conferences, workshops and similar scientific meetings, volumes published in this series will constitute a comprehensive up-to-date source of reference on a field or subfield of relevance in contemporary physics. Proposals must include the following:

- name, place and date of the scientific meeting
- a link to the committees (local organization, international advisors etc.)
- scientific description of the meeting
- list of invited/plenary speakers
- an estimate of the planned proceedings book parameters (number of pages/articles, requested number of bulk copies, submission deadline).

Please contact:

For Americas and Europe: Dr. Zachary Evenson; zachary.evenson@springer.com
For Asia, Australia and New Zealand: Dr. Loyola DSilva; loyola.dsilva@springer.com

Erma Yulihastin · Prayitno Abadi ·
Peberlin Sitompul · Wendi Harjupa
Editors

Proceedings of the International Conference on Radioscience, Equatorial Atmospheric Science and Environment and Humanosphere Science, 2021

Springer

Editors
Erma Yulihastin
National Research and Innovation Agency
Bandung, Indonesia

Prayitno Abadi
National Research and Innovation Agency
Bandung, Indonesia

Peberlin Sitompul
National Research and Innovation Agency
Bandung, Indonesia

Wendi Harjupa
National Research and Innovation Agency
Bandung, Indonesia

ISSN 0930-8989 ISSN 1867-4941 (electronic)
Springer Proceedings in Physics
ISBN 978-981-19-0307-6 ISBN 978-981-19-0308-3 (eBook)
https://doi.org/10.1007/978-981-19-0308-3

This Springer imprint is published by the registered company Springer Nature Singapore Pte Ltd.
The registered company address is: 152 Beach Road, #21-01/04 Gateway East, Singapore 189721, Singapore

Preface

The human environment is rapidly changing with complicated interactions, threatening human existence and human living. To establish the Sustainable Humanosphere, international collaboration and expansion of Humanosphere Science on a global scale is important.

In 2016, Research Institute for Sustainable Humanosphere (RISH) launched a program called the Humanosphere Asia Research Node (ARN) to strengthen the hub functions of international collaborative research and fostering talented people who expand the field of Humanosphere Sciences internationally. ARN integrates various facilities such as Equatorial Atmosphere Radar (EAR) and human networks in ASEAN region and Japan for consolidating the international collaborative research on "Sustainable Humanosphere."

The Equatorial Atmosphere Radar (EAR) is a VHF atmospheric radar located in Kototabang, West Sumatra, Indonesia. It has been operated by collaboration between RISH and National Institute of Aeronautics and Space of Indonesia (LAPAN) since 2001. RISH has conducted a collaborative research program by using the EAR and its related facilities since 2005 to enhance scientific research activity conducted with the EAR and associated facilities. To exchange information on a wide range of research results from the EAR and related facilities during 20 years, and to further strengthen the international network, RISH and LAPAN organize the symposium. This symposium is a joint program of LAPAN-Kyoto University with the title "The 6th Asia Research Node Symposium on Humanosphere Science," and "INternational Conference on Radioscience, Equatorial Atmospheric Science and Environment (INCREASE)." Due to the COVID-19 outbreak and travel restrictions, the symposium will be held online.

The symposium aims to share the concept and recent advances of Humanosphere in scientific and technological advances such as atmospheric science, environmental science, space science, material science, plant science, entomology, microbiology, ecology, forestry, wood science, chemistry, biochemistry, molecular biology, architecture, electromagnetic engineering, agriculture and other related sciences contributing to creating "Sustainable Humanosphere."

"The 6th Asia Research Node Symposium on Humanosphere Science," and "INternational Conference on Radioscience, Equatorial Atmospheric Science and Environment (INCREASE)" symposium was organized from the office of Research Center for Climate and Amospheric, LAPAN, on September 20–21, 2021, in Bandung, Indonesia, and held virtually considering high cases of COVID-19 around the globe.

Nevertheless, the committee received an overwhelming response in term of number of registered participants. The total participant attending this virtual symposium reach 114 people. The participants were from Indonesia, Japan, India, USA, Egypt, Europe, China and Australia.

We decided that the speakers could only present their work in oral for 10 m. After three or two speakers delivered their research, we accommodated question and answer sessions for 10 m for these speakers. Furthermore, we facilitate the manuscripts to be published at a journal with peer-reviewed process by experts in the field. One manuscript is reviewed by two reviewers and the handling editor decided whether a manuscript can be accepted or rejected mainly based on the reviewer's comments. In this symposium, 182 manuscripts submitted, 80 manuscripts were proceeded to publication process and suitable for scientific publication.

We want to thank participants and speakers who attend and support the 6th Asia Research Node Symposium and INCREASE. Also, we thank to scientific committee and reviewers for their precious time, and lastly many thanks to the local organizing committee for the hard work so that the 6th Asia Research Node Symposium and INCREASE can be done successfully. See you all in the next ARN and INCREASE Symposium.

Bandung, Indonesia

Dr. Erma Yulihastin
Editor in Chief of INCREASE
Proceeding 2021

Contents

Contributors

'Adany Fildzah Research Center for Chemistry, National Research and Innovation Agency, South Tangerang, Indonesia

Abdillah Muhammad R. Atmospheric Science Research Group, Faculty of Earth Sciences and Technology, Institut Teknologi, Bandung, Indonesia

Adetya Emmanuel Research Center for Climate and Atmosphere, National Research and Innovation Agency (BRIN), Bandung, Indonesia

Aditiya Arif Geodesy and Geodynamic Control Net Center, Bogor, Indonesia

Agustina Nova Department of Informatics, Sekolah Tinggi Teknologi Bandung, Bandung, Indonesia

Ahyuni Department of Geography, State University of Padang, Padang, Indonesia

Al Viandari Nourma Department of Soil Science and Land Resources, Faculty of Agriculture, IPB University, Dramaga, Bogor, Indonesia;
Indonesian Agricultural Environment Research Institute, Pati, Indonesia

Amelia Jihan Putri Meteorology Department, Bandung Institute of Technology, Bandung, Indonesia

Amor Giant Department of Geography, State University of Padang, Padang, Indonesia;
Department of Geography, Universitas Negeri Padang, Padang, Indonesia

Amri Iful Department of Electronics Engineering, Politeknik Jambi, Jambi, Indonesia

Andarini Dita Fatria Research Center for Climate and Atmosphere, National Research and Innovation Agency (BRIN), Bandung, Indonesia;
Department of Physics, Diponegoro University, Semarang, Indonesia

Anggina Stevani Center for Aerospace Policy Studies, LAPAN, BRIN, Jakarta, Indonesia;
National Research and Innovation Agency (BRIN), Jakarta, Indonesia

Anita Sita Heris Research Center for Biomass and Bioproducts, National Research and Innovation Agency (BRIN), Cibinong, Indonesia

Anugrah Ranie Dwi Detailed Spatial Planning of Environmental Carrying Capacity Area - Region 1, Ministry of Agrarian Affairs and Spatial Planning/National Land Agency, Jakarta, Indonesia

Aries Tanti Dyah Research Center for Climate and Atmosphere, National Research and Innovation Agency (BRIN), Bandung, Indonesia

Arisa Deasy Research Center for Geotechnology, Indonesian Institute of Sciences (LIPI), Bandung, Indonesia;
Research Center of Geotechnology, Indonesian Institute of Sciences (LIPI), Jakarta, Indonesia

Armer Aldi Nanda Department of Soil Science, Faculty of Agriculture, Andalas University, Padang, Indonesia

Bahar Afrizal Agam Atmospheric and Space Observation Office, National Research and Innovation Agency (BRIN), Agam, Indonesia

Bakti Hendra Geotechnology Research Center, National Research and Innovation Agency, Bandung, Indonesia

Balbo Fandy Department of Meteorology, Bandung Institute of Technology, Bandung, Indonesia;
Department of Environmental Engineering, Bandung Institute of Technology, Bandung, Indonesia

Bangsawan L. Gadjah Mada University, Yogyakarta, Indonesia;
Indonesia Agency for Meteorology Climatology and Geophysics, Jakarta, Indonesia

Birastri Wirid School of Atmospheric and Planetary Sciences, Sumatera Institute of Technology (ITERA), Lampung Selatan, Indonesia

Bunterngsook Benjarat BIOTEC-JGSEE Integrative Biorefinery Laboratory, National Center for Genetic Engineering and Biotechnology (BIOTEC), Science Park, National Science and Technology Development Agency (NSTDA), Pathumthani, Thailand

Cahyaningtyas Amatul Firdausya Nur Astronomy Study Program, Institut Teknologi Bandung, Bandung, Indonesia

Cahyono Waluyo Eko Research Center for Climate and Atmosphere, National Research and Innovation Agency (BRIN), Bandung, Indonesia

Champreda Verawat BIOTEC-JGSEE Integrative Biorefinery Laboratory, National Center for Genetic Engineering and Biotechnology (BIOTEC), Science

Park, National Science and Technology Development Agency (NSTDA), Pathumthani, Thailand

Chandra Indra Engineering Physics, School of Electrical Engineering, Telkom University, Bandung, Indonesia

Cholianawati Nani Research Center for Climate and Atmosphere, National Research and Innovation Agency (BRIN), Bandung, Indonesia

Dani Tiar Space Research Center, BRIN, Bandung, Indonesia

Dear Varuliantor Research Center for Space, National Research and Innovation Agency (BRIN), Bandung, Indonesia

Diwyacitta Karunika Research Center for Remote Sensing, National Research and Innovation Agency, Jakarta, Indonesia

Djamil Yudha Setiawan Research Center for Climate and Atmosphere, National Research and Innovation Agency (BRIN), Bandung, Indonesia

Dupe Zadrach Ledoufij Meteorology Department, Bandung Institute of Technology, Bandung, Indonesia

Ednofri National Research and Innovation Agency (BRIN), Agam, Indonesia

Ermawar Riksfardini A. Research Center for Biomass and Bioproducts, National Research and Innovation Agency (BRIN), Bogor, Indonesia

Fahrezi M. Zuvan Maulana Faculty of Agricultural Technology, Universitas Brawijaya, Malang, Indonesia

Fathrio Ibnu Research Center for Climate and Atmosphere, National Research and Innovation Agency (BRIN), Bandung, Indonesia

Fatria Dita Research Center for Climate and Atmosphere, National Research and Innovation Agency (BRIN), Bandung, Indonesia

Fatriasari Widya Research Center for Biomass and Bioproducts, National Research and Innovation Agency (BRIN), Cibinong, Indonesia

Faturahman Agri Research Center for Space, National Research and Innovation Agency (BRIN), Bandung, Indonesia

Fiantis Dian Department of Soil Science, Faculty of Agriculture, Andalas University, Padang, Indonesia

Filawati Siska BPAA Pasuruan, Pasuruan, Indonesia

Firdaus Permadhi Yan Meteorological, Climatology, and Geophysical Agency (BMKG), Jakarta, Indonesia

Fitria Research Center for Biomass and Bioproducts, National Research and Innovation Agency (BRIN), Cibinong, Indonesia;

Department of Biological Systems Engineering, Washington State University, Richland, WA, USA

Friska Vira Department of Physics, Faculty of Mathematics and Natural Sciences, Universitas Andalas, Padang, Indonesia

Gabriel Djoko Sihono Department of Industrial Engineering, Faculty of Engineering, Universitas Indonesia, Depok, Indonesia

Gammamerdianti Research Center for Climate and Atmosphere, National Research and Innovation Agency (BRIN), Bandung, Indonesia

Gemilang Wisnu Arya Research Institute for Coastal Resource and Vulnerability, Ministry of Marine Affairs and Fisheries, Padang, Indonesia

Gernowo Rahmat Research Center for Atmospheric Science and Technology, Indonesia National Agency of Research and Innovation (BRIN), Bandung, Indonesia

Ghosh Sayanta G.B. Pant, National Institute of Himalayan Environment, Himachal Regional Centre, Mohal-Kullu, Himachal Pradesh, India

Ginting Frisa Irawan Department of Soil Science, Faculty of Agriculture, Andalas University, Padang, Indonesia

Gouda K. C. CSIR Fourth Paradigm Institute, NAL Belur Campus, Bangalore, India

Gusnita Dessy Research Center for Climate and Atmosphere, National Research and Innovation Agency (BRIN), Bandung, Indonesia

Gustiandi Budhi Research Center for Remote Sensing, National Research and Innovation Agency, Jakarta, Indonesia

Hadi Tri Wahyu Atmospheric Science Research Group, Bandung Institute of Technology (ITB), Bandung, Indonesia

Hakim Arif Nur Rocket Technology Center, National Research and Innovation Agency (BRIN), Bogor, Indonesia

Hamdi Saipul Research Center for Climate and Atmosphere, National Research and Innovation Agency (BRIN), Bandung, Indonesia

Hamidi Muhammad Department of Physics, Andalas University, Padang, Indonesia

Handayani Lina Research Center for Geological Disaster, Research Organization for Earth and Marine Sciences, National Research and Innovation Agency (BRIN), Bandung, Indonesia

Hanif Muhammad Department of Geo-Informatics, Khon Kaen University, Khon Kaen, Thailand;
Research Center for Geological Disaster, Research Organization for Earth Sciences and Marine Sciences, National Research and Innovation Agency (BRIN), Bandung, Indonesia

Hanum Silvi Oktavia BPAA Pasuruan, Pasuruan, Indonesia

Harjana Teguh Research Center for Climate and Atmosphere, National Research and Innovation Agency (BRIN), Bandung, Indonesia

Harjupa Wendi Research Center for Climate and Atmosphere, National Research and Innovation Agency (BRIN), Bandung, Indonesia

Hatmaja Rahaden Bagas Research Center for Climate and Atmosphere, National Research and Innovation Agency (BRIN), Bandung, Indonesia

Hermawan Eddy Research Center for Climate and Atmosphere, National Research and Innovation Agency (BRIN), Bandung, Indonesia

Hermiati Euis Research Center for Biomass and Bioproducts, National Research and Innovation Agency (BRIN), Cibinong, West Java, Indonesia;
Research Center for Biomass and Bioproducts, National Research and Innovation Agency (BRIN), Bogor, Indonesia

Hidayat Fakhrijal Rizki Department of Soil Science, Faculty of Agriculture, Andalas University, Padang, Indonesia

Hidayat Rahmat Department of Geophysics and Meteorology, IPB University, Bogor, Indonesia

Hidayat Rizki Athorik Department of Geography, Universitas Negeri Padang, Padang, Indonesia;
Department of Geography, State University of Padang, Padang, Indonesia

Hidayat Rizki Atthoriq Department of Geography, State University of Padang, Padang, Indonesia

Hidayat Taufiq Department of Astronomy, InstitutTeknologiBandung, FMIPA, Bandung, Indonesia

Husin Asnawi Research Center for Space, National Research and Innovation Agency (BRIN), Bandung, Indonesia

Ihsan Candra Nur Research Center for Climate and Atmosphere, National Research and Innovation Agency (BRIN), Bandung, Indonesia

Indra Fauziah Fangia Meteorogical Study Program, Bandung Institute of Technology, Bandung, Indonesia

Indradjad Andy Research Center for Remote Sensing, National Research and Innovation Agency, Jakarta, Indonesia

Indrawati Asri Research Center for Climate and Atmosphere, National Research and Innovation Agency (BRIN), Bandung, Indonesia

Indrayani Poppy Department Architecture, University of Fajar, Makassar, Indonesia

Iswanto Apri Heri Department of Forest Product, Faculty of Forestry, Universitas Sumatera Utara, Medan, North Sumatera, Indonesia;
JATI-Sumatran Forestry Analysis Study Center, Universitas Sumatera Utara, Medan, North Sumatera, Indonesia

Jannatunnisa Luthfiyah Meteorological Study Program, Bandung Institute of Technology, Bandung, Indonesia

Jatmiko R. H. Gadjah Mada University, Yogyakarta, Indonesia

Jiyo Research Center for Space, National Research and Innovation Agency (BRIN), Bandung, Indonesia

Juaeni Ina Research Center for Climate and Atmosphere, National Research and Innovation Agency (BRIN), Bandung, Indonesia

Jumanissaba Aldiatama Department of Geophysics and Meteorology, IPB University, Bogor, Indonesia

Jumianti Nining Department of Physics, Faculty of Mathematics and Natural Science, Universitas Andalas, Padang, Indonesia

Juniarti Juniarti Department of Soil Science, Faculty of Agriculture, Andalas University, Padang, Indonesia

Komala Ninong Research Center for Climate and Atmosphere, National Research and Innovation Agency (BRIN), Bandung, Indonesia

Kombara Prawira Yudha Research Center for Climate and Atmosphere, National Research and Innovation Agency (BRIN), Bandung, Indonesia

Krismon Aldi Department of Meteorology, Faculty of Earth Sciences and Technology, Institut Teknologi, Bandung, Indonesia

Kuniyal Jagdish Chandra G.B. Pant National Institute of Himalayan Environment, Kosi-Katarmal, Almora, Uttrakhand, India

Lasmono Farid Research Center for Climate and Atmosphere, National Research and Innovation Agency (BRIN), Bandung, Indonesia

Lata Renu G.B. Pant, National Institute of Himalayan Environment, Himachal Regional Centre, Mohal-Kullu, Himachal Pradesh, India

Lianawati Adetya Research Center for Biomass and Bioproducts, National Research and Innovation Agency (BRIN), Bogor, Indonesia

Lubis Muhammad Adly Rahandi Research Center for Biomass and Bioproducts, National Research and Innovation Agency (BRIN), Cibinong, West Java, Indonesia

Lubis Retno Leodita Department of Soil Science, Faculty of Agriculture, Andalas University, Padang, Indonesia

Lubis Sandro W. Rice University, Houston, USA

Marlina Resti Research Center for Biomass and Bioproducts, National Research and Innovation Agency, Cibinong, Indonesia

Maryadi Edy Research Center for Climate and Atmosphere, National Research and Innovation Agency (BRIN), Bandung, Indonesia

Marzuki Marzuki Department of Physics, Faculty of Mathematics and Natural Science, Universitas Andalas, Padang, Indonesia

Maulana Tyo Department of Physics, Diponegoro University, Semarang, Indonesia

Monica Fadilla Department of Physics, Faculty of Mathematics and Natural Sciences, Universitas Andalas, Padang, Indonesia

Muhammad Fadhlil R. School of Geography, Earth, and Atmospheric Sciences, The University of Melbourne, Melbourne, Australia

Mumtahana Farahhati Department of Astronomy, InstitutTeknologiBandung, FMIPA, Bandung, Indonesia;
Space Research Center, National Research and Innovation Agency (BRIN), Bandung, Indonesia

Munthe Christine Cecylia Research Center for Atmospheric Science and Technology, National Research and Innovation Agency (BRIN), Bandung, Indonesia

Nafiisyanti Aisya Research Center for Climate and Atmosphere, National Research and Innovation Agency (BRIN), Bandung, Indonesia

Namigo Elistia L. Department of Physics, Andalas University, Padang, Indonesia

Nauval Fadli Research Center for Climate and Atmosphere, National Research and Innovation Agency (BRIN), Bandung, Indonesia

Neflia Space Research Center, National Research and Innovation Agency, Bandung, Indonesia

Nihei Naoto Faculty of Food and Agricultural Sciences, Fukushima University, Fukushima, Japan

Ningrum Riska Surya Research Center for Biomass and Bioproducts, National Research and Innovation Agency (BRIN), Cibinong, West Java, Indonesia;
Research Center for Biomass and Bioproducts, National Research and Innovation Agency (BRIN), Bogor, Indonesia

Noersomadi Research Center for Climate and Atmosphere, National Research and Innovation Agency (BRIN), Bandung, Indonesia

Norarat Rattanaporn Rajamangala University of Technology Lanna, Chiang Rai, Thailand

Nugroho Nunung Puji Balai Penelitian dan Pengembangan Teknologi Pengelolaan Daerah Aliran Sungai, Kartasura, Surakarta, Indonesia

Nurfindarti Erti Regional Development Planning Agency of Serang City, Serang, Indonesia

Nurhamiyah Yeyen Research Center for Biomass and Bioproducts, National Research and Innovation Agency, Cibinong, Indonesia

Nurjani E. Gadjah Mada University, Yogyakarta, Indonesia

Nurlatifah Amalia Research Center for Climate and Atmosphere, National Research and Innovation Agency (BRIN), Bandung, Indonesia

Nurokhman Research Center for Climate and Atmosphere, National Research and Innovation Agency (BRIN), Bandung, Indonesia

Oktaviani Maulida Research Center for Applied Microbiology, National Research and Innovation Agency (BRIN), Cibinong, West Java, Indonesia;
Research Center for Biomass and Bioproducts, National Research and Innovation Agency (BRIN), Cibinong, Indonesia

Pebralia Jesi Department of Physics, Faculty of Science and Technology, Jambi University, Jambi, Indonesia

Perwitasari Intan National Research and Innovation Agency (BRIN), Jakarta, Indonesia

Pradipta Rezy Institute for Scientific Research, Boston College, Newton, USA

Pramasari Dwi Ajias Research Center for Biomass and Bioproducts, National Research and Innovation Agency (BRIN), Cibinong, West Java, Indonesia;
Research Center for Biomass and Bioproducts, National Research and Innovation Agency (BRIN), Bogor, Indonesia

Pramudia Kadiman Erfitra Research Center for Atmospheric Science and Technology, National Research and Innovation Agency (BRIN), Bandung, Indonesia

Prasasti Indah Research Center for Remote Sensing, National Research and Innovation Agency (BRIN), Jakarta, Indonesia

Prasetiyo Kurnia Wiji Research Center for Biomaterials, Indonesian Institute of Sciences, Bogor, Indonesia

Pratama Ridho Research Center for Climate and Atmosphere, National Research and Innovation Agency (BRIN), Bandung, Indonesia

Pratikasari Resa Agency for Meteorology Climatology and Geophysics, Jakarta, Indonesia

Prayogo Luhur Moekti Faculty of Fisheries and Marine, University of PGRI Ronggolawe Tuban, Tuban, Indonesia

Priyatikanto Rhorom Space Research Center, BRIN, Bandung, Indonesia

Pulunggono Heru Bagus Department of Soil Science and Land Resources, Faculty of Agriculture, IPB University, Dramaga, Bogor, Indonesia

Purwaningsih Anis Research Center for Climate and Atmosphere, National Research and Innovation Agency (BRIN), Jakarta, Indonesia

Purwati Sri Department of Industrial Engineering, Faculty of Engineering, Universitas Indonesia, Depok, Indonesia

Putranto Muhammad Fadhlan Research Center for Climate and Atmosphere, National Research and Innovation Agency (BRIN), Bandung, Indonesia

Putri Gerhana Puannandra Sumedang Space and Atmospheric Observatory, BRIN, Sumedang, Indonesia

Radiana Atep Research Center for Climate and Atmosphere, National Research and Innovation Agency (BRIN), Bandung, Indonesia

Rahayu Soni Aulia Research Center for Climate and Atmosphere, National Research and Innovation Agency (BRIN), Bandung, Indonesia

Rahayuningsih Mulyorini Department of Agroindustrial Technology, Bogor Agricultural University, Bogor, Indonesia

Rahmat Ali Reseach Center for Limnology and Water Resources, National Research and Innovation Agency, Jakarta, Indonesia

Rahmatia Fahmi BPAA Agam, National Research and Innovation Agency (BRIN), Bukittinggi, Indonesia

Rajmi Saftia Laila Department of Soil Science, Faculty of Agriculture, Andalas University, Padang, Indonesia

Refindo Galang P. Department of Physics, Andalas University, Padang, Indonesia

Respati Muhamad R. Department of Geophysics and Meteorology, IPB University, Bogor, Indonesia

Riawan Edi Atmospheric Science Research Group, Faculty of Earth Science and Technology, Bandung Institute of Technology (ITB), Bandung, Indonesia

Risyanto Research Center for Climate and Atmosphere, National Research and Innovation Agency (BRIN), Bandung, Indonesia

Rizal Syahril Bina Darma University of Palembang, Palembang, Indonesia

Rodhibilah Fadia Idzni Biology Department, Faculty of Mathematics and Science, Padjajaran University, Bandung, West Java, Indonesia

Rosyidah Arlif Nabilatur Postgraduate Program in Earth Science, Institut Teknologi Bandung, Bandung, Indonesia

Ruhyana Nugrahana Fitria Regional Research and Development Planning Agency of Sumedang Regency, Sumedang, Indonesia

Rusmanansari Arifatus Hikmah Meteorology Department, Bandung Institute of Technology (ITB), Bandung, Indonesia

Saito Susumu Electronic Navigation Research Institute (ENRI), Tokyo, Japan

Sari Fahriya Puspita Research Center for Biomass and Bioproducts, National Research and Innovation Agency (BRIN), Cibinong, West Java, Indonesia

Sari Wilin Julian Research Center for Quantum Physics, National Research and Innovation Agency (BRIN), South Tangerang, Indonesia;
Research Center for Quantum Physics, National Research and Innovation Agency (BRIN), Bandung, Indonesia

Sasmitaloka Kirana Sanggrami Indonesian Center for Agricultural Postharvest Research and Development, Bogor, Indonesia

Satyawardhana Haries Research Center for Climate and Atmosphere, National Research and Innovation Agency (BRIN), Bandung, Indonesia

Saufina Elfira Research Center for Climate and Atmosphere, National Research and Innovation Agency (BRIN), Bandung, Indonesia

Setiahadi Bambang Indonesian National Institute of Aeronautics and Space (LAPAN), Pasuruan, Indonesia

Setiawan Fajar Reseach Center for Limnology and Water Resources, National Research and Innovation Agency, Jakarta, Indonesia

Setiawan Sonni Department of Geophysics and Meteorology, IPB University, Bogor, Indonesia

Setyawati Wiwiek Research Center for Climate and Atmosphere, National Research and Innovation Agency (BRIN), Bandung, Indonesia

Sholikhah Ni'matus Department of Physics, Faculty of Mathematics and Natural Sciences, Brawijaya University, Malang, Indonesia

Simanjuntak Yogi S. M. Department of Earth Science, Faculty of Earth Science and Technology, Bandung Institute of Technology (ITB), Bandung, Indonesia

Sinaga Edi Rikardo Department of Meteorology, Faculty of Earth Science and Technology, Bandung Institute of Technology (ITB), Bandung, Indonesia

Sjafrijon BPAA Agam, National Research and Innovation Agency (BRIN), Bukittinggi, Indonesia

Sofiati Iis Research Center for Climate and Atmosphere, National Research and Innovation Agency (BRIN), Bandung, Indonesia

Sondari Dewi Research Center for Biomass and Bioproducts, National Research and Innovation Agency (BRIN), Cibinong, West Java, Indonesia;
Research Center for Biomass and Bioproducts, National Research and Innovation Agency (BRIN), Bogor, Indonesia

Suaydhi Research Center for Climate and Atmosphere, National Research and Innovation Agency (BRIN), Bandung, Indonesia

Sudarmanto Research Center for Biomass and Bioproducts, National Research and Innovation Agency (BRIN), Cibinong, West Java, Indonesia

Sulistiyorini Desy Faculty of Health Science, Indonesia Maju University, Jakarta, Indonesia

Sumaryati Research Center for Climate and Atmosphere, National Research and Innovation Agency (BRIN), Bandung, Indonesia

Sunarti Titi Candra Department of Agroindustrial Technology, Bogor Agricultural University, Bogor, Indonesia

Sunarya Rachmat Research Center for Climate and Atmosphere, National Research and Innovation Agency (BRIN), Bandung, Indonesia

Suwardi Department of Soil Science and Land Resources, Faculty of Agriculture, IPB University, Dramaga, Bogor, Indonesia

Suwarman Rusmawan Atmospheric Science Research Group, Faculty of Earth Science and Technology, Bandung Institute of Technology (ITB), Bandung, Indonesia

Syafitri Nabilla F. Department of Physics, Andalas University, Padang, Indonesia

Syamani Firda Aulya Research Center for Biomass and Bioproducts, National Research and Innovation Agency, Cibinong, Indonesia

Syukri Ichsan Faishal Department of Soil Science, Faculty of Agriculture, Andalas University, Padang, Indonesia

Tanigaki Minoru Institute for Integrated Radiation and Nuclear Science, Kyoto University, Kyoto, Japan

Tank Albert Klein Met Office Hadley Centre for Climate Science and Services, Exeter, UK

Thakur Isha G.B. Pant, National Institute of Himalayan Environment, Himachal Regional Centre, Mohal-Kullu, Himachal Pradesh, India

Tiyas Diah A. Department of Geophysics and Meteorology, IPB University, Bogor, Indonesia

Toersilowati Laras Atmospheric Science Research Group, Faculty of Earth Science and Technology, Bandung, Indonesia

Tohari Adrin Research Center for Geological Disaster, Research Organization for Earth Sciences and Maritime, National Research and Innovation Agency (BRIN), Bandung, Indonesia

Triani Listi Restu Atmospheric Science Research Group, Faculty of Earth Science and Technology, Bandung, Indonesia

Trilaksono Nurjanna Joko Atmospheric Science Research Group, Faculty of Earth Sciences and Technology, Institut Teknologi Bandung, Bandung, Indonesia

Trismidianto Research Center for Climate and Atmosphere, National Research and Innovation Agency (BRIN), Bandung, Indonesia

Ueda Yoshikatsu Research Institute for Sustainable Humanosphere, Kyoto University, Gokasho, Uji, Kyoto, Japan

Utami Arika Indri Dyah Agency of Meteorological, Climatological, and Geophysics for Indonesia, Jakarta, Indonesia

Van Langenhove Herman Department of Green Chemistry and Technology, Faculty of Bioscience Engineering, Research Group Environmental Organic Chemistry and Technology (EnVOC), Ghent University, Gent, Belgium

Vila Jordi Wageningen University, Wageningen, The Netherlands

Walgraeve Christophe Department of Green Chemistry and Technology, Faculty of Bioscience Engineering, Research Group Environmental Organic Chemistry and Technology (EnVOC), Ghent University, Gent, Belgium

Wardhana Dadan Dani Research Center for Geological Disaster, Research Organization for Earth Sciences and Maritime, National Research and Innovation Agency (BRIN), Bandung, Indonesia

Watanabe Takashi Laboratory of Biomass Conversion, National Research and Innovation Agency (BRIN), Uji Campus, Gokasho, Uji, Kyoto, Japan

Wicaksana Sigit Kurniawan Jati Research Center for Climate and Atmosphere, National Research and Innovation Agency (BRIN), Bandung, Indonesia

Wicaksono Arif Research and Development Division, Regional Planning and Development Agency (BAPPEDA), Government of Bogor Municipality, Bogor City, Indonesia

Wihardjaka Anicetus Indonesian Agricultural Environment Research Institute, Pati, Indonesia

Winarko Anton Space Research Center, BRIN, Bandung, Indonesia

Wisha Ulung Jantama Research Institute for Coastal Resource and Vulnerability, Ministry of Marine Affairs and Fisheries, Padang, Indonesia

Wulandari Eka P. Research Center for Climate and Atmosphere, National Research and Innovation Agency (BRIN), Bandung, Indonesia

Xianfeng Wang Earth Observatory of Singapore, Nanyang Technological University, Singapore, Singapore;
Asian School of the Environment, Nanyang Technological University, Singapore, Singapore

Yenuar Tsany N. A. Department of Physics, Andalas University, Padang, Indonesia

Yohana Thesalonica Faculty of Agricultural Technology, Universitas Brawijaya, Malang, Indonesia

Yulanda Novika Department of Soil Science, Faculty of Agriculture, Andalas University, Padang, Indonesia

Yulfa Arie Department of Geography, State University of Padang, Padang, Indonesia;
Department of Geography, Universitas Negeri Padang, Padang, Indonesia

Yulihastin Erma Research Center for Climate and Atmosphere, National Research and Innovation Agency (BRIN), Bandung, Indonesia

Zulhakim Hazi Department of Soil Science, Faculty of Agriculture, Andalas University, Padang, Indonesia

Chapter 1
Reducing Sugars Production from Oil Palm Empty Fruit Bunches (OPEFB) by Combined Dilute Acids-Hydrothermal Pretreatment

Fahriya Puspita Sari, Fitria, Sita Heris Anita, Maulida Oktaviani, and Widya Fatriasari

Abstract OPEFB (oil palm empty fruit bunches) is one of the residues of oil-palm industry which can be utilized as bioethanol feedstock due to its high holocellulose content of about 60%. As with other bioethanol production from lignocellulosic biomass, OPEFB must be pretreated before subsequent enzymatic hydrolysis to break down or loosen the linkage of cellulose, hemicellulose, and lignin. Therefore, this study investigated the influence of three different acids (oxalic, maleic, and sulfuric acids), three concentrations (1, 3, 5%), and three heating durations (15, 30, 45 min) in an autoclave (121.1 °C) on the pretreatment of OPEFB and to evaluate sugar production after enzymatic hydrolysis of the pretreated OPEFB. The results show that maleic acid gave the highest reducing sugar yield of 19.78% at a 5% concentration for a 15 min heating duration. On comparison, 16.41 and 15.40% of reducing sugar yields were obtained from oxalic acid (5%, 15 min) and sulfuric acid (1%, 15 min) pretreated OPEFB, respectively. This study shows the promising result of using maleic acid for lignocellulosic pretreatment compared to oxalic and sulfuric acids.

1.1 Introduction

Lignocellulosic biomass is an intriguing alternative feedstock for the generation of biofuels, bio-based chemicals, and biomaterial precursors. Over the last two decades, lignocellulosic biomass such as sugarcane bagasse has grown in popularity [1], while bamboo [2], corn cob, rice husk, cassava peel [3] has been extensively studied for biofuel production. Oil palm empty fruit bunches (OPEFB) are the most researched

F. P. Sari (✉) · Fitria · S. H. Anita · M. Oktaviani · W. Fatriasari
Research Center for Biomass and Bioproducts, National Research and Innovation Agency (BRIN), Cibinong, Indonesia
e-mail: fahr007@brin.go.id

Fitria
Department of Biological Systems Engineering, Washington State University, Richland, WA, USA

E. Yulihastin et al. (eds.), *Proceedings of the International Conference on Radioscience, Equatorial Atmospheric Science and Environment and Humanosphere Science, 2021*, Springer Proceedings in Physics 275, https://doi.org/10.1007/978-981-19-0308-3_1

biomass for biofuel generation, among other parts or residues of oil palm biomass. Indonesia is the biggest oil palm producer in the world, followed by its neighboring country, Malaysia. Together, they supply nearly 90% of the global market with a predicted combined production of about 60 million tonnes this year, led by Indonesia with 43.5 million tonnes [4].

Indonesia's oil palm plantation area has continuously risen from 0.3 million hectares in 1980 to 12.7 million hectares in late 2018 [5, 6]. The amount of biomass residues has increased significantly in tandem with the expansion of palm oil plantations. According to Febriyanti et al. [6], one hectare of oil palm plantation yields approximately 1.5 tons of OPEFB. However, only 10% of these lignocellulosic biomass leftovers are utilized for power and steam in palm oil mills. The remainder, around 90%, is thrown off as waste [7]. On the other hand, lignocellulosic biomass is an attractive potential feedstock for biofuels, bio-based chemicals, and biomaterial precursors and has gained popularity during the previous two decades.

OPEFB, like the other lignocellulosic biomass, contains cellulose, hemicellulose, and lignin. OPEFB is made up of 43.00–46.48% cellulose, 12.70–21.53% hemicellulose, and 17.80–23.85% lignin. [8, 9]. Cellulose and hemicellulose can be hydrolyzed to sugars and fermented to create biofuels, including ethanol, butanol, and biogas. However, lignin protects cellulose and hemicellulose, resulting in a complex structure of biomass cell walls that contributes to its recalcitrance [10]. Pretreatment of OPEFB is required to open its recalcitrant structure and boost the digestibility of the cellulose and hemicellulose fractions from enzyme or chemical attack. Pretreatment methods for lignocellulosic biomass are classified into three types: chemical, physical or physicochemical, and biological [7, 10, 11].

The most generally described pretreatment method for OPEFB is chemical or biophysical pretreatment. The following methods, i.e. hydrothermal pretreatment [9], alkali pretreatment [12], organic acid-pressurized reactor pretreatment [13]; microwave oxalic acid pretreatment [14, 15], alkali fiber explosion [16], microwave maleic acid pretreatment [8], and liquid hot water-maleic acid [17, 18] were used to improve the digestibility of OPEFB. One of the most successful pretreatments in solubilizing hemicelluloses and rendering cellulose more accessible appeared to be acid pretreatment. Several acids, including inorganic acids such as nitric, hydrochloric, phosphoric, and sulfuric acids, and organic acids such as oxalic, maleic, and acetic acids, have been utilized in concentrated or diluted form in lignocellulosic pretreatment [19].

Concentrated acid pretreatment at low or mild temperature is used to obtain high sugar yields without enzymatic hydrolysis step. However, because this reaction is highly toxic and corrosive, acid-resistant reactors must be designed [10, 20]. Due to its simplicity, dilute acid is an alternative chemical pretreatment process suitable for hydrolysis of lignocellulosic biomass or to enhance enzymatic digestibility. Pretreatment with dilute acid is performed at a higher temperature for a shorter time or at a lower temperature for a more extended period [20]. The benefits of dilute acid pretreatment include high hemicellulose and cellulose hydrolysis reaction rates and reduced inhibitor generation [10].

Methods that combine physical changes and chemical reactions during processing are referred to as physicochemical pretreatment. Lignocellulosic biomass is treated at high temperatures or under pressure with a chemical substance, causing the refractory structure to be disrupted. One of the pretreatment methods that use no catalysts or chemicals other than water is called hydrothermal or autohydrolysis. The hydrothermal pretreatment of lignocellulosic material results in the dissolving and destruction of hemicellulose and lignin, making the cellulose more accessible. Hydrothermal pretreatment of *Miscanthus giganteus*, switchgrass, hybrid Pennisetum, and *Triarrhena lutarioriparia* greatly improved enzymatic hydrolysis, with cellulose conversion increasing by 40–50% [10].

There were only a few reports on the use of mild pretreatment to produce reducing sugar [21, 22] especially using hydrothermal pretreatment combined with dilute acid. In this present work, we performed pretreatment of OPEFB combined dilute acid and hydrothermal pretreatment to investigate the influence of the three different acids (oxalic, maleic, and sulfuric acids), three concentrations (1, 3, 5%), and three heating durations (15, 30, 45 min) at 121 °C of OPEFB and to assess sugar production after enzymatic hydrolysis of pretreated OPEFB.

1.2 Materials and Methods

1.2.1 Materials

OPEFB was sourced in Cibadak, Sukabumi, West Java, Indonesia. The OPEFB fibers were crushed into fine particles, sieved through a 40–60-mesh sieve, and stored in a sealed plastic bag. The OPEFB fibers used in this study were similar to those employed in our prior report [8]. It has 4.26% ash, 1.91% extractives, 17.8% acid-insoluble lignin (AIL), 43.0% cellulose, and 12.7% hemicellulose. All chemicals in this investigation were utilized in analytical grades, with no further purification.

1.2.2 Dilute Acids (DA) - Hydrothermal Pretreatment of OPEFB

The OFAT (One Factor at Time) technique was used for pretreatment at 121.1 °C. The studies were carried out in various acid concentrations of maleic acid (w/v), oxalic acid (w/v), and sulfuric acid (v/v) ranging from 1, 3, and 5% for 15, 30, and 45 min. To determine the optimal acid concentration, in an erlenmeyer flask, 10 g of ODW (Oven Dry Weight) OPEFB with a solid loading of 1:10 (g/ml) was immersed in 1, 3, and 5% concentrations of Maleic Acid (MA), Oxalic Acid (OA), and Sulfuric Acid (SA). The erlenmeyer flasks were tightly sealed before being heated in an autoclave (WiseClave, Wisd) at 121.1 °C in the liquid phase for 15 min. After cooling, the slurries were vacuum filtered using filter paper and washed with distilled water until neutral. The OPEFB pulp (solid fraction) was frozen for enzymatic hydrolysis. Following that, pretreatment with various heating durations (15, 30, and 45 min) is carried out using the optimal acid concentration determined in previous experiments using the same method. The brown compound was next tested in the liquid fraction. The solid recovery was calculated by Eq. (1.1).

$$\text{Solid recovery}\ (\%) = \left(\frac{100 - \text{MP}}{100} \times \frac{\text{WP}}{\text{WB}}\right) \times 100 \tag{1.1}$$

where MP is the moisture content of the pulp (%), WP is the weight of the pulp (g), and WB is the oven-dry weight of the initial biomass before pretreatment (g).

1.2.3 Combine Severity Factor (CSF)

In this study, determination of CSF was used to facilitate comparison between severity during combined DA-hydrothermal pretreatment and characteristics of the hydrolysate (pH) [9]. The CSF was calculated using Eq. (1.2) as planned by Chum et al. [23].

$$\text{Log CSF} = \log\left(\text{t exp exp}\left(\frac{\text{T} - 100}{14.75}\right)\right) - \text{pH} \tag{1.2}$$

where T is pretreatment temperature in °C, t is the heating duration in min, and the pH values were measured in the hydrolysates after pretreatment.

1.2.4 Enzymatic Hydrolysis

Cellulase enzyme (Meicellace from *Trichoderma viride*, Meiji Seika) was used to hydrolyze up to 100 mg of OPEFB pulp (ODW), with an enzyme loading of 40 FPU/g substrates. Enzymatic hydrolysis was carried out in a 50 mL centrifuge tube with 5 mL of 0.05 M citrate buffer, pH 4.8, 100 mL of 2% sodium azide, and 1 mL of the enzyme. Finally, the 0.05 M citrate buffer was added to get a total final weight of 10 g. The tubes were then firmly closed. The mixtures were then incubated at 50 °C for 72 h in an incubator shaker (Wisecube, WIS-30R) at 150 rpm. Enzymatic saccharification was carried out in duplicate. After enzymatic hydrolysis, the hydrolysate in the tube was chilled before being used for reducing sugar analysis.

1.2.5 *Reducing Sugar and Brown Compound Analysis*

According to Warrand and Janssen [24], the liquid fractions were examined for their brown compounds (quantified as absorbance at a wavelength of 490 nm) using UV Vis Spectrophotometer (Hitachi 2001), while reducing sugar concentrations was assessed using the DNS (Dinitrosalicylic Acid) method [25].

1.3 Results and Discussion

1.3.1 *The Influence of Acid Concentration on Solid Recovery and Reducing Sugar Yield*

The effect of MA, OA, and SA concentrations on OPEFB pretreatment was determined by varying the acid concentration 1, 3, and 5% for 15 min of heating. According to Pederesen et al. [21] the acidity of the pretreatment has the greatest effect on reducing sugar under mild conditions. Thus, the first step in this research was to determine the optimal acid concentration using three different acid concentrations and the shortest heating duration (15 min). Figure 1.1a shows the recovery of OPEFB solids after combined DA-hydrothermal pretreatment with different acid concentrations. Solid recovery decreased as acid concentration increased in MA, OA, and SA pretreatment. The decrease of OPEFB pulp after pretreatment was caused by the degradation of some polysaccharides and lignin in OPEFB during pretreatment. SA pretreatment gave the lowest solid recovery (59%) compared to MA and OA pretreatment. The more severe the pretreatment, the less OPEFB pulp recovered because more component degradation occurred [14].

Figure 1.1b shows that the pretreatment using MA, OA, and SA gave the highest reducing sugar yields of 19.78, 16.41, and 15.40%, respectively. MA and OA show the same pattern that increasing acid concentration will increase the reducing sugar yield. The highest reducing sugar yields in MA and OA pretreatment were obtained in the 5% of acid concentration. However, SA pretreatment was vice versa. Increasing sulfuric acid concentration resulted in decreasing reducing sugar yield. The concentration of 1% sulfuric acid gave the highest reducing sugar yield (15.40%). Sulfuric acid is a strong acid with a pKa value of −2.8 and will be ionized completely during pretreatment [26]. Therefore, sulfuric acid is more severe than MA and OA, which are weak acids with pKa values of 1.9 and 1.25, respectively. Not only higher than sulfuric acid on reducing sugar yield, MA and OA were also reported to have lower toxicity and are compatible for the subsequent fermentation process of the soluble fraction [27].

According to Yan et al. [22], dilute SA pretreatment under mild conditions (100–200 °C) is usually applied to disrupt the lignin-carbohydrate matrix. However, inhibitor products may be formed during the pretreatment process, rendering the process of enzymatic hydrolysis and yeast fermentation unfavorable. Additionally,

Fig. 1.1 Effects of acid concentration on solid recovery (**a**) and reducing sugar yield (**b**) using maleic acid, oxalic acid, and sulfuric acid for 15 min of heating duration

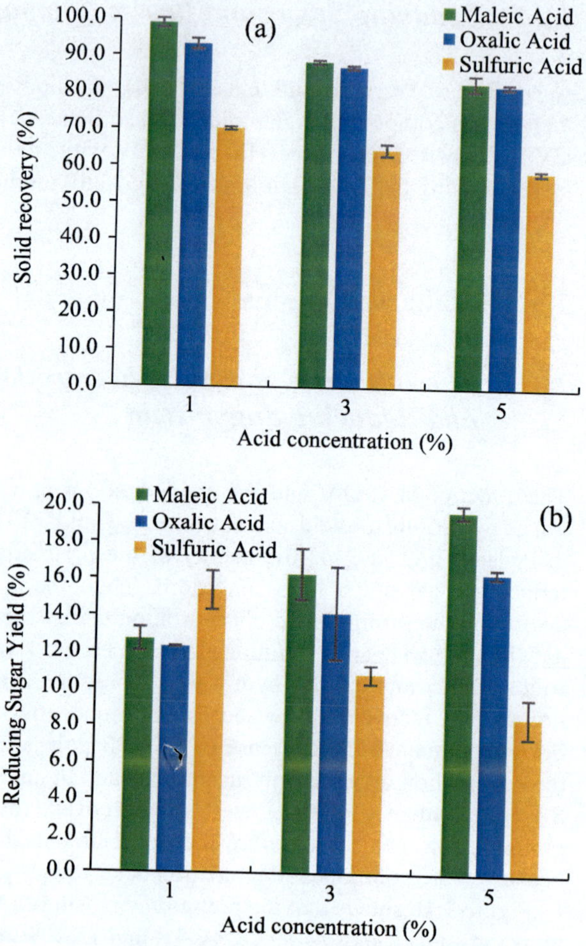

a higher SA concentration can result in more severe lignocellulose damage and can solubilize not only the hemicellulose fraction but also a portion of the glucan fraction that can be considered as cellulose [28]. Organic acids such as MA and OA can depolymerize hemicellulose selectively and stepwise control the acidity, allowing for more efficient hydrolysis of substrates over a wider temperature and pH range [22].

1.3.2 *The Influence of Heating Duration on Solid Recovery and Reducing Sugar Yield*

According to previous findings, the optimal acid concentration for pretreatment with MA and OA was 5%, while the optimal acid concentration for pretreatment with SA was 1%. Because the reducing sugar yield of MA and OA pretreatment was greater than that of SA pretreatment, a 5% acid concentration was used to determine the optimal heating duration. In this study, heating durations (15, 30, and 45 min) were carried out to explore the influence of heating duration in the production of reducing sugars at a constant acid concentration (5%) and heating temperature (121 °C). Heating duration is one of the main factors that influence the severity of pretreatment besides temperature and pH of pretreatment. The effect of heating duration on solid recovery and reducing sugar yield is shown in Fig. 1.2. A longer heating duration

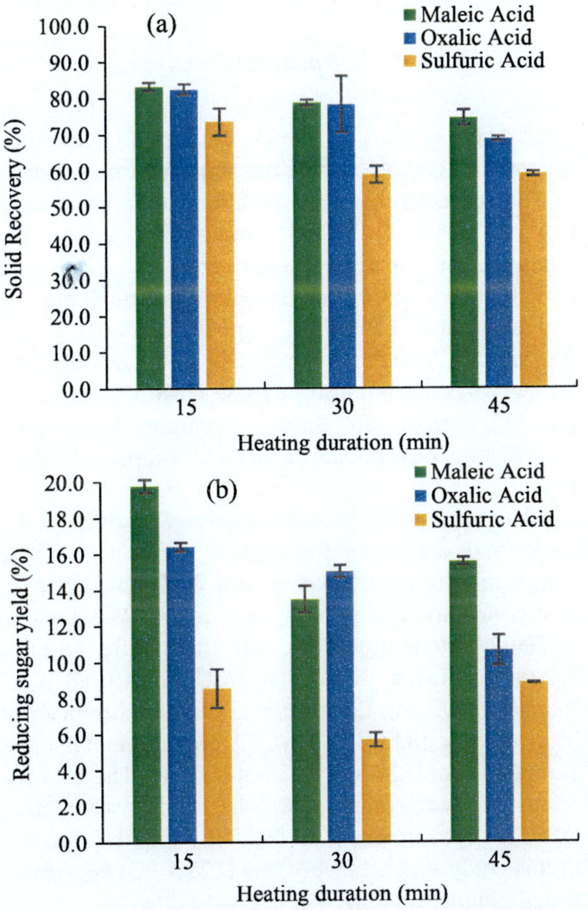

Fig. 1.2 Effects of heating duration on solid recovery (**a**) and reducing sugar yield (**b**) using 5% of maleic acid, oxalic acid, and sulfuric acid

from 15 to 45 min decreased the solid recovery in MA, OA, and SA (Fig. 1.2a). This can be attributed to the degradation of OPEFB components at a longer heating duration. Longer heating time not only degrades hemicellulose and lignin but also cellulose. The more cellulose is degraded and dissolved in the liquid fraction, the less reducing sugars after enzymatic hydrolysis (Fig. 1.2b).

In Fig. 1.2b, SA pretreatment gave the lowest reducing sugar yield of 5.68% for 30 min of heating. MA and SA pretreatments showed the same pattern that reducing sugar yield decrease when the duration of heating is 30 min and increases again at 45 min heating. The increase in heating duration causes degradation and solubility of chemical components in OPEFB, especially hemicellulose and lignin. Degradation of hemicellulose and lignin is beneficial during the enzymatic hydrolysis process because it can remove physical barriers and enhance the accessibility of enzymes to penetrate cellulose during hydrolysis [9, 29]. Based on these results, the heating duration of 15 min is the optimal condition for MA, OA, and SA pretreatment.

1.3.3 Brown Compound Analysis

The severe condition could initiate the decomposition of sugar monomers. Xylose and glucose can be decomposed into furfural and 5-hydroxymethylfurfural (HMF),respectively [30]. A HMF and furfural are undesirable secondary by-products that can negatively affect fermentation process [31]. The formation of brown compounds in the hydrolysate can be used as to indicate sugar degradation resulting from biomass hydrolysis. Browning reaction occurs from a non-enzymatic process which is mainly caused by caramelization and Maillard reaction. Both caramelization and Maillard reaction proceed simultaneously not only enhance brown color in the hydrolysates but also increase HMF and furfural formation and changes of pH [32, 33]. Furthermore, the brown compound formation can be caused by chemical modification on reducing sugars and compounds containing protein and amino acids [34].

The effect of acid concentration and heating duration on the absorbance of brown compounds is presented in Fig. 1.3. Figure 1.3a shows that the absorbance of brown compounds is relatively low and does not increase significantly with increasing acid concentrations in MA, OA, or SA. The absorbance of the brown compound in the OA pretreatment was constant, while the brown compound in the MA and SA pretreatments slightly increased from 0.18–0.20 and 0.13–0.24, respectively. Increasing heating duration also did not significantly affect the absorbance of the brown compound (Fig. 1.3b). SA pretreatment had the highest absorbance of brown compounds of 0.27 at 45 min of heating. The brown compounds in this study were relatively lower compared to other studies which use hydrothermal-maleic acid for OPEFB pretreatment [13, 18]. This may be due to the pretreatment temperature in this study was relatively low (121.1 °C) compared to other studies, which were usually higher than 170 °C.

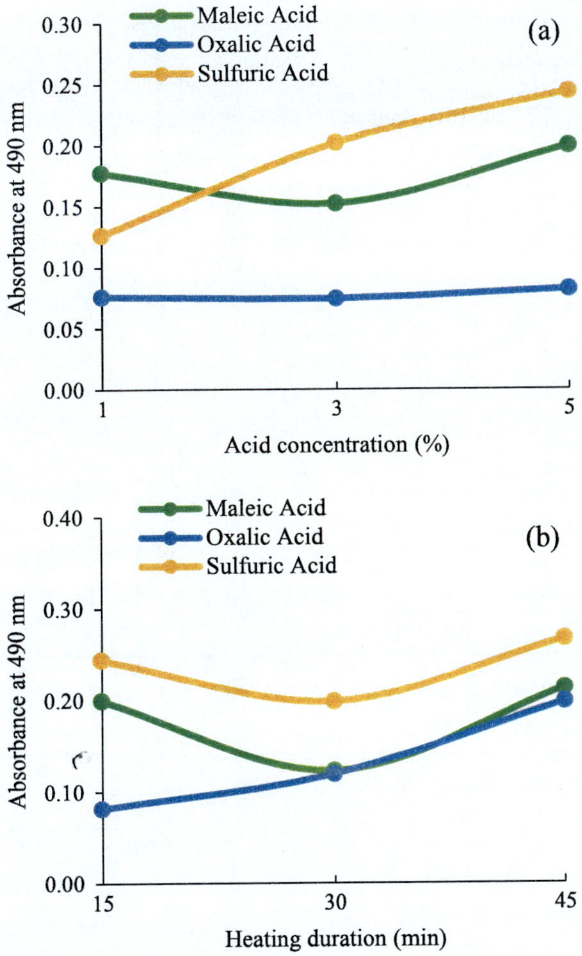

Fig. 1.3 The effects of acid concentration for 15 min of heating duration (**a**) and the effect heating duration using 5% of acid concentration (**b**) on the brown compound of OPEFB hydrolysates after combined DA-hydrothermal pretreatment at 121.1 °C

1.3.4 Effects of Combined Severity Factor on Pulp Recovery and Reducing Sugar Yield

The combined severity factors in this study were used to investigate the effect of heating duration and acid concentration on pulp recovery and reducing sugar yield. In Fig. 1.4a, the trend of pulp recovery decreases with increasing CSF. The more severe the pretreatment conditions, the more lignin, hemicellulose, and partial cellulose dissolved in the liquid fraction. This trend correlates with reducing sugar in higher CSF (Fig. 1.4b). The reducing sugars yield obtained after enzymatic hydrolysis showed that the pretreatment of MA, OA, and SA gave the highest yields at CSF of 1.0, 1.1, and 1.2, respectively. Meanwhile, CSF of SA pretreatment higher than 1.5 decreases the reducing sugar yield. As reaction severity increases, the

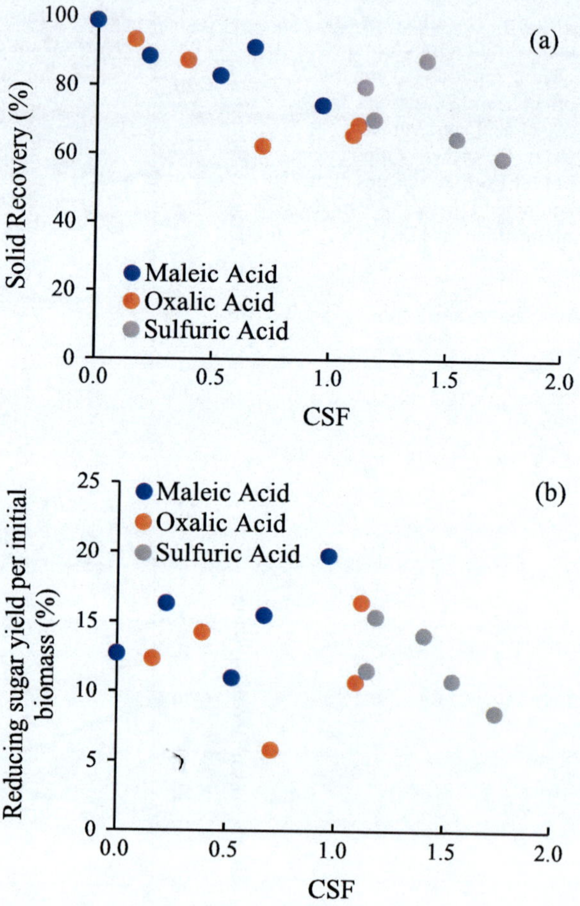

Fig. 1.4 Relationship between combine severity factor (CSF) and pulp recovery (**a**) and reducing sugar yield (**b**) of OPEFB after pretreatment

sugar monomers and oligomers will degrade into secondary degradation products such as HMF, furfural, formic acid, and other light organics [35]. The difference in CSF between MA, OA, and SA pretreatment was due to the parameters in the CSF comprised of heating duration, temperature, and pH, but not acid species. MA, OA, and SA have different acid dissociation constants, which will affect the severity during pretreatment [26]. An explanation of the effect of the acid dissociation constant of each acid species has been mentioned in discussion 3.1.

1.4 Conclusion

This study showed that pretreatment using combined maleic acid-hydrothermal pretreatment was a promising method to produce reducing sugars of OPEFB. A 5%

maleic acid combined with hydrothermal pretreatment for 15 min heating duration successfully disrupts the structure of OPEFB with high pulp recovery after pretreatment. Therefore, the reducing sugar yield after the enzymatic hydrolysis process can be achieved by 19.78% per initial biomass. The results of this study hopefully can be used as an alternative pretreatment of biomass, especially OPEFB. Further investigations are needed such as the chemical composition of pulp OPEFB after pretreatment and hydrolysate in liquid fraction to confirm the brown compound composition to get a deeper understanding of the hydrolysis and degradation process caused by the pretreatment. Analysis in structural changes such as morphological structure and crystallinity of OPEFB before and after pretreatment is also needed to give more insight into the severity of the pretreatment process.

Acknowledgements The authors gratefully acknowledge the Indonesian Institute of Sciences (LIPI) for funding this research through the DIPA of Research Center for Biomaterials LIPI in 2015 and Integrated Laboratory of Bioproduct (iLaB) for the facilities.

Authors Contribution All authors contributed equally as the main contributors to this work.

References

1. Suhardi, V.S.H., Prasai, B., Samaha, D., Boopathy, R.: Evaluation of pretreatment methods for lignocellulosic ethanol production from energy cane variety L 79-1002. Int. Biodeterior. Biodegrad. **85**, 683–687 (2013). https://doi.org/10.1016/j.ibiod.2013.03.021
2. Zhu, Y., Huang, J., Sun, S., Wu, A., Li, H.: Effect of dilute acid and alkali pretreatments on the catalytic performance of bamboo-derived carbonaceous magnetic solid acid. Catalysts **9** (2019). https://doi.org/10.3390/catal9030245
3. Awoyale, A.A., Lokhat, D.: Experimental determination of the effects of pretreatment on selected Nigerian lignocellulosic biomass in bioethanol production. Sci. Rep. **11**, 1–16 (2021). https://doi.org/10.1038/s41598-020-78105-8
4. Shahbandeh, M.: Global production volume palm oil by country 2020/2021. https://www-statista-com.hvhl.idm.oclc.org/statistics/856231/palm-oil-top-global-producers/%0A, https://www.statista.com/statistics/856231/palm-oil-top-global-producers/
5. BPS: Statistik kelapa sawit Indonesia 2018. Badan Pusat Statistik (2019)
6. Febriyanti, F., Fadila, N., Sanjaya, A.S., Bindar, Y., Irawan, A.: Pemanfaatan limbah tandan kosong kelapa sawit menjadi bio-char, bio-oil dan gas dengan metode pirolisis. J. Chemurg. **3**, 12 (2019). https://doi.org/10.30872/cmg.v3i2.3578
7. Rizal, N.F.A.A., Ibrahim, M.F., Zakaria, M.R., Abd-Aziz, S., Yee, P.L., Hassan, M.A.: Pretreatment of oil palm biomass for fermentable sugars production. Molecules **23**, 1–14 (2018). https://doi.org/10.3390/molecules23061381
8. Fatriasari, W., Raniya, R., Oktaviani, M., Hermiati, E.: The improvement of sugar and bioethanol production of oil palm empty fruit bunches (Elaeis guineensis Jacq) through microwave-assisted maleic acid pretreatment. BioResources **13**, 4378–4403 (2018). https://doi.org/10.15376/biores.13.2.4378-4403
9. Sari, F.P., Falah, F., Anita, S.H., Ramadhan, K.P., Laksana, R.P.B., Fatriasari, W., Hermiati, E.: Pretreatment of oil palm empty fruit bunch (Opefb) at bench-scale high temperature-pressure steam reactor for enhancement of enzymatic saccharification. Int. J. Renew. Energy Dev. **10**, 157–169 (2021). https://doi.org/10.14710/ijred.2021.32343

10. Jędrzejczyk, M., Soszka, E., Czapnik, M., Ruppert, A.M., Grams, J.: Physical and chemical pretreatment of lignocellulosic biomass (2019)
11. Baruah, J., Nath, B.K., Sharma, R., Kumar, S., Deka, R.C., Baruah, D.C., Kalita, E.: Recent trends in the pretreatment of lignocellulosic biomass for value-added products. Front. Energy Res. **6**, 1–19 (2018). https://doi.org/10.3389/fenrg.2018.00141
12. Duangwang, S., Sangwichien, C.: Optimizing alkali pretreatment of oil palm empty fruit bunch for ethanol production by application of response surface methodology. Adv. Mater. Res. **622**, 117–121 (2013). https://doi.org/10.4028/www.scientific.net/AMR.622-623.117
13. Sari, F.P., Solihat, N.N., Anita, S.H., Fitria, F., Hermiati, E.: Peningkatan produksi gula pereduksi dari tandan kosong kelapa sawit dengan praperlakuan asam organik pada reaktor bertekanan. REAKTOR (2017) https://doi.org/10.14710/reaktor.16.4.199-206
14. Anita, S.H., Fitria, Solihat, N.N., Sari, F.P., Risanto, L., Fatriasari, W., Hermiati, E.: Optimization of microwave-assisted oxalic acid pretreatment of oil palm empty fruit bunch for production of fermentable sugars. Waste Biomass Valorization **11** (2020). https://doi.org/10.1007/s12649-018-00566-w
15. Solihat N.N., Sari, F.P., Risanto, L., Anita, S.H.: Disruption of oil palm empty fruit bunches by microwave-assisted oxalic acid pretreatment methods. **49**, 244–257 (2017). https://doi.org/10.5614/j.math.fund.sci.2017.49.3.3
16. Waluyo, J., Anggraini, R.I.F., Kurniawan, H.H., Sudiyani, Y.: Physical chemical characterization of alkali pretreatment for oil palm empty fruit bunch. AIP Conference Proceedings 2024 (2018). https://doi.org/10.1063/1.5064358
17. Anita, S.H., Ashrianis, D.N., Fatriasari, W.: Pengaruh penambahan surfaktan pada fermentasi tandan kosong kelapa sawit dengan praperlakuan uap air panas bertekanan dengan katalis asam maleat. J. Ris. Kim. **12**, 65–74 (2021)
18. Fatriasari, W., Ulwan, W., Aminingsih, T., Puspita, F., Suryanegara, L., Heri, A., Ghozali, M., Nia, L., Hussin, M.H., Fudholi, A., Hermiati, E.: Optimization of maleic acid pretreatment of oil palm empty fruit bunches (OPEFB) using response surface methodology to produce reducing sugars. Ind. Crop. Prod. **171**, 113971 (2021). https://doi.org/10.1016/j.indcrop.2021.113971
19. Zhang, J., Bao, J.: Lignocellulose pretreatment using acid as catalyst. Handb. Biorefinery Res. Technol. 1–14 (2018). https://doi.org/10.1007/978-94-007-6724-9_3-1
20. Jung, Y.H., Kim, K.H.: Acidic Pretreatment. Elsevier B.V. (2015)
21. Pedersen, M., Johansen, K.S., Meyer, A.S.: Low temperature lignocellulose pretreatment: Effects and interactions of pretreatment pH are critical for maximizing enzymatic monosaccharide yields from wheat straw. Biotechnol. Biofuels. **4**, 11 (2011). https://doi.org/10.1186/1754-6834-4-11
22. Yan, Y., Zhang, C., Lin, Q., Wang, X., Cheng, B., Li, H., Ren, J.: Microwave-assisted oxalic acid pretreatment for the enhancing of enzyme hydrolysis in the production of xylose and arabinose from bagasse. Molecules **23** (2018). https://doi.org/10.3390/molecules23040862
23. Chum, H.L., Johnson, D.K., Black, S.K., Overend, R.P.: Pretreatment-Catalyst effects and the combined severity parameter. Appl. Biochem. Biotechnol. **24–25**, 1–14 (1990). https://doi.org/10.1007/BF02920229
24. Warrand, J., Janssen, H.: Controlled production of oligosaccharides from amylose by acid-hydrolysis under microwave treatment: comparison with conventional heating. **69**, 353–362 (2007). https://doi.org/10.1016/j.carbpol.2006.10.021
25. Lorenz, M.G.: Use of Dinitrosalicylic acid reagent for determination of reducing sugar. Anal. Chem. **31**, 426–428 (1959)
26. Qin, L., Liu, Z., Li, B., Dale, B.E., Yuan, Y.: Bioresource technology mass balance and transformation of corn stover by pretreatment with different dilute organic acids. Bioresour. Technol. **112**, 319–326 (2012). https://doi.org/10.1016/j.biortech.2012.02.134
27. Imman, S., Arnthong, J., Burapatana, V., Champreda, V., Laosiripojana, N.: Effects of acid and alkali promoters on compressed liquid hot water pretreatment of rice straw. Bioresour. Technol. **171**, 29–36 (2014). https://doi.org/10.1016/j.biortech.2014.08.022

28. Martínez-Patiño, J.C., Romero, I., Ruiz, E., Cara, C., Romero-García, J.M., Castro, E.: Design and optimization of sulfuric acid pretreatment of extracted olive tree biomass using response surface methodology. BioResources. **12**, 1779–1797 (2017). https://doi.org/10.15376/biores.12.1.1779-1797
29. Jeong, S., Lee, J.: Hydrothermal treatment. 61–74 (2015). https://doi.org/10.1016/B978-0-12-800080-9.00005-0
30. Palmqvist, E., Hahn-Hagerdal, B.: Fermentation of lignocellulosic hydrolysates . II : Inhibitors and mechanisms of inhibition mechanisms of inhibition. Bioresour. Technol. 74, 25–33 (2000). https://doi.org/10.1016/S0960-8524(99)00161-3
31. Saleem, M.E., Omar, R., Kamal, S.M.M., Biak, D.R.A.: Microwave-assisted pretreatment of lignocellulosic biomass: a review. J. Eng. Sci. Technol. **10**, 97–109 (2015)
32. Hermiati, E., Tsubaki, S., Azuma, J.I.: Cassava pulp hydrolysis under microwave irradiation with oxalic acid catalyst for ethanol production. J. Math. Fundam. Sci. **46**, 125–139 (2014). https://doi.org/10.5614/j.math.fund.sci.2014.46.2.2
33. Chen, S., Yang, D., Chen, H., Liu, S.: Effect of hot acidic fructose solution on caramelisation intermediates including colour, hydroxymethylfurfural and antioxidative activity changes. Food Chem. **114**, 582–588 (2009). https://doi.org/10.1016/j.foodchem.2008.09.089
34. Barclay, T., Ginic-markovic, M., Cooper, P.D., Petrovsky, N., Article, R.: The chemistry and sources of fructose and their effect on its utility and health implications. **3**, 67–82 (2012). https://ojs.abo.fi/ojs/index.php/jefc/article/view/145
35. Ruiz, H.A., Thomsen, M.H., Trajano, H.L.: Hydrothermal processing in biorefineries: production of bioethanol and high added-value compounds of second and third generation biomass (2017)

Chapter 2
The Decline of Surface Global Solar Radiation at Palembang City of Indonesia Correlated to the Smoke Events in 2019

Saipul Hamdi, Sumaryati, and Syahril Rizal

Abstract The decline of surface solar global radiation has been analyzed and reduced atmospheric visibility at Palembang, South Sumatra Province, Indonesia, during the smoke events in September-December 2019. The surface global solar radiation was measured at the campus of Bina Darma University at Palembang City using AWS integrated radiometer, and the visibility was observed in Talang Betutu, Palembang Airport. The smoke produced by forest fire surrounding Palembang caused declining in surface global solar intensity significantly. It also reduced atmospheric visibility in Palembang Airport down to 1.6 km, while in the atmospheric condition without smoke, the visibility was around 8.2 km. This condition made the sunrise in the morning as if delayed for some ten minutes. Among 86 days of data in this period, 46 days (53.3%) have lower surface solar radiation intensity. This study aims to get values of surface global solar radiation decline and the decrease in atmospheric visibility due to smoke events.

2.1 Introduction

Smoke is one of some factors that can decline solar radiation to reach the ground surface besides water vapor and dust. Latitude, altitude, and solar zenith angle are the other factors that contribute to declining solar radiation at surface level. Distance between solar and earth contributes to amount of solar radiation in the top of atmosphere. The smoke of forest fire decline not only solar radiation effectively by absorption and scattering processes [1] but also contains chemical compounds that endanger human health if only enter breathing system of human. Black carbon (BC) and organic carbon (OC) generated in forest fires are the aerosols and has a

S. Hamdi (✉) · Sumaryati
Research Center for Climate and Atmosphere, National Research and Innovation Agency (BRIN), Bandung, Indonesia
e-mail: saipulh@yahoo.com

S. Rizal
Bina Darma University of Palembang, Palembang, Indonesia

E. Yulihastin et al. (eds.), *Proceedings of the International Conference on Radioscience, Equatorial Atmospheric Science and Environment and Humanosphere Science, 2021*, Springer Proceedings in Physics 275, https://doi.org/10.1007/978-981-19-0308-3_2

wide impact on the global climate and local [2]. OC and BC also impact on human health [3] as carcinogenic [4, 5]. Smoke generated from forest and peatland fired in both Kalimantan and Sumatera is the primary source of smog transboundary in Southeast Asia and behave a significant contributor in increasing of aerosol content of atmospheric PM2.5 [6]. The smokes generated from forest fire and peatland fire in both Kalimantan and Sumatera contribute 55% of the smoke from all of the fire forests. Generally, the smoke and gaseous produced in the peatland fire in summer season are harmful to human life. Upper Respiratory Tract Infection or ISPA is one of serious impacts caused by the forest fire. Forest fired in surround area of Palembang, Indonesia, at last four months in 2019 has the profound impact to the environment at least decreasing visibility. Some area contributes to smoke event in Palembang i.e., residence of Banyuasin, Pampangan, Tulung Selapan, Pedamaran, Pemulutan, Cengal, Pematang, and Mesuji. In October 2019, at least 139 hotspots were recorded on Ogan Komering Ilir, and 67 hotspots in Banyuasin. Reducing visibility has a potential effect on the traffic accidents on roads and rivers. To anticipate the side impact of smokes Palembang City Local Government announced regulation for 500 schools (from kindergarten to high school) in the area impacted by peatland fire. The students must study from home and forbidden for both students and teachers come to school. The forest fire in South Sumatera used to happen in summer season period of July–November. This summer season is indicated in increasing in the surface temperature and even decreasing in humidity [7]. After field observation of forest fired in North Sumatra, it was understood that the smoke of the forest fired will accumulated and became denser after a period [8]. The smoke will cause the decline in visibility, indeed a decline in incoming solar radiation. The lowest visibility occurred on March 13, 14, and 15 of 2002; it was (0.6–0.7) km around the mid-day, and (1.2–2.5) km in around noon. The activities which has the direct impact on the decline of visibility is activity Polonia airport and piscatorial at Asahan and Labuhan Batu.

The solar radiation at top of the atmosphere is called a solar constant, and the value is 1368 W/m^2. The solar constant is not a constant value but depends on earth- sun distance and has periodicity. When passing the atmosphere, the solar radiation takes interact with atmospheric constituents, i.e., water vapor molecules, ozone, aerosol, and carbon dioxide, a part of solar radiation will be scattered and absorbed, and the rest part will reach the ground surface as incoming solar radiation or insolation. The solar spectrum, which is affected strongly by the stratospheric ozone layer, the ultraviolet B spectrum, with the result fluctuation of insolation can be an indicator of stratospheric ozone layer thickness. Except affected by scattering and absorption of atmospheric constituent insolation is also affected by topography, i.e., altitude and geography position. The daily pattern of incoming solar radiation is begun soon after the sun arise in the morning, and reaches maximum power at noon time, then will be end at sunset gradually. The orbital line of the earth circling the sun has one-year periodicity and impact to solar yearly intensity pattern, which has maximum and minimum in one year.

The smoke event in Palembang city and surround in September–December 2019 is caused by peatland fire in the area mentioned as before. Some factors that can be related to peatland fires, i.e., global warming [9]. It has been reported by IPCC (2007)

[10] that forest fire will happen over the world simultaneously to global warming. The IPCC Scenario SRES A2 indicates that increasing of fired area up to 500% will occured until the end of 21 century [11]. This study aims to analyze the decline of surface global solar radiation at Palembang at the smoke fog events caused by a forest fire in September-December 2019.

2.2 Data and Method

The intensity of solar radiation at surface level was measured with AWS integrated radiometer sensor in this study. We used AWS Vantage II Pro produced by Davis Instruments. This AWS was operated at the campus of The Bina Darma University of Palembang at 50 m height from the ground surface, and the sensors measured global solar intensity every second and recorded it to a data logger every 10 min. Furthermore, daily energy could be calculated from incoming solar radiation by integration to time. The duration of sunshine was calculated with a threshold of 120 W/m^2 while rainfall was used as secondary data. At meanwhile, the time for sunset and sunrise was found out from https://www.timeanddate.com/sun/ by using a filter of Palembang, and put the appropriate month/date in the calculation.

The assumption is the intensity of surface global solar radiation would reach more than 120 W/m^2 at 7 AM and decreased again to be less than 120 W/m^2 at 5 PM. The smoke used to appear at 4 PM and disappeared again after 8 AM during this event, and then, the threshold was chosen for 8 h to identify covering of smoke in daytime. Means, the days which have a duration of sunshine less than 8 h is suspected caused by smoke covering, even if it can be caused by a cloud covering, or caused by both. Refer to BMKG criteria about rainfall; the light rain is 5–20 mm/day, so that the days have rainfall more than 5 mm/day were not included in the discussion. This criterion was used for separating the days with rain especially suspected covered by smoke.

2.3 Results and Discussions

Forest fire occurs every year in the dry season in South Sumatra province, and the quantity increases when there is an El Nino phenomenon such as 2019. According to data issued by the Ministry of Environment and Forestry, the area of forest fires in the South Sumatra province as large as 8,785 ha in 2016, and it raised sharply in 2019 became 328,457 ha. The forest fire in 2019 has created an air environment polluted especially by smoke, from September until December.

Visibility is affected by air pollutants, particulate and gases, and weather parameters such as rainfall, humidity, and wind [12, 13]. But in the smoke event, particulate becomes very dominant in reducing visibility, as has been studied in Palangka Raya as an area affected by forest fires [14]. Based on observation of visibility as shown in Fig. 2.1, visibility in 2016 is relatively low all the year, with an annual average

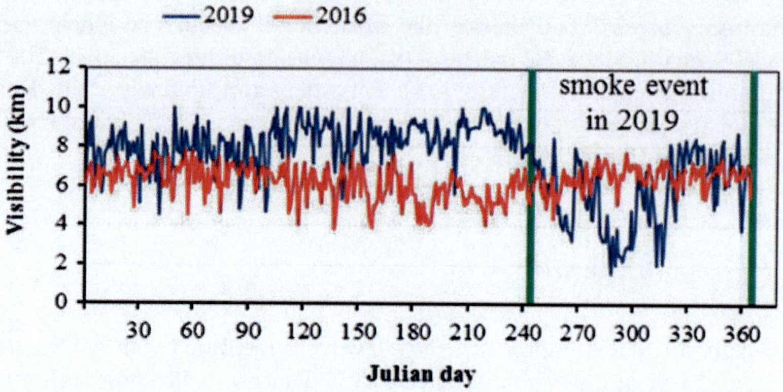

Fig. 2.1 Visibility in Palembang 2016 and 2019

of 6.3 km. It is supposed that the atmospheric environmental conditions in Palembang in 2016 are much polluted and get better in 2019. The pollutants are not only particulate but also certain gas pollutants SO_2, NO_2, O_3, and CO [12]. In 2019, under normal conditions the average visibility was 8.2 km, however in the smoke event, the average visibility drops to 5.7 km, even the lowest visibility is down to 1.6 km. Nevertheless, the annual average is still 7.4 km, which is higher than the 2016 annual average.

Figure 2.2 talk about the different of visibility around Ampera Bridge at a different time selected visually. Figure 2.2a shows the situation on January 19th, 2016, at 05:55,

Fig. 2.2 Ampera Bridge in Palembang City, taken on January 16th, 2016, at 05:55 AM (**a**) and on September 23rd, 2019, at 06:58 AM (**b**)

or 8 min before sunrise. At this time, the tower of Ampera Bridge was seen caused by the existence of water vapor and fog in the air was rare, relatively even so direct solar radiation not exposure yet. A different situation has happened shown in Fig. 2.2b, taken on the smoke of forest fired event on September 23rd, 2019. Currently, the sun was arisen at 05:50 or has exposure for 68 min after. The existence of forest-fired smoke causes to reduction the visibility, and this is correlated to the Tower of Ampera Bridge seen obscurely at a distance. Visibility is related to the visible spectrum of the sun and can be mean that solar radiation is affected strongly by forest fire smoke. On September 23rd, 2019, the air quality index at Palembang was 209 and increased to be 226 the next day.

The daily maximum intensity of surface global solar radiation at Palembang in 2019 is shown in Fig. 2.3. The horizontal axis is Julian Day (JD) or day number. The AWS solar radiation sensor passed the wavelength 300–1100 nm, or many parts of the visible spectrum (380–750 nm) and another part of both ultraviolet and infrared spectrum. Some collected data showed the very low intensity of solar radiation, for example, on JD 41 as amount as 379 W/m^2 or much less than the monthly average of February (about 1000 W/m^2). The 7.87 mm of rainfall has been recorded at 8:30–9:20 AM and 1:30–2:30 PM at JD-41. This rainfall is appropriate to the decline solar energy at the ground surface on JD-41 as amount as 162.17 Langley. This conformity can explain that water vapor and rain play an important role in decline solar radiation. Another atmospheric constituent plays in decline the solar radiation is the ozone layer. At meanwhile, the daily maximum intensity of solar radiation on JD 151-301 seems lower relatively compared to others periods of observation and can be explained by the position of the sun in the northern hemisphere. Change of distance between earth and sun will cause a change of intensity of solar irradiation reach top of the atmosphere.

Figure 2.4 shows comparability of maximum intensity of surface global solar radiation at Palembang in 2016 and 2019 on September-December, it correlates to forest fire smoke events in South Sumatera. The graph was also completed with time series of sunshine duration at the right vertical axis. September 1st is a coincidence

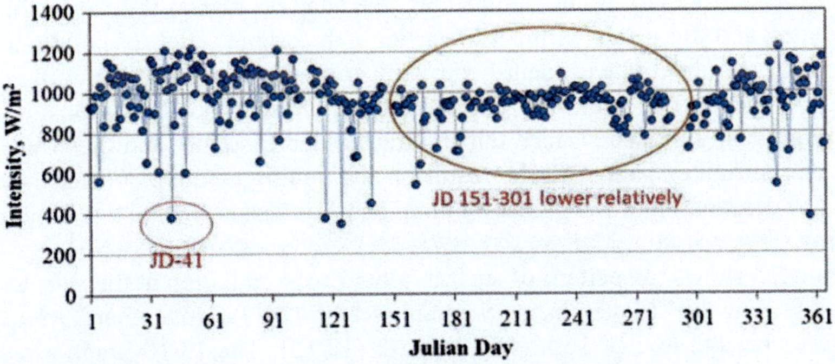

Fig. 2.3 The daily maximum intensity of global solar radiation in 2019

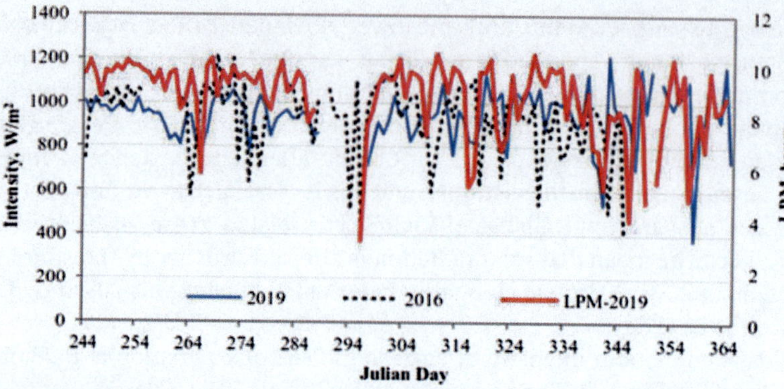

Fig. 2.4 The maximum intensity of surface global solar radiation at the smoke event in September-December 2019 and comparability at the same period in 2016. JD 274 = October 1st, JD 305 = November 1st, LPM = duration of sunshine

to JD = 244, and November 1st is a coincidence to JD = 305. Globally, the intensity of surface global solar radiation in 2019 is lower than in 2016. The amount of data in this period is 86 days of data, and there are 46 days of data (53.3% event) has a lower power. According to the measurement of solar radiation on the campus of Bina Darma University and method how to filter the data in this study there found 101 days of data September-December 2019. Some of data cannot be calculated caused by the technical reasons.

There are 85 days of data that have a duration of sunshine of more than 8 h and can be mean as inexistence of smoke covering on the mid-day at the region. On the other way, the number of days of data which have a duration of sunshine less than 8 h are 16 days of data and can be classified to be some classes of exposure. Indeed, it has been found one day of data in December has a duration of sunshine less than 5 h, and there was no rain also at that day. While there was no rain, the shorter duration of sunshine can be caused by continuous cover of clouds, indeed covering by mix of smoke and clouds. The impact of smoke in this event is more influence in the morning and early evening. In the morning, the sun was rose as if delayed for some ten minutes, and was disappearing as if earlier in the evening. Because of descending the planetary boundary layer identified by lower temperature in the morning causes smoke to concentrated intensely. The decreasing of solar radiation exposure in the evening gradually caused surface temperature to decrease too, and the atmosphere to became unstable again with the result of descending of the planetary boundary layer. The descending of the planetary boundary layer caused smoke to concentrate intensely.

Figure 2.5 shows the pattern of surface global solar radiation at the selected date on the smoke event at Palembang: September 24th (24/9), November 13rd (13/10) and December 8th (8/12), 16th (16/12), 18th (18/12). The 13/10 graph represents the profile of daily solar radiation generally. The intensity of the surface global solar radiation reaches lowest intensity on sunset and sunrise, and reaches the maximum

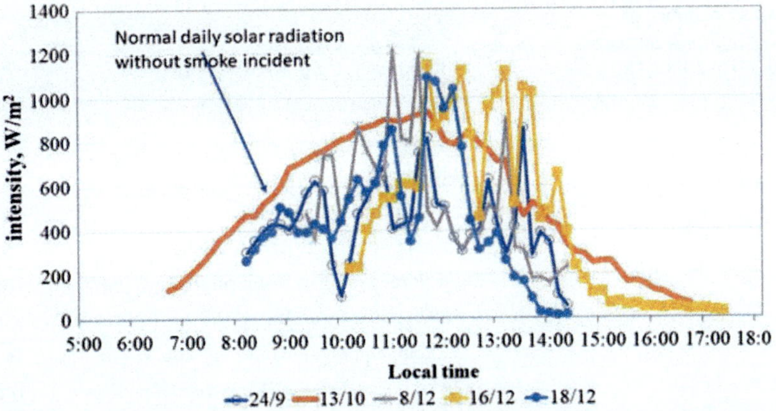

Fig. 2.5 The intensity of surface global solar radiation at Palembang on smoke forest fired event 2019. The sun rise at a range of time 05:37–05:59 and disappears at a range of time 17:50–18:12

power on sun culmination. The recorded intensity of solar radiation comes from both direct solar radiation and diffused solar radiation. By taking rough extrapolation of solar intensity, a threshold of 120 W/m^2 was reaching at around 06:30 AM. This threshold can be used for reference of determine how long it takes as if the delayed time for sunset and sunrise. In this forest-fired event, the sun rises on 24/9 and 18/12 were like as postpone for 40 min. A similar phenomenon occurred on 16/12 indeed took delayed more minutes up to 3 h. The smoke that covered the Palembang area on 16/12 in the morning was estimated very thick optically, but the smoke seems disappeared gradually after 10 AM. In this day, the surface global solar radiation reached a high value on maximum intensity after midday and can be related to inexistence of smoke and cloud cover. This delayed can be correlated to clean sky over Palembang and impacted by smoke in the morning only. The wind direction in November was dominantly easterly (14%), and wind speed was 1–11 knots [15].

At meanwhile, the intensity of surface global solar radiation on December 16th decreased rapidly to less than 120 W/m^2 at 15:00 or 2–3 h earlier than expected conditions. Indeed, the intensity of surface global solar radiation on December 18th decreased to be less than 120 W/m^2 close to 14:00 and down to zero rapidly. In these observed days, rainfall was not recorded, and reduce a duration of sunshine is assumed not caused by rain. Forest fired smoke contains black carbon (BC) and organic carbon (OC). A large amount of OC and BC, and a mix of some meteorological factors, impacts the solar radiation budget by scattering and absorption processes [16]. The solar radiation budget was impacted by forest fire event, and forest fire also impact air quality by enhancing some pollutant compounds. Forest fire also enhancing some chemical reactions which potentially depleting the atmospheric ozone layer [17]. Some of the PM2.5 particulates come from peatland fired is hygroscopic. The size of this particulate will be more prominent after binding water vapor molecules to attenuate the solar radiation effectively.

Table 2.1 shows the number of hotspots in South Sumatera Province period of 2015–2020 taken from LAPAN firehotspot website. Only 10 residences put in Table

Table 2.1 Number of hotspots in South Sumatera period 2015–2020 [18]

No	Regency	Hotspot
1	Kabupaten Ogan Komering Ilir	39,393
2	Kabupaten Musi Banyuasin	14,224
3	Kabupaten Banyuasin	8766
4	Kabupaten Muara Enim	3771
5	Kabupaten Musi Rawas	2942
6	Kabupaten Musi Rawas Utara	2522
7	Kabupaten Ogan Komering Ulu	1683
8	Kabupaten Ogan Ilir	1599
9	Kabupaten Ogan Komering Ulu Timur	1514
10	Kabupaten Penukal Abab Lematang Ilir	1462

2.1, which has most hotspots. The most hotspots found in Residence of Kayuagung as amount 39,393 hotspots. The other residences have smaller number of hotspots compare to Kayu Agung (Fig. 2.6).

In Fig. 2.5, Palembang City is marked by transparent arrow. Banyuasin is in northwest side of Palembang City, and Ogan Komering Ilir is East to South side of Palembang City. In South side of Palembang City there is Ogan Komering Ilir Residence. Subdistrict Pampangan, Tulung Selapan, Pedamaran, Cengal, and Mesuji are in Ogan Komering Ilir which contribute highly to smoke event in 2019.

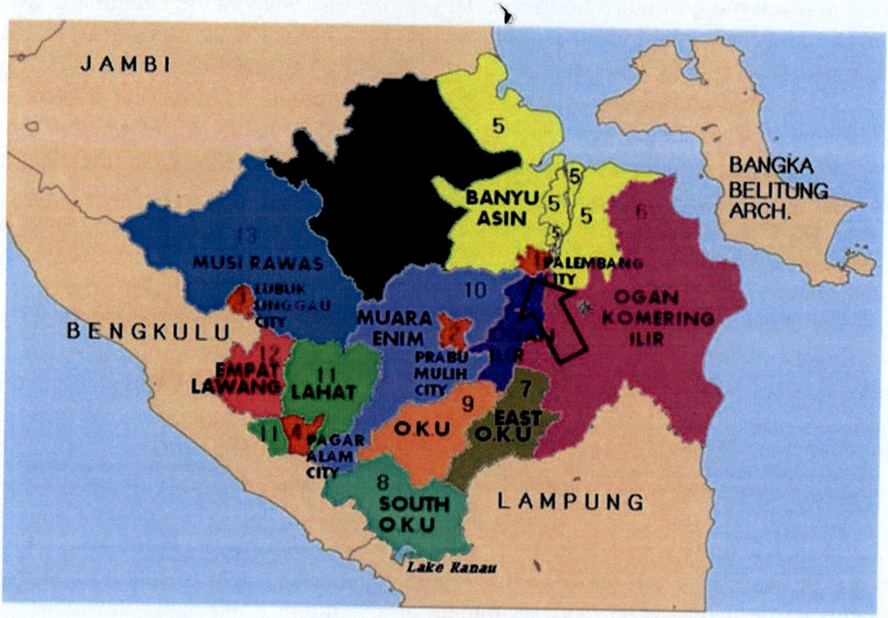

Fig. 2.6 The map of South Sumatera Province

2.4 Conclusions

At the forest fired event in September-December 2019, smoke from the surround Palembang area came to the inner city of Palembang and had a real impact on to the environment instantly. By visual, the smoke of forest fired reduced atmospheric visibility down to 1.6 km in the morning as if delayed sunrise for some ten minutes and sunset earlier. The smoke covering impacts to attenuation of global solar radiation at surface level by absorption and scattering processes; however, the decreasing intensity of the surface global solar radiation is not caused by smoke only. There are some factors contribute to decreasing of surface global solar radiation. Latitude, altitude, and solar zenith angle are the other factors that contribute to declining global solar radiation at surface level. Among 86 days of data of this period (September–December), there are 46 days (53.3%) have lower surface global solar radiation intensity.

Acknowledgements We wish to thank to the PSTA LAPAN and LPPM UBD, for facilitate this study, and the colleagues of the Atmospheric Environment Research Group for intensive discussion, and to the reviewers for the valuable comments and advice, which made it possible to elaborate the final version of this article.

References

1. Alonso-Blanco, E., Calvo, A.I., Pont, V., Mallet, M., Fraile, R., Castro, A.: Impact of biomass burning on aerosol size distribution, aerosol optical properties and associated radiative forcing. Aerosol Air Qual. Res. **006**, 708–724 (2014)
2. Healy, R.M., Wang, J.M., Jeong, C.-H., Lee, A.K.Y., Willis, M.D., Jaroudi, E., Zimmerman, N., Hilker, N., Murphy, M., Eckhardt, S., Stohl, A., Abbatt, J.P.D., Wenger, J.C., Evans, G.J.: Light-absorbing properties of ambient black carbon and brown carbon from fossil fuel and biomass burning sources. J. Geophys. Res. Atmos. **120**, 6619–6633 (2015)
3. Jacobson, L.D.S.V., Hacon, S.D.S., Castro, H.A.D., Ignotti, E., Artaxo, P., Saldiva, P.H.N., et al.: Acute effects of particulate matter and black carbon from seasonal fires on peak expiratory flow of school children in the Brazilian Amazon. PLoS ONE **9**(8), e104177 (2014)
4. Kaisovoja, T, Jalaya, P.I., Lamberg, H., Virn, A., Tapanainen, M., Torvela, E.T., Tapper, U., Sippula, O., Tissari, J., Hillamo, R., Hirvonen, M.-R., Jokinjemi, J.: Comparison of emissions and toxicology properties of fine particles from wood and oil boilers in small (20-25 kW) and medium (5019 MW) scale. Atmos. Environ. **77**, 193–201 (2013)
5. Garra, P., Maschowski, C., Liaud, C., Dieterlen, A., Trouve, G., Le Calve, S., Jaffrezo, J.L., Leyssens, G., Schonnenbecl, C., Kohler, S., Giere, R.: Fluoresence microscopy analysis of particulate matter from biomass burning: polyaromatic hydrocarbons as main contributors. Aerosol Sci. Technol. **49**(11), 1160–1169 (2015)
6. Fujii, Y., Iriana, W., Oda, M., Purigiwati, A., Tohno, S., Lestari, P., Mizohata, A., Huboyo, H.S.: Characteristics of carbonaceous aerosols emitted from Riau, Sumatera, Indonesia. Atmos. Environ. **87**, 167–169 (2014)
7. Reid, J.S., Hyer, E.J., Johnson, R.S., Holben, B.N., Yokelson, R.J., Zhang, J., Campbell, J.R., Christopher, S.A., DiGirolamo, L., Giglio, L., Holz, R.E., Kearney, C., Miettinen, J., Reid, E.A., Turk, F.J., Wang, J., Xian, P., Zhao, G., Balasubramanian, R., et al.: Observing and understanding the Southeast Asian aerosol system by remote sensing: an initial review and

analysis for the Seven Southeast Asian Studies (7SEAS) program. Atmos. Res. **122**, 403–468 (2013)
8. UPT-HB: Laporan Kegiatan Operasi udara Penipisan asap di Sumatera dan Kalimantan September–October 1997. BPPT, Jakarta (1997)
9. Carvalho, A., Flannigan, M., Logan, K., Gowman, L., Miranda, A.I., Borrego, C.: The impact of spatial resolution on area burned and fire occurrence projections in Portugal under climate change. Clim. Change **98**, 177–197 (2010)
10. IPCC: The physical science basis: climate change 2017. In: Contribution of Working Group I to the Fourth Assessment Report of the Intergovernmental Panel on Climate Change. Cambridge University Press, p. 996 (2007)
11. Nakicevonic, N., Alcamo, J., Davis, G., de Vries, B., Fenhann, J., Gaffin, S., Gregory, K., Grubler, A., Jung, T.Y., Kram, T., LaRovere, E., Michaelis, L., Mori, S., Morita, T., Pepper, W., Pitcher, H., Proce, L., Raihi, K., Roehrl, A., Rogner, H., Sankovski, A., Schlingser, M., Shukla, P., et al.: IPCC Spesial Report on Emissions Scenarios. Cambridge University Press, Cambdrige, UK and New York, USA (2000)
12. Du, K., Mu, K., Deng, J., Yuan, F.: Study on atmospheric visibility variations and the impacts of meteorological parameters using high temporal resolution data: an application of environmental Internet of Things in China. Int. J. Sust. Dev. World **30**(3), 238–247 (2013)
13. Wang, X., Zhang, R., Yu, W.: The effects of PM2.5 concentrations and relative humidity on atmospheric visibility in Beijing. JGR Atmosphere **124**(4), 2235–2259 (2019)
14. Sumaryati, Cholianawati, N., Indrawati, A.: The impact of forest fire on air-quality and visibility in Palangka Raya. J. Phys.: Theories Appl. **3**(1), 16–26 (2019)
15. BMKG: Buletin BMKG. In: Analisis hujan November 2019 dan prakiraan hujan Januari, Februari, dan Maret 2020 di Sumatera Selatan, vol. 14, no. Tahun XXXIII, December 2019
16. Bali, K., Mishra, A.K., Singh, S.: Impact of anomalous forest fire on aerosol radiative forcing and snow cover Himalayan region. Atmos. Environ. **150**, 264–275 (2017)
17. Carvalho, A., Monteiro, A., Flannigan, M., Solman, S., Miranda, A.I., Borrego, C.: Forest fires in a changing climate and their impacts on air quality. Atmos. Environ. **45**, 5545–5553 (2011)
18. Sari, N.A., Putra, R.A.: Analisis Statistik Deskriptif titik Panas di Wilayah Sumatera Selatan. In: Seminar Nasional Sains dan Teknologi Terapan, Surabaya (2020)

Chapter 3
Atmospheric Response to the Southern Java Upwelling Variability Associated with Positive Indian Ocean Dipole Event

Rahaden Bagas Hatmaja, Christine Cecylia Munthe, Erma Yulihastin, and Kadiman Erfitra Pramudia

Abstract Southern Java upwelling is closely related to the Indian Ocean Dipole (IOD) and El-Niño Southern Oscillation (ENSO) on interannual timescales. Recent studies stated that IOD tends to have a stronger correlation than ENSO in influencing the ocean-atmosphere dynamics in this region. This research investigated the atmospheric response to changes in sea surface temperature (SST) during more vigorous upwelling off the Southern Java coast. This was done based on the composite of oceanic and atmospheric processes in the upwelling system associated with the positive IOD events from 1993 to 2020. The data used is a reanalysis dataset from the European Centre for Medium-Range Weather Forecasts (ECMWF) with a spatial resolution of 0.25° × 0.25° from June to November. During positive IOD events, easterly wind anomaly generates upwelling Kelvin waves propagation along the equator to the coast off Sumatra-Java, thus strengthening the upwelling. The stronger upwelling, which transports cold water up to 1.48 Sv from deep water to the surface, contributes to reducing SST anomalies up to 0.9 °C. As the response of the SST changes due to strengthened upwelling, the wind stress reduced up to 0.006 N/m^2 and the near-surface air temperature by around 0.85 °C. Moreover, cold SST front during upwelling events also contributes to lowering the boundary layer height up to 27 m, which is suggested to increase the turbulence of the atmospheric lower layer.

3.1 Introduction

As part of the Maritime Continent and the Indian Ocean, the waters along the southern Java coast have complex oceanic and atmospheric characteristics, one of which is Southern Java upwelling [1–3]. The Southern Java Upwelling is a seasonal upwelling

R. B. Hatmaja (✉) · E. Yulihastin
Research Center for Climate and Atmosphere, National Research and Innovation Agency (BRIN), Bandung, Indonesia
e-mail: rahadenbagas@gmail.com

C. C. Munthe · K. E. Pramudia
Research Center for Atmospheric Science and Technology, National Research and Innovation Agency (BRIN), Bandung, Indonesia

E. Yulihastin et al. (eds.), *Proceedings of the International Conference on Radioscience, Equatorial Atmospheric Science and Environment and Humanosphere Science, 2021*, Springer Proceedings in Physics 275, https://doi.org/10.1007/978-981-19-0308-3_3

caused by Australia's south-east monsoon wind stress along the Java-Sumatra coast from June to August [1, 4]. Interannual variability of the Southern Java upwelling is closely related to El-Niño Southern Oscillation (ENSO) and Indian Ocean Dipole (IOD) [1, 5]. Recent study by Chen et.al. showed a significant correlation between wind stress (zonal and meridional) and the Dipole Mode Index (DMI) compared to the Nino 3 index, which indicates that there was a dominant contribution of IOD compared to ENSO on interannual variability of the Southern Java upwelling [6]. During positive IOD (pIOD) events, the upwelling transport along the southern coast of Java increased by 40–50% [7]. Stronger upwelling events resulted in cold SST anomalies lower than –1 °C extending west 90° E [5]. In contrast, this upwelling transport reduced dramatically to zero during negative IOD (nIOD) events. These findings suggest that upwelling can significantly impact SST variation, and therefore might be crucial in regulating regional and global climate.

Furthermore, earlier studies showed that upwelled cold water alternate SST and thus influence the atmosphere dynamics such as air temperature, marine boundary layer, and wind stress-SST coupling [8–14]. Upwelling was found to have a minor impact on regional atmospheric condition. However, it is considered locally over the area where upwelling is most common. These included 2 m drop in air temperature about 2 °C, as well as wind speed and a reduction in the mean boundary layer up to 0.25 m/s and 100 m, respectively [9]. The horizontal gradient in the ambient air temperature reflected the sea surface, but only in the lower atmosphere layer 150 m [10]. In addition, the marine boundary layer was typically defined by a strong temperature inversion and a shallow internal boundary layer over the cool upwelled water, with a 2 m air temperature typically at least 1 °C colder over the upwelling area [11].

Earlier studies about upwelling impact on the atmosphere as ocean–atmosphere interaction have been conducted in several places around the globe [12–19]. However, the studies conducted in the equatorial is still limited, particularly over the Southern Java upwelling system. In addition, due to the strengthened upwelling along the southern coast of Java during the pIOD event, concerns have been expressed about the atmospheric parameter i.e., wind stress, air temperature, and boundary layer height, over the upwelling system in response to changes in SST.

3.2 Data and Methods

3.2.1 Data

This study was conducted along the southern coast of Java Island, Indonesia, which is limited to the black box in Fig. 3.1. Oceanic parameter data used in this study is the monthly sea surface temperature reanalysis dataset from E.U. Copernicus Marine Service Information (CMEMS) [20]. The data were can be retrieved in https://marine.co-pernicus.eu/, with a spatial resolution of one-twelfth degree. Moreover, the atmospheric parameter data used are 10 m zonal and meridional wind, 2 m air temperature,

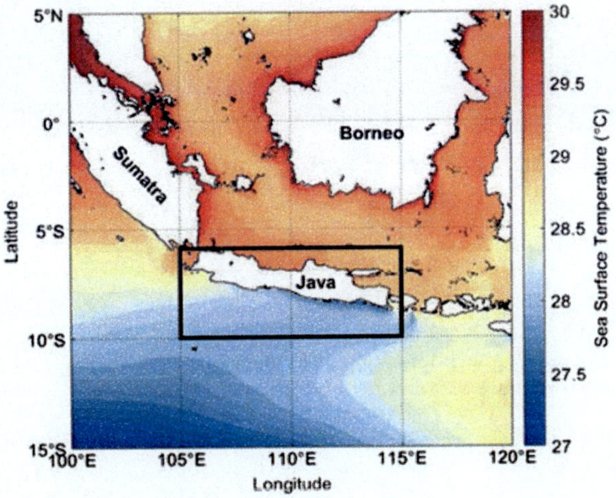

Fig. 3.1 The annual mean of sea surface temperature (SST) of the western Indonesia waters. The area of interest is specified in the black box

and boundary layer height, retrieved from the ERA5 monthly averaged reanalysis dataset by European Centre for Medium-Range Weather Forecasts (ECMWF) with a spatial resolution of a quarter degree. A detailed description of the ERA5 reanalysis products is provided in Copernicus Climate Change Service (C3S) Climate Data Store (CDS) [21]. All data used are 28 years from January 1993 to December 2020, which are considered as the era of assimilated data based on satellite observation. Hence, the reanalysis data used in this study are reliable in analyzing the ocean and atmosphere dynamics.

3.2.2 Methods

Southern Java upwelling is generated by wind stress is calculated from wind velocity by using Eq. 3.1 as follows [22].

$$\vec{\tau} = \rho_a \cdot Cd \cdot \left|\vec{V}\right| \cdot \vec{V} \tag{3.1}$$

where $\vec{\tau}$ is the wind stress vector, with τ_x and τ_y as the respective eastward and northward components, ρ_a is the density of air (1.225 kg/m^3), Cd is the drag coefficient (1.43×10^{-3}), and $\vec{V}$ is the wind component vector that consists of zonal component (u) and meridional component (v).

Furthermore, to obtain a comprehensive analysis of upwelling dynamics as wind stress and SST interaction, we calculated SST gradient (∇SST), crosswind SST gradient ($\nabla SST \times \vec{\tau}$), wind stress curl ($\nabla \times \vec{\tau}$), and Ekman pumping velocity (w_E) by using the following equations [12, 23, 24].

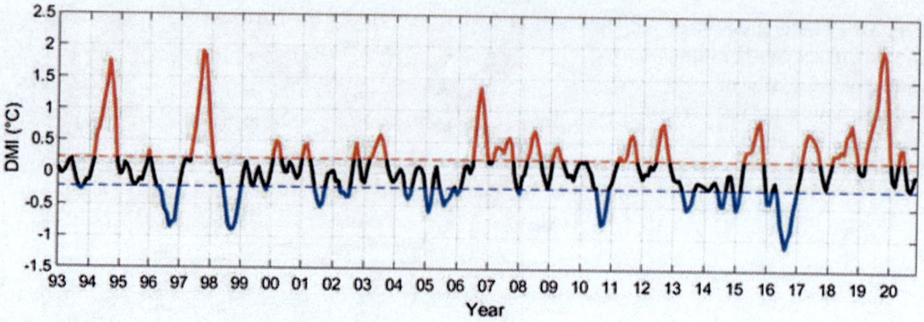

Fig. 3.2 Time series of Dipole Mode Index (DMI) during 1993–2020 period. The red (blue) dashed line denotes the threshold of positive (negative) events

$$\nabla SST = \frac{\partial SST}{\partial x} i + \frac{\partial SST}{\partial y} j \tag{3.2}$$

$$\nabla SST \times \vec{\tau} = \frac{\partial SST}{\partial x} \cdot \tau_y - \frac{\partial SST}{\partial y} \tau_x \tag{3.3}$$

$$\nabla \times \vec{\tau} = \frac{\partial \tau_y}{\partial x} - \frac{\partial \tau_x}{\partial y} \tag{3.4}$$

$$w_E = \frac{1}{\rho_w f} \nabla \times \vec{\tau} \tag{3.5}$$

where ρ_w is the density of seawater (1024 kg/m^3) and f is the Coriolis parameter.

Interannual variability associated with the pIOD event is analyzed by using composite analysis conducted in the anomaly field. The anomaly field is calculated as the deviation from the respective climatology based on 28 years data. The composite analysis is based on 12 events that occurred from 1993 to 2020 i.e., 1994, 1997, 2003, 2006, 2007, 2008, 2011, 2012, 2015, 2017, 2018, and 2019. These events are determined when the DMI or Dipole Mode Index from June to November is greater than 0.5 standard deviations or 0.24 °C for at least 3 consecutive months, see Fig. 3.2 [25].

3.3 Results and Discussion

3.3.1 The Oceanic and Atmospheric Feature of the Southern Coast of Java During Summer Monsoon

The wind patterns in western Indonesia are mostly influenced mainly by the Asian and Australian monsoons. From June to August, south-easterly winds blow parallel to the southern Java coastline, generating coastal upwelling (Fig. 3.3a). Coastal cooling occurs on the south-eastern cape of Java Island, where the SST can decrease up to 25 °C, then extend north-westward along the coast. The SST rises linearly from the coast to the ocean, with significant SST gradients mainly forming along the southern coast of Java (Fig. 3.3b). The coastal upwelling area is approximately represented by large SST gradients limited to 9° S. The SST gradient indicates a significant oceanic

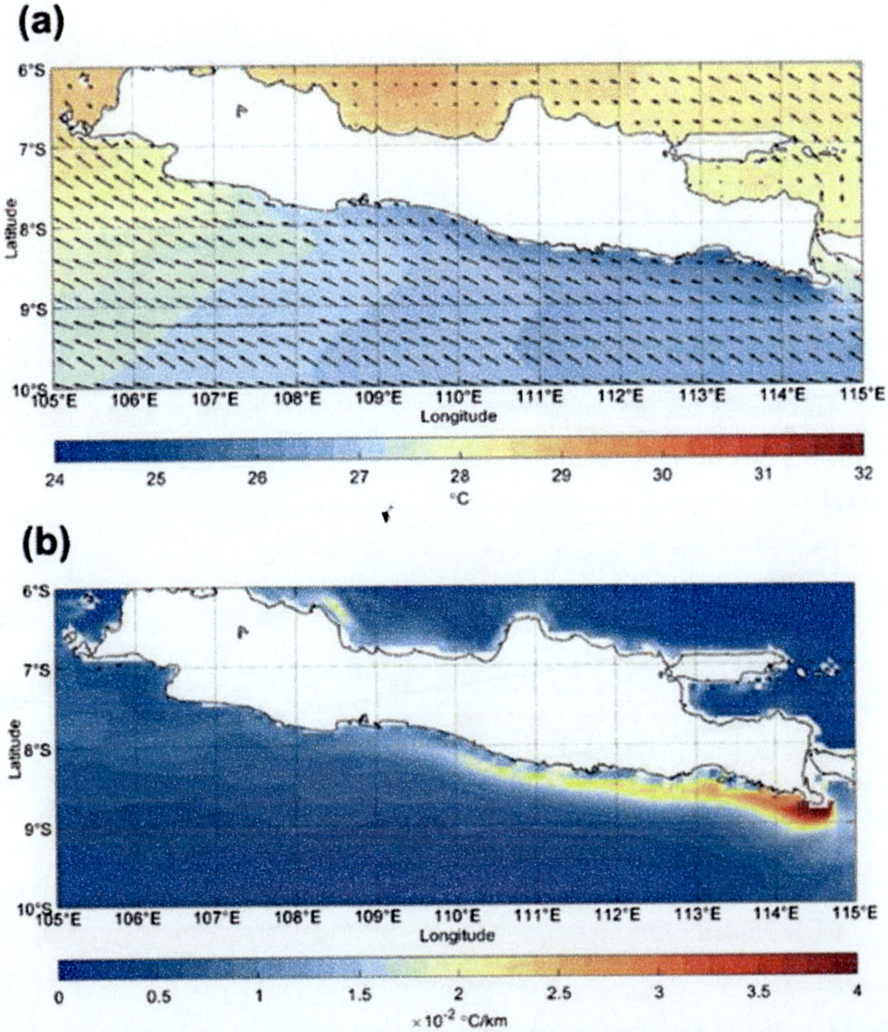

Fig. 3.3 Mean of **a** SST (shaded) overlaid with wind stress vector, and **b** SST gradient on the southern coast of Java Island, Indonesia, during summer monsoon in June to August based on 28 years data (1993–2020)

front, with a magnitude of SST gradient of more than 0.02 °C/km [16, 26, 27] along the coastline up to 60 km offshore.

Over the Southern Java upwelling system, the coupling between crosswind SST gradient and wind stress curl is observable (Fig. 3.4a). Strong crosswind SST gradients (greater than 0.02 °C/km) may be seen north-westward along the southern Java

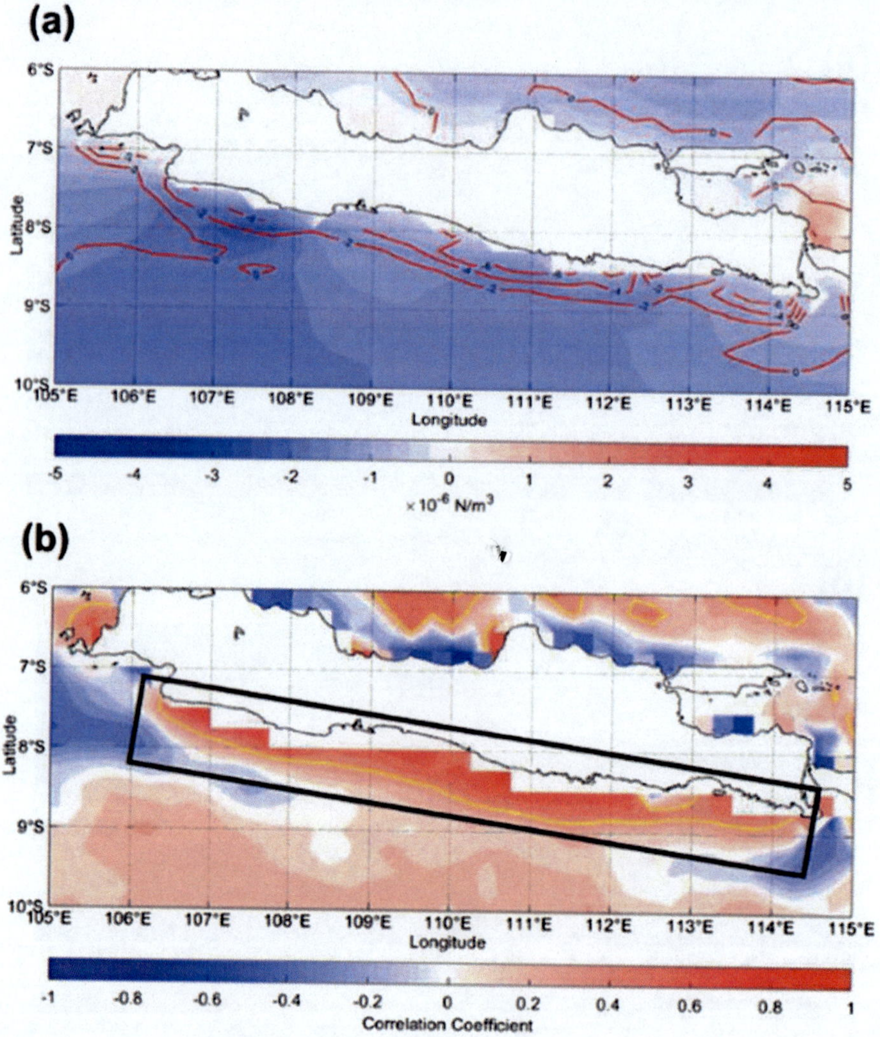

Fig. 3.4 **a** Mean of wind stress curl (shaded) overlaid with crosswind SST gradient (contoured, with the contour interval as 1×10^{-4} °C/km) on the southern coast of Java Island, Indonesia, during the summer monsoon in June to August based on 28 years data (1993 to 2020). **b** Correlation map between wind stress curl and crosswind SST gradient along the southern coast of Java, with 95% significance level denoted by yellow contour

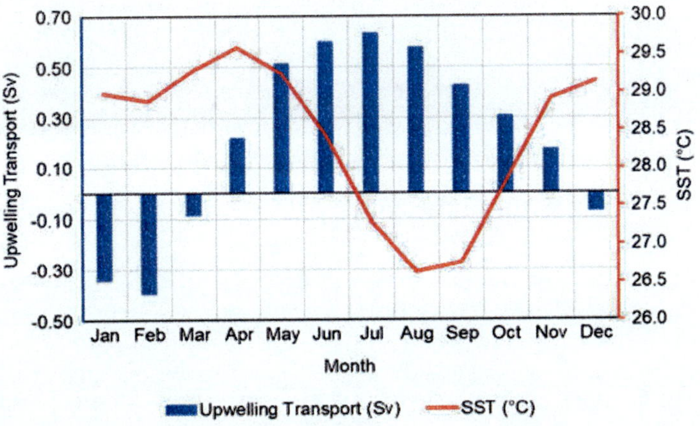

Fig. 3.5 Seasonal variation of area-averaged upwelling transport and SST along the southern coast of Java Island, Indonesia, which are calculated based on 28 years data (1993–2020)

coast up to 106° E, accompanied by a significant negative wind stress curl, and both decrease from the coast to offshore. The extent of the positive SST gradient and wind stress curl is roughly 60 km, which is compatible with the presence of a cold and thin front along the coast (Fig. 3.3b). The wind stress curl changes with the crosswind SST gradient, with a strong positive correlation between both about 60 km offshore (Fig. 3.4b).

The summer monsoon on the southern coast of Java plays a significant role in controlling coastal variability, mainly through Ekman pumping. As the south-east monsoon initiates in April, monthly upwelling transport shows a significant rise over the summer, followed by a drop in August (Fig. 3.5). A significantly large SST gradient is caused by the cold SST induced by coastal upwelling (Fig. 3.3). In the Southern Java upwelling system, negative wind stress curl generates Ekman pumping simultaneously. These results show that Ekman pumping is crucial for enhancing this upwelling system. In July, the upwelling transport via Ekman pumping peaks alternates the SST by one month lag (Fig. 3.5).

Furthermore, there is a significantly positive correlation between SST and wind stress, as seen in Fig. 3.6a. The wind is significantly weaker along the coast up to 80 km from the coastline than in the offshore region. The weak wind stress is located on the cool SST tongue. Decelerations in winds over cold SST might explain the inhomogeneous surface wind stress [18]. The SST increases by 1 °C from the coast to 80 km offshore, and the surface wind stress magnitude rises from 0.03 to 0.07 N/m^2 perpendicular to the coastline.

Moreover, near-surface air temperature and boundary layer height show similar characteristics in Fig. 3.6b. It is shown that air temperature as low as 24 °C along with the coast and marine boundary layer is approximately 400 m. Both parameters also increased along with the increasing SST from the coast to offshore.

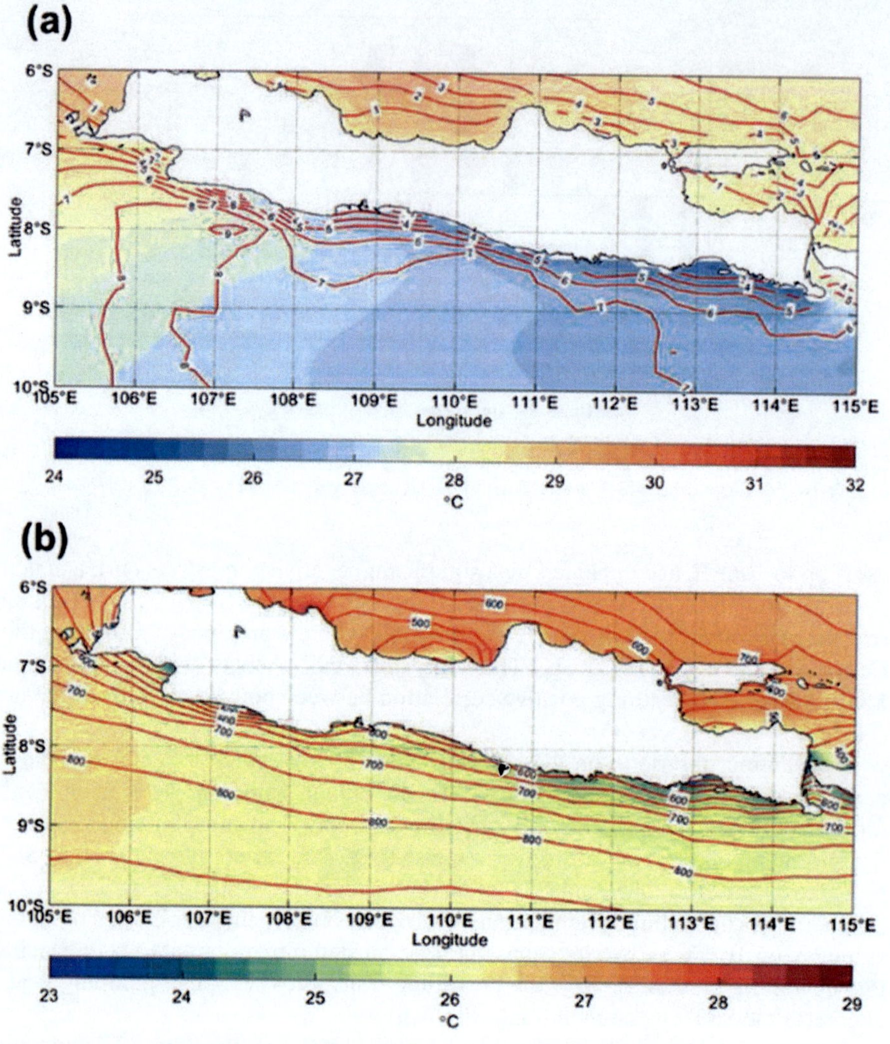

Fig. 3.6 Mean of **a** SST (shaded) overlaid with wind stress magnitude (contoured, with the contour interval as 1×10^{-2} N/m^2), and **b** near-surface air temperature (shaded) overlaid with boundary layer height (contoured, with the contour interval as 100 m) in the southern coast of Java Island, Indonesia, during the summer monsoon

3.3.2 Interannual Variability of Southern Java Upwelling During Positive Indian Ocean Dipole Event and the Atmospheric Response

Previous research found that pIOD events enhance upwelling transport in this region by 40 to 50%, controlled by IOD-related Wyrtki jet and thermocline transports from

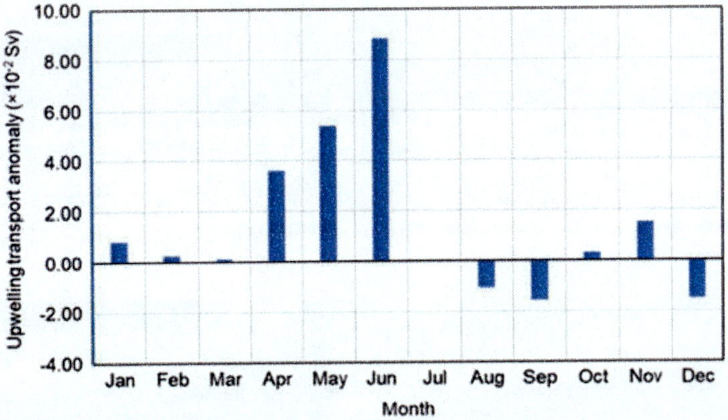

Fig. 3.7 Composite seasonal variation of area-averaged upwelling transport anomaly along the southern coast of Java Island, Indonesia, during the positive Indian Ocean Di-pole (pIOD) event

the Indian Ocean as remote equatorial forcing [7]. This study obtains the upwelling transport during the pIOD event from Ekman pumping induced by local wind stress. Figure 3.7 shows that the upwelling transport increased by 0.08 Sv in June. As the response of this strengthened upwelling, the cold SST anomaly and enhanced SST front occurs in the following season, i.e., the fall season in September to November, which peaks up to 0.9 °C, which coincided with the peak of pIOD event in September to October [28], see Fig. 3.8. This result suggests that the local forcing contributes to the earlier upwelling intensification and then reinforce by remote forcing from the equatorial Indian ocean.

As the response to cold SST anomaly rises in fall season, Fig. 3.9 a shows that wind stress reduced by around 4×10^{-3} to 6×10^{-3} N/m^2 along the coast where cold SST anomaly and significant SST gradient have occurred. In line with the decrease in SST along the Java southern coast, the air temperature near-surface and the height of the marine boundary layer also decreased up to 0.85 °C and 27 m, respectively. On the other hand, the spatial distribution of atmospheric parameters shows a distinctive large anomaly in the southwestern cape of Java Island, approximately 105 to 108° E. As this region is closer to the equator and as the part of an eastern pole of IOD, it is suggested that this region is also influenced by IOD-related equatorial wind anomaly, which increases due to the SST zonal gradient anomaly between the western and eastern pole of the Indian Ocean [25, 28].

3.4 Summary and Conclusions

In summer, the wind blows parallel south-easterly with the southern Java coastline, favorable for coastal upwelling. The mechanism within this upwelling system is when Ekman pumping induced by negative wind stress curl transports cold water,

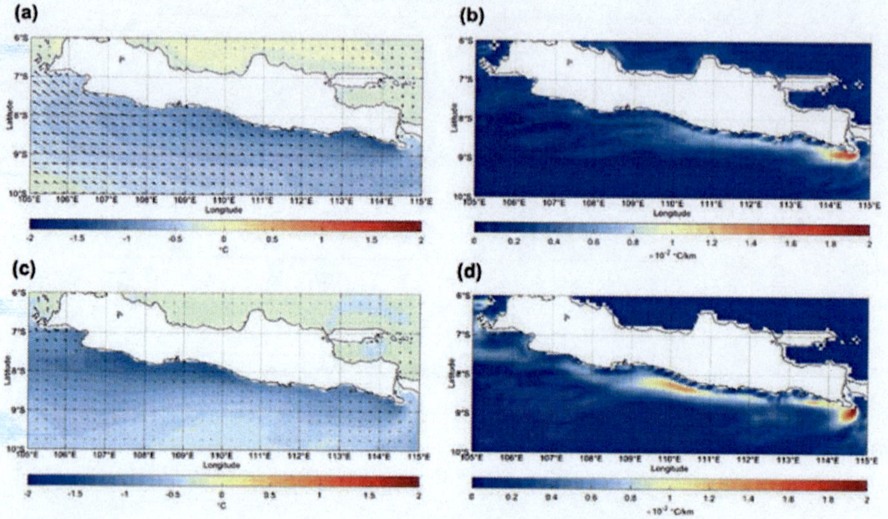

Fig. 3.8 Composite mean of summertime (June to August) **a** SST anomaly (shaded) overlaid with wind stress anomaly vector and **b** SST gradient anomaly on the southern coast of Java Island, Indonesia, during pIOD event. **c, d** are similar to (**a**) and (**b**) but in the fall season (September to November)

which peaks in July, then alternate the SST by one month lag. As the response of this mechanism, the wind stress is significantly weaker along the coast up to 80 km from the coast- line, where the cool SST tongue occurred. Moreover, near-surface air temperature and boundary layer height show similar characteristics, which reduce as low as 24 °C and up to approximately 400 m along the southern coast of Java. These atmospheric parameters increased along with the increasing SST from the coast to offshore.

During the pIOD events, the more vigorous upwelling transports cold deep-water up to 1.48 Sv to the surface in June, reducing SST anomaly up to 0.9 °C in the following season. As the response of cold SST anomaly due to the strengthened upwelling, the wind stress and the near-surface air temperature reduced up to 0.006 N/m^2 and 0.85 °C, respectively. Moreover, cold SST front during upwelling events also lowers the boundary layer height up to 27 m. On the other hand, a distinctive large anomaly of atmospheric parameters in the southwestern cape of Java Island, approximately 105–108° E, is suggested that this region is also influenced by IOD-related equatorial wind anomaly that increases due to SST zonal gradient anomaly between the western and eastern pole of Indian Ocean.

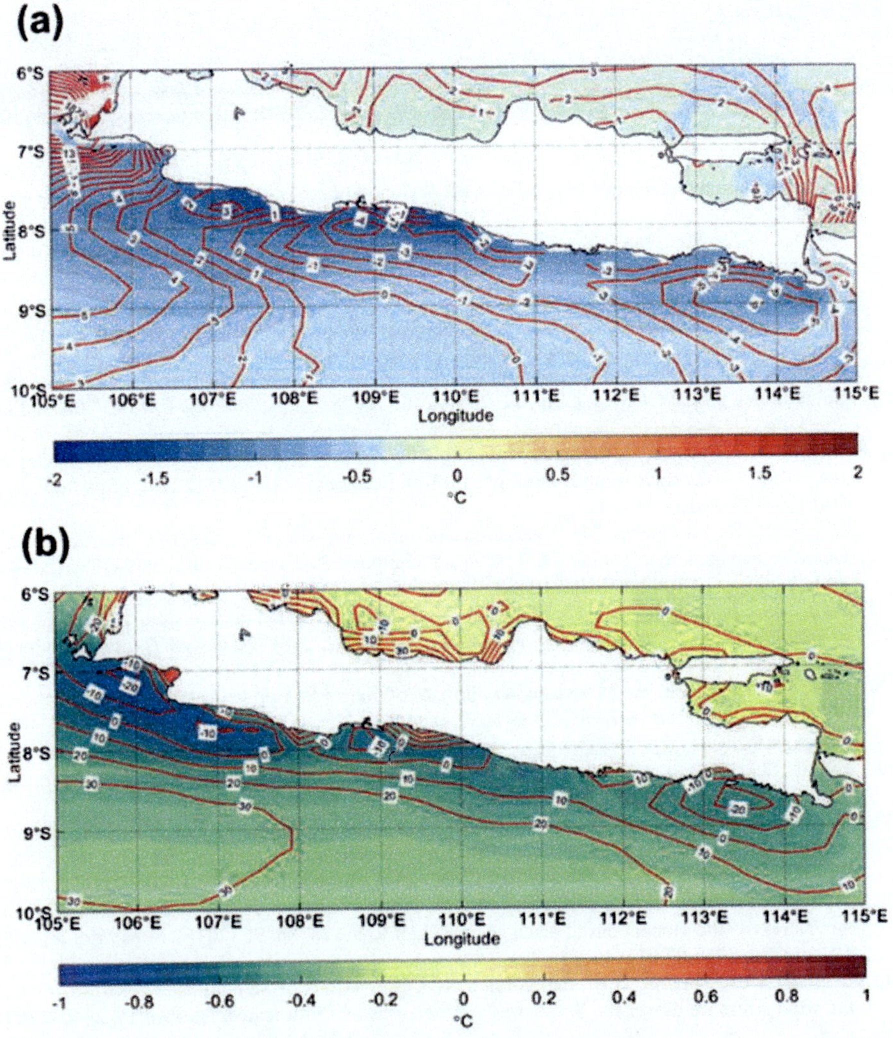

Fig. 3.9 The composite mean of **a** SST anomaly (shaded) overlaid with wind stress mag nitude anomaly (contoured, with the contour interval as 1×10^{-3} N/m^2), and **b** near-surface air temperature anomaly (shaded) overlaid with boundary layer height anomaly (con toured, with the contour interval as 10 m) in the southern coast of Java Island, Indonesia, in the fall season during the pIOD event

Acknowledgements This work was supported by the Indonesia Educational Endowment Fund (LPDP) through the Mandatory Productive Innovative Research program under National Research Priority [252/Menteri Ristek/Ka BRIN/E1/PRN/2020].

References

1. Susanto, R.D., Gordon, A.L., Zheng, Q.: Upwelling along the coasts of Java and Sumatra and its relation to ENSO. Geophys. Res. Lett. **28**, 1599–1602 (2001). https://doi.org/10.1029/2000GL011844
2. Chen, J., Chung, C.H.: Representation of global precipitation anomalies using four major climate patterns. Sci. China Technol. Sci. **58**, 927–934 (2015). https://doi.org/10.1007/s11431-015-5799-y
3. Delman, A.S., McClean, J.L., Sprintall, J., Talley, L.D., Bryan, F.O.: Process-specific contributions to anomalous java mixed layer cooling during positive IOD events. J. Geophys. Res.: Oceans **123**, 4153–4176 (2018). https://doi.org/10.1029/2017JC013749
4. Clarke AJ, Liu X (1993) Observations and dynamics of semiannual and annual sea levels near the eastern equatorial Indian ocean boundary. J. Phys. Oceanogr. 386–399
5. Du, Y., Qu, T., Meyers, G.: Interannual variability of sea surface temperature off Java and Sumatra in a global GCM. J. Clim. **21**, 2451–2465 (2008). https://doi.org/10.1175/2007JCLI1753.1
6. Chen, G., Han, W., Li, Y., Wang, D.: Interannual variability of equatorial Eastern Indian Ocean upwelling: local versus remote forcing*. J. Phys. Oceanogr. 789– 807 (2016). https://doi.org/10.1175/JPO-D-15-0117.s1
7. Nyadjro, E.S., McPhaden, M.J.: Variability of zonal currents in the eastern equatorial Indian Ocean on seasonal to interannual time scales. J. Geophys. Res: Oceans **119**, 7969–7986 (2014). https://doi.org/10.1002/2014JC010380
8. Perlin, N., Skyllingstad, E.D., Samelson, R.M., Barbour, P.L.: Numerical simulation of air-sea coupling during coastal upwelling. J. Phys. Oceanogr. **37**, 2081–2093 (2007). https://doi.org/10.1175/JPO3104.1
9. Sproson, D., Sahlée, E.: Modelling the impact of Baltic Sea upwelling on the atmospheric boundary layer. Tellus A: Dyn. Meteorol. Oceanogr. **66**, 24041 (2014). https://doi.org/10.3402/tellusa.v66.24041
10. Elliott, D.L., O'Brien, J.J.: Observational studies of the marine boundary layer over an upwelling region. Monthly Weather Rev. **105** (1977)
11. Bane, J.M., Levine, M.D., Samelson, R.M., Haines, S.M., Meaux, M.F., Perlin, N., Kosro, P.M., Boyd, T.: Atmospheric forcing of the Oregon coastal ocean during the 2001 upwelling season. J. Geophys. Res. C: Oceans **110**, 1–21 (2005). https://doi.org/10.1029/2004JC002653
12. Wang, Y., Castelao, R.M.: Variability in the coupling between sea surface temperature and wind stress in the global coastal ocean. Cont. Shelf Res. **125**, 88–96 (2016). https://doi.org/10.1016/J.CSR.2016.07.011
13. O'Neill, L.W., Chelton, D.B., Esbensen, S.K.: Observations of SST-induced perturbations of the wind stress field over the Southern Ocean on seasonal timescales. J. Clim. **16**, 2340–2354 (2003)
14. Chelton, D.B., Xie, S.P.: Coupled ocean-atmosphere interaction at oceanic mesoscales. Oceanography **23**, 54–69 (2010). https://doi.org/10.5670/oceanog.2010.05
15. O'Neill, L.W., Chelton, D.B., Esbensen, S.K.: The effects of SST-induced surface wind speed and direction gradients on midlatitude surface vorticity and divergence. J. Clim. **23**, 255–281 (2010). https://doi.org/10.1175/2009JCLI2613.1
16. Chelton, D.B., Schlax, M.G., Samelson, R.M.: Summertime coupling between sea surface temperature and wind stress in the California current system. J. Phys. Oceanogr. **37**, 495–517 (2007). https://doi.org/10.1175/JPO3025.1
17. Chen, H.H., Qi, Y., Wang, Y., Chai, F.: Seasonal variability of SST fronts and winds on the south-eastern continental shelf of Brazil. Ocean Dyn. **69**, 1387–1399 (2019). https://doi.org/10.1007/s10236-019-01310-1
18. Yu, Y., Wang, Y., Cao, L., Tang, R., Chai, F.: The ocean-atmosphere interaction over a summer upwelling system in the South China Sea. J. Mar. Syst. **208**, 103360 (2020). https://doi.org/10.1016/J.JMARSYS.2020.103360

19. Aguiar, A.L., Cirano, M., Marta-Almeida, M., Lessa, G.C., Valle-Levinson, A.: Upwelling processes along the South Equatorial Current bifurcation region and the Salvador Canyon (13°S), Brazil. Continental Shelf Res. **171**, 77–96 (2018). https://doi.org/10.1016/J.CSR.2018.10.001
20. Global Ocean Physics Reanalysis Products, E.U. Copernicus Marine Service Information. Available at:https://resources.marine.copernicus.eu/product-detail/GLOBAL_REANALYSIS_PHY_001_030. Accessed 20 Aug 2021
21. Hersbach, H., Bell, B., Berrisford, P., Biavati, G., Horányi, A., Muñoz Sabater, J., Nicolas, J., Peubey, C., Radu, R., Rozum, I., Schepers, D., Simmons, A., Soci, C., Dee, D., Thépaut, J-N.: ERA5 hourly data on single levels from 1979 to present. Copernicus Climate Change Service (C3S) Climate Data Store (CDS). Accessed 14 Apr 2021. https://doi.org/10.24381/cds.adbb2d47
22. Smith, R.L.: Upwelling. In: Oceanography and Marine Biology: An Annual Review, pp. 11–46 (1968)
23. Pickett, M.H., Paduan, J.D.: Ekman transport and pumping in the California Current based on the U.S. Navy's high-resolution atmospheric model (COAMPS). J. Geophys. Res.: Oceans **108** (2003) https://doi.org/10.1029/2003jc001902
24. Castelao, R.M., Barth, J.A.: Upwelling around Cabo Frio, Brazil: the importance of wind stress curl. Geophys. Res. Lett. **33** (2006). https://doi.org/10.1029/2005GL025182
25. Saji, N.H., Yamagata, T.: Structure of SST and surface wind variability during Indian Ocean dipole mode events: COADS observations *. J. Clim.: Am. Meteorol. Soc. **16**, 2735–2751 (2003)
26. Castelao, R.M.: Sea surface temperature and wind stress curl variability near a cape. J. Phys. Oceanogr. **42**, 2073–2087 (2012). https://doi.org/10.1175/JPO-D-11-0224.1
27. Castelao, R.M., Wang, Y.: Wind-driven variability in sea surface temperature front distribution in the California current system. J. Geophys. Res.: Oceans. **119**, 1861–1875 (2014). https://doi.org/10.1002/2013JC009531
28. Saji, N.H., Goswami, B.N., Vinayachandran, P.N., Yamagata T: A dipole mode in the tropical Indian Ocean (1999)

Chapter 4
Technological Properties of Formaldehyde Free Adhesives Based on Oxidized Starch Mixed with Different Crosslinkers for Plywood

Apri Heri Iswanto and Muhammad Adly Rahandi Lubis

Abstract Formaldehyde-free adhesive based on oxidized starch (OS) adhesives was formulated at different crosslinkers types and contents to replace urea–formaldehyde (UF) resins for plywood. Different types of crosslinker, i.e., blocked-pMDI (B-pMDI) and citric acid (CA), and levels of the crosslinkers (5.0, 7.5, 10%) were employed to tailor the adhesion performance of OS adhesives with H2O2/starch mole ratio of 0.5. Technological properties of OS adhesives such as primary properties, chemical and thermal properties were characterized using several methods. Adhesion performance of OS was evaluated in plywood panel according to the Japanese Agricultural Standard (JAS) 233:2003. FTIR spectroscopy revealed that the amide linkages were formed by reacting the OS with B-pMDI, and ester linkages were built by reacting OS with CA. These two linkages governed the adhesion performance of OS in plywood. OS/BpMDI adhesive generally had higher curing temperature values than the OS/CA. Moreover, greater content of the crosslinker accelerated the curing reaction. Tensile shear strength of plywood increased with higher crosslinker content from 0.61 MPa to 1.35 MPa, showing that crosslinker had a significant effect on enhancing the adhesion performance of OS adhesive by increasing crosslinking density and forming bigger network than without addition of crosslinker. Plywood bonded with OS/B-pMDI adhesive exhibited better adhesion than that of OS/CA bonded panel and without formaldehyde emission. Adding 10% crosslinker into OS resulted in plywood comparable to the UF bonded panel. This study suggested that OS adhesives mixed with B-pMDI can be used to produce plywood in the future.

A. H. Iswanto
Department of Forest Product, Faculty of Forestry, Universitas Sumatera Utara, Medan, North Sumatera 20155, Indonesia

JATI-Sumatran Forestry Analysis Study Center, Universitas Sumatera Utara, Medan, North Sumatera 20155, Indonesia

M. A. R. Lubis (✉)
Research Center for Biomass and Bioproducts, National Research and Innovation Agency (BRIN), Cibinong, West Java 16911, Indonesia
e-mail: marl@biomaterial.lipi.go.id

E. Yulihastin et al. (eds.), *Proceedings of the International Conference on Radioscience, Equatorial Atmospheric Science and Environment and Humanosphere Science, 2021*, Springer Proceedings in Physics 275, https://doi.org/10.1007/978-981-19-0308-3_4

4.1 Introduction

The sustainability of plywood manufacturing is depending on wood adhesives. The dominant adhesive used in plywood production is urea-formaldehyde (UF) resins, calculating around 30–40 million metric tons are utilized globally in 2020 [1]. UF resins adhesive is inexpensive, it has fast curing and has a good adhesion strength [2]. However, it requires elevate temperature to harden and to crosslink [3]. Moreover, UF resins are synthesized from formaldehyde which is volatile, hazardous and carcinogenic. This eventually can develop a sick building syndrome in indoor applications such as furniture materials [4].

Recently, the development of bio-based adhesives has become the main interest in the last decade because it is renewable, non-hazardous, and recyclable [5]. Starch has been used as a wood adhesive [6]. Still, it exhibits low water resistance, high viscosity and a slow curing of starch-based adhesives owing to the high affinity of hydroxyl groups for hydrogen bonding [7, 8]. Therefore, modification of starch is required prior its use for bonding wood.

Grafting and esterification have been attempted to modify starch, showing that an increased in water-resistance of the adhesive [9, 10]. Nevertheless, the high viscosity resulted from such modified adhesives, with viscosity over than 5000 mPa·s, restricts their use for wood adhesives. The range of viscosity used in the plywood production is considered between 100 and 500 mPa s [11, 12]. Oxidation can be used to lower the viscosity of starch to an applicable range. Oxidation reaction break down the glucosidic linkages, and alters the hydroxyl groups into carboxyl and aldehyde groups [13, 14].

To accelerate the curing of OS and to build a network structure, a crosslinker is usually mixed. Hydrogen bonding between starch molecules can enhance the properties of OS [15]. Poly-carboxylic acids, such as maleic acid and citric acid (CA), are commonly used as crosslinkers for OS [16, 17]. Among them, three carboxyl groups of CA are preferable for building ester linkages with hydroxyl groups of wood [18, 19].

In addition to the CA, a blocked polymeric methane diphenyl diisocyanate (B-pMDI) can be a crosslinker for OS owing to its high functionality [13]. The free isocyanate groups are discharged after a de-blocking reaction of B-pMDI [20]. Further, it can cross-link with aldehyde, carboxyl, or hydroxyl groups of the OS [13, 14]. The de-blocking temperature of B-pMDI should be lower than 100 °C, because the core temperature in the manufacturing of plywood during hot-pressing is between 100 and 120 °C [21]. Therefore, this work attempted to prepare a formaldehyde-free adhesive from OS and to enhance the adhesion of OS adhesives by incorporating with different types and levels of crosslinkers.

4.2 Materials and Methods

4.2.1 Materials

Corn starch with 28.5% of amylose and 18% moisture was used as starting material. H_2O_2 (30 wt%), $CuSO_4$ (99 wt%), $CH_3(CH_2)11SO_4Na$ (98 wt%), and $(NH_4)2S_2O_8$ (99 wt%) were used for the preparation of oxidized starch (OS). Polyvinyl alcohol (PVA, molecular weight: 22,000 g/mol) was added as an adhesion promoter. B-pMDI was prepared from liquid pMDI (±31% NCO), $NaHSO_4$ (58.5 wt%), and acetone (99.8 wt%) according to published methods. The B-pMDI and CA were utilized as a cross-linker for OS. UF resins with an F/U mole ratio of 1.0 and ammonium chloride (NH4Cl, 20 wt%) was used as a control adhesive [12, 21]. All chemicals were obtained from.

Merck (Darmstadt, Germany) and Sigma Aldrich (Missouri, United States). Sengon (*Paraserianthes falcataria* (L.) Nielsen) wood veneers, sizing 250 × 250 × 2 mm, was bought from local market (Cibinong, West Java, Indonesia) and was used to produce plywood.

4.2.2 Preparation of Oxidized Starch Adhesives

Briefly, around 100 g of starch and 150 mL of reverse osmosis water were homogenized under sonication for 30 min. The gelatinization of starch was done at temperature of 80 °C for 1 h. Around 0.5% of $CuSO_4$, 0.5% of (NH4)2S2O8, and 0.25% of CH3(CH2)11SO4Na were added to the gelatinized starch when the temperature reaching 50 °C. After that, 35.0 g of H2O2 was drop to prepare a 0.5 H2O2/starch mole ratio of OS. The temperature was maintained at 50 °C and the pH was controlled at 3 for 5 h under 500 rpm of agitation to complete the oxidation reaction.

As presented in Table 4.1, around 15% of PVA was added as an adhesion promoter. The mixture then was homogenized at 80 °C for 2 h. The OS adhesive was mixed with B-pMDI and CA at 5, 7.5 and 10%. The OS/B-pMDI and OS/CA mixtures then

Table 4.1 Formulation of oxidized starch adhesive mixed with B-pMDI and CA at different levels

Type of crosslinker	Level (%)	OS (g)	PVA (g)	Amount of crosslinker (g)
B-pMDI	5[a]	100.0	6.8	2.3
	7.5	100.0	6.8	3.4
	10	100.0	6.8	4.5
CA	5	100.0	6.8	2.3
	7.5	100.0	6.8	3.4
	10	100.0	6.8	4.5

[a]Based on the solids content of OS

were stirred at 80 °C for 10 min under 300 rpm of stirring. As a comparison, 1.0 F/U mole ratio UF resin with 61.2% of solids content, 240 mPa·s of viscosity, and 179 s of gel time was prepared accordingly [12, 21].

4.2.3 Characterization of Oxidized Starch Adhesives

The solids content of OS mixed with different types and levels of crosslinker was determined gravimetrically. Approximately 1 g of OS adhesive was dried for 3 h at 105 °C. Further, a gel time meter was used to determine the gel time of OS adhesive at different types and levels of crosslinker. The sample was heated on a hot-plate and the gel time was measured using gel time meter (Techne GT-6, Coleparmer, United States) at 100 °C. The viscosity of OS adhesive was measured at 25 °C using a spindle No. 27 at 60 rpm using a rheometer (Rheolab QC, Anton Paar, Austria). Prior to pH measurement of OS adhesive, the digital pH meter (Orion Star™ A211, Thermo Scientific, United States) was calibrated using a buffer solution having pH 4.0 and 10.0. Around 25 g of OS was used for the pH measurement.

An UATR–FTIR spectrometer (Spectrum Two, Perkin Elmer, United States) was used to observe the alteration in functional groups of OS mixed with different types and levels of crosslinker from 400 to 4000 cm^{-1}. Further, the crystal structure of the native starch and OS was determined using an X-ray diffractometer (XRD, Rigaku Miniflex II, Japan) with a CuKα radiation source of 0.15406 nm. The XRD patterns were obtained by scanning the samples in the range of 0°–50°. A differential scanning calorimetry (DSC 4000, Perkin Elmer, United States) was performed to analyse the peak temperatures (*Tp*) of the OS adhesives. Freeze-dried OS adhesives was heated from 30 to 200 °C at a heating rate of 10 °C/min under a nitrogen flow.

4.2.4 Plywood Manufacturing

Three-layer plywood panel, sizing 250 mm × 250 mm × 6 mm, were prepared using Sengon wood veneer. The OS adhesives was spread at a glue spread of 180 g/m^2 using double glue spread technique. Panel was pre-pressed for 20 min under 0.8 MPa of pressure at room temperature, and the panel was subsequently pressed at 120 °C for 10 min using same pressure. The panels then were kept at 27 ± 2 °C for 2 weeks before cutting and testing. Plywood bonded with UF resin adhesive was also prepared as a control [22].

4.2.5 Examination of Plywood Properties

Adhesion strength of OS bonded plywood was determined by means of tensile shear strength (TSS) according to JAS No. 233 [22]. Approximately twelve specimens sizing 25 × 80 × 6 mm were cut for TSS testing. Dry and wet TSS of plywood was examined using a UTM (50kN, Shimadzu, Japan) at a 2 mm/min of crosshead speed.

The FE values were examined by keeping ten samples sizing 150 × 50 × 6 mm in a desiccator for 24 h at 20 ± 2 °C. After that, around 25 mL of the solution was mixed with 25 mL of standard solution [22]. The mixtures were warmed for 10 min in a water bath at 65 ± 2 °C, and then were cooled to 25 ± 2 °C. UV–visible spectrophotometry (UV-1800, Shimadzu, Japan) was used to measure the FE at a wavelength of 412 nm.

4.3 Results and Discussion

4.3.1 Characteristics of Oxidized Starch Adhesive

Figure 4.1a reveals the presence of carbonyl of carboxyl and aldehyde groups in the OS at 1715 cm^{-1}, showing that starch has been de-polymerized and the oxidation alters the OH groups of the glucose unit to carboxyl and aldehyde groups using H2O2. Further, typical peaks at 3310, 2930, and 1640 cm^{-1} in both native starch and OS were attributed to OH, CH, and intramolecular OH bonds of starch polymer. There results are in agreement with published works [13, 14].

The XRD patterns revealed the change in the crystal structure of native starch after oxidation (Fig. 4.1b). The native starch had five typical peaks at diffraction angles (2θ) of 3.5°, 15.0°, 17.9°, 19.9°, and 22.8°, revealing an A-type structure of native starch [23]. The starch structure altered after the oxidation with H2O2,

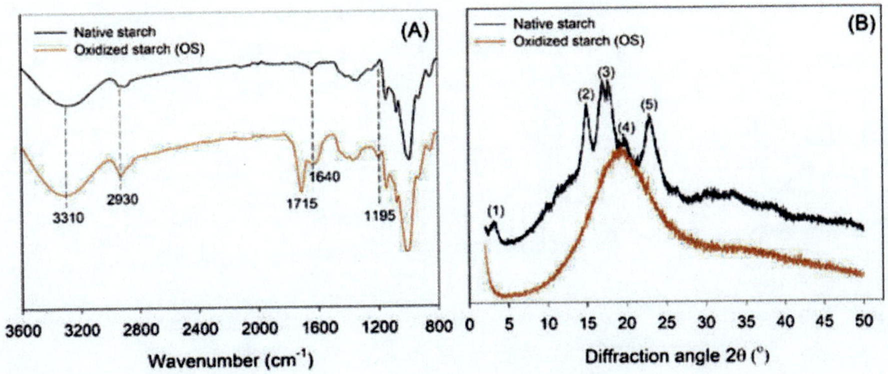

Fig. 4.1 Chemical properties of native and oxidized starch. **a** FTIR and **b** XRD

$$H_2O_2 + Cu^{2+} \rightarrow \cdot OH + OH^- + Cu^{3+}$$

$$\text{Starch-}CH_2OH + \cdot OH \rightarrow \text{Starch}\cdot CHOH + H_2O$$

$$\text{Starch}\cdot CHOH + Cu^{3+} \rightarrow \text{Starch-}CHO + H^+ + Cu^{2+}$$

$$\text{Starch}\cdot CHOH + H_2O_2 \rightarrow \text{Starch-}CHO + \cdot OH + H_2O$$

$$\text{Starch-}CHO + H_2O_2 + Cu^{2+} \rightarrow \text{Starch-}COOH + OH^- + Cu^{3+}$$

Scheme 4.1 Illustration of starch oxidation using H_2O_2 [24]

as revealed by XRD. The oxidation is breaking down the crystal structure of the starch by destroying the glucosidic linkages [24]. Smaller starch molecules with more amorphous structures are usually produced. As illustrated in Scheme 4.1 [24], two OH radicals were produced from the H2O2 by Cu^{2+} catalyst. The OH of the starch was then oxidized by radical OH groups and catalyst into aldehyde groups and released H_2O. Finally, the aldehyde groups were altered into carboxyl groups by the remaining H_2O_2.

The incorporation of crosslinkers into OS polymer alters the functional groups the OS-based adhesive. Mixing of OS with B-pMDI or CA created a new peak at 3258–3265 cm^{-1} (Fig. 4.2). This peak is attributed to the hydrogen bond between C = O of –NCO groups of B-pMDI or –COOH of CA due to the phase change from liquid to gel and the diffusion of OS molecules toward each other [25, 26]. B-pMDI was detected in the OS by the formation of C=C of aromatic pMDI at 1586 cm^{-1}. The intensity of C=O of –CHO and –COOH groups at 1720 cm^{-1} altered with increasing the amount of cross-linkers. Peaks at 2925 cm^{-1} corresponded to C–H of OS/B-pMDI and OS/CA. The peak of intermolecular hydrogen bonds in OS adhesive was found at 3300 cm^{-1}. The band assignments were in accordance to published works [13, 14].

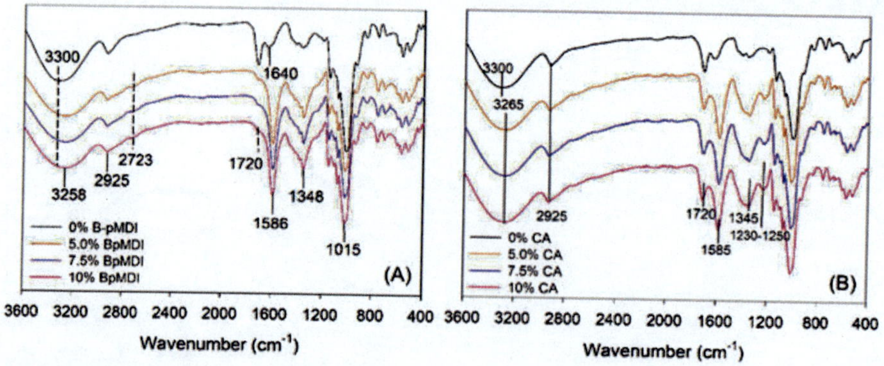

Fig. 4.2 FTIR spectra of oxidized starch mixed with crosslinker. **a** B-pMDI, **b** CA

4.3.2 *Oxidized Starch Adhesive Properties*

Pure OS adhesive with no addition of crosslinker has solids content of around 48.4%, a viscosity of 107.7 mPa s, a gel time of 582.5 s, and pH of 3.2 (Table 4.2). Incorporating of crosslinker generally resulted in higher solids content and viscosity, and lower the gel time compared to the pure OS. In particular, incorporation of B-pMDI in OS adhesive resulted in higher solids content and viscosity than the CA. This is probably because that the B-pMDI is a water dispersible type crosslinker [21], while CA is a water-soluble one [27]. Further, the gel time of OS/B-pMDI was lower compared to the OS/CA. The OS/B-pMDI was more viscous than the OS/CA, therefore resulting in a phase change from liquid to gel and diffusion of OS/B-pMDI molecules toward each other [25, 26]. As a comparison, the properties of UF resin adhesive was characterized. The OS adhesives mixed with different types and levels of crosslinker has lower solids content, viscosity and pH compared to UF resins. Thus, the gel time of OS adhesives are longer compared to the conventional UF resin. These findings were in accordance with the literature [13, 14].

Thermal behaviour of OS adhesives mixed with different types and levels of crosslinker is presented in Fig. 4.3. DSC analysis revealed two exothermic peak in OS adhesive mixed with different types and levels of crosslinker. The first exothermic

Table 4.2 Solids content, viscosity, gel time and pH of oxidized starch adhesive mixed with different crosslinkers

Properties	B-pMDI content (%)				CA content (%)			
	0	5	7.5	10	5	7.5	10	
Solids content (%)	48.4	50.1	51.9	53.2	49.7	50.8	51.9	61.2
Viscosity (mPa s)	107.7	197.5	215.9	238.5	115.2	122.6	135.9	240.0
Gel time (s)	582.5	547.2	525.8	512.3	580.5	573.7	565.8	179.0
pH	3.2	3.5	3.8	4.2	3.2	3.1	3.0	8.2

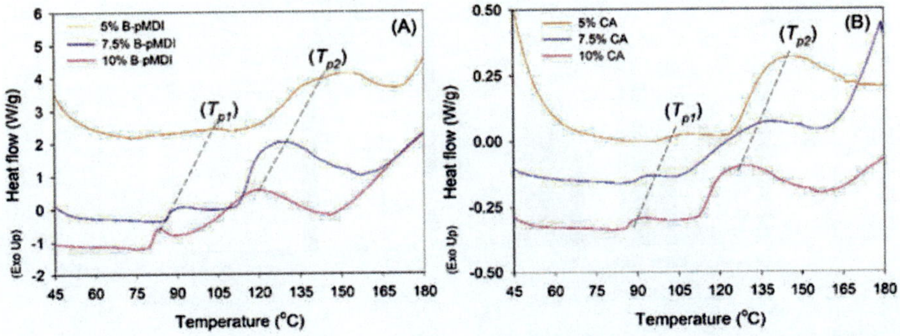

Fig. 4.3 DSC thermograms of oxidized starch adhesive with different crosslinkers. **a** OS/B-pMDI, and **b** OS/CA

peak (*Tp*1) at low temperature might be due to the alteration of the adhesive from liquid to gel due to the diffusion of molecules toward each other [25, 26]. The second exothermic peak (*Tp*2) at greater temperatures probably due to the cross-linking between either OS/B-pMDI or OS/CA. Incorporation of B-pMDI into OS resulted in lower *Tp*1 and *Tp*2 values than that of OS/CA. Further, increasing the content of crosslinker resulted in lower *Tp*1 and *Tp*2 values for both OS adhesives. At 5% content of crosslinker, both adhesives had *Tp*1 of approximately 103.5 °C. At 7.5% content, OS/B-pMDI had *Tp*1 of around 90.3 °C while OS/CA had about 92.7 °C. At 10% content, OS/B-pMDI had *Tp*1 of about 79.8 °C while OS/CA had about 90.1 °C. Similar trend has been found for *Tp*2, ranging from 120.3 to 145.6 °C for OS/B-pMDI and 127.4 to 146.2 °C for OS/CA. The core temperature of plywood is around 100–120 °C [21], therefore the OS adhesives mixed with B-pMDI and CA at different levels are compatible for plywood manufacturing.

4.3.3 *Performance of Plywood*

Plywood properties had been evaluated at different types and contents of crosslinker in the OS (Fig. 4.4). As a control, plywood panels bonded with pure OS had the TSS value of around 0.61 MPa, which did not fulfill the JAS standard No. 233 for plywood (>0.70 MPa) [22]. The TSS values were getting higher with greater level of crosslinker. The plywood properties met the minimum JAS standard of plywood. This result indicated that the incorporation of crosslinker enhanced the adhesion performance of the OS adhesives by forming a rigid network and increasing the cross-linking density than that of pure OS [13]. Plywood bonded with OS/B-pMDI adhesive had TSS value ranging from 0.95 to 1.35 MPa, while plywood bonded with OS/CA adhesive had TSS value from 0.92 to 0.98 MPa. This showed that OS/B-pMDI produced stronger plywood than that of OS/CA. This result showed that plywood

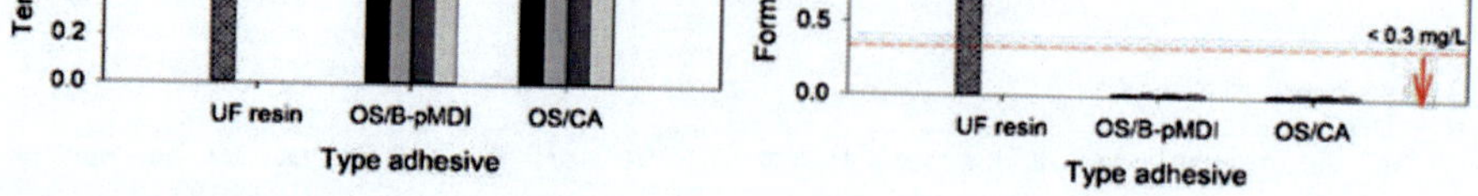

Fig. 4.4 Plywood properties bonded with OS adhesives. **a** Tensile shear strength, (B) formaldehyde emission

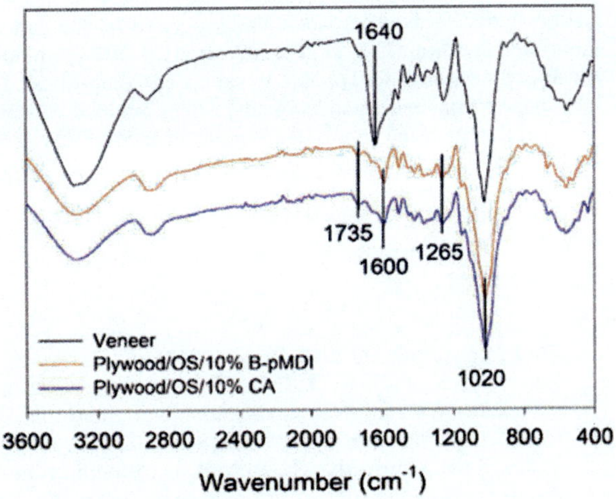

Fig. 4.5 FTIR spectra of wood veneer and plywood bonded with OS adhesive

bonded with both OS adhesives had comparable TSS value to UF bonded plywood panel.

The FE values of both OS/B-pMDI and OS/CA bonded plywood were around 0.01 mg/L. This result was hundred times lower than that of UF bonded plywood, which emitted around 1.9 mg/L of FE (Fig. 4.4b). This is obviously owing to that no formaldehyde was added in the formulation of the OS adhesives. Very low FE emitted from the OS bonded plywood might be originated from wood, called biogenic formaldehyde [28, 29]. This result indicated that OS can be utilized as an alternative to UF resins in the manufacture of plywood without emitted formaldehyde.

The change in functional groups of plywood bonded with OS adhesives was observed using FTIR spectroscopy (Fig. 4.5). The spectra revealed that ester linkages at 1730 cm^{-1} appeared in either OS/B-pMDI and OS/CA bonded plywood. This peak was not detected in the wood veneer. The adhesion of OS adhesives in the plywood might be enhanced by the formation of ester linkages, which was in agreement with published works [18, 19].

4.4 Summary and Conclusions

Technological properties of OS adhesives have been tuned with different types and levels of crosslinker to prepare plywood panels without formaldehyde emission. Amide linkages were formed in OS/B-pMDI, while the OS/CA formed ester linkages. The OS/B-pMDI bonded plywood produced better properties OS/CA bonded panel. Incorporating10% crosslinker into OS resulted in plywood with comparable properties to UF bonded panel. This study suggested that OS adhesives mixed with B-pMDI can be used to produce plywood without formaldehyde emission in the future.

Acknowledgements This work was supported by the Japan-ASEAN Science and Technology Innovation Platform (JASTIP) Fiscal Year 2020–2021 entitled 'Development of High Performance Bio-Based Adhesives for Durable Wood Based Composites". Authors are also grateful for the facilities support from integrated laboratory of bio-products (iLAB), Research Center for Biomaterials, National Research and Innovation Agency of Indonesia.

References

1. FAO: Forest Products 2019. Rome (2019)
2. Lubis, M.A.R., Park, B.-D.: Enhancing the performance of low molar ratio urea–formaldehyde resin adhesives via in-situ modification with intercalated nanoclay. J. Adhes. **97**, 1271–1290 (2021). https://doi.org/10.1080/00218464.2020.1753515
3. Yadav, S.M., Lubis, M.A.R., Park, B.-D.: Modification of nanoclay with different methods and its application in urea-formaldehyde bonded plywood panels. Wood Mater. Sci. Eng. 1–10 (2021). https://doi.org/10.1080/17480272.2021.1934897
4. Lubis, M.A.R., Park, B.D., Lee, S.M.: Microencapsulation of polymeric isocyanate for the modification of urea-formaldehyde resins. Int. J. Adhes. Adhes. **100**, 102599 (2020). https://doi.org/10.1016/j.ijadhadh.2020.102599
5. Aristri, M.A., Lubis, M.A.R., Yadav, S.M., Antov, P., Papadopoulos, A.N., Pizzi, A., Fatriasari, W., Ismayati, M., Iswanto, A.H.: Recent developments in lignin- and tannin-based non-isocyanate polyurethane resins for wood adhesives—A review. Appl. Sci. **11**, 4242 (2021). https://doi.org/10.3390/app11094242
6. Radley, J.A.: Adhesives from starch and dextrin. In: Industrial Uses of Starch and its Derivatives. pp. 1–50. Springer, Dordrecht (1976)
7. Arias, A., Feijoo, G., Moreira, M.T.: Evaluation of starch as an environmental-friendly bioresource for the development of wood bioadhesives. Molecules. 26 (2021). https://doi.org/10.3390/molecules26154526
8. Din, Z. ud, Chen, L., Xiong, H., Wang, Z., Ullah, I., Lei, W., Shi, D., Alam, M., Ullah, H., Khan, S.A.: Starch: an undisputed potential candidate and sustainable resource for the development of wood adhesive. Starch/Staerke **72** (2020). https://doi.org/10.1002/star.201900276
9. Selamat, M.E., Sulaiman, O., Hashim, R., Hiziroglu, S., Nadhari, W.N.A.W., Sulaiman, N.S., Razali, M.Z.: Measurement of some particleboard properties bonded with modified carboxymethyl starch of oil palm trunk. Meas. J. Int. Meas. Confed. **53**, 251–259 (2014). https://doi.org/10.1016/j.measurement.2014.04.001
10. Zia-ud-Din, Chen, L., Ullah, I., Wang, P.K., Javaid, A.B., Hu, C., Zhang, M., Ahamd, I., Xiong, H., Wang, Z.: Synthesis and characterization of starch-g-poly(vinyl acetate- co-butyl acrylate) bio-based adhesive for wood application. Int. J. Biol. Macromol. **114**, 1186–1193 (2018). https://doi.org/10.1016/j.ijbiomac.2018.03.178
11. Lubis, M.A.R., Yadav, S.M., Park, B.D.: Modification of Oxidized starch polymer with nanoclay for enhanced adhesion and free formaldehyde emission of plywood. J. Polym. Environ. **29**, 2993–3003 (2021). https://doi.org/10.1007/s10924-021-02101-w
12. Lubis, M.A.R., Park, B.-D.: Modification of urea-formaldehyde resin adhesives with oxidized starch using blocked pMDI for plywood. J. Adhes. Sci. Technol. **32**, 2667–2681 (2018). https://doi.org/10.1080/01694243.2018.1511075
13. Lubis, M.A.R., Park, B.D., Hong, M.K.: Tuning of adhesion and disintegration of oxidized starch adhesives for the recycling of medium density fiberboard. BioResources **15**, 5156–5178 (2020). https://doi.org/10.15376/biores.15.3.5156-5178
14. Lubis, M.A.R., Park, B., Hong, M.: Tailoring of oxidized starch's adhesion using crosslinker and adhesion promotor for the recycling of fiberboards. J. Appl. Polym. Sci. **136**, 47966 (2019). https://doi.org/10.1002/app.47966

15. Sridach, W., Jonjankiat, S., Wittaya, T.: Effect of citric acid, PVOH, and starch ratio on the properties of cross-linked poly(vinyl alcohol)/starch adhesives. J. Adhes. Sci. Technol. **27**, 1727–1738 (2013). https://doi.org/10.1080/01694243.2012.753394
16. Ghosh Dastidar, T., Netravali, A.: Cross-linked waxy maize starch-based "green" composites. ACS Sustain. Chem. Eng. **1**, 1537–1544 (2013). https://doi.org/10.1021/sc400113a
17. Ghosh Dastidar, T., Netravali, A.N.: "Green" crosslinking of native starches with malonic acid and their properties. Carbohyd. Polym. **90**, 1620–1628 (2012). https://doi.org/10.1016/j.carbpol.2012.07.041
18. Widyorini, R., Nugraha, P.A., Rahman, M.Z.A., Prayitno, T.A.: Bonding ability of a new adhesive composed of citric acid-sucrose for particleboard. BioResources **11**, 4526–4535 (2016). https://doi.org/10.15376/biores.11.2.4526-4535
19. Del Menezzi, C., Amirou, S., Pizzi, A., Xi, X., Delmotte, L.: Reactions with wood carbohydrates and lignin of citric acid as a bond promoter of wood veneer panels. Polymers (Basel) **10**, 833 (2018). https://doi.org/10.3390/polym10080833
20. Lubis, M.A.R., Park, B.-D., Lee, S.-M.: Performance of hybrid adhesives of blocked-pMDI/melamine-urea-formaldehyde resins for the surface lamination on plywood. J. Korean Wood Sci. Technol. **47**, 200–209 (2019). https://doi.org/10.5658/WOOD.2019.47.2.200
21. Lubis, M.A.R., Park, B.-D., Lee, S.-M.: Modification of urea-formaldehyde resin adhesives with blocked isocyanates using sodium bisulfite. Int. J. Adhes. Adhes. **73**, 118–124 (2017). https://doi.org/10.1016/j.ijadhadh.2016.12.001
22. Japanese Agricultural Standard (JAS) for Plywood: Japanese Agricultural Standard for Plywood (2003)
23. Buléon, A., Colonna, P., Planchot, V., Ball, S.: Starch granules: Structure and biosynthesis. Int. J. Biol. Macromol. **23**, 85–112 (1998). https://doi.org/10.1016/S0141-8130(98)00040-3
24. Parovuori, P., Hamunen, A., Forssell, P., Autio, K., Poutanen, K.: Oxidation of potato starch by hydrogen peroxide. Starch/Stärke. **47**, 19–23 (1995). https://doi.org/10.1002/star.19950470106
25. Zhang, S.D., Wang, X.L., Zhang, Y.R., Yang, K.K., Wang, Y.Z.: Preparation of a new dialdehyde starch derivative and investigation of its thermoplastic properties. J. Polym. Res. **17**, 439–446 (2010). https://doi.org/10.1007/s10965-009-9330-7
26. Upadhyay, S.: Chemical Kinetics and Reaction Dynamics. pp. 1–265. Anamaya Publisher, Springer. New York (2006)
27. Sutiawan, J., Hermawan, D., Massijaya, M.Y., Kusumah, S.S., Lubis, M.A.R., Marlina, R., Purnomo, D., Sulastiningsih, I.M.: Influence of different hot-pressing conditions on the performance of eco-friendly jabon plywood bonded with citric acid adhesive. Wood Mater. Sci. Eng. 1–10 (2021). https://doi.org/10.1080/17480272.2021.1884898
28. Tasooji, M., Wan, G., Lewis, G., Wise, H., Frazier, C.E.: Biogenic formaldehyde: content and heat generation in the wood of three tree species. ACS Sustain. Chem. Eng. **5**, 4243–4248 (2017). https://doi.org/10.1021/acssuschemeng.7b00240
29. Tasooji, M., Frazier, C.E.: Simple milligram-scale extraction of formaldehyde from wood. ACS Sustain. Chem. Eng. **4**, 5041–5045 (2016). https://doi.org/10.1021/acssuschemeng.6b01477

Chapter 5
Characteristic of Precipitable Water Vapor in 2017–2019 Over Bandung Estimated by GPS Data

Atep Radiana, Saipul Hamdi, Arif Aditiya, Fahmi Rahmatia, and Sjafrijon

Abstract In this paper, the average amount of water vapor in the atmosphere (PWV) from 2017 to 2019 has been calculated which, is derived from the GNSS data constellation using goGPS free software. The average PWV values for 2017, 2018, and 2019 were 38.9 mm, 34.7 mm, and 36.1 mm; respectively, and generally ranged between 11.4 and 56.3 mm. There is also a negative correlation between PWV and global solar energy of $r = -0.62$ and a gradient descent of -6.40 Ly/mm. The PWV daily profile shows that the smallest PWV occurs at midnight and the largest PWV occurs at the noon. The PWV value used is the average value calculated in the time range from 11 to 13 LT (UTC + 7).

5.1 Introduction

Water vapor plays a vital role in maintaining the life of living things on this planet earth it accounts for about 7% of the overall composition of the layers of the earth's atmosphere. One of the crucial functions of atmospheric water vapor is a storage place for latent energy that will support the environment if the water vapor condenses. This latent energy plays a role in atmospheric circulation, and becomes a source of energy in the occurrence of tropical disturbances such as tropical storms and tropical cyclones. Water vapor also affects the adiabatic heating and cooling process, so the vertical distribution of water vapor is important in maintaining atmospheric stability [14].

A. Radiana (✉) · S. Hamdi
Research Center for Climate and Atmosphere, National Research and Innovation Agency (BRIN), Bandung, Indonesia
e-mail: den_radiana@yahoo.co.id

A. Aditiya
Geodesy and Geodynamic Control Net Center, Bogor, Indonesia

F. Rahmatia · Sjafrijon
BPAA Agam, National Research and Innovation Agency (BRIN), Bukittinggi, Indonesia

E. Yulihastin et al. (eds.), *Proceedings of the International Conference on Radioscience, Equatorial Atmospheric Science and Environment and Humanosphere Science, 2021*, Springer Proceedings in Physics 275, https://doi.org/10.1007/978-981-19-0308-3_5

Precipitable Water Vapor (PWV) is the amount of water vapor content in the atmosphere when it is condensed in a specific column area at a specific time. Some of the factors controlling the daily cycle of PWV include broad-scale vertical movement of the atmosphere, the convergence of low-level humidity and its precipitation, and surface evapotranspiration. In weather forecasting, the results are strongly influenced by water vapor in the atmosphere because water vapor has a high variability over space and time. Changes in the composition of water vapor in the atmosphere will cause local weather changes.

Some of the techniques used to calculate PWV are radiosonde [11], microwave radiometer, lidar system [10], and GPS satellites [13]. Each technique has advantages and disadvantages and depends on the era. Radiosonde directly measures the water vapor content in the atmosphere at several altitudes. GPS satellites calculates the amount of water vapor in the atmosphere as compensation for refracting of the electromagnetic waves used. The lidar system calculates the amount of water vapor in the atmosphere based on the backscatter caused by the water vapor molecules.

In the Global Satellite Navigation System (GNSS), the accuracy will be determined by the tropospheric delay, which is very dependent on the amount of water vapor in the atmosphere. In a study in Egypt using GNSS data, Mousa et al. [12] reported a strong correlation between the zenith wet delay (ZWD) and PWV. The strong correlation indicates that ZWD is highly dependent on PWV. In this paper, we will calculate the total of water vapor (PWV) estimated by using the GNSS satellite and calculate the intensity of global solar radiation from 2017 to 2019.

5.2 Data and Method

In this paper, the PWV value is calculated from the Global Navigation Satellite System (GNSS) data collected by the Geospatial Information Agency (BIG) through a receiving system operated in Lembang. This data has been equipped with meteorology parameters in a separate file and processed using goGPS software, which is freely downloaded from the http://www.gogps-project.org page [7]. The PWV values used in this paper are the average values calculated over a 3-h time interval, i.e., between 11:00 and 13:00 WIB.

In the goGPS software, PWV is calculated according to the equation derived by Bevis et al. [2], namely

$$\mathrm{PWV} = \Pi \times \mathrm{ZWD} \tag{5.1}$$

ZTD or zenith total delay has two components, namely zenith wet delay (ZWD) and zenith hydrostatic delay (ZHD), so that ZWD can be defined as

$$\mathrm{ZWD} = \mathrm{ZTD} - \mathrm{ZHD} \tag{5.2}$$

Π is a constant issued by Askne and Nordeus [1] and is formulated in Eq. (5.3)

$$\rho\Pi = \frac{10^6}{\rho R_v\left[(k_3/T_m) + k_2'\right]} \tag{5.3}$$

ρ is the density of liquid water while R_v is the specific gas constant for water vapor and T_m is defined by Davis et al. [3] through Eq. (5.4)

$$T_m = (T + 273.15) \times 0.72 + 70.2 \tag{5.4}$$

The values of $k_2' = 22.1$ and $k_3 = 3.739$ are used, and have given satisfactory results in calculating PWV using GPS data [17].

The goGPS software is freeware and implements the Karman filter in its data processing. The file naming system used follows the RINEX file standard and has two data modes, namely stand-alone positioning with LSA, and relative positioning with EKF. Both are post-processing and real-time [7]. To support the analysis data on the global solar radiation power is used, which is recalculated into global energy. The power of global solar radiation was measured using an MS-802 pyranometer made by Eko Instruments (Japan) and recorded by the HIOKI 8400–20 data logger with a recording time interval of 10 min. The pyranometer is operated at the PSTA-LAPAN Bandung office (6°53′35.52″ S 107°35′12.16″ E), and calibrated every 3 years (Eko Instruments, Japan). The pyranometer pass solar radiation wavelength 285–3000 nm. At a meanwhile the GNSS receiver is operated in Lembang (6° 49′ 27.97″ S 107° 36′ 56.16′ E), and the distance between two locations is 17.2 km, Lembang is located in the north of the Bandung. The altitude of Lembang is 1329 m msl while the Bandung is located in a basin with an altitude of 740 m above sea level. GNSS equipment was calibrated by the manufacture and has been certified. The data is showed comply with standard for geodetic equipments.

5.3 Result and Discussion

The results of data processing using the goGPS software are freeware, illustrated in Fig. 5.1, namely the PWV values estimated using GNSS for three years 2017(1a), 2018(1b), and 2019(1c), while Fig. 5.2 is, Global Radiation Energy of the sun on the surface with the same year. Measurement data of global solar radiation intensity was measured in Bandung simultaneously using the MS-802 pyranometer. The horizontal axis is the Julian Day or Day Number.

Figure 5.1, is the PWV values estimated using GNSS for three years 2017(1a), 2018(1b), and 2019(1c).

The average PWV calculation results in 2017(1a) were 38.9 mm with 331 days of measured data, and 2018(1b) of 34.7 mm with 329 days of measured data, while in 2019(1c) of 36.1 mm with The measurement time is 362 data days, for 3 years the total water vapor in the atmosphere that has condensed in the column area is 1022 days (Table 5.1).

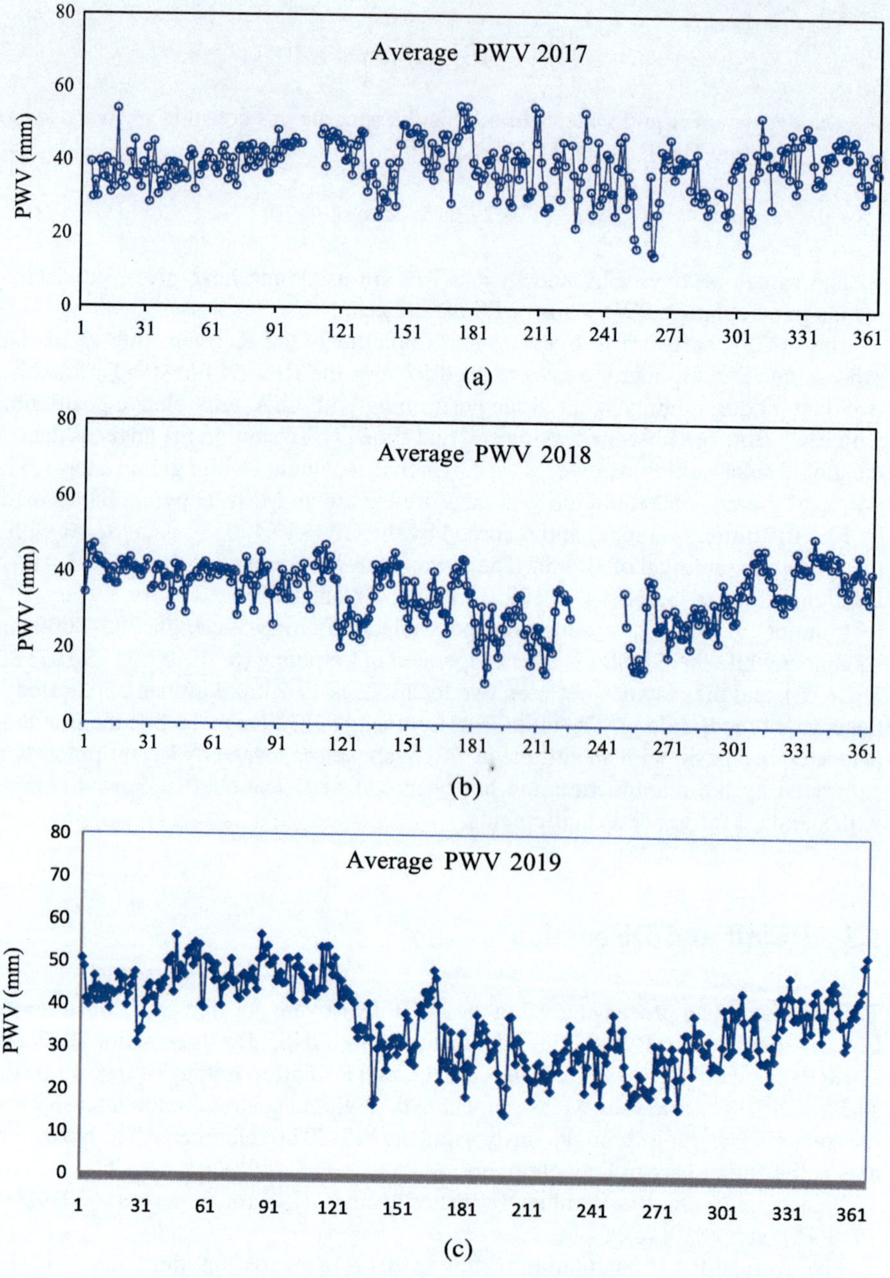

Fig. 5.1 **a** Graph of the average PWV in 2017, which is 38.9 mm. **b** Graph of the average PWV in 2018, which is 34.7 mm. **c** Graph of the average PWV in 2019, which is 36.1 mm

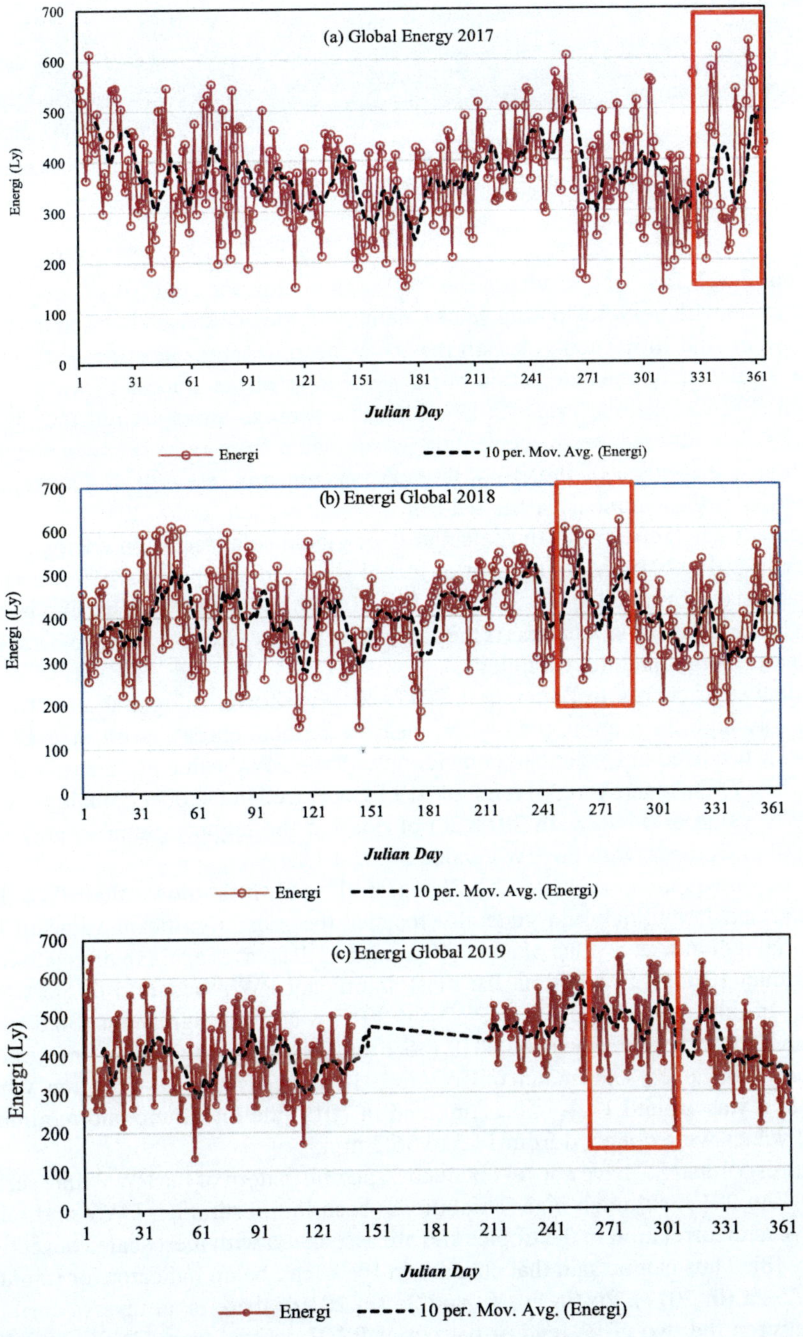

Fig. 5.2 **a** The highest of the calculation global solar energy 2017 occurred in December. **b** The highest global solar energy calculation results occur in October. **c** The highest global solar energy calculation results occur in October

Table 5.1 Statistics of PWV. Observations

Description	2017	2018	2019
Number of data days	331 days	329 days	362 days
Average annual PWV	38.8 mm	34.7 mm	36.1 mm
Standard deviation	7.0 mm	7.7 mm	9.6 mm
Highest PWV	55.4 mm	50.2 mm	56.3 mm
Lowest PWV	14.9 mm	11.4 mm	14.3 mm

From Fig. 5.1a–c, the results of the PWV calculation for three years occur in an actual cycle with very fluctuating values, namely in the dry season marked by a low PWV value and in the rainy season the PWV becomes high. For the distribution of PWV value data, the calculation of the percentage of the amount of water vapor content (PWV) for three years (2017–2019), the average interaction of the sample is 22.3%. The interaction of the sample is calculated from ratio between standard deviation and average of the PWV in each year. In this research, the average of interaction of the sample is calculated from 3 years of data (2017–2019).

Figure 5.2 is the result of the calculation of global solar radiation energy on the surface in 2017–2019. Measurement of global solar radiation intensity in Bandung was carried out simultaneously using the MS-802 pyranometer made by Eko Instrument Co. Ltd. The horizontal axis is the Julian Day or Day Number, while the vertical axis is the energy quantity (Langley).

Global solar energy in Bandung in 2017–2019 is shown in Fig. 5.2 and offers a reasonably high fluctuation, the highest measured global energy result in 2017(2a) of 634 Ly occurred in December, coinciding with the PWV value of 38.8 mm. Then the highest global energy in 2018(2b) of 619 Ly occurred in October coinciding with the PWV value of 34.7 Ly, in 2019(2c) of 648 Ly, the highest global energy that occurred in October, with the PWV value of 36, 1 Ly.

Based on research conducted by Zhang et al. [18], it is known that PWV has seasonal and latitudinal, and generally reaches the most significant value in the equatorial region with a value of about 40 mm, then decreases again in high latitudes to less than 10 mm. In addition, the most significant PWV occurred in the warm season. In this study, the average PWV in 2017 was most significant (38.8 mm) compared to 2018 (34.7 mm) and 2019 (36.1 mm), as well as the annual maximum and minimum values. The amount of PWV in 2017 ranged from 14.9–55.4 mm, while in 2018, it was around 11.4 – 50.2 mm, and in 2019, the maximum and minimum annual values were obtained from 14.3 to 56.3 mm.

Figures 5.1 and 5.2 have not yet obtained a specific pattern of the PWV time series for 2017 to 2019, although high variability is seen in the summer. PWV and rainfall have autocorrelation in the tropics and are associated with mesoscale convective activity [8]. Thus can be said that changes in PWV can be an indicator for rainfall. In Fig. 5.2a (in 2017), 2b (in 2018), and 2c (in 2019), there is an inverse correlation between the two. The trend or pattern of PWV derived from GNSS is highly dependent on the selection of the observed period [16]. In the study using the 20-year PWV time series, the opposite linearity could be clearly distinguished in the first ten years and the second ten years at several observation stations. Calculation of PWV in Bandung in 6–11 April 2001 using radiosonde data resulted in PWV ranging from 29.49 to 79.53 mm [5]. Compared with the amount of PWV from 2017 to 2019, the difference is quite large, with the most significant PWV value at the beginning of April only around 50 mm. Of course, this must take into the state of the season/weather between the two different observations.

Although Jin and Luo [9] offered cyclical or calculated PWV variability in the Canadian region, they even showed a solid daily of PWV in winter/summer [4].

The amount of global solar energy that arrives at the earth's surface can be related to the intensity of solar radiation. Based on research on the length of solar radiation in Bandung in 2014–2016, Hamdi [6] reports that the most extended duration of solar radiation occurs when the sun reaches the furthest distance from the earth (aphelion) and corresponds to the long cycle of daylight in one year. The farthest distance from the ground to the sun is reached on August 1, so that in the middle of the year, which coincides with the dry season, the earth's surface at the observation location experiences the most significant radiation. This will encourage evaporation at the surface and strong convection that lifts water vapor to the height of the upper troposphere. it promotes evaporation at the surface and strong convection that lifts water vapor to the heights of the upper troposphere.

The correlation between PWV and global solar radiation energy is shown in Fig. 5.3, with a coefficient of r = -–0.64 (medium correlation). This means that an increase in the amount of water vapor in the atmosphere (PWV) will cause a decrease in solar energy received by the earth's surface with a reduced coefficient of –6.40 Ly/mm (Ly = Langley. 1Ly = 1.36 kW m^{-2}). PWV affects solar irradiation through two processes, namely (1) direct absorption at several wavelengths, and (2) increasing the diameter of the aerosol causing scattering of solar irradiation into components of direct beam radiation and diffuse sky radiation. Water vapor in the atmosphere comes from evaporation (evaporation and evapotranspiration) of the surface and then condenses into clouds. The latent heat stored during evaporation will be rereleased when water vapor condenses, and this is a criticsl aspect of the global climate system

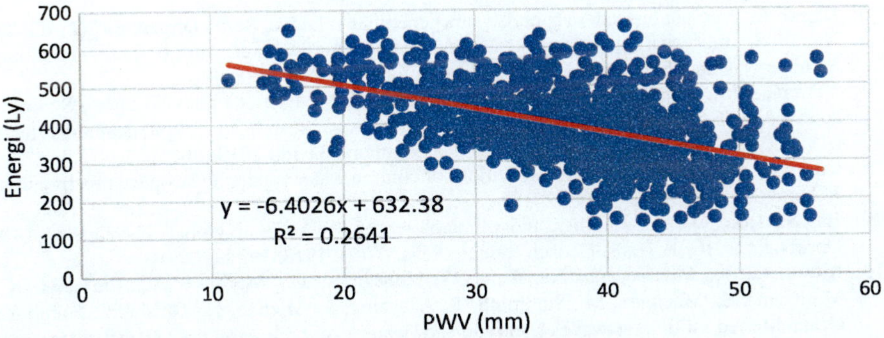

Fig. 5.3 Correlation between PWV and surface global radiant energy

[15]. Thus it becomes easy to understand that when the amount of water vapor in the atmosphere becomes large during the rainy season, the energy of solar radiation that arrives at the earth's surface becomes less, and vice versa during the dry season.

5.4 Conclusions

The negative correlation between PWV and global solar radiation energy of –0.64 indicates that there is an association in the moderate category. In the 2017–2019 periods, the amount of water vapor in the atmosphere calculated in the form of PWV shows a magnitude in the range of 11.4–56.3 mm. In dry season, the PWV is lower relatively compared to the PWV in rainy season. In the daily cycle, the PWV will be relatively larger at sunset and somewhat more minor at sunrise.

Acknowledgements We thank to the Head of LAPAN's Center for Atmospheric Science and Technology for facilitating this research, and the Geospatial Information Agency for providing GNSS data.

References

1. Askne, J., Nordius, H.: Estimation of tropospheric delay for microwaves from surface weather data. Radio Sci. **22**, 379–386 (1987)
2. Bevis, M., Businger, S., Chiswell, S., Herring, T.A., Anthes, R.A., Rocken, C., Ware, R.H.: GPS meteorology: mapping zenith wet delays onto precipitable water. J. Appl. Meteorol. **33**, 379–386 (1994)
3. Davis, J., Herring, T., Shapiro, I., Rogers, A., Elgered, G.: Geodesy by radio interferometry: effects of atmospheric modeling errors on estimates of baseline length. Radio Sci. **20**, 1593–1607 (1985)
4. Dai, A., Wang, J., Ware, R.H., Van Hove, T.: Diurnal variation in water vapor over North America and its implications for sampling errors in radiosonde humidity. J. Geophys. Res. **107**(D10), 4090 (2002). https://doi.org/10.1029/2001JD000642
5. Hamdi, S.: Precipitable water vapor dan cara penentuan nilainya. Berita Dirgantara **7**(2), 63–72.
6. Hamdi, S.: inieritas iradiasi permukaan dan lama penyinaran matahari di Bandung. Prosiding Seminar Nasional Sains Atmosfer 2017, Bandung, 26 Apr 2017 (2017)
7. Herrera, A.M., Suhandri, H.F., Realini, E., Reguzzoni, M., Clara de Lacy, M.: goGPS : open-source MATLAB software. Diakses dari. https://www.researchgate.net/publicapubl/278876018_goGPS_open-source_MATLAB_software. tanggal 31 Juli 2019 (2015)
8. Holloway, C., Neelin, J.: Temporal relations of column water vapour and tropical precipitation. J. Atmos. Sco. **67**, 1091–1105 (2010)
9. Jin, S., Luo, O.F.: Variability and climatology of PWV from global 13-years of GPS Observations. IEEE Trans. Geosci. Remote Sens. **47**(7), 1918–1924 (2009)
10. Kuwahara, T., Mizuno, A., Nagahama, T., Maezawa, H., Morihira, A., Toriyama, N., Murayama, S., Matsuura, M., Sugimoto, T., Asayama, S., Mizuno, N., Onishi, T., Fukui, Y.: Ground-based millimeter-wave observations of water vapor emission (183 GHz) at Atacama, Chile. Adv. Space Res. **42**(7), 1167–1171 (2008)
11. Lanzante, J.R., Klein, S.A., Seidelm, D.J.: Temporal homogenization of monthly radiosonde temperature data. Part II: Methodology. J. Climate **16**, 224–240 (2003)

12. Mousa, A., El-K., Aboualy, N., Sharaf, M., Zahra, H., Darrag, M.: Tropospheric wet delay estimation using GNSS: case study of a permanent network in Egypt. NRIAG J. Astron. Geophys. 76–86
13. Pramualsakdikul, S., Haas, R., Elgered, G., Scherneck, H.G.: Sensing of diurnal and semi-diurnal variability in the water vapour content in the tropics using GPS measurements. Meteorol. Appl. **14**, 403–412 (2007)
14. Prawirowardoyo, S.: Pendinginan dan pemanasan adiabatik, pp. 20–34. Meteorologi, Penerbit ITB Bandung (1996).
15. Trenberth, K.E., Jones, P.D., Ambenje, P., Bojariu, R., Easterling, D., Klein Tank, A., Parker, D., Rahimzadeh, F., Renwick, J.A., Rusticucci, M., Soden, B., Zhai, P.: Observations: surface and atmospheric climate change. In: Solomon, S., Qin, D., Manning, M., Chen, Z., Marquis, M., Averyt, K.B., Tignor, M., Miller, H.L. (eds.) Climate Change 2007: The Physical Science Basis. Contribution of Working Group I to the Fourth Assessment Report of the Intergovernmental Panel on Climate Change. Cambridge University Press, Cambridge, United Kingdom and New-York, NY, USA (2007)
16. Wang, X., Zhang, K., Wu, S., Li, Z., Cheng, Y., Yuan, H.: The correlation between GNSS-derived precipitable water vapor and sea surface temperature and its response to El Nino-Southern Oscillation. Remote Sens Environ **2016**, 1–12 (2018)
17. Yanxin, T., Lilong, L., Chaolong, Y.: Empirical model for measurement temperature and assessment of precipitable water vapor derived from GPS. Deodesy Geodyn **4**(4), 51–56 (2013). 10.3724.SP.J.1246.2013.04051
18. Zhang, Y., Xu, J., Yang, N., Lan, P.: Variability and trends in global precipitable water vapor retrieved from COSMIS radio occultation and radiosonde observations. Atmosphere **9**, 174 (2018). https://doi.org/10.3390/atmos9050174

Chapter 6
Convectively Coupled Equatorial Waves (CCEWs) Triggering Torrential Rainfall Events Over Sumatra, Indonesia

Muhamad R. Respati, Sandro W. Lubis, Sonni Setiawan, and Rahmat Hidayat

Abstract This study examines the influence of convectively coupled equatorial waves (CCEWs) on torrential rainfall events over Sumatra, Indonesia, during the rainy season (November to April). The analysis of the ERA-Interim reanalysis data and the gridded Asian Precipitation-Highly Resolved Observational Data Integration Toward Evaluation of Water Resources (APHRODITE) product from 1980 to 2007 showed that Kelvin waves have a major influence in the formation of the extreme rainfall event over Sumatra, compared to other equatorial waves such as equatorial Rossby (ER), mixed Rossby-gravity (MRG), and eastward inertio gravity (EIG) waves. In particular, Kelvin waves contribute up to +100% (−70%) change of extreme rainfall probability during their wet (dry) phase, followed by ER waves that contribute up to + 80% (−50%) probability change during their wet (dry) phase. The MRG and EIG waves have much less impact compared to the other waves. An examination of the characteristics of CCEWs reveals that the extreme rainfall modulation is consistent with the regions of active (suppressed) convection and convergence (divergence) induced by the equatorial waves over Sumatra, Indonesia.

6.1 Introduction

Sumatra is one of the five largest islands in Indonesia and is populated by more than 47 million people [1]. Rainfall is an important weather element to this island, as it plays a pivotal role in oil palm cultivation. Rainfall in Sumatra is very much influenced by the tropical climate modes of variability as it is located at the equator (Fig. 6.1). These modes of variability range from mesoscale to planetary-scale and from sub-seasonal to interannual timescale. On the intraseasonal time scale, Madden–Julian oscillation (MJO) strongly influences rainfall variability over Sumatra [2–4].

M. R. Respati (✉) · S. Setiawan · R. Hidayat
Department of Geophysics and Meteorology, IPB University, Bogor, Indonesia
e-mail: mreyhanrespati33@gmail.com

S. W. Lubis
Rice University, Houston, USA

E. Yulihastin et al. (eds.), *Proceedings of the International Conference on Radioscience, Equatorial Atmospheric Science and Environment and Humanosphere Science, 2021*, Springer Proceedings in Physics 275, https://doi.org/10.1007/978-981-19-0308-3_6

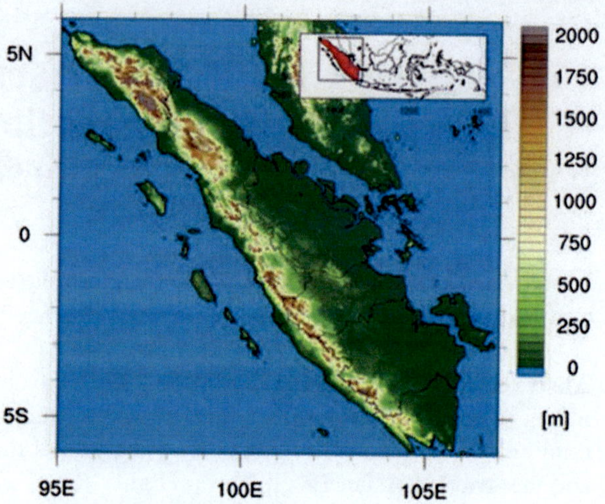

Fig. 6.1 Topographic map of Sumatra

Parts of regions in Sumatra are also affected by a seasonal variability: the monsoon [5, 6]. At longer timescales, there are El Nino-Southern Oscillation (ENSO) and Indian Ocean Dipole (IOD) that give impacts to interannual rainfall variability [7–10]. Despite clear evidence of the aforementioned phenomena on rainfall variability over Sumatra, the influence of equatorial waves that are coupled with the convection on rainfall variability over Sumatra remains elusive.

Convectively coupled equatorial waves (CCEWs) or equatorial planetary waves (EPW) are weather disturbances in the tropics triggered by the convection or latent heat release in the tropics [11]. CCEWs appear in the solutions of shallow water equations on equatorial β-plane for homogenous incompressible fluid where the Coriolis parameter is assumed to be proportional to latitude [12]. Generally, there are two wave modes found in the solutions: eastward or westward propagating inertio gravity waves and westward propagating Rossby waves. Two other wave modes, westward propagating mixed Rossby-gravity waves and eastward propagating Kelvin waves, are also found in particular solutions. There are four wave modes investigated in this study: Kelvin waves, equatorial Rossby (ER) waves, mixed Rossby-gravity (MRG) waves, and eastward inertio gravity (EIG) waves. The theoretical horizontal structures of those waves can be seen in Fig. 6.2.

Based on the horizontal structures, wave modes can be classified into symmetric waves and antisymmetric waves. Kelvin waves and ER waves are symmetric waves because their structures are the same in both hemispheres. MRG waves and EIG waves are antisymmetric waves because their structures in one hemisphere are the opposite of those in another hemisphere. In addition to that, based only on the structure of the wind field, wave modes can also be classified into more-rotational waves and more-divergent. waves [13]. ER and MRG are more-rotational waves, whereas Kelvin and EIG are more-divergent waves.

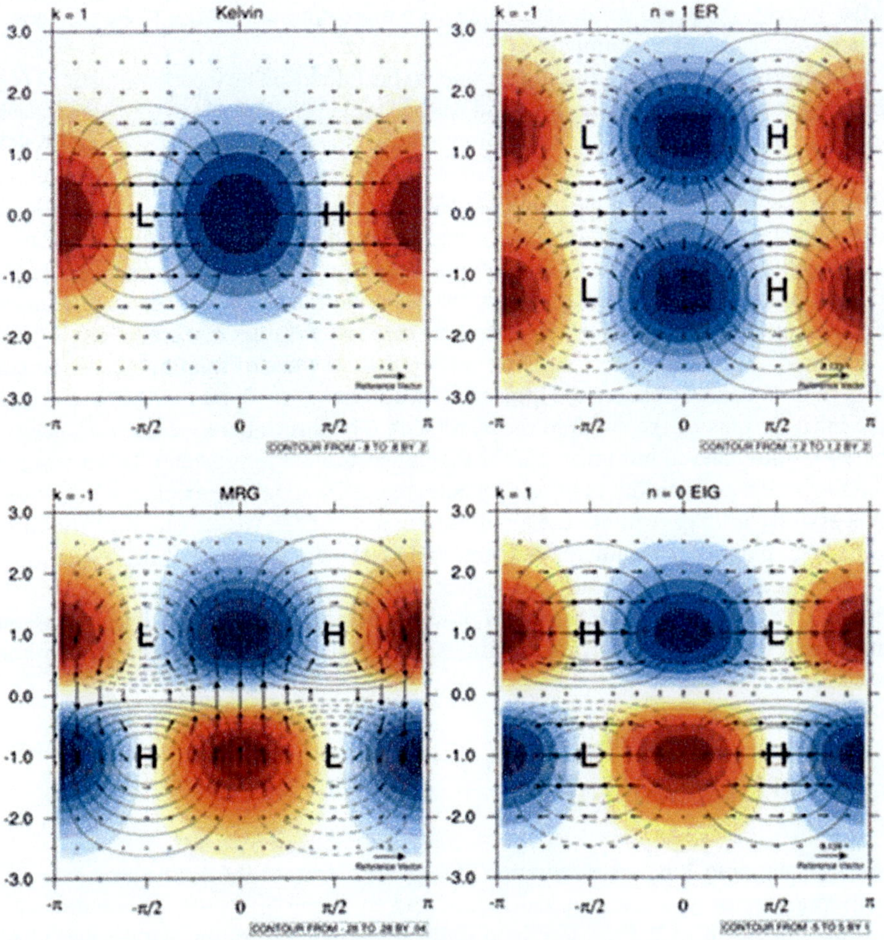

Fig. 6.2 Theoretical horizontal structures of Kelvin, ER, MRG, and EIG waves based on [12]. Color shadings are divergence of wind anomalies with blue (red) color representing convergence (divergence). Contour lines are geopotential height anomalies with solid (dashed) lines representing positive (negative) anomalies. Vectors are horizontal wind anomalies. Letters 'H' ('L') denote centers of high (low) pressure anomalies. All variables are nondimensionalized

Each wave mode has its distinct characteristics in its horizontal structures. Kelvin waves have zero meridional wind perturbations, while their zonal wind perturbations have peaks exactly at the equator and decay with latitudes. Centers of high or low pressure and centers of divergence or convergence of this wave mode are also exactly at the equator. ER waves have symmetric pairs of cyclonic or anticyclonic gyres with low or high pressure at the center. A prominent feature in the horizontal structures of MRG waves is the cross-equatorial flow with maximum meridional wind speed at the equator. The structure of the pressure field of EIG waves is the same as that of

MRG waves, while the wind disturbances are more similar to those of Kelvin waves where zonal winds are dominant.

Studies of CCEWs in Indonesia have so far been focused on the identification [14], the propagation [15], the horizontal and vertical structures [16, 17] of CCEWs, as well as their connection to the other low-frequency modes of variability such as ENSO, IOD and QBO [18–23], their behavior in the GCM model simulations [24, 25] and the development of the technique analysis [26]. However, only a few studies investigate the impacts of CCEWs on extreme rainfall in Indonesia [27–30]. Although Lubis and Jacobi have shown that CCEWs contribute about 16–20% of rainfall variability in the tropics [31], their regional-scale influence on rainfall variability in Indonesia remains unclear. Recent work by Ferrett et al. shows that CCEWs give impacts on daily rainfall in Sumatra to some extent [27]. Still, the use of satellite-based rainfall data in their analysis causes the impact of CCEWs over the landmass unclear due to the intrinsic bias of the satellite measurement. Therefore, here we use rain gauge-based rainfall data to minimize the error and use another approach to better capture the local impacts over the landmass of Sumatra. This local approach has been used successfully in the previous studies of the impact of CCEWs on rainfall variability over Java, Indonesia, Southern Vietnam, and West Africa [29, 32, 33].

The main goals of this study are to quantify the impacts of CCEWs on the distribution of daily rainfall and extreme rainfall over Sumatra during the rainy season and to investigate the underlying dynamical mechanisms of CCEW-induced extreme rainfall over Sumatra.

6.2 Data and Methods

We use outgoing longwave radiation (OLR) data from NOAA to investigate the convection activity in the tropical region and its variability [34]. This data has a spatial resolution of 2.5° × 2.5° and daily temporal resolution, with a period of 1 January 1979- 31 December 2018. Daily rainfall data are obtained from Asian Precipitation-Highly.

Resolved Observational Data Integration Toward Evaluation of Water Resources (APHRODITE) Monsoon Asia V1101 for the period of 1 January 1980–31 December 2007 [35]. APHRODITE rainfall is rain gauge measurements interpolated spatially onto latitude–longitude grids with 0.25° × 0.25° resolution. To analyze the influence of CCEWs on dynamic fields, we also use zonal wind, meridional wind, and geopotential height at 850 hPa from ECMWF ERA-Interim Reanalysis dataset with a spatial resolution of 0.75° × 0.75° and 6 hourly temporal resolution [36]. The dynamic field data are first averaged daily to match the temporal resolution of OLR and rainfall data. All data are processed using NCAR Command Language (NCL) software [37].

To isolate the activity of CCEWs, we filter the OLR data using two-dimensional fast Fourier transform (FFT) with the prescribed wavenumber and frequency as in Fig. 6.3 and Table 6.1. Before performing FFT, we remove the first three harmonics

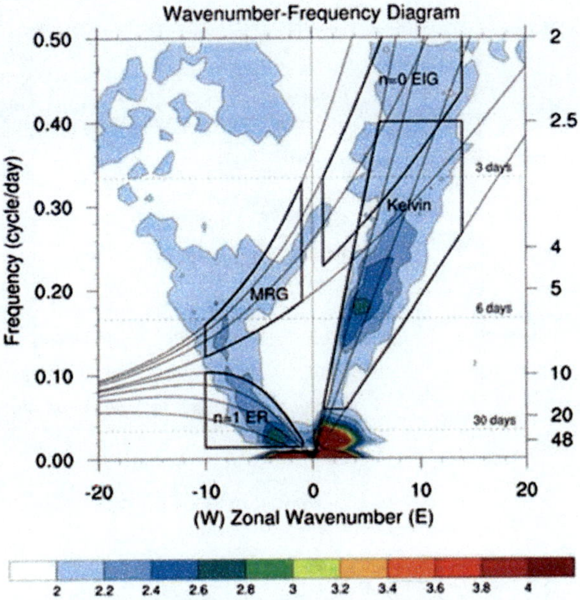

Fig. 6.3 Wavenumber-frequency diagram for OLR over 20° S–20° N. Thin black lines are dispersion curves for equivalent depth of 8, 12, 25, and 90 m. Thick black lines are limits of wavenumber and period used for each wave

Table 6.1 Wavenumber and period of each wave applied in the filtering technique [13, 31]

Wave Mode	Propagation	Wavenumber	Period (days)	Equivalent depth (m)
Kelvin	Eastward	1–14	2.5–17	8–90
Equatorial Rossby (ER)	Westward	1–10	9–72	8–90
Mixed Rossby-Gravity (MRG)	Westward	1–10	3–10	8–90
Eastward Inertio Gravity (EIG)	Eastward	1–14	1–5	12–50

and linear trends and taper the data to minimize spectral leakage [38]. The FFT is done using the 'kf_filter' function in NCL and is performed repeatedly for every latitude in the range of 20° S–20° N.

We divide the wave activity into eight consecutive phases based on the wave-filtered OLR anomaly and its temporal tendency. This method is first developed by Riley et al. in a study of MJO [39]. The determination of local wave phases is illustrated in Fig. 6.4. Wave-filtered OLR anomaly and its temporal tendency are averaged over a range of latitudes, 6° S–6° N for Sumatra, and are then standardized by dividing them with their respective global standard deviation. The standardized OLR anomaly and its tendency are plotted on the Hovmueller diagram in Fig. 6.4. Next, time series of the standardized anomaly and tendency at the reference longitude, 102° E, are plotted on a phase diagram with standardized anomaly as abscissa and standardized tendency as ordinate.

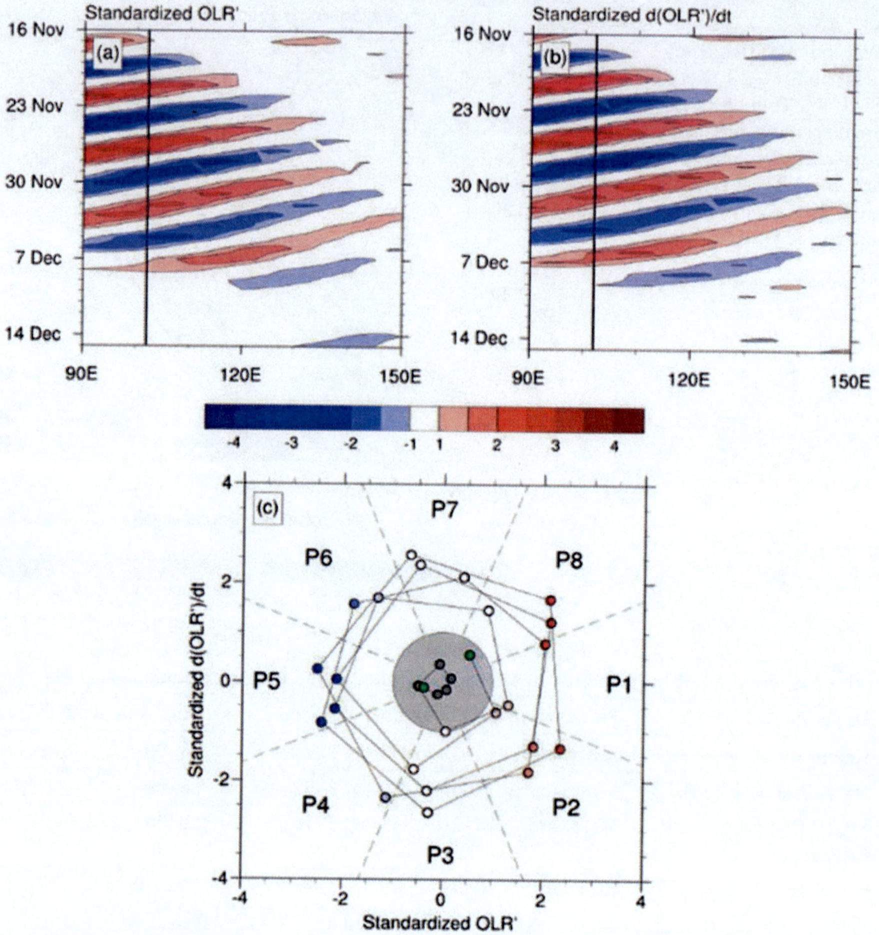

Fig. 6.4 Hovmueller diagram of **a** standardized OLR anomaly and **b** its temporal tendency and **c** the local wave phase diagram for MRG waves. The range of period used in this illustration is 16 November–15 December 1981. The vertical black line denotes reference longitude of 102° E. Green dots mark the start and the end of the period used for illustration

The phase diagram in Fig. 6.4 actually shows the life cycle of the waves on the reference longitude, starting from phase 1 to phase 8. Phase 1 is the driest phase because the OLR anomaly is maximum, and its tendency is zero. Dry phases continue until phase 2 before the transition phase occurs in phase 3 when the OLR anomaly switches signs from positive to negative. Wet phases start in phase 4, and the wettest condition is reached in phase 5 when the OLR anomaly is minimum, and its tendency is zero. OLR anomaly increases in phase 6, switches sign to positive in phase 7, and comes back to dry phases in phase 8. The amplitude of the wave activity is defined as the distance from origins in the phase diagram. In this study, only wave amplitude more than 1 is considered in the analysis.

Time-series of wave activity that has been divided into eight phases are used to create composites of daily rainfall and dynamic fields. Composite in a certain phase is the average of anomalies on days during the rainy season when the wave activity is in that phase. Anomaly is defined as the deviation from daily climatology. For dynamic field analysis, the anomaly is wave-filtered while the rainfall anomaly is not. Statistical significance of the anomaly is tested using two-tailed Student's t-test with 90% significance level.

Analysis of extreme rainfall is done by calculating the changes in extreme rainfall probability in a certain phase using

$$\Delta P_i = \frac{P_i(x \geq x_c) - P_{all}(x \geq x_c)}{P_{all}(x \geq x_c)} \tag{6.1}$$

where ΔP_i is the changes in extreme rainfall probability during phase *i*, $P_i(x \geq x_c)$ is the probability of rainfall during phase *i* exceeding threshold x_c, and $P_{all}(x \geq x_c)$ is the probability of rainfall during November–April exceeding threshold x_c [40]. The threshold used in this study is the 90th percentile of daily rainfall during the rainy season.

We also do a wet days analysis by calculating the changes in fraction of wet days in a certain phase using

$$\Delta f_i = \frac{n_i}{N_i} - \frac{n}{N} \tag{6.2}$$

where Δf_i is the changes in fraction of wet days during phase *i*, n_i is the number of wet days during phase *i*, N_i is the number of days during phase *i*, *n* is the number of wet days during the rainy season, and *N* is the number of days during the rainy season [27]. Here, wet days are defined as days with accumulated daily rainfall greater than 1 mm. In addition to changes in wet days fraction, we also calculate the changes in rainfall intensity on wet days. It is done the same way as creating composite of daily rainfall anomaly, but only wet days are included.

6.3 Results and Discussions

6.3.1 Climatological Condition

Indonesia is located in the monsoon region, where there are two dominant prevailing winds with nearly opposite directions [5]. During May–October, the prevailing wind is from the southeast, coming from a high pressure center over winter Australia (Fig. 6.5a).

The opposite condition is true during November–April when the wind is from the northeast before turning to the east at the equator (Fig. 6.5b). This change of prevailing winds is due to the change of position of the Sun in its annual cycle,

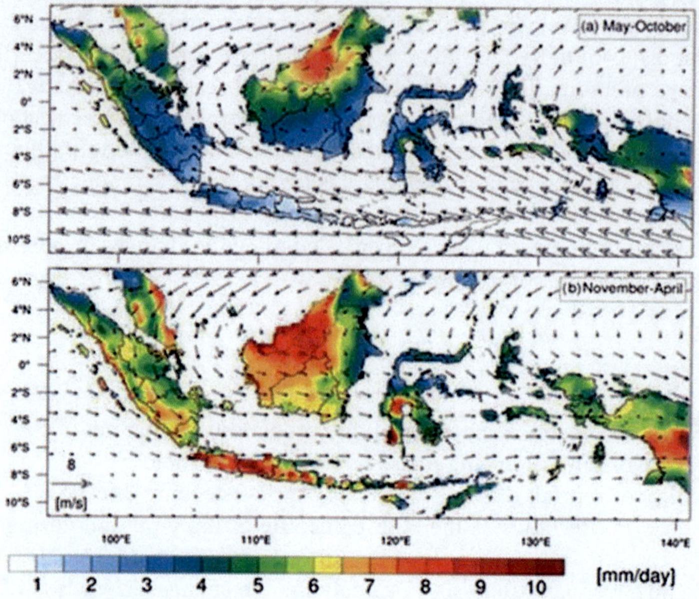

Fig. 6.5 Mean of daily rainfall (shading) and 850 hPa horizontal winds (vectors) during **a** May–October and **b** November–April

which is followed by the change of position of high/low pressure center around Indonesia [5].

The seasonal change of prevailing winds described above is accompanied by the change of rainfall in most parts of Indonesia (Fig. 6.5). Java Island and southern Celebes receive an amount of rainfall during November–April up to five times higher than during May- October. Southern parts of Sumatra also experience this increase in rainfall almost 100% from 3–4 mm/day to 7–8 mm/day. Due to the seasonal difference of rainfall amount, May–October is defined as dry season, and November–April is defined as rainy season although the difference is not significant in a few areas such as North Sumatra. The difference between the dry season and the rainy season is not only observed in mean rainfall, but also in rainfall variance (Fig. 6.6). An increase in rainfall variance means that the fluctuation of daily rainfall in the rainy season is greater than in the dry season.

6.3.2 *Waves Activity*

The strength of waves activity is inferred from the variance of wave-filtered OLR anomalies as in Fig. 6.7. The variance shows how fluctuative the perturbation on convection is due to each wave mode's activity. Kelvin waves have strong activity along the equatorial band, precisely on the intertropical convergence zone (ITCZ)

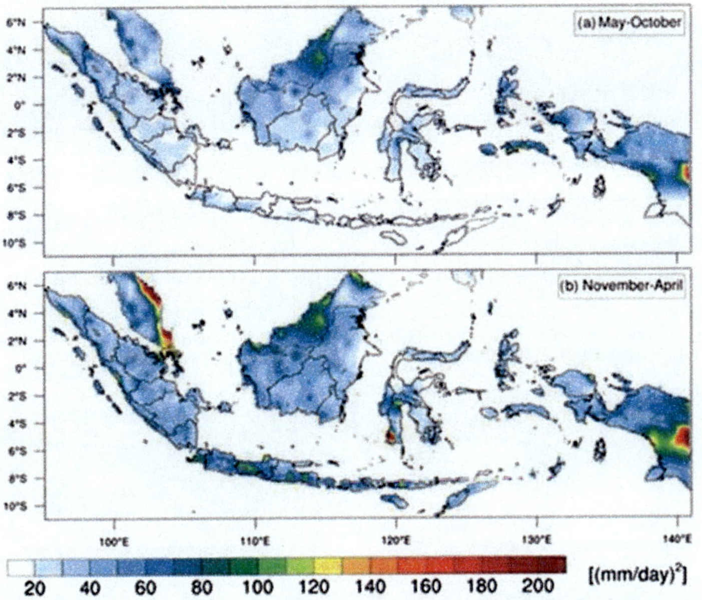

Fig. 6.6 The variance of daily rainfall during **a** May–October and **b** November–April

[11, 41]. On the other hand, the other three wave modes have peaks of activity off the equator, as predicted by the theoretical horizontal structures. ER waves have maximum activity at latitude of 12°–15° while the activity of MRG and EIG waves is maximum at latitude of 8°–10°. In general, the activity of all wave modes follows the seasonal position of the Sun: the maximum strength is displaced southward, and a bit weaker, during November–April. Over Sumatra, the strongest wave mode is Kelvin waves with variance of 120–140 W^2/m^4, whereas ER waves are at the second place, followed by MRG and EIG.

6.3.3 Modulation of Daily Rainfall

Modulation of Kelvin waves on daily rainfall during the rainy season in Sumatra is shown in Fig. 6.8. Daily rainfall is increased (decreased) during the wet (dry) phases, as indicated by positive (negative) rainfall anomalies. Wet phases start in phase 3 when a significant increase of rainfall is observed only in Aceh, western North Sumatra, and parts of West Sumatra. In phase 4, rainfall anomalies move to the east and get stronger, e.g., up to +3 mm/day along the west coast. The increase of daily rainfall is significant all over Sumatra in phase 5, but wanes in phase 6 as the transition phase starts. In phase 7, parts of Aceh and North Sumatra already experience a decrease of rainfall while positive anomalies are still observed in the eastern coast of South Sumatra and Lampung. In phase 8, negative anomalies are

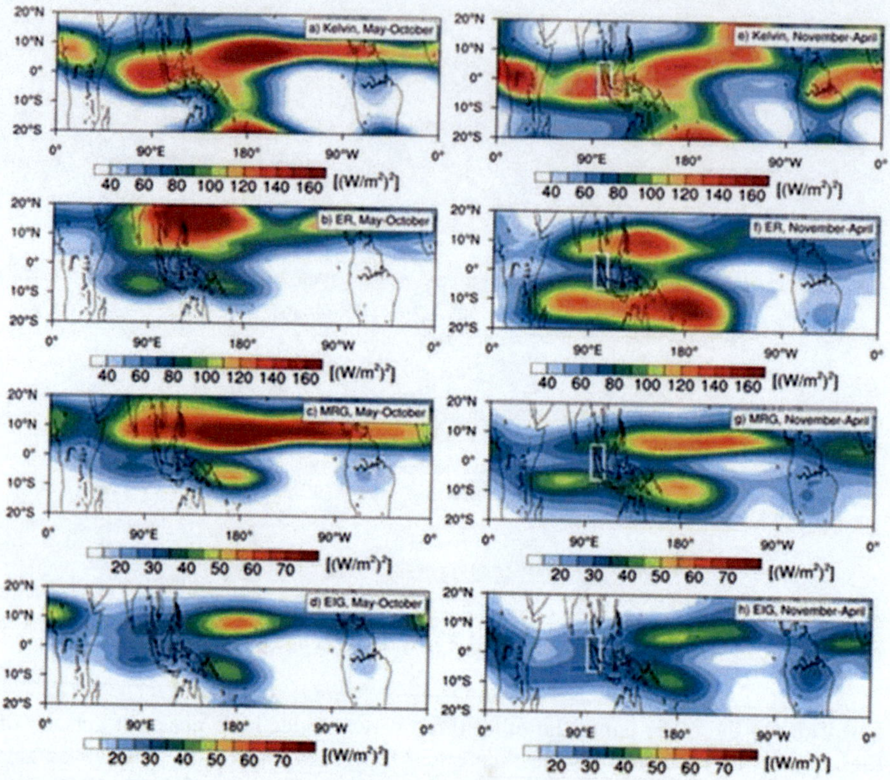

Fig. 6.7 The variance of wave-filtered OLR anomalies for **a**, **e** Kelvin, **b**, **f** ER, **c**, **g** MRG, and **d**, **h** EIG waves during **a–d** May–October and **e–h** November–April. White boxes on **e–h** denote Sumatra

significant over most of the island, indicating the start of the dry phases. Dry phases occur until phase 2, with a maximum decrease of rainfall up to −2 mm/day spotted in the west coast. The eastward propagation of increase or decrease of rainfall is consistent with the propagation of Kelvin waves.

Modulation of daily rainfall by ER waves (Fig. 6.9) has a different pattern and is slightly weaker compared to that by Kelvin waves. Wet phases start in phase 5 when there is a significant increase of rainfall only in the middle of the island. The wet phases last until phase 7, but the rainfall increase is never significant in the southern part of the island. The maximum positive anomaly is +2 mm/day or about 30–50% of the climatology and occurs on the northeast coast of Aceh and along the Bukit Barisan mountains. In phase 8, rainfall increase is only observed on the northern tip of the island, indicating the end of wet phases as the waves have now propagated westward. The modulation pattern during the dry phases, which last from phase 1 to phase 3, is very similar to that during the wet phases, where the maximum decrease of rainfall is also about −2 mm/day.

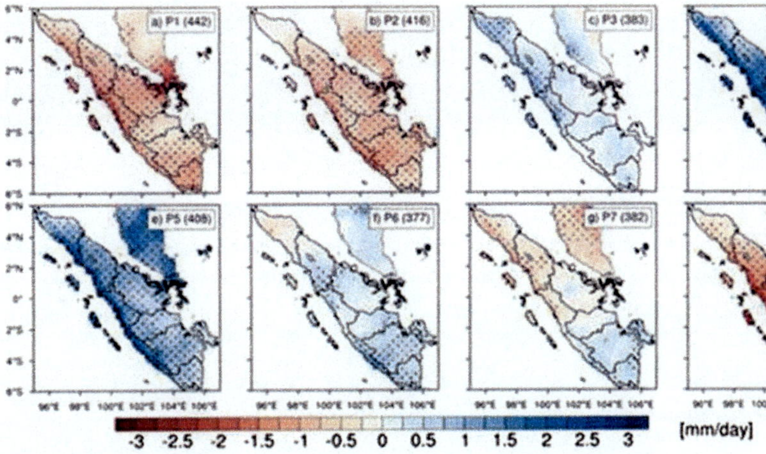

Fig. 6.8 Composite of daily rainfall anomalies per phase of Kelvin waves during the rainy season. Stipples denote areas with statistically significant anomalies at 90% level. Number of days in a certain phase when the waves activity is strong is shown in the top-right bracket

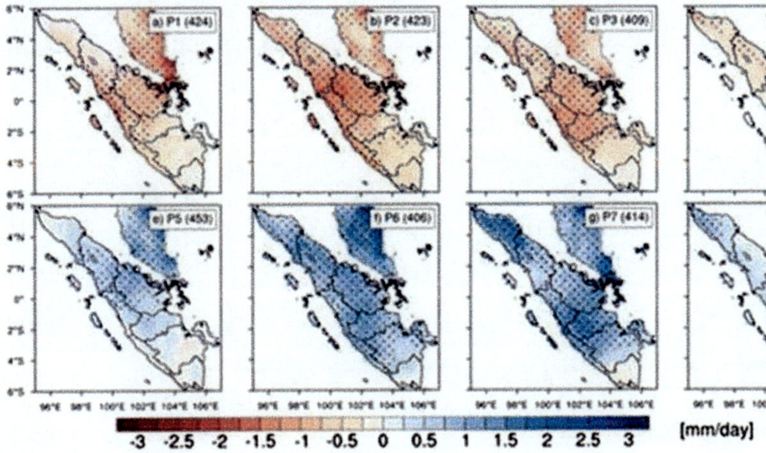

Fig. 6.9 As in Fig. 6.8, but for ER waves

MRG and EIG waves have weaker modulation amplitude (Figs. 6.10 and 6.11) than Kelvin and ER waves. The wet phases of MRG are clearly observed only in phase 5 and phase 6. The greatest increase of rainfall occurs in phase 6 and is only up to +1.5 mm/day, or half of that of Kelvin waves. The modulation during dry phases is significant over larger areas than during wet phases but with about the same amplitude, e.g., up to −1.5 mm/day in its driest condition. On the other hand, EIG waves are the weakest among the four wave modes investigated in this study. There is no obvious peak of wet phases for this wave mode. In its driest phase, only

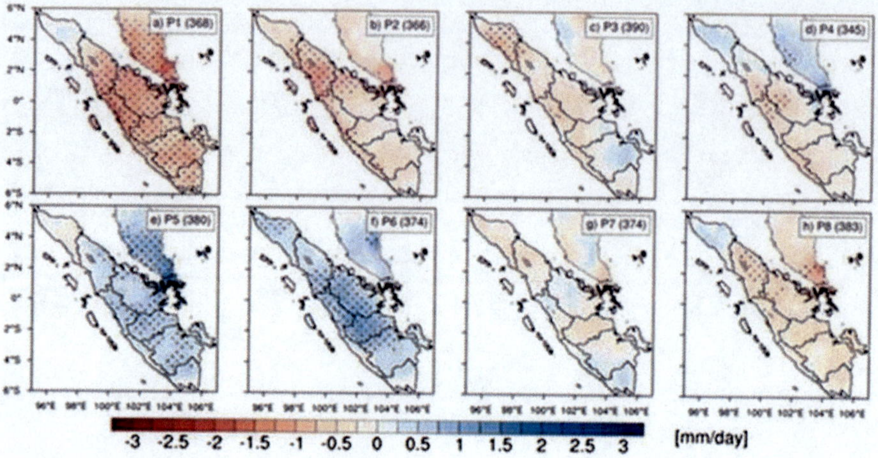

Fig. 6.10 As in Fig. 6.8, but for MRG waves

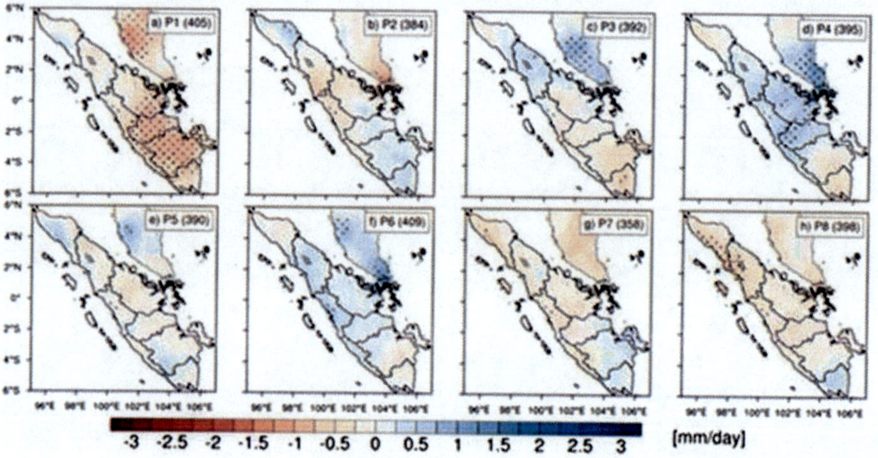

Fig. 6.11 As in Fig. 6.8, but for EIG waves

some parts of the eastern coast experience a significant decrease of rainfall, which is only about −1 mm/day. Propagation of this wave mode is also not clearly observed.

Figures 6.8, 6.9, 6.10 and 6.11 show modulations of daily rainfall, each by only one wave mode. In fact, there can be two or more waves with strong activity in a day that superpose each other. The wave superposition can be constructive or destructive. Figure 6.12 gives examples of constructive superposition of Kelvin waves and three other wave modes. CCEWs superposition is almost linear where the modulation amplitude by the combined waves is about the sum of the strengths of the superposing waves, similar to the results in the study of [29]. This interaction among waves needs

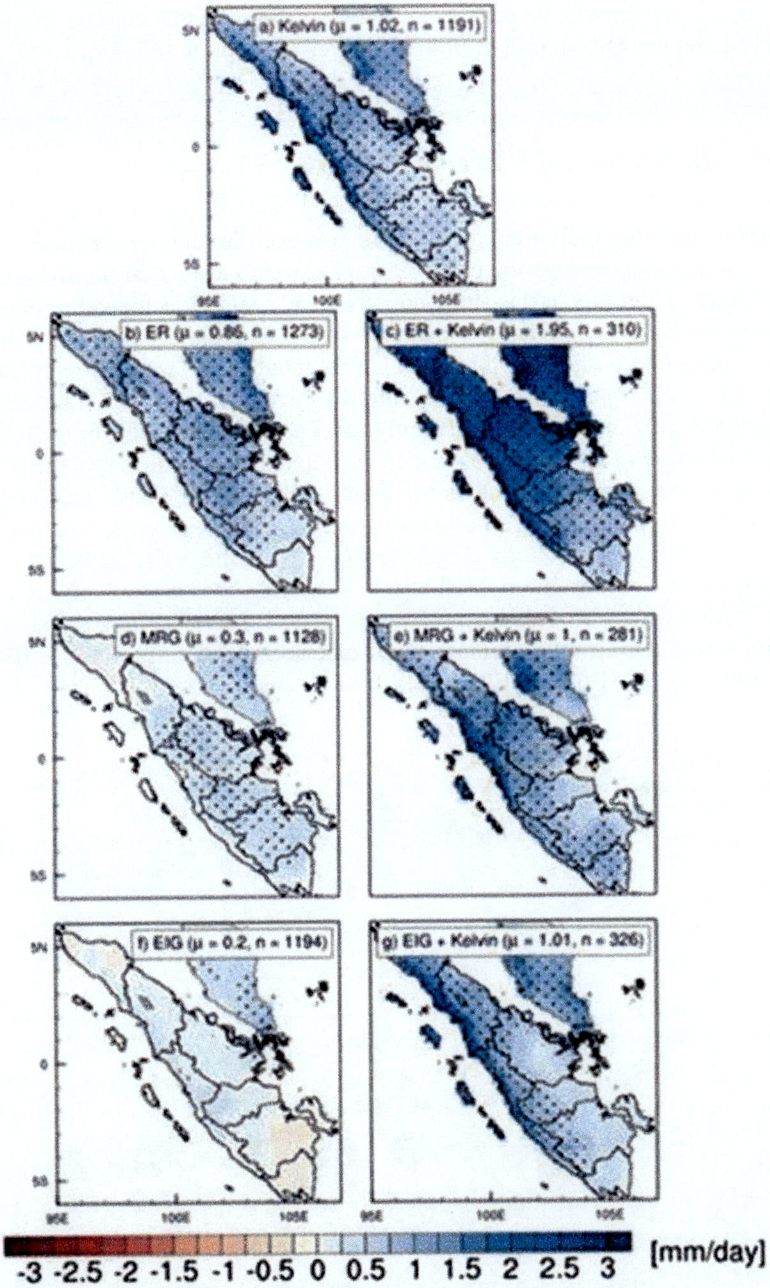

Fig. 6.12 Composite of daily rainfall anomalies during the wet phases of **a** Kelvin, **b** ER, **d** MRG, and **f** EIG waves and during the superposition between **c** ER and Kelvin, **e** MRG and Kelvin, and **g** EIG and Kelvin waves. Stipples denote areas with statistically significant anomalies at 90% level. Numbers on the top right corners are the areal average of daily rainfall anomalies (μ) and number of days (n)

to be further investigated because previous studies showed that it is possible that two waves having the same phase result in a weaker modulation [29, 33].

6.3.4 *Modulation of Extreme Rainfall*

Besides modulating daily rainfall, CCEWs also exert influence on extreme rainfall by increasing (decreasing) the probability of extreme events during the wet (dry) phases. Extreme events in this study are when daily rainfall is greater than the 90th percentile of daily rainfall during November–April, as explained in the methods. The extreme threshold is shown in Fig. 6.13 and is about 10–20 mm/day over Sumatra. Modulation of extreme rainfall has a pattern similar to that of daily rainfall in that the Kelvin waves have the strongest amplitude. The probability of extreme rainfall in the wettest phase of Kelvin waves is increased by up to + 100% or, in other words, the extreme events are two times more likely to occur (Fig. 6.14). This maximum increase is observed on the west coast of North Sumatra in phase 4. Meanwhile, the decrease of extreme probability is only up to −70% in the driest phase.

ER waves cause extreme rainfall events to be 1.8 times more likely to occur in phase 7 (Fig. 6.15). Modulation of ER waves during the dry phases has weaker amplitude with a maximum decrease of extreme probability only about −50%. Two other wave modes, MRG and EIG waves, have an even weaker influence on extreme

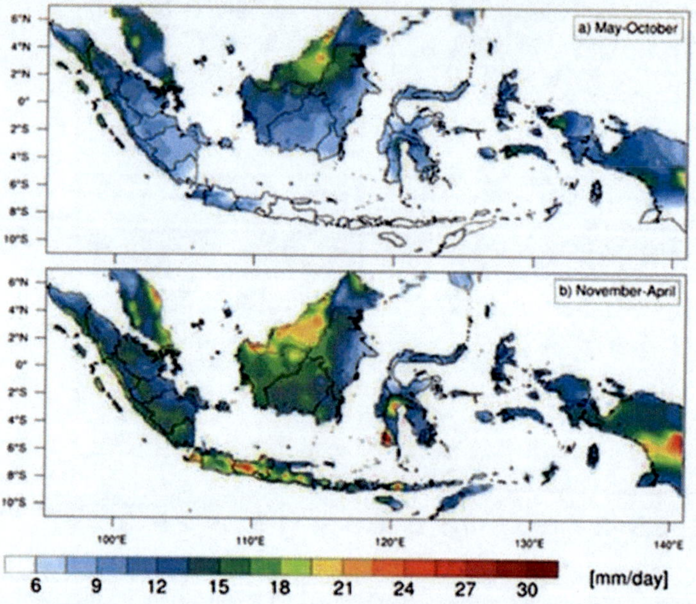

Fig. 6.13 90th percentile values of daily rainfall during **a** May–October and **b** November–April

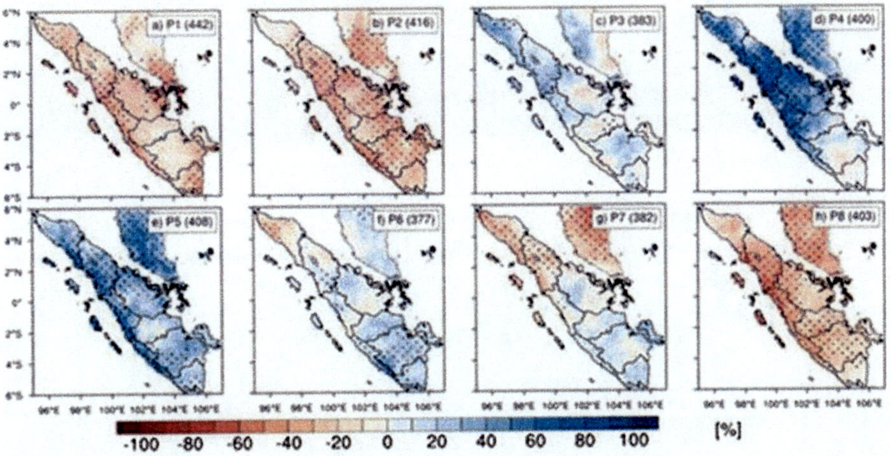

Fig. 6.14 Changes in probability of extreme rainfall per phase of Kelvin waves during the rainy season. Stipples denote areas with statistically significant changes at 90% level

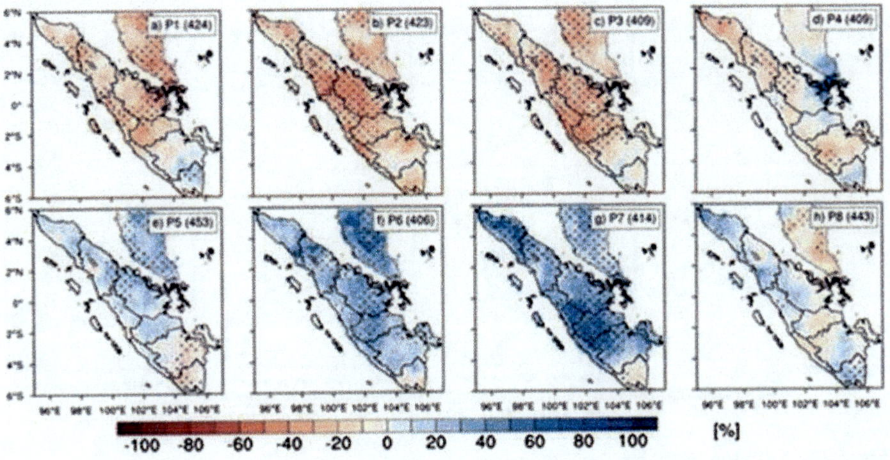

Fig. 6.15 As in Fig. 6.14, but for ER waves

rainfall. MRG waves increase the probability of extreme rainfall by only up to +40%, whereas during the dry phases the probability is decreased by up to −50% (Fig. 6.16). Meanwhile, the amplitude of modulation of EIG waves (Fig. 6.17) is only a quarter of that of Kelvin waves, with a maximum increase (decrease) of extreme probability only +20% (−30%).

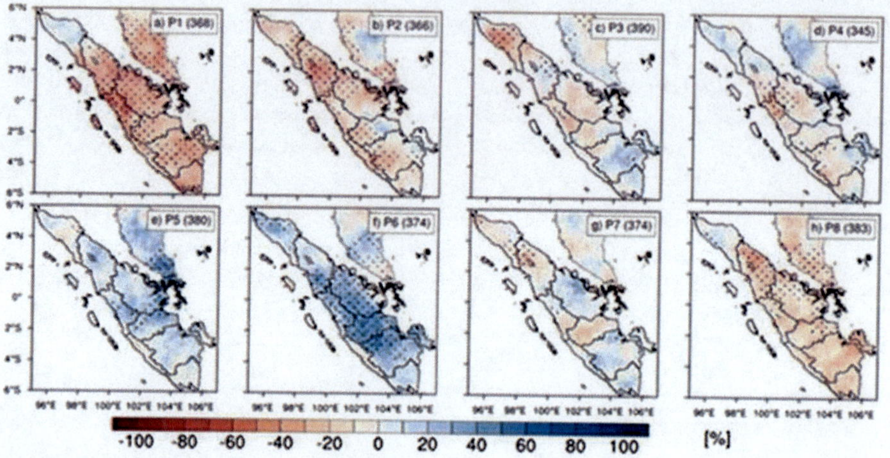

Fig. 6.16 As in Fig. 6.14, but for MRG waves

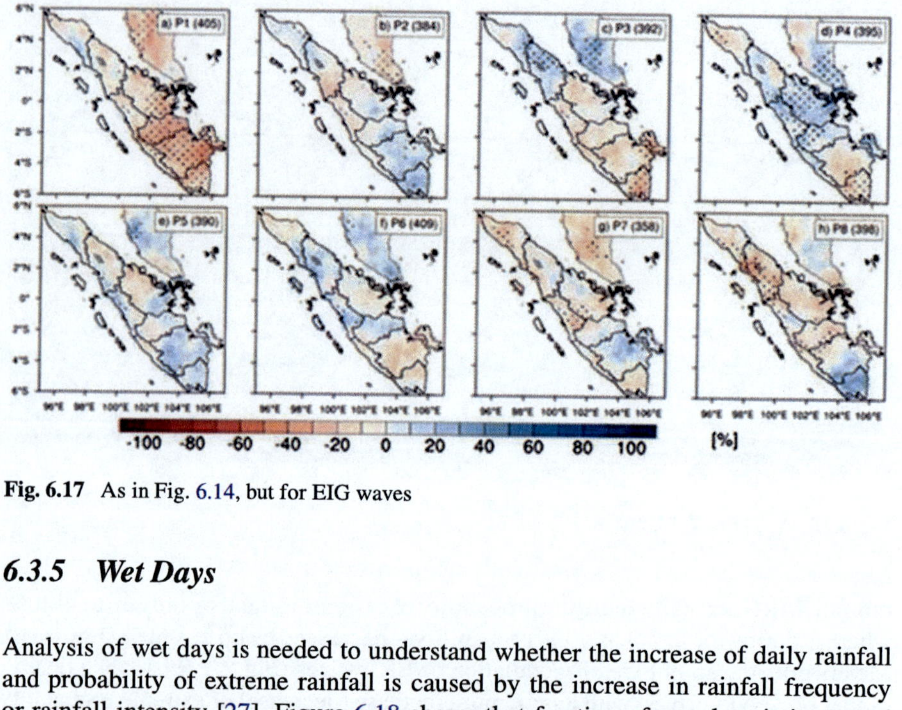

Fig. 6.17 As in Fig. 6.14, but for EIG waves

6.3.5 Wet Days

Analysis of wet days is needed to understand whether the increase of daily rainfall and probability of extreme rainfall is caused by the increase in rainfall frequency or rainfall intensity [27]. Figure 6.18 shows that fraction of wet days is increased (decreased) during the wet (dry) phases of Kelvin waves. Moreover, Fig. 6.19 also shows that the rainfall intensity on wet days is modulated with patterns almost identical to the daily rainfall. Analysis on the other three wave modes yields similar results (not shown). Those results lead to the conclusion that the increase of daily rainfall

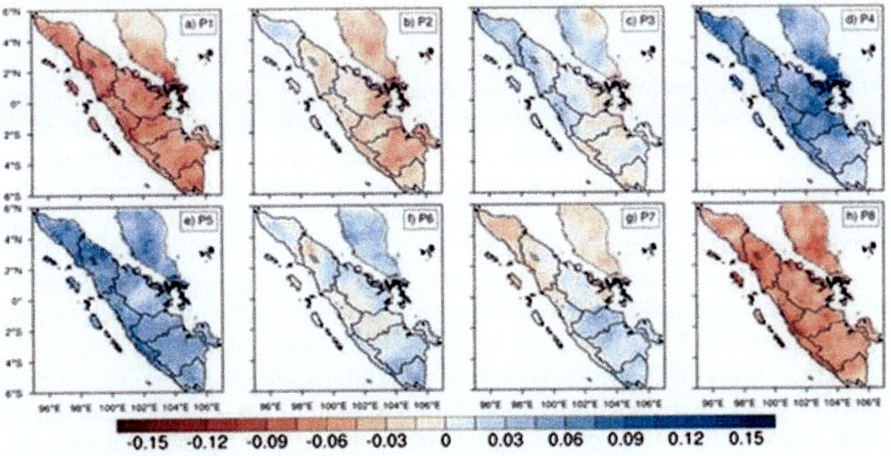

Fig. 6.18 Changes in fraction of wet days per phase of Kelvin waves during the rainy season

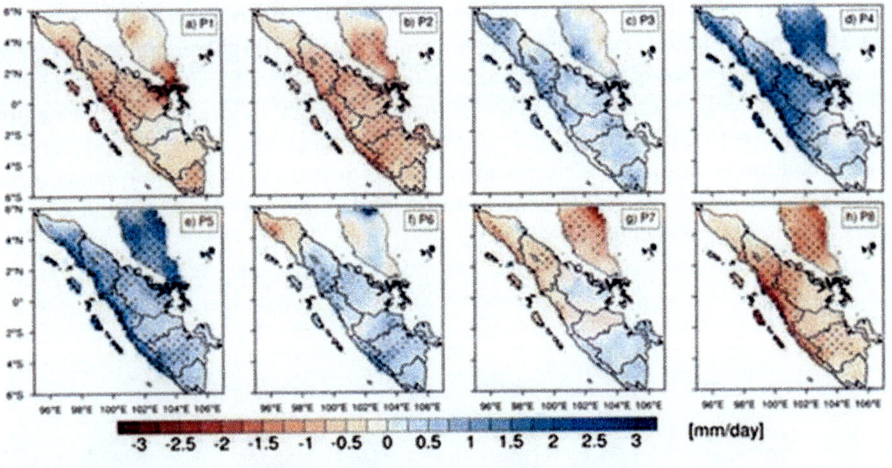

Fig. 6.19 Changes in rainfall intensity on wet days per phase of Kelvin waves during the rainy season. Stipples denote areas with statistically significant changes at 90% level

and probability of extreme rainfall is caused by both the increase in frequency of wet days and the increase in rainfall intensity on those wet days. This is in line with the results in the study of Ferrett et al. [27].

6.3.6 Mechanism of Modulation

The composite of anomalies of OLR and dynamic fields (horizontal winds and geopotential height) for Kelvin waves (Fig. 6.20) shows consistency with the theoretical horizontal structure in Fig. 6.2. Kelvin waves induce perturbations in the horizontal wind field in the form of zonal wind with maximum speed at the equator. Easterly (westerly) anomalies are observed in the region of anomalous low (high) pressure, which is also maximum at the equator. In the region where easterly and westerly winds meet, or region of convergence, OLR anomalies are negative. Conversely, positive OLR anomalies are located in the region of wind divergence. These negative (positive) OLR anomalies are right over Sumatra in phase 4 and 5 (8 and 1); at the same time, the daily rainfall is increased (decreased). Therefore, we conclude that Kelvin waves generate perturbations in the form of wind convergence (divergence) region during the wet (dry) phases where the convective activity is enhanced (suppressed) (Fig. 6.21).

Composite of dynamic field anomalies for ER waves (Fig. 6.21) also shows consistency with the theoretical horizontal structure. Perturbations by ER waves are in the form of pairs of low (high) pressure centers in both hemisphere encircled by cyclonic (anticyclonic) rotation. Between the two centers of low (high) pressure, there are strong westerly (easterly) wind anomalies at the quator where the flows from cyclonic (anticyclonic) cells meet. Unlike in the geopotential height field, symmetric structure of perturbation in OLR is not clearly observed. However, in general, the mechanism of modulation is similar to Kelvin waves that perturbations in dynamic fields form the region of convergence (divergence) during the wet (dry) phases where the activity of convection is enhanced (suppressed).

Horizontal structures of MRG and EIG waves are not clearly seen in geopotential height (not shown), so that we show horizontal winds and convergence in Figs. 6.22

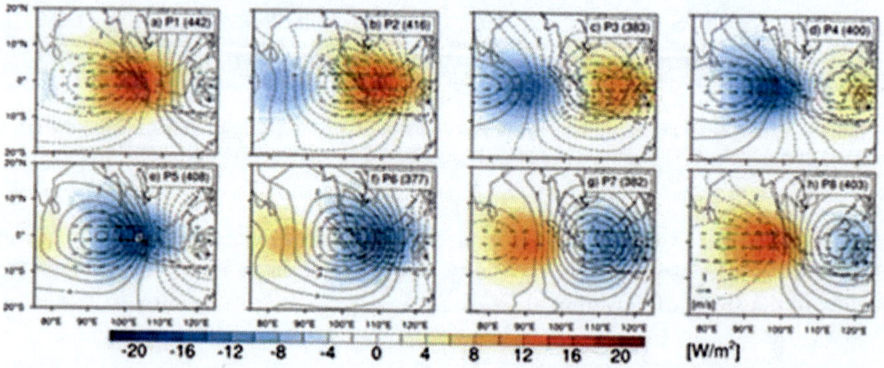

Fig. 6.20 Composite of Kelvin wave-filtered OLR (shading), geopotential height (contours), and horizontal winds (vectors) anomalies per phase of Kelvin waves during the rainy season. Only statistically significant OLR anomalies at 90% level are plotted. The contour interval for geopotential height anomalies is 0.2 m where solid (dashed) lines denote positive (negative) anomalies

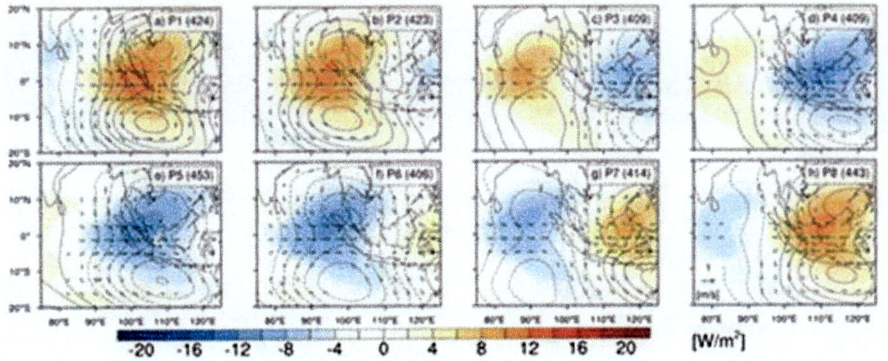

Fig. 6.21 As in Fig. 6.20, but for ER waves. The contour interval is 0.4 m

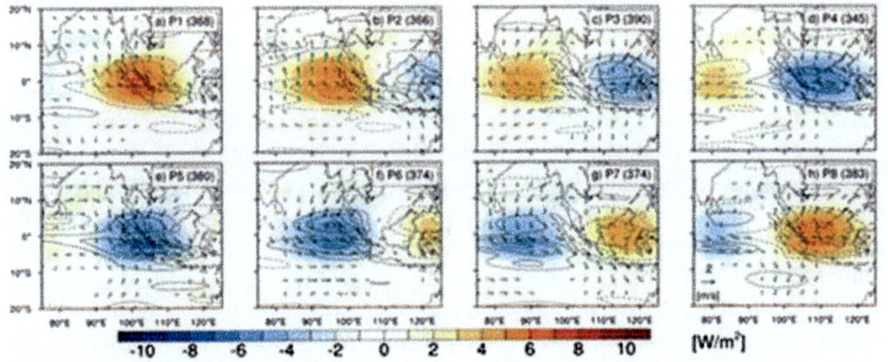

Fig. 6.22 Composite of MRG wave-filtered OLR (shading), convergence (contours), and horizontal winds (vectors) anomalies per phase of MRG waves during the rainy season. Only statistically significant OLR anomalies at 90% level are plotted. The contour interval for convergence anomalies is 0.4×10^{-7} s^{-1} where solid (dashed) lines denote convergence (divergence)

and 6.23. Cross-equatorial flow, a prominent feature of MRG waves, appears in Fig. 6.22 even though it is not pure meridional as in the theoretical structure. Meanwhile, perturbation by EIG waves in horizontal winds does not exhibit a clear structure. Anomalous winds in Fig. 6.23 resemble those of Kelvin waves but with stronger meridional components. This is possibly caused by the mixing of signals of those two wave modes, as their spectra intersect (Fig. 6.3) [42, 43]. However, regions of convergence (divergence) induced by MRG and EIG waves are collocated with active (suppressed) convective regions as indicated by negative (positive) OLR anomalies. This finding suggests a similar mechanism of modulation with the previous two wave modes.

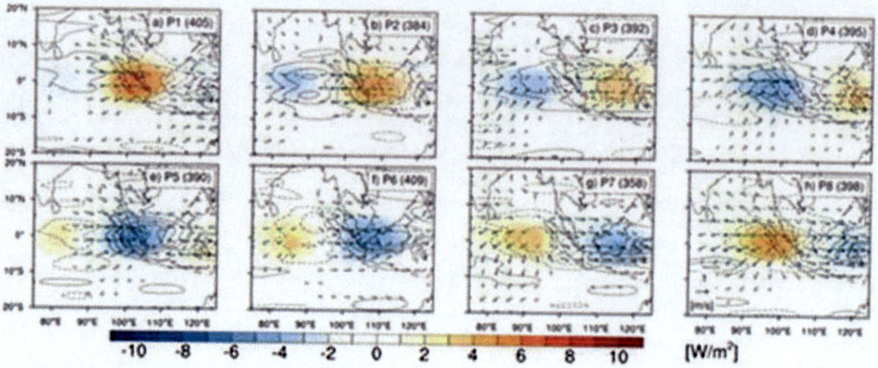

Fig. 6.23 As in Fig. 6.22, but for EIG waves

6.4 Conclusions

We have investigated the impact of CCEWs on extreme rainfall events on Sumatra during the rainy season (November to April) by using ERA-Interim reanalysis data and the gridded Asian Precipitation-Highly Resolved Observational Data Integration Toward Evaluation of Water Resources (APHRODITE) product from 1980 to 2007. Our results showed that CCEWs increase (decrease) the daily rainfall and the probability of extreme rainfall during their wet (dry) phases. It is found that the strongest influence of CCEWs comes from convectively coupled Kelvin waves, which increase the daily rainfall by up to 3 mm/day and contribute up to +100% (−70%) change of extreme rainfall probability during their wet (dry) phase. ER waves contribute up to +80% (−50%) probability change during their wet (dry) phase. On the other hand, the MRG and EIG waves have less impact compared to the other waves, consistent with their weaker activity over the region. Further analysis showed that the increase in daily rainfall and extreme rainfall events are related to the increase in frequency of wet days and the increase of rainfall intensity. Moreover, the modulation of extreme rainfall by CCEWs is linked to wave-induced circulation in the lower troposphere. In particular, CCEWs modulate horizontal winds and pressure centers such that regions of convergence are formed over Sumatra during the wet phases. In these convergence regions, the convective activity is enhanced and leads to the occurrence of extreme rainfall events.

Acknowledgements APHRODITE data are publicly available online from the Hirosaki University website at http://aphrodite.st.hirosakiu.ac.jp/download/.

Authors Contribution All authors contributed equally as the main contributors to this work.

References

1. Statistics Indonesia: Population of Indonesia by Province 1971, 1980, 1990, 1995, 2000, and 2010. https://www.bps.go.id/statictable/2009/02/20/1267/penduduk-indonesia-menurut-provinsi-1971-1980-1990-1995-2000-dan-2010.html. Accessed19 May 2020 (2012)
2. Zhang, C.: Madden-Julian oscillation. Rev. Geophys. **43**, 1–36 (2005)
3. Hidayat, R., Kizu, S.: Influence of the Madden-Julian oscillation on Indonesian rainfall variability in austral summer. Int. J. Climatol. **30**, 1816–1825 (2010)
4. Muhammad, F.R., Lubis, S.W., Setiawan, S.: Impacts of the Madden–Julian oscillation on precipitation extremes in Indonesia. Int. J. Climatol. **41**, 1970–1984 (2021)
5. McGregor, G.R., Nieuwolt, S.: Tropical climatology: an introduction to the climates of the low latitudes. Wiley, West Sussex (1998)
6. Aldrian, E., Susanto, R.D.: Identification of three dominant rainfall regions within Indonesia and their relationship to sea surface temperature. Int. J. Climatol. **23**, 1435–1452 (2003)
7. Philander, S.G.H.: El Nino Southern oscillation phenomena. Nature **302**, 295–301 (1987)
8. Saji, N.H., Goswami, B.N., Vinayachandran, P.N., Yamagata, T.: A dipole mode in the tropical Indian Ocean. Nature **401**, 360–363 (1999)
9. Nur'utami, M.N., Hidayat, R.: Influences of IOD and ENSO to Indonesian rainfall variability: role of atmosphere-ocean interaction in the Indo-Pacific sector. Procedia Environ. Sci. **33**, 196–203 (2016)
10. Supari, T.F., Salimun, E., Aldrian, E., Sopaheluwakan, A., Juneng, L.: ENSO modulation of seasonal rainfall and extremes in Indonesia. Clim. Dyn. **51**, 2559–2580 (2017)
11. Kiladis, G.N., Wheeler, M.C., Haertel, P.T., Straub, K.H., Roundy, P.E.: Convectively coupled equatorial waves. Rev. Geophys. **47**, 1–42 (2009)
12. Matsuno, T.: Quasi-geostrophic motions in the equatorial area. J. Meteorol. Soc. Jpn. **44**, 25–43 (1966)
13. Yasunaga, K., Mapes, B.: Differences between more divergent and more rotational types of convectively coupled equatorial waves. Part I: space-time spectral analyses. J. Atmos. Sci. **69**, 3–16 (2012)
14. Nababan, C.A.: Identifikasi gelombang Kelvin di lapisan tropopause Indonesia bagian barat dengan menggunakan data sounding NOAA. Thesis, IPB University (2009)
15. Ningrum, W.: Analisis penjalaran gelombang Kelvin di atas Kototabang berbasis data EAR (equatorial atmosphere radar). Thesis, IPB University
16. Lubis, S.W.: Analisis gelombang atmosfer ekuatorial planeter (EPW) di lapisan tropopause dan stratosfer bawah Indonesia dengan menggunakan data NCEP/NCAR Reanalysis I. Thesis, IPB University (2010)
17. Yuni, A.R.E., Lubis, S.W., Setiawan, S.: Vertical structure of convectively coupled equatorial waves (CCEWs) during boreal summer and winter. IOP Conf. Ser.: Earth Environ. Sci. **284**, 012010 (2019)
18. Fajary, F.R.: Karakteristik gelombang Kelvin ekuatorial atmosferik pada kondisi normal (1990), La Nina (1988–1989), dan El Nino (1997–1998). Thesis, IPB University (2013)
19. Alsepan, G., Lubis, S.W., Setiawan, S.: Analysis of the equatorial lower stratosphere Quasi-Biennial Oscillation (QBO) using ECMWF-Interim reanalysis data set. IOP Conf. Ser.: Earth Environ. Sci. **31**, 012032 (2016)
20. Fathullah, N.Z., Lubis, S.W., Setiawan, S.: Characteristics of Kelvin waves and mixed Rossby-gravity waves in opposite QBO phases. IOP Conf. Ser.: Earth Environ. Sci. **54**, 012032 (2017)
21. Kartika, Y.: Aktivitas gelombang Kelvin fase timuran dan fase baratan QBO dengan menggunakan data ketinggian geopotensial keluaran ECMWF- Interim. Thesis, IPB University (2017)
22. Kusuma, L., Lubis, S.W., Setiawan, S.: Unprecedented Quasi-Biennial Oscillation (QBO) disruption in 2015–2016: implications for tropical waves and ozone. IOP Conf. Ser.: Earth Environ. Sci. **284**, 012016 (2019)

23. Muhammad, F.R., Lubis, S.W., Tiarni, I., Setiawan, S.: Influence of the Indian Ocean Dipole (IOD) on convectively coupled Kelvin and mixed Rossby-gravity waves. IOP Conf. Ser.: Earth Environ. Sci. **284**, 012012 (2019)
24. Solihah, K.I., Lubis, S.W., Setiawan, S.: Influence of QBO on stratospheric Kelvin and mixed Rossby gravity waves in high-top CMIP5 models. IOP Conf. Ser.: Earth Environ. Sci. **149**, 012011 (2018)
25. Zakaria, D., Lubis, S.W., Setiawan, S.: Seasonal variability of convectively coupled equatorial waves (CCEWs) in recent high-top CMIP5 models. IOP Conf. Ser.: Earth Environ. Sci. **149**, 012030 (2018)
26. Mawarti, M.N., Lubis, S.W., Setiawan, S., Muhammad, F.R.: On the interpretation of EOF analysis of the convectively coupled equatorial waves. Proc. SPIE **11372**, 113720N (2019)
27. Ferrett, S., Yang, G.Y., Woolnough, S.J., Methven, J., Hodges, K., Holloway, C.E.: Linking extreme precipitation in Southeast Asia to equatorial waves. Quart. J. Roy. Meteorol. Soc. **146**, 665–684 (2020)
28. Baranowski DB, Flatau MK, Flatau PJ, Karnawati D, Barabasz K, Labuz M, Latos B, Schmidt JM, Paski JAI, Marzuki (2020) Social-media and newspaper reports reveal large-scale meteorological drivers of floods on Sumatra. Nat. Commun. **11**, 1–10
29. Lubis, S.W., Respati, M.R.: Impacts of convectively coupled equatorial waves on rainfall extremes in Java, Indonesia. Int. J. Climatol. **41**, 2418–2440 (2021)
30. Latos, B., Lefort, T., Flatau, M.K., Flatau, P.J., Permana, D.S., Baranowski, D.B., Paski, J.A.I., Makmur, E., Sulystyo, E., Peyrille, P., et al.: Equatorial waves triggering extreme rainfall and floods in Southwest Sulawesi, Indonesia. Mon. Weather Rev. **149**, 1381–1401 (2021)
31. Lubis, S.W., Jacobi, C.: The modulating influence of convectively coupled equatorial waves (CCEWs) on the variability of tropical precipitation. Int. J. Climatol. **35**, 1465–1483 (2015)
32. van der Linden, R., Fink, A.H., Pinto, J.G., Van, T.P., Kiladis, G.N.: Modulation of daily rainfall in Southern Vietnam by the Madden-Julian oscillation and convectively coupled equatorial waves. J. Clim. **29**, 5801–5820 (2016)
33. Schlueter, A., Fink, A.H., Knippertz, P., Vogel, P.: A systematic comparison of tropical waves over Northern Africa. Part I: influence on rainfall. J. Clim. **32**, 1501–1523 (2019)
34. Liebmann, B., Smith, C.A.: Description of a complete (interpolated) outgoing longwave radiation dataset. Bull. Am. Meteor. Soc. **77**, 1275–1277 (1996)
35. Yatagai, A., Kamiguchi, K., Arakawa, O., Hamada, A., Yasutomi, N., Kitoh, A.: APHRODITE: constructing a long-term daily gridded precipitation dataset for Asia based on a dense network of rain gauges. Bull. Am. Meteor. Soc. **93**, 1401–1415 (2012)
36. Dee, D.P., Uppala, S.M., Simmons, A.J., Berrisford, P., Poli, P., Kobayashi, S., Andrae, U., Balmaseda, M.A., Balsamo, G., Bauer, P., et al.: The ERA-Interim reanalysis: configuration and performance of the data assimilation system. Q. J. R. Meteorol. Soc. **137**, 553–597 (2011)
37. The NCAR Command Language (Version 6.6.2) [Software]: UCAR/NCAR/CISL/TDD, Boulder, Colorado (2019). https://doi.org/10.5065/D6WD3XH5
38. Wheeler, M., Kiladis, G.N.: Convectively coupled equatorial waves: analysis of clouds and temperature in the wavenumber-frequency domain. J. Atmos. Sci. **56**, 374–399 (1999)
39. Riley, E.M., Mapes, B.E., Tulich, S.N.: Clouds associated with the Madden-Julian oscillation: a new perspective from CloudSat. J. Atmos. Sci. **68**, 3032–3051 (2011)
40. Ren, H.L., Ren, P.: Impact of Madden-Julian oscillation upon winter extreme rainfall in southern China: observations and predictability in CFSv2. Atmosphere **8**, 1–14 (2017)
41. Roundy, P.E., Frank, W.M.: A climatology of waves in the equatorial region. J. Atmos. Sci. **61**, 2105–2132 (2004)
42. Yang, G.Y., Hoskins, B.J., Slingo, J.M.: Convectively coupled equatorial waves. Part III: synthesis structure and extratropical forcing. J. Atmos. Sci. **64**, 3438–3451 (2007)
43. Dias, J., Kiladis, G.N.: Influence of the basic state zonal flow on convectively coupled equatorial waves. Geophys. Res. Lett. **41**, 6904–6913 (2014)

Chapter 7
Kelvin Wave Activity in the UTLS Over the Maritime Continent from GPS RO Measurements

Diah A. Tiyas, Sandro W. Lubis, and Sonni Setiawan

Abstract Kelvin waves are one of the most dominant atmospheric waves in the tropical upper-troposphere and lower-stratosphere (UTLS). However, research related to the characteristics of Kelvin waves in the UTLS over the maritime continent is still limited. Here, we utilize long-term GPS RO measurements from January 2002 to December 2014 to study the characteristics of Kelvin waves over the maritime continent in the UTLS. The results show that the periodicity of the observed Kelvin wave over the Maritime Continent ranges from ~4–17 days and with the eastward zonal wavenumber of 1–4. The vertical wavelengths of the Kelvin wave are ranging from 2.8 to 4.9 km in the upper-troposphere, 2.8–7 km in the UTLS, and 4.9–12.1 km in the mid-stratosphere. It is also shown that the vertically propagating Kelvin wave has phase speed lines that tilt eastward with height and has a downward phase speed propagation in time. Furthermore, analysis of the Kelvin waves activity in the upper-troposphere reveals that propagation and phase speed of Kelvin waves are much slower than those in the UTLS and mid-stratosphere due to the convection. The seasonal analysis also indicates that Kelvin waves are stronger in the JJA and SON periods.

Acronym and Abbreviations

CHAMP	CHAllenging Minisatellite Payload
C/NOFS	Communications/Navigation Outage Forecasting System
FORMOSAT-3/COSMIC	FORMOsa SATellite mission-3/Constellation Observing System for Meteorology, Ionosphere, and Climate

D. A. Tiyas (✉) · S. Setiawan
Department of Geophysics and Meteorology, IPB University, Bogor, Indonesia
e-mail: diahhayu88@gmail.com

S. W. Lubis
Rice University, Houston, USA

E. Yulihastin et al. (eds.), *Proceedings of the International Conference on Radioscience, Equatorial Atmospheric Science and Environment and Humanosphere Science, 2021*, Springer Proceedings in Physics 275, https://doi.org/10.1007/978-981-19-0308-3_7

GRACE	GRavity And Climate Experiment
MetOp	Meteorological Operational satellite
SAC-C	Satelite de Aplicaciones Cientificas-C

7.1 Introduction

Kelvin waves are atmospheric disturbances in the tropics generated by convection or latent heat release and propagate eastward [15]. The equatorial Kelvin waves are considered to be one of the most dominant gravitational waves in the tropical stratosphere, especially in the dynamics of the middle atmosphere [30]. Kelvin waves with a period of 15–20 days, were first discovered in the Western Pacific at 20–80 mb [34]. Kelvin waves propagate eastward with a period of about 4.5–25 days and carry an upward westward momentum with a vertical wavelength of about 5–12 km [1, 3, 12, 21, 24, 31]. The meridional scale of the Kelvin waves ranges from 1300 to 1700 km [13]. Kelvin waves appear at an altitude of 16–60 km with different periods and wavelengths in the vertical direction [5].

Kelvin waves have significant impacts on the circulation and precipitation in the tropics. Several previous studies have shown that the Kelvin waves influence rainfall variability in the tropics with a contribution of up to 16–20% of the total intraseasonal rainfall variance [17]. In addition, Kelvin waves also increase the occurrence of extreme rainfall in the SE Asia and Java, Indonesia by up to 60% of the basic probability [8, 18]. Kelvin waves also play a pivotal role in the dynamics of the atmosphere as they transfer momentum and energy eastward from the troposphere to the stratosphere and even to the higher altitudes. Kelvin waves also contribute about 30–50% of the magnitude of the wind reversal associated with the Quasi-Biennial Oscillation (QBO), as well as 20–35% of the total wave force magnitude at a distance of 20–40 km [6].

GPS-RO (Global Positioning System Radio Occultation) has been used widely to study the atmospheric profiles in the UTLS over the maritime continent [28]. GPS-RO has a vertical resolution from 100 m in the troposphere [9] to 1.5 km in the stratosphere [16] and an accuracy of 0.7–1 K at 8–25 km [27]. The advantages of GPS-RO data include global coverage, long-term data stability, and high vertical resolution [16]. In addition, GPS-RO data is available in good coverage in space and time [2]. GPS-RO measures the atmospheric profile using transformation in electromagnetic wave signals from GPS with frequencies GPSL1 (1575.42 MHz) and L2 (1227.62 MHz) that pass through the atmosphere. Then the signal will be received by RO satellites such as CHAMP, GRACE, SAC-C, Formosat-3/COSMIC, MetOp, and C/NOFS satellites. This study uses a dry temperature atmospheric profile from measurements for analysis. This is because this study only investigated the temperature variability above 10 km where the humidity effect was small.

Various studies have investigated the activity of the Kelvin waves activity using GPS-RO in the tropics. In particular, observations of Kelvin wave activity have been carried out at UTLS over the Indian Ocean using GPS-RO with a period of more than 10 days, a vertical wavelength of 5–12 km, and a phase velocity of 75.6–92.6 km/hour [3]. Another study showed that Kelvin waves activity in the UTLS over the Indian Ocean has a phase speed of 8.9 m/s, a vertical wavelength of 5.6–6.2 km, and a period of 5–6 days [24]. Despite clear evidence of the Kelvin waves activity in the UTLS over the Indian Ocean, the characteristics of Kelvin waves over the maritime continent in the UTLS remain elusive. Here, we investigate the characteristics of Kelvin waves activity in the UTLS over the Maritime Continent by using 13-year GPS-RO data with a wider area coverage in the equator. We will also examine the behavior Kelvin waves activity over the maritime continent in the upper-troposphere, UTLS, and mid-stratosphere for different seasons.

7.2 Data and Methods

7.2.1 Data

Global Positioning System Radio Occultation (GPS-RO) data obtained from satellite measurements CHAMP (2002–2008), GRACE (2007–2014), SAC-C (2006–2011), Formosat-3/COSMIC (2006–2014) from The Wegener The Center for Climate and Global Change uses the Occultation Processing System (OPS) version 5.6 at longitudes 165°W–165°E. The data variable used is dry temperature. Dry temperature data is obtained from the GPS-RO observation bias with the assumption that water vapor can be ignored or zero. The dry temperature was used because the observations were carried out at 10 km and above, where humidity was negligible. The vertical resolution of GPS- RO is 100 m in the troposphere to 1.5 km in the stratosphere. While the level of accuracy is 0.7–1 K at a distance of 8 to 25 km. The data period used was January 1, 2002–December 31, 2014.

7.2.2 Methods

7.2.2.1 Anomalies and Detrending

The anomalies are computed by subtracting the daily climatological data and the detrending is estimated by removing the least-squares linear trend. Anomalies with positive values indicate the data is higher than the baseline, while negative anomalies indicate the data is lower than the baseline.

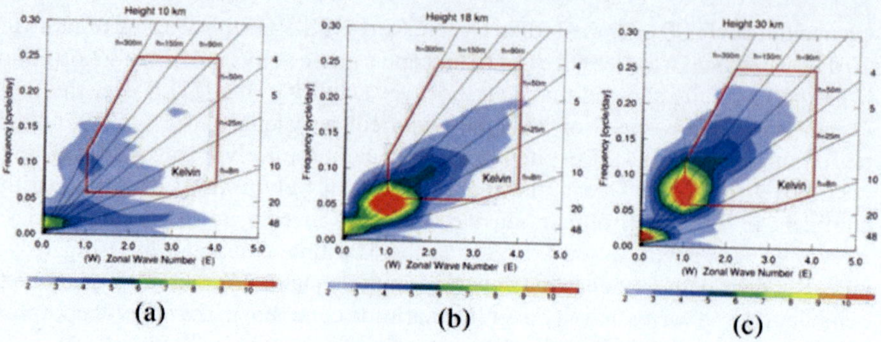

Fig. 7.1 Wavenumber-frequency diagram of temperature from GPS-RO measurement at **a** 10 km, **b** 18 km, and **c** 30 km calculated based on all calendar years

7.2.2.2 Space–Time Spectral Analysis (STSA)

The STSA method is used to isolate wave signals by decomposing the atmospheric field into zonal wave number (k)) and frequency (ω) using two-dimensional Fourier transform technique [11, 17, 20, 35]. There is no decomposition in symmetrical and asymmetrical components (e.g., see Figs. 7.1 and 7.2) in this STSA method [17]. The STSA procedure for each season is analyzed by grouping the data into 4 different seasons: December to February (DJF), March to May (MAM), June to August (JJA), and September to November (SON).

7.2.2.3 Wave Number and Frequency Filter

In this study, Kelvin waves are filtered from dry temperature anomalies with a period of 4–17 days with wave number k = + 1 to + 4 and at an equivalent depth of 8–300 m.

7.2.2.4 Calculation of Vertical Wavelength and Phase Speed

Vertical wavelength is calculated using vertical wave number, which is obtained from the following equation:

$$m^2 = \frac{N^2}{gh_e} - \frac{1}{4H^2} \tag{7.1}$$

where m indicates the vertical wave number, g indicates the acceleration of gravity and H indicates the wave height scale with a value of 8000 m. The Brunt–Väisälä frequency (N) indicates how much energy is required to exchange water parcels vertically. The Brunt–Väisälä frequency changes with altitude and depends on temperature (the gradient of potential temperature with altitude). Here, we assume the value of

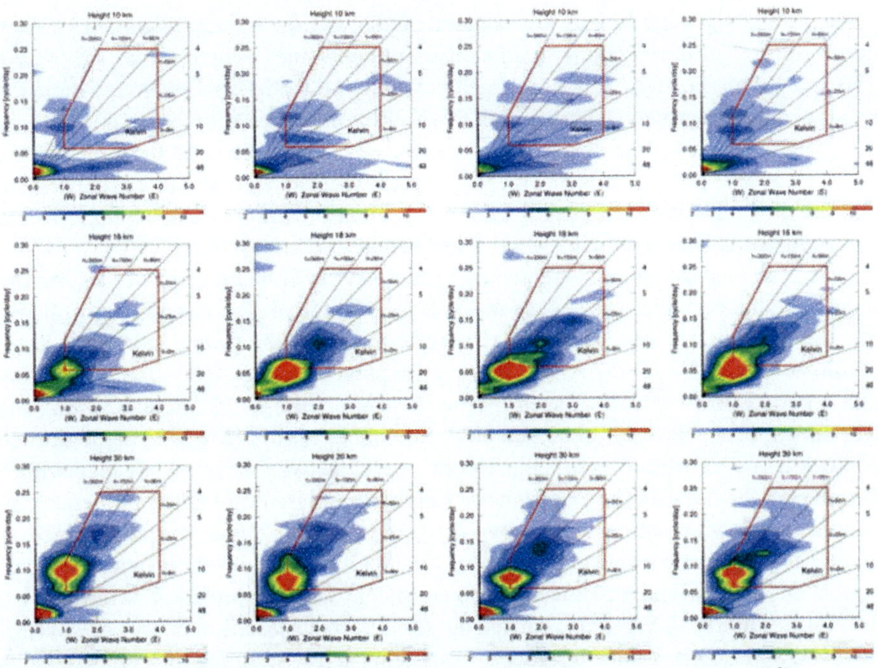

(a) (b) (c) (d)

Fig. 7.2 As in Fig. 7.1 but calculated for different seasons: **a** DJF, **b** MAM, **c** JJA, **d** SON

N = 0.02 s–1 is constant in the stratosphere as the variation in the UTLS is small. Vertical wavelength (λ) [14] and phase speed (c) [13] for Kelvin waves are calculated following this equation:

$$\lambda = \frac{2\pi}{m} \tag{7.2}$$

$$c = \sqrt{ghe} \tag{7.3}$$

7.3 Results and Discussions

7.3.1 Relational Dispersion Analysis of the Kelvin Waves

Figure 7.1 shows the wave spectrum from GPS-RO dry temperature data for 13 years (2002–2014). The positive sign of the wave number (k) indicates the Kelvin waves propagate from the west to the east. This direction of propagation is consistent with a general theory of the Kelvin waves based on the shallow water equation [13]. Each curve in Fig. 7.1 corresponds to an equivalent depth (he), which are 8, 25, 50, 90, 150 and 300 m. Equivalent depth is the thickness of the atmospheric layer where the wave occurs effectively. The strength of the Kelvin waves activity is indicated by the color contours on the diagram. Where the red color of the contour, indicates the Kelvin waves is getting stronger. Wave spectrum analysis with STSA was carried out at three different altitudes, namely 10 km to represent the upper-troposphere, 18 km for represent UTLS layer, and 30 km to represent mid-stratosphere.

The results show that the frequency of the wave will be greater with increasing height and the periodicity is getting faster. At an altitude of 30 km (mid stratosphere), the wave frequency reaches 0.06–0.23 cycles/day and the period is 4.2–17 days. While at an altitude of 18 km (UTLS) the frequency reaches 0.06–0.20 cycles/day and a period of 5–17 days. Based on the wave spectrum at 18 km and 30 km, the Kelvin waves are strongly observed at wave number $k = 1$. However, at an altitude of 10 km (upper-troposphere), the Kelvin waves activity is much weaker with a period of 6–17 days. In the troposphere, Kelvin waves are dominantly influenced by convective activity, which slows down their propagation. The lower phase speed of Kelvin activity in the troposphere is associated with strong convection and shallower equivalent depths. These results are also consistent with the previous studies [3, 19, 24]. Furthermore, STSA is also carried out for different seasons, namely DJF, MAM, JJA, and SON. This is done in order to comprehensively examine the characteristics of the Kelvin waves in the UTLS for different seasons.

The results of the seasonal analysis show that the Kelvin waves activity is stronger in the JJA and SON seasons. This is evident by analyzing their phase speed and the amplitude of the wave spectrum at each height. In particular, it is observed that the deeper the equivalent depth, the faster the phase speed, especially in JJA and SON. This is again consistent with previous studies [7, 33], which found that the Kelvin waves activity is stronger in the JJA and SON seasons. The difference in the characteristics of the Kelvin waves activity for different seasons is summarized in Table 7.2.

7.3.2 *Analysis of the Vertical Structure of Temperature*

The characteristics of the Kelvin waves can also be examined by observing the vertical structure of the temperature in the UTLS layers for different seasons (DJF, MAM, JJA, and SON). Here we also compare Kelvin waves activity at two different longitudes, which represent Western Indonesia (IBB) at 105°E and the Eastern Indonesia (IBT) at 135°E (Fig. 7.3). The comparison of the vertical profile of the temperature shows that at IBB, tropopause height is around 17.2 km during DJF and MAM. Meanwhile, during JJA and SON, the height of the tropopause is between 16.4 and 16.5 km. At IBT, the height of the tropopause during DJF and MAM range between 17.3 and 17.2 km. Meanwhile, during JJA and SON, the height of the tropopause is between

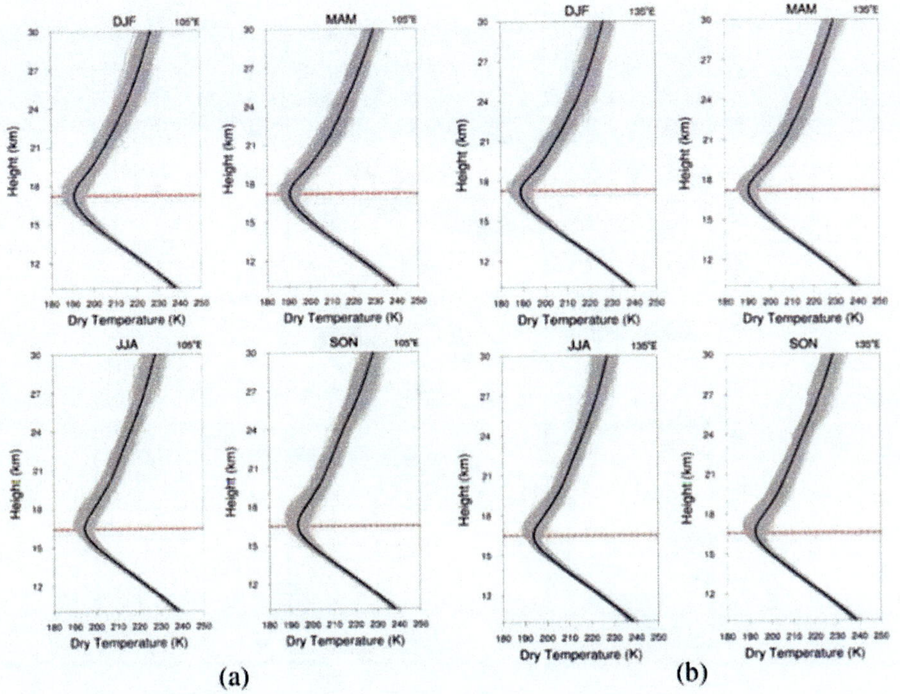

Fig. 7.3 Vertical profile of temperature from GPS-RO measurement (the red dotted line shows the tropopause layer) in **a** Western Indonesia (IBB) and **b** Eastern Indonesia (IBT)

16.5 and 16.6 km. This indicates that during DJF and MAM seasons, tropopause height is higher than during the JJA and SON seasons.

7.3.3 Analysis of the Vertical Propagation of Kelvin Waves

In this section, the analysis is focused in altitude of 15–30 km. This is because the spectrum of Kelvin wave in the upper-troposphere (about 10 km) are much weaker compared to the waves at the higher level (UTLS and mid-stratosphere). Analysis of the vertical structure of temperature-filtered Kelvin waves in the UTLS until mid-stratosphere shows that the vertically propagating Kelvin waves have phase speed lines that tilt eastward with height (Fig. 7.4). This is consistent with the general theory of the Kelvin waves and also consistent with the other previous studies [3, 23, 24, 29]. The results are also consistent with the characteristics of Kelvin waves previously reported [25, 34]. In addition, the observed Kelvin waves in Fig. 7.4 are dominant at an altitude of 17–27 km with the largest amplitude around the UTLS until mid-stratosphere. Analysis was conducted on 31 May 2002 to represent the

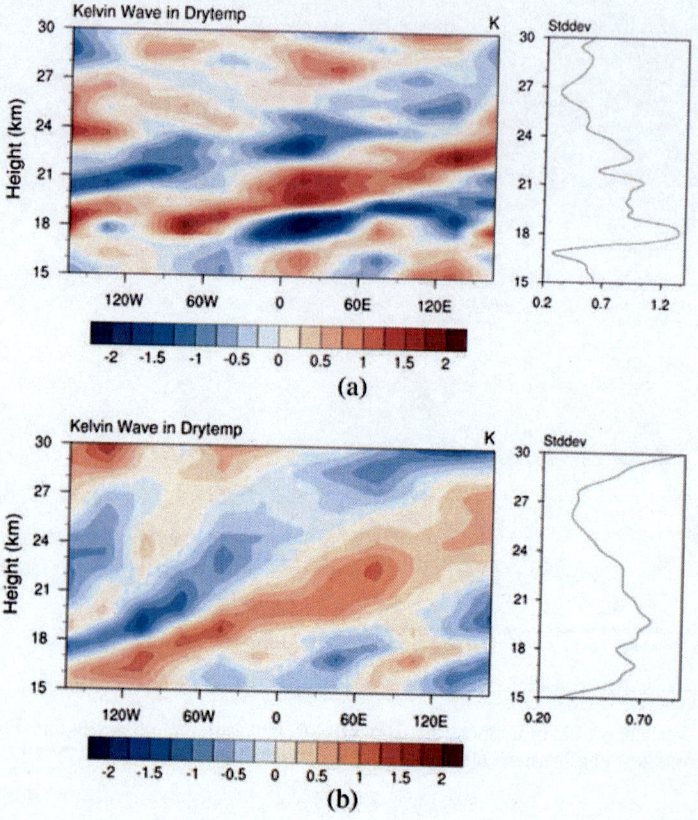

Fig. 7.4 Longitude-height section of the temperature-filtered Kelvin-waves anomaly at the equator: **a** 31 May 2002 and **b** 24 July 2007

MAM season and 24 July 2007 to represent the JJA season. The timing is based on the Kelvin waves activity, which was clearly observed at that time.

Figure 7.5 shows Kelvin waves activity that propagates downwards over time. Kelvin waves activity can clearly and strongly be seen in the JJA and SON seasons (Fig. 7.2). Kelvin waves activity from June to July in Fig. 7.5 shows strong Kelvin waves activity. This is consistent with the results of previous studies where Kelvin waves are intense in JJA and SON [7].

The Kelvin waves that are observed in Indonesia are the waves that are close to the source of the convective source, so this convection process will inhibit the propagation of the Kelvin waves in Indonesia [22]. Figure 7.6 also shows that the Kelvin waves begin to propagate below the UTLS height of ~15 km and extend to more than 30 km (mid-stratosphere). This is consistent with other research that conducted research on Kelvin waves at the same height [3]. In addition, these results are also consistent with a study that analyzed the characteristics of Kelvin waves in the tropical tropopause layer [10].

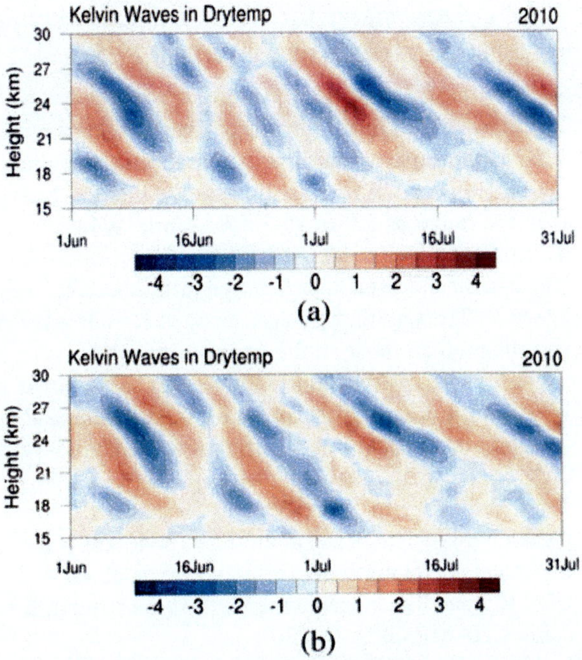

Fig. 7.5 Time-height section of the temperature-filtered Kelvin-wave anomaly at **a** 105° E and **b** 135° E

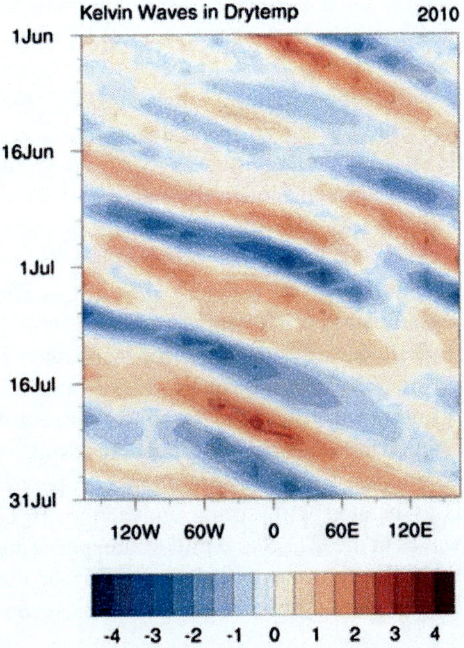

Fig. 7.6 Longitude-time section of the temperature-filtered Kelvin-waves anomaly at 18 km in the equator from June 1 to July 31, 2010

7.3.4 Analysis of Phase Speed and Vertical Wavelength of Kelvin Waves in the UTLS

The phase speed of Kelvin waves over the Maritime Continent in the UTLS is calculated as the square root of the gravitational acceleration multiplied by the equivalent depth. Based on Table 7.2, it is shown that the Kelvin waves phase speed at 10 km is in the range of 8.9 to 15.7 m/s. While at 18 km and 30 km, it ranges from 8.9 to 22.1 m/s and 15.7 to 38.3 m/s, respectively. Based on this result, the observed Kelvin waves over the Maritime Continent in the UTLS have a phase speed ranging from 8.9 to 38.3 m/s, with the phase speed of Kelvin waves in the mid-stratosphere is more significant than those in the upper-troposphere.

The vertical wavelengths of Kelvin waves over the Maritime Continent in the UTLS until mid-stratosphere are ranging from 2.8 to 12.1 km. The vertical wavelength is consistent with several previous studies such as research using Formosat-3/COSMIC with a vertical wavelength of 5–12 km [3] and analysis using the Long-Term Era- Interim Reanalysis in September 2008–February 2009 with a vertical wavelength of 12–18 km at altitude 10–40 km [4]. Also, a previous study found that a vertical wavelength of Kelvin waves is about 4–8 km with CHAMP and SAC-C GPS-RO satellite measurements [25], and found a vertical wavelength of 5–8 km using CHAMP GPS-RO [26].

Variations in the characteristics of the Kelvin waves for different seasons are not significant. The results of this study are consistent where the seasonal variations of Kelvin waves in the UTLS are not significant [7]. The only difference appears during the DJF season at an altitude of 18 km where the phase speed and vertical wavelength are in the lower range, namely 8.9–15.7 m/s and 2.8–4.9 km, while during MAM, JJA and SON the phase speed is 8.9–22.1 m/s and the vertical wavelength is 2.8–7 km. In addition, at an altitude of 30 km the phase speed and vertical wavelength of each season are constant in the range of 15.7–38.3 m/s and 4.9–12.2 km. In the 10 km layer, the Kelvin waves tend to be slower with a phase speed of 8.9–15.7 m/s and a shorter vertical wavelength of 2.8–4.9 km. Based on this result, it can be concluded that the observed Kelvin waves in this study are slow Kelvin waves based on the classification of Kelvin waves [32] (Table 7.1).

Kelvin waves in the upper-troposphere have different characteristics compared to the Kelvin waves in the UTLS and mid-stratosphere level. This characteristic difference is influenced by convection activity in the layer below. The characteristic of Kelvin waves in the upper-troposphere is the Kelvin waves coupled with the convection process. The lower phase speed of Kelvin activity in the upper troposphere is associated with strong convection. In comparison the characteristics of Kelvin waves in the UTLS and mid-stratosphere are the dry waves that are free or uncoupled with convection activity. A summary of the differences in the characteristics of the Kelvin waves for each season can be seen in Table 7.2.

Table 7.1 Brunt-Väisälä frequency, equivalent depth, phase speed, and vertical wavelength of Kelvin waves calculated from the dry temperature data of the GPS-RO measurement

$N(s^{-1})$				
Height (km)	DJF	MAM	JJA	SON
10	0.02	0.02	0.02	0.02
18	0.02	0.02	0.02	0.02
30	0.02	0.02	0.02	0.02
h_e (m)				
10	8–25	8–25	8–25	8–25
18	8–25	8–50	8–50	8–50
30	25–150	25–150	25–150	25–150
Phase speed (m/s)				
10	8.9–15.7	8.9–15.7	8.9–15.7	8.9–15.7
18	8.9–15.7	8.9–22.1	8.9–22.1	8.9–22.1
30	15.7–38.3	15.7–38.3	15.7–38.3	15.7–38.3
Vertical wavelength (λz) (km)				
10	2.8–4.9	2.8–4.9	2.8–4.9	2.8–4.9
18	2.8–4.9	2.8–7	2.8–7	2.8–7
30	4.9–12.1	4.9–12.1	4.9–12.1	4.9–12.1

Table 7.2 Characteristics of the Kelvin waves activity in the UTLS in the different seasons

		DJF	MAM	JJA	SON
	Vertical propagation	Downward propagating with time, and eastward phase tilts with height			
	Periodicity (days)		4–17		
10 km	Phase speed (m/s)		8.9–15.7		
	Zonal wave number		+ 1 until + 4		
	Equivalent depth (m)		8–25		
	Vertical wavelength (km)		2.8–4.9		
18 km	Phase speed (m/s)	8.9–15.7	8.9–22.1	8.9–22.1	8.9–22.1
	Zonal wave number		+ 1 until + 4		
	Equivalent depth (m)	8–25	8–50	8–50	8–50
	Vertical wavelength (km)	2.8–4.9	2.8–7	2.8–7	2.8–7
30 km	Phase speed (m/s)		15.7–38.3		
	Zonal wave number		+ 1 until + 4		
	Equivalent depth (m)		25–150		
	Vertical wavelength (km)		4.9–12.1		

7.4 Conclusions

We have investigated the characteristics of Kelvin waves in the UTLS until mid-stratosphere over the maritime continent by using long-term GPS-RO measurements from January 2002 to December 2014. The results indicate that the periodicity of observed Kelvin waves over the Maritime Continent range from ~4–17 days and with the eastward zonal wave number of 1–4. The dominant Kelvin waves activity is observed at an altitude of 15- 30 km. The phase speeds of Kelvin waves are in the range of 8.9–38.3 m/s. The vertical wavelengths of the Kelvin waves range from 2.8 to 4.9 km in the upper-troposphere, 2.8–7 km in the UTLS, and 4.9–12.1 km in the mid-stratosphere. It is also shown that the vertically propagating Kelvin waves have phase speed lines that tilt eastward with height and have a downward phase speed propagation from the stratosphere to the troposphere in time. In addition, Kelvin waves in the upper-troposphere have a slower phase speed compared to those in the UTLS and mid-stratosphere, due to coupling with convection. The seasonal analysis also indicates that Kelvin waves are stronger in the JJA and SON periods.

References

1. Andrews, D.G., Holton, J.R., Leovy, C.B.: Middle Atmospheric Dynamics. Academic Press (1987)
2. Anthes, R.A.: Exploring Earth's atmosphere with radio occultation: contributions to weather, climate and space weather. Atmos. Meas. Tech. **4**, 1077–1103 (2011). https://doi.org/10.5194/amt-4-1077-2011
3. Brahmanandam, P.S., Chu, Y.H., Liu, J.: Observations of equatorial Kelvin wave modes in FORMOSAT-3COSMIC GPS RO temperature profiles. J. Terr. Atmos. Oceanic Sci. **21**, 829–840 (2010). https://doi.org/10.3319/TAO.2010.01.06.01
4. Das, U., Pan, C.J.: Strong Kelvin wave activity observed during the westerly phase speed of QBO—a case study. Ann. Geophys. **31**, 581–590 (2013). https://doi.org/10.5194/angeo-31-581-2013
5. Dhaka, S.K., Krishna, M.B.V., Nagpal, O.P., Rao, R.R., Sasi, M.N., Sundaresan, S.: A study of equatorial waves in the Indian zone. J. Atmos. Terr. Phys. **57**, 1189–1202 (1995)
6. Ern, M., Preusse, P.: Quantification of the contribution of equatorial Kelvin waves to the QBO wind reversal in the stratosphere. Geophys. Res. Lett. **36**, 1–5 (2009)
7. Fathullah, N.Z., Lubis, S.W., Setiawan, S.: Characteristics of Kelvin waves and Mixed Rossby-Gravity waves in opposite QBO phase speeds. J. IOP Sci. (2017). https://doi.org/10.1088/1755-1315/54/1/012032
8. Ferret, S., Yang, G.Y., Woolnough, S.J., Methven, J., Hodges, K., Holloway, C.E.: Linking extreme precipitation in Southeast Asia to equatorial waves. Q. J. Royal Meteorol. Soc. **146**, 665–684 (2020). https://doi.org/10.1002/qj.3699
9. Gorbunov, M.E., Benzon, H.H., Jensen, A.S., Lohmann, M.S., Nielsen, A.S.: Comparative analysis of radio occultation processing approaches based on Fourier integral operators. Radio Sci. **39** (2004). https://doi.org/10.1029/2003RS002916
10. Harza, A., Lubis, S.W., Setiawan, S.: Activity of convective coupled equatorial wave in tropical tropopause layer in reanalysis and high-top CMIP5 models. IOP Conf. Ser.: Earth Environ. Sci. **149**, 012012 (2018). https://doi.org/10.1088/1755-1315/149/1/012012
11. Hayashi, Y.: Space-time spectral analysis and its applications to atmospheric waves. J. Meteorol. Soc. Jpn **60**, 156–171 (1982)

12. Holton, J.R., Alexander, M.J., Boehm, M.T.: Evidence for short vertical wavelength Kelvin waves in the department of energy-atmospheric radiation measurement Nauru99 radiosonde data. J. Geophys. Res. **106**, 20125–20129 (2001). https://doi.org/10.1029/2001JD900108
13. Holton, J.R.: An Introduction to Dynamic Meteorology, 4th edn. Elsevier, Washington (US) (2004)
14. Kawatani, Y., Sato, K., Dunkerton, T.J., Watanabe, S., Miyahara, S., Takashi, M.: The roles of equatorial trapped waves and internal inertia–gravity waves in driving the Quasi-Biennial oscillation. Part I: zonal mean wave forcing. J. Atmos. Sci. **67**, 963–980 (2010)
15. Kiladis, G.N., Wheeler, M.C., Haertel, P.T., Straub, K.H., Roundy, P.E.: Convectively coupled equatorial waves. Rev. Geophy. **47**(2), 1–42 (2009)
16. Kursinski, E.R., Hajj, G.A., Schofield, J.T., Linfield, R.P., Hardy, K.R.: Observing Earth's atmosphere with radio occultation measurements using the global positioning system. J. Geophys. Res. **102**, 23429–23465 (1997). https://doi.org/10.1029/97JD01569
17. Lubis, S.W., Jacobi, C.: The modulating influence of convectively coupled equatorial waves (CCEWs) on the variability of tropical precipitation. Int. J. Climatol. **35**, 1465–1483 (2015). https://doi.org/10.1002/joc.4069
18. Lubis, S.W., Respati, M.R.: Impacts of convectively coupled equatorial waves on rainfall extremes in Java, Indonesia. Int. J. Climatol. **41**(4) (2020). https://doi.org/10.1002/joc.6967
19. Lubis, S.W., Setiawan, S.: Identifikasi gelombang kelvin atmosfer di Indonesia berbasis data NCEP/NCAR Reanalysis I. Jurnal Fisika **10**, 71–82 (2010)
20. Mawarti, M.N., Lubis, S.W., Setiawan, S., Muhammad, F.R.: On the interpretation of EOF analysis of the convectively coupled equatorial waves. Proceedings of SPIE 11372, Sixth international symposium on LAPAN-IPB Satellite 113720N (2019). https://doi.org/10.1117/12.2541531
21. Mote, P.W., Dunkerton, T.J., Wu, D.: Kelvin waves in stratospheric temperature observed by the Microwave limb sounder. J. Geophys. Res. **107** (2002). https://doi.org/10.1029/2001JD001056
22. Nababan, C.A.: Identifikasi gelombang kelvin di lapisan tropopause Indonesia bagian barat menggunakan data sounding NOAA. Jurnal Sains dan Teknologi Modifikasi Cuaca **11**, 19–27 (2010)
23. Ningrum, W.: Analisis penjalaran gelombang Kelvin di atas Kototabang berbasis data EAR (equatorial atmosphere radar). Thesis, IPB University (2010)
24. Noersomadi, Hadi, T.W.: Downward propagating equatorial kelvin wave over the Eastern Indian Ocean as revealed from radiosonde and GPS radio occultation (CHAMP) data. Jurnal Matematika dan Sains. **15**, 39–45 (2010)
25. Randel, W.J., Wu, F.: Kelvin wave variability near the equatorial tropopause observed in GPS radio occultation measurements. J. Geophys. Res. **110**, 1–13 (2005). https://doi.org/10.1029/2004JD005006
26. Ratnam, M.V., Tsuda, T., Kozu, T., Mori, S.: Long-term behavior of the Kelvin waves revealed by CHAMP/GPS RO measurements and their effects on the tropopause structure. Ann. Geophys. **24**, 1355–2136 (2006)
27. Scherllin-Pirscher, B., Steiner, A.K., Kirchengatst, G., Kuo, Y.H., Foelsche, U.: Empirical analysis and modeling of errors of atmospheric profiles from GPS radio occultation. Atmos. Meas. Tech. **4**, 1875–1890 (2011). https://doi.org/10.5194/amt-4-875-2011
28. Scherllin-Pirscher, B., William, J.R., Joowan, K.: Tropical temperature variability and Kelvin-wave activity in the UTLS from GPS RO measurements. Atmos. Chem. Phys. **17**, 793–806 (2017)
29. Solihah, K.I., Lubis, S.W., Setiawan, S.: Influence of QBO on stratospheric Kelvin and mixed Rossby gravity waves in high-top CMIP5 models. IOP Conf. Ser.: Earth Environ. Sci. **149**, 012011 (2018). https://doi.org/10.1088/1755-1315/149/1/012011
30. Tsai, H.F., Tsuda, T., Hajj, G.A., Wickert, J., Aoyama, Y.: Equatorial Kelvin waves observed with GPS occultation measurements (CHAMP and SACC). J. Meteorol. Soc. Jpn **82**, 397–406 (2004). https://doi.org/10.2151/jmsj.2004.397

31. Tsuda, T., Murayama, Y., Wiryosumarto, H., Harijono, S.W.B., Kato, S.: Radiosonde observations of equatorial atmosphere dynamics over Indonesia 1. Equatorial waves and diurnal tides. J. Geophys. Res. **99**, 10491–10505 (1994). https://doi.org/10.1029/94JD00355
32. Yamazaki, Y., Miyoshi, Y., Xiong, C., Stolle, C., Soares, G., Yoshikawa, A.: Whole atmosphere model simulations of ultra-fast Kelvin wave effects in the ionosphere and thermosphere. J. Geophys. Res. Space Physics **125**, 1–25 (2020). https://doi.org/10.1029/2020JA027939
33. Yuni, A.R.E., Lubis, S.W., Setiawan, S.: Vertical structure of convectively coupled equatorial waves (CCEWs) during boreal summer and winter. IOP Conf. Ser.: Earth Environ. Sci. **284**, 012010 (2019). https://doi.org/10.1088/1755-1315/284/1/012010
34. Wallace, J.M., Kousky V.E.: Observational evidence of Kelvin waves in the tropical troposphere. J. Atmos. Sci. **25**, 900–907 (1968). https://doi.org/10.1175/15200469(1968)025<0900:OEOKWI>2.0.CO;2
35. Wheeler, M., Kiladis, G.N.: Convectively coupled equatorial waves analysis of clouds and temperature in the wavenumber–frequency domain. J. Atmos. Sci. **56**, 374–399 (1999). https://doi.org/10.1175/15200469(1999)056<0374:CCEWAO>2.0.CO;2

Chapter 8
The Influence of the QBO on the MJO-Related Rainfall Variability over the Maritime Continent

Aldiatama Jumanissaba, Sandro W. Lubis, and Sonni Setiawan

Abstract The stratospheric quasi-biennial oscillation (QBO) and tropospheric Madden–Julian oscillation (MJO) have a strong connection over the tropics during boreal winter. Observational analysis has shown that the MJO activity is relatively stronger during easterly QBO (EQBO) than during westerly QBO (WQBO). Using tropical Rainfall Measuring Mission Version 7 (TRMMV7) from 1998 to 2014, we investigate the impacts of the QBO phases on MJO-driven rainfall over the Maritime Continent from December to February. The results show that during EQBO (WQBO), daily rainfall anomaly associated with the wet phases of MJO is increased (decreased) by up to 3.3 mm/day (1.8 mm/day) especially over the eastern and southern regions of the Maritime Continent. Furthermore, during the EQBO the probability of extreme rainfall during the wet phases of MJO is significantly increased by up to 100% over the eastern and southern regions, with a stronger amplitude over the ocean than over the lands. MJO-related rainfall with respect to the different phase of MJO is associated with changes in cloud-radiative feedbacks due to the colder (warmer) tropopause temperature and upward (downward) vertical advection during QBOE (QBOW). These results suggest that QBO-MJO connection can alter large-scale convection and rainfall intensity and extremes over the Maritime Continent.

8.1 Introduction

Quasi-Biennial Oscillation (QBO) is characterized by an oscillation of the easterly and westerly winds in the stratosphere. It propagates downward into the troposphere with a rate of 1 km per month. QBO period varies between 24 and 30 months. The QBO amplitude reaches maximum (approximately 20 m/s) at the equator and decreases with latitude [1, 2]. QBO is formed as a result of the interaction between

A. Jumanissaba (✉) · S. Setiawan
Department of Geophysics and Meteorology, IPB University, Bogor, Indonesia
e-mail: aldiatamajs@gmail.com

S. W. Lubis
Rice University, Houston, USA

E. Yulihastin et al. (eds.), *Proceedings of the International Conference on Radioscience, Equatorial Atmospheric Science and Environment and Humanosphere Science, 2021*, Springer Proceedings in Physics 275, https://doi.org/10.1007/978-981-19-0308-3_8

the atmospheric waves and basic flow [3]. QBO can affect temperature changes and the total distribution of the ozone column in the tropical stratosphere [4, 5].

Over tropics, rainfall variability is influenced by various variability such as Madden–Julian Oscillation (MJO) [6], El-Nino and Southern Oscillation (ENSO), Indian Ocean Dipole (IOD), and Convectively Coupled Equatorial Waves (CCEWs) [7, 8] and diurnal cycles [9]. One that plays an important role in influencing intraseasonal rainfall over the maritime continent (MC) is MJO. Heavy rainfall occurs as a result of organized convection brings by the MJO, which propagates eastward from the Indian Ocean to the Pacific Ocean [10–12].

The correlation of the MJO Index to the QBO Index at various high levels shows a strong QBO-MJO relationship during boreal winter. The amplitude and frequency of MJO formation in the Indian Ocean are increased, as well as the number of more MJO days, during the easterly phase of QBO (EQBO). The amplitude and frequency are decreased, as well as fewer days, during the westerly phase of QBO (WQBO) [13–16]. Increased MJO activity in QBOE has possible effect in increasing intra-seasonal rainfall. This was proved by an increase in MJO-related rainfall anomaly during QBOE in East Asia [17]. However, the impact of QBO on MJO-related rainfall in the tropics, especially in the MC, is still in the studies.

The aims of this study are to analyze and quantify the easterly and westerly phase of QBO effect on MJO intra-seasonal rainfall, and distribution of rainfall extreme over MC during MJO active events. This study also investigates the connection mechanism between QBO, MJO and rainfall in the MC.

8.2 Data and Methods

To study the impact of QBO on MJO-related rainfall, we use re-analysis daily rainfall data from Tropical Rainfall Measuring Mission Version 7 (TRMMv7) with spatial resolution of 0.25° × 0.25° [18]. OLR dataset from NOAA [19] use as a proxy of MJO convection amplitude. We also use zonal wind at 50 hPa, vertical wind at 1000 to 200 hPa, air temperature at 100 hPa, and total column water vapor (TCWV) obtained from European Central Medium-range Weather Forecast (ECMWF) ERA-interim dataset [20]. And also, RMM Indices derived from www.bom.gov.au. The period we use is from 1 January 1998 to 31 December 2014 with daily resolution, except for zonal wind at 50 hPa. Our focus is on the time of boreal winter (December, January, and February) when the correlation of QBO-MJO is high [16].

8.2.1 QBO Index

Easterly and Westerly QBO is obtained using monthly zonal mean zonal wind anomaly at 50 hPa (U50) with 3 monthly moving averages and integrated over 10 °S–10 °N. Deviation from monthly climatology is used to obtain the anomaly [16].

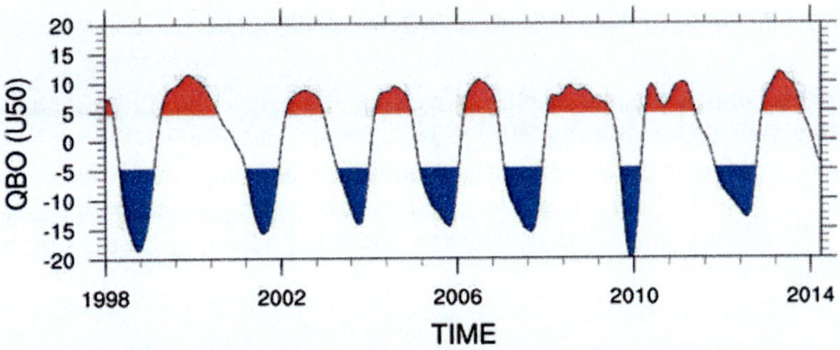

Fig. 8.1 U50 monthly anomaly used as QBO index. Easterly QBO phase are shaded in blue and westerly QBO phase are shaded in red

Easterly phase of QBO (EQBO) is identified as U50 below negative half its standard deviation, and westerly phase of QBO (WQBO) is identified as U50 over positive half its standard deviation. Blue and red shades show EQBO and WQBO in Fig. 8.1, respectively.

8.2.2 *MJO Index*

MJO Index used in this study is the real-time multivariate MJO (RMM) indices, which are calculated using the Empirical Orthogonal Function (EOF) technique based on zonal mean zonal winds at 850 and 200 hPa and OLR over 15 °S–15 °N. The two leading principal components of EOF are set as RMM Indices (RMM1 and RMM2) MJO active event is defined as MJO amplitude ($\sqrt{RMM1^2 + RMM2^2}$) greater than 1. MJO is divided into eight phases based on its enhanced convection area. Phase 1 in Africa, phase 2 and 3 in the Indian Ocean, phase 4 and 5 (also described as *wet phase*) in MC, phase 6 and 7 in the Western Pacific, and phase 8 in the western hemisphere [21].

8.2.3 *Band Pass Filter*

Butterworth band-pass filter technique is used to isolate intraseasonal signals of MJO with cut-off frequencies of 20–100 days [12, 22]. The filter is performed on daily anomaly data over 1998–2014.

8.2.4 Probability Change of Rainfall Extreme

We use probability change of rainfall extreme to analyze the distribution of MJO related rainfall extreme during QBO phases. We first separate MJO active events for each phase during boreal winter. After that, we calculate the cumulative probability (x) that exceeds the 95th percentile (x_c) of all days during boreal winter (P_{All}) and each MJO phase (P_{MJO}). The equation to calculate probability changes of extreme rainfall are:

$$\Delta P_{MJO} = \frac{P_{MJO}(x \geq x_c) - P_{All}(x \geq x_c)}{P_{All}(x \geq x_c)} \times 100 \quad (8.1)$$

where ΔP_{MJO} is the probability changes of extreme rainfall events for each phase of MJO [23].

8.2.5 Composite of Anomaly

To examine the impacts of QBO on MJO related rainfall and the connection between, we create composite by calculating the anomaly of Band-pass filtered data. We separate MJO active events, which are then divided into eight phases [21]. Finally, data are grouped by Easterly and Westerly QBO phases.

8.3 Results and Discussion

8.3.1 Intraseasonal Variations of OLR

Several studies have been using OLR as the main proxy of deep convection in the atmosphere [21, 24, 25]. Figure 8.2 shows that standard deviation of OLR-filtered MJO anomaly reaches its maximum at 22 W/m^2 over the southern part of the MC and is normally distributed along the equator. This indicates strong MJO amplitude during boreal winter in MC.

Comparison of MJO standard deviation OLR anomaly between EQBO and WQBO is shown in Fig. 8.3, which is used to examine the intraseasonal variations of MJO. Positive (negative) value occurs during EQBO (WQBO) reaching its maximum (minimum) approximately at 5 W/m^2 (−7 W/m^2) compare to all winter days over the southern part of the MC. A higher (lower) value of standard deviation OLR-filtered MJO anomaly leads to stronger (weaker) amplitude of MJO when EQBO (WQBO) occurs. Slower MJO eastward propagation and weaker barrier effect of MC causing more MJO events freely across the MC is a possible explanation of MJO deeper convection during EQBO [13, 15].

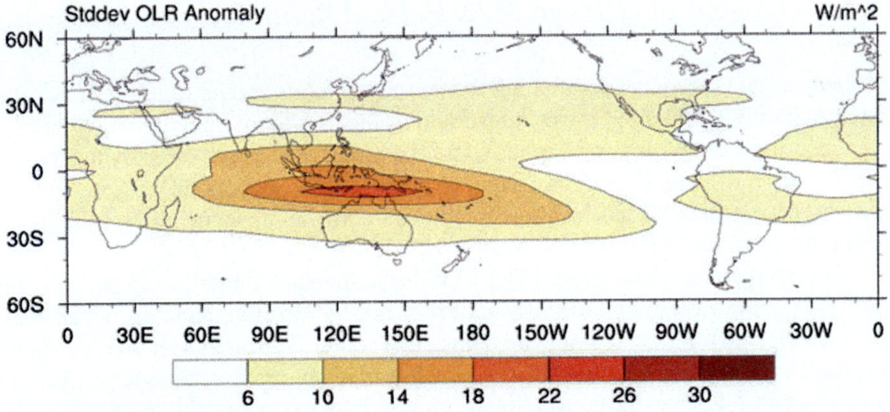

Fig. 8.2 Standard deviation of OLR-filtered MJO anomaly (W/m^2) during boreal winter

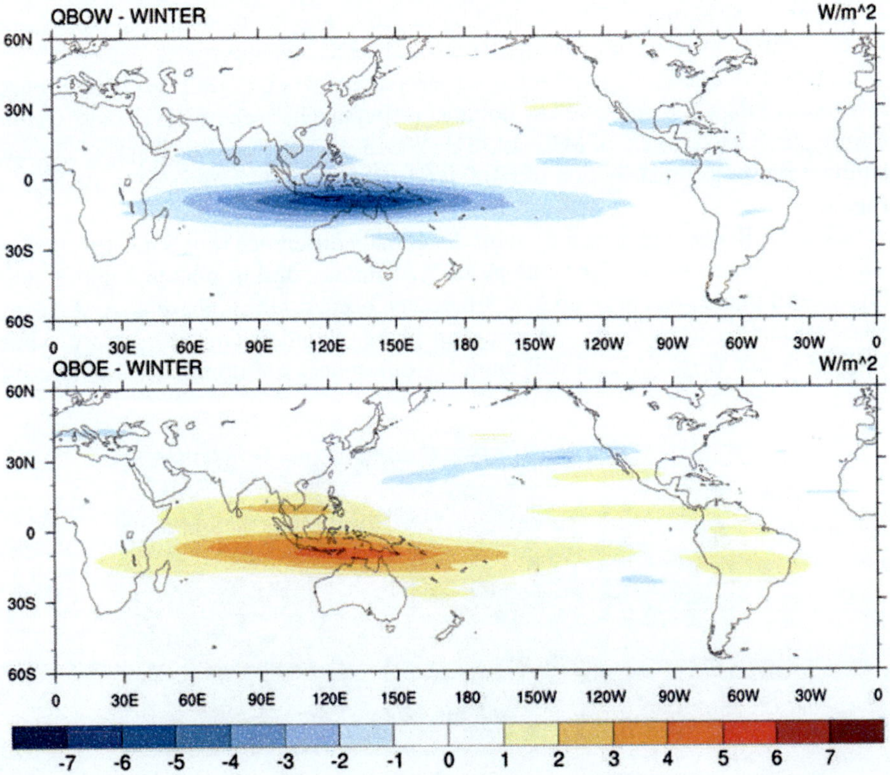

Fig. 8.3 Standard deviation OLR-filtered MJO anomaly difference (W/m^2) of WQBO (top) and EQBO (bottom) to all winter days during boreal winter

8.3.2 Impacts of QBO on MJO Related Rainfall

Boreal winter is associated with high formation of clouds and heavy rainfall over the MC. According to Fig. 8.4, Java and Borneo Island, Southern Sumatra and Sulawesi, and also the Central area of Papua have precipitation averages approximately at a range between 10 to 16 mm/day. Heavy rainfall in MC occurs due to intertropical convergence zone (ITCZ) location in the southern hemisphere and intraseasonal variability such as MJO and CCEWs [7, 9, 26].

Enhanced convection during MJO would also increase rainfall intensity over the MC [12] proved by eastward propagation of rainfall positive anomaly when MJO occurs (Fig. 8.5). Phase 1, since the convective center is located over Africa, shows negative anomaly coverage. Positive anomalies first emerge over the MC in phases 2 and 3. In phase 4, when the convection reaches its maximum, positive anomaly occurs with almost total coverage over the MC within a range of 4–6 mm/day. In phase 5 and phase 6, the convection center is shifting to the eastern part of the MC. Finally, in phases 7 and 8, the MC is covered by negative anomalies when the convection center location moves to the Pacific Ocean and western hemisphere.

Daily rainfall anomaly was increased (decreased) when EQBO (WQBO) occurs, as shown in Figs. 8.6 and 8.7. The contrast difference between them is most visible during phase 2 to phase 6 of MJO. EQBO (WQBO) has a clear (unclear) pattern and similar (dissimilar) distribution of rainfall positive anomaly compared to all days in Fig. 8.5.

When EQBO occurs (Fig. 8.6), rainfall anomaly difference shows mostly positive value. It reaches about +1.8 mm/day to 2.7 mm/day during phases 2 and 3. The biggest difference, reaching up to +3 mm/day, occurs during phase 4. And during phases 5 and 6, it resembles the values in phases 2 and 3. On the contrary, when WQBO occurs (Fig. 8.7), rainfall anomaly differences are dominated by negative

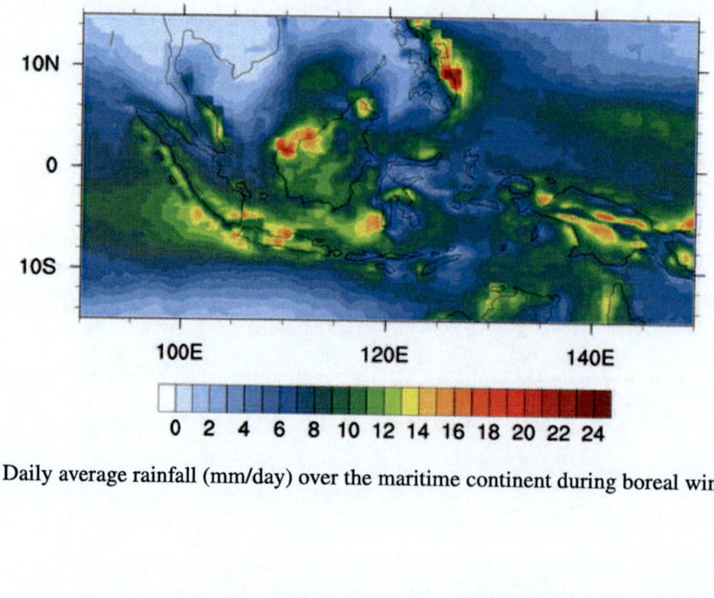

Fig. 8.4 Daily average rainfall (mm/day) over the maritime continent during boreal winter

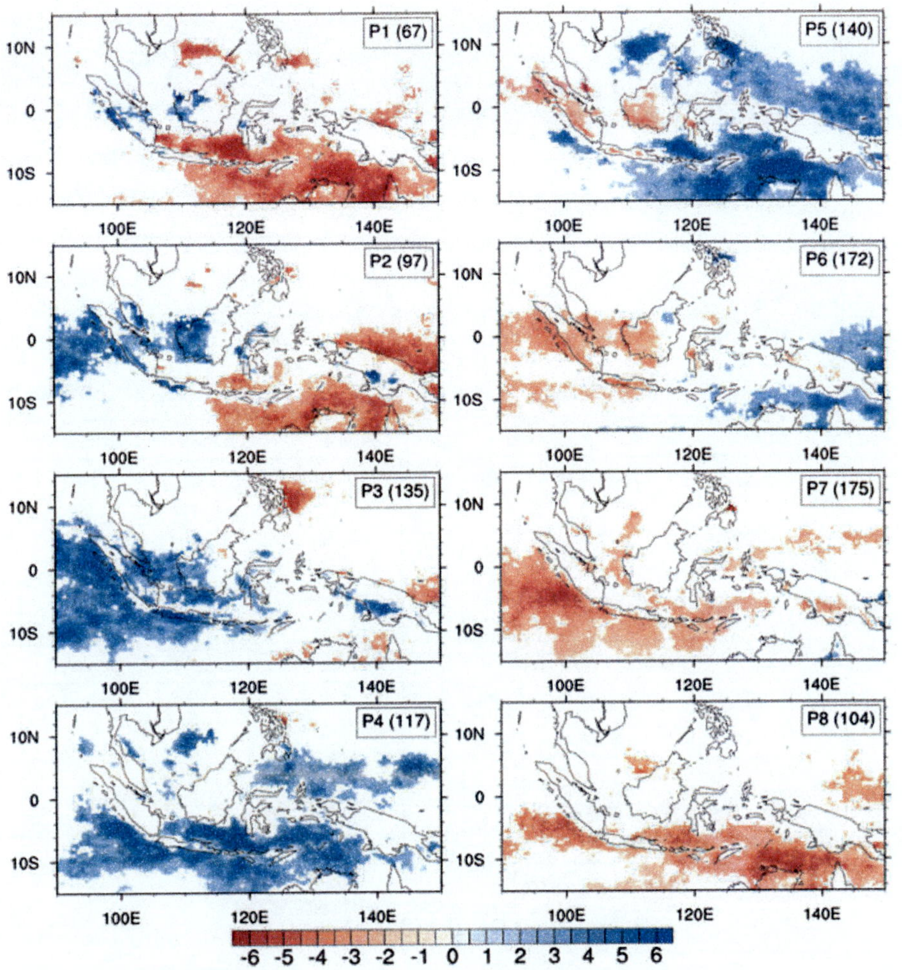

Fig. 8.5 Composites of MJO-daily rainfall anomaly (mm/day) during boreal winter. Only values that are statistically significant at the 95% level are shown

values. In phases 2 and 3, positive anomalous residues were seen and surrounded by unpatterned negative anomalies of −0.9 mm/day; however small areas of positive anomaly up to +1.8 mm/day are seen over Java Island in phase 3. Negative value reaching up from −0.9 to −1.8 mm/day over the southern part of the MC during phase 4. The negative anomaly shifts over the eastern part of the MC during phase 5, while no signs of positive or negative anomaly during phase 6.

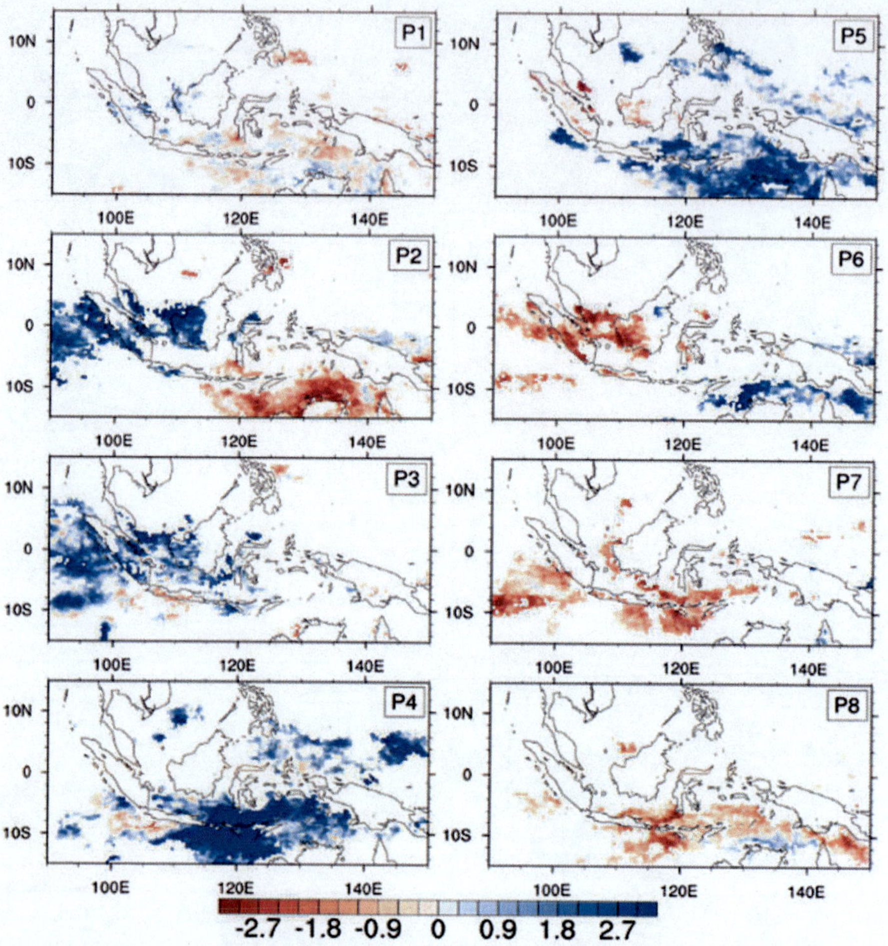

Fig. 8.6 Composites of daily rainfall anomaly difference (mm/day) of EQBO to all winter days during boreal winter. Only values that are statistically significant at the 95% level are shown

8.3.3 Impacts of QBO on MJO Rainfall Extreme

In this study, an extreme event is described as daily rainfall is greater than its 90th percentile during boreal winter, as explained in the previous section. According to the 90th percentile of daily rainfall in Fig. 8.8, extreme rainfall has possible values to reach two times higher than daily rainfall average. MJO modulated extreme rainfall in Fig. 8.9 shows an increase (decrease) in probability changes during phases 2, 3, 4, 5, and 6 (phases 1, 7, and 8). In phases 2 and 3, the probability change increased over the western part while in phases 5 and 6 over the eastern part of the MC with values in a range of 80–200%. The greatest increase is in phase 4 by up to 200%, mainly occurring at the seas.

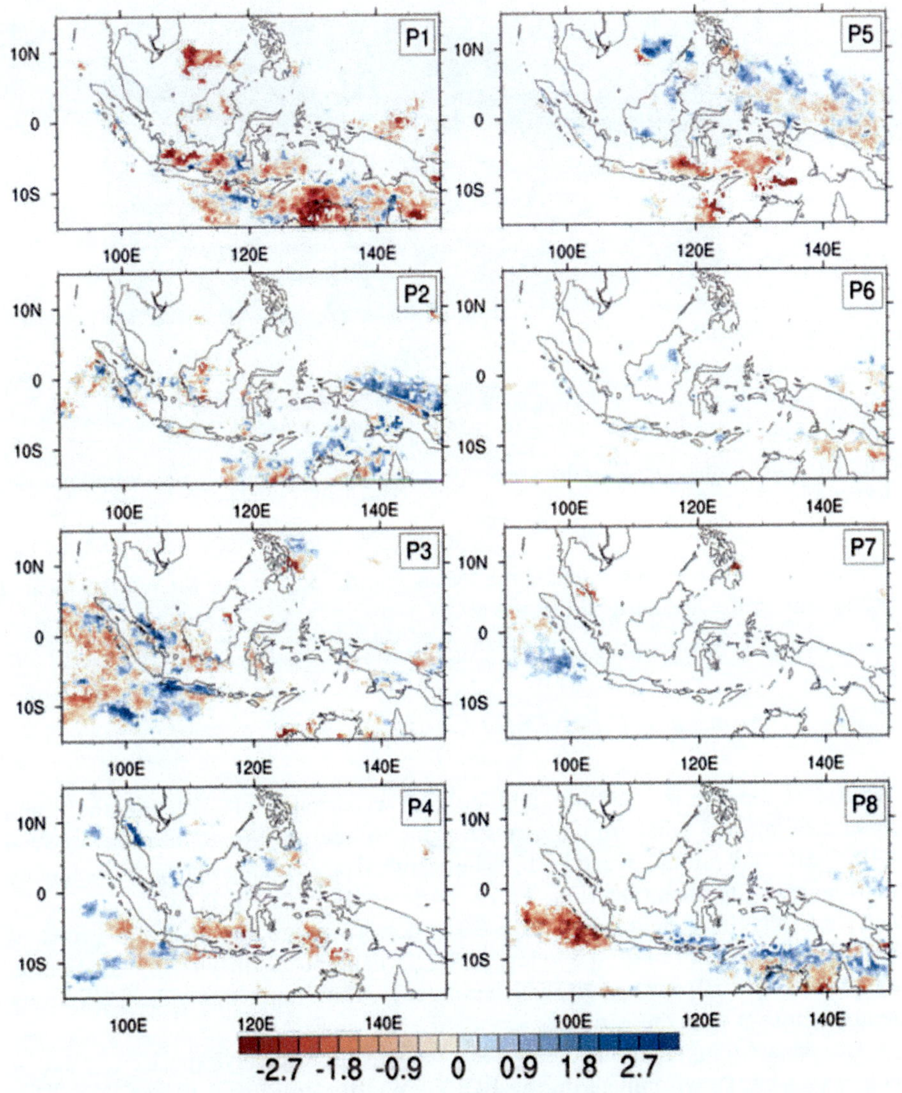

Fig. 8.7 Composites of daily rainfall anomaly difference (mm/day) of WQBO to all winter days during boreal winter. Only values that are statistically significant at the 95% level are shown

When EQBO occurs (Fig. 8.10), the probability change is widely distributed and increases up to two times higher (100%), in contrast to probability change in WQBO (Fig. 8.11) that has lower distribution and decreasing percentage to 70% in phase 4 and 5 over the southern part of MC. The EQBO probability change also observed has a similar pattern to Fig. 8.9. EQBO (WQBO) also has higher (lower) distribution over MC during phase 2 to phase 6 (phases 4 and 5 only). EQBO changes

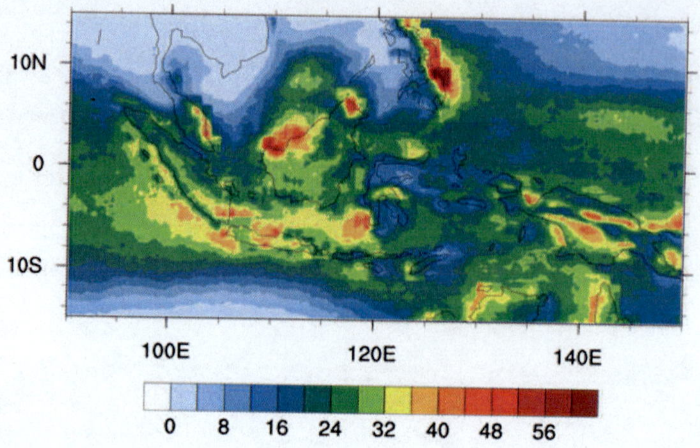

Fig. 8.8 Daily rainfall (mm/day) that exceeds 90th percentile over the maritime continent during boreal winter

in probability are mainly located at the seas across the MC, while for WQBO located at the seas on the northern part of the MC.

8.3.4 Dynamics of QBO-MJO Connections

We use tropopause temperature, total column water vapor (TCWV) anomaly, and integrated vertical wind from various heights to analyze the connection between QBO, MJO, and rainfall over the MC. The eastward propagation of negative anomaly of tropopause temperature (Fig. 8.12), positive anomaly of TCWV (Fig. 8.15), and vertically integrated wind (Fig. 8.18) correspond to the eastward propagation of positive daily rainfall anomaly (Fig. 8.6) and probability change of rainfall extremes (Fig. 8.10) during MJO active events. This harmonization indicates a strong relationship between the variables.

Tropopause temperature anomalies are shown with a negative (positive) value or colder (warmer) environment during EQBO (WQBO) in Figs. 8.13 and 8.14. This condition happened in MJO intensified activity over the MC in phase 2 to phase 6. QBO zonal wind anomaly and temperature anomaly satisfy the thermal wind equation. Wind shear anomaly of EQBO (WQBO) is associated with cold (warm) anomaly of lower stratospheric air temperature over the equator [2, 4, 5]. Tropopause cold temperature is linked to higher tropical tropopause height during QBOE [15]. This also leads to a better environment for organized and large-scale convection growth [27].

In accordance with the tropopause temperature, TCWV is believed as a factor in influencing convection growth. TCWV is described by ECMWF as total water vapor contained by the air from the surface to the top of the atmosphere. Hence, TCWV is

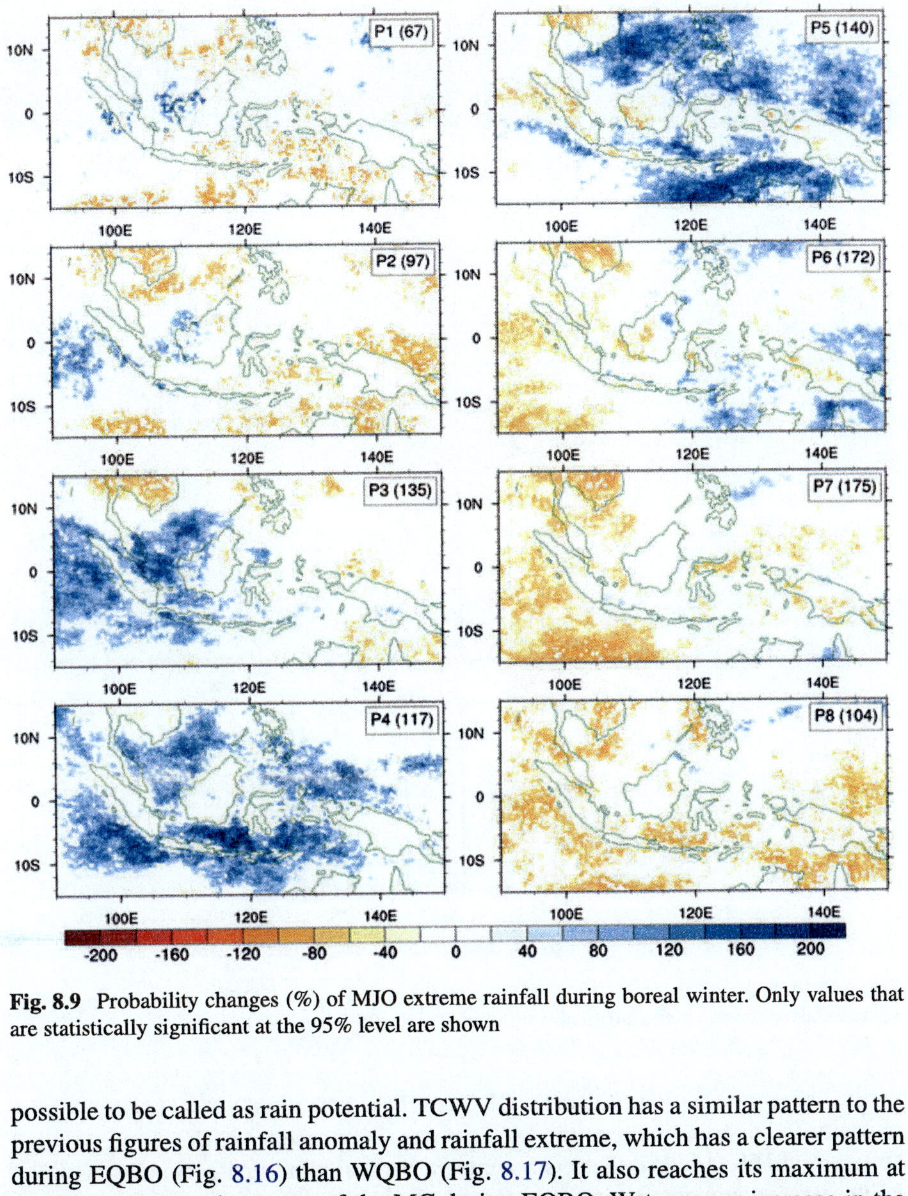

Fig. 8.9 Probability changes (%) of MJO extreme rainfall during boreal winter. Only values that are statistically significant at the 95% level are shown

possible to be called as rain potential. TCWV distribution has a similar pattern to the previous figures of rainfall anomaly and rainfall extreme, which has a clearer pattern during EQBO (Fig. 8.16) than WQBO (Fig. 8.17). It also reaches its maximum at phase 4 in the southern part of the MC during EQBO. Water vapor increase in the atmosphere would also increase daily rainfall anomaly [22].

Furthermore, the connection mechanism between QBO and MJO related rainfall can be explained more by the vertically integrated wind in various heights. In Figs. 8.18, 8.19 and 8.20, negative value shows upward vertical advection while positive value shows downward vertical advection [10]. Corresponding to the previous variables, vertically integrated wind also shows similar results. Upward (downward)

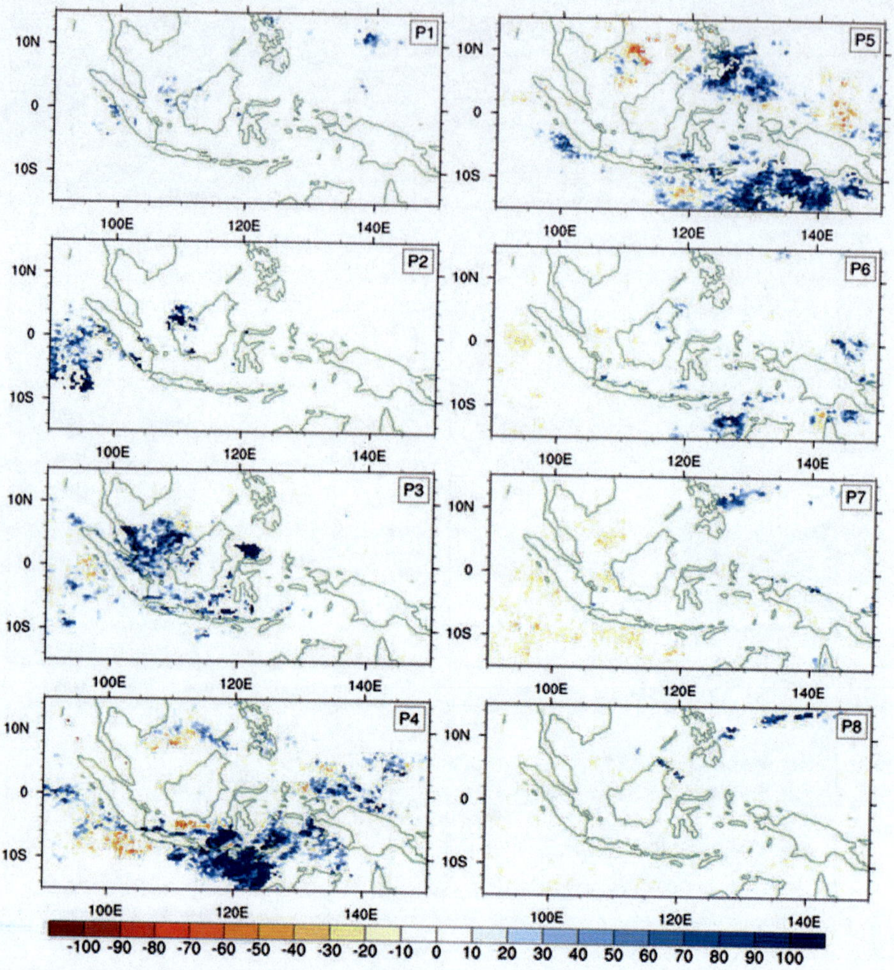

Fig. 8.10 Probability changes (%) of MJO extreme rainfall when EQBO occurs during boreal winter. Only values that are statistically significant at the 95% level are shown

vertical advection during EQBO (WQBO) in Figs. 8.19 and 8.20 has higher (lower) intensity and distribution over the MC in phase 2 to phase 6 of MJO. In the vertical structure of MJO [6, 25], there is an upward vertical wind that gets along with the MJO eastward propagation in the equator. This highly indicates that upward vertical advection has a strong link to the increase of daily rainfall anomaly and changes in the probability of extreme rainfall during EQBO.

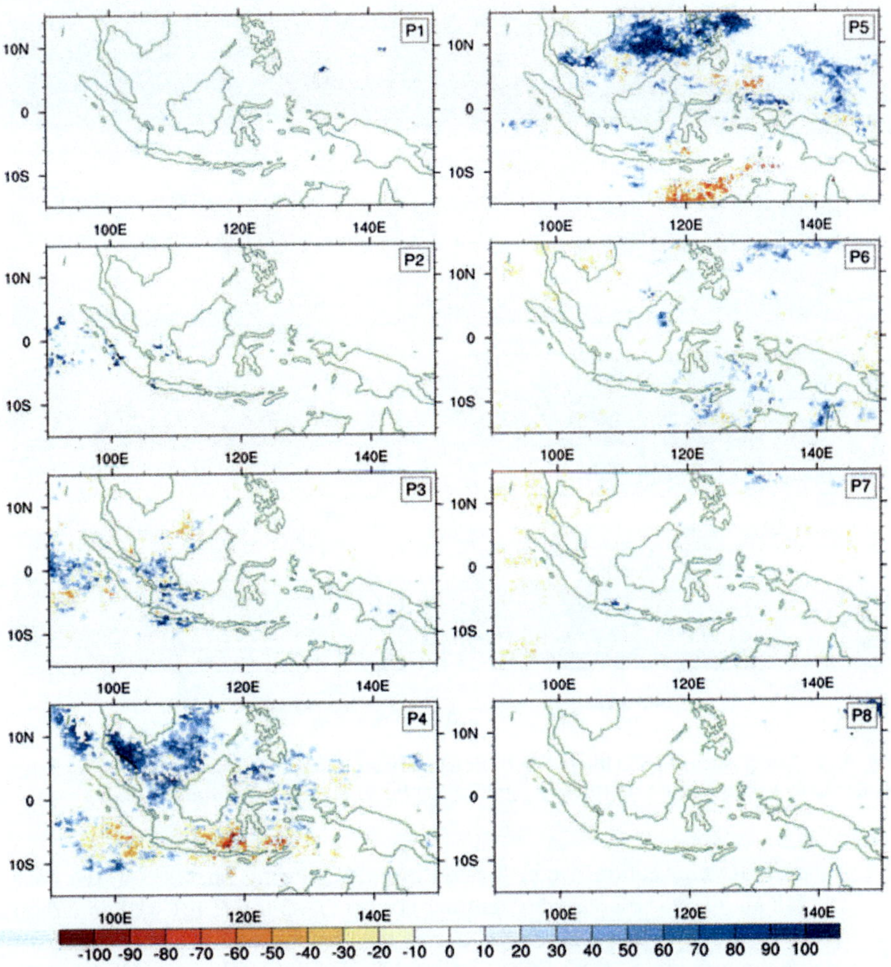

Fig. 8.11 Probability changes (%) of MJO extreme rainfall when WQBO occurs during boreal winter. Only values that are statistically significant at the 95% level are shown

8.4 Summary

We have investigated the impacts of the QBO on MJO rainfall anomaly and extreme rainfall in the maritime continent during boreal winter. We used daily rainfall data from the TRMMv7 satellite measurement and grouped them into different MJO and QBO phases. Our results show that different characteristics of MJO-related rainfall variability and extremes are evident between both different phases of QBO. The key results of our study are summarized below:

1. Stronger (weaker) amplitude of MJO occurs during EQBO (WQBO), relative to all winter days.

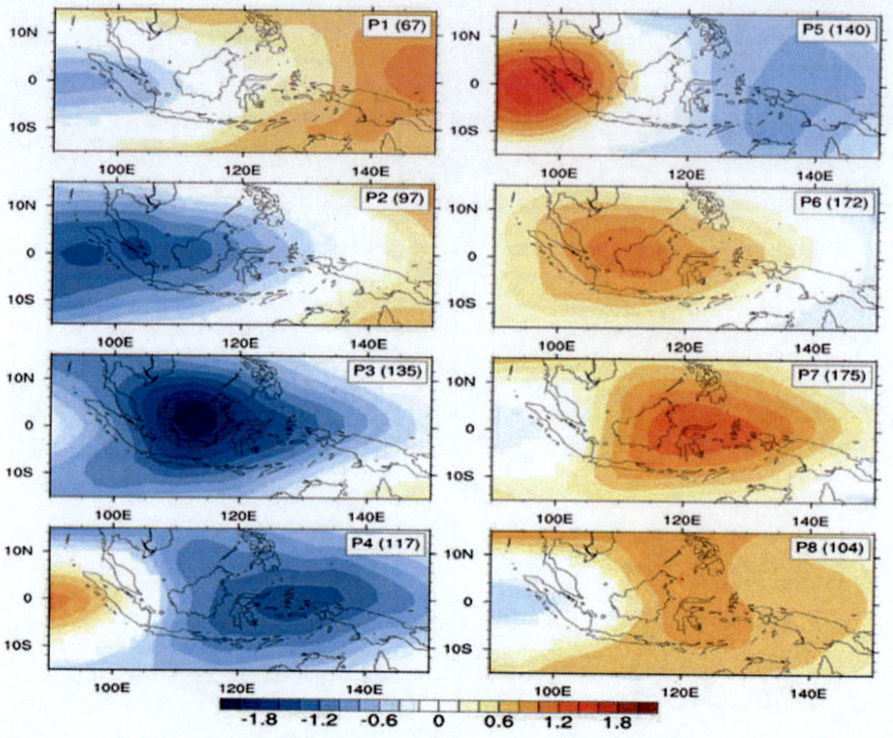

Fig. 8.12 Composites of MJO filtered daily air temperature anomaly (K) at 100 hPa during boreal winter. Only values that are statistically significant at the 95% level are shown

2. When MJO convection center located over MC during phases 3-4, the daily rainfall anomaly is increased (decreased) by up to +3 mm/day (−1.8 mm/day). The probability of extreme rainfall during the wet phase of MJO also increases (decreases) by up to 100% (70%) during EQBO (WQBO), especially over the eastern and southern regions of the MC.
3. Colder (warmer) tropopause temperature occurs in phase 2 to phase 6 of MJO during EQBO (WQBO), causing more (less) organized convection growth.
4. Similar to daily rainfall anomaly, TCWV is greater and more widely distributed during EQBO than during WQBO.
5. Upward (downward) vertical advection during EQBO (WQBO) indicates its high correlation to higher (lower) daily rainfall anomaly and increase (decrease) of changes in rainfall extreme probability.

Our results also suggest that (a) the colder (warmer) tropopause temperature, (b) higher (lower) distribution of total column water vapor, and (c) higher (lower) intensity of upward vertical advection during EQBO (WQBO), together favor to an increase (decrease) in daily rainfall anomaly and extreme probability distribution.

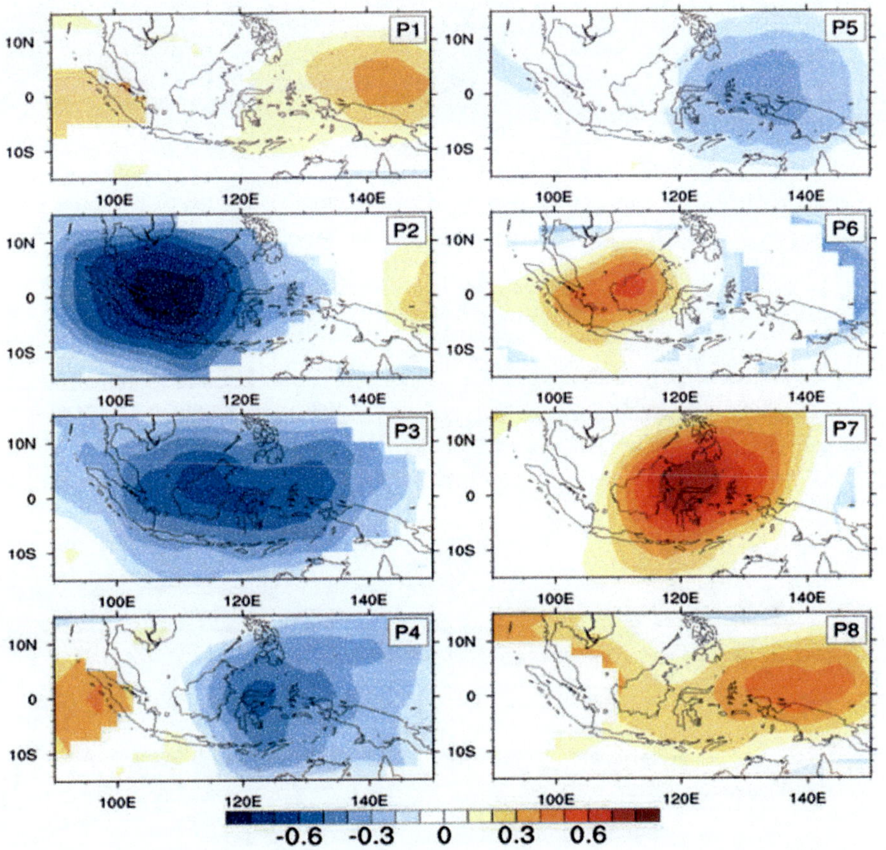

Fig. 8.13 Composite differences of MJO-filtered daily air temperature anomaly (K) at 100 hPa when EQBO occurs to all boreal winter days. Only values that are statistically significant at the 95% level are shown

Overall, we conclude that QBO has significantly influenced the MJO related rainfall variability and extreme events over the Maritime Continent.

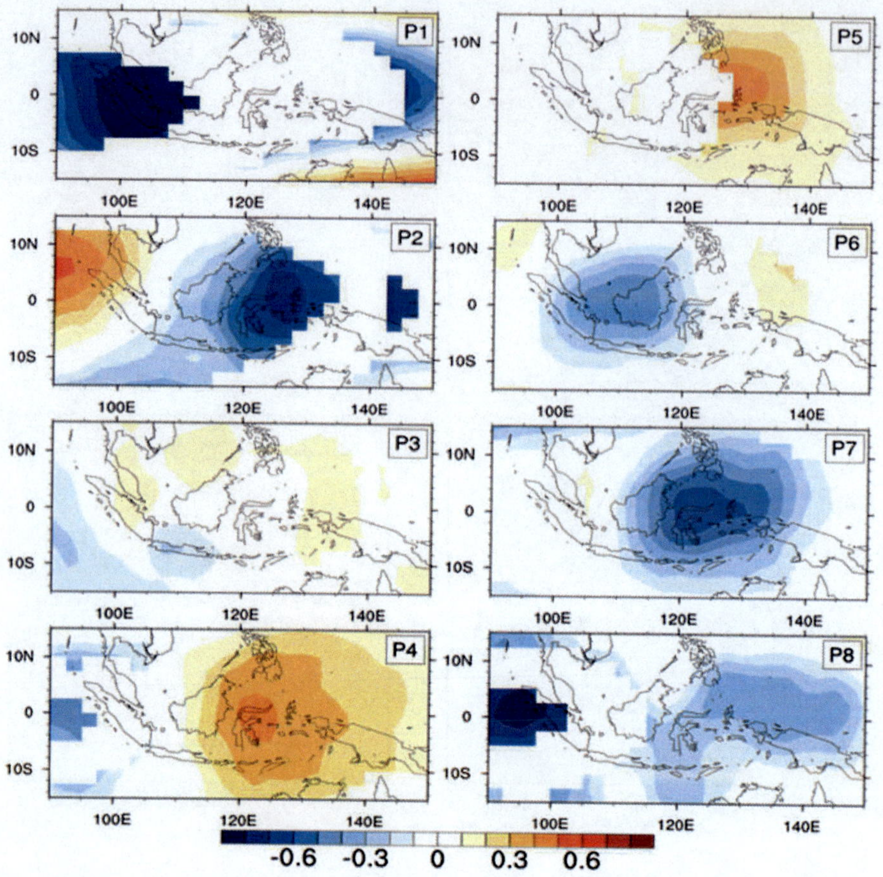

Fig. 8.14 Composite differences of MJO-filtered daily air temperature anomaly (K) at 100 hPa when WQBO occurs to all boreal winter days. Only values that are statistically significant at the 95% level are shown

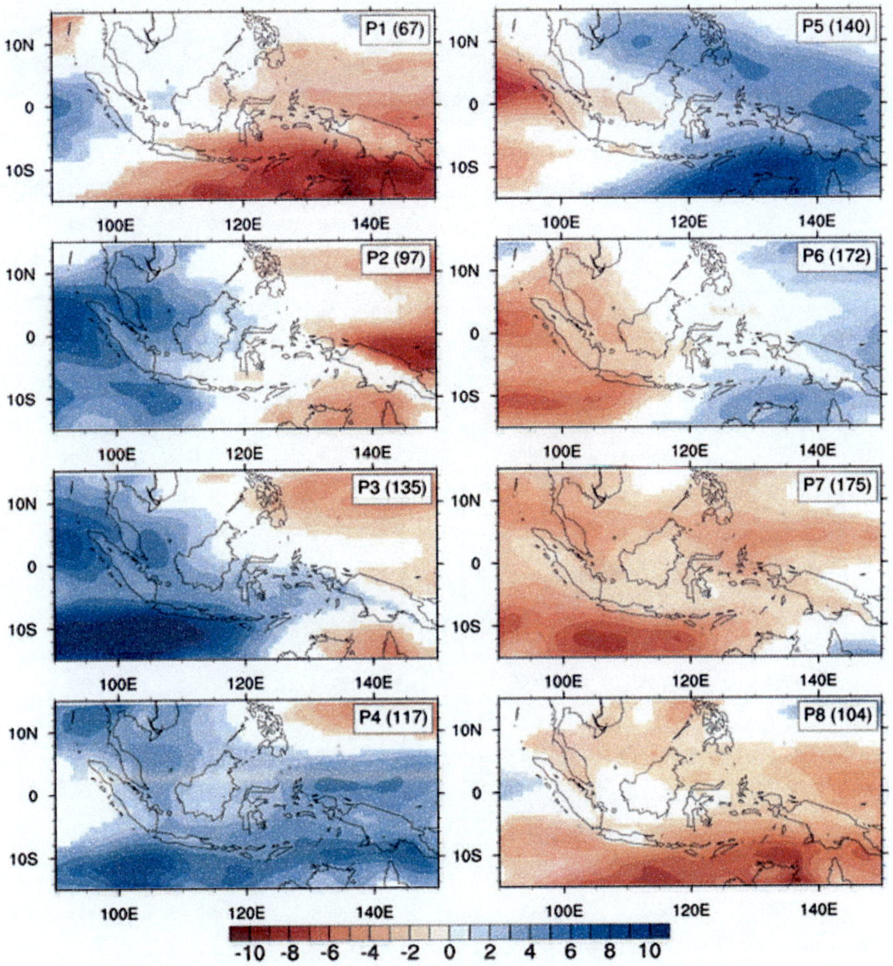

Fig. 8.15 Composites of MJO filtered daily total column water vapor anomaly (kg/m^2) during boreal winter. Only values that are statistically significant at the 95% level are shown

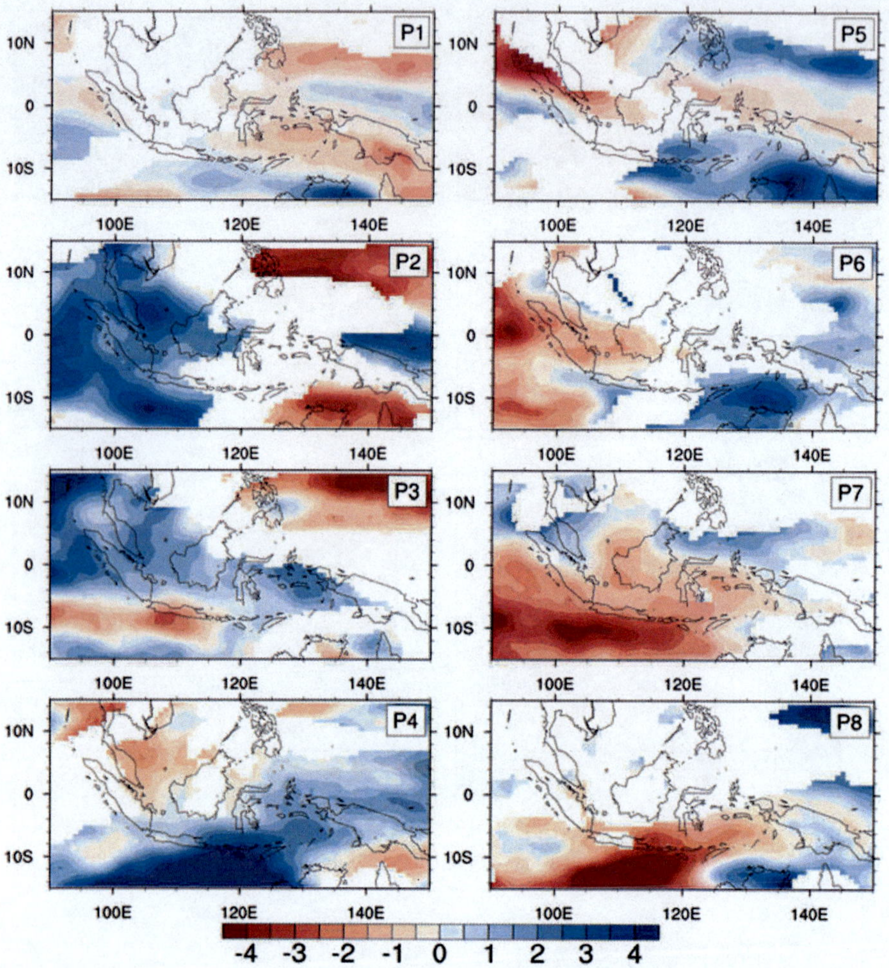

Fig. 8.16 Composite differences of MJO filtered daily total column water vapor anomaly (kg/m^2) when EQBO occurs to all boreal winter days. Only values that are statistically significant at the 95% level are shown

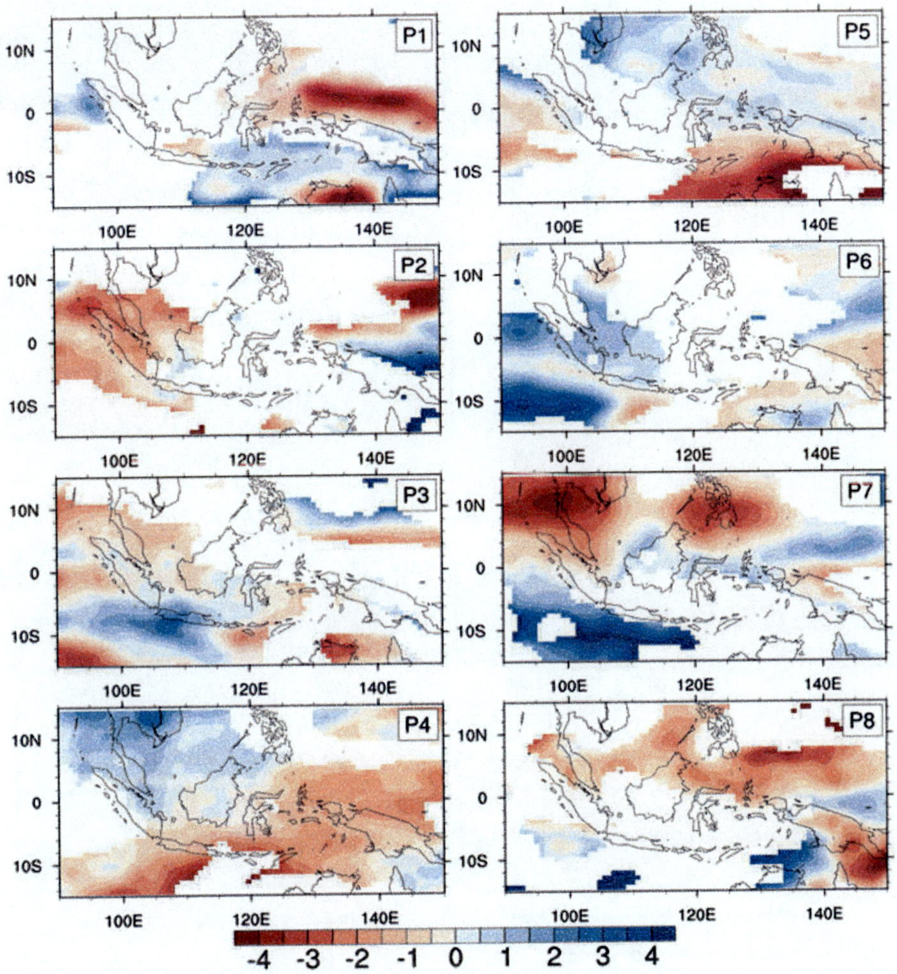

Fig. 8.17 Composite differences of MJO filtered daily total column water vapor anomaly (kg/m^2) when WQBO occurs to all boreal winter days. Only values that are statistically significant at the 95% level are shown

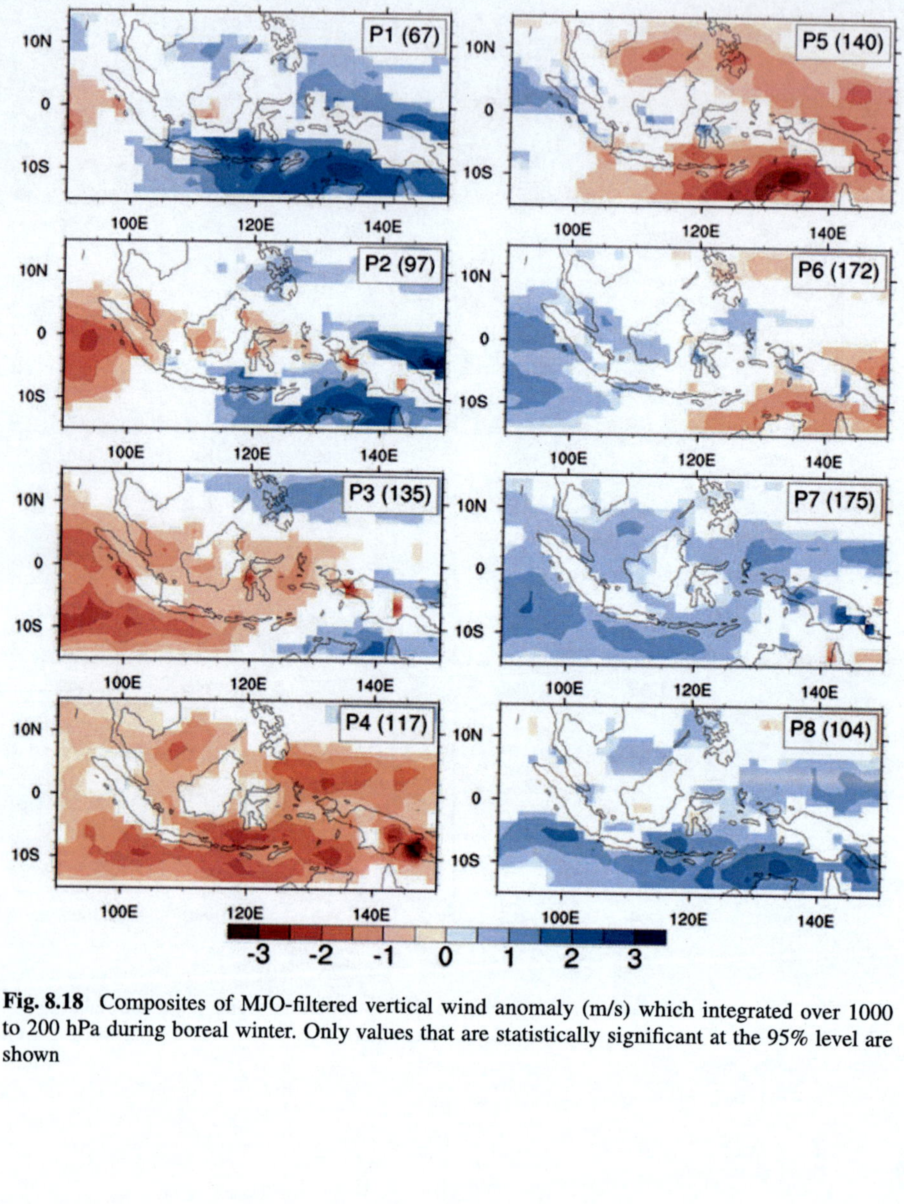

Fig. 8.18 Composites of MJO-filtered vertical wind anomaly (m/s) which integrated over 1000 to 200 hPa during boreal winter. Only values that are statistically significant at the 95% level are shown

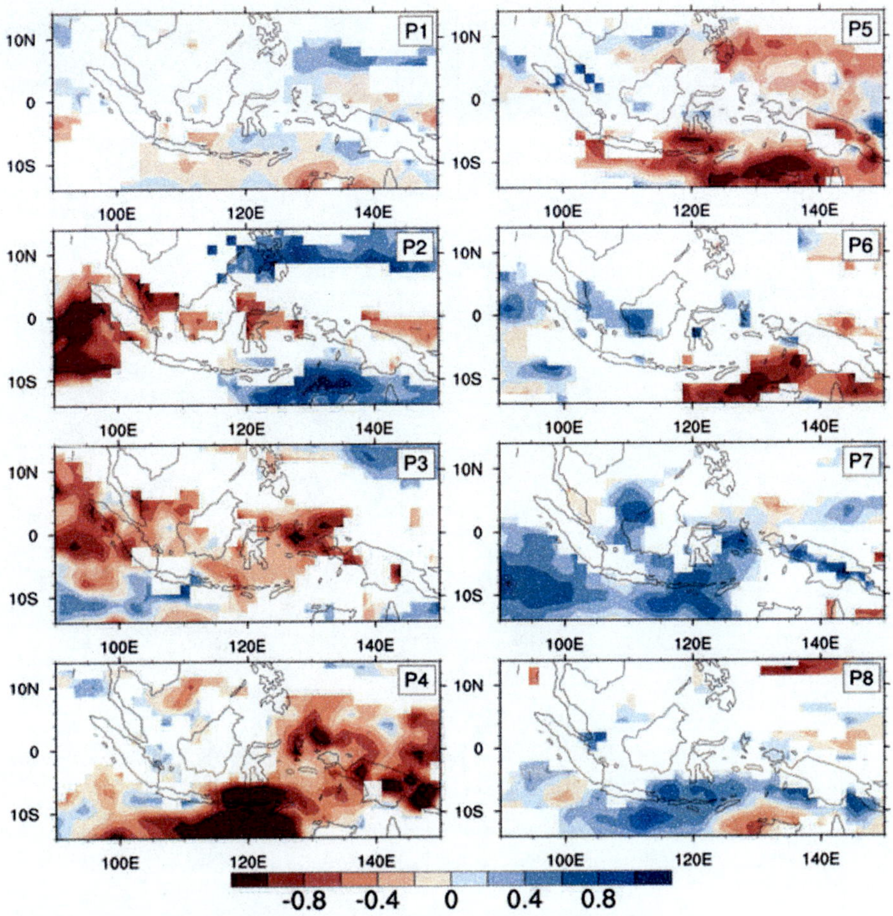

Fig. 8.19 Composite differences of MJO filtered vertical wind anomaly (m/s) which integrated over 1000 to 200 hPa when EQBO occurs to all boreal winter days. Only values that are statistically significant at the 95% level are shown

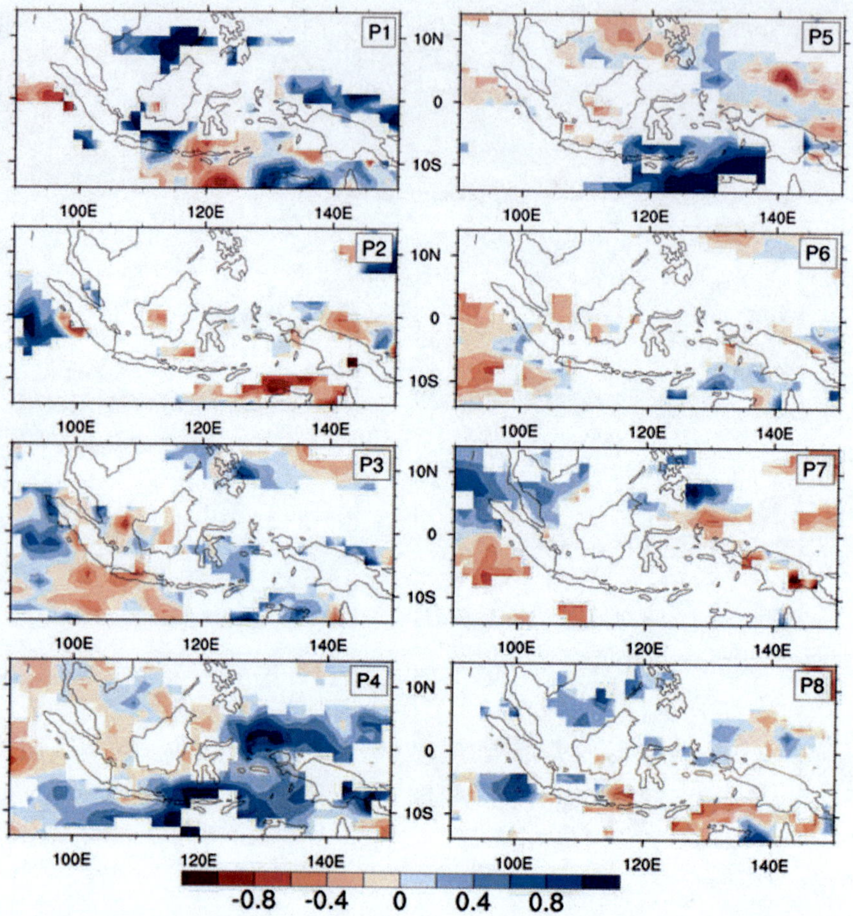

Fig. 8.20 Composite differences of MJO filtered vertical wind anomaly (m/s), which integrated over 1000 to 200 hPa when WQBO occurs to all boreal winter days. Only values that are statistically significant at the 95% level are shown

References

1. Baldwin, M.P., Gray, L.J., Dunkerton, T.J., Hamilton, K., Haynes, P.H., Holton, J.R., Alexander, M.J., Hirota, I., Horinouchi, T., Jones, D.B.A., Marquardt, C., Sato, K., Takahashi, M.: Stratopause 179–229 (2001)
2. Holton, J.R., Hakim, G.J.: An Introduction to Dynamic Meteorology, 5th edn (2012)
3. Lindzen, R.S., Holton, J.R.: A theory of the quasi-biennial oscillation. J. Atmos. Sci. **0469**, 1095–1107 (1972). https://doi.org/10.1175/1520-0469(1968)025%3c1095
4. Alsepan, G., Lubis, S.W., Setiawan, S.: Analysis of the equatorial lower stratosphere quasi-biennial oscillation (QBO) using ECMWF-interim reanalysis data set. IOP Conf. Ser. Earth Environ. Sci. **31** (2016). https://doi.org/10.1088/1755-1315/31/1/012032
5. Kusuma, L., Lubis, S.W., Setiawan, S.: Unprecedented quasi-biennial oscillation (QBO) disruption in 2015–2016: implications for tropical waves and ozone. IOP Conf. Ser. Earth Environ.

Sci. **284** (2019). https://doi.org/10.1088/1755-1315/284/1/012016
6. Madden, R.A., Julian, P.R.: Description of global-scale circulation cells in the tropics with a 40–50 day period. J. Atmos. Sci. **29**, 1109–1123 (1972). https://doi.org/10.1175/1520-0469(1972)029%3c1109:dogscc%3e2.0.co;2
7. Lubis, S.W., Jacobi, C.: The modulating influence of convectively coupled equatorial waves (CCEWs) on the variability of tropical precipitation. Int. J. Climatol. **35**, 1465–1483 (2015). https://doi.org/10.1002/joc.4069
8. Lubis, S.W., Respati, M.R.: Impacts of convectively coupled equatorial waves on rainfall extremes in Java, Indonesia. Int. J. Climatol. **41**, 2418–2440 (2021). https://doi.org/10.1002/joc.6967
9. Holton, J.R., Tan, H.-C.: The Influence of the Equatorial Quasi-Biennial Oscillation on the Global Circulation at 50 mb (1980)
10. Muhammad, F.R., Lubis, S.W., Setiawan, S.: Impacts of the Madden–Julian oscillation on precipitation extremes in Indonesia. Int. J. Climatol. **41**, 1970–1984 (2021). https://doi.org/10.1002/joc.6941
11. Xavier, P., Rahmat, R., Cheong, W.K., Wallace, E.: Influence of Madden-Julian oscillation on Southeast Asia rainfall extremes: observations and predictability. Geophys. Res. Lett. **41**, 4406–4412 (2014). https://doi.org/10.1002/2014GL060241
12. Hidayat, R., Kizu, S.: Influence of the Madden-Julian oscillation on Indonesian rainfall variability in austral summer. Int. J. Climatol. **30**, 1816–1825 (2010). https://doi.org/10.1002/joc.2005
13. Nishimoto, E., Yoden, S.: Influence of the stratospheric quasi-biennial oscillation on the Madden-Julian oscillation during austral summer. J. Atmos. Sci. **74**, 1105–1125 (2017). https://doi.org/10.1175/JAS-D-16-0205.1
14. Klotzbach, P., Abhik, S., Hendon, H.H., Bell, M., Lucas, C.G., Marshall, A., Oliver, E.C.J.: On the emerging relationship between the stratospheric quasi-biennial oscillation and the Madden-Julian oscillation. Sci. Rep. **9**, 1–9 (2019). https://doi.org/10.1038/s41598-019-40034-6
15. Zhang, C., Zhang, B.: QBO-MJO connection. J. Geophys. Res. Atmos. **123**, 2957–2967 (2018). https://doi.org/10.1002/2017JD028171
16. Yoo, C., Son, S.W.: Modulation of the boreal wintertime Madden-Julian oscillation by the stratospheric quasi-biennial oscillation. Geophys. Res. Lett. **43**, 1392–1398 (2016). https://doi.org/10.1002/2016GL067762
17. Kim, H., Son, S.W., Yoo, C.: QBO modulation of the MJO-related precipitation in East Asia. J. Geophys. Res. Atmos. **125**, 2–10 (2020). https://doi.org/10.1029/2019JD031929
18. Huffman, G.J., Adler, R.F., Bolvin, D.T., Nelkin, E.J., Code, N.G.: TRMM common-grid precipitation dataset status. Satell. Rainfall Appl. Surf. Hydrol. 1–19 (2008)
19. Liebmann, B., Catherine, S.A.: Liebmann OLR.pdf. Bull. Am. Meteorol. Soc. **77**, 1275–1277 (1996)
20. Dee, D.P., Uppala, S.M., Simmons, A.J., Berrisford, P., Poli, P., Kobayashi, S., Andrae, U., Balmaseda, M.A., Balsamo, G., Bauer, P., Bechtold, P., Beljaars, A.C.M., van de Berg, L., Bidlot, J., Bormann, N., Delsol, C., Dragani, R., Fuentes, M., Geer, A.J., Haimberger, L., Healy, S.B., Hersbach, H., Hólm, E.V., Isaksen, L., Kållberg, P., Köhler, M., Matricardi, M., Mcnally, A.P., Monge-Sanz, B.M., Morcrette, J.J., Park, B.K., Peubey, C., de Rosnay, P., Tavolato, C., Thépaut, J.N., Vitart, F.: The ERA-Interim reanalysis: configuration and performance of the data assimilation system. Q. J. R. Meteorol. Soc. **137**, 553–597 (2011). https://doi.org/10.1002/qj.828
21. Wheeler, M.C., Hendon, H.H.: An all-season real-time multivariate MJO index: development of an index for monitoring and prediction. Mon. Weather Rev. **132**, 1917–1932 (2004). https://doi.org/10.1175/1520-0493(2004)132%3c1917:AARMMI%3e2.0.CO;2
22. Fathurochman, I., Lubis, S.W., Setiawan, S.: Impact of Madden-Julian oscillation (MJO) on globa distribution of total water vapor and column ozone. IOP Conf. Ser. Earth Environ. Sci. **54**, 012034 (2017). https://doi.org/10.1088/1742-6596/755/1/011001
23. Ren, H.L., Ren, P.: Impact of Madden-Julian oscillation upon winter extreme rainfall in Southern China: observations and predictability in CFVs2. Atmosphere (Basel) **8** (2017). https://doi.org/10.3390/atmos8100192

24. Wheeler, M., Kiladis, G.N.: Convectively coupled equatorial waves: analysis of clouds and temperature in the wavenumber-frequency domain. J. Atmos. Sci. **56**, 374–399 (1999). https://doi.org/10.1175/1520-0469(1999)056%3c0374:CCEWAO%3e2.0.CO;2
25. Zhang, C.: Madden-Julian Oscillation, pp. 1–36 (2005). https://doi.org/10.1029/2004RG000158.1.INTRODUCTION
26. Yamanaka, M.D.: Physical climatology of Indonesian maritime continent: an outline to comprehend observational studies. Atmos. Res. **178–179**, 231–259 (2016). https://doi.org/10.1016/j.atmosres.2016.03.017
27. Son, S.W., Lim, Y., Yoo, C., Hendon, H.H., Kim, J.: Stratospheric control of the Madden-Julian oscillation. J. Clim. **30**, 1909–1922 (2017). https://doi.org/10.1175/JCLI-D-16-0620.1

Chapter 9
Water Quality Assessment for Detecting Submarine Groundwater Discharge (SGD) Pollution in the Coastal Area of Krakas Beach, North Lombok District, Indonesia

Wisnu Arya Gemilang, Ulung Jantama Wisha, and Hendra Bakti

Abstract One of the water resources in the coastal area is a submarine groundwater discharge (SGD) found in Krakas Beach, Lombok Utara, Indonesia. This spring discharge plays a role in providing high nutrient intake and balancing the nitrogen and other dissolved organic compounds within a coastal ecosystem. On the other hand, the SGD could be a freshwater resource for the local community and support marine tourism. The sustainable groundwater intake in the coastal area could cause coastal pollutions through eutrophication, temperature degradation and salinity shock to coral reefs. Thus, assessing water quality in the surrounding SGD area is crucial to determine the pollution level induced by freshwater contamination. The groundwater samples from the source of SGD and wells in the surrounding coastal area were collected and analyzed geochemically in the laboratory. We employed Alpha radioactivity analysis to detect the emergence of groundwater pollution. The ^{222}Rn of groundwater ranged from 339.3843.44 Bq/m^3, SGD 242.6–1626.15 Bq/m^3, and non-SGD area 39.93–89.42 Bq/m^3, respectively. We found that the nitrate concentration predominated the nutrient content of groundwater in the study area (approximately 23.93 mg/L). The high nitrate concentration in the groundwater was in line with the increasing nutrient evoked by the SGD (ranging from 1 to 3.3 mg/L), whereby the nitrate concentration exceeded the quality standard for marine biota. The N:P ratio was > 16 in the study area showing the imbalanced condition due to SGD. Additionally, the emergence of SGD tremendously declined the salinity, resulting in the low coral cover in the surrounding SGD area.

W. A. Gemilang (✉) · U. J. Wisha
Research Institute for Coastal Resource and Vulnerability, Ministry of Marine Affairs and Fisheries, Padang, Indonesia
e-mail: wisnu.gemilang@kkp.go.id

H. Bakti
Geotechnology Research Center, National Research and Innovation Agency, Bandung, Indonesia

E. Yulihastin et al. (eds.), *Proceedings of the International Conference on Radioscience, Equatorial Atmospheric Science and Environment and Humanosphere Science, 2021*, Springer Proceedings in Physics 275, https://doi.org/10.1007/978-981-19-0308-3_9

9.1 Introduction

Submarine groundwater discharge (SGD) is a seepage, emerging in the coastal area, seabed, and freshwater source in the sea [1]. SGD is a significant interaction between sea-groundwater that globally has been reported to bring several inorganic compounds (nutrient, heavy metals, carbon, and rare elements) toward the coastal area [2]. Many studies defined any environmental issues related to the primary productivity in the surrounding SGD sources, such as the red tides emergence due to the nutrient enrichment in the SGD area reported in Refs. [3, 4]. In several places, the nutrient intake from SGD is supposed to be higher than the river and surface runoff [5]. Furthermore, the varied chemical compositions of SGD impacts water quality and ecology degradation in the coastal area [6]. The excessive nutrient input in the SGD source could trigger local eutrophication [7].

Nitrate (N) and phosphate (P) concentrations of groundwater in the coastal area are often higher than in rivers [8]. This condition is related to the amount of N and P intake in the coastal zone through SGD sources [9]. The chemical concentration, primarily the nutrient originated from SGD, then exacerbates eutrophication tendency that eventually will induce several environmental issues by influencing the local vegetation, animal structures, and even the growth of harmful algal and red tides in the coastal area [10]. Therefore, evaluating the SGD potential effects on the N and P ratio in the coastal area is essential because this ratio could determine which nutrient is limiting the growth of phytoplankton and the efforts to figure out the environmental and ecological impacts in the coastal area.

Several related studies regarding nutrient supply implication of SGD had been conducted, such as in Bohai Bay China [11], East China Sea [12]. The detection of SGD in tropical areas is complicated due to the influence of external parameters [13]. Several meteorological factors influencing the SGD are rainfall intensity, humidity, evaporation, wind magnitude, and sunrays exposure [14]. SGD in Lombok Island is identified in the north and east regions [1, 15]. It comes with groundwater sources (submarine springs) that emerged around 2–10 m beneath the sea level. The water source pops out among coral reefs with a trough formation [16].

Previous studies commenced identifying SGD sites in Lombok Island since 2010 [1]. However, the assessment of water quality in the surrounding SGD sites is rarely conducted. Nutrient flux resulted from SGD must be compared to the other nutrient sources to determine the relative contribution of nutrients from SGD and its role for the future water quality and ecosystem degradation [17]. The unwanted SGD contaminant from groundwater will potentially pollute the environment near the coast. Hence, the assessment of SGD quality in North Lombok waters, particularly in Krakas Beach, is crucial to determine the influence of SGD on seawater quality. Therefore, this study aims to estimate the intake of nutrients from SGD in the North Lombok waters using *Radium Isotope Geochemical* detection and evaluate the respective contribution of nutrients to the surrounding ecosystem.

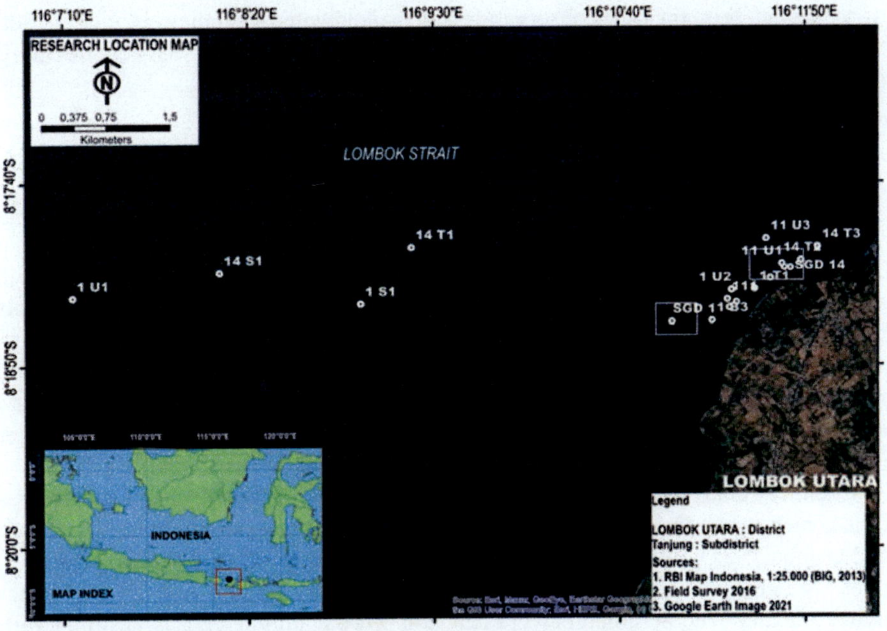

Fig. 9.1 Research location map modified from Google Earth image 2021

9.2 Materials and Methods

9.2.1 Research Location and Period

A study on SGD's characteristics and influence on water quality and ecosystem in North Lombok District, West Nusa Tenggara, Indonesia was conducted during March to December 2016. The research was focused on the coastal area of Krakas Beach. However, several stations were selected to compare the distribution of water quality parameters (Fig. 9.1). Generally, North Lombok waters has a common climate variability like the other areas in the West Nusa Tenggara, with rainy and dry seasons consecutively occurring every year [18]. A mountainous and hilly plateau dominates Lombok Island, and just a tiny part is a lowland.

9.2.2 Research Materials

Physical conditions of North Lombok area, such as morphology analysis, the rock type, hydrogeological, and hydro-chemical mapping toward the chosen elements in the field and water samples analysis in the laboratory, were measured and interpreted to identify the formation of the SGD. The hydrogeological mapping consisted of rock layer-brought water observation and groundwater level measurement. Additionally, the hydro-chemical survey consisted of salinity, Radon (^{222}Rn), conductivity, pH, and temperature measurements.

9.2.3 *Sampling and Analysis*

The survey was conducted by field-measuring and in situ samplings scattered in several locations in Lombok Island. In situ sampling had been done in 14 points near Krakas Beach (Lombok Utara), 10 points at Seruni Beach (Lombok Timur), and 2 points at Sekotong Beach (Lombok Barat) (Fig. 9.1). The field-measured parameters were pH, dissolved oxygen, turbidity, temperature, conductivity, salinity, and total dissolved solids, surveyed using Multi-Parameter Water Quality Meter model WQC-24 WMS-24 censors' type and TOA DKK Water Quality Checker.

Chemical analysis for the collected samples were conducted on several laboratories, such as the Geotechnology Research Center, Indonesian Institute of Science and Lombok Sea Cultivation Fisheries Center, Ministry of Marine Affairs and Fisheries. The analyses consisted of anion and cation elements, nutrients, and radioactivity. The analyzed parameters were Natrium (Na), Kalium (K), Magnesium (Mg), Calcium (Ca), Sulfate (SO_4), Chloride (Cl), Bicarbonate (HCO_3), Iron (Fe), Arsenic (As), Mangan (Mn), Silicate (SiO2), and calcium carbonate (CaCO3). The analyzed nutrient consisted of Ammonia (NH4-N), Nitrite (NO2-N), Nitrate (NO3-N), N total, P total, Orthophosphate (PO4-P). The analyzed radioactive element was Radon (^{222}Rn).

9.3 Result and Discussion

9.3.1 *Radon (^{222}Rn) Activity*

Surface water or seawater is supposed to have lower radon activity than groundwater. If the radon activity is restively higher and exceeds the standard radon for seawater, it hypothetically shows the presence of freshwater contamination near the coast. Several surveys at Krakas Beach and its surrounding were performed spatially for identifying radon activity.

Temporal radon measurement for seawater has been conducted at 14 observation points ($n = 14$), while the groundwater measurement was done at four local wells ($n = 4$). The temporal result of radon analysis varied considerably, ranging from 39.9 to 1266.2 and 339.2 to 3843.4 Bq/m^3 for seawater and groundwater, respectively. The results are shown in Table 9.1.

Generally, the average radon activity at SGD points was relatively 12 times higher compared to the seawater non-SGD. Meanwhile, the average radon activity was 0.6

Table 9.1 Statistical result of radon analysis in Krakas Beach

	Seawater non-SGD (Bq/m^3) (n = 4)	SGD (Bq/m^3) (n = 10)	Groundwater (Bq/m^3) (n = 4)
Max	89.42	1626.15	3843.44
Min	39.93	242.60	339.24
Average	63.89	788.06	1373.81

times higher compared to the well's groundwater. The value of radon activity in the wells at Seruni and Sekotong Beach varied considerably, ranging from 969.78 to 2726.48 Bq/m^3 and 2135.48 to 3928.84 Bq/m^3, respectively. These results were almost the same with Krakas Beach's observation.

SGD identification according to the trace element ^{222}Rn in several beaches and SGD sites in Lombok Island showed a relatively high concentration at Krakas Seruni and Sekotong Beach. This state was also caused by the relatively low conductivity in the surrounding SGD sites ranging from 26.6 to 30.7 mS/cm. In contrast, the conductivity was higher in the area without SGD contamination, ranging from 49.0 to 49.5 mS/cm.

Even though the groundwater mixing occurred, the SGD samples showed a saline characteristic (brine) with lower conductivity. In contrast, groundwater samples' conductivity at 25 m from the coastline ranged from 3.9 to 4.7 mS/cm. According to the previous study in Jakarta Beach [19], the inversely proportional correlation between ^{222}Rn and conductivity was observed. The higher the ^{222}Rn concentration, the lower the conductivity value for groundwater and vice versa.

Statistically, the correlations between ^{222}Rn and conductivity have a variation in R^2 value. The R^2 value for groundwater was 0.035, SGD 0.183, and the highest value was observed on seawater samples. These conditions reflect the groundwater supply from the land to the sea. The higher the conductivity value, the lower the value of ^{222}Rn (Fig. 9.2).

9.3.2 Nitrogen and Phosphorus Content Within Lombok Utara Groundwater

N and P concentration varied considerably depended on the input, soil type, and aquifer, aquifer permeability, the level of groundwater refillable, and climate [20]. The average nitrate concentration on groundwater in the coastal area of North Lombok ranged from 5.06–55.6 mg N/L, phosphate 0.29–0.62 mg P/L, Ammonia 0.01–1.33 mg/L, and nitrite averagely 0.01 mg/L (Table 9.2). Nitrate is a dominating N-compound in the aquifer compared to the reduced-N such as NH4 and dissolved organic nitrogen due to the groundwater mobility in the rock layers [21].

Groundwater's phosphate and nitrate concentrations in the study area were quite high. The relatively high phosphate concentration reflects the increased fertilizer and

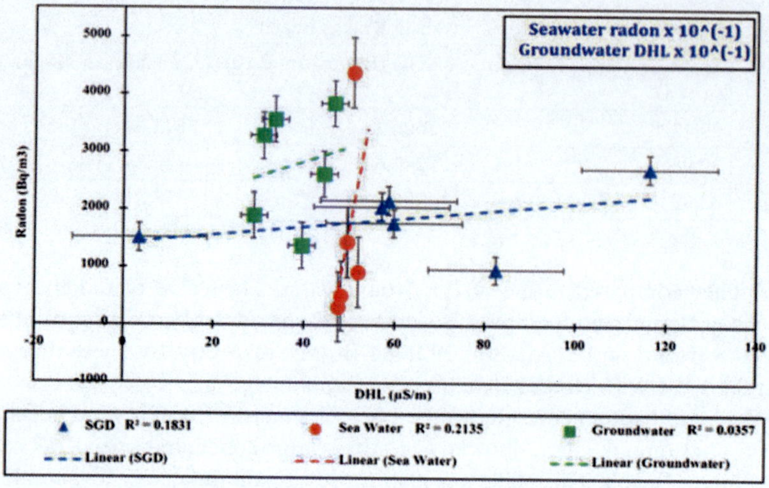

Fig. 9.2 Correlation between ^{222}Rn and conductivity to groundwater and SGD

Table 9.2 Nutrient concentration of groundwater in the North Lombok district

Parameter	Groundwater	Quality standard (PP No. 22, 2021)
Phosphate (mg/L)	0.29–0.62	0.2
Nitrate (mg/L)	5.06–55.6	10
Nitrite (mg/L)	0.01	0.06
Ammonia (mg/L)	0.01–1.33	0.1
Salinity (%)	0.01–0.06	–
pH	6.89–7.45	6–9

farm waste usage in the surrounding study area that filtrated within the soil layer. The absorption of phosphate into the soil and the characteristics of volcanic sediments in Lombok Island could influence the magnitude of PO_4 mobility on groundwater. The high nutrient content on groundwater in the coastal area of North Lombok District is related to the amalgamated geological and anthropogenic states and land use. Anthropogenic activities in the surrounding study area are categorized as settlement, agriculture, and husbandry because those three aspects come with 52%, 30%, and 4% of land use, respectively. The anthropogenic N and P source on the groundwater is commercial fertilizer, compost, and wastewater [22].

The flow of groundwater's nutrients is a significant path for contaminants entering the estuary and coastal area, mainly in areas with higher anthropogenic activity [23]. Although the magnitude of groundwater and living period can reduce the nutrient concentration before entering the river system [24], most nutrients can still enter the aquifer layer, eventually disembogued in the estuary [25]. The nitrate value within groundwater in the study area was 80% exceeding the quality standard established by PP No 22 in 2021. In contrast, nitrite, ammonia, and phosphate concentrations were under the standard (Table 9.2).

The other results of the physical parameters of groundwater are as follows: pH 6.89–7.27, DHL 112–1143 μS/cm; TDS 100–748 mg/L: the water type Na-Cl and Ca-HCOO3. Based on the type of water, it shows that the groundwater in the coastal area of North Lombok waters was characterized by the Na-Cl type indicating slightly saline water predomination. The type of Ca-HCO_3 showed the dominant groundwater interacting with limestone and dolomitic limestone. Based on the groundwater facies analysis correlated with the geological map of Lombok Island from the north to the south, the pattern of uncompacted groundwater flow experienced a hydro-geochemical evolution generated by facies Na-HCO_3. It ended up on the facies of Na-Cl that was situated in the coastal area. These groundwater facies characteristics are also related to the rock-arranged aquifer in the study area. Water samples that came from Alluvial Deposits (Qa) have a saline water characteristic.

9.3.3 Nitrogen and Phosphorus Content Within SGD in the North Lombok

Several previous studies have proven the SGD phenomena in the North Lombok coastal area [1, 15]. In this coastal zone, the value of N and P was highly influenced by the flow acceleration and redox features between seawater and groundwater. The high concentration of nutrients on groundwater indicates the tendency of higher nutrients on SGD sources. The phosphate concentration ranged from 0.03 to 0.12 mg P/L. The highest phosphate concentration was found at stations B1, 14T1, and 11U1 with 0.12, 0.07, and 0.06 mg P/L, respectively (Fig. 9.3).

A sufficiently low concentration of phosphate was observed in the surrounding SGD site with 0.04 mg P/L. Overall, the phosphate concentration in the North Lombok waters is categorized below the standard quality for marine biota. The phosphate near the coast ranged from 0.62 to 0.29 mg P/L, proving that land-source nutrient intake significantly triggered the nutrient concentration and accumulation. The farther the location from the source, the lower concentration of phosphate identified. This might be influenced by the water motion dynamics generated by oceanographic factors such as tides and long shore currents [26].

The highest concentration of nitrate was observed at stations 14T3, 14T2, and 1U1, with 1.6 up to 3 mg/L. The lowest value was identified at station SGD14 (1.2 mg/L), 1 U3 and 14 T1 (1.1 mg/L). The highest nitrate value was positioned around 0.7 km from the SGD site, while the lowest was around 3.1 km from the SGD site. Overall, the nitrate concentration was more dominant than ammonia, nitrite, and phosphate (Fig. 9.3). The main source of nitrate is river runoff, erosion, agricultural and settlement wastes [26]. Generally, the concentration of nitrate on groundwater in the coastal area ranged from 5.06 to 55.60 mg/L. It was higher than nitrate in the SGD site and the sea. Sea currents play a significant role in distributing dissolved

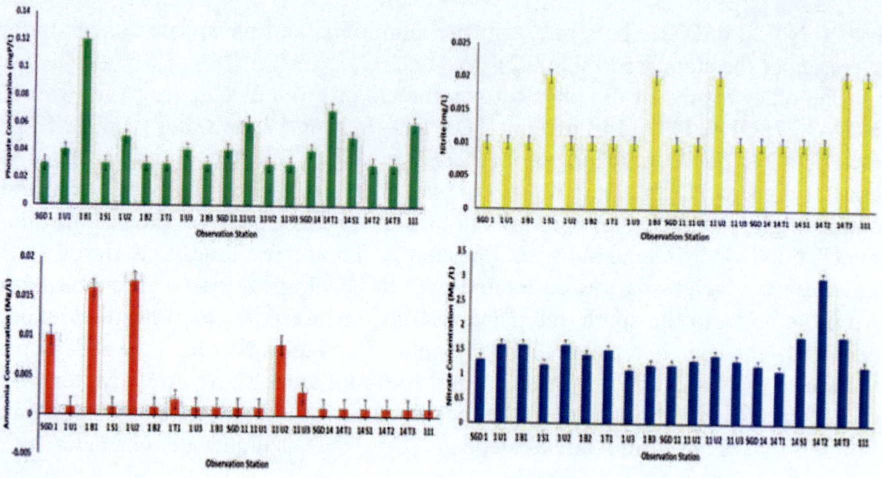

Fig. 9.3 Comparison between nitrate (blue), ammonia (red), phosphate (green) and nitrite (yellow) concentration in North Lombok waters

nutrients [27]. Due to the bottom materials consisted of rubbles and coarse sediment and the weak current features, the resuspension of the sediment-contained nutrient is impossible to occur. That is why the nutrient concentration in the surface water is lower than the groundwater sources.

Ammonia concentration is the lowest detected in the study area ranging from 0.009 to 0.01 mg/L. The highest concentration was found at stations SGD1, 1U2, and 1B1. According to the quality standard No. 22 in 2021, the ammonia concentration was below the standard for marine biota. The deeper the water, the more the compound can be found because of nutrient deposition in the surface bottom [28]. Moreover, the low concentration of nitrate, ammonia, and phosphate in the surface water is caused by the phytoplankton activity that intensively uses the nutrient to produce energy (photosynthesis) [29] (Table 9.3)

Nitrite concentration was relatively uniform, ranging from 0.01 to 0.02 mg/L. The highest concentration was observed in the station that was far away from the SGD site. The comparison average of nutrient composition assimilating with common

Table 9.3 Summary of the nutrient concentration monitoring results from North Lombok waters compared with the Government Regulation of Indonesia (PP RI) number 22 in 2021 (Standard quality for water and marine biota)

No	Dissolved nutrients	Standard quality for marine biota (mg/L)	North Lombok waters (mg/L)
1	Nitrate	0.008	1.1–3
2	Nitrite	0.06	0.02–0.01
3	Ammonia	0.3	0.017–0.001
4	Phosphate	0.015	0.12–0.03

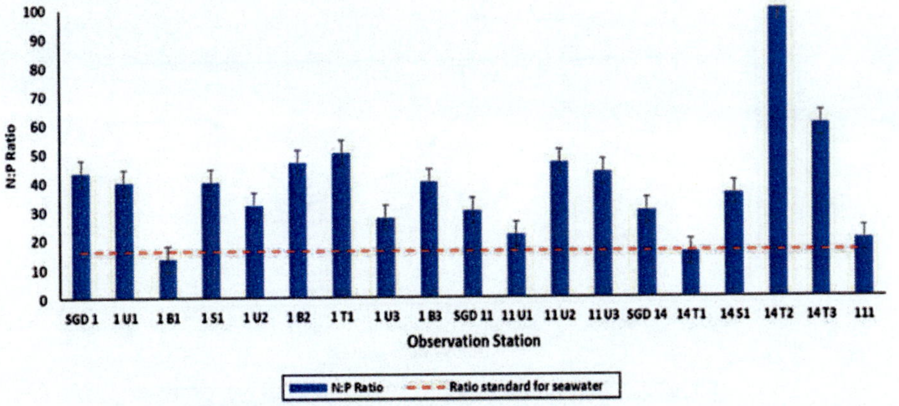

Fig. 9.4 Comparison between total dissolved Nitrogen (N) and dissolved inorganic Phosphate (P) ratios in North Lombok Waters

algae is called the Redfield ratio [30]. The N:P ratio varied considerably (Fig. 9.4). The average ratio was 38.7, with the highest value of 100 and 60. Phytoplankton is in the state of N- limited if the ratio is > 16, and it is P-limited if the ratio is < 16 [31]. We found that the highest value of the N:P ratio was identified at stations 14T2 and 1T1, approximately 500 m from the SGD site.

The lower N:P ratios were found at stations 1B1 (13.3) and 14T1 (15.7), categorized as N-limited. This state shows that the P concentration is higher than N. The natural ratio needed for supporting marine life is N:P = 16. The nitrate portion is supposed to be higher 16 times than the phosphate.

The freshwater intake in the coastal area triggers the decline in salinity that becomes the initial condition inducing coral mortality [32]. The disruption of low salinity causes the less-developed physiology control system of corals. This state will worsen if nutrient pollution occurs, impacting the increase in coral mortality [33]. Generally, the salinity in the SGD site ranged from 33–13.6%, while it was higher in the open sea (32.9–34.2%). Compared to the quality standard for marine biota, the salinity values in the study area are generally under the required standard (Fig. 9.5).

The salinity distribution was correlated with the previous study [34], whereby the concentration is sufficiently low. Due to the low salinity, the biota could not live in the surrounding SGD, such as the *Acrophore* sp. That is very sensitive to salinity and temperature shocks, only 1.06% *Acrophore* could live in, while the other species were found to adapt to the extreme condition (Fig. 9.6). Overall, the coral cover in the slightly outlying area from the SGD site was higher than the coral cover within the SGD site [29].

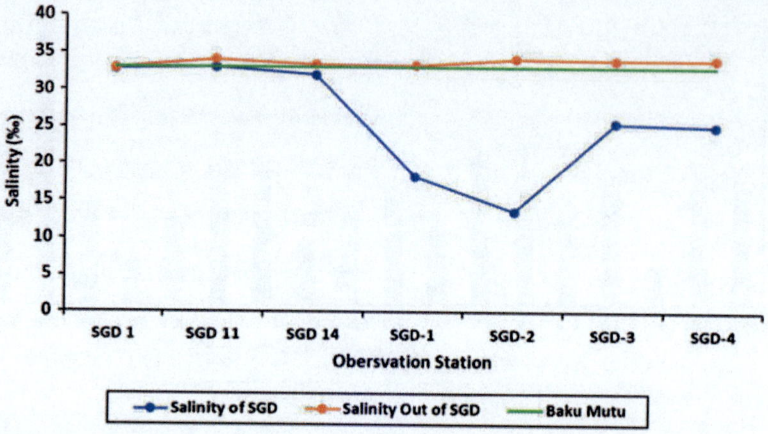

Fig. 9.5 The comparison of salinity profile in the SGD site and the slightly outlaying area

Fig. 9.6 The coral states at the station SGD1 and its surrounding. *Source* Field survey, 2016

9.4 Conclusions

The radon activity in the SGD site is 12-times higher than in the seawater-non-SGD area. The saline water features are identified in the surrounding SGD. However, the conductivity tends to be lower than in the non-SGD area. Most nitrate concentrations show excessive concentration compared to the quality standard. The nitrate is more dominating the water in the surrounding SGD. In the area around 0.5 km from the SGD site, nutrient enrichment is observed whereby the N-limited P is dominant. The SGD source is characterized by relatively low salinity and is under the quality standard allowed for marine biota. Salinity shock is potentially impacting the surrounding ecosystem. Additionally, the imbalanced nutrient states also allow the potency of eutrophication.

Acknowledgements Acknowledgment and gratitude are given to Research Institute for Coastal Resources and Vulnerability (RICRV) under the Ministry of Marine Affairs and Fisheries for research funding in 2016 with grant number of 2373.002.001.013 and to those who have helped during field survey.

Authors Contribution WAG and UJW are the main contributors to this work.

References

1. Lubis, R.F., Bakti, H., Suriadarma, A.: Submarine groundwater discharge (SGD) in Indonesia. RISET Geologi dan Pertambangan 57–62 (2011)
2. Johannesson, K.H., Palmore, C.D., Fackrell, J., Prouty, N.G., Swarzenski, P.W., Chevis, D.A., ... Burdige, D.J.: Rare earth element behavior during groundwater–seawater mixing along the Kona Coast of Hawaii. Geochimica Cosmochimica Acta 229–258 (2017)
3. Blanco, A.C., Watanabe, A., Nadaoka, K., Motooka, S., Herrera, E.C., Yamamoto, T.: Estimation of nearshore groundwater discharge and its potential effects on a fringing coral reef. Marine Pollut Bull 770–785 (2011)
4. Lee, Y.W., Kim, G.: Linking groundwater-borne nutrients and dinoflagellate red-tide outbreaks in the Southern sea of Korea using a Ra tracer. Estuarine, Coastal and Shelf Science 309–317 (2007)
5. Null, K.A., Dimova, N.T., Knee, K.L., Esser, B.K., Swarzenski, P.W., Singleton, M.J., ... Paytan, A.: Submarine groundwater discharge-derived nutrient loads to San Francisco Bay: implications to future ecosystem changes. Estuaries and Coasts 1299–1315 (2012)
6. Knee, K.L., Street, J.H., Grossman, E.E., Boehm, A.B., Paytan, A.: Nutrient inputs to the coastal ocean from submarine groundwater discharge in a groundwater-dominated system: relation to land use (Kona coast, Hawaii, USA). Limnol. Oceanogr. 1105–1122 (2010)
7. Moore, W.S.: The effect of submarine groundwater discharge on the ocean. Ann. Rev. Marine Sci. 59–88 (2010)
8. Slomp, C.P., Van Cappellen, P.: Nutrient inputs to the coastal ocean through submarine groundwater discharge: controls and potential impact. J. Hydrology 64–86 (2004)
9. Wang, X., Li, H., Zheng, C., Yang, J., Zhang, Y., Zhang, M., ... Zhang, X.: Submarine groundwater discharge as an important nutrient source influencing nutrient structure in coastal water of Daya Bay, China. Geochimica Cosmochimica Acta 52–65 (2018)
10. Chen, X., Cukrov, N., Santos, I.R., Rodellas, V., Cukrov, N., Du, J.: Karstic submarine groundwater discharge into the Mediterranean: radon-based nutrient fluxes in an anchialine cave and a basin-wide upscaling. Geochimica Cosmochimica Acta 467–484 (2020)
11. Wang, Q., Li, H., Zhang, Y., Wang, X., Xiao, K., Zhang, X., ... Dan, S.F.: Submarine groundwater discharge and its implication for nutrient budgets in the western Bohai Bay, China. J. Environ. Radioact. 212 (2020)
12. Peng, T., Zhu, Z., Du, J., Liu, J.: Effects of nutrient-rich submarine groundwater discharge on marine aquaculture: a case in Lianjiang, East China Sea. Sci. Total Environ. 786 (2021)
13. Moosdorf, N., Stieglitz, T., Waska, H., Dürr, H.H., Hartmann, J.: Submarine groundwater discharge from tropical islands: a review. Grundwasser 53–67 (2015)
14. Arief, M., Adawiah, S. W., Parwati, E., Hamzah, R., Prayogo, T. Pengembangan Model Ekstraksi Suhu Permukaan Laut Menggunakan Data Satelit Landsat 8 Studi Kasus: Teluk Lampung (Development model of sea surface temperature extraction using Landsat-8 satellite data, case study: Lampung Bay). Jurnal Penginderaan Jauh Pengolahan Data Citra Digital, 12 (2015).
15. Bakti, H., Lubis, R.F., Delinom, R., Naily, W.: Identify on submarine ground water discharge (SGD) on the alluvial coast of North Lombok, West Nusa Tenggara. Jurnal Lingkungan Bencana Geologi 133–149 (2012)

16. Lestiana, H., Sukristiyanti, S., Bakti, H., Lubis, R.F.: Pemanfaatan band termal citra landsat untuk identifikasi keluaran airtanah lepas pantai (kalp) di pantai utara Lombok. RISET Geologi Pertambangan 65–75 (2017)
17. Kress, N., Gertman, I., Herut, B.: Temporal evolution of physical and chemical characteristics of the water column in the Easternmost Levantine basin (Eastern Mediterranean Sea) from 2002 to 2010. J. Mar. Syst. **135**, 6–13 (2014)
18. As-Syakur, A.R.: Evaluasi zona agroklimat dari klasifikasi Schimidt-Ferguson menggunakan aplikasi Sistem Informasi Geografi (SIG). Jurnal PIJAR **4** (2012)
19. Burnett, W.C., Chanyotha, S., Wattayakorn, G., Taniguchi, M., Umezawa, Y., Ishitobi, T.: Underground sources of nutrient contamination to surface waters in Bangkok, Thailand. Sci. Total Environ. 3198–3207 (2009)
20. Nolan, B.T., Stoner, J.D.: Nutrients in groundwaters of the conterminous United States, 1992–1995. Environ. Sci. Technol. 1156–1165 (2000)
21. Cole, M.L., Kroeger, K.D., McClelland, J.W., Valiela, I.: Effects of watershed land use on nitrogen concentrations and δ15 nitrogen in groundwater. Biogeochemistry 199–215 (2006)
22. Voutchkova, D.D., Schullehner, J., Rasmussen, P., Hansen, B.: A high- resolution nitrate vulnerability assessment of sandy aquifers (DRASTIC-N). J. Environ. Manage. 111330 (2021)
23. Taniguchi, M., Burnett, W.C., Dulaiova, H., Siringan, F., Foronda, J., Wattayakorn, G., ... Ishitobi, T.: Groundwater discharge as an important land-sea pathway into Manila Bay, Philippines. J. Coast. Res. 15–24 (2008)
24. Michael, H.A., Mulligan, A.E., Harvey, C.F.: Seasonal oscillations in water exchange between aquifers and the coastal ocean. Nature 1145–1148 (2005)
25. Adyasari, D., Oehler, T., Afiati, N., Moosdorf, N.: Groundwater nutrient inputs into an urbanized tropical estuary system in Indonesia. Sci. Total Environ. 1066–1079 (2018)
26. Fujimoto, N., Sudo, R., Sugiura, N., Inamori, Y.: Nutrient-limited growth of microcystis aeruginosa and phormidium tenue and competition under various N:P supply ratios and temperatures. Limnol. Oceanogr. 250–256 (1997)
27. Geider, R.J., La Roche, J.: Redfield revisited: variability of C [ratio] N [ratio] P in marine microalgae and its biochemical basis. Eur. J. Phycol. **37**(1), 1–17 (2002)
28. LNCS Homepage: http://www.springer.com/lncs. Last accessed 21 Nov 2016
29. Simanjuntak, M.: Sea water quality observed from nutrient aspect, dissolved Oxygen and Ph in the Banggai waters, Central Sulawesi. Jurnal Ilmu Teknologi Kelautan Tropis **4**(2) (2012)
30. Makatita, J.R., Susanto, A.B., Mangimbulude, J.C.: Kajian zat hara fosfat dan nitrat pada air dan sedimen padang lamun Pulau Tujuh Seram Utara Barat Maluku Tengah. In: Seminar Nasional FMIPA-UT, vol. 23 (2014)
31. Utami, T.M.R., Maslukah, L., Yusuf, M.: Sebaran Nitrat (NO3) dan Fosfat (PO4) Di Perairan Karangsong Kabupaten Indramayu. Buletin Oseanografi Marina 31–37 (2016)
32. Kerswell, A. P., Jones, R.J.: Effects of hypo-osmosis on the coral Stylophora pistillata: nature and cause of low-salinity bleaching[1]. Mar. Ecol. Prog. Ser. 145–154 (2003)
33. Fabricius, K.E.: Effects of terrestrial runoff on the ecology of corals and coral reefs: review and synthesis. Mar. Pollut. Bull. 125–146 (2005)
34. Johan, O., Kusumah, G., Wisha, U.J.: Kondisi Terumbu Karang di kawasan KALP Pantai Krakas, Lombok Utara. Jurnal Segara **13**(3) (2017)

Chapter 10
The Influence of Boreal Summer Madden–Julian Oscillation on Precipitation Extremes in Indonesia

Fadhlil R. Muhammad, Sandro W. Lubis, and Sonni Setiawan

Abstract This study examines the influence of Madden–Julian Oscillation (MJO) on Indonesian precipitation during the extended boreal summer (May–September). The MJO is one of the dominant intra-seasonal variabilities that influence the extreme precipitation in the tropics, especially in Indonesia. Here, the episodes of intense precipitation (95th percentiles) during active MJO phases from 1998 to 2015 are evaluated, using the daily precipitation datasets from the gridded Asian Precipitation–Highly Resolved Observational Data Integration towards Evaluation of Water Resources (APHRODITE) and several station rain gauges. The boreal summer MJO influences the likelihood of extreme precipitation, especially in the west and north parts of Indonesia. The west part experiences an increase in the probability of extreme precipitation by up to 55 and 80% during phases 2 and 3, respectively. Moreover, the extreme precipitation probability in the north part increases by up to 40–70% during phases 2–4. On the other hand, the influence of MJO is relatively weak in the south and east parts of Indonesia. The contrast of the precipitation response between the north and south parts of Indonesia is consistent with the northward movement of boreal summer MJO.

10.1 Introduction

The Madden–Julian Oscillation is one of the most dominant modes of variability in the tropics on the intraseasonal timescale. The MJO is characterized by the eastward propagation of large cloud clusters over the tropical region with a period of

F. R. Muhammad (✉)
School of Geography, Earth, and Atmospheric Sciences, The University of Melbourne, Melbourne, Australia
e-mail: fadhlilrizki@student.unimelb.edu.au

S. W. Lubis
Rice University, Houston, USA

S. Setiawan
Department of Geophysics and Meteorology, IPB University, Bogor, Indonesia

E. Yulihastin et al. (eds.), *Proceedings of the International Conference on Radioscience, Equatorial Atmospheric Science and Environment and Humanosphere Science, 2021*, Springer Proceedings in Physics 275, https://doi.org/10.1007/978-981-19-0308-3_10

30–60 days [1–4]. A typical case of MJO is comprised of a pair of wet and dry phases associated with enhanced and suppressed convection, respectively [3]. The propagation of MJO can be divided into eight phases, with each phase has a period of 4–8 days [5]. The MJO propagation can be described as follows:

- Phase 1 is located over the western Indian Ocean and Africa.
- Phases 2 and 3 are located over the Indian Ocean.
- Phases 4 and 5 are located over the Maritime Continent, including Indonesia.
- Phases 6 and 7 are located over the western Pacific.
- Finally, the MJO ends its propagation in phase 8 over the Pacific Ocean.

The MJO can enhance or suppress the convection over the region, depending on its phase and region. Over the Maritime Continent, including Indonesia, a typical boreal winter MJO enhances the convection during phases 2–4 and suppresses the convection during phases 6–8 [6, 7]. This enhanced convection results in an increase in precipitation over the region. Furthermore, it has also been shown that the combination between large-scale convection, moisture convergence, and upward moisture advection associated with the boreal winter MJO can increase the likelihood of extreme precipitation events over Indonesia by up to 70% [7].

While the boreal winter MJO has been extensively studied, its counterpart, the boreal summer MJO, is still insufficiently researched. The main difference between boreal winter and boreal summer MJO is its propagation [3, 8]. The boreal winter MJO often thought of as the "dominant" MJO, exhibits a more southward tilt of propagation when it passes the Maritime Continent. On the other hand, the boreal summer MJO shows more northward tilt of the propagation when passing the Maritime Continent.

Because of the difference in propagation, the impact of boreal winter and summer MJO should be different, especially over Indonesia. Therefore, in this study, we will investigate these three main questions:

1. What are the impacts of boreal summer MJO on precipitation extremes in Indonesia?
2. Does the impact differ from boreal winter MJO? If so, in what location does it differs?
3. What are the underlying dynamics behind the impact?

10.2 Data and Methods

In this study, we focus on the extended boreal summer season (May–September). We use daily gridded precipitation datasets from Asian Precipitation–Highly Resolved Observational Data Integration Towards Evaluation of Water Resources (APHRODITE) with the period of 1998–2015 and a spatial resolution of 0.25° × 0.25° [9]. Furthermore, we also use 63 high-quality daily rain-gauge data from BMKG from 1987 to 2017.

To understand the dynamics behind the impact, we use bandpass-filtered daily anomalies with frequency cut-offs of 20–100 days [3]. Then, we calculate anomaly composites for each phase of the MJO for several variables such as outgoing long-wave radiation (OLR), vertically integrated moisture flux convergence (VIMFC), and vertical moisture advection. The VIMFC and vertical moisture advection are calculated using relative humidity, air temperature, zonal, meridional, and vertical wind at various pressure levels (i.e., 1000–100 hPa) [7]. The OLR datasets are obtained from NOAA and are used as a proxy for convection [10]. The other variables are obtained from NCEP-DOE Reanalysis II [11]. Both datasets consist of daily data from 1998 to 2015 with a spatial resolution of 2.5° × 2.5°.

In this study, the boreal summer MJO is defined using the Real-time Multivariate MJO Index (RMM) [5]. Prior to the analysis, the index is selected, so it only contains data from May to August. Then, the days with an MJO index of more than 1 are categorized as active MJO events, while those with less than 1 are categorized as weak MJO events.

We use probability composites to investigate the changes in the likelihood of extreme precipitation [7, 12]. The likelihood of precipitation exceeding the threshold (95th percentile, Fig. 10.1) is calculated for each active phase of the MJO. Then, the changes in the likelihood can be written as follows [7]:

$$\Delta P_{\mathrm{MJO}} = \frac{P_{\mathrm{MJO}}(x \geq x_c) - P_{\mathrm{summer}}(x \geq x_c)}{P_{\mathrm{summer}}(x \geq x_c)} \times 100\% \quad (10.1)$$

where ΔP is the percentage changes in the precipitation extremes likelihood, P_{MJO} is the likelihood of non-zero precipitation exceeding the threshold (x_c) for each active phase of the MJO, and P_{summer} is the same as P_{MJO}, except for all boreal summer days. The ΔP is calculated at each grid and station. Moreover, we also calculate

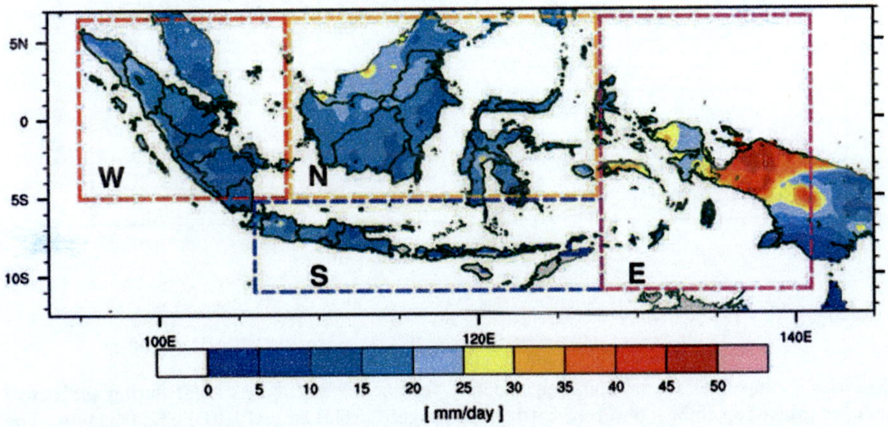

Fig. 10.1 The 95th percentile of precipitation during boreal summer (May–September). Red, orange, blue, and pink boxes represent the western, northern, southern, and eastern parts of Indonesia, respectively. Grey shades denote missing values

the area average of ΔP for each region in Indonesia (Fig. 10.1). The significance of anomaly and probability composites in this study are tested using the bootstrap test method with 1,000 synthetic composites [13].

10.3 Results and Discussion

10.3.1 MJO Impact on Regional Precipitation Anomalies

First, we analyze the impact of boreal summer MJO on the precipitation anomalies over Indonesia. Figure 10.2 shows the filtered anomaly composites during each phase of the MJO. The increase/decrease of precipitation occurs mainly over Sumatra, Borneo, Sulawesi, Molucca, and west of Papua islands by up to 3 mm/day. In general,

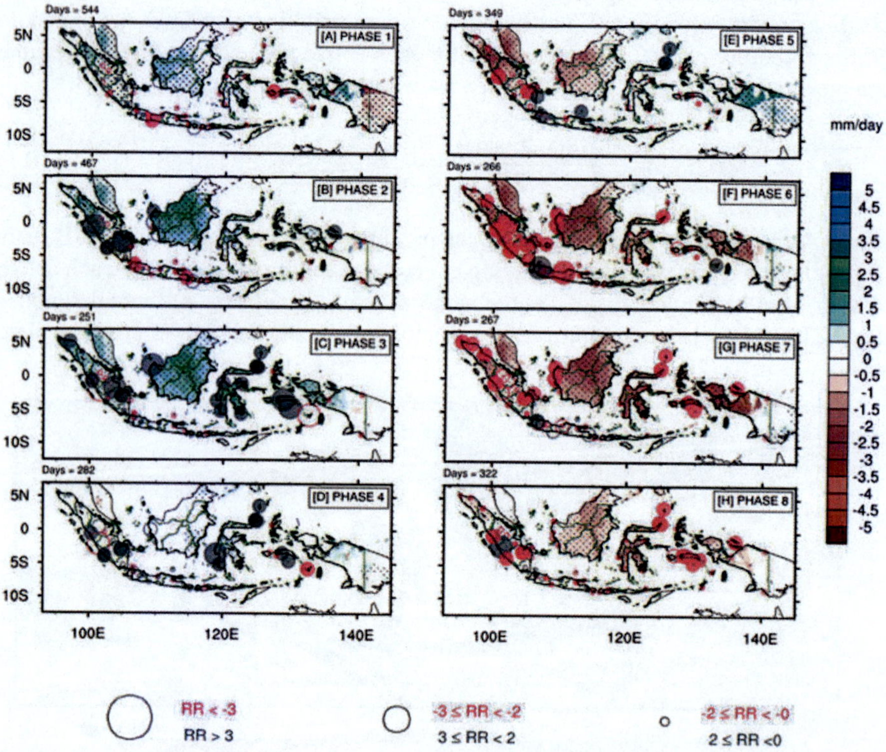

Fig. 10.2 Composites of precipitation anomaly for each MJO phase (A-H) during the boreal summer (May–September) observed by the rain-gauges (circle) and APHRODITE (shading). The filled circle and dots indicate the values that exceeded the 95% confidence level for the rain-gauges and APHRODITE, respectively

the results show that boreal summer MJO increases precipitation over the Indonesian region during phases 2–3 and decreases precipitation during phases 6–8.

From this result, we found that the impact on precipitation differs from that of boreal winter MJO. During boreal winter MJO, the precipitation affects most of the region in Indonesia, including Java [7]. On the other hand, the boreal summer MJO influences mostly on the northern region of Indonesia, with a very small influence over the southern region (i.e., Java, Bali, and Nusa Tenggara islands).

10.3.2 MJO Impact on Precipitation Extremes

In this section, we investigate the impact on the precipitation extremes. Figure 10.3 shows the changes in the likelihood of precipitation that exceeds the 95th percentile threshold (Fig. 10.1). Based on the figure, we conclude that MJO increases the

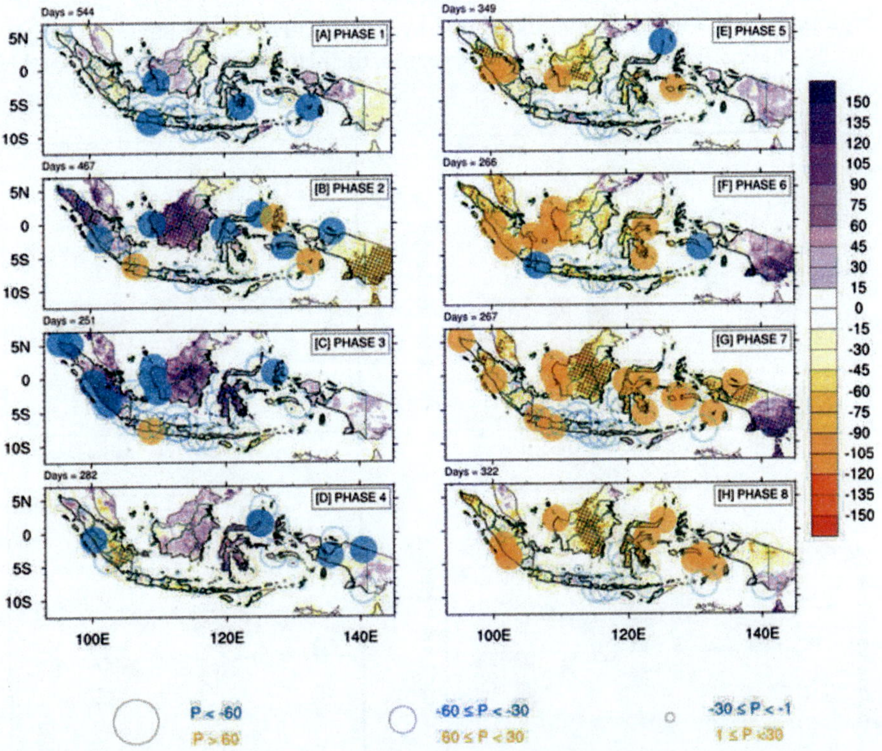

Fig. 10.3 Percentage changes in the probability of extreme precipitation events observed by the rain-gauges (circle) and APHRODITE (shading) data for each phase of the MJO (A-H). Filled circle and hatching indicate the values that exceeded the 95% confidence level for the rain-gauges and APHRODITE, respectively

likelihood of precipitation extremes in Indonesia during phases 2–3 and decreases the likelihood during phases 6–8. The increase in the likelihood is around 60–150% over the Sumatra, Borneo, and Sulawesi islands. Alternatively, during phases 6–8, the MJO decreases the likelihood by around 30–60%.

The strongest impact is seen during phase 3 of the MJO over central Borneo, central Sumatra, and most Sulawesi (Fig. 10.3). During this phase, the likelihood of precipitation extremes increased by up to 150%. The rain-gauge observation corroborates the result, which shows a high increase in the likelihood (>60%), particularly over Sumatra and Borneo. In contrast, a very small impact is found over the southern region of Indonesia. The difference between APHRODITE and rain-gauge observation could be due to the local effects (i.e., topography and diurnal cycle) [14–16]. Nevertheless, this result is generally consistent with the northward tilt of the boreal summer MJO propagation.

To better understand the regional differences of the impact, we calculate the area average of the impact based on geographical locations as in Fig. 10.1. Figure 10.4 shows that the boreal summer MJO increases the likelihood of precipitation extremes over Indonesia's western and northern regions (Fig. 10.4a and b). Over the western region, the MJO increases the likelihood by around 50% during phase 2 and 85% during phase 3. While over the northern region, the MJO increases the likelihood by

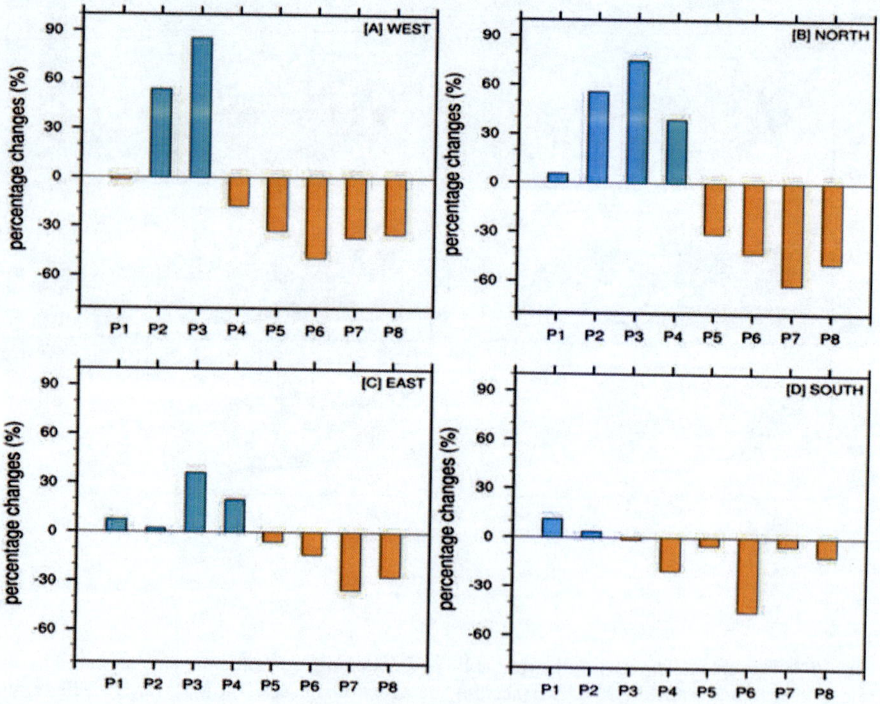

Fig. 10.4 Area-average of the probability composites for different regions in Indonesia

approximately 55, 75, and 35% during phases 2, 3, and 4, respectively. The eastern region of Indonesia is also affected by the boreal summer MJO, albeit in a smaller magnitude. The MJO increases the likelihood of precipitation extremes over the eastern region by around 20–30% during phases 3 and 4. On the contrary, the impact on the southern region is very small, only around up to 10% of the increase in the likelihood.

In general, the impact of the boreal summer MJO on precipitation extremes is the strongest during phases 2–3, affecting most of the western and northern part of Indonesia, similar to that of precipitation anomalies. The location of the impact differs from that of boreal winter MJO. The boreal winter MJO increases the likelihood of precipitation extremes over most of the regions of Indonesia [7], while the boreal summer MJO only strongly impacts the western and northern regions of Indonesia.

10.3.3 The Dynamical Links Between the Boreal Summer MJO and Precipitation Extremes

There are two important factors that are responsible for the increase in the likelihood of precipitation extremes, specifically the VIMFC [7, 17, 18] and vertical moisture advection [7, 19–21]. In this section, we investigate the role of VIMFC and vertical moisture advection anomalies associated with MJO on the precipitation extremes. The VIMFC is calculated as follows [7, 22]:

$$\mathrm{VIMFC} = -\frac{1}{g}\int_{1000\,\mathrm{hPa}}^{100\,\mathrm{hPa}}\left[-\left(\vec{\nabla}.\left(q\mathbf{u}_{H}\right)\right)\right]dp \tag{10.2}$$

where g is gravity (9.8 ms^{-2}), $\mathbf{u}_H$ is the horizontal wind, and q is specific humidity (kg/kg). Additionally, the vertical moisture advection can be written as:

$$\text{vert. qadv} = -\omega\frac{\partial q}{\partial p} \tag{10.3}$$

where ω is the vertical component of wind and p is pressure level. A large-scale moisture convergence from 1000–100 hPa and vertical moisture advection due to boreal winter MJO triggers the occurrence of precipitation extremes in Indonesia [7].

Figure 10.5 shows the filtered anomaly composites of VIMFC and OLR. Based on the figure, during phase 2, the MJO activity enhances the convection over most of Indonesia's western and northern regions (Fig. 10.5b). This enhanced convection is collocated with enhanced moisture convergence. Furthermore, during phase 3, the enhanced convection and convergence due to MJO are getting stronger. This result is consistent with the increase of precipitation extremes' likelihood. As expected, the boreal summer MJO propagation is more tilted to the north instead of to the south like boreal winter MJO. However, the northward tilt is not observed until after

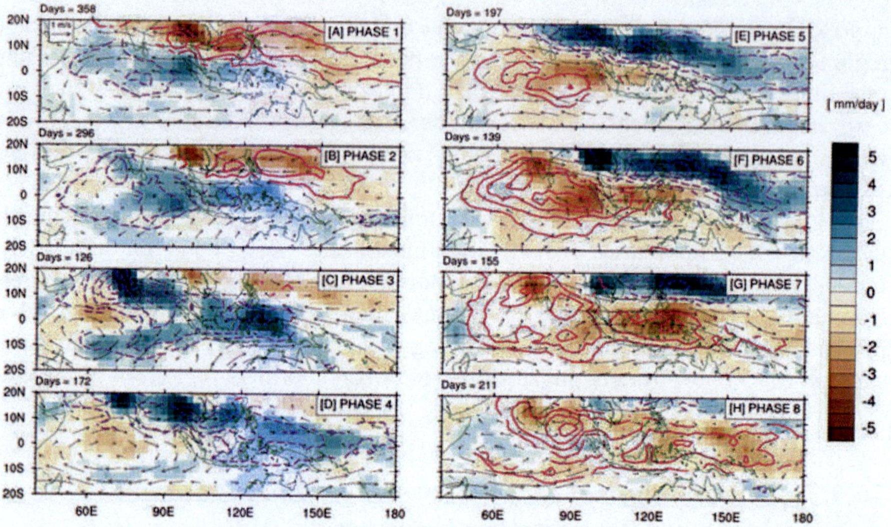

Fig. 10.5 Anomaly composites of VIMFC (shading, mm day^{-1}), OLR (contour, Wm^{-2}), and 850-hPa wind (ms^{-1}). Dashed blue (Solid Red) contour denotes the negative (positive) anomaly of OLR. The contour is from −50 to 50 with an interval of 5. The values shown are significant at a 95% confidence level

phase 4. This is consistent with the boreal summer MJO's impact on precipitation (Figs. 10.3 and 10.4), which only affected the northern part of Indonesia. In contrast, the MJO activity suppresses the convection and moisture convergence over most of the Indonesian regions during phases 6–8, resulting in a decrease in the likelihood of precipitation extremes (Fig. 10.6).

The enhanced (suppressed) convection and moisture convergence are consistent with the upward (downward) moisture advection due to MJO. The upward moisture advection is collocated with the enhanced convergence, resulting in an increase in precipitation and the precipitation extremes' likelihood. Similarly, the downward moisture advection is collocated with the suppressed convergence, causing a decrease in precipitation and the likelihood of the extremes. The strongest upward moisture advection is observed during phase 3 over 70°–90ºE, particularly over the 850–300 hPa, which is consistent with the strongest impact on precipitation. Moreover, throughout all phases, the upward moisture advection is mostly seen at the middle to the upper-level troposphere (i.e., 700–300 hPa). While the result is generally similar to that of boreal winter MJO, the boreal summer MJO exhibits less visible lower level (i.e., 1000–850 hPa) upward moisture advection compared to that of boreal winter MJO [7]. The lower-level advection favors a suitable environment for convection, resulting in a heavier rainfall [7, 21, 23].

In summary, the precipitation anomaly and the likelihood of precipitation extremes are directly associated with the VIMFC and vertical moisture advection due to MJO.

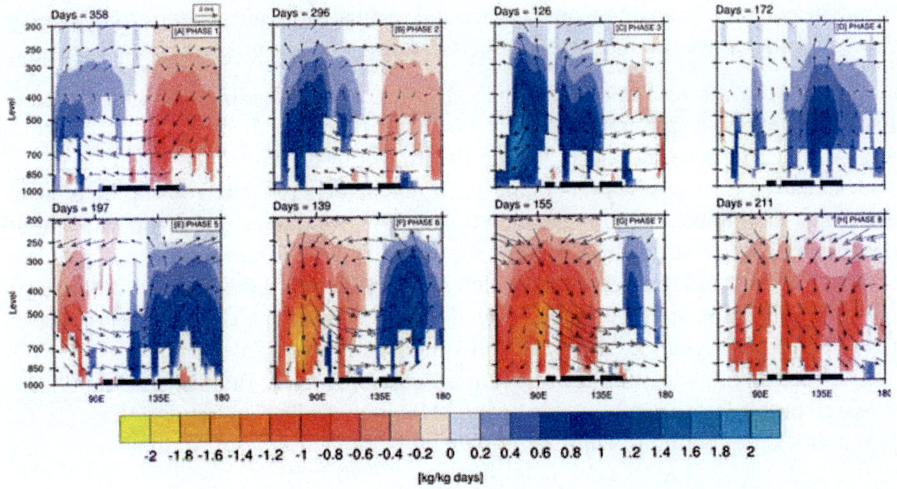

Fig. 10.6 Anomaly composites of vertical moisture advection (shading) and wind (vector) averaged over 11°N–11°S for each phase of MJO (A–H). Shaded values are significant at a 95% confidence level

The enhanced (suppressed) moisture convergence and upward (downward) moisture advection contribute strongly to the increased (decreased) precipitation and the likelihood of precipitation extremes in Indonesia.

10.4 Conclusions

The impact of boreal summer MJO on precipitation extremes and its dynamical links are investigated. We found that the boreal summer MJO significantly modulates the precipitation in Indonesia, particularly during phases 2 and 3. To sum up, we conclude the results as follows:

1. The impact of the boreal summer MJO is the strongest over the western and northern parts of Indonesia, with moderate impact over the eastern region. In contrast, the impact on the southern region is very small due to the northward tilt of boreal summer MJO propagation.
2. Phase 2 and 3 of the boreal summer MJO have the strongest impact on the Indonesian region. Both phases can increase the likelihood of precipitation extremes by up to 150% over the western and northern parts and around 30% over the eastern part of Indonesia.
3. On the contrary, the suppressed phase of the boreal summer MJO (phases 6–8) affects most of the Indonesian region, including the southern part. The suppressed phase of the MJO decreases the precipitation and the likelihood of precipitation extremes in Indonesia.

4. The enhanced moisture convergence and vertical moisture advection are directly responsible for the increase in precipitation and extreme precipitation events.

To sum up, the results suggest that MJO has a considerable impact on precipitation extremes during boreal summer in Indonesia, specifically over the western and northern parts of Indonesia. While the impact of boreal summer MJO is not as widespread as its boreal winter counterparts, the percentage changes of precipitation extremes' likelihood are high, reaching up to 150%. Moreover, due to the widespread decrease of precipitation during MJO's suppressed phase, future study is still required to better understand this response, which may be related to the complex interaction between MJO and other type of tropical disturbances, such as equatorial waves [24–26], Indian Ocean Dipole [27] and summer monsoon (also known as BSISO) [28, 29] in modulating summertime rainfall variability over Indonesia. Overall, this research highlights the importance of the MJO in providing an important source for probabilistic prediction of summertime precipitation extremes in Indonesia.

References

1. Madden, R.A., Julian, P.R.: Observations of the 40–50-day tropical oscillation—a review. Mon. Weather Rev. **122**, 814–837 (1994). https://doi.org/10.1175/1520-0493(1994)122%3c0814:OOTDTO%3e2.0.CO;2
2. Nakazawa, T.: Intraseasonal variations of OLR in the tropics during the FGGE year. J. Meteor. Soc. Jpn. **64**, 17–34 (1986)
3. Waliser, D., Sperber, K., Hendon, H., Kim, D., Maloney, E., Wheeler, M., Weickmann, K., Zhang, C., Donner, L., Gottschalck, J., Higgins, W., Kang, I.S., Legler, D., Moncrieff, M., Schubert, S., Stern, W., Vitart, F., Wang, B., Wang, W., Woolnough, S.: MJO simulation diagnostics. J. Clim. **22**, 3006–3030 (2009). https://doi.org/10.1175/2008JCLI2731.1
4. Weickmann, K.M., Lussky, G.R., Kutzbach, J.E.: Intraseasonal (30–60 Day) fluctuations of outgoing longwave radiation and 250 Mb stream function during Northern Winter. Mon. Weather Rev. **113**, 941–961 (1985)
5. Wheeler, M.C., Hendon, H.H.: An all-season real-time multivariate MJO index: development of an index for monitoring and prediction. Mon. Weather Rev. **132**, 1917–1932 (2004). https://doi.org/10.1175/1520-0493(2004)132%3c1917:AARMMI%3e2.0.CO;2
6. Hidayat, R., Kizu, S.: Influence of the Madden-Julian oscillation on Indonesian rainfall variability in austral summer. Int. J. Climatol. **30**, 1816–1825 (2010). https://doi.org/10.1002/joc.2005
7. Muhammad, F.R., Lubis, S.W., Setiawan, S.: Impacts of the Madden–Julian oscillation on precipitation extremes in Indonesia. Int. J. Climatol. **41**, 1970–1984 (2021). https://doi.org/10.1002/joc.6941
8. Zhang, C., Dong, M.: Seasonality in the Madden–Julian oscillation. J. Clim. **17**, 3169–3180 (2004)
9. Yatagai, A., Kamiguchi, K., Arakawa, O., Hamada, A., Yasutomi, N., Kitoh, A.: APHRODITE: constructing a long-term daily gridded precipitation dataset for Asia based on a dense network of rain gauges. Bull. Am. Meteor. Soc. **93**, 1401–1415 (2012). https://doi.org/10.1175/BAMS-D-11-00122.1
10. Liebmann, B., Smith, C.A.: Description of a complete (interpolated) outgoing longwave radiation dataset. Bull. Am. Meteor. Soc. **77**, 1275–1277 (1996)

11. Kanamitsu, M., Ebisuzaki, W., Woollen, J., Yang, S.-K., Hnilo, J.J., Fiorino, M., Potter, G.L.: NCEP-DOE AMIP-II Renalalysys (R-2). Bull. Am. Meteor. Soc. **83**, 1631–1643 (2002). https://doi.org/10.1175/BAMS-83-11
12. Wheeler, M.C., Hendon, H.H., Cleland, S., Meinke, H., Donald, A.: Impacts of the Madden-Julian oscillation on Australian rainfall and circulation. J. Clim. **22**, 1482–1498 (2009). https://doi.org/10.1175/2008JCLI2595.1
13. Wilks, D.S.: Statistical methods in the atmospheric sciences. Elsevier, Cambridge, US (2006)
14. Peatman, S.C., Matthews, A.J., Stevens, D.P.: Propagation of the Madden-Julian oscillation through the maritime continent and scale interaction with the diurnal cycle of precipitation. Q. J. R. Meteorol. Soc. **140**, 814–825 (2013). https://doi.org/10.1002/qj.2161
15. Hidayat, R., Kizu, S.: Influence of the Madden-Julian oscillation on Indonesian rainfall variability in Austral summer. Int. J. Climatol. **30**, 1816–1825 (2009). https://doi.org/10.1002/joc.2005
16. Lestari, S., King, A., Vincent, C., Karoly, D., Protat, A.: Seasonal dependence of rainfall extremes in and around Jakarta, Indonesia. Weather Climate Extremes **24**, 100202 (2019). https://doi.org/10.1016/j.wace.2019.100202
17. Adames, Á.F.: Precipitation budget of the Madden–Julian oscillation. J. Atmos. Sci. **74**, 1799–1817 (2017). https://doi.org/10.1175/jas-d-16-0242.1
18. Kirshbaum, D.J., Adler, B., Kalthoff, N., Barthlott, C., Serafin, S.: Moist orographic convection: physical mechanisms and links to surface-exchange processes. Atmosphere **9**, 1–26 (2018). https://doi.org/10.3390/atmos9030080
19. Adames, Á.F., Wallace, J.M.: Three-dimensional structure and evolution of the MJO and its relation to the mean flow. J. Atmos. Sci. **71**, 2007–2026 (2014). https://doi.org/10.1175/JAS-D-13-0254.1
20. Ren, H., Ren, P.: Impact of Madden–Julian oscillation upon winter extreme rainfall in Southern China: observations and predictability in CFSv2. Atmosphere **8**, 192–216 (2017). https://doi.org/10.3390/atmos8100192
21. Benedict, J.J., Randall, D.A.: Observed characteristics of the MJO relative to maximum rainfall. J. Atmos. Sci. **64**, 2332–2354 (2007). https://doi.org/10.1175/jas3968.1
22. van Zomeren, J., van Delden, A.: Vertically integrated moisture flux convergence as a predictor of thunderstorms. Atmos. Res. **83**, 435–445 (2007). https://doi.org/10.1016/j.atmosres.2005.08.015
23. Xavier, P., Rahmat, R., Cheong, W.K., Wallace, E.: Influence of Madden-Julian oscillation on Southeast Asia rainfall extremes: observations and predictability. Geophys. Res. Lett. **41**, 4406–4412 (2014). https://doi.org/10.1002/2014GL060241
24. Lubis, S.W., Jacobi, C.: The modulating influence of convectively coupled equatorial waves (CCEWs) on the variability of tropical precipitation. Int. J. Climatol. **35**, 1465–1483 (2015). https://doi.org/10.1002/joc.4069
25. Ferrett, S., Yang, G.Y., Woolnough, S.J., Methven, J., Hodges, K., Holloway, C.E.: Linking extreme precipitation in Southeast Asia to equatorial waves. Q. J. R. Meteorol. Soc. **146**, 665–684 (2020). https://doi.org/10.1002/qj.3699
26. Lubis, S.W., Respati, M.R.: Impacts of convectively coupled equatorial waves on rainfall extremes in Java, Indonesia. Int. J. Climatol. **41**, 2418–2440 (2021). https://doi.org/10.1002/joc.6967
27. Muhammad, F.R., Lubis, S.W., Tiarni, I., Setiawan, S.: Influence of the Indian Ocean Dipole (IOD) on convectively coupled Kelvin and mixed Rossby-gravity waves. IOP Conf. Ser. Earth Environ. Sci. **284**, 12012 (2019). https://doi.org/10.1088/1755-1315/284/1/012012
28. Lee, J.Y., Wang, B., Wheeler, M.C., Fu, X., Waliser, D.E., Kang, I.S.: Real-time multivariate indices for the boreal summer intraseasonal oscillation over the Asian summer monsoon region. Clim. Dyn. **40**, 493–509 (2013). https://doi.org/10.1007/s00382-012-1544-4
29. Jiang, X., Li, T., Wang, B.: Structures and mechanisms of the northward propagating boreal summer intraseasonal oscillation. J. Clim. **17**, 1022–1039 (2004). https://doi.org/10.1175/JCLI3861.1

Chapter 11
Smoke Propagation During Fire in Kalimantan and Sumatra in 2015 and 2019

Sumaryati, Dita Fatria Andarini, Nani Cholianawati, and Asri Indrawati

Abstract The forest and land fire events in Indonesia reached a peak in 2015 and 2019 since the last decade associated with strong and moderate El Nino phenomena. Those events have been produced a high concentration of smoke. This study aims to understand the propagation patterns of smoke from Kalimantan and Sumatra during those cases and their influence on the aerosol optical thickness (AOT) and visibility. The hotspots data observed by Terra, Aqua, and SNPP satellites were used to represent the forest and land fires, while the AOT data was obtained from the SNPP-VIIRS. Furthermore, the forward trajectory of the Hysplit model was applied to investigate the smoke propagation using matrix sources with a spatial resolution of $1° \times 1°$. The result shows that the background wind influenced the smoke propagation that led to transboundary smoke occurrences in September. Moreover, the events of the forest and land fires affected the higher value of AOT and lower visibility. The AOT distributions have a similar pattern to the smoke propagation. However, the visibility pattern does not perform the consistency with the propagation.

11.1 Introduction

The forest and land fire that occurs over several Indonesian regions almost every year can affect the environment and health sectors. In general, the intensity of those events is caused by both internal and external factors, which is the weather is the most important factor [1]. According to the historical events of forest fires in Indonesia, El Niño phenomena that affected a prolonged dry season is closely led to the forest fires. For instance, the Ministry of Environment and Forestry reported that 2,611,411 ha of forest fires occurred in 2015 related to strong El Nino, and a moderate El Nino in 2019 led to 1,592,010 ha of burned area. Interestingly, during those cases, around 75%

Sumaryati (✉) · D. F. Andarini · N. Cholianawati · A. Indrawati
Research Center for Climate and Atmosphere, National Research and Innovation Agency (BRIN), Bandung, Indonesia
e-mail: sumaryati@lapan.go.id

E. Yulihastin et al. (eds.), *Proceedings of the International Conference on Radioscience, Equatorial Atmospheric Science and Environment and Humanosphere Science, 2021*, Springer Proceedings in Physics 275, https://doi.org/10.1007/978-981-19-0308-3_11

of forest and land fire (vegetation fire) areas occurred in Sumatera and Kalimantan Islands which are 2,006,360 ha and 1,190,685 ha, respectively.

Besides the high intensity of fires, forest and land fires in the peatland area are also addressed as another crucial issue in Kalimantan and Sumatra. Biomass burning on peatlands occurs both in the surface and underground [2], where the underground biomass consists of plant roots and peatland. However, underground fires in peatlands are difficult to extinguish because the fires are deep in the ground, and thus, the underground burning produces much more smoke. Several studies compared the peat fire and savanna fire, where the results showed that more smoke is detected in the peat area than in the savanna [3, 4]. Those smoke fires consist of a complex mixture resulting from water vapor, particles, and gases as fire products.

Smoke is one of the most important compositions produced by a forest and land fire that has attracted the attention of governments, researchers, and the public. In detail, smoke is a visible air pollutant that can be observed easily without instruments, especially for a higher density smoke. Furthermore, the particle content in forest fire smoke causes several chronic health effects such as cardiovascular and lung cancer, and also acute diseases high concentrations (e.g. cough and respiratory infections) [5, 6]. People are more concerned about those chronic effects because they can suffer immediately to themselves. Whereas, many other vital problems led by the smoke fires, for instance, transboundary pollution, closure of learning activities in schools, cancellation of flights due to low visibility, and public health.

More importantly, the smoke contributes to decreasing the visibility and increasing the value of aerosol optical thickness (AOT), thus it will affect the characteristics of atmospheric optic [7, 8]. In addition, the particle of smoke on the surface area has also an effect on human health and visibility. If the transport of smoke exceeds the Planetary Boundary Layer (PBL), it will be affected the weather condition since the smoke particles act as the cloud condensation nuclei. Therefore, it can absorb and reflect the solar radiation which determines the radiation balance.

This study aims to understand the smoke propagation resulting from forest and land fires in Sumatra and Kalimantan and their impact on the changes of visibility and AOT during the highest cases in 2015 and 2019 associated with El Nino events. Moreover, the forward trajectory of the Hysplit model will be applied to identify the propagation of smoke. Several previous studies have implemented the Hysplit model using a single source trajectory to describe the smoke dispersion in Sumatra and Kalimantan [9]. Meanwhile, this study used a type of matrix source trajectory in the Hysplit model instead of a single source.

11.2 Data and Method

This study focuses on the forest fire events in 2015 and 2019 over Kalimantan and Sumatera that associated with strong and moderate El Nino events. Firstly, the forest burned areas were reported by the Indonesian Ministry of Environment and

Forestry that could be accessed through http://sipongi.menlhk.go.id/hotspot/luas_kebakaran. Meanwhile, the location of forest area were determined by the hotspots from Terra, Aqua, and SNPP satellites. Those datas were provided by the LAPAN and are available on the website (http://modis-catalog.lapan.go.id/monitoring/hotspot/index). Furthermore, the hotspot was selected based on the confidence level of more than 80%.

One of the most important particulates that could be seen during the vegetation fire events is smoke. Thus, this study identified the propagation of smoke by implementing the Hysplit model. The Hybrid Single-Particle Lagrangian Integrated Trajectory (Hysplit) was a trajectory model developed by Air Resources Laboratory of the National Oceanic and Atmospheric Administration (NOAA) to calculate and analyze simple air mass trajectories, complex transport, dispersion, as well as diffusion of atmospheric pollutants [10]. This model applied a hybrid diffusion model between Langrangian method for advection and diffusion calculation, and the concentration calculation used the Eularian approach [11]. This study run the Hysplit model interactively on the website by accessing https://www.ready.noaa.gov/HYSPLIT_traj.php.

The matrix sources were used to calculate the propagation of smoke based on the distribution of hotspot locations as shown in Fig. 11.1. The red, blue, and green dots represented the fires or sources of smoke with a spatial resolution of 1° × 1° (Fig. 11.1). In detail, two days forward trajectories were computed during the vegetation fire events, which are 5th, 15th, and 25th of September and October in 2015 and 2019 respectively. To obtain the diurnal variation of the smoke propagation, the starting time of each day was in the morning (06.00 Western Indonesian Time/WIT), noon (12.00 WIT), in the afternoon (18.00 WIT), and midnight 24.00 (WIT). For further analysis in understanding the propagation pattern, this study also utilized the daily wind component data from NCEP-DOE Reanalysis 2 during May, September, and October in 2015 and 2019.

To understand the relationship between visibility and AOT, this study used the visibility data from the observation stations over study area (6°S–10°N and 96°E–122°E)

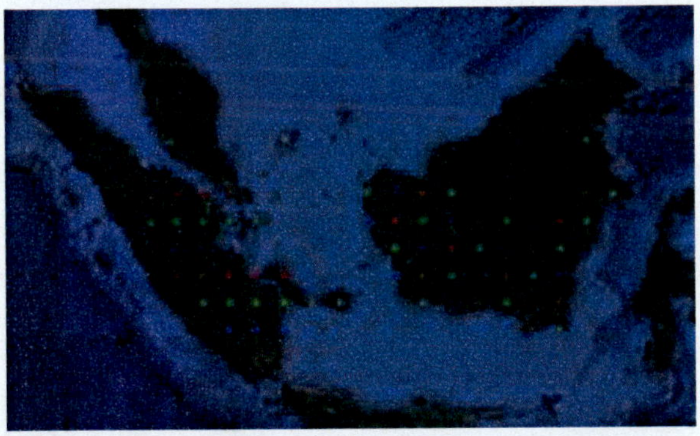

Fig. 11.1 Matrix source of forest fire

that obtained through online (https://www.ogimet.com/home.phtml.en). Meanwhile, the AOT was analyzed from the sensor of The Visible/Infrared Imager Radiometer Suite (VIIRS) in satellite of the Suomi National Polar-orbiting Partnership (SNPP). This dataset has the daily temporal resolution with a spatial of 0.25° × 0.25° of horizontal resolution, and accessible through ftp://ftp.star.nesdis.noaa.gov/pub/smcd/VIIRS_Aerosol/viirs_aod_gridded/idps/snpp/edraot550/. AOT was one of the products of VIIRS sensor in SNPP that has validated using another instrument which was MODIS and the in situ stations from Aerosol Robotic Network (AERONET) [12, 13]. Finally, the comparison of AOT and visibility was identified in two different conditions during the non-fire and forest/land fire events.

11.3 Result and Discussion

11.3.1 Forest Fire Events

During the last six-year (2014–2019), the biggest forest and land fire events in Kalimantan and Sumatra occurred in 2015 and 2019 as reported by the Indonesia Ministry of Environment and Forestry. It can be seen in Fig. 11.2, about 2.0 million ha of forest and land were burned during the 2015's event, while the fire case in 2019 has burned around 1.25 million ha. Those conditions were associated with the El Nino events, which were 2015 has strong El Nino and moderate category of El Nino happened in 2019. Even though the human factors, especially the economic problem, affected forest fires, however weather factors dominantly led to more prominent forest/land fire events. Therefore, the prolonged dry season in Kalimantan and Sumatra triggered by the El Niño phenomenon becomes a concern in discussing vegetation fires in those islands.

In general, the fire events can be determined by the number of hotspots as it showed a good correlation between the burned area and the number of hotspots that have more than 80% of confidence level [4]. Figure 11.3 depicts the number of hotspots each month based on three satellites (Aqua, Terra, and SNPP) in 2015 and

Fig. 11.2 Forest burned area

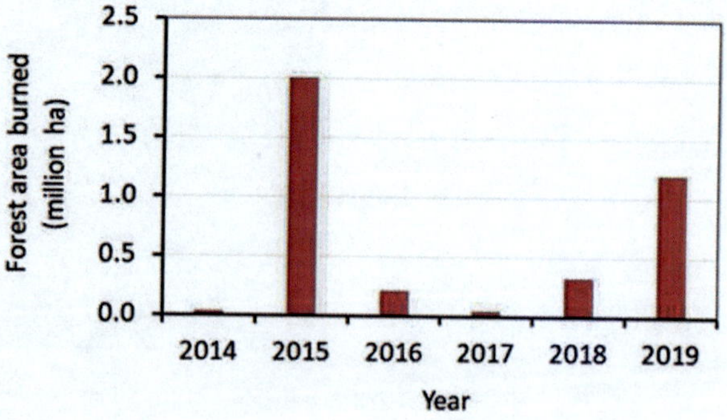

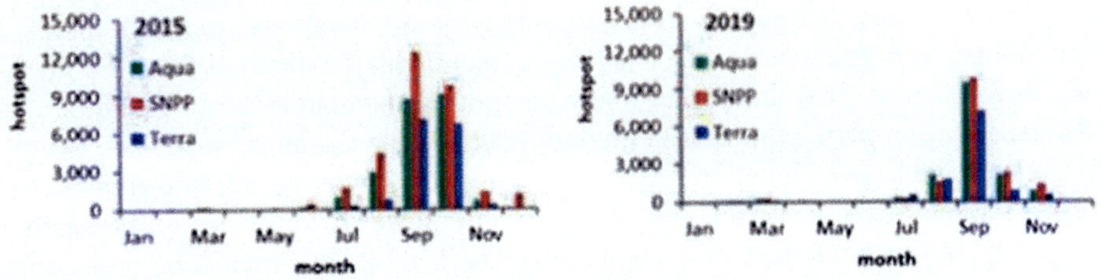

Fig. 11.3 The monthly hotspot in Sumatera and Kalimantan

2019. All satellites showed a similar pattern of hotspot numbers, where the hotspots emerged from July to November during those periods. Interestingly, the number of hotspots reached a peak in September and October of 2015, while the largest number of hotspots occurred in September 2019.

The majority of the regions in Kalimantan and Sumatera have the monsoonal rainfall type, while the remaining area is categorized as equatorial type [14]. Less amount of rainfall is received through June to August in both types, where this condition was sufficient to have dry forest biomass. More importantly, this dry season can be longer when an El Niño phenomenon was coming. It perhaps extends the dry season to October, November, or December, depending on the strength of El Niño. Eventually, those long periods of dry season drive pretty dry biomass conditions that can lead to intensively vegetation fires. As shown in Fig. 11.3, the number of hotspots in 2015 is higher than those in 2019, due to the different conditions of the El Nino phenomenon, which were strong and moderate El Nino in 2015 and 2019, respectively.

Spatially, the distribution of hotspot locations detected by Aqua, Terra, and SNPP satellites during September–October in 2015 and 2019 are shown in Fig. 11.4. In Sumatra, most of the hotspots are located in the southern land and eastern coastal area. Meanwhile, hotspots are found in almost areas in Kalimantan except several parts in the center and north region of the island. Specifically, SNPP and Aqua detected a higher number of hotspot locations than the Terra satellite.

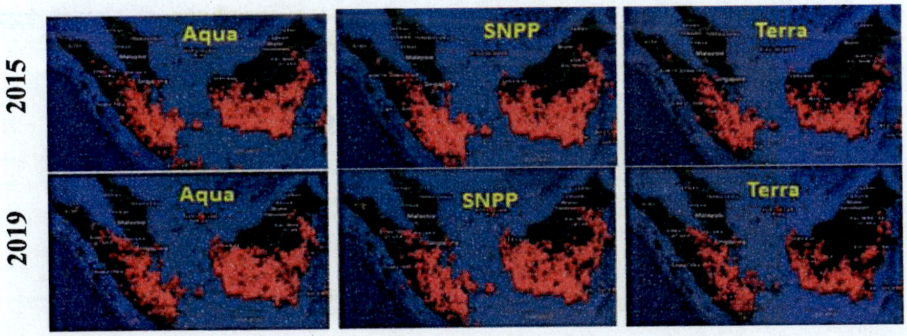

Fig. 11.4 The hotspot locations during forest and land fire events in 2015 and 2019

Based on the hotspot locations detected by the satellites, most of them are distributed in peatland areas. This is in agreement with the previous research studied by Prayoga et al. [15] that found many hotspots on peatlands during the fire cases. In general, complete combustion occurs if there is an adequate supply of oxygen. However, biomass burning on peatlands occurs not only in the surface area but also underground of the peatland area [2]. Due to the lower layer of soil in the peatland area having less amount of oxygen, it affects an incomplete combustion process that produces a huge amount of smoke instead of CO2. The previous study revealed that forest fires in Indonesia, which mostly occur on Sumatra and Kalimantan islands generate a higher concentration of smoke than savanna fires in Africa [3]. In addition, the smoke of forest fires consists of a complex composition of fine particles, water vapor, and gases released from biomass burning.

11.3.2 Smoke Propagation

According to the spatial distribution of the hotspot in Fig. 11.4, this study used the matrix sources with a horizontal resolution of 1° × 1° in simulating the Hysplit model. The output trajectory of the Hysplit model is generally used to describe a long-range transport air pollution and is one of the tools for calculating and simulating the dispersion of forest fires [16, 17]. Thus, the forward trajectory method was applied to simulate the transport of smoke from each source as represented by blue, green, and red dots in Fig. 11.5. The simulation was conducted during the peak fire events which was September 2015 in four different times to define the influence of the diurnal cycle.

The two days forward trajectory of smoke showed a similar pattern each day, however, the location of hotspots influence the different direction of the propagation.

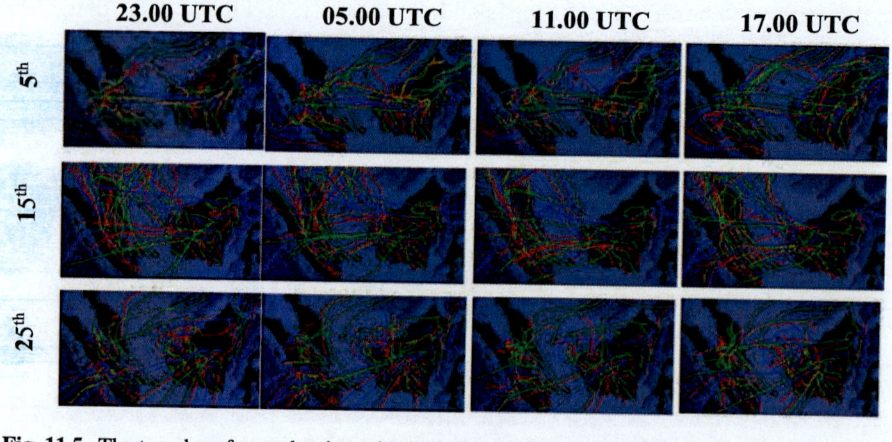

Fig. 11.5 The two days forward trajectories based on Hysplit model in September

The result of the Hysplit model also indicated that the smoke propagation has reached a distance of more than 100 km to the north of Sumatra and Kalimantan, thus it can be categorized as long-range transport. As a result, this long-range transport causes the transboundary pollution phenomenon in Singapore and Malaysia. In several previous studies that have been discussed the long-range transport issues dominantly analyzed the particulate as an object-focused study [18–20]. Particulate matter is a common name, while the specific names are called depending on its source, namely smoke for forest fires source, dust for sandstorms, volcanic ash for a volcano eruption.

It also can be seen that there were no significant differences in the smoke propagation in the diurnal trajectory based on the Hysplit model (Fig. 11.5). In Sumatra, the trajectory generally propagates to the north and turns to the northeast in the next two days. Interestingly, the path of trajectory in several areas of Kalimantan goes to the north, while others propagate to the west. The global circulation and a complex topography in Kalimantan and Sumatra especially the Bukit Barisan Mountains influenced the path [21, 22]. The Bukit Barisan Mountains located along the west of the island of Sumatra can be a barrier to the smoke propagation in Sumatra.

In order to compare the trajectory during the forest fire events in 2015 and 2019, this study also simulated the two days forward trajectory in the peak month of events as shown in Fig. 11.6. The smoke trajectories during October both in 2015 and 2019 have a shorter and irregular path in different levels than those in September. Regarding the topography barrier, the Bukit Barisan Mountains do not significantly affect the trajectory from several source locations of smoke in Kalimantan. The smoke emitted by fires in Kalimantan can cross the Bukit Barisan Mountains because the position of the smoke has reached the planetary boundary layer.

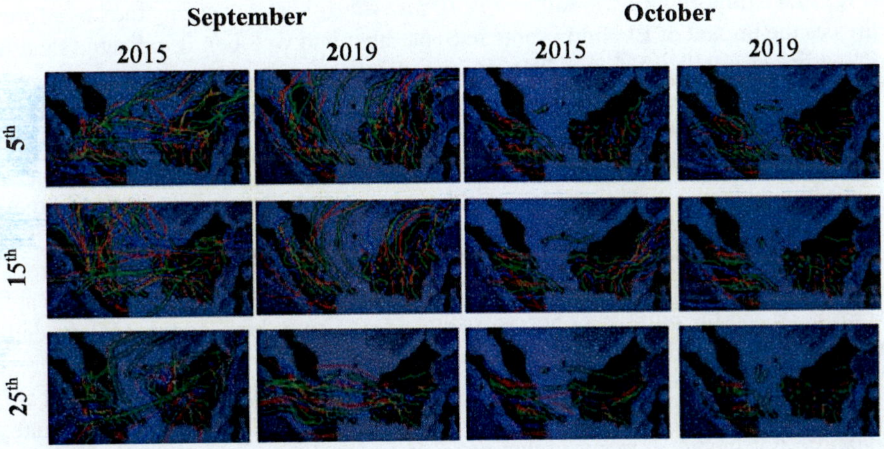

Fig. 11.6 The two days forward trajectories using matrix sources of Hysplit model in the peak period of forest and land fire events

11.3.3 The Aerosol Optical Thickness (AOT) and Visibility

Most of the forest fires in Kalimantan and Sumatra mainly occur in peatlands that can lead to incomplete combustion during the biomass burning process [11, 12]. This condition happens due to the insufficient supply of oxygen in the lower soil of peatland areas. Consequently, it causes the carbon content in the burned biomass to form many incomplete combustion products namely CO, HC, and particulate. Among those compounds, particulate is categorized as one of the smoke components. Hence, the fire events during the study period here produce a lot of smoke compared to the savanna fire [3, 4, 23]. Further, the smoke perhaps causes transboundary pollution, especially in the neighboring countries such as Singapore and Malaysia.

Besides the dense smoke give negative impacts on human health especially in respiratory, cardiovascular, lung cancer, as well as premature mortality [6, 24], it also changes the atmospheric optical characteristics. The atmospheric optical changes generally can be identified by the visibility and atmospheric optical thickness (AOT) or aerosol optical depth (AOD). Ground-based monitoring and satellite observation have been applied to analyze the fluctuation of AOT caused by forest fire events [25]. As an example of the research in Central Kalimantan that showed a significant reduction in visibility from an average of about 8 km during the normal condition (no fire event) to less than 1 km when detected haze from the forest fires [7]. Thus, to compare the AOT concentration in both conditions, this study used the month of May as a normal condition and forest fire events showed in the period of September and October as can be seen in Fig. 11.7.

The AOT value in Kalimantan and Sumatra was around 0.4 in May 2015 and 2019 when there were no forest fire cases, and then called as the background AOT (Fig. 11.8). Importantly, the value of AOT rose fourfold to 1.6 during the large forest fires as the impact of El Niño events in September and October. It indicated that the forest fire events in 2015 and 2019 produced dense smoke. This result is similar to the previous research studied by Zielinski that the AOT value in Europe increased during forest fire events [8].

According to the spatial distribution of AOT in Fig. 11.8, the higher value is concentrated in the hotspot areas and their surroundings, especially in the northern part, including the Karimata Strait, Singapore, Malaysia, and the South China Sea. Hence, it showed a similar pattern between the increase in AOT and smoke propagation paths analyzed by the forward trajectory model. Moreover, the burned area (Fig. 11.2) and the number of hotspots (Fig. 11.4) also influenced the AOT value in which the higher value spread over the larger areas in 2015 than those in 2019.

Similarly, the smoke of forest fires can be identified by the visibility parameter as it causes the reduction of visibility, which endangers transportation sectors especially aviation. For instance, many flights were delayed when the visibility in Palangka Raya, Central Kalimantan, as an affected area of smoke from a forest fire, reached less than 1 km [7]. It is supported by the previous result that represented a good correlation between visibility and PM10 [26], where PM10 is an essential compound of the smoke composition. Therefore, the decreased visibility can be assessed as one

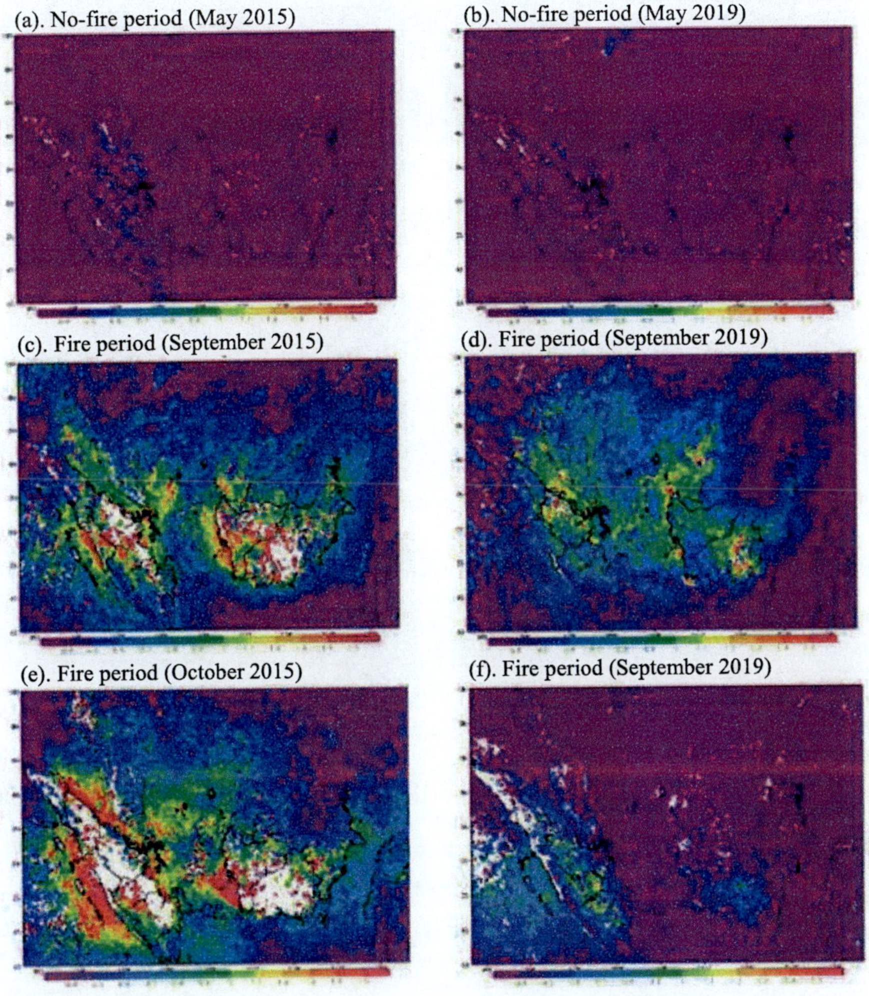

Fig. 11.7 The AOT value during the no-fire period (**a**, **b**), and fire period (**c–f**)

indication of the impact of smoke forest fires. To determine the different conditions of visibility during the fire and no fire periods, this study applied the same method with the analysis of AOT. September and October are chosen as the fire period, while the no-fire period of visibility is represented during May that classified as the background visibility (Fig. 11.9).

Figure 11.8 generally depicts that the visibility pattern during the forest and land fires in September and October has lower visibilities than those in May as denoted by a darker color. The lower value of visibilities in the forest and land periods are concentrated in the hotspots areas. However, it is difficult to determine the consistency of those patterns with the smoke trajectories as analyzed in the previous sections. It

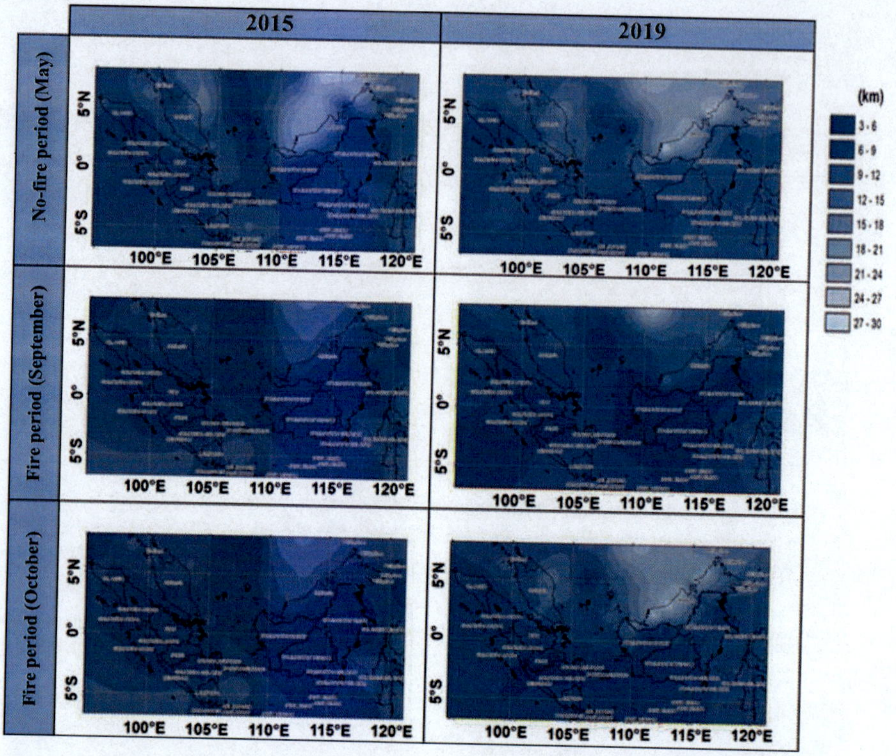

Fig. 11.8 The visibility during the fire period (September and October) and no-fire period (May)

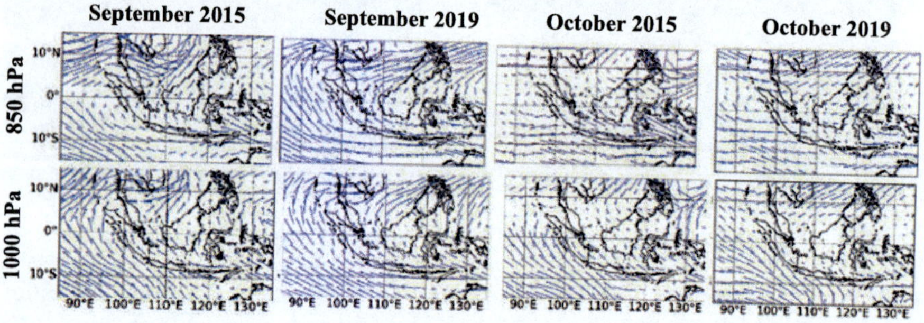

Fig. 11.9 Wind Vector during the forest and land fire events

is supposed that the used visibility data has a coarse spatial resolution since those data are available only at the airports in the study area.

The background wind during the fire events influenced the pattern and distance of the smoke trajectories, which affects also the AOT change pattern and visibility. In general, the smoke propagation during the forest and land fire events in 2015 and 2019 is concentrated in the surface level to around 850 hPa. Figure 11.9 shows that the wind dominantly blows from Australia that has a high pressure across the equator to Asia. This condition is associated with the Australian Winter Monsoon that normally occurs through June–September [27]. Hence, the background wind due to the Australian Monsoon drives the rapid propagation of the smoke from Sumatra and Kalimantan to the north and then turns eastward after crossing the equator. Consequently, the forest and land fire events in Sumatra lead to transboundary pollution over Peninsular Malaysia and Singapore.

A different wind pattern occurs in October as can be seen in Fig. 11.7. Even though the easterly wind caused by the Australian Winter Monsoon still exists, the northerly winds appear during this period in which is associated with the onset of the Asian Winter Monsoon. This onset indicates the sudden reversal of wind field [25]. Therefore, the seasonal transition of wind pattern influenced the smoke propagation blow to the west. Furthermore, this condition also leads to the path pattern of the forward trajectories, where the smoke has a closer path (Fig. 11.6). As a result, it reduced transboundary pollution in Malaysia and Singapore.

11.4 Conclusion

The propagation of smoke from forest and land fires during September–October in 2015 and 2019 are investigated by the Hysplit Model to perform two days forward trajectory using matrix sources. The results show that the trajectories dominantly prop- agate further to the north and turn to the northeast in September, thus it leads to a high potency of the smoke transboundary over Malaysia and Singapore. In contrast, the smoke trajectory in October tends to propagate in surrounding hotspot areas with irregular paths. This condition is influenced by the background wind in which the abrupt reversal of wind field occurred during the onset of the Asian Winter Monsoon. However, the local topography especially the existence of the Bukit Barisan Mountains has no dominant effect on the smoke trajectory in Kalimantan since it propagates above the Planetary Boundary Layer.

In addition, the analysis of AOT showed a consistent pattern with the smoke propagation. The AOT value significantly increased fourfold from 0.4 in normal conditions to 1.6 during the forest and land fire events. Those higher values of AOT are distributed in hotspots areas. On the other hand, even though the visibility parameter shows a lower value in forest and land fire events, the coarse horizontal spatial of the data leads to difficulties in clearly identifying the consistent pattern with the smoke trajectory. Therefore, this study recommends utilizing the higher resolution of visibility data for further analysis regarding the influence of smoke propagation on visibility.

Acknowledgements We want to thank the support of The National Institute of Aeronautics and Space and our colleagues in the Center for Atmospheric Science and Technology for discussing and suggestions. We also thank the agencies that have supported the data

References

1. Cahyono, A.A., Warsito, S.P., Andayani, A., Darwanto, D.H.: Faktor-faktor yang Mempengaruhi Kebakaran Hutan di Indonesia dan Implikasi Kebijakan. Jurnal Sylvia Lestari **3**(1), 103–112 (2015)
2. Usup, A., Hashimoto, Y., Takahashi, H., Hayasaka, H.: Combustion and thermal characteristic of peat fire tropical Peatland in Central Kalimantan, Indonesia. Tropic **14**(1), 1–19 (2004)
3. Lee, H., Jeong, S.-J., Kalashnikova, O., Tosca, M., Kim, S., Kug, J.-S.: Characteristic of Wildfire-Induced Aerosol Emissions from the Maritime Continent Peatland and central African Dry Savannah and CALIPSO Aerosol Product, vol. 123, pp. 3116–3125 (2018)
4. Sumaryati, Aandarini, D.F., Cholianawati, N., Indrawati, A.: Forest Fire in East Nusa Tenggara during 2015–2019: Comparison to Forest Fire in Kalimantan and Sumatera, in progress publishing
5. Black, C., Tesfaigzi, Y., Bassein, J.A., Miller, L.A.: Wildfire smoke exposure and human health: significant gaps in research for a growing public health issue. Environ. Toxicol. Pharmacol. **55**, 186–195 (2017)
6. Aguillera, R., Corringham, T., Gershunov, A., Benmarhinia, T.: Wildfire smoke impacts respiratory health mode than fine particles from other sources: observational evidence from Southern California. Nat. Commun. 1493 (2021)
7. Sumaryati, Cholianawati, N., Indrawati, A.: J. Phys: Theor. Appl. 3(1), 16–26 (2019)
8. Zielinski, T., Petelski, T., Strzalkowska, A., Pakszys, P., Przemyslaw: Impact of wild forest fires in Eastern Europe on aerosol composition and particle optical properties. Oceanology **58**(1), 13–24 (2016)
9. Khairullah, S.E., Makmur, E.E.S.: Trajectory and concentration PM10 on forest and vegetation peat-fire HYSPLIT model outputs and observations (Period: September– October 2015). IOP Conf. Ser.: Earth Environ. Sci. **58**, 1–13 (2017)
10. Draxler, R.R., Stunder, B., Rolph, G., Stein, A., Taylor, A.: HYSPLIT4 user's guide version 4.9 (2014)
11. Rolph, G., Stein, A., Stunder, B.: Real-time environmental applications and display system: READY. Environ. Model. Softw. **95**, 210–228 (2017)
12. National Aeronautics and Space Administration: Joint Polar Satellite System (JPSS) VIIRS Aerosol Optical Thickness (AOT) and Particle Size Parameter Algorithm Theoretical Basis Document (ATBD). Goddard Space Flight Center Greenbelt, Maryland. p 116 (2015)
13. Bhaskaran, S.: Applications of satellite data for aerosol optical depth (AOD) retrievals and validation with AERONET data. Atmos. Climate Sci. **1**, 61–67 (2011). https://doi.org/10.4236/acs.2011.12007
14. Aldrian, E., Susanto, R.D.: Identification of three dominant rainfall regions within Indonesia and their relationship to sea surface temperature. Int. J. Climatol. **23**, 1435–1452 (2003)
15. Prayoga, M.B.R., Yananto, A., Kusumo D.A.: Jurnal Sains Teknologi Modifikasi Cuaca **18**, 17–24 (2017)
16. Stein, A.F., Draxler, R.R., Rolph, G.D., Stunder, B.J.B., Cohen, M.D., Ngan, F.: Bull. Am. Meteorol. Soc. 2059–2077 (2015)
17. Goodrick, S.L., Achtemeier, G.L., Larkin, N.K., Liu, Y., Strand, T.M.: Modelling smoke transport from fires: a review. Int. J. Wildland Fire **22**(1), 1–13 (2013)
18. Aili, A., Xu, H., Kasim, T., Abulikemu, A.: Origin and transport pathway of dust storm and its contribution to particulate air pollution in Northeast edge of Taklimakan Desert, China. Atmosphere **12**, 113 (2021)
19. Begum, B.A., Biswas, S.K., Pandit, G.G., Saradhi, I.V., Waheed, S., Siddique, N., Seneviratne, M.C.S., Cohen, D.D., Markowitz, A., Hopke, P.K.: Long-range transport of soil dust and smoke pollution in the South Asian Region. Atmos. Pollut. Res. **2**, 151–157 (2011)
20. Oh, H.R., Ho, C.-H., Kim, J., Chen, D., Lee, S., Choi, Y.-S., Chang, L.-S., Song, C.-K.: Long-range transport of air pollutants originating in China: a possible major cause of multi-day high-PM10 episodes during cold season in Seoul, Korea. Atmos. Environ. **109**, 23–30 (2015)

21. Carvalho, A.C., Carvalho, A., Gelpi, I., Barreiro, M., Borrego, C., Miranda, A.I., Péres-Muñuzuri, V.: Influence of topography and land use on pollutants dispersion in the Atlantic coast of Iberian Peninsula. Atmos. Environ. **40**, 3969–3982 (2006)
22. Giovanni, L., Ferrero, E., Karl, T., Rotach, M.W., Staquet, C., Castelli, S.T., Zardi, D.: Atmospheric pollutant dispersion over complex terrain: challenges and needs for improving air quality measurements and modelling. Atmosphere **11**, 646 (2020). https://doi.org/10.3390/atmos11060646
23. Hayasaka, H., et al.: Peat-fire-related air pollution in Central Kalimantan, Indonesia. Environ. Pollut. **195**, 257–266 (2014)
24. Uda, S.K., Hein, L., Atmoko, D.: Assessing to health impacts of peatland firer: a case study for Central Kalimantan, Indonesia. Environ. Sci. Pollut. Res. **26**, 31315–31327 (2019)
25. Yuang, X., Zhao, C., Yang, Y., Yan, X., Fan, H.: Statistical aerosol properties associated with fire from 2002 to 2019 along with a case analysis in 2019 over Australia. Atmos. Chem. Phys. (2020). https://doi.org/10.5194/acp-2020-1139
26. Indrawati, A., Andarini, D.F., Cholianawati, N., Sumaryati: Analysis PM10 and Visibility During Forest Fire in Palangka Raya. IOP (in progress publishing)
27. Rajendran, K., Kitoh, A., Yukimoto, S.: South and East Asian summer monsoon climate and variation in the MRI coupled model (MRI-CGCM2). J. Clim. **17**, 763–782 (2004)
28. Duan, Y., Liu, H., Yu, W., Liu, L., Yang, G., Liu, B.: The onset of the Indonesian-Australian summer monsoon triggered by the first-branch eastward-propagating Madden-Julian oscillation. Am. Meteorol. Soc. **32**, 5453–5470 (2019) https://doi.org/10.1175/JCLI-D-18-0513.1

Chapter 12
Dynamic Mechanical Properties of Acetylated Cassava Starch-Chitosan Based Film, With Addition of Sorghum Micro-Fibrillated Cellulose

Yeyen Nurhamiyah, Resti Marlina, and Firda Aulya Syamani

Abstract The solubility of starch-based bioplastic films in water limits its function for varied applications. The incorporation of acetyl functional group and chitosan was conducted to improve the physical properties of starch-based bio-plastic film. Moreover, the mechanical properties of acetylated starch-chitosan-based film still need improvement for application. The addition of cellulose microfibrils from sorghum stalk fibers is expected to increase the mechanical properties of starch-based films. In the manufacture of films, chitosan was dissolved in glacial acetic acid (stirred at 250 rpm, 70 °C, 2 h), then acetylated starch was added (stirred at 250 rpm, 70 °C, 1 h), until a gel was formed. In this study, sorghum micro-fibrillated cellulose (MFC) with the weight ratio of 1–5% were added into acetylated cassava starch-chitosan (ACS-CS) based films. The resulting solution was poured onto a flexy glass, dried at room temperature for three days. The film capacity to persist water was observed by contact angle analysis. Functional groups formed in films were analyzed using Fourier Transform Infrared (FTIR). A dynamic mechanical analyzer (DMA) was operated to analyze the film's dynamic mechanical properties by heat changes. The research result shows that the addition of MFC increases modulus independently of the MFC content. Based on DMA analysis, the acetylated starch-chitosan film with 1% sorghum MFC was more elastic than other films in this study.

12.1 Introduction

Starch-based bioplastics have attracted significant attention since it shows great potential to substitute fossil-based plastic [1]. In addition, starch is a cheap and readily available source of plastic. The applications of starch-based plastics range from food packaging, agricultural, medical, and pharmaceutical [2]. Starch, which is a natural polymer containing amylose and amylopectin, has been used as a raw material for

Y. Nurhamiyah · R. Marlina · F. A. Syamani (✉)
Research Center for Biomass and Bioproducts, National Research and Innovation Agency, Jalan Raya Bogor Km. 46, Cibinong, Jawa Barat 16911, Indonesia
e-mail: fird003@brin.go.id

E. Yulihastin et al. (eds.), *Proceedings of the International Conference on Radioscience, Equatorial Atmospheric Science and Environment and Humanosphere Science, 2021*, Springer Proceedings in Physics 275, https://doi.org/10.1007/978-981-19-0308-3_12

bioplastics film production. Amylopectin comprises a short glucose chain of 10–60 units, in linear and branched-chain, interconnected by α (1–4) and α (1–6) glycosidic bonds. Amylose consists of an unbranched linear chain of glucose monomers interconnected by α (1–4) glycosidic bonds [3]. However, one of the significant drawbacks of starch-based bioplastic is that they have poor properties related to water such as swelling, moisture content, high water absorption, and low mechanical properties [4]. Chemical modification of starch, by acetylation, replaces the hydroxyl groups on the starch's glucose monomer with acetyl groups and increases their hydrophobicity [5]. The formation of acetyl functional group in starch was reported to enhance the hydrophobicity and tensile strength of film [6].

The other way to overcome the limitation of starch-based bioplastic is by using chitosan as a combination of ingredients in the formulation. It is because chitosan is more hydrophobic with additional antimicrobial properties [7]. Chitosan is recognized as a non-toxic, renewable, biodegradable, and biocompatible material. It is reported that the mixture of chitosan and acetylated starch resulted in excellent mechanical and barrier properties [8]. However, for varied application, acetylated starch-chitosan (ACS/CS) bioplastic film, its mechanical properties need to be improved.

The reinforcement in starch-based film can be in the form of pulp or cellulose fibers [9], cellulose microcrystalline [10], and cellulose microfibril [11]. The addition of oil palm trunk cellulose fibers into cassava starch improved starch-based film tensile strength and elongation compared with the addition of oil palm pulp fibers [9]. A similar trend was displayed by the incorporating of cellulose nanocrystal (CNC) into corn starch-based film [10]. In addition, the significant enhancement of thermoplastic starch (TPS) mechanical properties was achieved by introducing micro-fibrillated cellulose (MFC). The yield strength was improved by ~50%, whereas the stiffness by ~250% upon adding 20 wt% MFC compared to the native TPS [11].

In this study, cassava starch was modified with acetic anhydride and mix with chitosan to reduce the hydrophilic property to improve starch-based film characteristics. Furthermore, sorghum micro-fibrillated cellulose was used to enhance the mechanical properties of the starch-based film. The film characteristics were analyzed by its contact angle with water, functional groups, and dynamic mechanical properties.

12.2 Material and Methods

12.2.1 Material

Commercial cassava starch (brand: Gunung Agung) was purchased from the market. Acetic acid glacial, acetate anhydride, technical grade sodium hydroxide, and sodium chlorite (25% in solution) were obtained from Merck. All chemicals were used without further purification. Sorghum fibers (*Sorghum bicolor* (L.) Moench var

Lepeng) were obtained from the experimental field at LIPI Cibinong Science Center. Industrial grade shrimp chitosan was provided by PT Biotech Surindo.

12.2.2 *Modification of Cassava Starch*

Chemical modification of cassava starch was conducted by acetylation reaction. As much as 75 g cassava starch was reacted with 135 mL acetic acid glacial solution and 138 mL acetic anhydride in beaker glass by stirring at 250 rpm and heating to a temperature of 40 °C. Subsequently, 12.45 mL acetic acid glacial and 1.05 mL of concentrated sulfuric acid as a catalyst were added to the solution, and then the reaction was continued for 120 min at room temperature. The acetylation reaction was conducted in a fume hood. At the end of the reaction, cold-distilled-water (50 mL) was added gently to precipitate acetylated starch. The acid solution was removed by filtering with filter paper. Acetylated starch was then oven-dried at 80 °C for two hours. After drying, the acetylated starch was mashed and sieved with 80 mesh-screen to obtain acetylated starch powder.

12.2.3 *Preparation of Sorghum Micro-Fibrillated Cellulose*

Sorghum fiber was obtained from sorghum stalk, by processing in drum chipper and ring-flaker. Sorghum fiber (600 g) was reacted with 10% (w/v) Na OH (10 L) in the digester at temperature of 100 °C, and pressure of 760 mm Hg, for two hours, in order to remove lignin and other impurities. The resulting sorghum pulp was removed from the digester and rinse under running water to remove residual NaOH.

Afterward, the bleaching process was conducted. Sorghum pulp in 4.8 L of distilled water was reacted with 120 mL of 25% sodium chlorite (NaClO2) and 10 mL of 100% glacial acetic acid in the digester for three hours, at ~~a~~ temperature of 80 °C and pressure of 355 mm Hg. After the reaction was completed, the bleached pulp of sorghum was washed with running water to eliminate the residual bleaching solution.

The sorghum bleached pulp (1%) in 2 L distilled water was ground with Masuko Super Fine Grinder in three steps of fibrillation to obtain micro-fibrillated cellulose. The fibrillation steps were with gap 3 for five cycles, followed by fibrillation with gap 5 for five cycles and finally fibrillation with gap 7 for three cycles.

12.2.4 *Fabrication of Films*

The fabrication of bioplastic film was based on our previous work. Briefly, chitosan was dissolved in glacial acetic acid with a constant speed of 250 rpm, at 70 °C, for two

Table 12.1 Biocomposite film formulation

Code	Chitosan (g)	Starch (g)	Sorghum MFC (g)
B1	1.833	3.667	0.000
B2	1.815	3.630	0.055
B3	1.797	3.593	0.110
B4	1.778	3.557	0.165
B5	1.760	3.520	0.220
B6	1.742	3.483	0.275

hours. After that, acetylated-starch was inserted into chitosan solution, and the stirring and heating were continued for another hour until the starch was gelatinized. To this mixture, glycerol (10% wt) was added and stirred for 15 min until homogenized. The ratio of chitosan, starch, and glycerol was kept at 1:2:0.3, respectively. Finally, into starch-chitosan-glycerol mixture solution, sorghum micro-fibrillated cellulose (MFC) with the weight ratio of 1%, 2%, 3%, 4%, and 5% was added (Table 12.1) and stirred for another 15 min at the same speed and temperature. The mixture solution was then poured onto a flexi glass mold with the dimension of 20 cm × 20 cm. The sample was cooled to room temperature for three days until it formed a film sheet.

12.2.5 Characterization of Films

Contact Angle Analysis The characteristics of water droplets was determined by dropping distilled water (0.1 mL) on the film surface. The distilled-water-droplet on the film surface was recorded for 1 min with a Digital Microscope (Dino-lite Basic AM2111). The contact angle at the 60th second was measured using Image J software.

Functional Groups Analysis Fourier Transform Infra-Red (FTIR) analysis was performed on Perkin Elmer (Spectrum Two) with attenuated total reflectance (ATR) mode. For the ATR technique, the bioplastic film (1 cm × 1 cm) was s placed on top of a diamond crystal, and exposed to an infrared beam, across a range of wavelengths (wave- number range 4000 to 400 cm^{-1}), that passes through the diamond crystal forming an evanescent wave that penetrates and reflects off the surface off the sample. With a resolution of 1 cm^{-1}, the spectrums were recorded by an average of 16 scans.

Dynamic Mechanical Analysis The dynamic mechanical analysis was employed on Perkin Elmer DMA 8000 in tension mode. The sample with the dimension of 30 mm × 10 mm was hold for 1 min at 20 °C before being heated from 20 °C to 120 °C with a heating rate of 5 °C/min at 1 Hz frequency.

12.3 Result and Discussion

12.3.1 Water Contact Angle on Film Surface

Contact angle analysis was carried out to determine the characteristics of water droplets on the surface of the starch-chitosan based-film which was added with varied content of sorghum MFC. From Fig. 12.1a, we can observe that B1 film made from acetylated-starch and chitosan, shows the highest degree of contact angle with water. This shows that acetylated starch and chitosan successfully produced the film that resistant to water. On the other hand, the addition of sorghum micro-fibrillated cellulose (Fig. 12.1b–f) cause decreasing contact angle and water droplet more easily wetting film surface. This because sorghum MFC contains hydroxyl groups which that can react with water.

Figure 1g show characteristics of the water droplet in commercial cassava starch-based film surface. Low contact angle indicates that water droplet can easily be spreading on the surface of commercial cassava starch-based film. Conversely, commercial wrapping film made from low-density polyethylene (LDPE) show resistance to water droplet (Fig. 12.1h). LDPE is hydrophobic material with low wettability [12] and relatively inert because of its hydrophobic chains and high molecular weight [13].

Based on the research results, the performance of film made from blended acetylated cassava starch and chitosan was comparable with commercial wrapping film made from LDPE on the performance of holding back water droplets. However, the addition of sorghum micro-fibrillated into acetylated starch-chitosan film tend to reduce the water contact angle. Therefore, to maintain the physical performance and enhance the mechanical properties of the starch-based film, the application of

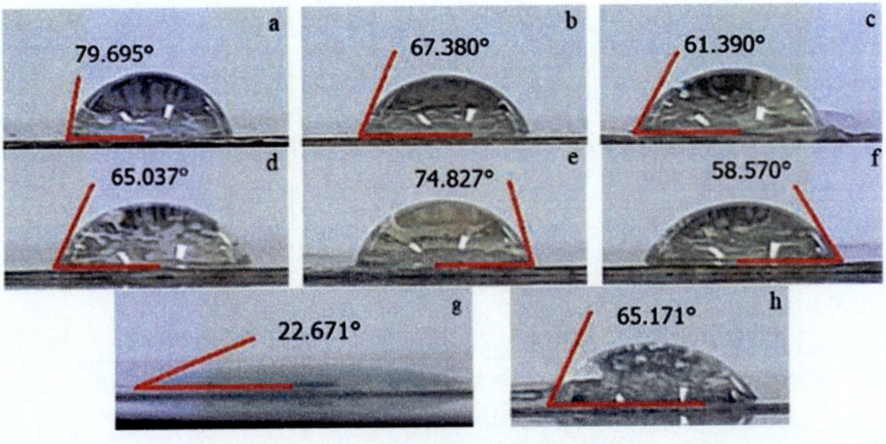

Fig. 12.1 Water contact angle on films surface: **a** B1 film, **b** B2 film, **c** B3 film, **d** B4 film, **e** B5 film, **f** B6 film, **g** commercial starch-based bioplastic film, **h** LDPE wrapping film

cellulose nanocrystal should be considered. Xia et al., reported that the addition of 0.3% cellulose nanocrystal into oxidized-acetylated corn starch could improve film tensile stress up to 18 Mpa and showed high contact angle (113°), compared with the film made from parent corn starch [14].

12.3.2 *Functional Groups in films*

ATR-FTIR is used to observing the possible interaction between film components. The FTIR spectra of the neat Acetylated Starch-Chitosan (ACS-CS) and ACS-CS film with the addition of various sorghum MFC contents are presented in Fig. 12.2. FTIR spectra of acetylated starch displayed at 1745, 1408, 1324, and 1242 cm^{-1}, associated with C = O stretching of the acetate group, CH3 anti-symmetric vibration, CH3 symmetric vibration, and carbonyl C-O stretching, respectively[15]. The wavenumber of chitosan is detected at 1642 and 1558 cm^{-1} which are assigned to amide I and amide II [16].

The characteristic of FTIR spectra on sorghum MFC was detected at 3282 cm^{-1} which was associated with O–H stretching (Table 12.2). The increase of sorghum MFC content in the film significantly increases the intensity of this wavenumber which could be related to the rise of hydrogen bonding in cellulose. A peak at 2936 cm^{-1} is related to C-H stretching vibration in cellulose and hemicellulose [17]. Peak at around 3200 cm^{-1} related to OH groups in water molecule. Film with higher content of acetylated sorghum MFC, tends to absorb lower water molecule, so that the

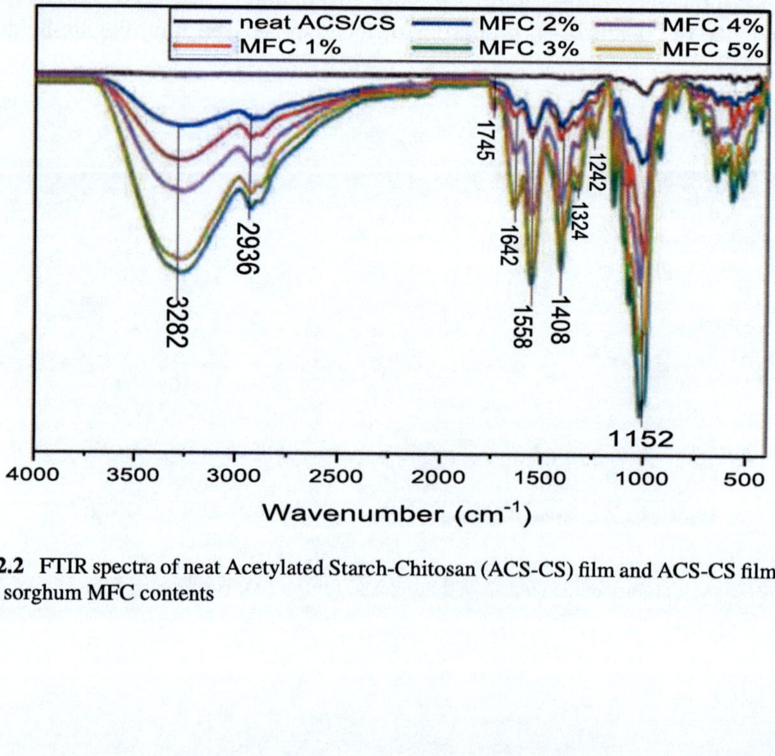

Fig. 12.2 FTIR spectra of neat Acetylated Starch-Chitosan (ACS-CS) film and ACS-CS films with varied sorghum MFC contents

Table 12.2 Functional groups in films

Bonding type	Wavenumber (cm^{-1})					
	B1 film	B2 film	B3 film	B4 film	B5 film	B6 film
O–H stretching	3282	3272	3270	3270	3268	3268
C-H stretching	2936	2930	2924	2930	2926	2926
C = O stretching in acetyl groups	–	–	–	–	1746	1740
C = C stretching in alkene	1642	1642	1636	1636	1634	1644
Deformation C-H from CH2	1558	1556	1548	1554	1552	1554
CH3 antisymmetric vibration	1408	1410	1408	1408	1408	1408
Saccharides	1152	1150	1152	1152	1152	1152
Amorphous structure	996	996	1016	1018	1022	1018

peak at around 3200 cm^{-1} shifted to lower wavenumber. This result was confirmed with the result of water contact angle analysis, as mentioned in previous section. Peak at around 2900 cm^{-1} related to CH groups. Film with higher content of acetylated sorghum MFC, contained higher CH groups, so that the peak at around 2900 cm^{-1} shifted to lower wavenumber.

12.3.3 Film Dynamic Mechanical Analysis

Storage modulus (E') is associated with film stiffness and elastic response. Figure 12.3 display the storage modulus as a function of temperature for neat ACS-CS film and ACS-CS film with the addition of varied sorghum MFC content. The addition of sorghum MFC increases the film storage modulus independently of the

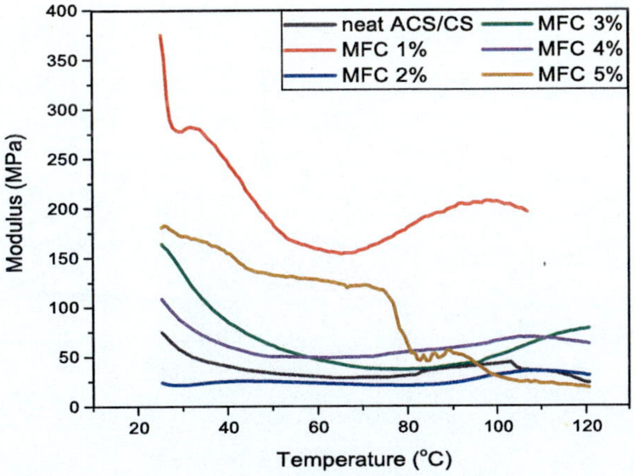

Fig. 12.3 The storage modulus of neat ACS-CS film and ACS-CS films with varied sorghum MFC contents

sorghum MFC content. The 1% sorghum MFC content in the film results the highest storage modulus. While an addition 2% sorghum MFC into film causes the lower storage modulus than its neat ACS-CS film (Fig. 12.3). It might be because the sorghum MFC suspension was more likely to be inhomogeneous as they include cellulose microfiber and microfiber bundles [18]. In addition, sorghum MFC surface consists of a high density of hydroxyl groups which tend to agglomerate [19]. Thus, higher sorghum MFC content led to a higher chance of the sorghum MFC to agglomerates in the films and create inefficient load transfer between sorghum MFC and film matrix.

Loss modulus (E") is related to the viscous response of the film, which is displayed in Fig. 12.4. The sorghum MFC integration effected the film loss modulus in the same trend as the film storage modulus. The film with 1% sorghum MFC shows the highest loss modulus, followed by the film with sorghum MFC content of 5%, 3%, and 4%. Al four films have a higher loss modulus than the neat ACS-CS film. Meanwhile, the film with 2% of sorghum MFC content shows a slight decrease of loss modulus than the neat ACS-CS film.

Loss factor (tan δ) is associated with the molecular motion and defined as tan δ = E"/E'. Figure 12.5 shows the loss factor graphs of neat ACS-CS film and ACS-CS film with the addition of varied sorghum MFC content. In dynamic mechanical analysis, tan delta is the ratio between loss modulus and storage modulus. The high value of tan delta indicates that the material is stiff. On the other hand, the low value of tan delta suggest that the material is elastic. The film made from acetylated starch-Chitosan with the addition of 1% sorghum MFC is the most elastic compare with other films. Addition of sorghum MFC into acetylated starch-chitosan increases the film stiffness. This because sorghum MFC was hydrolyzed when dispersed in acid solution to become microcrystalline cellulose.

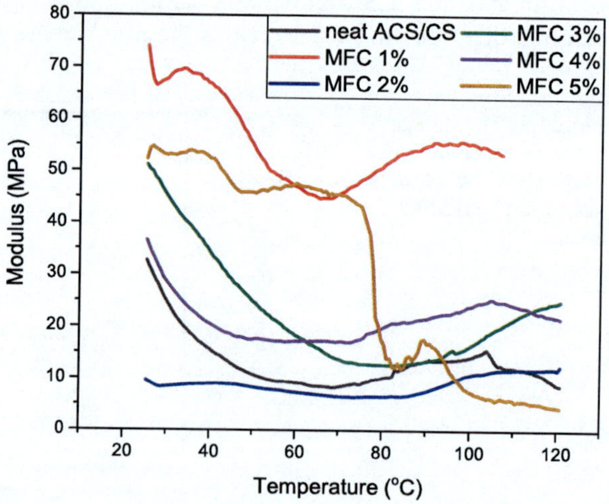

Fig. 12.4 The loss modulus of neat ACS/CS and its MFC composites

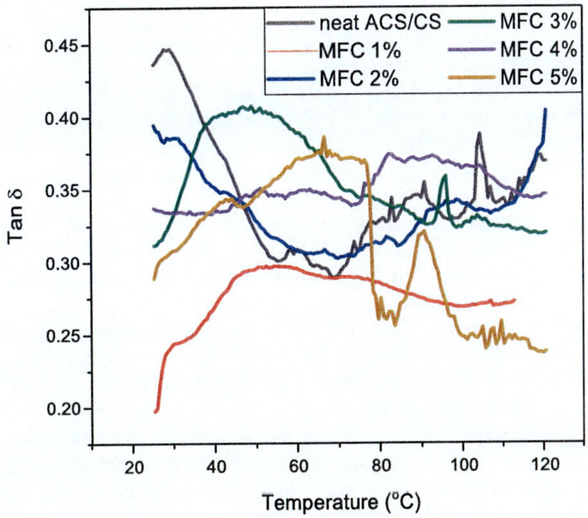

Fig. 12.5 The loss factor of neat ACS-CS film and ACS-CS film with the addition of varied sorghum MFC content

12.4 Conclusion

The acetylated cassava starch-chitosan-based film shows the best resistance to water due to the higher contact angle (74.695°) compared with other films in this study. The absorbance peak at 1746 cm^{-1} and 1408 cm^{-1} in FTIR spectra related to C = O and C-H in the acetyl group, respectively, proved that acetylated starch had been produced. Integration of sorghum micro-fibrillated cellulose (MFC) to blend the acetylated starch- chitosan film has been successfully prepared to enhance its mechanical properties. The result shows that the addition of MFC to the ACS-CS film enhances the storage modulus and loss modulus insignificantly to the sorghum MFC content. Film of acetylated starch-Chitosan (2:1) with the addition of 1% sorghum MFC is the most elastic compare with other films in this study.

Acknowledgements The authors would like to thank Japan International Corporation Agency (JICA) and Japan Science and Technology Agency (JST) for funding this research through SATREPS project with title of "The project for producing biomass energy and material through revegetation of alang-alang (*Imperata cylindrica*) fields". The authors also acknowledge the facilities, scientific and technical support form Advanced Characterization Laboratories Cibinong – Integrated Laboratory of Bioproduct, National Re- search and Innovation Agency through E- Layanan Sains, Badan Riset dan Inovasi Nasional.

References

1. do Val Siqueira, L., Arias, C.I.L.F., Maniglia, B.C., Tadini, C.C.: Starch-based biodegradable plastics: methods of production, challenges and future perspectives. Curr. Opin. Food Sci. **38**, 122–130 (2021)
2. Lumdubwong, N.: Applications of Starch-Based Films in Food Packaging. Elsevier (2019)
3. Sharma, A., Satyanarayana, T.: Microbial acid-stable α-amylases: characteristics, genetic engineering and applications. Process Biochem. **48**, 201–211 (2013)
4. Castillo, L.A., López, O.V, García, M.A., Barbosa, S.E. Villar, M.A.: Crystalline morphology of thermoplastic starch/talc nanocomposites induced by thermal processing. Heliyon **5**, e01877 (2019)
5. Colussi, R., Pinto, V.Z., El Halal, S.L.M., Biduski, B., Prietto, L., Castilhos, D.D., da Rosa Zavareze, E., Dias, A.R.G.: Acetylated rice starches films with different levels of amylose: Mechanical, water vapor barrier, thermal, and biodegradability properties. Food Chem. **221**, 1614–1620 (2016)
6. El Halal, S.L.M., Colussi, R., Biduski, B., do Evangelho, J.A., Bruni, G.P., Antunes, M.D., Dias, A.R.G., da Rosa Zavareze, E.: Morphological, mechanical, barrier and properties of films based on acetylated starch and cellulose from barley. J. Sci. Food Agric. **97**, 411–419 (2017)
7. Luchese, C.L., Pavoni, J.M.F., dos Santos, N.Z., Quines, L.K., Pollo, L.D., Spada, J.C.: Effect of chitosan addition on the properties of films prepared with corn and cassava starches. J. Food Sci. Technol. **55**(8), 2963–2973 (2018)
8. Tessaro, I.C., Escamilla-García, M., Reyes-Basurto, A., García-Almendárez, B.E., Hernández-Hernández, E., Calderón-Domínguez, G., Rossi-Márquez, G., Regalado-González, C.: Modified starch-chitosan edible films: physicochemical and mechanical characterization. Coatings **7**(12), 224 (2017)
9. Syamani, F.A., Nurjayanti, Pramasari, D.J., Kusumaningrum, W.B., Kusumah, S.S., Masruchin, N., Ermawati, R., Supeni, G., Cahyaningtyas, A.A.: Characteristics of bioplastic made from cassava starch filled with fibers from oil palm trunk at various amount. IOP Conf. Ser. Earth Environ. Sci. **439**, 012035 (2020)
10. Alves, J.S., dos Reis, K.C., Menezes, E.G.T., Pereira, F.V., Pereira, J.: Effect of cellulose nanocrystals and gelatin in corn starch plasticized films. Carbohyd. Polym. **115**, 215–222 (2015)
11. Lendvai, L., Karger-Kocsis, J., Kmetty, A., Drakopoulos, S.X.: Production and characterization of microfibrillated cellulose-reinforced thermoplastic starch composites. J. Appl. Polym. Sci. (2015). https://doi.org/10.1002/app.42397
12. Briassoulis, D., Giannoulis, A.: Evaluation of the functionality of bio-based food packaging films. Polym. Testing **69**, 39–51 (2018)
13. Suresh, B., Maruthamuthu, S., Kannan, M., Chandramohan, A.: Mechanical and surface properties of low-density polyethylene film modified by photo-oxidation. Polym. J. **43**, 398–406 (2011)
14. Xia, Y., Yang, G., Chen, J., Lin, Z.: Cellulose nanocrystal assisted dual modification of starch and subsequent polyvinyl alcohol blends. BioResources **14**(3), 7041–7055 (2019)
15. Teodoro, A.P., Mali, S., Romero, N., de Carvalho, G.M.: Cassava starch films containing acetylated starch nanoparticles as reinforcement: physical and mechanical characterization. Carbohyd. Polym. **126**, 9–16 (2015)
16. Branca, C., D'Angelo, G., Crupi, C., Khouzami, K., Rifici, S., Ruello, G., Wanderlingh, U.: Role of the OH and NH vibrational groups in polysaccharide-nanocomposite interactions: a FTIR-ATR study on chitosan and chitosan/clay films. Polymer **99**, 614–622 (2016)
17. Nguyen, H.D., Mai, T.T.T., Nguyen, N.B., Dang, T.D., Le, P., Dang, T.T., Tran, V.M.: A novel method for preparing microfibrillated cellulose from bamboo fibers. Adv. Nat. Sci. Nanosci. Nanotechnol. **4**(1), 015016 (2013)

18. Siró, I., Plackett, D.: Microfibrillated cellulose and new nanocomposite materials: a review. Cellulose **17**, 459–494 (2010)
19. Zimmermann, T., Pöhler, E., Geiger, T.: Cellulose fibrils for polymer reinforcement. Adv. Eng. Mater. **9**, 754–761 (2004)

Chapter 13
Characteristics of Acid Deposition in Urban and Sub-Urban Area

Asri Indrawati, Dyah Aries Tanti, Nurokhman, Indra Chandra, Atep Radiana, Wiwiek Setyawati, and Sumaryati

Abstract This study was conducted to examine the characteristics of acid deposition in Cipedes and Telkom University as an urban area and Tanjung Sari as a suburban area. The data used were wet deposition and dry deposition data from 2019 to 2020. Data processing was done using time series analysis. Calculation of the Fractional Acidity (FA), Neutralization Factor (NF), and the PMF 5 model were used to trace pollutant sources in urban and suburban areas. The results showed that the concentration of anions and cations for the Cipedes area was higher than at Telkom University and Tanjung Sari. The pH value Telkom Univ has a tendency more acidic than Cipedes and Tanjung Sari because the anion concentration contributing 52% of the total concentrations in Telkom Univ. The results of dry deposition using the passive sampling method showed that the concentrations of NO_2 and O_3 were the dominant gas concentrations in Cipedes, Telkom Univ, and Tanjung Sari. Fractional acidity shows that the acidity of rainwater in Cipedes, Telkom Univ, and Tanjung Sari can be neutralizing with the dominant neutralizing components were NH_4^+, Na^+, and Ca^{2+}. From the PMF model, this study found that local emission sources and long-distance emissions can affect the characteristics of acid deposition in urban and suburban areas. Local emissions such as traffic activity, industrial and biomass burning, and sea salt from ocean sources for long-range transport emissions.

13.1 Introduction

The chemical composition dissolved in rainwater was important information to determine the air quality of some areas [1]. The chemical composition of rainwater depends on the parameters of gases or particulates in the atmosphere emitted by

A. Indrawati (✉) · D. Aries Tanti · Nurokhman · A. Radiana · W. Setyawati · Sumaryati
Research Center for Climate and Atmosphere, National Research and Innovation Agency (BRIN), Bandung, Indonesia
e-mail: asri004@brin.go.id

I. Chandra
Engineering Physics, School of Electrical Engineering, Telkom University, Bandung, Indonesia

E. Yulihastin et al. (eds.), *Proceedings of the International Conference on Radioscience, Equatorial Atmospheric Science and Environment and Humanosphere Science, 2021*, Springer Proceedings in Physics 275, https://doi.org/10.1007/978-981-19-0308-3_13

natural sources, anthropogenic activities, and long-range transport pollutants. The acidity of rainwater and the amount of ionic concentration of rainwater is related to the chemical and physical processes during the incorporation process in rainwater [2].

Spatial differences directly affect the chemical nature of the acid deposition. Acid deposition characteristics for coastal areas have a different character from urban areas. Changes in the concentration of anions and cations in acid deposition are related to the source of the air mass that carries pollutants from the place of origin to the place where it rains. Characteristics anion and cation concentrations use to determine the main source of acid deposition [3]. Transboundary Air Pollution has occurred in Japan and Korea showed that the local contribution from acid deposition was only about 30%, while 70% by Transboundary Air Pollution (TAP). The analysis showed that the local pollutant sources in urban areas are due to industrial, power generation, and residential factors. The impact of TAP in the wet deposition of SO_4^{2-} and NO_3^- is higher than TAP from the concentration of PM2.5 [4].

Positive Matrix Factorization (PMF) was a method for identifying sources of air pollution [5, 6]. PMF only requires the input of the measurement concentration and the uncertainty value of the sample measurement. The result is a profile of pollutant sources divided into several factors [7]. This study was to examine the characteristics of acid deposition in urban and suburban, the effect of topography in Bandung and also identifies the pollution sources in urban and sub-urban using PMF 5.

13.2 Data Dan Method

13.2.1 Location and Data

The research location for the Urban was Cipedes and Telkom University, while the Suburban area was Tanjung Sari. A sampling of wet deposition and dry deposition was carried out during 2019–2020. Wet deposition sampling we carried out using an automatic rain sampler and rain gauge every rainy event. In wet deposition sampling, QA/QC was calculated based on the equilibrium between anion and cation concentrations with pH and conductivity values, according to the guidelines of EANET [8]. In dry deposition sampling, the passive sampling method used was the Ferm type was adopted by CSRIO Australia [9]. Sampling duration is carried out monthly (one month one sample) during 2019–2020. Monthly blank samples at each location we carried out to ensure the quality of the data produced (Fig. 1).

Fig. 13.1 Sampling location, **a** Cipedes, **b** Telkom Univ, **c** Tanjung Sari

13.2.2 Method

Calculation of Fractional Acidity and Neutralization Factor

Calculation of Fractional Acidity in wet deposition using the equations [10]:

$$FA = \frac{[H^{+}]}{[NO_3^{-}] + [nss - SO_4^{2-}]}$$

where:

FA	Fractional Acidity.
$[H^+]$	Ion concentration of H^+
$[NO_3]$	Ion concentration of NO_3^-
$[nss\text{-}SO_4^{2-}]$	Ion concentration of $nss\text{-}SO_4^{2-}$

Calculation of Neutralization Factor (NF) in wet deposition using the equations [11]:

$$NF_{X^{+}} = \frac{[X^{+}]}{[NO_3^{-}] + 2[nss - SO_4^{2-}]}$$

$$NF_{X^{2+}} = \frac{[X^{2+}]}{2[NO_3^{-}] + [nss - SO_4^{2-}]}$$

where:

NF	Neutralization factor
[X +]	Ion concentration of NH_4^+, Na^+, K^+
[X2 +]	Ion concentration of Ca^{2+} dan Mg^{2+}
[NO3]	Ion concentration of NO_3^-
[nss-SO42-]	Ion concentration of $nss\text{-}SO_4^{2-}$

PMF 5 Model

The PMF model from the US EPA was used to identify pollutant sources that contributed to the acidity of rainwater and the amount of ion concentration dissolved in rainwater. Based on the ionic concentration and uncertainty, the PMF model provides source profile and source contribution for each sample and the detailed descriptions in the EPA PMF 5.0 guide [12]. In this study, 147 samples of rainwater for Cipedes, 77 samples of rainwater for Telkom Univ, 106 samples of rainwater for Tanjung Sari, and ten parameters were used as input data for the PMF model.

13.3 Results and Discussion

13.3.1 Wet and Dry Deposition

The acidity of rainwater (pH of rainwater) depends on the amount of anion and cation concentrations of pollutants in the atmosphere dissolved in rainwater. The low pH value of rainwater indicates that the concentration of anions consisting of SO_4^{2-}, NO_3^-, Cl^- and nss-SO_4^{2-} is more dominant than the concentration of cations NH_4^+, Na^+, K^+, Mg^{2+}, Ca^{2+}, and nss-Ca^{2+}. Conductivity describes the amount of concentration of anions and cations dissolved in rainwater. Thus, levels of ionic concentrations in rainwater affect the pH value and conductivity of rainwater.

Based on Fig. 13.2, the acidity value of rainwater for Telkom Univ was lower than Cipedes and Tanjung Sari, with anion concentration contributing 52% of the total concentration. The dominant anion concentrations are nss-SO_4^{2-} and SO_4^{2-} with average concentrations from 2019 to 2021, namely 85.81 mol/L and 70.58 mol/L. Meanwhile, the pH value for both Cipedes and Tanjung Sari was more alkali, with an average pH value between 6.1 and 6.4. This condition occurs because the concentrations of anions and cations tended to have the same value, thus making the pH value close to the alkaline value. Then the tendency of the pH value to be more alkali in Cipedes due to the more dominant cation (neutralizing) concentration with a contribution of 53.3% of the total concentration. The Ammonium ion was a dominant concentration in Cipedes with an average concentration was 83 mol/L.

Load concentration for Cipedes and Telkom Univ shows higher than Tanjung Sari. In the urban area, many pollutants originating from anthropogenic activities such as transportation, industry, and housing emitting pollutants in urban areas. From Fig. 13.1, the ion concentrations in 2020 for Cipedes and Telkom Univ are lower than in 2019, related to the impact of the lockdown times due to the covid 19 pandemic. From July until October 2019, there is a peaks concentration of anion and cation, but in 2020 there is no significant peak in July–October. From this condition, it is clear that transportation and traffic emission is the most source contribution for pollutants emission in the urban area (Fig. 13.2).

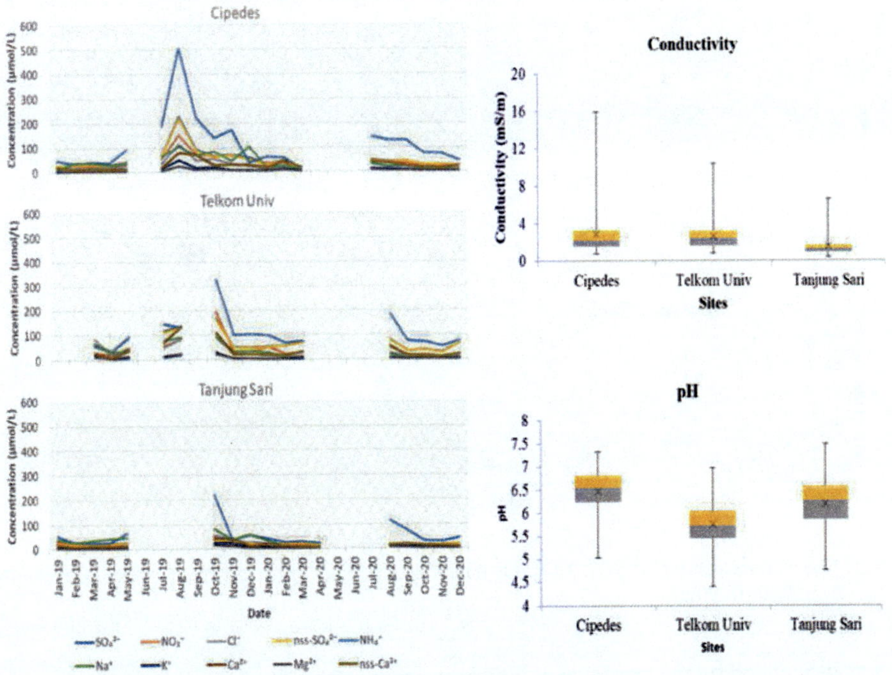

Fig. 13.2 Ion concentration, pH value, conductivity measured in Cipedes, Telkom Univ and Tanjung Sari from 2019 to 2020

Dry deposition measurements were carried out using the passive sampling method. This method is widely used in research on monitoring the concentration of gases such as SO_2, NO_2, O_3, and NH_3 [13–15]. Based on Fig. 13.3, the concentration of NO_2 for the Cipedes, Telkom Univ, and Tanjung Sari is more dominant than other gas concentrations. NO_2 emissions come from motor vehicles and other anthropogenic activities such as biomass burning [16]. Concentration of O_3 in Tanjung Sari is almost the same as the O_3 concentration in the urban area. This condition occurs related to VOCs sources from plant emissions. Biogenic volatile organic compounds (BVOCs) emitted by plants are isoprene, monoterpenes and higher terpenoids. The combination reaction of BVOC and NOx as a precursors of ozone in Tanjung Sari. NO_x concentration could be found in the exhaust gases from cars and transported over large distances by wind from the urban to suburban or rural area [17–19]. Many plants are surrounding the monitoring site in Tanjung Sari. Besides that, many industries near Tanjung Sari can contribute to VOCs emission as precursors of the formating O_3 in the ambient air. Several studies that have been conducted regarding the comparison of O_3 concentrations in urban and sub-urban areas have also found the same thing. Characteristics of O_3 concentration in urban and sub-urban areas also depend on O_3 production and its sensitivity to NO_x and VOC_s. [20]. Based on the results of the passive sampler, it appears that the formation of O_3 in Tanjung Sari

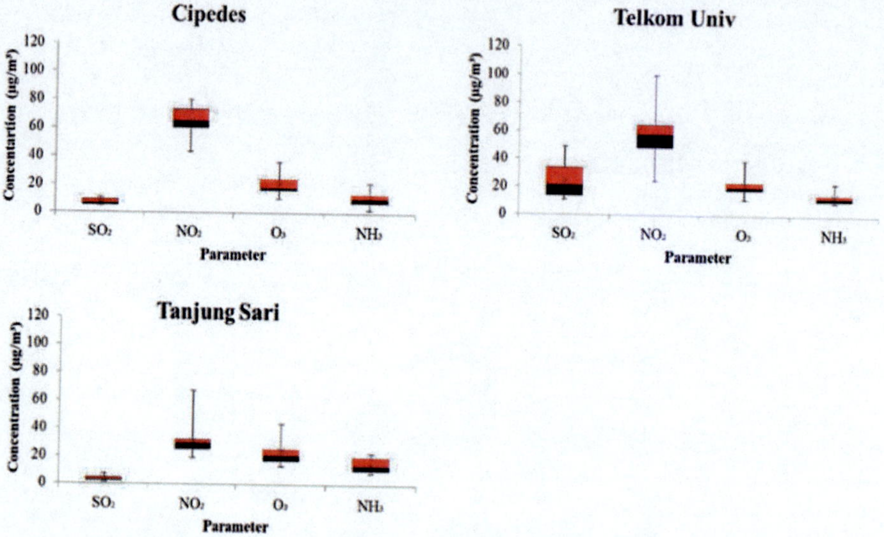

Fig. 13.3 Concentration of SO_2, NO_2, O_3 dan NH_3 gases in Cipedes, Telkom Univ and Tanjung Sari from 2019 to 2020

was influenced by the presence of VOCs, more sensitive to VOCs than NO_x because, for the same levels of O_3, NO_2 level was lower in Tanjung Sari.

The highest load concentration of dry deposition at Telkom Univ compared to other locations is due to many local emission sources from vehicle traffic, industry, residential areas around Telkom Univ and also from transports pollutants [19]. In addition, the results from the TAPM model in Fig. 13.4 show that the Telkom Univ area is the center of the basin point of Bandung. These topographic conditions can cause an accumulation of pollutant gases in Telkom Univ areas. Based on the research

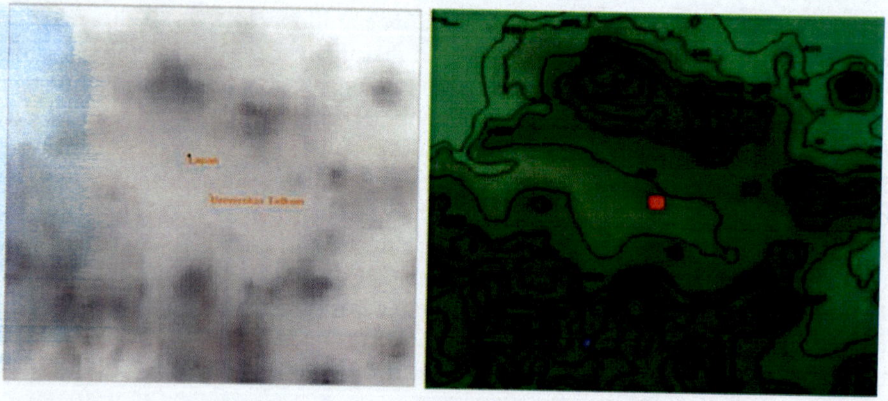

Fig. 13.4 TAPM result for topography at Telkom Univ

that has been done, the topography is an essential factor that influences the regional distribution of air pollutions because this condition is related to the accumulation of pollutant gases in an area [21]. Accumulation of pollutant gases also related to the inversion layer. The inversion layer in the Bandung Basin occurs persistently at night at the surface level and begins to rise in the morning. The layer inversion breaks at 10.00 AM for January and 12.00 AM for December. Break of temperature inversion occurs during the day. This inversion event correlates with the deposition and dilution of pollutants from the area surrounding the Bandung Basin [22].

13.3.2 Fractional Acidity and Neutralization Factor

Strong acid originating from anthropogenic emissions (H_2SO_4 and HNO_3) is the major impact that affects the acidity of wet deposition. Sulfate can be produced from anthropogenic sources and the ocean in the form of sea salt, so that in the calculation of Fractional Acidity (FA) nss-SO_4^{2-} was used. The neutralization process occurs when $Ca(OH)_2$ and $Mg(OH)_2$ react with H_2SO_4 and HNO_3 in the condensation cloud. Under the cloud, SO_2 adsorbs suspended particulates containing Ca^{2+} and Mg^{2+} then reacts with ammonia to produce neutralization [7].

The nss-SO_4^{2-} and NO_3^- species in wet deposition are the two main acid components. Therefore, FA can be used to evaluate the degree of acidity neutralization of wet deposition. If the FA value is equal to 1, wet deposition cannot neutralize. The acidity of rainwater is totally from H_2SO_4 and HNO_3. If the FA value is less than 1,

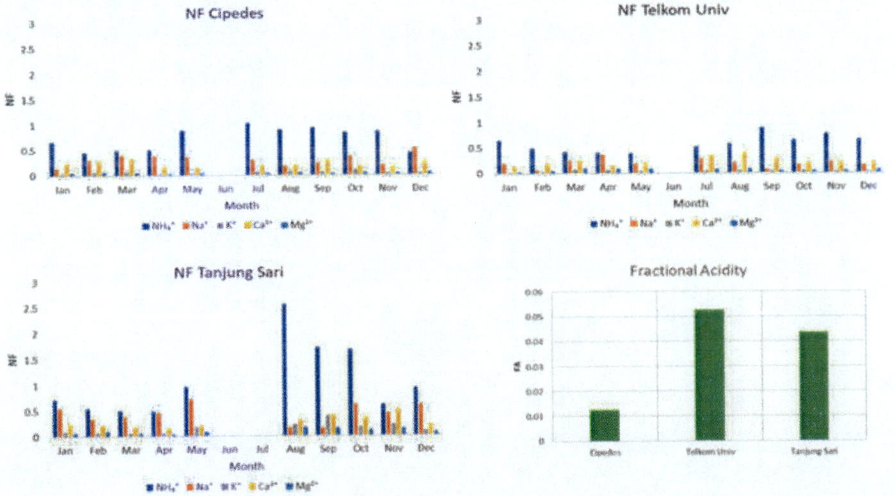

Fig. 13.5 Fractional acidity and neutralization factor for Cipedes, Telkom Univ and Tanjung Sari

the acidity of rainwater could be neutralized by NH_4^+, Ca^{2+}, K^+, and Mg^{2+}. Calculation of neutralization factor (NF) can show the ability of cations to neutralize free acids in rainwater.

In Fig. 13.5, the calculation of Fractional Acidity showed that Cipedes, Telkom University, and Tanjung Sari have the same characteristics of fractional acidity. The Fractional Acidity value is less than 1, so they could be neutralized by alkaline spices. The annual fractional acidity values in Cipedes, Telkom Univ, and Tanjung Sari were 0.01, 0.05, and 0.04, indicating that 99%, 95%, and 96% of the acidity in rainwater was neutralized by alkaline. Monthly data processing from 2019 to 2020 for neutralization factors in three locations shows that for the Cipedes, Telkom Univ, and Tanjung Sari, the dominant neutralizing factors were NH_4^+, Na^+, and Ca^{2+}.

13.3.3 Source Contribution PMF

In this study, the PMF model was used to determine pollutant sources that affect the chemical concentration of rainwater in Cipedes, Telkom Univ, and Tanjung Sari. The results showed for Cipedes, the identification of pollutants was represented by five factors, Telkom University by six factors, and Tanjung represented by four factors.

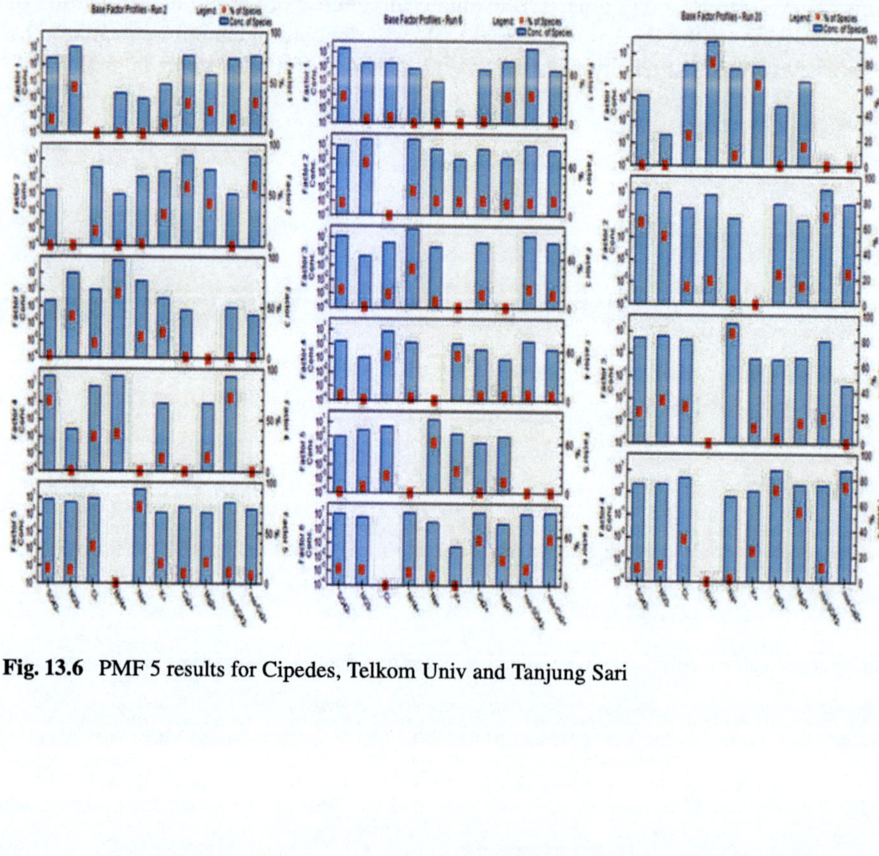

Fig. 13.6 PMF 5 results for Cipedes, Telkom Univ and Tanjung Sari

Based on the results of running PMF 5 in Fig. 13.6, the identification of Cipedes pollutant sources is represented by five factors, namely for factor 1 is traffic activity with a dominant concentration of NO_3^- (45.2%), Ca^{2+} (29.5%), and Mg^{2+} (22.5%). 6%). Factor 2 is a source of dust or soil (crustal source) with dominant concentrations of nss-Ca^{2+} (60.2%), Ca^{2+} (58.4%), and Mg^{2+} (41.8%). The third factor is a secondary aerosol ammonium nitrate, with a dominant concentration of NH_4^+ (63.6%), NO_3^- (41.5%) and K^+ (25.8%). The fourth factor was secondary aerosol ammonium sulfate with a dominant concentration of nss-SO_4^{2-} (72.4%), SO_4^{2-} (68.8%), and NH_4^+ (36.0%). The fifth factor is the source of sea salt with dominant concentrations of Na^+ (75.7%), Cl^- (36.5%) and Mg^{2+} (21.5%).

Identification of pollutant sources for Telkom Univ is represented by 6 factors, namely Factor 1 is a source of biomass burning with a dominant concentration of nss-SO_4^{2-} (33.6%), Mg^{2+} (32.7%) and SO_4^{2-} (32.7%). Factor 2 is vehicular traffic with dominant concentrations of NO_3^- (65.3%), NH_4^+ (30.6%), Na^+ (18.8%) and Ca^{2+} (18.3%). Factor 3 is a secondary source of ammonium sulfate aerosol with dominant concentrations of NH_4^+ (48.3%), nss-SO_4^{2-} (22.7%), SO_4^{2-} (22.3%) and Cl_- (17.2%). Factor 4 is industry with dominant concentration of K^+ (54,7%) and Cl^- (54.7%). Factor 5 is the source of sea salt with the dominant concentrations are Na^+ (61.4%), K^+ (27.6%) and Cl^- (21.3%). Factor 6 is the source of soil dust with the dominant concentration being nss-Ca^{2+} (57%), Ca^{2+} (56%), Mg^{2+} (31.9%).

Tanjung Sari, the identification of pollutants is represented by 4 factors, namely Factor 1 is an industrial source with a dominant concentration of NH_4^+ (81.3%), K^+ (63.6%) and Cl^- (23.5%). Factor 2 is the source of fossil fuel combustion with dominant concentrations of nss-SO_4^{2-} (69.3%), SO_4^{2-} (64.4%) and NO_3^- (53.2%). Factor 3 is coal combustion with a dominant concentration of Na^+ (86.4%), NO_3^- (33.9%), Cl^- (29%), SO_4^{2-} (24.1%). Factor 4 was a crustal source with dominant concentrations of nss-Ca^{2+} (74.9%) Ca^{2+} (72%) and Mg^{2+} (54.8%).

The effect of sea salt in Cipedes and Telkom Univ because due to the long-range transport pollutants. The positions of Cipedes and Telkom Univ are very far from the coast or the ocean. From the PMF 5 model, for the Cipedes and Telkom University locations, the dominant sea salt factor occurred on August 15, 2019. We ran HYSPLIT backward trajectory for 24 h with two altitude variations, namely 100, 300, 500 m, and 1000, 1500, 2000 m above the surface. This altitude variation is due to the influence of topography in Bandung. In Fig. 13.7, results from HYSPLIT using a backward trajectory shows the effect of sea salt sources arriving at Cipedes and Telkom Univ. This result shows the long range transport pollution from the ocean to Cipedes and Telkom Univ.

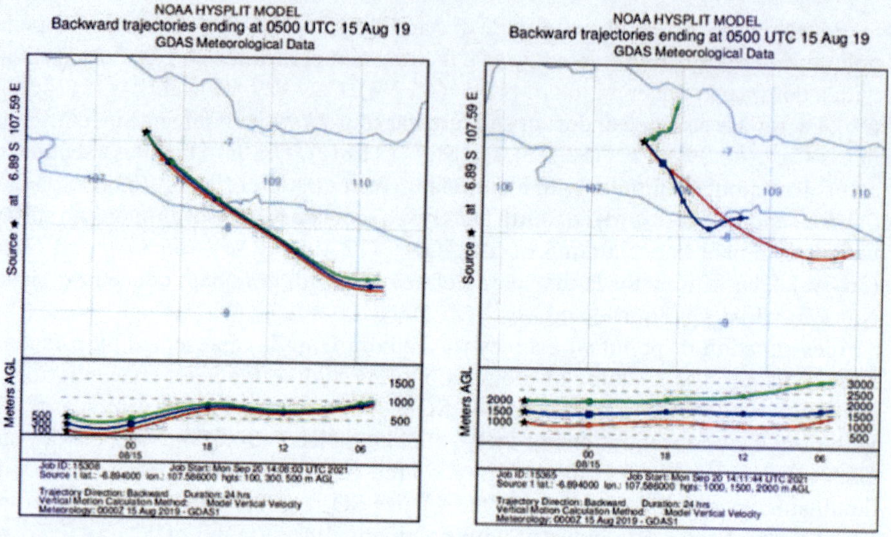

Fig. 13.7 Hysplit backward trajectory

13.4 Conclusions

Acid deposition for urban and sub-urban, Cipedes, Telkom Univ, and Tanjung Sari, have different characteristics. In urban areas, the concentration of pollutants from both wet deposition and dry deposition is greater than the concentration in suburban areas. For the pH value, the Telkom Univ area has slightly more acidic than Cipedes and Tanjung Sari. In addition to emission sources, topography also contributes to the regional distribution of pollutants, resulting in different characteristics between urban areas, Cipedes and Telkom Univ. Fractional acidity shows that both urban and suburban areas have a fractional acidity value <1. It is means the acidity of rainwater neutralized with alkaline neutralizing compounds. Based on PMF 5, the results show that local emission sources and long-distance emissions can affect the characteristics of acid deposition in urban and sub-urban areas.

13.5 Contributor Statements

1st author (Asri Indrawati): brainstorming ideas so that this research can arise, analysis wet deposition sample in laboratory, processing data, analyzing data, and writing papers.

2nd author (Dyah Aries Tanti): provided data sampling and analysis dry deposition sample in laboratory.

3rd author (Nurokhman): provided data sampling for Tanjung Sari.

4th author (Indra Chandra): brainstorming idea for data analysis and provided data sampling for Telkom Univ.

5th author (Atep Radiana): sampling wet deposition for Cipedes.

6th author (Wiwiek Setyawati): brainstorming ideas for statistical analysis.

7th author (Sumaryati): brainstorming ideas for TAPM and Hysplit.

References

1. Zunckle, M., Saizar, C., Zarauz, J.: Rainwater composition in northeast Uruguay. Atmos. Environ. **37**, 1601–1611 (2003)
2. Herra, J., Rodriguez, S., Baez, A.P.: Chemical composition of bulk precipitation in the metropolitan area of Costa Rica. Cent. Am. Atmos. Res. **94**, 151–160 (2009)
3. Chen, H.Y., Hsu, L.F., S. Z. Huang dan L. Zheng,: Assessment of the components and sources of acid deposition in Northeast Asia: a case study of the coastal and metropolitan cities in Northern Taiwan. J. Atmos. **11**, 983 (2020)
4. Yim, S.H.L., Gu, Y., Shapiro, M.A., Stephens, B.: Air quality and acid deposition impact of local emissions and trans boundary air pollution in Japan and South Korea. Atmos. Chem. Phy. **19**, 13309–13323 (2019)
5. Prevez, S., Dubey, N., Watson, J.H., Chow, J., Pervez, Y.: Impact of different household fuel use on source apportionment results of house-indoor RPM in Central India. Aerosol Air Qual. Res. **12**(1), 49–60 (2012)
6. Tiwari, S., A. Ranade dan D. Singh,: Study of Chemical species in rainwater at Ballia, a rural environment in eastern Uttar Pradesh, India. Indian J. Radio Space Phys. **35**, 35–41 (2016)
7. Rao, P.S.P., Tiwari, S., Matwale, J.L., Pervez, S., Tunved, P., Safai, P.D., Srivastava, A.K., Bisht, D.S., Singh, S., Hopke, P.K.: Sources of chemical species in rainwater during monsoon and nonmonsoonal periods over two mega cities in India and dominant source region of secondary aerosols. Atmos. Environ. **146**, 90–99 (2016)
8. EANET (2010) Technical manual for wet deposition monitoring in East Asia
9. Powell (2007) Chemlab wet chemistry measurement, work instructional manual for NATA accredited laboratories, CMAR-NATA-WC-2.0, 2nd edn
10. Balasubramanian, R., Victor, T., Chun, N.: Chemical and statistical analysis of precipitation in Singapore. Water Air Soil Pollut. **130**, 451–456 (2001)
11. Kulshrestha, U.C., Sarkar, A.K., Srivastava, S.S., Parashar, D.C.: Investigation into atmospheric deposition through precipitation studies at New Delhi (India). Atmos. Environ. **30**, 4149–4154 (1996)
12. Norris, G., Duall, R., Brown S., Bai, S.: EPA positive matrix factorization (PMF) 5.0 fundamentals and user guide. U.S. Environ. Prot. Agency Off. Res. Develop. Washington, D.C. EPA/600/R-14/108 (2014)
13. Indrawati, A., Tanti, D.A., Cholianawati, N., Sofyan, A., Cahyono, W.E.: Comparison of SO_2 and NO_2 Gas levels in ambient air between the passive method and active method (Case study: Jakarta City). Jurnal Teknologi Lingkungan **22**(1), 111–120 (2021)
14. Adon, M., Galy-Lacaux, C., Yoboue, V., Delon, C., Lacaux, J.P., Castera, P., Gardrat, E., Pienaar, J., Al Ourabi, H., Laouali, D., Diop, B., Sigha-Nkamdjou, L., Akpo, A., Tathy, J.P., Lavenu, F., Mougin, E.: Long term measurements of sulfur dioxide, nitrogen dioxide, ammonia, nitric acid and ozone in Africa using passive samplers. Atmos. Chem. Phys. **10**, 7467–7487 (2010)
15. Leiva G, M.A., Gonzales, B., Vargas, D., Toro, R., Morales S R.G.E.: Estimating the uncertainty in the atmospheric ammonia concentration in an urban area by Ogawa passive samplers. Microchem. J. **110**, 340–349 (2013)
16. Zainordin, N.S., N.A. Ramli, Elbayoumi, M. (2017) Distribution and temporal behaviour of O_3 and NO_2 near selected schools in Seberang Perai. Pulai Pinang Parit Buntar, Perak, Malaysia, Sains Malaysia **46**(2), 197–207

17. Paoletti, E., de Marco, A., Beddows, D.C.S., Harrison, R.M., Manning, W.J.: Ozone levels in European and USA cities are increasing more than at rural sites, while peak values are decreasing. Environ. Pollut. **192**, 295–299 (2014)
18. Sicard, P., Serra, R., Rossello, P.: Spatiotemporal trends in ground-level ozone concentrations and metrics in France over the time period 1999–2012. Environ. Res. **149**, 122–144 (2016)
19. Fitzky, A.C., Sanden, H., Karl, T., Fares, S., Calfapietra, C., Grote, R., Saunier, A., Rewald, B.: The interplay between ozone and urban vegetation-bvoc emissions, ozone deposition and tree ecophysiologgy. Front For Global Change. **2** (2019). https://doi.org/10.3389/ffgc.2019.00050
20. Huang, D., Li, Q., Wang, X., Li, G., Sun, L., He, B., Zhang, L., Zhang C.: Characteristics and trends of ambient ozone and nitrogen oxides at urban, suburban, and rural sites from 2011 to 2017 in Shenzhen, China. J. Sustain. (2018)
21. Zhang, Z., Xu, X., Qiao, L., et al.: Numerical simulations of the effects of regional topography on haze pollution in Beijing. Sci. Rep. **8**, 5504 (2018). https://doi.org/10.1038/s41598-018-23880-8
22. Sumaryati, Indrawati A., Tanti D.A.: Temperature gradient condition on the photochemical smog dilution in Bandung basin area. J Environ Technol **21**(2) (2020)

Chapter 14
Evaluation of CCAM Seasonal Prediction by GSMaP Satellite Rainfall Data in Indonesia

Haries Satyawardhana, Erma Yulihastin, Gammamerdianti, Candra Nur Ihsan, and Eka P. Wulandari

Abstract The Indonesia Maritime Continent (90–150 E, 12 S–12 N), so-called IMC, consists of a complex sea-land composition with the seasonal rainfall characteristics determined by the Asian-Australian monsoon system. As a result, seasonal rainfall has high variability in both time and spatial scales. Importantly, seasonal prediction is needed to support the agriculture sector, which mainly impacts the Indonesian economy, even though it was more challenging to predict. This study compares the dry and wet season experiments from Conformal Cubic Atmospheric Model (CCAM) output to GSMaP satellite rainfall data. The rainfall reconstruction is carried out based on the maximum correlation of members (11 results of seasonal prediction) with satellite rainfall. The results show that the predicted rainfall for the next eight months with the current updated predictive SST is well-quantitatively captured according to the GSMaP satellite rainfall data but estimates the season onset faster than GSMaP. Additionally, the correlation between model-simulated and satellite data exhibits high spatial variability and a quite different result between IMC's northern and southern parts. Rainfall reconstruction shows that the spatial pattern of predicted monthly rainfall in the dry and wet seasons is similar to the GSMaP rainfall pattern.

14.1 Introduction

As a maritime continent located between two oceans (Pacific and Indian Ocean) and between two continents (Asia and Australia), Indonesia has passed by monsoon wind, a periodic seasonal wind movement. The Asian Monsoon moves from the Asian continent to Australia by bringing wet air through Indonesia, causing the rainy season in Indonesia [1]. On the other hand, the Australian Monsoon moves from the Australian continent to the Asian continent by bringing dry air through Indonesia, resulting in a dry season in Indonesia. Besides the monsoon, Indonesia is

H. Satyawardhana (✉) · E. Yulihastin · Gammamerdianti · C. N. Ihsan · E. P. Wulandari
Research Center for Climate and Atmosphere, National Research and Innovation Agency (BRIN), Bandung, Indonesia
e-mail: hari031@brin.go.id

E. Yulihastin et al. (eds.), *Proceedings of the International Conference on Radioscience, Equatorial Atmospheric Science and Environment and Humanosphere Science, 2021*, Springer Proceedings in Physics 275, https://doi.org/10.1007/978-981-19-0308-3_14

also traversed by global climate variability phenomena such as the Madden Julian Oscillation (MJO), Indian Ocean Dipole Mode (IODM), and the El-Nino Southern Oscillation (ENSO) [2, 3]. This global phenomenon can affect the variability of rainfall and seasons in Indonesia, both temporally and spatially.

Seasonal prediction provides information about atmospheric conditions, especially temperature and rainfall, on a time scale of one or several seasons ahead. The requirement for seasonal predictions is increasing along with increasing public awareness of the effect of seasonal variability on important sectors such as agriculture, water resource management, and hydro meteorological disasters (floods and droughts), which of course have an impact on the economic sector [4]. In these sectors, stakeholders and decision-makers need seasonal forecasts to assist them in planning [5]. Therefore, it is necessary to have more accurate season predictions, and the way is by producing high-resolution seasonal predictions.

In 1991, the first seasonal predictions that combine the near real-time ocean modeling and analyses based on coupled ocean–atmosphere general circulation models were done [6]. For years, weather and climate prediction have been gradually evolving. Not only from the observation aspects but also technological advances have occurred at major operational centers worldwide [7]. As a result, weather and climate prediction information has become an essential issue in many sectors such as agriculture and food security, health, energy, water management, resource allocation, and disaster risk reduction and aim for decision-making purposes [8].

The National Oceanographic and Atmospheric Administration (NOAA) developed the North American Multi-Model Ensemble (NMME) experiment in 2011. They accomplished hind cast and real-time seasonal prediction using a multi-model approach with varying ensemble schemes to enhance the forecast quality [9]. Advance in real- time seasonal forecasts, the International Research Institute for Climate and Society (IRI), Columbia University, also provides monthly forecast products for the next six- month prediction by using the output of NMME.

This study uses GSMaP rainfall data as comparison data for predictive rainfall. Based on timescale analysis, GSMaP data has climate variability owned by observational data on the three rainfall patterns in Indonesia (local, equatorial, and monsoon patterns). Therefore, the spatial, temporal, and timescale characteristics of GSMaP data can be used to analyze climate variability in tropical areas such as monsoons, ITCZ, and MJO [10].

The purpose of this study is to test the accuracy of the Conformal Cubic Atmospheric Model (CCAM) rainfall prediction with a spatial resolution of 32 km and daily temporal resolution records from 2016 to 2017 for simulation of the dry season and rainy season using the GSMaP rainfall data to verify the CCAM seasonal prediction in IMC. The result will describe the rainfall simulation with a maximum correlation to obtain accurate rainfall results for each region in the Indonesian maritime continent.

14.2 Data and Methods

14.2.1 Data

This study uses data from CCAM processing output and GSMaP rainfall data. CCAM is an atmospheric model that was developed by Commonwealth Scientific and Industrial Research Organization (CSIRO), which was the development of the Division of Atmospheric Research Limited Area Model (DARLAM) in 1994 [11]. CCAM data has a final output with 32 km spatial resolution and a daily temporal resolution, although processed monthly. It has parameters other than rainfall, such as wind data, surface temperature, and others [12]. The running process using this model generating prediction data by conducting two experiments. The two experiments are experiment with initial conditions from GFS data on 1 December 2016, representing the wet season experiment (later on called WSE); meanwhile, 1 June 2017 initial condition represents the dry season experiment (later on called DSE). Each experiment consists of 11 simulations based on 11 different predictive SST data. Sea surface temperature (SST) prediction data are used from the ensemble results by The Predictive Ocean Atmosphere Model for Australia (POAMA), which will then using as input in CCAM. POAMA is a state-of-the-art seasonal to inter annual seasonal forecast system based on a coupled ocean/atmosphere model developed in a joint project involving the Bureau of Meteorology Research Centre (BMRC) and CSIRO Marine Research [8]. CCAM can be used to generate predictive atmospheric data with SST data guidance. Figure 14.1 shows the schematic regarding the simulation of season predictions carried out in this study.

The comparison data for the rainfall parameter is GSMaP rainfall data. GSMaP rainfall data used is GSMaP_mvk which has a spatial resolution of 0.1° and hourly

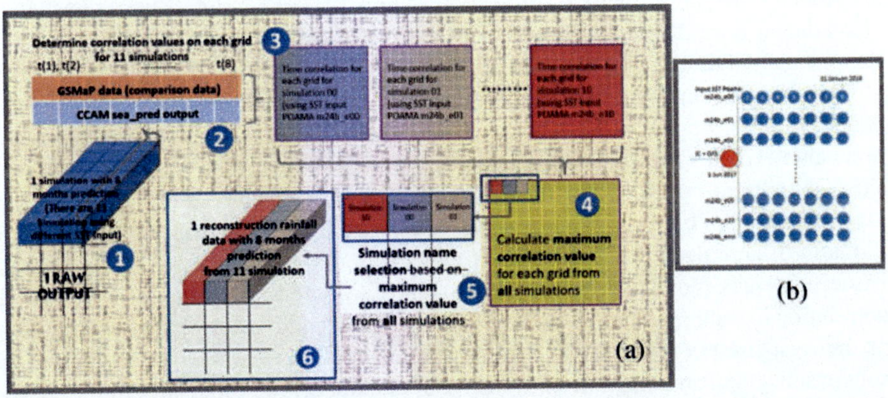

Fig. 14.1 **a** Research flow, and **b** Season prediction simulation scheme using CCAM with initial conditions GFS and SST prediction from POAMA. The date in the picture only illustrates the prediction scheme, not the date used for this study

and daily resolution. Before the comparative analysis, the two data (CCAM output and GSMaP) require adjusting in spatial and temporal resolution [13].

14.2.2 Methods

The research area coverage Maritime Continent, especially in the Indonesia region. The season predictions using CCAM provide up to 8 months of rainfall with 0.28° (32 km) spatial resolution and daily temporal resolution. The daily temporal resolution output process to monthly rainfall before the analysis. The CCAM output parameter that will be analyzed was rainfall.

The workflow for post-processing CCAM output and simulation illustration in this study describes in sequence with Fig. 14.1. CCAM running process are use two initial conditions that given 11 different SST inputs. The full simulations in this research carried out 22 simulations divided into two experiments based on initial conditions. The following steps were calculating the correlation coefficient between rainfall prediction (CCAM output) and GSMaP. From the 11 temporal correlation data, the highest (maximum) correlation value is calculated in each grid to produce maximum correlation value data used as the basis for rainfall reconstruction. The reconstruction result will describe the rainfall in the simulation that has a maximum correlation.

14.3 Results and Discussion

This research aimed to compare the predicted rainfall and the satellite rainfall using the spatial correlation method. Figures 14.2 and 14.3 show the spatial correlation of predicted rainfall for each member in the initial conditions of 1 June 2017 (dry season) and 1 December 2016 (wet season) with GSMaP rainfall.

The positive (negative) correlation value indicates that the predicted rainfall-series pattern follows (opposite) the satellite rainfall-series pattern. The high (low) correlation value explains that the expected rainfall series pattern was very similar or opposite (random) compared with the satellite rainfall-series pattern. Figure 14.2 shows the distribution of high correlations in the southern part of Indonesia, including south Sumatra, Lampung, Java, Bali, Nusa Tenggara, south and southeast Sulawesi, and western Borneo, mostly Papua. There are some differences in the value of the spatial correlation in each member, even though it is small. Negative correlation values were on the west coast of Sumatra, Central Borneo, South Sulawesi, and the west coast of southern Papua. It was refer to complexity of equator region especially In Indonesia which have the large source of humidity in the mountainous archipelagos which have strong local factor makes several model have lower predictability.

Most of simulations results produce a high correlation in the south Indonesia. It means that the rainfall prediction for eight months, both in the initial conditions of the

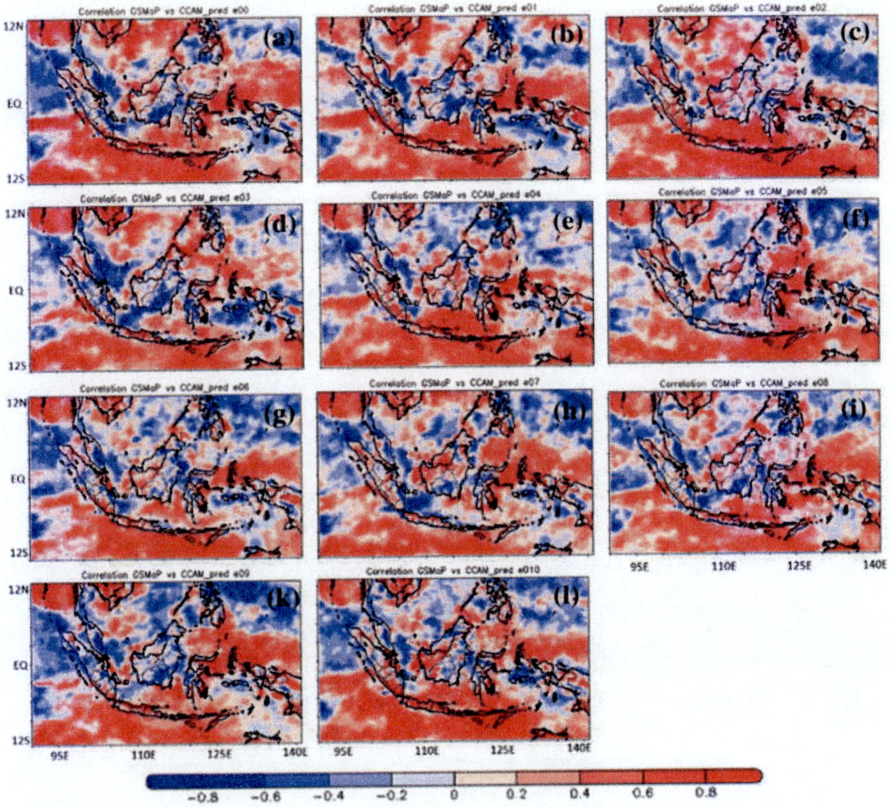

Fig. 14.2 Correlation of CCAM predictions for eight months and GSMaP rainfall in DSE using **a** e00, **b** e01, **c** e02, **d** e03, **e** e04, **f** e05, **g** e06, **h** e07, **i** e08, **j** e09, and **k** e10 SST predictions

dry and wet seasons, was well predicted in this region. Meanwhile, each simulation has different correlations in other areas, such as Sumatra, Borneo, Sulawesi, and Papua. The regions where has lower correlation values in each of these simulations means that regions only have a few simulation options.

Figure 14.4 shows the combination of maximum correlation values in each grid of member simulations.

The results of calculating the maximum correlation values between predicted rainfall and satellites in each grid show most regions in Indonesia have the highest correlation values in the range of 0.4–1. This value shows that from 11 simulations, both DSE and WSE provide a reasonably good correlation between predicted and satellite rainfall. However, a few areas only have a maximum correlation value below 0.2 and negative values. In the DSE, the lowest value of the maximum correlation coefficient was found in areas around the equator, such as in the central Sumatra, south Borneo, and central Sulawesi. In contrast, in the WSE, central Sumatra and west Borneo (along the Malaysia border) have the lowest maximum correlation

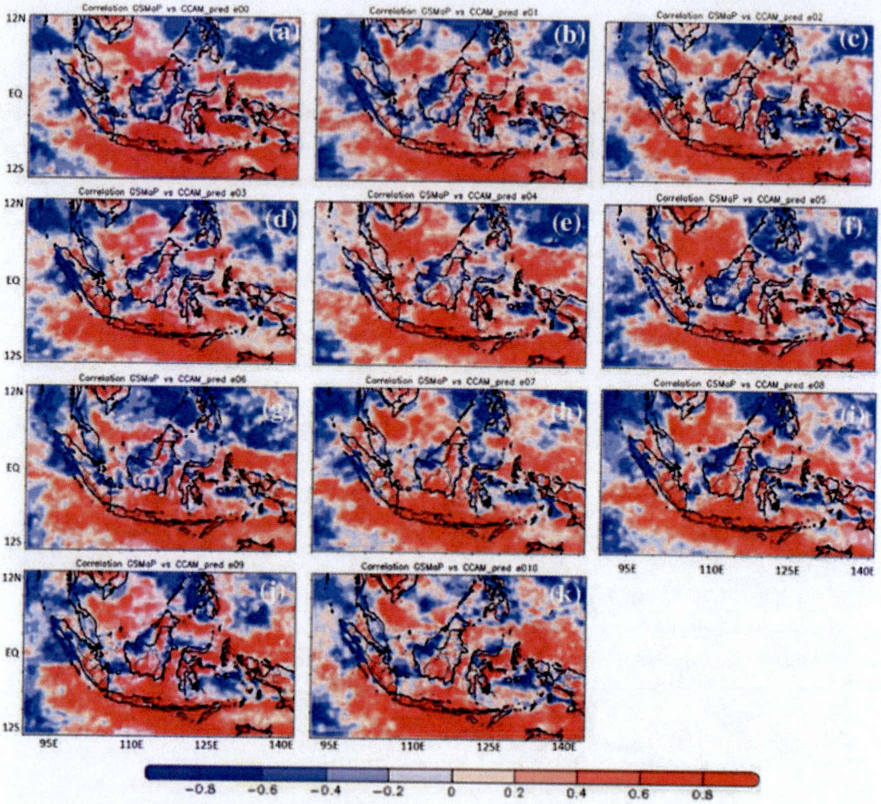

Fig. 14.3 Same as Fig. 14.2, but in WSE

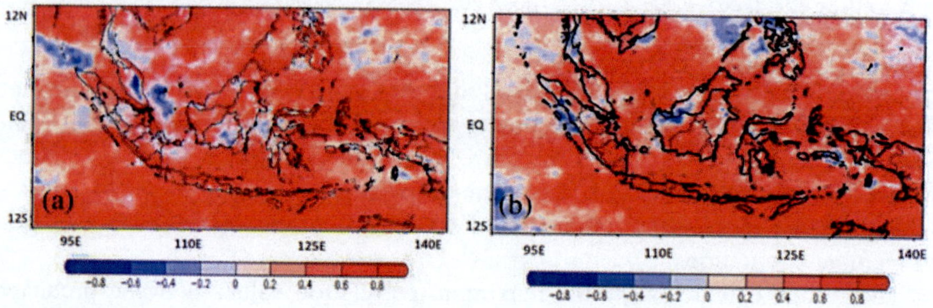

Fig. 14.4 Maximum correlation between CCAM rainfall prediction across all members and GSMaP for: **a** DSE and **b** WSE

value. This value shows that this CCAM seasonal prediction was not predictable for these areas, even though using the results from 11 simulations. The area with low predictability in both experiments is central Sumatra. This is similar to the research of The GFDL Global Atmospheric Model Development Team using the Geophysical Fluid Dynamics Laboratory Global Atmospheric Model (GFDLAM2), which showed a deficit in the annual mean of rainfall in the west of Sumatra [14]. In addition, another GCM, the Max Planck Institute AGCM (ECHAM5), also showed a negative rainfall bias [15].

Figure 14.5 is the result of calculating percentage area which has maximum correlation value. Maximum correlation coefficient value of the simulation using SST predictions e02 (e06) has the most significant area (smallest) in the experiment with IC 1 June 2017 with a coverage of 16% (5%) of the entire grid of research area. Meanwhile, in the IC experiment on 1 December 2016, the simulation with SST predicted e04 (e01) which has the most prominent (smallest) area with a coverage value of 15% (4%) of the entire grid in the research area. The simulation results of reconstructed rainfall based on the maximum correlation value in the DSE are shown in Fig. 14.6, while the WSE is in Fig. 14.7.

The comparison between the predicted rainfall reconstruction for the next eight months based on the simulation with the maximum correlation coefficient value with the GSMaP rainfall (Fig. 14.6) shows that, in general, the spatial pattern of predicted monthly rainfall movements in the dry season can already follow the satellite rainfall pattern. However, in some areas, the predicted rainfall area still looks under-estimated and tends to unpredict above 600 mm rainfall, as shown by satellite rainfall.

In addition to the spatial pattern of monthly rainfall, the change from the dry season to the wet season (wet season onset), especially in the southern part of Indonesia (Java, Bali, and Nusa Tenggara), experienced slight differences. Based on GSMaP satellite rainfall data, the southern Indonesia experiences the onset of the wet season

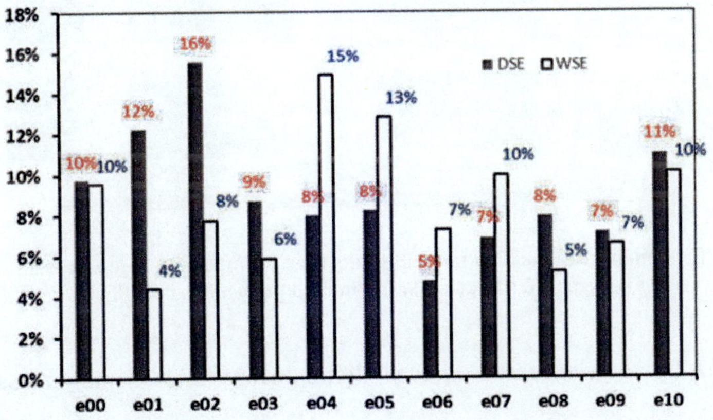

Fig. 14.5 Percentage area of the maximum correlation values with different predictive SST for dry and wet season experiments

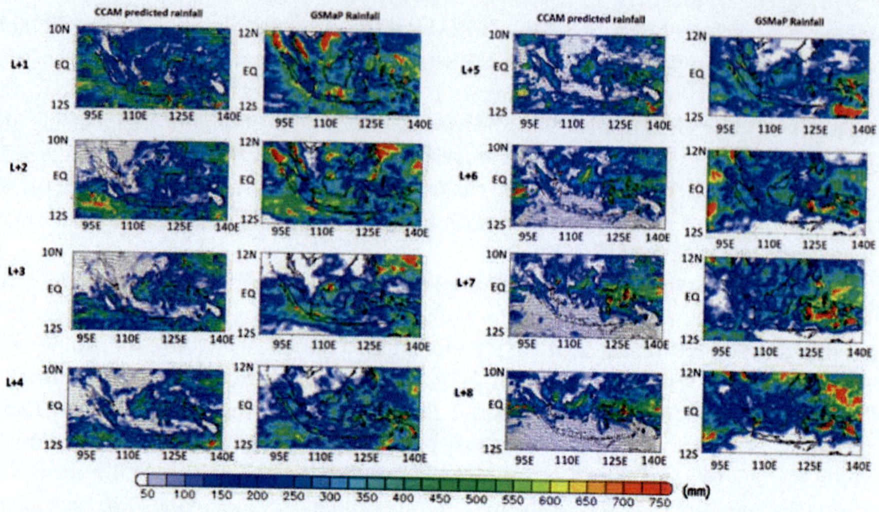

Fig. 14.6 The combined rainfall simulation results from all members were selected based on the maximum correlation between CCAM and GSMaP rainfall predictions in DSE

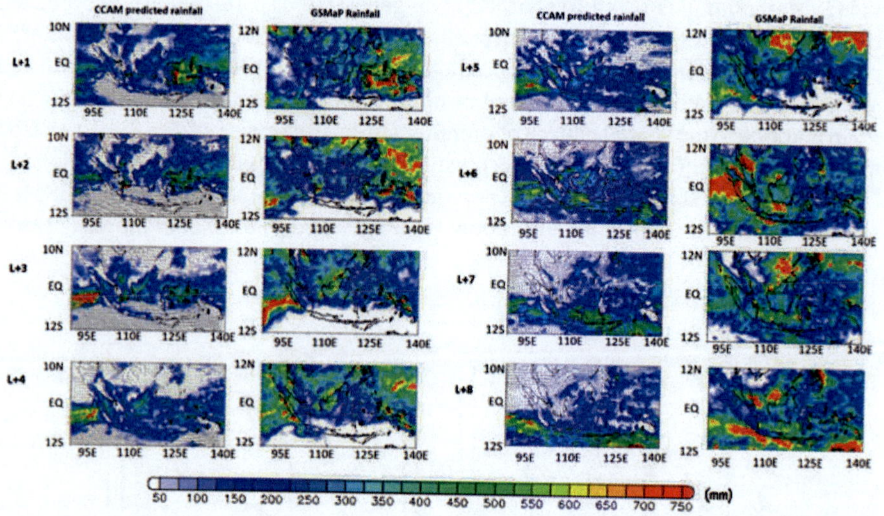

Fig. 14.7 The combined simulation rainfall results from all members were selected based on the highest correlation between CCAM and GSMaP rainfall predictions in WSE

in October, but the wet season onset prediction occur faster, namely in September 2017.

Figure 14.7 shows that the monthly rainfall in December 2016, the closest prediction result to the initial conditions (Lead +1 month/L + 1), has a similar spatial pattern to GSMaP. However, CCAM output cannot show high rainfall intensity in a few areas, such as South China Sea and Malacca. Excessive rainfall in these areas was not even predictable at Lead +2. The transition from the wet to the dry season in southern Indonesia occurs more quickly. It occurs uniformly and simultaneously, as seen in the May 2017 predictions, even though the GSMaP rainfall has only shown the dry season onset in May and July gradually by east to west propagation.

The results can be further advanced by increasing the horizontal resolution will reduce the dry bias of precipitation in the western Indian and western Pacific Ocean [15]. In addition, the increasing resolution will give an actual surface condition such as detailed coastline parameters, soil conditions, vegetation, and topography [16, 17].

Prediction results of the season onset and the intensity of seasonal rainfall for the next 6–8 months can be displayed spatially and time series through the Alpha-version of the Decision Support System (DSS) named KAMAJAYA using district-scale spatial resolution.

14.4 Summary and Conclusions

Rainfall simulation for the next eight months prediction using Conformal Cubic Atmospheric Model (CCAM) output with 11 Sea Surface Temperature (SST) input, get varied results for each simulation member. CCAM seasonal prediction was highly correlated with GSMaP rainfall in the southern part of Indonesia, while few areas in northwestern have a low correlation. Two experiments represent dry and wet seasons using 22 simulations shows the maximum correlation is dominated by the r-value > 0.6, except for the area around the equator and northwestern Indonesia. The best ensembles for simulating season predictions based on maximum correlation get varying results, both spatially and temporally. Reconstructing the rainfall prediction based on maximum correlation shows that the spatial pattern of predicted monthly rainfall in the dry and wet seasons is similar to the GSMaP rainfall pattern. However, this method was not reduce the bias between this two data, because basically, reconstructing rainfall prediction based on the correlation was only fixing the pattern of rainfall. Also this research still have limitation on onset determination because rainfall prediction estimates the onset faster than GSMaP. Therefore, it needs further research and testing in different years and long-term data in both seasons (wet and dry seasons).

Acknowledgements The authors expressed our gratitude for the GSMaP data support provided by the Japan Aerospace Exploration Agency (JAXA), POAMA's SST data provided by the Australian Bureau of Meteorology (BOM), and in this research. Also authors want to show our gratitude to CSIRO for developing CCAM that we can use to improve our capability in atmospheric modeling. This research was supported by the Indonesia Educational Endowment Fund (LPDP) through the Mandatory Productive Innovative Research program under National Research Priority [252/Menteri Ristek/Ka BRIN/E1/PRN/2020].

References

1. Dayem, K.E., Noone, D.C., Molnar, P.: Tropical western Pacific warm pool and maritime continent precipitation rates and their contrasting relationships with the Walker Circulation. J. Geophys. Res. Atmos. **112**(6), 1–12 (2007). https://doi.org/10.1029/2006JD007870
2. Wu, C.H., Hsu, H.H.: Topographic influence on the MJO in the maritime continent. J. Clim. **22**(20), 5433–5448 (2009). https://doi.org/10.1175/2009JCLI2825.1
3. Harijono, S.W.B.: Analisis Dinamika Atmosfer di Bagian Utara Ekuator Sumatera pada Saat Peristiwa El-Nino dan Dipole Mode Positif Terjadi Bersamaan. J. Sains Dirgant. **5**(2), 130–148 (2008). http://jurnal.lapan.go.id/index.php/jurnal_sains/article/view/341
4. Schepen, A., Wang, Q.J., Robertson, D.E.: Seasonal forecasts of Australian rainfall through calibration and bridging of coupled GCM outputs. Mon. Weather Rev. **142**(5), 1758–1770 (2014). https://doi.org/10.1175/MWR-D-13-00248.1
5. Ratri, D.N., Whan, K., Schmeits, M.: A comparative verification of raw and bias- corrected ECMWF seasonal ensemble precipitation reforecasts in Java (Indonesia). J. Appl. Meteorol. Climatol. **58**(8), 1709–1723 (2019). https://doi.org/10.1175/JAMC-D-18-0210.1
6. Kirtman, B., et al.: The North American multimodel ensemble: phase-1 seasonal- to-interannual prediction; phase-2 toward developing intraseasonal prediction. Bull. Am. Meteo. Soc. **95**, 585–601 (2014)
7. Ji, M., Kumar, A., Leetmaa, A.: A multiseason climate forecast system at the National Meteorological Center. Bull. Am. Meteo. Soc. **75**, 569–577 (1994)
8. Wang, G., et al.: POAMA: an Australian ocean-atmosphere model for climate prediction. Bull. Am. Meteorol. Soc. **1**, 4559–4563 (2004)
9. WMO. Guidance on Operational Practices for Objective Seasonal Forecasting. Chair, Publications Board, World Meteorological Organization (WMO), ISBN 978-92-63- 11246-9 (2020)
10. Fitria, M.: Validasi Data Curah Hujan Global Satellitc Mapping of Precipitation (Gsmap) Pada Tiga Pola Hujan di Indonesia. Bachelor Degree Thesis on IPB University (2016)
11. McGregor, J.L.: CCAM: geometric aspects and dynamical formulation. CSIRO Atmospheric Research Technical Paper No. 70 (2005)
12. Thatcher, M., McGregor, J., Dix, M., Katzfey, J.: A new approach for coupled regional climate modeling using more than 10,000 cores. In: Denzer, R., Argent, R.M., Schimak, G., Hřebíček, J. (eds.) Environmental Software Systems. Infrastructures, Services and Applications. ISESS 2015. IFIP Advances in Information and Communication Technology, vol. 448. Springer, Cham (2015). https://doi.org/10.1007/978-3-319–15994–2_61
13. Kachi, M., Kubota, T., Aonashi, K., Ushio, T., Shige, S., and Yamamoto, M.: Recent improvements in the global satellite mapping of precipitation (GSMaP), In: 2014 IEEE Geoscience and Remote Sensing Symposium, pp 3762–3765. https://doi.org/10.1109/IGARSS.2014.6947302
The GFDL Global Atmospheric Model Development Team.: The new GFDL global atmosphere and land model AM2-LM2: evaluation with prescribed SST simulations. J. Climate **17**(24), 4641–4673 (2004)
14. Wild, M., Roeckner, E.: Radiative fluxes in the ECHAM5 General Circulation Model. J. Climate **19**, 3792–3809 (2006)
15. Schiemann, R., Demory, M.E., Mizielinski, M., Roberts, M., Shaffrey, L., Strachan, J., Vidale, P.L. (2013) The sensitivity of the tropical circulation and Maritime continent precipitation to climate model resolution. Clim. Dyn. 42. https://doi.org/10.1007/s00382-013-1997-0
16. Strachan, J.: Understanding and modeling the climate of the Maritime Continent. Phd thesis, University of Reading (2007)
17. Bauer, P., Thopre, A., Brunet, G.: The quiet revolution of numerical weather prediction. Nature **525**, 47–55 (2015)

Chapter 15
Solar Wind Speed Time-Series Forecasting Based on Long Short-Term Memory (LSTM) Neural Network Model

Tiar Dani, Rhorom Priyatikanto, Anton Winarko, and Gerhana Puannandra Putri

Abstract Changes in solar wind speed is one of the earliest manifestations of space weather disturbances driven by solar flare, Coronal Mass Ejection (CME), or coronal hole. An increase in solar wind speed can cause geomagnetic disturbances indicated by either Dst or Kp indices. This study proposed a Long Short-Term Memory (LSTM) Neural Network model to produce hourly forecasts of solar wind speed for the next 24 h during solar minimum. The model uses input from solar wind speed observed by Advanced Composition Explorer (ACE) satellite in Lagrangian-1 (L1) orbit as internal input and the area of the geoeffective coronal hole from Solar Dynamic Observatory (SDO)/Atmospheric Imaging Assembly (AIA) at 193 Angstrom satellite detected by DeLuna as an external input. The model has an average RMSE (Root Mean Square Error) value of 40.04 km/s. We also conducted validation analysis using the forecast product and compared it with solar wind speed data observed by Deep Space Climate Observatory (DSCOVR) satellite from October 28, 2019, to May 9, 2021. We found the forecast results 77% correct, 11% overestimate, and 12% underestimate, with a best-fit correlation of 0.85.

15.1 Introduction

Solar wind dynamics around the Earth are strongly influenced by solar activities. Flare and CME are the dominant factors increasing solar wind speed during the maximum solar cycle phase. On the contrary, the coronal hole becomes the more domi- nant factor during the minimum solar cycle. The solar wind speed can reach twice its normal speed when a coronal hole appears on the solar disk. The increasing speed is even more significant when solar flare or Coronal Mass Ejection occurs. An earlier work [1] obtained a positive correlation with a lag time of 3.17 days

T. Dani (✉) · R. Priyatikanto · A. Winarko
Space Research Center, BRIN, Bandung, Indonesia
e-mail: tiar.dani@brin.go.id

G. P. Putri
Sumedang Space and Atmospheric Observatory, BRIN, Sumedang, Indonesia

E. Yulihastin et al. (eds.), *Proceedings of the International Conference on Radioscience, Equatorial Atmospheric Science and Environment and Humanosphere Science, 2021*, Springer Proceedings in Physics 275, https://doi.org/10.1007/978-981-19-0308-3_15

between the geoeffective coronal hole area, and the solar wind speed observed by the ACE satellite orbiting the Lagrange-1 (L1). Another study [2] obtained a lag time of 3.5 days.

There are several methods for forecasting solar wind speed. One method [3] used magnetohydrodynamic (MHD)-based physical model to forecast the solar wind speed. The model is called WSA-ENLIL and is used by the Space Weather Prediction Center (SWPC) of the National Oceanic and Atmospheric Administration (NOAA). Another method [4] forecasted solar wind speed using empirical correlation of coronal hole area and High-Speed Streams (HSS) that occurred on near-Earth four days later. Alternatively, one may utilize machine learning (ML) for forecasting, which is now widely used in various fields of science. For example, there were an approach [5] using a hybrid Artificial Neural Network (ANN) method, and another [6] using Deep Learning called WindNet to forecast the solar wind speed.

The Long Short-Term Memory neural network [7] is one approach for Deep Learning with Recurrent Neural Network (RNN) configurations. LSTM generally consists of cells, input gates, output gates and forget gates as its components. LSTM is very suitable for classifying, processing, and making a forecast based on time series data. In this study, we utilize LSTM method to forecast the solar wind speed time series 24 h in advance. Our LSTM model uses the geoeffective area of the coronal hole as an external input and solar wind speed as internal input.

15.2 Data and Methods

15.2.1 Data

Hourly averaged geoeffective coronal hole (GCH) area data were obtained from De- Luna detection [1] using SDO/AIA193 [8, 8] images from NASA/SDO, and the AIA science teams (https://sdo.gsfc.nasa.gov). The image has 512×512 pixels resolution and 12 min temporal resolution. Hourly solar wind speed level 2 data were obtained from ACE satellite at L1 [10] using the Solar Wind Electron Proton and Alpha Monitor (SWEPAM) instrument [11] (http://www.srl.caltech.edu/ACE/ASC/level2/lvl2DATA_SWEPAM.html). The data ranged from 2016 to 2018, which covered the period of solar minimum. We also used solar wind data from the DSCVR Satellites obtained from the NCEI DSCOVR teams (https://www.ngdc.noaa.gov/dscovr/portal/index.html#) from 2019 to 2021 as validation data.

Figure 15.1 shows the geoeffective coronal hole (GCH) area and solar wind speed (vSW) during 2016–2018. This data range was selected to represent the minimum phase of solar cycle where changes in the solar wind are caused by coronal holes. Changes in the GCH area have a sinusoidal pattern, with the largest area reaching about 12% of the total solar disk area at the end of 2018. The sinusoidal pattern in the GCH area shows periodicities of 13.5 days and 27 days. The same periodicity was also found in vSW with a delay of around 3.17 days relative to the GCH area [1].

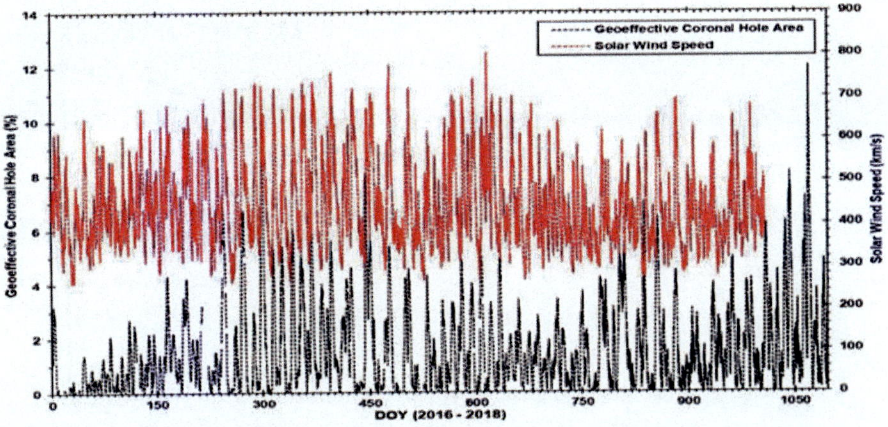

Fig. 15.1 Hourly time series of geoeffective coronal hole area and solar wind speed from 2016 to 2018

15.2.2 Methods

GCH areas are detected and calculated by DeLuNa from the SDO/AIA193 images. DeLuNa defined the GCH area located at ±30 heliographic latitude and −15–70° heliographic longitude and expressed in percent (%) of the total solar disk area. For the detailed detection method and GCH area calculation, see [1]. We averaged the 12 min resolution GCH area data and the 1 min solar wind speed data from the ACE satellite into hourly data.

The LSTM model was developed using the Python programming language by utilizing Keras which is a high-level API for artificial neural networks. Keras is a Deep Learning library that uses TensorFlow as a backend [12]. The LSTM Sequence to Sequence scenario was then used to forecast the solar wind speed. A lag times of 3.17 days was applied to GCH area data as suggested by [1]. Output of the LSTM model is the hourly solar wind speed for the next 24 h. The model was tested to obtain the average RMSE (Root Mean Square Error) and the hourly RMSE value. The data from 2016 to 2017 were used for training and the data from 2018 were used for testing.

15.3 Results

We obtained the optimum LSTM configuration to be 4 h of data input, two LSTM layers with 200 and 100 nodes, respectively, and one fully connected layer with 50 nodes. Figure 15.2 shows that the first hour of the forecast has an RMSE of 10.5 km/s. The RMSE reaches maximum value of 60.7 km/s at the 24th hour forecast. The consequence of the multi-step ahead forecast is that error will accumulate at each

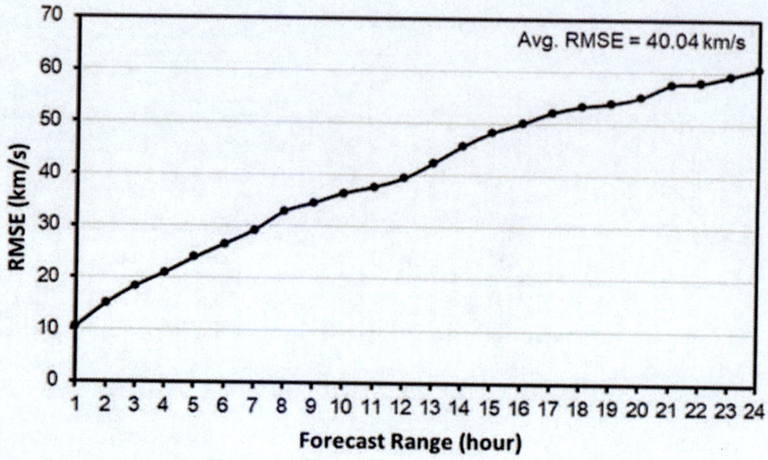

Fig. 15.2 Hourly RMSE values of the LSTM model for the 24 h forecast range

forecast step and lead to significant increase in the RMSE value. We obtained the average RMSE value to be 40.04 km/s for the 24 h interval as a whole.

Figure 15.3 shows some examples of solar wind speed forecasting in 2019, where the solar wind speed forecast results into three categories, i.e. correct, underestimated, and overestimated forecast. The chart shows the solar wind speed observation data from Deep Space Climate Observatory (DSCOVR) satellite (black line) [13], solar wind speed forecast in 24 h (red dotted) and forecast in the next 24 h (blue dotted). The RMSE value of the forecast in each hour are shown as the grey area.

Figure 15.4 shows the accuracy of solar wind forecasting during the period 28 October 2019 to 9 May 2021. The correct, overestimated, and underestimated forecast is designated when the difference between forecasted and observed solar wind speed is within ±40.04 km/s, more than 40.04 km/s, and less than 40.04, respectively. These designations were based on the average RMSE value (= 40.04 km/s) of the model. We obtained a best-fit correlation coefficient of 0.85, with 77%, 11%, and 12% in the correct, overestimated, and underestimated categories, respectively. The largest devia- tion value of the overestimated forecast is 205.85 km/s, with an average of 60.11 km/s. As for the underestimated forecast, the largest deviation value is 315.65 km/s, with an average of 83.53 km/s.

15.4 Conclusion

We have constructed a model to forecast solar wind speed based on the LSTM neural network. The model uses four hours of solar wind speed and coronal hole area data as input to produce hourly forecast of solar wind speed for the next 24 h. As a comparison, other models use four days of solar wind speed data and 3 days lag time

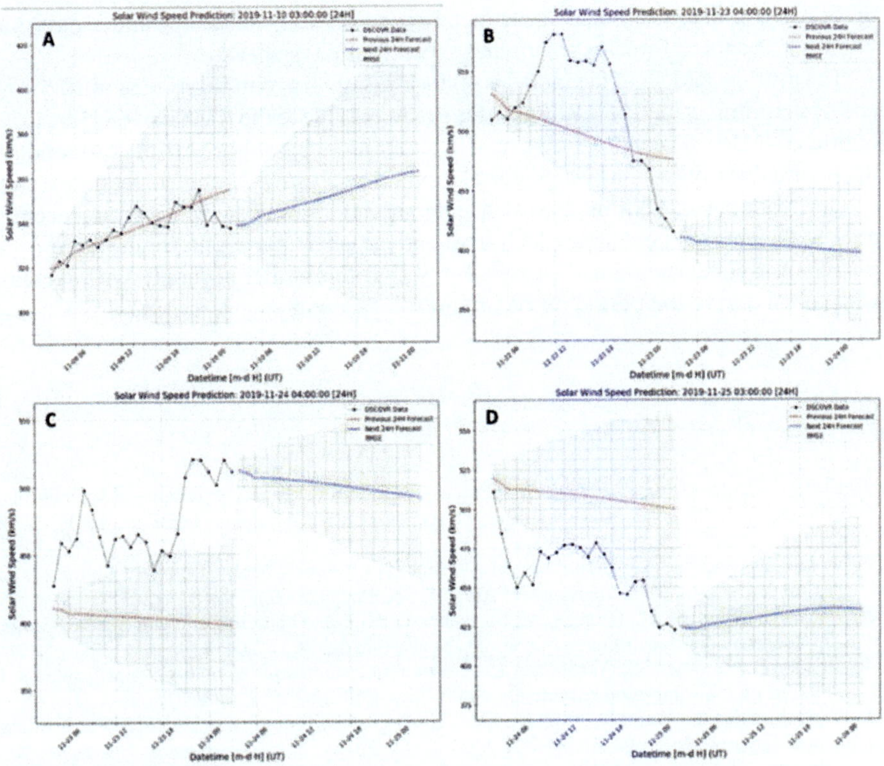

Fig. 15.3 Sample results of solar wind speed forecasting in 2019, illustrating correct forecast (**a**) and (**b**), underestimate forecast (**c**), and overestimate forecast (**d**)

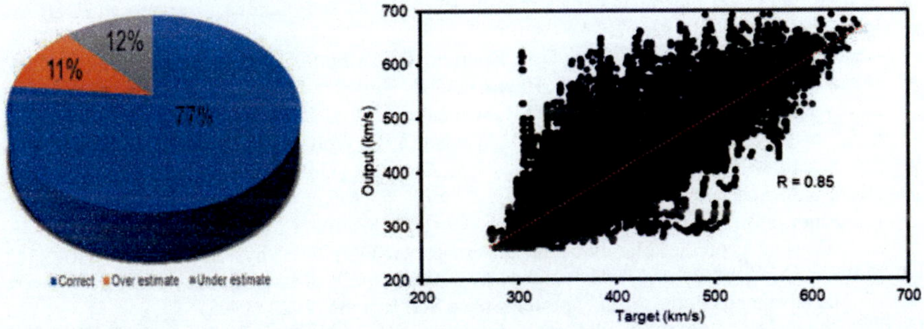

Fig. 15.4 Percentage of model validation for three categories, i.e. correct, overestimated, and underestimated (left) and correlation plot between observation and forecasted solar wind speed (right) from 28 October 2019 to 9 May 2021

on SDO/AIA data to forecast solar wind speed one day ahead [6] and 1 day solar wind speed data to forecast solar wind speed four days ahead [5].

The average RMSE of the model is 40.04 km/s, with a minimum value of 10.5 km/s and a maximum value of 60.7 km/s. This is comparable to the RMSE of 68 km/s using hybrid ANN [5]. The validation analysis of our model yields a best-fit correlation of 0.85 while [6] obtained a best-fit correlation of 0.55 using WindNet Deep Learning.

Our work has some limitations since the model training dataset is taking account only of the coronal hole area which frequently appears during minimum solar cycle phase. Consequently, there is a possibility that the model accuracy will decrease when used during the period of maximum solar activity.

References

1. Dani, T., Priyatikanto, P., Winarko, A., Putri, G.P.: Relationship and periodicity of solar coronal hole area with the solar wind speed and geomagnetic activity. J. Sains Dirgantara **17**(1), 27–38 (2019)
2. Rotter, T., Veronig, A.M., Temmer, M., Vršnak, B.: Relation between coronal hole areas on the sun and the solarwind parameters at 1 AU. Sol. Phys. **281**(2), 793–813 (2012)
3. Pizzo, V., Millward, G., Parsons, A., Biesecker, D.,Hill, S., Odstrcil, D.: Wang-Sheeley- Arge-Enlil cone model transitions to operations. Space Weather **9**(3) (2011)
4. Vršnak, B., Temmer, M., Veronig, A.M.: Coronal holes and solar wind high-speed streams: I. Forecasting the solar wind parameters. Solar Phys. **240**(2), 315–330 (2007)
5. Yang, Y., Shen, F., Yang, Z., Feng, X.: Prediction of solar wind speed at 1 AU using an artificial neural network. Space Weather **16**(9), 1227–1244 (2018)
6. Upendran, V., Cheung, M.C.M., Hanasoge, S., Krishnamurthi, G.: Solar wind prediction using deep learning. Space Weather **18**(9) (2020)
7. Hochreiter, S., Schmidhuber, J.: Long Short-Term Memory. Neural Comput. **9**(8), 1735–1780 (1997)
8. Pesnell, W.D., Thompson, B.J., Chamberlin, P.C.: The solar dynamics observatory (SDO). Sol. Phys. **275**(1–2), 3–15 (2012)
9. Lemen, J.R., Title, A.M., Akin, D.J., Boerner, P.F., Chou, C., Drake, J.F., Duncan, D.W., Edwards, C.G., Friedlaender, F.M., Heyman, G.F., Hurlburt, N.E., Katz, N.L., Kushner, G.D., Levay, M., Lindgren, R.W., Mathur, D.P., McFeaters, E.L., Mitchell, S., Rehse, R.A., Schrijver, C.J., Larry, A., Stern, R.A., Tarbell, T.D., Wuelser, J.-P., Wolfson, C.J., Yanari, C., Bookbinder, J.A., Cheimets, P.N., Caldwell, D., Deluca, E.E., Gates, R., Golub, L., Park, S., Podgorski, W.A., Bush, R.I., Scherrer, P.H., Gummin, M.A., Smith, P., Auker, G., Jerram, P., Pool, P., Soufli, R., Windt, D.L., Beardsley, S., Clapp, M., Lang, J., Waltham, N.: The atmospheric imaging assembly (AIA) on the solar dynamics observatory (SDO). Sol. Phys. **275**(1–2), 17–40 (2012)
10. Stone, E.C., Frandsen, A.M., Mewaldt, R.A., Christian, E.R., Margolies, D., Ormes, J.F., Snow, F.: The advanced composition explorer. Space Sci. Rev. **86**, 1–22 (1998)
11. McComas, D.J., Bame, S.J., Barker, P., Feldman, W.C., Phillips, J.L., Riley, P., Griffee, J.W.: Solar wind electron proton alpha monitor (SWEPAM) for the advanced composition explorer. Space Sci. Rev. **86**, 563–612 (1998)
12. Abadi, M., Barham, P., Chen, J., Chen, Z., Davis, A., Dean, J., Devin, M., Ghemawat, S., Irving, G., Isard, M., Kudlur, M., Levenberg, J., Monga, R., Moore, S., Murray, D.G., Steiner, B., Tucker, P., Vasudevan, V., Warden, P., Wicke, M., Yu, Y., Zheng, X.: TensorFlow: A system for large-scale machine learning. eprint arXiv:1605.08695 (2016)
13. NOAA Space Weather Prediction Center: Deep Space Climate Observatory (DSCOVR). NOAA National centers for environmental information. Dataset (2016). http://doi.org/https://doi.org/10.7289/V51Z42F7

Chapter 16
On the Mechanism of Anomalously Wet and Cold Weather Over Java in June

Suaydhi

Abstract Java is known to have very distinct rainy and dry seasons. The month of June usually is already in the period or the beginning of the dry season. The monthly rainfall average in June is usually below the annual average. The June surface temperature in Java is also cooler than that in the rainy season. However, some anomalies are observed during this month in some years with a relatively large amount of rainfall and colder weather than the June climatological average. Data from the fifth generation of ECMWF Atmospheric Reanalysis of the global climate (ERA5) is used to investigate the wet and cold episodes over Java in June. The analysis data from the Global Data Assimilation System (GDAS) are used to drive the HYSPLIT model to simulate the air mass trajectories. Hourly and six-hourly data of June rainfall, surface temperature, horizontal wind, and other atmospheric parameters between 2010 and 2021 are used in this research. It is found that the moisture source for Java rainfall in June mostly comes from the regions to the east of Java. Moisture from the Indian Ocean only contributes to the precipitation over Java when the MJO is active without positive Indian Ocean Dipole nor El Nino. The surface temperature will be colder over Java if a significant portion of the air mass comes from the high altitude of the tropical region or the high latitude regions.

16.1 Introduction

Java covers an area of about 150,000 square kilometers or less than 8% of the total land area of Indonesia. It is located in the southwestern part of Indonesia and has a relatively long and narrow shape, spanning from the west to the east with a slight tilt towards the southeast on the eastern part of the island. The latitude and longitude of the island are 6.0° S to 8.2° S and 105.4° E to 114.4° E, respectively. It comprises six provinces (Banten, West Java, Central Java, East Java, Special Capital Region

Suaydhi (✉)
Research Center for Climate and Atmosphere, National Research and Innovation Agency (BRIN), Bandung, Indonesia
e-mail: suaydhi@lapan.go.id

E. Yulihastin et al. (eds.), *Proceedings of the International Conference on Radioscience, Equatorial Atmospheric Science and Environment and Humanosphere Science, 2021*, Springer Proceedings in Physics 275, https://doi.org/10.1007/978-981-19-0308-3_16

of Jakarta, and Special Region of Yogyakarta). It is the most populous region of Indonesia, with around 56% of the total population of Indonesia residing in Java [1]. It is also the most developed region of Indonesia, with most business and economic activities conducted in Java. Despite that, Java is still the largest producer of rice in Indonesia, the staple food for most Indonesian people. The total rice production in Java alone is much greater than that of the rest of Indonesia [2]. It shows that Java remains one of the most important centers for food production in Indonesia.

Rice farming requires a lot of water, especially during the growing stage. The planning for planting season depends on the availability of water during the first two months. This planning is very crucial at the beginning of the dry season. In this case, June could be used to indicate whether the rest of the dry season will have less or more rainfall than the climatological average [3]. If there is still a good amount of rain during June, then starting another rice planting season would be very beneficial in increasing rice production. Otherwise, it would be a waste of money and time to start a new planting season. D'Arrigo and Wilson [4] demonstrate that a simple empirical forecast model of drought for Java using the August combination of DMI and Nino-3.4 SST significantly improves agricultural productivity. Increasing the preparedness for the dry spell is very important as it might become more severe and frequent in the future due to climate change [5].

The rainfall variability during the dry season over Java has long been studied [6, 7]. Its linkage to the sea surface temperature (SST) over the Indian Ocean via the Indian Ocean Dipole Mode [8] and over the Pacific Ocean via the El Nino Southern Oscillation (ENSO) can be found in many studies [3, 4, 9–11]. D'Arrigo et al. [9] and D'Arrigo et al. [10] relate the influence of Indian Ocean Dipole (IOD) and ENSO on Java monsoon drought by examining the tree ring records. Hatmaja et al. [11] show that the increase of rainfall during the September to November (SON) period in West Java is related to the negative IOD. A similar case is also found by Sahu et al. [12], where the extreme low stream flow events of Citarum River in West Java are associated with the positive IOD events during the June to August period.

Unlike the rainfall variability that has been studied extensively, the variability of surface temperature over Java has received far less attention. It might be that surface temperature in the tropical region is considered not to have any significant impact on daily life. Surface temperature is primarily affected by the amount of solar radiation received, which is mainly related to the apparent position of the Sun on the Earth's surface. The amount of solar radiation in the tropical region does not vary significantly compared with that in the subtropical or high latitude regions. Nevertheless, the surface temperature over Java in June can vary a few degrees from year to year. The surface temperature variation in the mountainous areas could be more significant than that in the low-lying coastal areas [13]. By studying the variation of surface temperature in a particular month (in this case in June), other factors that affect it can be separated from that of the apparent position of the Sun to the Earth's surface.

The weather of a particular area is influenced by air movement. The surface temperature and the probability of rainfall of that area are controlled by the air mass flowing to that area. The aim of this study is to elucidate the weather anomalies over

Java in June. Java usually undergoes dry season from June onwards. Any weather anomalies observed in June would also be detected during the subsequent months of the dry season [3]. The mechanism of the anomalous weather over Java can be studied by tracking the air mass trajectory. In this article, the air mass transport to Java Island is investigated using the Hybrid Single-Particle Lagrangian Integrated Trajectory (HYSPLIT) model for June between 2010 and 2021.

16.2 Data and Method

16.2.1 Data

The current study uses the fifth generation of the atmospheric reanalysis data from the European Centre Medium-Range Weather Forecast (ECMWF Reanalysis v5), better known as ERA5 reanalysis data [14]. The data used in the analysis include rainfall, surface temperature, air temperature at several pressure levels, relative humidity, and horizontal winds. The ERA5 hourly precipitation and surface temperature data at a resolution of $0.25° \times 0.25°$ are used to determine the weather characteristics of June between 2010 and 2021 around the island of Java. These data are obtained from the Copernicus Climate Data Store (CDS).

To the HYSPLIT trajectory model, six-hourly reanalysis data of the Global Data Assimilation System (GDAS) at a resolution of $0.5° \times 0.5°$ are used. These data consist of geopotential height, air temperature, specific humidity, and three-dimensional winds. The HYSPLIT mode is run via the TrajStat plugin from MeteoInfo weather mapping software [15] in the Air Resources Laboratory (ARL) compatible format.

16.2.2 Method

The trajectory calculation is carried out using the NOAA HYSPLIT model [16], which is run through the TrajStat plugin of the MeteoInfo software. In the trajectory calculation, a particle is assumed to follow the wind passively. The back-trajectory algorithm is applied to 17 grid points inside the target area. All parcels were integrated backward in time until 240 h (10 days). Ten days is the average period of residence of water vapor in the atmosphere [17]. A height of 10 m above ground level (agl) at the target area is used to analyze the air parcel source that contributes to the surface temperature at the target area, and a height of 1400 m agl is used to analyze the moisture source that contributes to the rainfall falling into the target area. In the analysis, the trajectories going into the 17 grid points inside the target area are averaged to obtain the general characteristics of various oair mass sources flowing into Java Island.

16.3 Results and Discussion

Table 16.1 shows the variation of rainfall and surface temperature anomalies during June over Java between 2010 and 2021. The symbols −, 0, and + in the table indicate negative, neutral, and positive states of IOD and ENSO, respectively. It can be seen from the table that anomalously wet Junes occurred in 2010, 2013, 2016, and 2021, and the anomalously dry Junes in 2015 and 2019. The net amount of rainfall anomalies can be summed together for the whole month. It is not the case for surface temperature anomalies. The highest warm surface temperature anomaly (dTmax) must be separated from the lowest cold surface temperature anomaly (dTmin). The anomalously warm Junes are indicated by the high values of dTmax or the high values of dTmin, which are 2010, 2014, 2016, and 2019. The anomalously cold Junes are indicated by the low values of dTmax or the low values of dTmin, which are 2011, 2012, and 2019. High (low) values here refer to the large positive (negative) differences between surface temperature of a particular year and the climatological mean of the surface temperature. The anomalously wet Junes coincide with the high number of wet days, i.e., 24 days in 2010, 23 days in 2013, 18 days in 2016, and 19 days in 2021. There is a relatively low number of hot days during the anomalously cold Junes (13 days in 2011, 17 days in 2012, and 16 days in 2019).

Figure 16.1 shows the Hovmoller diagram of the anomalously cold June of 2011 and the anomalously wet June of 2013. The anomalies were calculated with respect to the 1981– 2010 climatology. The Java region (105.5° E–114.5° E) is highlighted. It indicates that June can be very wet and can be cold. The propagation of cold and/or wet air shown in Fig. 16.1 is mostly from the east, except when the MJO is active. Figure 16.2 shows the time- height plot of air temperature and relative humidity for June 2011 and 2013. Relative humidity greater than 90% indicates a wet day. In early June 2011 (Fig. 16.2a), it was a wet and relatively warm day. In the middle of June 2011 (around 16 and 17 June), the weather was dry and relatively cold. Those events are consistent with the surface temperature anomaly shown in Fig. 16.1a, positive anomaly of surface temperature (warm) at the beginning and negative anomaly (cold) in the middle of June 2011. Figure 16.2b shows that the surface temperature was relatively cold during the wet days and warm during dry days. Hence, the weather can be wet and warm, dry and cold, wet and cold, or dry and warm on the same day.

From several cases of wet and warm days between 2010 and 2021 (not shown), the wet and warm days occur when there are shallow convective clouds [18]. According to [19], the top of the shallow cumulus clouds is around 2 km (800 mb), and these clouds appear to be isolated in time and space. It is indicated by the top of the high relative humidity (>80%) layers being lower than 800 mb (Fig. 16.2a). Shallow convection is dominated by vertical moisture transport that moistens the lower troposphere [19]. The wet and cold days occur when there are deep convective clouds (Fig. 16.2b). The high relative humidity (>80%) layer can be higher than the 650 mb pressure level (the pressure level in Fig. 16.2 is not fully shown to make the air temperature line

Table 16.1 Variation of rainfall and surface temperature anomalies over Java for June 2010–2021 (see text for details)

Years	2010	2011	2012	2013	2014	2015	2016	2017	2018	2019	2020	2021
IOD	0	0	0	–	0	+	–	+	0	+	0	–
ENSO	–	0	0	0	0	+	0	0	0	+	0	0
Rain	124	−36	−36	152	36	−57	79	35	−32	−86	−34	66
Wet day	24	8	8	23	16	5	18	13	7	1	8	19
dTmax	1.58	0.73	1.02	1.09	1.57	1.18	1.64	1.10	1.26	0.88	1.10	1.61
dTmin	−0.45	−0.79	−1.11	−0.37	−0.17	−0.35	−0.41	−0.26	−0.71	−0.90	0.22	−0.66
Hot days	27	13	17	23	28	21	28	27	28	16	30	22

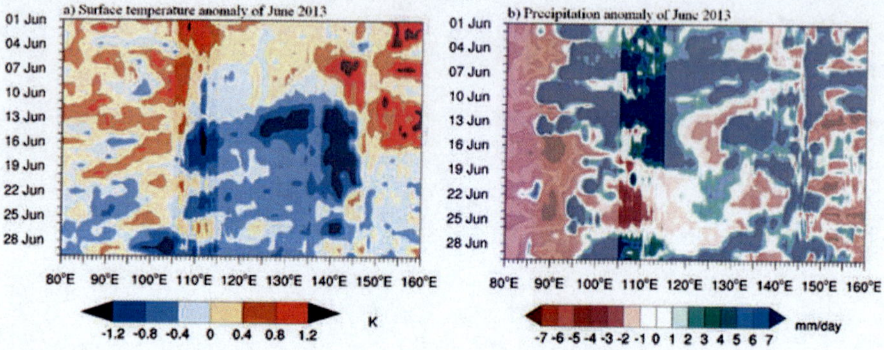

Fig. 16.1 Hovmoller diagram of **a** the anomalously cold June of 2011 and **b** the anomalously wet June of 2013

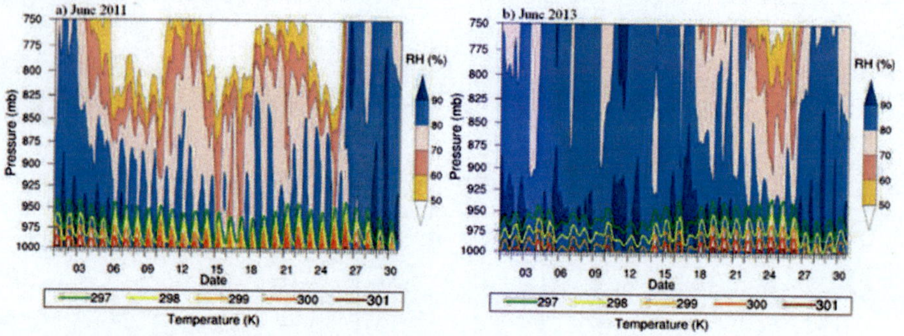

Fig. 16.2 Time-height distribution of air temperature (lines) overlaid on relative humidity (contours) from 1000 to 750 mb pressure levels for **a** June 2011 and **b** June 2013

plots more visible). When the relative humidity is mostly below 80%, the weather is usually warm.

The moisture transport pathways (or channels) for June 2013 are shown in Fig. 16.3. These pathways are obtained from the average of all back trajectories departing from the predefined grid points over the island of Java (Fig. 16.3a). The relative contribution of each trajectory is shown in numbers and indicated by the thickness of the lines. The major contributor for the water vapor during June 2013 is from the southeastern part of the Indian Ocean south of Java, accounting for around 25.3% of all trajectories. Air parcels from the south of Java enter the target area at the southwest boundaries. The average specific humidity of air parcels increases from 10 to 14 g/kg, indicating a significant portion of the moisture is absorbed along the pathways. The other two trajectories are coming from the Indian Ocean. The first is from the eastern part of the Indian Ocean, and the second is from the western part, accounting for around 11.5% and 18%, respectively. These two trajectories enter the target area at the west boundaries. The trajectories from the Indian Ocean as a moisture source for the target area only exist when the MJO is active. The other

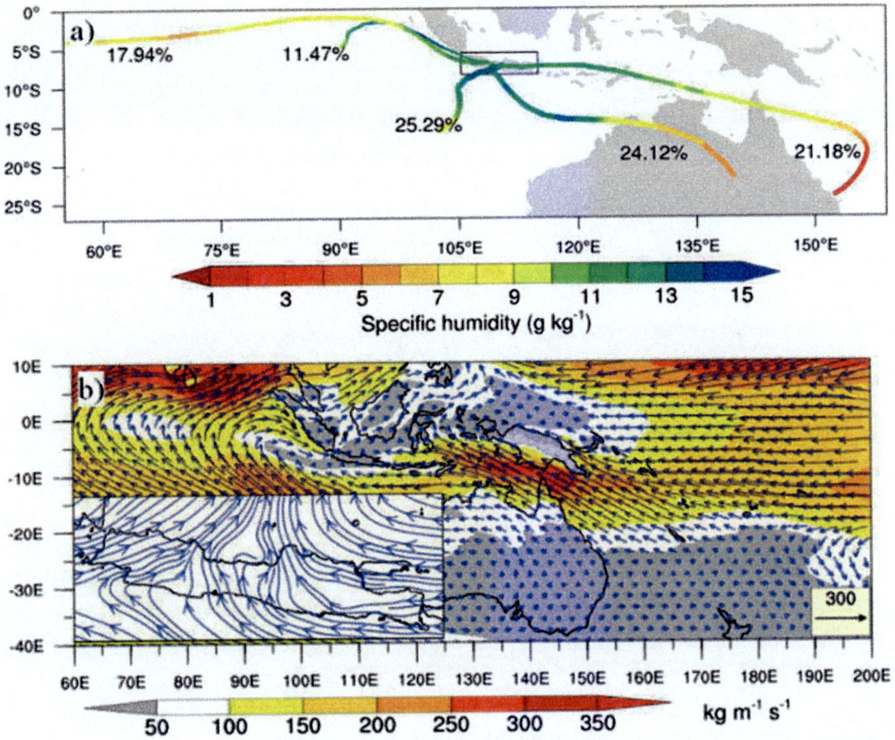

Fig. 16.3 Characteristics of moisture transport and moisture source for the wet days of June 2013 in Java. **a** Moisture transport channels based on average trajectories during the wet days of June 2013. Colors on the pathways indicate the average specific humidity of air parcels along the trajectories (g kg^{-1}). The thickness of the pathways represents the percentage of trajectories, also indicated in numbers. **b** The average of vertically integrated atmospheric water vapor transport during June 2013 (vector; kg m^{-1} s^{-1}) and shading represent the amount of water vapor transport

moisture source that triggers rainfall in the target area comes from the east. The air parcel trajectories in June 2016 are similar to the June 2013 case, because MJO was also active in June 2016. The moisture source for the anomalously wet Junes of 2010 and 2021, when the MJO is in the suppressed phase, the moisture source is only from the east.

To elucidate the pathways of the air parcels that enter the target area, Fig. 16.3b shows the average of the vertically integrated water vapor transport of the lower troposphere (from surface to 850 mb) during June 2013, with the higher resolution of the moisture transport shown in the inset. It shows that the air parcels enter the Java Island from the west/southwest and the east/southeast boundaries. The shading in Fig. 16.3b represents the amount of water vapor transport. The area with the higher moisture transport is more likely to be the moisture source than the lower one.

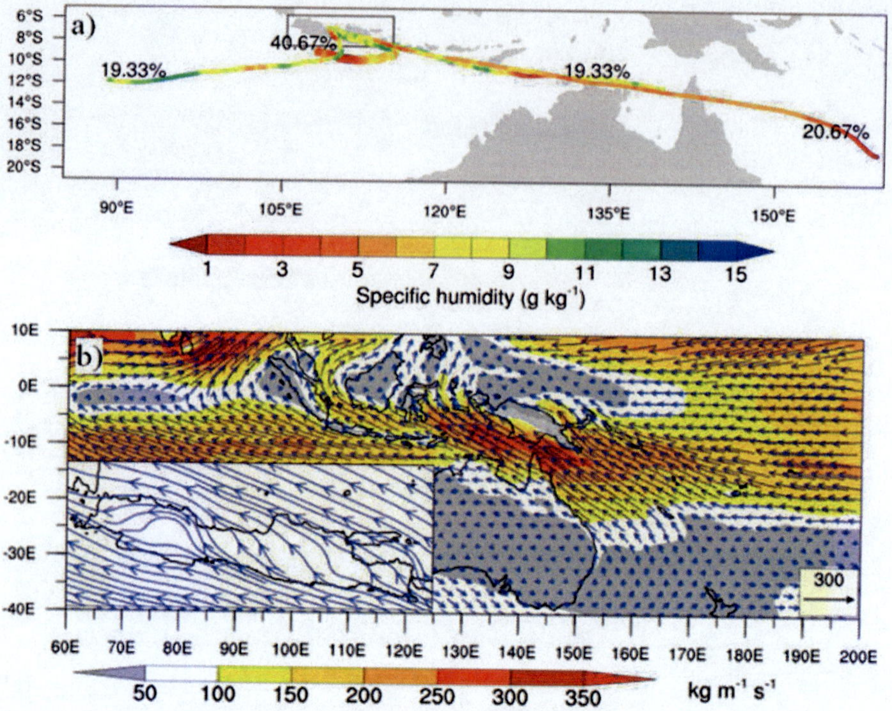

Fig. 16.4 Similar to Fig. 16.3, except for the dry days of June 2019

For comparison with the anomalously wet Junes, Fig. 16.4 shows the trajectories of the moisture transport pathways (Fig. 16.4a) and the average of the vertically integrated water vapor transport of the lower troposphere for June 2019. The trajectories of the air parcels enter the target area from the southeast boundaries (Fig. 16.4a). It shows that the specific humidity is very low along the trajectories. The average of the vertically integrated water vapor transport of the lower troposphere during June 2013 (Fig. 16.4b) shows that the air parcels enter the Java Island from the southeast boundaries. The southeasterly trajectories are the main features during the anomalously dry Junes. These are different from the easterly trajectories, which are some of the main features of the anomalously wet Junes.

Figure 16.5 shows the 3-dimensional trajectories of air parcels arriving at the near-surface of the target area. During the dry and cold days at 16 and 18 of June 2011 (shown in Fig. 16.2a), the trajectories in Fig. 16.5a show that the air parcels come from high altitudes in the high latitude region to arrive at the near-surface just around the target time. Because the air is cold, the relative humidity is relatively low, and no precipitation occurs. At the end of June 2011 (between the 28th and the 30th), the surface temperature was relatively cold, and there were deep convective clouds (indicated by the deep layers of high relative humidity shown in Fig. 16.2a). The air parcels come from high latitudes in the tropical region (Fig. 16.5a). The cold air from

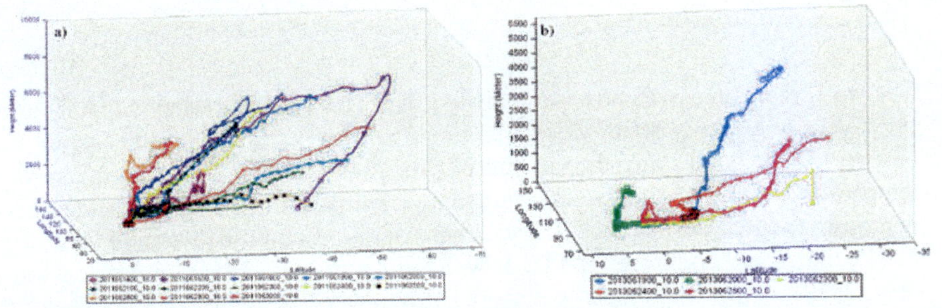

Fig. 16.5 The 3-dimensional trajectories of air parcels arriving at the near-surface of the target areas, **a** during the cold days of June 2011 and **b** during the warm days of June 2013

high altitude goes near the surface, and the deep convective clouds are formed above it. This phenomenon is similar to a cold front in the subtropical regions. The dry and warm weather (20–25 June 2011) can be explained as follows. The air parcels come from the high latitude regions and goes down near the surface of the tropical region several days before reaching the target area. These air parcels get warm as they pass through the tropical region, but do not bring enough moisture.

The weather on 19 and 20 June 2013 is slightly wet (see Figs. 16.1b and 16.2b). The rainfall that occurred on those dates might come from the shallow convective clouds (Fig. 16.2b). Figure 16.5b shows that the air parcels travel near the surface from the tropical region or the subtropical region, but they arrive near the surface of the tropical region long before they reach the target area. It is confirmed by Fig. 16.6 that shows the main trajectories of the warm air parcels. The shallow convective clouds bring less rainfall than the deep convective clouds (see Fig. 16.1b).

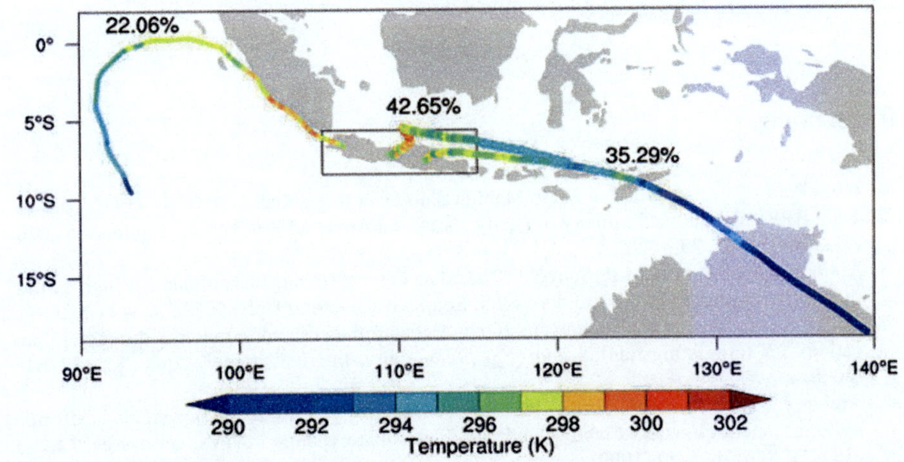

Fig. 16.6 The trajectories of the air parcels arriving near the surface of the target area for June 2013

16.4 Conclusion

The weather in June over Java is not always dry. This paper has shown that it varies from year to year. In some years, June can be very wet, and in other years can be very dry. It is not only the amount of precipitation that varies, but the surface temperature over Java also shows significant variation. From the analysis, it has been demonstrated that during June, the weather over Java could be wet and cool (e.g., 2021), or dry and cool (e.g., 2011), or wet and warm (e.g., 2013), or dry and warm (e.g., 2019).

The rainfall variability over Java during June is influenced by the common climate indices, such as ENSO, IOD, and MJO. The moisture source for the anomalously wet June usually comes from the east, although there is a negative IOD event, such as in 2021. The moisture from the Indian Ocean only contributes when the MJO is in active phases, such as in 2013 and 2016. The anomalously dry Junes occur during the positive IOD and the positive ENSO (El Nino) events, such as in 2015 and 2019. During the anomalously dry Junes, the trajectories of the air parcels arriving at Java are mainly southeasterly. While during from the anomalously wet Junes, the air parcel trajectories are easterly, and there are additional westerly/southwesterly trajectories during MJO active phase.

The surface temperature over Java in June can be anomalously cold when the air parcels that arrive near the surface of the target time come from the high altitudes in the tropical region or from the high latitude region. However, when the air parcels arrive near the surface of the tropical region long before they reach the target area, the June surface temperature will be warm. This is because the air parcels have travelled through the warm environment of the tropical region.

If the climate indices of ENSO, IOD, and MJO could be predicted accurately in advance, they would be very useful for the agricultural sector, or any other sectors that need climate information.

References

1. BPS. Potret Sensus Penduduk 2020: Menuju Satu Data Kependudukan. BPS- RI (2020)
2. BPS, BPPT. Executive Summary of Paddy Harvested Area and Production in Indonesia 2020. Jakarta: BPS-RI, 28 p (2020)
3. Yulihastin, E., Putranto, M.F., Suaydhi.: The effect of local forcing on anomalously high rainfall during the dry season in Java, Indonesia. J. Southwest Jiaotong Univ. **56**(3), 32–42 (2021)
4. D'Arrigo, R., Wilson, R.: El Nino and Indian Ocean influences on Indonesian drought: implications for forecasting rainfall and crop productivity. Int. J. Climatol. **2008**(28), 611–616 (2008)
5. Amien, I., Redjekiningrum, P., Kartiwa, B., Estiningtyas, W.: Simulated rice yields as affected by inter annual climate variability and possible climate change in Java. Clim. Res. **12**(2–3 SPEC. ISS. 6), 145–52 (1999)
6. Hackert, E.C., Hastenrath, S.: Mechanisms of Java rainfall anomalies. Mon. Weather Rev. **114**, 745–757 (1986)

7. Hastenrath, S.: Predictability of Java monsoon rainfall anomalies. Clim. Appl. Meteorol. **26**, 133–141 (1987)
8. Saji, N.H., Goswami, B.N., Vinayachandran, P.N., Yamagata, T.: A dipole mode in the tropical Indian Ocean. Nature **401**, 360–363 (1999)
9. D'Arrigo, R., Wilson, R., Palmer, J., Krusic, P., Curtis, A., Sakulich, J., et al.: Monsoon drought over Java, Indonesia, during the past two centuries. Geophys. Res. Lett. **33**(4) (2006)
10. D'Arrigo, R., Allan, R., Wilson, R., Palmer, J., Sakulich, J., Smerdon, J.E., et al.: Pacific and Indian Ocean climate signals in a tree-ring record of Java monsoon drought. Int. J. Climatol. **2008**(28), 1889–1901 (2008)
11. Hatmaja, R.B., Rusmanansari, A.H., Radjawane, I.M.: The dynamics of negative Indian Ocean dipole (nIOD) and its relation to the anomalous high rainfall in West Java Province, Indonesia. IOP Conf. Se.r Earth Environ. Sci. **303**(1) (2019)
12. Sahu, N., Behera, S.K., Yamashiki, Y., Takara, K., Yamagata, T.: IOD and ENSO impacts on the extreme stream-flows of Citarum river in Indonesia. Clim. Dyn. **39**(7–8), 1673–1680 (2012)
13. González, S., Garreaud, R.: Spatial variability of near-surface temperature over the coastal mountains in southern Chile (38° S). Meteorol. Atmos. Phys. **131**(1), 89–104 (2019)
14. Hersbach, H., Bell, B., Berrisford, P., Hirahara, S., Horányi, A., Muñoz-Sabater, J., et al.: The ERA5 global reanalysis. Q J R Meteorol. Soc. **146**(730), 1999–2049 (2020)
15. Wang, Y.Q.: An open source software suite for multi-dimensional meteorological data computation and visualisation. J Open Res. Softw. **7**(1), 1–9 (2019)
16. Draxler, R.R., Hess, G.D.: An overview of the HYSPLIT_4 modelling system for trajectories, dispersion and deposition. Aust. Meteorol. Mag. **47**(4), 295–308 (1998)
17. Numaguti, A.: Origin and recycling processes of precipitating water over the Eurasian continent: experiments using an atmospheric general circulation model. J. Geophys. Res. Atmos. **104**(D2), 1957–1972 (1999)
18. Bellenger, H., Yoneyama, K., Katsumata, M., Nishizawa, T., Yasunaga, K., Shirooka, R.: Observation of moisture tendencies related to shallow convection. J. Atmos. Sci. **72**(2), 641–659 (2015)
19. Masunaga, H.: A satellite study of tropical moist convection and environmental variability : a moisture and thermal budget analysis. J. Atmos. Sci. **70**, 2443–2466 (2013)

Chapter 17
Outstanding Issues in the Future Rainfall Projection Over Indonesia

Yudha Setiawan Djamil, Tri Wahyu Hadi, Wirid Birastri, and Wang Xianfeng

Abstract Towards the end of the twenty-first century, Earth climate has been projected by various general circulation models (GCMs) organized by the Climate Model Intercomparison Project phase-5 (CMIP5). However, multi-model projections of the future rainfall over Indonesia have a low level of confidence as shown by the broad range of its uncertainties. Several efforts and initiatives to improve the confidence level of the future climate projections over Indonesia, such as model selection, multi-model ensemble, and dynamical downscaling, were engaged recently. However, despite all the significant improvements, several issues remain, such as suppression of the smaller scale rainfall trend (e.g., Java) by the ensemble projections of the well-performed models. Another problem is the opposite sign of rainfall trend between the GCM simulation and its downscaled product. Engaging on a high-resolution climate model well-verified by most of the multi-scale features of the climate system made available to date (e.g., diurnal, paleoclimate, etc.) is one of the suggestions to further improve the spatial representation of the region and minimize the multi-model uncertainties, which then eventually improve the confidence level of the future rainfall projection over the region.

Y. S. Djamil (✉)
Research Center for Climate and Atmosphere, National Research and Innovation Agency (BRIN), Bandung, Indonesia
e-mail: yudh006@brin.go.id

T. W. Hadi
Atmospheric Science Research Group, Bandung Institute of Technology (ITB), Jl. Ganesa 10, Bandung 40132, Indonesia

W. Birastri
School of Atmospheric and Planetary Sciences, Sumatera Institute of Technology (ITERA), Jl. Terusan Ryacudu, Way Hui, Kecamatan Jati Agung, Lampung Selatan 35365, Indonesia

W. Xianfeng
Earth Observatory of Singapore, Nanyang Technological University, Singapore 639798, Singapore

Asian School of the Environment, Nanyang Technological University, Singapore 639798, Singapore

E. Yulihastin et al. (eds.), *Proceedings of the International Conference on Radioscience, Equatorial Atmospheric Science and Environment and Humanosphere Science, 2021*, Springer Proceedings in Physics 275, https://doi.org/10.1007/978-981-19-0308-3_17

17.1 Introduction

Global warming has been declared as a man-made catastrophe by the Intergovernmental Panel on Climate Change (IPCC). IPCC reported that global warming is mainly driven by the rising concentration of the anthropogenic greenhouse gasses. The warming then increases the occurrences of extreme weather events over many regions across the globe. Indonesia, as the fourth largest population in the world, has also been declared as one of the most vulnerable countries to experience the disastrous impact of global warming, especially due to the increase of extreme rainfall events [18]. However, future rainfall over Indonesia, as projected by various General Circulation Models (GCMs) organized by the Climate Model Intercomparison Project phase-5 (CMIP5), remains unclear. A previous study showed that there is a low agreement in the future rainfall projections by various GCMs over most regions in Indonesia, which is considered as part of the multi-model uncertainties [9]. On the other hand, it is well-known that most of the GCMs under-represent the terrestrial complexity of Indonesia due to their coarse grid resolution. These are the two main issues, among others, that prevent further analyses and application of the climate projections over Indonesia. Nevertheless, future climate projection with an acceptable level of confidence is urgently needed by the nation to develop a long-term mitigation strategy to face a potential climate crisis in the near future.

Initiatives and efforts to minimize multi-model uncertainties and to better represent the terrestrial complexity of Indonesia in the climate model have been taken progressively within the last decade. Solutions for both issues were approached mainly by conducting model selection based on their performances and downscaling them over the region of interest. The performance of a GCM in a certain area was evaluated by verifying the regional signatures of several large-scale events such as monsoon and El Nino-Southern Oscillation (ENSO) from their modern-day climate simulations [3, 13, 16]. This evaluation method is chosen mainly to compromise the coarse grid system used in most GCMs. Subsequently, efforts to downscale the output of the GCMs from the CMIP5 over Indonesia by using various Regional Climate Models (RCMs) has been done by the Centre for Climate Research Singapore (CCRS) and also by a regional initiative known as the Coordinated Regional Climate Downscaling Experiment (CORDEX) for Southeast Asia [12, 14].

Further improvement of model performances was also expected by calculating the ensemble-mean of the GCMs output [3]. The ensemble-mean is also useful to compile and display multi-model performances and behaviors such as those reported by the IPCC. Nowadays, calculating the ensemble-mean has become a common practice in multi-model analyses. However, ensemble-mean may suppress small-scale features resolved by the higher resolution model. Thus, the suppression is counterproductive for Indonesia since significant weather and climate events mainly occur in smaller scales following the terrestrial complexity of the region [15]. Therefore, the ensemble-mean potentially worsens the representation of climate over Indonesia in the GCM since it involves coarse grid models. A clearer description of this issue is shown in the next section of this paper by illustrating the process of calculating the ensemble-mean of the rainfall change over western Indonesia between future projections and present-day simulations of three GCMs from the CMIP5.

Outstanding issues are also identified in the downscaled product of the future rainfall change projected by the GCMs from the CMIP5. Future rainfall over regions in Indonesia that are better represented by the GCM disagreed with its downscaled products [8, 14]. Identification of these issues in model downscaling are discussed in the later section. Discussions on some possible solutions to the highlighted issues are presented as the closing remark of this review paper.

17.2 Outstanding Issues in a Multi-Model Ensemble

Multi-model analyses compile output from various GCMs regardless of their spatial resolutions since all models were eventually standardized into a uniform grid size comparable to the designated observational dataset [3]. Thus, the standardized grid size is mainly decided in compromise with the meteorological/climatological events used as the model evaluation criteria [13]. Eventually, the coarse grid resolution is commonly chosen as the standard grid size on par with the large-scale climate events used as the model evaluation criteria [8]. Therefore, contributions from high and low- resolution models are expected in the calculation of a multi-model ensemble.

Coarse grid models, by definition, under-represent small-scales features (e.g., intricate coastlines, topography, etc.). Thus, domination of coarse grid models in the calculation of the ensemble-mean of several equally weighted models is expected to suppress small-scales climate features resolved by the higher resolution models regardless of their pre-evaluated performances. This possible drawback is illustrated in this paper by calculating the ensemble-mean of the future rainfall change simulated by three well-performed GCMs from the CMIP5 as declared in [10]. The three models consist of two coarser (ACCESS1.0 and MPI-ESM-MR) and one finer (CCSM4) grid size model relative to each other (Table 17.1). The three models are also chosen by Kang et al. [8] to be downscaled over the western Maritime Continent as part of the CORDEX project. This illustration also highlights a possible misinterpretation of the rainfall change feature over the region following the suppression of the small-scales features.

In this illustration, small-scales features such as land-sea contrast on the change of the annual rainfall over the Java region were projected by the higher resolution model (CCSM4) (Fig. 17.1a). While the coarser grid models (ACCESS1.0 and MPI-ESM-

Table 17.1 Horizontal resolution of the atmospheric component of the three selected global climate models used in Kang et al. [8]. Information in the table is adapted from Kang et al. [8]

Model name (references)	Atmospheric resolution
CCSM4 [6]	0.9° × 1.25°
MPI-ESM-MR [7]	T63 (~1.875°)
ACCESS 1.0 [1]	1.25° × 1.875°

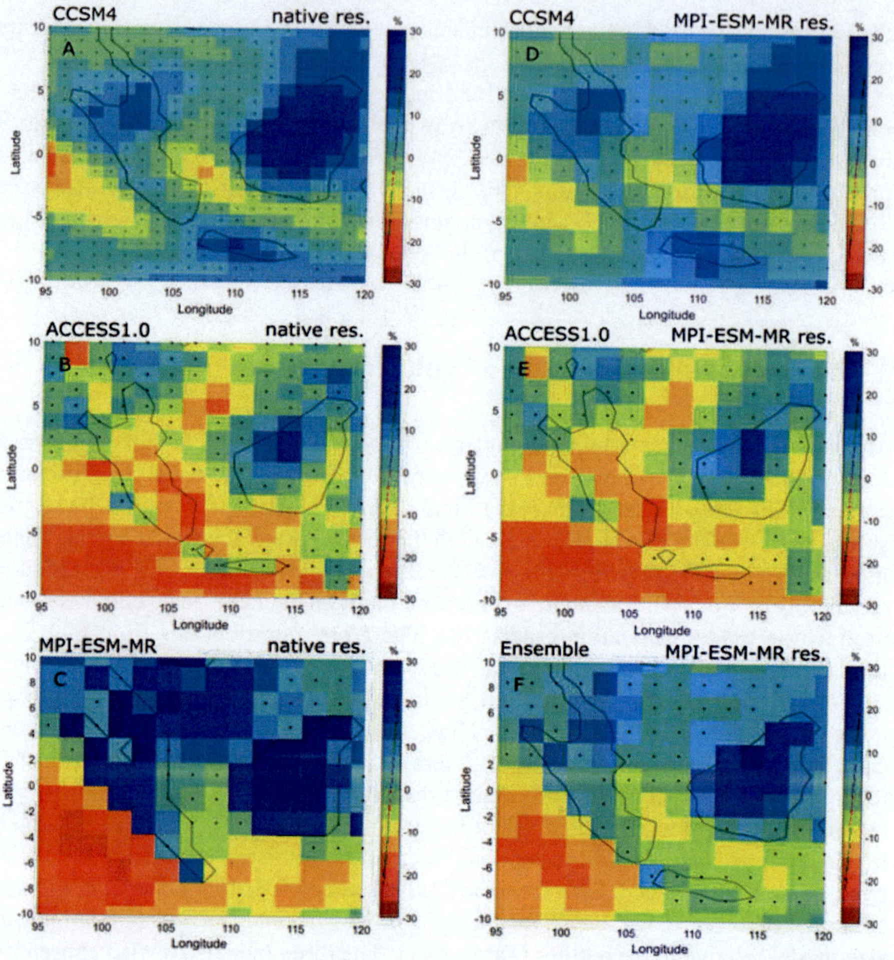

Fig. 17.1 Annual rainfall of the RCP8.5 (2050–2100) minus historical (1850–1900) simulations for the three GCMs (**a**, **b** and **c**) (CCSM4, ACCESS 1.0 and MPI-ESM-MR) from the CMIP5. **d** and **e** are the same with **a** and **b** but re-gridded based on the grid system of MPI-ESM-MR. **f** is the ensemble-mean of the three GCMs. Dotted areas are not significant at a 95% confidence level based on Welch's t-test. Coastlines are the atmospheric land fraction used by each model (in **f**, plotted coastlines are the land fraction used by the CCSM4) at 50% value

MR) projected drier conditions over the Java region homogeneously (Fig. 17.1b and c). Drier conditions over southern Indonesia, as shown by all three models, are likely to occur as changes in the oceanic rainfall since Java Island is under-represented in the coarser grid models (Fig. 17.1a, b, c, d and e). The ensemble-mean of the three models shows a wide-spread drier condition over the southern Indonesia following the domination of the coarser grid models (Fig. 17.1d, e, and f). A possible misinterpretation may arise when one perceived an oceanic rainfall as terrestrial events in the ensemble-mean since the plot is overlaid or the sea is concealed by a higher resolution coastlines which has a well-represented Java Island as purposely illustrated in Fig. 17.1f.

The visual misinterpretation is preventable by ensuring consistency between the grid size of the data and the coastlines plotted in the figure (e.g., Fig. 17.1f should plot coastlines as in Fig. 17.1c). In spite of that, a comprehensive solution can be achieved by limiting the contribution of the coarser grid models in the multi-model analyses, which will also prevent the misinterpretation from perplexing its downscaled product.

17.3 Outstanding Issues in the Downscaling of Rainfall Projections

Downscaling technique is a feasible approach to improve the representation of weather and climate over regions with a terrestrial complexity (e.g., Indonesia) undermined by most GCMs [10]. Efforts to develop downscaling techniques for a tropical atmosphere were mainly focused on the improvement of model performance in simulating high-resolution rainfall [5]. Reduction in rainfall biases by the ensemble-mean of the downscaled product in comparison with its GCMs forcing, as conducted by the CORDEX project, was claimed as an improvement of the model performance [8, 14]. However, despite the bias correction conducted by the CORDEX project, there is an opposite trend shown by the future rainfall projection over certain areas in Southeast Asia between the downscaled product and its GCM forcing. This issue is displayed clearly by Figs. 7, 8 and 9 in [8] and Figs. 10 and 11 in [17].

The disagreement between the downscaled product and its GCM forcing appears in several locations, such as Borneo, Sumatra, and Indochina/Malay Peninsula (see Figs. 7, 8 and 9 in [8] and Figs. 10 and 11 in [17]). For example, significantly higher rainfall over Borneo and Indochina/Malay Peninsula projected by the GCMs during boreal summer were downscaled oppositely, especially for the RCP8.5 simulations towards the end of the twenty-first century (see Fig. 8 in [8]). The disagreement in boreal summer is shown as a wider area of a suppressed rising air over the Maritime Continent. The expanded area of a suppressed rising air is suggested due to the enhancement of local circulations under a more realistic coastlines and sharper topography used in the RCM [8]. In spite of that, it remains unclear if the above explanation is also applied for the disagreements occur in another seasons and locations.

Another possible explanation for the disagreement is related to the downscaled product itself which may diverge far from its large-scale forcing [2]. For example, some RCMs show tendency to overly moistens the atmosphere as a result of excessive evaporation [5]. CCRS used the divergence of the downscaled product from their GCM forcing as an evaluation tool to further select models [12]. The downscaled product of the future rainfall projection will have a higher confidence level if its long-term trend is significantly larger than its divergence from its GCM forcing. Thus, a comparison between the trend and the divergence of the downscaled product can also be used as a model evaluation criteria.

17.4 Discussions

Outstanding issues in the future rainfall projection over Indonesia highlighted in this paper are potentially solvable with our current understanding of the climate and 'state-of-the-art' climate model. Climate simulation with a high spatial resolution is the main approach to improve the representation of Indonesia in the model. The desired resolution to properly represent the region is already achieved by the downscaled products of the GCMs from CMIP5. However, discrepancies between the downscaled product and its GCM forcing for the future climate projection remain to be resolved.

Various method to improve consistency between GCM and its downscaled product such as 'nudging' can always be developed and implemented [5, 10]. Alternatively, climate simulation with a high spatial resolution over a particular area can also be achieved by using a GCM with a variable-resolution grid system. This intricate grid system refined its spatial resolution towards the area of interest (e.g., from ~160 km to ~1 km) [11]. This technique provides several advantages; firstly, it will be less expensive than simulating a GCM in a similarly higher resolution setting. Secondly, it has the potential to replace the downscaling technique once their technical efficiency is comparable. Therefore, systematic studies to explore and improve the performance of variable-resolution GCMs over Indonesia should be engaged.

Level of confidence of the future climate projection can be improve by expanding the spectrum of weather and climate events used as the model evaluation criteria. Higher frequency modes such as diurnal and sub-seasonal events can be incorporated to evaluate model performance following the use of finer grid models (e.g., downscaled products, variable-resolution GCMs) [5]. In addition to the present-day climate verification, models should also able to simulate climate change under various scenarios such as those revealed by the paleoclimate studies. For example, changes of rainfall and ocean salinity in the Indian Ocean basin (including Indonesia) simulated by several GCMs under the last glacial maximum (~20.000 years ago) scenarios have been verified with the proxy datasets [4]. Eventually, future climate projected by high-resolution models (e.g., downscaled products, variable-resolution GCMs) well- performed in simulating multi-scale weather and climate events over Indonesia, including paleoclimate scenarios, should be prioritized.

In the last decade, CORDEX project and CCRS spearheaded regional efforts to improve the confidence level of the future climate projection over Southeast Asia, including Indonesia. Significant progress to better represent regional complexity in the future rainfall projection has been achieved. However, outstanding issues remain and must be solved by the next round of research. High- resolution models specifically fine-tuned to simulate weather and climate over Indonesia should be developed. The climate modeling community has yet to take greater challenges under the ever-increasing risk of an unavoidable climate crisis gradually experienced by the nation.

Acknowledgements We thank Wong Minn Lin for managing the CMIP5 model output dataset in the computational facility of the Earth Observatory of Singapore.

Authors contributions Yudha Setiawan Djamil acquired and analyzed GCMs simulations output from the CMIP5 online database. Yudha Setiawan Djamil wrote the manuscript, thus acts as the main contributor. Author member: Tri Wahyu Hadi contributes to the discussion of the main idea, suggested solutions, and also corrections and comments on the manuscript. Wirid Birastri contributes to the discussion of the issues in multi-model analyses, and also corrections and comments on the manuscript. Wang Xianfeng contributes to the discussion of the main idea, corrections and comments on the manuscript, and the approval to access the computational facility in the Earth Observatory of Singapore.

References

1. Bi, D., Dix, M., Marsland, S.J., O'Farrell, S., Rashid, H., Uotila, P., Hirst, A.C., Kowalczyk, E., Golebiewski, M., Sullivan, A., Yan, H.: The ACCESS coupled model: description, control climate and evaluation. Aust. Meteorol. Oceanogr. J. **63**(1), 41–64 (2013)
2. Bowden, J.H., Otte, T.L., Nolte, C.G., Otte, M.J.: Examining interior grid nudging techniques using two-way nesting in the WRF model for regional climate modeling. J. Clim. **25**(8), 2805–2823 (2012)
3. Chhin, R., Yoden, S.: Ranking CMIP5 GCMs for model ensemble selection on regional scale: case study of the Indochina Region. J. Geophys. Res. Atmos **123**(17), 8949–8974 (2018)
4. DiNezio, P.N., Tierney, J.E.: The effect of sea level on glacial Indo-Pacific climate. Nat. Geosci. **6**(6), 485–491 (2013)
5. Fonseca, R., Koh, T.Y., Teo, C.K.: Multi-scale interactions in a high-resolution tropical-belt experiment and observations. Clim. Dyn. **52**(5), 3503–3532 (2019)
6. Gent, P.R., Danabasoglu, G., Donner, L.J., Holland, M.M., Hunke, E.C., Jayne, S.R., Lawrence, D.M., Neale, R.B., Rasch, P.J., Vertenstein, M., Worley, P.H.: The community climate system model version 4. J. Clim. **24**(19), 4973–4991 (2011)
7. Giorgetta, M.A., Jungclaus, J., Reick, C.H., Legutke, S., Bader, J., Böttinger, M., Brovkin, V., Crueger, T., Esch, M., Fieg, K., Glushak, K.: Climate and carbon cycle changes from 1850 to 2100 in MPI-ESM simulations for the coupled model intercomparison project phase 5. J. Adv. Model. Earth Syst. **5**(3), 572–597 (2013)
8. Kang, S., Im, E.S., Eltahir, E.A.: Future climate change enhances rainfall seasonality in a regional model of western Maritime Continent. Clim. Dyn. **52**(1), 747–764 (2018)
9. Knutti, R., Sedláček, J.: Robustness and uncertainties in the new CMIP5 climate model projections. Nat. Clim. Chang. **3**(4), 369–373 (2013)
10. Lo, J.C.F., Yang, Z.L., Pielke Sr, R.A.: Assessment of three dynamical climate downscaling methods using the Weather Research and Forecasting (WRF) model. J. Geophys. Res. Atmos **113**(D9) (2008)
11. Lui, Y.S., Tam, C.Y., Tse, L.K.S., Ng, K.K., Leung, W.N., Cheung, C.C.: Evaluation of a customized variable-resolution global model and its application for high-resolution weather forecasts in East Asia. Earth Space Sci **7**(7) (2020)
12. Marzin, C., Rahmat, R., Bernie, D., Bricheno, L., Buonomo, E., Calvert, D., Cannaby, H., Chan, S., Chattopadhyay, M., Cheong, W.K., Hassim, M.E.: Singapore's second national climate change study - phase 1. Centre for Climate Research Singapore (2015)
13. McSweeney, C.F., Jones, R.G., Lee, R.W., Rowell, D.P.: Selecting CMIP5 GCMs for downscaling over multiple regions. Clim. Dyn. **44**(11), 3237–3260 (2015)
14. Ngo-Duc, T., Tangang, F.T., Santisirisomboon, J., Cruz, F., Trinh-Tuan, L., Nguyen-Xuan, T., Phan-Van, T., Juneng, L., Narisma, G., Singhruck, P., Gunawan, D.: Performance evaluation of

RegCM4 in simulating extreme rainfall and temperature indices over the CORDEX-Southeast Asia region. Int. J. Climatol. **37**(3), 1634–1647 (2017)
15. Qian, J.H.: Why precipitation is mostly concentrated over islands in the Maritime Continent. J. Atmos. Sci. **65**(4), 1428–1441 (2008)
16. Siew, J.H., Tangang, F.T., Juneng, L.: Evaluation of CMIP5 coupled atmosphere– ocean general circulation models and projection of the Southeast Asian winter monsoon in the 21st century. Int. J. Climatol. **34**(9), 2872–2884 (2014)
17. Tangang, F., Chung, J.X., Juneng, L., Salimun, E., Ngai, S.T., Jamaluddin, A.F., Mohd, M.S.F., Cruz, F., Narisma, G., Santisirisomboon, J., Ngo-Duc, T.: Projected future changes in rainfall in Southeast Asia based on CORDEX–SEA multi-model simulations. Clim. Dyn. **55**, 1247–1267 (2020)
18. Yusuf, A.A., Francisco, H.: Climate change vulnerability mapping for Southeast Asia. Economy and Environment Program for Southeast Asia (EEPSEA) (2009)

Chapter 18
Atmospheric Residual Layer Height Variations on Clear and Rainy Days Based on High-Resolution Radiosonde and Global Positioning System Radio Occultation Data

Resa Pratikasari, Nurjanna Joko Trilaksono, and Noersomadi

Abstract Local climate conditions, rainfall patterns, topography, and atmospheric circulation are factors that influence the structure and variation of the residual layer. The residual layer is an atmospheric layer that appears in the nighttime, this layer is remaining from the mixed layer, which is formed in the daytime. In this study, we analyzed the variation of the residual layer on clear and rainy days using residual layer height as a parameter. The residual layer heights were estimated by extracting and calculating refractivity (N) from high vertical resolution radiosonde and Global Positioning System Radio Occultation (GPS-RO) data using vertical gradient methods. The results show that the residual layer is more varied during rainy days. In particular, the residual layer would be (not) formed if the rain occurs in the afternoon until before (after) sunset, which in turn is related to the presence (absence) of capping inversion. In the clear day case, at the beginning of the night, the residual layer height would be as height as the mixed layer in the afternoon. Our further analysis found that the mixed layer height in the following day was not affected by the residual layer in the previous night. The variation of the mixed layer height on the following day is more determined by the weather condition at that time.

R. Pratikasari (✉)
Agency for Meteorology Climatology and Geophysics, Jakarta, Indonesia
e-mail: resa.pratikasari@bmkg.go.id

N. J. Trilaksono
Atmospheric Science Research Group, Faculty of Earth Sciences and Technology, Institut Teknologi Bandung, Bandung, Indonesia

Noersomadi
Research Center for Climate and Atmosphere, National Research and Innovation Agency (BRIN), Bandung, Indonesia

E. Yulihastin et al. (eds.), *Proceedings of the International Conference on Radioscience, Equatorial Atmospheric Science and Environment and Humanosphere Science, 2021*, Springer Proceedings in Physics 275, https://doi.org/10.1007/978-981-19-0308-3_18

18.1 Introduction

At the bottom of the atmosphere, there was a layer that was affected by the surface directly with a thickness of about 500–5000 m, this layer was called Atmospheric Boundary Layer (ABL). ABL consist of three main components, namely Mixed Layer (ML), Stable Boundary Layer (SBL), and Residual Layer (RL). The mixed layer was an unstable layer that occurs during the daytime, while the stable boundary layer was a stable layer that occurs at nighttime, and on the other hand, the residual layer was a neutral layer that has the same composition/tracer as the mixed layer, which appears at the nighttime (above the stable boundary layer). According to by definition, the RL was not part of ABL because this layer was not affected directly by the surface. However, in the analysis of ABL, mixed layer, stable boundary layer, and residual layer were inseparable components [1].

Since RL was an inseparable component of ABL, the factors that influence the RL dynamics were the same as those that affect ABL, such as local climate conditions, rainfall patterns, diurnal weather, and atmospheric circulations [2–8]. In addition to those factors, the RL and ABL dynamics in Indonesia which is located in tropics were also influenced by location (either near or far from waters), Inter-Tropical Convergence Zone (ITCZ), and climate variability such as ENSO, MJO, etc. [9–16].

Most studies on ABL focus on the ML dynamics, meanwhile the RL dynamics were still less discussed. However, because of the RL has the same composition/tracer as ML and provides favorable thermal conditions for the growth and development of ML during the day, RL can also be used to describe ABL characteristics [1, 17–19], which mean that the RL variation would significantly affect the ML variation. In this study, we would like to examine the characteristics and the dynamics of RL in clear and rainy days conditions using RL height as a parameter. The case study would take place in Pontianak, Indonesia.

18.2 Data and Methods

18.2.1 Data

The data that was used to analyze the RL height variations were high-resolution radiosonde and GPS-RO data. Radiosonde and GPS-RO data contained RL tracer including, temperatures, relative humidity, air pressure, and refractivity. The high-resolution radiosonde data used was derived from Supadio Meteorological Station, Pontianak, which was issued twice a day at 7:00 and 19:00 LT. The data has a vertical resolution of ~5 m but has a fairly low temporal resolution. Reasonably, in this study, we used GPS-RO data as a complement which has a better temporal resolution. The GPS-RO data yield a vertical observation atmospheric profile whose penetration could reach the surface, so it's suitable for residual layer analysis. The GPS-RO data used was derived from the Constellation Observing System for Meteorology Ionosphere and Climate 2 (COSMIC-2) satellite system, which has a vertical resolution of ~50 m [20–22]. The GPS-RO data was expected to complement in estimating the RL height at the time when radiosonde observations were not available because these data passing through the Pontianak area is 0–5 profile per day.

In addition to radiosonde and GPS-RO data, this study also used Meteorology Aviation Report (METAR) data to identify hourly weather conditions. METAR is aviation meteorological observation information that is issued every 30 min (for takeoff and landing purposes). METAR has contained weather condition information so that we could identify the weather changes accurately at any time.

18.2.2 Estimated Residual and Mixed Layer Height

In this study, the residual layer height was denoted as Z_{RL}, meanwhile, the mixed layer height was denoted as Z_{ML}. The method used to estimate Z_{RL}/Z_{ML} was minimum/maximum gradient vertical methods using virtual potential temperature (θ_v), mixing ratio (r), and refractivity (N) variables [1, 23, 24]. Z_{RL}/Z_{ML} was estimated at 800 m < altitude <5000 m, if Z_{RL}/Z_{ML} was found to be less than 800 m or more than 5000 m, the data were considered empty data [7, 14, 25].

Minimum/maximum gradient vertical method was a method to estimate Z_{RL}/Z_{ML} based on minimum/maximum changes of variables (u) to the changes of the height (du/dz) (Fig. 18.1), which is obtained using the three-point central difference formula method with the following equation:

$$\left(\frac{\partial u}{\partial z}\right)_i \approx \frac{u_{i+1} - u_{i-1}}{2\Delta z} \tag{18.1}$$

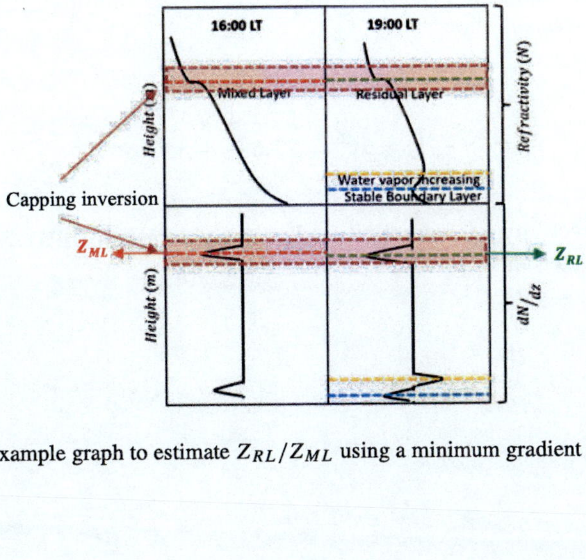

Fig. 18.1 The example graph to estimate Z_{RL}/Z_{ML} using a minimum gradient vertical methods

Virtual potential temperature (θ_v), mixing ratio (r), and refractivity (N) variables were tracers of residual and mixed layer height. Estimated Z_{RL}/Z_{ML} based on those variables, were obtained by the following equation:

1. Virtual Potential Temperature (θ_v)

   Estimating Z_{RL}/Z_{ML} use θ_v variables based on the vertical gradient method identified the θ_v maximum change to the height ($d\theta_v/dz$ maximum). The virtual potential temperature equation was as follow:

$$\theta_v = \theta(1 + 0.61.r) \tag{18.2}$$

where θ is the potential temperature in dry air and r is the water vapor mixing ratio (g/kg) [1, 23].

2. Mixing ratio (r)

   Estimating Z_{RL}/Z_{ML} use r variable based on the vertical gradient method identified the r minimum change to the height (dr/dz minimum). The mixing ratio equation was as follow:

$$r = \frac{\mu}{1-\mu} x 10^3 \tag{18.3}$$

where μ is specific humidity [1, 5, 23].

3. Refractivity (N)

   Estimating Z_{RL}/Z_{ML} use N variable based on the vertical gradient method identified the N minimum change to the height (dr/dz minimum). The refractivity equation was as follow:

$$N = (n-1) \times 10^6 = 77.6\frac{P}{T} + 3.73 \times 10^5 \frac{e}{T^2} \tag{18.4}$$

where N is refractivity, n is refractivity index, P is air pressure (hPa), T is temperature (K), and e is water vapor pressure (hPa) [23, 24, 26].

18.3 Results and Discussion

18.3.1 Residual Layer Height Estimation

Z_{RL} and Z_{ML} were identified as an inversion layer (stable layer), which is the boundary with the free atmosphere, or commonly called capping inversion [1]. There were various methods to identify the capping inversion, one of which is using the vertical profile of atmospheric layer obtained from radiosonde and GPS-RO observation [1, 5–8, 15, 23, 24, 26].

Estimating the inversion layer/capping inversion was to identify a significant increase/decrease in vertical profile gradient in radiosonde and GPS-RO data. The profile used to determine the increase/decrease in capping inversion is a variable based on the temperature and humidity [1, 24, 27]. In this study, the vertical profiles of variables that were used to identify capping inversion are the virtual potential temperature (θ_v), mixing ratio (r), and refractivity (N) using the vertical gradient methods (Fig. 18.2). Those three variables are considered suitable for analyzing Z_{RL} and Z_{ML} in Indonesia, because those variables contained temperature and water vapor as dominant components [24].

Z_{RL} and Z_{ML} based on virtual potential temperature (θ_v), mixing ratio (r), and refractivity (N) were denoted as $Z_{RL\theta_v}$ and $Z_{ML\theta_v}$, Z_{RLr} and Z_{MLr}, Z_{RLN} and Z_{MLN}. When estimating $Z_{RL\theta_v}$ and $Z_{RL\theta_v}$ were identified as maximum gradient, it based on the inversion layer definition that is temperature increase with height which indicates the stable layer. On the other hand, when estimating Z_{RLr}, Z_{MLr}, Z_{RLN}, and Z_{MLN} were identified as the minimum gradients because they show a water vapor significant decrease which indicates the presence of a semipermeable layer that could not be penetrated by water vapor. This semipermeable layer is capping inversion.

Generally, estimating $Z_{RL\theta v}$, Z_{RLr}, and Z_{RLN} yield nearly the same value, but there were differences in the water vapor composition in each variable (Eq. 18.2 and 18.4) that result in slightly different values. The correlation between Z_{RLr}, and Z_{RLN} was quite good, it is ~0.95, while the correlation between $Z_{RL\theta v}$ vs Z_{RLr} and $Z_{RL\theta v}$ vs Z_{RLN} is ~0.68. Those results occur due to water vapor contribution to the refractivity greater than the virtual potential temperature. The water vapor contribution of the refractivity variable is 0.7–2 times, while its contribution to the virtual potential temperature variable is 0.6 times [24].

Because of the refractivity has a good correlation with the mixing ratio and its suitable for studies in Indonesia, the following analysis Z_{RL} and Z_{ML} were estimated based on the refractivity that was denoted as Z_{RLN} and Z_{MLN}. In addition, the refractivity was also available in GPS-RO data, so it is easier to use radiosonde and GPS-RO simultaneously.

Estimating Z_{RLN} should meet some criteria to get an expected Z_{RLN} value. First, Z_{RLN} was estimated at an altitude of 800–5000 m, this is to minimize the stable boundary layer detection. Second, the Z_{RLN} was identified as the minimum vertical gradient after the maximum vertical gradient above the stable boundary layer (second minimum vertical gradient) (See Fig. 18.1). Third, there was a maximum vertical

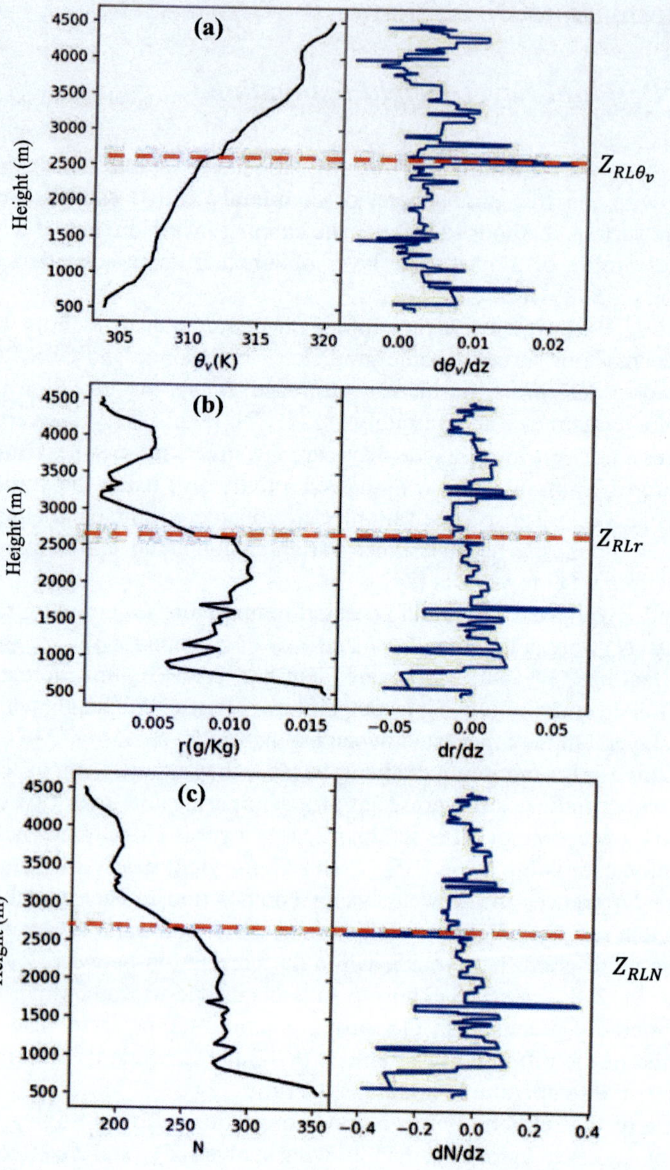

Fig. 18.2 Vertical profile and gradient of; **a** virtual potential temperature, **b** mixing ratio, **c** refractivity based on radiosonde data

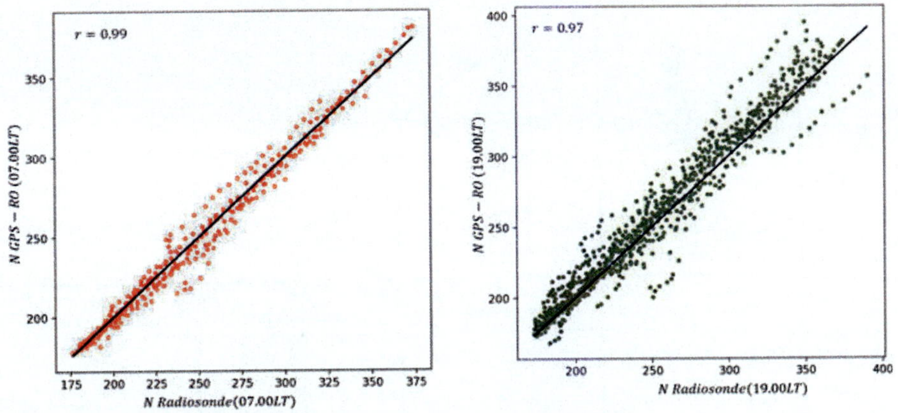

Fig. 18.3 The refractivity correlation between radiosonde and GPS-RO

gradient above the stable boundary layer that represented a water vapor concentration increasing. Fourth, the stable boundary layer is identified as the minimum vertical gradient near the surface (first minimum vertical gradient). And Fifth, the residual layer at night is neutral so there was no change in refractivity to altitude ($dN/dz \approx 0$).

18.3.2 The Comparison of Radiosonde and GPS-RO Data

The comparison of dN/dz profiles between the radiosonde and GPS-RO data shows a similar figure, even though the radiosonde has denser data compared to GPS-RO. Due to the radiosonde having a higher vertical resolution, the radiosonde can estimate Z_{RLN} value up to a single digit in meters. The minimum vertical gradient is also easier detected using radiosonde, while when using the GPS-RO it's a bit difficult. The GPS-RO has a lower resolution which causes it hard to distinguish between the residual layer and the stable boundary layer.

We validate radiosonde and GPS-RO data to ensure those data can be used simultaneously. The data that we validate is refractivity at the same level. The results show a good correlation (See Fig. 18.3) between radiosonde and GPS-RO that indicate those data can be used together.

18.3.3 Residual Layer Height Variation on Clear and Rainy Days

Residual Layer Height Variation on Clear Days. Figure 18.4 shows the growth and development profile of residual and mixed layers. When the weather conditions

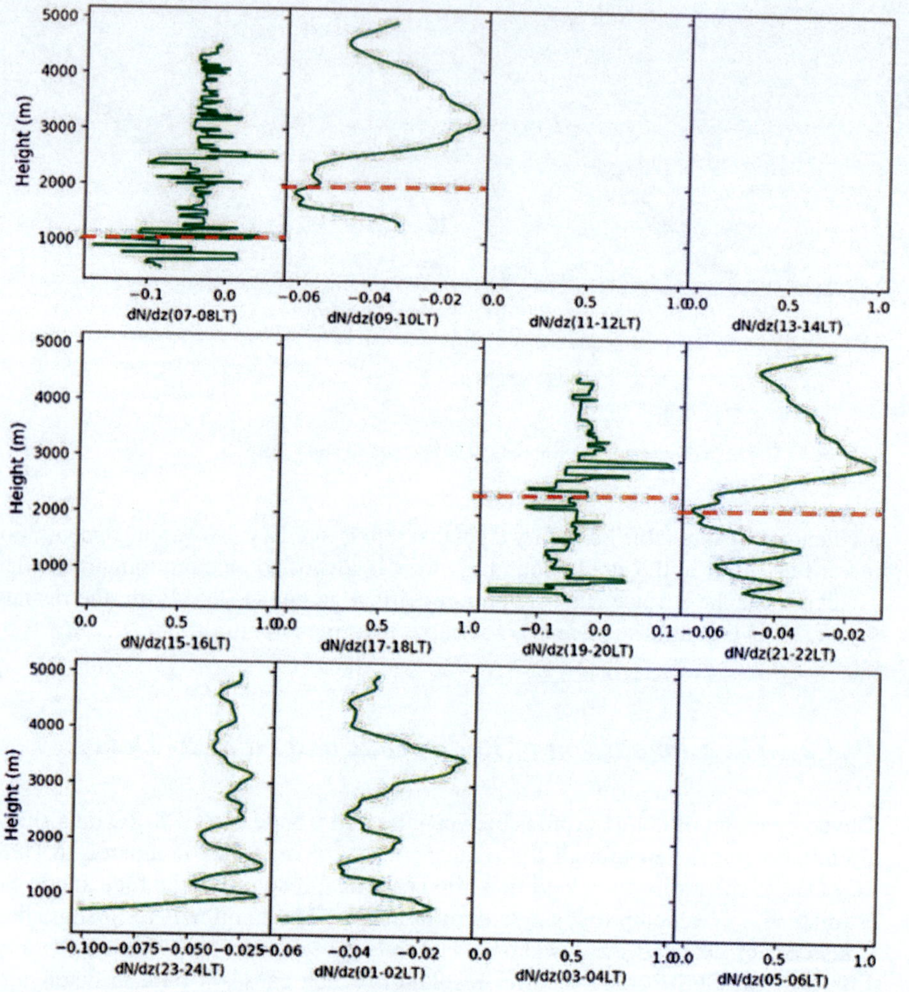

Fig. 18.4 Z_{RLN} and Z_{MLN} variations on clear days

are clear, the mixed layer would reach its peak in the afternoon. The mixed layer height in the afternoon would be the same as the residual layer height in the early evening. The mixed layer and residual layer on clear days are limited by a capping inversion. In the capping inversion, there is a significant water vapor decrease concentration because the water vapor cannot penetrate the capping inversion, which is a stable layer.

The residual layer height is maximum in the early evening, and it would decrease until near sunrise. The decrease was caused by the absence of solar radiation which is an external force that is triggered the layer to grow and develop.

Residual Layer Height Variation on Rainy Days. On the rainy days, ABL growth and development were obstructed by the rain clouds causing the surface temperature to be colder. The colder surface temperature causes an inversion in the ABL, which causes the air parcel to tend to be stable and cannot expand. On rainy days or there are a lot of rain clouds, Z_{RLN} and Z_{MLN} are identified as the cloud base height, which means the value of Z_{RLN} and Z_{MLN} are identified as the maximum gradient of refractivity because there was a significant increase of water vapor concentration in the rain clouds. On rainy days, the residual layer growth and development are divided into two conditions. First, when the rain does not cross the sunset, and second when the rain crosses through the sunset.

The Residual Layer Height Variation on the Day when The Rain does not Cross Sunset. When the rain stops before sunset, and the weather is relatively clear in the afternoon because the rain clouds have disappeared (Fig. 18.5), solar energy can reach thc surface and warm it. This condition triggered the air parcel to lift even though it only reaches a relatively low altitude. Due to there being a lift, a capping inversion was formed in the afternoon. In this condition, the residual layer would appear even at a relatively low altitude compared to when the weather is clear.

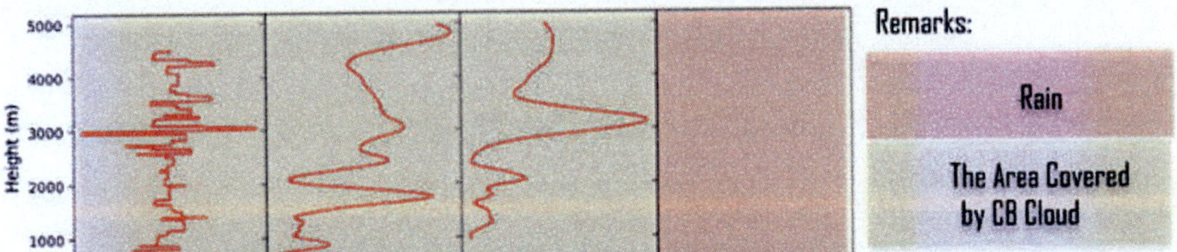

Fig. 18.5 Z_{RLN} and Z_{MLN} variations on the rainy days when the rain does not cross sunset

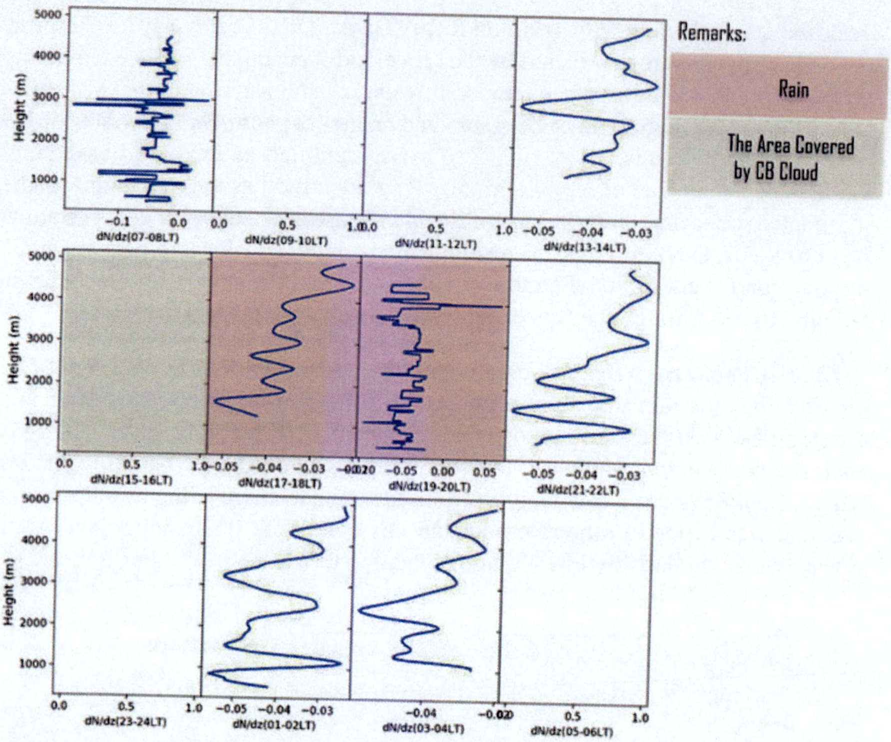

Fig. 18.6 Z_{RLN} and Z_{MLN} variations on the rainy days when the rain crosses sunset

The Residual Layer Height Variation on the Day when The Rain Crosses Sunset. Figure 18.6 show that the mixed layer rises to the maximum in the morning before the rain clouds were formed (strong convective). After the rain clouds were formed, the cloud would block the solar radiation causing the mixed layer height to decrease significantly. When the rain has occurred, the mixed layer and residual layer height are relatively low because they are identified as cloud base height. At that time, there was no inversion layer (stable boundary layer). And after the rain has stopped at nighttime, the surface temperature becomes colder, which causes the near-surface inversion to strengthen so that a stable boundary layer was formed. In this condition, no residual layer is formed because the temperature above the stable boundary layer is relatively homogeneous which causes the second capping inversion would not to form. The water vapor concentration in the air has decayed into rain so there is no significant water vapor decrease was detected above the stable boundary layer. It can be concluded that when the rain occurs at night, the residual would not appear.

After rain at nighttime, the stable boundary layer that was formed near the surface traps fog in the early morning until sunrise. This fog is formed from air-containing water vapor touching the cold surface of the earth. At daybreak, fog absorbs heat quickly causing the surface temperature to heat up faster. That condition causes the mixed layer would grow and develop rapidly in the morning and result in foggy weather.

18.3.4 The Relationship Between Residual and Mixed Layer Height Variation

Z_{RLN} variation is highly dependent on the weather condition in the afternoon especially, before sunset, it's proven by the correlation between Z_{MLN} at 16:00 LT and Z_{RLN} at 19:00 LT is 0.81. Near Sunset, the lift begins to weaken and disappear, leaving particulates, aerosols, and water vapor flying freely in the air. The condition of the atmospheric boundary layer was close to neutral, so the particles flew freely in all directions [1]. If in the afternoon the weather is so clear, the surface temperature is high, and the relative humidity is low, the Z_{MLN} would be high, as well as Z_{RLN}. Meanwhile, if the rain occurs in the afternoon, there are two possibilities; first, the residual would appear at a low altitude (the rain stop before sunset), or second, the residual layer would not appear (the rain stop after sunset).

In contrast, in the morning after sunrise, the Z_{MLN} growth and development was not affected by Z_{RLN}, it's shown by the correlation results between Z_{RLN} at 5:00 LT and Z_{MLN} at 7:00 LT is 0.1. Even though the residual layer provided energy for the air parcel to lift adiabatically [1, 19], that does not make the residual layer influence the mixed layer directly. Z_{MLN} growth and development is more influenced by weather conditions at that time dominantly. If at the time the weather is clear, Z_{MLN} would high and when the weather is rainy, Z_{MLN} would relatively low.

18.4 Conclusion

Estimating Z_{RL} is easier to use variables with more water vapor composition such as mixing ratio (r) and refractivity (N), because the inversion of the capping inversion tends to be weak at nighttime. The Z_{RL} is easier to estimate using radiosonde data because it is not affected by weather conditions, while GPS-RO data during rainy days could not penetrate to the surface. However, there are only two radiosonde data available in one day, so that GPS-RO data is required. The Z_{RL} variation is strongly affected by weather conditions in the afternoon, especially near sunset. If in the afternoon the weather is clear, Z_{ML} would be high as well as Z_{RL}. On the other hand, if the rains occur in the afternoon and stop before sunset, Z_{RL} would appear even at the low altitude. Meanwhile, if the rain stops after sunset, Z_{RL} would not appear because no capping inversion is formed. At the nighttime, Z_{RL} would decrease until break day. The following morning, Z_{RL} variation did not affect Z_{ML} variation even though the residual layer provides energy for the air parcel to lift adiabatically. The Z_{ML} growth and development influenced by the heating of solar radiation at the time.

References

1. Stull, R.B.: An introduction to boundary layer meteorology. Kluwer Academic Publisher (1988)
2. Ahrens, C.D.: Meteorology today: an introduction to weather, climate, and the environment, 9th edn. CA: Brooks/Cole, CengageLearning, Belmont (2009)
3. Arya, S.P.: Introduction to micrometeorology, 2nd edn. Academic Press, San Diego (2001)
4. Kaimal, J.C., Finnigan, J.J.: Atmospheric boundary layer flows their structure and measurement. Oxford University Press, New York (1994)
5. Liu, S., Liang, X.-Z.: Observed diurnal cycle climatology of planetary boundary layer height. J. Clim. **23**(21), 5790–5809 (2010)
6. Seidel, D.J., Zhang, Y., Beljaars, A., Golaz, J.C., Jacobson, A.R., Medeiros, B.: Climatology of the planetary boundary layer over the continental United States and Europe: Boundary Layer Climatology: U.S.and EUROPE. J. Geophys. Res. Atmos. **117**(D17), n/a-n/a (2012)
7. Wang, X.Y., Wang, K.C.: Estimation of atmospheric mixing layer height from radiosonde data. Atmos. Meas. Tech. **7**(6), 1701–1709 (2014)
8. Zhang, Y., Seidel, D.J., Zhang, S.: Trends in planetary boundary layer height over Europe. J. Clim. **26**(24), 10071–10076 (2013)
9. Ekayatni, S.W.: Hubungan Distribusi Aerosol dengan Ketinggian Planetary Boundary Layer (Studi Kasus: Riau and Sekitarnya). Bogor Agriculture University (2015)
10. Fitriani, V., Bey, A., June, T.: Estimasi Ketinggian Planetary Boundary Layer Indonesia Menggunakan Data Ecmwf Reanalysis Era-Interm. Jurnal Meteoroologi and Geofisika **18**(1), 21–31 (2017)
11. Hadi, T.W., Tsuda, T., Hashiguchi, H., Fukao, S.: Tropical sea-breeze circulation and related atmospheric phenomena observed with L-band boundary layer radar in Indonesia. J. Meteorol. Soc. Jpn **78**(2), 123–140 (2000)
12. Hashiguchi, H., et al.: Observations of the planetary boundary layer over equatorial Indonesia with an L band clear-air Doppler radar: Initial results. Radio Sci. **30**(4), 1043–1054 (1995)
13. Hermayana, F.F.: Kajian Atmosfer Bawah Wilayah Tropis And Subtropis. Bogor Agriculture Institute, Bogor (2012)
14. Pinandito, M., et al.: Mie scattering lidar observation of aero. Environ. Sci. **13**(2), 2015–2216 (2000)
15. Pratikasari, R.: Kajian teoritis and empiris distribusi spasial and temporal parameter-parameter atmospheric boundary layer. Bogor Agriculture University (2011)
16. Sugimoto, N., Matsui, I., Shimizu, A., Pinandito, M., Sugondo, S.: Climatological characteristics of cloud distribution and planetary boundary layer structure in Jakarta, Indonesia revealed by lidar observation. Geophys. Res. Lett. **27**(18), 2909–2912 (2000)
17. Cuesta, J., Marsham, J.H., Parker, D.J., Flamant, C.: Dynamical mechanisms controlling the vertical redistribution of dust and the thermodynamic structure of the West Saharan atmospheric boundary layer during summer. Atmos. Sci. Lett. **10**(1), 34–42 (2009)
18. Pahlow, M., Kleissl, J., Parlange, M.B.: Atmospheric boundary-layer structure observed during a haze event due to forest-fire smoke. Bound. Layer Meteorol. **114**(1), 53–70 (2005)
19. Qiang, Z., Sheng, W.: A study of the atmospheric boundary layer structure during a clear day in the arid region of Northwest China. Acta Meteor. Sin. **23**(3), 327–337 (2009)
20. Anthes, R.A., et al.: The COSMIC/FORMOSAT-3 mission: early results. Bull. Amer. Meteor. Soc. **89**(3), 313–334 (2008)
21. Kursinski, E.R., Hajj, G.A., Schofield, J.T., Linfield, R.P., Hardy, K.R.: Observing Earth's atmosphere with radio occultation measurements using the global positioning system. J. Geophys. Res. **102**(D19), 23429–23465 (1997)
22. Yue, X., et al.: Space weather observations by GNSS radio occultation: from FORMOSAT-3/COSMIC to FORMOSAT-7/COSMIC-2. Space Weather, **12**, 616–621 (2014)
23. Ao, C.O., et al.: Planetary boundary layer heights from GPS radio occultation refractivity and humidity profiles: PBL HEIGHT FROM GPS RO. J. Geophys. Res. **117**(D16), 1–18 (2012)

24. Basha, G., Ratnam, M.V.: Identification of atmospheric boundary layer height over a tropical station using high-resolution radiosonde refractivity profiles: comparison with GPS radio occultation measurements. J. Geophys. Res. **114**(D16), 1–11 (2009)
25. von Engeln, A., Teixeira, J.: A planetary boundary layer height climatology derived from ECMWF reanalysis data. J. Clim. **26**(17), 6575–6590 (2013)
26. Guo, P., Kuo, Y.-H., Sokolovskiy, S.V., Lenschow, D.H.: Estimating atmospheric boundary layer depth using COSMIC radio occultation data. J. Atmos. Sci. **68**(8), 1703–1713 (2011)
27. Garrat, J.R.: The atmospheric boundary layer, 1st edn. Cambridge University Press (1992)

Chapter 19
The Response of Rainfall in Sumatera to Indian Ocean Dipole Phenomenon

Fahmi Rahmatia, Listi Restu Triani, and Amalia Nurlatifah

Abstract Sumatera Island is one of the islands in Indonesia whose area has various rainfall patterns and is located directly adjacent to the eastern Indian Ocean. This causes rainfall in this island to be influenced by many things, one of that is the Indian Ocean Dipole (IOD). This study aims to see how the response of rainfall in Sumatera to the Indian Ocean Dipole phenomenon. The data used in this study is TRMM data with 0.25° spatial resolution and monthly temporal resolution. The Principal Component Analysis method is used to see how the spatial response of rainfall in Sumatera to the Indian Ocean Dipole phenomenon. While the Fast Fourier Transform (FFT) method is used to see what dominant phenomena affect rainfall conditions in the 10 capitals of the Province of Sumatera. The result of PC2 from PCA results on TRMM rainfall in Sumatera states that the majority area of Sumatera has a negative response to the IOD phenomenon. This means when the IOD has a positive phase, most areas in Sumatera will experience a lack of rainfall. Vice versa, when the IOD has a negative phase, then rainfall in Sumatera tends to increase. The FFT results for all provincial capitals in Sumatera state that the majority of Sumatera is affected by seasonal phenomena such as monsoons and interannual phenomena such as IOD. These two results prove that rainfall in Sumatera is heavily influenced by the IOD phenomenon.

F. Rahmatia (✉)
BPAA Agam, National Research and Innovation Agency (BRIN), Bukittinggi, Indonesia
e-mail: fahmi.rahmatia@brin.go.id

L. R. Triani
Atmospheric Science Research Group, Faculty of Earth Science and Technology, Bandung, Indonesia

A. Nurlatifah
Research Center for Climate and Atmosphere, National Research and Innovation Agency (BRIN), Bandung, Indonesia

E. Yulihastin et al. (eds.), *Proceedings of the International Conference on Radioscience, Equatorial Atmospheric Science and Environment and Humanosphere Science, 2021*, Springer Proceedings in Physics 275, https://doi.org/10.1007/978-981-19-0308-3_19

19.1 Introduction

The Indonesian Maritime Continent (BMI) is an archipelagic country located between two continents (Asia and Australia) and two oceans (India and the Pacific). This condition causes this area to have climatic characteristics that are different from other equatorial countries and make Indonesia have dynamic, unique, and complex weather climate characteristics.

Hamada et al. [1] stated that the Indonesian Maritime Continent has the most active convection process in the world (deep convection). This convection process is also influenced by the composition of the Indonesian Maritime Continent, two-thirds of which is ocean. It is also believed to be one of the factors that influence the weather and climate in BMI. In addition, Indonesia is also the meeting place of the zone circulation called the Walker Circulation and the meridional circulation called the Hadley Circulation.

One result that is closely related to the Walker Cycle is the ENSO and IOD phenomena. ENSO is a symptom of a deviation (anomaly) in the surface temperature of the Pacific Ocean on the west coast of Ecuador and Peru which is higher than the normal average. While the Indian Ocean Dipole (IOD) itself is a physical phenomenon of the interaction of the sea and the atmosphere in the equatorial Indian Ocean region which is characterized by a negative anomaly of sea surface temperature in the western part of the Indian Ocean [2]. This phenomenon is believed to be one of the strongest controls on global climate and Indonesia.

IOD is an ocean–atmosphere phenomenon at the equator of the Indian Ocean that affects the climate in Indonesia and other countries around the Indian Ocean basin [2].

As the name implies, the IOD is characterized by an anomaly of sea surface temperature or SST (Sea Surface Temperature) between the 'two poles' of the Indian Ocean, namely the western Indian Ocean (50° East Longitude-70° East Longitude, 10° South Latitude -10° North Latitude) and the southeast (90° East Longitude-110° East Longitude, 10°LS-0°LS). The difference in SST anomalies between these two regions is known as the Dipole Mode Index (DMI), and is used to measure the strength of the IOD itself. The period when the DMI is positive is generally referred to as the positive IOD period (IOD+) and the period when the DMI is negative is called the negative IOD period (IOD−).

In periods of positive IOD, the waters in the southeast Indian Ocean are generally "colder" (lower than the average temperature) while the waters in the western Indian Ocean will be "warmer" (higher than the average temperature) [2]. As a result, convection will shift from the east of the Indian Ocean to the west and bring a lot of rain to the eastern part of the African continent. On the other hand, the eastern Indian Ocean region (such as Indonesia) will experience minimal rainfall.

The characteristic period of a negative IOD is the opposite of a positive IOD. SST in the southeastern part of the Indian Ocean will be warmer while in the west it will be colder. This causes convection to move to the east and causes a lack of rainfall in eastern Africa while rainfall in Indonesia, especially western Indonesia, which borders the Indian Ocean will increase [2].

Sumatera island has unique climate characteristics both locally and regionally. This is because the area has mountains that stretch from north to south and is surrounded by the Indian Ocean, Malacca Strait, Java Sea, Karimata Strait and close to the South China Sea [3]. Topographic factors of regional weather systems have an important role in the amount and spatial pattern of rainfall in an area [4].

Rainfall in Sumatera has 2 types of rainfall, namely the equatorial type and the monsoon type. The equatorial rainfall type has a bimodal monthly rainfall distribution with two peaks of the maximum rainy season occurring almost throughout the year which usually occurs in March and October [5]. On the island of Sumatera, it occurs in the central and northern parts of Sumatera. The monsoon rainfall type has unimodal characteristics (one peak of the rainy season) where the dry season occurs in June, July, and August. Wet months occur in December, January and February. While the remaining six months is a transition period. The area that has a monsoon rain type for the island of Sumatera is in the southern part of Sumatera [3].

The IOD index is defined as the difference between the SST anomaly in the western (50–70° E, 10° S–10° N) and eastern (90–110° E, 10° S–Eq) tropical Indian Ocean. The dipole on the SST anomaly is accompanied by abundant rainfall in the western part of the Indian Ocean-East Africa and rare rainfall in the eastern part of the Indian-Indonesian Ocean. This is similar to the Bjerknes-type air-sea interaction in the tropical Pacific [6].

Therefore, research is needed that describes how the response of rainfall on the island of Sumatera to the Indian Ocean Dipole phenomenon.

19.2 Data Dan Method

19.2.1 Data

TRMM Rainfall Data

Rainfall data measured by the Tropical Rainfall Measuring Mission (TRMM) satellite can be used in Indonesia because the focus of TRMM is to measure rain in the tropics [7, 8]. This TRMM satellite data has provided data nationally since 1998 in the form of daily and 3 h rainfall. The use of TRMM data in the form of daily rainfall in Indonesia has been widely used, such as landslide early warning systems in Pacitan and Brebes, grouping of rainfall patterns in Indonesia [9], and evaluation of rainfall forecasts for the area around the Makassar Strait [10]. The TRMM that used in this research is TRMM 3B43 which has monthly temporal resolution and 0.25° spatial resolution.

Dipole Mode Index (DMI) Data

DMI data shows the strength of the Indian Ocean Dipole (IOD). This data is obtained from JAMSTEC. The DMI calculation is obtained from the difference in sea surface temperature (SST) from the western Indian Ocean on the East Coast of Africa (50° E–70° E, 10° S–10° N) and the eastern Indian Ocean on the West Coast of Sumatera (90° E–110° E, 10° S–Equator).

The calculation of the DMI value is obtained in the following way:

$$DMI = SMLA - SMLB$$

where:

SML A Sea surface temperature in the western Indian Ocean region (50° E–70° E, 10° S–10° N).

SML B Sea surface temperature in the southeastern part of the Indian Ocean (90° E–110° E, 10° S-0° S).

19.2.2 Method

PCA Calculation

The Principal Component is a linear combination of the vector p of the random variables Xi, …, Xp. Let matrix X = [X1,…., Xp] has a covariance matrix. In this case it is a positive and symmetric definite matrix with eigenvalues 1 2 … p > 0 and calling the corresponding eigenvectors for each 1 2 … p > 0 is e1, …, ep which are orthogonal to each other, by looking for a linear combination, namely:

$$Y_i = e_i^T X = e_{11}X_1 + e_{21}X_{2+\ldots+}e_{pi}X_{p,} \quad \text{i} = 1, 2, \ldots \text{p}$$

With this selection,

$$Var(Y_i) = e_i^T \Sigma e_i = \lambda_1 \quad \text{i} = 1, 2, p$$

$$Cov(Y_i, Y_k) = e_i^T \Sigma e_i = 0, i \neq k$$

The variance describing PC i is defined as

$$\frac{\lambda_i}{\lambda_1 + \lambda_2 + \ldots + \lambda_p}, k = 1, \ldots, p.$$

The value of e_{ki} represents a measure of the importance of the k-th variable to the i-th PC. Specifically, e_{ki} reveals the correlation between the Y_i component and the X_i variable. This is explained by using the correlation coefficient between the Y_i component and the X_i variable is:

$$\rho_{Y_i X_i} = \frac{e_{ki}\sqrt{\lambda_1}}{\sqrt{\sigma_{kk}}}$$

where σ_{kk} is the standard deviation of the k-th variable [11].

Analysis of the Relationship Between Two Data

Pearson Correlation

Pearson correlation is a correlation measure used to measure the strength and direction of a linear relationship between two variables. Two variables are said to be correlated if a change in one variable is accompanied by a change in the other variable, either in the same direction or in the opposite direction. The correlation coefficient only measures the strength of the linear relationship and not the non-linear relationship. It should also be remembered that the absence of a strong linear relationship between the variables does not mean that there is no causality, cause-effect relationship.

The Pearson correlation coefficient is defined as follows [12]:

$$r = \frac{n\sum XY - (\sum X)(\sum Y)}{\sqrt{\left[n\sum X^2 - (\sum X)^2\right]\left[n\sum Y^2 - (\sum Y^2)\right]}}$$

where r represents the correlation coefficient, X and Y represent the variable for which you want to know the linear relationship.

The meaning of the value of r:

- positive r indicates a unidirectional relationship. If the value of X increases, then the value of Y also increases.
- r is negative, indicating an inverse relationship. If the value of X increases, then the value of Y decreases.
- a large r value indicates a strong relationship. X and Y have a related linear relationship.
- a small r value indicates a weak relationship. X and Y have a less linear relationship.

Fast Fourier Transform

FFT is an algorithm that calculates the Discrete Fourier Transform (DFT) of a sequence, or its reciprocal (IDFT). FFT converts a signal from its original domain (often space or time) to a representation in the frequency domain and vice versa. DFT is obtained by decomposing the sequence of values into different frequency components [13]. Or in short, "Fast Fourier Transform" (FFT) is an important measurement method in audio and acoustic measurement science. It converts the signal into its individual spectral components and thus provides frequency information about the signal. FFT is used for fault analysis, quality control, and condition monitoring of machines or systems.

19.3 Results and Discussion

19.3.1 The Major Signal in Rainfall Data

Based on the fast Fourier transform analysis, it is known that the biggest signal that affects rainfall data in Sumatera is a signal with a period of 1 year (Table 19.1). This happened in Banda Aceh, Medan, Padang, Pangkal Pinang, Palembang, Bandar Lampung, and Bengkulu. It is known that this one-year or seasonal phenomenon that affects rainfall in Sumatera is the monsoon phenomenon. This is supported by Hermawan's research in 2007 which states that the majority of rainfall in Sumatera is heavily influenced by the monsoon climate [14].

The second strongest phenomenon that affects rainfall in the majority of regions in Sumatera is the 3–6 year phenomenon or the intra-seasonal phenomenon (Table 19.1). To estimate or predict what this phenomenon is, further analysis of rainfall in Sumatera is carried out using the PCA (Principal Component Analysis) and Pearson correlation methods.

19.3.2 Rainfall Pattern in Sumatera with PCA

The spatial variability of the rainfall data on the island of Sumatera is shown in Fig. 19.1. This variability is sought using the Principal Component Analysis (PCA) method. In the PCA plot of rainfall on the island of Sumatera, there are several principal modes. PC 1 accounted for 40% of the total variance, while PC 2 accounted for about 13% of the total variance.

Table 19.1 Peak power and period of rainfall data using PCA 1998–2017

No	Location	Power 1	Period 1 (Year)	Power 2	Period 2 (Year)
1	Banda Aceh	105,014.8	1.02	57,197.09	0.97
2	Medan	365,848.3	1.02	117,950.2	21.3
3	Padang	199,944.34	1.02	91,379.48	3.56
4	Pekan Baru	134,079.2	21.3	87,874.9	3.56
5	Batam	79,142.42	21.3	77,464.47	4.27
6	Jambi	96,893.05	21.3	74,187.06	5.33
7	Pangkal Pinang	215,494.7	1.02	70,288.82	3.05
8	Palembang	461,664	1.02	160,365.1	0.97
9	Bengkulu	389,317.3	1.02	162,006.4	0.97
10	Bandar Lampung	1,064,492.2	1.02	286,243.5	0.97

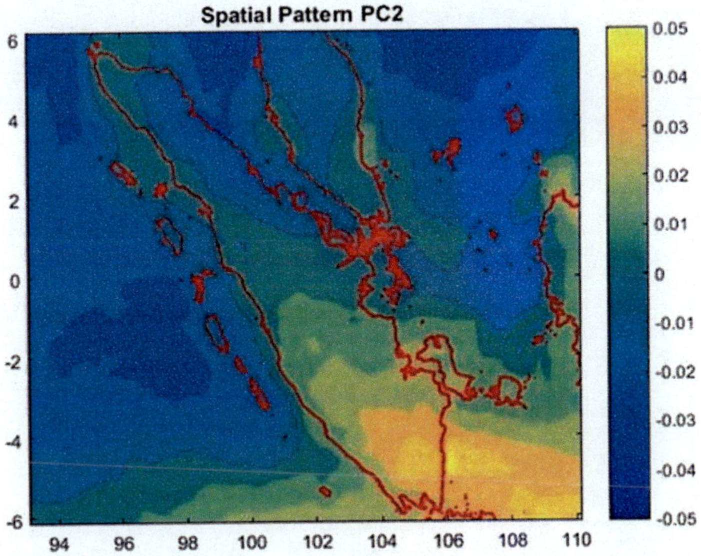

Fig. 19.1 PC2 of rainfall's spatial pattern in Sumatera

From the calculation results, the PC1 pattern shows overall uniformity on the island of Sumatera, which tends to be similar to the seasonal pattern (monsoon). This is also supported by the results of the FFT calculations which are presented in Table 19.1. In the PC2 pattern, which shows the non-uniformity, especially in the northern and southern parts of the island of Sumatera, it can be seen that the southern region has the opposite pattern to the northern region. Where the southern part is affected by PC2 is stronger than the northern part of the island of Sumatera. However, it is not yet known what phenomena are represented by PC2, which based on FFT refers to internal phenomena. So to make sure, see the correlation between DMI and ONI with rainfall on the island of Sumatera.

The PC2 pattern which is principal and represents 13% of the total variance is also similar to the map of the correlation between DMI and rainfall on the island of Sumatera presented in Fig. 19.2. Thus, this pattern of IOD influence on rainfall is related to PC2 which explains about 13% of the overall variance.

19.3.3 Linear Relationship Between Rainfall and Indian Ocean Dipole

Calculation of the correlation between DMI and ONI with rainfall on the island of Sumatera shows that the ONI-rainfall correlation is smaller than the DMI-rainfall correlation. So it is suspected that what affects rainfall on the island of Sumatera after the monsoon is the Indian Ocean Dipole.

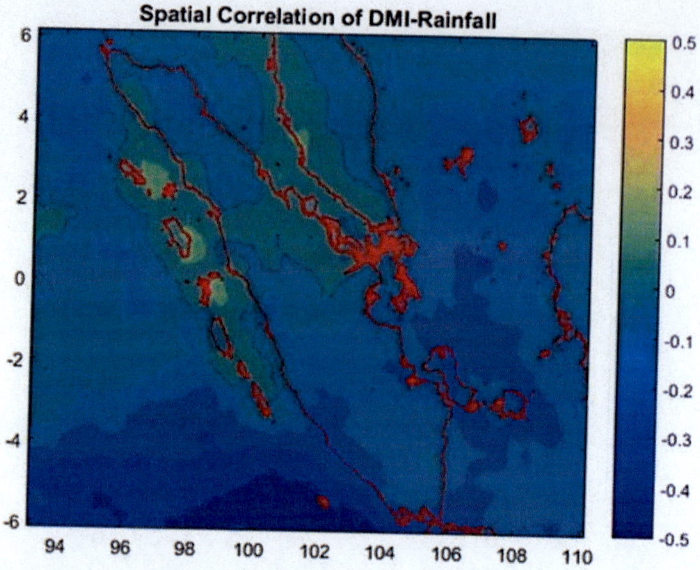

Fig. 19.2 Spatial correlation between DMI data and rainfall data

Based on the calculation of the Pearson correlation between DMI data and rainfall in Sumatera, the strongest and negative correlation tends to occur in the southern area of Sumatera (Fig. 19.2). Meanwhile, other areas on the mainland of Sumatera tend to experience negative correlations, although with a smaller strength than the southern regions of Sumatera. This raises the hypothesis that the effect of IOD is strongest in the southern area of Sumatera. To prove it, an analysis of the calculation of rainfall anomalies was carried out at the time of positive and negative IOD events in the southern area of Sumatera.

In almost all areas in mainland Sumatera, DMI and rainfall data tend to be negatively correlated, which means that when the DMI condition is positive, it tends to be followed by a decrease in rainfall and vice versa when the DMI is negative, tends to be followed by an increase in rainfall in Sumatera.

Analysis of the relationship between the incidence of IOD and rainfall in Sumatera was also carried out by calculating the correlation between DMI data and normalized rainfall data. The calculation results are in line with the depiction of the spatial correlation map presented in Fig. 19.2. The calculations show the strongest correlation between DMI data and normalized rainfall data occurs in most cities in southern Sumatera (Table 19.2). The strongest correlations occurred in Pangkal Pinang, Palembang, Bengkulu, Bandar Lampung, and Jambi.

Most of the correlations belong to a weak negative correlation ($-0.5 < r < 0$). This is probably because the data used in calculating this correlation are all data in various conditions (strong IOD, weak IOD, and neutral conditions). If the data is filtered to only IOD event data, it is likely that the correlation coefficient will strengthen.

Table 19.2 Table of correlation between DMI and normalized rainfall data in 10 Regions

No	Location	R (correlation)
1	Banda Aceh	−0.1591
2	Medan	−0.0808
3	Padang	−0.0435
4	Pekan Baru	0.1098
5	Batam	−0.0576
6	Jambi	−0.2146
7	Pangkal Pinang	−0.3134
8	Palembang	−0.2462
9	Bengkulu	−0.2822
10	Bandar Lampung	−0.2182

Based on the depiction of the PC2 map, correlation map, and correlation table, it can be seen that in general, the effect of IOD on rainfall conditions in Sumatra is inversely proportional. That is, a positive IOD tends to make in a reduction on rainfall in Sumatera and a negative IOD tends to result in an increase on rainfall in Sumatera. However, the PC2 map shows a different pattern between northern Sumatra and southern Sumatra. The correlation map also shows that the southern part of Sumatra tends to be more strongly correlated with DMI than the northern region. This shows that there is a tendency for the response of rainfall in southern Sumatra to the rainfall data to be stronger than the northern region.

19.3.4 Sumatera Island Rainfall Conditions During IOD

IOD Positive. In general, the impact of positive IOD on rainfall conditions in Sumatera according to previous research is that there tends to be a reduction in rainfall intensity [15]. In the case of 1998–2017, there were several positive IOD events. These incidents included 2002, 2007, 2008, 2011, 2012, 2015, and 2016.

In 2002, there was a positive IOD which caused the area south of Sumatera to tend to experience a negative anomaly of rainfall, which means that in that year the southern part of Sumatera experienced a reduction in rainfall (Fig. 19.3a). The same thing also happened in 2007 where the area of Southern Sumatera experienced the highest reduction in rainfall compared to other regions (Fig. 19.3b). Similar events occurred in the IOD positive events that occurred in 2008, 2011, 2012, and 2015 wherein these events a decrease in rainfall occurred in Sumatera (Fig. 19.3c, d, e and f). Of all cases of positive IOD, the most reduction in rainfall occurred in the southern part of Sumatera. Reductions in the central and southern regions of Sumatera tend to occur but with a weaker magnitude or strength compared to the southern regions.

To better understand the effect of positive IOD on rainfall conditions on the island of Sumatera, a composite analysis of rainfall anomalies when a positive IOD occurs

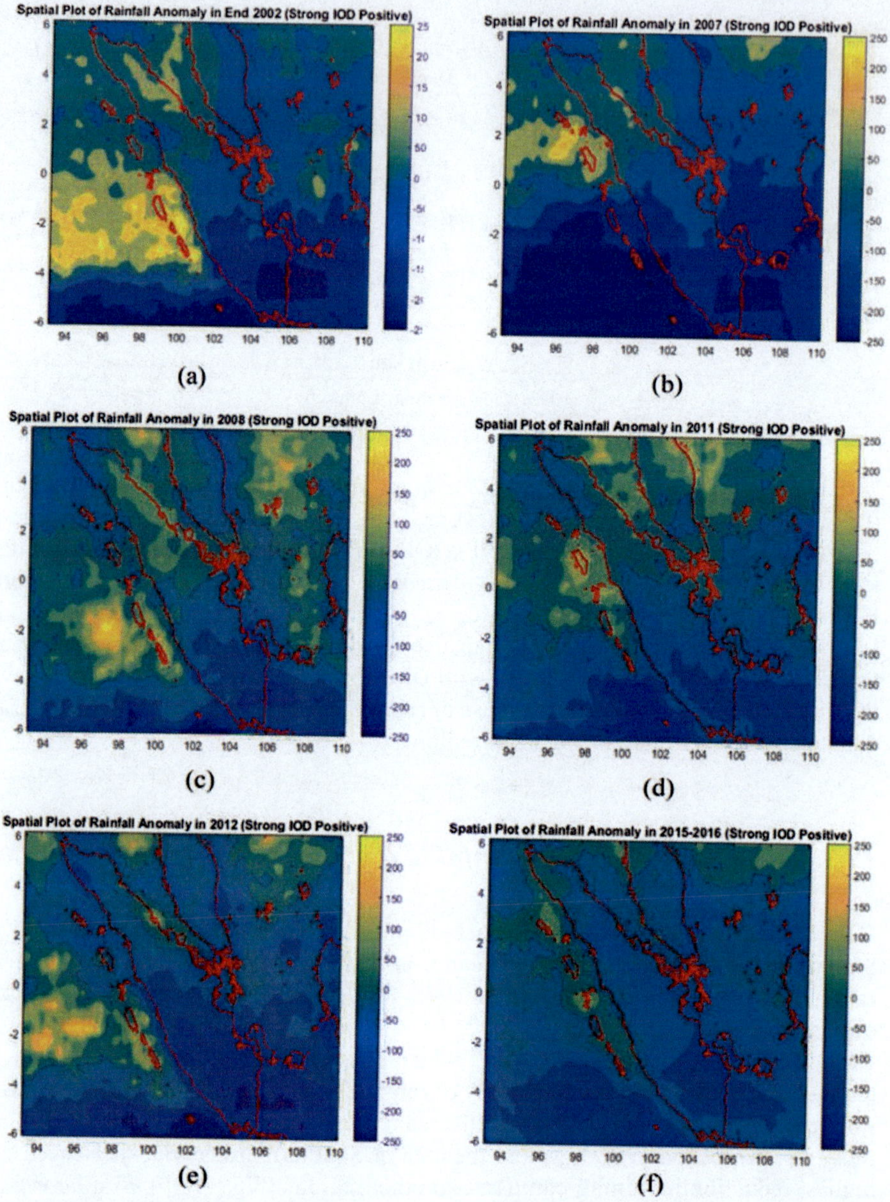

Fig. 19.3 Spatial plot of Rainfall Anomaly in **a** 2002-end, **b** 2007, **c** 2008, **d** 2011, **e** 2012, and **f** 2015–2016

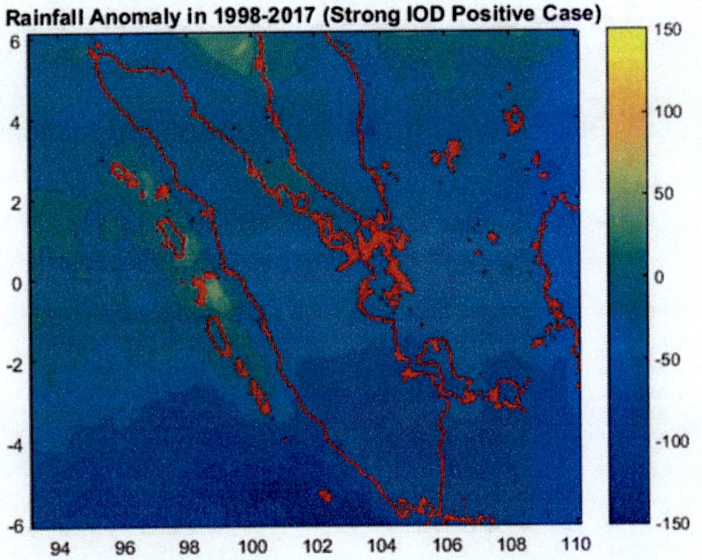

Fig. 19.4 Composite of Rainfall Anomaly in 1998–2017 (on IOD Positive event)

is carried out which is presented in the plot of Fig. 19.4. In Fig. 19.4 it can be seen that due to the occurrence of positive IOD, a lot of rainfall reduction occurs in Southern Sumatera. Meanwhile, in the central and northern parts, there tends to be a reduction in rainfall, although not as much as in the southern part of Sumatera (Fig. 19.4).

IOD Negative. The negative IOD condition causes additional rainfall in the majority of Indonesia, especially in Sumatera [15]. This statement is supported by several facts about the incidence of negative IOD at the end of 1998, 2000, 2001, and 2005. It can be seen that in those years most of Sumatera experienced an increase in rainfall (Fig. 19.5). Just like the positive IOD incident, the negative IOD impact was also strongest in the southern area of Sumatera where when the IOD was negative, this area experienced the most additional rainfall compared to other areas.

A more in-depth understanding of how the negative IOD phenomenon affects rainfall conditions on the island of Sumatera is presented in Fig. 19.6. Figure 19.6 shows a composite of rainfall anomalies during negative IOD events where it is seen that almost the entire Sumatera region experiences an increase in rainfall during this condition. This means that it can be concluded that the incidence of negative IOD resulted in an increase in rainfall in almost the entire Sumatera region.

19.4 Conclusions

Based on calculations using FFT, rainfall in 10 capital cities on the island of Sumatera is influenced by seasonal phenomena (monsoons) and inter-annual phenomena. The

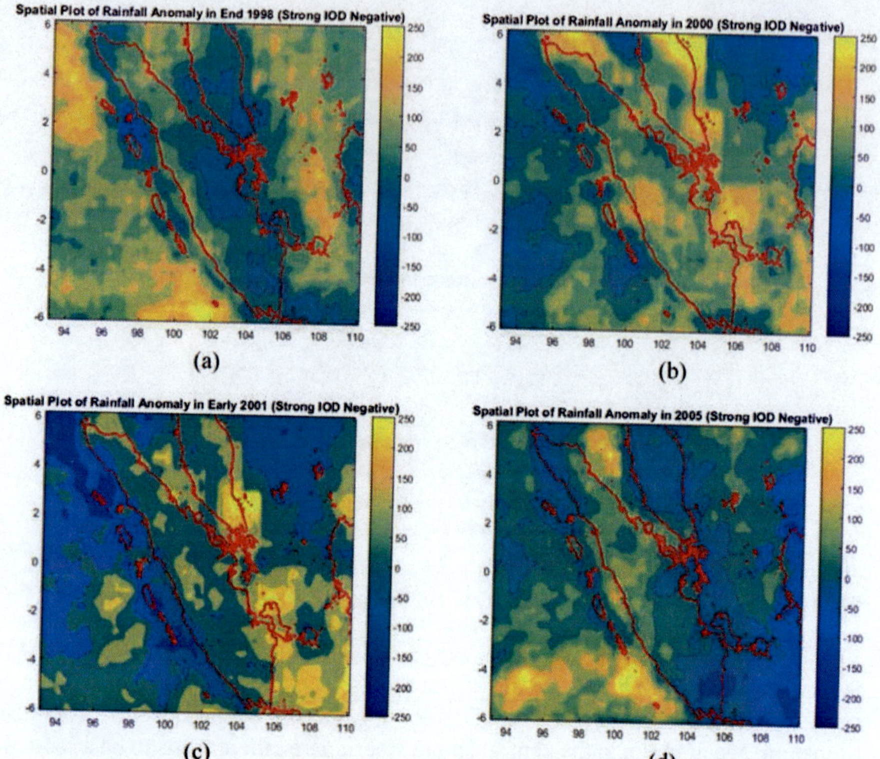

Fig. 19.5 Spatial plot of Rainfall Anomaly in **a** 1998, **b** 2000, **c** 2001, dan **d** 2005

plot of the entire Sumatera Island with PCA shows that PC2 which explains 13% of the overall variance is an interannual phenomenon that is estimated to be IOD. The spatial variability in PC2 shows that the southern area has a different pattern from the upper part of Sumatera. This is in line with the correlation map between DMI and rainfall which shows that the southern region of Sumatera Island has the strongest correlation with DMI. IOD exerted the greatest influence in the southern region of Sumatera. This is indicated by the strong correlation between DMI and rainfall data in this area. The response of rainfall in southern Sumatra to IOD is stronger than in the north. This is explained by the stronger correlation between DMI and rainfall in the south Sumatera region than the north.

19.5 Contributor Statements

1st author (Fahmi Rahmatia): brainstorming ideas so that this research can arise, processing data, analyzing, and writing papers.

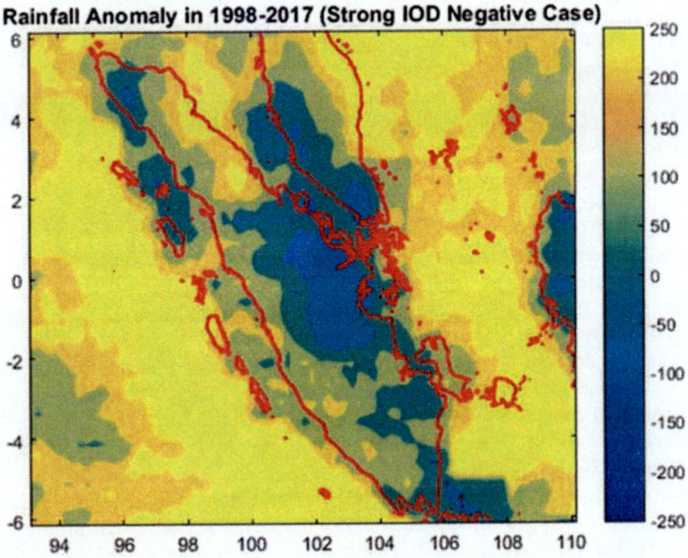

Fig. 19.6 Composite of Rainfall Anomaly in 1998–2017 (on IOD Negative event)

2nd author (Listi Restu Triani): brainstorming ideas so that this research can arise, collecting data, analyzing, and writing papers.

3rd author (Amalia Nurlatifah): brainstorming ideas so that this research can arise, collecting data, analyzing, and writing papers.

All authors provided feedback, discussed results and contributed to the final manuscript.

References

1. Hamada, J.I., Mori, S., Yamanaka, M.D., Kubota, H., Haryoko, U., Lestari, S., Syamsudin, F.: Interannual rainfall variability over northwestern jawa and its relation to Indian ocean dipole and El Nino-Southern oscillation events. SOLA **8**, 069–072 (2012)
2. Hameed, S., Goswami, B.N., Vinayachandran, P.N., Yamagata, T.: A dipole mode in the tropical Indian Ocean. Nature **401**(6751), 360–363 (1999)
3. Gara, M.I., Dwiridal, L., Nugroho, S.: Analysis of the characteristics of the return period of rainfall using the Iwai Kadoya method in the local Area of West Sumatera. Pillar Phys. 47–52 (2019)
4. Enyew, B., Steeneveld, G.: Analysing the impact of topography on precipitation and flooding on the. J. Geol. Geosci. 173 (2014)
5. Aldrian, E., Dwi Susanto, R.: Identification of three dominant rainfall regions within Indonesia and their relationship to sea surface temperature. Int. J. Climatol. 23, 1435–1452 (2003)
6. Bjerknes, J.: Atmospheric teleconnections from the equatorial Pacific. J. Phys. Oceanography, 163–172 (1969)

7. Huffman, G., Adler, R.F., Bolvin, D.T., Gu, G.: The TRMM multisatellite precipitation analysis (TMPA): Quasi-Global, multiyear, combined-sensor precipitation estimates at fine scales. J. Hydrometeorology **8**(1) (2007)
8. Schumacher, C., Houze, R.A.: Stratiform rain in the tropics as seen by the TRMM precipitation radar. J. Clim. **16**(11), 1739–1756 (2003)
9. Kuswanto, H., Naufal, A.: Evaluation of performance of drought prediction in Indonesia based on TRMM and MERRA-2 using machine learning methods. Methods **6**, 1238–1251 (2019)
10. Arno, G.: Distribution of accuracy of TRMM daily rainfall in Makassar strait. Forum Geografi. **32**(1), 38–52 (2018)
11. Johnson, R.A., Wichern, D.W.: Applied multivariate statistical analysis. New Jersey p 07458 Prentice Hall (2002)
12. Von Storch, H., Zwiers, F.W.: Statistical analysis in climate research. Cambridge University Press, p 484 (1999)
13. Heideman, M., Johnson, D., Burrus, C.: Gauss and the history of the fast fourier transform. IEEE ASSP Mag. **1**(4), 14–21 (1984)
14. Hermawan, E.: Penggunaan Fast Fourier Transform Dalam Analisis Kenormalan Curah Hujan Di Sumatera Barat dan Selatan Khususnya Saat Kejadian Dipole Mode. Jurnal Meteorologi dan Geofisika (2007)
15. Nurutami, M.N., Hidayat, R.: Influences of IOD and ENSO to Indonesian rainfall variability: role of atmosphere-ocean interaction in the indo-pacific sector. Procedia. Environ. Sci. **33**, 196–203 (2016)

Chapter 20
Analysis and Verification of Fire Danger Rating System (FDRS) Parameters in Land and Forest Fire in West Kalimantan in 2019 and Its Relationship with Hotspots and Rainfall

Jihan Putri Amelia, Zadrach Ledoufij Dupe, and Indah Prasasti

Abstract Land and forest fire in 2019 adversely affected Indonesia, one of them occurred in West Kalimantan. The government has been trying to develop an early warning system that is FDRS. However, the utilization of FDRS still requires verification on fire events in the field, especially in West Kalimantan, and the study of its relationship to the distribution of hotspots and rainfall. The study aims to analyze and verify FDRS parameters in land and forest fire events in West Kalimantan in 2019 and its relationship to hotspots and rainfall. The data used in the study were FDRS data from LAPAN, climate data from meteorological observations, hotspot data from MODIS Terra-Aqua, and location of the burned area from the Ministry of Environment and Forestry for verification. The analysis method in this study used correlation analysis (r) and coefficient of determination (R^2). The results showed that the parameter values of FFMC, DC, ISI, and FWI in the area of forest and land fires ranged from 48 to 89, 6 to 733, 0 to 5, 0 to 23. The FDRS parameter is strongly influenced by rainfall conditions 21 days before the fire event ($\overline{r} = -0.54$ to -0.66) and the number of days with light rainfall before the fire event ($\overline{r} = 0.54$–0.61). About 27.85% of fire events are affected by rainfall conditions 21 days before the fire event, which is the lower the accumulation of rainfall 21 days before the fire event, the potential for fire events will increase. About 28.51% of fire events are affected by the number of days with light rainfall (<20 mm/day) before the fire event; the longer of the day with rainfall <20 mm, the greater the potential for fire events. The number of hotspots ($\overline{r} = 0.62$–0.68) and the burned area ($\overline{r} = 0.64$–0.74) have possibility can increase along with the classification of parameter FDRS. The results also showed that about 70.48% of fire events could be identified based on the number of hotspots, and FWI has an excellent ability to be used as an early warning indicator of land and forest fire in West Kalimantan.

J. P. Amelia (✉) · Z. L. Dupe
Meteorology Department, Bandung Institute of Technology, Bandung 40132, Indonesia
e-mail: ameliajihanputri@gmail.com

I. Prasasti
Research Center for Remote Sensing, National Research and Innovation Agency (BRIN), Jakarta, Indonesia

E. Yulihastin et al. (eds.), *Proceedings of the International Conference on Radioscience, Equatorial Atmospheric Science and Environment and Humanosphere Science, 2021*, Springer Proceedings in Physics 275, https://doi.org/10.1007/978-981-19-0308-3_20

20.1 Introduction

Land and forest fires are some of the environmental issues that can impact the economic sector in various countries. It is also considered a threat to sustainable development due to its direct effect on ecosystems and increase carbon emissions that could further affect the global climate. Based on data from Global Wetlands (2019), Indonesia is the second country that has a peatland area of 22.5 million hectares (ha) after Brazil [10]. The period of land and forest fire in Indonesia usually occurs in the dry season, namely August, September, and October, as well as the transition period. In 2019, the fire was also related to the El Niño Southern Oscillation (ENSO), which caused the fire to get worse. This happened because, during the ENSO period, there was a significant reduction in the amount of rainfall in several regions in Indonesia [3].

In 2019, massive land and forest fire occurred in Indonesia, especially in west Kalimantan. Based on data from the Ministry of Environment and Forestry in 2019, the land and forest fire area in West Kalimantan in 2019 amounted to 151,919 ha [17] and BNPB (2019) recorded the most significant number of hotspots in West Kalimantan with 384 hotspots on September 16, 2019 [6].

One of the efforts to mitigate land and forest fire in the development of early warning information systems through the Fire Danger Rating System (FDRS), which was first developed by Canada which was used for policy in making decisions related to land and forest fire management. FDRS is a rating used to determine the rate of land/forest fire events in a region based on weather conditions (rainfall, relative humidity, air temperature, and wind speed), vegetation, and others, which affect fire behavior. FDRS consists of six parameters, namely Fine Fuel Moisture Code (FFMC), Duff Moisture Code (DMC), Drought Code (DC), Initial Spread Index (ISI), Build-Up Index (BUI), and Fire Weather Index (FWI). FFMC is used to describe the potential level of the ignitability of a fire, DMC for the potential flammability of fuels, DC for the potential for drought and smoke, ISI for the difficulty of fire control, BUI for fuel availability index, and FWI for fire weather index [7]. However, LAPAN only uses four parameters in FDRS monitoring activities, namely: FFMC, DC, ISI, and FWI.

The FWI parameters of FDRS perform well in detecting fire [12]. The same result was also showed that there is a linear relationship between FWI and fire events [5]. The results of Vetrita et al., study also showed that the parameters FFMC ($r = 0.68$) and DC ($r = 0.88$) estimated using satellite data had an excellent correlation with those obtained from observation data. However, the values of FFMC and DC from satellite data were higher (overestimate), especially in the dry season [20]. Nevertheless, FFMC and DC parameters derived from satellite data are quite well used for monitoring and early warning of land and forest fire. Meanwhile, Prasasti et al., found that DC parameters correlate quite well with burned area in the Central Kalimantan [14].

FDRS parameters usually use point data of the station and to be spatial information requires interpolation techniques [11]. However, this can lead to errors, especially if the station's point distance is very far away and does not represent the topography of the region [13]. The differences in interpolation techniques used will also result in different spatial information [11]. This corresponds with Baihaqi's research, which said that the resulting FWI value tends to be overestimated if utilize observation data [2]. One way to overcome the limitations of observation data in this research is to use prediction data and remote sensing data that have wide enough coverage to be cost-effective and efficient. Since 2005, LAPAN has developed FDRS that leverages remote sensing data to overcome the limitations of observation data. However, the result of FDRS parameters is still not widely verified to land and forest fire events and are studied the relationship of hotspot and rainfall monitoring data. In a previous study, Prasasti et al., uses NOAA as input data for FDRS calculations [14]. Based on that, the author sees a gap between research on FDRS, starting from the data used and its verification.

Therefore, the novelty in this study will use FDRS data from LAPAN that utilizes Himawari-8 satellite data and Weather Research Forecasting (WRF) model output. The aims of this study also analyze and verify the value of FDRS in land and forest fire area in 2019 in West Kalimantan to verify the capabilities of FDRS methods that have been developed by LAPAN in land and forest fire detection. In addition, it analyzes the relationship of FDRS parameters with hotspot and rainfall data to predict potential land and forest fires.

20.2 Data and Methods

20.2.1 Data

The data used in this study is FDRS parameter data obtained from PUSFATJA-LAPAN with a temporal resolution of 24 h and spatial 4 km. In addition, the data used as supporting data are meteorological observation data, hotspot data (from MODIS Terra Aqua), and burned area data (from The Ministry of Environment and Forestry) for verification.

20.2.2 Extraction of FDRS in Burned Area

Extraction of FDRS values was performed on the burned area through the following stages. The first stage is to equate the resolution of the FDRS data grid with the smallest size of the burned area. Then, the next step is FDRS data overlay with the location of the burned area, then identified the value of each FDRS parameter in each burned area.

20.2.3 Number of Hotspots with Radius 2 km from Burned Area

The number of hotspots in the burned area with a radius of 2 km is calculated from the center of the burned area. The first stage is to create a shapefile buffer for the burned area with a radius of 2 km from the center of the burned area. Then, the hotspot data overlay on the burned area shapefile data with a buffer area of 2 km and extract the number of hotspots that go into the burn area buffer. Next, calculate the number of hotspots by the month and input them in Microsoft Excel.

20.2.4 Extraction and Determination of Parameters to Be Analyzed

Extraction is carried out against FDRS parameters, hotspots, rainfall, number of days with light rainfall (<20 mm/day) in the burned area. The first thing is to calculate the average, minimum, and maximum values of the FDRS parameter values located in the burned area and the number of hotspots within a radius of 2 km from the burned area. Furthermore, it calculates the accumulation of rainfall 7 days before the fire (CH7), rainfall 14 days before the fire event (CH14), rainfall 21 days before the fire event (CH21), and the number of days with light rainfall (CHR) before the fire event. The fire events in this study were the location, date, and area of the burned area. The analysis was conducted by utilizing correlation method (r) and coefficient of determination (R^2) to analyze FDRS's ability to detect land and forest fire events in West Kalimantan in 2019.

Correlation

The correlation value (r) was used to measure the strength of relationships between parameters used in the study. In this study, the correlation used to measure the strength of the relationship between CH7, CH14, CH21, CHR, the number of hotspots within a radius of 2 km of the burned area (independent variable) against the area of the burned area and the FDRS parameter (dependent variable). In addition, a correlation was also used to measure the relationship between FDRS parameters (independent variables) and the burned area (dependent variables).

$$r_{xy} = \frac{\sum(x_i - \overline{x}) - (y_i - \overline{y})}{\sqrt{\sum(x_i - \overline{x})^2 \sum(y_i - \overline{y})^2}} \tag{20.1}$$

where,

x : Independent variables
$\overline{x}$: x mean

Table 20.1 Correlation coefficient interpretation (Sugiyono, 2010 in Safitri [16])

Coefficient interval	Level of relationship
0.800–1.000	Very strong
0.600–0.799	Strong
0.400–0.599	Moderate
0.200–0.399	Weak
0.000–0.199	Very weak

y : Dependent variables
$\overline{y}$: y mean (Table 20.1).

Coefficient of Determination

The coefficient of determination (R^2) is used to describe the percentage of a diversity of dependent changes that can be explained by independent change. The greater the R^2 value (close to 1), the greater the influence that independent variables (CH7, CH14, CH21, CHR, hotspot) have on dependent variables (FDRS Parameters and Burned Area).

20.3 Results and Discussions

20.3.1 Hotspot Distribution Patterns in West Kalimantan in 2019 and Their Relationship to Rainfall and Humidity

The spatial distribution of hotspots in the period 2019 in West Kalimantan is shown in Fig. 20.1. The accumulation of these hotspots obtained from the Terra-Aqua MODIS satellite from low to high confidence levels. The extraction results showed that the number of hotspots in West Kalimantan in 2019 reached 13,587 hotspots. Out of 13,587 hotspots, the Ketapang regency has more hotspots than any other district, amounting to 5826 throughout 2019. Other areas are Sintang 1490 hotspots, Sanggau 1078 hotspots, Kubu Raya 999 hotspots, Kapuas Hulu 932 hotspots, Kayong Utara 772 hotspots, Melawi 653 hotspots, Landak 465 hotspots, Sekadau 348 hotspots, Pontianak 340 hotspots, Sambas and Bengkayang 305 hotspots, Singkawang 46 hotspots, and the lowest number of hotspots was in Pontianak as many as 28 hotspots.

The number of hotspots fluctuated from January to June, then began to increase in July to as many as 384 hotspots and continued to increase until September reached the peak with the highest number of hotspots, that is 8447 hotspots. After entering October, hotspots fluctuate (up and down) and run into a drastic decrease in December with the number of hotspots as many as 15 hotspots. Thus, it appears that the fire period in 2019 occurred around the MAM and JJA—SON (Fig. 20.2).

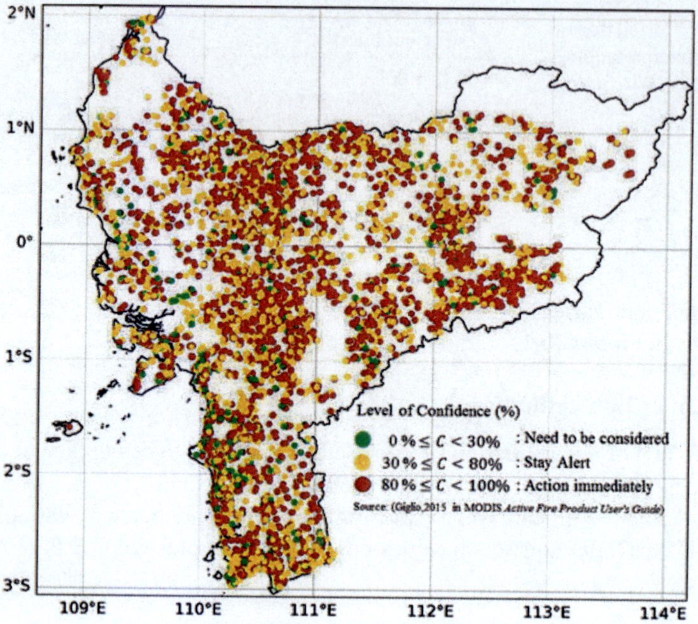

Fig. 20.1 Hotspot spatial distribution in West Kalimantan in 2019

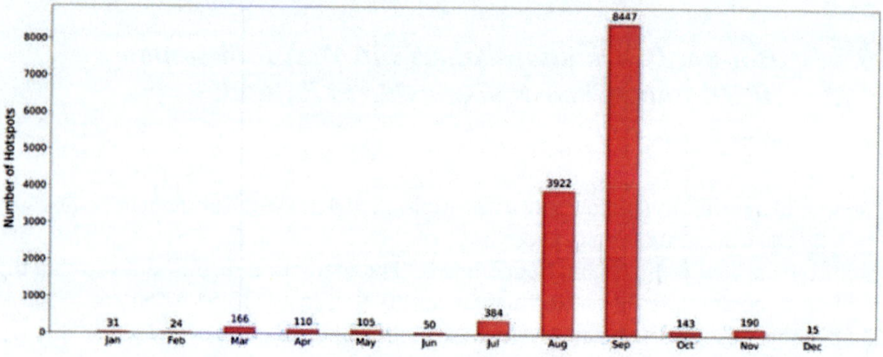

Fig. 20.2 Hotspot distribution in West Kalimantan in 2019

Meanwhile, Fig. 20.3 shows the relationship between hotspot monthly and relative humidity distribution patterns (Fig. 20.3a) and rainfall (Fig. 20.3b) from several stations in the West Kalimantan. Based on Fig. 20.3, in the period March to May, the number of hotspots is less than in June to August, because March to May is the transition month between the rainy to the dry season. Then, the number of hotspots increased dramatically in September, which appeared to coincide with low humidity conditions in that month and a decrease in rainfall. This condition triggers an increase in the intensity of forest fire. Then, the number of hotspots tends to continue to

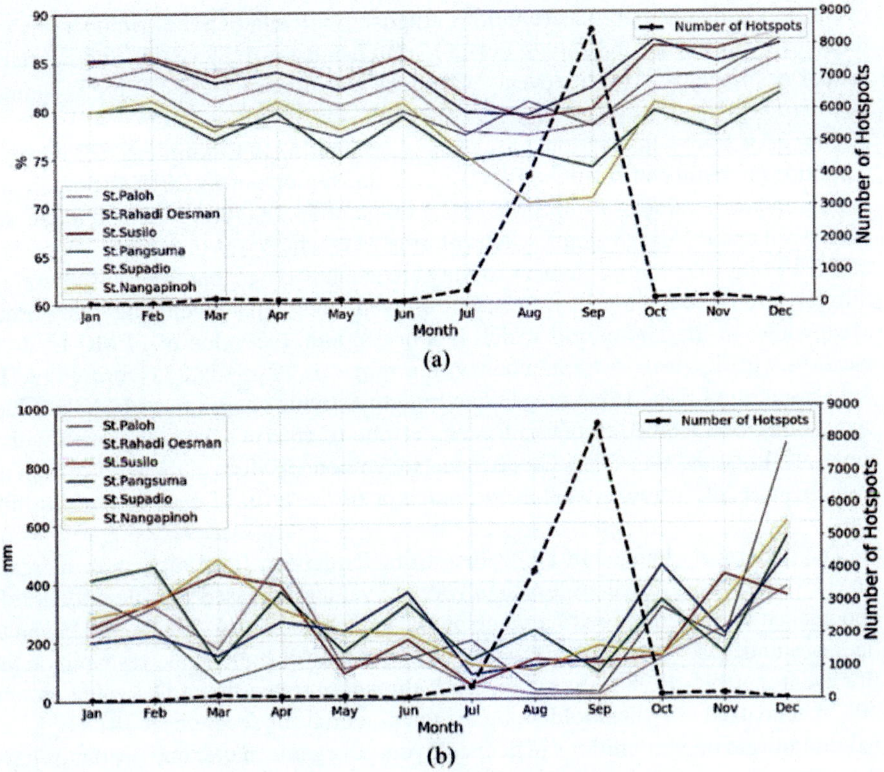

Fig. 20.3 Hotspot distribution with **a** relative Humidity and **b** rainfall in West Kalimantan in 2019

decrease from October to December. These conditions are also seen along with increased humidity and rainfall. This shows that the distribution of hotspots in the West Kalimantan region is greatly affected by humidity and rainfall conditions. The results of Syaufina et al., showed that rainfall would affect the condition of water content (moisture) of organic fuels on peatlands [18]. Tosic et al., found that relative humidity and rainfall are the best predictors used for forest fire prediction models. In addition, Tosic et al., also stated the intensity and size of the area affected by the fire depend on the value of meteorological elements [19].

20.3.2 FDRS Parameters in Burned Area

In this discussion, the value of FDRS parameters extracted from the location or based on a map of the fire area that occurred in West Kalimantan from January to December 2019. The map of the fire area was obtained from the Ministry of Environment and Forestry. During this period, there were 55 events of land and forest fire, the most events in October 2019 (10 events). In August, September, and November each of the 8 fire events. May and July (4 events), March and April (3 events), and February, June, December each two fire events. The fewest number of fires occurred in January 2019 (1 event).

Fig. 20.4 shows fluctuations in average, maximum, and minimum FFMC value conditions in burned areas in the West Kalimantan region during 2019. From January to February, the value of FFMC was in the range of 53.23–70.97. Then, it decreased until April and increased again, when entering May which was in the range of 59.55–68.25. FFMC values from January to May are at a low range, but fire is also burning. Similarly, in December there were still fires even though there were only two events. Meanwhile, in the JJA period which is a dry month, the value of FFMC tends to increase with its peak in September with a range of 79.65–89.23. During the JJA dry season, the FFMC value range from high to extreme conditions. FFMC's values again decreased from October to December due to entering the wet season with a range of 48.66–83.94. From October to November, FFMC values tend to remain high. The FFMC value is used as an indicator of the level of ease of burning fine fuel.

The pattern of changes in DC values from January to June appears to increase little by little, with DC values less than 100. DC values increased significantly in July and peaked in September with a range of DC values of 572.64–733.28 and began to decrease until December 2019 (Fig. 20.5). The results of Field et al., research on the fire event in 1994–1998 in Sumatra and Kalimantan stated that a DC value around 400, can be used as a threshold value for the potential for severe haze [8]. High DC values indicate drier conditions in deeper layers of organic matter and peatlands have the potential to increase the potential of deep-burning and upper layer (smoldering) fire. Thus, high DC values from August to November 2019 in the West Kalimantan region are in danger of drought and smoke. The DC value decreased dramatically when entering December, which was in the range of 6.94–7.59. This is due to the increasing rainfall.

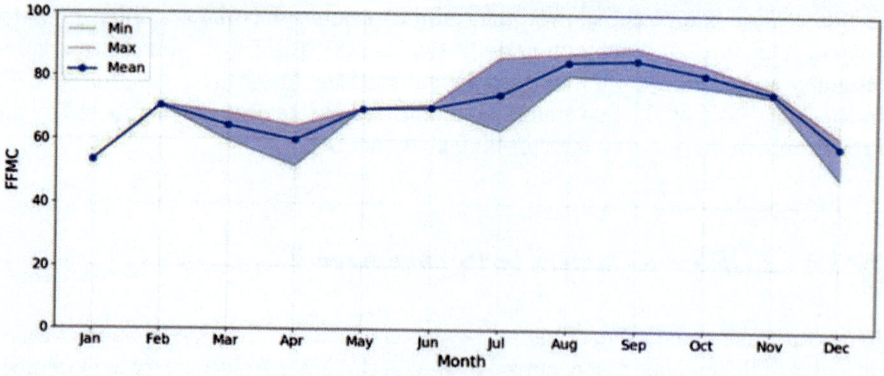

Fig. 20.4 FFMC time series in West Kalimantan in 2019

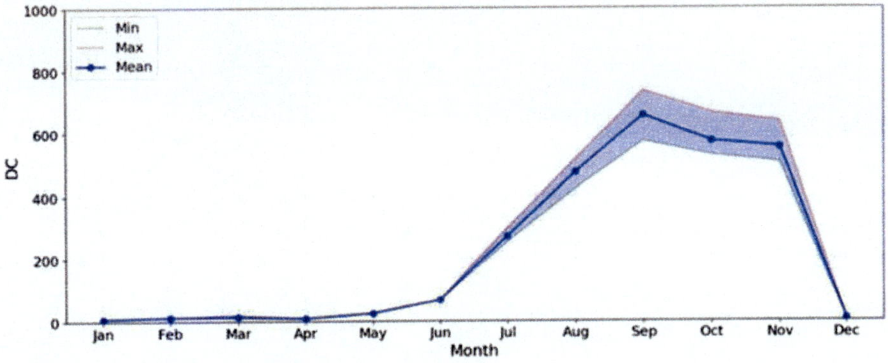

Fig. 20.5 DC time series in West Kalimantan in 2019

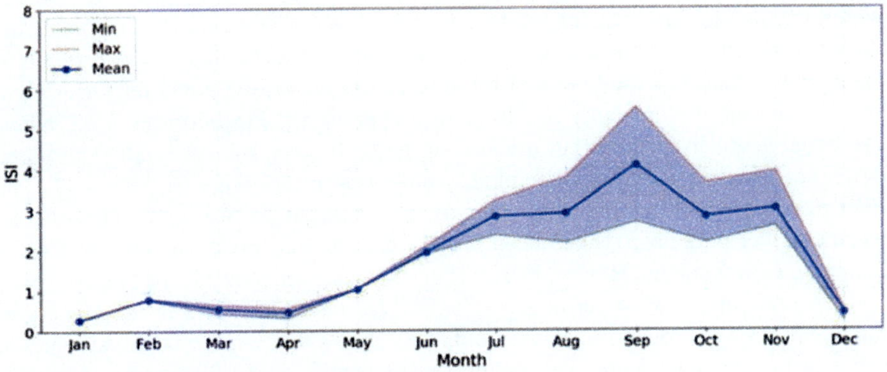

Fig. 20.6 ISI time series in West Kalimantan in 2019

Figure 20.6 shows a pattern of ISI value changes that illustrate the potential ease of spreading fire. From January to May, it was seen that ISI values were in the range of 0.27–0.79, which means its potential for low fire spread. A significant increase in ISI value has occurred since June. Furthermore, ISI values continued to increase until they reached a peak in September with a value of 2.69–5.59 and began to decrease again when entering October to December ISI values. In December, the ISI value was in the range of 0.22–0.49, which means the potential spread of fire decreased. ISI values are used as an indicator of the level of ease of fire spread. Even the intensity of grassland fires can be estimated using ISI parameters only. This is because grasslands are easier to dry simultaneously and do not find a fuel that includes medium or heavy burn ease [7].

Based on Fig. 20.7 shows that the FWI value from January to May is in the range of 0.09–0.31 or in safe fire weather conditions which means if there is a fire, then in that month the fire is easily extinguished by itself. When entering the dry season, the FWI value increased sharply with a peak in September with a value of 11.09–23.14,

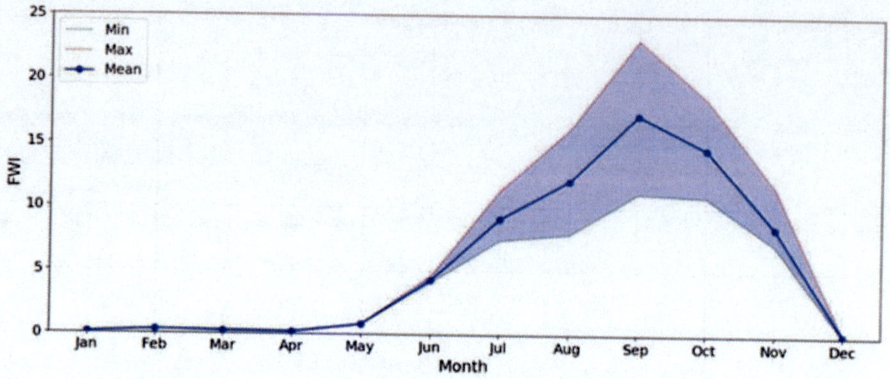

Fig. 20.7 FWI time series in West Kalimantan in 2019

which means the potential for weather is very easy for forest fires to occur. If the values of FFMC, DC, and ISI are in extreme conditions, then the fire is difficult to extinguish because of the potential for ease of fire ignite, the potential for drought and smoke, and the difficulty of fire control also higher. Furthermore, FWI values decreased again in October but are still in the high category. FWI values dropped dramatically when entering December, which was in the range of 0.07–0.17. The FWI parameter is a common indicator used in monitoring fire potential based on weather. FWI parameters have been widely used in several countries, such as in America, New Zealand, Russia, Fiji, and in some countries in Europe [7].

In the January to May period, FFMC, DC, ISI, and FWI were in the low (safe) range, but fires still occurred in the field. This shows that a low FDRS value is not necessarily safe against fire. This condition is because almost 99% of fire events in Indonesia are caused by human behavior and climate factors are only a trigger.

20.3.3 Relationship of FDRS Parameters with Hotspot, Rainfall, and Burned Area

This study also analyzed the pattern of relationships between FDRS parameters and hotspots, rainfall, number of days with light rainfall (number of days with rainfall <20 mm/day), and the burned area (Figs. 20.8, 20.9, 20.10, and 20.11).

Based on Figs. 20.8, 20.9, 20.10, and 20.11, it can be seen that the highest FFMC, DC, ISI and, FWI values coincide with the condition of the highest number of hotspots (Figs. 20.8a, 20.9a, 20.10a, and 20.11a). The maximum values for FFMC, DC, ISI, and FWI also occur when the number of days with the lightest rainfall is 17–18 days from July to September (Figs. 20.8c, 20.9c, 20.10c, and 20.11c). In addition, increased the values of FFMC, DC, ISI, and FWI are strongly affected by the condition of decreasing the amount of rainfall, when rainfall 7 days, 14 days, and 21 days before the fire event were significantly reduced, so the potential for easy ignition

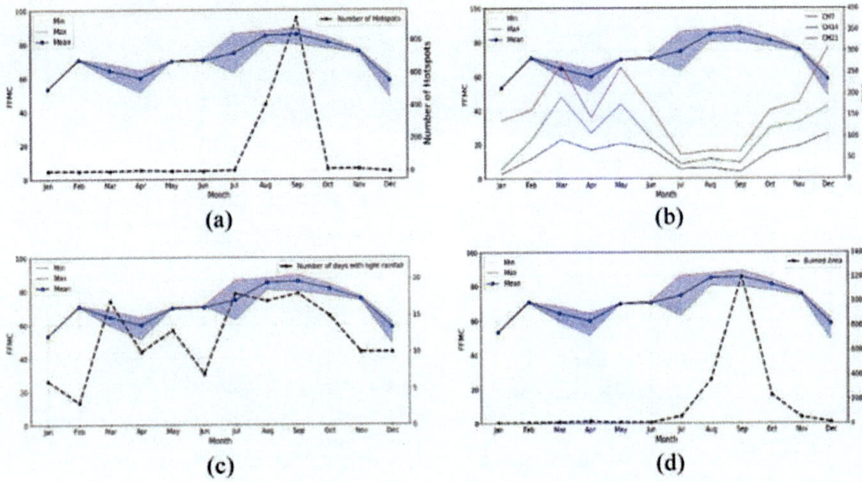

Fig. 20.8 Time series between FFMC with **a** number of hotspot, **b** CH, **c** CHR, and **d** burned area

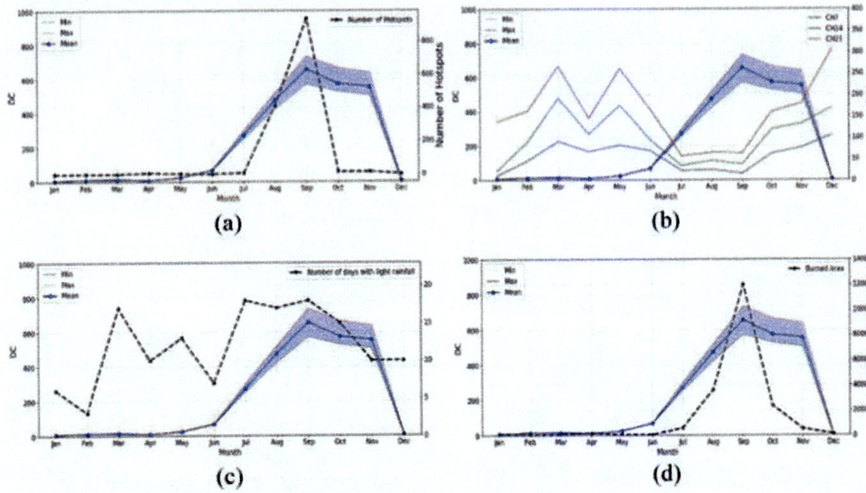

Fig. 20.9 Time series between DC with **a** number of hotspot, **b** CH, **c** CHR, and **d** burned area

(FFMC), drought and smoke (DC), potential fire control difficulty (ISI), and fire weather (FWI) will slowly increase (May to July period). However, an increase in the amount of rainfall does not necessarily reduce the value of the FDRS parameter (October to December period), which means that there is a lag time of about 1 month after the rainfall increases. This pattern is very clear as shown in Figs. 20.8b, 20.9b, 20.10b, and 20.11b. This condition caused the burned area to increase slowly and was seen most widely in September 2019 (Figs. 8d, 20.9d, 20.10b, and 20.11b).

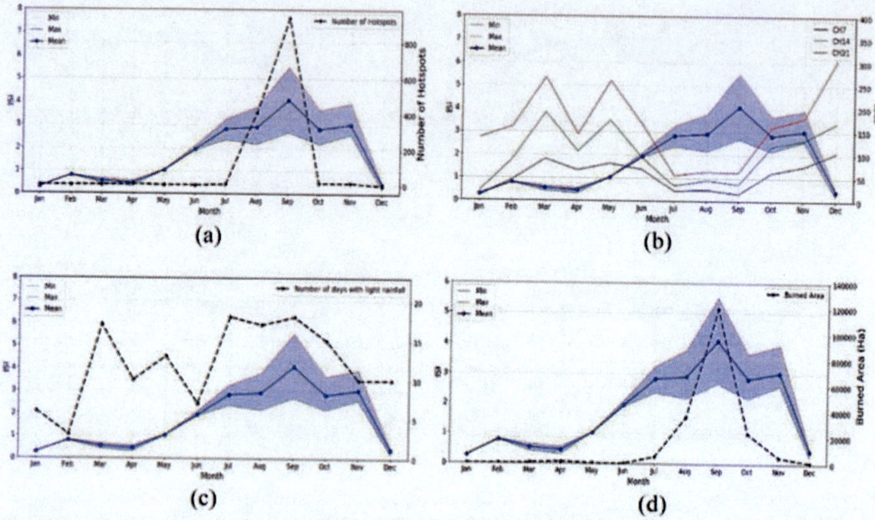

Fig. 20.10 Time series between ISI with **a** number of hotspot, **b** CH, **c** CHR, and **d** burned area

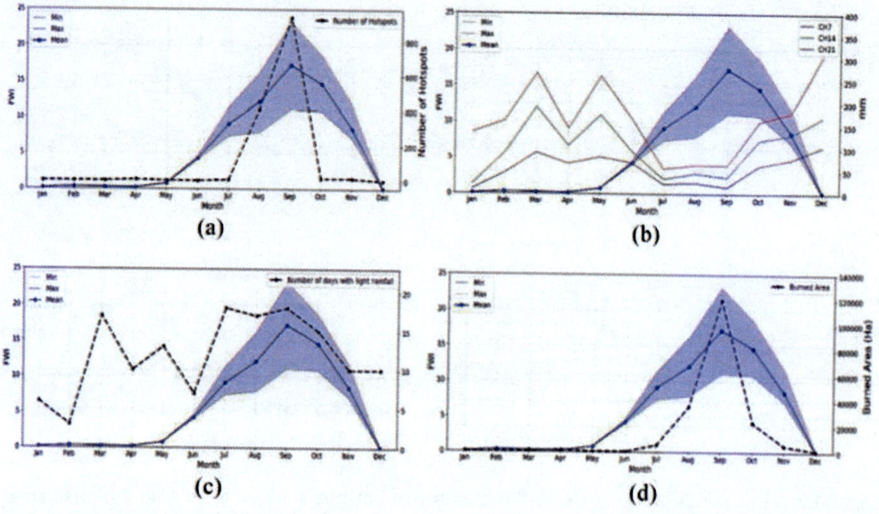

Fig. 20.11 Time series between FWI with **a** number of hotspot, **b** CH, **c** CHR, and **d** burned area

The results of this study also show that a decrease in the level of air humidity also affects increasing the number of hotspots and expanding the burned area (Fig. 20.12). In addition, an increase in the number of hotspots also increases the burned area (Fig. 20.13a), a decrease in rainfall 7 days, 14 days, and 21 days before a fire and an increase in the number of days with light rainfall (<20 mm/day) also has the potential to increase the number of hotspots (Fig. 20.13b and c).

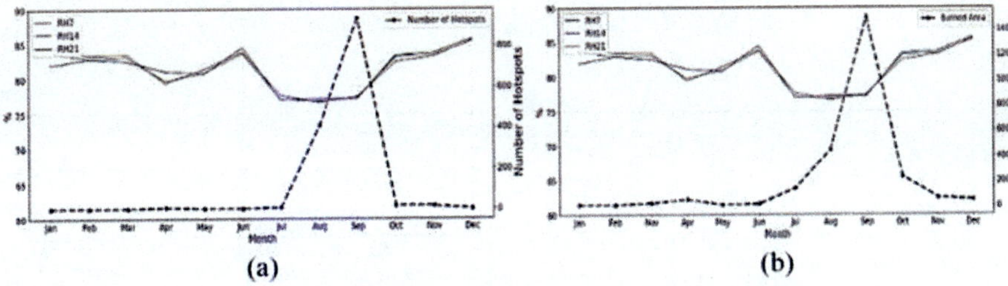

Fig. 20.12 Time series between relative humidity with **a** hotspot and **b** burned area

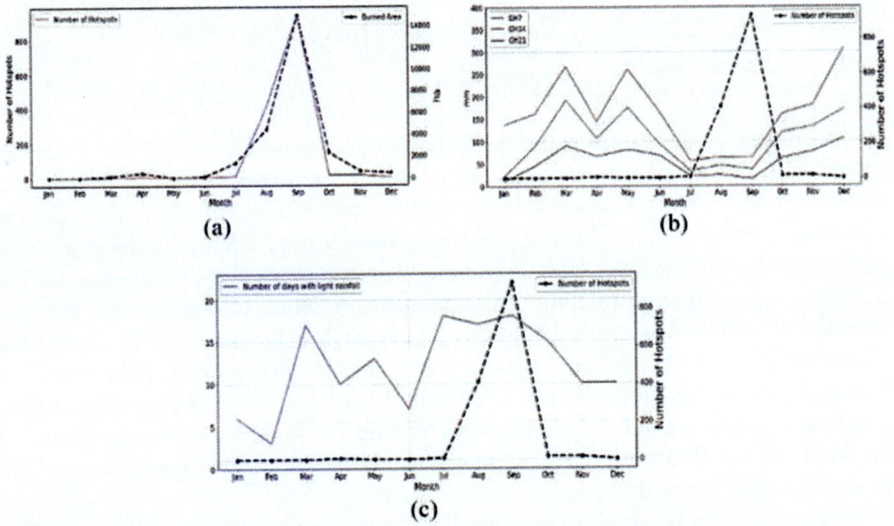

Fig. 20.13 Time series between number of hotspot with **a** burned area, **b** CH7, CH14, CH21, and **c** CHR

The burned area also appears to increase with decreasing rainfall 7 days, 14 days, and 21 days before the fire and with a longer period of light rainfall days (Fig. 20.14). The FDRS parameter in the West Kalimantan region in 2019 has a tendency of high values in the dry season period, with a peak in September with FFMC >79, DC >572, ISI >2, and FWI >11. In addition, the burned area reached 14,995.84 ha. The critical value of this FDRS parameter indicates the possibility of a high fire igniting because the condition is in a drought and fire hazard. This critical value should serve as an early warning against fire hazards and the local government start to increase awareness and alertness for land and forest fires disasters.

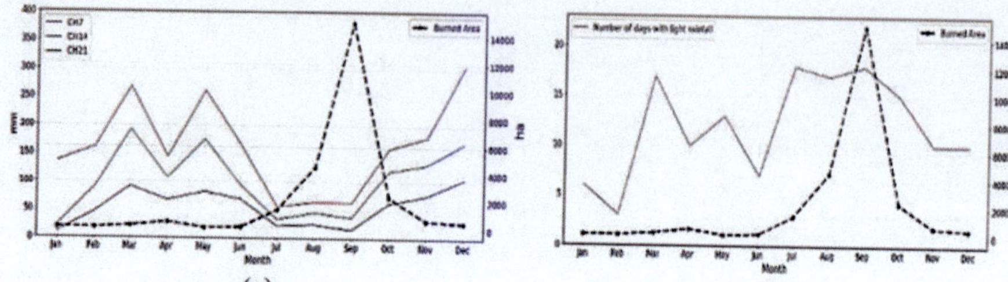

(a) (b)

Fig. 20.14 Time series between burned area with **a** CH7, CH14, CH21 and, **b** CHR

20.3.4 Relationship of Rainfall and Hotspot to FDRS Parameters

Based on the Table 20.2 results of the correlation between the FDRS parameter values with the accumulation of rainfall 7 days before the fire event (CH7) and the accumulation of rainfall 14 days before the fire event (CH14) that indicates a low to moderate correlation. Meanwhile, the correlation of the FDRS parameter with the accumulation of rainfall 21 days before the fire event (CH21) is classified as having a moderate to strong correlation with a strong correlation on the average value of FFMC, ISI, and FWI. Thus, if the rainfall 21 days before the fire event decreases, the FDRS parameter will be higher and the risk of fire potential increases. The results also found that the correlation between CH21 and FWI was relatively high. This shows that lower the amount of rainfall 21 days before the fire event, so the weather potential for fire hazard (FWI) was increased (negative correlation), and about 42.31% of fire events are affected by this condition.

The correlation between FDRS and CHR has a low to strong correlation. The highest correlation value was obtained with the FWI parameter with a coefficient of determination of 40.66% (R^2). It shows that about 40.66% of the FWI parameters are influenced by the condition of the length of the day of light rainfall. If the length of the day of light rain is longer, the potential for the fire weather index will be higher (positive correlation).

Meanwhile, the correlation of FDRS parameter values with the number of hotspots shows a moderate to strong correlation. The correlation value with the maximum,

Table 20.2 Correlation of rainfall, hotspot to FDRS parameters

	CH7	CH14	CH21	CHR	Hotspot
FFMC	−0.35 to −0.39	−0.28 to −0.44	−0.52 to −0.54	0.31–0.66	0.54–0.62
DC	−0.35 to −0.36	−0.32 to −0.44	−0.52 to −0.53	0.53–0.54	0.60–0.62
ISI	−0.40 to −0.44	−0.43 to −0.44	−0.52 to −0.66	0.46–0.56	0.50–0.69
FWI	−0.43 to −0.47	−0.43 to −0.45	−0.52 to −0.65	0.58–0.62	0.58–0.71
Hotspot	−0.51	−0.44	−0.52	−0.48	

Table 20.3 Coefficient of determination of rainfall, hotspot to FDRS parameters

	CH7	CH14	CH21	CHR	Hotspot
FFMC	13.33	10.63	35.08	31.53	65.54
DC	13.32	11.3	30.16	32.28	56.06
ISI	19.45	20.48	44.17	32.58	55.68
FWI	21.88	20.3	42.31	40.66	60.31
Hotspot	58.35	38.11	34.90	33.06	

minimum, and average FWI in burned areas in West Kalimantan shows the highest correlation among other FDRS parameters. An increase in FWI conditions, 60.31% will have the potential to increase the number of hotspots. It means that increasing FDRS parameters will have the opportunity to increase more than 50% of the number of hotspots.

The correlation between the number of hotspots and the CHR is quite strong and 33.06% (R^2) of the CHR variance will affect the increase in the number of hotspots. If seen from the graph of the relationship pattern of the CHR with hotspots (Fig. 20.13c), it shows that in the period from March to September the number of days with light rainfall is getting longer. This condition certainly causes the fuel to become drier and will increase the potential for easy fire; this is aligned with the theory of Chen et al. [4]. In addition, during this period there was an equinox phenomenon that occurred twice a year, namely on March 20 and September 23 [1]; this event caused the air temperature to increase so that it became one of the factors that triggered the ease of fuel to burn [15]. Then, if we look at the hotspot relationship with CH7, CH14, CH21, it produces a moderate negative correlation. Based on the correlation results, it shows that CH21 has the highest correlation among other CH with the number of hotspots.

This indicates that if the rainfall 21 days before the fire event decrease, the number of hotspots is predicted to increase with a diversity level of around 34.90% (R^2) (Table 20.3).

20.3.5 *Relationship of Rainfall, Hotspot, and FDRS Parameters to Burned Area*

Based on Table 20.4, if the rainfall 21 days before the fire decreases, the chance of fire spreading will increase and 27.85% (R^2) the variance of the burned area affected by

Table 20.4 Correlation and coefficient of determination of rainfall, hotspots, FDRS on fire event

	CH7	CH14	CH21	CHR	Hotspot	FFMC	DC	ISI	FWI
r	−0.49	−0.44	−0.53	0.52	0.98	0.53–0.64	0.65–0.68	0.55–0.73	0.65–0.76
R^2(%)	23.34	18.94	27.85	28.51	70.48	41.03	45.57	47.94	55.24

CH21. Then, the correlation of CHR with the burned area has a moderate relationship and 28.51% (R^2) the variance of the burned area is influenced by CHR, which means that the longer the number of days with light rainfall (CHR) the potential to expand the fire area.

Hotspots do not necessarily describe forest fires, but according to Ivan et al., hotspots that are numerous and clustered indicate a strong indication of fire occurrences in the area, so that hotspot conditions like this can be used as an indicator of forest fires [9]. The results of this study indicate that the greater the number of hotspots, the higher the potential for the burned area with a correlation of 0.98 and about 70.48% of the extent of the burned area can be identified based on a large number of hotspots.

Most of burned area with FDRS parameter values showed a strong relationship and FWI had the highest correlation. About 55.24% (R^2) burned area variance can be identified by FWI. This shows that FWI can be used as an excellent fire early warning indicator.

Meanwhile, the correlation results with each tested variable have a moderate to a very strong relationship with the burned area. The results show that the burned area will increase as the amount of rainfall before the fire decreases (CH21), the number of days with light rainfall (CHR) increases, the number of hotspots, and FDRS parameters, especially FWI. Thus, the FDRS parameters, especially FWI, can be used as a fire early warning system, while the CH21 and CHR parameters can be used as predictive parameters for forest/land fires in West Kalimantan. The implementation was applied using the results of predictions of rainfall for the next 3 months and the potential for rain less than 50 mm.

20.4 Conclusion

Land and forest fire in West Kalimantan in 2019, around 27.85% was affected by rainfall conditions before the fire event ($r = -0.53$ and $R^2 = 27.85\%$), about 28.51% by the number of days with light rainfall ($r = 0.52$ and $R^2 = 28.51\%$), and about 70.48% can identified based on the number of hotspots ($r = 0.98$ and $R^2 = 70.48\%$). In addition, the FDRS parameter is affected quite strongly by the rainfall conditions 21 days before the fire event ($\overline{r} = -0.54$ to -0.66) and the number of days with light rainfall ($\overline{r} = 0.54$–0.61). The higher and the extreme class of each FDRS parameter, the number of hotspots will increase ($\overline{r} = 0.62$–0.68) and the burned area will also potentially increase ($\overline{r} = 0.64$–0.74). These results indicate that FDRS; especially FWI, has a good ability to be used as an early warning system for potential land and forest fire that occurred in West Kalimantan in 2019.

Acknowledgements The author would like to thank Remote Sensing Application Center, Indonesian National Institute of Aeronautics and Space for the support of FDRS data, especially to Mrs.

Farikhotul Chusnayah, S.T., M.Eng, the Ministry of Environment and Forestry, in particular Mr. Radian Bagiyono, S. Hut., M.For and Mr. Judin Putranto, for support of the map of the fire event area.

Authors Contribution Jihan Putri Amelia (main contributor: data processor, analyst, writing); Zadrach Ledoufij Dupe (member contributor: drafter, advisor, and consultant); Indah Prasasti (member contributor: drafter, analyst, and writing).

References

1. Aflahah, E., Hidayati, R., Hidayat, R., Alfahmi, F.: The restoration of hotspots as an indicator of forest fires in kalimantan based on climate factors. JPSL **9**(2), 405–418 (2019). https://doi.org/10.29244/jpsl.9.2.405-418
2. Baihaqi, A.: Use of Canadian Forest Fire Weather Index Method in Determining the Level of Forest Fire Vulnerability (Study Case: West Kalimantan). Bandung, Bandung Institute of Technology (2014)
3. Bangsawan, L., Pandjaitan, B.S.: Identification of forest fires in West Papua using Himawari-8 Satellite. In: Proceedings Workshop Operational Weather Satellite, 2 Dec 2015
4. Chen, F., Fan, Z., Niu, S., Zheng, J.: The Influence of Precipitation and Consecutive Dry Days on Burned Areas in Yunnan Province. Advances in Meteorology, Hindawi Publishing Corporation, Southwestern China (2014)
5. Cloppet, E., Regimbeau M.: Fire weather index: from high-resolution climatology to climate change impact study. In: International Conference on current knowledge of Climate Change Impacts on Agriculture and Forestry in Europe, 3–6 May 2011. COST- WMO, Topolcianky. [Accessed Aug 10, 2021]
6. CNN Indonesia: Comparing Land and Forest Fire in 2015 and 2019 (2019). Accessed on November 9, 2020, from https://www.cnnindonesia.com/teknologi/20190918104533-199-431485/membandingkan-karhutla-di-indonesia-pada-2015-dan-2019.
7. De Groot, W.J., Field, R.D., Brady, M.A., Roswintiarti, O., Mohamad, M.: Development of the Indonesian and Malaysian fire danger rating systems. Mitig. Adapt. Strat. Glob. Change. **12**, 165–180 (2006). https://doi.org/10.1007/s11027-006-9043-8
8. Field, R.D., Wang, Y., Roswintiarti, O., Guswanto: A drought-based predictor of recent haze events in western Indonesia. Atmos. Environ. **38**, 1869–1878 (2004)
9. Ivan, P.A., Ifran, D.I., Muhnandar.: Vegetation Fires in Indonesia: The Interpretation of NOAA Derived Hot-Spot Data. Palembang, Center for Forest Inventory and Mapping Region II and the Regional Office for Forestry and Plantations (1999)
10. Katadata Publication Team.: Indonesia's Second Largest Peat Area in the World (2019). Accessed on November 9, 2020, from https://katadata.co.id/timpublikasikatadata/infografik/5e9a519433cb1/luas-gambutindonesia-terbesar-kedua-di-dunia.
11. Khomarudin, M., Roswintiarti, O., Tjahjaningsih, A.: Estimated Weather Elements to Support Land/Forest Fire Danger Rating System with MODIS data. Jakarta, LAPAN (2005)
12. Lawler, M.: Analysis of the American National Fire Danger Rating System (NFDRS) vs the Canadian Fire Weather Index System (FWI) for the Superior National Forest. District Fire Management Officer: U.S Forest Service (2004)
13. Narasimhan, B., Srinivasan, R.: Determination of Regional Scale Evapotranspiration of Texas from NOAA-AVHRR Satellite. Final Report Submitted to Texas Water Resources Institute. March. 5. Texas, USA (2002)
14. Prasasti, I.: Utilization of READY-ARL NOAA Data and CMORPH for Land and Forest Fire Risk Model Development in Central Kalimantan. Bogor, Bogor Agricultural Institute (2012)

15. Pyne, S.J.: Introduction to Wildland Fire: Fire Management in the United States. John Willey and Sons, New York (1984)
16. Safitri, W. R.: Pearson correlation analysis in determining the relationship between dengue hemorrhagic fever incidence with population density in Surabaya City in 2012–2014. J. Nurs. Sci., **2**(2) (2016). ISSN 2477-4391
17. SiPongi Karhutla Monitoring System: Recapitulation of Land and Forest Fires (Ha) Per Province in Indonesia in 2015–2020. Jakarta, PKHL Directorate of the Ministry of Environment and Forestry. Accessed on November 9, 2020 (2015), from http://sipongi.menlhk.go.id/hotspot/luas_kebakaran
18. Syaufina, L., Nuruddin, A.A., Basharuddin, J., See, L.F., Yusof, M.R.M.: The Effects of climatic variations on peat swamp forest condition and peat combustibility. J. Tropical Forest Manage. **5**(1), 1–14 (2004)
19. Tošić, I., Mladjan, D., Gavrilov, M., Živanović, S., Radaković, M., Putniković, S., Petrović, P., Mistridželović, I., Marković, S.: Potential influence of meteorological variables on forest fire risk in Serbia during the period 2000–2017. Open Geosci. **11**, 414–425 (2019). Published by De Gruyter. https://doi.org/10.1515/geo-2019-0033
20. Vetrita, Y., Prasasti, I., Haryani, N.S., Priyatna, M., Khomarudin, M.R.: Drought and fine fuel moisture code evaluation: an early warning system for forest/land fire using remote sensing approach. Int. J. Remote Sens. Earth Sci. **9**(2), 140–147 (2012)

Chapter 21
Exploring an Extension of Space Situational Awareness in Southeast Asian Region Utilizing EAR Observation Data

Afrizal Bahar, Varuliantor Dear, Asnawi Husin, Agri Faturahman, Jiyo, and Rezy Pradipta

Abstract The concept of Space Situational Awareness (SSA) refers to the ability to detect and track threats, enabling appropriate mitigation measures aimed at protecting space assets. In this paper, we explored a possible extension of this concept by considering asset protection and connectivity to satellites in orbit by users under various circumstances. Turbulence in the ionospheric layer at low latitude can degrade connectivity to satellites (GNSS and SATCOM). Field Aligned Irregularities (FAI) from such turbulence are often detected by Equatorial Atmosphere Radar (EAR) when the radar beam is pointed perpendicular to magnetic field lines. EAR has been operational in Kototabang (0.20° S, 100.32° E; 10.36° S dip latitude), with beam steering capability up to 30 degrees from the zenith direction. At ionospheric altitude, EAR direct horizontal coverage is around ~200 km. However, considering the physical properties of Equatorial Plasma Bubbles (EPB), we can extend the effective "viewing area". The north–south elongation of EPB implies an additional ~4000 km coverage from EAR meridionally. The eastward drift of EPB implies a further ~2000 km extension of coverage zonally. In this scheme, EAR effectively "guesses" the location of FAI when they are not directly in the field of view. Our study reconstructed the EPB geometry in three dimensions (3-D) based on EAR FAI data and the above assumptions. The reconstructed 3D EPB geometry may be useful to many types of users, such as those in the maritime, aviation, or aerospace sectors operating in the Southeast Asian region, extending the SSA concept more widely.

A. Bahar (✉)
Agam Atmospheric and Space Observation Office, National Research and Innovation Agency (BRIN), Agam, Indonesia
e-mail: afri003@brin.go.id

V. Dear · A. Husin · A. Faturahman · Jiyo
Research Center for Space, National Research and Innovation Agency (BRIN), Bandung, Indonesia

R. Pradipta
Institute for Scientific Research, Boston College, Newton, USA

E. Yulihastin et al. (eds.), *Proceedings of the International Conference on Radioscience, Equatorial Atmospheric Science and Environment and Humanosphere Science, 2021*, Springer Proceedings in Physics 275, https://doi.org/10.1007/978-981-19-0308-3_21

21.1 Introduction

Traditionally, the term Space Situational Awareness (SSA) refers to the ability to detect and keep track of threats and implement appropriate mitigation measures aimed at protecting space and ground assets. For example, refer to the definition of SSA by Marcus Holzinger [1], PricewaterhouseCoopers (PwC) [2], and the United States Government Accountability Office (GAO) [3]. In this work, we want to explore an extension of the SSA concept. We noted that the idea of Space Situational Awareness (SSA) generally refers to the ability to detect and keep track of threats, emphasizing enabling appropriate mitigation measures aimed at protecting space assets. In this paper, we explore the possibility of extending this concept by focusing on safeguarding space assets and facilitating users' connectivity to satellites in orbit under various geospace conditions.

The Equatorial Atmosphere Radar (EAR) operates at 47.0 MHz with peak powers of 100 kW. It is located at Kototabang (0.20° S, 100.32° E; 10.36° S dip latitude), Bukittinggi, West Sumatra. EAR has an active phased-array antenna system, similar to that of the MU radar in Japan. It has a circular antenna array approximately 110 m in diameter consisting of 560 three-element Yagi antennas. EAR system configuration allows the antenna beam to be steered up to 30 ° from the zenith direction [4].

The primary scientific purpose of the EAR is to study the atmosphere from 1.5 to 20 km altitude, primarily the investigations of wind, waves, and turbulence in the neutral atmosphere. On the other hand, the EAR is also designed to observe the upper atmosphere (the ionosphere) with an altitude of more than 90 km in directions perpendicular to the geomagnetic field direction. Therefore, the EAR can operate as a coherent scatter radar to study field-aligned irregularities (FAI) in the ionospheric E and F regions. For upper atmospheric (ionospheric) observations, the primary study for EAR is the investigation characteristics and signature of that FAI. For detecting the FAI, the EAR views a fan-shaped azimuth-range sector (see Fig. 21.1). This azimuth scan covers approximately 600 km in the zonal direction at 500 km altitude [4].

Equatorial plasma bubbles (EPBs) is a depletions in plasma density of the ionosphere which known as the Equatorial ionospheric irregularities caused by the condition of the Rayleigh–Taylor instability (RTI). EPB affecting to the communication and satellitebased navigation system. EPB disrupt the radio communication, navigation, and geopositioning. EPB will impacting the performance and reliability of space borne and ground based electronic systems. The information of EPB occurrences will help the user to mitigate its impact to the related technology.

21.2 Methodology

Field Aligned Irregularities (FAI) indicate equatorial plasma bubbles (EPB) in the lowlatitude ionosphere. To successfully detect FAI, EAR beams must be directed

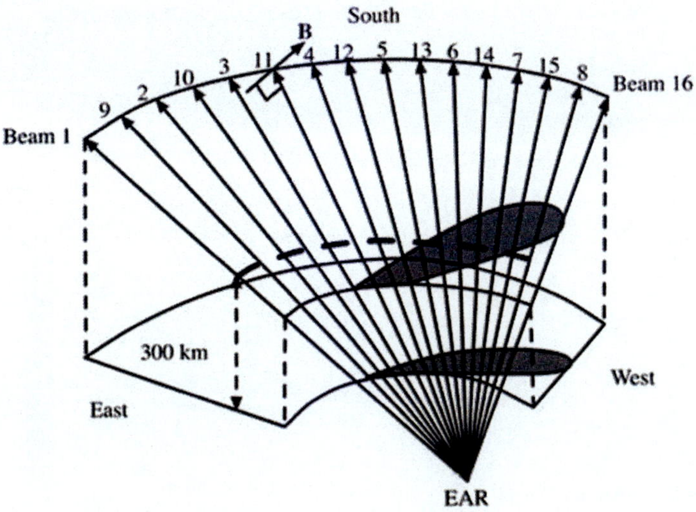

Fig. 21.1 An illustration of EAR fan-beam configuration for detecting FAI [4]

perpendicular to the geomagnetic lines of force. Based on its beam steering alone, the coverage of the EAR field-of-view is about 200 km West–East. However, EPB has unique physical properties that can deduce the actual area affected by EPB.

This expansion of adequate coverage (North–South, West–East, 3D geometry) is valuable information that can be incorporated into regional SSA capabilities. Although EAR only detects a small part of the EPB, by utilizing the unique physical properties of EPB, we can deduce the actual EPB geometry in 3D, which is significantly larger than the part detected by the EAR beam. In terms of technological implications, EPB can disrupt GNSS and SATCOM radio signals due to scintillation.

21.3 Result and Discussion

Figure 21.2 shows an example of FAI detected by the EAR fan-beam on the night of 1 March 2011. Here, the FAI appeared as high signal-to-noise (SNR) values within the radar beam. The EAR fan-beam only captured a slice of the entire EPB structures. The next step is to turn this single-slice FAI into a 3D structure that is elongated north–south, representing the proper geometry of EPB that gave rise to the observed FAI in the EAR observation data on that night.

The dipole field line Eq. (21.1) can be used for the north–south mapping of the FAI structures [5]. In the dipole approximation, the altitude h of a field line at different magnetic latitudes λ can be expressed as

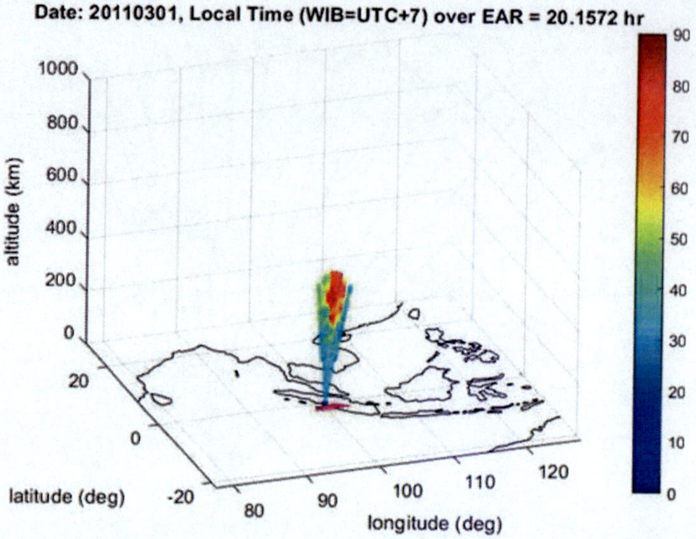

Fig. 21.2 The standard fan beam of EAR detecting the presence of some FAI over the Kototabang area

$$h = (R_{eq} + h_A) * \cos^2 \lambda - R_{eq} \tag{21.1}$$

where *Req* is the equatorial Earth's radius and h_A is the apex altitude of the field line at $\lambda = 0$ degrees. The resulting 3D point cloud from this geomagnetic field mapping procedure is shown in Fig. 21.3.

Usually, FAI/EPB starts growing around sunset time (18:00–19:00 LT). After that, FAI/EPB will drift eastward at speed close to 100 m/s. An hour or two past midnight, FAI/EPB will usually disappear. FAI/EPB observed over Kototabang probably came from an area to the west of Kotatabang (adopting a discussion presented by Fukao, URSI 2005). Then, the FAI/EPB will continue to sweep toward the area east of Kotatabang. An intuitive way to visualize this drift motion is by using a keogram plot, shown in Fig. 21.4 for the EAR FAI data on 1 March 2011. This means that the proper area where FAI/EPB exists during the entire night can be deduced from the EAR observation data, and this area is many times larger than the field-of-view of EAR itself.

For the practical purpose of representing the EPB movement from west to east, a surface mesh (derived from the 3D point cloud distribution) is used in the data visualization. In the 3D data visualization, the constructed surface mesh can be shifted along the east–west direction according to the observed velocity of the FAI structures and the time of observations during the night. This is illustrated diagrammatically in Fig. 21.5.

Figure 21.6 is an example of the full 3D geometry reconstruction of EPB detected in the area around EAR Kototabang on 1 March 2011 starting at 18:5 LT until 01:00

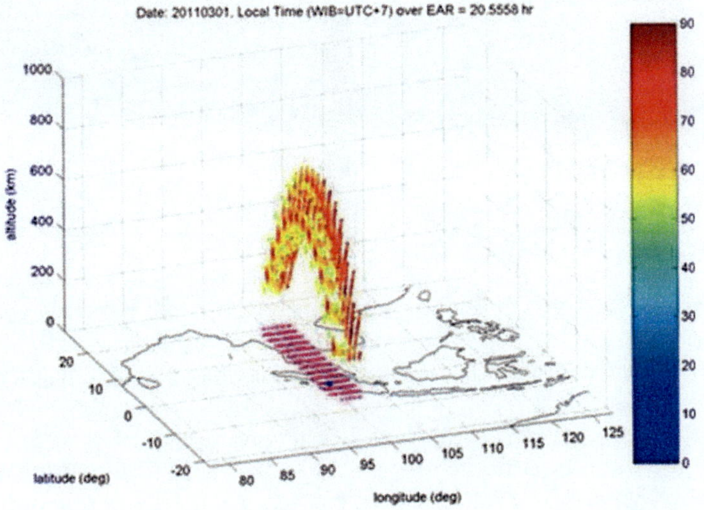

Fig. 21.3 Spatial distribution of FAI mapped north–south along the geomagnetic field lines

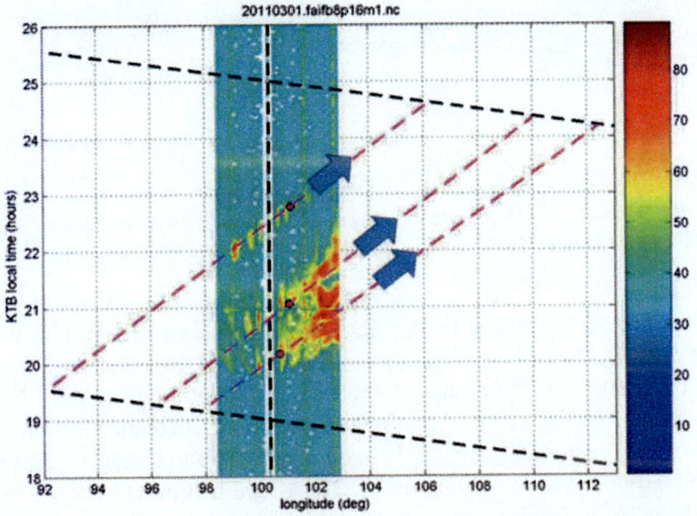

Fig. 21.4 Keogram data plot of power SNR recorded by EAR, which allows for a drift speed estimation

LT on the next day. Here we can see the projected movement of the EPB position, from the west side of Kototabang at longitude 90° E to its final position at longitude 115° E. In this case, the north–south extent of the EPB structure is from latitude −

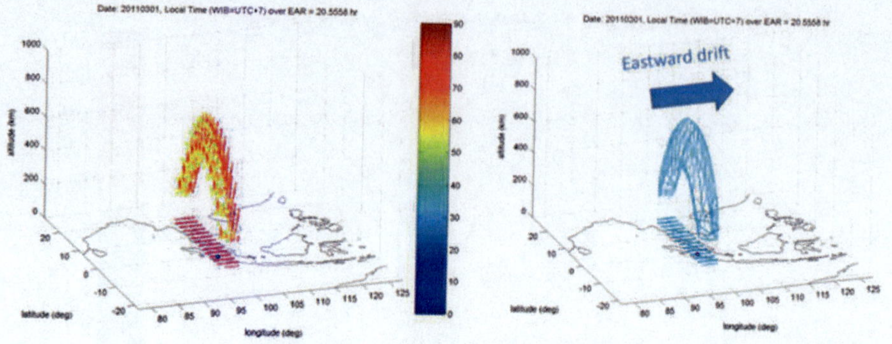

Fig. 21.5 FAI/EPB drift information reconstruction results using a 3D visualization format

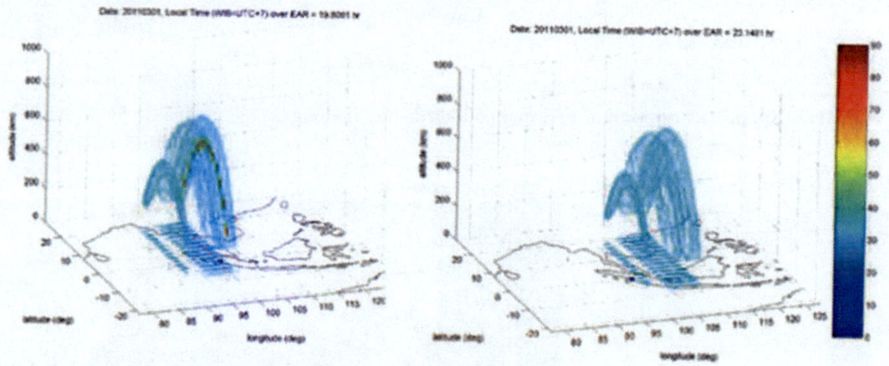

Fig. 21.6 Example of the 3D geometry and zonal drift motion reconstruction of EPB that were detected in the area around EAR Kototabang on March 1, 2011

5° S to latitude 20° N. Based on the 3D geometry reconstruction and the west-to-east drift of EPB, the effective coverage area that EAR can provide (for awareness information of EPB) is 2500 km × 2500 km = 6,250,000 km^2.

When implemented as an operational system, this extended SSA concept has several potential customers, especially in the Southeast Asian region. This region has a relatively large economy with a total GDP of 3.08 trillion US dollars in 2020 [6]. The Southeast Asian region is also quite active in maritime transport, civil aviation traffic, and geological explorations on land and offshore. This trend is illustrated in Figs. 21.7 and 21.8 based on news articles from the Australian Broadcasting Corporation and Vox.

Various sectors of the regional economic activity, maritime shipping, and geological explorations in the Indian Ocean and Southeast Asia require GNSS and SATCOM services. Therefore, we can expect that many people (especially users of GNSS and SATCOM) may want to know if their connectivity to those satellite-based services is interrupted due to FAI or EPB.

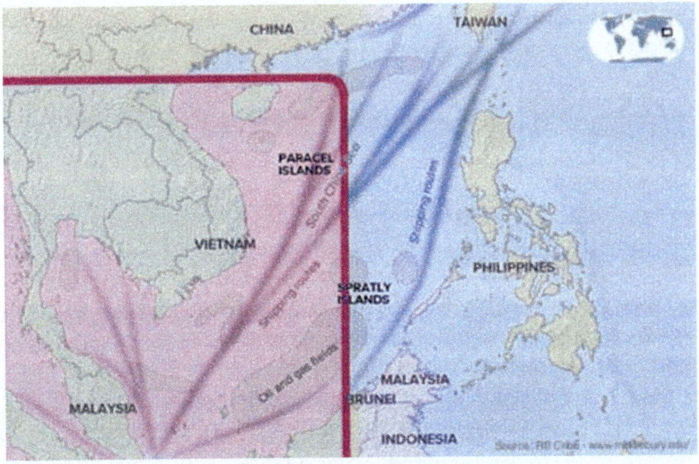

Fig. 21.7 Strategic shipping routes and geological exploration sites in the Southeast Asian region, which would be within the service area of EAR Kototabang under the extended SSA concept (*Source* Australian Broadcasting Corporation, https://www.abc.net.au)

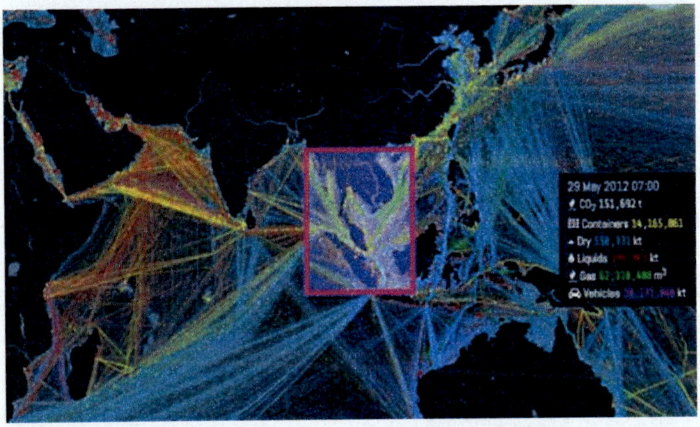

Fig. 21.8 Congested maritime shipping lanes in the Indian Ocean and Southeast Asian region, pointing to the technological importance of satellite-based navigation and communication channels using GNSS and SATCOM (*Source* Vox, https://www.vox.com)

Given the congested maritime shipping lane in the Indian Ocean and Southeast Asia, the role of satellite-based navigation using GNSS and communication channels via SATCOM is very important. A disruption in satellite connectivity to those involved in the maritime transport sector will significantly affect the economic lifeline of many in this region. An extended concept of SSA that includes a concern about the users' connectivity to various satellite services by providing information on the regions in space that may cause disruptions, would therefore be a valuable goal to pursue and achieve. "Area that is effectively covered by the EPB geometry and drift reconstruction is highlighted by the rectangle".

21.4 Conclusion

In general, although EAR only detects a small part of EPB that may be present in lowlatitude ionosphere over Southeast Asia, by using our knowledge on physical properties of EPB, we can deduce the full 3D geometry of the EPB which is much larger than what can be directly detected by the EAR beam. We demonstrate that a comprehensive 3D reconstruction of the EPB geometry and drift motion is possible, effectively covering an area of approximately 6,250,000 km^2 in the Southeast Asian region. A full 3D reconstruction of FAI/EPB geometry using EAR data can be a valuable information to be included in an extended concept of SSA (i.e. taking into account not only protection of space assets but also users' connectivity). This conceptual development has a potential to be implemented and utilized in ASEAN region.

Acknowledgements The Equatorial Atmosphere Radar (EAR) belongs to the Research Institute for Sustainable Humanosphere (RISH) of Kyoto University. The EAR was operated by RISH and the National Institute of Aeronautics and Space of Indonesia (LAPAN). We would like to thank RISH for providing and sharing this EAR data.

References

1. Prof. Marcus J. Holzinger Hompage, http://holzinger.gatech.edu/?q=SSA. Accessed 17 Sept 2021
2. Campbell, E.: Spratly Islands: Foreign Correspondent visits remote reef flashpoint where Filipino marines hold out against Chinese navy, Australian Broadcasting Corp., URL https://www.abc.net.au/news/2014-05-20/oil-key-in-spratly-islands-dispute-south-china-sea/5463080 (2015). Accessed 17 Sept 2021
3. United States Government Accountability Office (GAO), Space Situational Awareness: Status of Efforts and Planned Budgets, URL https://www.gao.gov/products/gao-16-6r. Accessed 19 Sept 2021
4. Fukao, S., Ozawa, Y., Yokoyama, T., Yamamoto, M., Tsunoda, R.T.: First observations of the spatial structure of F region 3-m-scale field-aligned irregularities with the Equatorial Atmosphere Radar in Indonesia. J. Geophys. Res. **109**, A02304 (2004). https://doi.org/10.1029/2003JA010096
5. VanZandt, T.E., Clark, W.L., Warnock, J.W.: Magnetic apex coordinates: a magnetic coordinate system for the ionospheric F2 layer. J. Geophys. Res. **77**, 2406–2411 (1972)
6. Gross domestic product (GDP) of the ASEAN countries from 2011 to 2021. https://www.statista.com/statistics/796245/gdp-of-the-asean-countries/. Accessed 17 September 2021
7. Plumer, B.: This is an incredible visualization of the world's shipping routes, Vox, URL https://www.vox.com/2016/4/25/11503152/shipping-routes-map (2017). Accessed 19 Sept 2021

Chapter 22
The Comparison of Z-R Relation Methods on Convective and Stratiform Rain of Quantitative Precipitation Estimation (QPE) in East Java Area

L. Bangsawan, R. H. Jatmiko, and E. Nurjani

Abstract QPE can be calculated using the Z-R relation algorithm by empirical equations between weather radar reflectivity and rainfall. The application of various Z-R relations has been widely carried out in Indonesia. However, there is still little research on the most suitable ones to be applied, especially regarding the convective and stratiform rain types. This study aims to compare several Z-R relations, namely Rosenfeld Tropical, Fujiwara, and Arida et al., to determine the most effective algorithm in some areas of East Java. SRI product is accumulated into PAC and RIH to be validated at each rain gauge location during season periods and by considering the radius of the weather radar. The results show that the best Z-R relation algorithm is Arida et al., with the highest correlation value 0.76 and the lowest RMSE 0.82 mm/hr, followed by Fujiwara then Rosenfeld Tropical with the highest correlation values 0.71 and 0.67 and the lowest RMSE 1.2 mm/hr and 2.09 mm/hr. The correlation tends to be higher during DJF as the peak period of the rainy season, and RMSE tends to be lower in the stratiform rain. The strongest correlation and the smallest RMSE occur in the closest radius within 50 km from the weather radar location.

22.1 Introduction

Weather radar can estimate rainfall in certain areas, which is known as quantitative precipitation estimation (QPE) [1]. QPE measurement is instrumental, especially in areas that do not have a rain gauge, so that rainfall can be estimated both in high spatial and temporal resolution [2, 3]. QPE is based on the relationship between radar reflectivity obtained from the meteorological object backscatters and the rainfall

L. Bangsawan (✉) · R. H. Jatmiko · E. Nurjani
Gadjah Mada University, Yogyakarta 55281, Indonesia
e-mail: laode.bangsawan@bmkg.go.id

L. Bangsawan
Indonesia Agency for Meteorology Climatology and Geophysics, Jakarta 10610, Indonesia

E. Yulihastin et al. (eds.), *Proceedings of the International Conference on Radioscience, Equatorial Atmospheric Science and Environment and Humanosphere Science, 2021*, Springer Proceedings in Physics 275, https://doi.org/10.1007/978-981-19-0308-3_22

produced. Researchers have studied many algorithms by creating various algorithms derived from empirical equations between the object's reflectivity recorded by radar and the resulting rain referred to as the Z-R relation [4–6]. Z represents for reflectivity quantified by dBZ and R represents for rain rate quantified with mm/hr. Fujiwara conducted research on the Z-R relationship algorithm in several different areas that are geographically located in mid-latitude by using radar data in Florida, Illinois, Tokyo, and Darmstadt to determine the occurrence of rain and formulating Z-R relationships using the least square method. Rosenfeld, Wolf, and Atlas (1992) in Darwin, which is one of the tropical climate areas in northern Australia, built a Z-R relation model from radar data located in the region through a matching method by considering the climatic conditions, rainfall types, and parameters on weather radar. Arida, Renggono, and Dupe built a Z-R relation algorithm model using weather radar reflectivity data and automatic rain gauges in Soroako, South Sulawesi Province, Indonesia. They also used the least square method to determine the coefficient of the Z-R relation equation by taking into account the type of convective rain characterized by a rainfall intensity $\geq$10 mm/hr and stratiform rain identified by rainfall intensity $<$10 mm/hr [7, 8].

There is still little research on the most suitable Z-R relation to be applied in East Java, Indonesia. The study of QPE produced by Z-R relation to the type of rain, season period, and radius of the weather radar is also rare to be found. This study aims to compare several Z-R relations, namely Rosenfeld Tropical, Fujiwara, and Arida et al., to determine the most effective Z-R relation algorithm in determining QPE in some areas of East Java. The type of rain, namely convective and stratiform, and the effect of the peak rainy season period, namely December, January, February (DJF) and the transition season, namely March, April, and May (MAM) as well as the radius of the weather radar would be considered on the QPE performance resulted from the Z-R relation.

22.2 Study Area

The study area covers a part of East Java, Indonesia, as shown in the following picture.

Figure 22.1 shows the location of the Sidoarjo C-band weather radar, owned by the Indonesia Agency for Meteorology Climatology and Geophysics (BMKG), and 14 rain gauges spread over the research area. The general specification of the weather radar is presented in the following Table 22.1.

The issue sometimes faced by rain gauges is the incompleteness of the data due to the disconnection of the communication networks or the mechanical and electrical troubles that affect the measurement results [9]. Therefore, the rain gauges used in this study were selected by purposive sampling that considering the completeness of the quantity and quality of data during the study period. On the other hand, the scan results of the weather radar are strongly affected by the surrounding topographical

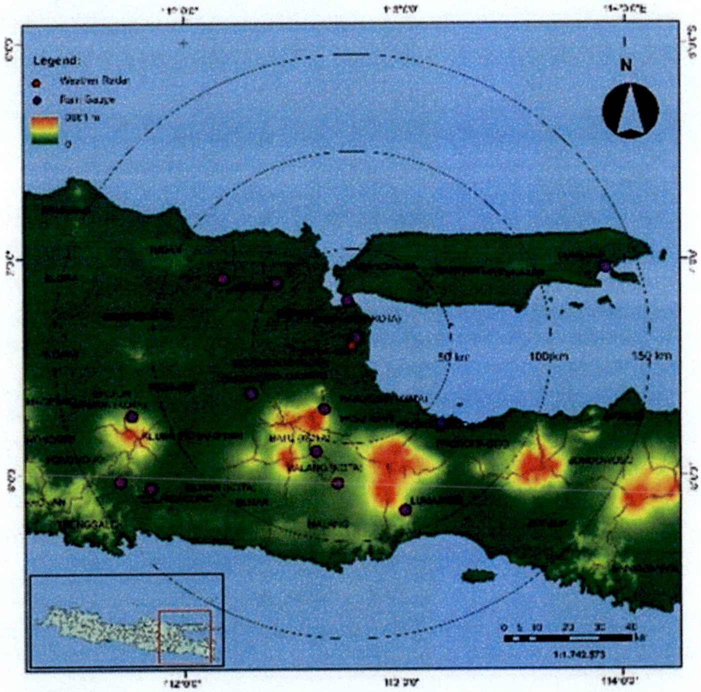

Fig. 22.1 The location of Sidoarjo Weather Radar with 150 km maximum radius of radar observation and the rain gauges location

Table 22.1 General specification of the weather radar

Parameter	Value
Manufacturer	Gematronik radar
Installation year	2006
Latitude	7° 24′ 36” S
Longitude	112° 45′ 36” E
Altitude	3 m
Tower height	23 m
Frequency	5.6–5.65 GHz
PRF	250–2400 Hz
Pulse width	0.5–2 μs
QPE Z-R relation	Rosenfeld tropical ($Z = 250R^{1,2}$)
	Fujiwara ($Z = 205R^{1,48}$)
	Arida et al. ($Z = 96R^{2,066}$)

conditions so that the sampling in areas that are blocked by mountains, mainly behind the mountainous region in the southern part of the research area, is minimized to avoid the beam blockage [10].

22.3 Data and Method

The data needed in this research are as follows:

a. Location (latitude and longitude) of the Sidoarjo C band weather radar and rain gauges.
b. Raw data reflectivity of the Sidoarjo C band weather radar with 500 $\times$ 500 m spatial resolution and 10 min temporal resolution in December 2019–May 2020.
c. Hourly observation rainfall data from a rain gauge in December 2019–May 2020.
d. The dominant cloud base height from synoptic weather observation data in December 2019–May 2020.
e. The dominant freezing level from upper-air weather observation data (open source radiosonde data from http://weather.uwyo.edu/upperair/sounding.html) in December 2019–May 2020.

Pre-processing and processing of weather radar data use the RainDart Rainbow software, which has several hydrological products to create QPE. The stages carried out in this research are:

a. Apply attenuation correction towards raw data to eliminate the effect of beam attenuation due to hydrometeorological objects in the form of rain. Attenuation correction for weather radar raw data at the preprocessing stage is quite sensitive in eliminating the attenuation effect [11, 12].
b. Apply the Z-R relation algorithms of the Rosenfeld Tropical, Fujiwara, and Arida et al., to the corrected raw data based on the Eq. (22.1) to produce the QPE Surface Rainfall Intensity (SRI) product, which is the main hydrological product with the exact temporal resolution as the raw data. The input data for QPE SRI is the dominant cloud base height value to be filled in the parameter height above ground as the lowest level of data processed by QPE. The highest level processed by QPE is the dominant freezing level value to be filled in the max height above the mean sea level (MSL) parameter. The height of the cloud base is part of the macrophysical conditions that contribute directly to the occurrence of rain [13, 14].
c. Produce a PAC product, which is a derivative of the SRI product, to produce an hourly rainfall accumulation map according to Eq. (22.2) and (22.3) then produce a RIH product by inputting rain gauge location data to obtain hourly QPE then grouped into convective type with rainfall intensity $\geq$10 mm/hr and stratiform with rainfall intensity <10 mm/hr [6–8]. Furthermore, the grouping is carried out again based on the peak period of the DJF rainy season and the

MAM transition season, considering the monsoon rain pattern in the East Java area [15].

d. Classifying rain gauge rainfall at each rain gauge location at a radius of 50, 100, and 150 km from the location of the weather radar because it is essential to see the significant effect of distance on QPE [16]. The classification was also carried out based on convective and stratiform rain types as well as DJF and MAM periods to be validated with QPE Z-R relation algorithm Rosenfeld Tropical, Fujiwara, and Arida et al.

The backscatter converted by the radar system into reflectivity values can be an indicator in determining the QPE of weather radar [17] through the Z-R relation with the following formula:

$$Z = aR^b \tag{22.1}$$

where Z is the reflectivity of the weather radar, R is the estimated rainfall, a and b are the coefficients of the Z-R relation. QPE generated from Eq. (22.1) is from SRI hydrological products that have 10 min temporal resolution, while PAC products generate hourly QPE from SRI products based on the following formula:

$$A = \sum_{i=1}^{N-1} A_i \tag{22.2}$$

The PAC calculation considers neighboring pixels in accordance with the size of the number of pixels, which in this study used a pixel size of 3 × 3. N is the number of neighboring pixels and Ai is the sum of rainfall in step i which is formulated by the following equation:

$$A_i = (t_i - t_{i-1}) \cdot \frac{R_i + R_{i-1}}{2} \tag{22.3}$$

where t_i is the ith time and R_i is the ith amount of rainfall. The calculation process is carried out according to one-hour intervals which are repeated according to the time series and to produce an accumulation of QPE every hour.

Validation is carried out when there are convective and stratiform rain events according to hourly rain data recorded on the rain gauges. Correlation analysis and RMSE were used to determine the strength of the relationship and error between QPE and observational rainfall. The correlation value of + 1 indicates a very strong positive relationship, −1 a very strong negative relationship, and 0 there is no relationship, with a scale range of every 0.2 from very weak, weak, moderate, strong, to very strong [3, 18]. There is no specific value interval to measure the amount of RMSE, but it is known that the greater the RMSE value, the greater the error of the QPE [10, 19].

22.4 Results and Discussion

22.4.1 *QPE for PAC Products with Rosenfeld Tropical, Fujiwara, and Arida Et Al.'s Z-R Relation Algorithms*

QPE for PAC products can be carried out after producing QPE for SRI products because the PAC products made every hour are derivatives of QPE for SRI products whose calculations are refer to Eq. (22.2) and (22.3). The following are the settings for the RainDart Rainbow application in processing weather radar data with the PAC product of the Z- R relation algorithm Rosenfeld Tropical, Fujiwara, and Arida et al., (Fig. 22.2):

The setting in the interval section is determined every hour (1 h) with a look backward mode so that the accumulation of rain every hour is calculated at the beginning of the hour. Thus, the QPE PAC products is the hourly composite of QPE SRI products [20]. The time offset is set at 00:00, indicating that there is no time zone difference from UTC. In this case, UTC is used to match the time of meteorological operational rain measurement observations. Repeat time 01:00 h indicates the QPE PAC recording time repeats every hour. The following is a QPE map produced by these algorithms:

Figure 22.3 shows that almost the entire study area was covered by rain on May 28th, 2020, at 11:00 UTC, with rainfall intensity that was generally higher in the southeast to the west and lower in the east, central, and north. The Rosenfeld Tropical algorithm produces a QPE PAC map with the highest rainfall intensity, then the Fujiwara and last Arida et al., which creates a QPE PAC map with the lowest rainfall intensity. This is because in the Z-R relation as in the Eq. (22.1), the a coefficient value of the Rosenfeld Tropical algorithm is much greater than the a coefficient value of the Fujiwara and Arida et al. Although it is known that the b coefficient of the Rosenfeld Tropical algorithm is smaller than Fujiwara and Arida et al., this shows that in the exponential relationship of the three Z-R relations, the influence of the a

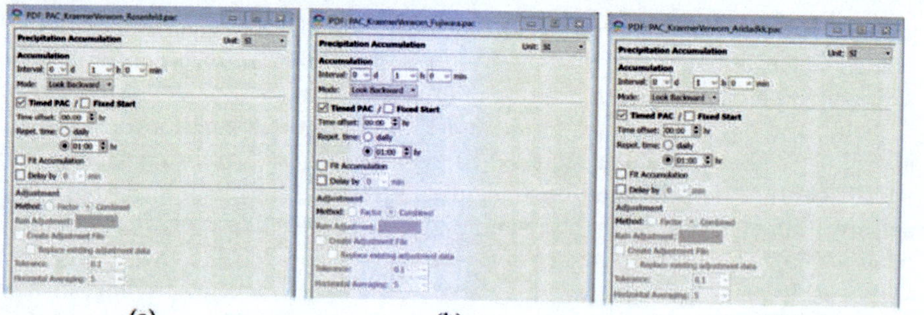

(a) (b) (c)

Fig. 22.2 The RainDart Rainbow setting for QPE product PAC with Z-R relation algorithms: **a** Rosenfeld Tropical, **b** Fujiwara, and **c** Arida et al.

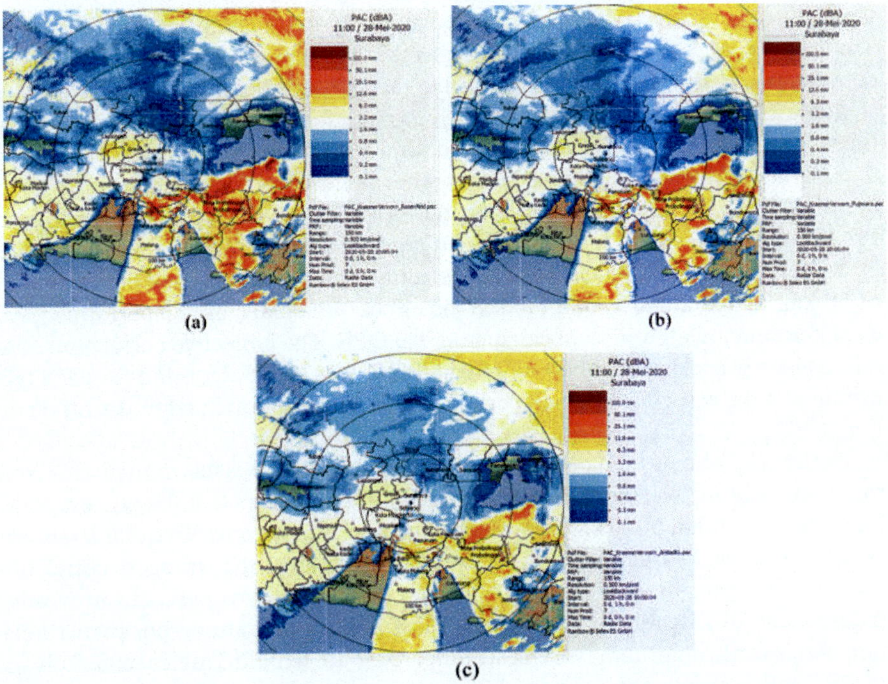

Fig. 22.3 Maps of QPE for PAC products with Z-R relation algorithms: **a** Rosenfeld Tropical, **b** Fujiwara, and **c** Arida et al.

coefficient is more dominant compared to the b coefficient value. This relationship causes the greater the reflectivity value of the weather radar, the greater the resulting QPE value [4–6].

Weather radar beams in the southern part are blocked (beam blocking) so that these areas cannot display QPE. Beam blocking due to the mountainous terrain in the south part of East Java caused weather radar elevation scans to be blocked at elevations of 0.5°, 1.5°, and 2.4°, which prevented weather radars from obtaining data [10]. The QPE RIH product is used to provide information on the QPE data at rain gauge locations according to Fig. 22.1 so that it can then be analyzed and validated based on the type of rain and the period of the season that occurs.

22.4.2 Correlation Comparison of QPE with Z-R Relation Algorithm Rosenfeld Tropical, Fujiwara, and Arida et al.

To examine the QPE correlation at each Z-R relation algorithm towards rain type and the season period, several comparison charts are made. The following table compares the highest correlation values of observed rainfall with QPE Z-R relation Rosenfeld Tropical, Fujiwara, and Arida et al., from the 14 locations of rain gauges for the convective and stratiform rain types during the DJF and MAM periods.

Table 22.2 shows that the best correlation values for convective and stratiform rain types from the applied Z-R relation algorithm tend to be higher in the DJF period than in the MAM period. This is because the condition of water vapor is greater in the peak period of the DJF rain compared to the rainy to dry transition period of MAM so that it tends to form clouds and produce rain that is more easily detected by weather radar to make QPE [21]. The comparison is also carried out at a radius of 50 km, 100 km, and 150 km to see the effect of distance on the performance of each Z-R relation algorithm in calculating the QPE. The collective correlation of all rain gauges was calculated based on the QPE Z-R relation algorithm for each type of rain and the associated season period. The results are presented in the following Fig. 22.4.

Figure 22.4 shows the correlation curve changes as the distance increases from the location of the weather radar. In Fig. 22.4a–d, it appears that the highest correlation occurs within 50 km of the weather radar. Both in convective and stratiform rain types as well as during the DJF and MAM periods, the strongest correlation between rain gauge rainfall and QPE occurs within a 50 km radius from weather radar where all Z-R relation algorithms show a weak to a strong positive correlation. Arida et al., algorithm is the best compared to Rosenfeld Tropical and Fujiwara in QPE with a correlation value of 0.75 for the convective rain type during the DJF period (Fig. 22.4a), 0.71 for the convective rain during the MAM period (Fig. 22.4b), 0.76 for the stratiform rain type during the DJF period (Fig. 22.4c), and 0.68 for the stratiform rain type during the MAM period (Fig. 22.4d).

The trendline shows a decrease in the correlation value with increasing distance. Arida et al., algorithm is always in a higher position than the others. The best correlation between QPE and rainfall observations occurs at locations close to the radar within a radius of 50 km, which is a strong positive relationship, while at locations far from the radar within a radius of >50 and >100 km the correlation tends to be weak and even negative [22]. This indicates that the performance of Arida et al., algorithm for determining QPE is the best as the distance to the weather radar changes. The quality of the weather radar QPE usually becomes signicantly reduced within a radius of more than 100 km [23].

Table 22.2 Highest correlation values comparison of observed rainfall with QPE Z-R Relation Rosenfeld Tropical, Fujiwara and Arida et al.

Z-R Relation	Season period	Rain type convective	Stratiform
Arida et al	DJF	0,75	0,76
	MAM	0,71	0,68
Fujiwara	DJF	0,60	0,71
	MAM	0,62	0,60
Rossenfeld tropical	DJF	0,55	0,67
	MAM	0,59	0,58

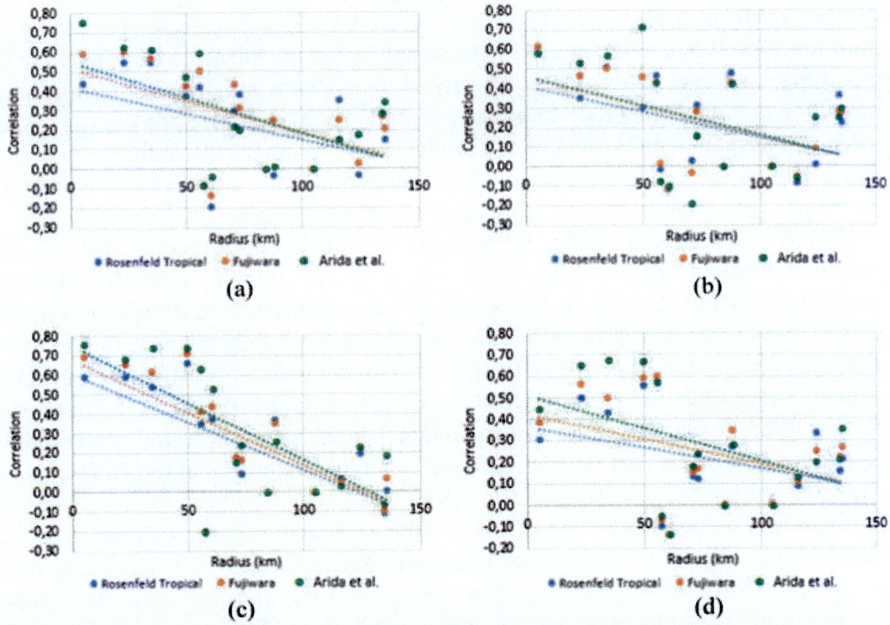

Fig. 22.4 QPE correlation values comparison of the Rosenfeld Tropical, Fujiwara, and Arida et al., algorithms with rainfall observations at a radius of 50 km, 100 km, and 150 km in the convective rain type during **a** DJF and **b** MAM and stratiform rain type during **c** DJF and **d** MAM

22.4.3 RMSE Comparison of QPE with Z-R Relation Algorithm Rosenfeld Tropical, Fujiwara, and Arida et al.

The following table compares the RMSE value of observed rainfall with the QPE Z-R relation Rosenfeld Tropical, Fujiwara, and Arida et al., from the 14 rain gauge locations in the convective and stratiform rain types for the DJF and MAM periods:

Table 22.3 shows that the RMSE value tends to be lower in the stratiform type of rain because the error of the QPE data from the observational rainfall is smaller

Table 22.3 Smallest RMSE values comparison of observed rainfall with QPE Z-R Relation Rosenfeld Tropical, Fujiwara and Arida et al.

Z-R Relation	Season period	Rain type convective	Stratiform
Arida et al	DJF	5.66	1.21
	MAM	8.65	0.82
Fujiwara	DJF	14.34	1.52
	MAM	15.74	1.20
Rosenfeld tropical	DJF	19.84	2.54
	MAM	17.89	2.09

in the stratiform type of rain than the convective type. It can also be caused by the threshold for the convective rain type, which is greater than the threshold for the stratiform rain type. Thus, the possibility of the QPE error value for the convective rain type is higher than the stratiform rain type [6–8]. Massive convective rainfall events in the DJF and MAM periods produce high weather radar reflectivity data to produce high QPE according to the Z-R relation algorithm used. It has the potential to result in the large errors of the observed rainfall [21, 24].

Based on the season period, the peak period of DJF rainfall has an RMSE value that is not much different from the transition period from rainy to dry MAM in both convective and stratiform rain types. It means that the QPE error values for the convective and stratiform rainfall types are not much different from the observations during the peak rainfall period of the DJF and the transition of MAM, as can also be seen in Table 22.2. Through this RMSE assessment, it is known that the rain types have a different pattern that is more dominant than the seasonal period, which shows that the aspect of the rain types is more influential in determining the QPE error on the observed rainfall.

Figure 22.5 shows the error curve of QPE concerning observed rainfall as the distance increases from the location of the weather radar. In Fig. 22.5a–d., the error of QPE to the observed rainfall occurs along with the distance from the weather radar. Within 50 km from the weather radar, the highest QPE error is 150.98 mm/hr

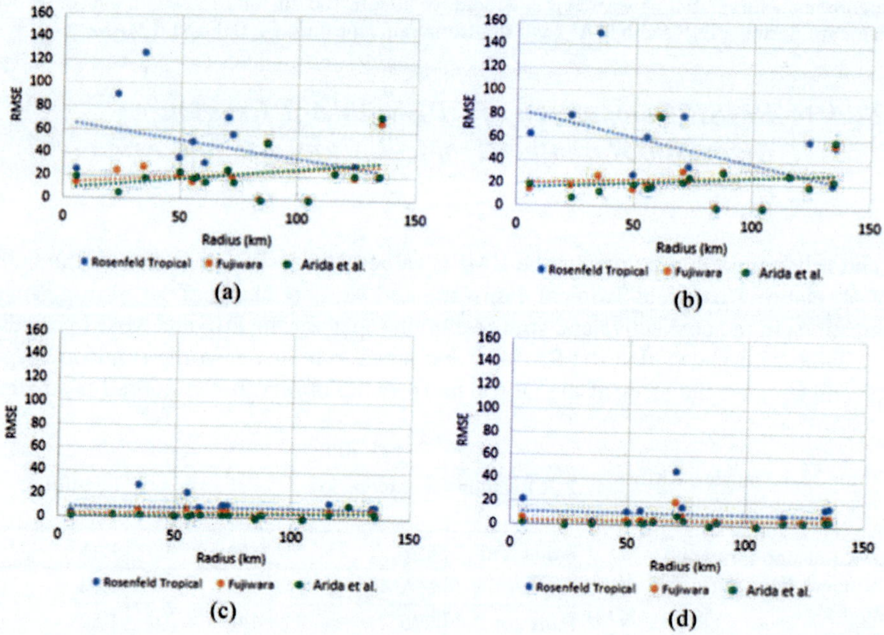

Fig. 22.5 QPE RMSE values comparison of the Rosenfeld Tropical, Fujiwara, and Arida et al., algorithms with rainfall observations at a radius of 50 km, 100 km, and 150 km in the convective rain type during **a** DJF and **b** MAM and stratiform rain type during **c** DJF and **d** MAM

experienced by the Z-R Rosenfeld Tropical relation in all types of rain and seasons. The smallest error in the same distance is encountered by Arida et al., also in all types of rain and seasons, with the smallest error of 0.82 mm/hr, which occurs in the stratiform rain type during the MAM period.

In general, the error of the weather radar QPE to the observed rainfall is the smallest within 50 km from the radar location and gets larger at distances > 50 km and >100 km [3, 22, 23]. The trendline shows that the change in the RMSE value tends to increase towards the distance from the weather radar on the Arida et al., algorithm and Fujiwara in all types of rain and season. The Arida et al., algorithm is always in a lower position than Fujiwara. Different things are presented by the Rossenfeld Tropical algorithm that indicates a large error at close range from the radar and decreases with increasing distance. However, It does not improve the Rossenfeld Tropical QPE performance because the farther from the weather radar the higher elevation of the weather radar, so it cannot accurately describe rainfall estimates [25, 26].

22.5 Conclusions

During the study period, the best Z-R relation algorithm for the convective rain type was the Arida et al. Its correlation values and RMSE are 0.75 and 5.66 mm/hr, which occurred during the DJF period. Likewise, the best Z-R relation in the stratiform rain type is the Arida et al., algorithm with a correlation value of 0.76, which also occurs in the DJF period and an RMSE value of 0.82 mm/hr, which arises in the MAM period. In both convective and stratiform rain types, the most robust relationship and the slightest error occurred within the closest radius of 50 km from the location of the weather radar. The correlation tends to be higher in the DJF period due to the vast availability of water vapor and atmospheric instability in the peak period of the rainy season, which tends to form clouds and produce rain that is easily detected by weather radar to produce QPE. RMSE tends to be lower in the stratiform rain type because the error of the QPE data to the observed rainfall is smaller in that type of rain.

Acknowledgements The author feels grateful to the weather radar subdivision and online service database of BMKG who has provided raw data reflectivity and rain gauge observation data in this study. High appreciation is also given to the laboratory administrator of the remote sensing department of Gadjah Mada University for giving allowance to utilize the high specification PC to run the data.

References

1. Nicol, J.C., Austin, G.L.: Attenuation correction constraint for single- polarisation weather radar. Meteorol. Appl. **10**(4), 345–354 (2003)

2. Qu, J.J., Gao, W., Kafatos, M., Murphy, R.E., Salomonson, V.V.: Earth science satellite remote sensing. Sci. Instruments, **1** (2006)
3. Giarno, M.P. Hadi, S. Suprayogi, Murti, S.H.: Daily quantitative precipitation estimates use weather radar reflectivity in South Sulawesi. IOP Conf. Ser. Earth Environ. Sci. **256**(1), 0–11 (2019)
4. Fujiwara, M.: Raindrop-size distribution from individual storms. J. Atmos. Sci. **22**(5), 585–591 (1965)
5. Rosenfeld, D., Wolff, D.B., Atlas, D.: General probability-matched relations between radar reflectivity and rain rate. J. Appl. Meteorol. **32**(1), 50–72 (1993)
6. Arida, V., Renggono, F., Dupe, Z.L.: Radar reflectivity factor relation with rainfall intensity for C-band weather radar in Soroako, Sulawesi Selatan. J. Sci. Weather Modif. Technol., **13** (2012)
7. Penide, G., Protat, A., Kumar, V.V., May, P.T.: Comparison of two convective/stratiform precipitation classification techniques: radar reflectivity texture versus drop size distribution-based approach. J. Atmos. Ocean. Technol. **30**(12), 2788–2797 (2013)
8. Tenório, R.S., Moraes, M.C.D.S., Kwon, B.H.: Raindrop distribution in the Eastern Coast of Northeastern Brazil using disdrometer data. Rev. Bras. Meteorol. **25**(4), 415–426 (2010)
9. Habib, E., Krajewski, W.F., Kruger, A.: Sampling errors of tipping-bucket rain gauge measurements. J. Hydrol. Eng. **6**(2), 159–166 (2001)
10. Ali, A., Deranadyan, G., Hairuly Umam, I.: An enhancement to the quantitative precipitation estimation using radar-gauge merging. Int. J. Remote Sens. Earth Sci. **17**(1), 65 (2020)
11. Bringi, V.N., Chandrasekar, V.: Polarimetric Doppler Weather Radar. Cambdridge University Press (2004)
12. Rico-Ramirez, M.A.: Adaptive attenuation correction techniques for C-band polarimetric weather radars. IEEE Trans. Geosci. Remote Sens., **50**(12) (2012)
13. Lamer, K., Naud, C.M., Booth, J.F.: Relationships between precipitation properties and large-scale conditions during subsidence at the Eastern North Atlantic observatory. J. Geophys. Res. Atmos. **125**(7), 1–16 (2020)
14. Mróz, K., Battaglia, A., Kneifel, S., Von Terzi, L., Karrer, M., Ori, D.: Linking rain into ice microphysics across the melting layer in stratiform rain: a closure study. Atmos. Meas. Tech. **14**(1), 511–529 (2021)
15. Aldrian, E., Dwi Susanto, R.: Identification of three dominant rainfall regions within Indonesia and their relationship to sea surface temperature. Int. J. Climatol. **23**(12), 1435–1452 (2003)
16. Zhang, G.: Weather Radar Polarimetry. CRC Press, New York, US (2016)
17. Zhang, P., Liu, X., Li, Z., Zhou, Z., Song, K., Yang, P.: Attenuation correction of weather radar reflectivity with arbitrary oriented microwave link. Adv. Meteorol. **2001**, 2017 (2017)
18. Ratner, B.: The correlation coefficient: Its values range between 1/1, or do they. J. Targeting, Meas. Anal. Mark., **17**(2), 139–142 (2009)
19. Meyer, T.: Root mean square error compared to, and contrasted with, standard deviation. Surv. L. Inf. Sci. **72**(3), 107–108 (2012)
20. Szturc, J., Ordka, K., Jurczyk, A.: Quality control algorithms applied on weather radar reflectivity data. Doppler Radar Obs.—Weather Radar, Wind Profiler, Ionos. Radar, Other Adv. Appl., (2012)
21. Blenkinsop, S., Lewis, E., Chan, S.C., Fowler, H.J.: Quality-control of an hourly rainfall dataset and climatology of extremes for the UK. Int. J. Climatol. **37**(2), 722–740 (2017)
22. Gourley, J.J., Tabary, P., Parent du Chatelet, J.: Data quality of the Meteo-France C-band polarimetric radar. J. Atmos. Ocean. Technol. **23**(10), 1340–1356 (2006)
23. Bordoy, R., Bech, J., Rigo, T., Pineda, N.: Analysis of a method for radar rainfall estimation considering the freezing level height. Tethys, J. Weather Clim. West. Mediterr., no. January (2010)
24. Ramsauer, T., Weiß, T., Marzahn, P.: Comparison of the GPM IMERG final precipitation product to RADOLAN weather radar data over the topographically and climatically diverse Germany. Remote Sens., **10**(12) (2018)

25. Hosseini, S.H., et al.: Evaluation of a new X-band weather radar for operational use in south Sweden. Water Sci. Technol. **81**(8), 1623–1635 (2020)
26. Neuper, M., Ehret, U.: Quantitative precipitation estimation with weather radar using a data- and information-based approach. Hydrol. Earth Syst. Sci. **23**(9), 3711–3733 (2019)

Chapter 23
Ionospheric Observation Using Equatorial Atmosphere Radar (EAR) Kototabang for the 26 December 2019 Annular Solar Eclipse Research

Agri Faturahman, Varuliantor Dear, Jiyo, Afrizal Bahar, Asnawi Husin, and Rezy Pradipta

Abstract Solar eclipse passage can cause some change in the ionosphere, which makes it a valuable event to be studied using multi-instruments. On 26 December 2019, an annular solar eclipse passed in the vicinity of Equatorial Atmosphere Radar (EAR) at Kototabang station. We utilize the EAR's capability to detect field-aligned irregularities (FAI) to discover changes in the ionosphere during the eclipse. Other instruments also supported the EAR observation. The results of EAR observations show some appearance of FAI (strong echo power backscatter) at the altitude of ionospheric E layers (90–100 km) during the solar eclipse. Traces of strong sporadic E layers (with spread) were also observed on ionograms from the FMCW ionosonde at the exact location, confirming the EAR observation. The appearance of FAI and these sporadic E layers were likely to be related to the solar eclipse event. The results of several other studies also supported this hypothesis. On the eclipse day, we found that the ordinarily occurring FAI at 150 km altitude disappeared; meanwhile, on 25 and 27 December 2019, the 150 km altitude FAI was observed. We suspect that the absence of FAI at the height of 150 km was related to the solar eclipse phenomena. Another interesting finding is a spread echo backscatter around noon on 25 December 2019, one day before the eclipse.

A. Faturahman (✉) · V. Dear · Jiyo · A. Husin
Research Center for Space, National Research and Innovation Agency (BRIN), Bandung, Indonesia
e-mail: agri.faturahman@brin.go.id

A. Bahar
Agam Atmospheric and Space Observation Office, National Research and Innovation Agency (BRIN), Agam, Indonesia

R. Pradipta
Institute for Scientific Research, Boston College, Newton, USA

E. Yulihastin et al. (eds.), *Proceedings of the International Conference on Radioscience, Equatorial Atmospheric Science and Environment and Humanosphere Science, 2021*, Springer Proceedings in Physics 275, https://doi.org/10.1007/978-981-19-0308-3_23

23.1 Introduction

The annular solar eclipse has occurred on 26 December 2019. The annular phase of this solar eclipse is visible from some parts of West Asia, South Asia, and Southeast Asia, including Indonesia. In the Indonesian region, the path of this eclipse annularity crossed some areas, especially Sumatra and Kalimantan (Borneo) island, as shown in Fig. 23.1. This eclipse path is also coincidentally close to the Equatorial Atmosphere Radar (EAR) sites at Kototabang, Bukit Tinggi, West Sumatra. A solar eclipse provides a unique opportunity to investigate the dynamics in the ionosphere as a response to the sudden changes of solar radiation. The effects of a solar eclipse on the ionosphere and its dynamics are still interesting as practical research topics in ionospheric studies [5]. This study aims to investigate the impact of this eclipse on the dynamics of ionospheric changes, especially in the E region (90–100 km), using EAR data.

23.2 Observation Methods

Equatorial Atmosphere Radar (EAR) is one of the doppler radars. It is built near the equator, Bukittinggi, West Sumatra, located at 865 m above sea level. EAR operates at a frequency of 47 MHz with a peak output power of 100 kW [6]. The EAR diameter is approximately 110 m. It uses a circular antenna array consisting of 560 three-element Yagi antennas. The scientific purpose of the EAR is to study the dynamics

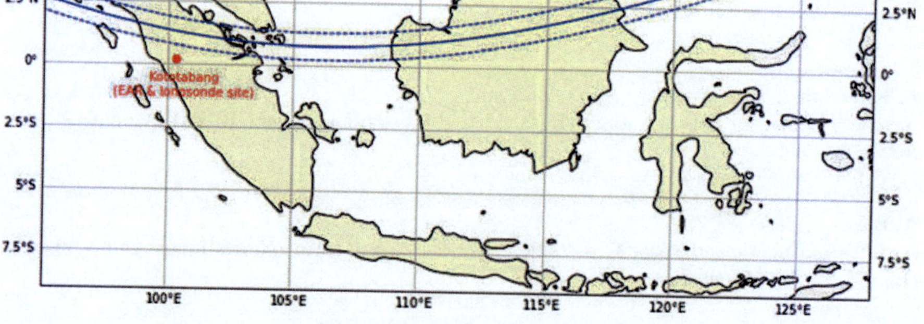

Fig. 23.1 Annular solar eclipse path over Indonesian region

and electrodynamic coupling processes of the equatorial atmospheres (troposphere and lower stratosphere).

Furthermore, EAR can also observe echoes from ionospheric irregularities at an altitude of more than 90 km (upper atmosphere). The data to be analyzed to investigate the upper atmosphere echo from ionospheric irregularity, namely field-aligned irregularities (FAI), is the range time-intensity (RTI) plot data. FAI is obtained from signal-to-noise ratio (SNR), Doppler speed, and the spectral widths of the E and F regions of the ionosphere.

For this study, observations were made from 25 to 27 December 2019. During this observation period, the EAR operated under similar setting conditions, as shown in Fig. 23.2. Two beam directions were used during the observation project: beam 1 with an azimuth angle of 150.1°, a zenith angle of 21.4°, and beam 2 with an azimuth angle of 180°, a zenith angle of 18.8°. Above Kototabang, the initial phase of the eclipse begins at 03:18 UT (10:18 LT), the peak of the eclipse, with a maximum obscuration of 91.50% at 05:11 UT (12:11 LT), and the eclipse ends at 07:08 UT (14:08 LT). Our investigation focused on the E layer of the ionosphere and events at an altitude range of 90–150 km.

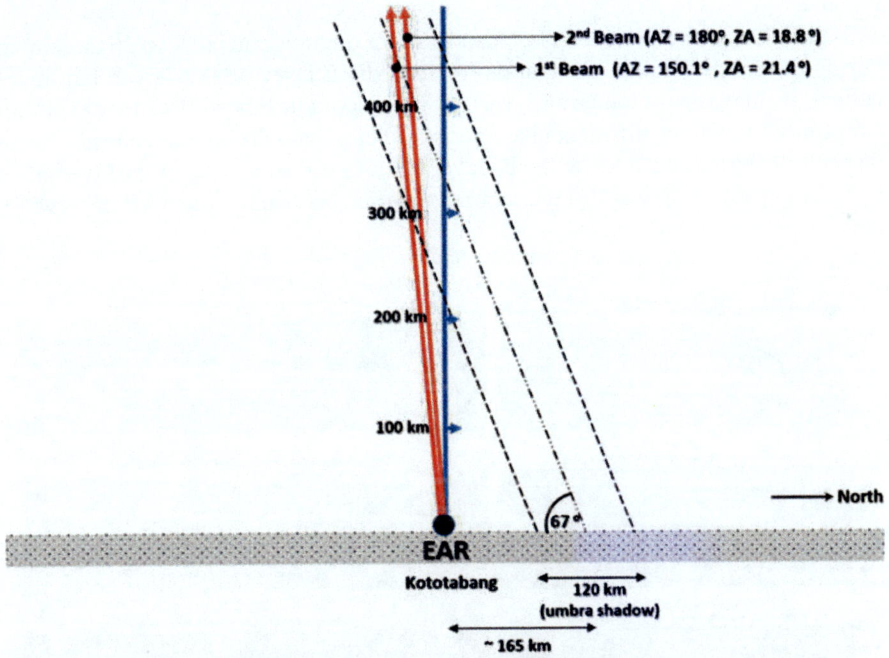

Fig. 23.2 EAR observation setting on 25–27 December 2019

23.3 Observation Results

Figure 23.3 shows the RTI backscatter power (SNR) and Doppler velocity plot for EAR observation results 25–27 December 2019. Observation data on 25 and 27 December were selected as an appropriate comparison for the observation on the eclipse day. The strong echo power backscatter, SNR value, indicated the presence of FAI. FAI in the ionospheric E layer was observed at noon 26 December, on the day of the eclipse, and on 27 December. These results show that FAI in the E layer region of the ionosphere during eclipse has unique characteristics. On 26 December, before the eclipse peak, an echo appeared higher above the lower echo for approximately 1 h (11:40 LT-12:40 LT). This echo backscatter then disappears and reappears 30 min later until the end of the eclipse (13:00 LT–14:20 LT).

Meanwhile, in the RTI doppler velocity plot, we observed FAI in the E layer on 26 and 27 December. The exciting feature is the signature of signal amplitude that is not constant for the echo around 100 km towards the end of the eclipse. There is another identified phenomenon from Fig. 23.3. On 25 and 27 December, an echo backscatter usually occurs was observed at an altitude of 150 km, indicating FAI's presence. However, on the eclipse day, 26 December, no echo backscatter was observed at that altitude.

Figure 23.4 combines the RTI plot and a series of ionogram plots on 26 December, during the eclipse period. On that day, a sporadic E layer was observed before the eclipse. In this series of ionograms, we follow the sporadic E layer (Es) trace evolution over time, and we see different characteristics before and during the eclipse. During this eclipse period, the trace of the Es layer tends to spread and cover the layer above it. Traces of strong spread Es layers from ionogram confirming the EAR observation data.

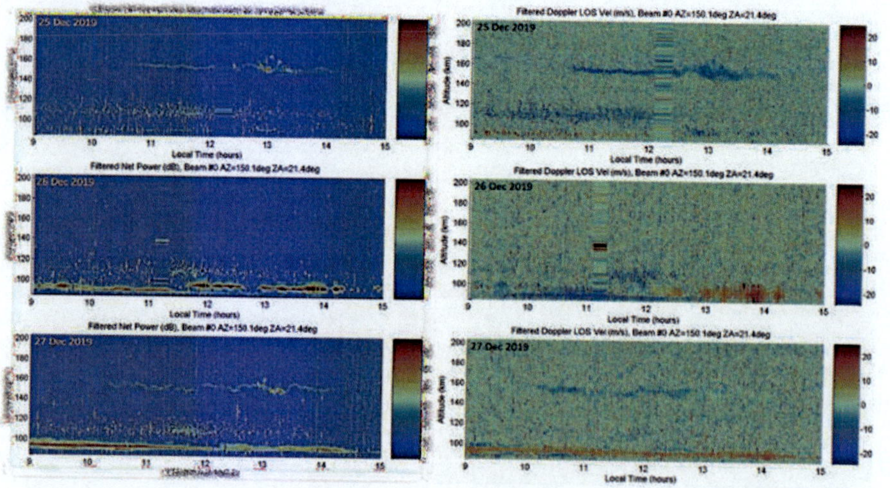

Fig. 23.3 Range time-intensity SNR (Signal to Noise Ratio) and doppler velocity plot

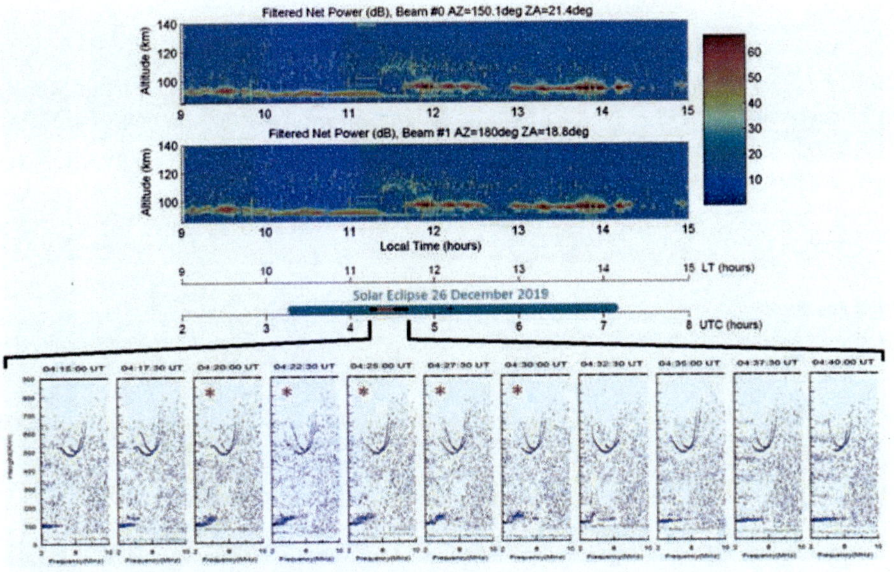

Fig. 23.4 Range time-intensity SNR and Ionogram time series plot

Figure 23.5 accidentally found another exciting finding outside the FAI occurrence in the E region during the solar eclipse. The burst of echoes presence from ~400 km altitude shortly after 11:00 LT on 25 December is a possible indication of FAI in the F region during daytime. FAI occurrence in the F region is an infrequent event in the day. We found it accidentally because EAR is operating in the daylight during the solar eclipse observation project.

23.4 Discussion

FAI features in ionospheric E layers on 26 and 27 December are consistent with several other past observations [4, 10]. On 26 and 27 December, we observed FAI events with continuous and decreasing echoes characteristics. That pattern during the daytime is a common feature in several FAI observations, particularly for low latitude regions.

On the other hand, on 26 December, there is a unique feature: additional echoes above the lower echoes before the eclipse peak. This result is almost identical to the observations made during the solar eclipse of 11 August 1999 in India [7]. We suspect that another echo appears due to specific mechanisms; the eclipse triggers the possibility. The ionospheric E layer irregularities might be due to the mechanism of gradient drift instability. However, combinations of several agents may play a simultaneous role, not just a single agent.

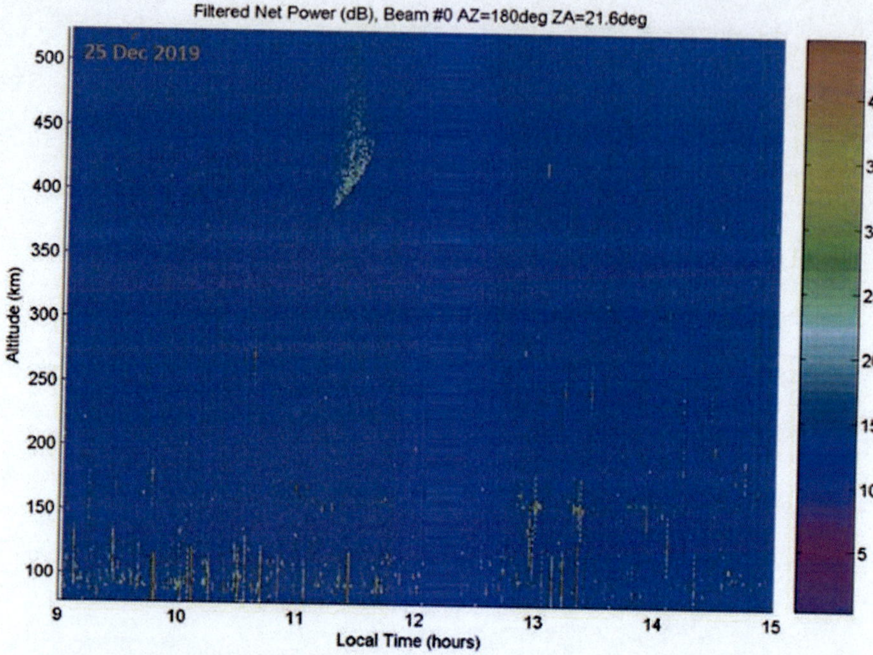

Fig. 23.5 Range time-intensity SNR plot for 25 December 2019

In addition, we also use ionosonde data for comparison to EAR data. From the ionosonde data (ionogram), we can see the evolution of sporadic E layer traces. The Es layer has different characteristics before and during the eclipse. The traces of that Es layer appear to be spreading during the eclipse. From the EAR observations, Doppler velocity fluctuated significantly around an altitude of 100 km during the eclipse. This fluctuation is more visible at the end phase of the eclipse with a wave-like structure.

The previous studies showed that solar eclipses profoundly affect the ionospheric E region and sporadic E layer. When the solar eclipse occurs, the cooling effect of the lunar shadow is considered to form meridional wind shear at the altitude of Es layers. The wind shear generally enhanced the electron density in the Es layer during the eclipse period [1]. Other studies show that during the eclipse period, atmospheric gravity waves (AGW) can be formed. The AGW induced by the solar eclipse suspect to deform the sporadic E layer and produce moving wave-like structures. The tilted plasma concentration in the wave-like structures reflected radio waves from ionosonde to form the spread Es in the recorded ionograms [3]. Some further investigation is needed if we want to ensure AGW was created during the eclipse.

On the eclipse day, 26 December, no echo backscatter was observed at an altitude of 150 km, while on 25 and 27 December, that echo backscatter was identified in the RTI plot. We suspect that 150 km FAI did not occur because, at the time of the eclipse, specific conditions for FAI to appear at that altitude were not met. Meanwhile, on comparison days, 25 and 27 December, the mechanism is possible for the FAI to occur, so that echo backscatter was detected. The results of previous observations related to 150 km FAI during the 15 January 2010 solar eclipse in India concluded almost the same thing [9]. According to their observations, 150 km echo backscatter is still observed during the eclipse, but only appears at the beginning and the end of the eclipse phase. Until now, the mechanism of the formation of 150 km FAI is still not fully understood. Many factors influence its appearance. One of them is the dependence on the solar EUV flux, which is also complex [8].

The burst of echoes from ~400 km altitude shortly after 11:00 LT was identified on 25 December 2019, one day before the eclipse. It is a possible FAI occurrence in the ionospheric F region during the daytime. FAI in the F region is an infrequent event in the day. We found it because EAR is operating during the solar eclipse observation project. Some hypotheses from the previous study [2] stated that FAI occurrence in the daytime is the remnant of postsunset/postmidnight plasma bubbles. However, we need further study regarding the peculiar appearance of FAI in this F layer at noon, but the mode of operation required for such investigation is not yet known.

23.5 Conclusion

The annular solar eclipse event at the end of 2019 coincidentally passed close to the EAR Kototabang site. So we took advantage to investigate the effect of that solar eclipse on the dynamics of the ionosphere. The EAR observed FAI on the eclipse day, especially at the ionospheric E layer (90–100 km). This FAI was identified in the RTI plot of echo backscatter signal (SNR) and doppler velocity. The FAI at altitude 100 km was also confirmed by ionosonde data, where the sporadic E layers traces (spread Es) was observed during the eclipse. Spread Es indicates an irregularity in the E region ionosphere. The backscatter echoes at an altitude of 150 km, which is usually occurred, disappeared during an eclipse on 26 December 2019. Accidentally, we found another interesting finding. The burst of echoes presence from ~ 400 km altitude on 25 December is a possible indication of FAI in the F region during daytime.

Acknowledgements The Equatorial Atmosphere Radar (EAR) belongs to the Research Institute for Sustainable Humanosphere (RISH), Kyoto University. The EAR was operated under international collaboration between RISH and the National Institute of Aeronautics and Space of Indonesia (LAPAN). We would like to thank RISH for providing and sharing this EAR data.

References

1. Chen, G., Zhao, Z., Zhou, C., Yang, G., Zhang, Y.: Solar eclipse effects of 22 July 2009 on sporadic-E. Ann. Geophys. **28**(2), 353–357 (2010). https://doi.org/10.5194/angeo-28-353-2010
2. Chen, G., Jin, H., Yan, J., Zhang, S., Li, G., Yokoyama, T., Yang, G., Yan, C., Wu, C., Wang, J., Zhong, D., Li, Y., Wang, Z.: Low-latitude daytime F region irregularities observed in two

geomagnetically quiet days by the Hainan coherent scatter phased array radar (HCOPAR). J. Geophys. Res. Space Phys. **122**(2), 2645–2654 (2017). https://doi.org/10.1002/2016JA023628
3. Chen, G., Zhao, Z., Zhang, Y., Yang, G., Zhou, C., Huang, S., Li, T., Li, N., Sun, H.: Gravity waves and spread Es observed during the solar eclipse of 22 July 2009. J. Geophys. Res. Space Physics **116**(9), 1–7 (2011). https://doi.org/10.1029/2011JA016720
4. Choudhary, R.K., Mahajam, K.K., Singh, S., Kumar, K.: First VHF radar observations of tropical latitude E-region field aligned irregularities. Geophysical Research Letterse **23**(25), 3683–3686 (1996)
5. Dear, V., Husin, A., Anggarani, S., Harjosuwito, J., Pradipta, R.: Ionospheric effects during the total solar eclipse over Southeast Asia-Pacific on 9 March 2016: part 1 vertical movement of plasma layer and reduction in electron plasma density. J. Geophys. Res. Space Phys. **125**(5), 1–15 (2020). https://doi.org/10.1029/2019JA026708
6. Fukao, S., Hashiguchi, H., Yamamoto, M., Tsuda, T.: Equatorial atmosphere radar (EAR): system description and first results. Radio Sci. **38**(3) (2003). https://doi.org/10.1029/2002RS002767
7. Patra, A.K., Choudhary, R.K., St-Maurice, J.P.: Solar eclipse-induced E-region plasma irregularities observed by the Gadanki radar. Geophys. Res. Lett. **36**(13), 3–7 (2009). https://doi.org/10.1029/2009GL038669
8. Patra,A.K., Pavan Chaitanya, P., St. Maurice, J.P., Otsuka, Y., Yokoyama, T., Yamamoto, M.: The solar flux dependence of ionospheric 150 km radar echoes and implications. Geophys. Res. Lett., **44**(22), 11,257–11,264 (2017). https://doi.org/10.1002/2017GL074678
9. Patra, A.K., Pavan Chaitanya, P., Tiwari, D.: Characteristics of 150 km echoes linked with solar eclipse and their implications to the echoing phenomenon. J. Geophys. Res. Space Phys. **116**(A5), 1–9 (2011). https://doi.org/10.1029/2010ja016258
10. Patra, A.K., Sripathi, S., Rao, P.B., Choudhary, R.K.: Gadanki radar observations of daytime E region echoes and structures extending down to 87 km. Ann. Geophys. **24**(7), 1861–1869 (2006). https://doi.org/10.5194/angeo-24-1861-2006

Chapter 24
Atmospheric Methane Variability During Upwelling Events in the Southern Coast of Java, Indonesia

Rahaden Bagas Hatmaja, Fildzah 'Adany, Wilin Julian Sari, and Prawira Yudha Kombara

Abstract As a greenhouse gas with a radiative forcing of 0.48 ± 0.05 W/m^2, methane can trap heat in the atmosphere better than carbon dioxide. Methane is produced naturally from the ocean, especially during coastal upwelling events due to the effect of the biogeochemical process. This study investigates the contribution of the seasonal coastal upwelling events towards atmospheric methane variability in the southern coast of Java, Indonesia, from 2011 to 2020. The result shows that Atmospheric methane concentrations along the southern coast of Java have an annual cycle with strong seasonal variability that peaks around boreal autumn with a concentration around 1800–1820 ppb/volume each year. Seasonal upwelling begins in May and peaks in July. Atmospheric methane concentrations are significant in July and peak in September. Hence there is a two-month lag between upwelling intensification and seasonal variations in plankton and atmospheric methane concentrations above the upwelling system. In addition, the interannual variability of atmospheric methane concentration is related to Indian Ocean Dipole that coincides with increased upwelling intensity and phytoplankton abundance.

24.1 Introduction

Methane (CH_4) is the second-highest greenhouses gas emitted to the atmosphere, after carbon dioxide (CO_2), but it has a global warming potential (GWP) 25 times

R. B. Hatmaja (✉) · P. Y. Kombara
Research Center for Climate and Atmosphere, National Research and Innovation Agency (BRIN), Bandung, Indonesia
e-mail: rahadenbagas@gmail.com

F. 'Adany
Research Center for Chemistry, National Research and Innovation Agency (BRIN), South Tangerang, Indonesia

W. J. Sari
Research Center for Quantum Physics, National Research and Innovation Agency (BRIN), South Tangerang, Indonesia

E. Yulihastin et al. (eds.), *Proceedings of the International Conference on Radioscience, Equatorial Atmospheric Science and Environment and Humanosphere Science, 2021*, Springer Proceedings in Physics 275, https://doi.org/10.1007/978-981-19-0308-3_24

higher than CO_2 [1, 2]. Methane emissions come from anthropogenic such as landfills, agriculture, farming activities, and natural from the ocean. The global ocean emits about 5×10^6–25×10^6 ton/year of methane, contributing about 5–13% of the total emissions from natural sources [3].

The oceanic mechanism that increases methane emissions is coastal upwelling [4]. Coastal upwelling plays a vital role in transporting cold and nutrient-rich deep-water to the surface, thus increasing phytoplankton abundance, which plays critical controls over biogeochemical CH_4 formation processes [5]. Moreover, previous studies have confirmed that the marine sediment produced CH_4 and transported it to the ocean's surface through coastal upwelling [6].

As a maritime continent, Indonesia has numerous upwelling spots that are important to be investigated, one of which is Southern Java upwelling system. Southern Java upwelling system is a seasonal upwelling that occurs during the boreal summer monsoon in June to August. This coastal upwelling is generated by south-easterly monsoon wind stress along the coast [7]. Consequently, the chlorophyll-a concentration increased due to the elevated phytoplankton in boreal summer [8]. Moreover, Java Island, as the most populated island in Indonesia, has contributed to emitting greenhouse gases from its anthropogenic activities.

Increasing methane concentrations affects the equilibrium anthropogenic and natural-sourced methane [2]. Yet, unfortunately, there are limited studies on oceanic methane. Therefore, it is necessary to investigate the contribution of the Southern Java coastal upwelling events towards atmospheric methane variability in the southern coast of Java.

24.2 Data and Methods

24.2.1 Data

This study used the atmospheric methane concentration, surface wind, and phytoplankton data over ten years from January 2011 to December 2020. The level-3 monthly atmospheric methane concentration data at 1000 hPa were obtained from the Atmospheric Infrared Sounder (AIRS) onboard satellite Aqua), with a spatial resolution of 1°. These data are accessible through the website of Giovanni by NASA (https://giovanni.gsfc.nasa.gov/gionanni). Moreover, the surface wind data with a spatial resolution of 0.25° for the same period were obtained from the ERA5 reanalysis products developed by the (ECMWF) [9]. The wind data were utilized to estimate the upwelling transport via Ekman pumping. Furthermore, the monthly phytoplankton data expressed as carbon in seawater is from the PISCES biogeochemical model reanalysis dataset from E.U. Copernicus Marine Service Information (CMEMS). The data were can be retrieved in https://marine.copernicus.eu/, with a spatial resolution of 0.25°. The analysis of this study is focused on the black box area in Fig. 24.1, which are limited to 7°–10° S and 105°–114° E.

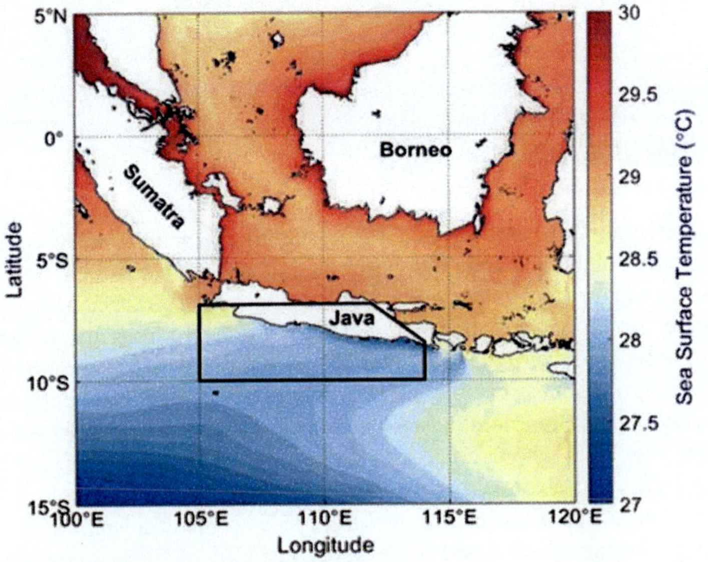

Fig. 24.1 Mean sea surface temperature (SST) of western Indonesia seas, where upwelling spot is indicated by cold SST distribution. Southern Java upwelling as the region of interest is limited in the black box

24.2.2 Methods

The estimation of the upwelling transport via Ekman pumping is derived from the wind velocity by following this procedure [10, 11]:

a. Firstly, we calculated the wind stress by using the following Eq. 24.1.

$$\vec{\tau} = \rho_a \cdot Cd \cdot \left|\vec{V}\right| \cdot \vec{V} \tag{24.1}$$

where, $\vec{\tau}$ is the wind stress vector, with τ_x and τ_y as the respective eastward and northward components, ρa is the density of air (1.225 kg/m^3), Cd is the drag coefficient (1.43×10^{-3}). $\vec{V}$ is the wind component vector that consists of zonal component (u) and meridional component (v).

b. Then, we calculated the wind stress curl by using Eq. 24.2 as follows.

$$\nabla \times \vec{\tau} = \frac{\partial \tau_y}{\partial x} - \frac{\partial \tau_x}{\partial y} \tag{24.2}$$

c. Lastly, the Ekman pumping velocity was obtained by dividing the wind stress curl with the density of seawater (ρw), which assumed to be constant at 1024 kg/m^3 and Coriolis parameter (f) [10]

$$w_E = \frac{1}{\rho_w f} \nabla \times \vec{\tau} \tag{24.3}$$

24.3 Results and Discussion

Time-series plots from area-averaged atmospheric methane concentration along the southern coast of Java Island, Indonesia, which are limited to 7°–10° S and 105°–114° E, for ten years from January 2011 to December 2020 shown in Fig. 24.2. This figure shows that the atmospheric methane concentration has an annual cycle with strong seasonal variabilities that peak around the boreal autumn season with a concentration of about 1800 to 1820 ppb/volume each year. The concentration then drops below 1780 ppb/volume in winter. This cycle is further confirmed by spectral analysis in Fig. 24.3, where the Fourier spectrum has a distinct peak around one year at the 95% significance level.

The investigation of seasonal upwelling's role on the atmospheric methane variability is further analyzed based on seasonal variation of Ekman pumping velocity, plankton concentration, and atmospheric methane concentration. This investigation was carried out during the boreal summer (JJA) and boreal autumn (SON) seasons. Based on the spatial distribution of respective parameters in Fig. 24.4, it is highly likely that the summer monsoonal wind stress generates Ekman pumping along the southern coast of Java Island and weakened in the following season [12, 13].

In contrast, the plankton and atmospheric methane concentration show a significant increase in the boreal autumn season. This finding suggested a lag time between upwelling transport and plankton concentration as well as atmospheric methane concentration along the southern coast of Java Island. The previous studies also showed the same result: coastal upwelling points such as the Arabian Sea NW and SW, off the Canarias Islands, and Namibia (SW Africa) are identified in locations with increased concentrations and emissions of CH_4 [4, 13–16].

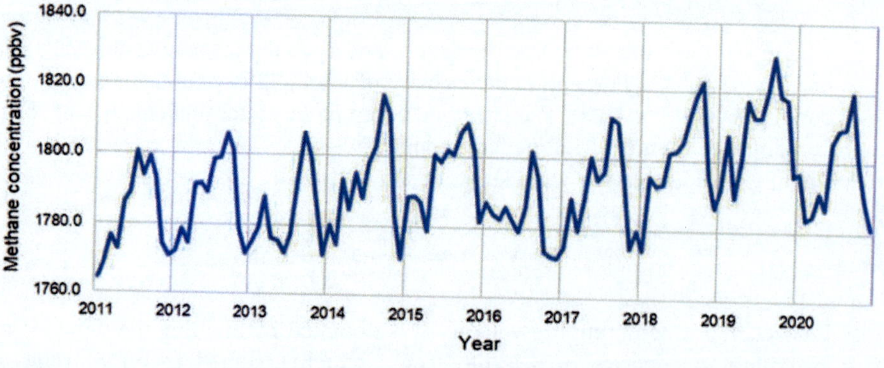

Fig. 24.2 Time series of area-averaged atmospheric methane concentration along the southern coast of Java Island, Indonesia, for ten years from January 2011 to December 2020

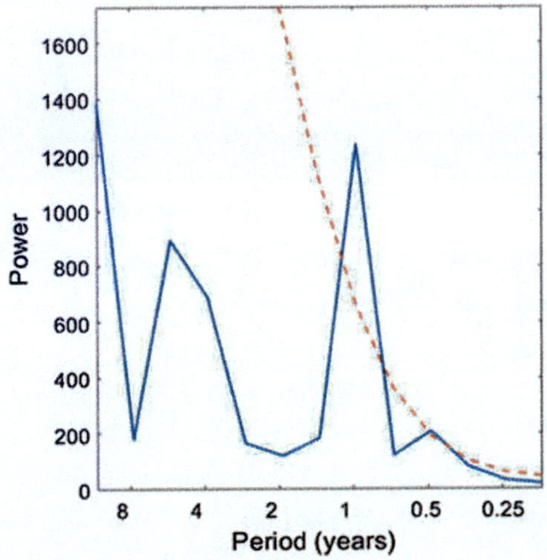

Fig. 24.3 Spectral analysis of the area-averaged methane concentration along the southern coast of Java Island, Indonesia. The Red dashed line indicates a 95% significance level

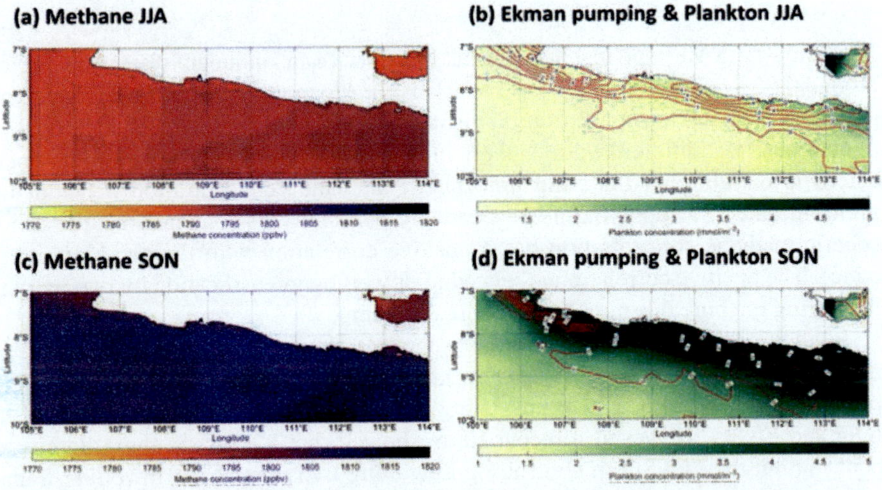

Fig. 24.4 Spatial distribution of **a** methane concentration and **b** plankton concentration (shaded) overlaid with Ekman pumping (contour) in the boreal summer season. **c** and **d** are similar to **a** and **b** but in boreal autumn season. The contour interval is 1×10^{-5} m/s

In order to confirm the delay between upwelling transport, plankton abundance, and atmospheric methane concentration, each parameter is normalized to the respective standard deviation. Upwelling transport induced by Ekman pumping was obtained by integrating the Ekman pumping velocity over the region of interest. Based on Fig. 24.5, it can be seen that the upwelling transport along the southern coast of

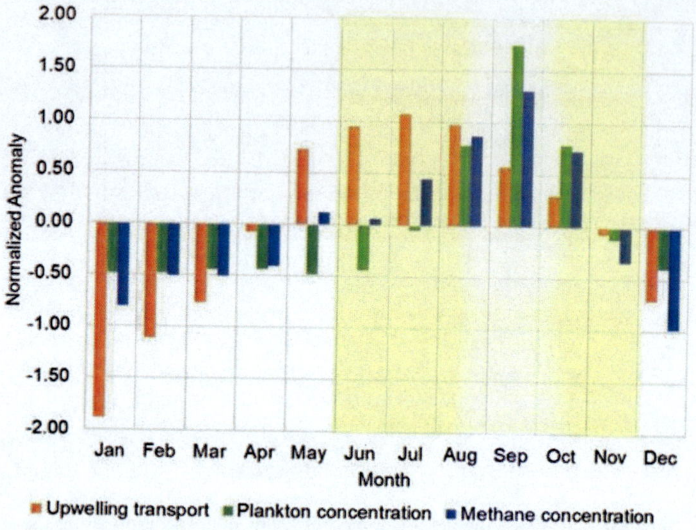

Fig. 24.5 Annual cycle of upwelling transport, plankton concentration, and atmospheric methane concentration averaged along the southern coast of Java Island, Indonesia

Java initiated in May and reached its peak in July with an amount of water transported around 0.66 Sv (1 Sverdrup (Sv) = 1×10^6 m^3/s). Along with the increased upwelling transport, the plankton concentration rose significantly in July and reached its peak in September. In addition, the atmospheric methane concentration shows similar characteristic as well. Plankton's and atmospheric methane's concentration are about 4.2 $mmol/m^3$ and 1811 ppb/volume, respectively. These results conclude that the atmospheric methane concentration has a positive correlation with the plankton abundance. The positive correlation is related to upwelling intensification by two months lag during the boreal summer to the boreal autumn.

In addition to the seasonal variability, the interannual variability in atmospheric methane concentration over the Sothern Java upwelling was further investigated and presented in Fig. 24.6. Upwelling along the southern coast of Java is related to IOD and ENSO. However, a recent study showed that IOD has a more dominant contribution than ENSO in controlling upwelling transport [17]. Upwelling transport increases by 40–50% during positive IOD events and reduces effectively to zero during negative IOD events, influenced by IOD-related Wyrtki jet and thermocline transports from the Indian Ocean as remote equatorial forcing. Therefore, this study focuses on the relation between IOD phenomena and atmospheric methane interannual variability, particularly in 2016 and 2019, when the negative and positive IOD occurs.

During the negative IOD event in 2016, the summertime upwelling transport was the lowest in 2011–2020, as low as 0.46 Sv. The weakening upwelling results in the

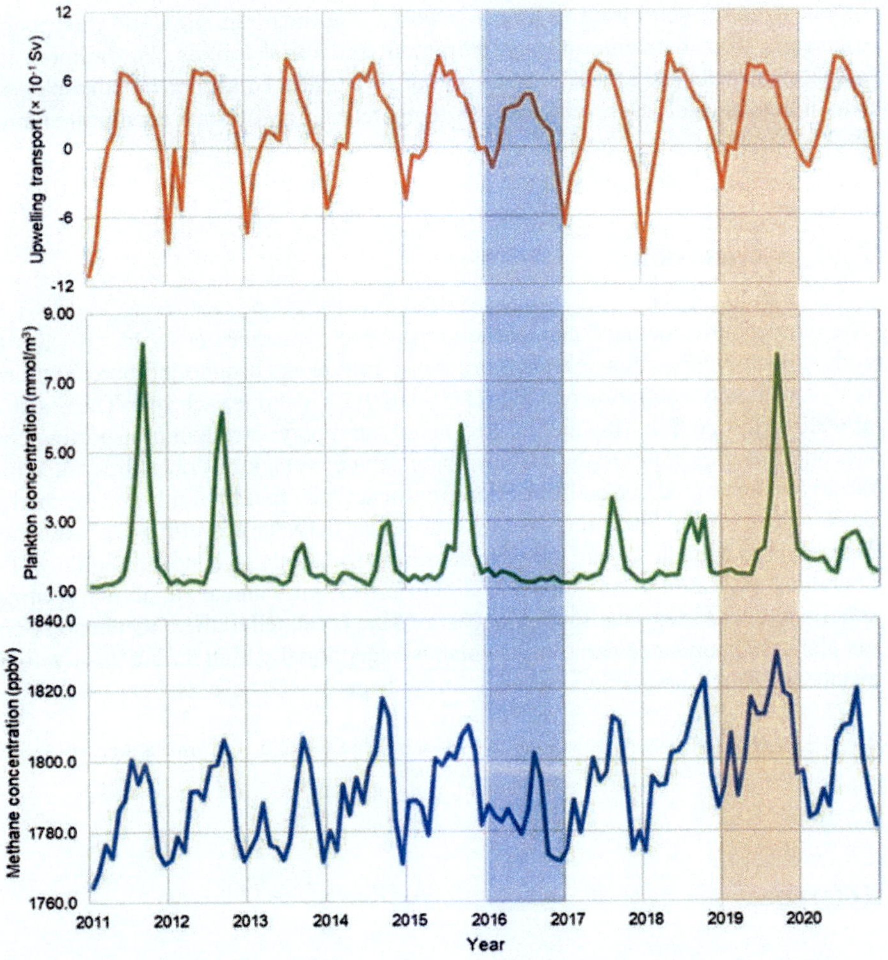

Fig. 24.6 Time series of area-averaged upwelling transport (top panel), plankton concentration (middle panel), and atmospheric methane concentration (bottom panel) along the southern coast of Java Island, Indonesia, for ten years from January 2011 to December 2020. Blue shaded indicates negative Indian Ocean Dipole event, and orange shaded indicates positive Indian Ocean Dipole event

most negligible plankton concentration in this year. Consequently, the atmospheric methane concentration was decreasing compared to the previous year.

On the other hand, during the positive IOD event in 2019, marked by orange shade, the plankton concentration was extremely high, up to 7.73 mmol/m^3, and the atmospheric methane concentration peaked at about 1830 ppb/volume occurred in September. However, there is no significant upwelling transport intensification via Ekman pumping as local forcing. This phenomenon indicated that the plankton bloom is controlled by upwelling transport via Ekman pumping and thermocline transports from the Indian Ocean as remote equatorial forcing. Furthermore, the plankton growth along the southern coast of Java is also controlled by other physical and biological processes, such as light, temperature, run-off, that need further study in the future.

24.4 Conclusion

The atmospheric methane concentration along the southern coast of Java has an annual cycle with strong seasonal variability that peaks around the boreal autumn season with a concentration of about 1800–1820 ppb/volume each year. The seasonal upwelling initiated in May and peaked in July and plays a role in transporting cold and nutrient- rich water to the surface, increasing the plankton concentration. While the atmospheric methane concentration significantly increased in July and reached its peak in September. Hence, there is a two-month lag between upwelling intensification and plankton abundance and increased atmospheric methane concentration over the upwelling system in its seasonal variation. Further, interannual variability of atmospheric methane along the southern coast of Java is related to the IOD phenomenon via plankton abundance variability, which is exceptionally high during positive IOD events and shows no plankton growth during negative IOD events.

Acknowledgements The authors gratefully acknowledge the E.U. Copernicus Marine Service Information and Giovanni for providing satellite data.

References

1. Ciais, P., Sabine, C., Bala, G., et al.: The physical science basis. Contribution of working group I to the fifth assessment report of the intergovernmental panel on climate change. Chang IPCC Clim., 465–570 (2013)
2. Farías, L., Tenorio, S., Sanzana, K., Faundez, J.: Temporal methane variability in the water column of an area of seasonal coastal upwelling: a study based on a 12 year time series. Prog Oceanogr **195**, 102589 (2021). https://doi.org/10.1016/j.pocean.2021.102589
3. Saunois, M., Bousquet, P., Poulter, B., et al.: The global methane budget 2000–2012. Earth Syst. Sci. Data **8**, 697–751 (2016). https://doi.org/10.5194/essd-8-697-2016
4. Kock, A., Gebhardt, S., Bange, H.W.: Methane emissions from the upwelling area off Mauritania (NW Africa). Biogeosciences **5**, 1119–1125 (2008). https://doi.org/10.5194/bg-5-1119-2008
5. Florez-Leiva, L., Tarifeño, E., Cornejo, M., et al.: High production of nitrous oxide (N_2O), methane (CH_4) and dimethylsulphoniopropionate (DMSP) in a massive marine phytoplankton culture. Biogeosciences Discuss **2010**, 6705–6723 (2010). https://doi.org/10.5194/bgd-7-6705-2010
6. James, R.H., Bousquet, P., Bussmann, I., et al.: Effects of climate change on methane emissions from seafloor sediments in the Arctic Ocean: a review. Limnol. Oceanogr. **61**, S283–S299 (2016). https://doi.org/10.1002/lno.10307

7. Steinke, S., Prange, M., Feist, C., et al.: Upwelling variability off southern Indonesia over the past two millennia. Geophys. Res. Lett. **41**, 7684–7693 (2014). https://doi.org/10.1002/2014GL061450
8. Forryan, A., Naveira Garabato, A.C., Vic, C., et al.: Galápagos upwelling driven by localized wind–front interactions. Sci Rep **11**, 1277 (2021). https://doi.org/10.1038/s41598-020-80609-2
9. Hersbach, H., Bell, B., Berrisford, P., et al.: The ERA5 global reanalysis. Q J. R. Meteorol. Soc. **146**, 1999–2049 (2020). https://doi.org/10.1002/qj.3803
10. Smith, R.L.: Upwelling. In: Barnes, H. (ed.) Oceanography and Marine Biology An Annual Review, Vol. 6, pp. 11–, George Allen and Unwin Ltd (1968)
11. Castelao, R.M., Barth, J.A.: Upwelling around Cabo Frio, Brazil: The importance of wind stress curl. Geophys. Res. Lett. **33** (2006). https://doi.org/10.1029/2005GL025182
12. Gülzow, W., Rehder, G., Schneider v. Deimling, J., et al.: One year of continuous measurements constraining methane emissions from the Baltic Sea to the atmosphere using a ship of opportunity. Biogeosciences **10**, 81–99 (2013). https://doi.org/10.5194/bg-10-81-2013
13. Rehder, G., Collier, R.W., Heeschen, K., et al.: Enhanced marine CH_4 emissions to the atmosphere off Oregon caused by coastal upwelling. Global Biogeochem. Cycles **16**, 2–11 (2002). https://doi.org/10.1029/2000GB001391
14. Monteiro, P.M.S., van der Plas, A., Mohrholz, V., et al.: Variability of natural hypoxia and methane in a coastal upwelling system: oceanic physics or shelf biology? Geophys. Res. Lett. **33** (2006). https://doi.org/10.1029/2006GL026234
15. Yadav, I.C., Devi, N.L.: Biomass burning, regional air quality, and climate change. In: Encyclopedia of Environmental Health. Elsevier, pp 386–391 (2019)
16. Sudheesh, V., Gupta, G.V.M., Naqvi, S.W.A.: Massive methane loss during seasonal hypoxia/anoxia in the nearshore waters of Southeastern Arabian Sea. Front. Mar. Sci. **7**, 324 (2020)
17. Nyadjro, E.S., McPhaden, M.J.: Variability of zonal currents in the eastern equatorial Indian Ocean on seasonal to interannual time scales. J. Geophys. Res. **119**, 7969–7986 (2014). https://doi.org/10.1002/2014JC010380

Chapter 25
Preliminary Study on the Utilization of Sugarcane Trash and Corncob for Xylo-Oligosaccharides and Xylose Production Through Dilute Acid Hydrolysis

Thesalonica Yohana, M. Zuvan Maulana Fahrezi, Adetya Lianawati, Riska Surya Ningrum, Dwi Ajias Pramasari, Riksfardini A. Ermawar, Dewi Sondari, and Euis Hermiati

Abstract Sugarcane trash (ST) and corncob (CC) are abundant sources of lignocellulosic biomass with the potential to produce xylo-oligosaccharides (XOS) and xylose in the biorefinery process. XOS are oligomers of sugars that consist of xylose units with DP ranging from 2–10 linked by β-(1,4)-bond and beneficial to use as prebiotics in food and pharmaceutical industries. In addition, xylose is a xylan monomer that can be used as a sweetener. Some organic acids have been reported as suitable catalysts for xylan extraction. In this study, the dilute organic acids, i.e., acetic acid and oxalic acid, were used for hydrolyzing xylan-rich biomass (ST and CC) to produce XOS and xylose. Distilled water was used as a control treatment to be compared with dilute acid hydrolysis. The pretreatment was conducted by heating the biomass in an autoclave at 121 °C for 30, 45, and 60 min with a solid/liquid ratio (1:10). The result shows that the highest XOS concentration from ST and CC was 4.74 g/ L after hydrolysis using 2% oxalic acid for 45 min and 1.31 g/L after hydrolysis using 2% acetic acid for 60 min, respectively. Meanwhile, the highest xylose concentration obtained from ST and CC was 10.04 g/L after hydrolysis using 2% oxalic acid for 45 min and 13.94 g/L after hydrolysis using 2% oxalic acid for 60 min, respectively. The results of this preliminary study confirm that acid hydrolysis has the potential to be used to produce of XOS and xylose from xylan-rich ST and CC.

T. Yohana · M. Z. M. Fahrezi
Faculty of Agricultural Technology, Universitas Brawijaya, Malang, Indonesia

A. Lianawati · R. S. Ningrum · D. A. Pramasari (✉) · R. A. Ermawar · D. Sondari · E. Hermiati
Research Center for Biomass and Bioproducts, National Research and Innovation Agency (BRIN), Bogor, Indonesia
e-mail: dwi.ajias.pramasari@brin.go.id

E. Yulihastin et al. (eds.), *Proceedings of the International Conference on Radioscience, Equatorial Atmospheric Science and Environment and Humanosphere Science, 2021*, Springer Proceedings in Physics 275, https://doi.org/10.1007/978-981-19-0308-3_25

25.1 Introduction

Hemicellulose is the primary raw material for xylose and other variety of xylo-sugars such as xylo-oligosaccharides (XOS). Guo et al. [1] explained that CC consists of hemicellulose in large proportion, particularly macromolecule xylan. Another biomass, ST, is also rich in hemicellulose that reaches 20–35% [2]. Therefore, it seems promising to utilize hemicellulose of ST and CC for a source of value-added products, such as sweetener, prebiotic, polymers, composites, and pharmaceutical applications [3].

For the industry, the dilute acid hydrolysis method is commonly used to solubilize sugars contained in hemicellulose. Because of its low cost and high efficiency, sulfuric acid is the most commonly used for hemicellulose extraction. However, there are some disadvantages of using sulfuric acid, because it causes equipment corrosion and generation of inhibitors. For this reason, other acids, such as acetic acid and oxalic acid, have been used as alternatives for acid hydrolysis. Moderate temperatures during pretreatment (140-190°C) can effectively remove and recover most of the hemicellulose as dissolved sugars, and the addition of acid can increase the yield of product [4]. At lower temperatures ($T<160$°C), acids hydrolyze the hemicellulose into soluble oligomers, such as XOS, then to sugar monomers, such as xylose. Finally, the sugar monomers could be degraded into decomposed products, such as furfural or 5-hydroxymethyl furfural. During the reaction process, the oligomers produced by a random attack by acid on the hemicellulose chains with varying DP (degree of polymerization) [5–8].

In this study, we investigated the use of organic acids in the hydrolysis of ST and CC to produce XOS and xylose. Acetic acid and oxalic acid were used in the acid hydrolysis. Distilled water was also used as a control treatment to compare with dilute acid. The effects of dilute acid and hydrolysis duration on the concentrations of XOS and xylose, as well as pulp or solid residue properties, were investigated. This work is a preliminary study for further development of the utilization of ST and CC for the production of hemicellulose-based products as a part of the biorefinery concept.

25.2 Materials and Methods

25.2.1 Materials

ST was obtained from PT PG Rajawali II, Subang, West Java. CC was obtained from Sumedang, West Java. ST was cut into small pieces (5–7 cm), then was crushed into small particles using a ring flaker. Meanwhile, the CC was directly crushed in a ring flaker. The feedstocks were sun-dried, and then they were milled into smaller particles using a hammer mill and disc mill. After that, they were sieved to get materials having a particle size between 40 to 60 mesh (250–420 μm). These raw materials were stored in an air-tight plastic bag at room temperature before use.

25.2.2 *Methods*

Determination of pH and Chemical Composition. The pH of initial biomass suspension and pretreatment hydrolysate was measured by pH meter. Extractive and ash content of untreated and pretreated ST and CC were determined by following the Technical Association of Pulp and Paper Industry (TAPPI) Standard [9]. Acid-insoluble lignin (AIL) and Acid-soluble lignin (ASL) were determined following the National Renewable Energy Laboratory LAP-003[10]. The holocellulose and α-cellulose were determined according to Wise et al. and Rowell [11, 12].

Dilute Acid Hydrolysis. Dilute acid hydrolysis was conducted at 121°C, 1 atm for 30, 45, 60 min in an autoclave. Organic acids (oxalic and acetic acids) were diluted to a concentration of 2 % (w/v) acid solution. After that, 20 g oven-dry weight of ST or CC was suspended in the acid with a solid-liquid ratio of 1:10 (w/v). Hydrolysis with distilled water was conducted as the control treatment to be compared with the diluted acid hydrolysis. After the hydrolysis was completed, the samples were cooled in an ice bath. The liquid and solid fractions were separated using vacuum pump filtration. The solid fraction was washed with water several times until neutral and was stored at 5°C in a refrigerator for further analysis. Meanwhile, the liquid fraction was analyzed for its reducing sugar, xylose, and XOS concentration.

Determination of Reducing Sugars Concentration, XOS, and Xylose Content. Total reducing sugars were determined using 3,5-dinitrosalicylic acid (DNS) assay [13] with D-glucose as a standard. The absorbance at 540 nm was measured using spectrophotometer UV-Vis (UV-1800, Shimadzu Scientific Instruments Inc, Kyoto, Japan). Xylose content was determined using a D-xylose assay kit (Megazyme, Wicklow, Ireland). The absorbance of the assay at 340 nm was measured by a microplate reader (TECAN Infinite® 200 Pro). To determine XOS content, the hydrolysate was hydrolyzed with 4% H2SO4 at 121 °C for 1 h, and the xylose content of this assay was also determined [14]. The XOS was quantified by the difference between xylose concentration before and after the acid hydrolysis of the hydrolysate.

Statistical Analysis. The experimental results were analyzed using SPSS (Statistical Package for Social Science) as analysis of variance (ANOVA) at a significance level of $\alpha = 0.05$. Three independent variables (raw material, hydrolysis time, and medium type) were investigated for their significance. Fisher's Least Significant Difference (LSD) was used to determine the significant differences within the variable levels.

Table 25.1 Chemical composition of biomass

Components	CC	PCC	ST	PST
Ash content	3.13 ± 0.06	0.74 ± 0.02	8.41 ± 0.70	6.95 ± 0.09
Extractive in EtOH-Benzene	8.95 ± 2.06	25.14 ± 0.51	7.16 ± 2.02	14.62 ± 2.63
Lignin				
AIL	14.34 ± 0.47	23.35 ± 0.17	12.17 ± 0.11	17.56 ± 1.66
ASL	8.06 ± 0.21	3.00 ± 0.11	4.95 ± 0.10	2.74 ± 0.01
Holocellulose	58.95 ± 0.21	54.56 ± 0.09	66.10 ± 1.46	60.51 ± 0.55
α-cellulose	27.02 ± 0.79	42.44 ± 0.37	36.37 ± 1.08	44.83 ± 0.85
Hemicellulose	31.93 ± 0.58	12.11 ± 0.46	29.73 ± 0.39	15.68 ± 0.30

Note CC: Initial Corncob; PCC: Pretreated corncob (oxalic acid 2% for 60 min); ST: Initial Sugarcane trash; PST: Pretreated sugarcane trash (oxalic acid 2% for 60 min)

25.3 Results and Discussion

25.3.1 *Chemical Composition of Biomass*

The chemical compositions of ST and CC are presented in Table 25.1. The ST contains higher ash and α-cellulose than does the CC, while the CC contains higher AIL, ASL, and hemicellulose than does the ST. From this composition, it can be predicted that CC is more potential for producing hemicellulose-based products, such as XOS or xylose, than ST. The α-cellulose and AIL content of ST used in this study is comparable with those reported by Hermiati et al. [6], while the hemicellulose content is lower and extractive and ash content is higher than those in the same report. However, similar ash and hemicellulose content is reported by Pippo et al. [15]. The hemicellulose and AIL content of CC used in this study is comparable with that previously reported [16, 17]. In contrast, the α-cellulose content is lower and ash content is higher than those in the same reports.

25.3.2 *The pH of Biomass Suspension*

The pH of biomass suspension before and after hydrolysis is shown in Table 25.2. The pH values of the hydrolysates obtained from hydrolysis using oxalic acid and water are lower than those of the initial biomass suspensions. In contrast, those obtained from the hydrolysis using acetic acid are vice versa. The decrease of pH in oxalic acid and water hydrolysates are probably due to the cleavage of some acetyl groups in the hemicellulose that generates some organic acids, such as acetic acid, during the hydrolysis [16]. The use of some hydrogen ions from acetic acid during hemicellulose hydrolysis is probably responsible for the increase of pH in acetic acid hydrolysates [16].

Table 25.2 pH of biomass suspension

Hydrolysis		pH (Corncob)		pH (Sugarcane trash)	
		Before	After	Before	After
Oxalic acid	30 min	1.40 ± 0.42	1.12 ± 0.01	1.40 ± 0.42	1.13 ± 0.01
	45 min	1.40 ± 0.42	1.11 ± 0.01	1.40 ± 0.42	1.15 ± 0.01
	60 min	1.40 ± 0.42	1.10 ± 0.01	1.40 ± 0.42	1.08 ± 0.00
Acetic acid	30 min	2.80 ± 0.24	3.37 ± 0.01	2.80 ± 0.24	3.60 ± 0.03
	45 min	2.80 ± 0.24	3.31 ± 0.01	2.80 ± 0.24	3.61 ± 0.01
	60 min	2.80 ± 0.24	3.26 ± 0.01	2.80 ± 0.24	3.53 ± 0.04
Water	30 min	7.19 ± 0.58	4.48 ± 0.01	7.19 ± 0.58	5.99 ± 0.04
	45 min	7.19 ± 0.58	4.40 ± 0.01	7.19 ± 0.58	5.73 ± 0.05
	60 min	7.19 ± 0.58	3.58 ± 0.01	7.19 ± 0.58	4.68 ± 0.05

25.3.3 Production of Reducing Sugars from The Hydrolysis

Total reducing sugars concentrations obtained in the acid hydrolysates are shown in Fig. 25.1. Hydrolysis with oxalic acid revealed the highest reducing sugar concentrations from ST and CC, which were 35.11 and 45.31 g/L, respectively. The highest reducing sugars from ST was obtained after 45 min hydrolysis, while that from CC was after 60 min of hydrolysis. Biomass types, hydrolysis time, and medium types significantly affect reducing sugars production in the hydrolysates.

CC hydrolysates have significantly higher reducing sugar concentrations than do ST hydrolysates. Reducing sugars concentrations obtained from hydrolysis using acetic acid and distilled water are similar and do not quite different. These reducing sugars are significantly lower than those produced from hydrolysis using oxalic acid. Chen et al. [18] explained that acid hydrolysis with acetic acid recovered lower

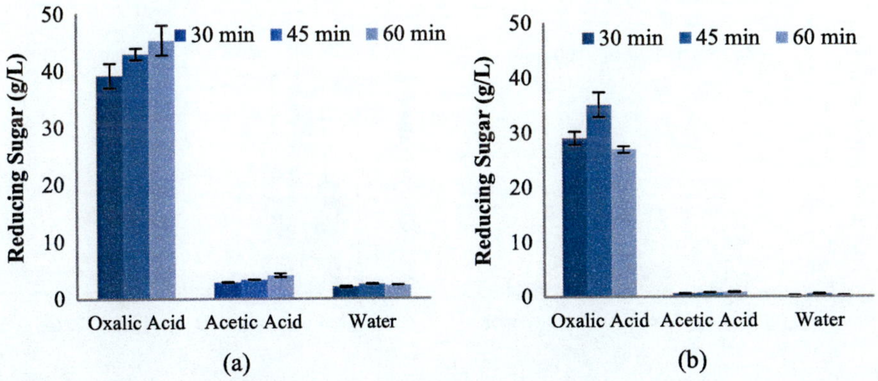

Fig. 25.1 Reducing sugar content (g/L) of CC **a** and ST **b** hydrolysate obtained after hydrolysis in various media and hydrolysis time

reducing sugar in hydrolysate than that other organic acids. This means that acetic acid has a low ability to remove xylan from the biomass. Each level of hydrolysis time resulted in significantly different reducing sugars concentration in the hydrolysates.

25.3.4 Production of XOS and Xylose from the Hydrolysis

Fig. 25.2a shows that the highest XOS concentration from CC (1.31 g/L) was obtained from hydrolysis using acetic acid, while the lowest XOS concentration was from hydrolysis using oxalic acid. On the contrary, Fig. 25.2b shows that the highest XOS concentration from ST (4.74 g/L) was obtained from hydrolysis using oxalic acid, and the lowest (0.13 g/L) was from hydrolysis using water. These are probably due to the difference in biomass characteristics and the strength of the acids. Statistical analysis also shows that biomass types and medium types used in hydrolysis significantly affect the production of XOS and xylose.

The structure of xylan, including the degree of polymerization, branching, and the amount of acetyl group, as well as the lignin carbohydrate complex in the two feedstocks, could be different, which supports an easier degradation of xylan in the CC. Lower ash content in CC (Table 25.1) might also cause an easier degradation of CC. The strength of an acid, shown by the pKa value, also affect the hydrolysis of xylan. The pKa is the negative log base ten of the Ka value (acid dissociation constant). The lower the value of pKa, the stronger the acid and the greater its ability to donate its protons. Oxalic acid is a strong acid with pKa1 2.25 and pKa2 4.28, while acetic acid is a weak acid with pKa 4.76. The use of weak acid in the hydrolysis of CC could still produce XOS. During the hydrolysis of acetic acid, hydronium ions cleave the interior and/or exterior bonds of the hemicellulose, so that XOS is dissolved [5]. Meanwhile, when strong acid was used, the xylan in CC mainly was converted directly to xylose. This is supported by the higher xylose concentration resulted from

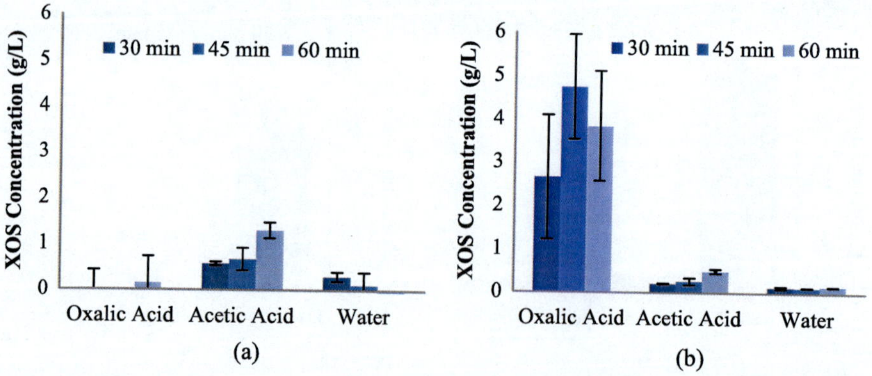

Fig. 25.2 XOS concentration (g/L) of CC **a** and ST **b** hydrolysate in various solvents and hydrolysis time

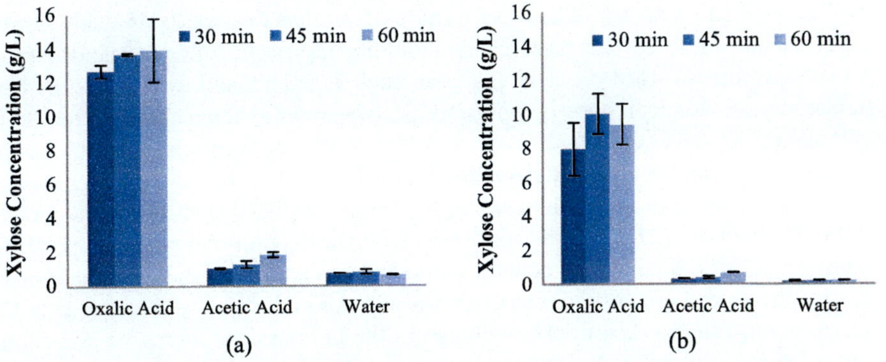

Fig. 25.3 Xylose concentration (g/L) of CC **a** and ST **b** hydrolysate in various solvents and hydrolysis time

CC (Fig. 25.3a). On the other hand, ST that has higher ash content needs stronger acid to hydrolyze its xylan, so that the use of oxalic acid with an appropriate duration of hydrolysis was able to produce XOS as well as xylose. Acid concentration and reaction time affect the degradation of XOS into monomeric sugars [17]. However, based on our study, hydrolysis time did not have significant effect on the production of XOS. Hydrolysis of wheat straw using four organic acids that has pKa 3.75–4.75 shows that the yield of XOS increases with an increase of pKa so that acetic acid that has the highest pKa could produce the highest XOS, which was about 8.5 g/L [18]. Nevertheless, the temperature and acid concentration used (170 °C and 5%, respectively) are much higher than those used in our study. The use of water for hydrolysis of biomass in our study could not produce a significant amount of XOS, which is on the contrary to that reported by Jang et al. [19]. This probably due to different temperatures used, 121 °C in our study, and 180-190 °C in Jang et al. [19].

Xylose concentration obtained from this experiment is presented in Fig. 25.3. The highest xylose concentration from CC (13.94 g/L) was obtained from hydrolysis using oxalic acid for 60 min. On the other hand, the highest concentration of xylose from ST (10.03 g/L) was obtained from hydrolysis using oxalic acid for 45 min. An extension of reaction time could increase the xylose concentration in CC, which is similar to the research conducted by Jang [19] using sweet sorghum bagasse. On the contrary, prolonged hydrolysis until 60 min decreases the xylose concentration in ST hydrolysate. However, statistical analysis shows that the difference in xylose production from each level of hydrolysis time is not significant. The low xylose content obtained from water and acetic acid hydrolysis of CC and ST indicates that using those mediums for hydrolysis at 121 °C is not adequate. Hydrogen ion concentration is directly correlated with the hydrolysis reaction. Therefore, the lower the pKa value of the acid, the more influential the hydrolysis process. Oxalic acid has a lower pKa value that promotes hydrolysis reaction efficiently [20].

Hydrolysis using diluted acid causes structural changes in cellulose, lignin, and hemicellulose [17]. The cellulose content increased from 27.02% to 42.44% in CC and from 36.37% to 44.83% in ST (Table 25.1) due to hydrolysis using oxalic acid. According to several researchers, cellulose increased following acid hydrolysis due to the dicarboxylic acid characteristics of oxalic acid, which has the best selectivity for hemicellulose hydrolysis [18]. Meanwhile, hemicellulose from both biomasses decreases after hydrolysis. It indicates that hemicellulose, which has an amorphous structure, is easily degraded into oligosaccharides or monomer sugars. A higher decrease of hemicellulose might produce higher reducing sugars content in the hydrolysate [21, 22]. It is confirmed with the data on reducing sugars in this study (Fig 25.1). Meanwhile, cellulose structure has a high crystallinity that makes it very strong and not easy to break. As a result, the majority of the cellulose in the biomass has remained, and its percentage in the biomass has increased due to the loss of hemicellulose. Cellulose and lignin content increased due to the removal of other components in the samples [23].

25.4 Conclusion

The xylan-rich biomass CC and ST have been utilized to obtained both XOS and xylose using the dilute acid hydrolysis process. CC hydrolyzed with 2% oxalic acid produced the highest concentration of xylose. However, a 2% acid concentration of oxalic acid could degrade XOS into monomeric sugars. Acetic acid could produce a higher XOS concentration from CC because its pKa value was higher than the other acid. On the other hand, ST hydrolyzed with 2% oxalic acid produced the highest yield of XOS. The findings of this study revealed that acid hydrolysis has the potential to produce XOS and xylose from xylan-rich ST and CC.

Acknowledgements This study was funded through Life Sciences DIPA Project 2021. The authors gratefully acknowledge Integrated Laboratory of Bioproducts, National Research and Innovation Agency Cibinong for its facilities, scientific and technical support.

References

1. Guo, L., Hu, Y., Wu, L., Liang, C., Zhang, W.: The green hydrolysis technology of hemicellulose in corncob by the repeated use of hydrolysate. Chinese J. Chem. Eng. **26**, 191–195 (2018)
2. Moodley, P., Kana, E.B.G.: Optimization of xylose and glucose production from sugarcane leaves (Saccharum officinarum) using hybrid pretreatment techniques and assessment for hydrogen generation at semi-pilot scale. Int. J. Hydrogen Energy. **40**, 3859–3867 (2015)
3. Brienzo, M., Siqueira, A.F., Milagres, A.M.F.: Search for optimum conditions of sugarcane bagasse hemicellulose extraction. Biochem. Eng. J. **46**, 199–204 (2009)
4. Shahbazi, A., Zhang, B.: Dilute and concentrated acid hydrolysis of lignocellulosic biomass. Bioalcohol Prod. Biochem. Convers. Lignocellul. Biomass. 143–158 (2010).
5. Zhang, H., Xu, Y., Yu, S.: Co-production of functional xylooligosaccharides and fermentable sugars from corncob with effective acetic acid prehydrolysis. Bioresour. Technol. **234**, 343–349 (2017)

6. Hermiati, E., Laksana, R.P.B., Fatriasari, W., Kholida, L.N., Thontowi, A., Yopi, Arnieyanto, D.R., Champreda, V., Watanabe, T.: Microwave-assisted acid pretreatment for enhancing enzymatic saccharification of sugarcane trash. Biomass Convers. Biorefinery. (2020).
7. Lee, Y.Y., Iyer, P., Torget, R..: Dilute-Acid Hydrolysis of Lignocellulosic Biomass. Adv. Biochem. Eng. Biotechnol. 65, (1999).
8. Lin, Q., Li, H., Ren, J., Deng, A., Li, W., Liu, C., Sun, R.: Production of xylooligosaccharides by microwave-induced, organic acid-catalyzed hydrolysis of different xylan-type hemicelluloses: Optimization by response surface methodology. Carbohydr. Polym. **157**, 214–225 (2017)
9. TAPPI: [TAPPI] Techincal Association of The Pulp and Paper Industry. , Atlanta (1996).
10. Ehrman, T., Ruiz, R.O., Templeton, D., Adney, B., Hsu, D., Hayward, T.K., Combs, N.S., Schmidt, S.L., Philippidis, G.P., Brown, L., Torget, R.W., Posey-Eddy, F., Okafor, J., Roberson, C.: Chemical analysis and testing laboratory analytical procedures. NREL- Protocols (l. 1–18). 1–186 (1998)
11. Wise, L., Murphy, E., D'Addieco, A.: Chlorite holocellulose, its fractionation and bearing on summative wood analysis and studies on the hemicelluloses. Pap Trade J. **122**, 11–19 (1946)
12. Rowell, R.M.: Handbook of wood chemistry and wood composites, second edition. , Florida (2012).
13. Miller, G.L.: Use of dinitrosalicylic acid reagent for determination of reducing sugar. Anal. Chem. **31**, 426–428 (1959)
14. Surek, E., Buyukkileci, A.O.: Production of xylooligosaccharides by autohydrolysis of hazelnut (Corylus avellana L.) shell. Carbohydr. Polym. 174, 565–571 (2017).
15. Pippo, W.A., Luengo, C.A., Alberteris, L.A.M., Garzone, P., Cornacchia, G.: Energy recovery from sugarcane-trash in the light of 2nd generation biofuels. Part 1: Current situation and environmental aspects. Waste and Biomass Valorization. 2, 1–16 (2011).
16. Yang, R., Xu, S., Wang, Z., Yang, W.: Aqueous extraction of corncob xylan and production of xylooligosaccharides. LWT - Food Sci. Technol. **38**, 677–682 (2005)
17. Wan, L., Gao, Z., Wu, B., Cao, F., Jiang, M., Wei, P., Jia, H.: Hydrolysis of corncob hemicellulose by solid acid sulfated zirconia and its evaluation in xylitol production. Appl. Biochem. Biotechnol. **193**, 205–217 (2021)
18. Chen, Y., Yan, Z., Liang, L., Ran, M., Wu, T., Wang, B., Zou, X., Zhao, M., Fang, G., Shen, K.: Comparative evaluation of organic acid pretreatment of eucalyptus for kraft dissolving pulp production. Materials (Basel). 13, (2020).
19. Jang, S.K., Jung, C.D., Yu, J.H., Kim, H.: Environmentally friendly approach for the production of glucose and high-purity xylooligosaccharides from edible biomass byproducts. Appl. Sci. **10**, 1–12 (2020)
20. Bensah, E.C., Mensah, M.: Chemical pretreatment methods for the production of cellulosic ethanol: Technologies and innovations. Int. J. Chem. Eng. 2013, (2013).
21. Ogunbayo, A.O., Olanipekun, O.O., Babatunde, D.E.: Effect of Pre-treatment method on the Hydrolysis of Corn cob and Sawdust. Anadolu Univ. J. Sci. Technol. Appl. Sci. Eng. 17, 795–811 (2016).
22. Aboagye, D., Banadda, N., Kambugu, R., Seay, J., Kiggundu, N., Zziwa, A., Kabenge, I.: Glucose recovery from different corn stover fractions using dilute acid and alkaline pretreatment techniques. J. Ecol. Environ. **41**, 1–11 (2017)
23. Lee, J.W., Jeffries, T.W.: Efficiencies of acid catalysts in the hydrolysis of lignocellulosic biomass over a range of combined severity factors. Bioresour. Technol. **102**, 5884–5890 (2011)

Chapter 26
Influence of Local Wind Respect to the $PM_{2.5}$ Dispersion in Bandung Basin

Prawira Yudha Kombara and Nani Cholianawati

Abstract The $PM_{2.5}$ dispersion simulation using WRFChem has been conducted. WRFChem is a numerical model that consists of the weather model coupled with atmospheric chemistry. The simulations were performed a few days during the dry season and rainy season at the Bandung Basin. As known, Bandung basin is one of the areas with complex topography in Indonesia. The meteorological factors affect the dispersion of $PM_{2.5}$ in the air, one of which is wind. Based on the numerical model simulation, we achieved a result that the Bandung Basin area has local wind called anabatic wind and katabatic wind. The anabatic wind occurs during the daytime, while the katabatic wind occurs at night time. This study aims to describe the influence of local wind patterns with respect to the dispersion of air pollutants at Bandung Basin. The result of this study shows that there is the difference in the dispersion of $PM_{2.5}$ between the dry season and the wet season. In the dry season, $PM_{2.5}$ is more concentrated equally in the Bandung basin, especially in the morning and afternoon. In contrast, in the noon and afternoon, the enhancement of $PM_{2.5}$ concentration occurs in urban areas. Then in the wet season, $PM_{2.5}$ tends to accumulate in the southeastern of Bandung basin, precisely in the morning and evening. In contrast, in the noon and afternoon, there is no concentration accumulation in urban areas such as in the dry season. This is due to the difference in local wind patterns between the dry season and the wet season. In addition, other meteorological factors such as surface potential temperature and mean sea level pressure were also explored to describe more the influence of the local wind with respect on $PM_{2.5}$ dispersion.

26.1 Introduction

Air pollution has become a serious problem in big cities in Indonesia, including Bandung city. Bandung has a topography that is surrounded by mountains with

P. Y. Kombara (✉) · N. Cholianawati
Research Center for Climate and Atmosphere, National Research and Innovation Agency (BRIN), Bandung, Indonesia
e-mail: praw005@brin.go.id

E. Yulihastin et al. (eds.), *Proceedings of the International Conference on Radioscience, Equatorial Atmospheric Science and Environment and Humanosphere Science, 2021*, Springer Proceedings in Physics 275, https://doi.org/10.1007/978-981-19-0308-3_26

the result that it resembles a bowl or basin. Therefore, it is often referred to as the Bandung basin. Like other big cities, the Bandung Basin also experiences air pollution problems. The increasing number of residents makes the level of mobility and the use of private vehicles increase too. The impact of this makes the air quality in the Bandung Basin worse. According to Fitri et al., one of the pollutants that have to become a concern is $PM_{2.5}$ because it has microscopic size and can penetrate the circulatory system and the protective lining of the lungs [4].

One of the steps to overcome the problem of air pollution in the Bandung Basin is to study the dispersion pattern of $PM_{2.5}$. The dispersion of pollutants is strongly influenced by meteorological parameters such as wind, temperature, humidity, and atmospheric stability [8]. The Bandung Basin has local wind patterns called anabatic winds and katabatic winds. Anabatic winds blow from the basin towards the top of the hill that occur during the day. Then the katabatic winds are the opposite of the anabatic wind, which blow down the slope towards the valley floor and occur at night [7]. The local wind has a different pattern between the dry season and the wet season. As in the wet season based on Kombara et al., the anabatic wind pattern is divided into three patterns depending on the convective condition. The three patterns are weak convection (WC), strong convection (SC), and strong convection accompanied by rain events (SCR).

Analyzing how local wind affects $PM_{2.5}$ dispersion can use dynamic modeling tools. One of the dynamic models that can be used is WRFChem. WRFChem is a mesoscale dynamic model that combines atmospheric models with chemical models that include calculations of atmospheric chemical processes, photolysis, deposition, dispersion and emission mixing, aerosol interactions, and radiation [1, 3, 5]. This model combines topographic, atmospheric, and emission data to produce a simulation of the dispersion and concentration of emissions in an area at a time. This model uses several atmospheric and chemical parameters such as convection, microphysics, boundary layer, radiation, RADM, GOCART, MOZART, and so on. Therefore, in this study, we used WRFChem to simulate the dispersion of $PM_{2.5}$ in the Bandung Basin during the wet and dry seasons and then analyzed the effect of local winds on the dispersion pattern of $PM_{2.5}$.

26.2 Data and Method

This study using some data such as:

1. Final Reanalysis (FNL), this data is used as initial meteorological conditions to run the WRFChem model. This data can be downloaded on the page: http://rda.ucar.edu/datasets/ds083.2. The data has a spatial resolution of 0.25^0 and an hourly temporal.
2. Surface observation data from the Automatic Weather Station (AWS) at one location. The location is the PRTA-BRIN Bandung office. The data is used from 2nd to 5th July 2019. Surface observation data used for verifying the simulation results of the WRFChem model.

3. Global emission data, the data used consists of anthropogenic and biogenic emission data. Anthropogenic emission data used is EDGAR-HTAP data. This data is global anthropogenic emission data with a resolution of 0.1×0.1 degrees in 2010 for BC, OC, CO, NOx, SO2, NH3 emission. Then for biogenic emissions using trace species from the terrestrial ecosystems that be computed by the Model of Emission of Gases and Aerosols from Nature (MEGAN) [5].

26.2.1 The WRFChem Model

To describe the dispersion of pollutant emissions, a dynamic model can be used. One example is using the WRFChem model. WRFChem is a combined model of atmospheric modeling with atmospheric chemistry. The distribution of pollutant emissions is influenced by various factors, one of which is meteorological factors such as wind, temperature, atmospheric stability, etc. In addition, there are also topographic factors and atmospheric photochemical reaction factors. There are many groups has used the WRFChem model to simulate and predict the distribution of pollutant emissions in an area over a certain of period time [3, 5, 6, 9, 10].

To get higher resolution of the model result, the downscaling method is used. The downscaling method was done in WRF Preprocessing System (WPS) part. There are several parts in WRFChem model such as WPS, the chemistry module, WRF main code and post processing part. The WPS is the first part and do preparation the initial input for the model, include the downscaling method. Downscaling was done follow two domains resolution are defined. Two domains are defined in this study as shown in Fig. 26.1, with first domain covers the part of Java and Sumatra region and second domain covers Bandung and its surroundings. First domain has 5 km spatial resolution and 1 km spatial resolution for second domain. The simulation was conducted in the dry and wet seasons. There are four days samples in each season.

Furthermore, for simulation settings in this study are shown in Table 26.1. Table 26.1 shows several schemes of parameterization that be used in this study.

26.2.2 The WRFChem Evaluation

To evaluate the output of WRFChem model in this study, two simple methods are used. Those are Root Mean Square Error (RMSE) and Coefficient Correlation. For the equation of RMSE is shown in the first equation:

$$RMSE = \sqrt{\sum_{i=1}^{n} \left(X_{obs,i} - X_{\mathrm{mod}\, el,i}\right)^2} \tag{26.1}$$

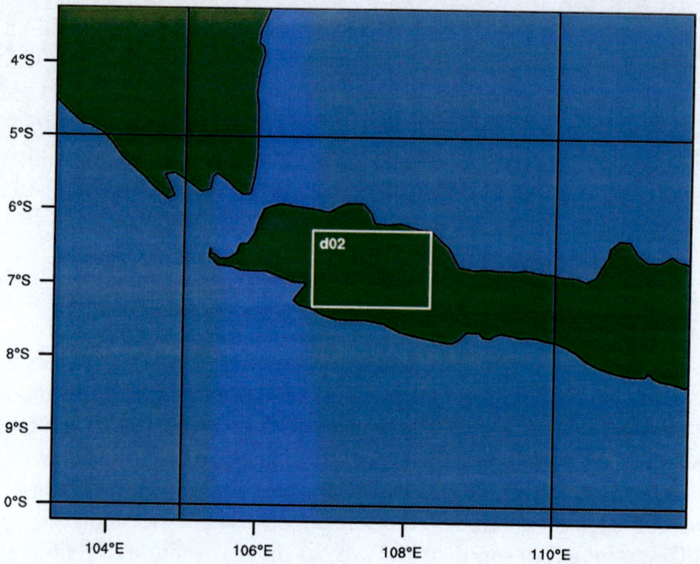

Fig. 26.1 Two domains design's in this study

Table 26.1 The simulation settings

Parameterization	Domain 1	Domain 2
Cumulus	Grell 3	–
Microphysics	Purdue Lin	Purdue Lin
Boundary layer	YSU	YSU
Radiation (shortwave and longwave)	RRTMG	RRTMG
Surface	NOAH MP	NOAH MP
Chemical	CBMZ 4 Bins	CBMZ 4 Bins

and for Coefficient Correlation (CC) equation:

$$CC = \frac{\sum(X - \overline{X})(Y - \overline{Y})}{\sqrt{\sum(X - \overline{X})^2 \sum(Y - \overline{Y})^2}} \tag{26.2}$$

Several parameters of the WRFChem simulation were computed with both equations, such as surface temperature and surface pressure, to evaluate the simulation result.

26.2.3 Exploratory Data Analysis (EDA)

Before describing the influence of local circulation with respect to $PM_{2.5}$ in the Bandung basin, the local circulation has to be identified. To identify the local circulation, the EDA method is used in this study, such as composite and anomaly. Several parameters of the WRFChem simulation such as potential temperature, mean sea level pressure (MSLP), and meridional wind are computed its composite and anomaly.

26.3 Result and Analysis

26.3.1 Meteorological Parameters Evaluation

The meteorological parameters play an essential role in the dispersion of PM2.5 pollutants. Therefore, the meteorological parameters of the simulation results must be evaluated first. The parameters were surface temperature, and surface pressure. As previously mentioned, the two parameters were evaluated by the RMSE method and the correlation coefficient. The RMSE results and the correlation coefficient are shown in Fig. 26.2.

The evaluation results shown in Fig. 26.2 are divided into two, Fig. 2a and b for the dry season and Fig. 2c and d for the wet season. The simulation results in the dry season show good the correlation values for the surface temperature and MSLP parameters reaching 0.98 and 0.76. The RMSE value for the surface temperature parameter is 0.81 however the MSLP parameter is still quite large, with a value of 4.7. If we compare the time-series graph of the two parameters, the results are pretty good because the simulation results can follow the pattern of the observation data. Then the simulation results in the wet season are good enough. The correlation value for the two parameters is still above 0.5 even though the RMSE value is still large enough. The simulation results of surface temperature in the wet season show an underestimate while for the MSLP parameter shows an overestimate.

26.3.2 An Identification of Local Wind in Bandung Basin

Identifying local wind in the Bandung basin based on the simulation result use the EDA method. Here, there are three main parameters that can be used to identify local wind. Those are potential temperature (theta), MSLP, and meridional wind (v wind). By using the EDA method, the anomaly of those parameters is computed.

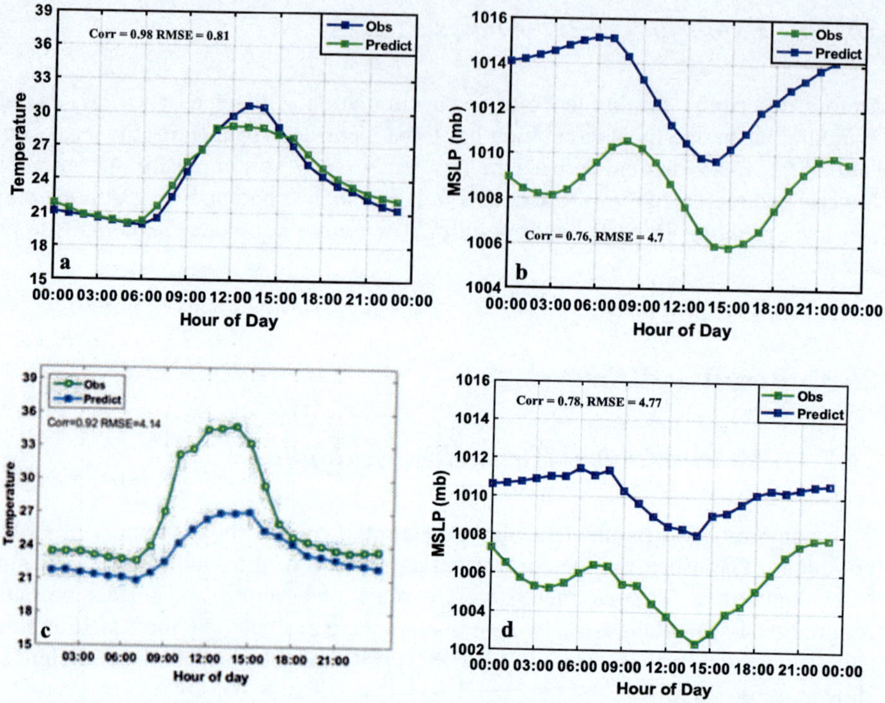

Fig. 26.2 The evaluation of simulation result, **a** surface temperature in the dry season, **b** MSLP in the dry season, **c** surface temperature in the wet season, **d** MSLP in the wet season

- Dry Season

The spatial plots of theta anomaly and MSLP in the dry season are shown in Fig. 26.3. Starting at 09 am local time, positive theta anomalies occurred in the northern and southern regions of the Bandung Basin. The positive theta anomaly makes the two areas warmer than the surrounding area and creates a low-pressure center. This is confirmed by the MSLP anomaly spatial plot. Both regions have negative MSLP anomalies indicating low pressure centers. The existence of low-pressure centers in these two areas makes the wind move both sides of the basin towards the top of the mountain so that at that time, the anabatic winds begin to be active.

At 12 pm local time, the theta anomaly turned negative, but the MSLP anomaly still showed negative values on the north and south sides of the basin so that the wind was still moving on both sides, but the movement to the north side tends to be stronger. Then at 5 pm local time, the wind direction on the north side turned towards the center of the basin, indicating the regime changed to a katabatic wind.

- Wet Season

The spatial plots of theta anomaly and MSLP in the wet season are shown in Fig. 26.4. As in the dry season, starting at 09 am local time, positive theta anomalies occurred in

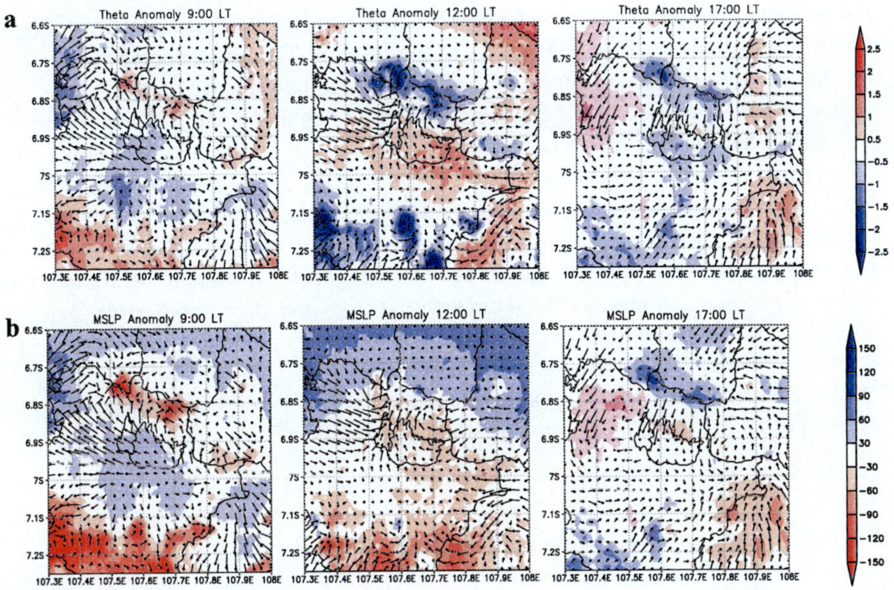

Fig. 26.3 Theta anomaly **a** and MSLP anomaly **b** in the dry season from 09, 12, 17 local time (left to right)

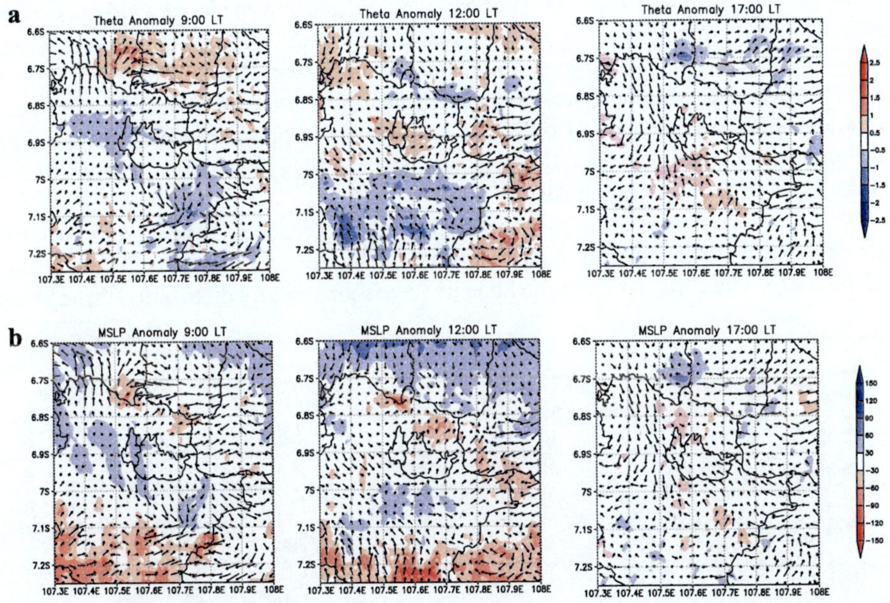

Fig. 26.4 Identical with Fig. 26.3 (for wet season)

the northern and southern areas of the Bandung Basin. It creates a low-pressure center that is confirmed by the MSLP anomaly spatial plot. The existence of low-pressure centers in these two areas makes the anabatic winds begin to be active.

At 12 pm local time, the theta anomaly turned negative as in the dry season, and the MSLP anomaly still shows negative values on the north and south sides of the basin so that the wind was still moving on both sides. However, the anabatic winds heading southward from the Bandung Basin tend to be stronger. It is confirmed by the more negative MSLP anomaly value in the southern region of the basin, resulting in a more significant pressure gradient and stronger wind blowing to the southern region. Based on Kombara et al., in the wet season, when strong convection category, anabatic winds tend to blow more strongly to the south. Such a pattern is caused by stronger convective activity in the southern region [7]. Then at 5 pm local time, the anabatic winds changed to the katabatic winds. It is confirmed by the wind direction on the north side turned towards the center of the basin.

26.3.3 $PM_{2.5}$ Dispersion in Bandung Basin

- Dry Season

The dispersion of $PM_{2.5}$ pollutants in the Bandung basin during the dry season is shown in Fig. 26.5. The dispersion of $PM_{2.5}$ seems to follow the local wind movement pattern for this dry season. At 09 am local time (Fig. 26.5 left), the anabatic winds began to be active, and there was a movement of air masses to both sides of the basin. At that time, the pollutants were concentrated in the north and south of the basin. In addition to the presence of pollutant sources where pollutants are concentrated, the local wind gives an impact on the dispersion of $PM_{2.5}$ pollutants to the north and south sides of the Bandung basin at 09 am local time.

By 12 pm local time (Fig. 26.5 center), the heating reaches its peak, and the air is already mixing. The anabatic wind is still active but tends to move to the north side of the basin. The $PM_{2.5}$ at this time appears to be more evenly distributed throughout

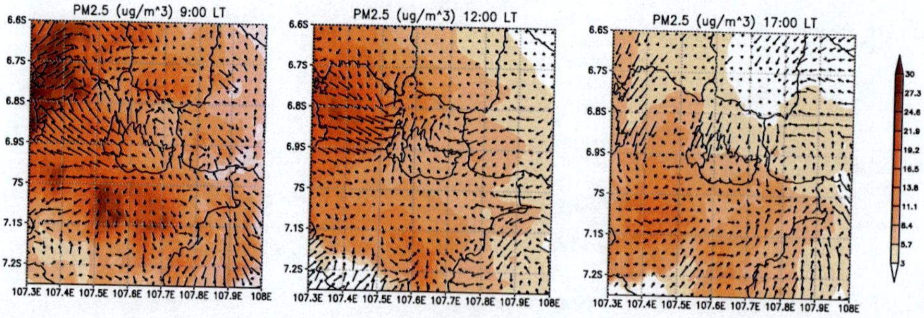

Fig. 26.5 $PM_{2.5}$ dispersion in the dry season from 09, 12, and 17 local time (left to right)

the Bandung Basin area. However, the $PM_{2.5}$ is more concentrated in the south, northwest, and north, although not as high as at 09 am.

In the afternoon, at 05 pm (Fig. 26.5 right), local wind circulation reversed direction, mainly on the north side of the basin. The wind direction changes towards the center and the south side of the Bandung basin, indicating that the anabatic regime has changed to the katabatic regime. This change of wind regime causes pollutants to move and accumulate in the southern region of the Bandung basin.

- Wet Season

As in Figs. 26.5 and 26.6 shows the dispersion of $PM_{2.5}$ pollutants in the wet season. At 09 am local time, the anabatic winds begin to be active with movements to the northern and southern of the Bandung basin. In the wet season, the $PM_{2.5}$ is seen to be more concentrated in the southern region of the basin. This difference is thought to be because the anabatic wind blowing to the south is stronger than that blowing to the north. As explained earlier, this can be visually confirmed by looking at the MSLP anomaly in Fig. 26.4b (left). The MSLP anomaly area is more negative in the southern region so that the pressure gradient is stronger and causes the wind that blows to the south to be stronger as well and transport the pollutants to there. Therefore, more pollutants are concentrated in the southern–southeastern area of the basin.

At 12 pm local time (Fig. 26.6 center), anabatic winds are still blowing to the north and south of the Bandung basin. However, the concentration of $PM_{2.5}$ pollutants in the southern-southeastern region is slightly reduced but still dominate. Unlike in the dry season, where $PM_{2.5}$ is evenly distributed in almost all areas of the Bandung Basin, in the wet season, $PM_{2.5}$ remains more concentrated in the southeastern-eastern area of the basin.

The wind circulation in the Bandung basin is not only influenced by local circulation but also by global circulation such as monsoons. In the wet season, there is a winter Asian monsoon which brings a lot of water vapor and causes a wet or rainy season in the Bandung basin. Pratama and Sofyan, did a study to simulate the PM_{10} dispersion in Bandung basin use the WRFChem model. They got the result that in the wet season, PM_{10} tends to more concentrated in the southeastern region of Bandung

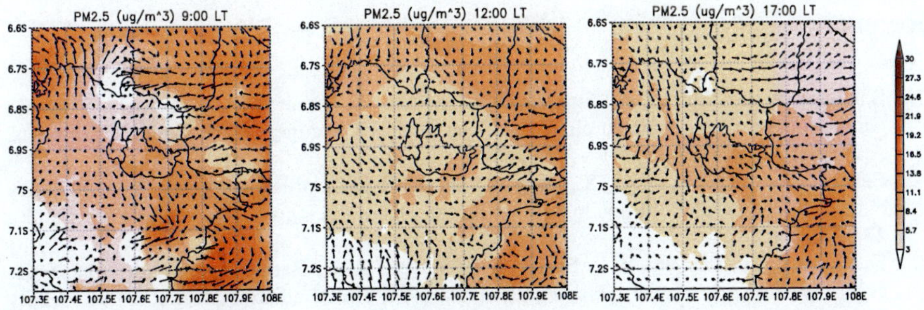

Fig. 26.6 Identical with Fig. 26.5 for the wet season

basin. The identical result was found in this study too. In the wet season, the $PM_{2.5}$ is more concentrated in the southeastern-eastern of the basin. The previous study stated that there is the role of monsoon circulation. Besides, it is also influenced by the complex topography of the Bandung basin. Therefore, the presence of convective activity in the wet season that triggers convective rain makes the concentration of $PM_{2.5}$ in the wet season not as large as in the dry season because the atmosphere is cleaned by rainwater [9]. In addition, pollutants dispersion is influenced by combination between local emission and meteorological condition. Meteorological condition is divided become two, there are local scale and global scale. The local scale is more dominant when the global scale is inactive or in weak condition [2].

At 05 pm local time (Fig. 26.6 right), cooling begins to occur and the anabatic winds begin to change to katabatic winds. $PM_{2.5}$ is slightly shifted to urban areas due to the wind that blows towards the city or the basin's center. Even in the katabatic wind regime $PM_{2.5}$ is still dominantly concentrated in the southeastern area of the basin.

26.4 Conclusion

Based on this research, we can conclude that the WRFChem model can simulate $PM_{2.5}$ dispersion in the Bandung basin well. Furthermore, the difference local wind pattern in Bandung basin among the dry and the wet season can be solved so that the difference $PM_{2.5}$ dispersion can be known too. Local wind pattern gives influence respect to $PM_{2.5}$ dispersion in Bandung basin in both seasons. In the dry season, $PM_{2.5}$ is concentrated in the basin more evenly at morning and afternoon than the wet season. In the wet season, the $PM_{2.5}$ is more concentrated in southeastern and eastern of Bandung basin especially at noon and afternoon.

Acknowledgements The authors gratefully acknowledge to LAPAN-BRIN has provided the data of surface observation from Automatic Weather Station (AWS) for model evaluation.

References

1. Baklanov, A., Schlünzen, K., Suppan, P., Baldasano, J., Brunner, D., Aksoyoglu, S., et al.: Online coupled regional meteorology chemistry models in Europe: current status and prospects. Atmos. Chem. Phys. **14**(1), 317–398 (2014)
2. Bei, N., Zhao, L., Wu, J., Li, X., Feng, T., Li, G.: Impacts of sea-land and mountain-valley circulations on the air pollution in Beijing-Tianjin-Hebei (BTH): a case study. Environ. Pollut., **234** (2018) (2017)
3. Fast, J.D., Gustafson Jr., W.I., Easter, R.C., Zaveri, R.A., Barnard, J.C., Chapman, E.G., et al.,: Evolution of ozone, particulates, and aerosol direct radiative forcing in the vicinity of Houston using a fully coupled meteorology-chemistry-aerosol model. J. Geophys. Res.: Atmos **111**(D21) (2006)

4. Fitri, D.W., Afifah, N., Anggarani, S.M.D., Chamidah, N.: Prediction concentration of $PM_{2.5}$ in Surabaya using ordinary Kriging Method. AIP Conference Proceedings **2329**, 060030 (2021)
5. Grell, G.A., Peckham, S.E., Schmitz, R., McKeen, S.A., Frost, G., Skamarock, W.C., et al.: Fully coupled "online" chemistry within the WRF model. Atmos. Environ. **39**(37), 6957–6975 (2005)
6. Kedia, S., Cherian, R., Islam, S., Das, S.K., Kaginalkar, A.: Regional simulation of aerosol radiative effects and their influence on rainfall over India using WRFChem model. Atmos. Res. **182**(2016), 232–242 (2016)
7. Kombara, P.Y., Junnaedhi, I.D.G.A., Riawan, E.: Characteristic of anabatic wind in Bandung basin observed by AWS. IOP Conference Series: Earth and Environmental Science 2019
8. Perez, I.A., Garcia, M.A., Sanchez, M.L., Pardo, N., Fernandez-Duque, B.: Key points in air pollution meteorology. Int. J. Environ. Res. Public Health **2020**(17), 8349 (2020)
9. Sofyan, APA.: PM10 Air pollution dispersion analysis in bandung city using wrfchem data assimilation. Jurnal Teknik Lingkungan **26**(1)
10. Zhang, Q., Tong, P., Liu, M., Lin, H., et al.: A WRF-Chem model-based future vehicle emission control policy simulation and assessment for the Beijing-Tianjin-Hebei region, China. J. Environ. Manage., (2020)

Chapter 27
Correlation Between Radiation Measurement on the Field Slopes Using KURAMA-II(Kyoto University RAdiation MApping System) and Environmental Radioactivity in the Soil Depth Direction

Yoshikatsu Ueda, Naoto Nihei, Rattanaporn Norarat, and Minoru Tanigaki

Abstract Using KURAMA-II (Kyoto University Radiation Measurement System), we investigated the environmental radioactivity after decontamination in the rice fields of Iitate Village, Fukushima Prefecture. We compared the cross-correlation between the measurement results of KURAMA-II and those of actual soil sampling to confirm the measurement error, especially in the sloping area around the rice field. The results showed an excellent agreement between the soil sample of 5 cm surface layer even in slope.

27.1 Introduction

Tohoku region in Japan was suffered from the triple disaster of the earthquake, the tsunami and Fukushima Daiichi Nuclear Power Plant accident on March 11, 2011. More than 15,000 people lost their lives, and over 2500 are still officially reported as missing, while a further 6000 suffered injuries. In total, over 470,000 people were evacuated from their homes because of radioactive leakage from the power plant. There were about four types of radioactive materials that were diffused: Cs-134,

Y. Ueda (✉)
Research Institute for Sustainable Humanosphere, Kyoto University, Gokasho, Uji 611-0011, Kyoto, Japan
e-mail: yueda@rish.kyoto-u.ac.jp

N. Nihei
Faculty of Food and Agricultural Sciences, Fukushima University, Fukushima, Japan

R. Norarat
Rajamangala University of Technology Lanna, Chiang Rai, Thailand

M. Tanigaki
Institute for Integrated Radiation and Nuclear Science, Kyoto University, Kyoto, Japan

E. Yulihastin et al. (eds.), *Proceedings of the International Conference on Radioscience, Equatorial Atmospheric Science and Environment and Humanosphere Science, 2021*, Springer Proceedings in Physics 275, https://doi.org/10.1007/978-981-19-0308-3_27

Cs-137, I-131, and Sr-90. Iodine-131, which has a short half-life, had no effect in a long time, and Sr-90 was not a problem because the amount diffused was small in Fukushima. Since radioactive cesium(Cs-134, Cs-137) has a large impact on the environment, research on its decontamination attracted much attention. Since 2011, the Japanese government promoted decontamination works, such as contaminated surface soil removal and re-covering the area with uncontaminated soil to reduce and block the radiation. Meanwhile, continuous monitoring of radioactive cesium in soil and air dose is required to evaluate the fate of radiocesium and establish strategies for remediation and management of the contaminated land. In 2011, to measure radiocerium that has fallen to the ground surface, Kyoto University Research Reactor Institute (now Institute for Integrated Radiation and Nuclear Science) had developed a system that can measure the density of radioactive cesium on the ground surface in situ using simple measurements [1]. This system can be used to measure the density of radiocesium contamination on the surface of the ground as you move around, either on a vehicle or on a motorcycle, and the data is stored on a cloud server via the Internet, so that the measured data can be analyzed immediately [2, 3].

In the observation with the walking survey KURAMA-II (Fig. 27.1), the method of measuring the ground surface dose rate is based on the use of two CsI sensors. The lower CsI sensor is set up to detect radiation only in the lower direction by covering the surrounding area (upper and lateral directions) with lead except for the lower direction. From the radiation intensity of the entire space measured by the upper sensor, only the contamination density on the ground surface is detected. Walking surveys can measure the radiation intensity every second, which is a great tool, especially for the observation of narrow roads and rivers where cars and motorcycles cannot enter. On the other hand, there is a problem with the height of the observer, the posture of the walker, and a large number of errors in the measurement of non-flat areas. We confirm this data error due to posture and on the slope. We have checked what kind of error is caused by various walking.

We have measured the surface contamination density of agricultural land in Fukushima prefecture using the KURAMA-II system. In particular, in mountainous areas such as Iitate Village, the influence of the environment around the field (irrigation rivers, etc.) and the slope of the mountain next to the field can be more significant. In mountainous forest areas, there are many places where decontamination work has not been done yet, and there is a possibility of radiocesium leakage. In particular,

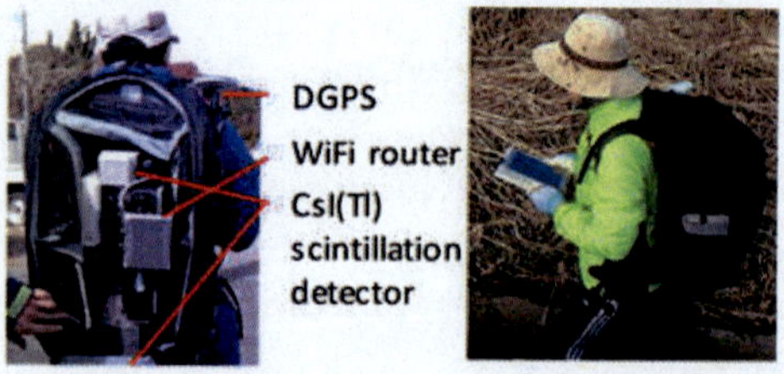

Fig. 27.1 KURAMA-II system

we have been trying to determine the effect of KURAMA-II measurements on the accuracy of surface contamination densities on this slope. In this paper, we report on these results.

27.2 Experimental

27.2.1 Experimental Setup

For the slope measurements by KURAMA-II, we used a field in Iitate Village, Fukushima Prefecture (Fig. 27.2), to carry out the measurements. Specifically, we conducted walking survey measurements in the rice fields and areas with some slopes, and collected more measurement data, especially in the slopes around the irrigation rivers next to the rice fields. The measurements were conducted on December 15, 2017 and March 17, 2018. To simultaneously compare the results of measurements by KURAMA-II with actual soil sampling, soil sampling of the top and bottom of the slope (December 15, 2017) and also the middle part of the slope was additionally conducted (March 17, 2018). During this sampling, in order to check the concentration of radiocesium in the soil in the direction of depth from the ground surface, the concentrations in the soil were measured [4] in three levels of depth: 0–5 cm, 5–10 cm, and 10–15 cm starting from the ground surface (0 cm).

The results of the walking sampling by KURAMA-II are shown in the color map, and the soil sampling points are shown in Fig. 27.3: 8 points (upper and lower) from 17-A to H for the 2017 sample, and 8 points (upper, middle, and lower) from 18-A to 18-H for the 2018 sample. We also measured the concentration of radiocesium in the walking survey of KURAMA-II. The ground surface contamination density was measured from the following formula for calculating the ground surface contamination density [2, 3].

Fig. 27.2 Iitate village

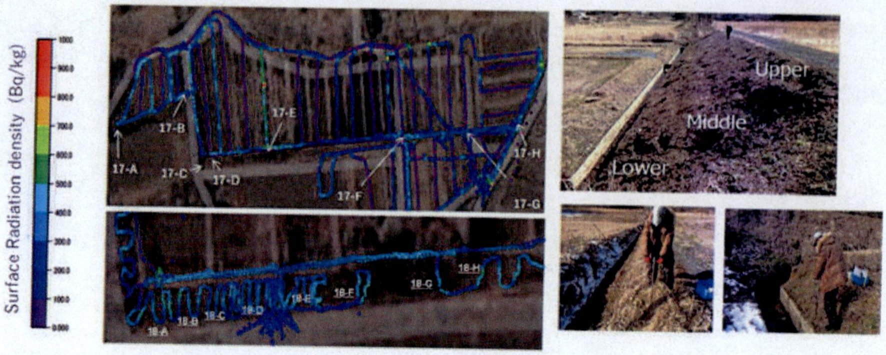

Fig. 27.3 Survey results by KURAMA-II and soil sampling points

$$\mathrm{C_{KURAMAII}} = \left(\mathrm{D_{Lower}} - \mathrm{D_{Upper}}\right) \times 4000 + 48$$

$$\mathrm{C_{KURAMAII}} \ldots SurfaceContaminationDensity\ (Bq/kg)$$

$$\mathrm{D_{Upper}} \ldots Doserate(Upper)$$

$$\mathrm{D_{Lower}} \ldots Doserate(Lower)$$

To compare the surface contamination density of KURAMA-II ($\mathrm{C_{KURAMAII}}$) with that of soil samples, an x–y plot (y-axis: KURAMA-II, x-axis: soil samples) was prepared as a correlation chart. samples) were prepared. In particular, for soil samples, the correlation by the depth and the overall average was charted, and the correlation was evaluated by Coefficient of determination (COD).

27.3 Results and Discussion

27.3.1 Concentration of Radiocesium in Sampled Soil

The results for radiocesium concentrations in soil sampled in 2017 and 2018 are shown in Fig. 27.4.

The trend in sampling at both years was that the concentration of radiocesium in the soil decreased with depth, although there were some exceptions (17-H). This indicates that radiocesium does not penetrate very deep into the soil, but only up to about 5 cm in the surface layer. Regarding the comparison of cesium concentrations at different locations on the upper, middle, and lower slopes, radiocesium concentrations were higher in the middle and lower parts of the slope in the surface layer

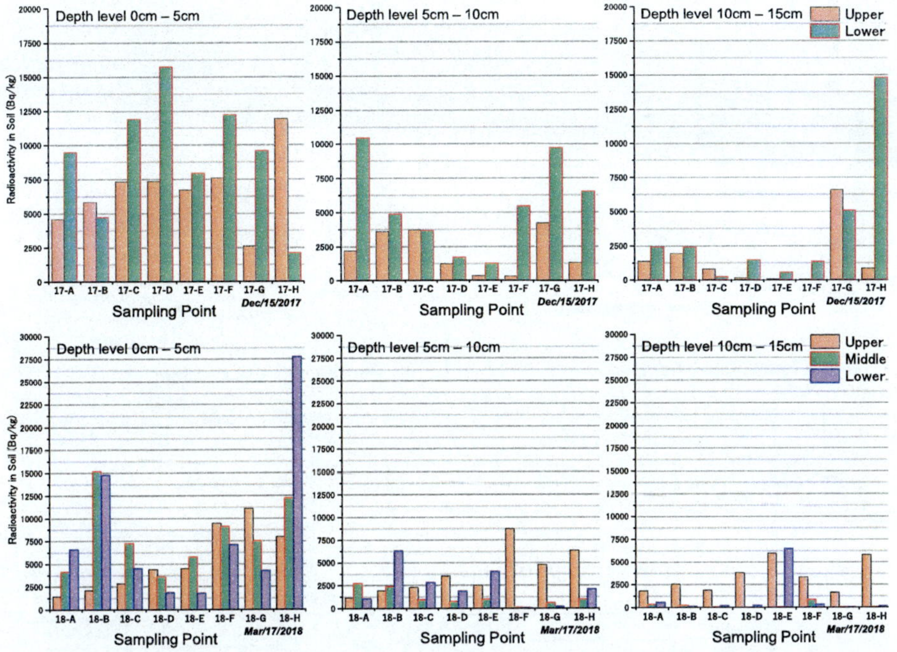

Fig. 27.4 Radiation densities of soil samples by depth (upper graph: 2017, lower graph: 2018 and Upper(upper slope), Middle(middle slope), Lower(lower slope))

(0–5 cm, left graphs). This result is also considered reasonable, since radiocesium may move from the top to the bottom of a slope due to external influences such as rain. Regarding the distribution of concentration in the lower layer (10–15 cm) in 2018, the fact that only the concentration of radiocesium in the upper part of the slope was detected at many points suggests that the points where soil samples were taken in 2018 were places where there was still minimal human movement. Soil erosion was kept to a minimum. In addition, there was a constant transfer of radiocesium from the mountain next to the slope, which may have resulted in a certain amount of radiocesium in the deep soil.

The results of the correlation between KURAMA-II and the soil samples are then shown in Figs. 27.5 and 27.6.

The COD shows a high correlation between the overall mean and the 0–5 cm surface layer in Fig. 27.5. In particular, the correlation with the surface layer is excellent, indicating that KURAMA-II can measure the surface contamination density well. On the other hand, there is a sevenfold difference between the results and the actual measurement of the soil, so it is necessary to conduct calibration sampling to match the precise measurement. It was also found that there was a gradual increase in variability below a depth of 5 cm. The accuracy of KURAMA-II was not very good for underground contamination concentrations because of the principle of measuring the surface contamination density from the air.

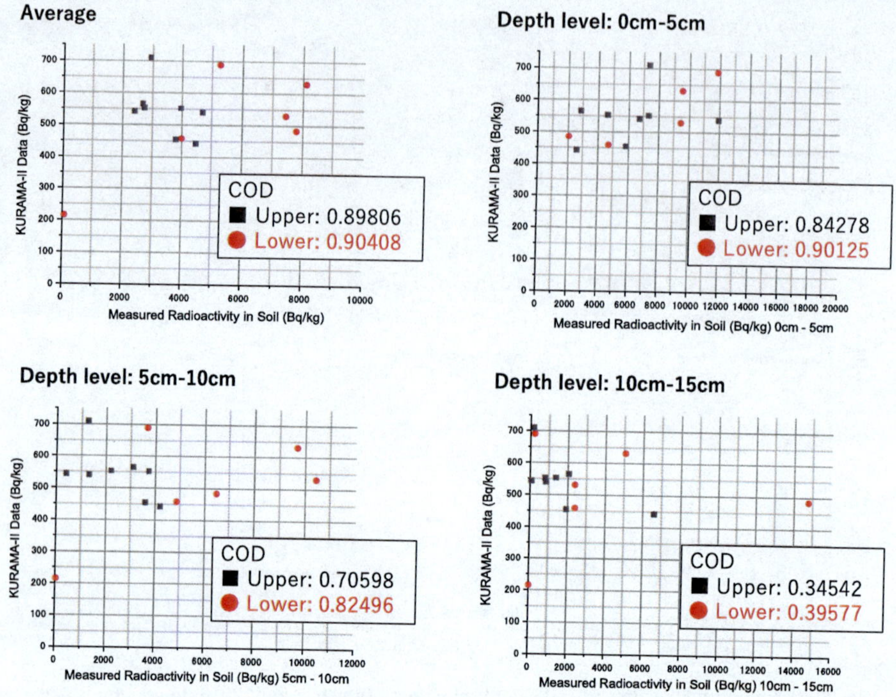

Fig. 27.5 Comparison of radiation density of soil samples with measurements by KURAMA-II (2017)

The sample results for 2018 are shown in Fig. 27.6. The trend of the graphs is similar to that of 2017, with excellent correlations, especially with soil samples in the surface layer 0–5 cm. On the other hand, there is a poor correlation in the middle and lower part of the slope. These results may be due to the fact that the 2017 slope was measured on a relatively gentle slope with a low height, while the 2018 slope, as shown in the photo (Fig. 27.3), is about 5 m high and is easily affected by the sides of the slope, resulting in a low correlation. However, even on a slope with a certain degree of steepness and height, the surface contamination density measured by KURAMA-II was sufficient.

27.4 Conclusion

In this study, we confirmed the difference between the surface contamination density measurement by KURAMA-II and the actual soil sampling, especially in the slope area of a rural field. Comparison of correlation plots showed that the correlation was very high, especially in the surface layer of 0–5 cm. KURAMA-II was able to measure the surface contamination density accurately. On the other hand, there was

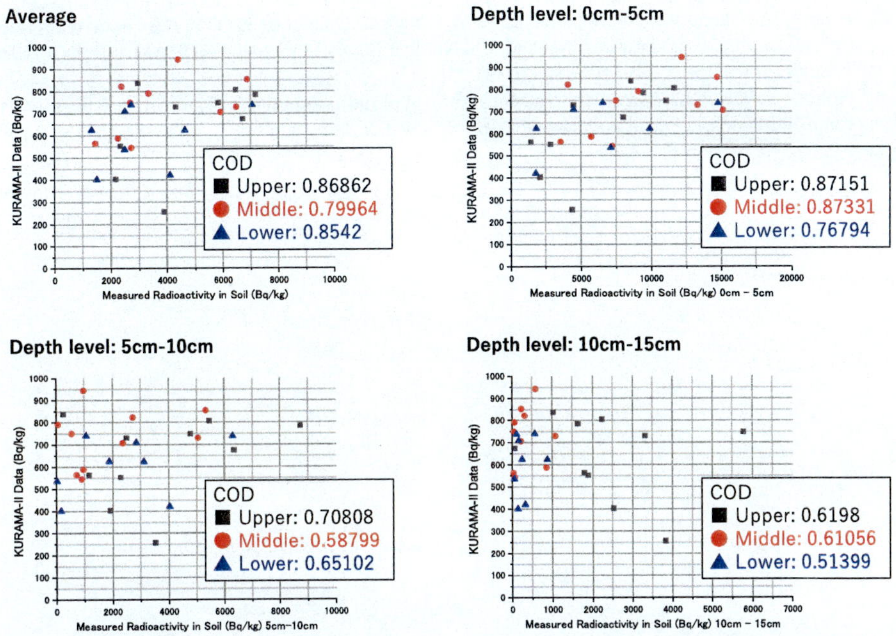

Fig. 27.6 Comparison of radiation density of soil samples with measurements by KURAMA-II (2018)

a sevenfold difference between the measured and measured densities on the slope, indicating that the initial measurement is essential to calibrate the difference. In the future, we would like to collect calibration data for more accurate measurement and use this system to measure the surface contamination density in the soil more accurately, even if it is simple.

Acknowledgements This research is supported by Humanosphere Mission research by Research Institute for Sustainable Humanosphere, Kyoto University, and by Research Units for Exploring Future Horizons, Kyoto University.

References

1. Tanigaki, M., Okumura, R., Takamiya, K., Sato, N., Yoshino, H., Yoshinaga, H., Kobayashi, Y., Uehara, A., Yamana, H.: Development of KURAMA-II and its operation in Fukushima. Nucl. Instrum. Methods Phys. Res. Sect. A. **781**, 57–64 (2015)
2. Andoh, M., Mikami, S., Tsuda, S., Yoshida, T., Matsuda, N., Saito, K.: Decreasing trend of ambient dose equivalent rates over a wide area in eastern Japan until 2016 evaluated by car-borne surveys using KURAMA systems. J Environ Radioact **210**, 105813 (2019)

3. Andoh, M., Yamamoto, H., Kanno, T., Saito, K.: Measurement of ambient dose equivalent rates by walk survey around Fukushima Dai-ichi Nuclear Power Plant using KURAMA-II until 2016. J. Environ. Radioact. **210**, 105812 (2019)
4. Nobori, T., Tanoi, K., Nakanishi, T.: Measurement of radiocesium in soil and crops using NaI(Tl) scintillation counter with autosampler. Jpn. J. Soil Sci. Fertilizer **84**(3), 182–186 (2013)

Chapter 28
Chemical Characteristics of Rainwater in Bandung and Koto Tabang

Fandy Balbo, Rusmawan Suwarman, and Arika Indri Dyah Utami

Abstract We conducted air quality research by reviewing rainwater's chemical composition in Bandung and Koto Tabang, with a sampling period from December 2013 to November 2014. The principal ions in the chemical composition of rainwater consist of Ca^{2+}, Mg^{2+}, Na^{+}, K^{+}, NH_4^{+}, Cl^{-}, SO_4^{2-}, and NO_3^{-}. We analyze the data using statistical methods, Acid–Base Equilibrium, Enrichment Factor, Factor Analysis, and Back Trajectory. The statistical techniques are used to analyze the development of major ions in seasonal time patterns and the effect of precipitation on the ion's concentration. Acid–base equilibrium analysis is used to interpret the process of neutralizing the acidity of rainwater. Hereinafter, the enrichment factor, analysis factor, and backward trajectories are used to analyze the distribution of rainwater element sources. This study shows that the low concentrations are associated with high rainfall events (dilution effect). In September–November, high concentrations of SO_4^{2-} and NO_3^{-} were identified due to anthropogenic activity under low rainfall conditions. The pH of rainwater in low conditions at both locations is identified as acidic rainwater (pH 4.51 and 4.97). The neutralization process with alkaline is quite effective. However, the pH value is still low, caused by the higher concentration of acidity and not comparable to the alkaline. The Investigation of source apportionment with Enrichment Factor, Analysis Factor, and Backward Trajectories result, show Bandung's pollutants come from local anthropogenic sources, while in Koto Tabang, it is indicated to long-distance transport pollutants.

F. Balbo (✉)
Department of Meteorology, Bandung Institute of Technology, Bandung, Indonesia
e-mail: fandy.balbo@gmail.com

Department of Environmental Engineering, Bandung Institute of Technology, Bandung, Indonesia

R. Suwarman
Atmospheric Science Research Group, Faculty of Earth Science and Technology, Bandung Institute of Technology (ITB), Bandung, Indonesia

A. I. D. Utami
Agency of Meteorological, Climatological, and Geophysics for Indonesia, Jakarta, Indonesia

E. Yulihastin et al. (eds.), *Proceedings of the International Conference on Radioscience, Equatorial Atmospheric Science and Environment and Humanosphere Science, 2021*, Springer Proceedings in Physics 275, https://doi.org/10.1007/978-981-19-0308-3_28

28.1 Introduction

Acid rain is a long-term effect due to a high level of air pollution when precipitation dissolves accumulated gases and particulates in the atmosphere. Acid rain is defined as a pH below 5.65 caused by high sulfates and nitrates dissolved in rainwater. Interaction of precipitation and pollutants influences the chemical characteristics and composition of rainwater. Rainwater composition keeps records of air quality and air pollution conditions in the sampling area.

A previous study in Hyderabad, India, show that rainwater in urban areas has experienced twice acid rain events [1]. In another study in West Jordan, on the chemical composition and metal content of rainwater, the concentration of Ca^{2+} was higher than that of other elements, indicating that there was a neutralization process (acid-base neutralization) [2]. Local anthropogenic sources influence rainwater composition as vehicles, industrial plants, and agricultural activities [2]. A study that collected the rainwater sample at Pune India as an urban site and at Sinhagad, India, as a rural site, compared urban and rural sites samples and found higher concentrations of SO_4^{2-} in rural areas [3]. The concentration of SO_4^{2-} is associated with Ca^{2+} concentration, indicates that elements are from Africa, carried with air masses, or we call it a long-range transport pollutant phenomenon. This study [3] informed that there is an opportunity for acid rain in rural sites.

This research aims to determine the composition of rainwater in Bandung (urban site) and Koto Tabang (rural area) to determine air quality and air pollution in these two locations. Bandung, the capital city of West Java province in Indonesia, was chosen as an urban area with high anthropogenic activity, especially on weekends. In comparison, Koto Tabang is a village in West Sumatra, Indonesia, selected as a rural area because it is adequately far from urban and industrial areas. Furthermore, Koto Tabang is the location of the global atmospheric monitoring point by the Agency of Meteorological, Climatological, and Geophysics—Global Atmospheric Watch station.

28.2 Study Area

We use the sample from two sites, Bandung city as an urban area and Koto Tabang as a rural area. In Bandung city, we collect the sample data from the Agency of Meteorological Climatological and Geophysics for Bandung with the point coordinate is 6.883S 107.583E. We choose Bandung city as an urban area because it has a 2.458.503 population with a 167.3 km^2. Bandung city is the capital of West Java Province, which is the most populous province in Indonesia. Bandung city, as a central government for West Java province, has high anthropogenic activity. In addition, Bandung city is also surrounded by industrial areas in the south and east of the city of Bandung. Meanwhile, Koto Tabang is a village in West Sumatra's province, located at coordinates 0.222S 100. 323E. Koto Tabang was chosen as a rural area because this area is far from urban areas (112 km from the provincial capital of West

Sumatra). The nearest small town is the town of Bukittinggi, with a distance of 18 km. The sampling area in Koto Tabang is surrounded by forest and little agricultural area. In addition, Koto Tabang is the location of the Global Atmospheric Watch station, which is the only global atmospheric observation station in Indonesia. In Fig. 28.1, we can see the sampling locations at both sites and show the difference between Bandung as an urban area and Koto Tabang as a rural area.

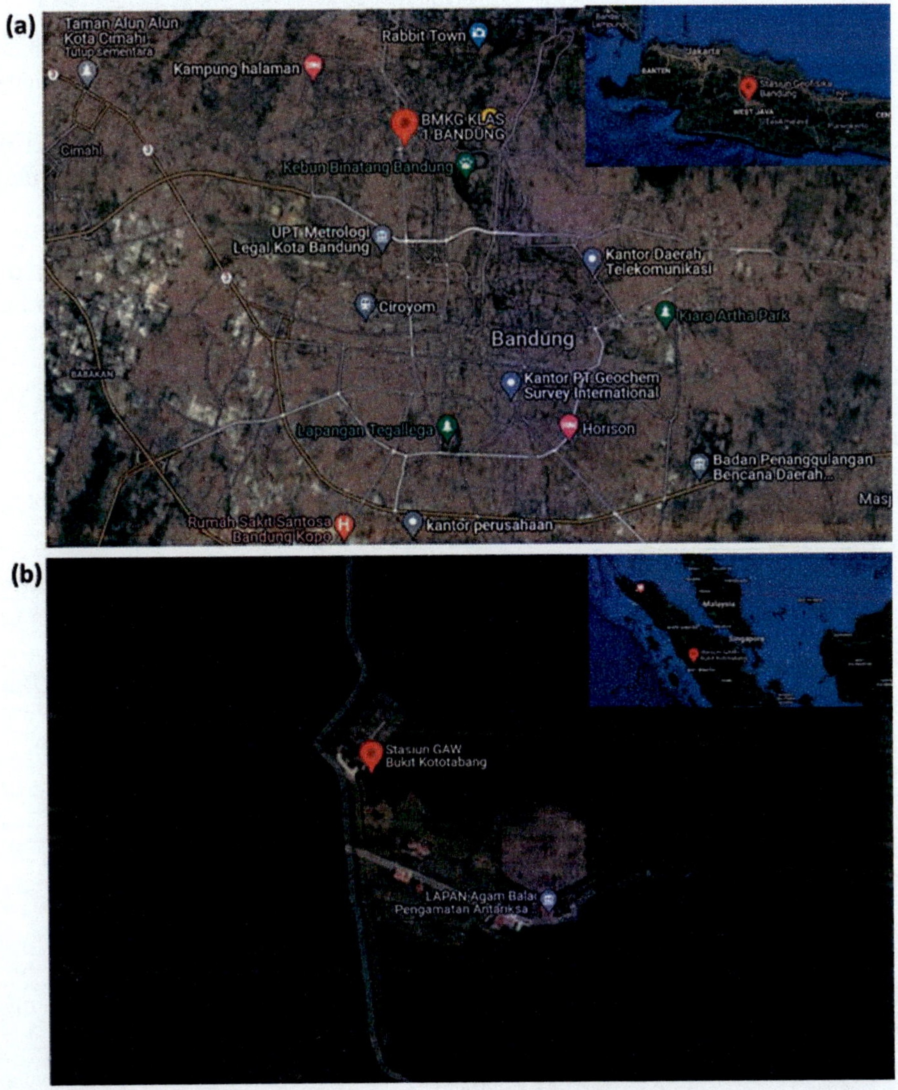

Fig. 28.1 Maps of sampling locations, **a** Bandung City, and **b** Koto Tabang. *Source* https://maps.google.com/

28.3 Data and Methods

We use the weekly concentration of rainwater composition (in μg/L units) from the Indonesian Meteorology, Climatology, and Geophysics Agency from December 2013–November 2014. The chemical elements of rainwater discussed in this study are Ca^{2+}, Mg^{2+}, Na^{+}, K^{+}, NH_4^{+}, Cl^{-}, SO_4^{2-}, NO_3^{-}. and rainwater pH. Chemical analysis of rainwater uses three main methods: elemental concentration variations, rainwater acidity analysis, and identification of elemental sources. First, we will analyze variations in elemental concentrations, which are compiled every three months, in December–January–February (DJF), March–April–May (MAM), June–July–August (JJA), and September–October–November. (SON), to analyze the development of major ions in seasonal time patterns. Next, we analyze the effect of precipitation on the concentration of the elements to illustrate the dilution effect process.

Second, the analysis of the acidity of rainwater using three methods, i.e., pH distribution and time series, the neutralization factor, acidity contribution, and acidity fractionation. Acid fractionation is calculated by Eq. (28.1) [4]. The neutralization factor (NF) calculated by Eq. (28.2) [1].

$$FA = \frac{[H^{+}]}{([NO_3^{-}] + [nssSO_4^{2-}])} \tag{28.1}$$

$$NF_x = \frac{[X]}{([NO_3^{-}] + [nssSO_4^{2-}])} \tag{28.2}$$

where X in Eq. (28.2) is the concentration of the alkaline. Third, we will identify sources of rainwater elements using an enrichment factor and factor analysis to determine the source of rainwater elements in the sampling area. Identifying the source of ions with by finding the value of enrichment using Eq. (28.3) [5] and Eq. (28.4) [4].

$$EF_{seawater} = \frac{(X/Na^{+})_{rainwater}}{(X/Na^{+})_{seawater}} \tag{28.3}$$

$$EF_{crust} = \frac{(X/Ca^{2+})_{rainwater}}{(X/Ca^{2+})_{crust}} \tag{28.4}$$

where X is the concentration of the ion to be identified. Furthermore, knowing the percentage of source contribution in the rainwater element can be calculated using Eq. (28.6) [4], Eq. (28.7) [4], and Eq. (28.8) [4].

$$\%SSF = \frac{(X/Na^{+})_{seawater}}{(X/Na^{+})_{rainwater}} \times 100\% \tag{28.5}$$

$$\%CF = \frac{(X/Ca^{2+})_{crust}}{(X/Ca^{2+})_{rainwater}} \times 100\% \tag{28.6}$$

$$\%AF = 100\% - (\%SSF + \%CF) \tag{28.7}$$

As a discussion, the HYSPLIT Backward Trajectory model is used to analyze the direction and movement of air masses towards the sampling location to determine the transport of pollutants. We use the HYSPLIT Backward Trajectories online version from http://ready.arl.noaa.gov/HYSPLIT.php.

28.4 Results

28.4.1 Variation Analysis

The scope of this research is a preliminary study, so we present four groups of data to see a correlation between rain and elemental concentration. Seasonal variation in three months elements of rainwater and accumulation rainfall is shown in Figure 28.2. In general, elemental concentration is low when rainfall is high and vice versa. In Bandung (Fig. 28.2a), SO_4^{2-} and NO_3^- concentrations are higher than other elements. It is due to the high levels of SO_2 and NO_2 pollutants from anthropogenic activities (vehicles and industrial activities). The high concentration of SO_4^{2-} and NO_3^- indicates an urban area with high anthropogenic activity. Furthermore, we know the anthropogenic activity does not recognize seasonal patterns, so the pollutants increase and accumulate in the atmosphere.

We concern about the element concentration scale. Bandung has an element concentration that reaches 9000 μg/L. Meanwhile, in Koto Tabang, the maximum concentration is 2000 μg/L. Hitherto, there are no quality standards or regulations regarding limits or levels of rainwater. This study slightly challenges identifying whether the concentration of SO_4^{2-} and NO_3^- in Koto Tabang was high or low. However, to determine the condition of rainwater in a polluted state, it can be identified through the measured pH value in the category of acid rain ($pH < 5.65$) or in average rainwater ($pH \geq 5.65$).

Further, we concern about an association between elemental concentration and rain- fall events. Figure 28.3 show a relationship between elemental concentration and precipitation condition. The dilution effect is less pronounced during small rainfall events than during high rainfall events, influencing gas and particle in the atmosphere [6]. The chemical composition of rainwater depends on the lifetime of particles in the air. During the low rainfall season, the particles in the air can last for a long time, so the particles will be accumulated and lead to high concentrations. As a result, the particle scavenging process in low rainfall causing higher solute concentrations. When more rain events, particle scavenging from the atmosphere will become more

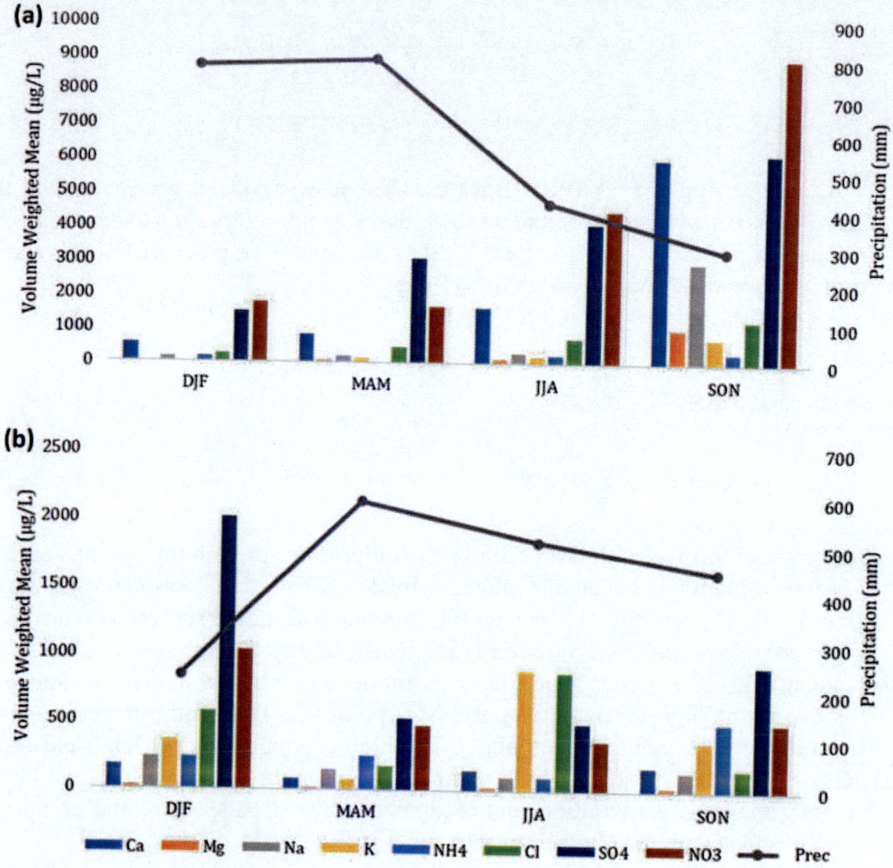

Fig. 28.2 Seasonal variation of rainwater composition **a** Bandung and **b** Koto Tabang

frequent, so solute concentration is low. It states the applicability of the dilution effect process by rainwater to the element concentration (Fig. 28.3).

28.4.2 Acidity Analysis

The chemical composition of rainwater is generally composed of acids and alkalines dissolved in rainwater. The existence of an acid and an alkaline provides information about the neutralization process between both. In theory, we understand that if the acid and base concentrations are nearly balanced, they will make the solution at a state around normal pH. However, if the acid concentration is higher than the base concentration, the solution will become acidic and vice versa. Therefore, this theory

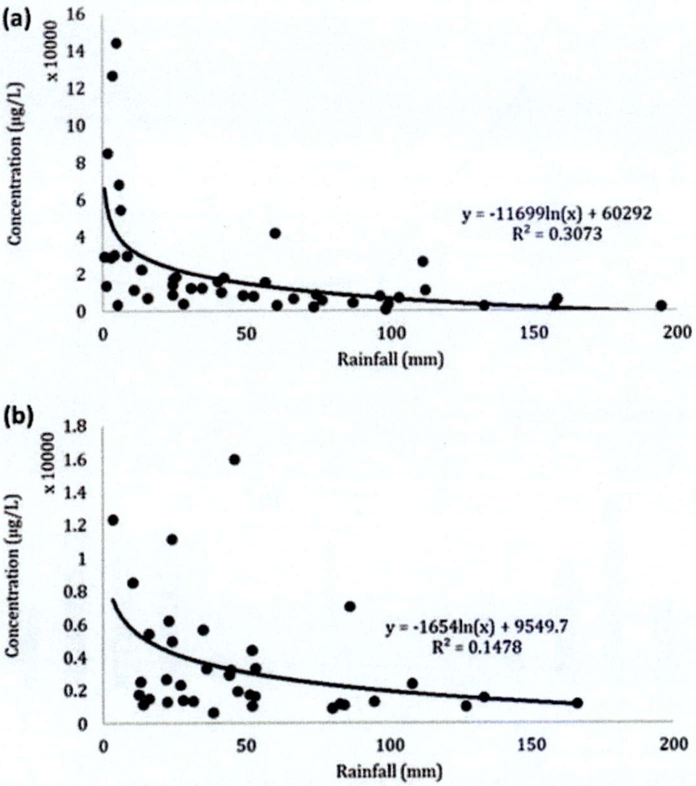

Fig. 28.3 Scatterplot of total elements concentration to rainfall **a** Bandung and **b** Koto Tabang

applies to analyze of rainwater acidity, where rainwater acts as a solvent for pollutants in the atmosphere (Fig. 28.4).

In investigating the rainwater neutralization process, several aspects are considered. Acid fractionation (AF) can provide an information neutralization process that occurs effectively or vice versa. If AF = 1, it is considered that rainwater acidity generated by SO_4^{2-} and NO_3^- will not be neutralized at all [4]. In this study, the AF value for Bandung and Koto Tabang is less than 0.01. The AF value less than 0.01 indicates that the neutralization process by alkaline is 99% effective [4]. In Bandung, the neutralization process by Ca^{2+} is faster than Mg^{2+}, K^+, and NH_4^+. Meanwhile, in Koto Tabang, Ca^{2+} and NH_4^+ neutralize the acidity of rainwater faster than $Mg^{2+}$and K^+. The neutralization process looks adequate and complete in neutralizing acids by alkaline in rainwater. However, this is contrast proportional to the pH conditions in the two sites. Figure 28.5 shows the pH distribution for Bandung (Fig. 28.5a) and Koto Tabang (Fig. 28.5b).

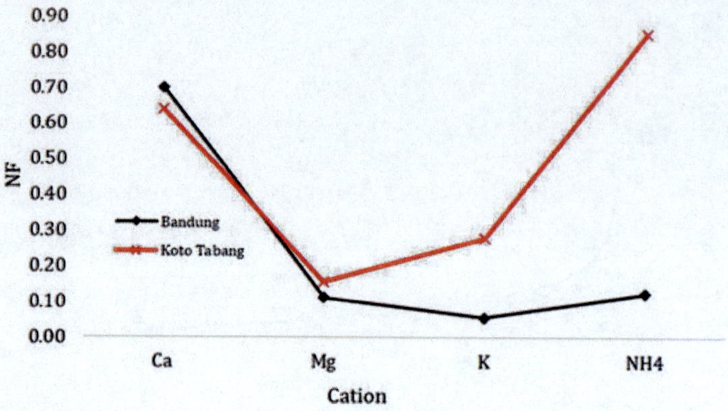

Fig. 28.4 Neutralization factor

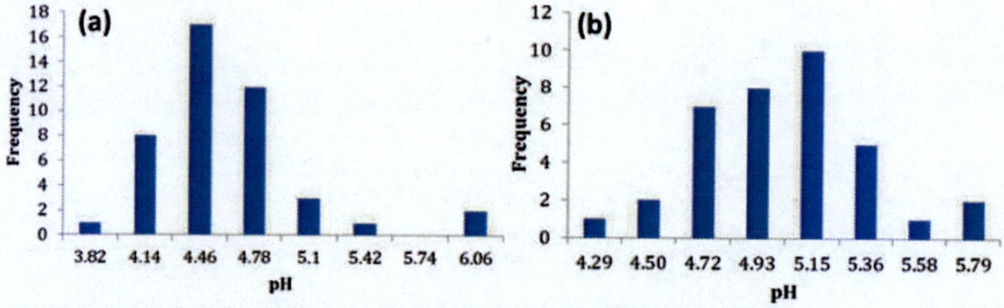

Fig. 28.5 Distribution frequency of pH in **a** Bandung and **b** Koto Tabang

Figure 28.5 provides information in both regions that acid rain occurs throughout the year. Bandung and Koto Tabang only have twice events with rain pH above 5.65. In Bandung, the degree of acidity of rainwater ranges from pH 4.00 to 4.78. From December 2013—November 2014, there was acid rain with a pH of 3.82. Meanwhile, in Koto Tabang, the pH value ranges from 4.72 to 5.36. AF and NF have provided information about a practical and fast neutralization process by Ca^{2+} and NH_4^+ ions. However, the pH value still shows the number of occurrences of acid rain. If we back to the Figure 28.2, the amounts between alkaline and acidic elements are not balanced. Allegedly, the concentration of acid tends to be very abundant compared to the alkaline concentration, causing the solution to have a degree of acidity that tends to be more acidic.

28.4.3 Sources Identification

The two previous results show that the abundance of high acid elements dissolved in rainwater, even though the neutralization process has been running well and effectively. The high acid concentration in rainwater indicates high concentrations of acidic contaminants, such as around the sampling area. Bandung as an urban area, we can assume that these pollutants come from local sources in Bandung. However, it is different from Koto Tabang, a rural area that should not have emission sources around the sampling area. In this third subsection, we will further investigate the source of the rainwater element. Studies of [6–12] state that there are three main sources of gases and particles found in rainwater is the crust, sea, and anthropogenic. The most pollution intensity will be shown in the relative contributions of terrestrial and anthropogenic [4, 6].

We use the Enrichment Factor (EF) to identify the elemental source of rainwater [4, 13–15] which is based on the ratio of elements found between ions accumulated in the atmosphere or rainwater, compared to a similar ratio for the reference element. The study of [16] stated that Na is an element used as a reference element that is pure from the sea. In another study of [14], for elements originating from the continental crust, Al and Ca elements are used because they are natural sources for almost unchanged lithophilic elements. In this study, referring to both studies, Na is used as a reference element for sources from the ocean, and Ca is used as a reference element for sources from the earth's crust. So that in calculating the Enrichment Factor using Eqs. 28.4 and 28.5.

Tables 28.1 and 28.2 show the value of the Enrichment factor for Bandung and Koto Tabang. If the Enrichment Factor (EF) values closes to 1, the element declared associate with the reference element, so have the same source as the reference element. In Bandung and Koto Tabang, the value of Enrichment factor for $Mg^{2+}$and Cl^- has a value close to 1, indicating that the element is associated with Na^+, Mg^{2+} and Cl^- and can be categorized as elements derived from seawater. On the other hand, Mg^{2+} is also associated with Ca^{2+} and K^+, providing information that both

Table 28.1 Enrichment factor relative to seawater

EFs	Ca^{2+}/Na^+	Mg^{2+}/Na^+	K^+/Na^+	Cl^-/Na^+	SO_4^{2-}/Na^+	NO_3/Na^+
Bandung	62.69	1.50	15.58	0.85	46.62	281,412.91
Koto Tabang	21.70	0.64	132.20	2.57	47.72	183,812.47

Table 28.2 Enrichment factor relative to the crust

EFc	Mg^{2+}/Ca^{2+}	K^+/Ca^{2+}	Cl^-/Ca^{2+}	SO_4^{2-}/Ca^+	NO_3/Ca^+
Bandung	0.22	0.25	114.93	107.87	971.66
Koto Tabang	0.27	6.04	1007.70	318.95	1833.24

Table 28.3 Sources of contribution rainwater elements

Bandung				Koto Tabang		
	%SSF	%CF	%AF	%SSF	%CF	%AF
Ca^{2+}	1.60	98.40		4.61	95.39	
Mg^{2+}	66.71	33.29		73.63	26.37	
K^+	6.42	93.58		0.76	99.24	
Cl^-	81.96	0.87	17.17	91.11	0.10	8.79
SO_4^{2-}	2.15	0.93	96.93	2.10	0.31	97.59
NO_3		0.10	99.90		0.05	99.95

originate from the crust. Meanwhile, SO_4^{2-} and NO_3^- have a tremendous EF value, so categorized as originating from the other sources, as anthropogenic sources.

Table 28.3 shows the percentage of source contributions for Bandung and Koto Tabang. SSF (Sea Salt Fraction) represents elements of contribution from the sea. CF (Crust Fraction) represents elements of contribution from the crust, and AF (Anthropogenic Fraction) represents elements from anthropogenic activity. The value percentage of Mg^{2+} and Cl^- is high in SSF, so elements come from the sea (sea spray). Ca^{2+}, Mg^{2+} and K^+ high contribution percentage in CF. The elements come from the crust. SO_4^{2-} and NO_3^- high contribution percentage in AAF, the elements come from an anthropogenic source.

Factor analysis is used to identify the source of elements of rainwater using the statistical method. The source from the sea and the crust hereafter will be referred to as a natural source. In Bandung, factor 1 members are Ca^{2+}, Mg^{2+}, Na^+, K^+, Cl^-, and SO_4^{2-}. The SO_4^{2-} in factor 1 associated with Mg^{2+}, Na^+, and Cl^- indicates that SO_4^{2-} originates from a long-range source. We named Factor 1 as a factor explain Natural Source (Seawater and Crust). Factor 2 members are NO_3^- and NH_4^+. Factor 2 explain that NO_3^- and NH_4^+ origins from local anthropogenic source (transportation, industry and burning biomass). SO_4^{2-} has a moderate relationship (value 0.587) with NO_3^- and NH_4^+ that explain SO_4^{2-} also origins from local anthropogenic activities. In this case, the acidity of rainwater in Bandung is caused by SO_4^{2-} which origins from local anthropogenic and other distant sources.

Koto Tabang has two factors loading. All of the elements are members of factor 1, except NH_4^+. Factor 1 explains about 70% variances. SO_4^{2-} and NO_3^- associate with alkaline element with significant value. That indicates that SO_4^{2-} and NO_3^- in Koto Tabang come from natural source same alkaline elements. We can conclude that pollutants in Koto Tabang are due to the phenomenon of long-range transport pollutants. The SO_4^{2-} and NO_3^- are associated with Ca^{2+} in from of $CaSO_4$ and $Ca(NO_3)_2$, Mg^{2+} in form $MgSO_4$ and $Mg(NO_3)_2$, K^+ in form KNO_3, and Na^+ in form Na_2SO_4 and $NaNO_3$. Factor 1 explains about 'natural source.' Factor 2 explained 13% variances with NH_4^+ as a single member. NH_4^+ indicated as an element from local anthropogenic come from biomass burning. Land fires around sampling locations and

Table 28.4 Factor loading from analysis factor result

Variable	Bandung		Koto Tabang	
	Factor 1	Factor 2	Factor 1	Factor 2
Ca^{2+}	0.886	0.434	0.919	0.021
Mg^{2+}	0.816	0.123	0.891	0.291
Na^{+}	0.871	0.046	0.907	0.059
K^{+}	0.903	0.408	0.831	−0.348
NH_4^+	0.423	0.780	−0.029	0.891
Cl^{-}	0.736	0.605	0.923	−0.051
SO_4^{2-}	0.745	0.587	0.886	−0.274
NO_3	0.028	0.906	0.863	−0.326
Variance (%)	71	13	70	13

agriculture activities give a contribution to NH_4^+ concentration. Factor 2 explains about 'local anthropogenic source'. Table 28.4 shows the results of factor analysis.

28.5 Discussion

We use the Backward trajectory to determine the possibility of the origin of air masses assumed to bring pollutants from a location to the sampling point. The backward trajectories were modified based on the HYSPLIT Model access via the NOAA ARL READY Website. Administration trajectories with 72 h running time are computed for 100, 300, and 500 m above ground level (AGL). We take backward trajectory results that represent the conditions in each season.

Figure 28.6 shows backward trajectories in Bandung. Figure 28.6a explains long-range transport pollutants from a source to a sampling point. That identified trajectory through Jakarta, Bekasi, and Karawang areas. That three areas are industrial. Figure 28.6b, c explain that air masses from southeast Bandung are an industrial area. Figure 28.6b, c explain anthropogenic local source contribution in Bandung. Figure 28.6d shown air masses from southeast and south come from the Hindia Ocean. On the date indicated of high pH (6.07). High ion alkaline concentration can be neutralization acidity.

Figure 28.7 shown the backward trajectory in the Koto Tabang site. In Koto Tabang is not visible local trajectory. In Fig. 28.7a, b, the air mass moves from the northeast. The initial area of the trajectory contains an oil industry, and the air mass may carry pollutants from industrial emissions, thus indicating the phenomenon of long-distance transportation pollutants. Furthermore, Figure 28.7c shows the contribution from the ocean in the rainwater element, indicated by the high concentrations of

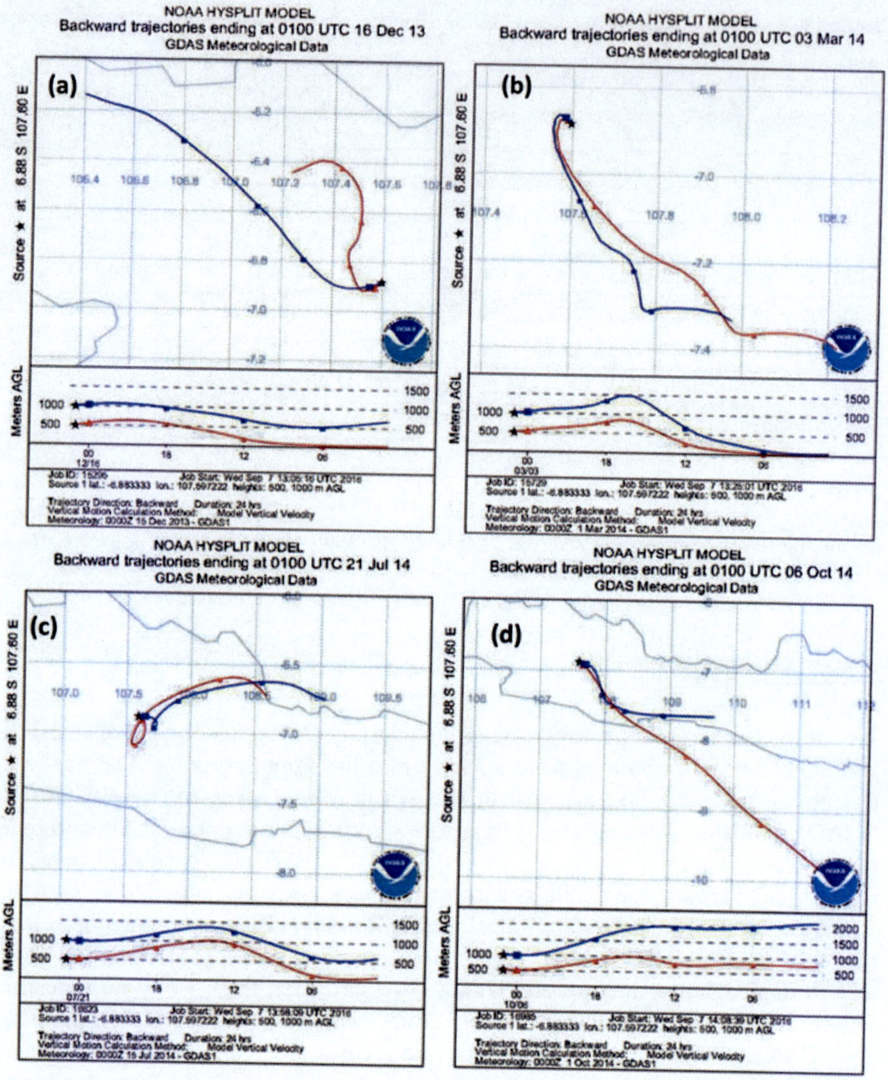

Fig. 28.6 Backward trajectory in Bandung, **a** 16 December 2013 (northwest), **b** 3 March 2013 (southeast), **c** 21 July 2014 (northeast), dan **d** 6 October 2014 (southeast). The Greenline is trajectory end in 500 m levels, the blue line is trajectory end in 300 m levels and, the red line is trajectory end in 100 m levels

Mg^{2+}, Na^{+}, and Cl^{-} at the same time. Figure 28.7d shows that the pollutant comes from the eastern region. The contaminants that reach the sampling point are thought to have come from Riau and the contribution of land fires carried by air masses.

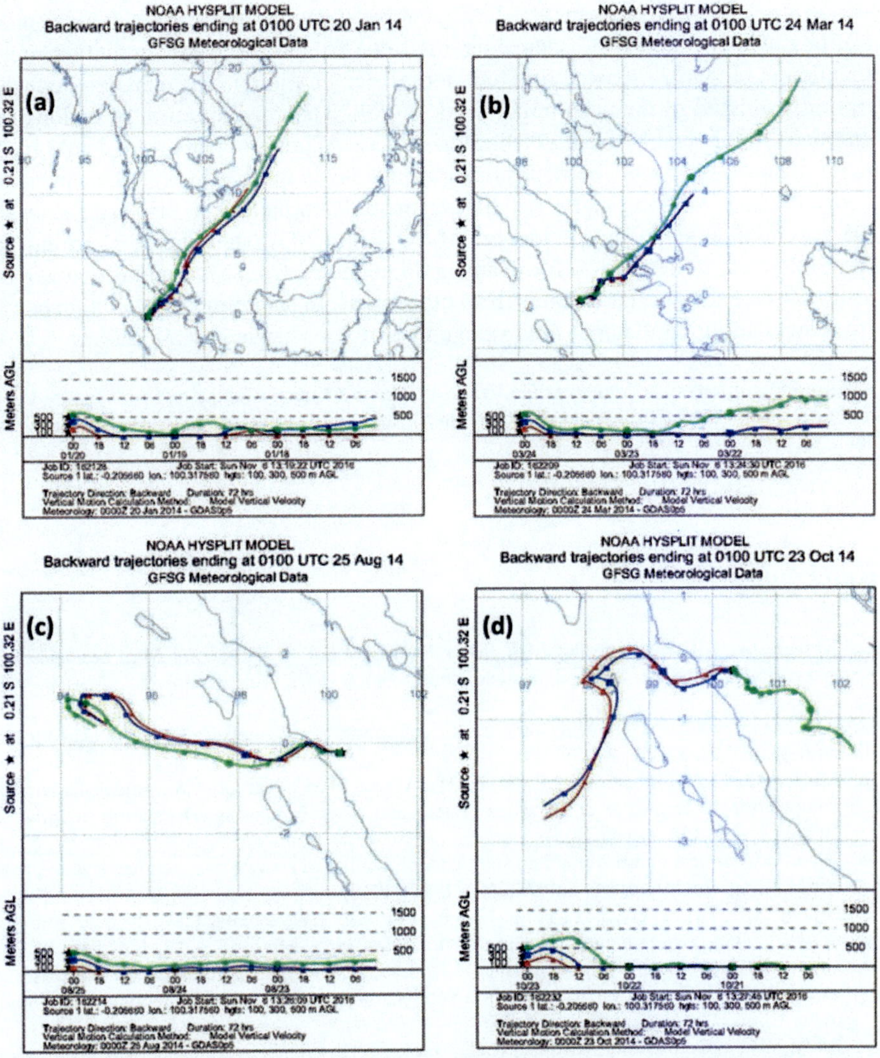

Fig. 28.7 Backward trajectory in Koto Tabang. **a** 20 January 2014 (northeast), **b** 24 March 2014 (northeast), **c** 25 August 2014 (west), dan **d** 13 October 2014 (east). The Greenline is trajectory end in 500 m levels, the blue line is trajectory end in 300 m levels and, the red line is trajectory end in 100 m levels

28.6 Conclusion

Chemical analysis of rainwater in Bandung as an urban area and Koto Tabang as a rural area uses three main methods: variation of elemental concentration, analysis of rainwater acidity, and identification of elemental sources. Seasonal conditions affect variations in rainwater elements. In September–October–November, high concentrations of SO_4^{2-} and NO_3^- were identified due to anthropogenic activity patterns that do not recognize seasons and low rainfall accumulation. The elemental concentration is related to the accumulation of rainfall. The concentration of cations and anions is influenced by rainfall (dilution effect). Alkali elements have a role in the neutralization process. The neutralization process is quite effective. However, the concentration of acidity that is still higher and not comparable to the base causes the pH to remain acidic. The abundances of SO_4^{2-} and NO_3^- were identical around the sampling area. Pollutants in Koto Tabang do not come from local anthropogenic but long-distance transport pollutants from other areas. In Bandung, the pollutant comes from local and long-distance anthropogenic activities carried by air masses.

Acknowledgements The author would like to thanks to Climate Change and Air Quality Laboratory, Agency of Meteorological, Climatological, and Geophysics for Indonesia for providing data.

References

1. Kulshrestha, U.C., Kulshrestha, M.J., Sekar, R., Sastry, G.S.R., Vairamani, M.: Chemical characteristics of rainwater at an urban site of south-central India. Atmos. Environ. **37**, 3019–3026 (2003)
2. Al-Khashman, O.A.: Chemical characteristics of rainwater collected at a western site of Jordan. Atmos. Res. **91**, 53–61 (2009)
3. Momin, G.A., Ali, K., Rao, P.S.P., Safai, P.D., Chate, D.M., Praveen, P.S.: Study of chemical composition of rainwater at an urban (Pune) and a rural (Sin-hagad) location in India. J. Geophys. Res. **110**, 1–10 (2005)
4. Xiao, J.: Chemical composition and source identification of rainwater constituents at an urban site in Xi'an. Environ. Earth Sci. **75**(209), 1–12 (2016)
5. Huang, X.F., Li, X., He, L.Y., Feng, N., Hu, M., Niu, Y.W., Zheng, L.W.: 5- Year study of rainwater chemistry in a coastal mega-city in South China. Atmos. Res. **97**, 185–193 (2010)
6. Huang, D.Y., Xu, Y.G., Peng, P., Zhang, H.H., Lan, J.B.: Chemical composition and seasonal variation of acid deposition in Guangzhou South China: comparison with precipitation in other major Chinese cities. Environ. Poll. **157**, 35–41 (2009)
7. Kelly, V.R., Lovett, G.M., Weathers, K.C., Likens, G.E.: Trends in atmospheric concentration and deposition compared to regional and local pollutant emission at a rural site in southeastern New York, USA. Atmos. Environ. **36**, 1569–1575 (2002)
8. Terada, H., Ueda, H., Wang, Z.F.: Trend of acid rain and neutralization by yellow sand in east Asia—a numerical study. Atmos. Environ. **36**, 503–509 (2002)
9. Hu, G.P., Balasubramanian, R., Wu, C.D.: Chemical characterization of rain- water at Singapore. Chemosphere **51**, 747–755 (2003)
10. Laudon, H., Westling, O., Bergquist, A., Bishop, K.: Episodic acidification in northern Sweden: a regional assessment of the anthropogenic component. J. Hydrol. **297**, 162–173 (2004)
11. Rogora, M., Mosello, R., Marchetto, A.: Long-term trends in the chemistry of atmospheric deposition in Northwestern Italy: the role of increasing Saharan dust deposition. Tellus **56B**, 426–434 (2004)
12. Das, R., Das, S.N., Misra, V.N.: Chemical composition of rainwater and dust fall at Bhubaneswar in the east coast of India. Atmos. Environ. **39**, 5908–5916 (2005)

13. Zhang, M., Wang, S., Wu, F., Yuan, X., Zhang, Y.: Chemical compositions of wet precipitation and anthropogenic influences at a developing urban site in southeastern China. Atmos. Res. **84**, 311–322 (2007)
14. Cao, Y.Z., Wang, S.Y., Zhang, G., Luo, J.Y., Lu, S.Y.: Chemical characteristics of wet precipitation at an urban site of Guangzhou, South China. Atmos. Res. **94**, 462–469 (2009)
15. Wang, H., Han, G.L.: Chemical composition of rainwater and anthropogenic influences in Chengdu, Southwest China. Atmos. Res. **99**, 190–196 (2011)
16. Samara, C., Tsitouridou, R., Balafoutis, C.H.: Chemical composition of rain in Thessaloniki, Greece, in relation to meteorological conditions. Atmos. Environ. **26B**(3), 359–367 (1992)

Chapter 29
Evaluation of the Surface Wind Speed, Shear of Wind Speed, Shear of Wind Direction and Richardson Number at Soekarno-Hatta Airport Using Wyoming Radiosonde Data

Ina Juaeni, Elfira Saufina, Ridho Pratama, Dita Fatria, Wendi Harjupa, and Ibnu Fathrio

Abstract Radiosonde data, which includes surface wind speed (SWS), temperature, and wind profiles used in this study are data at Soekarno-Hatta airport (Cengkareng) in Jakarta. The data period is January to June 2020 used to evaluate the shear of wind direction, shear of wind speed, and Richardson number as the atmospheric parameters that are significantly noticed in flight. In aviation, the Richardson number is used as a rough measure of air turbulence, which values below unity indicating significant turbulence. The surface wind speed can be evaluated directly from the observation data. In contrast, the shears of wind direction and wind speed are determined by calculating the differences in every meter in a vertical direction (from the surface to 10 km). The result shows that the minimum and maximum wind speeds are 0 and 20 ms^{-1}, respectively. During this period, there was an event where the surface wind speed exceeded the threshold value of safe surface wind speed for flight (15 ms^{-1}). Moreover, both the shears of wind speed and wind direction are far below the threshold value. Based on the analysis of turbulence using the Richardson number, the percentage of turbulence events is about 53%. This situation occurs under stable conditions characterized by the CAPE < 1000 J kg^{-1} (no-convective activity) and unstable conditions (CAPE > 1000 J kg^{-1}). Furthermore, there was no significant difference in results related to Southern Oscillation Index (SOI) phase difference and Madden–Julian Oscillation (MJO) amplitude difference. The correlation test shows that there is no correlation between SOI and MJO amplitude to the aviation parameters.

I. Juaeni (✉) · E. Saufina · R. Pratama · D. Fatria · W. Harjupa · I. Fathrio
Research Center for Climate and Atmosphere, National Research and Innovation Agency (BRIN), Bandung, Indonesia
e-mail: inaj001@brin.go.id

E. Yulihastin et al. (eds.), *Proceedings of the International Conference on Radioscience, Equatorial Atmospheric Science and Environment and Humanosphere Science, 2021*, Springer Proceedings in Physics 275, https://doi.org/10.1007/978-981-19-0308-3_29

29.1 Introduction

LAPAN's Strategic Plan 2020–2024 are prepared by researching meteorological parameters related to aviation safety. This strategic plan is realized in several studies. This research is one of those studies. Based on Boeing [1], airplane accidents account for 13% of all aircraft accidents. Of the 181 aircraft accidents from 2017 to 2021, 6 events were related to turbulence encounters, and 12 events were related to controlled flight into terrain (Fig. 29.1).

When the aircraft used are having a sophisticated navigation system, flown by trained pilots, flight safety is determined by the weather [3–6]. The Etihad Airways plane accident on May 4th, 2016, one of which was a plane accident due to weather; the plane experienced severe shaking due to turbulence related to convection activity [7].

In a stable atmosphere, daytime surface winds sometimes have high speeds with varying directions. If this occurs at the airport area, it will disrupt the take-off or landing process of the plane. According to [5], accidents on the runway have a reasonably large percentage (11.11%). Other data show that, in Indonesia, from 2004 to 2018, the percentage of weather-related aircraft accidents was 23%, which occurred at the airports. The rest, 77% happened on the way [6]. International data [4] shows that 1998 of 2832 aircraft accidents occurred on the way, 90 had no known cause , 153 due to engine failure, 71 due to fuel, four due to fire and smoke, and 576

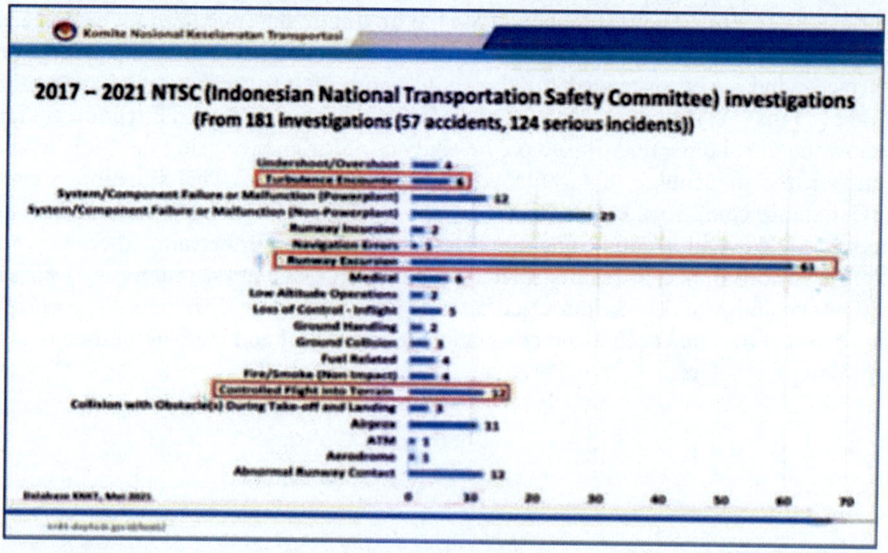

Fig. 29.1 Indonesian National Transportation Safety Committee (NTSC) investigations (Original title: Investigasi KNKT 2017–2021) [2]

occurred at the airport (landing, ground handling, take-off). In addition to the wind, surface parameters that need to be considered because of their impact on a flight are visibility, runway reach, cloud base height, dew point temperature, and air pressure [8].

The atmospheric parameters related to aviation safety are wind, and turbulence, fog and visibility, aerosol and dust, ceiling, rain, and snow in intensity, and rate, icing, ice microphysics, convection and intensity of precipitation, microburst, hail, and lightning [9–11]. The research on wind, and turbulence, will be presented in this paper. In brief, the purpose of this study is to evaluate the surface wind speed, shear of wind speed, shear of wind direction, and turbulence. We also assess the relationship of these parameters with the El Niño atmospheric phenomenon represented by Southern Oscillation Index (SOI) and Madden Julian Oscillation (MJO) characterized by the real-time multivariate MJO (RMM) index amplitude.

In this study, the atmospheric profiles are shown from the surface to an altitude of about 10 km, which is the height commonly used for commercial aircraft [12]. The selected location is Soekarno-Hatta airport which has very high flight traffics.

This article is divided into four sections. The first section is the introduction. The second section describes the data and the method used. The third section contains the results and finally the conclusions in the fourth section.

29.2 Data and Method

This research used daily radiosonde data, which has one or two data a day. The location is Soekarno Hatta airport (Cengkareng), with the selected period is from January to June 2020. It has been mentioned that the parameters evaluated are surface wind speed (SWS), shear of wind speed (SoWS), shear of wind direction (SoWD), and the Richardson number as a sign for turbulence. Those parameters are calculated from radiosonde data which includes wind speed profile, wind direction profile and potential temperature profile. The type and the source of data can be seen in Table 29.1.

Surface winds in a stable atmosphere during the day are powerful and random compared to the night. This condition is essential to know, especially for the purpose of taking off or landing the aircraft. Several accidents have been recorded as being caused by sudden surface winds blowing across the airport runway.

Table 29.1 Source of data

No.	Type	Source
1	Radiosonde data	http://weather.uwyo.edu/upperair/sounding.html
2	SOI	https://data.longpaddock.qld.gov.au/SeasonalClimateOutlook/SouthernOscillationIndex/SOIDataFiles/DailySOI1933-1992Base.txt/
3	MJO index	http://www.bom.gov.au/climate/mjo/graphics/rmm.74toRealtime.txt

Wind shear is the change of wind (direction or speed) between two points. If the change occurs in a vertical direction, it is called vertical wind shear. Meanwhile, if it happens at two points in the horizontal direction, it is called horizontal shear. In addition, surface wind speed is extracted directly from radiosonde data, while the shear of wind is calculated by Eq. 29.2. Wind direction shear is determined with a similar equation, but wind speed is replaced by wind direction.

For the horizontal direction the equation is used:

$$w_{sh} = \frac{(u_2 - u_1)}{(x_2 - x_1)} \tag{29.1}$$

with

w_{sh} = horizontal wind shear

u_2–u_1 = difference of wind speed between point 2 and point 1 (horizontal)

x_2–x_1 = distance between point 2 and point 1 (horizontal).

For the vertical direction used Eq. (29.2):

$$w_{sv} = \frac{(w_2 - w_1)}{(z_2 - z_1)} \tag{29.2}$$

with

w_{sv} = vertical wind shear

w_2–w_1 = difference of wind speed between point 2 and point 1 (vertical)

z_2–z_1 = altitude distance.

Convective Available Potential Energy (CAPE) is the maximum buoyancy of an undiluted air parcel, related to the potential updraft strength of thunderstorms. On a thermodynamic diagram, this is called positive area and can be seen as the region between the lifted parcel process curve and the environmental sounding, from the parcel's level of free convection to its level of neutral buoyancy. CAPE may be expressed as follows:

$$CAPE = \int_{p_f}^{p_n} R_d(T_{vp} - T_{ve})d \ln p \tag{29.3}$$

where T_{vp} is the virtual temperature of a lifted parcel moving upward moist adiabatically from the level of free convection to the level of neutral buoyancy, T_{ve} is the virtual temperature of the environment, R_d is the specific gas constant for dry air, p_f is the pressure at the level of free convection (LFC), and p_n is the pressure at the level of neutral buoyancy (Equilibrium Level; EL). The value depends on whether the moist-adiabatic process is considered to be reversible or irreversible (conventionally irreversible, or a pseudo adiabatic process in which condensed water immediately falls out of the parcel) and whether the latent heat of freezing is considered (conventionally not). It is assumed that the environment is in hydrostatic balance

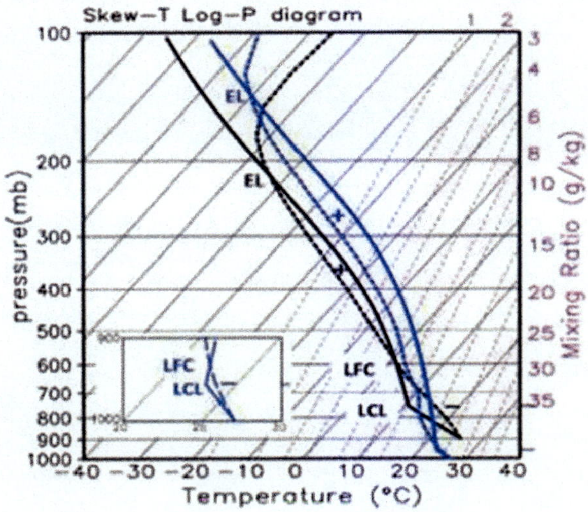

Fig. 29.2 CAPE is a positive area between LFC and EL [13]

and that the pressure of the parcel is the same as that of the environment. Virtual temperature is used for the parcel and environment to account for the effect of moisture on air density (https://glossary.ametsoc.org/wiki/Convective_available_potential_energy) (Fig. 29.2).

Richardson number evaluated using the following equation:

$$R_i = \left(\frac{9.81}{\theta}\frac{d\theta}{dh}\right) / \left(\frac{du}{dh}\right)^2 \tag{29.4}$$

R_i = Richardson number, g = 9.81; gravitational acceleration (m/s^2), θ = potential temperature, du = difference of wind speed and direction between z_2 and z_1, dh = difference of height between z_2 and z_1.

In this study, the vertical shear will be determined by Eq. (29.2) and the profile evaluated is from surface to 10 km [14]. If the wind shear increase, the potential for air transportation hazards increase too. Surface wind and wind shear fall into the category of hazards if they have reached or exceeded the threshold value. This study used threshold values from National Oceanic and Atmospheric Administration NOAA [15], as shown in Table 29.2.

29.3 Results and Discussion

It was starting with the first result. The minimum and maximum surface wind speeds (SWS) are 0 and 20 ms^{-1}, respectively. It can be seen in Fig. 29.3, during the research period, only one incident was found where the surface wind speed exceeded the

Table 29.2 The threshold for aviation parameters

No.	Parameter	Threshold
1	Surface horizontal wind speed	<15 m/s
2	Horizontal wind speed for small planes	<15 m/s
	Horizontal wind speed for large aircraft	<23 m/s
3	Shear of wind direction	<90°/m
4	Shear of horizontal wind speed	<2.5/s
5	Richardson number	0 < Ri < 0.25

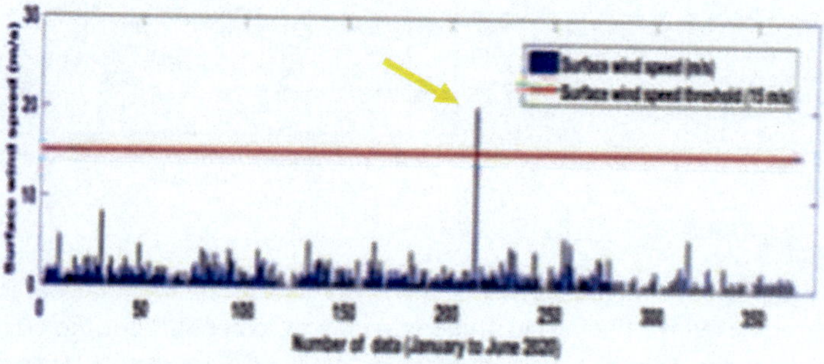

Fig. 29.3 Surface wind speed at Soekarno-Hatta airport

threshold value as marked in the yellow arrow. It occurred on April 14th, 2020, at 00:00 Z (07:00 LT). Even though the surface wind speed does not seem too vulnerable, but we must be aware of it because can suddenly appear. Additionally, there are no reports of aircraft accidents recorded during the research period [16].

Wind shear is a sudden change in wind direction and speed. Strong wind shear disrupts the stability of the aircraft, will cause the airplane to change direction and speed suddenly. If the wind shear is accompanied by turbulence, the aircraft may shake violently. Considering the impact on air transportation safety, measurement and research related to surface wind and wind shear at airports become important [9, 17, 18]. The second result is the SoWS. As shown in Fig. 29.4. SoWS range between 0 and 0.1. It is far below the threshold value. The maximum SoWD in Fig. 29.5 reached 40 deg m^{-1}, which only occurred once (yellow arrow). The mean shear of wind direction is less than 20 deg m^{-1}; it is far below the threshold. The shear of the wind direction threshold is 90 deg m^{-1}. Thus, SoWS and SoWD are not harmful to the flight. However, it does not rule out the possibility of it happening suddenly.

Based on the analysis of turbulence using the Richardson number, the percentage of turbulence events is about 53%, as shown in Fig. 29.6. Therefore, turbulence is relatively frequent at Soekarno-Hatta airport. This result is following the KNKT data mentioned in the introduction section. Generally, it occurred at the surface (coded 1) and at a particular height (coded 2). Turbulence occurs at very varied altitudes (77, 102, 300, 301, 775, 900, 1501, 1509, 1511, 1515, 1520, 1524, 3171, 3301, 5362,

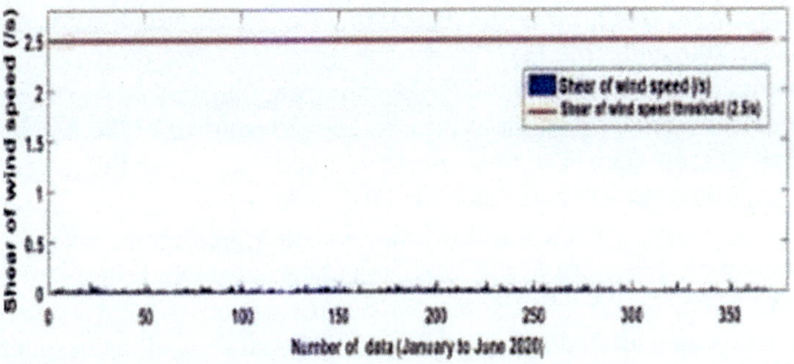

Fig. 29.4 SoWS at Soekarno-Hatta airport

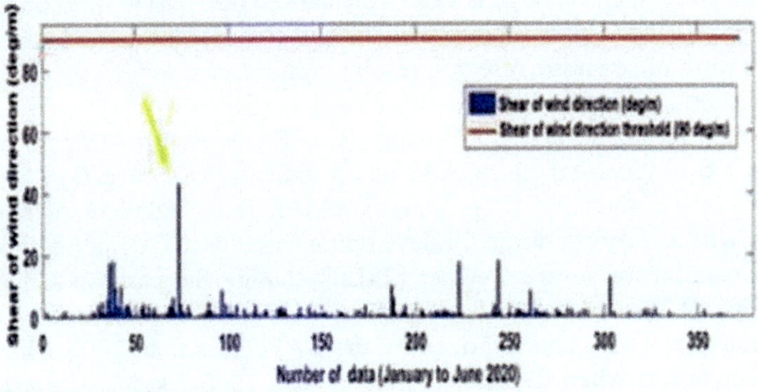

Fig. 29.5 SoWD at Soekarno-Hatta airport

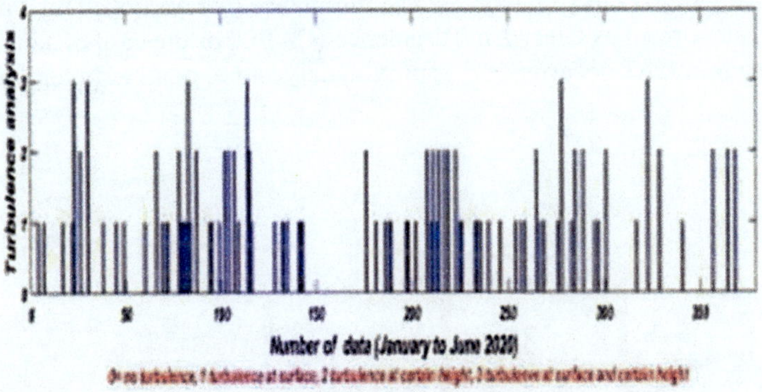

Fig. 29.6 Turbulence analysis based on Ri at Soekarno-Hatta airport

5395, 5708, 5812, 6005, 6024, 6310, 7590, 7793, 8999, 9019, 9179, 9384, 9514, 9673–9720, 9700, and 9740 m).

Turbulence is instability triggered by shear, strong shear evokes strong turbulence. Wind shear converts laminar flow into an eddy (turbulent). The height that is of great concern in air transportation related to wind shear is at the lower level (less than 500 m) when the aircraft takes off [19]. It is similarly stated in [20] that the improvement of data accuracy and aviation weather predictions currently depends on the accuracy of determining and modeling lower-level wind shear and turbulence. At the same time, the best measurement or observation method to date is in situ observations by aircraft and compliance with radiosonde as done in this research is the second-best method [21].

Aeronautical meteorological research is very important to improve flight safety but unfortunately, such research is still rarely carried out. Discussions on wind shear mostly appear in the form of reference books or technical reports, and it is difficult to find in the form of scientific research results in Indonesia, especially at local scales or at the airports.

Convective Available Potential Energy (CAPE) be used to indicate convection activity at the observation site. CAPE as an indicator of convection activity was carried out by [22–26] and [27]. Nonzero CAPE is an indicator of atmospheric instability, a necessary condition for developing cumulus and cumulonimbus clouds with the potential for severe weather [28]. By taking the case on January 1st to 31th in Fig. 29.7., it is found that turbulence occurred both in stable conditions (no-convective activity) characterized by the CAPE less than 1000 J kg^{-1} and in unstable conditions when CAPE was greater than 1000 J kg^{-1}. The examples of the occurrence of turbulence at low CAPE (<400 J kg^{-1}) are on January 25th and January 30th. While, on January 4th, there was no turbulence even though the CAPE was relatively high (2296.71 J kg^{-1}). The turbulence that occurs along with a small CAPE is categorized as Clear Air Turbulence (CAT). For the case of January 2020, the frequency of CAT occurrence is approximately 54.6% of all turbulence incidents.

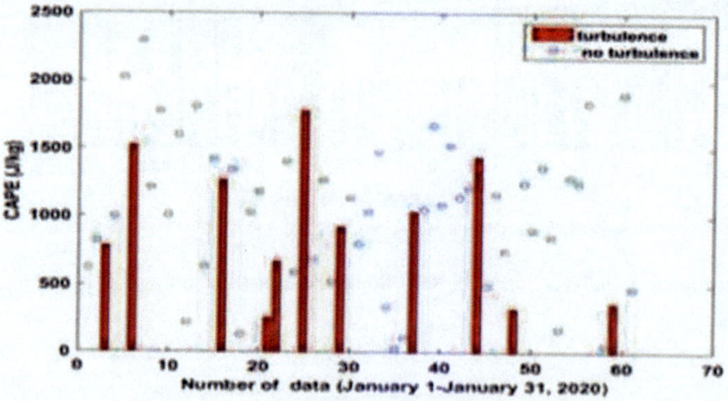

Fig. 29.7 Comparison of CAPE pattern and turbulence existing

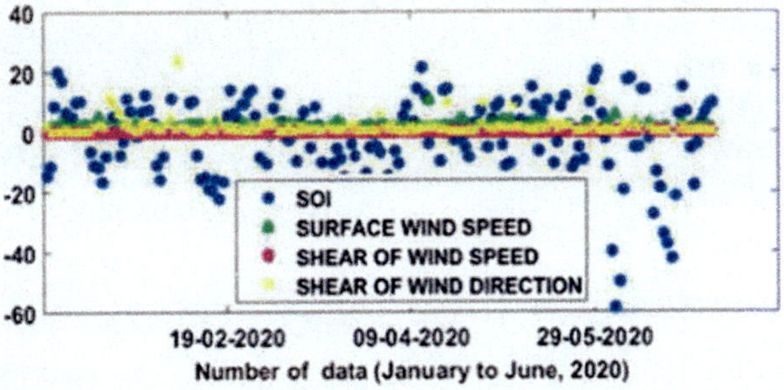

Fig. 29.8 Comparison of SOI pattern and aviation meteorology parameters patterns at Soekarno-Hatta airport

CAT is more dangerous than bad weather turbulence, because there are no signs like clouds, rain, or lightning.

The comparison between the SOI pattern and the aviation meteorology parameters patterns is shown in Fig. 29.8. The aviation meteorology parameters pattern has a different pattern with the SOI. Similarly, the MJO amplitude pattern also shows no similarity to the aviation parameters pattern (SWS, SoWS, SoWD, turbulence), as shown in Fig. 29.9.

Furthermore, statistical tests were carried out to evaluate the relationship between these parameters. Table 29.3 shows the calculation correlation (r_c) and table-based on the coefficient (r_t), between aviation meteorology parameters and SOI and MJO amplitude. The calculated r_c is smaller than $r_{t,}$ for all cases, which means no correlation between El Niño and MJO to the SWS, SoWS, SoWD, and turbulence. Provisionally, this is due to the weak intensity of El Niño and MJO.

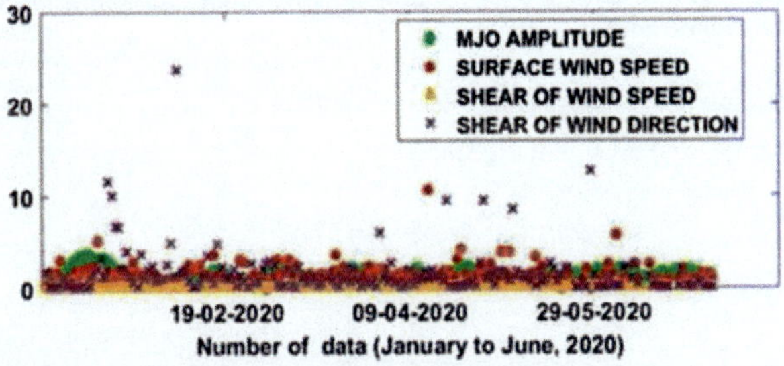

Fig. 29.9 Comparison of MJO amplitude pattern and aviation meteorology parameters patterns at Soekarno-Hatta airport

Table 29.3 The statistical results

Calculated r_c ($\alpha = 0.05$)					
SOI versus SWS	SOI versus SoWS	SOI versus SoWD	MJO versus SoWD	MJO versus SoWS	MJO versus SWS
−0.093	0.077	0.099	0.041	0.082	0.084
Table base $r_t = 0.14$					

r_c = r based on calculation
r_t = r by table
$r_c < r_t$: no correlation

29.4 Conclusion

The surface wind speed, shear of wind speed and, shear of wind direction at Soekarno-Hatta airport show the variation with time. However, the maximum value is still below the threshold value. Based on this result, it can be declared no high-level threat of these parameters to the flight. The frequency of occurrence of turbulence is 53% which occurs both in clear and cloudy or rainy conditions. This shows that the occurrence of turbulence does not depend on the weather. Case studies to evaluate the relationship of aviation meteorology parameters to atmospheric phenomena (El Niño and MJO) show that no correlation between El Niño and MJO to the SWS, SoWS, SoWD, and turbulence. This is the preliminary results which can be used as a base information to develop the real-time system information for aviation. In the future we need to have data from many airports or sampling area, and longer data period to draw a complete conclusion. Relationship studies also need to be carried out with local atmospheric parameters.

Acknowledgements Thank LAPAN for funding this research and appreciation to the research group for valuable suggestions and comments.

Authors Contribution All authors contributed equally as the main contributors to this work.

References

1. Boeing: Statistical summary of commercial jet airplane accidents, worldwide operations 1959–2016 (2017). www.boeing.com
2. KNKT: Aircraft crashes in Indonesia and relationship with atmosphere condition (Original title: Penyebab kecelakaan penerbangan di Indonesia dan kaitannya dengan kondisi atmosfer) (2021). http://knkt/dephub/go.id/knkt/, FGD-LAPAN
3. International Civil Aviation Organization (ICAO): Annex 13 Aircraft Accident and Incident Investigation Tenth Edition. International Civil Aviation Organization, Montreal, International Society of Air Safety Investigators (ISASI), Positions on Air safety Investigation Issues, 3rd edn. 24 Agustus 2003 (2010)

4. International Civil Aviation Organization (ICAO): Guidelines for Investigator Training and Education, First Edition, Agustus 1999. International Civil Aviation Organization (ICAO) dan Commercial Aviation Safety (2011)
5. KNKT: Type of aircraft incidents in 2017. www.knkt.go.id (2020)
6. Republika: Aircraft accidents/incidents for last 15 years (Original title: Kecelakaan pesawat 15 tahun terakhir). www.republika.co.id (2020)
7. Narayana Muthia Sani dan Rahmat Gernowo: Turbulence analysis on the accident of the Etihad Airways EY-474 on May 4th, 2016 using Weather Research and Forecasting method (Original title: Analisis turbulensi pada pesawat Etihad Airways EY-474 tanggal 4 Mei 2016, dengan metode Weather Research and Forecasting). Youngster Phys J. ISSN: 2302-7371, 07 1, 34–39 (2018)
8. Gultepe, I., Pearson, G., Milbrandt, J.A., Hansen, B., Platnick, S., Taylor, P., et al.: The fog remote sensing and modeling (FRAM) field project. Bull. Am. Meteor. Soc. **90**, 341–359 (2009)
9. Gultepe, I., Sharman, R., Williams, P.D., Zhou, B., Ellrod, G., Minnis, P., Trier, S., Griffin, S., Yum, S.S., Gharabaghi, B., Feltz, W., Tamimi, M., Pu, Z., Storer, L.N., Kneringer, P., Weston, M.J., Chuang, H., Thobois, L., Dimri, A.P., Dietz, S.J., Gutember, B., França, M.V., Almeida, F.L.: Albuquerque Neto: a review of high impact weather for aviation meteorology. Pure Appl. Geophys., **176**, 1869–1921 (2019)
10. Kumaran, Y., Nagarajan, S.: Aviation meteorology. Int. J. Latest Technol. Eng., Manag. Appl. Sci. **VI**(VIIIS), 90–94 (2017)
11. Page, C.: Understanding Aviation Meteorology and Weather Hazards with Ground-Based Observations, SN 978-3-642-12967-4, 161–173 (2010)
12. Smith, J.W.: Temperature distribution and the layer of minimum temperature. J. of Appl. Met. **2**, 655–667 (1963)
13. Chen J., Dai, A., Zhang, Y., Rasmussen, K.L.: Changes in convective available potential energy and convective inhibition under global warming. J. of Clim. **33**(6), 2025–2050 (2020)
14. National Weather Service (NOAA): Pilot's Guide for Aviation Weather. https://www.weather.gov/ (2020)
15. List of accidents and incidents commercial aircraft. http://aviation-safety.net. Accessed Sept 2021
16. Besson, C., Dolfi-Bouteyre, A., Canat, G., Cézard, N., Augère, B., Durécu, A., Lombard, L., Valla, M., Hallermeyer, A.: Doppler LIDAR Developments for Aeronautics. J. Aerosp. Lab. **12** (2016)
17. Luce, H., Nakamura, T., Yamamoto, M.K., Yamamoto, M., Fukao, S.: MU radar and lidar observations of clear-air turbulence underneath cirrus. Mon. Weather Rev. **138**, 438–452 (2010)
18. World Meteorological Organization: Aerodome reports and forecast. A Users Handbook, WMO No. 782 (2019)
19. Chun, H., Kim, J., Lee, D., Kim, S., Strahan, M., Pettegrew, B., et al.: Research collaborations for better predictions of aviation weather hazards. Bull. Am. Meteorol. Soc. **98**, ES103–ES107 (2017). https://doi.org/10.1175/BAMS-D-17-0010.1
20. Benjamin, S.G., Jamison, B.D., Moninger, W.R., Sahm, S.R., Schwartz, B.E., Schlatter, T.W.: Relative short-range forecast impact from aircraft, profiler, radiosonde, VAD, GPS-PW, METAR, and mesonet observations via the RUC hourly assimilation cycle. Mon. Weather Rev. **138**, 1319–1343 (2010)
21. Monkam, D.: Convective available potential energy (CAPE) in Northern Africa and tropical Atlantic and study of its connections with rainfall in Central and West Africa during Summer 1985. Atmos. Res. **62**(1–2), 125–147 (2002)
22. Adams, D.K., Souza, E.P.: CAPE and convective events in the Southwest during the North American monsoon. Mon. Wea. Rev. **137**, 83–98 (2009)
23. Williams, E.: Comments on "thunderstorms above frontal surfaces in environments without positive CAPE. Part I: a climatology". Mon. Wea. Rev. **119**, 2511–2513 (1991)
24. Williams, E., Renno, N.: An analysis of the conditional instability of the tropical atmosphere. Mon. Wea. Rev. **121**, 21–36 (1993)

25. Adams, D.K., Rennó, N.O.: Remarks on Arakawa–Schubert's quasi-equilibrium theory. J. Atmos. Sci. **60**, 178–181 (2003)
26. McBride, J., Frank, W.: Relationships between stability and monsoon convection. J. Atmos. Sci. **56**, 24–36 (1999)
27. Yano, J., Chaboureau, J., Guichard, F.: A generalization of CAPE into potential-energy convertibility. Quart. J. Roy. Meteor. Soc. 861–875 (2005)
28. Moncrieff, M.W., Miller, M.J.: The dynamics and simulation of tropical cumulonimbus and squall lines. Q. J. R. Meteorol. Soc. **120**, 432, 373–394 (1976)

Chapter 30
Interpolation Methods Evaluation on Arbitrary Straight Line of Flight Weather Data

Aisya Nafiisyanti

Abstract The JATAYU Decision Support System (DSS) for air transportation safety implements a feature that displays data along the flight path (currently, the flight path is a straight line from departure to arrival airport's coordinates). To obtain the data, the interpolation method is used to provide more detailed information. There are two interpolation methods tested in this study, namely Bicubic interpolation and Nearest Neighbour interpolation. The current resolution used in the decision support system is 5 km, while the produced interpolation data should ideally have a higher resolution. Data with 2.5 and 1 km resolution are generated to act as comparison data. The case study areas are the islands of Java, Bali, and Nusa Tenggara. Based on the test results using Mean Absolute Scaled Error and Normalized Root Mean Squared Error, both methods show almost the same interpolation results. However, Bicubic interpolation gave slightly better results than Nearest Neighbour interpolation.

30.1 Introduction

The number of flights in Indonesia is rising from year to year (before the Coronavirus pandemic). As recorded on the official website of the Indonesian Central Bureau of Statistics, there has been an increase in the number of passengers over the past five years at the main airports in Indonesia [1]. In addition, there is also the addition of new airports at several locations, both international and domestic airports. Even during the pandemic time, some vital flights remain operated despite applying restrictions for the passenger's safety.

Indonesia itself is a country located in an archipelagic region where its water is 2:3 of the whole country, which causes its atmosphere to have variation hydrometeorological phenomenon's. Several atmospheric phenomenon conditions affect the airspace where flight activities take place. Since weather condition affects flight, it is

A. Nafiisyanti (✉)
Research Center for Climate and Atmosphere, National Research and Innovation Agency (BRIN), Bandung, Indonesia
e-mail: aisya.nafi@gmail.com

E. Yulihastin et al. (eds.), *Proceedings of the International Conference on Radioscience, Equatorial Atmospheric Science and Environment and Humanosphere Science, 2021*, Springer Proceedings in Physics 275, https://doi.org/10.1007/978-981-19-0308-3_30

crucial to have a supporting system that is reliable to the aviation parties to perform safety flight activities.

One way to establish a supporting system for safe flight is by providing weather condition data that consist of parameters that affect flight. So that, the data can be one of several deciding factors to execute, postpone or cancel a flight activity.

Generally, flight data is the data standardized into METAR format. The data source can come from permanent observation tools such as radiosonde, which is point/station-wise. This data is obtained from flying a balloon sonde in a certain area or airfields. There is also data derived from the forecast results of TAF (Terminal Aerodrome Forecast). Another approach is to use spatial data that can be obtained from satellite or model simulations.

In a flight, a line is formed from two points. The starting point and the ending point are the representation of the departure airport coordinates and the arrival airport coordinates, respectively. In reality, the line formed is not always straightforward. However, to facilitate this research, the line formed is assumed as a straight line. Then, the processed data is the data traversed by the straight line.

This paper focuses on data generated from the simulation model examination. In the weather model, the user can set the resolution of the data. The higher resolution, the more computational facilities are required. This is because more grids must be generated. In addition, if the simulation applies the nesting method, additional processors are needed to process the area of the child domain(s). Due to the limited number of computing facilities, the resulting data has not yet reached the expected resolution. Therefore, the interpolation method is used.

Some of the commonly used methods are Binary interpolation, Bicubic interpolation, Nearest Neighbour. Each of these methods is widely used for image/pictures rescaling. However, these methods are also capable of rescaling non-image spatial data [2, 3].

This project aims to evaluate the ability of the interpolation method towards spatial flight data obtained from the weather model simulation results, in this case, Weather Research Forecast.

30.2 Methods

30.2.1 Bicubic Interpolation

This method is an extension to cubic interpolation, which is capable of two-dimensional processing data [4]. It has smoother edges compared to bilinear interpolation. In a python programming language, this method is grouped into spline interpolation, which produces smoother results [5]. The formula is represented in Eq. 30.1.

$$p(x, y) = \sum_{i=0}^{3}\sum_{j=0}^{3} a_{ij}x^i y^j \tag{30.1}$$

If assumed (0, 0), (0, 1), (1, 0), and (1, 1) are four corners with functions f and the derivatives f_x, f_y, f_{xy} as known values.

30.2.2 Nearest Neighbour Interpolation

Nearest Neighbour is the simplest interpolation method that assigns the unknown grid value using the nearest neighbouring grid value [6]. The advantage of this method is that it keeps the original value, and the value obtained does not go through the interpolation process [7]. In addition, because there is no calculation, this method tends to consume faster computational time than other methods.

30.2.3 MASE

Mean Absolute Scaled Error (MASE) is an improvement of the error calculation formula to overcome the shortcoming of Mean Absolute Percentage Error (MAPE). As known in MAPE, there is division by zero possibility. Thus, it does not suitable for the data study since some of the collected data has 0 value [8]. Besides, it does not depend on data scale (invariance), so that it can be used on data sets with a large or small range of values. It also acts symmetrically where the penalization for positive error prediction is equal to negative error prediction. Likewise, the penalization in a large forecast is equivalent to a small forecast. The formula is represented in Eq. 30.2.

$$\text{MASE} = \frac{|e_i|}{\frac{1}{T-1}\sum_{t=2}^{t}|Y_t - Y_{t-1}|} \tag{30.2}$$

where e is the forecast error in a given period, and i is the number of forecast data. Given Y_i is the actual value and F_i is the forecast value in the i period, then $e_i = Y_i - F_i$. Meanwhile, t is the time of a particular actual value, and T is the amount of total time.

30.2.4 NRMSE

Normalized Root Mean Squared Error (NRMSE) is an improvement for traditional Root Mean Squared Error. The normalization aims to handle large range values of a data set or sets with outlier data values. The normalization used in this study is the difference between the maximum and minimum values of prediction data. The method is widely used in various research [1, 2]. The formula is expressed in Eq. 30.3.

$$\text{NRMSE} = \frac{\sqrt{\frac{\sum_{t=1}^{T}(Y_t - F_t)^2}{T}}}{F_{\max} - F_{\min}} \tag{30.3}$$

where T is the amount of data in a period, Y_t and F_t are the observation and prediction data on a particular time, respectively. Meanwhile, $F_{\max}$ and $F_{\min}$ are the maximum and minimum values of the prediction data.

Both MASE and NRMSE has the following condition: the closer the error value to 0, the more similar the objects.

30.2.5 Data

The data used in this study is the simulation result of the WRF weather model with −10° to −5.5° latitude and 101.9° to 121.8° longitude settings, which represent the islands of Java, Bali, and Nusa Tenggara. The simulation represents flight weather parameters on June 30th, 2021, at 00, 01, and 02 UTC. The generated data are Convective Available Potential Energy (CAPE), CIN, Low-Level Turbulence (LLTURB), Low-Level Wind Shear (LLWS), total accumulated sleet (ICE), cloud ceiling (CLOUD CEIL), Mean Sea Level Pressure (MSLP), and visibility (VIS).

Each generated parameters data is in 2-dimensional spatial form with three different resolutions: 0.05°, 0.025°, and 0.01°. The data is produced in JSON format. The 0.05° resolution data is then compared to 0.025° and 0.01° resolution data. The aim is to test the ability of interpolation methods that are applied to the low-res data.

The study applied three scenarios for testing. Three straight lines: vertical, diagonal, and horizontal are arranged with coordinate ranges that did not exceed the coordinates of the original data. Each line is represented by two points. The first point represents the departure airport, and the second point represents the arrival airport. Each point is randomly chosen. The three straight-lines scenarios are then applied to all parameters to see the trend or pattern.

30.3 Result and Analysis

Each parameter is tested except cloud ceiling (CLOUD_CEIL) and total accumulated sleet (ICE) because the simulation model produced zero value data at the generated time. Since the initial clock of the simulation is a clock without precedent data input, which results in zero data, the simulation results at 00 UTC are not used.

Figure 30.1 is a charts collection of the comparison results for data traversed by a straight line in the domains with a resolution of 0.05°, 0.025°, and 0.01°. In some tests,

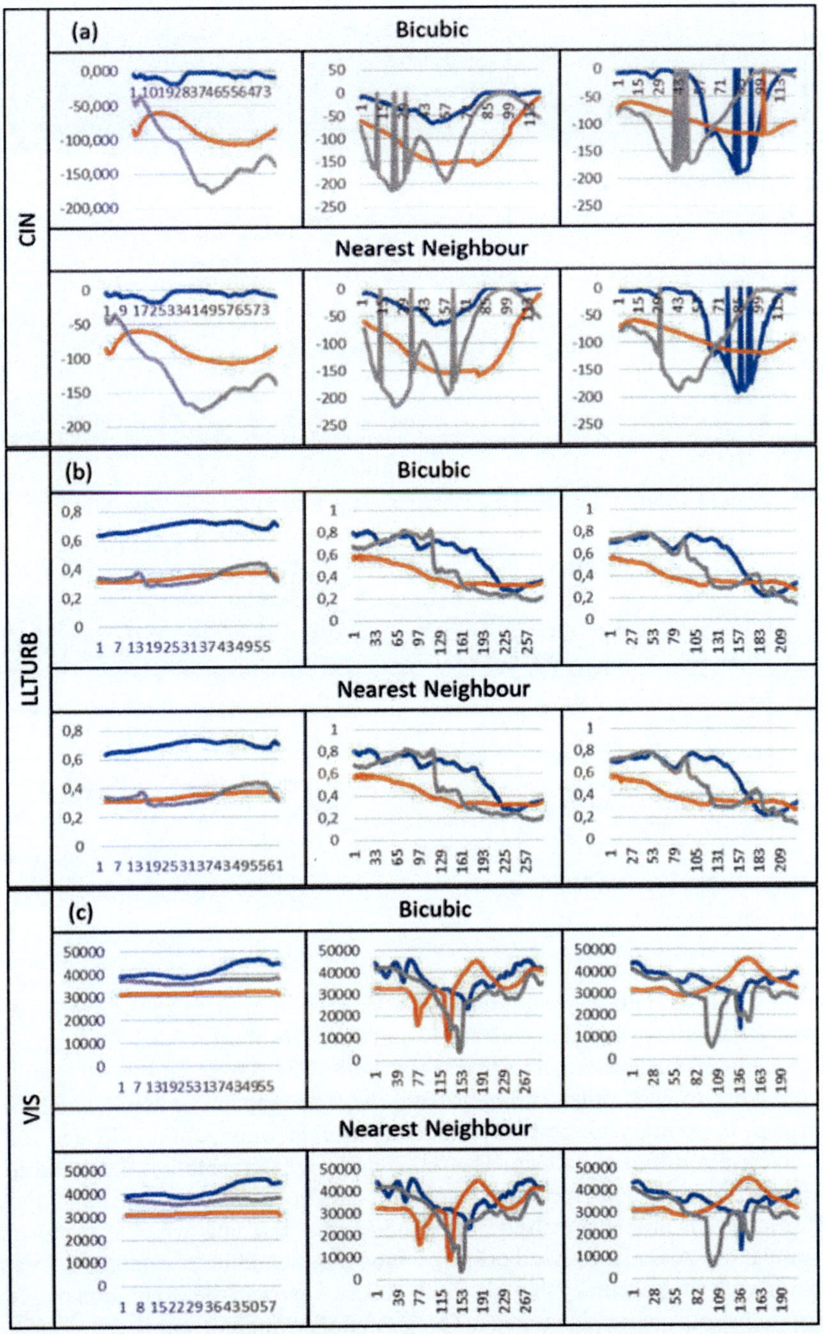

Fig. 30.1 Comparison between interpolation results using Bicubic and Nearest Neighbour methods of 0.05°, 0.025°, and 0.01° resolution data of each flight weather data parameter. Orange, grey, and blue lines represent interpolation results of 0.05°, 0.025°, and 0.01° resolution data, respectively

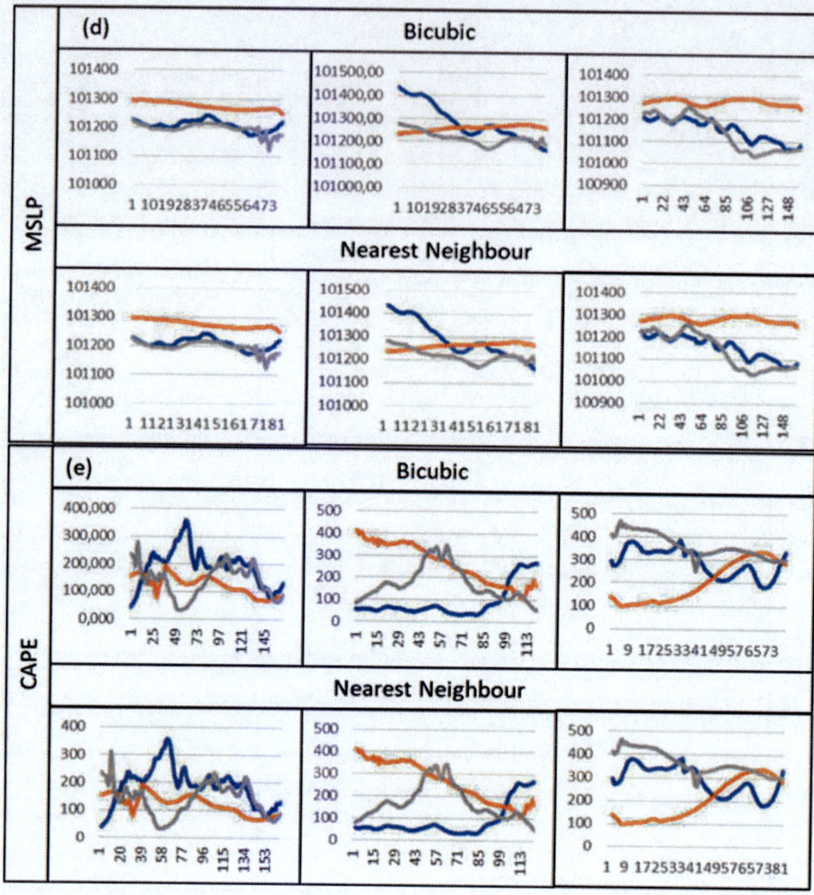

Fig. 30.1 (continued)

especially in tests with vertical lines, the 0.05° resolution data appears to have big differences when compared to the 0.025° and 0.01° resolution data. Take examples of CIN and LLTURB data with vertical test lines. The 0.025° and 0.01° resolution data are close to each other, different from the 0.05° resolution data, which has a larger gap. In contrary to the two parameters aforementioned, the VIS and MSLP seem to have a different pattern. The 0.05° and 0.025° resolution data tend to be more similar; meanwhile, the 0.01° resolution data is outcasted.

This pattern is probably caused by the character of the data, which is the result of the simulation. As known that there is no model which able to simulate a weather condition with such accuracy that can reach 100%. However, the simulation may/may not capture atmospheric phenomena. This probability might occur in the simulation results used in this study. In simulations with higher resolutions (0.025° and 0.01°), the model possibly was able to capture the local phenomenon that is remote and in a small area represented in a sole grid. Whereas in a low-resolution simulation, the phenomenon is averaged with the surrounding conditions that are still on the same grid, so that, the value that represents the phenomenon is influenced by the value in the area that does not experience certain events. This causes the 0.025° resolution data at some tests to have a lesser difference towards 0.01° resolution data rather than the 0.05° resolution data and vice versa.

As shown in the figure, there are not many differences between the interpolation results with Bicubic and Nearest Neighbour methods. This is seen from the comparison of each chart of the same data and line types. More details are provided in the comparison table for measuring error values using MASE and NRMSE (Table 30.1).

The difference between Bicubic and Nearest Neighbour error values using MASE and NRMSE are maximum 46,626 and 0,736,626, and minimum 0039 and 0,000,121 respectively. It can be concluded that the two methods at some tests have significant differences and do not have a quite considerable difference at the other tests. So that, a more detailed comparison is made. However Bicubic interpolation has smaller error values than Nearest Neighbour interpolation (see the Bicubic columns, the number of marked yellow columns represent Nearest Neighbour is 27, meanwhile the number of non-coloured columns represent Bicubic interpolation is 33). This means Bicubic methods performed slightly better. This is following [9], which states that the results of Bicubic interpolation are better than other methods.

Table 30.1 Error-values collection for Bicubic and Nearest Neighbour Methods using MASE and NRMSE

		Bicubic				Nearest Neighbor			
		MASE		NRMSE		MASE		NRMSE	
		0.01°	0.025°	0.01°	0.025°	0.01°	0.025°	0.01°	0.025°
CIN	V	57,900	39,390	1,746	0,882	57,939	38,822	1,747	0,882
	H	39,566	4,977	0,629	0,427	40,292	5,549	0,634	0,441
	D	22,718	7,369	0,557	0,535	69,344	17,444	1,108	0,546
LLTURB	V	181,532	45,427	5,626	2,290	181,806	45,553	5,635	2,296
	H	66,176	30,753	1,523	0,580	65,773	30,088	0,786	0,272
	D	97,298	18,326	0,854	0,270	84,178	16,064	0,839	0,272
VIS	V	253,104	55,682	9,481	2,018	253,798	56,148	9,487	2,031
	H	21,365	19,961	0,276	0,057	21,524	20,053	0,277	0,057
	D	42,654	13,832	0,498	0,269	42,322	10,517	0,501	0,264
MSLP	V	96,614	3,274	1,626	0,197	96,525	3,347	1,626	0,200
	H	85,308	19,770	2,065	0,839	84,678	19,502	2,056	0,834
	D	136,217	13,447	3,079	0,203	116,575	12,825	3,071	0,203
CAPE	V	26,205	11,344	0,743	0,472	25,972	10,316	0,726	0,419
	H	39,802	18,968	0,746	0,521	39,594	18,634	0,748	0,525
	D	32,782	12,570	0,698	0,415	33,024	13,857	0,702	0,441

V—vertical, H—horizontal, D—diagonal

According to the results of the MASE and NRMSE calculations, the interpolation value of 0.05° resolution data against 0.025° resolution data has a smaller error than 0.01° resolution data. This indicates that the interpolation can work well with a resolution that is not too far apart. Interpolation result towards data with two times higher resolution than the initial resolution is still relevant. The significant difference in the case of 0.05° resolution data to 0.01° resolution data may occur for the reasons previously mentioned (the data used as comparisons are simulation data that probably capture local phenomena).

30.4 Conclusion

A study to review interpolation results using Bicubic and Nearest Neighbour for arbitrary straight-line data has been conducted. Both Bicubic and Nearest Neighbour methods show almost the same interpolation results. However, Bicubic interpolation performed slightly better than Nearest Neighbour interpolation. Since the simulation data was used as a comparison, there are differences between actual and comparison data because the higher resolution data probably captured local phenomena. Interpolation cannot be performed to obtain values at a much higher resolution than the original data. Data interpolation as much as half of the original data resolution is still relevant.

Acknowledgements The author would like to thank Dr Teguh Harjana and Dr Ibnu Fathrio for the support and constructive discussions during the study. She also thanks Atmospheric Science and Technology Research Center in the Indonesian National Research and Innovation Agency for providing infrastructure to perform model simulation.

References

1. bps.go.id. 2021. Badan Pusat Statistik. [online] Available at: https://www.bps.go.id/linkTableDinamis/view/id/812. Accessed 14 Sept 2021
2. Khomchenko, A., Sidenko, I.: Comput. Sci. Eng. **1**, 29 (2020)
3. Kalideen, M., Osmanoglu, M., Tugrul, B.: Physical Science and Engineering. 61st edn, pp. 1–11. Ankara: Communications Faculty of Sciences University of Ankara Series A2-A3 (2019)
4. Ashraf, I., Hur, S., Park, Y.: IEEE Access **5**, 8250 (2017)
5. Agrapart, Q.A., Batailly, A.: Polytechnique Montréal, pp. 3–46. Montreal, Quebec, 2020
6. Lepot, M., Aubin, J.-B., Clemens, F.: Water **9**, 796 (2017)
7. Lyra, G.B., Correia, T.P., de Oliveira-Júnior, J.F., Zeri, M.: Theoret. Appl. Climatol. **134**, 955 (2017)
8. Chen, C., Twycross, J., Garibaldi, J.M.: PLOS ONE **12** (2017)
9. Gao, S., Gruev, V.: Opt. Express **19**, 26161 (2011)

Chapter 31
Performance Analysis of Platinum Wire Temperature Sensor for Radiosonde

Soni Aulia Rahayu, Rachmat Sunarya, Edy Maryadi, Laras Toersilowati, Listi Restu Triani, and Christine Cecylia Munthe

Abstract At present, atmospheric observations are needed, both horizontally and vertically for a particular area. Observation of the atmosphere vertically can be done using a radiosonde. Radiosonde has many uses including helping to provide weather information in flight and also for researching vertical atmospheric conditions in the territory of Indonesia. One of the sensors carried by the radiosonde is the temperature sensor. So far, radiosonde sensors depend on imports from abroad. This research will focus on designing a platinum wire-based temperature sensor. The platinum wire used is platinum wire with a diameter of 0.025 mm with a resistance of 99.99% or $\pm 10\ \Omega$. The platinum wire is wound into a zero Ohm resistor with the number of turns according to the required specifications. After going through laboratory testing, then the platinum wire-based temperature sensor is carried out field testing using a reference radiosonde by following the provisions of the World Meteorological Organization (WMO). The results obtained show that the response time of the platinum sensor can follow the reference data. The correlation value obtained is 0.99 for temperature measurements from +27 °C to −83.7 °C.

31.1 Introduction

Vertical observation of the atmosphere is very much needed for the territory of Indonesia. There are many uses for vertical atmospheric observations, one of which

S. A. Rahayu (✉) · R. Sunarya · E. Maryadi
Research Center for Climate and Atmosphere, National Research and Innovation Agency (BRIN), Bandung, Indonesia
e-mail: soni.aulia.rahayu@brin.go.id

L. Toersilowati · L. R. Triani
Atmospheric Science Research Group, Faculty of Earth Science and Technology, Bandung, Indonesia

C. C. Munthe
Research Center for Atmospheric Science and Technology, National Research and Innovation Agency (BRIN), Bandung, Indonesia

E. Yulihastin et al. (eds.), *Proceedings of the International Conference on Radioscience, Equatorial Atmospheric Science and Environment and Humanosphere Science, 2021*, Springer Proceedings in Physics 275, https://doi.org/10.1007/978-981-19-0308-3_31

is for the world of aviation, aviation requires vertical weather conditions to anticipate weather changes during flight. One of the equipment used for vertical atmospheric observation is radiosonde [1].

Radiosonde is equipment that is flown using atmospheric balloon rides to a certain height depending on needs [2]. One that is measured by radiosonde is vertical temperature. Vertical temperature measurement is very important for further research, because until now the use of radiosonde is still imported from abroad. In this research, the development of a temperature sensor for vertical atmospheric observations using platinum wire will be carried out.

Platinum wire in principle includes a Resistance Temperature Detector (RTD) [3]. where this RTD is part of a very small technological device which is now known as the Micro-Electro-Mechanical System (MEMS) [4]. MEMS are often used as micro-scale machines consisting of components measuring from 1 to 100 μm [5].

RTD is a tool used to determine the value of a temperature using a sensitive element of pure platinum, copper, or nickel wire, which provides a limited resistance value for each temperature in its temperature range [6, 7]. The hotter the object, the greater or higher the value of its electrical resistance, and vice versa.

This research is focused on the development of a platinum wire-based temperature sensor. This research is a follow-up study from previous research. In previous studies, radiosondes from the development of PSTA LAPAN were given the symbol R-Sonde using a temperature sensor that was already on the market [8]. The results obtained are still not valid after reaching the lowest temperature. Therefore, it takes innovation and development of advanced temperature sensors so that vertical temperature data can be obtained validly according to reference data. In this research, platinum wire on the market is used as the primary material for making temperature sensors using MEMS technology. So that platinum wire which was just an ordinary wire can become a temperature sensor that can measure temperature at specific locations and low temperatures.

31.2 Data and Method

In this research, the material used is a platinum wire with a diameter of 0.025 mm and a resistance of 99.99% or ±10 Ω. The block diagram of this research can be seen in Fig. 31.1.

In principle, platinum wire is wound into a zero Ohm resistor with several turns of about 50–200 turns according to the required specifications and the diameter of the wire used. The smaller the diameter of the wire, the more turns of wire are needed, and vice versa [9].

This platinum wire-based temperature sensor is developed using the principle of Resistance Temperature Detector (RTD) [2]. This sensor is a passive sensor that requires a converter to read the data. This converter functions to change the amount of resistance produced by the platinum wire to a digital scale so that it can be read by measuring instruments in the form of temperature data. At room temperature

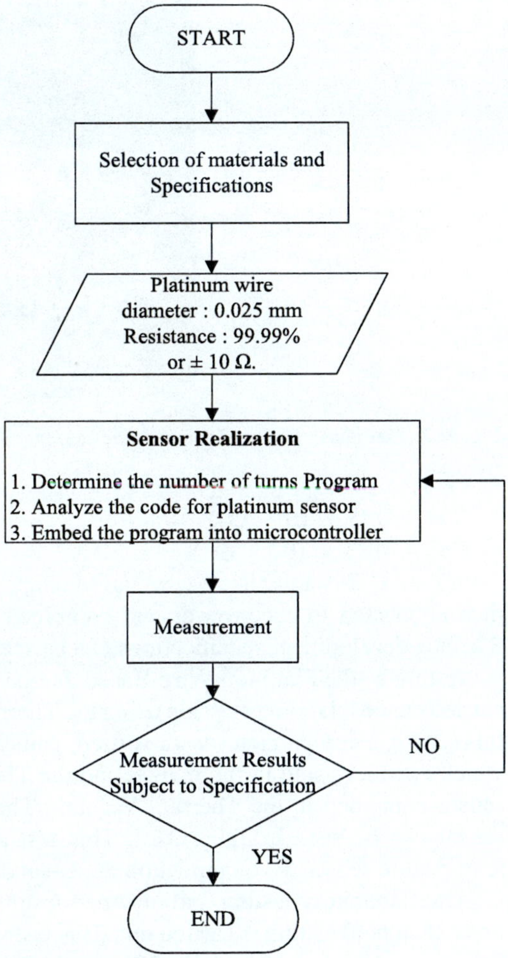

Fig. 31.1 Block diagram of platinum wire design

(20–25 °C), the temperature sensor resistance is about 10 Ω [10]. This resistance value can vary according to the temperature difference. To calculate it, we can use Eqs. (31.1) and (31.2) [11].

$$R_t = R_0\left[1 + At + Bt^2 + C(t - 100)t^3\right],\ -200\,^\circ\mathrm{C} < t < 0\,^\circ\mathrm{C} \tag{31.1}$$

$$R_t = R_0\left(1 + At + Bt^2\right),\ 0\,^\circ\mathrm{C} < 850\,^\circ\mathrm{C} \tag{31.2}$$

where
R_t = Resistance value at t°C.
R_0 = Resistance value at 0 °C.
$A = 3.9083 \times 10^{-3}\ {}^\circ\mathrm{C}^{-1}$.

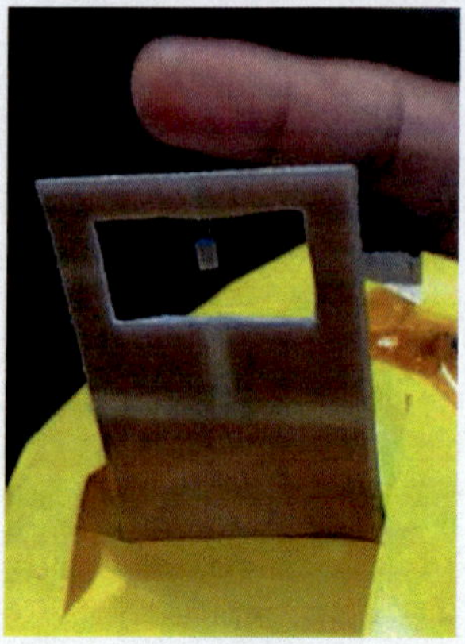

Fig. 31.2 Platinum wire-based temperature sensor

$B = -5775 \times 10^{-7}$ °C^{-2}.
$C = -4183 \times 10^{-12}$ °C^{-3}.

In Fig. 31.2 is a platinum wire-based temperature sensor that has been designed, then connected to a converter and combined with the development of R-Sonde. R-sonde development specifications can be seen in Table 31.1.

Testing of the Platinum Wire-Based Temperature Sensor in the early stages was carried out on a laboratory scale using the Thermal Vacuum HVT 400 ILS. In testing this sensor, a comparison sensor is used, namely the DS18b20 Temperature Sensor which is widely sold in the market and the Thermal-Ribbon™ Sensor S651 series sensor contained in the Thermal Vacuum. The specifications of each comparison sensor can be seen in Table 31.2. This test aims to see the performance of the temperature sensor development on the R-sonde that has been developed.

After laboratory testing and obtained results according to the desired specifications, then field testing is carried out. This test was carried out using an atmospheric balloon on October 16 and 17, 2019 at the Pameungpeuk Atmospheric and Technology Test Center, Garut Regency, West Java with coordinates 007° 38′ 46″ S, 107° 41′ 15″ E at an altitude of ±7 m above sea level. This field test uses a reference

Table 31.1 Specifications of R-sonde

Specification	R-sonde
Frequency	433 MHz
Accuracy	±5 °C
Dimension	9 × 5 × 3 (mm)
Output power	100 mW
Modulation	GFSK
Baud rate	1200 bps
Temperature range	−250 °C to +900 °C

Table 31.2 Sensor specifications in laboratory testing [12]

Specification	Sensor DS18b20	Platinum wire based temperature sensor	Thermal-Ribbon™ sensor series S651
Range	−55 °C to + 125 °C	−250 °C to + 900 °C	−200 °C to + 200 °C
Accuracy (°C)	±5	±0.01	±0.15

Table 31.3 Sensor specifications in field testing [14–16]

Specification	Platinum wire based temperature sensor	Thermocouple	MPL3115A2	Sensor DS18b20	Vaisala RS41
Range	−250 °C to + 900 °C	−270 °C to + 1800 °C	−40 °C to + 85 °C	−55 °C to + 125 °C	−90 °C to + 90 °C
Response time (°C)	±0.01	±0.1	±1	±5	±0.01

radiosonde in accordance with the provisions of the World Meteorological Organization (WMO) [13] as well as several temperature sensors on the market. In this test, an R-sonde with several temperature sensors and a reference radiosonde is flown in parallel on an atmospheric balloon. This is so that all temperature data received by the receiver are at the same height. The temperature sensors that are flown are listed in Table 31.3.

31.3 Results and Discussion

The development of this platinum wire-based temperature sensor uses platinum wire with a diameter of 0.0025 mm. After the process of designing the platinum wire into a sensor that can measure temperature, then the sensor is combined in the R-Sonde development circuit with a microcontroller, Tx radio, GPS, Tx antenna and others as shown in Fig. 31.3.

To determine the performance of the temperature sensor that has been designed, laboratory scale testing is carried out. This test was carried out at the LAPAN Satellite Technology Center on 18–19 September 2019 using a Thermal Vacuum HVT 400 ILS with specifications Temp Range: −70 to +150 °C Vacuum Limit: 5×10^{-2} mbar. Testing the R-Sonde with a platinum wire-based temperature sensor was carried out using a comparison sensor, namely the Thermal-Ribbon™ Sensor series S651 (sensor on Thermal Vacuum) and the DS18b20 sensor on the market (see Fig. 31.4).

This test is set at 5 °C intervals for 5 min and held for 5 min so that it is stable at a certain point and the changes can be seen. Starting from a room temperature of around +25 °C, then decreasing every 5 °C until a temperature of −70 °C, then rising again to a temperature of +30 °C and finished.

Fig. 31.3 R-Sonde with platinum wire-based temperature sensor

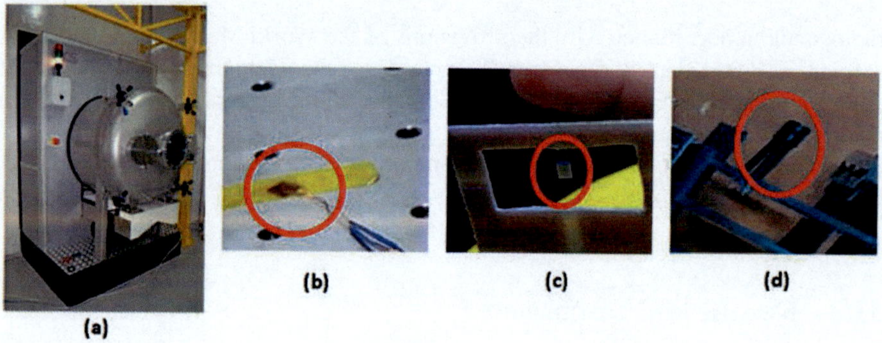

Fig. 31.4 Thermal vacuum (**a**), Sensor seri S651 (**b**), sensor platinum (**c**) and sensor DS18b20 (**d**)

On the first day of testing on September 18, 2019, testing was carried out using 2 reference sensors in the thermal vacuum with the S651 series (blue and gray color), platinum sensors and DS18b20 sensors. From Fig. 31.5 it can be seen that the lowest temperature measured for the platinum sensor (orange color): −38 °C and the DS18b20 sensor (yellow color) cannot reach 0 °C. Where the Y axis is the measured °C temperature value and the X axis is the data starting from the 1st data (the first measured data) to the 2507th data (the last data measured from the chamber). It can be seen that the platinum sensor can follow the pattern from the reference data, although the response time is slightly longer than the reference sensor. While the DS18b20 sensor cannot measure 0 °C temperature.

On the second day of testing on September 19, 2019, it was seen that the two sensors tested experienced an increase in performance as shown in Fig. 31.6. Where the Y axis is the measured °C temperature value and the X axis is the data starting

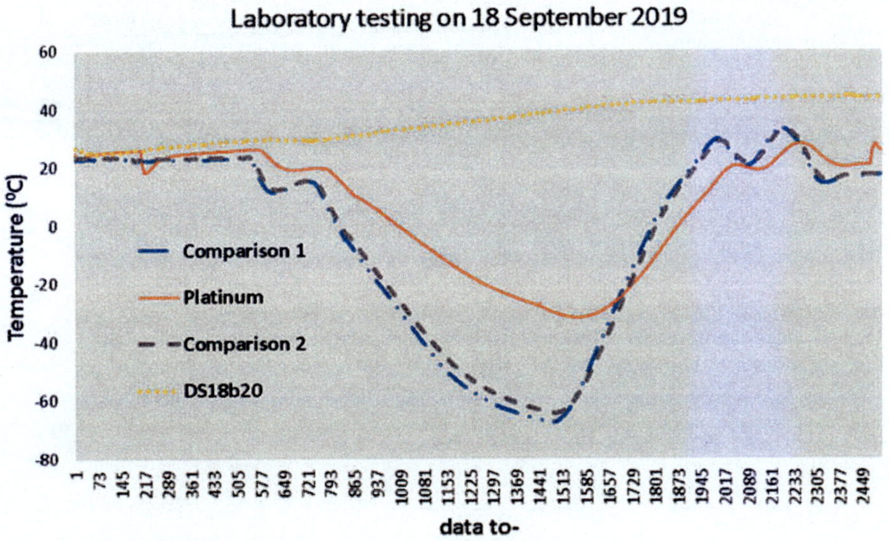

Fig. 31.5 First day of laboratory testing

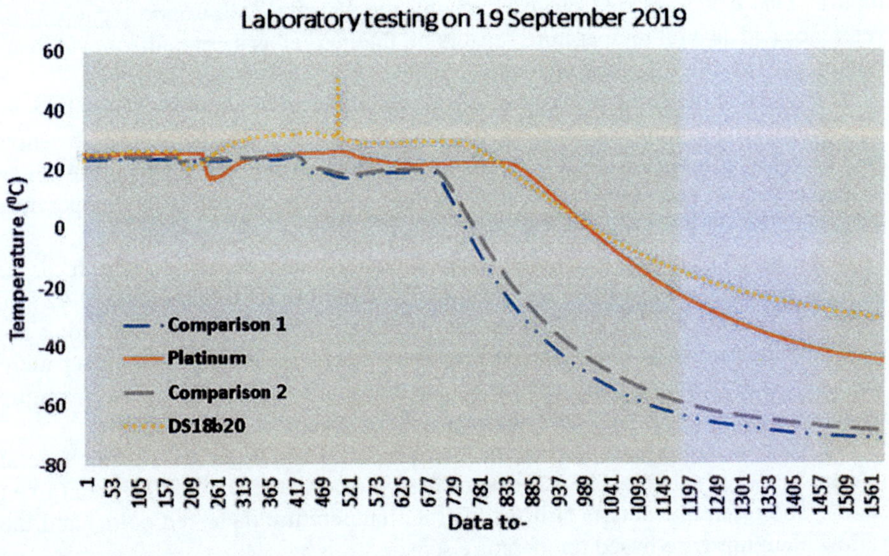

Fig. 31.6 Second day of laboratory testing

from the 1st data (the first measured data) to the 2507th data (the last data measured from the chamber). This second day of testing could not be carried out until the end because the battery ran out of power so it could not send data to the receiving system. The lowest temperature measured on the platinum sensor:—58 °C, and the lowest temperature on the DS18b20 sensor:—35 °C.

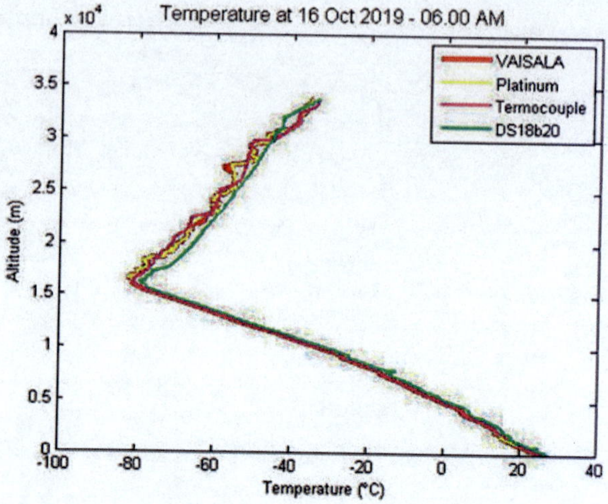

Fig. 31.7 The results of the first day of field testing

From the results of laboratory tests, further testing is carried out in the field. This aims to see the performance of the platinum wire-based temperature sensor in actual atmospheric conditions, whether it is in accordance with the required specifications or not. This test is carried out by carrying the Vaisala Radiosonde payload as a reference and several temperature sensors on the market as a comparison which are combined with the R-sonde payload.

In Fig. 31.7, the red color is the Vaisala data, the yellow color is the platinum wire-based temperature sensor data, the magenta color is the thermocouple sensor and the green color is the DS18b20 sensor data. Where the vertical axis is the height in meters (m) up to a height of 40 km, while the horizontal axis is the temperature data unit °C with a range of −100 to 40 °C.

In testing on October 16, 2019, the maximum altitude reached 33.5 km. It can be seen from several graphs that they have the same pattern up to a height of 17.2 km. After an altitude of 17.2 km (lowest temperature −83.7 °C) it starts to look not following the pattern from the Vaisala reference data. The most following the pattern from the Vaisala temperature sensor data is the platinum wire-based temperature sensor.

The second day of testing was on October 17, 2019 (Fig. 31.8) using several sensors, namely the MPL3115A2 sensor (blue color) and the DS18b20 sensor (green color). The reference data is still the Vaisala temperature data (red color) and the yellow platinum wire-based temperature sensor.

The maximum altitude on the second day is lower than on the first day, which is 32.7 km. The MPL3115A2 temperature sensor does not work properly, it can be seen from the data obtained that it is constant at 24 °C. For platinum wire-based temperature sensors, the pattern still follows the pattern of the reference data. For the DS18b20 sensor, the pattern is still the same as the previous day at an altitude below 17 km the pattern follows the reference data pattern, but at an altitude above 17 km the data received starts to be inaccurate.

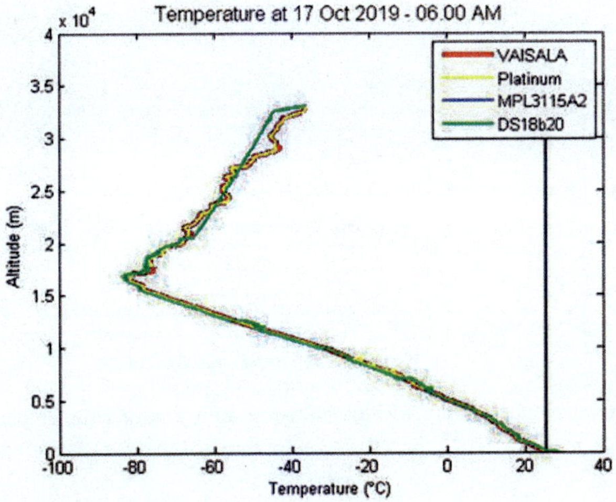

Fig. 31.8 The results of the second day of field testing

From several sensors, it can be seen that there is a bias towards the Vaisala reference temperature sensor data for each sensor starting from the platinum wire-based temperature sensor, MPL3115A2, DS18b20 and thermocouple (Fig. 31.9). Overall the bias of each sensor to the reference sensor is not large. It starts to show a large bias value after an altitude of 17 km as described in the profile graph above. Seen from the smallest to the largest bias, respectively, namely the platinum wire-based temperature sensor, MPL3115A2, DS18b20 and thermocouple.

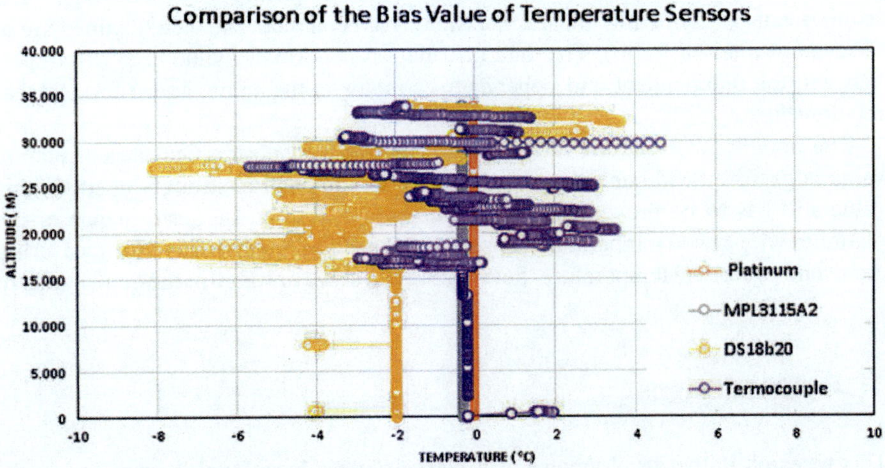

Fig. 31.9 The bias value of each sensor to the reference sensor

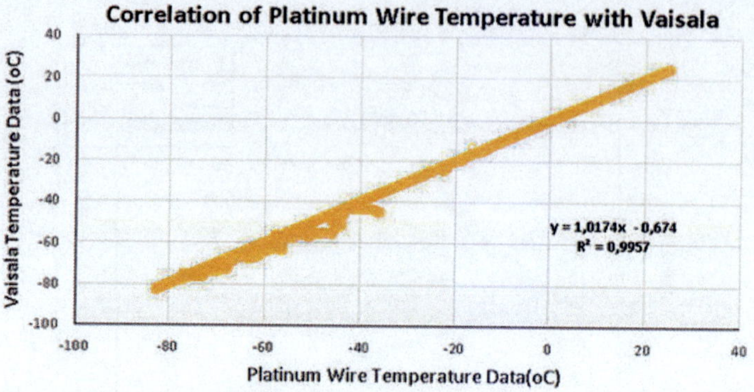

Fig. 31.10 Correlation of platinum wire-based temperature sensor with temperature sensor from Vaisala

From the data obtained, the correlation plot between the platinum wire-based temperature sensor and the Vaisala temperature sensor is obtained as shown in Fig. 31.10. Where the correlation value is obtained close to 1, which means that the correlation value is positive. Although the correlation value obtained is strongly positive, this value still has to be proven by more field testing. So that the performance of the platinum wire-based temperature sensor can be used as an alternative for measuring vertical atmospheric temperature. Field testing can serve to stabilize the data obtained.

The correlation plot between the platinum wire-based temperature sensor and the Vaisala temperature sensor is obtained from the data obtained, as shown in Fig. 31.9. The author uses simple linear regression to determine the correlation, where the relationship between the independent and dependent variables will be sought. The requirements for selecting linear regression have been met. The data obtained are of the type of interval (ratio). The data distribution is normal, the number of samples between the independent and dependent variables is the same, and it has a linear relationship.

The obtained correlation value is close to 1, which means that the correlation value is positive. Although the correlation value obtained is strongly positive, this value still has to be proven by more field testing. So that the performance of the platinum wire-based temperature sensor can be used as an alternative for measuring atmospheric vertical temperature, field testing can serve to stabilize the data obtained.

31.4 Conclusions

This research is the development of a platinum wire-based temperature sensor so that vertical temperature data can be obtained validly according to reference data. The platinum wire used is a wire with a diameter of 0.025 mm. The platinum wire is wound into a zero Ohm resistor with the number of turns of about 50–200 turns according to the required specifications and the diameter of the wire used.

This platinum wire-based temperature sensor is coupled with the R-Sonde range of payload systems. This aims to see the performance of the sensor when measuring temperature vertically. Tests are carried out in the laboratory and in the field. Laboratory scale testing obtained good results up to −58 °C.

Field testing was carried out in South Garut. The results obtained that the value of the platinum wire-based temperature sensor is closest to the Vaisala temperature data pattern which is the reference data with the smallest bias and the correlation is close to 1. So that the performance of the platinum wire-based temperature sensor can be used as an alternative sensor for measuring vertical atmospheric temperature. It's just that field testing is needed to stabilize the data obtained. This research can be continued by comparing several temperature sensors on the market with platinum wire-based temperature sensors that have been designed in this research, to obtain a platinum wire-based temperature sensor that has reliable performance like the manufacturer's production.

Acknowledgements The author would like to thank the Center for Atmospheric Science and Technology (PSTA) team and from the Garut Space and Atmospheric Technology Test and Observation Center who cannot be mentioned one by one, who have collaborated and assisted in this research.

Contributors Statements
All authors have contributed equally.

References

1. Lehtinen, R., Suryo, P., Jauhiainen, H.: Comparison of Two Radiosonde Pressure Measurements: Pressure Sensor Versus GPS-Derived Pressure. Vaisala (2015)
2. Shi, Y., Luo, Y., Zhao, W., Shang, C., Wang, Y., Chen, Y.: A radiosonde using a humidity sensor array with a platinum resistance heater and multi-sensor data fusion. Sensors **13**, 8977–8996 (2013). https://doi.org/10.3390/s130708977
3. Santos, E.J.P., Vasconcelos, I.B.: RTD based smart temperature sensor: process development and circuit design. In: IEEE International Conference Microelectronics., 11–14 (2008)
4. Hernowo, D.G.O., Triyana, K.: Development of Electromechanical Sensor Prototype Based on Strain Gauge Principle using Poly(3,4-thylenedioxythiophene): Poly (styrene sulfonic acid) (PEDOT: PSS) (Original title: Pengembangan Prototipe Sensor Elektromekanik Berbasis Prinsip Strain Gauge menggunakan Poly (3,4-thylenedioxythiophene): Poly (styrene sulfonic acid) (PEDOT: PSS)). Jurnal Fisika Indonesia No: 49, XVII (2013). ISSN:1410-2994
5. Tingting, L., Ming, Q., Qingan, H.: Design of a new type of MEMS temperature sensor. J Funct Mater Devices **14**(01), 223–226 (2016)
6. Lourenco, M.J., Serra, J.M., Nunes, M.R., Vallera, A.M., Castro, C.A.: Thin-film characterization for high-temperature applications. Int. J. Thermophys. **19**, 1253–1265 (1998)
7. Sohi, A.N: A multifunctional MEMS pressure and temperature sensor for harsh environment applications. A thesis presented to the University of Waterloo Ontario, Canada (2013)
8. Rahayu, S.A., Sunarya, R., Maryadi, E., Nugroho, G.A., Triani, L.R.: Analysis of radiosonde temperature profile measurement in Pamengpeuk. J. Phys.: Conf. Ser., The 2019 Conference

on Fundamental and Applied Science for Advanced Technology, IOP Publishing, 1373 012003 (2019). https://doi.org/10.1088/1742-6596/1373/1/012003
9. Farnell: A guide to temperature sensor design. 15th April 2018. https://sg.element14.com/a-guide-to-temperature-sensor-design, last accessed on Juli 2019
10. Veeramani, P., Vimala Juliet, A., Sam Jebakumar, J., Jagadish, R.: Design and fabrication of temperature sensor for weather monitoring system using micro electro mechanical system technology. Orient. J. Chem. ISSN: 0970-020 X, 2510-2516 (2018)
11. Trivedi, R., Mathur, G., Mathur, A.: A Survey on platinum temperature sensor. Int. J. Soft Comput. Eng. (IJSCE) 1(NCAI2011) (2011)
12. Thermal-Ribbons: https://www.distrelec.biz/Web/Downloads/a_/10/ytts103a_10.pdf, last accessed 2019/10/12
13. Nash, J.: Measurement of upper-air pressure, temperature and humidity. World Meteorol. Organ. (2015). http://library.wmo.int/opac, last accessed 2019/10/12
14. General Technical Informations for Resistance Thermometers (RTs): Messtechnik Schaffhausen GmbH, https://www.mts.ch/documents/Technical_informations_Resistance_Thermometers.pdf, last accessed 2019/10/12.
15. MAX31865 Datasheet: http://pdf1.alldatasheet.com/datasheet-pdf/view/492628/MAXIM/MAX31865.html, last accessed 2019/10/12
16. Vaisala: Radiosonde RS41 (2017). https://www.vaisala.com/en/products/instruments-sensors-and-other-measurement-devices/soundings-products/rs41, last accessed 2019/10/12

Chapter 32
Feasibility Test of Using Disposable Baby Diaper Waste as Raw Material for Fiberboard

Sri Purwati, Djoko Sihono Gabriel, and Kurnia Wiji Prasetiyo

Abstract The high number of births in Indonesia causes the waste of disposable baby diapers to be higher. Some people have used disposable baby diaper waste as a more valuable item, but this is still not enough to overcome the problems caused by disposable baby diaper waste. This study aims to increase the value of disposable baby diaper waste by utilizing it as a raw material for fiberboard. Experiments were carried out to produce fiberboard made from disposable baby diaper waste combined with citric acid as an adhesive in various compositions and pressing times. The fiberboard is tested on several parameters based on the National Standardization Agency of Indonesia standard requirements. From the results of the fiberboard testing, all treatments met the National Standardization Agency of Indonesia's legal requirements. The best treatment chosen was fiberboard with a compressed time of 15 min, and the composition of raw materials and adhesives was 10:2.

32.1 Introduction

In recent years population growth has continued to increase. Indonesia has a relatively high birth rate, the number of births in Indonesia in 2019 was 4,800,700 people [1]. The high birth rate in Indonesia also increases the use of disposable diapers. In 2020, the use of disposable baby diapers in Indonesia reached 254.04 million kg, an increase of 4.44% compared to the previous year [2].

Disposable baby diaper waste is made of super-absorbent plastic and gel, making it difficult to decompose from the soil [3]. Disposable baby diaper waste can also spread infectious diseases when disposed of in landfills [4]. The impact of disposing of disposable baby diapers into the environment is a big problem for the government,

S. Purwati (✉) · D. S. Gabriel
Department of Industrial Engineering, Faculty of Engineering, Universitas Indonesia, Depok, Indonesia
e-mail: ananda.sripurwati@gmail.com

K. W. Prasetiyo
Research Center for Biomaterials, Indonesian Institute of Sciences, Bogor, Indonesia

E. Yulihastin et al. (eds.), *Proceedings of the International Conference on Radioscience, Equatorial Atmospheric Science and Environment and Humanosphere Science, 2021*, Springer Proceedings in Physics 275, https://doi.org/10.1007/978-981-19-0308-3_32

society, and the preservation of the water ecosystem [5]. Therefore, we need an alternative to further optimize disposable diaper waste into items that have added value to reduce the problem of disposable baby diaper waste.

Several alternatives have been made to utilize the waste of disposable baby diapers. Some researchers use disposable baby diaper waste as a planting medium [6], compost [7–9], and pulp [10]. Several practitioners, individuals, and organizations have used baby diaper waste, such as Lestari Ningsih using disposable baby diaper waste into paper [11], Prabang Setyono using disposable baby diaper waste as raw material for acoustic panels [12], and PT. Softex Indonesia has been recycling the waste of disposable baby diapers into bricks, fertilizer, and planting media [13]. Some people have used disposable baby diaper waste as a more valuable item, but this is still not enough to overcome the problems caused by disposable baby diaper waste [14]. Therefore, it is essential to find alternatives in utilizing the waste of disposable baby diapers.

Fiberboard is a panel produced from compression of wood fibers or other lignocellulosic materials with the primary bond derived from the relevant material, especially lignin or other materials, especially adhesives, to obtain particular properties [15]. Disposable baby diaper waste is the main ingredient for making fiberboard because it contains tissue, paper, and pulp, which are derivatives of the wood fiber.

32.2 Materials and Methods

32.2.1 Materials

The material used in this work is fiber pulp which consists of organic materials (paper, pulp, and tissue) and inorganic materials (non-woven polypropylene) from baby diaper waste obtained from a waste treatment company. Meanwhile, citric acid is used as an adhesive on fiberboard. Citric acid is a natural adhesive and is cheaper than other adhesives. Citric acid has the advantage that it is easily soluble in water, poses no threat to human health, and is more environmentally friendly [16].

32.2.2 Experimental Design

The experimental design aims to take the form of the experiment to be carried out following the existing problems, to obtain as accurate information as possible from a phenomenon that occurs with minimal cost or experimental effort [17]. The experimental design process begins with selecting the Design of Experiment (DOE) method that best suits the problems raised in the research. The technique is the Design of Experiment Factorial Design method.

Table 32.1 Experimental design using factorial design

Hot-pressing time (min)	Composition									
	Baby diapers waste: citric acid = 10:2					Fiber pulp from disposable baby diapers waste: citric acid = 10:3				
15	1A1	1A2	1A3	1A4	1A5	1B1	1B2	1B3	1B4	1B5
25	2A1	2A2	2A3	2A4	2A5	2B1	2B2	2B3	2B4	2B5

The second step is the initial sample planning. Initial sample planning is done by determining the factors, levels, and response variables. Determination of factors and levels for each factor is based on literature study and consultation with researchers and technicians working at the LIPI Biomaterials Research Center in Cibinong. From the results of discussions and literature studies in previous studies, the factors used in this study were the length of pressing time and the composition of baby diaper waste with adhesive. There are two levels on the hot-pressing time factor is 15 and 25 min. And there are two levels in the composition factor of baby diaper waste with adhesive, namely 10:2 and 10:3. The response variables used in this research are the physical and mechanical properties of the fiberboard: Density, Moisture Content, Water Absorption, Thickness Swelling, Modulus of Elasticity (MOE), and Modulus of Rupture (MOR) tests.

The third step is the determination of treatment and replication. Treatment is a set of experimental conditions to be used in experimental units within the selected scope. The treatment in this research is a combination of factors and levels (Table 32.1).

Based on Table 32.1, the treatment in this work is as follows:

- 1A: Fiberboard with a hot-pressing time of 15 min and composition of fiber pulp from baby diaper waste and adhesive is 10:2.
- 1B: Fiberboard with a hot-pressing time of 15 min and composition of fiber pulp from baby diaper waste and adhesive is 10:3.
- 2A: Fiberboard with a hot-pressing time of 25 min and composition of fiber pulp from baby diaper waste and adhesive of 10:2.
- 2B: Fiberboard with a hot-pressing time of 25 min and composition of fiber pulp from baby diaper waste and adhesive of 10:3.

Each treatment received five repetitions with a total of 20 experiments. The fourth step is to determine the order of the experiments through randomization to avoid bias in the experimental results. Randomization process using Ms. Excel (Table 32.2).

32.2.3 Production of Fiberboard

The initial process in the production of fiberboard is material preparation. Material preparation is carried out in two stages, namely calculating material requirements

Table 32.2 Experiment sequence according to randomization using ms. excel

Experiments sequences	Treatment	Experiments sequences	Treatment	Experiments sequences	Treatment
1	2A3	8	2A2	15	2B2
2	1B4	9	2A1	16	1A3
3	2B3	10	1A4	17	1A1
4	1B3	11	2A5	18	1A2
5	2B1	12	2B4	19	1B2
6	1B1	13	2B5	20	1A5
7	1B5	14	2A4		

and milling fiber pulp from baby diaper waste so that the size is smaller and uniform (Fig. 32.1a). The size of the fiberboard sample to be made is 25 cm × 25 cm × 0.9 cm, and the target density is 0.8 g/cm^3, so the total weight of the fiberboard sample is 450 g. In treatments 1A and 2A, the materials needed to make fiberboard samples were 436.4 g of fiber pulp from baby diaper waste, 75 g of citric acid, and 65 ml of water. In treatments 1B and 2B, the materials needed to make fiberboard samples were 403 g of fiber pulp waste from baby diaper waste, 112.5 g of citric acid, and 97 ml of water.

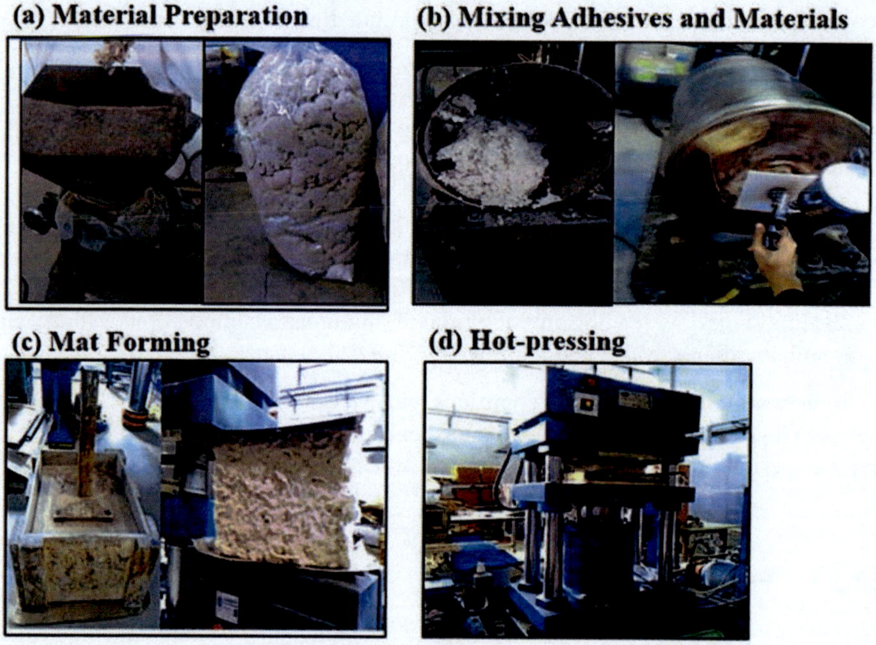

Fig. 32.1 Fiber-board production process

Fig. 32.2 Fiber-board from the hot-pressing process

The second process is mixing raw materials and adhesives using a blender drum and spray gun (Fig. 32.1b). The third process is mat-forming (Fig. 32.1c). The third process is mat-forming (Fig. 32.1c) by molding a mixture of raw materials and adhesives on a wooden mold measuring 25 cm × 25 cm. The last process in the production of fiberboard is the hot-pressing process (Fig. 32.1d). For the thickness to be the same, both sides of the plate were supported with 0.9 cm thick iron rods and then hot pressed at 200 °C for 15 and 25 min (Fig. 32.2).

32.2.4 Fiberboard Testing

Fiberboard testing based on SNI 01-4449-2006 standard is testing the physical and mechanical properties of fiberboard. The physical properties test consisted of Density, Moisture Content, Water Absorption, and Swelling Thickness tests. In contrast, the mechanical properties testing consists of the Modulus of Elasticity (MOE) and the Modulus of Rupture (MOR).

32.3 Results and Discussion

32.3.1 Physical and Mechanical Properties of the Fiberboard

All sample manufacturing and testing activities are conducted in the laboratory at Research Center for Biomaterials, Indonesian Institute of Sciences, Cibinong. In MOE and MOR tests, expert technicians test samples with a Universal Testing

Machine (UTM). Fiberboard testing from disposable baby diaper waste in this study contained four treatments, each of which had five replications so that the total sample was 20. And compare the average results of five replications in each treatment with the Indonesian National Standard (SNI). The following is a graph of the test results, with the *X*-axis as the treatment and the *Y*-axis being the average test results of 5 replications for each treatment.

The fiberboard density of the test results is 0.707–0.763 g/cm^3. Treatment 1B is the highest density, and treatment 1A is the lowest density. SNI requirements for density test on fiberboard are 0.4–0.84 g/cm^3. All treatments followed the SNI density standard (Fig. 32.3). Based on the test results, medium-density fiberboard is a type of fiberboard from disposable baby diapers.

Moisture content shows the water content in the fiberboard when it is in balance with the surrounding environment. The test results state that the moisture content of the fiberboard is 4.36–5.83%. Treatment 1B had the highest water content, and treatment 2A had the lowest water content. The SNI requirement for testing the fiberboard moisture content is $\leq$13%. All treatments followed the SNI moisture content standard (Fig. 32.4). The amount of adhesive solution affects the water content. The greater the sticky solution, the higher the water content of the fiberboard, so that the fiberboard treated with 1B and 2B had a higher moisture content than the 1A and 2A treatments.

There is no water absorption test in the SNI provisions. Still, for fiberboards in general, a water absorption test is needed to determine whether the fiberboard made is

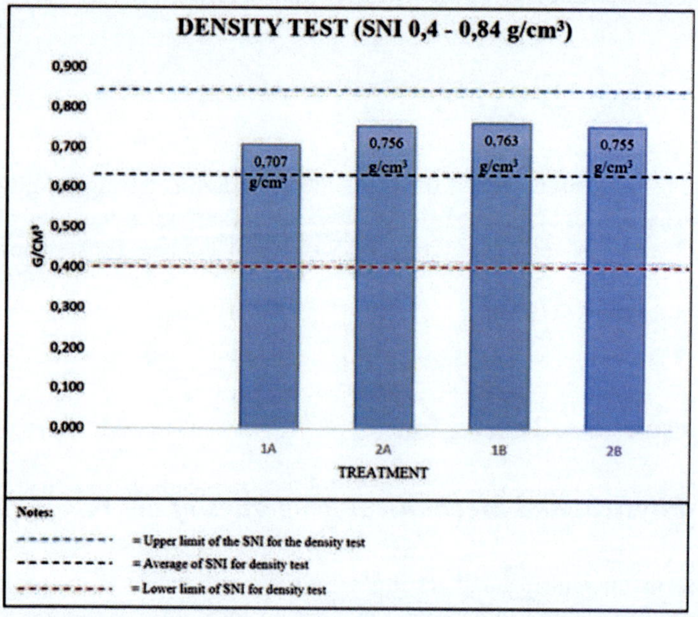

Fig. 32.3 Comparison of density test results with SNI

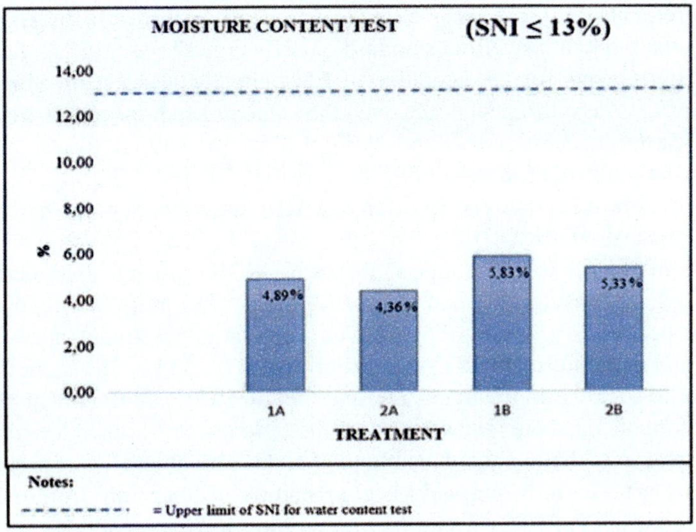

Fig. 32.4 Comparison of moisture content test results with SNI

resistant to water. The smaller the water absorption test value, the better. Fiberboard with the best test value was fiberboard with treatment 1B (Fig. 32.5).

The test results of thickness swelling are 4.17–10.93%. Treatment 1A had the highest thickness swelling, and treatment 2 B had the lowest thickness swelling.

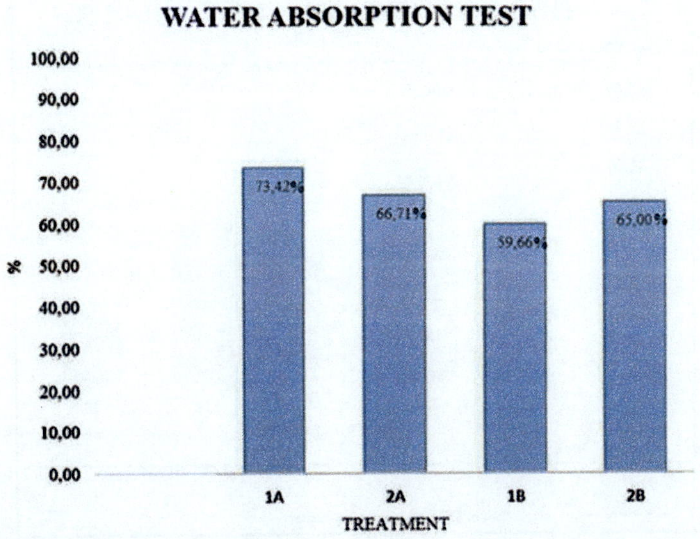

Fig. 32.5 Water absorption test results

SNI requirements for thickness swelling test on fiberboard are <17%. All treatments following the thickness swelling standard of SNI (Fig. 32.6).

SNI requirements for the Modulus of Elasticity (MOE) test on fiberboard are ≥2500 Kgf/cm^2. The test results obtained the value of Modulus of Elasticity (MOE) of fiber-board between 6140.08 and 8703.21 Kgf/cm^2 so that all treatments were following the standard of Modulus of Elasticity (MOE) of SNI (Fig. 32.7). Treatment 1A had the highest Modulus of Elasticity (MOE), and treatment 2B had the lowest Modulus of Elasticity (MOE).

SNI requirements for Modulus of Rupture (MOR) test on fiberboard are ≥30 Kgf/cm^2. The test results obtained that the Modulus of Rupture (MOR) on the fiber-board was between 49.96 and 65.95 Kgf/cm^2 so that all treatments were following the Modulus of Rupture (MOR) standard of SNI (Fig. 32.8). The highest modulus of rupture (MOR) was in treatment 1A, and the lowest was in treatment 1B.

All treatments meet the requirements of the Indonesian National Standard (SNI). The best treatment that is most feasible to choose is treatment 1A (fiberboard with a hot-pressing time of 15 min and the composition of fiber pulp from baby diaper waste and adhesives of 10:2). Treatment 1A requires the lowest adhesive cost, which is as much as 20%, and the hot-pressing time in treatment 1A is also faster, which is 15 min. In addition, the Modulus of Rupture (MOR) test from treatment 1A was the highest compared to the Modulus of Rupture (MOR) from other treatments, which was 65.95 Kgf/cm^2. The Modulus of Rupture (MOR) test is one of the mechanical properties of the board that shows the strength of the board to withstand the loads acting on the board.

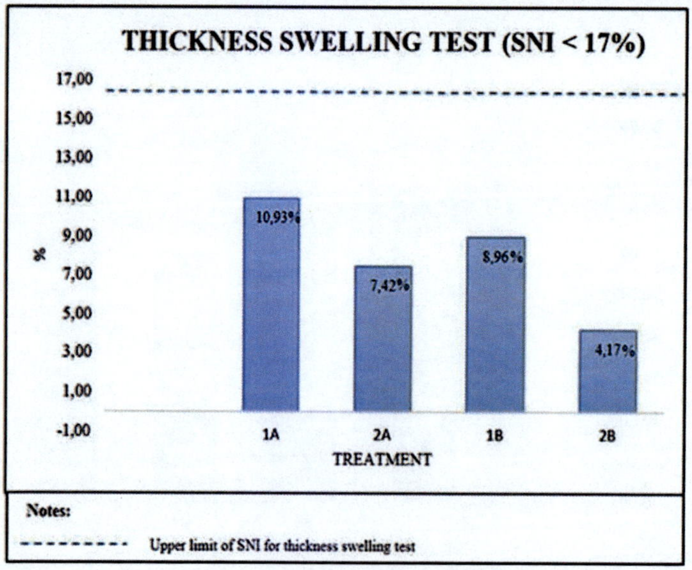

Fig. 32.6 Comparison of thickness swelling test results with SNI

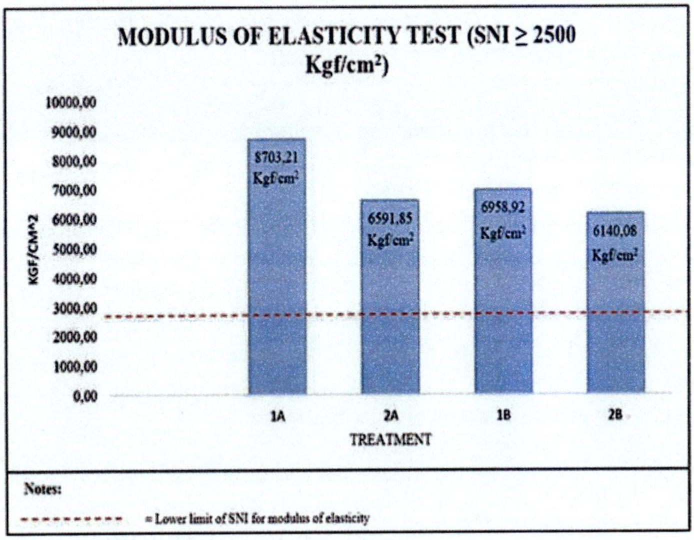

Fig. 32.7 Comparison of modulus of elasticity (MOE) test results with SNI

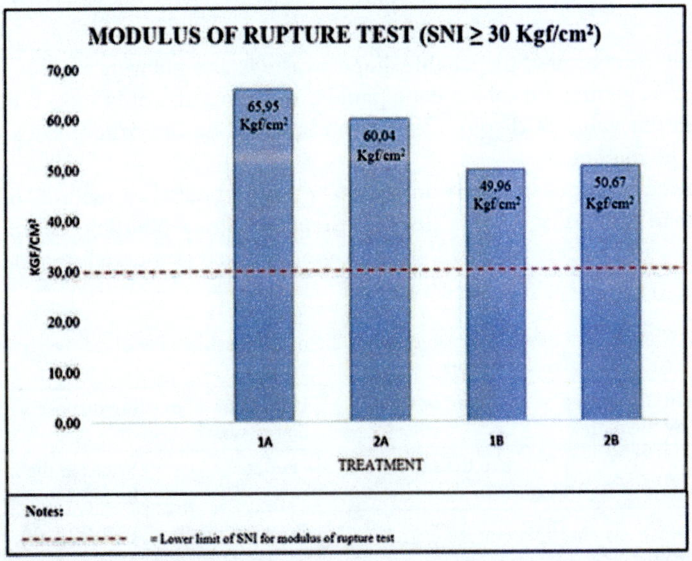

Fig. 32.8 Comparison of modulus of rupture (MOR) test results with SNI

32.3.2 Prospects of Utilizing Disposable Baby Diaper Waste as Raw Material for Fiberboard

The results of testing the physical and mechanical properties of fiberboard from disposable baby diaper waste are that all treatments produce fiberboard according to the Indonesian National Standard (SNI) attract the interest of board manufacturers to produce fiberboard from disposable baby diaper waste. Fiberboard from baby diaper waste produces a density between 0.707 and 0.763 g/cm^3, which is included medium-density fiberboard (MDF). Medium-density fiberboard (MDF) is currently in demand by many consumers as a substitute for plywood and is usually used as a raw material for making furniture. Thus, if manufacturers produce fiberboard from baby diaper waste materialize, then there will be valuable benefits from disposable baby diaper waste as a raw material for fiberboard.

32.3.3 Comparison of Increased Value of Disposable Baby Diaper Waste as Fiberboard Raw Material with Current Utilization

The current utilization of disposable diaper waste is as a planting medium, compost, pulp, paper, raw material for acoustic panels, and a mixture of bricks. Table 32.3 the increase in the value of disposable diapers is based on the selling price of similar goods on the market.

Utilization of disposable baby diaper waste as fiberboard raw material has a higher increase than the use of disposable baby diaper waste as planting media, compost, pulp, paper, and bricks, but has a lower increase when compared to acoustic panel

Table 32.3 Comparison of the increase in the value of disposable baby diaper waste based on the selling price of similar items in the market

Utilization of disposable baby diaper waste developed by the author		Utilization of existing disposable baby diaper waste	
Output	Prices in the market per kg	Output	Prices in the market per kg
Medium density fiberboard with 9 mm thickness	IDR 20,270.27	Growing media	IDR 6500.00
		Compost	IDR 6000.00
		Pulp	IDR 3454.54
		Paper	IDR 1500.00
		Acoustic Panel	IDR 24,750.00
		Brick	IDR 1000.00

raw material. However, disposable baby diapers as raw materials for fiberboard are more promising in the market because the market demand for fiberboard is higher than for acoustic panels.

32.4 Conclusion

Baby diaper waste, which so far has only been considered to have no value, turns out to have the prospect of becoming a product that has a higher value by using it as a raw material for fiberboard. All treatments in this work produce fiberboard following the National Standardization Agency of Indonesia standard requirements in density, moisture content, swelling thickness, modulus of elasticity (MOE), and modulus of rupture (MOR) test. Fiberboard with a hot pressing time of 15 minutes and the composition of fiber pulp from baby diaper waste and 10:2 adhesive is the best and most appropriate fiberboard. Because it requires the lowest adhesive cost, the fastest pressing time, and is the strongest in bearing load compared to other treatments. The results of the water absorption test on the fiberboard from disposable baby diaper waste are quite high. In a further study, biodeterioration tests on fiberboard production from disposable baby diaper waste could be considered to determine their response against fungi.

References

1. Knoema: Indonesia Number of Births in 2019: Monday, 31 November 2020, Jakarta (2020)
2. Statista: Baby diapers Indonesia: Volume: Friday, 25 September 2020, Jakarta (2020)
3. Chełstowska, P.M., Susdorf, A.K., Wasylka, J.P.: Environmental problems and health risks with disposable baby diapers: monitoring of toxic compounds by application of analytical techniques and need of education. TRAC. 143 (2021)
4. APHA: Health and environmental hazards of disposable diapers: Sunday, 02 February 2021, Jakarta (2021)
5. Sholikhah, M., Zunariyah, S.: Gerakan Ecoton dalam Upaya Pemulihan Sungai Brantas. J. Dev. Soc. Change **2** (2019)
6. Nawawi, M.I., Azizah, M.N., Andini, T.: Pemanfaatan Limbah Popok Bayi Sebagai Alternatif Media Tanam. Universitas Sebelas Maret, Solo (2017)
7. Colón, J., Ruggieri, L., Sa´nchez, A., Gonza´lez, A., Puig, I.: Possibilities of composting disposable. Waste Manag. Res. **29**, 249–259 (2010)
8. Colo'n, J., et al.: Performance of compostable baby used diapers in the composting process with the organic fraction of municipal solid waste. Waste Manage. **33**, 1097–1103 (2013)
9. Zulfikar, Aditama, W., Nasrullah: Decomposition process of disposable baby diapers in organic waste with Takakura method. Int. J. Sci. Healthc. Res. **4** (2019)
10. Ichiura, H., Nakaoka, H., Konishi, T.: Recycling disposable diaper waste pulp after dehydrating the superabsorbent polymer through oxidation using ozone. J. Clean. Prod. **276** (2020)
11. Media Indonesia: Yunita Lestari Ningsih Membuat Kertas dari Popok Bekas: Friday, 06 November 2020, Jakarta (2020)
12. Cristianto, E.: Dr. Prabang Manfaatkan Limbah Popok Sekali Pakai untuk Bahan Baku Panel Akustik: Sunday, 01 November 2020, Jakarta (2020)

13. Kurniawan, D.: Limbah Popok Bisa Diolah Jadi Batu Bata Hingga Media Tanam: Monday, 02 November 2020, Jakarta (2020)
14. Kurnia, D., Murdaningsih, D.: Limbah Popok Bisa Diolah Jadi Batu Bata Hingga Media Tanam: Monday, 02 November 2020, Jakarta (2020)
15. BSN: SNI 01-4449-2006 Papan Serat. Jakarta: Badan Standarisasi Nasional (2006)
16. Umemura, K., et al.: Application of citric acid as natural adhesive for wood. J. Appl. Polym. Sci. (2011)
17. Montgomery, D.C.: Design and Analysis of Experiments. John Wiley & Sons, New York (2012)

Chapter 33
Intraseasonal and Interannual Variability of Cumulonimbus Cloud Over the Maritime Continent

Suaydhi and Gammamerdianti

Abstract The distribution of cumulonimbus clouds over the Maritime Continent is analyzed over different atmospheric conditions to investigate the influence of El Niño, La Niña, and MJO. The data used here are infrared brightness from the Himawari family satellites from 2000 to 2016, rainfall data from GSMaP, and ERA Interim reanalysis data. In this paper, the cloud types are identified using the 2-D threshold-based cloud type classification method in which a threshold of 180 K is used as cloud-top minimum value. The results show that Cumulonimbus (Cb) and Mature Cumulonimbus (MCb) are more affected by El Niño, La Niña, and MJO phenomena than other cloud types. When the MJO is active, Cb and MCb clouds are mostly concentrated in the vicinity of Sumatra, especially during El Niño. A clear sky is observed over the middle part of Indonesia during El Niño combined with active MJO. Cb and MCb clouds are more evenly distributed around Indonesia when there is MJO during normal or La Niña conditions. The cloud distribution changes during the MJO event because it modifies and/or strengthens the low-level convergence in the atmosphere.

33.1 Introduction

The Maritime Continent region consists of more than 17,000 islands and is interspersed by warm seas. This condition makes the region one of the dominant heat sources for global atmospheric circulation. Ramage [1] was the first to point out the importance of the Maritime Continent to the global climate. This heat source brings about a lot of convection clouds over the region. Although clouds in the atmosphere are differentiated according to their shape and height, they play similar roles in the global atmospheric circulation. Convective clouds are important in the vertical heat

Suaydhi (✉) · Gammamerdianti
Research Center for Climate and Atmosphere, National Research and Innovation Agency (BRIN), Bandung, Indonesia
e-mail: suaydhi@brin.go.id

E. Yulihastin et al. (eds.), *Proceedings of the International Conference on Radioscience, Equatorial Atmospheric Science and Environment and Humanosphere Science, 2021*, Springer Proceedings in Physics 275, https://doi.org/10.1007/978-981-19-0308-3_33

distribution in the troposphere and radiation budget [2], and cirrus clouds influence the thermal structure of the atmosphere [3].

Research on clouds over the Indonesian region is usually limited to cloud covers, such as [4, 5]. However, there is only a few studies on cloud types conducted over the region [6, 7]. Temporal and spatial distribution patterns of clouds are more easily identified, if observed cloud types are known. Risyanto and Sipayung [6] used the ISCCP (International Satellite Cloud Climatology Project) to classify cloud types over Indonesia using MODIS Terra/Aqua data. On the other hand, Lasmono et al. [7] investigated the diurnal variation of Cumulonimbus (Cb) and Cirrus (Ci) clouds using cloud identification method developed by Suseno and Yamada [8]. Cb and Ci clouds have different diurnal variations over Java Island, Java Sea, and Borneo Island. The peak of the Cb cloud cycle occurs when the Ci cloud is at its minimum and vice versa [7].

Similar to precipitation, the variation in the cloud distribution is influenced by the state of the atmosphere. Eastman et al. [9] show that El Nino events increase the cloud cover over Central Pacific and decrease cloud cover over Indonesia. Different cloud types may have different behavior during El Nino, La Nina, and MJO (Madden–Julian Oscillation) events. This paper is aimed at investigating those effects over the Maritime Continent. The following section discusses the data and method used in this research. Section 33.3 presents the results and the discussion. It ends with a conclusion in Sect. 33.4.

33.2 Data and Method

33.2.1 Data

Infrared data from Himawari family satellites [Himawari-8, Multi-functional Transport Satellites (MTSAT, both MTSAT1R and MTSAT2), GOES9, and GMS (Geostationary Meteorological Satellite)] are used to identify cloud types. These data have a spatial resolution of 5 km and a temporal resolution of 1 h. ECMWF (European Centre Medium-Range Weather Forecast) Reanalysis or ERA Interim data [10] are used to describe the meteorological conditions. The 0.25° spatial resolution and 6-h temporal resolution data are used in this paper. The GSMaP precipitation data are used to validate the Cumulonimbus (Cb) cloud identification. The precipitation data have a resolution of 0.1° spatially and 1 h temporally. Other data used in this work are the El Nino and La Nina events combined with the Madden–Julian Oscillation (MJO) event. This combination is used to study the effect of MJO on the atmospheric circulation during normal, El Nino, or La Nina conditions over the Maritime Continent. Only MJO event in phase 3 or 4 is used for this purpose.

Himawari-8 and its predecessors are geostationary satellites positioned above the equator around 145°E longitude at a height of around 35,800 km. The fixed position of these satellite families make them ideal for studying the variability of various cloud types above the Maritime Continent. Two channels of infrared data (IR1 and IR2) are used to derive the cloud types using a two-dimensional threshold diagram (2D-THR) method developed by Suseno and Yamada [8]. Seventeen years of data from 2000 until 2016 are used in this work.

33.2.2 *Method*

The 2D-THR method classifies clouds into six types: cumulonimbus (Cb), mature cumulonimbus (MCb), low-level cloud (LC), mid-level cloud (MC), thin cirrus (TnCi), and thick cirrus (TkCi). This method combines a threshold value from the brightness temperature of infrared channel 1 (TIR1) and a threshold value from the difference between the brightness temperature of the infrared channel 1 and the infrared channel 2 (ΔTIR1-IR2). For example, the minimum threshold value of TIR1 for Cb cloud is 200 K, and the maximum value of TIR1 is 227 K. The threshold values of ΔTIR1-IR2 for Cb clouds are −2.0 and 2.0 K. But, according to Chen and Houze Jr. [11], the cloud top temperature can be lower than 200 K, i.e., 180 K. Hence, the minimum threshold value of TIR1 for the Cb cloud is modified to 180 K in this work. Further discussion will be presented in the next section.

33.3 Results and Discussion

Figure 33.1 shows Cb cloud and precipitation event over Java on 11th December 2014 at 9 o'clock local time (LT). The upper panel of Fig. 33.1 shows the Cb cloud identified using a minimum threshold value of 200 K. This Cb cloud seems to have some unidentified area inside it. When the 180 K threshold value is used, the Cb cloud is almost perfectly filled (middle panel of Fig. 33.1). Cb cloud with 180 K minimum threshold value resembles much more closely to the precipitation event (bottom panel of Fig. 33.1) than the 200 K threshold value Cb cloud.

Suseno and Yamada [8] state that their method of cloud identification is also valid for the tropical region. However, Fig. 33.1 shows that there is a shortcoming found in their approach. The Cb cloud over Indonesia identified using their method is hollow inside the cloud. The existence of hollow part inside the Cb cloud is probably because their approach was developed over Japan, where the height of the troposphere is lower than that over the Maritime Continent. Hence, the Cb cloud over Japan is shallower and has a lower cloud top temperature than over the Maritime Continent. The use of 180 K as a threshold value for the cloud temperature is in accordance with the finding of Chen and Houze Jr. [11] over the tropical Pacific region that the top temperature of Cb cloud can reach as low as 180 K. The rest of the results will use Cb cloud with a minimum threshold value of 180 K.

The monthly phase of the Cb cloud over the Maritime Continent for three different atmospheric states (normal, La Nina, and El Nino) is shown in Fig. 33.2. Figure 33.2a

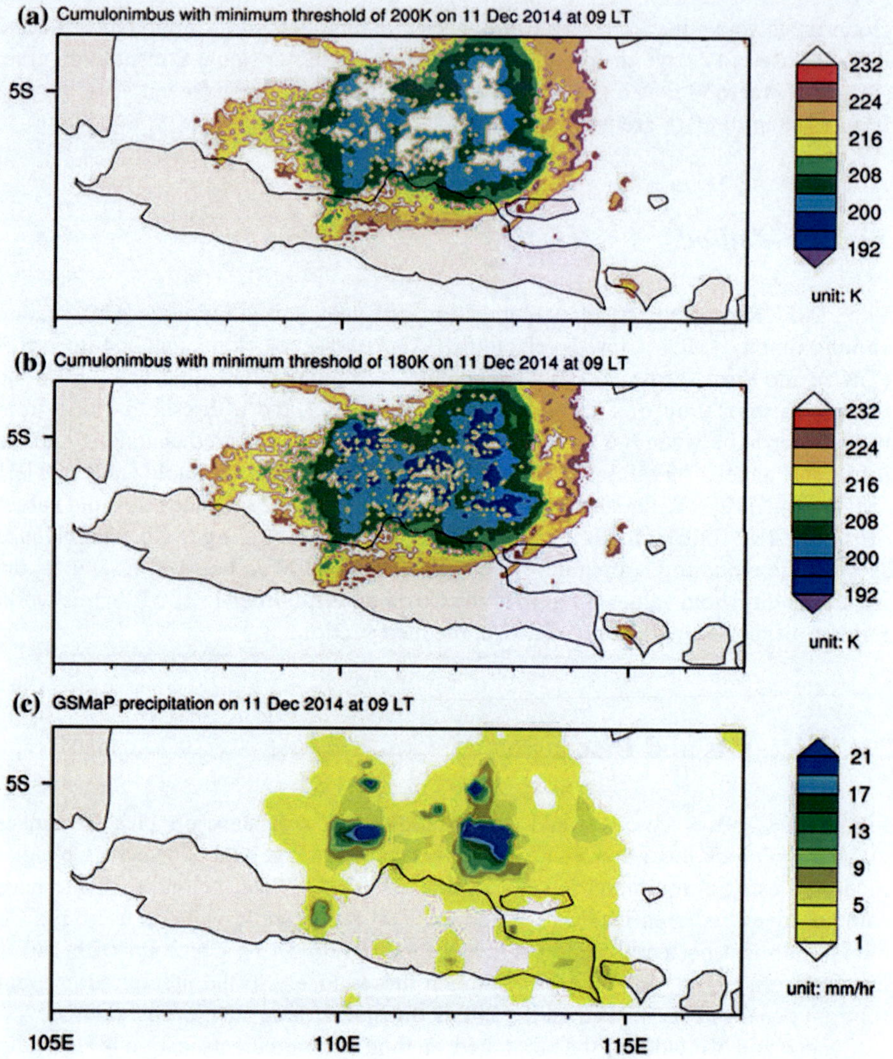

Fig. 33.1 Comparison of Cb cloud obtained with a threshold value of 200 K (**a**) and 180 K (**b**) for the minimum temperature, and precipitation from GSMap data on 11th December 2014 at 9 LT (**c**) precipitation event corresponds to the same date and hour as Cb cloud depicted above

shows the Cb cloud pattern during normal years. It shows that most of the Maritime Continent has peak Cb cloud cover in January, especially over the sea. The island of Sumatra mainly has three different regimes, the upper third has a peak of Cb cloud cover in April, the middle third around the equator has a peak in November and the lower third has a peak in December. Java and Lesser Sunda Islands (Nusa Tenggara) mainly have a peak of Cb cloud cover in January. The peaks of the Cb cloud cover over Borneo occur in five different months over different parts of the island. Celebes

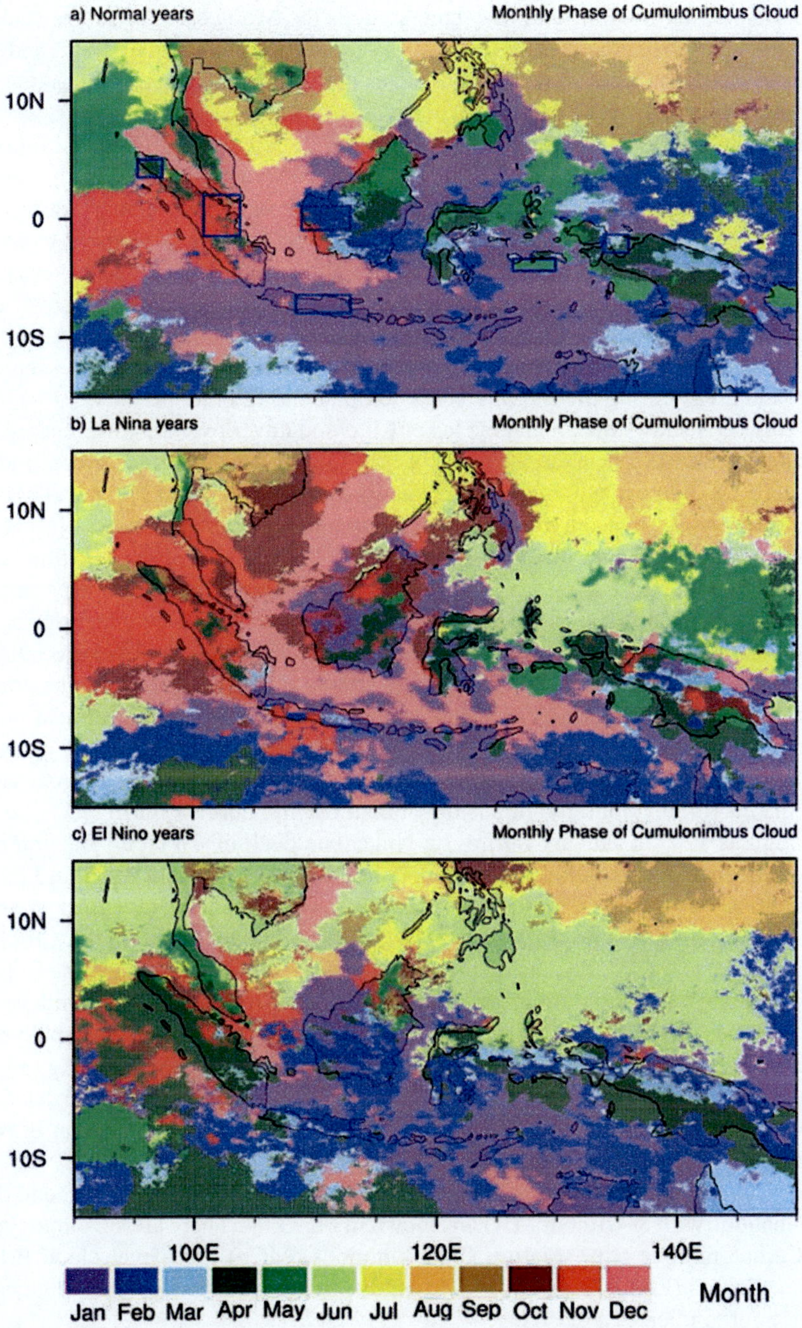

Fig. 33.2 The monthly phase of Cb clouds during normal (**a**), La Nina (**b**), and El Nino years (**c**) for a period between 2000 and 2016. The six blue boxes (**a**) are selected for areal average time series

has peaks mainly in April and May, and so does the Moluccas. Papua has peaks in May to January from the west to the east, in sequence. The Karimata Strait and Java Sea have peaks in December. The blue boxes in this Fig. indicate six areas that will be analyzed further.

Figure 33.2b shows Cb cloud cover during the La Nina years. The distribution of the Cb clouds during the La Nina event is much different from that of the normal years. For example, the Cb cloud over Java mainly occurs in January during the normal years, but it spreads over five different months from November to March during the La Nina years. A completely different pattern of Cb cloud cover is also seen during the El Nino years, as shown in Fig. 33.2c. The Cb cloud cover over Sumatra that dominantly occurs in November during the normal and the La Nina years peaks in April during the El Nino years. Different patterns are also seen over other land areas during the El Nino years compared with other periods.

Figure 33.3 shows the annual cycles of Cb cloud cover over six different regions indicated by blue boxes in the top panel of Fig. 33.2. In this Fig., blue, green, and red lines indicate annual cycles of Cb cloud fraction during normal, La Nina, and El Nino years, respectively. In Aceh (Fig. 33.3a), the maximum Cb cloud fraction is just below 6% in April during the normal years, a little lower than that the during El Nino years, but it can exceed 10% during the La Nina years in September to November. A similar feature is also observed over eastern Sumatra, but the peak is November during the normal and La Nina years and in April during the El Nino event (Fig. 33.3b). A very contrasting annual cycle is observed over Java (Fig. 33.3c). During the normal years, the Cb cloud fraction can reach around 7.5% in January, but almost no Cb clouds are observed in the dry season (August and September). The annual cycles of Cb cloud during El Nino and La Nina years are similar to that of the normal years. Still, the maximum cloud fraction is only about 6% in February during El Nino and is as high as 12% in January during La Nina. The peak of Cb cloud fraction over western Borneo occurs in February at about 5% during the normal years, in January and February at about 3.5% during the El Nino years, and in November at about 9.5% during the La Nina years (Fig. 33.3d). Over southern Maluku, the Cb cloud fraction reaches 4% in May during the normal years, 3% in April during the El Nino years, and 7.5% in May during the La Nina years (Fig. 33.3e). A maximum of 5% of Cb cloud is observed over northwestern Papua in March during the normal years, 4% in March during El Nino, and about 8.5% in April during La Nina (Fig. 33.3f). Those six areas show that more (less) Cb clouds are observed during La Nina (El Nino) events. These results agree with the finding of Eastman et al. [9] that El Nino decreases the cloud cover over Indonesia.

The cloud distribution during normal, La Niña, and El Niño conditions, and their combinations with or without MJO are shown in Fig. 33.4. There are six cloud types, i.e., Cumulonimbus (Cb), Mature Cumulonimbus (MCb), mid-level cloud (Mid), low-level cloud (Low), thick cirrus (TkCi), dan thin cirrus (TnCi). The white color in that figure indicates a clear sky.

During the El Nino condition without MJO on 1st December 2009 (Fig. 33.4a), the Cb clouds were seen over the northern part of Papua with cirrus clouds covering most of the Indonesian region. However, when MJO was active on 17th September

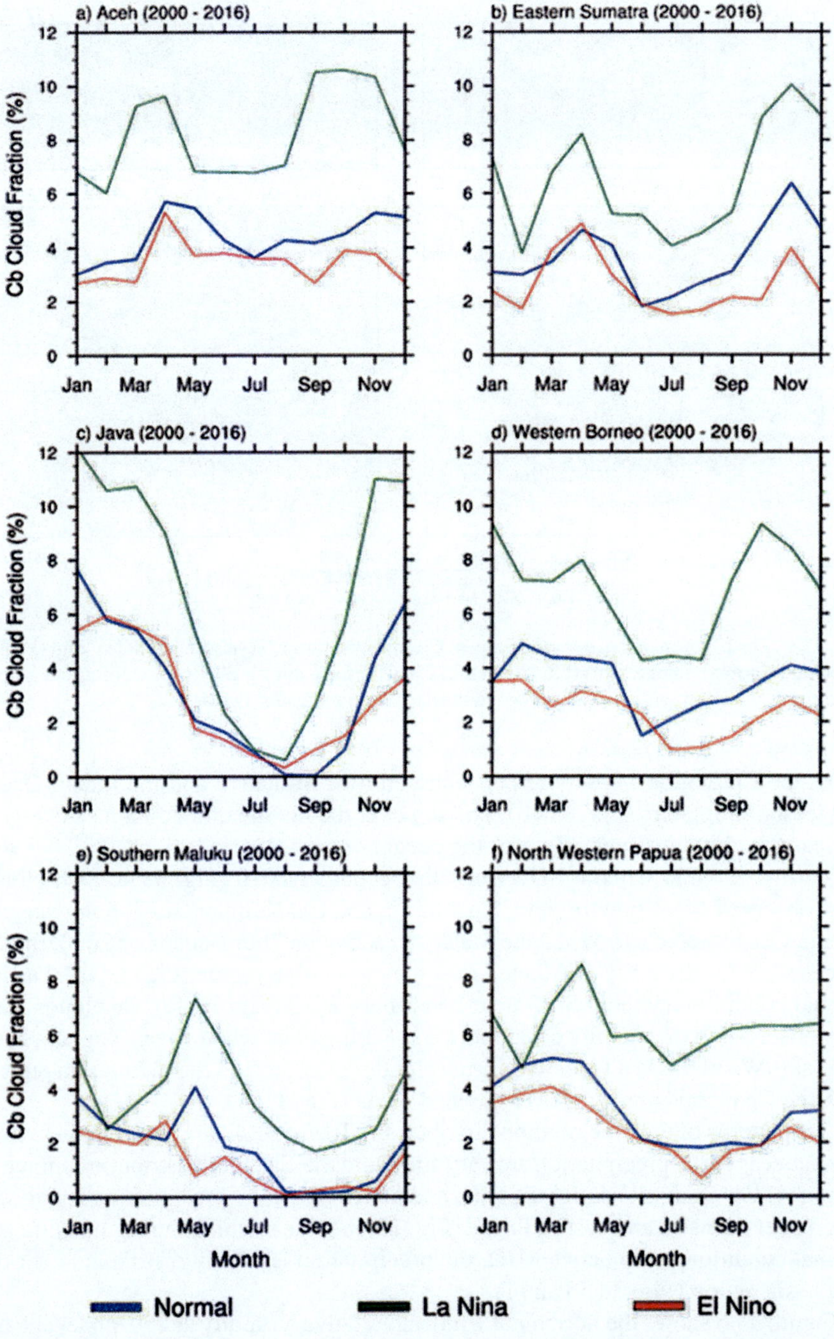

Fig. 33.3 Time series of the six areas depicted in Fig. 33.2a for normal, La Nina and El Nino years for (**a**) Aceh, (**b**) eastern Sumatera, (**c**) Java, (**d**) western Borneo, (**e**) southern Maluku, and (**f**) north-western Papua

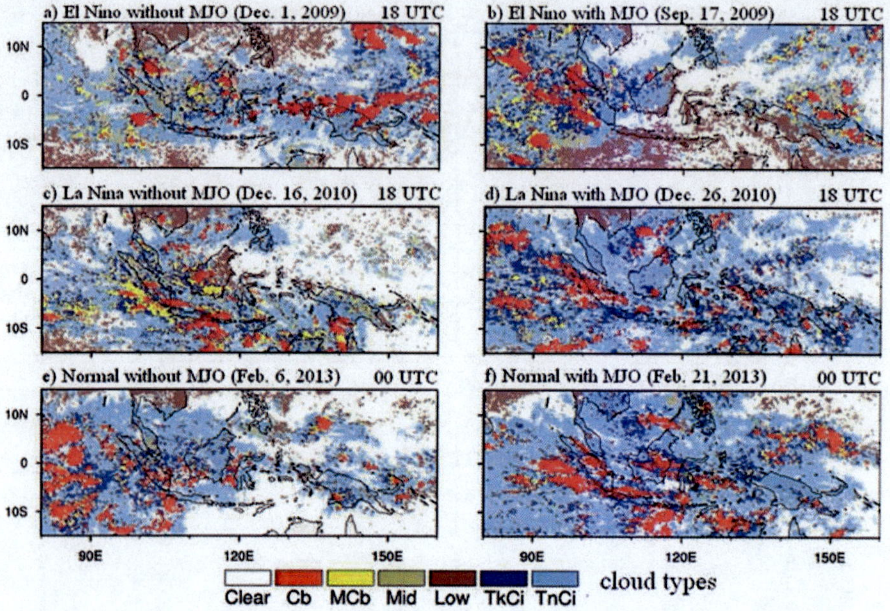

Fig. 33.4 Distribution of various cloud types: Cb (red), mature Cb (yellow), middle cloud (khaki), low cloud (brown), thick Cirrus (blue), and thin Cirrus (cyan), during El Nino, La Nina, and normal conditions, with and without MJO over Indonesia. White indicates a clear sky

2009, the Cb clouds were more concentrated over Sumatera and the Indian Ocean just west of Sumatera (Fig. 33.4b). The sky over the middle of the Indonesian region is clear (no clouds), especially over the ocean.

During La Niña without MJO on 16th December 2010 (Fig. 33.4c), Cb clouds were observed over Sumatra, Java, Nusa Tenggara, and Kalimantan. The sky over the Pacific Ocean was clear. When the MJO was active on 26th December 2010, the Cb clouds were more evenly distributed over the Indonesian region (Fig. 33.4d). During normal conditions without MJO on 6th February 2013 (Fig. 33.4e), Cb clouds were only observed in the vicinity of Sumatra and Kalimantan, while Nusa Tenggara has a clear sky. When the MJO is active, cirrus clouds cover most of the Indonesian region, with the Cb clouds observed over the land areas (Fig. 33.4f).

The patterns of the precipitation distribution TRMM 3B42 are shown in Fig. 33.5. The patterns of the precipitation are very similar to the Cb cloud distribution shown in Fig. 33.4. For example, during El Niño and active MJO, the precipitation is concentrated over Sumatra and east of Papua (Fig. 33.5b). Meanwhile, during La Niña and normal conditions with active MJO, the precipitation is evenly distributed over the Indonesia region (Fig. 33.5d and f).

Figure 33.6 shows the horizontal wind and relative humidity at 850 mb level from ERA Interim dataset. During El Nino and suppressed MJO on 1st December 2009, the low-level wind flows from the middle of Indonesia towards the Pacific Ocean

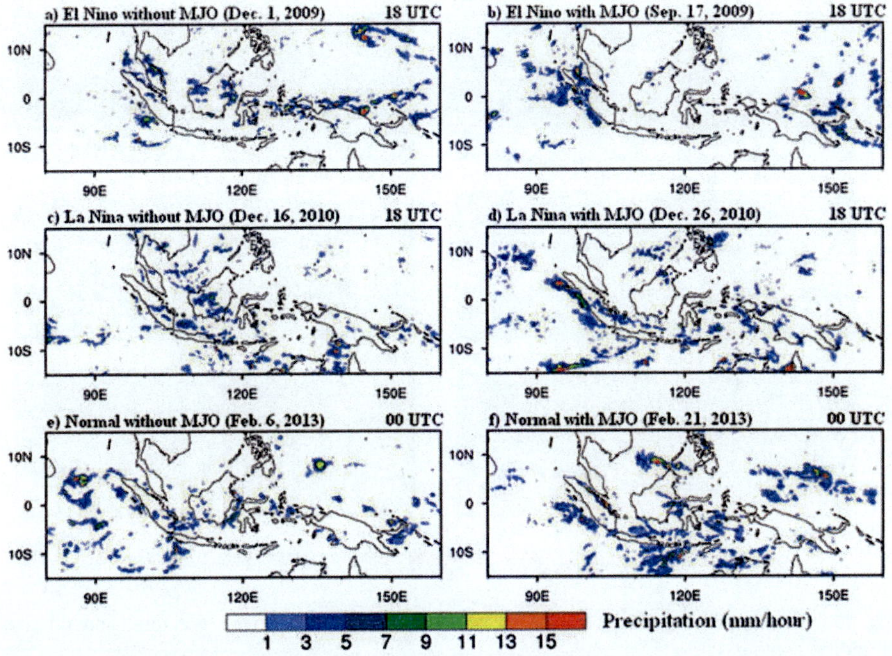

Fig. 33.5 The patterns of precipitation distribution from TRMM 3B42 (in mm/hour) during El Nino (top), La Nina (middle) and normal (bottom) conditions, without MJO (left) and with MJO (right)

(Fig. 33.6a). The air over this region is temporarily void, and it is immediately occupied by the dry incoming wind from Australia. The effect of active MJO on 17th September 2009 is indicated by the strengthening of the low-level wind from the Indian Ocean towards Sumatra (Fig. 33.6b). The center of the low-level convergence is getting stronger during active MJO than during the suppressed MJO (Fig. 33.6a). The air from Australia is drawn towards the center of the convergence, resulting in low-level airflow from the Pacific Ocean towards eastern Indonesia despite El Nino (Fig. 33.6b).

During La Niña and suppressed MJO on 16th December 2010 (Fig. 33.6c), the low-level wind blows towards Indonesia from the Indian and the Pacific Oceans. When the MJO is active on 26th December 2010 (Fig. 33.6d), the low-level wind gets stronger, and a vortex is formed just north of Kalimantan (i.e., Borneo Vortex). A similar vortex is also observed during the normal conditions with active MJO on 21st February 2013 (Fig. 33.6f). When the MJO is suppressed during the normal condition (6th February 2013), the convergence center is located to the west of Aceh (Fig. 33.6e).

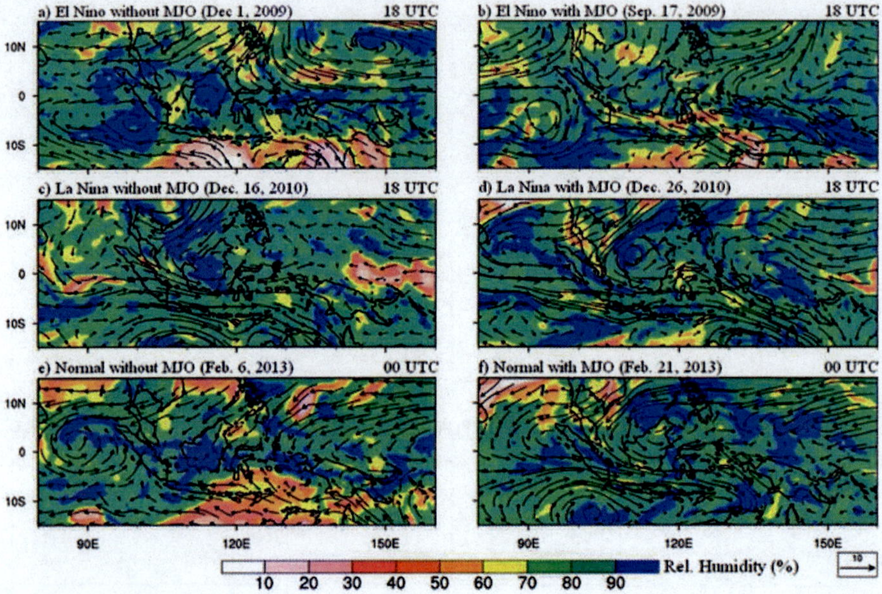

Fig. 33.6 Similar to Fig. 33.5, except for low-level wind convergence (vectors) and relative humidity (contours)

33.4 Conclusion

This paper has discussed the variation of the distribution of cloud types over Indonesia during El Nino/La Nina in combination with the active/suppressed phase of MJO. There are six combinations of these phenomena which influence the cloud distribution over Indonesia. The 850 mb wind and relative humidity data were used to analyze the atmospheric circulation around Indonesia.

The centers of low-level convergence center are located over the southwest of Sumatra during El Niño, to the west of Aceh during the normal conditions, and north of Sulawesi La Niña. These low-level convergence centers move when MJO is active. During El Niño, the low-level convergence over the southwest of Sumatra strengthens when the MJO is in active phase. This condition causes a lower density of air in the middle of the Indonesian region, which is subsequently filled by the dry air from Australia. Consequently, a clear sky over this area is observed. The Cb and MCb clouds are concentrated over Sumatra. During the normal and La Niña conditions, MJO triggers an intense convergence center north Kalimantan. This convergence center attracts moisture from both the Indian and Pacific Oceans to the Indonesian region, and Cb and MCb clouds form all over Indonesia.

References

1. Ramage, C.S.: Role of a tropical "Maritime Continent" in the atmospheric circulation. Mon. Wea. Rev. **96**(June), 365–370 (1968)
2. Inoue, T., Satoh, M., Miura, H., Mapes, B.E.: Characteristics of cloud size of deep convection simulated by a global cloud resolving model over the Western Tropical Pacific. J. Meteorol. Soc. Jpn. **86A**, 1–15 (2008)
3. Gu, Y., Liou, K.N., Ou, S.C., Fovell, R.: Cirrus cloud simulations using WRF with improved radiation parameterization and increased vertical resolution. J. Geophys. Res. Atmos. **116**(6), 1–14 (2011)
4. Houze, Jr, R,A., Geotis, S.G., Marks, Jr. F.D., West, A.K.: Winter Monsoon Convection in the Vicinity of Borneo. Part I: structure and time variation of the clouds and precipitation. Mon. Wea. Rev. **109**(8):1595–1614 (1981)
5. Gastellu-Etchegorry, J.P.: Cloud cover distribution in Indonesia. Int. J. Remote Sens. **9**(7), 1267–1276 (1988)
6. Risyanto, S.S.B.: Analisis tipe awan ISCCP menggunakan data MODIS Terra/Aqua. In: Prosiding Seminar Sains Atmosfer 2013. 2013
7. Lasmono, F, Suaydhi, Nafiisyanti, A.: Cloud diurnal variation above Indonesia on 2014. In: Prosiding Seminar Nasional Sains Atmosfer 2015, pp. 76–80
8. Suseno, D.P.Y., Yamada, T.J.: Two-dimensional, threshold-based cloud type classification using MTSAT data. Remote Sens. Lett. **3**, 737–746 (2012)
9. Eastman, R., Warren, S.G., Hahn, C.J.: Variations in cloud cover and cloud types over the ocean from surface observations, 1954–2008. J. Clim. **24**, 5914–5934 (2011)
10. Berrisford, P., Dee, D., Fielding, K., Fuentes, M., Kallberg, P., Kobayashi, S, et al.: The ERA-Interim archive. ERA Rep Ser [Internet]. **1**(1), 1–16 (2009). Available from: http://www.ecmwf.int/publications/library/do/references/list/782009
11. Chen, S.S., Houze, R.A., Jr.: Diurnal variation and life cycle of deep convective systems over the tropical Pacific warm pool. Q. J. R. Meteorol. Soc. **123**, 357–388 (1997)

Chapter 34
Impact of Plasma Depletion on the Occurrence of Scintillation in the Minimum Years Phase of Solar Cycle 23

Ednofri and Susumu Saito

Abstract Plasma depletion as part of space weather has a role in scintillation events that can impact satellite communication systems, satellite-based navigation, and radio communications. Therefore, it is now essential to understand how space weather events play a role in scintillation events. This study uses an All-Sky Imager (ASI), ionosonde, and GPS Scintillation Receivers were used in Kototabang (0.2° S, 100.3° E, −10.4° magnetic latitude) Indonesia. Plasma depletion events in the form of Equatorial Plasma Bubble (EPB) and Equatorial Spread F (ESF) will be observed from a day-to-day occurrence during minimum solar activity (2008–2009), then studied based on the emergence of post-sunset, midnight, and post-midnight as well as relation to scintillation. Of the four days of EPB emergence during minimum solar activity, there were more EPB occurrences around midnight than in the post-sunset period, and no EPB occurrences were found for post-midnight. EPB that appears after sunset is immediately followed by scintillation events, while the emergence at midnight takes several hours before scintillation occurs. At the peak of minimum solar activity (2009), the emergence of EPB was not followed by scintillation events either at post-sunset or midnight. This indicates that plasma depletion and scintillation at the peak of the year of minimum solar activity do not always coincide. We found a specific case where scintillation was not observed in an EPB but in the next adjacent EPB. It cannot be explained by the simple decay of irregularities at later local times.

Ednofri (✉)
National Research and Innovation Agency (BRIN), Agam, Indonesia
e-mail: ednofri@brin.go.id

S. Saito
Electronic Navigation Research Institute (ENRI), Tokyo, Japan

E. Yulihastin et al. (eds.), *Proceedings of the International Conference on Radioscience, Equatorial Atmospheric Science and Environment and Humanosphere Science, 2021*, Springer Proceedings in Physics 275, https://doi.org/10.1007/978-981-19-0308-3_34

34.1 Introduction

Large-scale plasma depletion in the equatorial ionosphere or plasma bubbles is an area where plasma density is much lower than the plasma density of the environment and lies along the line of the earth's magnetic field [1]. Large-scale plasma irregularities, known explicitly as equatorial plasma bubbles (EPBs), are associated with equatorial spread-F (ESF). EPBs associated with ESF is produced through the generalized Rayleigh–Taylor (RT) instability. Small-scale irregularities possibly generated by secondary instability processes can cause severe ionospheric scintillations, and affect satellite-based navigation and communication systems [2, 3]. It may result in considerable communication hazards, and thus is of great practical interest [4].

In the navigation sector, the reliability of the Global Navigation Satellite System (GNSS) services is seriously demanded for the presence of the EPB which causes deviations of tens of meters or more from the correct position [5]. The study conducted by Juan [6] over a certain number of days using a solar maximum year ground-based receiver (2014–2015) found that the amplitude scintillation has no effect at high latitudes for GNSS navigation accuracy, but has an effect at low latitudes resulting in worsening of navigation. The many applications of satellite signals have been used in many applications in today's society, the importance of predicting the emergence of EPB/ESF and ionospheric scintillation, and efforts to reduce the losses they cause are more urgent than ever. Although several attempts to predict GNSS scintillation at low latitudes have been made [7–9], a final result remains to be delivered.

Plasma bubbles, localized plasma density depletions in the ionosphere at equatorial and low-latitude regions, have been studied with various radio and optical techniques [10–13]. Although the EPB has been studied over the last several decades, it still attracts attention because of its strong influence on radio waves of trans-ionospheric communication. In fact, the connection between the EPB incident and the outage of the UHF satellite with the helicopter communication link in Afghanistan on March 4, 2002, has been reported by Kelly [14]. Some aspects of observation, generally statistically, such as morphology, seasonal behavior, latitude-long variations, the general features of EPBs, and their medium-term variability have been the subject of much research in the past and have been studied reasonably well, the factors controlling the day-to-day and short-term variability of EPB and the occurrence of scintillation are still unclear [15–17]. However, prediction and day-to-day variability of the impact of EPB on scintillation is still a challenging problem even today, especially in the solar minimum period.

Since the intensity of amplitude scintillation is proportional to the amplitude of plasma density perturbations, the GPS scintillation cannot be observed when the plasma density in the F region is low after midnight during minimum solar conditions [18]. Scintillation occurrence near and after midnights is known to decrease because ionospheric irregularities with smaller scale sizes disappear more quickly [19]. Equatorial scintillations during a high solar activity period are be sufficiently intense to disable many communications and navigation systems [20]. Under quiet geomagnetic conditions, the equatorial scintillation activity develops in the post-sunset hours. It typically lasts for 4–5 h until midnight, sometimes extending for a few hours into the post-midnight period. It presents strong seasonal and longitudinal variations that can be roughly explained by the alignment of the sunset terminator with the magnetic meridian [21, 22].

It is important to understand the role of space-weather events on scintillations. Therefore, this study was conducted to determine the impact of the occurrence of EPB on scintillation at any given time. To study the occurrence of EPB, an all-sky imager (ASI) is used to detect large-scale airglow depletion as two-dimensional images. Three closely-spaced single-frequency (1.5754 GHz) GPS scintillation receivers (Astech G12) are used to detect EPBs as scintillations of the GPS signals and measure their drift velocities. If we assume that the irregularities are located at an altitude of 350 km, we can observe the scintillation occurrence within the area of approximately 500 km around the receiver. The spatial correlation between the ionospheric scintillation and the large-scalc EPBs structure seen in ASI images in some cases for the selected day was studied at post-sunset, midnight, and post-midnight, and efforts were made to discuss the results under different atmospheric coupling processes associated with it. The supporting data from another instrument like ionosonde were used as well.

34.2 Data and Methods

The propagation of EPBs in the form of a two-dimensional image obtained from an All-Sky Imager (ASI) mounted on Kototabang (0.2° S, 100.3° E, −10.4° magnetic latitude) Indonesia. ASI has been installed in Kototabang since October 2002 as part of the Optical Mesosphere Thermosphere Imagers (OMTIs) [23]. This equipment uses a fisheye lens ($f = 24$ mm), a cooled CCD camera, and five pieces of optical filters with pass-bands that match the different emission processes from different heights. The wavelength used in this study was 630.0 nm (typically emitted at altitudes around 250 km) to observe the appearance of the plasma bubble in the form of a two-dimensional image.

To obtain information on the occurrence of ESF and its relationship to plasma bubbles, this study used an ionosonde of the Frequency Modulated Continuous Wave (FMCW) type installed by the National Institute of Information and Communications Technology Japan to detect the vertical profile of the electron density in the bottom-side ionosphere as well as the plasma bubbles occurrences as spread echoes on ionograms. The ionosondes continuously transmit radio waves from 2 to 30 MHz and receive echoes from the ionosphere to provide the bottom side plasma density profile every five minutes.

GPS scintillation receivers have been installed at the exact location as the ASI and ionosonde. The sampling rate of the GPS receivers is 20 Hz. The S4 index introduced by Yeh and Liu [24] is taken as an indicator of scintillation intensity. The S4 index is estimated every one minute. Figure 34.1 shows the location of Kototabang.

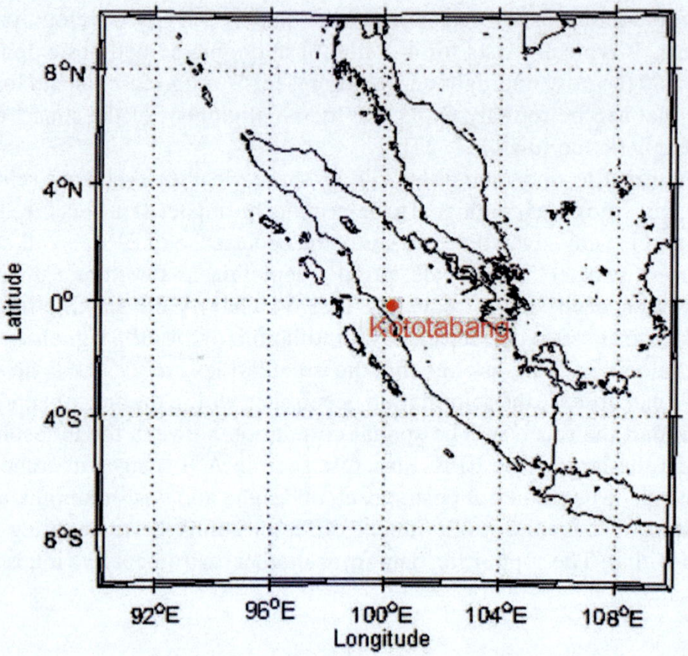

Fig. 34.1 Location of ASI, ionosonde, and GPS scintillation receivers at Kototabang station

Plasma depletion events were observed daily during minimum solar activity (2008–2009) and then studied based on the occurrence of post-sunset, midnight, and post-midnight. Due to the tropical and mountainous climate around Sumatra Island, the sky over Kototabang tends to be cloudy and rainy. Figure 34.2 shows some sample images of airglow at 630.0 nm above Kototabang at night with observations of different sky conditions; clear sky, cloudy, rainy, and clear sky with the occurrence of EPB. The event of EPB is characterized by a dark structure (marked by an arrow) that extends from north to south. Geographic North is at the top, and geographic east is to the right of the center of the image.

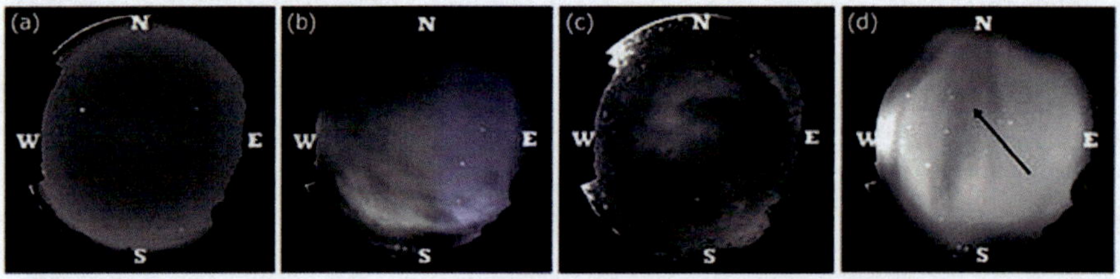

Fig. 34.2 Airglow images of the sky condition were observed with a wavelength of 630.0 nm, **a** clear sky with many stars, **b** cloudy, **c** rainy, and **d** clear sky with lots of stars and seen the occurrence of EPB

During 2008–2009, there were four days of EPB occurrences, three days in 2008 (05 February, 26 March, and 19 December) and one day in 2009 (17 August). We identified the occurrence of EPB in three events, namely post-sunset between the hours of 19:00-21:00 LT (local time is seven hours ahead of the universal time), midnight between the hours of 22:00-00:00 LT, and post-midnight between the hours of 01:00-03:00 LT. We identified the midnight EPB events in following two cases:

(1) an EPB does not appear in post-sunset but occurs at midnight only; (2) an EPB appears in midnight with a break after EPB occurrences in the post-sunset. Appearance of a post-midnight EPB is identified in the similar way. The spatial correlation between the large-scale EPBs structure seen in ASI images, the ionospheric scintillation, and ESF at post-sunset, midnight, and post-midnight in some cases for the selected day were also studied.

34.3 Results and Discussion

34.3.1 05 February 2008 (Midnight Occurrence)

Figure 34.3a shows the development and movement of EPBs from the 630 nm image on the night of 05-02-2008 at midnight (shown with a red arrow). On that night, the ASI observations begin at 12:33:30 UT. Signs of the initial occurrence of EPB can be seen at 15:57:15 UT when very weak depletions are observed. Increased intensity of 630 nm and also movement from west to east are also seen from the image. The occurrence of EPB is proper in the middle of the image at 18:17:02 UT. After that, it weakens and disappears. On the same night, as shown in Fig. 34.3b, the ESF has observed at 17:00:00 UT even though F-layer is not spread completely.

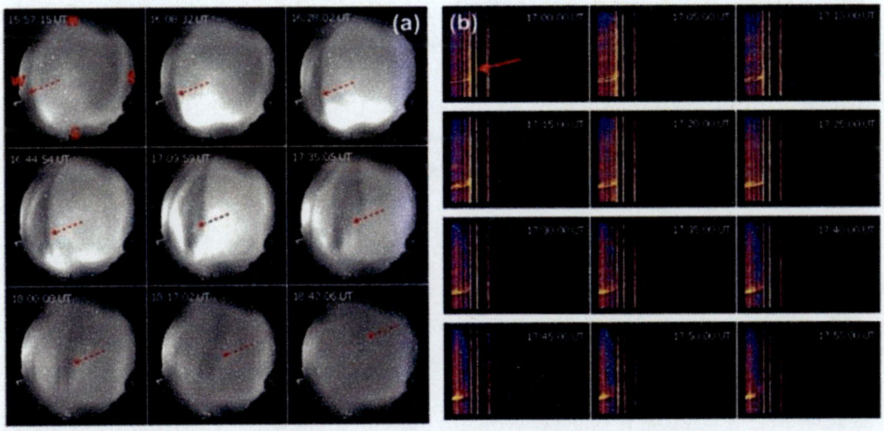

Fig. 34.3 The sequence of events **a** EPBs from ASI and **b** ESF from ionosonde observed at night of 05-02-2008

The complete ESF occurs at 17:35:00 UT, where we can no longer determine the frequency and elevation values of the F-layer. The comparison of Fig. 34.3a, b shows that the initial occurrence of ESF was observed around ~60 min since the onset of the EPB that appeared at midnight. The delay in ESF emergence after the initial EPB event during the solar minimum led to higher ESF occurrence rates at midnight or post-midnight. This is in agreement with the ESF observations in the American, Pacific, and Southeast Asian longitudinal sectors conducted by Li [25], showing a higher occurrence rate of irregularities at midnight/post-midnight than post-sunset during solar minimum.

The appearance of these structures is intriguing since the amplitude of the pre-reversal enhancement (PRE) is extremely reduced during these conditions. In addition to the PRE, it is also believed that plasma fluctuations in the equatorial bottom side F-region could be responsible for the observed variability in ESF occurrence [26]. These plasma perturbations would provide initial perturbations (seed waves) that are amplified by the instability process and develop into the large-scale plasma depletions associated with ESF. It has been suggested that, when the initial perturbations have significant amplitudes, ESF can develop even under weak/absent PRE conditions [27, 28].

Because these observations use different measurement techniques, the areas measured by each technique are also different. ASI is used to observe the whole sky, while the ionosonde inside the beam (width depending on frequency, about 45-degree radius for 8 MHz). To see the correlation of EPB and ESF, the data obtained on 05-02-2008 were mapped on a geographic map assuming an emission altitude of 250 km. Dots on the map show the ionospheric pierce point (IPPs) of GPS satellites tracked by the GPS receiver. The signal transmitted from the satellite to the receiver on earth penetrates the ionospheric shell (top of the F-layer), the intersection between the line of sight and this shell is called IPP. The EPB seen in the All-Sky Image is only a trace of plasma depletion near the height of the F peak [29]. The dark region in the center of the map is the EPBs (marked with a red arrow), moving eastward.

The spatial relationship between EPB and ESF on the map shows that when the EPB enters the field of view of the ionosonde (indicated by the blue circle), as shown in Fig. 34.4a, b, then ESF events are observed on the ionogram shown in Fig. 34.3b. However, in Fig. 34.4a, although it can be seen that the EPB has entered the field of view of the ionosonde, the ESF has not yet occurred. This can be understood that to have ESF or multiple traces, we need numerous radio wave paths that satisfy the reflection condition. If the curvature of the ionosphere is not much, radio waves transmitted in more off-zenith angles will not be reflected back to the ionosonde.

The keogram plots of EPB occurrence and corresponding observed scintillation for the night time on 05-02-2008 are shown in Fig. 34.5a, b, respectively. EPBs propagating eastward can be regarded as dark regions and indicated by oblique ovals. To reduce the multipath effects of GPS signals, the data with elevation angles less than 30° were not used in this study.

From Fig. 34.5a, b, it can be seen for 05-02-2008 when EPB that appears at 15:57:15 UT (midnight), is not immediately followed by scintillation, where the scintillation occurs more than two hours after that (shown with a red box). This

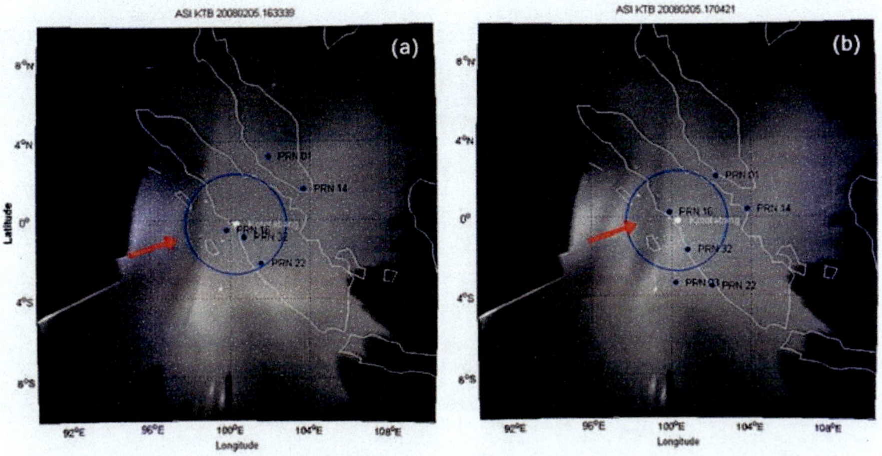

Fig. 34.4 Movement of the EPB to the field of view of the ionosonde (05-02-2008)

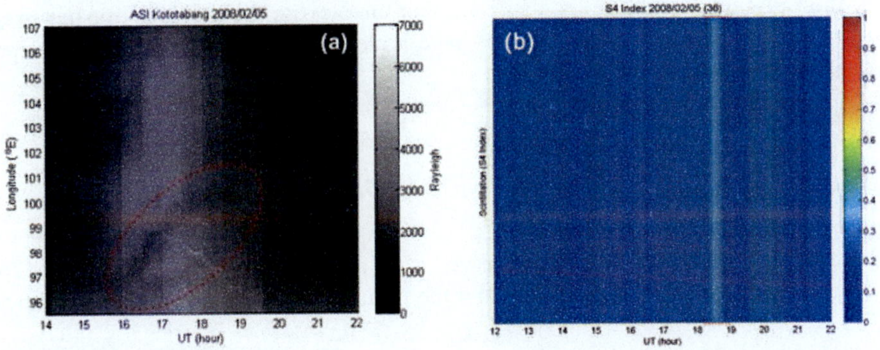

Fig. 34.5 Intensity variations of 630.0 nm airglow in the west–east meridian (**a**) and scintillation occurrence (**b**) on 05-02-2008

indicates that the occurrence of midnight EPB, takes longer to affect the ionosphere layer during approaching minimum solar conditions. And it is seen that the time of scintillation arises when the EPB begins to develop fully or is in the middle of the image at 18:17:02 UT in Fig. 34.3a.

By performing the same analysis as shown in Fig. 34.4, the spatial correlation between EPB and scintillation can be seen in Fig. 34.6. The size and color of the dots represent the magnitude of the scintillation (S4 index). As the S4 index enhances, the dot enlarges, and the color changes from blue to green, yellow, and further to red.

On 05-02-2008, PRN16 encountered an EPB (Fig. 34.6a), but S4 was not enhanced. On the other hand, in Fig. 34.6b, S4 on PRN20 was enhanced when it encountered another EPB at a slightly later local time. This is puzzling because one

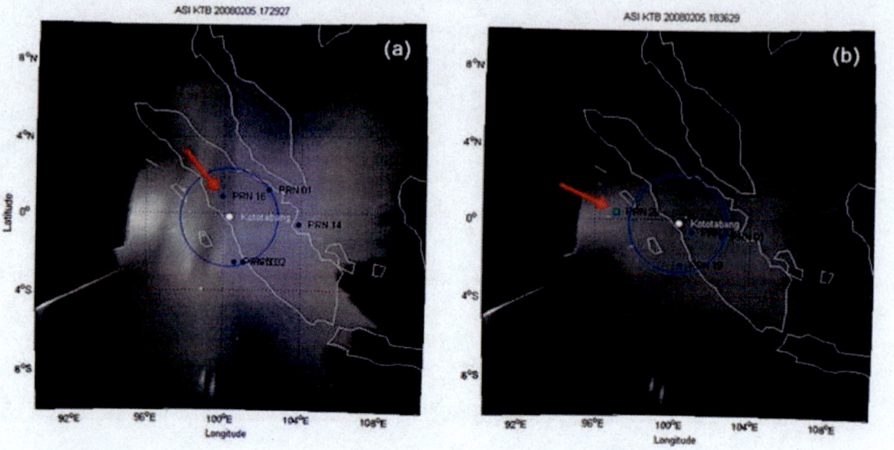

Fig. 34.6 Same as Fig. 4 but one PRN is right above EPB (05-02-2008)

of the EPBs has irregularities, while another one next to the previous one has irregularities. At the moment, we do not have a clear explanation and left as an open question.

34.3.2 26 March 2008 (Post-sunset Occurrence)

Figure 34.7a is the EPBs event observed after the post-sunset of 26-03-2008. The observations begin at 12:12:15 UT, and the initial appearance of the EPBs is shown

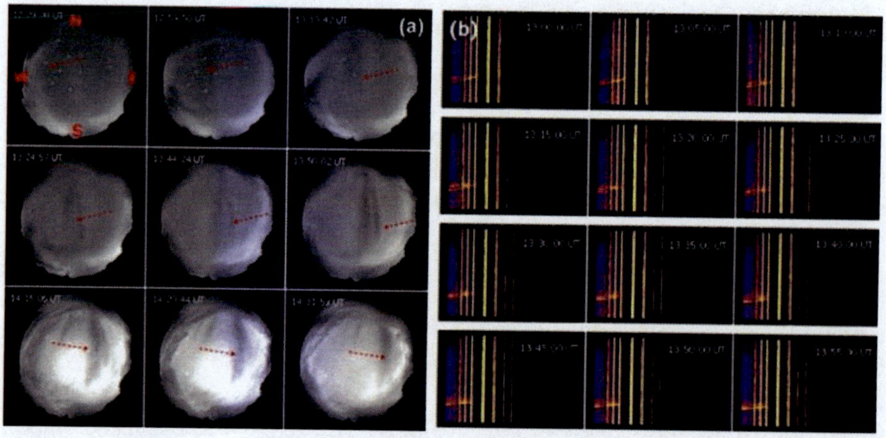

Fig. 34.7 The sequence of events **a** EPBs from ASI and **b** ESF from ionosonde observed at night of 26-03-2008

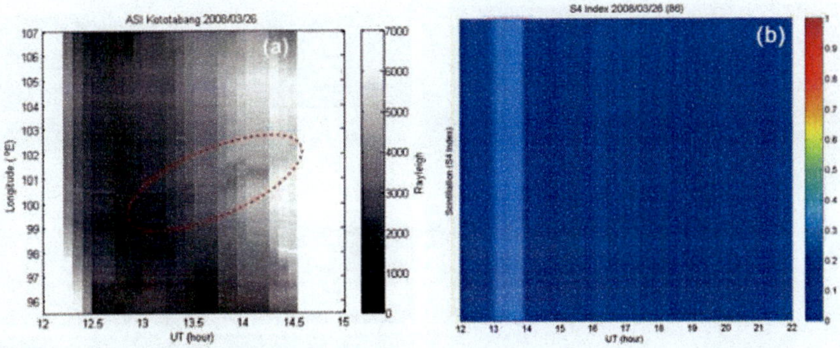

Fig. 34.8 Intensity variations of 630.0 nm airglow in the west–east meridian (**a**) and scintillation occurrence (**b**) on 26-03-2008

on the image at 12:29:08 UT. Observation stopped at 14:31:59 UT due to technical errors from the equipment even though EPB was still appearing. All EPBs are seen moving eastwards, and the EPBs in the middle look very strong (deeper depletions), whereas the visible on the east side look weak (weak depletions). Although not yet complete, from Fig. 34.7b, the ESF at night was observed at 13:00:00 UT or about 30 min after the initial appearance of the EPB.

With the same analysis as Fig. 34.5a, b, the keogram plots of EPB occurrence and corresponding observed of scintillation for the night time on 26-03-2008 are shown in Fig. 34.8a, b, respectively. It can be seen from Fig. 34.8a, b for 26-03-2008 when EPB that appears after post-sunset is immediately followed the incidence of scintillation. Even the appearance of scintillation almost coincides with the time of the ESF. However, by comparing Fig. 34.8a, b, although airglow depletions were still present after 14:00:00 UT, scintillation was no longer observed. This is a similar case observed by Saito [30] where 3 m scale irregularities disappeared after midnight, even though there were airglow depletions. This may indicate that drivers of the secondary instability was weak in the approaching minimum solar conditions. These facts are generally consistent with previous equatorial scintillation observations.

34.3.3 19 December 2008 (Midnight Occurrence)

The last occurrence of the EPB observed in 2008 was on 19-12-2008 midnight, as shown in Fig. 34.9a. Observation of ASI that night began at 12:12:15 UT, and the initial occurrence of EPB was observed at 15:47:10 UT. Observation stopped at 17:05:30 UT due to technical errors of the equipment even though EPB was still appearing. At 16:48:37, UT EPB looks fully developed and is in the middle of the image. In Fig. 34.9b, the occurrence of the ESF that night was observed at 16:35:00 UT. It takes about 50 min for the ESF to occur after the initial occurrence of the EPB.

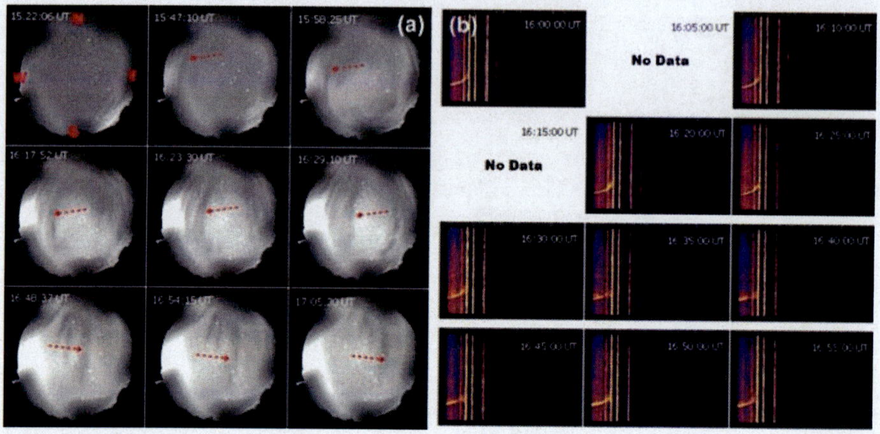

Fig. 34.9 The sequence of events **a** EPBs from ASI and **b** ESF from ionosonde observed at night of 19-12-2008

This event is almost the same as shown in Fig. 34.3a, b above for the case of EPB events that occur at midnight.

The occurrence of EPB on 19-12-2008 midnight shown in Fig. 34.10a was not followed by the occurrence of scintillation in Fig. 34.10b. Using the same format as Fig. 34.6, we found PRN22 and PRN31 encountered EPBs, and S4 did not enhance for both satellites. Our statement is the same for this condition as previously mentioned in Fig. 34.6. Scintillation occurrence near and after midnights is known to decrease because ionospheric irregularities with smaller scale sizes disappear more quickly [19]. Plasma density irregularities of various size scales are generated within the plasma bubbles, but due to plasma diffusion the smaller scale irregularities disappear earlier than the larger scale irregularities. In fact, Otsuka [31] shows that the ionospheric scintillation over Kototabang quickly disappears after midnight.

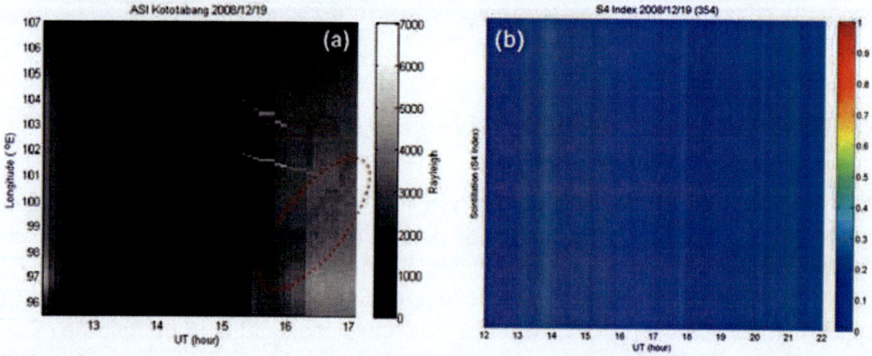

Fig. 34.10 Intensity variations of 630.0 nm airglow in the west–east meridian (**a**) and scintillation occurrence (**b**) on 19-12-2008

The ionospheric irregularities causing scintillations are under weak scintillation theory, physically dependent on the root mean square of the plasma density fluctuations and the square root of the thickness of the scattering medium. As reported by Huang [32], scintillations become weak in later local time, as the ambient plasma density is high at an early local time after sunset and becomes very low near dawn. The amplitude of plasma density perturbations also decreases with local time at night. Sripati [33] has suggested that the Rayleigh–Taylor instability that initiates a post-midnight plasma bubble during the solar minimum period does not extend to irregularities on a size scale of several hundred meters resulting in the absence of GPS scintillation.

Unlike the post-sunset EPB and ionospheric scintillation which is positively correlated with solar activity and has similar seasonal patterns for maximum and minimum solar activity, the emergence of post-midnight EPB and scintillation has complex behavior. EPB after midnight does not necessarily cause ionospheric scintillation, which is dependent on changes in total electron density [32].

34.3.4 17 August 2009 (Post-sunset and Midnight Occurrence)

In the minimum peak of solar activity (2009), we found the occurrence of EPB on 17-08-2009. On this night, EPB appeared twice, during post-sunset and midnight, as shown in Fig. 34.11a. Observation of ASI that night began at 12:12:15 UT, and the first appearance of EPB was observed at 14:01:13 UT (post-sunset). EPB looks fully developed and centered on the image at 15:27:43 UT. The EPB continues to move eastward and eventually weakens and disappears after 16:59:49 UT. At 17:05:27 UT (midnight), the start of a new EPB reappearance is not a continuation of the previous EPB. During solar minimum years, some post-midnight EPBs are not continuations of post-sunset EPBs but are initially generated during midnight or post-midnight [18]. In fact, under geomagnetic active conditions, EPB can be generated post-midnight due to the contribution of the disturbed electric field during a storm [34]. The existence of EPB was seen in the middle of the image at 18:37:34 UT, and after that, it continued to weaken and disappear. On the same night, the initial appearance of the ESF shown in Fig. 34.11b was observed at 14:30:00 UT or about 30 min after the initial appearance of the EPB (14:01:13 UT). While the initial appearance of the second ESF was observed at 18:10:00 UT or about 60 min after the second EPB appeared (17:05:27 UT).

Similar to Fig. 34.10a,b on 19-12-2008, the appearance of EPB on 17-08-2009 shown in Fig. 34.12a was not followed by the scintillation event in Fig. 34.12b. Even though that night, there were two EPB events, one of which appeared post-sunset. This fact may indicate that in the Kototabang region, during the peak of minimum solar activity, the weak EPBs do not constantly evolve into the fully developed plumes with the associated small-scale irregularities that generate GPS scintillations. This

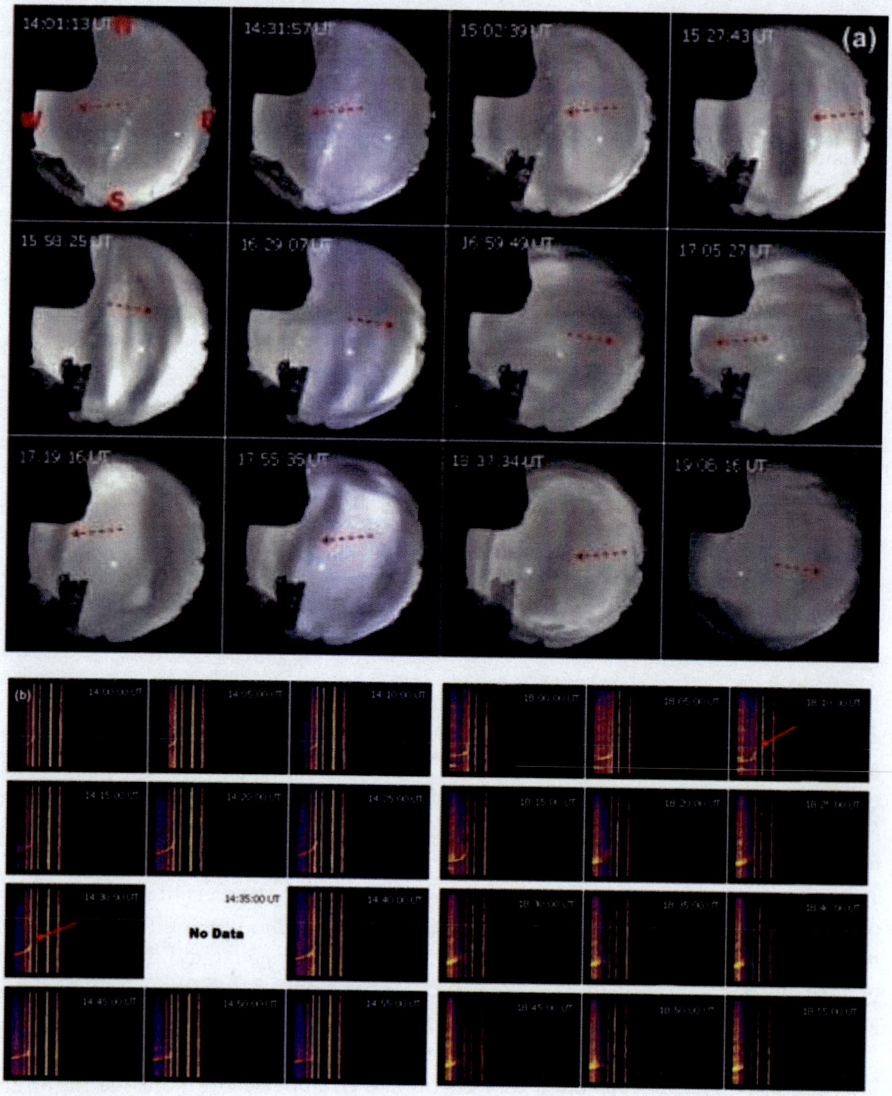

Fig. 34.11 The sequence of events **a** EPBs from ASI and **b** ESF from ionosonde observed at night of 17-08-2009

also indicates that EPB growth does not develop at a certain height. The occurrence of scintillation at any location away from the magnetic equator depends on the altitude at which the EPB develops below E $\times$ B [35]. The geometry of the peak height and beam path for the current location (Kototabang; 0.2° S, 100.3° E, $-10.4°$ magnetic latitude) indicates that the EPB must grow above the magnetic equator to produce scintillation at this location.

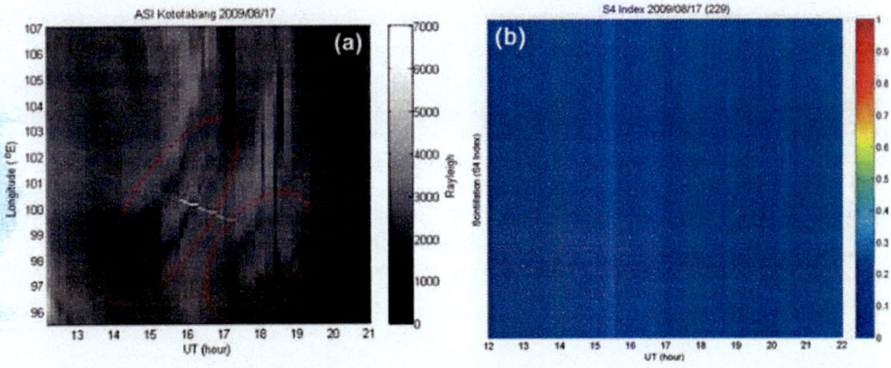

Fig. 34.12 Intensity variations of 630.0 nm airglow in the west–east meridian (**a**) and scintillation occurrence (**b**) on 17-08-2009

Another possibility is that the total integrated density fluctuation (which can be limited by the background plasma density) along the GPS ray path is inadequate to generate signal scintillations at the L-band frequencies, although plasma bubbles are present. The EPBs during this time is too low for significant scintillations to exist at 1.5754 GHz, even though the underlying EPBs structure is still present.

34.4 Conclusion

From the four days of EPB events during minimum solar activity (2008–2009), we found more EPB midnight occurrences than post-sunset and none of EPB occurrences for post-midnight. The correlation between the occurrence of EPB and ESF was obtained from observations using ASI and ionosonde. Existing data show that the average ESF is observed between about 30 min for post-sunset and about 60 min for post-midnight from when EPBs appear in all-sky images. The observation of EPBs can be explained by slower eastward plasma drift velocity at later local times. In 2008 when approaching minimum solar activity, the post-sunset EPB event was immediately followed by the occurrence of scintillation. In contrast, the EPB midnight event, scintillation took more than two hours to occur. However, all occurrences of scintillation are classified as weak scintillation (S4 < 0.4). At the end of 2008 and when it entered the bottom of minimum solar activity (2009), the occurrence of EPB was not always followed by the occurrence of scintillation for both post-sunset and midnight.

We found a specific case where scintillation was not observed in an EPB but in the next adjacent EPB. It cannot be explained by the simple decay of irregularities at altering local times. It shows that plasma depletion and scintillation when entering the peak of a minimum year of solar activity does not always coincide.

Using only a few days of observational data because, given the limitations of observations from EPBs that can be interrupted by local weather conditions and moon phases, we admit our results cannot fully represent the effects of plasma depletion on general scintillation during minimum solar activity.

Acknowledgements The observation of ASI and GPS receivers at Kototabang station was carried out in collaboration with the Institute for Space-Earth Environmental Research (ISEE), Nagoya University, Japan, and the National Institute of Aeronautics and Space (LAPAN), Indonesia. The ionosonde observation at Kototabang was carried out in cooperation with NICT, Japan, as part of the SEALION project with LAPAN. The data for ASI and GPS receivers at Kototabang are available at ISEE, Nagoya University. The data for ionosonde at Kototabang is available at NICT. We want to thank anonymous reviewers for their valuable comments enabling us to improve our manuscript.

References

1. Fagundes, P.R., et al.: Observations of day-to-day variability in precursor signatures to equatorial F-region plasma depletions. Ann. Geophys. **17**, 1053–1063 (1999)
2. Bhattacharyya, A.: Chaotic behaviour of ionospheric turbulence from scintillation measurements. Geophys. Res. Lett. (1990). https://doi.org/10.1029/GL017i006p00733
3. Wernik, A.W., Secan, J.A., Fremouw, E.J.: Ionospheric irregularities and scintillation. Adv. Space Res. **31**(4), 971–981 (2003)
4. Zheng, J.C, Huang, Z.G, Wu, Z., Hao, X.Q., Ji, J.L., Ning, B.Q., Hu, L.H., Zhao, X.K., Li, G.Z.: Onset location of the scintillation-producing spread-F plume over Sanya. Sci. China Earth Sci. **59**, 1692–1699 (2016). https://doi.org/10.1007/s11430-015-0265-5
5. Kintner, P.M., Kil, H., Beach, T.L., de Paula, E.R.: Fading timescales associated with GPS signals and potential consequences. Radio Sci. **36**(4), 731–743 (2001)
6. Juan, J.M., Sanz, J., Gonzalez-Casado, G., Garcia, A.R., Camps, A., Riba, J., Barbosa, J., Blanch, E., Altadill, D., Pérez, R.O.: Feasibility of precise navigation in high and low latitude regions under scintillation conditions. J. Space Wea. Space Clim. **8**(A05) (2018). https://doi.org/10.1051/swsc/2017047
7. Rezende, L.F.C., de Paula, E.R., Stephany, S., et al.: Survey and prediction of the ionospheric scintillation using data mining techniques. Space Weather **8**, S06D09 (2010). https://doi.org/10.1029/2009sw000532
8. Alfonsi, L., Wernik, A.W., Materassi, M., Spogli, L.: Modelling ionospheric scintillation under the crest of the equatorial anomaly. Adv Space Res **60**(8), 1698–1707 (2017)
9. Grzesiak, M., Cesaroni, C., Spogli, L., De Franceschi, G., Romano, V.: Regional short-term forecasting of ionospheric TEC and scintillation. Radio Sci **53**(10), 1254–1268 (2018)
10. Hysell, D.L.: An overview and synthesis of plasma irregularities in equatorial spread F. J Atmos Sol-Terr Phys **62**(12), 1037–1056 (2000)
11. Makela, J.J.: A review of imaging low-latitude ionospheric irregularity processes. J Atmos Sol-Terr Phys **68**(13), 1441–1458 (2006)
12. Kelley MC, Makela J.J, de La Beaujardière O, Retterer J.: Convective ionospheric storms: a review. Rev Geophys 49: RG 2003, doi:https://doi.org/10.1029/2010RG000340, (2011)
13. Abdu, M.A.: Equatorial spread F development and quiet time variability under solar minimum conditions. Indian J. Radio Space Phys. **42**, 168–183 (2012)
14. Kelly, M.A., Comberiate, J.M., Miller, E.S., Paxton, L.J.: Progress toward forecasting of space weather effects on UHF SATCOM after operation Anaconda. Space Weather **12**, 601–611 (2014). https://doi.org/10.1002/2014SW001081
15. Abdu, M.A.: Outstanding problems in the equatorial ionosphere thermosphere electrodynamics relevant to spread F. J. Atmos. Solar-Terr. Phys. **63**(9), 869–884 (2001)
16. Abdu, M.A.: Day-to-day and short-term variabilities in the equatorial plasma bubble/spread F irregularity seeding and development. Prog. Earth Planet Sci. **6**, 11 (2019). https://doi.org/10.1186/s40645-019-0258-1

17. Yamamoto, M., Otsuka, Y., Jin, H., et al.: Relationship between day-to-day variability of equatorial plasma bubble activity from GPS scintillation and atmospheric properties from Ground-to-topside model of Atmosphere and Ionosphere for Aeronomy (GAIA) assimilation. Prog. Earth Planet Sci. **5**, 26 (2018). https://doi.org/10.1186/s40645-018-0184-7
18. Otsuka, Y.: Review of the generation mechanisms of post-midnight irregularities in the equatorial and low-latitude ionosphere. Prog. Earth Planet Sci. **5**, 57 (2018). https://doi.org/10.1186/s40645-018-0212-7
19. Basu, S., Basu, S., Aarons, J., McClure, J.P., Cousins, M.D.: On the coexistence of kilometer- and meter-scale irregularities in the nighttime equatorial F region. J. Geophys. Res.: Space Phys. **83**(A9), 4219–4226 (1978). https://doi.org/10.1029/JA083iA09p04219
20. Groves, K.M., Basu, S., Weber, E.J., Smitham, M., Kuenzler, H., Valladares, C.E., Sheehan, R., McKenzie, E., Secan, J.A., Ning, P., McNeill, W.J., Moonan, D.W., Kendra, M.J.: Equatorial scintillation and systems support. Radio Sci. **32**, 2047–2064 (1997)
21. Abdu, M.A., Bittencourt, J.A., Batista, I.S.: Magnetic declination control of the equatorial F region dynamo electric field development and spread F. J. Geophys. Res. **86**(A13), 11, 443–11,446 (1981). https://doi.org/10.1029/JA086iA13p11443
22. Tsunoda, R.T.: Control of the seasonal and longitudinal occurrence of equatorial scintillations by the longitudinal gradient in integrated E region Pedersen conductivity. J. Geophys. Res. **90**(A1), 447–456 (1985). https://doi.org/10.1029/JA090iA01p00447
23. Shiokawa, K., Katoh, Y., Satoh, M., Ejiri, M.K., Ogawa, T., Nakamura, T., Tsuda, T., Wiens, R.H.: Development of optical mesosphere thermosphere imagers (OMTI). Earth Planets Space **51**, 887–896 (1999)
24. Yeh, K.C., Liu, C.-H.: Radio wave scintillations in the ionosphere. Proc. IEEE **70**(4), 324–360 (1982). https://doi.org/10.1109/proc.1982.12313
25. Li, G., Ning, B., Abdu, M.A., et al.: On the occurrence of postmidnight equatorial F region irregularities during the June solstice. J Geophys. Res. **116**, A04318 (2011). https://doi.org/10.1029/2010JA016056
26. Tsunoda, R.T., Yamamoto, M., Tsugawa, T., Hoang, T.L., Tulasi Ram, S., Thampi, S.V., Chau, H.D., Nagatsuma, T.: On seeding, large-scale wave structure, equatorial spread F, and scintillations over Vietnam. Geophys. Res. Lett. **38**(20), L20102 (2011). https://doi.org/10.1029/2011GL049173
27. Abdu, M.A., Alam Kherani, E., Batista, I.S., de Paula, E.R., Fritts, D.C., Sobral, J.H.A.: Gravity wave initiation of equatorial spread F/plasma bubble irregularities based on observational data from the SpreadFEx campaign, Annales Geophys. **27**(7), 2607–2622 (2009). https://doi.org/10.5194/angeo-27-2607-2009
28. Tsunoda, R.T.: On seeding equatorial spread F during solstices. Geophys. Res. Lett. **37**(5), L05102 (2010). https://doi.org/10.1029/2010GL042576
29. Kil, H., Paxton, L.J., Oh, S.-J.: Global bubble distribution seen from ROCSAT-1 and its association with the evening prereversal enhancement, J. Geophys. Res. **114**, A06307 (2009). https://doi.org/10.1029/2008JA013672
30. Saito, S., Fukao, S., Yamamoto, M., Otsuka, Y., Maruyama, T.: Decay of 3-m-scale ionospheric irregularities associated with a plasma bubble observed with the equatorial atmosphere radar. J. Geophys. Res.: Space Phys. **113**(A11), A11318 (2008). https://doi.org/10.1029/2008JA013118
31. Otsuka, Y., Shiokawa, K., Ogawa, T.: Equatorial ionospheric scintillations and zonal irregularity drifts observed with closely-spaced GPS receivers in Indonesia. J. Meteorol. Soc. Jpn. **84A**, 343–351 (2006). https://doi.org/10.2151/jmsj.84A.343
32. Huang, C.-S., De La Beaujardiere, O., Roddy, P.A., Hunton, D.E., Liu, J.Y., Chen, S.P.: Occurrence probability and amplitude of equatorial ionospheric irregularities associated with plasma bubbles during low and moderate solar activities (2008–2012). J. Geophys. Res.: Space Phys. **119**(2), 1186–1199 (2014). https://doi.org/10.1002/2013ja019212
33. Sripathi, S., Bose, S., Patra, A.K., Pant, T.K., Kakad, B., Bhattacharyya, A.: Simultaneous observations of ESF irregularities over Indian region using radar and GPS. Ann Geophys **26**, 3197–3213 (2008). https://doi.org/10.5194/angeo-26-3197-2008

34. Fukao, S, Ozawa, Y, Yamamoto, M, Tsunoda, RT.: Altitude extended equatorial spread F observed near sunrise terminator over Indonesia. Geophys. Res. Lett. **30**(22), 2137 (2003). https://doi.org/10.1029/2003G. L0183 83
35. Kumar, S., Gwal, A.K.: VHF ionospheric scintillations near the equatorial anomaly crest: solar and magnetic activity effects. J. Atmos. Solar Terr. Phys. **62**(3), 157–167 (2000)

Chapter 35
Effect of Precipitation Time on the Physicochemical Properties of Modified Sago Starch

Riska Surya Ningrum, Fadia Idzni Rodhibilah, Dewi Sondari, Sudarmanto, and Dwi Ajias Pramasari

Abstract Sago starch is offered as an alternative diet to rice and is mainly consumed by people in the eastern region of Indonesia. This starch content is high carbohydrate and fiber with a low glycemic index. Sago starch is broadly used in the food industry, textile, paper, pharmaceuticals, biotechnology, and poultry industries as a thickener, stabilizer, and gelling agent. However, it has some disadvantages, such as being over sticky, high viscosity, no resistance to acid, and unclear properties causing un-optimal utilization of the starch. Therefore, the effect of ethanol precipitation time (E1 = 1 h, E2 = 2 h, E3 = 3 h, E4 = 4 h, and E5 = 5 h) on the starch physicochemical properties (moisture, amylose content, solubility, chemical functional group, particle size, morphological, swelling power, thermal behavior, and crystallinity) has been investigated in this study. The properties of modified sago starch were analyzed by gravimetric method, spectroscopy, Fourier Transform Infrared (FTIR), Particle Size Analyzer (PSA), Field Emission Scanning Electron Microscope (FE-SEM), Different Scanning Calorimetric (DSC), and X-Ray Diffractometer (XRD). This study showed precipitation time affects differently to each physicochemical property. The 2 h (E2) precipitation is the optimum time to improve the physicochemical properties of sago starch. This treatment can increase thermal behavior and crystallinity, followed by decreasing the amylose content, moisture content, water-solubility, swelling power, and hydroxyl-group, which affect the hydrophobicity and reduce the particle size.

R. S. Ningrum (✉) · D. Sondari · Sudarmanto · D. A. Pramasari
Research Center for Biomass and Bioproducts, National Research and Innovation Agency (BRIN), Cibinong, West Java, Indonesia
e-mail: riska.surya.ningrum@brin.go.id

F. I. Rodhibilah
Biology Department, Faculty of Mathematics and Science, Padjajaran University, Bandung, West Java, Indonesia

E. Yulihastin et al. (eds.), *Proceedings of the International Conference on Radioscience, Equatorial Atmospheric Science and Environment and Humanosphere Science, 2021*, Springer Proceedings in Physics 275, https://doi.org/10.1007/978-981-19-0308-3_35

35.1 Introduction

Sago starch is a starch extracted from sago plant stem (*Metroxylan* Sp.) processing. Native sago has a high viscosity, is over sticky, is not resistant to acid, and has unclear properties [1]. Native sago is needed to be modified therefore, it can be used in the food industry, textile, paper, pharmaceutical, biotechnology, and poultry industries as a thickener, stabilizer, and gelling agent [2]. The aims of starch modification process are to repair or enhance the sago's properties, such as tensile strength, elongation at break, and water absorption [3]. The modification process can be done using chemical, physic, biology, and combination of them. There are some studies about the modification of sago starch:

According to Table 35.1, the characteristics of modified sago depend on the type of modification process. The precipitation method is a simple, fast, and easy

Table 35.1 Type of starch modification process

Classification	Type of modification	Agent	Modified starch properties	References
Chemistry	Oxidation	H_2O_2	Modified sago obtaining from acetylation method is the most hydrophobic and elastic, and also has the highest of Tensile Strength than other	[4]
	Acetylation	Acetate anhydrous		
	Crosslink	Sodium acetate		
	Precipitation	Ethanol		
	Microemulsion	Cyclohexane, ethanol, and surfactant	Modified sago has smaller particle sizes of 83 nm	[5]
	Acid hydrolysis	HCl	Modified sago is more soluble in water and lower gelling point	[6]
	Acid hydrolysis	Lactic acid	The solubility, swelling power, and baking expansion of modified sago was increase significantly	[7]
Enzymatic	Hydrolysis	A-amylase, glucoamylase pululanase (dextrozyme)	Dextrozyme produced more reducing sugar yield	[8]
Physic	Ball milling and Ultrasonic		The particle size of modified sago become smaller and the surface area was increase	[9]
	Hydrolysis	High Shear Homogenization (HSH)	The particle size of modified sago is 118.6 nm	[10]

Table 35.2 Type of modified starch using precipitation method

Type of starch	Precipitate agent	Modified starch properties	References
Waxy maize starch acetate	Acetone	Nanospheres more hydrophobic	[12]
Rice flour	Ethanol	The result is a dry, porous, pregelatinized powder with the particle size 75 μm (flour-derived powder) and 41.6 μm (starch-derived powder)	[13]
Arrowroot	NaOH, n-butanol, ethanol	Particle size range 78.6–538.7 nm, consisting of high porosities	[14]
Waxy maize	Ethanol 99%, surfactant (SDS and Tween 80)	Particle size range 100–200 nm	[15]
Wheat	NaOH, urea, ethanol 98%	Particle size range 50–70 nm, yield of 92 ± 5%	[2]

modification type process for starch [11]. Table 35.2 is shown that there are some studies that examine the modification of starch using the precipitation method.

Precipitation is one modification technique to occur the modified starch with smaller particle size [16]. Winarti et al. [14] used ethanol as the non-solvent on precipitation method. They occurred a high porosity of the starch, therefore, can be used as reinforce and filler in the nanocomposite materials, and also as carrier matrix for the active ingredients that is less soluble in water [12]. In addition, to obtain the smaller modified starch, usually in nanoparticle size, there are some pretreatment examined before precipitation process, such as hydrolyzed the starch using acid then neutralized using alkali [14], autoclaved the starch [13], reacted starch with sodium hydroxide and urea [2], dissolution by DMSO and adding the surfactant to the starch solution [15].

Different from other previous research, in this study, the modification process has been done without any pretreatment due to knowing the effect of precipitation time on the modification process using ethanol on the physicochemical properties of modified sago starch. That physicochemical properties include moisture content, water-solubility, swelling power, particle size, functional group, morphology, thermal stability, and crystallinity.

35.2 Materials and Methods

35.2.1 Materials

Sago starch, ethanol 96% (Merck), and distilled water were used in this modified sago starch process.

35.2.2 Methods

Preparation of modified sago starch using ethanol precipitation method [17]. Sago starch dissolved with distilled water (1:19), heated at 50 °C while stirred at 350 rpm for 1 h, then cooling down at room temperature. Added ethanol 96% as dropwise into it (the volume of ethanol is equal with distilled water), while stirring at 700 rpm for particular time (1, 2, 3, 4, and 5 h). We separated the solvent (ethanol) using centrifugation. Keep and dry the precipitate (modified sago starch) using a freeze dryer (Labconco).

Physicochemical analysis. The moisture content of modified sago starch was determined by the gravimetrical method that measure water removal from the initial mass [18]. Amylose content was analyzed by the spectroscopy method at a wavelength of 625 nm and using amylose solution as the standard [4]. The solubility of modified sago starch was determined at 70 °C using the previous method [4]. Moisture content, amylose content, and solubility test were carried out two times. The functional groups of modified sago were analyzed using FTIR Spectrum (Perkin Elmer) at a wavenumber of 4000–400 cm^{-1} and attenuated total reflectance (UATR) technique. The particle size of modified sago was determined by Particle Size Analyzer (PSA). The morphological samples were analyzed using Field Emission Scanning Electron Microscope (FE-SEM Thermo Scientific—Quattro S) on the 816× of magnification. The thermal behavior of modified sago was monitored using Differential Scanning Calorimetric (DSC 4000 Per-kin-Elmer) with the range of temperature is 20–120 °C and the rate is 10 °C/min. The crystallinity of modified sago was analyzed using an X-ray diffractometer (XRD-6100 Shimadzu) with a Cu target at 40 kV at a diffraction angle ranging from $2\theta = 20°$ to 45°.

35.3 Result and Discussion

35.3.1 Preparation of Modified Sago Starch Using Ethanol Precipitation Method

There are 15–30% amylose and 70–80% amylopectin in the sago starch [7], which both of compounds has many hydroxyl groups that can be formed the intermolecular hydrogen bonding. The heating and stirring process of the starch solution will be broke the intermolecular bond and entering the water molecule in the starch granule. Water changed to an amorphous region form, which is amylose and branches of amylopectin, called swelling power [19]. Greater water amount entered in the starch granule will increase swelling power. If the heat temperature rises, gelatinization will happen. The initiation of the gelatinization process is characterized by the presence of leaching amylose, therefore, gel is formed, and the viscosity is increased [20]. The maximum gelatinization can be characterized by the breakdown of the starch granules. However, in this study, the heating process has been done in the lower gelatinization temperature, so the full gelatinization did not happen yet. For the evidence, we can see the morphology test result (Fig. 35.2), it showed that not all the granule was broken.

After heating and stirring process for 1 h, the sago starch solution was cooled down to room temperature. Retrogradation or structure rearrangement of amylose and amylopectin will be happened in this state [21]. When the water bound to the starch molecule is released slowly, the hydrogen bond will be formed again between amylose-amylose, amylose-amylopectin, and amylopectin-amylopectin. Amylose retrogradation is faster than amylopectin retrogradation.

After that, the ethanol precipitation process is carried out using a variations of precipitation time, there are 1 h (E1), 2 h (E2), 3 h (E3), 4 h (E4), and 5 h (E5). In the precipitation process, the transition from a dissolved state to the solid-state will be happened after addition ethanol as non-solvent [15]. Amylose that comes out of granule will be precipitated after adding dropwise of ethanol. Modification using the non-solvent precipitation method will affect the particle size and shape of modified sago [11]. Some factors influencing the successful of this method are the technique used, the concentration of polymer (starch), the ratio of polymer non-solvent, and the type of non-solvent used [12, 15]. After precipitation, ethanol was separated from the starch solution by centrifugation, then dried using a freeze dryer.

The reaction between sago starch and ethanol is a physical reaction because there are no new bonds are formed in the spectra of FTIR (Fig. 35.1). Spectra FTIR shows that all the samples have similar absorption of functional group. However, that absorption intensity of the functional group is different (Table 35.3).

Fig. 35.1 Spectrum FTIR of modified sago starch

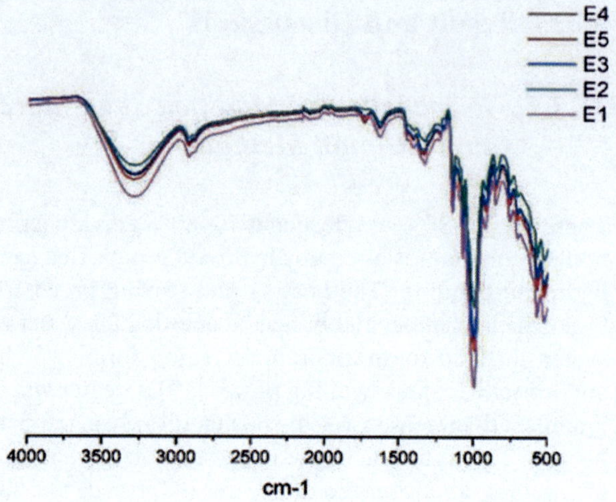

Table 35.3 The functional group containing in the modified sago starch

Wavenumber (cm^{-1})	Functional group
3297–3289	– OH stretching
2932–2928	– C–H stretching
1417–1415	– CH3 group
1338–1336	– CH2-group
1149–1076	C–O group

Fig. 35.2 Particle size (μm) of modified sago

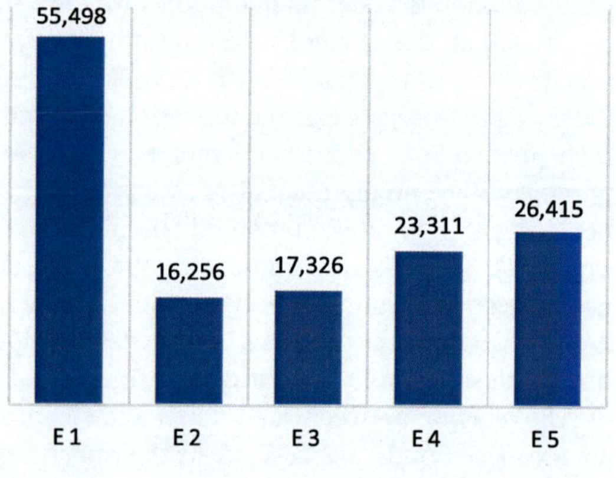

35.3.2 *The Effect of Precipitation Time on the Characteristics of Modified Sago Starch: Size, Morphology, and Composition*

The particle size of modified sago starch can be determined using PSA. Figure 35.2 shows that on the precipitation two until five hours (E2 until E5), the particle size of sago starch was increased with increasing time of precipitation. It was happened due to the sago particles will be aggregated first before being precipitated by ethanol. However, when sago starch is precipitated for 1 h (E1), the particle size is the highest one due to the amount of ethanol added dropwise has not been able to bind and precipitate the sago particles; therefore, many sago particles are aggregated. The particle size of modified sago is 16.256–55.498 μm. E1 has the largest particle size than others, and E2 has the smallest particle size. Although both use ethanol as non-solvent precipitation, the particle size of modified sago is bigger than waxy maize, which is 0.2–0.3 μm [15] and wheat starch, which is 50–70 nm [2]. Hence, it can be known that precipitation method without any pretreatment such as dissolution by DMSO and adding the surfactant to the starch solution [15] or reacted starch with sodium hydroxide and urea [2], cannot produce the nano-sized of modified starch, including sago starch.

Morphology and shape of modified sago starch can be determined by FESEM analysis. Figure 35.3 shown that shape of native sago starch is spherical and its surface is smooth. But after modification process, some granule was changed into an irregular shape and the surface became rough. It was indicated that the heating process and precipitation using ethanol could be destroyed the starch granule, therefore, affected the particle size and shape of sago starch [13]. Figure 35.3 also show that the particle size of E2 is smaller and the damage of that granule less than other samples. It caused

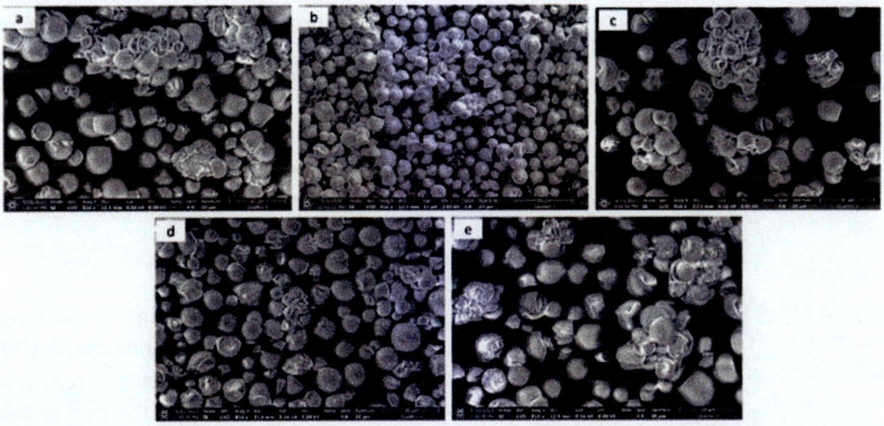

Fig. 35.3 Morphological analysis of modified sago using FESEM **a** E1, **b** E2, **c** E3, **d** E4 and **e** E5 on the 816× of magnification

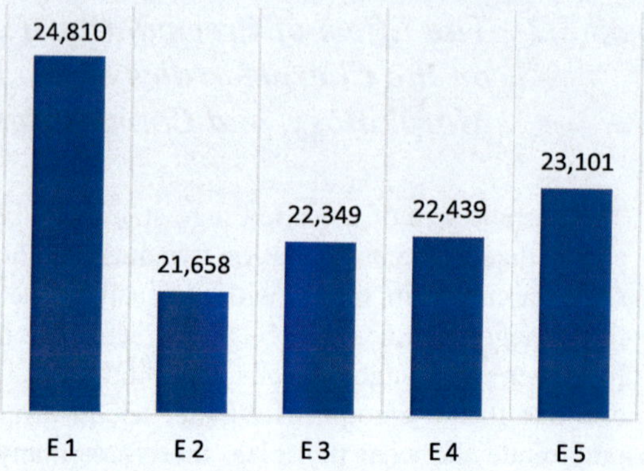

Fig. 35.4 The amylose content (%) of modified sago

that E2 has the best thermal stability than others (Table 35.2) therefore, it did not fully gelatinized.

Starch is a polysaccharides containing of two major polymers, which is amylose and amylopectin. Amylose is a polymer of low molecular weight, consist of linear chains bound by α-1,4 glycosides links, amorphous, and more soluble in hot water. At the same time amylopectin has a higher molecular weight, consist of linear chains bound by α-1,4 glycosides links and highly branches chains formed by α-1,6 glycosides links, crystalline, and less soluble in hot water [22]. High amylose content in the starch will be affected its physicochemical.

In this study, the amylose content of modified sago is a ranges from 21.66–24.81% (Fig. 35.4). E2 has the lowest amylose content, and E1 has the highest amylose content than other modified sago. It is probably happens because of the fastest precipitation time (1 h), able to precipitate the amylose molecule which leaches during the gelatinization process. While the longest precipitation time (5 h) also able to precipitate the starch (E5) due to a long time allows the amylose in granule starch to have retrograded completely, therefore, the precipitated is high amylose of sago starch. Amylose content of this modified sago is smaller than sago from [10], containing 28.93% amylose and 71.07% amylopectin.

35.3.3 The Effect of Precipitation Time on the Crystallinity of Modified Sago Starch

Crystallinity properties of modified sago was determined by Differential scanning calorimeter (DSC) and X-ray Diffraction (XRD) analysis. In this test, the initiation of gelatinization is known as T-onset and the position where the endothermic reaction occurs at the maximum is known as T-peak. The higher of T-onset, the higher the temperature required to make a gelatinized sample. E2 has the highest T-onset than other modified sago (Table 35.4) because of it has the smallest particle size. Smaller granules will gelatinize at high temperature than larger granules [21].

Table 35.4 Thermal properties of native and modified sago

Samples	Tonset (°C)	Tpeak (°C)	Enthalpy (J/g)
E1	79.22	95.87	148.723
E2	94.75	103.15	−34.7
E3	82.77	90.72	67.61
E4	81.58	88.83	34.441
E5	80.58	88.59	28.396

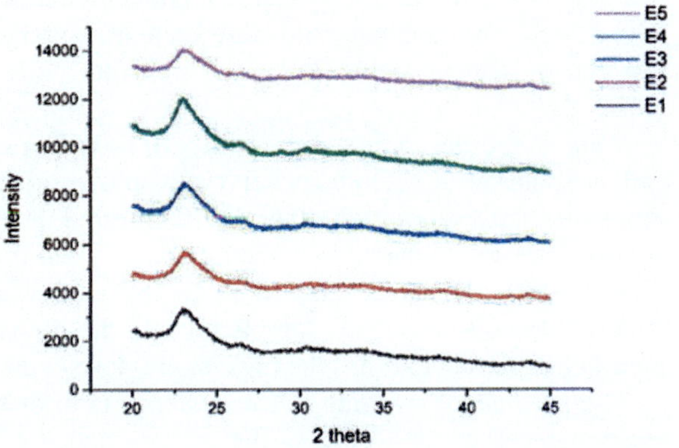

Fig. 35.5 Diffractogram of XRD

The crystallinity of modified sago can be seen on the diffractogram (Fig. 35.5), where E2 has the highest and E1 has the lowest crystallinity. This is because E2 has the lowest and E1 has the highest amylose content. Besides that, according to morphology analysis, it is known that E2 has less damage of the granules, which indicated E2 did not gelatinize entirely yet because it has the best thermal stability than other modified sago. According to Fig. 35.5, the modified sago has 2θ values of 23° on the XRD diffraction peaks. It is indicated that this modified sago starch has A-type structures. There are three types of crystalline structure in starch, which are A-type, B-type, and C-type [23].

35.3.4 The Effect of Precipitation Time on the Hydrophobicity of Modified Sago Starch

The hydrophobicity of modified sago was determined using FTIR analysis, moisture content, water solubility, and swelling power. According to FTIR spectra (Fig. 35.1), it can be known that each modified sago has a different intensity of hydroxyl group (–OH) uptake. That intensity shows the amount of hydroxyl group in the sample. The more hydroxyl group, the polarity and ability to form the hydrogen bond were greater; therefore, water solubility is also increased (more hydrophilic). E2 has the lowest

Table 35.5 Characteristic of modified sago starch

Analysis	Modified sago				
	E1	E2	E3	E4	E5
Moisture content (%)	8.637	6.632	7.294	8.240	8.405
Water solubility (%)	74.437	46.909	49.954	50.136	52.495
Swelling power (%)	43.382	27.678	33.322	34.629	42.711

intensity of hydroxyl group uptake. It concluded that E2 was the most hydrophobic than others. The results of moisture content, water solubility, and swelling power test can be seen in Table 35.5. The moisture content of modified sago is about 6.632–8.637%.

Water solubility indicates the ability of modified sago to dissolve in water. Basically, all starch is soluble in hot water and insoluble in cold water. In a starch molecules, amylose is more soluble than amylopectin due to molecular mass of amylose being less than amylopectin, and crystallinity of amylopectin being higher than amylose [22]. According to Table 35.5, the water solubility of modified sago is about 46.909–74.437%. Sample E1 has the highest water solubility than other modified sago due to it has the highest amylose content.

The purpose of swelling power analysis is to determine the power of starch to expand. Swelling is a process that occurs because the presence of excess water and heat. When a starch is dissolved into water and heating, the intermolecular hydrogen bond will be broken and replaced by new bonds between starch molecules and water [19]. Water molecules is easier bonding with hydroxyl group of amylose than amylopectin, due to amylose structure is amorphous while amylopectin is crystalline. The bond between hydroxyl group of amylose and water caused the starch granule swelled, and the more water molecule entered the granule, the particle size of the granule became greater. Hence, the more amylose content of sago starch, the greater swelling power. Factors affecting the swelling power are temperature, long-chain, distribution of molecular weight, amylose and amylopectin content [24]. The swelling power of modified sago is about 27.678–43.382%. Modified sago E2 has the lowest swelling power, and E1 has the highest one.

35.4 Conclusion

Precipitation time on the modification of sago starch using ethanol is affect differently by each physicochemical properties of modified sago. The shortest (1 h) and the longest (5 h) precipitation times were caused a decrease in the thermal stability, crystallinity, and hydrophobicity of modified sago. Sample E1 had the highest value on moisture content (8.63%), water solubility (74.43%), swelling power (43.38%), amylose content (24.81%), and particle size (55.49%) but it had lowest thermal behavior (79.22 °C). In contrast, precipitation for 2 h can be obtained modified sago

(E2) which had the lowest on moisture content (6.63%), water solubility (46.9%), swelling power (27.67%), amylose content (21.65%), and particle size (16.25%), but it also had the highest thermal stability, crystallinity, and hydrophobicity.

Acknowledgements The authors would like to thank for the Ministry of Research, Technology and Higher Education for Program Insentif Riset Sistem Inovasi Nasional (INSINAS) 2018 and for the facilities, scientific and technical support from Advanced Characterisation Laboratories Cibinong Integrated Laboratory of Bioproduct (iLab), Research Center for Bio-material, National Research and Innovation Agency (BRIN).

References

1. Okazaki, M.: The structure and characteristics of sago starch. In: Sago Palm. 247–259 (2018)
2. Momenpoor, B., Danafar, F., Bakhtiari, F.: Size controlled preparation of starch nanoparticles from wheat through precipitation at low temperature **56**, 131–141 (2019). https://doi.org/10.4028/www.scientific.net/JNanoR.56.131
3. Xie, F., Pollet, E., Halley, P.J., Avérous, L.: Starch-based nano-biocomposites. Prog. Polym. Sci. **39** (2013). https://doi.org/10.1016/j.progpolymsci.2013.05.002
4. Sondari, D.: Modification of sago starch for edible coating. IOP Conf. Ser. Mater. Sci. Eng. (2019). https://doi.org/10.1088/1757-899X/543/1/012013
5. Chin, S.F., Azman, A., Pang, S.C.: Size controlled synthesis of starch nanoparticles by a microemulsion method. J. Nanomater. (2014)
6. Abdorreza, M.N., Abd Karim, A.: Mechanical, barrier, physicochemical, and heat seal properties of starch films filled with nanoparticles. J Nano Res **25**, 90–100 (2012). https://doi.org/10.4028/www.scientific.net/JNanoR.25.90
7. Sumardiono, S., Rakhmawati, R.B., Pudjiastuti, I.: Physicochemical and rheological properties of sago (MetroxylonSagu) starch modified with lactic acid hydrolysis and UV rotary drying. AJChE **18**, 41–53 (2018)
8. Azmi, A., Malek, M.I., Puad, N.I.: A review on acid and enzymatic hydrolyses of sago starch. Int. Food Res. J. **24**, 265–273 (2017)
9. Suki, F.M.M., Azura, A.R., Azahari, B.: Effect of ball milled and ultrasonic sago starch dispersion on sago starch filled natural rubber latex (SSNRL) films. Procedia Chem. **19**, 782–787 (2016). https://doi.org/10.1016/j.proche.2016.03.085
10. Maryam, A.K., Novelina, E.: Production of sago (Metroxylon Sp.) starch nanoparticles using hydrolysis-high shear homogenization (HSH) method. IOP Conf. Ser. Earth Enviromental Sci. **258** (2019). https://doi.org/10.1088/1755-1315/258/1/012046
11. Chin, S.F., Pang, S.C., Tay, S.H.: Size controlled synthesis of starch nanoparticles by a simple nanoprecipitation method. Carbohydr. Polym. **86**, 1817–1819 (2011). https://doi.org/10.1016/j.carbpol.2011.07.012
12. Tan, Y., Xu, K., Li, L., Liu, C., Song, C., Wang, P.: Fabrication of size-controlled starch-based nanospheres by nanoprecipitation. Appl. Mater. Interfaces **1**, 1–4 (2009). https://doi.org/10.1021/am900054f
13. Patindol, J., Shih, F., Ingber, B., Champagne, E., Boue, S.: Porous rice powder from the precipitation of gelatinized flour or starch paste with ethanol Porous rice powder from the precipitation of gelatinized flour or starch paste with ethanol *. Starch **00**, 1–8 (2012). https://doi.org/10.1002/star.201200134
14. Winarti, C., Sunanti, T.C., Mangunwidjaja, D., Richana, N.: Preparation of arrowroot starch nanoparticles by butanol-complex precipitation, and its application as bioactive encapsulation matrix. Int. Food Res. J. **21**, 2207–2213 (2014)

15. Saari, H., Fuentes, C., Sjöö, M., Rayner, M., Wahlgren, M.: Production of starch nanoparticles by dissolution and non-solvent precipitation for use in food-grade Pickering emulsions. Carbohydr. Polym. **157**, 558–566 (2017). https://doi.org/10.1016/j.carbpol.2016.10.003
16. Kim, H., Park, S.S., Lim, S.: Preparation, characterization and utilization of starch nanoparticles. Colloids Surfaces B Biointerfaces. (2014). https://doi.org/10.1016/j.colsurfb.2014.11.011
17. Ningrum, R.S., Sondari, D., Amanda, P., Widyaningrum, B.A., Burhani, D., Akbar, F., Sampora, Y.: Properties of edible film from modified sago starch precipitated by butanol. J. Sains Mater. Indones. **21**, 164–169 (2020)
18. Pelissari, F.M., Andrade-Mahecha, M.M., Sobral, P.J., do A., Menegalli, F.C.: Food hydrocolloids comparative study on the properties of flour and starch films of plantain bananas (Musa paradisiaca). Food Hydrocoll. **30**, 681–690 (2013). https://doi.org/10.1016/j.foodhyd.2012.08.007
19. Vamadevan, V., Bertoft, E.: Food hydrocolloids observations on the impact of amylopectin and amylose structure on the swelling of starch granules. Food Hydrocoll. **103** (2020). https://doi.org/10.1016/j.foodhyd.2020.105663
20. Guo, P., Yu, J., Copeland, L., Wang, S., Wang, S.: Mechanisms of starch gelatinization during heating of wheat flour and its effect on in vitro starch digestibility. Food Hydrocoll. **82**, 370–378 (2018). https://doi.org/10.1016/j.foodhyd.2018.04.012
21. Tako, M., Tamaki, Y., Teruya, T., Takeda, Y.: The principles of starch gelatinization and retrogradation. Food Nutr. Sci. **05**, 280–291 (2014). https://doi.org/10.4236/fns.2014.53035
22. Cornejo-ramírez, Y.I., Martínez-cruz, O., Toro-Sanchez, C.L.D., Wong-corral, F.J., Borboa-flores, J., Cinco-moroyoqui, F.J.: The structural characteristics of starches and their functional properties. CyTA—J. Food. **16**, 1003–1017 (2018). https://doi.org/10.1080/19476337.2018.1518343
23. Dome, K., Podgorbunskikh, E., Bychkov, A., Lomovsky, O.: Changes in the crystallinity degree of starch having different types of crystal structure after mechanical pretreatment. Polymers (Basel) **12**, 1–12 (2020)
24. Lin, C.-L., Lin, J.-H., Lin, J.-J., Chang, Y.-H.: Properties of high-swelling native starch treated by heat-moisture treatment with different holding times and iterations. Molecules **25**, 1–12 (2020)

Chapter 36
Air Quality in the Bandung Basin of Indonesia as Measured by Passive Sampler

Wiwiek Setyawati, Dyah Aries Tanti, and Asri Indrawati

Abstract The Greater Bandung or Bandung Basin is a metropolitan area surrounding Bandung City, the Capital of The West Java Province. It is a basin-shaped region surrounded by mountains and has more than 8.5 million population. Therefore, air quality degradation will significantly impact the regions' sustainability. Air pollution monitoring is required to get information regarding air quality and its seasonal variability. We carried out a monthly sampling of ambient air using the Ferm-type passive gas sampler method adopted by CSIRO Australia at eleven monitoring points from 2014 to 2018. Air pollutant concentration values in urban site were determined by averaging concentration in Cikadut, Cipedes, Kopo, Kebon Kalapa, Martadinata Soreang and Padalarang, while in rural were in Ciparay, Lembang, Tanjung Wangi and Dago. The air quality parameters measured were NO_2, SO_2, and O_3. The five-year mean NO_2 and SO_2 concentrations in urban (71.87 ± 35.36 and 6.72 ± 1.84 $\mu g/m^3$) were relatively higher than in rural (35.06 ± 14.60 and 3.79 ± 1.27 $\mu g/m^3$) sites. On the contrary, the mean O_3 concentration was relatively more elevated in rural (21.86 ± 9.23 $\mu g/m^3$) than in urban (17.60 ± 6.55 $\mu g/m^3$) sites. The composition in ambient air from high to low was $NO_2 > O_3 > SO_2$. The seasonal mean NO_2 and O_3 concentrations in the Bandung Basin were maximum during the transitional dry to wet season (SON) and minimum during the wet season (DJF). We reported that the annual mean NO_2 concentrations in the urban site during the 2014–2018 period were above Indonesian National Air Quality Standards (NAAQS).

36.1 Introduction

Nitrogen dioxide (NO_2), sulfur dioxide (SO_2), and ozone (O_3) are some of the target species because of their adverse effects on human health and vegetation [1–4]. NO_2 and SO_2 are primary pollutants emitted from fossil fuel combustion, mostly from

W. Setyawati (✉) · D. Aries Tanti · A. Indrawati
Research Center for Climate and Atmosphere, National Research and Innovation Agency (BRIN), Bandung, Indonesia
e-mail: wiwieksetyawati21@gmail.com

E. Yulihastin et al. (eds.), *Proceedings of the International Conference on Radioscience, Equatorial Atmospheric Science and Environment and Humanosphere Science, 2021*, Springer Proceedings in Physics 275, https://doi.org/10.1007/978-981-19-0308-3_36

transportation and industries [5, 6]. These two species might lead to acid rain, which is very harmful to the environment [7]. On the contrary, O_3 is the secondary pollutant produced from its precursors, especially nitrogen oxides (NO_x) and volatile organic compounds (VOC), in the presence of sunlight [6, 8].

The Greater Bandung is a basin-shaped region surrounded by volcanic highlands, and the centre has the lowest altitude of about 665 m [9]. The inversion layer, developed in the valley because colder air is trapped below warmer air, can limit the vertical air pollution movement from the surface to the upper atmosphere [10]. This phenomenon can worsen air quality in the Bandung Basin, especially during the night.

The Greater Bandung or Bandung Basin is a metropolitan area surrounding Bandung City, the Capital of The West Java Province. It is an important economic centre in the Western part of Indonesia and the third Mega-City after Jakarta and Surabaya, with more than 8.5 million populations. Based on the population census in 2010, The population density was about 2342/km^2. Population growth of 4.7% resulted in a 43.5% increase in motorized vehicles registered in Bandung from 2014 to 2018 [11, 12]. Moreover, the existence of the Cipularang Toll Road, connecting the metropolitan Jakarta to Bandung, also accelerated the number of motor vehicles entering Bandung areas. Furthermore, many attractive places have increased the visit of foreign and domestic tourists by 0.13% [11, 12]. There was also a massive increase in the number of industries in the area by 144% [11, 12].

To evaluate the state of ambient air in the urban and rural areas of Bandung Basin, episodic simultaneous measurements of NO_2, SO_2 and O_3 were performed between January 2014 to December 2018 using passive samplers. This method is widely used worldwide to determine air quality, particularly within international frameworks, such as the acid deposition Monitoring Network in East Asia (EANET) [7] and the World Meteorological Organization (WMO) [13].

36.2 Methods

36.2.1 Sampling Location

The Bandung basin is located in the West Java Province of Indonesia. It is a big intra-montane basin with its lowest altitude of 665 m and is surrounded by up to 2400-m volcanic highlands [9]. Table 36.1. shows the annual mean of meteorological characteristics of the Bandung Basin from 2014 to 2018. It had high mean humidity (74.5–80%) and wind speed variabilities (perennial ground wind speed of 1.54–4.2 m/s). The temperature was about 23.4–23.8 °C, and the sunshine duration was approximately 52–67%. The precipitation event was 17.4–25 days. The rainfall was 182.4–295.8 mm.

The Acid Deposition Monitoring Network in East Asia (EANET) classifies acid deposition monitoring sites into urban, rural, and remote areas [14]. We collected

Table 36.1 The annual mean of meteorological data in Bandung Basin [11, 12, 14–16]

Parameter	Year				
	2014	2015	2016	2017	2018
T (°C)	23.4	23.56	23.8	23.52	23.6
RH (%)	77.17	74.50	80	77.38	75.09
Sunshine duration (%)	60.3	67	52	56.25	62
Precipitation event (day)	18.8	17.4	25	19.75	17.5
Rainfall (mm)	198.8	184.74	295.8	191.84	182.4
Wind speed (m/s)	1.54	2.15	2.06	2.4	4.2

samples from eleven monitoring sites in the Bandung basin region. They are Cikadut (CIK), Cipedes (CIP), Kopo (KOP), Kebon Kalapa (KEB), Martadinata (MAR), Soreang (SOR), Padalarang (PAD), Ciparay (CIP), Dago (DAG), Lembang (LEM), and Tanjung Wangi (TJW), as shown in Fig. 36.1. The seven former sites were classified as urban, while the four were rural areas, as indicated in Table 36.2.

The Bandung Basin region has monsoonal type seasons and is influenced by The SouthEast monsoon/Australian monsoon, as shown in Fig. 36.2 [15]. The wet season occurs from December to February (DJF), while the dry from June to July (JJA). The wet to dry transitional season occurs from March to May (MAM), and the dry to wet season is from September to November (SON). The seasonal mean is analyzed according to these four seasons: DJF, MAM, JJA, and SON.

Fig. 36.1 Eleven monitoring sites in the Bandung Basin regions

Table 36.2 Description of each monitoring site used in the study

No	Location	Coordinates		Altitude (meter)	Description
		Latitude	Longitude		
1	Cikadut (CKT)	−6.890000 S	107.670000 E	746	Urban
2	Cipedes (CPS)	−6.895171 S	107.588370 E	746	Urban
3	Kopo (KOP)	−6.976744 S	107.568077 E	672	Urban
4	Kebon Kalapa (KBK)	−6.926263 S	107.607889 E	703	Urban
5	Martadinata (MAR)	−6.915637 S	107.629971 E	698	Urban
6	Soreang (SOR)	−7.010000 S	107.530000 E	720	Urban
7	Padalarang (PAD)	−6.831000 S	107.469000 E	697	Urban
8	Ciparay (CPY)	−7.037500 S	107.710000 E	690	Rural
9	Lembang (LEM)	−6.828000 S	107.630000 E	1239	Rural
10	Tanjung Wangi (TJW)	−6.948000 S	107.497000 E	803	Rural
11	Dago (DAG)	−6.868733 S	107.623849 E	842	Rural

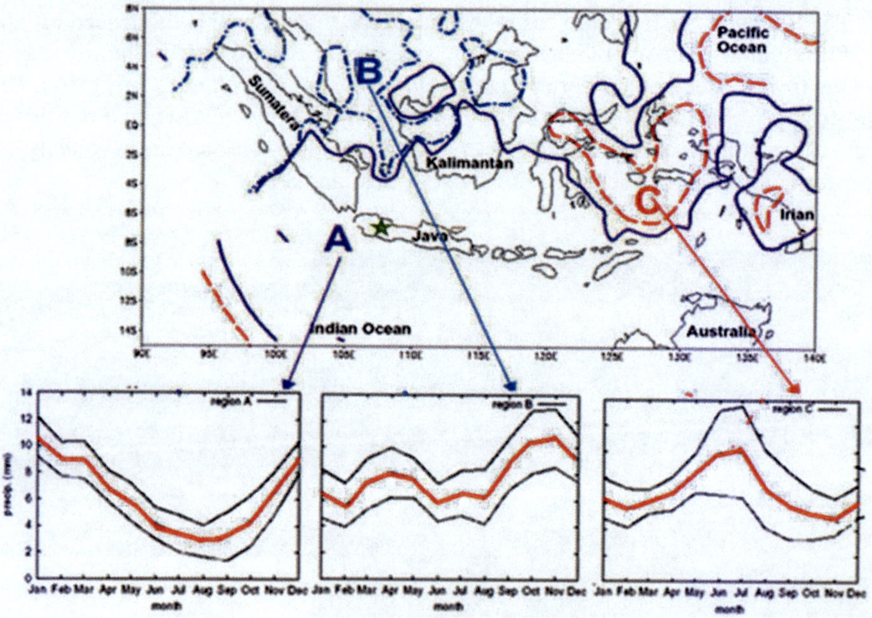

Fig. 36.2 Climatology classification in Indonesia. Bandung basin region is indicated by the symbol green star [15]

36.2.2 Sampling and Analysis Methods

The ambient air sample is collected using a Ferm-type passive gas sampler adopted by CSIRO, Australia [16], as shown in Fig. 36.3. The passive sampler method was chosen because of its ease of use, analytical sensitivity, target compounds, and sample duration [17]. The passive gas sampler consisted of four Teflon filters (Advantech hydrophobic PTFE membrane filter, 1.0 μm pore size, and 25 mm diameter). These Teflon filters used to absorb gas NO_2, SO_2, and O_3 were impregnated with absorbant reagents before deployment, as indicated in Table 36.3. A field blank sample for every set of filters prepared was provided to determine any contaminants present before installation. The passive gas samplers holders were installed about 1.5 m above ground in an area free of any obstacles and were exposed to ambient air for one month.

The sampling duration was one month, and the sampling period was from January 2014 to December 2018. So there are 60 (sixty) data for each parameter from every monitoring site. The monthly mean concentration in the urban site was calculated as the mean concentration in Cikadut, Cipedes, Kopo, Kebon Kalapa, Martadinata Soreang, and Padalarang, while in rural was in Ciparay, Lembang, Tanjung Wangi, and Dago.

The reaction to trap $SO_{2(g)}$, $NO_{2(g)}$ and $O_{3(g)}$ on the filters can be expressed in Eqs. (36.1)–(36.3) as follows:

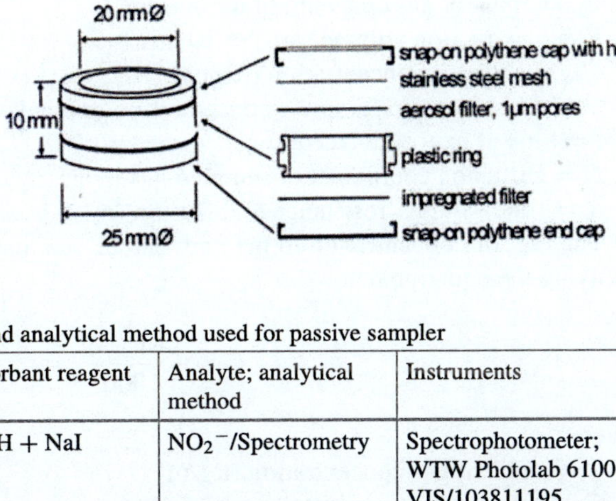

Fig. 36.3 Schema of Ferm-type passive gas sampler [18]

Table 36.3 Reagent, analyte, and analytical method used for passive sampler

No	Gases measured	Absorbant reagent	Analyte; analytical method	Instruments
1	NO_2	NaOH + NaI	NO_2^-/Spectrometry	Spectrophotometer; WTW Photolab 6100 VIS/103811195
2	SO_2	NaOH	SO_4^{2-}/Ion Chromatography	Ion Chromatography; Dionex ICS-1500/04070620
3	O_3	$K_2CO_3 + NaNO_2$	NO_3^-/Ion Chromatography	Ion Chromatography; DIONEX ICS-1500/04070620

Table 36.4 The molecular weight, diffusion coefficient, and total air resistance used in the calculation of ambient concentration [18]

Gases measured	Molecular weight (g/mol)	Diffusion coefficient in air (m^2/s)	Total sampler resistance to diffusion (m^{-1})
NO_2	46.01	1.54×10^{-5}	38.64
SO_2	64.06	1.32×10^{-5} at 20 °C	38.64
O_3	48.00	1.64×10^{-5} at 25 °C	36.98

$$SO_2 + 2OH^- \rightarrow 2H^+ + SO_4^{2-} \tag{36.1}$$

$$2NO_2 + 3I^- \rightarrow 2NO_2^- + I_3^- \tag{36.2}$$

$$O_3 + NaNO_2 \rightarrow NaNO_3 + O_2 \tag{36.3}$$

The ambient concentration of NO_2, SO_2, and O_3 in nmol/m³ can be calculated by the following equation:

$$C_0 = \frac{EVxC_{aq}xR}{txD} \tag{36.4}$$

where:

C_o = ambient gas concentration (nmol/m³).
EV = extraction volume (mL) = 10 mL.
C_{aq} = aqueous concentration (µmol/L).
= concentration in sample—concentration in blank.
t = time of exposure (second).
D = Diffusion coefficient of sampled gas in air (m²/s) as shown in Table 36.4.
R = Total sampler resistance to diffusion (m^{-1}) as shown in Table 36.4.

The unit of concentration in the ambient air can also be converted to µg/m³ by applying the equation below:

$$C = \frac{C_0xMW}{1000} \tag{36.5}$$

where:

C = ambient gas concentration (µg/m³).
C_o = ambient gas concentration (nmol/m³).
MW = molecular weight (g/mol) as shown in Table 36.4.

36.3 Result and Discussion

36.3.1 Monthly Linear Trend

The monthly NO_2, SO_2, and O_3 concentrations in urban and rural sites exhibit similar patterns, as shown in Fig. 36.4. The monthly NO_2 and SO_2 concentrations in urban areas were more outstanding than in rural sites, while O_3 was vice versa. The five-year

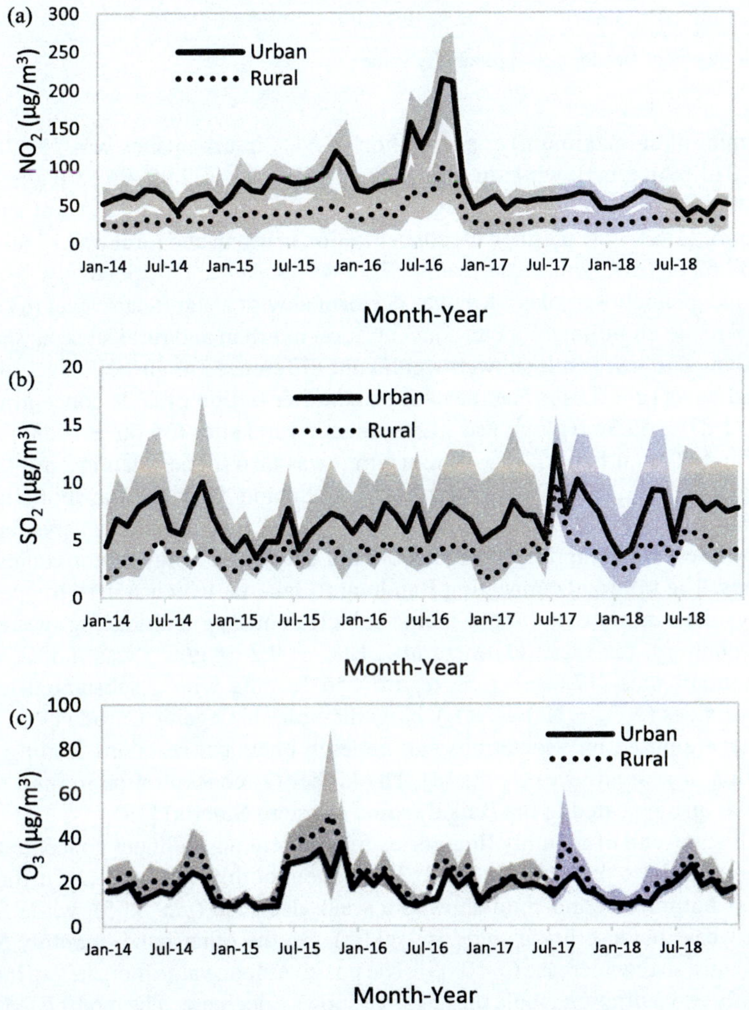

Fig. 36.4 Time series for monthly mean **a** NO_2, **b** SO_2, and **c** O_3 in urban and rural sites from January 2014 to December 2018. The black line represents urban, while the black dotted line rural area. The shaded grey area is the standard deviation of the monthly mean

Table 36.5 The independent t-test result compares air pollutant concentration in urban and rural sites of Bandung Basin

Gases measured	Regional type	Mean ± standard deviation (μg/m^3)	df	*t*	*p*
NO_2	Urban	71.87 ± 35.36	78.553	7.453	0.000
	Rural	35.06 ± 14.60			
SO_2	Urban	6.72 ± 1.82	105.324	10.229	0.000
	Rural	3.79 ± 1.27			
O_3	Urban	17.60 ± 21.86	106.403	−2.916	0.000
	Rural	21.86 ± 9.23			

Note df—degree of freedom, *p*—probability value

mean (minimum–maximum) concentration of NO_2 in urban sites was 71.87 μg/m^3 (35.64–213.69 μg/m^3), while in rural was 35.06 μg/m^3 (22.99–96.88 μg/m^3). SO_2 concentration in urban was 6.72 μg/m^3 (3.37–12.97 μg/m^3) and in rural sites was 3.79 μg/m^3 (1.63–9.81 μg/m^3). O_3 concentration in urban and rural was 17.60 μg/m^3 (8.80–37.62 μg/m^3), and 21.86 μg/m^3 (7.92–49.74 μg/m^3), respectively.

The independent samples t-test was performed with a significant level (α) = 0.05 to compare the air pollutant mean concentration in urban and rural sites, as shown in Table 36.5. The test result showed significant differences in air pollutants in urban and rural areas ($p < 0.01$). The mean ± standard deviation of NO_2 concentration in urban (71.87 ± 35.36 μg/m^3) was higher than in rural sites (35.06 ± 14.60 μg/m^3). The mean SO_2 in urban (6.72 ± 1.82 μg/m^3) was also higher than in rural (3.79 ± 1.27 μg/m^3). NO_2 and SO_2 are the primary pollutants from the transportation and industrial sectors [19]. Bandung metropolitan areas (i.e., Bandung City, Bandung Regency, West Bandung Regency, and Cimahi City) have many motor vehicles and industries. The toll road connecting Bandung to Jakarta, which started to operate in 2005, resulted in more and more motor vehicles entering and leaving the regions. On the contrary, the mean O_3 in rural (21.86 ± 9.23 μg/m^3) was more elevated than in urban sites (17.60 ± 6.55 μg/m^3). In an area with a substantial level of nitrogen oxides ($NO_x = NO + NO_2$), biogenic Volatile Organic Carbon (VOC) such as isoprene emitted by vegetations can undergo chemical reactions leading to the production of ground-level O_3 [8, 18]. The higher O_3 concentration in rural than in urban was also reported in the Baikal region (Eastern Siberia) [19].

The linear trend of monthly time series for ambient air pollutant concentration in urban and rural are presented in Table 36.6. The monthly NO_2 and O_3 in Bandung Basin for both urban and rural showed a weak decrease ($p > 0.05$), while SO_2 in urban a weak increase linear trend ($p > 0.05$). On the other hand, monthly SO_2 in rural shown a strong increase ($p < 0.05$). The positive slope value indicates an increase in monthly concentration, while the negative shows a decrease. The monthly NO_2 rate of decline in urban sites (−0.291 μg m^{-3} month^{-1}) was about 3.5 times that in rural (−0.084 μg m^{-3} month^{-1}), but for the monthly O_3 in rural (−0.065 μg m^{-3} month^{-1}) was about 15 times faster than in urban areas (−0.013 μg m^{-3} month^{-1}). The rate

Table 36.6 The linear trend of gaseous air pollutants in urban and rural of Bandung Basin between January 2014 to December 2018

Parameter	Type	Slope (μg/m^3 per month)	*p*	R2
NO_2	Urban	−0.291	0.291	0.073
	Rural	−0.084	0.445	0.010
SO_2	Urban	0.022	0.108	0.044
	Rural	0.019	0.041	0.070
O_3	Urban	−0.013	0.789	0.001
	Rural	−0.065	0.350	0.015

Note p—probability; *R2*—coefficient of determination

of growth for the monthly SO_2 in urban (0.022 μg m^{-3} month^{-}1) was about 1.16 faster than in rural sites (0.019 μg m^{-3} month^{-1}). Since 2017, Indonesia has adopted the EURO 4 standard fuel to reduce sulfur emissions from motor vehicles [20]. The factors that affect the ambient air pollutant concentration are the source strength, meteorological conditions (such as wind speed and wind direction, relative humidity, temperature, and sunlight), chemical reactions in the atmosphere, and the altitude [4–6, 21, 22] Biogenic volatile organic compounds (VOC), emitted by vegetation primarily found in a rural site, are essential for O_3 formation since the alkyl peroxyl radical (ROO^+) that form during the oxidation of VOCs convert NO to NO_2 from which O_3 is formed [6].

By focusing on these three gaseous pollutants, the percentage of five-year mean NO_2, SO_2, and O_3 concentrations in urban and rural were shown in Fig. 36.5. NO_2 dominated with 75% and 58%, while for SO_2, there were 7% and 6% in urban and rural sites, respectively. O_3 was 18% and 36% in urban and rural, respectively. Therefore, high to low concentrations in urban and rural areas are $NO_2 > O_3 > SO_2$.

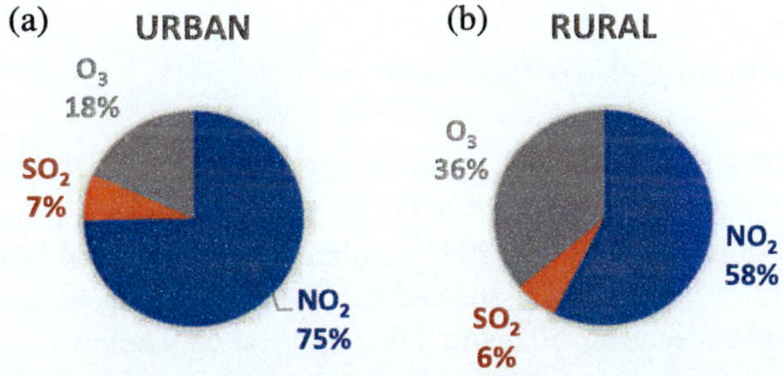

Fig. 36.5 Composition of air pollutants in **a** urban and **b** rural areas

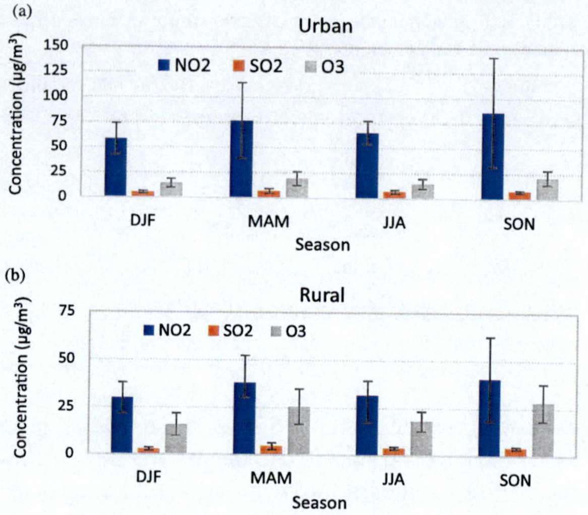

Fig. 36.6 Seasonal means in **a** urban and **b** rural areas in Bandung Basin. The top of the bar indicates the mean value, while the error bar is the standard deviation value

36.3.2 Seasonal Mean

Figure 36.6 shows the seasonal mean NO_2, SO_2, and O_3 in urban and rural sites. The maximum NO_2, SO_2, and O_3 concentrations in urban were found during the transitional dry to wet season (SON), while the minimum concentrations were during the wet season (DJF). The maximum O_3 in a coastal Mega-City of Jakarta was also recorded during September to November, and minimum during December to March [18]. NO_2 and O_3 maximum and minimum were found in the rural site during the transitional dry to wet season (SON) and wet season (DJF). Minimum SO_2 in the rural area was also found in the wet season (DJF), but the maximum was in the transitional wet to dry season (MAM).

36.3.3 Annual Mean

The Ministry of Environment and Forestry has published the National Ambient Air Quality Standard (NAAQS) for nitrogen dioxide (NO_2), sulfur dioxide (SO_2), and ozone (O_3) [1]. The 2021 regulation revises the NAAQS previously released in 1999 [23]. Figure 36.7a. showed that the annual NO_2 concentrations in urban areas were all exceeded (>50 μg/m^3), but for rural well below the NAAQS (<50 μg/m^3). On the contrary, annual mean SO_2 and O_3 concentrations in urban (<45 μg/m^3) and rural sites (<35 μg/m^3) were all below the NAAQS (Fig. 36.7b, c).

Fig. 36.7 The annual mean **a** NO_2, **b** SO_2, and **c** O_3 in Bandung Basin from 2014 to 2018. The top of the bar is Q3 (75% quartile), and the bottom is Q1 (25% quartile) values. The top error bar indicates the maximum, while the bottom minimum values. The bottom of the grey bar is the median. The orange dots represent the mean values. The red dotted lines are the value of the standard annual concentration in NAAQS

(a)

(b)

(c)

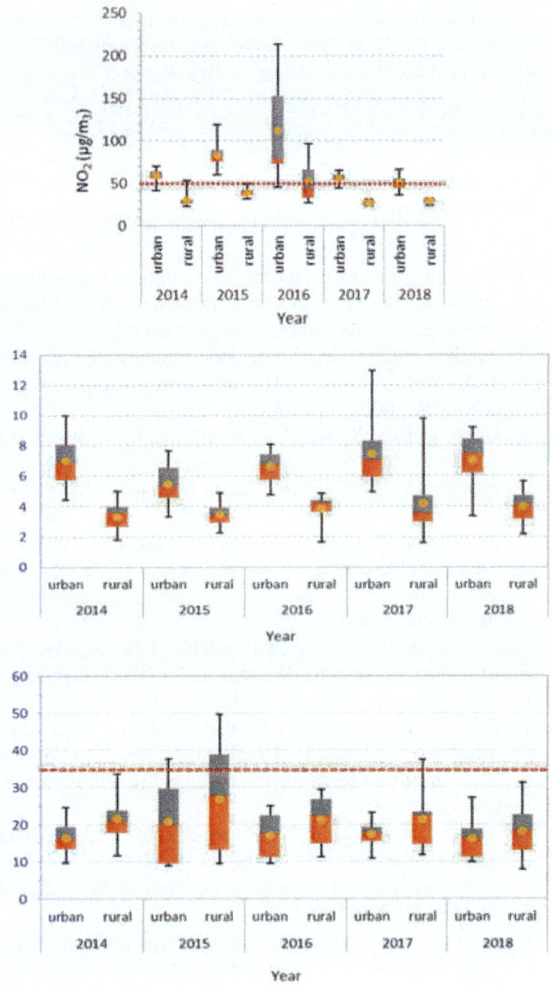

36.4 Conclusion

The five-year mean showed higher NO_2 and SO_2 concentrations in urban than rural sites. On the contrary, O_3 concentration was relatively more elevated in rural than in urban areas. The composition in ambient air from high to low was $NO_2 > O_3 > SO_2$. The seasonal mean NO_2 and O_3 concentrations in the Bandung Basin were maximum during transitional dry to wet season (SON) and minimum during wet season (DJF) periods. We reported that the annual mean NO_2 concentrations in urban from 2014 to 2018 were above NAAQS in Indonesia.

Acknowledgements We thank The Chemistry Laboratory in The Center for Atmospheric Research and Technology for providing the air quality data for the study.

Author Contributions Wiwiek Setyawati performed the data processing and analysis and wrote the manuscript; Dyah Aries Tanti conducted sampling in the field and testing in the laboratory; Asri Indrawati contributed to data quality assurances.

References

1. The Ministry of Environment and Forestry: The Government Regulation of The Republic of Indonesia No. 22/2021 about Implementation of Environmental Protection and Management. Indonesian Government, Jakarta (2021)
2. Markovic, D.M., Markovic, D.A., Jovanovic, A., Lazic, L., Mijic, Z.: Determination of O_3, NO_2, SO_2, CO, and PM10 measured in Belgrade Urban Area. Environ. Assess. Monit. **145**, 349–359 (2008)
3. Chen, J., Jian, X., Shi, C., Liu, R., Lu, R., Zhang, L.: Association between gaseous pollutants and emergency ambulance dispatches for asthma in Chengdu, China: a time-stratified case-crossover study. Environ. Health Prev. Med. **24**(20), 1–10 (2019)
4. Economopoulou, A.A., Economopoulos, A.P.: Air pollution in Athens basin and health risk assessment. Environ. Monit. Assess. **80**, 277–299 (2002)
5. Yamada, E., Kimura, M., Tomozawa, K., Fuse, Y., Yamada, T.: Behaviour and origin of atmospheric NO_2, SO_2, and O_3 in Mountains around the Kyoto Basin. Bull. Chem. Soc. Jpn. **73**, 1291–1297 (2000)
6. Ngoasheng, M., Beukes, P.J., van Zyl, P.G., Swartz, J.-S., Loate, V., Krisjan, P., Pambani, S.M., Kulmala, M., Vakkari, V., Laakso, L.: Assessing SO_2, NO_2 and O_3 in rural areas of the North West Province. Clean Air J. **31**(1), 1–14 (2021)
7. EANET homepage, https://www.eanet.asia, last accessed 11 February 2021.
8. Churkina, G., Kuik, F., Bonn, B., Lauer, A., Grote, R., Tomiak, K., Butler, T.M.: Effect of VOC emissions from vegetation on air quality in Berlin. Environ. Sci. Technol. **51**, 6120–6130 (2017)
9. Gumilar, I., Abidin, H.Z., Hutasoit, L.M., Hakim, D.M., Sidiq T.P., Andreas, H.: Land subsidence in Bandung Basin and its possible caused factors. In: 3rd International Symposium on Earthquake and Disaster Mitigation on Procedia Earth and Planetary Science, pp. 47–62. Elsevier, Bandung (2015)
10. Sumaryati, Indrawati, A., Tanti, D.A.: Temperature gradient condition on the photochemical smog dilution in Bandung Basin Area. Jurnal teknologi Lingkungan **21**(2), 219–226 (2020)
11. Statistics Indonesia: Jawa Barat Province in Figures, 1st edn, Statistics of Jawa Barat Province, Bandung (2015)
12. Statistics Indonesia: Jawa Barat Province in Figures, 1st edn, Statistics of Jawa Barat Province, Bandung (2019)
13. Charmichael, G.R.: Report on Passive Samplers for Atmospheric Chemistry No. 122: Measurement and Their Role in GAW, 1st edn, USA, WMO-GAW (1997)
14. Statistics Indonesia: Jawa Barat in Figures, 1st edn, Statistics of Jawa Barat Province, Bandung (2016)
15. Statistics Indonesia: Jawa Barat in Figures, 1st edition, Statistics of Jawa Barat Province, Bandung (2017)
16. Statistics Indonesia: Jawa Barat in Figures, 1st edn, Statistics of Jawa Barat Province, Bandung (2018)
17. EANET: Technical Manual for Wet Deposition Monitoring in East Asia, Japan, 1st edn. Network Centre EANET (2010)
18. Powell, J.:Work Instruction Manual for Accredited Laboratories, 2 edn. CSIRO (2007)
19. Aldrian, E., Susanto, R.D.: Identification of three dominant rainfall regions within Indonesia and their relationship to sea surface temperature. Int. J. Climatol. **23**, 1435–1452 (2003)

20. EPA: Passive Samplers for Investigation of Air Quality: Method Description, Implementation, and Comparison to Alternative Sampling Methods, EPA, USA (2014)
21. Ning, G., Wang, S., Ma, M., Ni, C., Shang, Z., Wang, J., Li, J.: Characteristics of air pollution in different zones of Sichuan Basin, China. Sci. Total Environ. **612**, 975–984 (2017)
22. Liu, Y., Zhou, Y., Lu, J.: Exploring the relationship between air pollution and meteorological conditions in China under environmental governance. Sci. Rep. **10**(14518), 1–11 (2020)
23. The Ministry of Environment and Forestry: Regulation No P.20/MENLHK/SETJEN/KUM.1/3/2017 about A Quality Standard for Exhaust Emission from New Type Motor Vehicles category M, N and O, 1st edn, KLHK (2017)
24. Khuriganova, O.I., Obolkin, V.A., Golobokova, L.P., Bukin, Y.S., Khodzher, T.V.: Passive sampling as a low-cost method for monitoring air pollutants in the Baikal Region (Eastern Siberia). Atmosphere **10**(470), 1–13 (2019)
25. Suhadi, D.R., Awang, M., Hassan, M.N., Abdullah R., Muda, A.H.: Review of photochemical smog pollution in Jakarta Metropolitan, Indonesia. Am. J. Environ. Sci. **1**(2), 110–118 (2005)
26. The Ministry of Environment and Forestry: The Government Regulation of The Republic of Indonesia No. 41/1999 about Implementation of Environmental Protection and Management, The Ministry for Environment and Forestry, Jakarta (1999)

Chapter 37
A Low-Cost and Practical Method for Determining Raindrop Size from Spray Nozzle by Using a Mobile Phone Camera

Jesi Pebralia and Iful Amri

Abstract This study proposed a new method to determine the size of the raindrops from the spray nozzle. The technique is a combination of conventional methods using flour and digital image processing. The instruments in this method consist of a mobile phone camera used to make a digital image, a ruler used to create a scale on the digital image, and graphic design software used to organize the digital image, as well as an image processing software ImageMIF to determine the size of the raindrops. This method was tested on a rain simulator to evaluate the raindrop's size distribution. It was done by taking an image of spherical grains on the flour hit by raindrops. Thus, this method is efficient because a mobile phone could save time. In addition, it will make the experiment cost to be relatively inexpensive. Also, measurement accuracy could be achieved because the raindrops whose diameter is smaller than 0.5 mm could be measured by implementing this method. Moreover, this method could be an alternating method to solve the disadvantage of the previous method. Thus, this technique may be much helpful for researchers to ascertain a quantitative relationship between the drop size distribution and other variables from the rainwater or the spray nozzle.

37.1 Introduction

The nozzle has any functions; those are in the artificial rain [1–4], in the horticulture [5, 6], the soil erosion [7–9] and the landslide [1, 10, 11]. The nozzle is chosen because of its character that can break down the fluid flow to the droplet or spray so that it resembles rainwater. Rainwater characteristics such as size, shape, velocity,

J. Pebralia
Department of Physics, Faculty of Science and Technology, Jambi University, Jalan Raya Jambi—Ma. Bulian, Muaro, Jambi 36361, Indonesia

I. Amri (✉)
Department of Electronics Engineering, Politeknik Jambi, Jalan Lingkar Barat II, Jambi 36129, Indonesia
e-mail: iful.amri@yahoo.com

E. Yulihastin et al. (eds.), *Proceedings of the International Conference on Radioscience, Equatorial Atmospheric Science and Environment and Humanosphere Science, 2021*, Springer Proceedings in Physics 275, https://doi.org/10.1007/978-981-19-0308-3_37

kinetic energy, and droplet size distribution are essential for scientific, commercial, and industrial applications [5]. Or even the measurement of the drop size distribution generated by the sprinkler nozzle can be used to express the relationship quantitatively between the drop size distribution and the operating pressure and the nozzle size [5]. In addition, the measurement of the drop size can also be used to evaluate the impact of water droplets on the soil.

By using a disdrometer, a direct and accurate measurement of the drop size can be performed. However, the equipment required for such a procedure is very sophisticated and expensive. One of the most accessible methods for measuring the discharge of droplet sizes is by using flour [12, 13]. This method uses flour to cover the raindrops from a nozzle and a digital caliper to measure the diameter of droplets in the flour. It is done by measuring the existing droplet on the flour surface. By applying a direct measurement of this method, it would take much time because the flour has to be dried in the oven first before it is measured. Also, the digital caliper is particularly susceptible to contact with the droplet, so it will break the system and make the data obtained to be inaccurate.

Meanwhile, the size of droplets can be measured by using digital image processing [5]. The result of measurement using this method is accurate, but it uses a reasonably expensive cost because it must use a high-speed camera, such as SLR. Therefore, in this study, we combine the two methods so that the data obtained from measurement is accurate and also low-cost for the experiment. The proposed method employs the flour for the droplet collector media of the rain form nozzle. Then, it will be used a phone cell camera to get a digital image. A phone cell camera is a practical solution in this method because in the digital era, almost everyone has it. Additionally, by using this method, the drying stage after the flour is showered can also be avoided. Thus, it will be saving temporary time.

In this research, there are two primary objectives that will be achieved. First, to propose a new method for determining the size of the raindrops from the spray nozzle by using a phone cell camera. Using this method, the measurement accuracy is very high, efficient, and the experimental cost is relatively inexpensive. Second, to show measurement data about raindrop size on rainfall simulator which is one of our work.

37.2 Materials and Methods

The experiment method used in this study was adapted from several previous studies [1, 2, 4, 7, 14], so obtained a novelty method that can determine raindrop size from spray nozzle by using phone cell camera. In which, by using this new method, the measurement accuracy can be reached, practical, and the experimental cost is relatively inexpensive. In this study, the procedure of the experiment is illustrated in Fig. 37.1. The first step was started by preparing a 0.5 m^2 platform to collect water droplets from a nozzle of the rain simulator. Then, the platform was filled with flour whose thickness is 3 mm. The purpose is when raindrops hit it, the sprinkles will not evaporate and become a spherical grain [12]. Then, the ruler is installed on one side

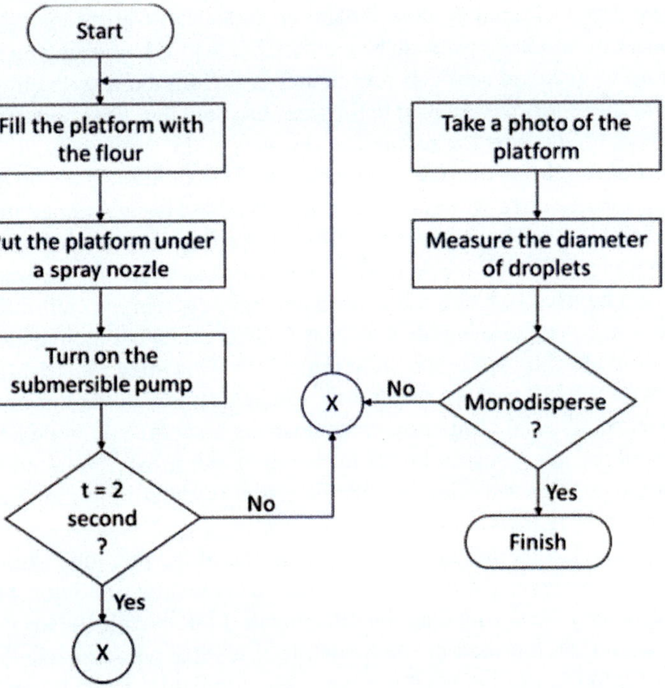

Fig. 37.1 The measurement process of raindrop size distribution

of the platform. It was used to make a reference line for scaling grains of rainwater on the photo when the calibration process was done in the image processing software. The installation process of raindrop size distribution is shown in Fig. 37.2a.

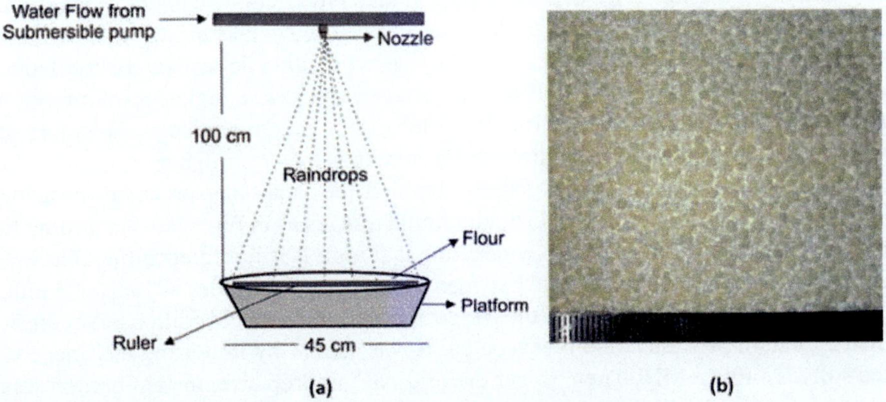

Fig. 37.2 **a** Experiment design, **b** the raindrop morphology on the flour

In the next step, after all the components and materials were prepared, the measurement process of raindrops size can be executed. It begins by turning on a submersible water pump to make an artificial rain that comes from the nozzle on the rain simulator. Then, the platform is placed in the rain for about 2 s. The short duration (1–2 s) aims to avoid duplicating the number of droplets at the same point [13, 15, 16], and the number of raindrops that fall to the surface of the flour depends on the type of the rain. Then, the platform which has been hit by rainwater is scanned using a mobile phone camera, so it is obtained a digital image of spherical grains of rainwater. While taking the image, the tick labels of the ruler which is in the image must be visible. Those would be used to make a line used as a reference in the calibration process.

The next step is measuring the diameter of droplets by using imageMIF software. The digital image obtained is shown in Fig. 37.2b. This image was captured by using Redmi 3 with 13 MP camera resolution. ImageMIF is software used to determine the diameter of material of fiber on the nanoscale. Generally, it could be found in the laboratory which has research topics in that area, but it may be found in a different name. Moreover, this software is very effective to determine the diameter of the spherical grain of rainwater.

In this research, this method was tested on one of the previous works, it is a rain simulator used to trigger ground movement on the landslide simulator. This simulator consists of two nozzles, which are labeled nozzle 1 and nozzle 2. Nozzle 1 was used to cover rain on the top area of the landslide simulator while nozzle 2 was used to cover rain on the bottom area of the landslide simulator. Thus, it could determine the performance for every nozzle by evaluating the raindrop size distribution of the spray nozzles on the rain simulator.

37.3 Results and Discussion

In this study, a digital image of a raindrop was taken by a mobile camera. However, it is not all digital cameras of the mobile phone can be used to take digital images of the raindrop because the quality of the digital image will highly depend on the resolution of each camera because it will affect the data accuracy. The higher resolution of a mobile phone camera, the better quality of the digital image produced. Therefore, it would be better when the resolution of the camera is 5 MP or higher.

Furthermore, after the digital image is obtained, the data is then processed by using a vector and designing software to attain digital images as in Fig. 37.3. According to that picture, there is a black line whose line is 5 mm. It is a reference line that was created by drawing a straight line between tick labels of the ruler as long as 5 mm. Thus, it was obtained the scale of the image appropriate to real conditions. After that, the calibration process could be executed. It was started by importing the image to the software ImageMIF. Then, to get the original raindrop size, the calibration was done by equating the reference line in the image of Fig. 37.3 to the reference line in the ImageMIF software.

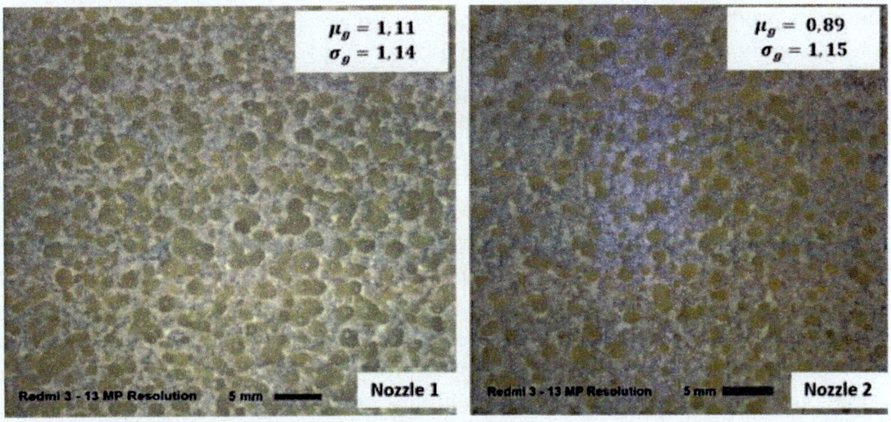

Fig. 37.3 The digital image of the raindrop on the flour

After equating the reference line has been finished, the raindrop size measurement process could be performed. To determine raindrop size, it could be done by measuring the diameter of every spherical grain hit by a raindrop. It is carried out by means of drawing transverse lines on each circle of spherical grain. This step should be repeated until the total desired data was achieved. After that, the obtained data could be saved to be processed and analyzed. In this study, the number of raindrops that could be measured was 250 spherical grains. The measurement data of the diameter of the raindrop size are shown in Fig. 37.4a, b.

Figure 37.4 shows that the smallest diameter of spherical grain hit by raindrops on spray nozzle 1 is 0.463 mm, and its largest diameter is 2.63 mm so it could be declared the average raindrop size on spray nozzle 1 is 1.17 mm. While on spray nozzle 2, the smallest diameter of spherical grain hit by raindrops is 0.408 mm, and its biggest diameter is 1.78 mm, as well as the average raindrop size on spray nozzle 2 is 0.93 mm. Thus, based on these data, it could be concluded that the diameter distribution of

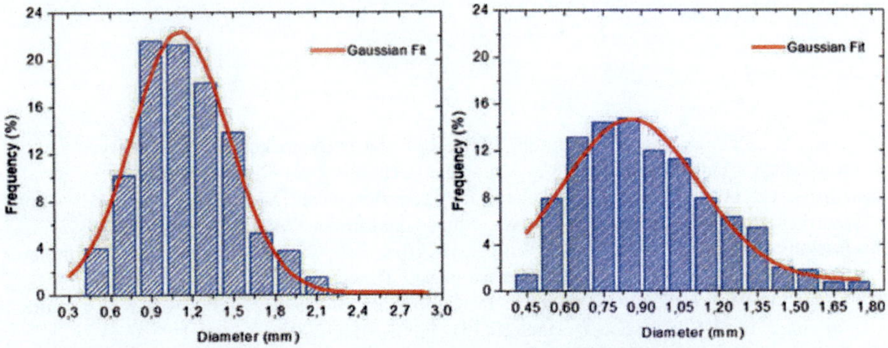

Fig. 37.4 The distribution of the raindrop size from the rain simulator

raindrops produced by the rain simulator referred to a Gaussian distribution pattern for both spray nozzle as shown in Fig. 37.4.

Moreover, to determine the uniformity of the raindrop from the spray nozzle, then the geometry means, and deviation standard should be known. Based on the data in Fig. 37.4, the geometry means for spray nozzle 1 is 1.11, with a deviation standard of geometry is 1.14, whereas, for spray nozzle 2, it is obtained the geometry mean 0.89, with a standard deviation of geometry 1.15. Based on the results, the standard geometry values for both spray nozzles were less than 1.2. Thus, it can be concluded that the raindrop size from the rain simulator indicates a monodisperse characteristic or uniform [17].

On the other hand, raindrops that have a diameter below 0.5 mm probably exist in the experiment. However, these raindrops are very difficult to detect and measure by using a digital caliper [12] because it is highly susceptible to contact with droplets that can make the droplet to be broken. Therefore, this method will be an alternating method to solve this disadvantage. By using this method, the droplet which has a diameter below 0.5 mm can be measured. According to the result measurement data, the smallest droplet diameter obtained is 0.408 mm produced by a spray nozzle 2.

37.4 Conclusions

A simple method implemented by using a mobile phone camera could be implemented to determine the raindrop size from a spray nozzle. This method consists of a mobile phone camera used to make a digital image, a ruler used to create a scale on the digital image, and graphic design software used to organize the digital image, as well as an image processing software to determine the raindrop size. This method is very practical so that will save time, and the accuracy of measurement is high, as well as the experiment cost is relatively inexpensive. Moreover, the method may be alternating to determine the relationship between the drop size distribution and other physical variables from the rainwater or the spray nozzle.

References

1. Lora, M., Camporese, M., Salandin, P.: Design and performance of a nozzle-type rainfall simulator for landslide triggering experiments. CATENA **140**, 77–89 (2016)
2. Aksoy, H., et al.: A rainfall simulator for laboratory-scale assessment of rainfall-runoff-sediment transport processes over a two-dimensional flume. CATENA **98**, 63–72 (2012)
3. Iserloh, T., Fister, W., Seeger, M., Willger, H., Ries, J.B.: A small portable rainfall simulator for reproducible experiments on soil erosion. Soil Tillage Res. **124**, 131–137 (2012)
4. Yu, B., Ciesiolka, C.A.A., Langford, P.: Calibration of an oscillating nozzle-type rainfall simulator. Earth Surf. Process. Land. **28**(13), 1483–1490 (2003)
5. Sudheer, K.P., Panda, R.K.: Digital image processing for determining drop sizes from irrigation spray nozzles. Agric. Water Manag. **45**(2), 159–167 (2000)

6. Salvador, R., Bautista-Capetillo, C., Burguete, J., Zapata, N., Serreta, A., Playán, E.: A photographic method for drop characterization in agricultural sprinklers. Irrig. Sci. **27**(4), 307–317 (2009)
7. Sukhanovski, Y.P., Ollesch, G., Khan, K.Y., Meissner, R.: A new index for rainfall erosivity on a physical basis. J. Plant Nutr. Soil Sci. Fur Pflanzenernahrung Und Bodenkd. **65**(1), 51–57 (2002)
8. Arnaez, J., Lasanta, T., Ruiz-Flaño, P., Ortigosa, L.: Factors affecting runoff and erosion under simulated rainfall in Mediterranean vineyards. Soil Tillage Res. **93**(2), 324–334 (2007)
9. Fornis, R.L., Vermeulen, H.R., Nieuwenhuis, J.D.: Kinetic energy-rainfall intensity relationship for Cental Cebu, Philippines for soil erosion studies. J. Hydrol. **300**, 20–32, 5AD (2005)
10. Hong, Y., Adler, R.F., Huffman, G.: An experimental global prediction system for rainfall-triggered landslides using satellite remote sensing and geospatial datasets. IEEE Trans. Geosci. Remote Sens. **45**(6), 1671–1680 (2007)
11. Iverson, R.M.: Landslide triggering by rainfall infiltration. Water Resour. Res. **36**(7), 1897–1910 (2000)
12. Mazon, J., Viñas, M.: A low-cost experiment for determining raindrop size distribution. Weather **68**(2), 44–49 (2013)
13. Pérez-Latorre, F.J., de Castro, L., Delgado, A.: A comparison of two variable intensity rainfall simulators for runoff studies. Soil Tillage Res. **107**(1), 11–16 (2010)
14. Iserloh, T., et al.: European small portable rainfall simulators: a comparison of rainfall characteristics. CATENA **110**, 100–112 (2013)
15. Júnior, S.F.S., Siqueira, E.Q.: Development and calibration of a rainfall simulator for urban hydrology research. In: 12th International Conference Urban Drainage, 11–16 Sept. 2011, no. September, p. 8 (2011)
16. Regmi, T.P., Thompson, A.L.: Rainfall simulator for laboratory studies. Appl. Eng. Agric. **16**(6), 641–647 (2000)
17. Fromen, C.A., et al.: Synthesis and characterization of monodisperse uniformly shaped respirable aerosols Catherine. AIChE J. **504**, 3–194 (2013)

Chapter 38
Biomonitoring of Polycyclic Aromatic Hydrocarbons in the Ambient Air Using Plants: A Review

Desy Sulistiyorini, Christophe Walgraeve, and Herman Van Langenhove

Abstract One of the groups of pollutants in the atmosphere is Polycyclic Aromatic Hydrocarbon (PAHs). PAHs compounds are currently extensively studied due to their harm to human and ecosystem health. Biomonitoring using plants was currently introduced as an alternative to monitoring pollutants through active air sampling to minimize the bias effect of the short-term active air sampling. Biomonitoring is also considered an effective technique to be applied in developing countries for the reason that it can avoid the high cost of instrumental monitoring. Previous studies from the last decade were analyzed to gain insight into current practices, progress, and challenges. In addition, content analysis was employed to systematically characterize and classify the existing biomonitoring application. Therefore, the emphasis in this review will be placed on the use of bioaccumulation and biomarker responses in the plants as monitoring tools for PAH concentrations for the ecosystem and its limitations.

38.1 Introduction

There is a growing attention toward air pollution recently due to its impacts on human health and damages to the environment. The air pollutants have significantly increased primarily due to the increase of anthropogenic activities [1]. The air pollution source can be from the natural origin or anthropogenic sources. The air pollution source from anthropogenic origin impose harmful effects which are associated with

D. Sulistiyorini (✉)
Faculty of Health Science, Indonesia Maju University, Jalan Harapan No. 50, 12610 Jakarta, Indonesia
e-mail: desy.sulistiyorini@yahoo.com

C. Walgraeve · H. Van Langenhove
Department of Green Chemistry and Technology, Faculty of Bioscience Engineering, Research Group Environmental Organic Chemistry and Technology (EnVOC), Ghent University, Coupure Links 653, 9000 Gent, Belgium

E. Yulihastin et al. (eds.), *Proceedings of the International Conference on Radioscience, Equatorial Atmospheric Science and Environment and Humanosphere Science, 2021*, Springer Proceedings in Physics 275, https://doi.org/10.1007/978-981-19-0308-3_38

emission of particulate matters which comprises heavy metals, black carbon, polycyclic aromatic hydrocarbons and other substances that present in the atmospheric environment [2] An earlier study indicated that the air pollution sources arose from the combustion process, mainly related to the traffic and vehicle-related emissions. Nevertheless, the other sources such as from industrial discharge and resuspension of road dust also can influence the air quality [3].

The effect of air pollution on the environment is apparent, especially in the plant. Initially, it can be observed the changes in the foliar element concentrations, in which the situation can be worsened in a long time by the injury of the plant tis- sues, even the death of the plant [4]. The impacts of air pollution on the human health were also massively studied. Both carcinogenic effects and noncarcinogenic effects are linked to the exposure of outdoor air pollution according to International Agency for Research on Cancer (IARC). Outdoor air pollution is categorized as group 1 which is carcinogenic to human. Whereas the noncarcinogenic effects are manifested in different ranges of magnitude of the effect, from subclinical physiological abnormalities to a fatality [5]. Previous studies suggested acute and chronic exposure to air pollution associated with the greater occurrence of cardiovascular disease and mortality [6–9] and associated with respiratory disease risk [10, 11].

38.2 Polycyclic Aromatic Hydrocarbon Compounds (PAHs) in the Air

38.2.1 Source of PAHs Emissions

One of a group of pollutants in the atmosphere is Polycyclic Aromatic Hydrocarbon (PAHs). Although PAHs exist in various environments, the atmosphere is the primary means of PAHs dispersal. It receives the PAHs environmental load which leads to the ubiquity of PAHs in the environment. PAHs compounds currently are extensively studied due to their harm to the human and ecosystem health [12, 13] that will be discussed in the later section. Once emitted, PAHs will go through different environmental fate. The dispersion of PAHs occurs in the atmospheric environment as well as the terrestrial environment [12].

PAHs are not produced chemically for industrial activities. However, PAHs are commercially used as intermediaries in the manufacture of pharmaceutical products, agricultural products, photographic products, thermosetting plastics, materials for lubrication, and other chemical industries. The general use of PAH in the manufacturing process is in the production of explosives, fibres, dyes, plastics, pesticides, pharmaceuticals, agrochemicals, pigments, resins, and thermoset plastics. Other industries also utilize PAHs such as in the asphalt used for the road construction, in the field of electronics, plastics, and liquid crystals [12, 14].

PAHs originate both from natural and anthropogenic activities, although the anthropogenic activities play a role as the predominant sources. PAHs can be released during natural processes such as the volcano eruption, open burning, and natural loss of coal and petroleum deposits. The other sources of PAHs are anthropogenic activities. Based on the source types of PAHs, the anthropogenic sources can be categorized to large point sources, dispersed sources, and other sources. The examples of large point sources are industrial processes and incinerators (petroleum refineries, coke and aluminium production, coal gasification and liquefying plants, coal tar pitch and asphalt production, etc.) The dispersed sources may come from motor vehicle exhausts, jet aircraft exhaust, cigarette smoking, wood-burning stoves, and agricultural sources (combustion of agricultural products). The other anthropogenic sources are from petroleum product spills, sewage sludge, and tar and creosote-containing waste materials [12].

Based on the formation process, PAHs can be categorized into pyrogenic PAHs, petrogenic PAHs, and biological PAHs. The pyrolysis process, which is the burning of organic material with insufficient or in the absence of oxygen, will generate PAHs so-called *pyrogenic* PAHs which commonly exist at a higher concentration in the urban area. The other type of PAHs, called *petrogenic* PAHs, released along with the use, storage, and transport of the crude oil and its derivative products which commonly related with oil spillage and leakage and incomplete combustion the crude oil and its derivatives. The last category is *biological* PAHs which are released from biological activities in nature [12, 15].

The source of PAHs can be identified and examined based on different keys, such as the temperature of the formation and the number of rings. The number of rings also can be used to distinguish between petrogenic and pyrogenic PAHs. The five-member rings PAHs are more commonly exist as petrogenic PAHs than as pyrogenic PAHs due to the extensive time of petroleum hydrocarbon formation which is a favourable condition in the marshalling of the rings. On the other hand, the PAHs from pyrolytic sources are rapidly converted into more stable six-member rings PAHs [12]. Whereas PAHs with three aromatic rings found to be a major fraction of the atmospheric pollutant load, for instance, phenanthrene and alkylated derivatives exist in abundant quantities due to the combustion and evaporation of fossil fuel [16].

Motorized vehicles emissions have become increasingly recognised as the contributor of the air pollution and shown to be associated with health problem and mortality based on the proximity to the roads. It was suggested in the previous studies that the primary source of PAHs exposure was from vehicle exhaust based on the calculated PAHs diagnostic ratios [17]. The previous study also suggested that PAHs are generated from vehicular emission gasoline and diesel driven vehicles [18].

PAHs are found in greater concentration in the urban areas than in rural areas, because of the proximity to the source of the PAHs in urban areas. The concentration of PAHs in an urban area was found to be higher than in remote sites due to the increase of vehicular emission and small dispersion of the atmospheric pollutants, and it was well described that the ambient air nearby congested road is highly polluted with PAHs. It was reported in the previous study that the background concentration levels of some representative PAHs compound in the air was 0.02–1.2 ng/m^3 in rural areas and 0.15–19.3 ng/m^3 in the urban areas [12, 17, 19–21]. Recent findings revealed that there was increasing concentration gradient of PAHs along the rural–urban gradient

[22] and urban areas yielded the highest average PAHs concentration, followed by rural, and finally by remote sites [23].

38.2.2 Physicochemical Characteristics, Transport, and Fate of PAHs

PAHs are compounds which are ubiquitous and environmentally persistent and therefore, they are considered as persistent organic pollutants (POPs). PAHs compounds consist of hydrogen and carbon atoms and comprised of two or more benzene rings that are bound together in linear, cluster, or angular (as it can be seen in Fig. 38.1 arrangements with a pair of carbon atom shared between rings in their molecules [12, 13, 21, 24]. Generally, PAHs compounds have characteristics such as high melting and boiling points, low vapour pressure, and very low aqueous solubility [21]. The fate of PAHs in different compartments in the environment is determined by various characteristics of PAHs that is described in the Table 38.1 that will be explained further in this section.

The first characteristic of PAHs is having variability of the molecular weight. PAHs compounds which are categorized as a low molecular weight PAHs (with two or three rings) normally occur in the atmospheric environment in the vapour phase.

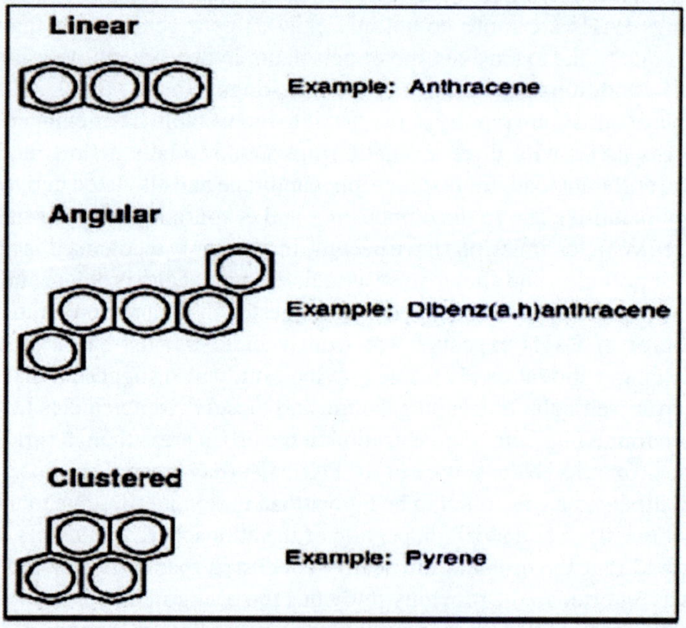

Fig. 38.1 The molecular arrange ments of the PAHs compounds [12]

Table 38.1 PAHs and its chemical and physical properties

	Molecular formula	Molar mass (g)[a]	Boiling Point (°C)[f]	Melting Point (°C)[f]	Henry's Law constant (Pa m^3/mol) at 25 °C[a]	Log KOW at 25 °Ca	Log KOA at 25 °Ca, g	Water Solubility (mg/L) at 25 °C[a]	IARC group[d,e]	Structure	No of Rings[b,c]
Naphthalene	$C_{10}H_{18}$	128.17	218	80.2	42.6	3.37	5.1	31	2B	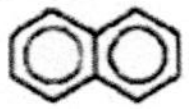	2
Acenaphthylene	$C_{12}H8$	152.19	280	93	12.7	4	6.36	16	n.e	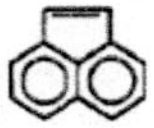	3
Acenaphthene	$C_{12}H_{10}$	154.21	279	95	18.5	3.92	6.3	3.8	3		3
Fluorene	$C_{13}H_{10}$	166.22	295	116	9.8	4.18	6.7	1.9	3		3
Phenanthrene	$C_{14}H_{10}$	178.23	340	100.5	4.3	4.46	7.5	1.1	3	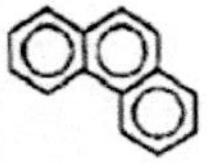	3
Anthracene	$C_{14}H_{10}$	178.23	342	216.4	4.45	4.49	7.3	0.04	3	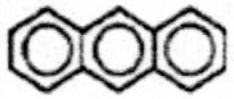	3

(continued)

Table 38.1 (continued)

	Molecular formula	Molar mass (g)[a]	Boiling Point (°C)[f]	Melting Point (°C)[f]	Henry's Law constant (Pa m^3/mol) at 25 °C[a]	Log KOW at 25 °Ca	Log KOA at 25 °Ca, g	Water Solubility (mg/L) at 25 °C[a]	IARC group[d,e]	Structure	No of Rings[b,c]
Fluoranthene	$C_{16}H_{10}$	202.25	375	108.8	1.96	8.9	8.6	0.2	3	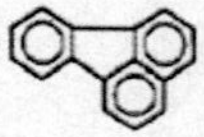	4
Pyrene	$C_{16}H_{10}$	202.25	150.4	393	1.7	8.8	8.6	0.13	3	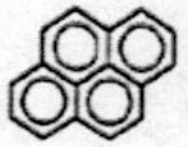	4
Benzo[a]anthracene	$C_{18}H_{12}$	228.29	400	160.7	1.22	5.8	9.5	0.011	2B	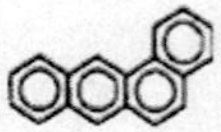	4
Chrysene	$C_{18}H_{12}$	228.29	448	253.8	0.53	5.73	10.4	0.0019	2B	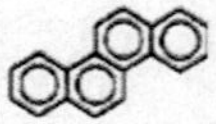	4
Benzo[b]fluoranthene	$C_{20}H_{12}$	252.31	481	168.3	0.05	5.78	11.3	0.0015	2B	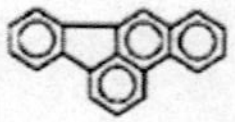	5
Benzo[k]fluoranthene	$C_{20}H_{12}$	252.31	480	215.7	0.04	6.5	11.2	0.0008	2B	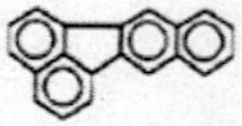	5
Benzo[a]pyrene	$C_{20}H_{12}$	252.31	496	178.1	0.07	6.35	10.8	0.0015	1	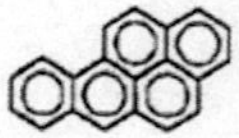	5

(continued)

Table 38.1 (continued)

	Molecular formula	Molar mass (g)[a]	Boiling Point (°C)[f]	Melting Point (°C)[f]	Henry's Law constant (Pa m^3/mol) at 25 °C[a]	Log KOW at 25 °Ca	Log KOA at 25 °Ca, g	Water Solubility (mg/L) at 25 °C[a]	IARC group[d,e]	Structure	No of Rings[b,c]
Dibenzo[a,h]anthracene	C_{24}H14	278.35	524	266.6	0.0075	6.5	11.8	0.0005	2A	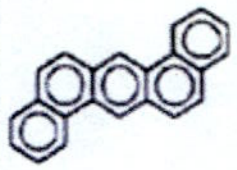	6
Benzo[g,h,i]perylene	$C_{22}H_{12}$	276.33	500	277	0.031	6.63	12.5	0.00014	3		6
Indeno[1,2,3-c-d] py-rene	$C_{22}H_{12}$	276.33	536	163.6	0.029	6.7	12.4	0.00019	2B	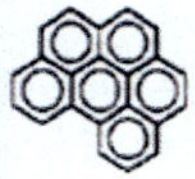	6

[a]From [25]; [b]From [12]; [c]From [27]; [d]From [28]; [e]IARC (International Agency for Research on Cancer) classification from [28] : group 1 = carcinogenic to humans, group 2A = probably carcinogenic to human, group 2B = possibly carcinogenic to humans, group 3 = not classifiable as to carcinogenicity to humans, n.e. = not evaluated; [f]From [29]; [g]From previous study suggested that Kp is closely related to KOA, which is the fourth parameter. The octanol/air partitioning coefficient (KOA) is dependent on the molecular weight of the compounds and the temperature. The lower the value of KOA, the PAHs will tend to exist in the vapour phase [26, 30]. Therefore, based on Tables 38.1 and 38.2, naphthalene, the PAHs compound with the lowest KOA will tend to be found in the vapour phase, and the tendency to be found in the vapour phase is lower for the PAHs compounds with higher KOA

Table 38.2 Categories of PAHs based on the partitioning between gas and particles [26]

PAHs categories	No. of rings	Log KOA	% gas	% particles	Means of deposition
Light PAHs	2–3	<8.5	80–100	s	Gas deposi tion
Intermediate PAHs	4	8.5–11	20–100	0–80	Gas and particle deposition
Heavy PAHs	>4	>11	0–20	80–100	Particle deposition

On the other hand, PAHs compounds with an intermediate molecular weight (with four rings) are partitioned between vapour and particulate phase which depends on different parameters. Whereas PAHs compounds that are categorized as multi-ringed PAHs with heavy molecular weight (with five rings) are bound to particles. Generally, PAHs which are lighter than pyrene normally exist in the gaseous phase, whereas the PAHs that heavier than benzo[a]anthracene and chrysene bind to particles [12, 17].

The second characteristic of PAHs is the variability of the vapour pressure and the aqueous solubility of PAHs. PAHs compounds have low vapour pressure and aqueous solubility [12, 21]. The vapour pressure tends to decrease as the molecular weight increase. The relationship between the vapour pressure and the molecular weight of PAHs was studied (See Fig. 38.2). PAHs compound with low vapour pressure tends to be associated with the vapour phase and less likely to be deposited on the surface. On the other hand, PAHs with high vapour pressure will tend to be associated with the solid phase. This relationship will create a variability in the relative distribution or partitioning of PAHs in the air [12, 17].

The third characteristic of PAHs is that they have low aqueous solubility and hydrophobic or lipophilic characteristics, and therefore PAHs are very soluble in organic solvents. The aqueous solubility of PAHs is decreased as with the addition of one extra ring as shown in Fig. 38.2. PAHs with their low aqueous solubility and their low vapour pressure make them more readily bound to the particulate matter than PAHs with higher vapour pressure, such as benzene [12, 21, 25]. Consequently, the wet deposition through rain and fog is rare. PAHs mostly exist in dry deposition through the particles and gaseous deposition which both will undergo atmospheric transport [12, 26].

The previous study suggested that Kp is closely related to KOA, which is the fourth parameter. The octanol/air partitioning coefficient (KOA) is dependent on the molecular weight of the compounds and the temperature. The lower the value of KOA, the PAHs will tend to exist in the vapour phase [26, 31]. Therefore, based on Tables 38.1 and 38.2, naphthalene, the PAHs compound with the lowest KOA will tend to be found in the vapour phase, and the tendency to be found in the vapour phase is lower for the PAHs compounds with higher KOA.

The fifth parameter is the octanol–water partitioning coefficient (Kow). Kow is related to the solubility of PAHs in water. The aqueous solubility of PAHs will decrease, and the adsorption to the particulate phase will increase as the KOw value increases [12]. From Fig. 38.3, it can be found that the value of KOw tends to

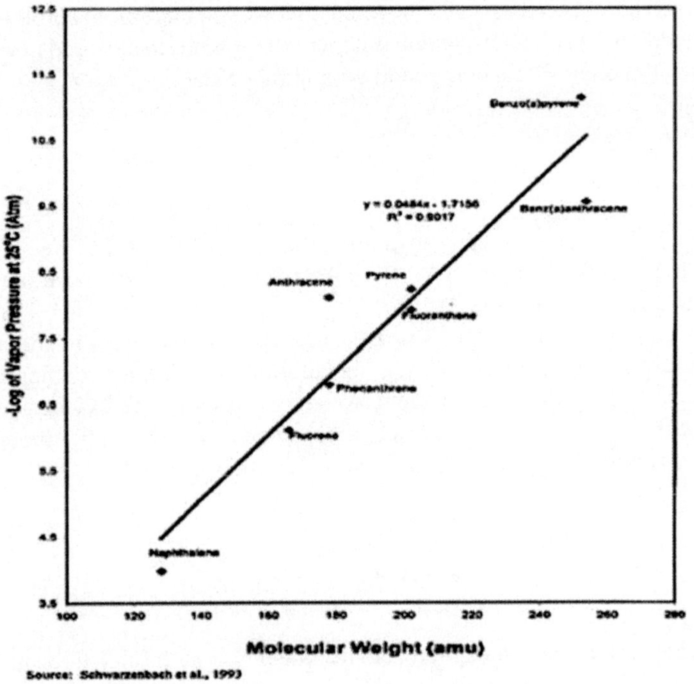

Fig. 38.2 The relationship between vapour pressure and molecular weight of PAHs [12]

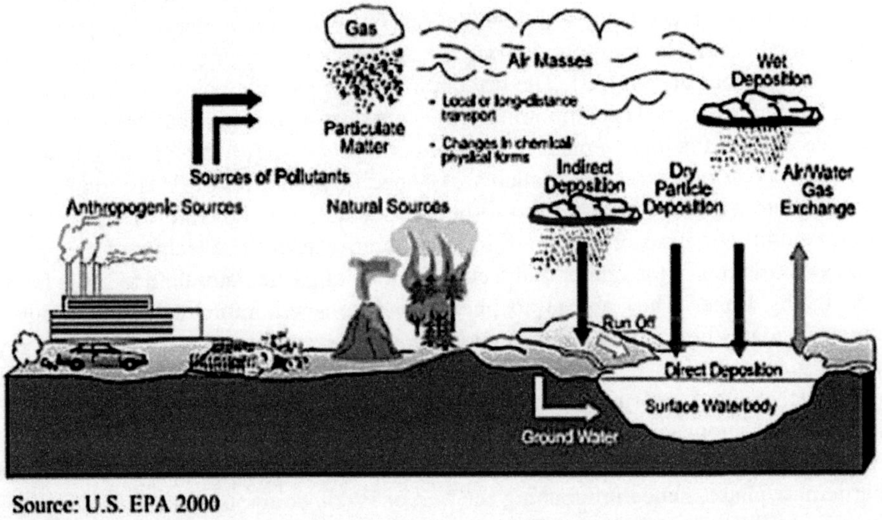

Fig. 38.3 The fate and removal of PAHs compound in the atmospheric environment [12]

increase the number of rings and the molecular weight increase. Therefore, the PAHs compound with lowest KOw, which is naphthalene, tends to be found in vapour phase more than the other PAHs compound with higher Kow.

$$K_{OA} = \frac{K_{OW}}{K_H}$$

where KOW is octanol–water partitioning coefficient, and KH is Henry's constant. Based on the parameter of KOA, the PAHs are categorized into three categories which are described in Table 38.2.

From Table 38.2, it can be observed that the fate of PAHs in the atmosphere is influenced by the number of rings, octanol air partitioning coefficient which is proportionally related to the octanol–water partitioning coefficient and inversely related with Henry's constant. Hence, the removal of PAH from atmospheric envi ronment strongly depends on those factors [12, 26, 32].

38.2.3 *The Fate of PAHs in the Environment*

PAHs removal from the atmospheric environment occurs through three mechanisms, which is the removal through dry deposition, wet deposition, plant uptake, and degradation [12, 26].

For particle-bound PAHs, the removal mechanisms occur with the process of dry deposition, wet deposition, and degradation. Deposition is a significant path- way of removal of PAHs compound from the atmosphere which is driven by gravity force or washout of rain [33]. The dry deposition or particle bound of PAHs occurs with the settling of PAHs compounds which are adsorbed to particulates to the earth surface without the involvement of precipitation process. The rate of this deposition depends on various parameters, such as the specific properties of PAHs compounds, the characteristics and types of the sorbent particulates, and the atmospheric conditions (stability of atmospheric condition, and temperature). The weight of the particulates contributes to the rate of their settling. Heavier particulates tend to settle faster than the light ones. The stable atmospheric conditions will improve the settlement of the particulates better than the unstable atmospheric conditions which have energy that is adequate to prevent the settling of the particulates. The temperature also plays a role in the rate of settlement by influencing the partitioning of the PAHs compounds in between vapour and solid or particulate phase. The higher the temperature, the more tendency of the PAHs compound to be found in the vapour phase than in the particulate phase, hence influencing the rate of PAHs compound settlement via dry deposition [12]. Particle-bound PAHs are also removed through the wet deposition mechanisms which occur along with the precipitation. The PAHs compounds that are adsorbed to the particulates will be eliminated through scrubbing process in the wet deposition process. It was reported in the previous study that the PAHs compounds which are adsorbed to particulates are more easily removed through wet deposition than PAHs compounds in the vapour phase [12].

For PAHs compounds which exist in the vapour or gaseous phase, the removal oc curs through deposition and chemical reaction. The first mechanism is the PAHs removed in their vapour phase in which they adsorbed to aerosol particles, while the second one is the elimination of their deposition on the of vegetation, water, and soil [24, 34]. The deposition of PAHs compound depends on the gas to particle conversion [24]. Gaseous PAHs compounds that are found in the vapour phase will be removed through dissolution in wet deposition process [12]. However, the removal of PAHs from the vapour phase is almost independent to the proximity of the site to the emission sources. Therefore, they may be found in the remote sites and dominate the total mixture of PAHs in a variable environment such as in rural, suburban, and urban areas. Gaseous PAHs can be found on the site which is far away from the sources because they can undergo long range transport in the atmosphere after being released from the source. The source of gaseous PAHs can be from primary sources, such as from combustion process in the indus try and vehicle exhaust, and from secondary sources, such as from the volatilisation of PAHs from soil or plants [26].

PAHs in the atmosphere also will undergo a chemical reaction with reactive and oxidative species and can undergo photoreaction. PAHs compound in the atmosphere may undergo chemical reaction, for instance, gaseous and particle-bound PAHs are prone to chemical transformation by the other compounds, such as ozone, hydroxyl, and nitrate radicals which will form PAHs derivative compounds such as nitro-PAHs and oxygenated PAHs [26, 35]. PAHs compounds can also undergo degradation in the atmospheric environment via photo-degradation. The photodegradation rate of PAHs compounds depends on the structure of the PAHs compounds. Normally, the PAHs compounds with linear, 2-rings, and some clustered PAHs are more easily photodegraded than angular PAHs (see Table 38.2, such as phenanthrene and dibenzo[a,h]anthracene, because they are more stable in structure. The photodegradation tends to occur easily to the PAHs compounds that exist in the vapour phase or aqueous phase. The chemical degradation of PAHs compounds depends on different parameters, such as the molecular weight and the structure of the PAHs compounds, the physical state, temperature, and the strength of the oxidizing agents [12].

PAHs compounds are lipophilic which can accumulate in the lipid tissue of plants and animals. On the other hand, PAHs will not have a tendency to build up in a plant tissue which has high water content. Therefore, the waxy surface of vegetables and fruits can concentrate the low molecular mass of PAHs mainly through surface adsorption. The surface of plants, such as peel, outer leaves, normally has a greater concentration of low molecular weight PAHs than inner tissues of the plants. The careful washing will contribute to the removal of PAHs up to 50% of the total PAHs. PAHs compounds that are bound to the particulate matters are easily washed off from the surface, whereas the PAHs compounds which are in the waxy layer are less efficiently removed [12].

38.2.4 The Health Effects and Environmental Impacts of Atmospheric PAHs

PAHs become a great concern because of the ubiquitous occurrence and their potential toxic effects on the human and ecosystem health [15]. The exposure of PAHs to general population occurs through the breathing of ambient and indoor air, smoking cigarettes, and breathing smoke from open fireplaces. The exposure of PAHs compounds to human may lead to teratogenic, mutagenic, carcinogenic effect and influence the immune system [12, 21, 36]

The effect of PAHs on human health can occur in the short term (acute effects) or long term (chronic effects. These effects are manifested on human health differently based on the routes of exposure, the concentration of the PAHs, and the relative toxicity of the PAHs itself. Other factors, such as pre-existing health status and age may contribute to the health effects. PAHs are hydrophobic compounds, which is highly soluble in lipid. Therefore their bioavailability after inhalation is significant [12, 21].

The carcinogenicity of PAHs has been evaluated by multiple organization such as United States Environmental Agency (U.S. EPA), International Agency for Research on Cancer (IARC) and the United States Agency for Toxic Substances and Disease Registry (ATSDR) [28]. EPA indicated 16 compounds of PAHs which are frequently occurred in the environmental monitoring samples, namely acenaphthene, acenaphthylene, anthracene, fluoranthene, fluorene, naphthalene, phenanthrene, pyrene, benzo[a]anthracene, benzo[b]fluoranthene, benzo[k]fluoranthene, benzo[ghi]perylene, benzo[a]pyrene, chrysene, dibenzo[a,h]anthracene, and indeno[1,2,3-cd]pyrene [17, 28, 37]. However, the carcinogenic effect of an individual PAHs is not exactly similar. As it is mentioned in Table 38.1, IARC has categorized PAHs compound into group 1, 2A, or 2B which means known, possibly, or probably carcinogenic to human. Ben-zo[a]pyrene belongs to group 1 which means it is a carcinogenic compound, whereas other PAHs compounds, such as naphthalene, chrysene, ben-zo[a]anthracene, benzo[k]fluoranthene, and benzo[b]fluoranthene which belong to group 2B, which means they are probably carcinogen to human. Meanwhile, United States Environmental Protection Agency (US EPA) has categorized seven PAHs compounds, which is benzo[a]anthracene, benzo[a]pyrene, benzo[b]fluoranthene, benzo[k]fluoranthene, chrysene, dibenzo[a,h]anthracene, and indeno[1,2,3-cd]pyrene as probable human carcinogens. An excess risk of lung cancer is the most significant health outcome that arises from the inhalation exposure to PAHs [12].

The genotoxicity effects resulting from exposure to PAHs have been reported from the previous studies. PAHs are metabolized enzymatically resulting in different metabolites, which some are reactive. PAHs will undergo metabolic activation into diol-epoxides, which will bind covalently to the DNA and lead to carcinogenic effects. The mutagenic effect of PAHs needs first an enzymatic activation. It was studied mainly using benzo[a]pyrene compound. It was found that benzo[a]pyrene primarily binds at exocyclic amino groups of purines which induce base-pairs substitutions, which is G > T transversions, in in vivo transgenic mice and the Ha-ras oncogene in mouse skin tumours. The similar transversions were analyzed in the case of p53 mutations in lung cancers due to tobacco smoke [12, 21, 35].

It was also reported in a previous study that immunosuppression is the most frequent effect reported after the exposure of PAHs. Atmospheric or topic exposure via in vitro systems was reported resulting in immune-potentiation in the literature. However, the precise mechanism by which PAHs induce immunosuppressant is still not clearly established. The effect of PAHs on the immune system has been considered at a systemic level in previous studies, regardless their routes of expo- sure. It was suggested that the immunosuppressant effect might relate to the mechanism in which PAHs induce cancer [12]. Immunological changes due to PAHs exposure was also associated with allergic asthma and related to increased asthma severity [22].

PAHs considered organic xenobiotics that is persistent in the environment [37]. Concerning the toxicity to the environment, the exposure of PAHs compounds in the atmosphere will harm the plants. The direct exposure to the leaves will lead to damages of the leaves with symptoms such as necrosis, collapsing mesophyll, bronzing and silvering of the leaves surfaces. PAHs compounds in the atmosphere together with the UV radiation will also induce the leaves damage through the production of reactive oxygen species which causing photosensitization and the production of daughter components which are more toxic than PAHs compounds themselves as the parent compounds, such as quinones, benzoic acids, and phenols which pose high water solubility, availability, and more reactivity [26].

PAHs compound concentration in the atmospheric environment, therefore, needs to be controlled with regulated standards to avoid the impacts to the human health and environment. World Health Organization (WHO) established a guidance in 2003 for the unit risk of lung cancer of benzo[a]pyrene with the value 87×10^{-6} ng/m^3 for a lifetime exposure [12]. In European legislation, an individual com- pound of PAHs that is regulated is limited to benzo[a]pyrene in the fraction of particulate matter (including inorganic salts, crustal material, and elementary carbon) [38].

38.3 Biomonitoring of PAHs

Biomonitoring is the use of bio-organisms to assess the atmospheric condition by monitoring the changes within the organism itself, either behavioral changes or morphological changes, because of the exposure to certain compounds, whether it is organic or inorganic compound, to compare the susceptibility of individual sites and investigate spatial distribution toward atmospheric pollution [4, 26, 39]. Biomonitoring can provide an assessment of the integrated uptake of the environmental exposure to the organisms through all exposure routes since the compounds are bioavailable and more likely to be assimilated and concentrated in the issues [17]. The biomonitoring using plant was recently introduced as an alternative to the monitoring of pollutants using active air sampling to minimise the bias effect of the short-term active air sampling [16]. The use of plants as biomarkers began in the 1960s due to their cuticle which contains lipid that is capable of accumulating persistent organic pollutants (POPs) [40]. The biomonitoring methods also have been carried out in the several previous studies to determine the concentration of certain inorganic pollutants in the atmospheric environments such as heavy metals and trace elements, and airborne fluoride [1, 4, 24, 31, 41].

Biomonitoring of PAHs has been done in recent years worldwide in North America [22, 42], South America [24, 43], Europe [23, 38, 44–46], Asia [20, 37, 47], and Africa [48]. The species which were used for biomonitoring of PAHs varied from lichen, mosses, trees, needles, bromeliads, hedge plant, and grass cultures [24, 38].

38.3.1 Atmospheric Deposition of PAHs and PAHs Uptake by Vascular Plants

The deposition of the pollutants, for instance, PAHs, in the atmospheric environment to the terrestrial environments partially will be accumulated on the plant foliage. Biomonitoring using the plant as passive samplers is to determine the PAHs content that is accumulated in the surface area of the leaves. The accumulation of particle-bound PAHs occurs when its deposition accumulated in the waxy layer of the leaves, whereas the uptake of PAHs in the gaseous phase occurs via stomata [17, 24]. The concentration of the targeted pollutants in plants can represent the combined levels of the PAHs in the surrounding air over time which can be useful in identifying the point source of the PAHs emissions [24, 37].

The effects of PAHs uptake to the plants vary with different types of PAHs, the concentration level of PAHs in the ambient, duration of exposure, and characteristics of the plant itself [26]. The leaf properties, such as the leaf features and the leaf components contribute to the accumulation of PAHs. The leaf features such as the surface area, cuticular waxes, hairs, and some stomata contribute to the accumulation and uptake of the PAHs. Hairy leaves tend to have more surface area to which enhance the capturing of PAHs from the air and better retention of the particulates. The waxy layer and lipids in the cuticular and epicuticular will increase the accumulation of PAHs compounds as they have lipophilic characteristic [4, 17]. The roughness and integrity of the cuticle were reported in the previous study to influence the adhesion of particle on the leaves surface [4].

The uptake of PAHs compounds from the environments in the vascular plants occurs directly and indirectly. The direct uptake happens through foliar uptake from the atmospheric environments to the leaves, whereas the indirect uptake hap- pens from the air and the passing through soils and the roots of the plants (see Fig. 38.4). In the case, if the plants grow in unpolluted soils, the direct uptake be- comes the dominating pathway of the entering of PAHs compounds to the plants [26]

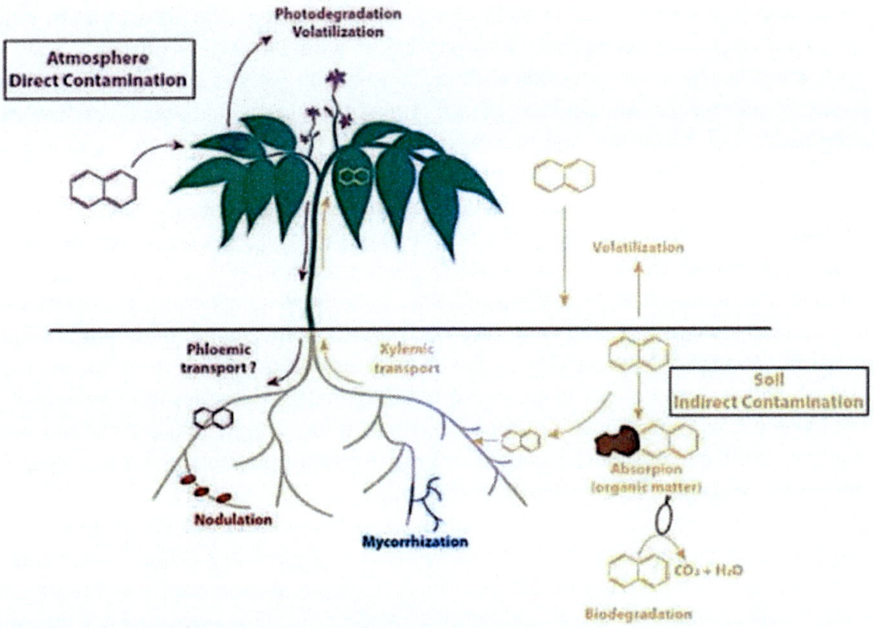

Fig. 38.4 PAHs exposure pathways for in situ vascular plants [26]

The direct uptake of the PAHs compound from the air to the plants determined by three processes, which are the diffusion process, kinetically limited gaseous deposition, and wet and dry particle-bound deposition (see Fig. 38.4). Those processes are dependent on different factors, such as the concentration of PAHs in the atmosphere, environmental conditions, and plant features. As for plant features, it has been noted that leaf surface, hairiness, and the composition of cuticles influenced the efficiency of intercepting and retaining of PAHs compounds. More retention of PAHs compounds in hairy leaves and leaves with more surface areas. The cuticle with waxy layer and cutin poses hydrophobic characteristics which help to improve the PAHs retention capacity [26].

Once the PAHs compounds are retained within the leaves, they will undergo two different pathways. The first mechanism is the adsorption to the cuticle and the second mechanism is the absorption to the inner tissues of the leaves reaching the cytoplasm of the epidermal cells [26, 49]. The adsorption to the cuticles happens with the facilitating waxy layers and hydrophobic characteristic of the leaves which is dependent on the temperature of the environment [26]. It has been noted from the previous studies that PAHs compounds with low molecular weight tend to undergo adsorption to the cuticle easier than the high molecular weight PAHs compounds, for instance PAHs in the vapor phase tend to undergo this mechanism to undergo direct uptake via stomata or via outer cuticular [26, 44]. The previous study suggested a mechanism in which a light PAHs compounds will undergo the absorption to inner tissues, such as in epidermis and mesophyll cells within 1–2 days [26]

On the other hand, compounds with high molecular weight are easily washed off since they will be accumulated and quickly transferred to the leaf wax layer. The will be embedded in the cuticle and they will be slowly desorbed from the particles and will diffuse to the cuticle [26]. A previous PAHs biomonitoring study using six tree species implied that airborne particulates are likely the sources PAHs for cuticles, whereas gaseous PAHs are likely the source PAHs for inner tissues probably due to the difficulty of heavy PAHs to migrate into inner tissues [47].

Other than plant leaves, needles also have a waxy surface which composed mainly from fatty acids (methyl esters), polyesters, paraffin, and secondary alcohols which probably enhance the effectivity in the entrapment of organic pollutants, such as PAHs. It was reported in numerous previous study that pine needles have been recognized as a useful biomonitoring tool of PAHs as it has a lipid-rich cuticle and pore structure which makes them have a high affinity for vapor phase semi-volatile organic compounds and can intercept particles [23, 40, 50].

Several studies also mentioned the tree bark as a biomonitor of PAHs. Due to the porous surface of tree bark, it is well recognized as a good biosorbent material and passive samplers for air pollutants [43, 50, 51]. However, the bark was mentioned in the previous study having an almost inert surface in the presence of organic and organic substances. The accumulation of air contaminant to the bark is mainly by the deposition from the atmosphere onto the outer layer [50].

38.3.2 Type of Plants Used in Biomonitoring of PAHs

Both non-vascular and vascular plants have been used in the previous studies to perform biomonitoring of PAHs. For the non-vascular plant, especially bryophytes because they are distributed widely and have a simple mechanism of PAHs uptake by means of diffusion through the process of exchange surface with the air. However, vascular plants are more commonly used to perform biomonitoring of PAHs due to their adaptability to the atmospheric environment and PAHs [26].

In the vascular plant, gymnosperm and angiosperm species were used to perform biomonitoring to PAHs compounds. The previous studies compared the PAHs concentration recovered in urban and rural areas. It has been noted that the results were highly variable between different plants used for biomonitoring. The plant species that belong to gymnosperm which was selected for biomonitoring mainly from the genus of *Pinus*, such as *P. nigra, P. sylvetris, P. pinaster, P. massoniana, and P. pinea* with the accumulation of PAHs compound in their nee- dles within year-round. On the other hand, angiosperm species that were used for biomonitoring such as deciduous trees (*Acer, Quercus, Corylus, Ficus, Fraxinus genus*) cereals (*Z. mais*), vegetables *(P. sativum, B. oleacera, L. sativa), and grassland (F. ovina, H. lanatus, T. repens, and L. perenne*) [20, 26].

38.3.3 Advantages and Disadvantages of Biomonitoring PAHs Using Plants

Biomonitoring using plant sample offers a continuous monitoring of PAHs for an extended period [23] since they are accumulated in the leaves for their whole lifetime [34, 39] regardless of the influence of the atmospheric situation [34].

Biomonitoring especially favourable due to its abundance and availability of the samples, reliability as a mean of passive monitoring, and it is convenient be- cause of less use of labour and less time-consuming [22]. Biomonitoring is also considered easy in term of sampling with no involvement of complex and high- cost technical sampling equipment [39]. Those reasons make the biomonitoring of pollutant depositions, (i.e. PAHs) favourable as an alternative to the conventional methods using air sampler, especially in the areas where complex and expensive air sampling equipment are unavailable. Biomonitoring is also considered a convenient, available, and reliable passive monitoring of PAHs which is effective to be applied in the developing countries since it can avoid the high cost of instrumental monitoring [22, 24, 37, 38].

By choosing biomonitoring, it is possible to obtain detailed view about various concentration level within a region which is not feasible to be attained using high volume sampler (HVS). Besides, it can also provide additional information regard ing outdoor air quality resulting in a time-weighted average signal over an extended period, which is longer if it is compared to HVS (24 h) [38]. As a mean of passive monitoring or passive samplers, biomonitoring can also analyze temporal accumulation of PAHs compounds to minimize the bias from short-term active sampling, although previous study suggested that passive sampling using high volume samplers is also capable of reflecting spatial and temporal fluctuation of PAHs concentration [52–54]. Another previous study combined the result from data from active using HVS (24 h) and passive sampling using pine needles to identify the origin of PAHs contamination source, and it was found that both data complement each other [42]

Nevertheless, biomonitoring also has drawbacks in the application. The biomonitoring of PAH with plant samples was suggested to be an economical and efficient system, but regarding the evaluation of the PAHs accumulation in the plants, it is critical to determine the accumulation of the PAHs during the plants growth (i.e. the growth of needles or leaves) and the behavior of the PAHs in the plants [55].

Previous study suggested the difficulties in establishing the clear-cut quantitative relationship between concentration of pollutants in the air and leaf [17, 37] although it was reported in the previous studies that accumulation of PAHs in leaves was sensitive to the changes in air concentration [22]. Previous studies have generated data about PAHs accumulation in plants. However, there are difficulties in comparing those data due to their heterogeneity, such as the difference in units used and different PAHs compounds monitored between previous studies [26]

It was reported in the previous study that the accumulation of PAHs in the leaves in the biomonitoring had similar pattern to those found in the passive air samplers (PAS) in the dry season, however in the wet season less significant correlation between the total PAHs from biomonitoring and PAS compared to those in the dry season, presumably due to lower concentration found in the air during wet season and washout process [24, 33].

A study compared the responsiveness of biomonitoring method to PAS in determining the differences between the sites as classification criteria in a principal component analysis (PCA). PCA aimed to group variables with similar analysis into factors, which can be interpreted either as an emission source or as chemical interaction. The study tested different system of equations, and it was found that with data from PAS at concentration less than 3000 ng sampler^{-1}, biomonitoring did not have a good response and hence did not allow determining the differences between the sites (i.e. city center, neighborhood in the proximity of industrial sites, etc.), but a good fit of the data to a linear regression was observed with the use of PAS data from the sites with concentration more than 3000 ng sampler [33]

38.4 Sampling Campaign

Different types of plants were collected in different previous studies, considering the ubiquity of the plant in the region under study [38], such as evergreen species of hedge pant *Taxus bacata* [38], pine needles of pine trees from different species such as *Pinus jeffreyi* [22], *Pinus massoniana* L. [37], *Pinus nigra* [16, 42], *Pinus pinea* [23], *Pinus brutia* [56], *Pinus strobus* [42], evergreen giant shrubs leaves *Calotropis gigantean* which is hairy and broad [54], holm oak *Quercus ilex* [34, 49, 55, 57], Italian ryegrass cultures *Lolium multiforum* [24], epiphytic plant *Tillandsia capillaris* [33], epiphytic lichens *Pseudevernia furfuracea* [46, 58, 59] and *Parmelia sulcata* [46], moss samples *Hylocomium splendens* [19], *Hypnum cupressiforme*, and *Sphagnum capillifolium* [46]. The selection of the plant types used for biomonitoring study was performed with several considerations, such as the distribution of the plants within the area under study and the ability to retain the air contaminants [55] as it was reported in a study which used 11 different plant species for PAHs analysis resulting in various total PAHs concentration which was assumed due to differences in surface roughness of the leaves and stems which leading to the increasing PAHs retainment potential [48]. Other criteria make pine needles were more frequently used in previous studies as biomonitoring tools because the needles from 1 to 3 years old are found in the same branch so that their age are easily determined [52]. The independence from soil makes epiphytic plants, such as *Tillandsia capillaris* favourable to be used as biomonitoring tools because it is clearly independent of soil condition. Therefore it is easier to establish patterns for biomonitoring studies [33].

Different sources of PAHs compounds as air pollutant were studied, such as traffic-related PAHs emission with different variables such as road type (freeway, major arterial, minor arterial, major collector, and minor collector), road density, traffic density, and different level of traffic [22, 33, 34, 38, 43], and industrial-related PAHs, such as emission from coal-fired thermal power plants [54], different field of industries, such as petrochemical industry, ceramic industry, metal industry (iron and steel company, metal-mechanical industry) [24, 37], cement industry [24].

A different method of collecting the plants varied in the previous studies. Removing the part of the plants by hand-picking was described in previous studies [38, 55], with the use of gloves was mentioned in the earlier study [34, 38]. It was also mentioned that such care was taken to minimize the contact with leaf surface [55] and needle surface [37]. Another previous study collected the plants and subsequently exposed in net bags at 3 m above the ground at similar height as the PAS was located [24, 33].

Different part of the plants which were collected performs biomonitoring in previous studies. For instance, the needles of pine trees were collected considering the position toward the source of PAHs emission in the street, which can be parallel or perpendicular, to avoid bias from branch-to branch variation [22]. The height of the plants in which the leaves, needles, or other parts of the plants were collected in previous studies also considered, such as at breast height (approximately 1.4 m), 1–2 m above ground, 2 m above the ground [55], 1.5–2.5 m above the ground [22, 23, 54].

The selection of the location where the samples were collected was performed considering different factors. The wind direction was considered in a previous study selecting the sites in which biomonitoring was conducted (downwind, up-wind, and perpendicular to the wind direction) [54], while in another study the selection of the location where the needles were collected at four directions [22, 37] (south, north, east, west) in the outer part and middle part [37]. The proximity to the emission source, such as roads, was also considered in selecting the sampling location [23].

The age of the leaves and needles used for biomonitoring was also considered in the previous studies in term of accumulation of PAHs [23, 37, 55]. Previous biomonitoring studies of PAHs using pine needles showed higher accumulation of PAHs in the third- year needles than in two-year needles [23] and higher concentration in the needles collected in the previous year than needles collected in the current year [37]. Another biomonitoring study using plant leaves, carefully selected the leaves started from bud breaks, then continued to collect the leaves every two months in the same branch in the period approximately 16 months to observe the dynamics of PAHs accumulation during that period [55]. Besides the age of the leaves and needles, the age and size of the trees were also considered in the previous study [37] (Table 38.3).

38.5 Sample Preparation and Sample Extraction

Previous studies demonstrated different methods of sample preparation, sample extraction, and clean up. Sampling preparation was performed before the extraction process. Different treatments to the samples were conducted in previous studies before the extraction process. It was mentioned in a previous study, after the sample collection, the samples were immediately stored in the refrigerator at temperature

Table 38.3 Sampling campaign

Location	Main source of emission	Species type	Plant part location	Amount	Age	Additional in-formation about sampling method	References
Guangdong Province, South China	Emission from petrochemical, ceramic, iron, and steel industry	Pine needles: *Pinus massoniana* L	Minor branches situ ated at the outer part of the middle canopy at the four directions (south, north, east, west)	NA	1. Five mature trees with the similar size and age 2. Picking needles with different age from a current year and previous year	Needles were taken care to minimize hand contact	[37]
Warsaw	Motor vehicle exhaust	Mosses: *Hylocomium splendens*	Taking the mosses in a small forest clearing or on the open heath	NA	NA	NA	[19]
Province of Cordoba, Spain	(1) Vehicular traffic (2) Metal and mechanical industry (3) Cement industry (4) Indoor and out-door gran ite floors and a company of liquid fuel distribution	Epiphytic species leaves: *Tillandsia capilaris*	Plants were collected from the tree trunks from unpolluted sites and then placed in a bag 3 m above ground level	8–10 plants	NA	Sampling was conducted at different times of the year (March 21st- June 21st) and (September 1st-December 1st)	[33]

(continued)

Table 38.3 (continued)

Location	Main source of emission	Species type	Plant part location	Amount	Age	Additional in- formation about sampling method	References
Ghent, Belgium	Traffic	Hedge plant leaves: *Taxus baccata*	NA	NA	NA	The hedge plants were handpicked us ing gloves	[38]
Palermo, Itay and coun tryside Cefalù	Traffic	Holm oak leaves: *Quercus ilex*	Leaves collected at dif ferent points in the same trees at about 2–3 m above the ground	50–100 g	NA	Leaves collected using rubber gloves	[34]
Portugal and Spain	NA	Pine needles and barks: *Pinus pinaster* and *Pinus pinea*	Needles were chosen from bottom branches of trees and re- moved in one piece Barks were collected at the most external layer at 0.5–1.5 m above the ground	NA	NA	NA	[50]

4 °C and stored with avoidance to the exposure of sunlight and then rapidly carried to the laboratory to be frozen [34]. In another study, it was mentioned that the samples were cut first, then freeze dried, and kept frozen at a temperature of −20 °C or −21 °C [34, 38].

It was mentioned that the washing process to the plant samples was performed before extraction [19], whereas in other previous studies the washing process was omitted, and also there were studies performed using both washed and unwashed samples [37, 57]. A previous study compared the PAHs concentration in washed and unwashed samples with the result that water-washing significantly reduce the PAHs concentration in needle sample [37], while other study revealed that there was no difference in terms of PAHs concentration between washed and unwashed leaves, although the difference of PAHs concentration between urban and rural area was more pronounced in unwashed leaves [57].

A special procedure was performed in the sample preparation, for instance, a waxy-layer stripping was performed by 10 min sonication using dichloromethane as a solvent, and then the solvent is evaporated from the samples before the extraction [22]. Other special procedures were performed with the sample from evergreen hedge plant before extraction process was performed by freezing the sample with liquid nitrogen (Air Liquide) before homogenizing the sample by grinding [38].

The sample the homogenized by chopping [22] or grinding to reduce the size using mortar [38] before the performing of extraction. The amount of the samples which are used in the extraction differ from one to another study with dry weight was considered [34, 38]. The dry weight was determined with the correlation with the determination of water content of the samples, for instance by drying the samples at 105 °C [34] or 110 °C [38] for one night [34].

The extraction methods were performed in different techniques. The analysis methods of the samples which are obtained from vegetation have been developed in the last 20 years based on the multiple residue extraction methods [40]. First method that was commonly used is ultrasonic extraction (USE) with different kind of solvents or solvent mixtures, such as dichloromethane (DCM) [54], a mixture of n-hexane and acetone (2 mL; 1:1, v/v) [33]. The second method was Soxhlet extraction method using DCM in 48 h [37]. The third method was pressurized liquid extraction (PLE) using an accelerated solvent extractor (ASE) which was manufactured by Dionex with different solvents were used in previous studies [22]. A detailed procedure in PLE extraction methods was developed in the previous study using DCM as the extraction solvent. The different temperature of extraction (50, 100, 150, and 200 °C) and different amount of Florisil powder used (0, 1, 3, and 7.5 g) to evaluate the most optimum procedure in terms of recovery and matrix effect [34]. The fourth method of extraction was microwave-assisted extraction (MAE) method performed with a modified version of the WP700P17-3 domestic oven from Electric Co [50]. The fifth method of extraction was shake- flask extraction in which sample and solvent placed on a shaker for several hours shaking [19].

After extraction procedure, the evaporation procedure was also performed, although this procedure was also skipped or not mentioned in another previous study. The evaporation procedure was performed to reduce the volume of the extract which is normally performed using the flow of nitrogen gas, for instance in the Turbovap LV evaporator using evaporator at 25 °C until a certain volume was achieved [34].

A previous study compared two different extraction methods, which were MAE and USE. Both methods were tested in pine needles, and bark samples are resulting in a very similar recovery for both needles and bark, with a value of recovery between 70 and 130% in most of the cases [50].

PLE extraction was considered having a higher recovery than the other methods, such as Soxhlet extraction and USE due to its aggressiveness in terms of extraction capabilities, therefore the chromatogram presented more sample interferences, intricating the identification and quantification of the target compounds which was due to the high temperature used (150 °C) [40]. PLE was also considered as a fast and automated extraction method with low volume used in comparison with USE and Soxhlet extraction method [34]. However, ultrasonic extraction method was more easily available and more cost effective in term of the equipment cost [40].

Other than extraction methods, the selection of extraction solvent or solvent mixtures was mentioned to be an important parameter in term of recovery of the compounds. This is due to the certain characteristics of the compound with different polarity and volatility of the compounds. Solvents or solvents mixtures are selected particularly to widen the spectrum of the polarity of the extraction. The other reason behind the selection of solvents and solvents mixtures was the boiling point. The lower the boiling point, it will be more preferred to be used due to the faster evaporation process, for instance, dichloromethane (DCM). Various solvents and solvent mixtures were reported to be used in previous studies such as a mixture of hexane:acetone (1:1), hexane, hexane:dichloromethane (1:1), etc. [50] (Table 38.4).

Sample Clean-Up

After sample extraction and evaporation, some different clean-up procedures were performed to the sample to eliminate unwanted compounds, such as lipid content [37, 40] and elemental sulphur [37]. However, it was mentioned that clean-up process was omitted in the previous study because the analysis using selected ion monitoring mode (SIM) allowed direct analysis by gas chromatography the concentrated extract [34].

The first methods that were commonly used in clean up procedure were fractionation using gel permeation chromatography (GPC) column. The second method involved traditional glass column, for example, X-4083 (1 cm I.D.) glass column which was filled with silica or alumina [40]. The third method was clean-up using SPE (solid-phase extraction) cartridges packed with Florisil [38, 40], silica or alumina [40]. In another study, it was mentioned that a filtration process using Sartorius filters (0.22 μm, Goettingen, Germany) was performed after the extraction process [33].

From previous studies, the clean-up procedure using SPE cartridges involved the conditioning process in the initial step. The conditioning process was performed before the loading of the sample using solvent or solvent mixtures, for instance the mixture (1:1) of 50 mL hexane and dichloromethane [40], a mixture of 18 mL of DCM: cyclohexane [38]. The next step was the loading of the samples, followed by the elution with solvent or solvent mixtures. For instance, in the previous

Table 38.4 Sample preparation and extraction

Sample preparation description	Extraction methods	Solvents usage	Amount of sample	Evaporation methods	Additional information	References
(1) Using washed and un- washed needles (2) Samples were cut into 1-cm long sections, freeze dried, kept frozen −20 °C until analysis (3) Needle samples mixed with 5 g anhydrous sodium sulphate (4) Internal standard of mixture (200 μg/L, 5 μL) of acenaphthene-d10, phenanthrene-d10, chrysene-d12, and perylene-d12 was added	Soxhlet extraction; 48 h	Dichloro- methane; volume: N/A	5 g	Using evaporator; temperature: N/A	Removal of elemental Sulphur influence, 5 g of activated copper granules were added to the collection flask	[37]

(continued)

Table 38.4 (continued)

Sample preparation description	Extraction methods	Solvents usage	Amount of sample	Evaporation methods	Additional information	References
(1) The sample was hand cleaned, washed with distilled water and air dried at the room temperature and packed to a plastic sleeve net to form spherical moss bag (2) The sample placed in a glass with 50% 2 N KOH and 20 ng/mL (3) An internal standard of 9-phenyl anthracene (RdH) was added	Shake-flask extraction; 5 h	Cyclohexane; 200 mL	6 g	(1) First evaporation: Rotary evaporator at temperature 40 °C to dryness (2) Perform second evapora tion to dryness	(1) After extraction fin ished, the upper layer was separated, washed twice with 2 mL of 2 N sulfuric acid and 2 mL of saturated sodium chloride, dried over anhydrous sodium sulphate before first evaporation (2) Extract from first evaporation was transferred into flask with addition of small amount of ethanol before second evaporation	[19]
The samples were homogenized and placed in the extraction cell along with n-hexane and acetone	Ultrasonic extraction (USE), 20 min	n-hexane and acetone; (2 mL; 1:1, v/v)	0.3 g	Evaporation in the steady stream of nitrogen	The extracts filtered with Sartorius filters (0.22 μm) before evaporation	[33]

(continued)

Table 38.4 (continued)

Sample preparation description	Extraction methods	Solvents usage	Amount of sample	Evaporation methods	Additional information	References
(1) A cellulose filter was inserted in the bottom of clean extraction cell and then different amount of Florisil powder (0, 1, 3, 7.5 g), and topped with a cellulose filter again (2) Leaves sample was mixed with 3.5 g diatomaceous earth	Pressurized liquid extraction (PLE)	Dichloro- methane	3 g	Evaporation using Turbovap LV evaporator at 25 °C until 0.5 mL	Before evaporation, the extract was filtered through a column containing 10 g of sodium sulphate and pre-rinsed with dichloro-methane and water to re- move impurities	[38]
Leaves were defrosted and mixed with equal quantities of sodium sulphate anhydrous	USE, 20 min, repeated three times	DCM, 40 mL	5 g	Evaporation using rotary evaporator with thermostatic bath at 35 ± 5 °C until final volume 10 mL	Before evaporation, the ex- tracts were filtered through a clean Pasteur pipette with solvent-rinsed glass wool and pre-cleaned anhydrous sodium sulphate	[34]
(1) MAE: Needle (cut into 1 cm bits), bark (ground into < 1 cm^2 bits with pestle and mortar (2) USE: similar to MAE (see point 1)	(1) Microwave-assisted ex- traction (MAE), 30 min (2) USE, 10 min, repeat ed two more times	(1) For MAE: Mix- ture hex- ane:D CM (1:1), 90 mL (2) For USE: hexane: DCM (1:1), 30 mL	2 g	Evapo- rated to almost dryness us- ing Buchi R- rotavapor (until volume 0.5 mL approximately)	NA	[50]

study, a 50 mL of solvent mixture (1:1) hexane-dichloromethane and a 50 mL of di- chloromethane was used for the elution process [40]. An additional step was mentioned in a study before the clean-up procedure was performed, which was the filtration using 10 g sodium sulphate in the form of anhydrous powder for water removal purpose in a column that was pre-rinsed using DCM [38].

The traditional glass column was also used in the cleaning up process, for instance, the traditional gas column of X-4083 that was filled with 5 g of Florisil or silica or alumina. The conditioning process was also performed prior to the loading of the samples using solvent o solvent mixtures. The column then was activated in the temperature of 400 °C for 12 h and deactivated using 1.2% of ultrapure water and 1 cm high of ultrapure water. After that, the loading of the sample was performed followed by elution process using solvent or solvent mixtures [40].

It is understood that the recovery of the PAHs compounds can be higher due to coelution from the matrix and it can be lower due to inefficient ionization. A coelution with other matrix compound was presumed in a previous study resulting in a higher recovery of PAHs compounds due to their vulnerability in coeluting with the matrix compound which cannot be eliminated selectively by GC–MS. An additional cleaning-up procedure was suggested in the previous study as a solution for the higher recoveries of PAHs compound using different methods, such as gel permeation chromatography, saponification, or additional SPE clean up, despite the risk of additional losses [40].

A second evaporation was mentioned in the previous study after the clean-up procedure was performed [38, 40], until the final drop, followed by reconstitution process, for instance by adding 1 mL of DCM [35], a mixture of acetonitrile (ACN) and water (0.4 mL; 6:4, v/v) [33].

In comparison between different types of the column used for cleaning up in a previous study, it was suggested that different sorbent materials used in the cartridges (Florisil, silica, or alumina) and different extraction methods used provided different recovery results. For needles samples, extraction using ultrasonic extraction followed by elution using SPE cartridge packed with alumina was chosen as methods which yield the preferable recoveries. The recoveries were compared be- tween different extraction methods, for instance, a recovery ranging between 65 and 102% using Soxhlet extraction with relative standard deviation (RSDs) below 7.5% and 72 and 100% using ultrasonic extraction (RSDs below 8%), followed by PLE extraction with the recovery ranging between 70 and 130%. The range of recovery between 70 and 130% is acceptable according to EPA. In the previous study, the disposable SPE cartridge was chosen over a traditional glass column for the easy use and time-saving purposes [40] (Table 38.5).

Sample Analysis

The analysis methods used in previous studies were different in the equipment and techniques. The equipment used to analyze PAHs compound mainly were gas chromatography (GC) coupled with a mass spectrometry (MS) or a high- resolution mass spectrometry (HRMS), although for PAHs analysis high performance liquid chromatography (HPLC) was also used [24, 33, 37, 38, 40],

Table 38.5 Cleaning-up methods

Cleaning-up methods	Solvent used	Additional evaporation	Additional information	References
(1) To remove elemental sulphur: elution using chromatographic column packed with silica/alumina (2:1, v/v) with the addition of anhydrous sodium sulphate (2) To eliminate lipid: using gel permeation chromatography (GPC) column	(1) To remove elemental sulphur: using 60 mL of mixture dichloromethane (1) To remove lipid: using 80 mL of a mixture of dichloromethane and *n*- hexane (1:1 v/v)	(1) Elution from the chromatographic column was evaporated in the evaporator; solvent changed to 1 mL hexane (2) Elution from GPC column was vacuum evaporated to 1 mL and finally 0.2 mL	Removal of elemental sulphur, 5 g of activated copper granules was added in the collection flasks	[37]
Silica column (6 mL SPE tube, packed with 1.2 g silica gel.)	Elution with 10 mL of cyclohexane	Elution from silica column was evaporated	After evaporation, the residue was re-dissolved in 1 mL of methanol then filtered into an auto sampler vial	[19]
The extracts filtered with Sartorius filters (0.22 μm) be- fore evaporation	N/A	N/A	N/A	[33]
Using SPE cartridges packed with 2 g Florisil which was conditioned with 10 mL dichloromethane (DCM):cyclohexane (1:1; v/v)	Elution with 18 mL DCM:cyclohexane (1:1; v/v)	Evaporation using Turbovap LV evaporator at 25 °C until final drop	N/A	[38]
The extracts were filtered through a clean Pasteur pipette with solvent-rinsed glass wool and pre-cleaned anhydrous sodium sulphate	N/A	N/A	N/A	[34]

(continued)

Table 38.5 (continued)

Cleaning-up methods	Solvent used	Additional evaporation	Additional information	References
For MAE and USE: Using SPE alumina cartridges (5 g, 25 mL) preconditioned with 50 mL hexane:DCM (1:1)	For MAE and USE: Elution with 50 mL hexane:DCM (1:1), followed by adding 50 mL DCM	Evaporation using a rotary evaporator to volume 0.5 mL with nitrogen flow	Two fractions were collected into the same pear-shaped flask and transferred to 2 mL glass vials	[50]

HPLC was performed in the analyzing process in the previous study by injecting a certain amount of reconstituted sample in the liquid chromatography column, for instance, Vydac 201 TP C18, 250 mm × 2.1 mm; 5 mm (Deerfield, IL, USA) with gradient elution program with an initial mobile phase at 60:40 (v/v) of ACN:water, then it changed linearly to 100% ACN at 18 min, and then changed again to the initial phase 60:40 after 3 min. The peaks of the sample were identified based on the retention times of the standard compounds [24, 33].

The analysis of PAHs compounds using GC was described in several previous studies. A certain amount of extract, for instance, one microliter [37, 38, 54], was injected to perform chromatographic separation in the capillary column. Different types and dimension of the capillary column were mentioned in the previous study, for example, A Restek Rxi-17Sil MS column with 30 m × 0.25 mm ID × 0.25 μm film thickness was used [38]. A DB-5 column of 30 m × 0.25 mm I.D was also used in another study to perform separation [40].

A detailed procedure was explained in a previous study using the high-resolution mass spectrometer (HRMS). After chromatographic separation, the compounds were subjected to electron ionization to be detected in the mass spectrometer. It was mentioned in a previous study that a 70 eV electron ionization were subjected to the compound with a temperature of the ion source was set at 270 °C and with multiple ion detection (MID) modes was applied in the mass spectrometer [38]. Another study used selection ion monitoring (SIM) mode [37, 43].

Peak detection and integration were performed in the previous studies using different software. PAHs identification in the analysis using GC–MS was performed with different software, such as Masslab 1.4, Excalibur 1.2 (from Finnigan, Waltham, MA, USA) [40], XCalibur software (Thermo Finnigan, version 2.07) spectrometer [38] by internal standard method using deuterated PAHs compounds (Ratola et al. 2006). Whereas PAHs identification in the analysis using HPLC was performed based on the retention times and UV spectrum and quantified by their respective peaks areas using Chromeleon (Dionex) software [19] (Table 38.6).

Concentration of PAHs

Several previous studies reported that between 16 EPA PAHs compounds, it was reported in several biomonitoring studies that PAHs with three aromatic rings

Table 38.6 Sample analysis

Equipment	Internal standard used	Re-constitution	Volume injected	Column	Program	References
GC–MS	1 µg of hexamethylbenzene was added as internal standard before performing instrumental analysis	NA	1 µL	NA	(1) Selected ion monitoring (SIM) mode (2) Flow: 1.2 mL/min of helium (3) 60 °C for 3 min (4) Raised to 200 °C at rate of 2 °C/min (5) Increased to 260 °C at 4 °C/min with 10 min hold time 6)260 °C at 2 °C/min with 10 min hold time	[37]
HPLC	0.123 to 3.3 µg of PAHs (PAHs mix, Fluka)	After evaporation, the sample was redissolved in 1 mL of methanol and then filtered into an autosampler vial	20 µL	STH 585 column, four channel UV Di- ode Array Detector UVD 340S with Pro- Sphere 300 PAH, 5 µm, 150 × 4.6 mm column	(1) Temperature 25 °C (2) Flux 2 mL/min (3) Equilibration 15 min (4) 0–3 min 50% water, 50% acetonitrile; 3–23 min 100% acetonitrile; 23–30 min 100% acetonitrile	[19]

(continued)

Table 38.6 (continued)

Equipment	Internal standard used	Re-constitution	Volume injected	Column	Program	References
HPLC	Standard solutios (EPA 525 PAH Mix B) purchased from Supelco contains acenaphthene, acenaphthylene, anthracene, fluoran-thene, fluorene, naphthalene, phenanthrene, pyrene, benzo[a]anthracene, ben-zo[b]fluoranthene, benzo[k]fluoranthene, benzo[ghi]perylene, benzo[a]pyrene, chrys-ene, dibenzo[a,h]anthracene, and in-deno[1,2,3-cd]pyrene	The residue was dissolved in acetonitrile and water 0.4 mL; 6:4, v/v	25 μL	Vydac 201TP C18, 250 mm × 2.1 mm x; 5 mm	(1) Temperature 28 °C (2) Flux 0.3 mL/min (3) Initial mobile phase (60:40; v/v) acetonitrile:water; 18 min 100% acetonitrile; after 3 min change back to initial phase (60:40) acetonitrile:water	[33]
GC-HRMS	7 perdeuterated PAHs (± 200 μg/mL in tolu- ene) including benzo[a]anthracene-d12, benzo[g,h,i]perylene-d12, benzo[a]pyrene- d10, phenanthrene–d10, pyrene–d10	1.) After evaporation, the sample was redissolved in 1 mL of DCM 2.) After reconstitution, the final extract was held in an ultrasonic bath for 5–10 s and filtered over 13/0.2 RC into GC vial	1 μL	Restek Rxi-17Sil MS column with 30 m × 0.25 mmID × 0.25 μm	(1) Multiple ion detection (MID) mode (2) Initial temperature 70 °C for 2.5 min (3) Raised to 320 °C at rate of 10 °C/min, held for 15 min	[38]

(continued)

Table 38.6 (continued)

Equipment	Internal standard used	Re-constitution	Volume injected	Column	Program	References
GC–MS	Perdeuterated internal standards (±200 μg/mL in cyclohexane) including acenaph- thene-d10, chrysene-d12, phenanthrene-d10, perylene-d12	The residue was dissolved in 0.5 mL solution containing perdeuterated internal standards	1 μL	Equity-5 with 30 m × 0.25 I.D.; 0.5 μm fused silica capillary column, injection splitless mode	(1) Selected ion monitoring (SIM) mode (2) Flux 1.6 mL/min with helium as carrier gas (3) 40 °C for 2 min (4) Raised to 100 °C at rate of 40 °C/min (5) Increased to 200 °C at 10 °C/min (6) 325 °C at 30 °C/min for 8 min	[34]
GC–MS	A volume of 50 ng/mL anthracene-d10	The residue was dissolved in 1 mL of hexane with the addition of internal standards solution	2 μL	DB-5 column coated diphenyl-polydimethylsiloxane 30 m × 0.25 mm I.D; 0.25 μm, injection in splitless mode	(1) Electron impact mode (EI, 70 eV) (2) Flux 50 cm/s with helium as carrier gas (3) Initial temperature 60 °C for 1 min (4) Raised to 175 °C at rate of 6 °C/min held for 4 min (5) Increased to 235 °C at 3 °C/min (6) 300 °C at 8 °C/min for 5 min	[50]

composed a significant amount in the atmospheric pollutant load [16, 23]. Phenanthrene was suggested having the highest abundance in airborne concentration [22, 53, 60], except for naphthalene with varying concentration in term of spatial distribution which was affected by the presence of highways, railroads, and industrial, commercial zones [22, 53], hence, it was used as diagnostic tool in a study in a high trafficked region [53]. Phenanthrene in the atmosphere was mentioned in the previous study having the major source from fossil fuel combustion and fossil fuel evaporation [53].

PAHs with high molecular weight, such as benzo[b]fluoranthene, benzo[k]fluoranthene, benzo[ghi]perylene, benzo[a]pyrene, dibenzo[a,h]anthracene, and indeno[1,2,3-cd]pyrene and PAHs with medium molecular weight, such as anthracene, fluoranthene, phenanthrene, pyrene, benzo[a]anthracene, and chrysene was reported in the previous study to be found higher in the vicinity of the emis- sion source. On the other hand, PAHs with low molecular weight, such as acenaphthene, acenaphthylene, fluorene, naphthalene was found with no much variation observed among all the sites [54].

Several previous studies compared the abundance of the PAHs accumulated during exposure in the biomonitoring between wet season and dry season. It was found that phenanthrene and pyrene constituted the highest fraction in both sea- sons, with the total concentration PAHs was found higher in the dry season than a wet season [24, 33]. A study of dynamics of the accumulation of PAHs resulting that PAHs accumulation was found at higher concentration during the winter [55, 58, 61] (Tables 38.7 and 38.8).

Table 38.7 Total PAHs concentration

Location	Total PAHs concentration in leaf (ng/g, d.w.) (Details of the location information)	Type of location	References
Ceramic manufacturer steel and iron manufacturer petrochemical plant Remote clear sites	1927.20 1564.70 1143.86 1044.78	Industrial areas (petrochemical, ceramics, and iron and steel manufacturing) and remote clear sites	[37]
Warsaw range Łomianki (Control 1) National Park Kampino (Control 2)	828–3573 2373 244	Warsaw (in two arterial streets courses) Control areas 1 in suburban Lomianki control areas 2 in forest area (National Park Kampinos)	[19]

Table 38.8 Concentration of different PAHs compounds

References	[37]				[19]		
Site information	Petrochemical plant	Ceramic manufacturer	Iron and steel manufacturer	Remote clear sites	Warsaw range	Control C1 (mean)	Control C2 (mean)
PAHs (±SD) ng/g d.w							
Naphthalene	198.90 ± 77.54	141.5 ± 34.1	116.05 ± 21.6	320.1 ± 30.3	5–102	52	6
Acenaphthylene	31.8 ± 3.3	21.9 ± 0.2	9.2 ± 1.8	33.1 ± 3.6	6–35	19	2
Acenaphthene	49.8 ± 11.4	46.8 ± 7.9	50.8 ± 0.5	55.4 ± 6.5	6–74	33	7
Fluorene	45.9 ± 24.7	143 ± 82	69.4 ± 8.3	36.6 ± 8.4	17–92	49	10
Phenanthrene	157.5 ± 59.2	467 ± 246	252 ± 29.9	203.9 ± 32	133–702	387	39
Anthracene	35.1 ± 11.1	70.2 ± 35.9	76.9 ± 8.9	49.1 ± 11.4	9–62	35	2
Fluoranthene	134.7 ± 70.3	207 ± 100	198.5 ± 62.3	79.3 ± 18.4	197–875	527	35
Pyrene	265.6 ± 115	366 ± 172	182.2 ± 15.5	94.7 ± 27.4	147–900	504	39
Benzo[a]anthracene	116.1 ± 76.5	93.3 ± 49.0	150.1 ± 30.4	81.1 ± 30.3	25–188	91	8
Chrysene	61.5 ± 33.2	287 ± 152	182.6 ± 54.3	57.2 ± 21.5	17–460	191	22
Benzo[b]fluoranthene	83.8 ± 28.8	113.3 ± 3.7	93.7 ± 17.5	25.0 ± 6.1	48–309	164	23
Benzo[k]fluoranthene	73.7 ± 9.6	69.5 ± 13.1	79.0 ± 7.6	33.9 ± 7.0	5–129	56	14
Benzo[a]pyrene	44.0 ± 10.1	72.6 ± 22.6	53.5 ± 12.1	21.2 ± 11.8	24–205	61	7
Dibenzo[a,h]anthracene	N/A	47.6 ± 43.6	40.9 ± 10.8	10.9 ± 3.5	4–24	10	4
Benzo[g,h,i]perylene	N/A	126.9 ± 20.1	N/A	N/A	30–181	94	10
Indeno[1,2,3-c,d]pyrene	N/A	61.4 ± 4.4	37.7 ± 9.7	39.3 ± 114.8	28–213	100	15

References

1. Rodriguez, J.H., Weller, S.B., Wannaz, E.D., Klumpp, A., Pignata, M.L.: Air quality biomonitoring in agricultural areas nearby to urban and industrial emission sources in Córdoba province, Argentina, employing the bioindicator *Tillandsia capillaris*. Ecol. Indic. **11**(6), 1673–1680 (2011). https://doi.org/10.1016/j.ecolind.2011.04.015
2. Sæbø, A., Popek, R., Nawrot, B., Hanslin, H.M., Gawronska, H., Gawronski, S.W.: Plant species differences in particulate matter accumulation on leaf surfaces. Sci. Total Environ. **427–428**, 347–354 (2012). https://doi.org/10.1016/j.scitotenv.2012.03.084
3. Squizzato, S., Cazzaro, M., Innocente, E., Visin, F., Hopke, P.K., Rampazzo, G.: Urban air quality in a mid-size city—PM2.5 composition, sources and identification of impact areas: from local to long range contributions. Atmos. Res. **186**, 51–62 (2017). https://doi.org/10.1016/j.atmosres.2016.11.011
4. Petrova, S., Yurukova, L., Velcheva, I.: Possibilities of using deciduous tree species in trace element biomonitoring in an urban area (Plovdiv, Bulgaria). Atmos. Pollut. Res. **5**(2), 196–202 (2014). https://doi.org/10.5094/APR.2014.024
5. Megido, L., Suárez-Peña, B., Negral, L., Castrillón, L., Fernández-Nava, Y.: Suburban air quality: human health hazard assessment of potentially toxic elements in PM10. Chemosphere **177**, 284–291 (2017). https://doi.org/10.1016/j.chemosphere.2017.03.009
6. Franklin, B.A., Brook, R., Arden Pope, C.: Air pollution and cardiovascular disease. Curr. Probl. Cardiol. **40**(5), 207–238 (2015). https://doi.org/10.1016/j.cpcardiol.2015.01.003
7. Huang, F., et al.: Gaseous air pollution and the risk for stroke admissions: a case-crossover study in Beijing, China. Int. J. Environ. Res. Public Health **14**(2) (2017). https://doi.org/10.3390/ijerph14020189
8. Shrey, K., Suchit, A., Deepika, D., Shruti, K., Vibha, R.: Air pollutants: the key stages in the pathway towards the development of cardiovascular disorders. Environ. Toxicol. Pharmacol. **31**(1), 1–9 (2011). https://doi.org/10.1016/j.etap.2010.09.002
9. Simkhovich, B.Z., Kleinman, M.T., Kloner, R.A.: Air pollution and cardiovascular injury. epidemiology, toxicology, and mechanisms. J. Am. Coll. Cardiol. **52**(9), 719–726 (2008). https://doi.org/10.1016/j.jacc.2008.05.029
10. Pannullo, F., et al.: Quantifying the impact of current and future concentrations of air pollutants on respiratory disease risk in England. Environ. Heal. A Glob. Access Sci. Source **16**(1), 1–14 (2017). https://doi.org/10.1186/s12940-017-0237-1
11. Yin, P., et al.: Particulate air pollution and mortality in 38 of China's largest cities: time series analysis. BMJ **356**, 1–12 (2017). https://doi.org/10.1136/bmj.j667
12. Abdel-Shafy, H.I., Mansour, M.S.M.: A review on polycyclic aromatic hydrocarbons: source, environmental impact, effect on human health and remediation. Egypt. J. Pet. **25**(1), 107–123 (2016). https://doi.org/10.1016/j.ejpe.2015.03.011
13. Garrido, A., Jiménez-Guerrero, P., Ratola, N.: Levels, trends and health concerns of atmospheric PAHs in Europe. Atmos. Environ. **99**, 474–484 (2014). https://doi.org/10.1016/j.atmosenv.2014.10.011
14. Foan, L., et al.: Reconstructing historical trends of polycyclic aromatic hydrocarbon deposition in a remote area of Spain using herbarium moss material. Atmos. Environ. **44**(26), 3207–3214 (2010). https://doi.org/10.1016/j.atmosenv.2010.05.019
15. Shen, H., et al.: Global atmospheric emission of polycyclic aromatic hydrocarbons from 1960 to 2008 and future predictions. Environ. Sci. Technol. **47**, 6415–6424 (2013)
16. Lehndorff, E., Schwark, L.: Biomonitoring airborne parent and alkylated three-ring PAHs in the greater cologne conurbation II: regional distribution patterns. Environ. Pollut. **157**(5), 1706–1713 (2009). https://doi.org/10.1016/j.envpol.2008.12.006
17. Srogi, K.: Monitoring of environmental exposure to polycyclic aromatic hydrocarbons: a review. Environ. Chem. Lett. **5**(4), 169–195 (2007). https://doi.org/10.1007/s10311-007-0095-0

18. Sharma, H., Jain, V.K., Khan, Z.H.: Characterization and source identification of polycyclic aromatic hydrocarbons (PAHs) in the urban environment of Delhi. Chemosphere **66**(2), 302–310 (2007). https://doi.org/10.1016/j.chemosphere.2006.05.003
19. Orliński, R.: Multipoint moss passive samplers assessment of urban airborne polycyclic aromatic hydrocarbons: concentrations profile and distribution along Warsaw main streets. Chemosphere **48**(2), 181–186 (2002). https://doi.org/10.1016/S0045-6535(02)00062-0
20. Prajapati, S.K., Tripathi, B.D.: Biomonitoring seasonal variation of urban air polycyclic aromatic hydrocarbons (PAHs) using *Ficus benghalensis* leaves. Environ. Pollut. **151**(3), 543–548 (2008). https://doi.org/10.1016/j.envpol.2007.04.013
21. Rengarajan, T., Rajendran, P., Nandakumar, N., Lokeshkumar, B., Rajendran, P., Nishigaki, I.: Exposure to polycyclic aromatic hydrocarbons with special focus on cancer. Asian Pac. J. Trop. Biomed. **5**(3), 182–189 (2015). https://doi.org/10.1016/S2221-1691(15)30003-4
22. Noth, E.M., Katharine Hammond, S., Biging, G.S., Tager, I.B.: Mapping and modeling airborne urban phenanthrene distribution using vegetation biomonitoring. Atmos. Environ. **77**, 518–524 (2013). https://doi.org/10.1016/j.atmosenv.2013.05.056
23. Ratola, N., Alves, A., Psillakis, E.: Biomonitoring of polycyclic aromatic hydrocarbons contamination in the Island of crete using pine needles. Water. Air. Soil Pollut. **215**(1–4), 189–203 (2011). https://doi.org/10.1007/s11270-010-0469-y
24. Rodriguez, J.H., Wannaz, E.D., Franzaring, J., Klumpp, A., Fangmeier, A., Pignata, M.L.: Biomonitoring of airborne fluoride and polycyclic aromatic hydrocarbons in industrial areas of Cordoba, Argentina, using standardized grass cultures of *Lolium multiflorum*. Atmos. Pollut. Res. **6**(3), 444–453 (2015). https://doi.org/10.5094/APR.2015.049
25. Tobiszewski, M., Namieśnik, J.: PAH diagnostic ratios for the identification of pollution emission sources. Environ. Pollut. **162**, 110–119 (2012). https://doi.org/10.1016/j.envpol.2011.10.025
26. Desalme, D., Binet, P., Chiapusio, G.: Challenges in tracing the fate and effects of atmospheric polycyclic aromatic hydrocarbon deposition in vascular plants. Environ. Sci. Technol. **47**(9), 3967–3981 (2013). https://doi.org/10.1021/es304964b
27. Alam, M.S., Delgado-Saborit, J.M., Stark, C., Harrison, R.M.: Using atmospheric measurements of PAH and quinone compounds at roadside and urban background sites to assess sources and reactivity. Atmos. Environ. **77**, 24–35 (2013). https://doi.org/10.1016/j.atmosenv.2013.04.068
28. Zelinkova, Z., Wenzl, T.: The occurrence of 16 EPA PAHs in food—a review. Polycycl. Aromat. Compd. **35**(2–4), 248–284 (2015). https://doi.org/10.1080/10406638.2014.918550
29. Haritash, A.K., Kaushik, C.P.: Biodegradation aspects of polycyclic aromatic hydrocarbons (PAHs): a review. J. Hazard. Mater. **169**(1–3), 1–15 (2009). https://doi.org/10.1016/j.jhazmat.2009.03.137
30. Rodriguez, J.H., Wannaz, E.D., Salazar, M.J., Pignata, M.L., Fangmeier, A., Franzaring, J.: Accumulation of polycyclic aromatic hydrocarbons and heavy metals in the tree foliage of *Eucalyptus rostrata*, *Pinus radiata* and *Populus hybridus* in the vicinity of a large aluminium smelter in Argentina. Atmos. Environ. **55**, 35–42 (2012). https://doi.org/10.1016/j.atmosenv.2012.03.026
31. Wannaz, E.D., Carreras, H.A., Rodriguez, J.H., Pignata, M.L.: Use of biomonitors for the identification of heavy metals emission sources. Ecol. Indic. **20**, 163–169 (2012). https://doi.org/10.1016/j.ecolind.2012.02.022
32. Kipopoulou, A.M., Manoli, E., Samara, C.: Bioconcentration of polycyclic aromatic hydrocarbons in vegetables grown in an industrial area. Environ. Pollut. **106**, 369–380 (1999)
33. Wannaz, E.D., Abril, G.A., Rodriguez, J.H., Pignata, M.L.: Assessment of polycyclic aromatic hydrocarbons in industrial and urban areas using passive air samplers and leaves of *Tillandsia capillaris*. J. Environ. Chem. Eng. **1**(4), 1028–1035 (2013). https://doi.org/10.1016/j.jece.2013.08.012
34. Orecchio, S.: PAHs associated with the leaves of *Quercus ilex* L.: extraction, GC-MS analysis, distribution and sources. Assessment of air quality in the Palermo (Italy) area. Atmos. Environ. **41**(38), 8669–8680 (2007). https://doi.org/10.1016/j.atmosenv.2007.07.027

35. Walgraeve, C., Demeestere, K., Dewulf, J., Zimmermann, R., Van Langenhove, H.: Oxygenated polycyclic aromatic hydrocarbons in atmospheric particulate matter: molecular characterization and occurrence. Atmos. Environ. **44**(15), 1831–1846 (2010). https://doi.org/10.1016/j.atmosenv.2009.12.004
36. Lehndorff, E., Schwark, L.: Biomonitoring of air quality in the Cologne conurbation using pine needles as a passive sampler—Part II: polycyclic aromatic hydrocarbons (PAH). Atmos. Environ. **38**(23), 3793–3808 (2004). https://doi.org/10.1016/j.atmosenv.2004.03.065
37. Sun, F., Wen, D., Kuang, Y., Li, J., Li, J., Zuo, W.: Concentrations of heavy metals and polycyclic aromatic hydrocarbons in needles of Masson pine (*Pinus massoniana* L.) growing nearby different industrial sources. J. Environ. Sci. **22**(7), 1006–1013 (2010). https://doi.org/10.1016/S1001-0742(09)60211-4
38. Walgraeve, C., De Wispelaere, P., Van der Elst, F., Van Langenhove, H.: Development of an analytical method to determine oxy-PAHs and PAHs in *Taxus baccata* leaves. Anal. Bioanal. Chem. **409**(1), 335–347 (2017). https://doi.org/10.1007/s00216-016-0008-9
39. Wolterbeek, B.: Biomonitoring of trace element air pollution: principles, possibilities and perspectives. Environ. Pollut. **120**(1), 11–21 (2002). https://doi.org/10.1016/S0269-7491(02)00124-0
40. Ratola, N., Lacorte, S., Alves, A., Barceló, D.: Analysis of polycyclic aromatic hydrocarbons in pine needles by gas chromatography-mass spectrometry. Comparison of different extraction and clean-up procedures. J. Chromatogr. A **1114**(2), 198–204 (2006). https://doi.org/10.1016/j.chroma.2006.03.110
41. Bonanno, G.: *Ricinus communis* as an element biomonitor of atmospheric pollution in urban areas. Water. Air. Soil Pollut. **225**(2) (2014). https://doi.org/10.1007/s11270-013-1852-2
42. Tomashuk, T.A., Truong, T.M., Mantha, M., McGowin, A.E.: Atmospheric polycyclic aromatic hydrocarbon profiles and sources in pine needles and particulate matter in Dayton, Ohio, USA. Atmos. Environ. **51**, 196–202 (2012). https://doi.org/10.1016/j.atmosenv.2012.01.028
43. Pereira Netto, A.D., Barreto, R.P., Moreira, J.C., Arbilla, G.: Spatial distribution of polycyclic aromatic hydrocarbons in *Terminalia catappa* L. (Combretaceae) bark from a selected heavy road traffic area of Rio de Janeiro City, Brazil. J. Hazard. Mater. **142**(1–2), 389–396 (2007). https://doi.org/10.1016/j.jhazmat.2006.08.034
44. De Nicola, F., et al.: Improved biomonitoring of airborne contaminants by combined use of holm oak leaves and epiphytic moss. Chemosphere **92**(9), 1224–1230 (2013). https://doi.org/10.1016/j.chemosphere.2013.04.050
45. Lehndorff, E., Schwark, L.: Biomonitoring of air quality in the Cologne conurbation using pine needles as a passive sampler—Part III: major and trace elements. Atmos. Environ. **44**(24), 2822–2829 (2010). https://doi.org/10.1016/j.atmosenv.2010.04.052
46. Vingiani, S., De Nicola, F., Purvis, W.O.: Active Biomonitoring of Heavy Metals and PAHs with Mosses and Lichens : A Case Study in the Cities of Naples and London (2015). https://doi.org/10.1007/s11270-015-2504-5
47. Wang, Y.Q., Tao, S., Jiao, X.C., Coveney, R.M., Wu, S.P., Xing, B.S.: Polycyclic aromatic hydrocarbons in leaf cuticles and inner tissues of six species of trees in urban Beijing. Environ. Pollut. **151**(1), 158–164 (2008). https://doi.org/10.1016/j.envpol.2007.02.005
48. Sojinu, O.S., Sonibare, O.O., Ekundayo, O., Zeng, E.Y.: Biomonitoring potentials of polycyclic aromatic hydrocarbons (PAHs) by higher plants from an oil exploration site, Nigeria. J. Hazard. Mater. **184**(1–3), 759–764 (2010). https://doi.org/10.1016/j.jhazmat.2010.08.104
49. De Nicola, F., Lancellotti, C., Prati, M.V., Maisto, G., Alfani, A.: Biomonitoring of PAHs by using *Quercus ilex* leaves: source diagnostic and toxicity assessment. Atmos. Environ. **45**(7), 1428–1433 (2011). https://doi.org/10.1016/j.atmosenv.2010.12.022

50. Ratola, N., Lacorte, S., Barceló, D., Alves, A.: Microwave-assisted extraction and ultrasonic extraction to determine polycyclic aromatic hydrocarbons in needles and bark of *Pinus pinaster* Ait. and *Pinus pinea* L. by GC-MS. Talanta **77**(3), 1120–1128 (2009). https://doi.org/10.1016/j.talanta.2008.08.010
51. Orecchio, S., Gianguzza, A., Culotta, L.: Absorption of polycyclic aromatic hydrocarbons by Pinus bark: analytical method and use for environmental pollution monitoring in the palermo area (Sicily, Italy). Environ. Res. **107**(3), 371–379 (2008). https://doi.org/10.1016/j.envres.2008.02.010
52. Klánová, J., Čupr, P., Baráková, D., Šeda, Z., Anděl, P., Holoubek, I.: Can pine needles indicate trends in the air pollution levels at remote sites? Environ. Pollut. **157**(12), 3248–3254 (2009). https://doi.org/10.1016/j.envpol.2009.05.030
53. Lehndorff, E., Schwark, L.: Biomonitoring airborne parent and alkylated three-ring PAHs in the greater Cologne conurbation I: temporal accumulation patterns. Environ. Pollut. **157**(4), 1323–1331 (2009). https://doi.org/10.1016/j.envpol.2008.11.037
54. Sharma, A.P., Tripathi, B.D.: Assessment of atmospheric PAHs profile through *Calotropis gigantea* R.Br. leaves in the vicinity of an Indian coal-fired power plant. Environ. Monit. Assess. **149**(1–4), 477–482 (2009). https://doi.org/10.1007/s10661-008-0224-7
55. Alfani, A., De Nicola, F., Maisto, G., Prati, M.V.: Long-term PAH accumulation after bud break in *Quercus ilex* L. leaves in a polluted environment. Atmos. Environ. **39**(2), 307–314 (2005). https://doi.org/10.1016/j.atmosenv.2004.09.001
56. Amigo, J.M., Ratola, N., Alves, A.: Study of geographical trends of polycyclic aromatic hydrocarbons using pine needles. Atmos. Environ. **45**(33), 5988–5996 (2011). https://doi.org/10.1016/j.atmosenv.2011.07.058
57. De Nicola, F., Maisto, G., Prati, M.V., Alfani, A.: Leaf accumulation of trace elements and polycyclic aromatic hydrocarbons (PAHs) in *Quercus ilex* L. Environ. Pollut. **153**(2), 376–383 (2008). https://doi.org/10.1016/j.envpol.2007.08.008
58. Kodnik, D., Candotto Carniel, F., Licen, S., Tolloi, A., Barbieri, P., Tretiach, M.: Seasonal variations of PAHs content and distribution patterns in a mixed land use area: a case study in NE Italy with the transplanted lichen *Pseudevernia furfuracea*. Atmos. Environ. **113**, 255–263 (2015). https://doi.org/10.1016/j.atmosenv.2015.04.067
59. Nascimbene, J., et al.: Patterns of traffic polycyclic aromatic hydrocarbon pollution in mountain areas can be revealed by lichen biomonitoring: a case study in the Dolomites (Eastern Italian Alps). Sci. Total Environ. **475**, 90–96 (2014). https://doi.org/10.1016/j.scitotenv.2013.12.090
60. Ratola, N., Amigo, J.M., Oliveira, M.S.N., Araújo, R., Silva, J.A., Alves, A.: Differences between *Pinus pinea* and *Pinus pinaster* as bioindicators of polycyclic aromatic hydrocarbons. Environ. Exp. Bot. **72**(2), 339–347 (2011). https://doi.org/10.1016/j.envexpbot.2011.04.012
61. De Nicola, F., Maisto, G., Prati, M.V., Alfani, A.: Temporal variations in PAH concentrations in *Quercus ilex* L. (holm oak) leaves in an urban area. Chemosphere **61**(3), 432–440 (2005). https://doi.org/10.1016/j.chemosphere.2005.01.082

Chapter 39
Satellite Rainfall Estimation from Himawari-8 Multi Channels Observation Based on AWS Data Trained Machine Learning Methods

Farid Lasmono, Risyanto, Fadli Nauval, Elfira Saufina, Trismidianto, and Teguh Harjana

Abstract Rainfall estimation from geostationary satellites has the advantage of its relatively high temporal resolution. Therefore, the data can be used for supporting the early warning systems of hydrometeorological disasters. This study implemented machine learning methods to estimate rainfall using the multi-channels data from the Himawari-8 satellite at several locations in Indonesia, i.e., Bandung (West Java), Pasuruan (East Java), and Agam (West Sumatera). The ground-based rainfall measurements taken from installed Automatic Weather Stations (AWS) at the three sites were involved in providing reference values and labels of data training. The brightness temperature (BT) of 9 infrared channels of Himawari-8 and 36 combinations of their brightness temperature differences (BTD) were used as explanatory variables. We introduced the three steps of machine learning procedures: (i) separation of rain and no-rain event; (ii) classification of rainfall category; and (iii) estimation of rainfall rate. Generally, machine learning methods used in this study for rainfall retrieval are applicable for site-scale with acceptable accuracies. The Linear Discriminant Analysis (LDA) model has had the highest accuracy (Acc = 0.75, POD = 0.75, SR = 0.76, POFD = 0.24) compared to the other 8 classifiers models for detecting rain and no-rain event (step 1) and for classifying the categorical rainfall (step 2). The selected regression models for rain rate estimation, based on the highest correlation, included the Extra Trees Regressor (ETR), the Random Forest (RF), and the Multiple Linear Regression (MLR). Overall, we obtained a good coefficient of correlation from comparing the actual rainfall and the estimated from Himawari-8 ($R^2 = 0.92$). However, the generalization results of these selected models to be applied for larger spatial coverage areas has not been produced consistently. The lack of AWS data sites and the relatively short data period might be the issues. From the calculation of feature importance, we found that BTD of $BT_{9.6\mu} - BT_{13.3\mu}$ (Band12 and 16), which correlates to cloud top height, had the highest contribution for rainfall retrieval based on the LDA machine learning method.

F. Lasmono (✉) · Risyanto · F. Nauval · E. Saufina · Trismidianto · T. Harjana
Research Center for Climate and Atmosphere, National Research and Innovation Agency (BRIN), Bandung, Indonesia
e-mail: farid.lasmono@brin.go.id

E. Yulihastin et al. (eds.), *Proceedings of the International Conference on Radioscience, Equatorial Atmospheric Science and Environment and Humanosphere Science, 2021*, Springer Proceedings in Physics 275, https://doi.org/10.1007/978-981-19-0308-3_39

39.1 Introduction

Rainfall data are crucial in many hydrometeorology applications such as the Decision Support System (DSS) related to hydrometeorological disasters. A high-quality rainfall data, in terms of accuracy, frequent monitoring, spatial resolution, and significant area coverage are needed allowing the DSS to recommend the best possible action to mitigate the disasters. These requirements, however, sometimes cannot be fulfilled by only one kind of rainfall observation instrument. The satellite's infrared instruments, for example, have limitations on their indirect relationship with rain and only detect the top properties of clouds. Nevertheless, the geostationary satellites data still become a primary source of information for the DSS due to their advantage on relatively high temporal resolution which enables the instrument to nicely capture the cloud development and estimate the precipitation on the near real-time observation.

Himawari-8 is a made-in-Japan geostationary satellite that has operates since 2015, covering most of the Asia and the Australia, including Indonesia. The satellite is equipped with an Advance Himawari Imager (AHI) instrument, which can observe the Earth and atmosphere for every 10-min through the three visible (VIS), three near-infrared (NIR), and ten infrared (IR) bands [1]. Despite its difficulties in revealing the whole 3-dimension structure of clouds, this multi-channel sensor at least increases the possibility to yield a better rainfall retrieval result. On the other side, the techniques to retrieve rainfall from geo-satellite have been much developed, such as by machine learning methods. It is a method that allows a computer to make decisions or predictions based on data patterns and characteristics [2].

Many studies have introduced the application of machine learning for estimating rainfall. Hirose et al. [3] implemented a rain forest algorithm for Himawari-8 rainfall estimation and found that the method was also able to detect the warm type precipitation clouds for the case of heavy rainfall in Japan. Kuhnlein et al. [4, 5] also implemented said algorithm for Meteosat Second Generation (MSG) Spinning Enhanced Visible and Infrared Imager (SEVIRI) which produced good results. Tsay et al. [6] incorporated the Convolutional Auto-Encoder (CAE) deep learning method to the established CNN-PERSIANN algorithm for geo-satellite rainfall retrieval operated in Taiwan. For the Indonesia region case, Risyanto et al. [7] implemented the machine learning methods from Himawari-8 infrared trained by the GSMaP precipitation products. The study showed a decent accuracy result. However, the estimated rainfalls were provided in rain rate classification. The values of rain intensity have not yet been determined.

This study implemented the machine learning methods from the brightness temperature (BT) of IR channels of Himawari-8 to estimate rain rate based on the Automatic Weather Station (AWS) rainfall data training. Due to the lack of data availability, we used only three locations of AWS sites, consists of Bandung (West Java), Pasuruan (East Java), and Agam (West Sumatera). We proposed a 'three steps' machine learning scheme to minimize errors that appear higher when applying machine learning conversion directly from the brightness temperature to rain rate product. We then generalized the selected machine learning models to be implemented for large area scales. In this study, we also investigated the contribution of each predictor (BT and their differences/BTD of IR channels) in retrieving rainfall, so that the ML procedures can be conducted efficiently.

39.2 Data and Methods

The Himawari-8 BT of IR channels dataset was obtained from a HimawariCast receiver installed at Research Center of Atmospheric Science and Technology (PRSTA), LAPAN, Bandung, West Java, Indonesia. The rainfall data of the AWS instrument were extracted from three locations, i.e., Bandung, Pasuruan, and Agam site, to provide reference labels and rain rate values for data training. We established data training from 1 December 2019 to 29 February 2020 based on the sample balanced method, so that the number of rainy data samples was matched with non-rainy data samples, and represented enough all rainfall categories. Himawari-8 data were taken at the point of the AWS site, and the pixel size is about 4 $\times$ 4 km area. Both AWS and Himawari-8 data were matched temporally to become hourly data. The list of BT and BTDs (brightness temperature difference) as explanatory variables (predictors) from all possible combinations of IR channels is shown in Table 39.1. For the validation purpose, we separated 20% of the dataset randomly. In addition, the GSMaP data product was also used to verify the ML result for a larger spatial area.

The ML procedure in this study was conducted in three steps as depicted in Fig. 39.1. The first step of ML procedure aims to separate rain and no-rain events. The rain event detected from step 1 will be continued to the following procedure, that is, classification of categorical rainfall, consisting of light, medium, heavy, and

Table 39.1 Combination of the Himawari-8 observation band that is considered, as an explanatory variable. Numbers indicate observation center wavelength

μm	6.2	6.9	7.3	8.6	9.6	10.4	11.2	12.4	13.3
6.2	6.2	6.2–6.9	6.2–7.3	6.2–8.6	6.2–9.6	6.2–10.4	6.2–11.2	6.2–12.4	6.2–13.3
6.9		6.9	6.9–7.3	6.9–8.6	6.9–9.6	6.9–10.4	6.9–11.2	6.9–12.4	6.9–13.3
7.3			7.3	7.3–8.6	7.3–9.6	7.3–10.4	7.3–11.2	7.3–12.4	7.3–13.3
8.6				8.6	8.6–9.6	8.6–10.4	8.6–11.2	8.6–12.4	8.6–13.3
9.6					9.6	9.6–10.4	9.6–11.2	9.6–12.4	9.6–13.3
10.4						10.4	10.4–11.2	10.4–12.4	10.4–13.3
11.2							11.2	11.2–12.4	11.2–13.3
12.4								12.4	12.4–13.3
13.3									13.3

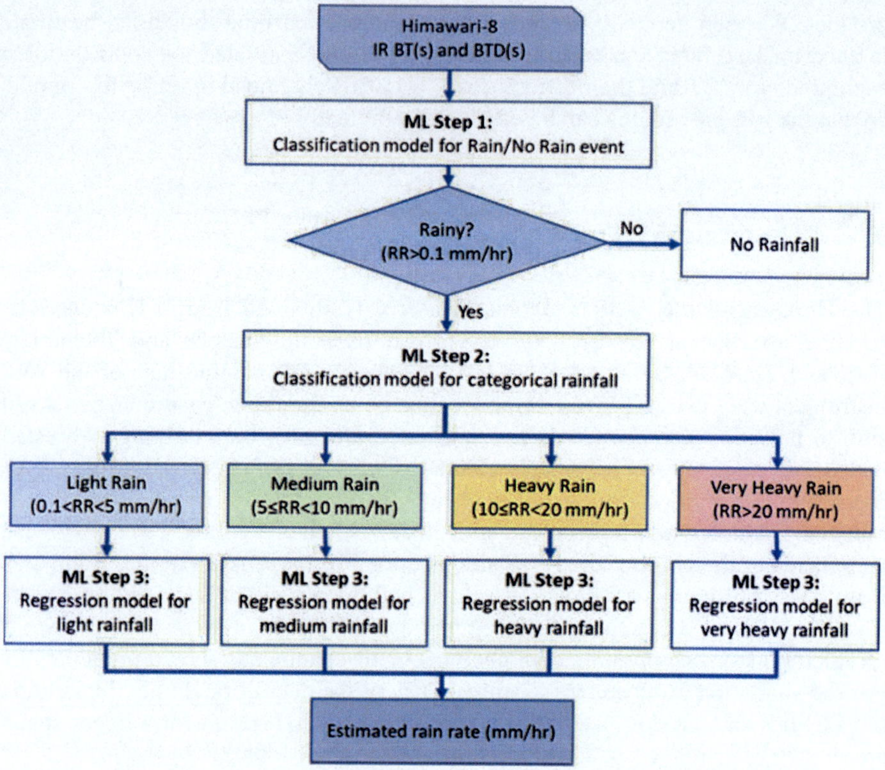

Fig. 39.1 Machine learning procedure for satellite rainfall estimation used in the study

very heavy rain. This rainfall category is based on Indonesia Meteorology, Climatology, and Geophysics Agency (BMKG) criteria [8], as shown in Table 39.2. The ML classification as implemented in steps 1 and 2 used nine different algorithms, i.e., Logistic Regression (LR), K-Neighbour Classifier (KNN), Gaussian Naive Bayes (NB), Linear Discriminant Analysis (LDA), Quadratic Linear Analysis (QDA), Classification and Regression Tree (CART), Random Forest Classifier (RF), Extra Trees Classifier (ETC) and Multi-Layer Perceptron (MLP). The third procedure of ML is to estimate the rain rate from each categorical rainfall by applying the regression ML

Table 39.2 Rain rate category based on BMKG criteria

Class/category	Rain rate intensity	
	Hourly (mm/h)	Daily (mm/day)
Light rainfall (drizzle)	0.1–4.9	0.1–19.9
Medium rainfall	5.0–9.9	20.0–49.9
Heavy rainfall	10.0–20	50.0–100
Very heavy rainfall	>20	>100

Table 39.3 Contingency table used for computation of categorical statistics

Model estimated	Observed data	
	Rain	No-rain
Rain	Hits (H)	False alarm (F)
No-rain	Misses (M)	Correct negative (C)

method. The algorithms include Linear Regression (LIN), Bayesian Ridge (BR), Support Vector Machine (SVM), K-Neighbors Regression (KNR), Decision Tree Regression (DTR), Gradient Boosting Regression (GBR), Random Forest Regression (RFR), Multiple Linear Regression (MLR) and Extra Trees Regression (ETR). All processes of ML in the study utilized the SK-learn python library package developed by Scikit Learn Team [9].

The use of several different algorithms for classification and regression in the study was conducted to select the best model for each ML step, based on their accuracy rank. The accuracy score (fraction of correct prediction) is defined as [8]:

$$accuracy(y, \hat{y}) = \frac{1}{n_{samples}} \sum_{i=1}^{n_{sample}-1} 1(\hat{y}_i = y_i), \tag{39.1}$$

where y is the estimated/predicted value of rainfall, $\hat{y}$ the actual rainfall condition, and n is the number of samples and i is -th sample. The accuracy ranges from 0 to 1, where 0 means all estimations/predictions are false, while 1 means all estimation/prediction are accurate. In addition, the categorical statistics score, including the Probability of Detection (POD), Success Ratio (SR), and Probability of False Detection (POFD), were also calculated, especially in classifying rain and no-rain event. The scores are calculated as follows [10], while the contingency table (Table 39.3) shows the categories included in the formula [11, 12]:

$$POD = H/(H + M) \tag{39.2}$$

$$SR = H/(H + F) \tag{39.3}$$

$$POFD = F/(F + C) \tag{39.4}$$

39.3 Results and Discussion

This section will describe the results and the discussion of the research with data and methods explained in the previous section.

39.3.1 Classification Model of Rain Event and Categorical Rainfall

The first ML procedure was performed to separate rain and no rain event. We found that the Linear Discriminant Analysis (LDA) model has had the highest accuracy score in this step. Figure 39.2a shows the accuracy score for each location. The mean accuracy score at Pasuruan (86%) is higher than at Bandung (66%) and Agam (63%). Satellite observation usually has a limitation on detecting local scale activities, thus it needs the relatively large scale of rain events to be sufficiently detected by satellite sensor. In the case of Pasuruan, this high accuracy means that satellite was able to distinguish the rain and no-rain event from the BTs and BTDs observation. It is possibly because the rain in Pasuruan site were large enough and most of them were the convective type precipitation, which is more detectable by the IR channels instruments. On the contrary, the lower accuracy scores in Bandung and Agam were probably due to the frequent local rain event and the occurrence of non-convective type precipitation, such as the rainfall that generated from the warm type clouds.

In the attempt for model generalization, we combined the data from the three sites and established another data training. The result accuracy for this merged scheme is shown in Fig. 39.2b. The LDA model gave the highest accuracy (75%) and SR score (76%), and the lowest POFD (24%). The MLP model, despite having an unusual low SR (16%), has had the highest POD (77%). However, the model also generated the lowest accuracy score and the highest POFD, which means the result of MLP model was not consistent. The LDA still the model to be selected for detecting rain and no-rain events.

The rainfall retrieval was then proceeded to the second step, which is the classification of categorical rainfall. This ML step 2 took a similar scheme as the first step that merged the three sites data. The result indicated that, again, the LDA model got the first rank based on an accuracy score (66%) as shown in Fig. 39.3. The mean accuracy for the ML step 2 was lower than the step 1, which is acceptable since the

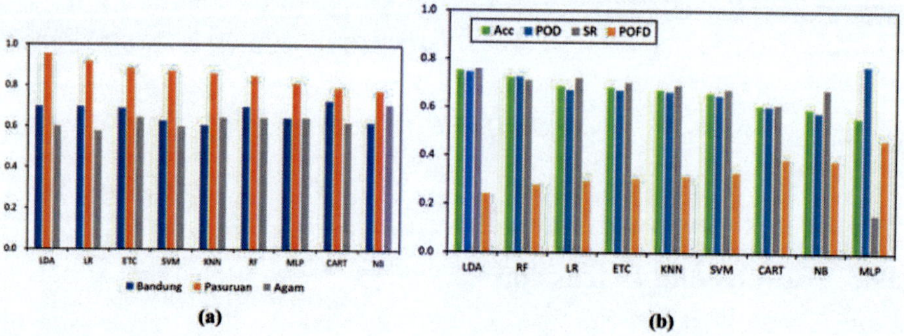

Fig. 39.2 **a** Accuracy score of classifier models used for detecting the event of rain or no-rain at the three sites. **b** Accuracy, POD, success ratio (SR), and POFD calculated from the classifier models used for detecting the event of rain or no-rain by merging the three sites data

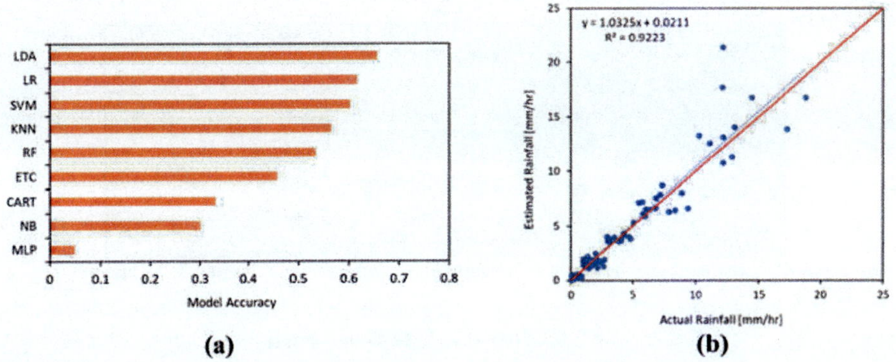

Fig. 39.3 **a** Accuracy calculated from the classifier models used for detecting categorical rainfall by merging the 3 sites data. **b** Comparison of actual estimated rainfall rate from machine learning based Himawari-8 at 3 sites

algorithms need to classify the dataset into 4 classes (categorical rainfall), compared to the step 1 that only has 2 classes (rain and no-rain).

The last ML procedure conducted in the study is to convert each rainfall category to rain rate estimation by utilizing regression models. As stated, 9 regression models were investigated in this step. The test of accuracy score showed that the ETR were the selected models for the light rainfall, followed by the RFR for medium rainfall, and the MLR for both heavy and very heavy rainfall category. Comparison plot of actual rain rate taken from the AWS data and the estimated rain rate from Himawari-8 ML is displayed in Fig. 39.3b. The comparison shows a good correlation ($R^2 = 0.92$) between the actual and estimated rain rate, which means the ML method used in the study generally can be applied for rainfall retrieval from BT and BTD of IR channels of Himawari-8. However, the plot also shows larger errors obtained on the heavy to very heavy rainfall category. This was probably due to the lack of samples of data training for the heavy and very heavy rainfall events available during the study period.

39.3.2 Generalization for Spatial Area Coverage

The result of ML rainfall retrieval aforementioned above is based on the AWS data trained from the available 3 sites. In the study we also conducted an attempt to generalize the selected models to be implemented to larger spatial area coverage. The LDA model would be used for both classification of rain/no-rain area and categorical rainfall separation. Figure 39.4a shows the result of the LDA model for rain/no-rain classification from Himawari-8 data at 1 January 2020, 00UTC (for reference, the GSMaP product at a similar time observation is displayed in Fig. 39.5a). It is clearly shown that the LDA model failed to deliver good pattern and distribution of rainfall

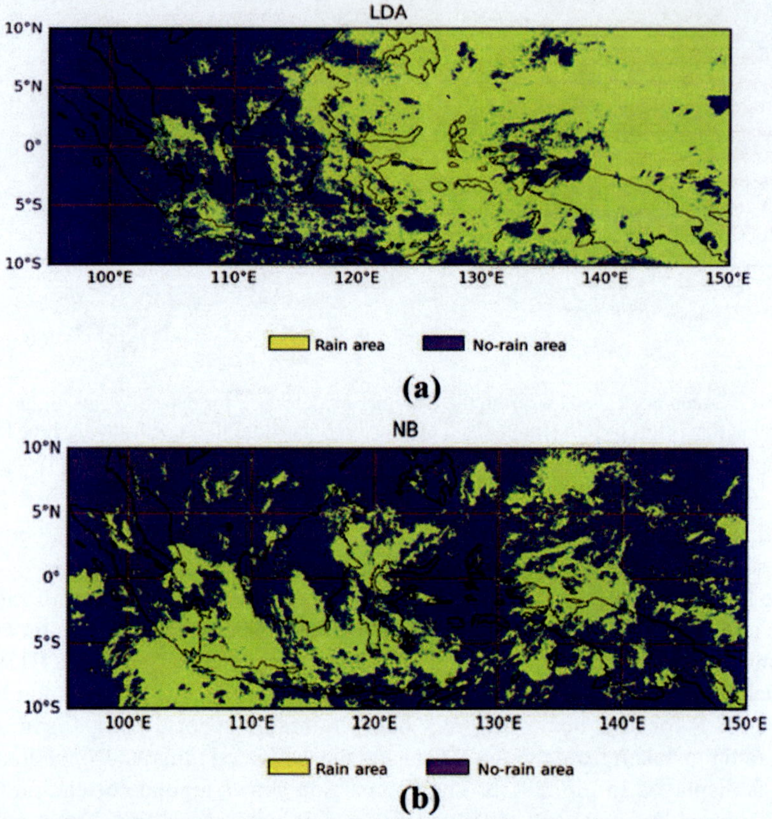

Fig. 39.4 Rain area classification generated from the Linear Discriminant Analysis (LDA) **a** and Gaussian Naïve Bayes (GNB). **b** Model implemented on Himawari-8 spatial data at 1 January 2020, 00 UTC

area in this case. We then performed another experiment that used the other classifier models. The best model that spatially nearly matched the GSMaP data was the Gaussian Naïve Bayes (GNB), shown in Fig. 39.4b. A similar result is also reported by Risyanto et al. [7], which found the GNB as the most suited model for separating the pixels from rain and no-rain. The GNB model, however, in this case mainly still produced overestimation on many rainfalls detected areas.

To further proceed the rain area to be converted to rain rate estimation, we conducted the ML procedure of step 2 and 3 by using the selected models, i.e., the LDA for categorical rainfall classification, and the regression models for each rainfall category (light, medium, heavy, and very heavy). The final result of this procedure is shown in Fig. 39.5b. We can conclude from the comparison with the GSMaP data, that the estimated Himawari-8 rainfall could not generate consistent rainfall pattern and distribution, especially when generating the classification of categorical rainfall. The LDA model established from the AWS data trained cannot be applied for the

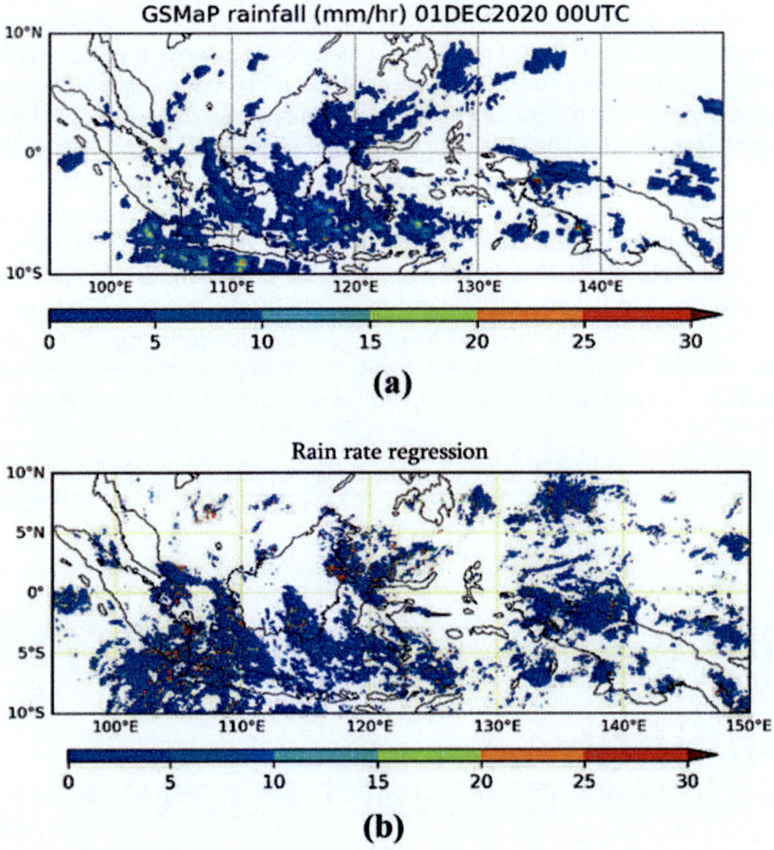

Fig. 39.5 **a** GSMaP rainfall at 1 January 2020, 00 UTC. **b** Machine learning estimated rainfall from Himawari-8 using the selected models at 1 January 2020, 00 UTC

case of a larger spatial scale area. Overall, the performance of ML models for the generalization case has not produced the good results as expected. This was possible because of the lack of the AWS sites used for data training in the study, as skillfully as the short period, only three months data, which limited the samples of rainfall events in each category, especially for heavy and very heavy rainfall.

39.3.3 Feature Importance Calculation

We investigated the contribution from each BT and BTD of Himawari-8 IR channels to the retrieved rainfall. For this purpose, we used the calculation of feature importance (FI) to find which explanatory variable has the most contribution. FI is a score assigned to the features of an ML model that defines how important is a part to the model's prediction. In this case, the features included all the 9 BT and 36 BTD of IR used in the study for proceeding the ML process. Since the first step of ML procedure, i.e., detecting rain/no rain event, is the initial crucial part of the retrieval, we calculated the FI score for this procedure. Figure 39.6a shows the result of the FI calculation for the LDA classification of rain/no-rain event. It indicates that B1216, which is BTD of Band 12 (9.6μ) and Band 16 (13.3μ) is the most crucial feature contributes to the rain event prediction. The BTD of 9.6μ and 13.3μ has been used as a proxy for cloud top height [13], so it was acceptable it has a good correlation to the predicted rainfall event. We also conducted the FI calculation for the GNB model, which was the selected model we used in the study for the generalization of the rain/no-rain area detection model. The result can be viewed in Fig. 39.6b. It shows that the BTD of Band 11 (8.6μ) and Band 15 (12.4μ) has the most contribution. Krebs et al. [14] and Strabala et al. [15] have used this BTD as a proxy for cloud

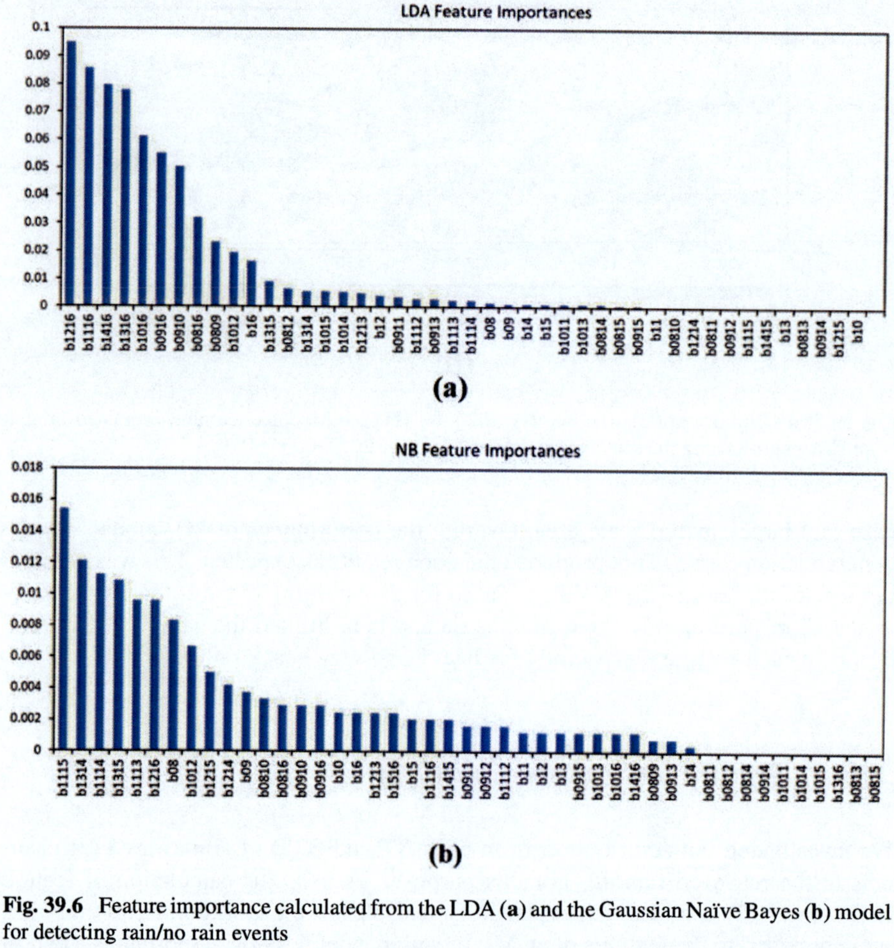

Fig. 39.6 Feature importance calculated from the LDA (**a**) and the Gaussian Naïve Bayes (**b**) model for detecting rain/no rain events

optical thickness retrieval. FI calculation from Fig. 39.6a, b indicates that the most contribution are coming from the BTD amongst the Band 11 to Band 16 (between 8.6μ and 13.3μ), which correlates to the cloud properties [15].

39.4 Conclusion

In general, machine leaning methods used in this study for rainfall retrieval is applicable for the site scale with acceptable accuracies. The Linier Discriminant Analysis (LDA) model has had the highest accuracy (Acc = 0.75, POD = 0.75, SR = 0.76, POFD = 0.24) compared to the other 8 classifiers models for detecting rain and no-rain event (step 1) and for classifying the categorical rainfall (step 2). The selected regression models for rain rate estimation, based on the highest correlation, included the Extra Trees Regressor (ETR), the Random Forest (RF), and the Multiple Linear Regression (MLR). A good coefficient of correlation has been obtained from the comparison of the actual rainfall and the estimated from Himawari-8 ($R^2 = 0.92$). We conducted an attempt to generalize these selected models to be implemented for larger spatial coverage area, which has not produced consistent results. This was probably due to the lack of AWS data sites and the relatively short data period used in the study. Based on the calculation of feature importance, we found that BTD(9.6μ–13.3μ), and BTD(8.6μ–12.4μ) had the highest contribution for rainfall retrieval based on LDA and GNB machine learning method.

Acknowledgements This work is fully supported by Research Center for Atmospheric Science and Technology of LAPAN-BRIN. We would like to thank local operators in Bandung, Pasuruan and Agam sites for maintaining the instrument and preparing the rainfall dataset. We also acknowledge the Scikit Learn team for developing excellent python machine learning modules.

References

1. Bessho, K., Date, K., Hayashi, M., Ikeda, A., Imai, T., Inoue, H., Kumagai, Y., Miyakawa, T., Murata, H., Ohno, T., Okuyama, A., Oyama, R., Sasaki, Y., Shimazu, Y., Shimoji, K., Sumida, Y., Suzuki, M., Taniguchi, H., Tsuchiyama, H., Uesawa, D., Yokota, H., Yoshida, R.: An introduction to Himawari-8/9—Japan's new-generation geostationary meteorological satellites. J. Met. Soc. Jpn **94**(2), 151–183 (2016)
2. Koza, J.R., Bennett, F.H., Andre, D., Keane, M.A.: Automated Design of Both the Topology and Sizing of Analog Electrical Circuits Using Genetic Programming. Artificial Intelligence in Design '96, pp. 151–170. Springer, Dordrecht (1996)
3. Hirose, H., Shige, S., Yamamoto, M.K., Higuchi, A.: High temporal rainfall estimations from Himawari-8 multiband observations using the random-forest machine-learning method. J. Met. Soc. Jpn,. **97**(3), 689–710 (2019)
4. Kühnlein, M., Appelhans, T.: Precipitation estimates from MSG SEVIRI daytime, nighttime, and twilight data with random forests. J. Appl. Meteorol. Climatol. **53**, 2457–2480 (2014). https://doi.org/10.1175/JAMC-D-14-0082.1

5. Kühnlein, M., Appelhans, T., Thies, B., Nauß, T.: Improving the accuracy of rainfall rates from optical satellite sensors with machine learning—a random forests-based approach applied to MSG SEVIRI. Rem. Sen. Environ. **141**, 129–143 (2014). https://doi.org/10.1016/j.rse.2013.10.026
6. Tsay, J., Kao, K., Chao, C., Chang, Y.: Deep learning for satellite rainfall retrieval using Himawari-8 multiple spectral channels. Preprints 2020, 2020100648 (2020). https://doi.org/10.20944/preprints202010.0648.v1
7. Risyanto, F., Lasmono, T., Harjana.: Himawari-8 rainfall estimation from infrared channels based on machine learning methods. In: AIP Conference Proceedings 2366, 050004 (2021). https://doi.org/10.1063/5.0060010
8. Gustari, I., Wahyu Hadi, T., Hadi, S., Renggono, F.: Akurasi Prediksi Curah Hujan Harian Operasional di Jabodetabek: Perbandingan dengan model WRF. J. Meteorol. Dan Geofisika. **13**(2), 119–130 (2012)
9. Pedregosa, F., Varoquaux, G., Gramfort, A., Michel, V., Thirion, B., Grisel, O., Blondel, M., Prettenhofer, P., Weiss, R., Dubourg, V., Vanderplas, J., Passos, A., Cournapeau, D., Brucher, M., Perrot, M., Duchesnay, E.: Scikit-learn: machine learning in python. J. Mach. Learn. Res. **12**, 2825–2830 (2011)
10. Lekula, M., Lubczynski, M.W., Shemang, E.M., Verhoef, W.: Validation of satellite-based rainfall in Kalahari. Phys. Chem. Earth **105**, 84–97 (2018)
11. Ebert, E.E.: Methods for verifying satellite precipitation estimates. In: Measuring Precipitation from Space: EURAINSAT and the Future, pp. 345–356 (2007)
12. Stanski, H.R., Wilson, L.J., Burrows, W.R.: Survey of Common Verification Methods in Meteorology. Research Report No.89-5 (1989)
13. Kwon, E., Sohn, B., Schmetz, J., Wats, P.: Intercomparison of height assignment methods for opaque clouds over the tropics. J. Atmos. Sci. **46**(1), 11–19 (2010)
14. Krebs, W., Mannstein, H., Bugliaro, L., Mayer, B.: A new day and night time meteosat second generation cirrus detection algorithm mecida. J. Atmos. Chem. Phys. **7**, 10933–10969 (2007)
15. Strabala, K.I., Ackerman, S.A., Menzel, W.P.: Cloud properties inferred from 8–12micron data. J. Appl. Meteorol. **33**(2), 212–229 (1994)

Chapter 40
Utilization of Agroindustrial by Product for Bioinsecticide Production

Kirana Sanggrami Sasmitaloka, Mulyorini Rahayuningsih, and Titi Candra Sunarti

Abstract Microbial bioinsecticides are product produced by microorganisms that can kill insect pests and disease-carrying vectors. *Bacillus thuringiensis* is one of the well-known bioinsecticide resources, and is utilized in organic farming to replace chemical insecticides. Substrate composition affects the characteristics of bioinsecticides products, especially in growth, toxicity, and potential effects. This research aimed to produce bioinsecticides from *Bacillus thuringiensis* using agroindustrial by-product as a substrate and its characteristics. The study was designed using a completely randomized design, consisting of two factors, namely the type *Bacillus thuringiensis* (*Bacillus thuringiensis* subsp. *berliner* and *Bacillus thuringiensis* subsp. *aizawai*) and kinds of a substrate (cassava dregs, pulp of coffee, starch fractions of iles-iles, and sago dregs), with four replication. The results showed that *Bacillus thuringiensis* could be used agroindustrial by-products as a carbon source in solid media cultivation. Cultivation of *Bacillus thuringiensis* subsp. *aizawai* using cassava dregs as a substrate can produce the highest toxicity with a VSC of 10.9 log CFU/g, LC50 value of 0.09 μg/ml, and potency of 9103 IU/mg, during cultivation of *Bacillus thuringiensis* subsp. *berliner* using the pulp of coffee as a substrate can produce the highest toxicity with VSC of 10.15 log CFU/g, LC50 value 0.08 μg/ml, and potency 10,476 IU/mg.

40.1 Introduction

Bioinsecticides are materials contain toxic compounds that can kill insect pests and disease-carrying vectors by plants or using organisms such as viruses, bacteria, and fungi. Generally, it is utilized in organic farming for the replacement of chemical insecticide as pest control. The most widely developed bacteria were *Bacillus*

K. S. Sasmitaloka (✉)
Indonesian Center for Agricultural Postharvest Research and Development, Bogor, Indonesia
e-mail: ks.sasmitaloka@gmail.com

M. Rahayuningsih · T. C. Sunarti
Department of Agroindustrial Technology, Bogor Agricultural University, Bogor, Indonesia

E. Yulihastin et al. (eds.), *Proceedings of the International Conference on Radioscience, Equatorial Atmospheric Science and Environment and Humanosphere Science, 2021*, Springer Proceedings in Physics 275, https://doi.org/10.1007/978-981-19-0308-3_40

thuringiensis because these bacteria can form δ-endotoxin, which is toxic to insect larvae. The effect of δ-endotoxin as a bioinsecticide can be seen if eaten by insects, so it is relatively safe against other insects that do not consume plant parts [1]. However, the use of bioinsecticide In Indonesia is still rarely because their market is still an imported product, so that the price is relatively high. This problem can be solved by producing bioinsecticides using the active compounds of *B. thuringiensis* and agro-industrial by-products as a substrate.

Agroindustrial by-product have components that are needed for the microorganism's growth, namely carbon and nitrogen. Carbon is necessary for providing energy [2], while nitrogen is needed for the formation of biomass, while minerals used for spore growth and the production of δ-endotoxin [3]. Several studies state that bioinsecticides can be produced using agroindustrial by-products as substrates such as cassava dregs [4], soybean meal [5], and coconut water [6]. Their composition affects the characteristics of bioinsecticides products.

Bioinsecticides can be produced in solid media cultivation due to the characteristics of bacteria that support this cultivation, specifically facultative aerobics, motile, spores in solid media cultivation are formed faster than liquid media cultivation [7]. The residual substrate can be used as a protector of *B. thuringiensis.* This research aimed to produce bioinsecticides from *B. thuringiensis* using agroindustrial by-product as a substrate and its characteristics.

40.2 Materials and Methods

40.2.1 Materials

The isolates used were *Bacillus thuringiensis* subsp. *aizawai* (*BtA*) and *Bacillus thuringiensis* subsp. *berliner* (*BtB*) obtained from Bogor Agricultural University Culture Center. Raw materials used were cassava dregs (*Manihot utilissima*) obtained from Cimahpar (Bogor, West Java), the pulp of coffee (*Coffea* sp) obtained from Tanjung Karang (Lampung), the starch of iles-iles (*Amorphophallus muelleri Blume*) obtained from Madiun (East Java), and sago dregs (*Metroxylon* sp) obtained from Cimahpar (Bogor, West Java). The growth mediums used was nutrient broth and nutrient agar. Minerals used are $MgSO_4.7H_2O$, $MnSO_4.H_2O$, $ZnSO_4.7H_2O$, $FeSO_4.7H_2O$, and $CaCO_3$. The main equipment used was autoclave, laminar, oven, incubator, and others.

40.2.2 Bioinsecticides Production

One culture loop of *B. thuringiensis* was inoculated in 50 ml of nutrient broth (NB) and incubated in a rotary shaking incubator at 150 rpm, 30 °C for 15 h. The carbon source used is industrial agricultural waste which is reduced in size to 20 mesh. The nitrogen source commonly used in the production of bioinsecticides is urea. Fermentation was carried out on solid media using Capalbo et al. [8] method. The weight of the carbon source substrate was 25 g with the initial pH of 6.8–7.2 and incubated for 96 h in an incubator at 30 °C. The study was designed using a C/N ratio of 7 on each cultivation medium, and the a_w value was 0.92 [9]. The mineral composition (trace elements) were added according to the method of Rahayuningsih et al. [4], namely $CaCO_3$ (1 g/l), $MgSO_4.7H_2O$ (3 g/l), $MnSO_4.7H_2O$ (0.02 g/l), $ZnSO_4.7H_2O$ (0.02 g /l), and $FeSO_4.7H_2O$ (0.02 g/l). The treatment consisted of two factors, namely the type *B. thuringiensis* (*Bacillus thuringiensis* subsp. *berliner* and *Bacillus thuringiensis* subsp. *aizawai*) and kinds of the substrate (cassava dregs, pulp of coffee, starch fractions of iles-iles, and sago dregs), with four replication.

40.2.3 Analysis

Bioinsecticides were analyzed for their performance, such as loss of fiber content [10], total cell count [11], total spore count [11], toxicity (LC50) [12], and potency [13]. Our research was conducted by Completely Random Design (CRD). All data reported as a mean of four replicates. The means was compared using the Duncan test with the significant level at 5% using SPSS software 21 version.

40.3 Results and Discussion

40.3.1 Loss of Fiber Content

The substrate used as carbon and nitrogen source during cultivation [14]. Loss of fiber content indicates that the isolate has consumed that substrate their growth so that the several substrate components such as carbon, nitrogen, water, oxygen, and minerals, are reduced. Based on Fig. 40.1, *BtA* have a loss of fiber content higher than *BtB*. The loss of fiber content in *BtA* was 5.20–7.33%, while the loss of fiber content in *BtB* was 4.36–5.78%.

The decrease in fiber content occurred because the cellulase enzyme produced by the isolate would hydrolyze the cellulose in the media into simpler components that were easily digested and used by the isolates for growth. The higher loss of fiber content on a substrate from the fermentation of *BtA* is cassava dregs (7.33%), while in *BtB* is the pulp of coffee (5.26%).

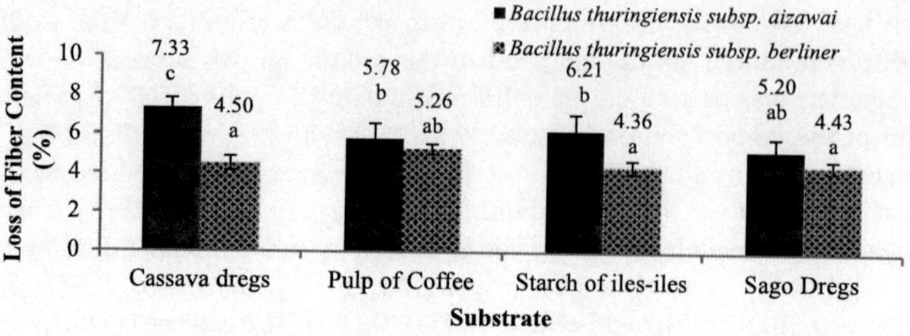

Fig. 40.1 Loss of fiber content of microbial bioinsecticide. Remark: numbers followed by the same letter on the bar chart are not significantly different based on Duncan's test 5%

40.3.2 Total Cell Count

The growth of isolates was measured by total cell count. The total cell count will be higher than the total spore count. It shows the total of vegetative cells and spores. The results showed that the isolates were able to grow on all carbon source substrates. Cultivation of bioinsecticides using *BtA* produced a high total cell count (10.26–11.68 log CFU/g), but the highest number of cells was produced on the pulp of coffee (11.3 log CFU/g). Cultivation using *BtB* also produced a high total cell count (10.86–12.45 log CFU/g), where the highest number was found on the pulp of coffee as a substrate (12.45 log CFU/g) (Fig. 40.2).

The pulp of coffee has low carbohydrate content but high glucose content [14]. Isolates are thought to prefer to consume simple sugars than starch and fiber. Isolates will consume complex substrates when simple substrates have been used up. In the initial growth phase, amylase has not yet been formed so that the isolates would utilize simple sugars for cell growth [15]. During the stationary phase, the simple sugars in the substrate begin to decrease, and complex substrates (starch and protein) would be hydrolyzed into simple sugars and amino acids.

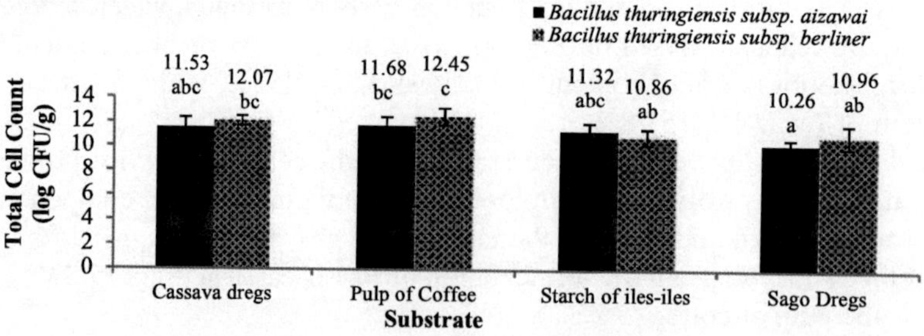

Fig. 40.2 Total cell count of microbial bioinsecticide. Remark: numbers followed by the same letter on the bar chart are not significantly different based on Duncan's test 5%

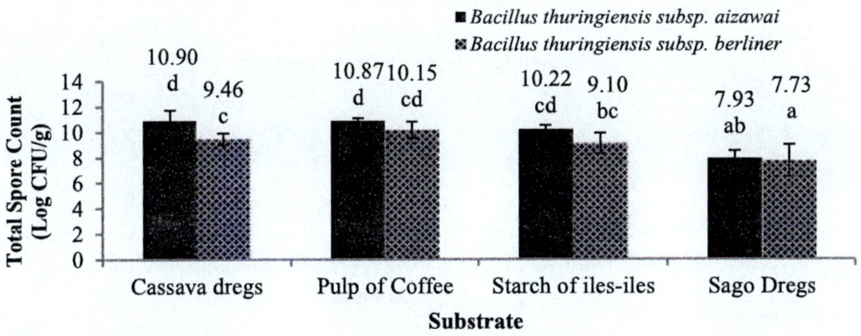

Fig. 40.3 Total spore count of microbial bioinsecticide. Remark: numbers followed by the same letter on the bar chart are not significantly different based on Duncan's test 5%

40.3.3 Total Spore Count

The bioinsecticide product is measured by total spore count. It affects toxicity. The greater of total spores count, the more protein crystals are produced, which will affect the toxicity of the bioinsecticide. Generally, spores are formed when the exponential phase ends, i.e., at the beginning of the stationary phase. Cells will continue to reproduce until one or more types of nutrients in the medium are reduced for bacterial growth [11]. In these conditions, spores and spore bodies that are contain one or more protein crystals are produced [16].

The growth of *B. thuringiensis* spores on the four carbon source substrates was generally well developed (7.73–10.90 log CFU/g). The cultivation of *BtA* produced a high total spore count (7.93–10.90 log CFU/g), but the highest cells was produced on cassava dregs (10.90 log CFU/g). Cultivation using *BtB* produced a lower total spore count (7.73–10.15 log CFU/g) than *BtA*, where the highest number was found on the pulp of coffee as a substrate (10.15 log CFU/g).

Cultivation of *BtA* on the pulp of coffee resulted in a higher total cells count than cassava dregs. The total spore count on the cultivation of *BtA* using cassava dregs was the highest. This is due to the cultivation of *BtA* on the cassava dregs as substrate produced more spores than vegetative cells count. The total cells count consists of the number of spores and vegetative cells count. However, the ratio between the number of spores and vegetative cells produced has not been ascertained yet (Fig. 40.3).

40.3.4 LC50

Bioinsecticides toxicity is generated by bioassay to determine how much material is needed to cause the death of the target insect. LC50 indicates the concentration of toxins that can kill 50% of the target insect. It describes the potential of protein crystals, which are insecticidal. The smaller of LC50 value, the more effective the product, which means the greater the degree of toxicity.

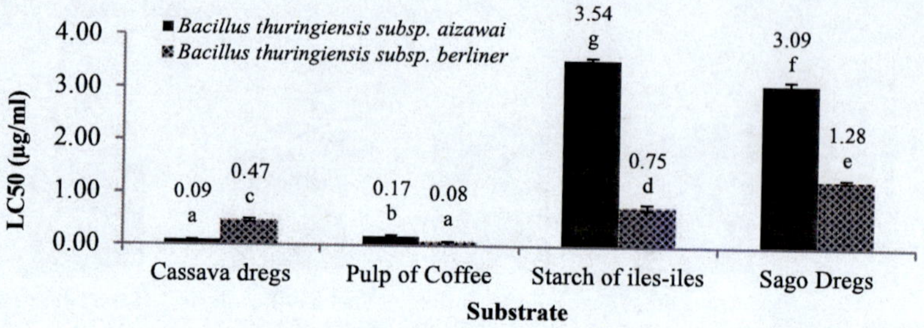

Fig. 40.4 LC50 of microbial bioinsecticide. Remark: numbers followed by the same letter on the bar chart are not significantly different based on Duncan's test 5%

BtA and *BtB* is very effective in controlling larvae of the Lepidoptera and Diptera [17]. This study used *Croccidolomia binotalis* (cabbage caterpillar) instar III to analyze the level of bioinsecticide toxicity of *B. thuringiensis*, because at this stage, the larvae are experiencing a large appetite, so the possibility to eat bioinsecticides attached to the leaves will also be more significant [18]. The results showed that the larvae of *C. binotalis* that died as a result of the treatment, the body became wrinkled, dry, and changed color to blackish brown. In addition, the rest of the larvae that did not die did not mean that they were still healthy, but rather that the larvae showed abnormal symptoms, such as being less active, becoming small or thin, and being slow or unable to progress to the next instar (Fig. 40.4).

Cultivation of *B. thuringiensis* produces bioinsecticide with LC50 around 0.08–3.54 µg/ml. The strain of isolates affects the toxicity of bioinsecticide. Strain *BtA* produces LC50 of 0.09–3.54 µg/ml, but the lowest LC50 is produced in cassava dregs as a substrate (0.09 µg/ml). Strain *BtB* produces LC50 of 0.08–1.28 µg/ml, but the lowest LC50 produced in the pulp of coffee as a substrate (0.08 µg/ml). This result was in line with data of loss of fiber and total spore count. Cultivation *BtA* using cassava dregs and *BtB* using the pulp of coffee that produces lowest LC50 value, also produces highest loss fiber content and total spore count.

40.3.5 Potency

An effective bioinsecticide product will have a small LC50 value and a large potential for bioinsecticide products. LC50 only shows the relative potency of the product because the potential of bioinsecticide products is expressed in international units (SI). *BtA* is very effective in controlling larvae of *Lepidoptera* and *Diptera* because it produces the Cry II gene [16], while *BtB* is very effective in controlling *Lepidoptera* because it produces the Cry I gene. According to Crespo et al. [17], Cry gene belonging to *BtB* includes CryI ab, Cry I ac, and Cry IF, which are toxic to Lepidoptera.

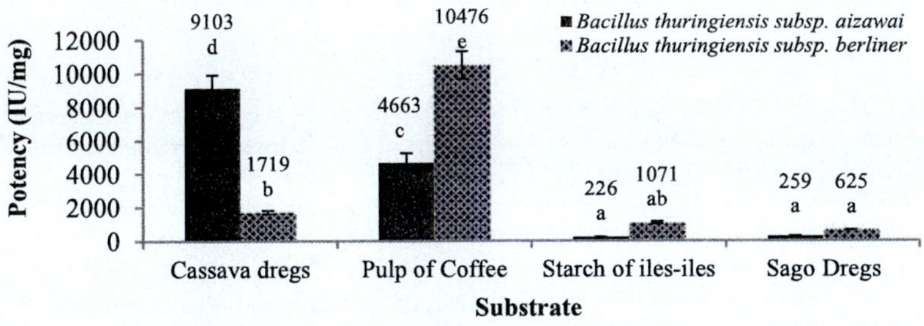

Fig. 40.5 The potency of microbial bioinsecticide. Remark: numbers followed by the same letter on the bar chart are not significantly different based on Duncan's test 5%

Protein crystals are insecticidal. However, the bioinsecticide produced is classified as an insecticide that works as a stomach poison. Bioinsecticide will kill the target insect if it is eaten and enters the insect's digestive organs [19]. In general, protein crystals are protoxin. However, proteolysis activity in the digestive systems of insects makes it turn into a toxin. The toxin will be hydrolyzed by the protease enzyme found in the insect's intestine into a toxin that is smaller toxin and easy to insert into the intestinal wall [15]. After insertion occurs, the insect's intestinal wall becomes damaged, and the insect becomes unable to eat and eventually dies.

Bioinsecticide produced in this research has a potency of 226–10,476 IU/mg. Cultivation of *BtA* has a potency of 226–9103 IU/ mg, lower than the potency of bioinsecticides produced by *BtB* (625–10,476 IU/mg). This result is same with other research, which stated that bioinsecticide with a low LC50 value would produce high potency. LC50 value of bioinsecticide by *BtA* using cassava dregs as a substrate is 0.09 μg/ml, but this product has high potency, which is 9103 IU/mg. LC50 value of bioinsecticide by *BtB* using a pulp of coffee as a substrate is 0.08 μg/ml, but this product has high potency, which is 10476 IU/mg. When compared with Bactopein (commercial bioinsecticide products which have an LC50 value of 0.05 μg/ml and a potency of 16,000 IU/mg), the bioinsecticides produced from the cultivation of *B. thuringiensis* has a lower potency than commercial bioinsecticide products (16,000 IU/mg) (Fig. 40.5).

40.4 Conclusion

Bacillus thuringiensis can produce cellulase and amylase enzymes, so it can be used agroindustrial by-products as a carbon source in solid media cultivation. Cultivation of *BtA* using cassava dregs as a substrate can produce the highest toxicity with a VSC of 10.9 log CFU/g, LC50 value 0.09 μg/ml, and potency 9103 IU/mg. Cultivation of *BtB* using the pulp of coffee as a substrate can produce the highest toxicity with VSC of 10.15 log CFU/g, LC50 value of 0.08 μg/ml, and potency of 10,476 IU/mg.

References

1. Li, H., Bouwer, G.: Evaluation of the synergistic activities of *Bacillus thuringiensis* Cry proteins against *Helicoverpa armigera* (Lepidoptera: Noctuidae). J. Invertebr. Pathol. **121**, 7–13 (2014)
2. Zhuang, L., Zhou, S., Wang, Y., Liu, Z., Xu, R.: Cost-effective production of *Bacillus thuringiensis* biopesticides by solid-state fermentation using wastewater sludge: effects of heavy metals. J. Bioresource Technol. **102**, 4820–4826 (2011)
3. Sanahuja, G., Banakar, R., Twyman, R.M., Capell, T., Christou, P.: *Bacillus thuringiensis*: a century of research, development and commercial applications. J. Plant Biotechnol **9**, 283–300 (2011)
4. Rahayuningsih, M., Syamsu, K., Dawis, A.A., Purnawati, R.: Penggandaan skala produksi bioinsektisida *Bacillus thuringiensis* var.*israelensis* untuk membasmi jentik nyamuk *Aedes aegypti*. J. Ilmu Pert. Indones. **12**(2), 123–130 (2007)
5. Shojaaddini, M., Saeid, M., Mahvash, K., Ali, A.T.: Development of a cost effective medium for production of *Bacillus thuringiensis* bioinsecticide using food barley. J. Plant Prot. Res. **50**(1), 9–14 (2010)
6. Susanti, L., Blondin, Ch.P.: Pengaruh ketinggian habitat kelapa (*Cocos nucifera*) terhadap pengembangbiakan *Bacillus thuringiensis* H-14 dan toksisitasnya terhadap jentik (*Anopheles aconitus*). Media Penelitian dan Pengembangan Kesehatan. **23**(1), 23–30 (2013)
7. Eswarapriya, B., Gopalsamy, B., Kameswari, B., Meera, R., Devi, P.: Insecticidal activity of *Bacillus thuringiensis* IBT—15 strain against *Plutella xylostella*. J. Pharm. Tech. Res. **2**(3), 2048–2053 (2010)
8. Capalbo, D.M.S., Valicente, Fh., Moraes, I.O., Pelizer, L.H.: Solid state fermentation of *Bacillus thuringiensis tolworthi* to control Fal Armyworm in Maize. J. Biotechnol. **4**, 112–125 (2001)
9. Foda, M.S., Fawkia, M., El-Beih, M., Maysa, E., Nora, N.A., El-Gamal.: Physiological formation of mosquitocidal toxin by a novel *Bacillus thuringiensis* isolate under solid state fermentation. J. Life Sci. **7**, 144–152 (2010)
10. AOAC.: Official Methods of Analytical of the Association of Official Analytical Chemist. AOAC, Washington (2006)
11. Marzban, R.: Investigation on the suitable isolate and medium for production of *Bacillus thuringiensis*. J. Biopest. **5**(2), 144–147 (2012)
12. Uhan, T.S., Sulastrini, I.: Efektivitas aplikasi kombinasi *Steinernema carpocapsae* dan biopestisida *Bacillus thuringiensis* terhadap mortalitas *Crocidolomia pavonana* F. pada tanaman kubis di rumah kaca. J. Hort. **18**(1), 38–45 (2008)
13. Pang, J.C., Chen, M.L., Ho, Y.C., Yang, C.Y., Tzeng, C.C., Kao, S.S., Tsen, H.Y.: Effect of fermentation conditions on the enterotoxigenicity, cytotoxicity and pesticidal activity of *Bacillus thuringiensis* strains isolated in Taiwan. J. Bioresour Technol. **101**, 1871–1876 (2010)
14. Luambano, N.D., Lopez, R.H.M., Kimenju, J.W., Powers, S.J., Narla, R.D., Wanjohi, W.J., Kerry, B.R.: Effect of temperature, pH, carbon and nitrogen ratios on the parasitic activity of *Pochoniachlamydosporia* on *Meloidogyne incognita*. J. Biol. Control **80**, 23–29 (2015)
15. Xu, C., Chinte, U., Chen, L., Yao, Q., Meng, Y., Zhou, D., Bi, L.J., Rose, J., Adang, M.J., Wang, B.C., Yu, Z., Sun, M.: Crystal structure of Cry51Aa1: a potential novel insecticidal aerolysin-type β-pore-forming toxin from *Bacillus thuringiensis*. Biochem. Biophys. Res. Commun. **462**, 184–189 (2015)
16. Somwatcharajit, R., Tiantad, I., Panbangred, W.: Coexpression of the silent cry 2Ab27 together with cry1 genes in *Bacillus thuringiensis* subsp. *aizawai* SP41 leads to formation of amorphous crystal toxin and enhanced toxicity against *Helicoverpa armigera*. J. Invertebr. Pathol. **116**, 48–55 (2014)
17. Crespo, A.L.B., Ana, R.S., Herbert, A.A.S., Eliseu, J.G.P., Juan, F., Blair, D.S.: Cross-resistance and mechanism of resistance to Cry1Ab toxin from Bacillus thuringiensis in a field-derived strain of European corn borer *Ostrinia nubilalis*. J. Inverteb Pathol. **107**, 185–192 (2011)

18. Li, H., Bouwer, G.: Toxicity of *Bacillus thuringiensis* cry proteins to Helicoverpaarmigera (Lepidoptera: Noctuidae) in South Africa J. Invertebr. Pathol. **109**, 110–116 (2012)
19. Yilmaz, S., Ayvaz, A., Akbulut, M., Azizoglu, U., Karaborklu, S.: A novel Bacillus thuringiensis strain and its pathogenicity against three important pest insects J. Stored Prod. Res. **51**, 33–40 (2012)

Chapter 41
Spatial Distribution of Nutrient Export from the Catchment Area of Lake Rawapening

Nunung Puji Nugroho

Abstract Eutrophication is a severe environmental problem faced by the lake ecosystems in Indonesia, including Lake Rawapening. It is a natural process in which the enrichment of aquatic systems by nutrients, especially by nitrogen (N) and phosphorus (P) compounds, causes uncontrolled growth of plants and algae, resulting in an imbalance between production and consumption of algae. Eutrophication has adverse impacts on the lake ecosystem, such as degrading water quality and decreasing biodiversity. Thus, reducing lake eutrophication was set as one of the super-priority programs of the Lake Rescue Movement. Information on the spatial distribution of nutrient export becomes essential to implement the program effectively. This study aims to estimate the magnitude and the spatial distribution of nutrient export from the catchment area of Lake Rawapening. The analysis was based on the Nutrient Retention sub-model of the InVEST (*Integrated Valuation of Environmental Services and Tradeoffs*). The results indicated that the total nutrient loads of N and P in the catchment area of Lake Rawapening were 2,263,276 kg/year and 428,697 kg/year, respectively. Total nutrient exports of N and P that entered Lake Rawapening were 847,899.59 kg/year and 153,145.44 kg/year, respectively, 37.5% and 35.7% of the total nutrient loads of N and P. Panjang Sub-sub watershed is the largest source of nutrient exports. Therefore, Panjang Sub-sub watershed should be selected as the focus area for implementing the reducing eutrophication program of Lake Rawapening.

41.1 Introduction

Indonesia has around 840 lakes, both large and small [1–3]. Lake ecosystems in principle consist of aquatic ecosystems (lake waters), riparian ecosystems (transitional zones), and terrestrial ecosystems of the catchment areas [1, 3]. Lake ecosystems

N. P. Nugroho (✉)
Balai Penelitian dan Pengembangan Teknologi Pengelolaan Daerah Aliran Sungai, Jl. Jend. Ahmad Yani Pabelan, P.O. Box 295, Kartasura, Surakarta, Indonesia
e-mail: npnugroho09@gmail.com

E. Yulihastin et al. (eds.), *Proceedings of the International Conference on Radioscience, Equatorial Atmospheric Science and Environment and Humanosphere Science, 2021*, Springer Proceedings in Physics 275, https://doi.org/10.1007/978-981-19-0308-3_41

serve many functions and benefits for humans and other living things. The functions and benefits of lakes can be observed in terms of direct benefits, ecological functions, production results, and their specificities, such as flood and drought control, transportation route, tourism, research and education, sediment anchoring, global climate control, water and energy supply, fisheries and agriculture support, as well as habitat for various flora and fauna [3].

Many aquatic ecosystems (e.g., lakes and reservoirs) worldwide suffer from eutrophication [4]. It is considered the most serious pollution problem that causes water quality degradation of aquatic ecosystems [4–6]. The origin of the word eutrophication comes from the Greek, i.e., *eu*, which means abundant, and *trophi*, which means nutrients [7]. In general, eutrophication is defined as a process in which the enrichment of aquatic systems by excess plant nutrients, especially nitrogen (N) and phosphorus (P) compounds, causes uncontrolled plant and algal growth, resulting in an imbalance between algae production and consumption [4, 6–9]. Eutrophication was initially a slow and natural process in the geological history of a lake with accumulated sediments and nutrients in water bodies. In the end, it was accelerated by human activities that triggered a high rate of nutrient inputs of point and non-point sources into the lake body [10–12]. Lake eutrophication can cause complex ecological impacts to lake ecosystems that hinder their utilizations [6, 13]. The effects of lake eutrophication include an increase in algal production and biomass (algae blooms), a decrease in biodiversity, a reduction in the amount of dissolved oxygen, blockage of waterways by aquatic weeds (e.g., water hyacinth), a decrease in quantity and quality of water, and a decrease in the economic value of the lake [10, 13]. The reduction of dissolved oxygen due to deoxygenation of water after the occurrence of algae blooms can kill fish massively [10, 13].

In Indonesia, lake ecosystems face various problems, both related to the survival of the biota, as well as related to the quantity and quality of the water [14], e.g., lake eutrophication. Those problems are mainly caused by the increase of human activities that pay less attention to sustainability [14]. Eutrophication level was one of the criteria used in determining the national priority lakes [3]. Considering that eutrophication is one of the main causes of lake ecosystem degradation, thus countermeasures of lake eutrophication become a super-priority program of Lake Rescue Movement [15–17]. To support the program's implementation and the formulation of appropriate policies and actions to reduce lake eutrophication, it is crucial to gather information on the magnitude and the spatial distribution of nutrient export at the catchment level. This study aims to estimate the magnitude and spatial distribution of nutrient export (N and P) from the catchment area of Lake Rawapening.

41.2 Material and Methods

41.2.1 Study Site

The study was conducted at the catchment area of Lake Rawapening, i.e., Rawapening Sub-watershed. Geographically, it is situated between 110°17′ and 110°30′ East longitude and 7°5′ and 7°25′ South latitude [13]. Lake Rawapening is a semi-natural lake [18, 19] and one of the fifteen priority lakes set at the 2009 Indonesia Lakes National Conference (INLC) I in Bali [20] and also listed in the Presidential Regulation Number 60 of 2021 concerning National Priority Lakes Rescue [21]. Lake Rawapening suffers from various environmental problems such as water quality degradation, biodiversity decline, excessive cover of water hyacinth (up to 70%), high erosion and sedimentation rate that leads to lake size diminishment [21]. The high rate of erosion and sedimentation was partly triggered by the increasing residential and industrial areas in the catchment area of Lake Rawapening [22, 23]. Lake Rawapening catchment, which is upstream of Tuntang Watershed, consists of nine sub-sub watersheds, namely: (1) Rengas, (2) Panjang, (3) Torong, (4) Parat, (5) Sraten, (6) Ringis, (7) Kedungringin, (8) Legi and (9) Galeh [20, 24]. In terms of government administration, Rawapening Sub-watershed covers seventy-two subdistricts/villages spread over ten districts in Semarang Regency and Salatiga City [13, 24] with a total area of about 27,130 ha [24]. The climate in Rawapening area is a humid tropical climate (Af-Am) based on Koppen climate classification [25], and the elevation ranges from 494 to 2,865 m above sea level (masl) [24]. The map of the research location is presented in Fig. 41.1.

As a tectonic-volcanic lake [18], Lake Rawapening was formed through the eruption of Ungaran Tua Volcano, where basalt lava flows from the eruption blocked Kali Pening in Tuntang area, which resulted in the submergence of Kali Pening valley [26, 27]. The valley eventually became a natural reservoir and was known as Lake Rawapening. The function of Lake Rawapening is as a barrier and reservoir for surface water flow, so that it can be utilized for agricultural irrigation, hydropower, raw water and water for industry, tidal paddy fields, inland fisheries, flood control in downstream areas, tourism, handicrafts, organic fertilizer from peat, and mushroom cultivation facilities [28, 29].

41.2.2 Data

This study is a GIS (geographic information system)—based study using spatial and nonspatial data collected from relevant institutions and published literature. Data needed in nutrient retention modeling using the Nutrient Retention sub-model are:

1. Digital elevation models (DEM): obtained from the Shuttle Radar Topographic Mission (SRTM-DEM) image with a spatial resolution of 30 m.

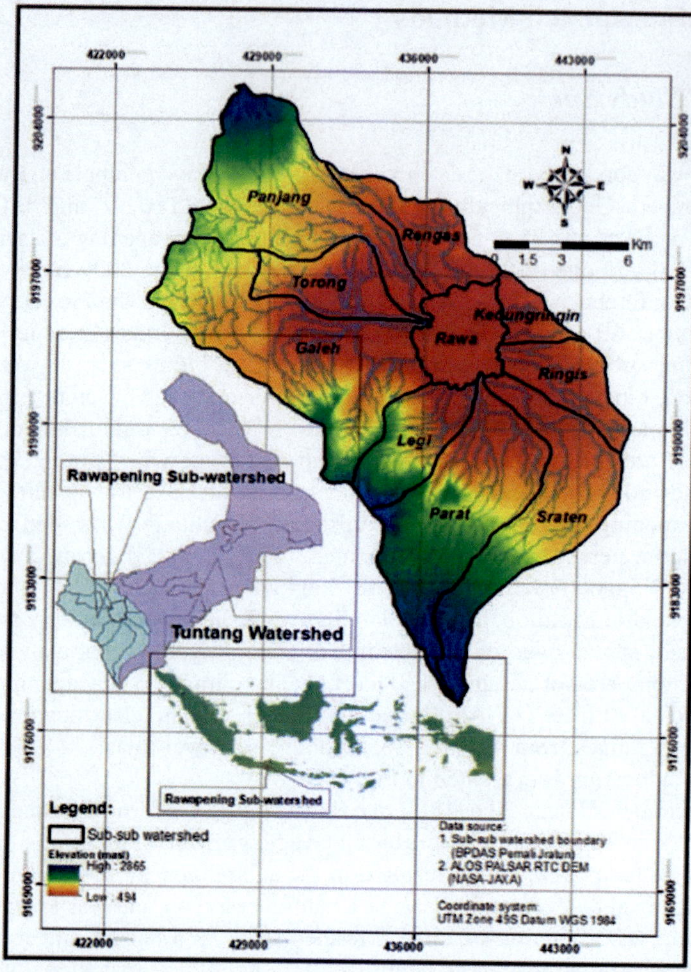

Fig. 41.1 Map of the study area depicting the elevation and the sub-sub watershed boundary (Rawapening Subwatershed)

2. Soil depth: based on the soil type map from BBWS (*Balai Besar Wilayah Sungai*) Pemali Juana and land system map (Regional Physical Planning Programme for Transmigration—RePPProT)
3. Annual rainfall: calculated from monthly rainfall data of 19 rainfall stations obtained from *PuSDATARu* (*Pusat Sumber Daya Air dan Tata Ruang*) Central Java (2008–2017)
4. Land cover/use: obtained from updated RBI (*Rupa Bumi Indonesia*) map from BIG (*Badan Informasi Geospasial*)
5. Watershed/sub-watershed boundaries: obtained from BPDASHL (*Balai Pengelolaan Daerah Aliran Sungai dan Hutan Lindung*) Pemali Jratun

6. A biophysical table containing information:
7. lucode: code (integer) for each land cover/use class.
8. lulc_desc: description/name of each land cover/use class.
9. root_depth: maximum root depth for vegetation cover/land use class (mm). For the non-vegetated land cover/land use class, the value is 1.
10. Kc: coefficient of plant evapotranspiration for each land cover/ use class, used to calculate potential evapotranspiration using plant physiological characteristics to modify the reference evapotranspiration (decimal value between 0 and 1.5).
11. load_n/load_p: nutrient load (N/P) for each land cover/use class (kg/ha/year) based on the coefficient of nutrient export.
12. eff_n/eff_p: vegetation filtering value per pixel for each land cover/use class (percent between 0 and 1).
13. Flow accumulation threshold: the minimum number of pixels that flow into a pixel to be considered part of the river network (default value is 1,000).

41.2.3 Data Analysis

The calculation of nutrient export from the catchment area of Lake Rawapening was carried out using the Nutrient Retention sub-model in the InVEST (Integrated Valuation of Environmental Services and Tradeoffs) model developed by the Natural Capital Project, Stanford University [30]. The Nutrient Retention sub-model calculates the amount of nutrient exported and retained, both at each pixel and in each watershed or sub-watershed. Calculations at the pixel scale make it possible to describe the heterogeneity of key driving factors in water yields, such as soil type, precipitation, and vegetation type. This model assumes that: (1) non-point sources of water pollution from nutrient exports can be reduced by vegetation acting as a filter and (2) water flows down slopes along natural flow paths [30].

The Nutrient Retention sub-model calculates the amount of nutrients exported and retained by various land covers through three stages. The first stage is to calculate the annual average runoff from each land cover class using the Water Yield sub-model. The second stage is to determine the quantity of nutrients exported and retained by each land cover class based on a look-up table containing the nutrient loading (export coefficients) and filtering capacity of each land cover type [30, 31]. The Nutrient Retention sub-model uses the annual average export coefficient developed by Reckhow et al. [32] from various field studies in the United States. However, the value of the export coefficient from local studies is prioritized [33, 34]. The nutrient loading value is adjusted (ALV) for differences in conditions between the location where the export coefficient was developed and the location where the model was applied by using a hydrological sensitivity score (HSS) as in Eq. (41.1) [30].

$$ALV_x = HSS_x \times pol_x \tag{41.1}$$

where ALV_x = *Adjusted Loading Value* at pixel x, HSS_x = *Hydrological Sensitivity Score* at pixel x, and pol_x = export coefficient at pixel x. HSS_x was calculated using Eq. (41.2).

$$HSS_x = \frac{\lambda_x}{\overline{\lambda_x}} \tag{41.2}$$

where λ_x = runoff index at pixel x, was calculated using Eq. (41.3), and $\overline{\lambda_x}$ = mean runoff index in the watershed of interest.

$$\lambda_x = \log\left(\sum_U Y_u\right) \tag{41.3}$$

where $\sum_U Y_u$ = sum of the water yield of pixels along the flow path above pixel x (including the water yield at pixel x).

The third step is to calculate the amount of nutrients exported and retained in each pixel. After the nutrient load of each pixel is determined, the amount of nutrient retained (nutrient retention) in each pixel can be calculated along with the surface flow that carries nutrients (pollutants) from upstream to the river (water body) according to the slope. The amount of retained nutrients is determined by the filtering capacity of the land cover type in each pixel. By following the flow of nutrient loads at each pixel to the water body, the model also tracks the amount of nutrients reaching the river/water body. Next, the model aggregates/accumulates the nutrient load that reaches the river from each pixel to get the loading value at the watershed/sub-watershed level.

41.3 Results and Discussion

41.3.1 The Magnitude of Nutrient Load, Export, and Retention from the Catchment Area of Lake Rawapening

Based on the results of the analysis using the InVEST model, the total nutrient loads of N and P in the catchment area of Lake Rawapening were 2,263,276 kg/year and 428,697 kg/year, respectively. Total estimated nutrient exports of N and P that entered Lake Rawapening were 847,899.59 kg/year (37.5% of the total) and 153,145.44 kg/year (35.7%), respectively. Panjang Sub-sub watershed was the largest source of nutrient exports, followed by Parat and Sraten Sub-sub watersheds. The contribution of each sub-sub watershed was 325,008 kg/year (38.3%), 185,024 kg/year (21.8%) and 102,100 kg/year (12.0%) of N nutrients and 48,789 kg/year (31.9%), 31,067 kg/year (20.3%) and 17,257 kg/year (11.3%) of P

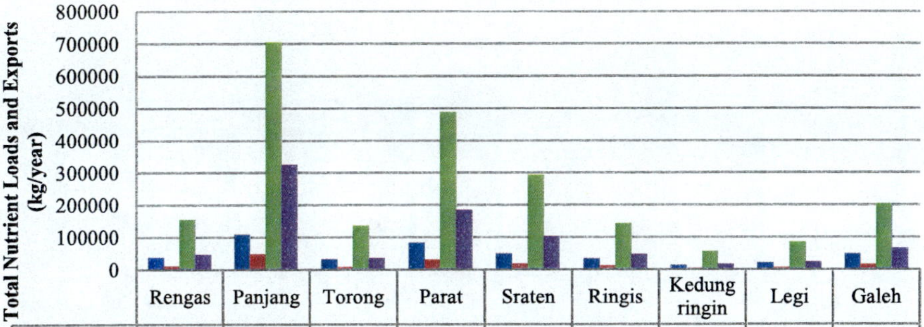

	Rengas	Panjang	Torong	Parat	Sraten	Ringis	Kedung ringin	Legi	Galeh
P_load_tot	37314.12	109878.2	32717.29	83629.24	49917.75	34327.71	13171.75	20083.18	47657.33
P_exp_tot	11289.79	48789.74	8597.60	31067.28	17257.13	11206.25	3924.68	5614.07	15398.89
N_load_tot	154627.1	704602.4	136146.5	489297.8	292437.4	142852.0	55108.96	84509.85	203693.6
N_exp_tot	46892.41	325008.4	36181.19	185024.5	102100.4	46761.83	16456.08	23766.85	65707.78

Fig. 41.2 Total nutrient loads and exports for each sub-sub watershed in the catchment area of Lake Rawapening

nutrients (Fig. 41.2). Piranti [13] indicated that human activities at the catchment areas determined the rate of nutrient exports. The high nutrient exports of N and P from those three sub-sub watersheds mainly came from the vegetable farmland, settlements, irrigated paddy fields, and rainfed paddy fields. Urbanization and fertilizer production contribute to the increase in N and P exports that cause lake eutrophication [4]. It is related to the excessive use of fertilizer in agricultural practices and the disposal of domestic waste.

Galeh Sub-sub watershed, as the largest sub-sub watershed, has a much lower export value of nutrients than Panjang Sub-sub watershed due to the dominance of mixed private forest cover, which has high retention capacity. Kedungringin Sub-sub watershed, as the smallest sub-sub watershed, is also the most minor contributor of nutrients, i.e., 16,456 kg/year (N) and 3,925 kg/year (P). The quantity of nutrient exports into lake bodies is strongly influenced by land cover conditions, climate, and topography [30]. In this case, land cover determines the magnitude of the nutrient loads and the ability to retain nutrients. In the catchment area of Lake Rawapening, the land cover class with the most significant nutrient loads is the vegetable farmland, and the forest has the largest nutrient retention ability. Climate, especially rainfall and temperature, determine the amount of water yield that becomes surface runoff. Topography, especially slope, determines the direction of runoff that carries nutrient loads.

The percentage of nutrient exports in each sub-sub watershed ranged from 26.58% (Torong Sub-sub watershed) to 46.13% (Panjang Sub-sub watershed) for N and between 26.28% (Torong Sub-sub watershed) to 44.40% (Panjang Sub-sub watershed) for P (Fig. 41.3). Thus, the existing land covers could retain about 54–74% of the total nutrient loads. The high ability of nutrient retention in Torong Sub-sub watershed was attributed to the dominant mixed private forest cover class, which has a retention value of 0.5. This value indicates that 50% of the total nutrient loads

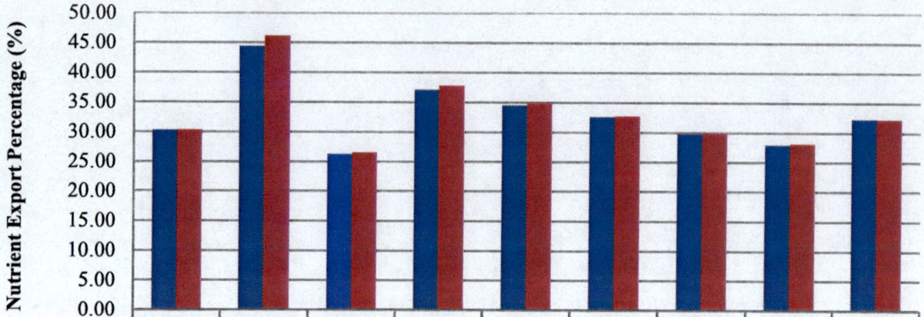

	Rengas	Panjang	Torong	Parat	Sraten	Ringis	Kedung ringin	Legi	Galeh
■ P_percent	30.26	44.40	26.28	37.15	34.57	32.64	29.80	27.95	32.31
■ N_percent	30.33	46.13	26.58	37.81	34.91	32.73	29.86	28.12	32.26

Fig. 41.3 Nutrient export percentages for each sub-sub watershed in the catchment area of Lake Rawapening

entering the mixed private forest will be retained. Meanwhile, the high export of nutrients in Panjang Sub-sub watershed was strongly influenced by the high nutrient loads of vegetable farmland, irrigated paddy fields, and rainfed paddy fields. Murtiono and Wuryanta [35] found that the high nutrient export that enters Lake Rawapening was caused by the application of chemical fertilizer in vegetable farmland, irrigated paddy fields and rainfed paddy fields. In addition to having high nutrient loads, those three land covers also have low retention capabilities, i.e., 0.2, 0.1, and 0.1, respectively. Therefore, almost all of the nutrient loads flowed to the downslope land cover to the lake body. A previous study found that the paddy field had the highest nutrient loads (Total P) in the catchment area of Lake Mrica [34]. The distribution of land cover classes in the Lake Rawapening catchment area is presented in Fig. 41.4.

41.3.2 Spatial Distribution of Nutrient Export from the Catchment Area of Lake Rawapening

Spatially, the distribution pattern of nutrient exports of N and P is similar to hotspot areas in vegetable farmland, irrigated paddy fields, and rainfed paddy fields (Fig. 41.5), as also observed by Murtiono and Wuryanta [35] due to high input of inorganic fertilizer. The nutrient export value of N ranges from 0 to 54.21 kg/pixel, while P ranges from 0 to 6.71 kg/pixel (the size of each pixel is 900 m^2). If converted to kg/ha, the nutrient export value ranges from 0 to 602.33 kg/ha for N and from 0 to 74.56 kg/ha for P. Spatial distribution of nutrient export from the catchment area of Lake Rawapening was dominated by low export values due to the domination of mixed private forest cover (around 7,900 ha), which has low nutrient loads and high retention capacity. The amount of nutrient exports is not only determined by the land

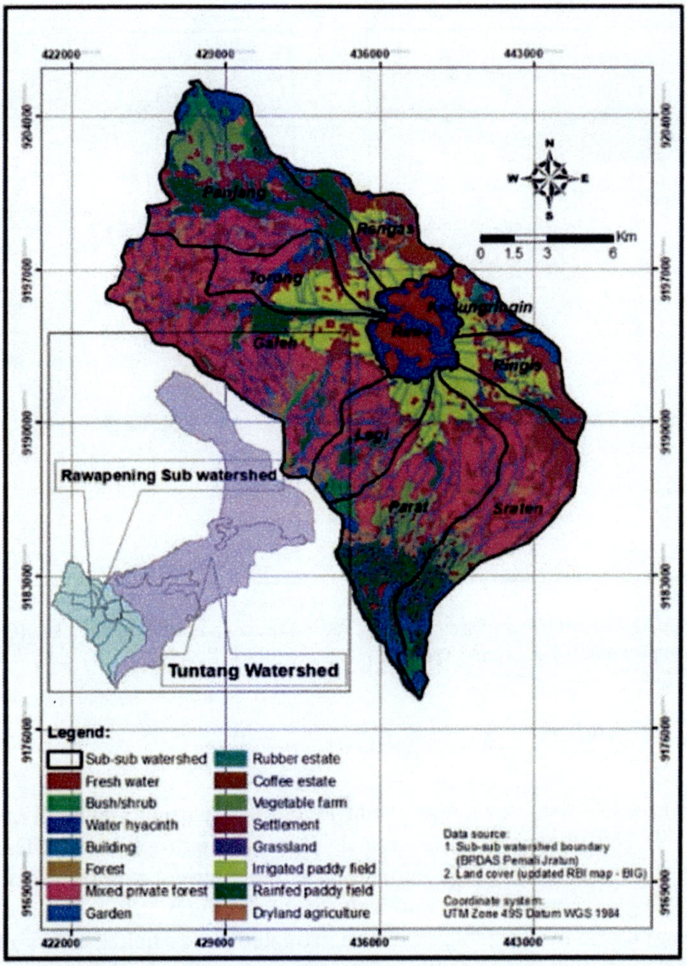

Fig. 41.4 The land cover/use classes in the catchment area of Lake Rawapening

cover type (which determines the amount of nutrient loads and retention capacity) but also by the proximity to the stream network, similar to the sediment export [36]. Sources of nutrient loads close to streams will have a relatively higher export value because the chances of being retained by land cover will be lower concerning the time travel of nutrient loads to stream. Thus, the nutrient loads will directly flow into the river to the lake body. In this regard, Lake Rawapening acts as a sink of sediment and nutrients that lead to eutrophication [37]. Redhead et al. [38] suggested that InVEST model can provide valuable information on the spatial distribution of nutrient exports to decision-makers, especially in terms of relative differences among catchments.

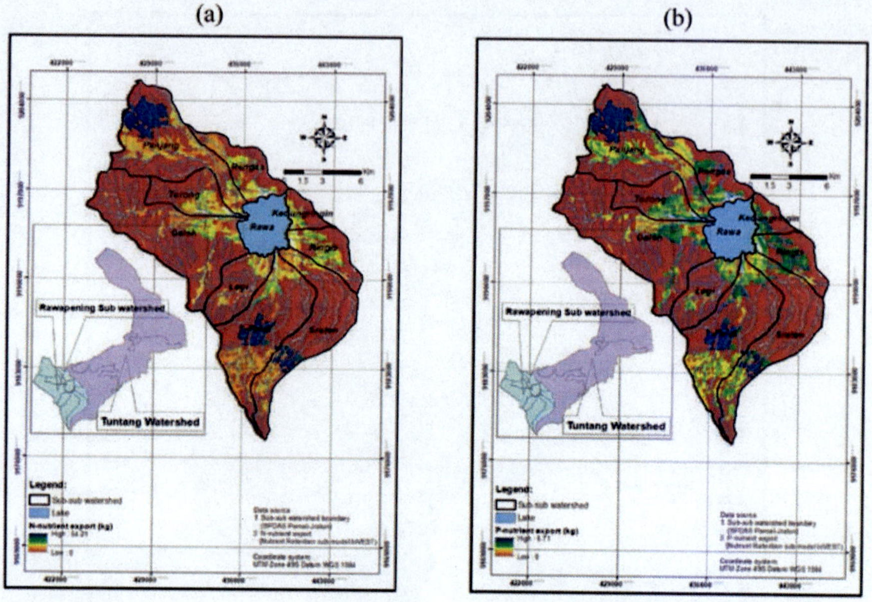

Fig. 41.5 Spatial distribution of nutrient exports in the catchment area of Lake Rawapening: **a** N-nutrient export and **b** P-nutrient export

41.4 Conclusions

Lake Rawapening receives a significant amount of nutrients annually, especially nitrogen (N) and phosphorous (P) that lead to lake eutrophication. The Nutrient Retention sub-model of InVEST estimated the total nutrient exports of N and P were 847,899.59 kg/year and 153,145.44 kg/year, respectively. Spatially, high nutrient exports were found in vegetable farmland, irrigated paddy fields, and rainfed paddy fields and close to streams. Eutrophication in Lake Rawapening has caused the accelerated growth of water hyacinth, the degradation of water quality, and the decrease in biodiversity that disrupt the utilization of the lake. Eventually, it will negatively impact the peoples who depend on the existence and services provided by Lake Rawapening. Therefore, lake rescue programs should be undertaken to overcome the eutrophication problem. Considering that vegetable farmland, irrigated paddy field, and rainfed paddy field contribute high nutrient export due to the application of inorganic fertilizer, it is crucial to provide agricultural extension services to farmers to raise their awareness related to wise-use of fertilizer, organic farming, and application of soil and water conservation measures related to nutrient retention. Government resources for countermeasures of lake eutrophication should be allocated to hotspot areas, such as Panjang Sub-sub watershed, which has high nutrient exports.

Acknowledgements The author would like to express sincere gratitude to the Ministry of Environment and Forestry through *Balai Penelitian dan Pengembangan Teknologi Pengelolaan Daerah*

Aliran Sungai for providing research funding. The author thanks Asep Hermawan and Santi Fandriani for their assistance in data collection.

References

1. Irianto, E.W., Triweko, R.W.: Eutrofikasi waduk dan danau: permasalahan, pemodelan dan upaya pengendalian. Pusat Penelitian dan Pengembangan Sumber Daya Air, Badan Penelitian dan Pengembangan Kementerian Pekerjaan Umum, Bandung (2011)
2. Haryani, G.S.: Kondisi danau di Indonesia dan strategi pengelolaannya. In: Prosiding pertemuan ilmiah tahunan Masyarakat Limnologi Indonesia I (Perkembangan Limnologi dalam Mendukung Pembangunan Berkelanjutan di Indonesia: Tantangan dan Harapan), Cibinong (pp. 1–19). Pusat Penelitian Limnologi (2013)
3. KLH: *Grand design* penyelamatan ekosistem danau Indonesia. Kementerian Lingkungan Hidup (KLH) Jakarta (2014)
4. Jorgensen, S., Tundisi, J.G., Tundisi, T.M.: Handbook of Inland Aquatic Ecosystem Management, vol. 5. CRC Press (2012)
5. Han, B., Reidy, A., Li, A.: Modeling nutrient release with compiled data in a typical Midwest watershed. Ecol. Ind. **121**, 107213 (2021). https://doi.org/10.1016/j.ecolind.2020.107213
6. Khan, M.N., Mohammad, F.: Eutrophication: challenges and solutions. In: Eutrophication: Causes, Consequences and Control, pp. 1–15. Springer (2014)
7. Cardoso, A., Free, G., Nõges, P.O.K., Poikane, S., Solheim, A.: Lake management criteria. In: Likens, G. (ed.) Encyclopedia of Inland Waters. Elsevier, Amsterdam, The Netherlands (2009)
8. Morsy, K., Morsy, A., Morsy, M., Thakeb, H.: Eutrophication of aquatic ecosystems: a viewpoint on the environmental impact of climate change. J. Environ. Sci. Eng. B **6**, 506–514 (2017). https://doi.org/10.17265/2162-5263/2017.10.002
9. Likens, G.E. (ed.).: Lake Ecosystem Ecology: A Global Perspective. Academic Press (2010)
10. Ansari, A.A., Gill, S.S., Lanza, G.R., Rast, W. (eds.): Eutrophication: Causes, Consequences and Control. Springer, Netherlands (2011)
11. Aoki, I.: Entropy Principle for the Development of Complex Biotic Systems. Elsevier, Amsterdam (2012)
12. Chislock, M.F., Doster, E., Zitomer, R.A., Wilson, A.E.: Eutrophication: causes, consequences, and controls in aquatic ecosystems. Nat. Educ. Knowl. **4**(4), 10 (2013)
13. Piranti, A.S.: Pengendalian eutrofikasi Danau Rawapening (Monograf). Penerbit Universitas Jenderal Soedirman, Purwokerto (2019)
14. Irianto, E.W., Triweko, R.W., Yudianto, D.: Pengembangan kriteria status mutu ekosistem danau sebagai bagian dari indikator pengelolaan terpadu wilayah sungai. J. Teknik Hidraulik **1**(1), 27–42 (2010). https://doi.org/10.32679/jth.v1i1.227
15. KLH: Gerakan Penyelamatan Danau (GERMADAN) Toba. Kementerian Lingkungan Hidup (KLH), Jakarta (2014)
16. KLH: Gerakan Penyelamatan Danau (GERMADAN) Tempe. Kementerian Lingkungan Hidup (KLH), Jakarta (2014)
17. KLH: Gerakan Penyelamatan Danau (Germadan) Danau Rawapening. Kementerian Lingkungan Hidup (KLH), Jakarta (2011)
18. Soeprobowati, T., Hadisusanto, S., Gell, P., Zawadzki, A.: The diatom stratigraphy of Rawapening Lake, implying eutrophication history. Am. J. Environ Sci. **8**(3), 334–344 (2012). https://doi.org/10.3844/ajessp.2012.334.344
19. Chulafak, G.A., Kushardono, D., Yulianto, F.: Utilization of multi-temporal Sentinel-1 satellite imagery for detecting aquatic vegetation change in Lake Rawapening, Central Java, Indonesia. Papers Appl. Geogr. **7**(3), 316–330 (2021). https://doi.org/10.1080/23754931.2021.1890193

20. Apriliyana, D.: Pengaruh perubahan penggunaan lahan SubDAS Rawapening terhadap erosi dan sedimentasi Danau Rawapening. J. Pembangunan Wilayah dan Kota **11**(1), 103–116 (2015). https://doi.org/10.14710/pwk.v11i1.8661
21. Republik Indonesia: Peraturan Presiden Nomor 60 Tahun 2021 tentang Penyelamatan Danau Prioritas Nasional. In: Lembaran Negara Republik Indonesia Tahun 2021 Nomor 143. Sekretariat Kabinet Republik Indonesia, Jakarta (2021)
22. Purboseno, S.: Evaluasi perubahan tata guna lahan sebagai upaya menjaga kerbelanjutan fungsi waduk. Media Komunikasi Teknik Sipil **19**(2), 167–174 (2013). https://doi.org/10.14710/mkts.v19i2.8426
23. Sanjoto, T.B., Sidiq, W., Nugraha, S.B.: Land cover change analysis to sedimentation rate of Rawapening Lake. Int. J. GEOMATE **18**(70), 294–301 (2020). https://doi.org/10.21660/2020.70.ICGeo50
24. Nugroho, N.P.: Estimasi hasil air dari daerah tangkapan air Danau Rawa Pening dengan menggunakan Model InVEST. Majalah Ilmiah Globë **19**(2), 157–166 (2017). https://doi.org/10.24895/MIG.2017.19-2.578
25. Soeprobowati, T.R.: Mitigasi danau eutrofik: Studi kasus Danau Rawapening. In: Prosiding Seminar Nasional Limnologi VI, pp. 36–48 (2012)
26. Machbub, B., Suwanto, A., Harahap, T.N., Manurung, H., Retnowati, I., Rachmiati, S., Rustadi, W.C.: Daya tampung beban pencemaran air dan zonasi Danau Rawa Pening. In: Kementerian Lingkungan Hidup, Jakarta, p. 80 (2012)
27. Soeprobowati, T.R.: Peta batimetri Danau Rawapening. . Bioma **14**(2), 78–84 (2012). https://doi.org/10.14710/bioma.14.2.78-84
28. KLH: Profil 15 Danau Prioritas Nasional 2010–2014. Kementerian Lingkungan Hidup (KLH), Jakarta (2011)
29. Soeprobowati, T.R., Suedy, S.W.A.: Status trofik Danau Rawapening dan solusi pengelolaannya. J. Sains Dan Matematika **18**(4), 158–169 (2010)
30. Sharp, R., Tallis, H.T., Ricketts, T., Guerry, A.D., Wood, S.A., Chaplin-Kramer, R., Nelson, E., Ennaanay, D., Wolny, S., Olwero, N., Vigerstol, K., Pennington, D., Mendoza, G., Aukema, J., Foster, J., Forrest, J., Cameron, D., Arkema, K., Lonsdorf, E., Kennedy, C., Verutes, G., Kim, C.K., Guannel, G., Papenfus, M., Toft, J., Marsik, M., Bernhardt, J., Griffin, R., Glowinski, K., Chaumont, N., Perelman, A., Lacayo, M., Mandle, L., Hamel, P., Vogl, A.L., Rogers, L., Bierbower, W.: InVEST 3.1.3 User's Guide. The Natural Capital Project, Stanford University, University of Minnesota, The Nature Conservancy, and World Wildlife Fund. Stanford (2015)
31. Sharps, K., Masante, D., Thomas, A., Jackson, B., Redhead, J., May, L., Prosser, H., Cosby, B., Emmett, B., Jones, L.: Comparing strengths and weaknesses of three ecosystem services modelling tools in a diverse UK river catchment. Sci. Total Environ. **584**, 118–130 (2017). https://doi.org/10.1016/j.scitotenv.2016.12.160
32. Reckhow, K.H., Beaulac, M.N., Simpson, J.T.: Modeling Phosphorus loading and lake response under uncertainty: a manual and compilation of export coefficients. In. U.S. Environmental Protection Agency, Washington D.C. (1980)
33. Bhagabati, N., Barano, T., Conte, M., Ennaanay, D., Hadian, O., McKenzie, E., Olwero, N., Rosenthal, A., Suparmoko, Shapiro, A., Tallis, H., Wolny, S.: Visi Hijau untuk Sumatera: Menggunakan informasi jasa ekosistem untuk membuat rekomendasi rencana peruntukan lahan di tingkat provinsi dan kabupaten. In: The Natural Capital Project (p. 152). WWF-US, dan WWF-Indonesia (2012)
34. Piranti, A.S., Soedarmadji, Waluyo, G., Suwardi: Transport nutrien penyebab eutrofikasi dari Daerah Tangkapan air Waduk Mrica Banjarnegara. Biosfera **32**(1), 66–73 (2015). https://doi.org/10.20884/1.mib.2015.32.1.297
35. Murtiono, U.H., Wuryanta, A.: Telaah eutrofikasi pada Waduk Alam Rawapening. In: Prosiding Seminar Nasional Geografi UMS 2016: Upaya Pengurangan Risiko Bencana Terkait Perubahan Iklim, Surakarta, pp. 170–181. Muhammadiyah University Press (2016)
36. Brooks, K.N., Ffolliott, P.F., Gregersen, H.M., DeBano, L.F.: Hydrology and the Management of Watersheds, 4th edn. Wiley, Oxford (2013)

37. Sulastri, Henny, C., Handoko, U.: Environmental condition and trophic status of Lake Rawa Pening in Central Java. Oseanologi dan Limnologi di Indonesia **1**(3), 23–38 (2016). https://doi.org/10.14203/oldi.2016.v1i3.20
38. Redhead, J.W., May, L., Oliver, T.H., Hamel, P., Sharp, R., Bullock, J.M.: National scale evaluation of the InVEST nutrient retention model in the United Kingdom. Sci. Total Environ. **610**, 666–677 (2018). https://doi.org/10.1016/j.scitotenv.2017.08.092

Chapter 42
Public Urban Park Quality Assessment Using Fuzzy C Means Classification of Land Surface Temperature and Social Function

Arif Wicaksono

Abstract This study describes the process to assess public urban park (PUP) quality in Bogor Municipality, Indonesia, using social function key performance indicator (KPI) and land surface temperature (LST) from Landsat 8. Eighteen PUPs located in high LST were analyzed to observe their social function based on KPI values. Three fuzzy linguistic classes of LST have been generated from the fuzzy c means (FCM) classification technique. A method called fuzzy class separation analysis is proposed in this study to select the most acceptable fuzzy separation among three fuzzy membership values ranges. This fuzzy class separation analysis used Jaccard Similarity Index (JSI) to identify the least intersected raster cells from three fuzzy images, namely (a) > 0.7–1, (b) > 0.75–1, and (c) > 0.8–1. The result from JSI shows that fuzzy LST image with > 0.75–1 membership value has the smallest number of intersected cells, and therefore high LST raster image was extracted from > 0.75 to 1. Within images from > 0.75 to 1 fuzzy membership, 30.99% raster cells were classified as high LST, 37.14% as medium LST, 27.27% as low LST, and 4.60% unclassified. High LST image was integrated with PUPs vector datasets using the extraction menu in ArcMap 10.3 and yielded in eighteen PUP locations. Based on social function KPI values in each PUP, seven PUP locations within high LST can be classified as compatible with KPI standard value (0.68–1), while eleven locations can be associated with less compatible (0.34–0.67). Therefore, seven PUP locations with compatible standards need proper maintenance to keep their social function quality. In contrast, the other eleven locations with less compatible standards need improvement to increase their social function quality.

A. Wicaksono (✉)
Research and Development Division, Regional Planning and Development Agency (BAPPEDA), Government of Bogor Municipality, Bogor City 16121, Indonesia
e-mail: arifsez1977@gmail.com

E. Yulihastin et al. (eds.), *Proceedings of the International Conference on Radioscience, Equatorial Atmospheric Science and Environment and Humanosphere Science, 2021*, Springer Proceedings in Physics 275, https://doi.org/10.1007/978-981-19-0308-3_42

42.1 Introduction

Great evapotranspiration in the vegetated areas is reducing the urban-heat island effect. Moreover, the areal extent and distribution urban green space affect the quality of life of the city's inhabitants through park aesthetic and social function [1]. Assessing the quality of public urban parks (PUP) has been an essential activity, particularly for urban park planners within a municipality government, with two objectives, namely to improve facilities for urban parks with poor quality, and to implement proper maintenance to keep better quality urban parks. Therefore, it is crucial to identify which public urban parks are located in high land surface temperature (LST), and to assess whether public urban parks in high LST locations meet appropriate social functions. This will strengthen the Sustainable Development Goals in terms of increasing urban quality of life from social perspectives, where public urban parks as parts of urban green space play essential roles. Previously, several indicators have been used to assess the urban park social function quality, such as key performance indicator (KPI) in Maccini Sombala Parks, South Sulawesi [2], and Bendosari Park, Salatiga [3]. In addition, assessment of appropriate urban park facilities suitable to improve social function has been performed using ratings for the social desirability of specific park elements [4].

In the related field of urban park social function, [5] used public awareness surveys to score each urban park based on recognition from the urban citizens. However, for a municipal government to conduct the methods mentioned earlier, public urban parks' social function quality measurement might be time-consuming since a lot of respondents are needed for detailed surveys. This is exacerbated by considering that some municipal governments have limited resources in urban parks management. Nowadays, Geographic Information System (GIS) and Remote Sensing technologies provide opportunities to cover a wide area of public urban parks within the city limit. Landsat 8 data as one of the remote sensing technologies offer multi-bands images with various spatial resolution. Its multi-bands can be processed to yield land surface temperature raster images. Meanwhile, vector datasets containing public urban park locations resulted from GIS can be integrated with land surface temperature from Landsat 8 to obtain PUP locations within specific classification of LST.

For a closer look, PUP locations within high LST are fascinating to observe especially whether facilities in these PUPs meet the quality standard to improve social function in surrounding neighborhoods. Areas that fall within high LST require classification of LST data using the fuzzy method to convert numerical values of LST into fuzzy linguistic values. Since the words ''low, medium, high" can be classified as fuzzy linguistic values in GIS, a fuzzy classification method is needed to classify LST data.

Fuzzy c means (FCM) [6] is one of the methods to classify numerical values into fuzzy linguistic classes. Hence, FCM can be used to classify LST in detecting high LST class, and then PUP locations within this high LST class can be assessed to observe its social function quality. Consequently, this paper aims at three objectives, namely to classify LST from Landsat 8 using FCM, to identify PUP locations in high LST class, and to assess the social function quality of PUPs located in high LST.

42.2 Materials and Methods

In this paper, bands 4, 5, and 10 from Landsat OLI/TIRS Collection 2 Level 1 were downloaded from the USGS Earth Explorer website (https://earthexplorer.usgs.gov/). Bands 4 and 5 were processed to generate normalized difference vegetation index (NDVI). The NDVI image was then further processed to obtain surface emissivity (ε), and input ε to generate LST in the form of raster maps with a spatial resolution of 30 m × 30 m. FCM clustering algorithm in R software version 3.6.3 [7] was used to process LST images and resulted three classes of LST, namely high LST, medium LST, and low LST. The overview research flowchart in this paper can be observed in Fig. 42.1.

A method called fuzzy class separation analysis is proposed in this study to select the most acceptable fuzzy separation among three fuzzy membership value ranges using Jaccard Similarity Index (JSI). JSI values used to identify the least intersected raster cells from three fuzzy images, namely (1) > 0.7–1, (2) > 0.75–1, and (3) >

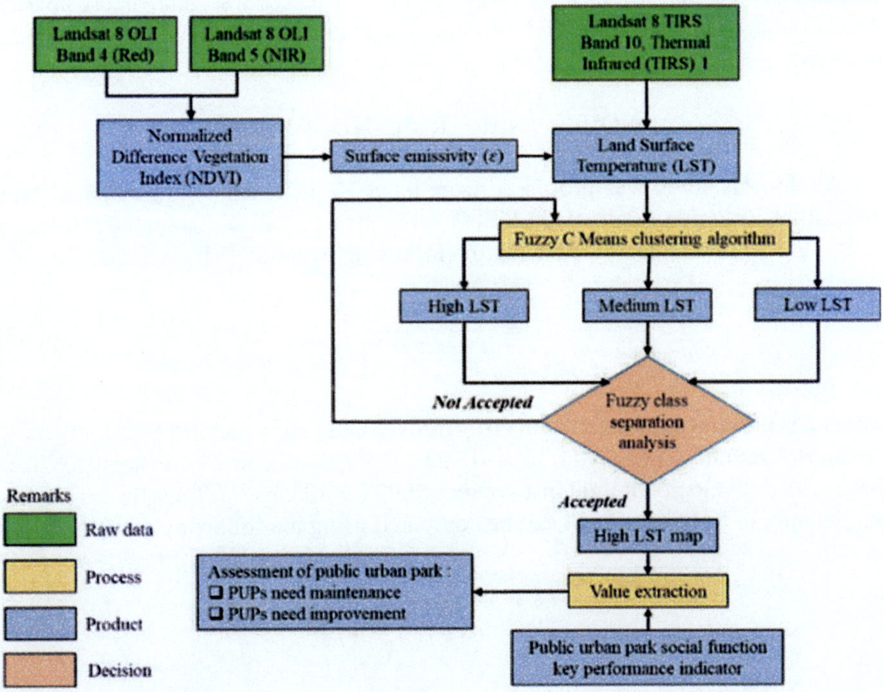

Fig. 42.1 Research flow chart

0.8–1. The least intersected cells from JSI indicated that separation in the image is the best and therefore selected for further analysis. For this analysis, only a high LST class was selected, and its raster image was used to select the PUP locations within high LST. Social function KPI values from eighteen selected PUP locations within high LST were surveyed and classified into two categories, namely suitable with KPI standard and less compatible.

42.2.1 *Study Area*

Bogor Municipality has six sub-districts and 68 villages. Based on GIS datasets downloaded from the Indonesian Geospatial Agency website (https://tanahair.indonesia.go.id/portal-web), Bogor Municipality Area is 112.5 km^2. The map of the study area is shown in Fig. 42.2.

42.2.2 *Land Surface Temperature (LST)*

At first, NDVI can be generated using near-infrared (NIR) band 5 of Landsat 8 and Red band 4 of Landsat 8 OLI. In this paper, NDVI image was generated using the ArcGIS image analyst menu. The formula to calculate NDVI can be observed in the following:

$$\text{NDVI} = (\text{NIR} - \text{Red})/(\text{NIR} + \text{Red}) \tag{42.1}$$

LST for this study was processed from band 10 of Landsat 8 TIRS dated May 24th, 2020. Emissivity-corrected LST (T_s) in each raster cell with a spatial resolution of 30 m × 30 m was generated using the following equation [8]:

$$T_S = \frac{T_B}{1 + \left(\frac{\lambda \sigma T_B}{hc}\right) \ln \varepsilon} \tag{42.2}$$

where λ is the effective wavelength (10.9 mm for band 10 in Landsat 8 data) [9], $\sigma =$ Stefan Boltzmann's constant (1.38 × 10–23 J K^{-1}), $h =$ Planck's constant (6.626 × 10–34 Js), $c =$ velocity of light in a vacuum (2.998 × 108 ms^{-1}). TB is the brightness temperature in Kelvin where it can be computed using the following equation [9]:

$$T_B = \frac{K_2}{ln\left(\left(\frac{K_1}{L_\lambda}\right) + 1\right)} \tag{42.3}$$

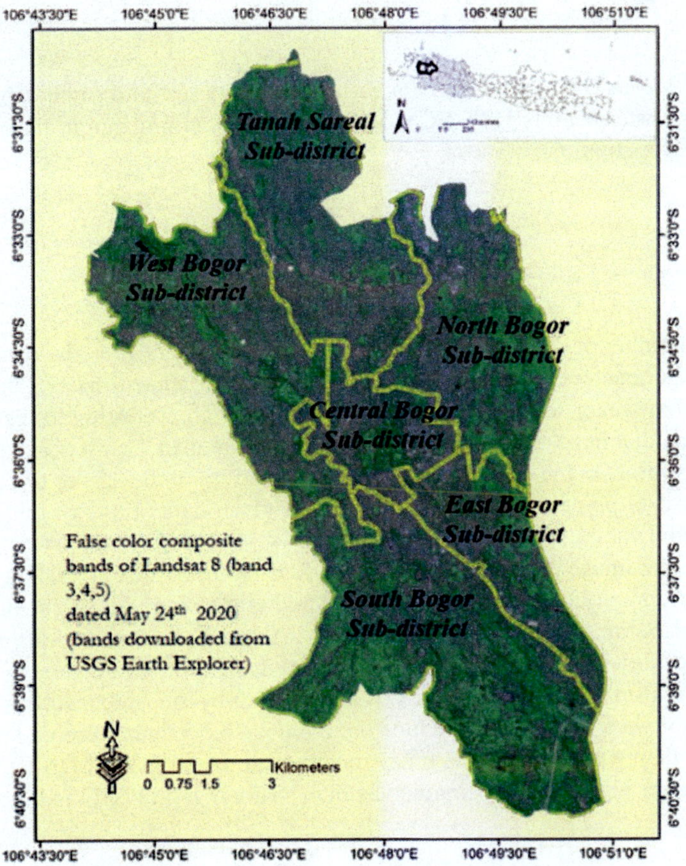

Fig. 42.2 Study area map

where $L_{\lambda} = 0.0003342*\text{DN} + 0.1$, K_1 and K_2 are calibrated constant for Landsat 8 K_1 is 774.89 and K_2 is 1321.08. The estimation of surface emissivity (ε) was performed using the NDVI threshold method [10] and fractional vegetation (F_v) in each pixel which is described in the following equation:

$$\varepsilon = 0.004\left(\frac{NDVI - NDVI_{\max}}{NDVI_{\max} - NDVI_{\min}}\right)^2 + 0.986 \quad (42.4)$$

For $\text{NDVI}_{\max}$, the maximum NDVI value is 0.5, where pixels represent healthy vegetation, while in $\text{NDVI}_{\min}$, 0.2 is the value of minimum NDVI where pixels represent bare soils [9].

42.2.3 Fuzzy C Means (FCM)

Fuzzy C Means (FCM) is a technique to classify data sets into smaller groups based on their cluster centers [11]. FCM is based on the minimization of the following objective function [6]:

$$J_m = \sum_{i=1}^{D} \sum_{j=1}^{N} \mu_{ij}^m x_i - c_j^2 \tag{42.5}$$

$D =$ the number of data points, $N =$ the number of clusters, $m =$ the fuzzy partition matrix exponent for controlling the degree of fuzzy overlap, with $m > 1$. Fuzzy overlap refers to how fuzzy the boundaries between clusters are, referring to the number of data points that have significant membership in more than 1 cluster, where $x_i =$ the ith data point, $c_j =$ the center of the jth cluster, and $\mu_{ij} =$ the degree of membership in the ith data point and the jth cluster. For a given data point, x_i, the sum of the membership values for all clusters is 1.

LST raster image was standardized using R software version 3.6.3. R is a free software environment for statistical computing and graphics, and it can be downloaded from https://cran.r-project.org/mirrors.html. After the standardization process, the FCM algorithm was applied to LST raster data using R software version 3.6.3. The FCM algorithm was described by [7], and its step-by-step procedure was applied using R software version 3.6.3. Once the membership values were yielded from R software, they were input to the fuzzy membership menu in ArcMap 10.3 software to form raster maps in fuzzy linguistic values, namely high LST, medium LST, and low LST.

42.2.4 Fuzzy Class Separation Analysis

For implementing fuzzy separation analysis, to select which fuzzy membership value is the best fuzzy classification of LST, a method called Jaccard Similarity Index (JSI) was applied. Hence, the smallest value of the JSI among three FCM results is chosen to be the best fuzzy LST classes. JSI for GIS works using the following formula [12]:

$$I_i = \frac{T_{11}}{T_{21} + T_{12} + T_{11}}. \tag{42.6}$$

where T_{21} is elements in B, T_{12} is elements in A, and T_{11} is elements in A n B.

42.2.5 Social Function Key Performance Indicator (KPI)

Assessment of the social function quality of PUPs located in the high LST class was performed using KPI. A form containing parameters for social function in each PUP location within the high LST class was sent electronically to four experts in Bogor Municipality's Housing Division. They were asked to answer questions for six social function parameters adapted from [3], namely accessibility, health support, security level, comfort level, social activity, and education.

Three scores category were given for experts when they had to answer the questions which specifically 1 = not suitable, 2 = less suitable, 3 = suitable. The results from each experts were combined to yield a KPI score for each PUP. KPI works using the following formula [2]:

$$KPI = \frac{actualvalue(1to3)}{standardvalue(3)} \tag{42.7}$$

For KPI classification [2], a PUP with a KPI value between 0 and 0.33 can be classified as incompatible with the standard, while a PUP with a KPI value between 0.34 and 0.67 can be classified as less compatible with the standard. Moreover, when a PUP has a KPI value between 0.68 and 1, therefore, it can be categorized as compatible with the standard.

42.3 Results

42.3.1 LST

NDVI image resulted from Eq. 42.1 is displayed in Fig. 42.3a, while the LST image is shown in Fig. 42.3b. It can be observed that in Fig. 42.3a, that the highest NDVI value is around 0.613, and the lowest is −0.416. The greener the color of the image indicates that its NDVI value is getting higher while the redder the color indicates its NDVI value is getting lower.

Interestingly, Bogor Municipality city center has a greener color which is concentrated in Bogor Botanical Garden. It can also be observed that in Fig. 42.3b, LST visually corresponds with the NDVI image in Fig. 42.3a. In Fig. 42.3b, the highest LST value is around 31.39 °C, and the lowest is around 22.72 °C.

42.3.2 Fuzzy Class Separation Analysis

Resulted c and σ values from the FCM algorithm using R software version 3.6.3 were inputted into ArcMap 10.3 fuzzy membership menu using gaussian function.

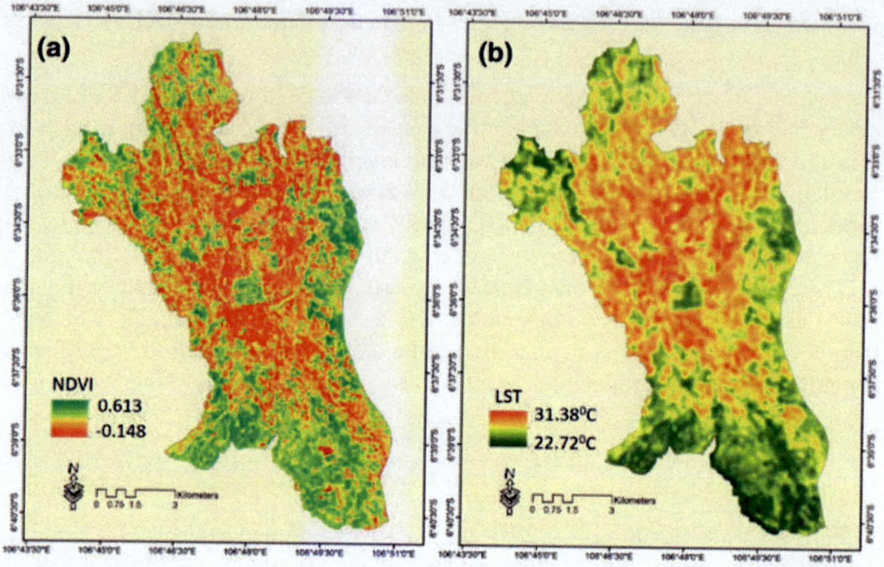

Fig. 42.3 **a** NDVI, **b** LST

In this paper, there are three fuzzy membership values to be compared using Jaccard Similarity Index (JSI), namely (a) > 0.7–1, (b) > 0.75–1, and (c) > 0.8–1. For detailed observation, the three images with different fuzzy membership values are displayed in Fig. 42.4.

When visually observed, raster cells in LST images with > 0.7–1 [figure (a–c)] and > 0.8–1 [figure (g–i)] have the tendency to much overlapped among high, medium, and low classes. It was seen from Fig. 42.4b, h, that yellow color almost covers the whole study area map. Not to mention that in Fig. 42.4a, g, raster cells with red color dominate the map. Therefore from a visual perspective, it can be summarized from Fig. 42.4 that LST image classes from > 0.75 to 1 fuzzy membership values are the best in terms of fuzzy class separation.

Quantitatively, when using JSI to select the best separable fuzzy class, it can be seen from Fig. 42.5 that overlapped raster cells between high LST and medium LST in > 0.7–1 values is the biggest among others. Interestingly, no overlapped cells when high LST and low LST were intersected, which are indicated by zero JSI value (see Fig. 42.5). Furthermore, the JSI values for the intersection between medium LST and low LST shown by orange color bar are always lower than other intersections (blue and grey colors). This indicates that medium LST raster cells resulted from the FCM algorithm tend to dominate the study area map. Thus, it corresponds with visual interpretation from Fig. 42.4b, h. Based on the results of fuzzy class separation analysis both using visual perspective and JSI, it can be summarized that the LST classes from > 0.75 to 1 fuzzy membership values are the best and selected for further analysis.

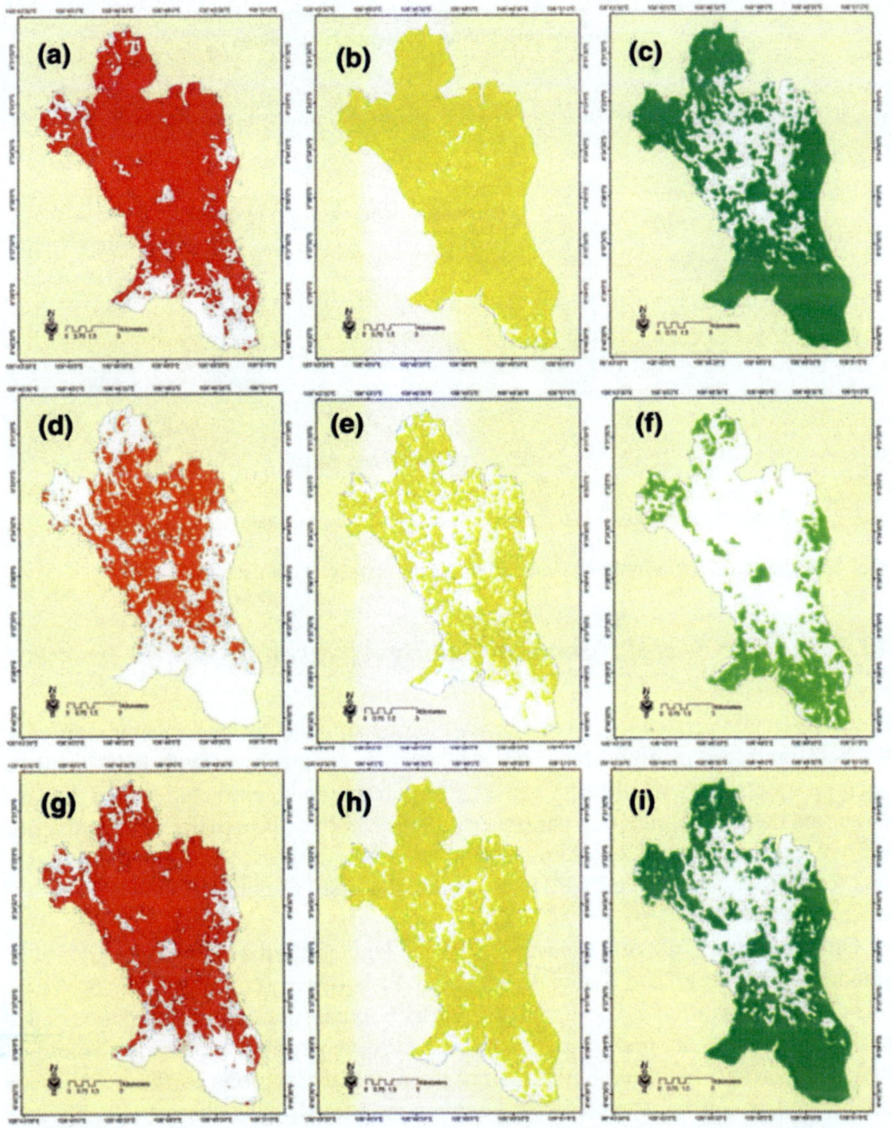

Fig. 42.4 **a** high LST (>0.7–1), **b** medium LST (>0.7–1), **c** low LST (>0.7–1), **d** high LST (>0.75–1), **e** medium LST (>0.75–1), **f** low LST (>0.75–1), **g** high LST (>0.8–1), **h** medium LST (>0.8–1), **i** low LST (>0.8–1)

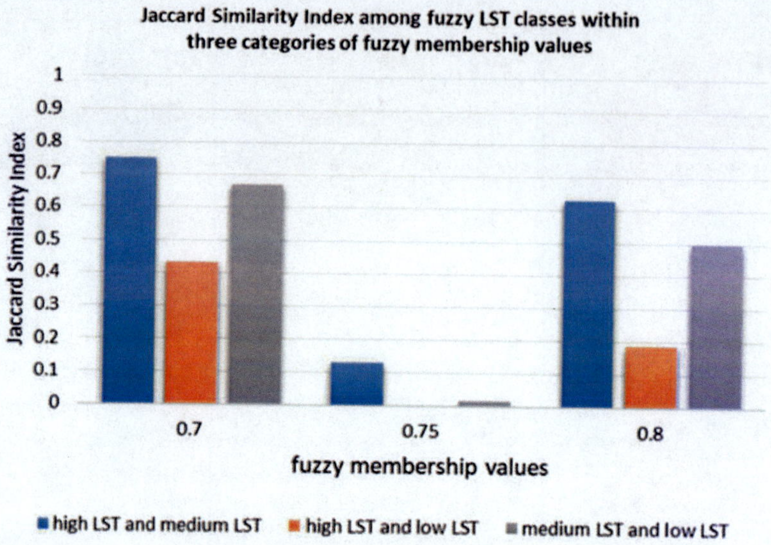

Fig. 42.5 Jaccard Similarity Index among LST fuzzy classes in three categories

42.3.3 *PUP Social Function Quality Assessment*

Since LST classes from > 0.75 to 1 fuzzy membership values have been identified as the best, therefore its high LST class is selected to be integrated with PUP locations vector datasets. As a result, eighteen PUP locations can be extracted within the LST class, and their social function parameters were scored by the experts. Social function KPI values were computed from experts' scoring, can be seen in Fig. 42.6. Based on their KPI values, seven PUP locations are classified as compatible with standard since they have KPI between 0.67 and 1.

On the contrary, the other eleven PUPs can be classified as less compatible with standard since they have a social function KPI between 0.66 and 0.34. Based on this classification, PUPs whom compatible with social function standard are needed to be properly maintained, so they continue to serve people located in surrounding high LST. When facilities built to strengthen social functions in these PUPs are maintained, it can be expected that people continuously visit these PUPs because they might fulfill their social needs such as meeting with each other.

On the other hand, for PUPs with less compatible with social function standards, improvement is urgently needed for facilities within the parks which fulfilling people's social needs. This improvement plan may include adding new facilities which strengthen social function such as children's play area, gazebo, and jogging track.

The spatial distribution of PUPs located within high LST can be observed in Fig. 42.7, where green color points represent PUPs with compatible social function

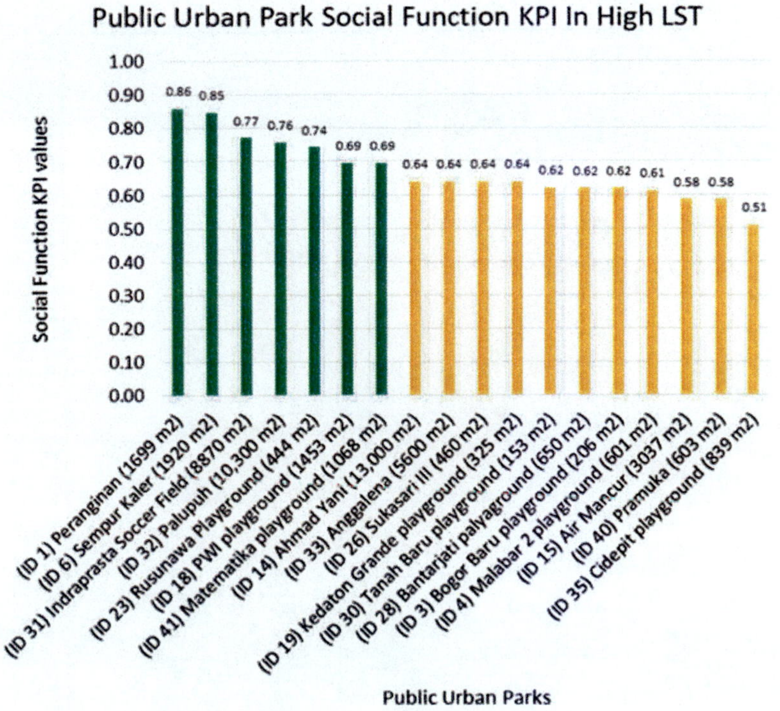

Fig. 42.6 PUP social function KPI values within high LST locations

standards and therefore need proper maintenance. In addition, PUPs with orange color indicated that they need improvement to increase their social function quality.

42.4 Conclusions

Based on the Fuzzy C Means method, this study results in 30.99% high LST class, 37.14% medium LST class, 27.27% low LST class, and 4.60% unclassified. Based on fuzzy class separation analysis using Jaccard Similarity Index, fuzzy membership value (>0.75–1) shows the best separation for LST classification using Fuzzy C Means when compared to fuzzy membership value (>0.8–1) and fuzzy membership value (>0.7–1). Based on social function KPI values, seven public urban park locations within high LST can be classified with compatible with KPI standard value (0.68–1), while eleven locations can be associated with less compatible with KPI standard (0.34–0.67). Therefore, seven public urban park locations with compatible standards need proper maintenance to keep their social function quality. In comparison, the other eleven locations with less compatible standards need improvement to increase their social function quality.

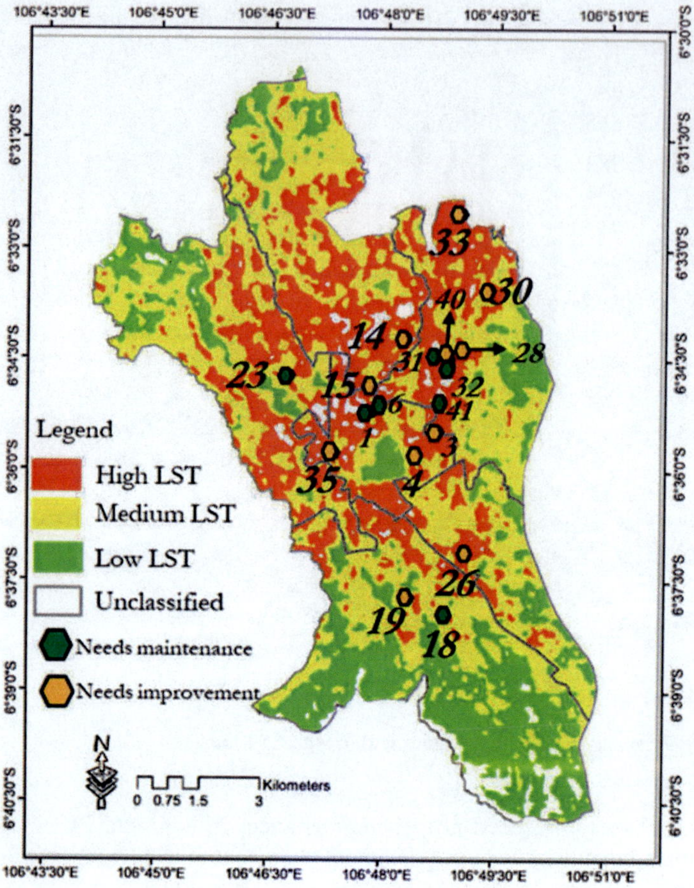

Fig. 42.7 PUP locations in high LST

References

1. Heidt, V., Neef, M.: Benefits of Urban green space for improving urban climate. In: Carreiro, M.M., Song, Y.C., Wu, J. (eds.) Ecology, Planning, and Management of Urban Forests. Springer, New York (2008). https://doi.org/10.1007/978-0-387-71425-7_6
2. Yanti, C.W.B., Nurfaida, N., Pramasarandy, A.K.A.: Evaluasi Nilai Fungsional dan Estetika Taman Maccini Sombala Kota Makassar sebagai Taman Hortikultura. Jurnal Arsitektur Lansekap **1**(2), 50–57 (2015)
3. Kristi, R., Simanjuntak, B.H.: Evaluation of the Aesthetic and Social Function of the Bendosari Park Salatiga City, Indonesian. J. Comput. Model. **1**(2), 87–92 (2018)
4. Daniels, B., Zaunbrecher, B.S., Paas, B., Ottermanns, R., Ziefle, M., Roß-Nickoll, M.: Assessment of urban green space structures and their quality from a multidimensional perspective. Sci. Total Environ. **615**, 1364–1378 (2018)
5. Zanon, D., Victoria, P.: A Model for Estimating Urban Park Visitation: Parks Victoria. Parks Victoria, Victoria (1998)

6. Bezdek, J.C.: Pattern Recognition with Fuzzy Objective Function Algorithms. Springer, New York (1981)
7. Cebeci, Z.: Partitioning Cluster Analysis Using Fuzzy C-Means. https://cran.r-project.org/web/packages/ppclust/vignettes/fcm.html
8. Weng, Q., Yang, S.: Managing the adverse thermal effects of urban development in a densely populated Chinese city. J. Environ. Manage. **70**(2), 145–156 (2004)
9. Guha, S., Govil, H., Dey, A., Gill, N.: Analytical study of land surface temperature with NDVI and NDBI using Landsat 8 OLI and TIRS data in Florence and Naples city, Italy. Eur. J. Remote Sens. **51**(1), 667–678 (2018)
10. Sobrino, J.A., Jiménez-Muñoz, J.C., Paolini, L.: Land surface temperature retrieval from LANDSAT TM 5. Remote Sens. Environ. **90**, 434–440 (2004)
11. Mathworks Homepage, https://www.mathworks.com/help/fuzzy/fuzzy-c-means-clustering.html
12. Jardón, E., Jiménez, E., Romero, M.: Spatial Markov Chains implemented in GIS. In 2018 International Conference on Computational Science and Computational Intelligence (CSCI), pp. 361–367

Chapter 43
Zone of Biodiversity Extinction in Coral Reef Ecosystem in Sumatra Island Using Climate Change Scenarios

Muhammad Hanif, Rizki Atthoriq Hidayat, Giant Amor, Luhur Moekti Prayogo, Ahyuni, and Arie Yulfa

Abstract Global climate change has a significant impact on the environment, especially on tropical ecosystems. Coral reef ecosystems mostly live in shallow and surface waters that heat up quickly, and they are adapted to live near the temperature tolerance limits for their lives. However, they are susceptible to environmental changes, especially to extreme changes in water temperature. This study aims to predict the extinction zone of coral reef ecosystems with climate change scenarios. The data used in this study is based on geospatial data obtained from remote sensing satellite imagery such as Aqua Modis from 2003 to 2020, Eco-region. For climate change scenarios that impact coral reef life, use one part of Habitat Suitability (HS), namely water temperature. Analysis The sea surface temperature (SST) was performed by transforming the SST from satellite from imagery, then interpolating the data with Inverse Distance Weighted (IDW) to complete the blank data for each image recording. The application of linear forecasting method is used to predict future sea surface temperature rise. The results of this study we found that sea surface temperatures tend to fluctuate, and we found that the increase in sea temperature from 2003 to 2020 was 1.11 °C from this data we created a baseline map by 18 years of average annual temperature, to predict data for the next 2070 years and this map grouped into four zones coral reef: disappear, bleaching, stress, and stable zones. The temperatures above 30.88 °C are the disappearance zone. The eco-regions included in the extinction area are the Andamans of western Sumatra and parts of the Sunda Malacca strait, which are covered by extinction because these temperatures are beyond tolerance for life. This scenario provides an overview of the priority locations for the conservation of aquatic ecosystems that need to be mitigated from the impacts of climate change.

R. A. Hidayat · G. Amor · Ahyuni · A. Yulfa
Department of Geography, State University of Padang, Padang 25171, Indonesia

L. M. Prayogo
Faculty of Fisheries and Marine, University of PGRI Ronggolawe Tuban, Tuban 62381, Indonesia

M. Hanif (✉)
Department of Geo-Informatics, Khon Kaen University, Khon Kaen 40002, Thailand
e-mail: muhammad.h@kkumail.com

E. Yulihastin et al. (eds.), *Proceedings of the International Conference on Radioscience, Equatorial Atmospheric Science and Environment and Humanosphere Science, 2021*, Springer Proceedings in Physics 275, https://doi.org/10.1007/978-981-19-0308-3_43

43.1 Introduction

Indonesia is the largest archipelagic country globally, with a coastline of more than 81,000 km and more than 17,508 islands. Indonesia also has extensive coral reefs and can protect the shoreline of the Indonesian archipelago [1]. Coral reefs are a collection of coral animals that form the structure of coral reefs on the seabed. Coral reef systems originate from calcium carbonate (CaCO3) deposits in the ocean produced by polyps [2]. Coral reef ecosystems are marine ecosystems located in tropical waters. This ecosystem has high productivity and biodiversity. Coral reefs in a healthy condition have many benefits for the environment and the community's economy. We know many advantages, namely increasing reef fish stocks, diving tourism activities, and protecting coastal ecosystems [1, 3]. The problem of coral reef damage that is quite popular is coral bleaching. Coral bleaching is an event that disrupts the symbiotic relationship between corals and coral endosymbiont algae. A study conducted in 2000 witnessed worldwide coral bleaching events, which, in turn, resulted in unprecedented coral mortality and destruction of coral reef ecosystems [4]. Coral bleaching is a response from groups of scleractinian, alcyonarian corals, clams, and anemones to changes in the environment in which they live, namely dramatic changes in water temperature causing the population of symbiodinium (zooxanthellae algae) to leave/decline, leaving coral tissue that causes the pale or white discharge [5, 6].

The IUCN Global Species Program in 2005 began conducting global assessments of marine species, with an estimated extinction risk for 20,000 species. The results of the current evaluation have been carried out on 16,000 species, including coral reefs. [8, 9]. Increasing threats from environmental dynamics result in a decrease in the number and distribution of biodiversity. 11]. A recent global assessment of the state of biodiversity by the Intergovernmental Science-Policy Platform on Biodiversity and Ecosystem Services (IPBES) estimates that one million species are threatened with extinction [12].

The purpose of this study is to predict the increase in sea surface temperature and predict the life zone of coral reef ecosystems that will experience threats due to rising sea surface temperatures in the future. This research is focused on observing the dynamics of sea surface temperature, predicting the increase in sea surface temperature (SST) in 2070, and threats to coral reef ecosystems. The basic approach to extinction zones uses the Habitat Suitability (HS) zone of coral reefs based on sea surface temperature from a group of ecoregions that will face extinction in the future. We did this research on the island of Sumatra, which has many stretches of coral reefs. The territory of Sumatra Island is directly adjacent to the Indian Ocean and the Malacca Strait.

43.2 Data and Methods

43.2.1 Data

This research uses two types of data, and the first is Aqua Modis Satellite (Moderate Resolution Imaging Spectro radiometer). This data is from 2003 to 2020, and this data are taken from the official website of NASA (The National Aeronautics and Space Administration) using the Google Earth Engine platform. The specifications of this data are 2330 km (cross-track) by 10 km (along-path at nadir) and infrared waves. Infrared waves are very effective in observing sea surface temperatures [17]. The second is the Eco-region groups this data in 2007. The use of eco-regions can make it easier to identify coral reef distribution groups based on eco-region units. The data was sourced from the results of mapping classified by previous researchers [41]. Other supporting data are land boundaries, administrative boundaries, which are sourced from Indonesia government http://tanahairindonesia.go.id.

43.2.2 Methods

This research must go through several stages, the stage of interpretation and initial data processing. At this stage, the SST transformation and data refinement will be carried out with IDW. After that, proceed to the second stage, namely the analysis of sea surface temperature predictions using a forecasting linear method calculation; this process is carried out with statistical equations to see trends in data development linearly and predict future data developments. The research procedure is as follows:

43.2.2.1 Analysis the Sea Surface Temperature (SST)

To determine the threat of extinction of coral reef ecosystems in the future, we use one of the indicators from The Habitat Suitability (HS). The Habitat Suitability (HS) model is the standard approach to predicting and describing species responses to global climate change is. The Habitat Suitability (HS) model uses several variables to support species living in ecosystems, the one variable is temperature. These criteria will explain how strong the relationship between species and climate is [13, 14]. Future coral reef observers argue that in the face of rising sea surface temperature (SST) is one consequence of global warming [16].

SST analysis is a technique of changing the pixel value information in digital satellite imagery, from the object brightness value to the water surface temperature value, which is part of the HS criteria as an indicator of the extinction of coral reef ecosystems in the future, following equation [19] [43]:

$$\mathrm{Tb} = \mathrm{c1}/\left(\mathrm{vi} * \left(\mathrm{alog}\left((\mathrm{c2})/\left(\mathrm{vi}^5 * \mathrm{B}\right) + 1\right)\right)\right) \tag{43.1}$$

Table 43.1 Coefficient for band 31 dan 32 [19]

Coefficient	≤ 0,7 K	> 0,7 K
C1	1,228,552	1,692,521
C2	0,9,576,555	0,95,558,419
C3	0,1,182,196	0,0,873,754
C4	1,774,631	1,199,584

where:

Tb sea surface temperature (°K)
c1, c2 constant of radiation, which is c1 is 1,1910659x108 [W m^{-2} sr^{-1} (μm -1)-4] and c2 is 1,438,833 × 104 [K μm].
vi central wave length
B Band 20, band 31 dan band 32

The SST analysis process is carried out with a cloud data computing device, namely the Google earth engine (GEE), we apply several scripts to obtain the data results (Table 43.1).

43.2.2.2 Inverse Distance Weighted (IDW)

This method assumes that each input point has a local effect and decreases with distance. IDW distance is an inverse method obtained from a mathematical equation. In this interpolation method, we adjust the relative influence of each sample point. The power value in the IDW interpolation is determined by the weight and distribution of the input points so that adjacent points produce a more detailed surface. The effect decreases as the distance from the input point increases. The value of each point will be used in the distance derivative function when calculating the interpolation [42].

At this stage, we convert each raster sheet resulting from the STT analysis into data points. When this data is of point type, we interpolate the data with the Q-GIS tool. The following two stages are carried out, first calculating each image sheet's average water temperature value. The average values from LST will develop into a statistical analysis of the linear forecasting method to find temperature predictions in 2070. The second stage is where all the results interpolation data overly by adding and dividing all the raster data to get the average of the sea surface temperature as a baseline for spatial data for future predictions.

43.2.2.3 Linear Trend (Semi Average)

To predict future sea surface temperature trends we use a linear trend technique. This forecasting method is part of decomposition techniques, this technique can be interpreted to break down time series data in a time series. This solution forms the basis of the decomposition technique and identifies each component of the data separately. This method is also known as the time series method. One part of this technique is a linear trend, and this can be interpreted in the time series that can be described in a plot close to a straight line the trend (semi average) method used to get the current average to be two parts. The steps are (1) dividing the existing data into two groups with the same amount. (2) Determine the base year in the middle of group I. (3) In each group, determine the X, semi total, and average values. (4) The total value of X in group I must be zero. The trend formula is calculated by the formula [22]:

$$Y = a + bX \tag{43.2}$$

where:

a average group I
b (average group II–average group I)/n
n the amount of data for each group
X the value is determined based on the base year.

The data used in this projection is data extracted from SST from satellite image analysis. The next is the last process is used the overlay technique is a process for overlapping spatial data. Where this overlay technique is a form of geometric data representation in the principle of slices [23, 24]. The overlay process has a minimum requirement that there are two types of data use. The data from the overlay analysis will produce new information, which is a combination of two data and represents unique characteristics and information [25]. In this study, what will be overlaid are a map of the 2070 sea surface temperature projection, a map of the distribution of coral reefs, and the eco-region.

43.3 Results and Discussion

43.3.1 Sea Surface Temperature Change

The results of processing Aqua Modis satellite imagery used the SST algorithm. We obtain the average annual temperatures. This yearly average temperature is obtained by calculating the data every month, and we do an average value-based assessment of each pixel of the raster data. Before we calculate the average value, in fact, a lot of invalid data has a condition (null) or loses information in an area, so this will affect the results of the analysis if no modifications or improvements are made to fill in the raster that has a value (null) or 0. Therefore, we convert raster to point, and we interpolate the cell values for each pixel so that we can find the average temperature in each image sheet. We do not use the cloud masking principle because each image sheet on a particular day has a difference in sea surface temperature. The consequences if used cloud masking, and mosaic the data that fills in the blank part of this image raster sheet is data from a different time recording, and different pixel value of sea temperature, so this will misleading for final calculation. So it is more appropriate to interpolate using the values of neighboring pixels on one image sheet at the same recording time to replace invalid value of the data. The principle of interpolation is also widely applied by climate observers to map rainfall because the interpolation model can represent the impression of other regions more precisely in conditions of limited space. In the image analysis sheet, we found the presence of invalid data so that, we lost information, so to overcome this, to improve each invalid data we interpolate the data with an interpolation approach model, this is done so that the analysis data in raster form can cover the entire area without losing information important in one spot (Fig. 43.1).

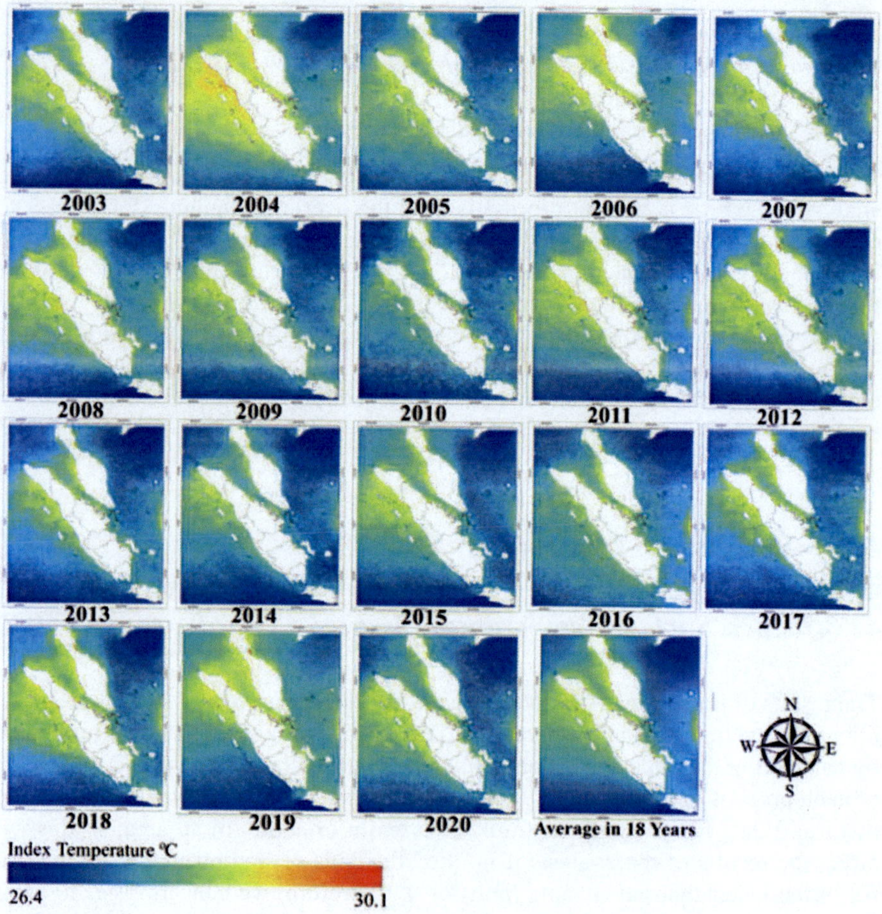

Fig. 43.1 Annual of imagery the sea surface temperature

Overall this diagram illustrates information on the fluctuations in the representation of sea surface temperature information. The sea surface temperature in this research location has increased, but this temperature rises and falls on the average annual temperature according to cycles and natural phenomena. The starting point of the data in the early years was at 29.44 °C, while there was a sharp increase in 2010 where the lamina phenomenon occurred with the annual average air temperature at 30.05 °C.

The temperature changes that look very significant also occurred in 2016. This year is an El Nino Southern Oscillation (ENSO) phenomenon occurred where the average sea surface temperature at the research location was at 30.18 °C. In 2016 the increase in sea surface temperature during the ENSO period caused the bleaching of coral reefs in many areas in Indonesia [26]. Whether or not the ENSO case has an impact on Indonesia's coral reef ecosystems, the El Nino climate anomaly that has occurred throughout 2015–2016 has caused mass deaths and significant coral bleaching worldwide [27] (Fig. 43.2 and 43.3).

We calculated the average temperature for 18 years on the raster map data as the basis for visualizing the projections. Furthermore, the main calculation we do is to forecast the sea surface temperature in 2070 with the statistical equation Forecast Linear Prediction, this approach is semi-average to forecast data more acceptable, where the results of linear forecasts in 2070 we find the average of normal sea surface temperature is at 30.88 °C. The condition of the future temperature is a precarious for the life of biodiversity, which is very vulnerable to hot air temperatures. The predicted temperature conditions in 2070 exceed the hot air temperature in the ENSO, and Lanina anomaly phenomena in the previous year. The threat of ecosystem extinction in the future is an assessment of the facts that have occurred previously during the

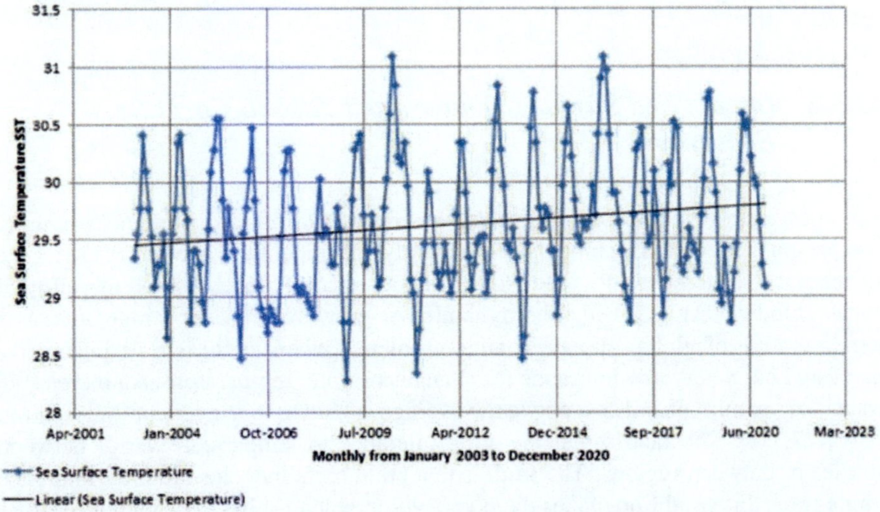

Fig. 43.2 Statistic monthly of the sea surface temperature from January 2003 to December 2020

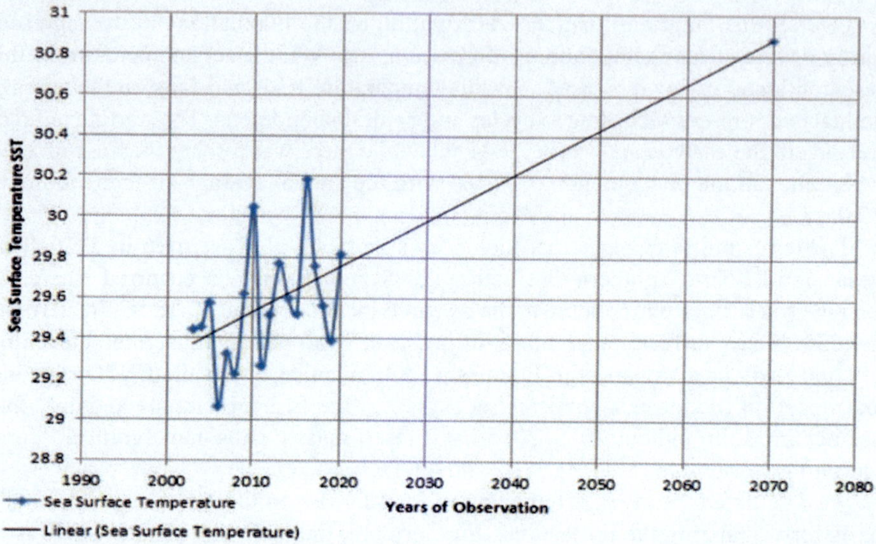

Fig. 43.3 The annual statistic of the sea surface temperature from 2003 to 2020, and prediction in 2070

ENSO phenomenon, where changes in sea surface temperature occur over a certain period and are not permanent. However, with the prediction model for sea surface temperature rise in 2070, this is an assessment of the temperature increase of the ocean, not in an ENSO indication. The increase in the usual average sea surface temperature in the future is believed to cause the extinction of coral reef ecosystems and biodiversity riches.

43.3.2 Impact Sea Surface Temperature Rise to Coral Reef Ecosystem

The ecosystem response of coral reefs to climatic and other disturbances is more complex than predicted by models based mainly on temperature anomalies [28]. Like an anomaly, sea temperatures are driving increased coral bleaching and mortality of marine biodiversity [29, 30]. The dynamics of growing seawater temperature is a trending issue of global climate change problems, where coral is a very sensitive fauna and has a very low tolerance for changes in water temperature. An increase in water temperature of 1–1.5 C above the average is known to trigger coral bleaching.

[6, 27, 31]. The fact is that the coral tolerance to temperature varies between species or between regions. The truth is that coral reefs live close to their tolerance limits zone, this condition makes them very vulnerable to shifts in water temperature limits [26]. The typical temperature ranges of habitat suitability (HS) for coral reefs

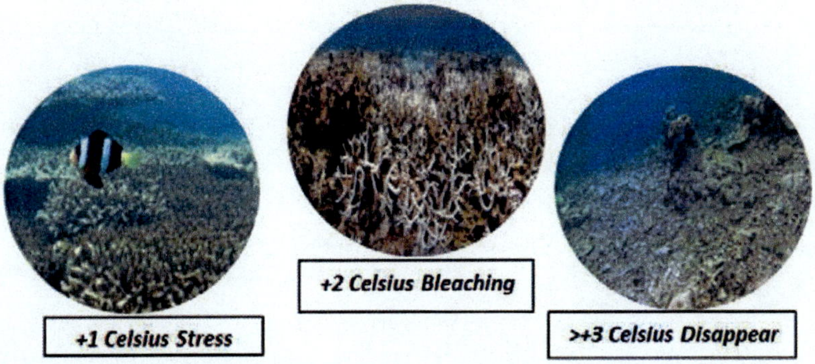

Fig. 43.4 Coral reefs respond to change sea surface temperature [29, 30]

are between 26 and 28 °C [32]. Coral reefs can also gain heat tolerance of 1–1.5 °C through shifting algae living in coral tissue, coral cover above 5% extended to 2065 [33]. Basic mainly of the principle of coral reef vulnerable to temperature changes, and character HS for coral reef. We decide to group the spatial map of sea surface temperature outside average temperature for HS prediction into three classes.

In Fig. 43.4, we observe differences in the environmental and physical conditions of coral reefs caused by changes in sea surface temperature. An increase of 1 °C in the temperature of the coral reef ecosystem from the HS temperature causes depression to occur where the appearance of coral reefs begins to fade. An increase in water temperature of more than 2 °C has an impact on coral reef bleaching. In contrast, the fatal condition that occurs when the rise in ocean surface temperature exceeds 3 degrees Celsius from the HS temperature causes the coral reef ecosystem is disappearing.

Figure 43.5 is the result of grouping sea surface temperature data for Sumatra Island in 2070. We grouped the data into interval classes to make it easier to describe spatial information. This group refers to the sensitivity level of coral reef ecosystems to changes in water temperature. The surface temperature zones that dominate Sumatran waters in 2070 are the bleaching zone at the number of 61.5%, the stress zone at the number of 36.5%, and the disappear zone at the number of 1.8%. From SST prediction in 2070 an initial approach can be drawn that in the future. It will be a lot of massive coral reef bleaching, which is the result of a linear change in the average temperature of the ocean surface from 2020 to 2070.

This temperature change causes the majority of coral reef bleaching on the island of Sumatra. This bleaching of coral reefs will affect the reduction in the amount of aquatic biodiversity. Some of them are believed to be extinct because the temperature of these waters is a constant temperature that will occur in the future, not a temporary change such as the ENSO phenomenon. Because, we discovered much literature was explained the temperature anomaly of temporary ocean surface temperature disturbances coral reefs, it will be able to recover if the water temperature in their environment returns to normal, even this will take up to three years. The forecasting

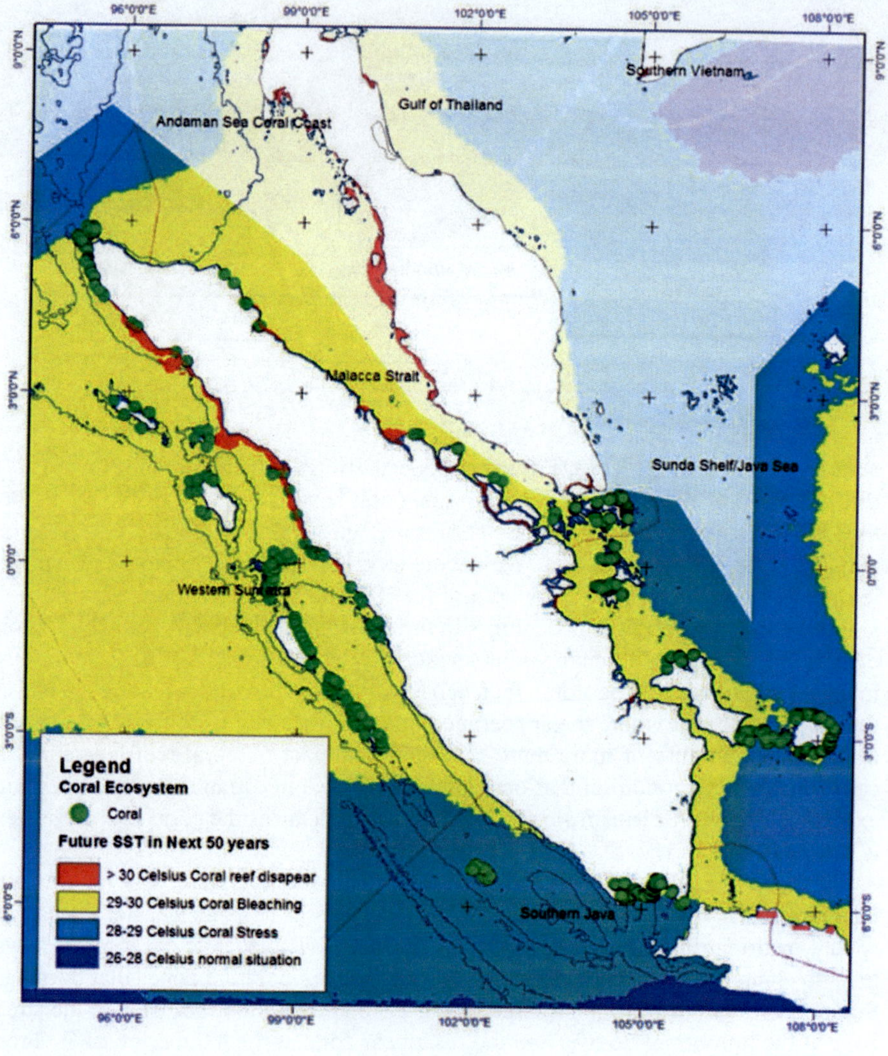

Fig. 43.5 Map of figure is the zone of changes in ocean surface temperature to the biodiversity extinction of coral reef ecosystems in 2070, and distribution existing spot of coral reef in Sumatra Island

scenario is that the average temperature of the waters in the future does not support coral reef HS, so the return of the water temperature to previous average is not a common thing to support habitat for habitat restoration.

Some studies explicitly combine species-area projections of species loss to incorporate climate change as a driver via models of changing global ecosystems [21]. The climate change responses suggested that 10 to 14% of species would become extinct [34]. Coral reef ecosystems have an extensive type of biodiversity compared to other marine ecosystem groups. So the occurrence of damage or extinction in coral reef ecosystems forms a linear relationship with the death of another biodiversity which is in symbiosis with coral reef ecosystems. The approach in this study in, which we assess the risk of coral reef damage caused by rising sea surface temperatures, and it able to describe future extinction zones. [26] The case of coral bleaching occurred in Indonesia's. In one time has caused the death of a group of coral landscapes in many locations, and this event has a very significant influence causing a reduction in the quantity of fish biodiversity that lives and is in symbiosis with coral reefs. Nevertheless, the results of the survey data show that the abundance of herbivorous fish has decreased while the biomass has increased but not significantly so that the recovery process is possible. Recovery of hard corals after disturbance is directly proportional to the rescue of reef fish communities [35]. Recovery of hard corals after bleaching can occur if there is no interference intervention during recovery [36]. In Bali, hard corals take three years to recover themselves [37]. (Table 43.2).

This pointed the need for methods that link spatial to explain spatial information [14]. The cause of climate change that impacts coral reefs does not only come from an increase in sea surface temperature. Many environmental variables contributed to coral reef, and ocean problems, the complex environmental analysis and management strategies need to be carried out to increase the resilience and resilience of coral reef communities to the impacts of climate change [38]. Management and policy actions to update climate change coping strategies on biodiversity can substantially reduce future climate change impacts [39].

Table 43.2 Ocean surface temperature zone that exceeds the tolerance limit of HS coral reef by eco-region group in 2070

Province	Eco-region	Temperature °C	Life zone	SST (%)
Andaman	Nicobar Islands	28–29	Stress	23
	Andaman Sea Coral Coast	28–29	Stress	39
		29–30	Bleaching	39
Sunda Shelf	Malacca Strait	> 30	Disappear	22
		28–29	Stress	39
		29–30	Bleaching	39
Java Transitional	Southern Java	28–29	Stress	50
		29–30	Bleaching	50
Sunda Shelf	Sunda Shelf/Java Sea	28–29	Stress	35
		29–30	Bleaching	35
		> 30	Disappear	30
Andaman	Western Sumatra	> 30	Disappear	34
		28–29	Stress	22
		29–30	Bleaching	44

43.4 Summary and Conclusions

From the results of this study, we can conclude that the dynamics of the SST sea surface temperature. This condition increases sharply during times of climatic anomalies such as the ENSO. Interpolation approach to perfect the invalid data from SST transformation is helps improve the calculation results and minimizes the problem of spatial information in modeling ocean surface temperatures. Linear trend projections provide of future sea surface temperatures with more realistic results because the primary approach of all available data is a basis for calculations. Here we found that the average normal sea surface temperature in the next fifty years will be at 30.88 °C. This future average temperature is not very suitable for the Habitat Suitability HS indicator of coral reef ecosystems. This prediction has not been able to describe the ENSO anomaly that will occur in fifty years. The grouping of future sea surface temperatures into these three zones shows that the majority of the territorial waters of the island of Sumatra will be in the bleaching zone, and several main spots are in the disappearance zone. Most of the coral ecosystem on the Andaman Western Sumatra eco-region group is covered by the bleaching zone. Here we emphasize that this research is still lacking because predictions of ecosystem extinction in the future are only based on future sea surface temperature increases, not involving other environmental factors. Future research needs to use another environmental approach for analysis coral reef extinction. It can be analyzed by spatially to represent this information on extinction more strongly.

Acknowledgements We say thanks to NOAA and USGS for providing open-source remote sensing data. Thanks to the committee of the INCREASE conference, which allow presenting idea, and article in this conference. We appreciate to all researcher team which supports this research program.

References

1. Aulia, Q,A., Sari N.W.P.: Coral Bleaching, Karang Hidup Atau Mati? Oseana **45**(2) (2020). p-ISSN 0216-1877. e-ISSN: 2714-7185. Available at https://oseana.lipi.go.id
2. Papu, A.: Kondisi tutupan terumbu karang Pulau Kapoposang, Kabupaten Pangkajene Kepulauan, Provinsi Sulawesi Selatan. Jurnal Ilmiah Sains (2011). https://https://doi.org/10.35799/jis.11.1.2011.32
3. Sari, N.W.P. (2016). Coral reef, penyerap atau penghasil carbon? Oseana, XL1(2): 32–40. Available at http://oseanografi.lipi.go.id
4. Fine, M., Oren, U., Loya, Y.: Bleaching effect on regeneration and resource translocation in the coral Oculina patagonica. Marine Ecol. Progress Ser. **234**, 119–125 (2002). https://https://doi.org/10.3354/meps234119
5. Hughes, T.P., Bellwood, D.R., Folke, C.S., McCook, L.J., Pandolfi. J.M.: No-take areas, herbivory and coral reef resilience. TRENDS Ecol. Evol. **22**, 1–3 (2006). https://https://doi.org/10.1016/j.tree.2006.10.009
6. Douglas, A.E.: Coral Bleaching—How and Why? Mar. Pollut. Bull. **46**, 385–392 (2003). https://doi.org/10.1016/S0025-326X(03)00037-7
7. IUCN: The IUCN Red List of Threatened Species. Version 2020-2 (2020). Available at http://www.iucnredlist.org

8. Carpenter, K.E., et al.: One-third of reef-building corals face elevated extinction risk from climate change and local impacts. Science **321**(80), 560–563. https://doi.org/10.1126/science.1159196
9. Miranda, R., et al.: Monitoring extinction risk and threats of the world's fishes based on the sampled red list index. Res. Square (2020). https://doi.org/10.21203/rs.3.rs-129206/v1
10. Tilman, D., et al.: Future threats to biodiversity and pathways to their prevention. Nature **546**, 73–81 [Page 15/23] (2017). https://doi.org/10.1038/nature22900
11. Almond, R.E.A., Grooten, M., Petersen, T.: Living planet report 2020—bending the curve of biodiversity loss. WWF (2020). Available at https://www.zsl.org
12. Ruckelshaus, M.H., et al.: The IPBES Global Assessment: Pathways to Action. Trends Ecol. Evol. **35**, 407–414 (2020). https://doi.org/10.1016/j.tree.2020.01.009
13. Pearson, R.G., Dawson, T.: Predicting the impact of climate change on species distribution: are bioclimate envelope models useful? Glob. Ecol. Biogeogr. **12**, 361–371 (2003). https://doi.org/10.1046/j.1466-822X.2003.00042
14. David, K.A., Akcakaya, H.R., Thuiller, W., Midgley G.F., Pearson R.G., Phillips S.J.: Predicting extinction risks under climate change: coupling stochastic population models with dynamic bioclimatic habitat models. J. R. Soc. Biol. Lett. Glob. Change Biol. (2008). http://dx.doi.org/https://doi.org/10.1098/rsbl.2008.0049
15. McClanahan, T.R., Baird, A.H., Marshall, P.A., Toscano, M.A.: Comparing bleaching and mortality responses of hard corals between southern Kenya and the Great Barrier Reef, Australia. Mar. Pollut. Bull. (2004). https://doi.org/10.1016/j.marpolbul.2003.08.024
16. Nolan, C., et al.: Past and future global transformation of terrestrial ecosystems under climate change. Science **361**, 920–923 (2018, August 31). Avalaible at http://science.sciencemag.org/. Science **361**(6405), 920–923. https://doi.org/10.1126/science.aan5360
17. Green, P.E., et al.: Remote Sensing Handbook for Tropical Coastal Management. United Nations Educational, Scientific and Cultural Organization (2000). ISBN 92-3-103736-6
18. Llewellyn-Jones, D.T., Minnett, P.J., Saunders, R.W., Závody, A.M.: Satellite multichannel infrared measurements of sea surface temperature of the N.E. Atlantic Ocean using AVHRR/2. Quart. J. R. Met. Soc. **110**, 613–631 (1984). https://doi.org/10.1002/qj.49711046504
19. Brown, O.B., Minnett, J.P.: MODIS infrared sea surface temperature algorithm algorithm theoretical basis document. University of Miami. Miami, FL 33149-1098 (1999). Under Contract Number NAS5-31361. Available at https://modis.gsfc.nasa.gov
20. Prabhakara, C., Dalu, G., Kunde, V.G.: estimation of sea surface temperature from remote sensing in the 11 μm to 13 μm window region. J. Geophys. Res. **79**(12), 1744–1749 (1974). https://doi.org/10.1029/JC079i033p05039
21. Pimm, S.L.: Biodiversity: climate change or habitat loss—which will kill more species? Curr. Biol. 18, R117–R119 (2008). https://doi.org/10.1016/j.cub.2007.11.055; (pmid: 18269905)
22. Yudaruddin, R.: Forecasting Untuk Kegiatan Ekonomi dan Bisnis. RV Pustaka Horizon. Samarinda Kalimantan Timur. **75119** (2019). ISBN 978-602-5431-48-7. Available at https://feb.unmul.ac.id
23. Hanif, M., et al.: Multi spectral satellite data to investigate land expansion and related micro climate change as threats to the environment. In: IOP Conference Series: Earth and Environmental Science. **303** (2019). https://doi.org/10.1088/1755-1315/303/1/012030
24. Hidayat, A.R., et al.: Modeling of Conservation Priority Zone for the Hel-meted Hornbill (Rhinoplax Vigil) In Silokek Geopark Area, West Sumatra (2020). Pre-prints (www.preprints.org). Posted: 27 September 2020. https://doi.org/10.20944/preprints202009.0676.v1
25. Hidayat, R.A., Hanif, M.: Spatial modeling of the threat of damage to the peatland ecosystem in the Mainland of Bengkalis regency, Riau Province. J. Soc. Polites. Spec. Issue Youth Peatland Conf. e-ISSN 2620-4975. https://doi.org/10.33541/sp.v21i3.2249
26. Setiawan, F., Muttaqin, A., Taringan, S.A., Muhidin, M., Hotmariyah, Sa-blin, A.: Pemutihan Karang Akibat Pemanasan Global Tahun 2016 Terhadap Ekosistem Terumbu Karang: Studi Kasus Di Twp Gili Matra (Gili Air, Gili Meno Dan Gili Trawangan) Provinsi Ntb. J. Fish. Mar. Sci. (2017). https://doi.org/10.21776/ub.jfmr.2017.001.02.1

27. Saptarini, D., dan Muzaki, F.K.: Kajian Jenis dan Bentuk Pertumbuhan Karang yang Rentan Mengalami Pemutihan di Perairan Selat Madura, Jawa Timur. Pros. Semin. JIWECC (2010)
28. McClanahan, T.R., Maina, J., Ateweberhan, M.: Regional coral responses to climate disturbances and warming is predicted by multivariate stress model and not temperature threshold metrics. Clim. Change **131**, 607–620 (2015). https://doi.org/10.1007/s10584-015-1399-x
29. Hoegh-Guldberg, O., et al.: Coral reefs under rapid climate change and ocean acidification. Science **318**(5857), 1737–1742 (2007). https://doi.org/10.1126/science.1152509
30. Herr, D., Grantly, G.R.: The Ocean and Climate Change Tools and Guidelines for Action (2009). Published by: IUCN, Gland, Switzerland. ISBN 978-2-8317-1201-7. books@iucn.org or www.iucn.org/publications
31. Baker, A.C., Glynn, P.W., Riegl, B.: Climate change and coral reef bleaching: an ecological assessment of long-term impact, recovery trends and future outlook. Estuar. Coast. Shelf Sci. **80**, 435–471 (2008). https://doi.org/10.1016/j.ecss.2008.09.003
32. LIPI.: Pemutihan Karang (Bleaching Coral) dan Kejadian bleaching tahun 2016 (2016). Available at http://coremap.or.id/berita/1172
33. Buddemeier, R.W., Lane, D.R., Martinich, J.A.: Modeling regional coral reef responses to global warming and changes in ocean chemistry: Caribbean case study. Clim. Change **109**, 375–397 (2011). https://doi.org/10.1007/s10584-011-0022-z
34. Maclean, M.I., Wilson, J.R.: Recent ecological responses to climate change support predictions of high extinction risk. Proc. Natl. Acad. Sci. U.S.A. **108**, 12337–21234 (2011). https://doi.org/10.1073/pnas.1017352108;pmid:21746924
35. Emslie, M.J., Cheal, A.J., Sweatman, H., Delean, S.: Recovery from dis-turbance of coral and reef fish communities on the Great Barrier Reef, Australia. Mar. Ecol. Prog. Ser. **371**, 177–190 (2008). https://doi.org/10.3354/meps07657
36. Pratchett, M.S, Wilson, S.K., Baird, A.H.: Declines in the abundance of Chaetodon butterflyfishes following extensive coral depletion. J. Fish Biol. **69**. 1269–1280 (2016). https://doi.org/10.1111/j.1095-8649.2006.01161.x
37. Simarangkir, O.R., Yulianda, F., Boer, M.: Community recovery of hard coral post bleaching event in Amed, Bali. J. Ilmu Pertanian (JIPI) **20**(2), 158–163 (2015). https://doi.org/10.18343/jipi.20.2.158
38. Wooldridge, S.A., Done, T.J., Thomas, C.R., et al.: Safeguarding coastal coral communities on the central Great Barrier Reef (Australia) against climate change: realizable local and global actions. Clim. Change **112**, 945–961 (2012). https://doi.org/10.1007/s10584-011-0229-z
39. Driscoll, D.A., Felton, A., Gibbons, P., et al.: Priorities in policy and management when existing biodiversity stressors interact with climate-change. Clim. Change **111**, 533–557 (2012). https://doi.org/10.1007/s10584-011-0170-1
40. Pimm, L.S., et al.: The biodiversity of species and their rates of extinction, distribution, and protection. Biodivers. Status. Sci. Mag. **344**(6187). https://doi.org/10.1126/science.1246752
41. Spalding, M.D., et al.: Marine ecoregions of the world: a bioregionalization of coastal and shelf areas. J. Biosci. **57**(7), 573 (2007, July/August). www.biosciencemag.org
42. Merwade, V.M., Maidment, D.R., Golff, J.A.: Anisotropic consideration while interpolating river channel bathymetry. J. Hydrol. **331**, 731–741 (2006)
43. Dwi Ayu, R.A, Sukojo, B.M., Jaelani, L.M.: Studi Perubahan Suhu Permukaan Laut Menggunakan Satelit Aqua Modis. J. Geoid **7**(1), 73–78 (2011, Februari)

Chapter 44
Comparison of Cloud Base Height in Kototabang from Ceilometer, Radiosonde and Himawari-8 Observations During 2016–2020

Ridho Pratama, Muhammad Fadhlan Putranto, Fahmi Rahmatia, Ina Juaeni, Risyanto, Wendi Harjupa, Fadli Nauval, and Marzuki Marzuki

Abstract Cloud data can be obtained through direct observation, and by using models on radiosonde or satellite data. Cloud observations in real time will be very helpful in predicting the potential for rain events in a place. Ceilometer CT25K can observe the cloud base height for three cloud layers up to 7.5 km. In this study, Ceilometer CT25K data was used to determine the average cloud base height in December January February (DJF), March April May (MAM), June July August (JJA), and September October November (SON) during 2016–2018. The average cloud base height in Kototabang is compared to the Lifted Condensation Level (LCL) of the Radiosonde. Furthermore, Himawari-8 Geostationary Satellite data was used to confirm the presence of the clouds. The result shows that there is a difference in the cloud base height of the Radiosonde (LCL) with the cloud base height from the Ceilometer. The Brightness Temperature (TB) value of Band 13 of the Himawari-8 Geostationary Satellite observation in the same period shows that the average cloud top temperature is ranging from 240 to 270 K.

44.1 Introduction

The movement of large clouds leading to Indonesia generates many higher rainfall-potential regions. For instance, the monsoon circulation in December, January and

R. Pratama (✉) · M. F. Putranto · I. Juaeni · Risyanto · W. Harjupa · F. Nauval
Research Center for Climate and Atmosphere, National Research and Innovation Agency (BRIN), Bandung, Indonesia
e-mail: ridh003@brin.go.id

F. Rahmatia
BPAA Agam, National Research and Innovation Agency (BRIN), Bukittinggi, Indonesia

M. Marzuki
Department of Physics, Faculty of Mathematics and Natural Science, Universitas Andalas, Padang, Indonesia

E. Yulihastin et al. (eds.), *Proceedings of the International Conference on Radioscience, Equatorial Atmospheric Science and Environment and Humanosphere Science, 2021*, Springer Proceedings in Physics 275, https://doi.org/10.1007/978-981-19-0308-3_44

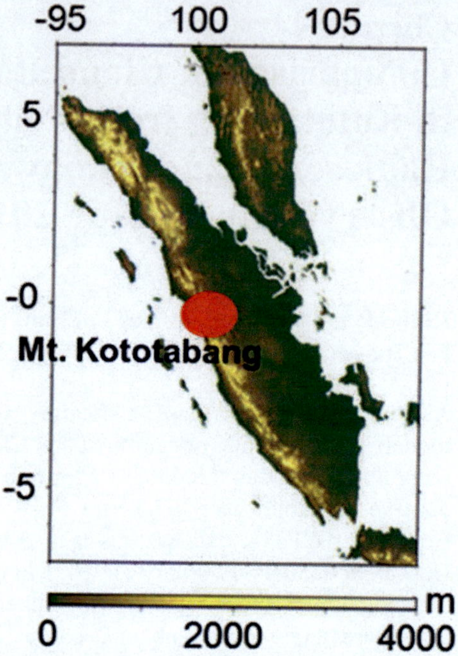

Fig. 44.1 Installed ceilometer location (Mt. Kototabang)

February (DJF) causes an increase in rainfall in western Indonesia [1]. In addition, the MJO circulation that moves from east to west, indicated by large-scale cloud movements, also produces abundant rainfall in Indonesia, in phases 2, 3, 4 and 5 in particular.

In order to observe the movement of atmospheric circulation in Indonesia, especially the equatorial region of Sumatra, an atmospheric observation facility has been built on the island of Sumatra, precisely in Agam Regency, West Sumatra (Fig. 44.1). In addition to the Equatorial Atmospheric Radar (EAR), X-Band Radar, Lidar, and others are also installed at this location.

Ceilometer, also installed At the EAR observation center, is a type of Lidar that uses a white light source. Ceilometers can notify about the height of the cloud base. Although there have been several studies related to clouds in Kototabang, the utilization of Ceilometer data is still terribly restricted. Therefore, this study aims to analyze the results of the Ceilometer data processing during 2016–2018. Ceilometers can provide continuous and objective observations with high accuracy and have been used to evaluate cloud vertical structures [2–4]. Cloud Base Height can also be determined by using satellite-data [4, 5]. Therefore, apart from the Ceilometer, data from the Himawari-8 and Radiosonde satellite observations would also be used to compare cloud characteristics for several cases.

44.2 Data and Methodology

The main data used in this study was the data on the base height of the clouds observed by the Ceilometer CT25K from 2016 to 2020. Ceilometer data was obtained from the Agam Atmospheric Observation Center, West Sumatra. Observational data from the Ceilometer is converted to ASCII data to separate the height of the cloud base level 1, level 2 and level 3. The resolution of the cloud base height by this Ceilometer is 50 ft with a temporal of 15 s.

To see cloud conditions in some cases, data on the height of the cloud tops observed by the Himawari-8 Satellite was used. The data for band 13 Himawari-8 was obtained from the JMA website (https://www.data.jma.go.jp/mscweb/data/himawari/). The spatial resolution of the Himawari-8 satellite in band 13 (infrared) is 4 km $\times$ 4 km per pixel with a temporal resolution of 10 min.

Type of radiosonde that we used is Vaisala RS41-SG, S/N: M2710588 with response time from 0.3 to 10 s. These are the resolutions of some of the parameters in the radiosonde, 0.01 °C (Temperature), 0.1% (RH), 0.01 hPa (Pressure), 0.1 gpm (Geopotential Height), 0.1 m/s (Wind Speed), and 0.1° (Wind Direction). From radiosonde data used LCL and CCL parameters. LCL can be calculated or determined graphically using standard thermodynamic diagrams such as the T-skew log P or the tephigram. The calculation to get the LCL is to use the equation built by Yin et al. [6] and continued by Romps [7]. In the case of this study, the determination of LCL and CCL has been integrated in the Vaisala workstation (WS), so that LCL and CCL are outputs of the Vaisala WS system, along with the atmospheric parameter profile measured by radiosonde (P, T, RH, wind).

The data from the observation of atmospheric parameters from the radiosonde was also utilized to try observing the basic difference in cloud heights from the estimation of the radiosonde and the results of direct observations from the Ceilometer. This study only analyzed the height of the cloud base from the first layer due to the lack of ceilometer observation data in the second and third layers.

44.3 Results

The daily average cloud base height is obtained by averaging the cloud base height from layer 1 ceilometer CT25K every hour from 2016 to 2020. The highest average cloud base height is achieved during midday (8 AM LT) and then continues decreases until late afternoon (5 PM LT) and rises again at night as shown in Fig. 44.2. This is in line with the results obtained by previous researchers using Radiosonde data, especially for single-layer clouds [8].

According to the basic theory, the height of the cloud base is higher in the morning than in the afternoon. Cloud growth is strongly influenced by adiabatic processes, namely solar heating and cooling from long-wave radiation [9]. When solar heating is very strong (during the day), it heats the air, reduces the relative humidity, increases

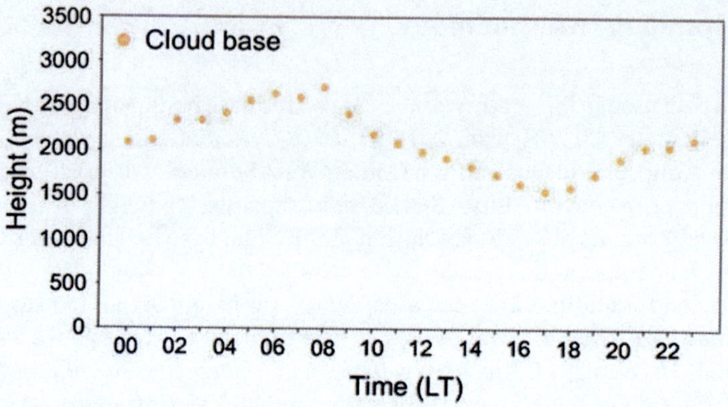

Fig. 44.2 Diurnal cloud base height

the evaporation [10] and increase atmospheric stability. As a result, cloud growth becomes slower so that its height is lower, including its base height. However, this condition will only occur in one- and two-layer clouds. For clouds with more layers, sunlight has been absorbed by the clouds in the upper layer, so the clouds in the layers below can grow to reach a higher altitude. This effect was dominant during the summer (dry) such as in the JJA and SON periods.

Figure 44.3 shows the average height of the cloud base in the DJF, MAM, JJA and SON periods during the day and night. It can be seen that in the MAM and JJA periods the average height of the cloud base during the day (2774 and 2274 m) is higher than at night (2737 and 2193 m) but the difference is not too big. Meanwhile, during DJF and SON, the average height during the day (2193 and 1818 m) is lower than at night (2644 and 2148 m). Figure 44.4 shows the average seasonal cloud base height every year (2016–2020).

Figure 44.5 (top panel) illustrates a comparison of direct observation of cloud base height by Ceilometer (green) and cloud height estimation using radiosonde observation data where the gray bar is the estimated cloud base height using the Lifted

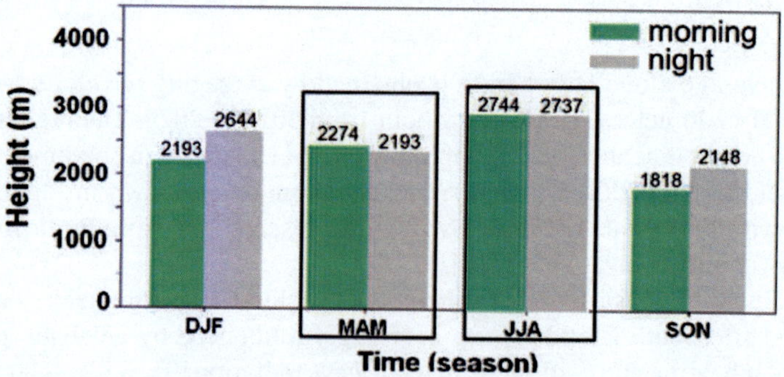

Fig. 44.3 Seasonal average cloud base height in 2016–2020

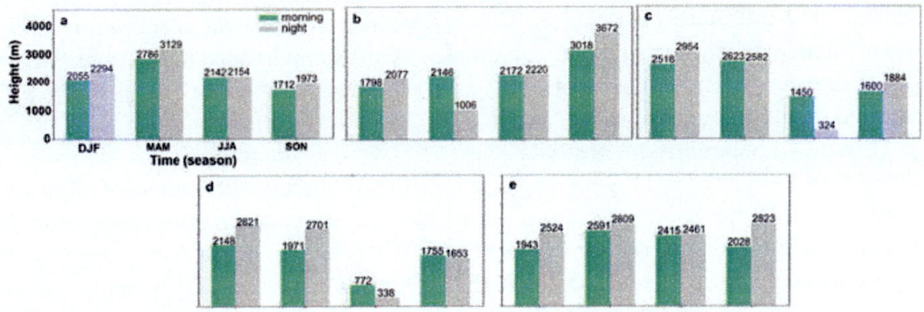

Fig. 44.4 Seasonal cloud base height in **a** 2016, **b** 2017, **c** 2018, **d** 2019, and **e** 2020

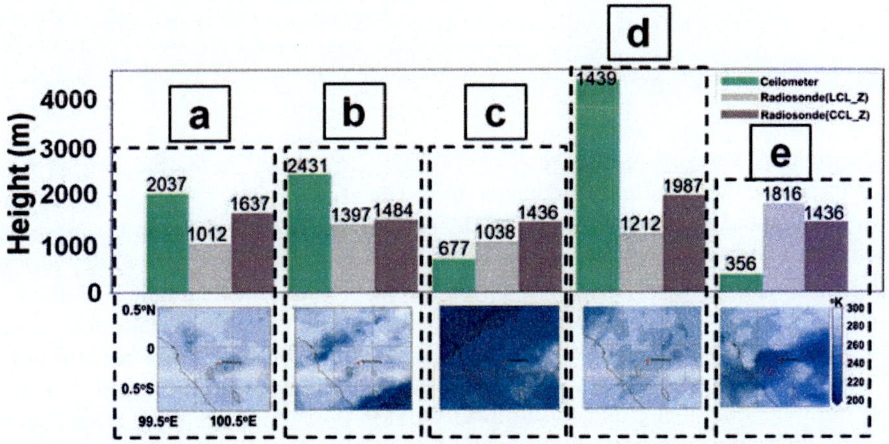

Fig. 44.5 Comparison of cloud base height observed by Ceilometer, Radiosonde and Himawari-8 **a** 1 September 2016 (07:00 LT), **b** 1 September 2016 (10:50 LT), **c** 1 September 2016 (15:00 LT) **d** 2 September 2016 (07:20 LT), and **e** 2 September 2016 (18.50 LT)

Cloud Level (LCL) approach and the brown bar is the estimation of the height of the cloud base using the Convective Cloud Level (CCL) approach. CCL is generated when there is strong surface heat causing an adiabatic increase in air mass and mixing occurs at the planetary boundary layer, so that the heat in the layer near the surface no longer follows the adiabatic lapse rate. The mixing layer will be thicker, LCL will be the top of the mixed layer. When this happens and the heat at the surface continues to cause a cloud base to form above the mixed layer and the level at which this occurs is called the CCL.

Figure 44.5 bottom panel is a spatial view of clouds in the Kototabang area. Figure 44.5 (top panel) shows that there was a difference in the value of observing the height of the cloud base directly and the height of the cloud base from the estimation of Radiosonde data. It was also observed that there was a contrast in the height of the LCL and CCL. The diversity in the height of the cloud base was probably due to the varied distance between the Ceilometer and Radiosonde observations. With this distance, the observed clouds might also be different as seen in the observations by Himawari-8. There was a difference in brightness temperature (BT) in the area around the observation site.

Figure 44.5 can also be seen that there are two data samples with a lower cloud height base (top panel). Figure 45.5c, e show that when the cloud base height was lower, from Himawari-8 data it was known that at the same time there were low BT values in almost all observation areas. From this it is known that at that time a low BT value was detected on a wide scale. With this limited case, it can be stated that lower the cloud base height, the lesser BT value. A low BT value is identical to the height of the cloud, where with a low BT the resulting cloud can turn into cumulonimbus clouds. This type of cloud usually produces heavier rain [11, 12].

44.4 Conclusions

The highest average Cloud Base Height (CBH) is observed in the morning (8 AM Local Time), then continues to decline until late afternoon (5 PM Local Time) and rises again at night. In the MAM and JJA periods, the average CBH in the morning is higher than in the evening, but the difference is insignificant. Meanwhile, during DJF and SON, the average altitude during the day is lower than at night. The diurnal and seasonal variations in CBH from the Ceilometer have a similar pattern to those based on Radiosonde data, as reported by a previous study. Lower CBH from the Ceilometer at Kototabang coincides with low TB data from the Himawari-8 observation, indicating that clouds with lower CBH in this region are thick clouds with a high cloud top. However, the current study is based on limited data, more data analysis is being carried out, and the results will be reported in another paper.

References

1. Lestari, S., King, A., Vincent, C., Karoly, D., Protat, A.: Seasonal dependence of rainfall extremes in and around Jakarta, Indonesia. Weather Clim. Extremes **24** (2019)
2. Costa-Surós, M., Calbó, J., González, J.A., Martin-Vide, J.: Behavior of cloud base height from ceilometer measurements. Atmos. Res. **127**, 64–76 (2013)
3. Vaishnav, R., Sharma, S.K., Shukla, K.K., Kumar, P.: A comprehensive statistical study of cloud base height using ceilometer over Western India. Adv. Space Res. **63**(5), 1708–1718 (2018)
4. Barrena, C.A., Vazquez, D.P., Arias, J.R., Pescador, J.T.: Macriscopic cloud properties in The WRF NWP model. An assessment using sky camera and ceilometer data. J. Geophys. Res. Atmosp. **120**, 10297–10312 (2015)
5. Yoo, H., Li, Z.: Evaluation of cloud properties in The NOAA/NCEP global forecast System using multiple satellite products. Clim. Dyn. **39**(12), 2769–2787 (2012)

6. Yin, J., Albertson, J.D., Rigby, J.R., Porporato, A.: Land and atmospheric controls on initiation and intensity of moist convection: CAPE dynamics and LCL crossings. Water Resour. Res. **51**(10), 8476–8493 (2015)
7. Romps, D.: Exact expression for the lifting condensation level. J. Atmos. Sci. (12), 3891–3900 (2017)
8. Primadona, H., Marzuki, Shafii, M.A., Hasiguchi, H.: Diurnal Varian in Cloud Vertical Structure over Sumatra from Radiosonde Observations During CPEA-I and CPEA-II Campaigns. In: AIP Conference Proceedings (2021)
9. Zhang, J., Chen, H., Li, Z., Fan, X., Peng, L., Yu, Y., Cribb, M.: Analysis of cloud layer structure in Shouxian, China using RS92 radiosonde aided by 95 GHz cloud radar. J. Geophys. Res. Atmos., 115 (2010)
10. Gebremariam, S., Li, S., Weldegaber, M.: Observed Correlation between Aerosol and cloud base height for low clouds at Baltimore and New York, United States. Atmosphere **9**, 143 (2018)
11. Houze, R.A.: Chapter 8—cumulonimbus and severe storms. Int. Geophys. **104** (2014). (Academic Press)
12. Cotton, W.R., Bryan, G., Van den Heever, S.C.: Chapter 8—cumulonimbus clouds and severe convective storms. Int. Geophys. **99**, 315–454 (2011). (Academic Press)

Chapter 45
Influence of Different Pretreatment Methods and Yeast Strains on Xylitol Production from Sugarcane Trash Hemicellulose Hydrolysate

Maulida Oktaviani, Fahriya Puspita Sari, Benjarat Bunterngsook, Euis Hermiati, Verawat Champreda, and Takashi Watanabe

Abstract The two main factors that influence the success of xylitol production from lignocellulosic biomass are as follows: (1) A pretreatment method that can increase cellulose digestibility, resulting in high sugar production and low concentration of by-products. (2) The use of microorganisms that can efficiently use sugar (especially xylose) and have a high tolerance to the inhibitors contained in hemicellulose hydrolysate. The present study aimed to study the influence of different pretreatment methods and yeast strains used in the xylitol production from sugarcane trash hemicellulose hydrolysate. The pretreatment was conducted using two methods: (1) microwave-maleic acid (1.8%) pretreatment at 180 °C for 5 min and (2) liquid hot water-sulfuric acid (1%) at 170 °C for 30 min. The hemicellulose hydrolysate produced from each pretreatment was then used for fermentation using four different yeast strains to produce xylitol. Fermentation was carried out at 30 °C and 200 rpm for

All authors contributed equally as the main contributors to this work.

M. Oktaviani (✉)
Research Center for Applied Microbiology, National Research and Innovation Agency (BRIN), Jl. Raya Bogor Km. 46, Cibinong, West Java 16911, Indonesia
e-mail: maulida.oktaviani@brin.go.id

F. P. Sari
Research Center for Biomass and Bioproducts, National Research and Innovation Agency (BRIN), Jl. Raya Bogor Km. 46, Cibinong, West Java 16911, Indonesia

E. Hermiati
Research Center for Biomass and Bioproducts, National Research and Innovation Agency (BRIN), Jl. Raya Bogor Km. 46, Cibinong, West Java 16911, Indonesia

T. Watanabe
Laboratory of Biomass Conversion, National Research and Innovation Agency (BRIN), Uji Campus, Gokasho, Uji 611-0011, Kyoto, Japan

B. Bunterngsook · V. Champreda
BIOTEC-JGSEE Integrative Biorefinery Laboratory, National Center for Genetic Engineering and Biotechnology (BIOTEC), Science Park, National Science and Technology Development Agency (NSTDA), Pathumthani 12120, Thailand

E. Yulihastin et al. (eds.), *Proceedings of the International Conference on Radioscience, Equatorial Atmospheric Science and Environment and Humanosphere Science, 2021*, Springer Proceedings in Physics 275, https://doi.org/10.1007/978-981-19-0308-3_45

72 h. *Kluyveromyces marxianus* TBRC 1524 gave the highest result at 24 h of fermentation, with 2.29 g/l xylitol production from 20.29 g/l xylose contained in sugarcane trash hemicellulose hydrolysate after microwave-maleic acid pretreatment. Unfortunately, no xylitol was produced by all the yeast strains after fermentation using liquid hot water-sulfuric acid pretreated sugarcane trash hemicellulose hydrolysate. It was concluded that microwave-maleic acid and *K. marxianus* TBRC 1524 could be a potential pretreatment method and a potential yeast strain to produce xylitol from sugarcane trash hemicellulose hydrolysate.

45.1 Introduction

Thailand and Indonesia are the two biggest sugarcane producers in the world. With a total harvested area of 1.8 Mha, Thailand produced 131 Mtons of sugarcane with an average yield of 71.4 tons/ha in 2019. The total harvested area of sugarcane in Indonesia is less than Thailand. With only 0.44 Mha of a total harvested area in 2019, Indonesia produces 29.1 Mtons of sugarcane, resulting in an average yield of 65.6 tons/ha [1]. Although mechanized sugarcane harvest was introduced in the late '90 s, most sugarcane plantations in both countries still adopt the traditional method of harvesting, such as burning sugarcane leaves to ease the cutting and transporting of sugarcane to factories [2, 3]. If unburned, farmers usually left this residue in an open land as a soil fertilizer, although this practice needs to be studied further to determine the impact on soil quality. However, the amount of this sugarcane trash (ST) accounted for 10–40% of the total biomass of sugarcane, depending on crop age, sugarcane variety, type of soil, and other factors [4, 5]. Sugarcane leaves contain cellulose (35–50%), hemicellulose (20–35%), and lignin (15–20%) [6]. This indicates that ST has a great potency for the production of value-added products such as bioethanol or bio-oil used in the energy sector, and also xylitol, arabitol, or other products used in the food industry.

Xylitol ($C_5H_{12}O_5$) is a five-carbon sugar alcohol with a molecular weight of 152.15 g/mol [7]. The demand for xylitol as a natural sweetener has steadily increased since it has a low glycemic index and does not raise insulin levels [8]. Conventional chemical methods using nickel catalysts to produce xylitol from pure xylose must be abandoned immediately due to high energy and cost-demanding [9]. On the other hand, biotechnological production of xylitol from xylose contained in lignocellulose hydrolysate has given a very promising alternative as its low cost, more sustainable and environmentally safe [10]. Various natural xylose-fermenting yeasts from different species, such as *Candida tropicalis, Candida guilliermondii, Debaryomyces nepalensis, Pichia stipitis*, and *Kluyveromyces marxianus*, have been investigated for their abilities to produce xylitol from lignocellulose biomass such as corn stover [11], sugarcane bagasse [12], and corn cob [13]. However, research on the ability of these species to produce xylitol from sugarcane trash has never been reported yet.

To date, dilute-sulfuric acid pretreatment using steam explosion or liquid hot water is still considered as the effective way for solubilizing the hemicellulose fraction of lignocellulose biomass. However, the use of sulfuric acid generates more by-products that can inhibit fermentation by microorganisms. On the other hand, a previous study using a combination of microwave-maleic acid has been proved as a promising strategy since it produces high xylose concentration. However, the use of organic acid is still questionable due to the high costs [14]. Therefore, in this study, we investigated the effects of different pretreatment methods, including liquid hot water-sulfuric acid dan microwave-maleic acid and different yeast strains on xylitol fermentation using ST hemicellulose hydrolysate.

45.2 Materials and Methods

45.2.1 Biomass Preparation and Composition Analysis

The sugarcane (*Saccharum officinarum* L.) trash used in this study was the same as that in our previous report [15]. It was collected from PT PG Rajawali II, Subang, West Java, Indonesia. Before being used, it was ground into small particles using a hammer mill and fractionated using sieves to obtain 40–60 mesh-sized powder. The ST consists of 35.54% cellulose, 35.54% hemicellulose, 17.44% lignin, 6.86% ash, and 2.25% extractives [15].

45.2.2 Sugarcane Trash Hemicellulose Hydrolysates Preparation

The hemicellulose hydrolysate of ST was obtained by acid hydrolysis using two methods: (1) Liquid hot water (LHW) with 1% of H_2SO_4 and without catalyst and (2) microwave digester with 1.8% of maleic acid.

45.2.2.1 Liquid Hot Water (LHW) Method

LHW pretreatment was performed in a 600 mL stainless steel high-pressure reactor equipped with a mixing system and a thermocouple for internal temperature measurement (Parr Reactor 4560, Parr Instrument, Moline, IL, USA). An amount of 3.0 g (oven-dry weight) ST was inserted into the vessel, then 1% of H_2S0_4 was added to reach a solid and liquid ratio of 1:10 (w/v). Subsequently, the nitrogen gas was flushed into the reactor for purging and adjusting the initial pressure to 40 bar with continuous stirring at 300 rpm. The samples were heated at 170 °C for 30 min. After the hydrolysis reaction was over, the vessel was immediately ice water-cooled for 30 min to release the heat. Then, the slurries were filtered using Whatman (No. 4) filter paper to separate insoluble solids. The pH of the hydrolysate was measured using a pH meter. The hydrolysate was collected in glass bottles and kept in a refrigerator at 10–11 °C for further experiments. LHW pretreatment without catalyst was carried out using distilled water with the same steps as pretreatment using H_2SO_4.

45.2.2.2 Microwave Irradiation Method

Microwave irradiation method was performed in a microwave irradiator equipped with a stirrer and thermocouple (Milestone START D, Milestone Inc., Bergamo, Italy). An amount of 3.0 g (oven-dry weight) ST was inserted into the Teflon® vessel, then 1.8% maleic acid was added to reach a solid and liquid ratio of 1:10 (w/v). Subsequently, the samples were degassed at a 20-bar vacuum for 5 min before the pretreatment. The samples were heated at 180 °C, 195 rpm for 5 min. After the hydrolysis reaction was over, the vessel was immediately ice water-cooled for 30 min to release the heat. Then, the slurries were filtered through filter paper using Whatman (No. 4) filter paper to separate insoluble solids. The pH of the hydrolysate was also measured using a pH meter. The hydrolysate was collected in glass bottles and kept in a refrigerator at 10–11 °C for further experiments.

45.2.3 Yeast Strains and Inoculum Preparation

This study was carried out using four yeast strains provided by Integrative Biorefinery Laboratory, National Center for Genetic Engineering and Biotechnology (BIOTEC), National Science and Technology Development Agency (NSTDA), Thailand, including *Scheffersomyces stipitis* TBRC 3582, *Debaryomyces nepalensis* TBRC 2102, *Candida intermedia* TBRC 61101, and *Kluyveromyces marxianus* TBRC 1524. The inoculum was prepared by growing the strains in the YPX medium containing 10 g/L yeast extract, 20 g/L peptone, 20 g/L xylose, 0.2 g/L $MgSO_4.7H_2O$, 5 g/L KH_2PO_4, and 1 g/L urea for 24 h at 30 °C, 150 rpm. The cells were harvested by centrifugation (4000 rpm, 4 °C, 10 min), then washed and resuspended in a sterilized 0.9% (w/v) NaCl solution.

45.2.4 Fermentation of Sugarcane Trash Hemicellulose Hydrolysate

Production of xylitol from ST hemicellulose hydrolysate was conducted in a closed system. YPX media containing 10 g/L yeast extract, 10 g/L peptone, and 20 g/L xylose was used as a fermentation control. Prior to fermentation, the pH of hydrolysate was adjusted to 5.0 by 10 N NaOH. The hydrolysates were enriched with 5 g/L yeast extract, 2 g/L KH_2PO_4 and 0.3 g/L $MgSO_4.7H_2O$, then sterilized through a 0.22 μm microfilter. Further, the yeast inoculum (10%, v/v) was added to a 250 mL Erlenmeyer flask containing 100 mL of prepared hydrolysate medium. Finally, the flasks were incubated on a rotary shaker (200 rpm) at 30 °C for 72 h. Samples of the fermentation broth (0.5 mL) were periodically withdrawn after 0 h, 24 h, 48 h, and 72 h. They were then centrifugated at 4000 rpm for 10 min to separate the supernatant and the cell pellet. The supernatant was used to determine xylose concentration, while the cells pellet was used to measure cells growth. All the experiments were carried out in duplicate, and the data are expressed as mean values.

45.2.5 *Analytical Methods*

Yeast cell concentration was determined by its turbidity at 600 nm (OD_{600nm}) using a microplate reader (VersaMax Plus, USA). Sugars (xylose, glucose, arabinose, mannose, and galactose) and inhibitory by-products (lactic acid, formic acid, acetic acid, levulinic acid, 5-HMF, and furfural) contained in sugarcane trash hydrolysate were analyzed by high-performance liquid chromatography (HPLC) (SPD-M10A, Shimadzu, Kyoto, Japan) equipped with a refractive index detector using an Aminex HPX 87H column (Bio-Rad, Hercules, CA) operating at 65°°C with 5 mM H_2SO_4 as the mobile phase at a flow rate of 0.5 mL/min. Xylose and Xylitol concentration contained in fermentation products were analyzed by Ultimate 3000 HPLC System (DionexCo., Sunnyvale, CA, USA) equipped with a refractive index detector using Coregel-87H3 column (Transgenomic, San Jose, CA, USA) operating at oven temperature 80 °C with 5 mM H_2SO_4 as the mobile phase at a flow rate of 0.5–0.6 mL/min. Xylitol yield (Yp/s) was calculated as the ratio between xylitol concentration and consumed xylose concentration. Xylitol productivity (Qp) was calculated as the ratio between xylitol concentration and fermentation time. The efficiency of xylose-to xylitol bioconversion was calculated as the ratio between the experimental Yp/s value (g/g) and the theoretical Yp/s value (1.01 g/g).

45.3 Results and Discussions

3.1. Chemical Compositions of Sugarcane Trash Hemicellulose Hydrolysate

The major component of biomass such as cellulose, lignin, and hemicellulose will be solubilized into liquid fractions during pretreatment. The effect of microwave-MA pretreatment and LHW with and without catalyst on sugars contained in the liquid fraction is presented in Table 45.1. In the hydrolysate form microwave-MA pretreatment, most of the hemicellulose was lost during pretreatment and solubilized in the liquid fraction in the form of xylose. In this study, microwave-MA pretreatment resulted in the highest concentration of xylose. The concentration of xylose reached 20.28 g/L, which was much higher than the LHW pretreatment with and

Table 45.1 Chemical compositions of sugarcane trash hemicellulose hydrolysate

Components (g/L)	Microwave—maleic acid (MA)	LHW—sulfuric acid catalyst	LHW—without catalyst
pH	1.74	3.42	3.89
Glucose	1.68	1.62	1.16
Xylose	20.29	7.41	3.60
Arabinose	2.85	1.76	1.46
Lactic acid	nd	1.18	nd
Formic acid	0.25	8.92	8.46
Acetic acid	0.77	5.93	6.82
5-HMF	nd	1.12	0.50
Furfural	nd	1.42	1.09

* nd = not detected

without a catalyst which only reached 7.41 and 3.60 g/L, respectively. It can be concluded that microwave-MA pretreatment is more effective to degrade the hemicellulose into xylose than LHW pretreatments. The glucose concentration obtained after pretreatment is almost the same in all pretreatment conditions in this research.

Table 45.1 also shows the effect of pretreatment on acidity in the liquid fraction. After pretreatment, microwave-MA pretreatment has the lowest pH than LHW with and without catalyst. Decomposition of cellulose and hemicellulose induced the formation of inhibitory by-products such as formic acid, acetic acid, and levulinic acid, furfural, HMF. Production of by-products suggested that some sugar monomers degraded to lower molecular weight compounds under severe conditions. LHW pretreatment with and without catalyst produce higher by-products concentration (acetic acid, 5-HMF, and furfural) than does microwave-MA pretreatment. The degradation of sugars can be occurred due to its long-duration exposure to high reaction temperature. LHW pretreatment has a longer duration of heating (30 min) than does microwave pretreatment (5 min), which can influence a higher concentration of by-product in the liquid fraction [16]. Pretreatment using microwave heating provides uniform, direct, and rapid heating; thus, the formation of by-products can be avoided [15]. Acetic, formic, and levulinic acids are weak acids that can inhibit the cell growth of microorganisms during the fermentation process and may reduce the yield or production of xylitol [17, 18].

45.3.1 *Effect of Different Media on Yeast Strains Growth During Fermentation*

To investigate the effect of pretreatment methods on yeast strains growth, four types of media were used as a fermentation broth including hemicellulose hydrolysate

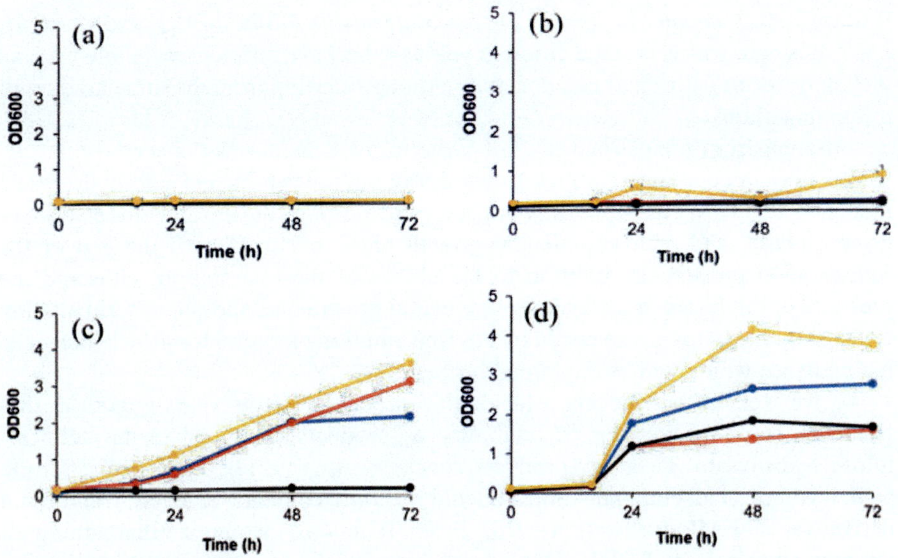

Fig. 45.1 Comparison of yeast growth curve on different media: hemicellulose hydrolysate from LHW-H_2SO_4 pretreatment (**a**), LHW-no catalyst (**b**), microwave-maleic acid pretreatment (**c**), and YPX media (**d**). *Blue* for *D. nepalensis* TBRC 2102; *red* for *S. stipitis* TBRC 3582; *black* for *C. intermedia* TBRC 61101; and yellow for *K. marxianus* TBRC 1524

from liquid hot water (LHW)-H_2SO_4 pretreatment, LHW-no catalyst pretreatment, microwave-maleic acid pretreatment, and yeast peptone xylose (YPX) media as a control medium. As shown in Fig. 45.1, each yeast strain developed different responses when they were grown in various media. None of the yeast strains showed growth on LHW-H_2SO_4 hemicellulose hydrolysate (Fig. 1a), meanwhile only *K. marxianus* TBRC 1524 was able to grow on LHW-no catalyst hemicellulose hydrolysate (Fig. 45.1b). In microwave-maleic acid hemicellulose hydrolysate, *K. marxianus* TBRC 1524, *S. stipitis* TBRC 3582, and *D. nepalensis* TBRC 2102 were able to grow; meanwhile *C. intermedia* TBRC 61101 did not show any growth response (Fig. 45.1c). As expected, all yeast strains were able to grow on YPX media since that media did not contain inhibitors that could inhibit the yeast growth (Fig. 1d).

As can be seen in Fig. 45.1d, the growth curve of all yeast strains showed a typical yeast growth curve during fermentation. Three strains, namely *K. marxianus* TBRC 1524, *D. nepalensis* TBRC 2102, and *C. intermedia* TBRC 61101, went through a phase of adaptation to YPX media for 16 h, then entered the exponential phase until 48 h of fermentation, and finally reached their stationary phase until the end of the fermentation process. The exception for strain *S. stipitis* TBRC 3582, this strain went the stationary phase faster than the three other strains (at 24 h of fermentation). Although it showed a similar growth pattern, each strain grew with a different cell density. Among all strains, *K. marxianus* TBRC 1524 showed the highest cell density, followed by *D. nepalensis* TBRC 2102, *C. intermedia* TBRC 61101, and *S. stipitis* 3582. It is noteworthy that the different yeast strains have different metabolisms, and they may develop different cell densities although starting from the same amount of initial inoculum.

Different from typical behavior of yeast in YPX media, we observed that two strains developed a different growth pattern in microwave-maleic acid hydrolysate media. As shown in Fig. 45.1c, *K. marxianus* and *S. stipitis* did not enter the stationary growth phase, and their exponential growth phase occurred until the end of the fermentation process. In addition to the ability of these strains to withstand the presence of inhibitors in media, the exponential growth until the end of fermentation is also very likely to be influenced by the consumption of xylitol for growth and cells maintenance which will be explained further.

As we well know, one main factor that probably affects yeast growth during fermentation is the toxicity of inhibitory by-products contained in the hemicellulose hydrolysate. These by-products consist of weak acids (acetic acid, formic acid, levulinic acid, etc.), phenolics (phenol, vanillin, coumaric acid, etc.), and furan derivatives (5-HMF, furfural) [19, 20]. From Table 45.1, we notice that among all pretreatment methods, both LHW pretreatment with H_2SO_4 catalyst or without catalyst produced more inhibitory by-products than did microwave-maleic acid pretreatment. The failure of microorganisms to grow on the LHW hydrolysates was probably due to the high concentration of by-products. However, the information on the degree of inhibition of each by-products to yeast growth was still limited. It is hard to determine which compound is the most toxic to the yeast cells, since we did not investigate the inhibitory effect caused by individual products in this research. Moreover, negative effects on yeast growth may also be caused by the total concentration of all by-products that were above the inhibitory threshold concentration tolerated by yeast [21]. In addition, the synergistic effects between these multiple by-products may enhance the toxicity of hydrolysate, thus reducing the performance of yeast cells.

Maintaining pH homeostatis is very critical for cell growth since it influences all metabolisms, including nutrients absorption and utilization, degradation of substrate, also a synthesis of protein and DNA. Weak acids, in the non-dissociated form, can enter the cell by passive diffusion, resulting in the acidification of the cell's cytoplasm. Reduction of intracellular pH will cause morphological changes in cells, prolonging the lag phase, damaging the integrity of the plasma membrane, thus leading to improper ion transportation. Adverse effects may be the deactivation of cellular enzymes and the damage of DNA [22–26].

In a recent study, about 5.93–6.82 g/L of acetic acid was contained in LHW hydrolysate. This concentration seems too high for the yeast strains, although the concentration of acetic acid between 1.5 and 9 g/L was typical in the non-concentrated hydrolysate [27]. According to López-Linares et al., [28], a concentration of acetic acids higher than 3 g/L reduced the performance of *D. hansenii* NRRL Y-7426 and *C. guilliermondii* FTI 20037 during xylitol fermentation of rapeseed straw hemicellulose hydrolysate. Vanmarcke et al., [29] also showed that the presence of 6.3 g/L acetic acids in lignocellulose hydrolysate inhibited the growth of yeast GSE16-T18 and

MD4 strains. Formic acid (pKa = 3.75) at lower concentration might have a higher toxic effect on cells growth and xylose fermentation than levulinic acid and acetic acid (pKa = 4.76) due to the smaller molecule size [30]. The concentrations of formic acid contained in both LHW hydrolysates in this study (8.46–8.92 g/L) were higher than that of the other lignocellulosic hydrolysate previously reported (0.1–7.4 g/L) [31–33]. Zhao et al., [34] reported that 2–4 g/L of formic acid contained in sugarcane bagasse hydrolysate decreased 25–40% biomass of *Rhodotorula toruloides* during fermentation.

Regarding the furan derivatives, there was no furfural, nor 5-HMF detected in the microwave-maleic acid hydrolysate. It may also be the possible reason why most of the yeast survived in the microwave-maleic acid hydrolysate. Furan compounds tend to be more toxic than formic and acetic acids since they can directly affect cell metabolism [20]. According to Senatham et al., [35], yeast growth was inhibited by the presence of 0.9–2 g/L furfural. Also, at least 0.5 g/L of HMF has been reported to be toxic for yeast cells [36]. The above findings indicate that to minimize the inhibition effects on yeast cells, the concentration of furan derivatives in hemicellulose hydrolysate should be as low as possible, preferably none.

When comparing yeast, it seems that *K. marxianus* TBRC 1524 showed the highest tolerance to the toxic compound present in both microwave-maleic acid and LHW-without catalyst hydrolysate (Fig. 45.2d). As mentioned before, this strain also showed the highest cell density among all strains used. *C. intermedia* TBRC

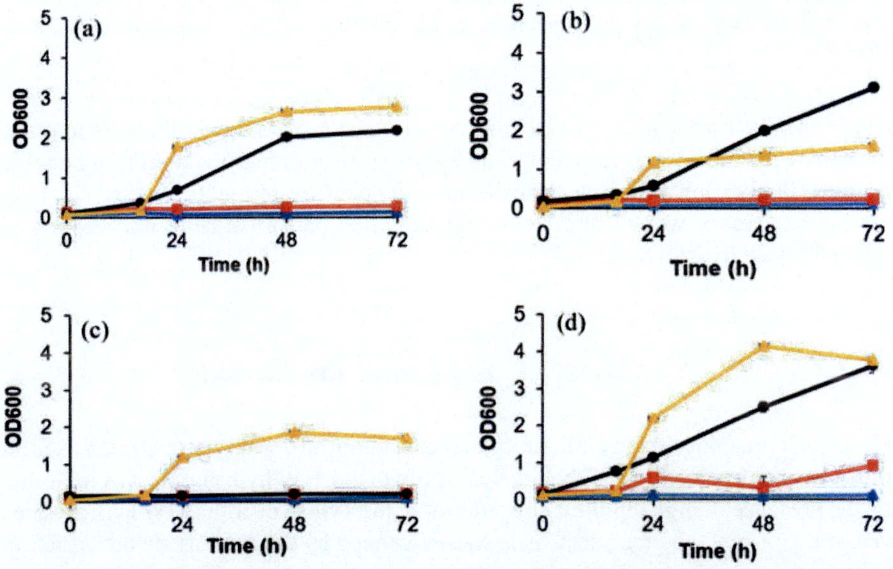

Fig. 45.2 Growth comparison of *D. nepalensis* TBRC 2102 (**a**), *S. stipitis* TBRC 3582 (**b**), *C. intermedia* TBRC 61101 (**c**), and *K. marxianus* TBRC 1524 (**d**) during fermentation using different media. *Blue* for LHW-H2SO4; *red* for LHW-no catalyst; *black* for Microwave-maleic acid; and yellow for YPX

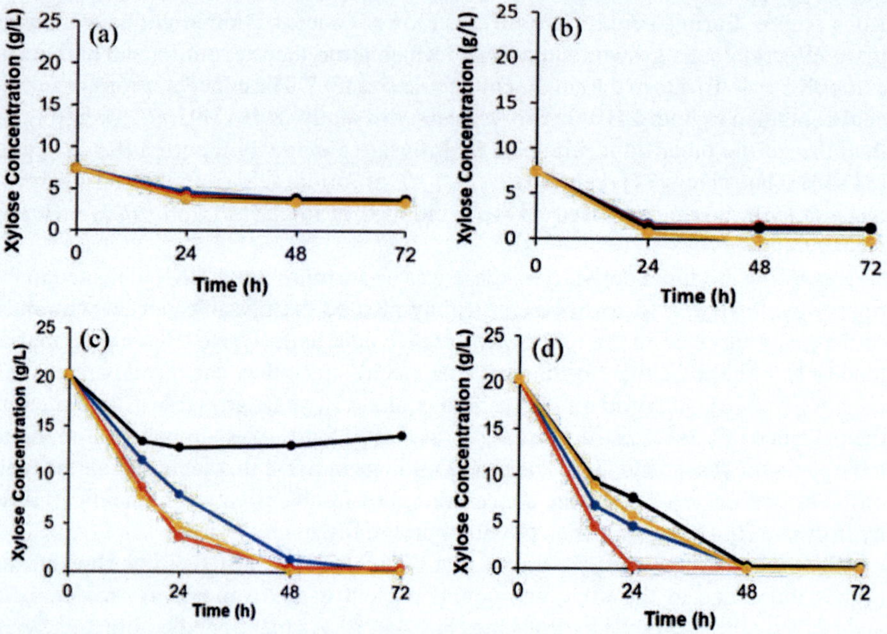

Fig. 45.3 Xylose consumption by four natural xylose-fermenting yeasts during fermentation using hemicellulose hydrolysate from LHW-H_2SO_4 pretreatment (**a**), LHW-no catalyst (**b**), microwave-maleic acid pretreatment (**c**), and YPX media (**d**): *blue* for *D. nepalensis* TBRC 2102; *red* for *S. stipitis* TBRC 3582; *black* for *C. intermedia* TBRC 61101; and yellow for *K. marxianus* TBRC 1524

61101 has the lowest tolerance to inhibitory by-products, since it did not show any viability even in microwave-maleic acid hydrolysate with minimal inhibitor concentrations (Fig. 45.3c). Thus, selecting the naturally pentose-fermenting yeast with high inhibitor tolerance would improve the fermentation performance of non-detoxified hemicellulose hydrolysate.

45.3.2 Xylose Consumption and Xylitol Production

Figure 45.3 displays the profile of xylose consumption for all yeast strains used in this research. Interestingly, although yeast growth in LHW hydrolysate was inhibited by the presence of high inhibitor concentration, the xylose in both LHW hydrolysates was still consumed by the yeast. This was evidenced by the decreased concentration of xylose during the first 24 h of fermentation (Figs. 45.3a, b). This phenomenon is possible in the hydrolysates that contain a high concentration of weak acids, since yeast requires additional energy to transport the protonated ion through the plasma membrane [21]. On the contrary, the decrease of xylose concentration on microwave-maleic acid hydrolysate and YPX media was most likely due to its utilization for growth and yeast metabolism [37]. In YPX media, xylose was consumed within 24–48 h of fermentation, which was faster than that in the microwave-maleic acid hydrolysate. In microwave hydrolysate, xylose was consumed within 48–72 h by three yeast strains except for *C. intermedia* TBRC 61101, which did not wholly consume xylose in the medium.

Since yeasts did not grow on and ferment xylose, there was no xylitol produced from both LHW hydrolysates. Figure 45.4 represents the xylose consumption, productivity, and yield during fermentation of microwave-maleic acid ST hydrolysate and YPX media. Maximum xylitol production from microwave-maleic acid hydrolysate varied from 1.1 to 2.3 g/L (from an initial xylose concentration of 20.3 g/L) with productivity between 0.03 and 0.1 g/L/h (Fig. 4a), while the yield ranged from 0.06 to 0.15 g xylitol/g xylose consumed or as high as 5.5–11.2% (Fig. 4b) of theoretical value after 48 h of fermentation. *K. marxianus* TBRC 1524 produced the highest xylitol concentration (2.3 g/L) with volumetric productivity of 0.096 g/L/h, yielding 0.15 g xylitol/g xylose consumed or as high as 11.2% of the theoretical value after 24 h of fermentation. Regarding the toxic compound contained in microwave hydrolysate, xylitol production by *K. marxianus* was not affected by

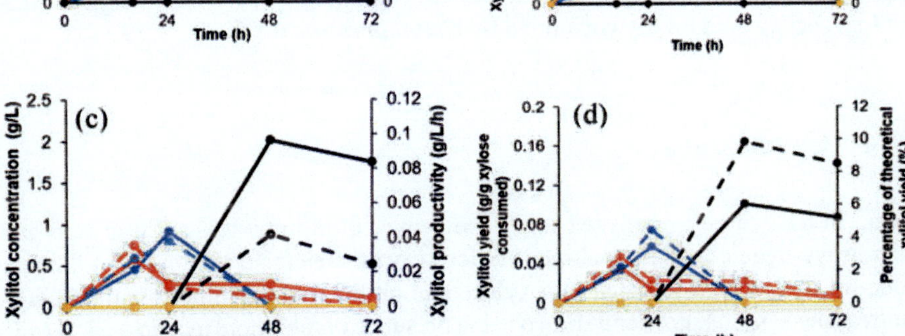

Fig. 45.4 Xylitol consumption, productivity, and yield during fermentation of microwave-maleic acid sugarcane trash hemicellulose hydrolysate (**a**, **b**) and YPX media (**c**, **d**) by four natural xylose-fermenting yeasts: *blue* for *D. nepalensis* TBRC 2102; *red* for *S. stipitis* TBRC 3582; *black* for *C. intermedia* TBRC 61101; and yellow for *K. marxianus* TBRC 1524. *Straight line* for xylitol concentration and yield; *dashed line* for xylitol productivity and the percentage of theoretical xylitol yield

the presence of inhibitors. This xylitol yield obtained was comparable and could be reached faster with the result reported by Dasgupta et al., [38], who got xylitol yield 0.16 g xylose/g xylose (initial xylose concentration of ~20 g/L) after 40 h of sugarcane bagasse hydrolysate fermentation by *K. marxianus* IIPE45. Although the xylitol concentration and yield from our result were lower than the result that reported by Albuquerque et al., [39] (6.76 g/L and 0.036 g/g), the volumetric productivity of *K. marxianus TBRC* 1524 in this study was higher than *K. marxianus* CCA510 in that study which produced 0.070 g xylitol/L/h after 96 h of fermentation of cashew apple bagasse hydrolysate. Moreover, it needs to be underlined that *K. marxianus* CCA510 did not produce xylitol before 24 h of fermentation, comparing with our result that as much as 0.81 g/L of xylitol was produced in 16 h of fermentation. It shows that the strain *K. marxianus* TBRC 1524 has the potency to grow in the microwave-maleic acid sugarcane trash hydrolysate, even without detoxification.

The result of fermentation in the YPX media was different from that in the microwave-maleic acid hydrolysate. The highest xylitol concentration obtained from fermentation in YPX media was produced by *C. intermedia* TBRC 61101 (2.01 g/L) with the highest volumetric productivity of 0.04 g/L/h (Fig. 4c), yielding 0.1 g xylitol/g xylose consumed or as high as 9.8% from theoretical value (Fig. 4d) after 48 h of fermentation. Interestingly, no xylitol was produced by *K. marxianus* TBRC 1524. This finding was probably related to the presence of certain compounds, such as acetic acid, in the ST hydrolysate from pretreatment with microwave-maleic acid, which might have created a favorable condition to enhance xylitol production by *K. marxianus* TBRC 1524 [40]. Cheng et al. [41] reported that the presence of acetic acid below 1 g/L favored xylitol production by *Candida guilliermondii*, while the concentration of acetic acid above 3 g/L was toxic to the cells. Rodrigues et al., [42] also showed that 1.6 g/L acetic acids was consumed by *Pichia stipitis* YS-30 during fermentation of corn stover hydrolysate. Another finding using microwave-maleic acid hydrolysate that contained a similar concentration of acetic acid showed that 0.77 g/L acetic acid totally consumed by *Candida tropicalis* InaCC Y799 [43].

45.4 Conclusions

Xylan in ST can be hydrolyzed to xylose using liquid hot water-sulfuric acid and microwave-maleic acid pretreatment. Microwave-maleic acid pretreatment is more effective to degrade the xylan into xylose and produce lower inhibitors than LHW pretreatment. Xylose concentration and the presence of inhibitors (acetic acid, formic acid, HMF, and furfural) in the hydrolysate strongly affected the cell's growth and xylitol production. LHW hydrolysates contained a high concentration of acetic acid (>5 g/L), formic acid (>8 g/L), 5-HMF (>1 g/L), and furfural (>1 g/L) which inhibited the cell's growth and made the yeasts unable to produce xylitol. The low concentration of xylose in the hemicellulose hydrolysate (<10 g/L) probably was another factor that inhibited the xylitol production. On the contrary, the presence of other compounds, such as acetic acid, in favorable concentration in the hemicellulose hydrolysate might enhance the xylitol production. The highest concentration of xylitol obtained by *K. marxianus* TBRC 1524 was 2.3 g/L, with a yield of 0.15 g xylitol/g xylose consumed using non detoxified microwave-maleic acid ST hydrolysate. It is concluded that *K. marxianus* TBRC 1524 is a potential yeast for producing xylitol from ST hemicellulose hydrolysate.

Acknowledgements The authors would like to thank Japan Science and Technology Agency (JST) for funding this research through the Japan-ASEAN Science, Technology and Innovation Platform (JASTIP) under the activity of the Strategic International Collaborative Research Program (SICORP). This work is a part of the JASTIP project "Utilization of Hemicellulose Fraction from Acid Pretreatment of Lignocellulose for Production of Xylitol" 2019–2020, funded by the Japan Science and Technology Agency and e-ASIA project "Integrated Biorefinery of Sugarcane Trash" 2019–2022.

The author also would like to thank Dr. Marisa Raita and Dr. Nopparat Suriyachai for supervising the sugarcane trash pretreatment using LHW pretreatment.

References

1. FAO: FAOSTAT, http://www.fao.org/faostat/en/#data/QCL. Last accessed 2021/05/11
2. Antonio, B.W., Lenço, P.C., Carvalho, D.J., Veiga, J.P.S.: The generation of residual biomass during the production of bio-ethanol from sugarcane, its characterization and its use in energy production. Renew. Sustain. Energy Rev. **29**, 589–603 (2014). https://doi.org/10.1016/j.rser.2013.08.056
3. Junpen, A., Pansuk, J., Garivait, S.: Estimation of reduced air emissions as a result of the implementation of the measure to reduce burned sugarcane in Thailand. Atmosphere. **11**, (2020). https://doi.org/10.3390/atmos11040366
4. Jutakanoke, R., Tolieng, V., Tanasupawat, S., Akaracharanya, A.: Ethanol production from sugarcane leaves by *Kluyveromyces marxianus* S1.17, a genome-shuffling mediated transformant. BioResources **12**, 1636–1646 (2017)
5. Jenjariyakosoln, S., Gheewala, S.H., Sajjakulnukit, B., Garivait, S.: Energy for sustainable development energy and GHG emission reduction potential of power generation from sugarcane residues in Thailand. Energy Sustain. Dev. **23**, 32–45 (2021). https://doi.org/10.1016/j.esd.2014.07.002
6. Moodley, P., Kana, E.B.G.: Optimization of xylose and glucose production from sugarcane leaves (*Saccharum officinarum*) using hybrid pretreatment techniques and assessment for hydrogen generation at semi-pilot scale. Int. J. Hydrogen Energy. **40**, 3859–3867 (2015). https://doi.org/10.1016/j.ijhydene.2015.01.087
7. Parajó, J.C., Domínguez, H., Domínguez, J.M.: Biotechnological production of xylitol. Part 2: Operation in culture media made with commercial sugars. Bioresour. Technol. **65**, 203–212 (1998). https://doi.org/10.1016/S0960-8524(98)00036-4
8. Dasgupta, D., Bandhu, S., Adhikari, D.K., Ghosh, D.: Challenges and prospects of xylitol production with whole cell bio-catalysis : a review. Microbiol. Res. **197**, 9–21 (2016). https://doi.org/10.1016/j.micres.2016.12.012
9. Eryasar, K., Karasu-Yalcin, S.: Evaluation of some lignocellulosic byproducts of food industry for microbial xylitol production by *Candida tropicalis*. 3 Biotech. 6, 1–7 (2016). https://doi.org/10.1007/s13205-016-0521-8
10. Albuquerque, T.L.D., da Silva, I.J., de Macedo, G.R., Rocha, M.V.P.: Biotechnological production of xylitol from lignocellulosic wastes: A review. Process Biochem. **49**, 1779–1789 (2014). https://doi.org/10.1016/j.procbio.2014.07.010

11. Hong, E., Kim, J., Rhie, S., Ha, S.J., Kim, J., Ryu, Y.: Optimization of dilute sulfuric acid pretreatment of corn stover for enhanced xylose recovery and xylitol production. Biotechnol. Bioprocess Eng. **21**, 612–619 (2016). https://doi.org/10.1007/s12257-016-0483-z
12. Silva, D.D.V., Felipe, M.G.A., Mancilha, I.M., Luchese, R.H., Silva, S.S.: Inhibitory effect of acetic acid on bioconversion of xylose in xylitol by *Candida guilliermondii* in sugarcane bagasse hydrolysate. Braz. J. Microbiol. **35**, 248–254 (2004)
13. Yewale, T., Panchwagh, S., Sawale, S., Jain, R., Dhamole, P.B.: Xylitol production from non-detoxified and non-sterile lignocellulosic hydrolysate using low-cost industrial media components. 3 Biotech. **7**, 1–9 (2017). https://doi.org/10.1007/s13205-017-0700-2
14. Hermiati, E., Oktaviani, M., Ermawar, R.A., Laksana, R.P.B., Kholida, L.N., Thontowi, A., Mardiana, S., Watanabe, T.: Optimization of xylose production from sugarcane trash by microwave-maleic acid hydrolysis. Reaktor **20**, 81–88 (2020). https://doi.org/10.14710/reaktor.20.2.81-88
15. Hermiati, E., Laksana, R.P.B., Fatriasari, W., Kholida, L.N., Thontowi, A., Yopi, Arnieyanto, D.R., Champreda, V., Watanabe, T.: Microwave-assisted acid pretreatment for enhancing enzymatic saccharification of sugarcane trash. Biomass Convers. Biorefinery (2020). https://doi.org/10.1007/s13399-020-00971-z
16. Sari, F.P., Falah, F., Anita, S.H., Ramadhan, K.P., Laksana, R.P.B., Fatriasari, W., Hermiati, E.: Pretreatment of oil palm empty fruit bunch (Opefb) at bench-scale high temperature-pressure steam reactor for enhancement of enzymatic saccharification. Int. J. Renew. Energy Dev. **10**, 157–169 (2020). https://doi.org/10.14710/ijred.2021.32343
17. Kim, Y., Ximenes, E., Mosier, N.S., Ladisch, M.R.: Soluble inhibitors/deactivators of cellulase enzymes from lignocellulosic biomass. Enzyme Microb. Technol. **48**, 408–415 (2011). https://doi.org/10.1016/j.enzmictec.2011.01.007
18. Palmqvist, E., Hahn-Hägerdal, B.: Fermentation of lignocellulosic hydrolysates. II: Inhibitors and mechanisms of inhibition. Bioresour. Technol. **74**, 25–33 (2000). https://doi.org/10.1016/S0960-8524(99)00161-3
19. Koppram, R., Albers, E., Olsson, L.: Evolutionary engineering strategies to enhance tolerance of xylose utilizing recombinant yeast to inhibitors derived from spruce biomass. Biotechnol. Biofuels. **5**, 1–13 (2012). https://doi.org/10.1186/1754-6834-5-32
20. Jung, Y.H., Kim, K.H.: Evaluation of the main inhibitors from lignocellulose pretreatment for enzymatic hydrolysis and yeast fermentation. BioResources. **12**, 9348–9356 (2017). https://doi.org/10.15376/biores.12.4.9348-9356
21. Suko, A.V., Bura, R.: Enhanced xylitol and ethanol yields by fermentation inhibitors in steam-pretreated lignocellulosic biomass. Ind. Biotechnol. **12**, 187–194 (2016). https://doi.org/10.1089/ind.2015.0026
22. Narayanan, V., Sànchez, V., Niel, E.W.J.V., Grauslund, M.F.G.: Adaptation to low pH and lignocellulosic inhibitors resulting in ethanolic fermentation and growth of *Saccharomyces cerevisiae*. AMB Express **6**, 1–13 (2016). https://doi.org/10.1186/s13568-016-0234-8
23. Kim, D.: Physico-chemical conversion of lignocellulose : inhibitor effects and detoxification strategies : a mini review. Molecules **23**, 1–21 (2018). https://doi.org/10.3390/molecules23020309
24. Zhang, K., Wells, P., Liang, Y., Love, J., Parker, D.A., Botella, C.: Effect of diluted hydrolysate as yeast propagation medium on ethanol production. Bioresour. Technol. **271**, 1–8 (2019). https://doi.org/10.1016/j.biortech.2018.09.080
25. Moysés, D.N., Reis, V.C.B., de Almeida, J.R.M., de Moraes, L.M.P., Torres, F.A.G.: Xylose fermentation by *Saccharomyces cerevisiae*: challenges and prospects. Int. J. Mol. Sci. **17**, 1–18 (2016). https://doi.org/10.3390/ijms17030207
26. Guan, N., Liu, L.: Microbial response to acid stress: mechanisms and applications. Appl. Microbiol. Biotechnol. **104**, 51–65 (2020). https://doi.org/10.1007/s00253-019-10226-1
27. Duarte, L.C., Carvalheiro, F., Neves, I., Girio, F.M.: Effects of aliphatic acids, furfural, and phenolic compounds on *Debaryomyces hansenii* CCMI 941. Appl. Biochem. Biotechnol. **121–124**, 413–425 (2005). https://doi.org/10.1385/ABAB:121:1-3:0413

28. López-Linares, J.C., Romero, I., Cara, C., Castro, E., Mussatto, S.I.: Xylitol production by *Debaryomyces hansenii* and *Candida guilliermondii* from rapeseed straw hemicellulosic hydrolysate. Bioresour. Technol. **247**, 736–743 (2018). https://doi.org/10.1016/j.biortech.2017.09.139
29. Vanmarcke, G., Demeke, M.M., Foulquié-Moreno, M.R., Thevelein, J.M.: Identification of the major fermentation inhibitors of recombinant 2G yeasts in diverse lignocellulose hydrolysates. Biotechnol. Biofuels. **14**, 1–15 (2021). https://doi.org/10.1186/s13068-021-01935-9
30. Thomsen, M.H., Thygesen, A., Thomsen, A.B.: Identification and characterization of fermentation inhibitors formed during hydrothermal treatment and following SSF of wheat straw. Appl. Microbiol. Biotechnol. **83**, 447–455 (2009). https://doi.org/10.1007/s00253-009-1867-1
31. Montipó, S., Pedroso, G.B., Bevilaqua, D.B., Prestes, O.D., Corona-González, R.I., Martins, A.F.: Building block lactic acid from rice husks and agave bagasse. Waste and Biomass Valorization. **7**, 1495–1507 (2016). https://doi.org/10.1007/s12649-016-9554-9
32. Borrega, M., Sixta, H.: Water prehydrolysis of birch wood chips and meal in batch and flow-through systems: a comparative evaluation. Ind. Eng. Chem. Res. **54**, 6075–6084 (2015). https://doi.org/10.1021/acs.iecr.5b00908
33. Samad, A., Zhang, J., Chen, D., Liang, Y.: Sophorolipid production from biomass hydrolysates. Appl. Biochem. Biotechnol. **175**, 2246–2257 (2015). https://doi.org/10.1007/s12010-014-1425-x
34. Zhao, X., Peng, F., Du, W., Liu, C., Liu, D.: Effects of some inhibitors on the growth and lipid accumulation of oleaginous yeast *Rhodosporidium toruloides* and preparation of biodiesel by enzymatic transesterification of the lipid. Bioprocess Biosyst. Eng. **35**, 993–1004 (2012). https://doi.org/10.1007/s00449-012-0684-6
35. Senatham, S., Chamduang, T., Kaewchingduang, Y., Thammasittirong, A., Srisodsuk, M., Elliston, A., Roberts, I.N., Waldron, K.W., Thammasittirong, S.N.R.: Enhanced xylose fermentation and hydrolysate inhibitor tolerance of *Scheffersomyces shehatae* for efficient ethanol production from non-detoxified lignocellulosic hydrolysate. Springerplus **5**, 1–8 (2016). https://doi.org/10.1186/s40064-016-2713-4
36. Bajwa, P.K., Pinel, D., Martin, V.J.J., Trevors, J.T., Lee, H.: Strain improvement of the pentose-fermenting yeast *Pichia stipitis* by genome shuffling. J. Microbiol. Methods. **81**, 179–186 (2010). https://doi.org/10.1016/j.mimet.2010.03.009
37. Sharma, N.K., Behera, S., Arora, R., Kumar, S.: Enhancement in xylose utilization using *Kluyveromyces marxianus* NIRE-K1 through evolutionary adaptation approach. Bioprocess Biosyst. Eng. **39**, 835–843 (2016). https://doi.org/10.1007/s00449-016-1563-3
38. Dasgupta, D., Bandhu, S., Adhikari, D.K., Ghosh, D.: Challenges and prospects of xylitol production with whole cell bio-catalysis : a review. Microbiol. Res. **197**, 9–21 (2017). https://doi.org/10.1016/j.micres.2016.12.012
39. de Albuquerque, T.L., Gomes, S.D.L., Marques, J.E., da Silva, I.J., Rocha, M.V.P.: Xylitol production from cashew apple bagasse by *Kluyveromyces marxianus* CCA510. Catal. Today. **255**, 33–40 (2015). https://doi.org/10.1016/j.cattod.2014.10.054
40. da Silva, D.D.V., de Arruda, P.V., Vicente, F.M.C.F., Sene, L., da Silva, S.S., das Graças de Almeida Felipe, M.: Evaluation of fermentative potential of *Kluyveromyces marxianus* ATCC 36907 in cellulosic and hemicellulosic sugarcane bagasse hydrolysates on xylitol and ethanol production. Ann. Microbiol. **65**, 687–694 (2015). https://doi.org/10.1007/s13213-014-0907-y
41. Cheng, K.K., Zhang, J.A., Ling, H.Z., Ping, W.X., Huang, W., Ge, J.P., Xu, J.M.: Optimization of pH and acetic acid concentration for bioconversion of hemicellulose from corncobs to xylitol by *Candida tropicalis*. Biochem. Eng. J. **43**, 203–207 (2009). https://doi.org/10.1016/j.bej.2008.09.012
42. Rodrigues, R.C.L.B., Kenealy, W.R., Jeffries, T.W.: Xylitol production from DEO hydrolysate of corn stover by *Pichia stipitis* YS-30. J. Ind. Microbiol. Biotechnol. **38**, 1649–1655 (2011). https://doi.org/10.1007/s10295-011-0953-4
43. Oktaviani, M.: peningkatan produksi xilitol dari hidrolisat hemiselulosa limbah daun tebu (*Saccharum officinarum* L.) melalui teknik adaptasi dan mobilisasi sel. (2019)

Chapter 46
Modulation of Diurnal Cycle of Rainfall Over the Western Maritime Continent Associated with Cold Surge and the Madden–Julian Oscillation

Aldi Krismon, Muhammad R. Abdillah, and Nurjanna Joko Trilaksono

Abstract The Maritime Continent exhibits strong diurnal cycle of rainfall (DCR) that largely contributes to total precipitation in the tropics and to global atmospheric teleconnection. The DCR was shown to be modulated under various background conditions such as those induced by cold surge (CS) in the South China and the active phase of Madden–Julian Oscillation (MJO). However, collaborative impacts of these phenomena on the DCR are not well understood. Here we analyze the relative contributions of CS and MJO (phase 3–5) in changing DCR characteristics over the western Maritime Continent during the recent climate (2009–2019). By using harmonic analysis, we find that in general CS and/or MJO do not affect the timing of diurnal peaks of rainfall but do cause significant anomalies in diurnal amplitude, clarifying that diurnal peak modulation greatly contributed to daily mean anomalies. Different DCR changes are observed in Java, Borneo, and Sumatra Islands and their adjacent seas. The most prominent pattern is found in southwestern Sumatra when CS and MJO occur simultaneously.

46.1 Introduction

The Maritime Continent (MC) consists of many islands scattered in the warm equatorial seas. Thermal contrasts between land and sea in MC are very pronounced. In afternoon, sea-breezes regularly occur, and land often experiences deep convective clouds and thunderstorms, which sometimes are enhanced by topographic effects [1–3]. Whereas in morning, the relatively weak land-breeze induces precipitation in the sea. The resulting diurnal circulation and precipitation contributes largely to the

A. Krismon
Department of Meteorology, Faculty of Earth Sciences and Technology, Institut Teknologi, Bandung, Indonesia

M. R. Abdillah (✉) · N. J. Trilaksono
Atmospheric Science Research Group, Faculty of Earth Sciences and Technology, Institut Teknologi, Bandung, Indonesia
e-mail: m.rais@itb.ac.id

E. Yulihastin et al. (eds.), *Proceedings of the International Conference on Radioscience, Equatorial Atmospheric Science and Environment and Humanosphere Science, 2021*, Springer Proceedings in Physics 275, https://doi.org/10.1007/978-981-19-0308-3_46

total annual precipitation in MC and even in the tropical belt [3]. The precipitation in MC also has impacts on remote areas in the extratropics since the deep convections release abundant latent heat that may trigger upper-level divergence and wave trains [4].

The diurnal cycle of rainfall (DCR) in MC is known to be affected by various modes of variability that change the activity of DCR [5–11]. Two prominent variabilities in boreal winter are cold surges [12–14] and the Madden–Julian Oscillation (MJO) [15, 16]. In a classical study by [5], they explained the changes in timing and amplitude of DCR in Borneo due to enhanced monsoonal flow associated with cold surges. [17] emphasized the important role of background flow anomalies that causing more early morning rain in north of Java while reducing afternoon rain in the inland, which appear to be attributed to the intermittent northerly surge events [18–20]. On the other hand, changes in DCR associated with MJO account for 80% of the MJO precipitation signal and the strongest anomalies appear leading the main envelope of MJO by several days [10, 11]. However, it is uncertain how the DCR response is when a cold surge concurrently appears.

The aim of current study is to investigate the combined effects of cold surges and MJO on DCR in MC. Previous studies relating to both events have shown that co-occurrences of cold surges and MJO lead to a much larger increase of synoptic precipitation anomalies [13, 14, 21, 22]. We are interested to find whether the impacts are due to changes in DCR or daily mean rainfall.

46.2 Data and Method

The period of study covers the recent 10 boreal winters (Dec-Feb 2009/10 to Dec-Feb 2018/19). We utilize the hourly precipitation data of GSMaP [23], hourly low-level circulation variables from ERA5 reanalysis [24], and daily MJO RMM index [16].

We use a definition of cold surge introduced by [13]. The indicators of cold surges are the strengthened northerly wind in the South China Sea (SCS) as in [21] and enhanced pressure in south of China to ensure the influence of cold air intrusion from inland Asia. A cold surge is identified when the northerly wind exceeds 8 m s^{-1} over SCS (5°–10°N, 107°–115°E) and mean sea level pressure (MSLP) exceeds 1020 hPa over south of China (18°–22°N; 105°–122°E). An MJO event is identified when the MJO phases are 3, 4, or 5 according to the RMM index with an amplitude ≥ 1.

We identify days that satisfy the above conditions and define three types of events: MJO, CS, and both MJO and CS. The first type indicates days that satisfy the MJO criteria but does not satisfy the cold surge criteria, vice versa for the second type. Both-type event is labelled when the criteria of MJO and cold surge are satisfied. One identified day is considered as one event. We find a total of 114 MJO days, 124 CS days, and 15 both MJO and CS days. Simple composites analyses are used to extract dominant synoptic and diurnal patterns of the three types of events. Harmonic analysis of diurnal cycle is used to quantify the timing and magnitude of DCR.

46.3 Result and Discussion

Firstly, we evaluate the composite anomaly features of daily precipitation and 925-hPa wind for MJO, CS, and both MJO and CS events (Fig. 46.1). The composites are generally consistent with previous studies [13, 14, 22], where the MJO and cold surges individually show enhanced precipitation in the western Maritime Continent and the impacts are greatly increased when they co-occur. MJO tends to bring more rainfall in the Indonesian seas and west of Sumatra associated with westerlies (Fig. 46.1a), which indicate enhanced westerly moisture fluxes from the Indian Ocean. Meanwhile, cold surge pattern is denoted by positive precipitation anomalies in downstream regions of the amplified northerly monsoon flow: southern SCS near Peninsular Malaysia and north of Borneo, and north of Java (Fig. 46.1b). Whereas the rainfall west of Sumatra is suppressed. The simultaneous events of MJO and CS bring heavier precipitation responses, especially in north of Java, east of Sumatra, and northwestern Borneo (Fig. 46.1c). A more mixed feature is observed. It may be due to fewer number of samples in the concurrent type.

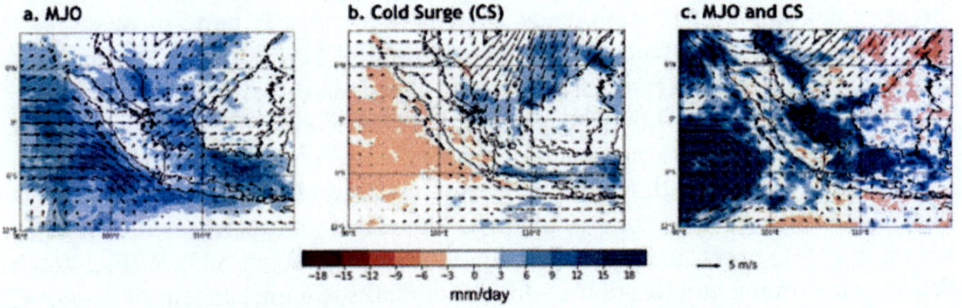

Fig. 46.1 Composite anomalies of daily precipitation (shaded) and 925-hPa wind field (vector) associated with MJO, CS, and both MJO and CS events

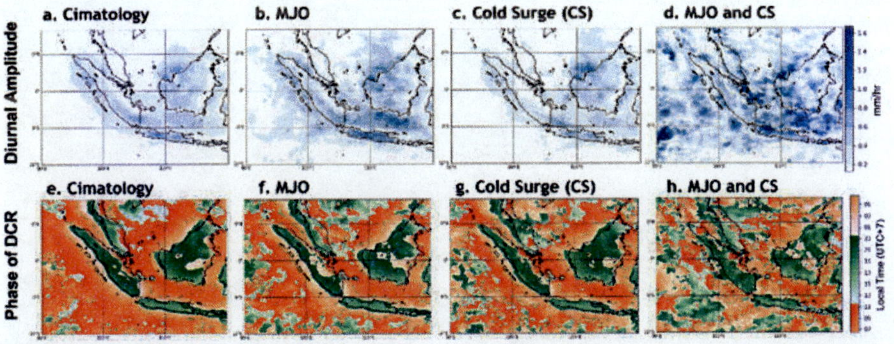

Fig. 46.2 Harmonic analysis of DCR. **a–d** are the amplitude of DCR and **e–h** are the phases of DCR shown by local time (UTC + 7). From left to right: climatological mean, MJO, CS, and both MJO and CS events

Figure 46.2a, e show a harmonic analysis of DCR from climatological mean data in boreal winter. The diurnal amplitude of precipitation spans over the islands and seas in the western Maritime Continent (Fig. 46.2a). Some prominent areas are Java Island, north of Java, southern Sumatra, and northwestern Borneo. During MJO event, the amplitude is enhanced in a large part of western Maritime Continent, consistent with the finding of [10] that utilized a radar-based only satellite precipitation dataset. Meanwhile, cold surge event increases the amplitude mostly in northwestern Borneo and north of western Java (Fig. 46.2b, c). Both MJO and CS event also show increases in DCR amplitude but with several enclosed peaks of strong DCR increases (Fig. 46.2d). In general, the increases of DCR amplitude are consistent with the main daily rainfall anomalies in Fig. 46.1, suggesting that the DCR modulation have large contributions to the overall precipitation anomalies associated with MJO and CS.

In climatological perspective, the timing of DCR clearly differs between land and sea, where land (sea) precipitation tends to appear in the afternoon (morning) (Fig. 46.2e). However, the overall changes in DCR phases associated with MJO, CS, and MJO and CS events are difficult to generalize and probably distinct between regions in the Maritime Continent (Fig. 46.2f–h). For example, during cold surges, the early morning precipitation tends to be more frequent in north of western Java (Fig. 46.2g). It is consistent with the recent result of [19, 20]. Changes in diurnal phases are further investigated in subsequent analyses.

To investigate in detail the changes in DCR and to allow the analysis of diurnal propagation of convections associated with the three events, we analysis three regions of interest (Fig. 46.3) with Hovmöller diagrams. Figures 46.4, 46.5, 46.6 and 46.7 show the results. In Java, the amplitude of diurnal rainfall increases over land and sea during MJO event, including the Indian Ocean side south of Java (Fig. 46.4b). Whereas the timing and location of diurnal rainfall seem unchanged. In case of CS

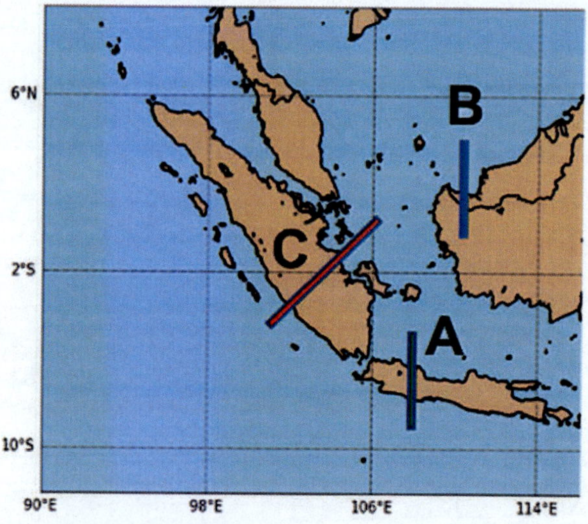

Fig. 46.3 Three selected regions for Hovmöller analyses in next figures. A, B, C labels indicate representatives of Java, Borneo, and Sumatra, respectively

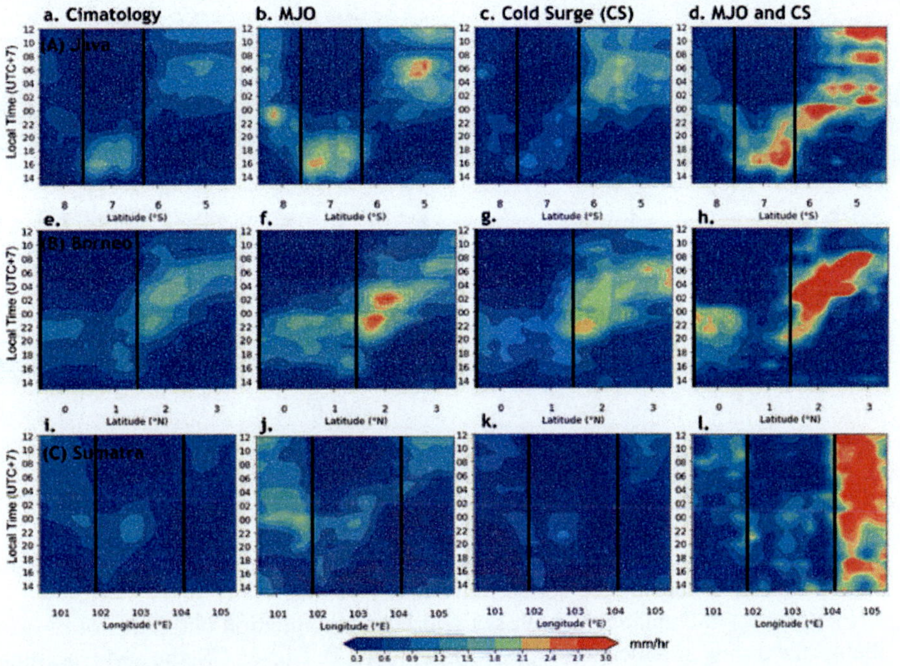

Fig. 46.4 Hovmöller plots of precipitation in three regions of interest shown in Fig. 46.3: **a–d** Java, **e–h** Borneo, and **i–l** Sumatra. From left to right: climatological mean, MJO, CS, and MJO and CS event. In Java and Borneo, latitude-time axes are chosen while in Sumatra, longitude-time axes are preferred

event, offshore precipitation in the early morning is enhanced and shifted towards the island (Fig. 46.4c). However, afternoon precipitation over land is reduced. The result agrees well with [17, 19, 20]. During the concurrent event, the amplitude is greatly enhanced, including early morning offshore rainfall and afternoon inland rainfall (Fig. 46.4d). Only slight increase is observed south of Java.

Physical mechanisms that control the changes in DCR are explored. Figure 46.5 show meridional wind and divergence features associated with the three events. In MJO case, no significant differences with climatology are observed (Fig. 46.5b, f) because the main flow are westerly, and the precipitation responses are more attributed to large scale convergences and instability [16]. Meanwhile, cold surges clearly induce stronger northerlies that amplify convergence zone near to the northern coast of Java in early morning (Fig. 46.5c, g), which are favorable for convections. In the inland, we observe that divergence is stronger and consequently lead to suppressed rainfall in the afternoon. For both MJO and CS event, the northerlies are rather weak (Fig. 46.5d), which is consistent with the result of [13]. However, we would have found stronger northerlies if the CS index was based on cross-equatorial surge index introduced by [22]. The convergence is much larger in the concurrent event,

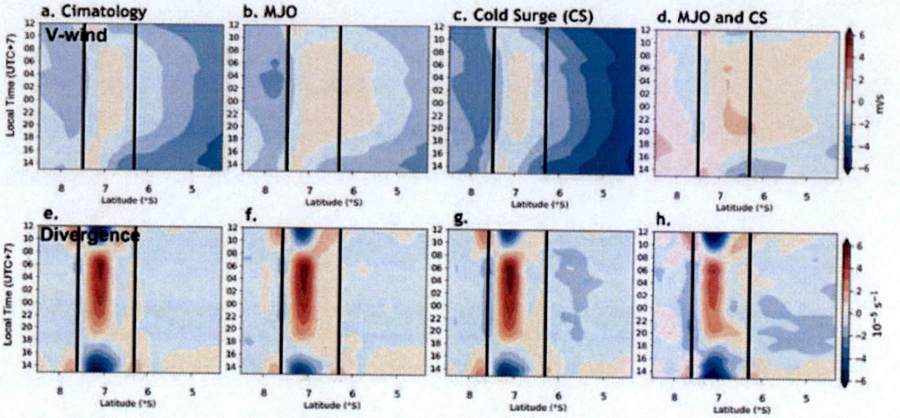

Fig. 46.5 Hovmöller plots of **a**–**d** meridional wind and **e**–**h** low-level divergence in Java. From left to right: climatological mean, MJO, CS, and MJO and CS event

including southern Java, and the divergence is reduced. These conditions lead to more precipitation during the concurrent event.

In Borneo, MJO enhances both sea and land precipitation but does not clearly change the timing and location of DCR (Fig. 46.4e, f). During CS, the early morning offshore precipitation is amplified and extended further to the north, but its peak is closer to the coastline of Borneo (Fig. 46.4g), indicating the impact of enhanced background flow like the previous case in Java. During the simultaneous events of MJO and CS, the early morning precipitation signal is very robust (Fig. 46.4h). It resembles a more organized system rather than that in Java. It is possibly because the definition of cold surge based on SCS region, which is closer to Borneo.

The diurnal evolution of wind and divergence in Borneo is shown in Fig. 46.6. The northerly wind in MJO is reduced, but strongly enhanced in CS and moderately enhanced in MJO and CS events. The convergence is enhanced in CS, which is confined in offshore from midnight to morning. Convergence due to CS and MJO are quite large and spanning a large area and more persistent.

In Sumatra, the afternoon rainfall in inland spreads out to the west and east of Sumatra (Fig. 46.4i), similar to that shown in [25]. During MJO event, the propagation of rainfall toward Indian Ocean is amplified (Fig. 46.4j). It is possibly because the westward gravity wave and gravity current more effectively interact with the westerlies background flow and lead to greater precipitation. In cold surge however, the rainfall in the land and west of Sumatra is suppressed but offshore rainfall in the east of Sumatra is increased at noon (Fig. 46.4k). During a concurrent event, very strong precipitation response appears in the east of Sumatra, which persists in the whole day (Fig. 46.4l). The associated circulation features are shown in Fig. 46.7. Here we use zonal wind in Fig. 46.7a–d. During MJO, it is evident that the westerlies are enhanced. Local scale convergence develops in the west of Sumatra. During cold surge, the apparent wind is westerlies. Some convergences are increased in the east of

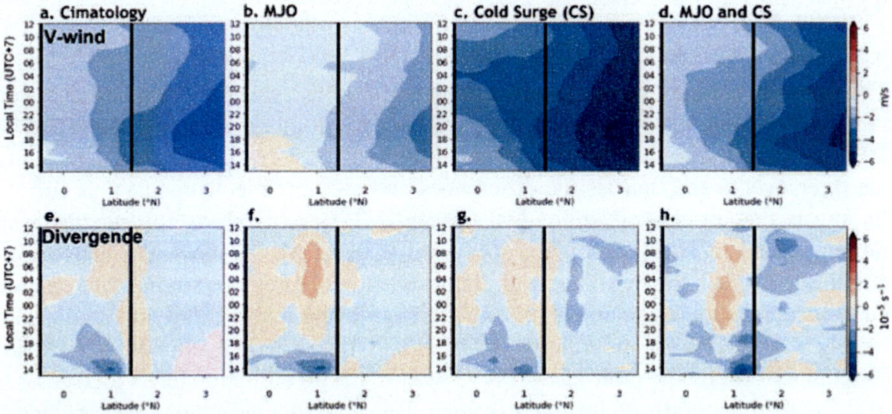

Fig. 46.6 Hovmöller plots of **a**–**d** meridional wind and **e**–**h** low-level divergence in Borneo. From left to right: climatological mean, MJO, CS, and MJO and CS event

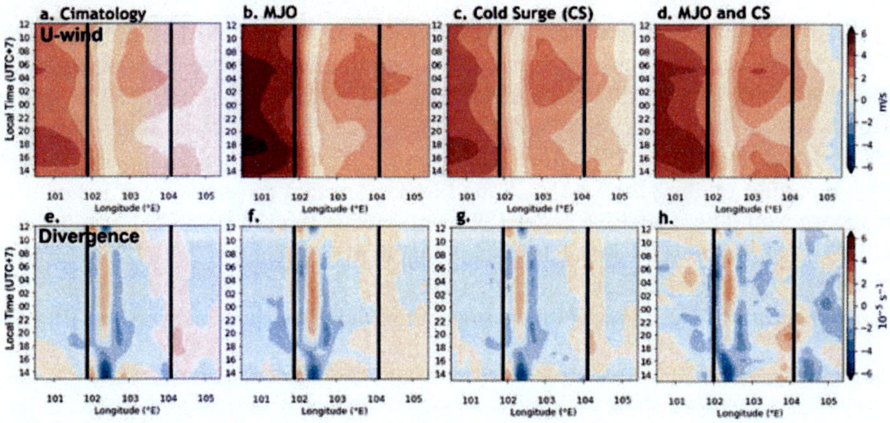

Fig. 46.7 Hovmöller plots of **a**–**d** zonal wind and **e**–**h** low-level divergence in Sumatra. From left to right: climatological mean, MJO, CS, and MJO and CS event

Sumatra, possibly due to convergence between the background westerly and northwesterly associated with cold surge (see spatial pattern in Fig. 46.1). Meanwhile, the concurrent MJO and CS event shows many parts of enhanced convergences in Sumatra and its vicinity, especially in east of Sumatra, although the convergence-divergence patterns are rather mixed probably due to low number of samples for composite analysis.

46.4 Conclusion

This is a preliminary study that investigates possible changes in DCR associated with MJO, CS, and combine MJO and CS events. From a systematic analysis over 10 years of boreal winter data, we show distinct DCR responses associated with the three events and discuss the differences and possible mechanisms. Our current finding is that in general, individual CS and MJO events show enhancements in the amplitude of DCR. For the co-occurred CS and MJO events, the increase in amplitude is much greater. The result is consistent with previous studies that utilized daily-mean rainfall anomalies [13, 14, 22], suggesting that the daily anomalies are largely attributed to the modulation of DCR. The timing and location of DCR do not differ much among the three cases. For CS cases, the precipitation tends to get closer to the coastlines north of Java and Borneo. But when the effect of MJO included, the shifting is not clear. The east of Sumatra experiences large increase in rainfall when MJO and CS occur simultaneously. The apparent responses of precipitation are partly explained through interactions between local and background wind flow as well as topography.

Some problems and unanswered questions need to be addressed in future studies. For example, the unbalanced number of samples among the three cases. It could be mitigated with a better definition of CS and MJO or a longer study period. In the Hovmöller analysis, this study only considers three transect line to represent Java, Borneo, and Sumatra DCR. More detailed investigation is needed to capture the whole picture relating the DCR, CS, and MJO. Another raised concern is how the response would be if a CENS index [22] is utilized instead of the SCS CS index. As CENS is defined closer to Java Island, the southward momentum of CENS is expected to be more pronounced than that associated with CS in SCS. Previous studies have shown results consistent with our finding but limited to western Java region [19, 26].

Acknowledgements This study is supported by PPMI-ITB Riset Kolaboratif FITB. We are very grateful to the reviewers who provide useful comments on the early version of this study.

References

1. Yang, G.-Y., Slingo, J.: The diurnal cycle in the tropics. Mon. Weather Rev **129**, 18 (2001)
2. Qian, J.-H.: Why precipitation is mostly concentrated over Islands in the Maritime Continent. J. Atmos. Sci **65**, 1428–1441 (2008). https://doi.org/10.1175/2007JAS2422.1
3. Yamanaka, M.D., Ogino, S.-Y., Wu, P.-M., Jun-Ichi, H., Mori, S., Matsumoto, J., Syamsudin, F.: Maritime Continent coastlines controlling Earth's climate. Prog. Earth Planet Sci **5**, 21 (2018). https://doi.org/10.1186/s40645-018-0174-9
4. Neale, R., Slingo, J.: The Maritime Continent and Its role in the global climate: a GCM study. J. Clim. **16**, 15 (2003)
5. Johnson, R.H., Priegnitz, D.L.: Winter monsoon convection in the vicinity of North Borneo. Part II: effects on large-scale fields. Mon. Weather Rev. **109**, 1615–1628 (1981). https://doi.org/10.1175/1520-0493(1981)109%3c1615:WMCITV%3e2.0.CO;2

6. Araki, R., Yamanaka, M.D., Murata, F., Hashiguchi, H., Oku, Y., Sribimawati, T., Kudsy, M., Renggono, F.: Seasonal and interannual variations of diurnal cycles of wind and cloud activity observed at Serpong, West Jawa, Indonesia. J. Meteorol. Soc. Jpn. **84A**, 171–194 (2006). https://doi.org/10.2151/jmsj.84A.171
7. Qian, J.-H., Robertson, A.W., Moron, V.: Interactions among ENSO, the Monsoon, and diurnal cycle in Rainfall Variability over Java, Indonesia. J. Atmos. Sci. **67**, 3509–3524 (2010). https://doi.org/10.1175/2010JAS3348.1
8. Rauniyar, S.P., Walsh, K.J.E.: Scale Interaction of the diurnal cycle of rainfall over the Maritime Continent and Australia: influence of the MJO. J. Clim. **24**, 325–348 (2011). https://doi.org/10.1175/2010JCLI3673.1
9. Oh, J.-H., Kim, K.-Y., Lim, G.-H.: Impact of MJO on the diurnal cycle of rainfall over the western Maritime Continent in the austral summer. Clim. Dyn. **38**, 1167–1180 (2012). https://doi.org/10.1007/s00382-011-1237-4
10. Peatman, S.C., Matthews, A.J., Stevens, D.P.: Propagation of the Madden-Julian oscillation through the Maritime Continent and scale interaction with the diurnal cycle of precipitation: MJO propagation and scale interaction with the diurnal cycle. QJR Meteorol. Soc. **140**, 814–825 (2014). https://doi.org/10.1002/qj.2161
11. Birch, C.E., Webster, S., Peatman, S.C., Parker, D.J., Matthews, A.J., Li, Y., Hassim, M.E.E.: Scale Interactions between the MJO and the Western Maritime Continent. J. Clim. **29**, 2471–2492 (2016). https://doi.org/10.1175/JCLI-D-15-0557.1
12. Zhang, Y., Sperber, K.R., Boyle, J.S.: Climatology and interannual variation of the East Asian winter monsoon: results from the 1979–95 NCEP/NCAR reanalysis. Mon. Weather Rev. **125**, 15 (1997)
13. Lim, S.Y., Marzin, C., Xavier, P., Chang, C.-P., Timbal, B.: Impacts of boreal winter monsoon cold surges and the interaction with MJO on Southeast Asia Rainfall. J. Clim. **30**, 4267–4281 (2017). https://doi.org/10.1175/JCLI-D-16-0546.1
14. Abdillah, M.R., Kanno, Y., Iwasaki, T., Matsumoto, J.: Cold surge pathways in East Asia and their tropical impacts. J. Clim. **34**, 157–170 (2021). https://doi.org/10.1175/JCLI-D-20-0552.1
15. Madden, R.A., Julian, P.R.: Detection of a 40–50 day oscillation in the zonal wind in the tropical pacific. J. Atmos. Sci. **28**, 702–708 (1971). https://doi.org/10.1175/1520-0469(1971)028%3c0702:DOADOI%3e2.0.CO;2
16. Wheeler, M.C., Hendon, H.H.: An all-season real-time multivariate MJO index: development of an index for monitoring and prediction. Mon. Weather Rev. **132**, 16 (2004)
17. Koseki, S., Koh, T.-Y., Teo, C.-K.: Effects of the cold tongue in the South China Sea on the monsoon, diurnal cycle and rainfall in the Maritime Continent. QJR Meteorol. Soc. **139**, 1566–1582 (2013). https://doi.org/10.1002/qj.2052
18. Trilaksono, N.J., Otsuka, S., Yoden, S.: A Time-Lagged ensemble simulation on the modulation of precipitation over West Java in January–February 2007. Mon Wea Rev **140**, 601–616 (2012). https://doi.org/10.1175/MWR-D-11-00094.1
19. Yulihastin, E., Wahyu Hadi, T., Sari Ningsih, N., Ridho Syahputra, M.: Early morning peaks in the diurnal cycle of precipitation over the northern coast of West Java and possible influencing factors. Ann. Geophys **38**, 231–242 (2020). https://doi.org/10.5194/angeo-38-231-2020
20. Yulihastin, E., Hadi, T.W., Abdillah, M.R., Fauziah, I.R., Ningsih, N.S.: Propagation of convective systems associated with early morning precipitation and different northerly background winds over Western Java. J. Meteorol. Soc. Jpn. Ser. II **100**, 99–113 (2022). https://doi.org/10.2151/jmsj.2022-005
21. Chang, C.-P., Harr, P.A., Chen, H.-J.: synoptic disturbances over the equatorial South China Sea and Western Maritime Continent during boreal winter. Mon. Weather Rev. **133**, 489–503 (2005). https://doi.org/10.1175/MWR-2868.1
22. Hattori, M., Mori, S., Matsumoto, J.: The cross-equatorial northerly surge over the Maritime Continent and Its relationship to precipitation patterns. JMSJ **89A**, 27–47 (2011). https://doi.org/10.2151/jmsj.2011-A02
23. Kubota, T., Aonashi, K., Ushio, T., Shige, S., Takayabu, Y.N., Kachi, M., Arai, Y., Tashima, T., Masaki, T., Kawamoto, N., Mega, T., Yamamoto, M.K., Hamada, A., Yamaji, M., Liu,

G., Oki, R.: Global satellite mapping of precipitation (GSMaP) products in the GPM era. In: Levizzani, V., Kidd, C., Kirschbaum, D.B., Kummerow, C.D., Nakamura, K., Turk, F.J. (eds.) Satellite Precipitation Measurement, vol. 1, pp. 355–373. Springer International Publishing, Cham (2020)

24. Hersbach, H., Bell, B., Berrisford, P., Hirahara, S., Horányi, A., Muñoz-Sabater, J., Nicolas, J., Peubey, C., Radu, R., Schepers, D., Simmons, A., Soci, C., Abdalla, S., Abellan, X., Balsamo, G., Bechtold, P., Biavati, G., Bidlot, J., Bonavita, M., Chiara, G.D., Dahlgren, P., Dee, D., Diamantakis, M., Dragani, R., Flemming, J., Forbes, R., Fuentes, M., Geer, A., Haimberger, L., Healy, S., Hogan, R.J., Hólm, E., Janisková, M., Keeley, S., Laloyaux, P., Lopez, P., Lupu, C., Radnoti, G., de Rosnay, P., Rozum, I., Vamborg, F., Villaume, S., Thépaut, J.-N.: The ERA5 global reanalysis. Q. J. R. Meteorol. Soc. **146**, 1999–2049 (2020). https://doi.org/10.1002/qj.3803
25. Mori, S., Jun-Ichi, H., Tauhid, Y.I., Yamanaka, M.D., Okamoto, N., Murata, F., Sakurai, N., Hashiguchi, H., Sribimawati, T.: Diurnal land-sea rainfall peak migration over Sumatera Island, Indonesian Maritime Continent, observed by TRMM satellite and intensive rawinsonde soundings. Mon. Weather Rev. **132**, 19 (2004)
26. Mori, S., Hamada, J.-I., Hattori, M., Wu, P.-M., Katsumata, M., Endo, N., Ichiyanagi, K., Hashiguchi, H., Arbain, A.A., Sulistyowati, R., Lestari, S., Syamsudin, F., Manik, T., Yamanaka, M.D.: Meridional march of diurnal rainfall over Jakarta, Indonesia, observed with a C-band Doppler radar: an overview of the HARIMAU2010 campaign. Prog. Earth Planet Sci **5**, 47 (2018). https://doi.org/10.1186/s40645-018-0202-9

Chapter 47
Comparative Analysis of Upwelling Characteristics in Northeast and Southwest of Indonesian Seas Area

Sigit Kurniawan Jati Wicaksana and Iis Sofiati

Abstract Information of the characteristics of upwelling in the Indonesian sea has a very important role to estimate the potential fishing areas. The purpose of this study is to identify upwelling based on the Ekman pumping value that occurs in northeastern and southwestern Indonesia. The data used in this study are surface wind, sea surface temperature, chlorophyll-a and geostrophic current. Where the last three data are used as supporting analysis data for the calculation results of Ekman pumping. The method used in this study is to calculate Ekman pumping based on surface wind data for a period of 25 years (1995–2019). From the results, it can be seen that the Sea Surface Temperature (SST) at the time of the December January February (DJF) in the southern sea of Java is warmer than its surroundings around 30 °C, and vice versa during the June July August (JJA) about 27 °C. Chlorophyll-a (0.9 mg/m^3), geostrophic currents (1.1 m/s), and Ekman pumping (4.8×10^{-5} m/s) have a fairly high intensity in the northern region of Java during DJF, while during JJA it occurs in southern Java. From the temporal analysis, it can be seen that the minimum intensity of chlorophyll a, geostrophic currents and Ekman pumping occurs during DJF and maximum occurs during JJA and vice versa for SST. This study supports previous studies, where the maximum Ekman pumping was followed by the minimum SST that occurred in both observation areas. However, chlorophyll-a and geostrophic currents in the southwest of Indonesia follow seasonal characteristics but not for the northeastern region.

47.1 Introduction

The upwelling occurs when the wind blows from the coast, allowing water masses from deeper layers to fill the void remains. This event causing them to be dragged south away from the coast [1]. The nutrients carried by upwelled water (such as

S. K. J. Wicaksana (✉) · I. Sofiati
Research Center for Climate and Atmosphere, National Research and Innovation Agency (BRIN), Bandung, Indonesia
e-mail: sigi020@brin.go.id

E. Yulihastin et al. (eds.), *Proceedings of the International Conference on Radioscience, Equatorial Atmospheric Science and Environment and Humanosphere Science, 2021*, Springer Proceedings in Physics 275, https://doi.org/10.1007/978-981-19-0308-3_47

phosphate and nitrate) from the lower layer to the upper layer or even to the surface layer of the ocean make upwelling regions important fishing grounds [2]. Ocean circulation and atmospheric interactions are extremely complicated in the Indonesia region [3]. Upwelling events and monsoon winds are two outcomes of the complex system. The strong winds play an important role in determining where and how long upwelling events occur everywhere. As a result of the monsoon system's physical effects on oceanographic factors, the upwelling phenomenon occurs, which supports the biological productivity of marine waters where upwelling occurs [4].

Coastal upwelling is a type of wind-driven upwelling. The wind's movement will cause a water mass transport known as Ekman transport [5]. Aside from Ekman transport, there is also an Ekman pumping, which plays an important role in the formation of upwelling on the coast.

Currently, there are actually many methods to detect upwelling [6–9], but in Indonesia most of the methods still use satellite-based remote sensing. However, most research has been conducted in a relatively short period of time and with a low spatial resolution [10]. Given the importance of upwelling and the scarcity of satellite data, other methods of determining upwelling are required, one of which is analyzed using Ekman transport and Ekman pumping values. Therefore, the objective of this paper is to understand the spatial and temporal evolution of upwelling, especially for coastal upwelling, as it can be used as a predictive tool, which is quite valuable for designing proper fishery management plans.

47.2 Data and Methods

47.2.1 Data

The study area is separated into two sections: southwestern Indonesia (8°N–4°S and 115°–150°E) and southwest Indonesia (4°–16°S and 100°–135°E) (see Fig. 47.1).

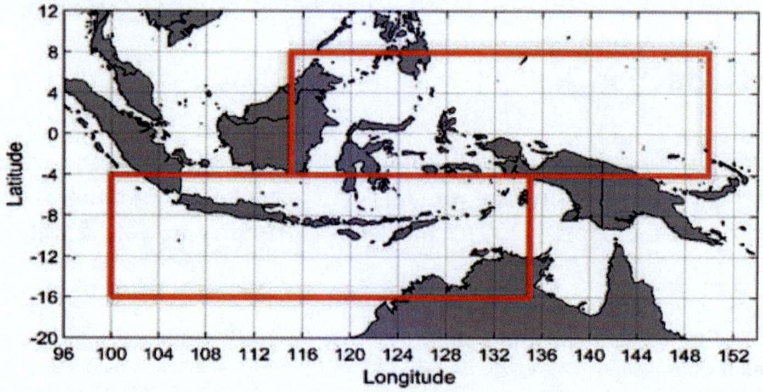

Fig. 47.1 Map of the research area

This was decided to represent open sea waters towards the Indian and Pacific Oceans. The material used in this study consists of main data and supporting data. The main data used are monthly surface wind (10 m) data were obtained from http://www.ecmwf.int and the supporting data are sea surface temperature (SST) from http://www.ecmwf.int, chlorophyll-a and geostrophic current data from https://marine.copernicus.eu. All of the above data have a spatial resolution of 0.25° × 0.25° and range from January 1995 to December 2019 (25 years).

47.2.2 Methods

In this study, the analysis of each variable (Ekman pumping, SST, chlorophyll-a and geostrophic currents) was carried out by seasonal analysis to determine the distribution pattern of each variable in monthly variations throughout the year (1995–2019) which aims to determine the detailed phenomena that occur every month both spatially and temporally.

The upwelling phenomenon can be quantified by calculating the Ekman pumping associated with Ekman transport. The initial step in determining Ekman transport and Ekman pumping is to compute wind stress and wind stress curl for each wind component. The value of Ekman transport for components parallel and perpendicular to the coast is estimated using wind stress. Equation 1 expresses the Ekman pumping calculation [11, 12]:

$$-\mathrm{W_E} = \frac{\partial M_{xE}}{\partial x} + \frac{\partial M_{yE}}{\partial y}. \tag{47.1}$$

where:

$\mathrm{W_E}$ Ekman pumping ($\mathrm{m^2s^{-1}}$)
$\mathrm{M_{xE}}$ Ekman transport parallel to the coast ($\mathrm{m^2s^{-1}}$)
$\mathrm{M_{yE}}$ Ekman transport upright to the coast ($\mathrm{m^2s^{-1}}$).

The Ekman pumping value with a negative sign WE < 0 indicates that there is a convergence of Ekman transport on the surface which moves the water mass towards the seabed, which is known as downwelling [12]. If the Ekman pumping value is positive, WE > 0 indicates that there is a divergence of Ekman transport on the surface that moves the water mass towards the sea surface, which is referred to as an upwelling process. Ekman transport and Ekman pumping values can be used to represent the movement of coastal upwelling events.

The next analysis used a composite technique, where the results of the monthly analysis were combined into three months and averaged. This technique applies to each parameter, during the observation period. In this study, the composite technique was used to observe seasonal changes in each component.

47.3 Results

The waters of northeastern and southwestern Indonesia are interesting to study due to their opposite spatial and temporal variability, as can be seen from the distribution of chlorophyll-a, SST and upwelling.

Figure 47.2 shows the spatial and temporal distribution of SST and surface winds in southwestern Indonesia. During the June July August (JJA) period, SST reached 27 °C and began to rise during the September October November (SON) period. Monthly wind movements show southwesterly winds with higher speeds in the southern waters of Java in the JJA period and begin to decline in the SON period. Meanwhile, during December January February (DJF), wind speed tends to decrease and reaches a minimum during the March April May (MAM) period.

Figure 47.3 illustrates the monthly climatology of chlorophyll-a in southwestern Indonesia. The concentration of chlorophyll-a is centered in the north of Java during the western monsoon (DJF) and MAM periods. Meanwhile, during the Australian monsoon (JJA) period, the concentration of chlorophyll-a increased to 0.84 mg/m^3 and continued to rise during the SON period. The concentration of chlorophyll-a

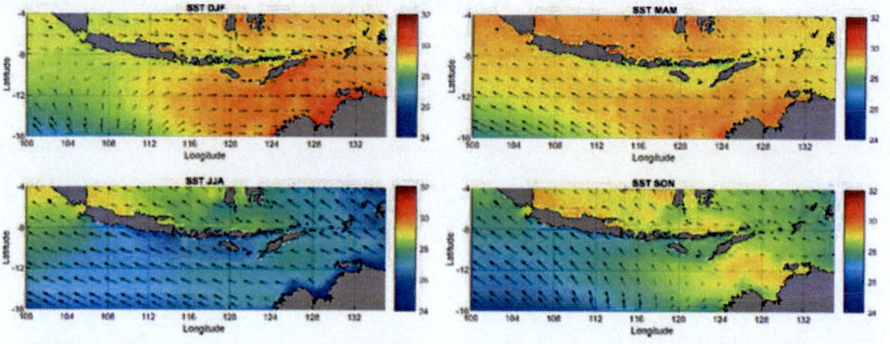

Fig. 47.2 Spatial distribution of seasonal SST and surface winds from 1995 to 2019 in southwestern Indonesia

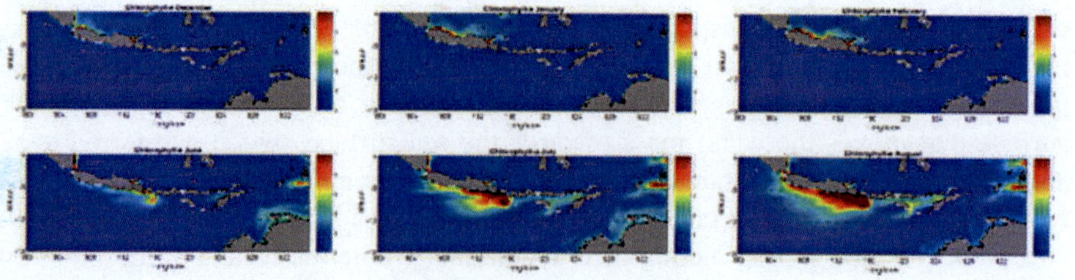

Fig. 47.3 Monthly climatology of Chlorophyll-a distribution from December to August (1995–2019)

was found to be higher and more widely distributed from the coast during the SON period than during the JJA period.

When the easterly wind blows strongly during the SON period, it acts as a wind-driven motion in the upwelling phenomenon during the Australian monsoon. The abundance of chlorophyll-a concentrations indicates that cold SST conditions brought the nutrient-rich water mass to the sea surface. This rising movement of water mass is also caused by layer stratification, which has a density difference in each layer because the temperature decreases with increasing water depth. This causes energy to be released, causing the water mass to move vertically. On the other hand, during DJF the abundance of chlorophyll-a concentration tends to decrease, where the sea surface temperature increases while the wind speed decreases.

Figure 47.4 shows the distribution of chlorophyll-a in Indonesia's northeast region. In general, the concentration of chlorophyll-a in this area is low. During the JJA period it has the highest concentration of chlorophyll-a and the vast distribution. This condition is related to the distribution of average SST during the JJA period, particularly in the Banda Sea, where SST reach 26–27 °C with strong easterly winds (Fig. 47.5). The concentration of chlorophyll-a continues to decline during the SON period and dropped significantly throughout the DJF to MAM period. Although the concentration of chlorophyll-a is stable (almost always all year round) in some areas near the coast of the island of Papua. SST begins to increase in the period from SON to MAM.

During the Asian monsoon, the waters in Indonesia will warm because it is influenced by warm currents and winds from the Asian continent. During the Australian monsoon, however, a mass flow of cold water from eastern Australia affects the waters of eastern Indonesia [13]. High concentrations of chlorophyll-a can be found in some areas during this time period. It tends to extend to the central and western parts of the Banda Sea in the east. Higher concentrations are found in the south of Java near Bali and along all of the coasts [3].

Geostrophic currents are used to detect and analyze the emergence of several phenomena on the high seas, including upwelling and downwelling [14]. Upwelling

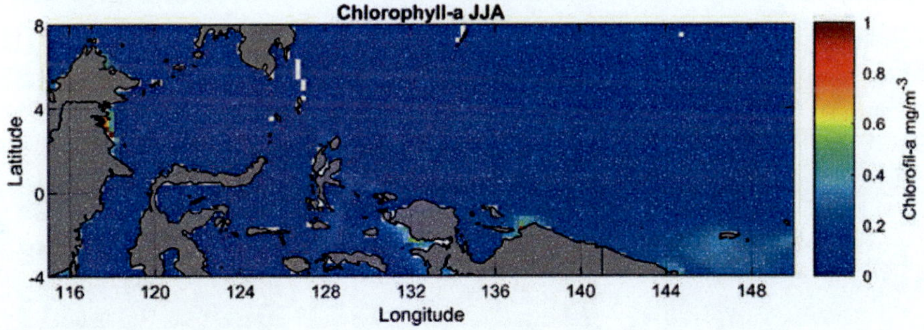

Fig. 47.4 Annual distribution of chlorophyll-a from 1995 to 2019 in the northeastern Indonesia

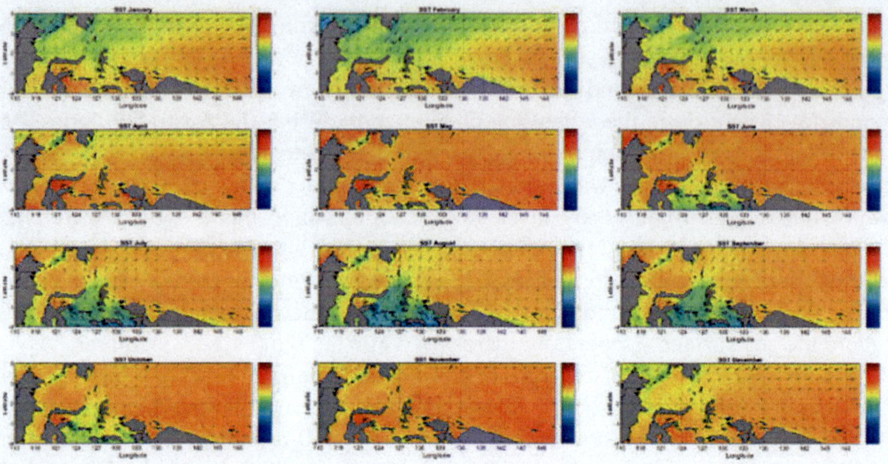

Fig. 47.5 Monthly climatology of SST and surface winds distribution from 1995 to 2019 in northeast Indonesia

(downwelling) events are thought to occur in areas with cyclonic Eddies (anticyclones) from geostrophic currents [15]. Figure 47.6 shows the distribution pattern of geostrophic currents analyzed on a monthly basis, which shows the movement of geostrophic currents following the pattern of monsoon movement. Starting in January, the intensity of geostrophic currents was quite high in the waters north of Sulawesi. Then it moves eastward to the waters of Maluku in March, this condition coincides with the occurrence of the Asian monsoon. Along with changes in wind movement starting from May to October, geostrophic currents move eastward with a fairly strong intensity occurring in the southern waters of Maluku and Papua, then

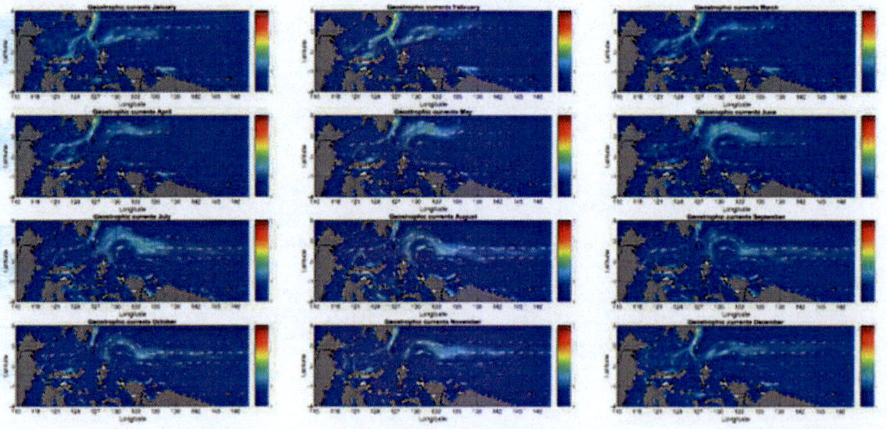

Fig. 47.6 Monthly climatology of geostrophic currents distribution from 1995 to 2019 in northeast Indonesia

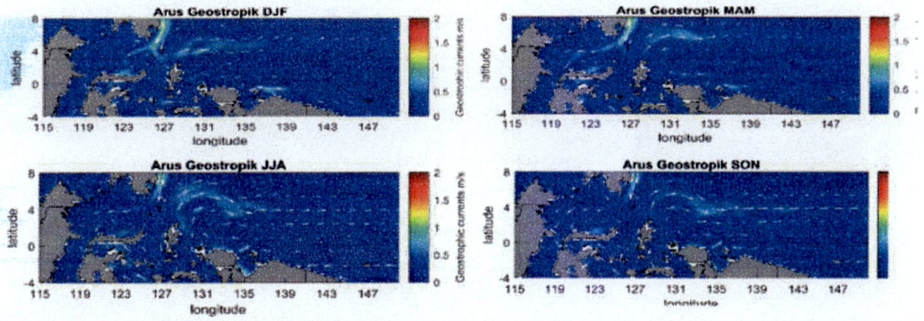

Fig. 47.7 Spatial distribution of Ekman pumping seasonally during 1995–2019 in northeast Indonesia

weaken and change direction again following the next monsoon. In the same area, a cyclonic event was seen which indicated an upwelling process. With the above results, this study supports previous research states that utilization of geostrophic currents can detect the potential fishing zone (ZPPI) [16].

Using the composite technique method, an analysis of the geostrophic current conditions that occurred in the waters of northeastern Indonesia was carried out (**Fig. 47.7**). Seasonal analysis for geostrophic currents shows a significant difference. The dominance of geostrophic currents from the west and from the north resulted in the confluence of geostrophic currents that occurred in most of the northern waters of Maluku and Papua from January to December. In June July August (JJA), geostrophic currents that move from the east with a small intensity, converge with geostrophic currents from the north with a fairly large intensity. Because it is blocked by land, the geostrophic current from the east turns, so that near the mainland this condition results in an increase in the intensity of the geostrophic current.

Based on the Ekman pumping calculations, Figs. 47.8 and 47.9 show the distribution of upwelling and downwelling in northeastern and southwestern Indonesia. Blue indicates upwelling and green indicates downwelling. In the northeastern region, around the equator there is an empty area, this happens because this calculation uses the Coriolis force, where the value of the Coriolis force will disappear the closer it is to the equator [17].

The average Ekman pumping in the Northeast (Fig. 47.8) reached 2,079 × 10–5 m/s during the JJA period, coinciding with the Australian monsoon. In August, Ekman pumping reached a peak of 3.33 × 10–5 m/s. The high pumping efficiency and abundance of chlorophyll-a during this time period indicate the presence of upwelling in the Banda Sea. The Australian monsoon's Ekman transport values (red arrows) appear to be higher than the Asian monsoon. During the Australian monsoon, Ekman's longshore transport pattern in the Banda Sea approaches the coast, whereas during the Asian monsoon, the pattern of Ekman transport tends to move offshore [18, 19].

In the southwest (Fig. 47.9), the JJA period has the highest Ekman pumping value of 2.240 × 10–5 m/s, while February has the lowest with a value of − 1530 ×

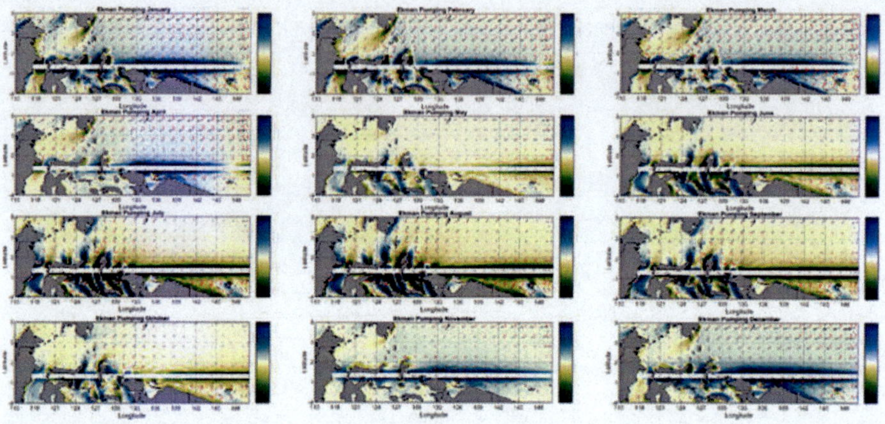

Fig. 47.8 Monthly climatology of Ekman pumping distribution from 1995 to 2019 in northeast Indonesia

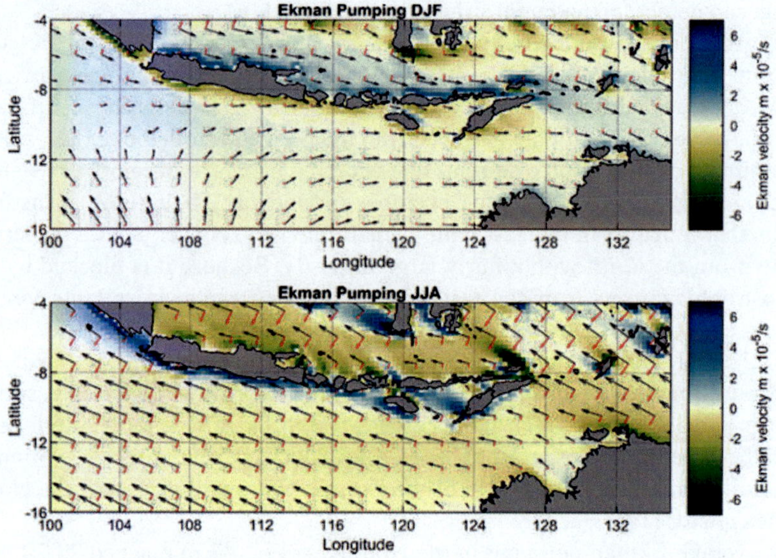

Fig. 47.9 Spatial distribution of Ekman Pumping average from 1995 to 2019 in southwestern Indonesia

10^{-5} m/s. Throughout the year, the upwelling phenomenon is quite strong, particularly on the island of Java. Upwelling occurred in the northern waters of Java during the DJF period, while downwelling occurred in the southern waters. The intensity then decreased during the MAM period, and the position of the upwelling shifted to the south of Java. This occurs due to a shift in wind direction from the Asian monsoon to the Australian monsoon.

The distribution of Ekman transport (red arrow) in the south of Java during the Australian monsoon is shown going towards the high seas, while the Ekman transport moves towards the coast during the Asian monsoon. During the Australian monsoon, this situation corresponds to a high concentration of chlorophyll-a and a lower SST. The investigation of Ekman pumping and Ekman transport reveals that the upwelling mechanism that occurs in southern Java is a coastal upwelling with substantial consequences for ecosystems and fisheries [20].

In general, the spatial distribution difference between the DJF and JJA periods is clearly visible in both regions. Upwelling occurs in the north of the island during the DJF period and vice versa during the JJA period. This occurs because, during MAM, Ekman's pumping position changes due to changes in wind direction associated with the transition from the Asian cold monsoon to the Australian cold monsoon.

The following Figs. 47.10 and 47.11 are monthly temporal variation for each variable in the study region. As can be seen in the graph, there is a similar pattern between

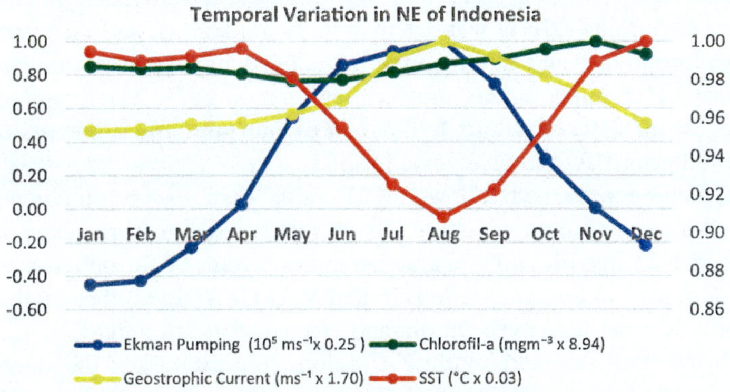

Fig. 47.10 Monthly temporal variation of Ekman pumping, SST, chlorophyll-a and geostrophic current in the northeastern Indonesia

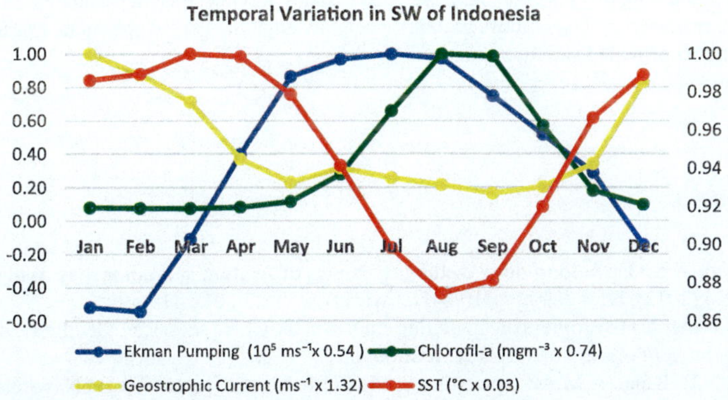

Fig. 47.11 Monthly temporal variation of Ekman pumping, SST, chlorophyll-a and Geostrophic current in the southwest Indonesia

the northeast and southwest regions. Where the SST value decreased followed by an increase in Ekman pumping and chlorophyll-a, and vice versa. In April, the Ekman pumping value began to increase and SST decreased in May, while chlorophyll-a only increased in July. This is in accordance with previous studies [21], where high upwelling does not directly affect the concentration of chlorophyll-a, but there is a time lag before the concentration of chlorophyll-a increases.

In general, upwelling events will periodically repeat in the south Australian monsoon (wind from the southeast). This is in accordance with previous research that upwelling in southern Java occurs during the southeast monsoon [22, 23].

47.4 Conclusion

Upwelling in Indonesia is affected by monsoon, it is evidenced by changes in the concentration and upwelling conditions that change seasonally. SST in the southern sea of Java during the DJF is warmer than its surroundings, and vice versa during the JJA. Chlorophyll-a and Ekman pumping is fairly intense in the northern region of Java during DJF, but not in the southern region during JJA. The temporal analysis shows that the intensity of chlorophyll-a, and Ekman pumping is lowest during DJF and highest during JJA, and vice versa for SST.

The upwelling area had a lower SST value than its surroundings and the chlorophyll-a concentration increased. The lowest average sea surface temperature is in August with a value of 25.09 °C and the highest average chlorophyll-a concentration is 0.78 mg/m^3 in September. In DJF and MAM, SST in northeastern Indonesia is noticeably warmer than in the north, and vice versa at JJA and SON.

In both the northeast and southwest regions, upwelling has the same pattern. Upwelling in the DJF period occurs in the north of the island and in the JJA period it occurs in the south of the island due to the change of monsoon.

Acknowledgements We would like to thank Dr. Mochamad Furqon Azis Ismail for providing a method for processing this research data. We wish to thank Reviewer(s) for comments and critics to improve the manuscript. The authors gratefully acknowledge the E.U. Copernicus Marine Service Information that provided satellite data.

References

1. Pranowo, W.S.: Upwelling-downwelling dynamics of Arafura and timor seas. Widyariset. **15**, 415–424 (2012). https://doi.org/10.14203/widyariset.15.2.2012.415-424
2. Hu, J., Wang, X.H.: Progress on upwelling studies in the China seas. Rev. Geophys. **54**, 653–673 (2016). https://doi.org/10.1002/2015RG000505
3. Purba, N.P., Khan, A.M.A.: Upwelling session in Indonesia waters. World News Nat. Sci. **25**, 72–83 (2019)

4. Ningsih, N.S., Rakhmaputeri, N., Harto, A.B.: Upwelling variability along the southern coast of Bali and in Nusa Tenggara waters. Ocean Sci. J. **48**, 49–57 (2013). https://doi.org/10.1007/s12601-013-0004-3
5. Barth, J.A., Menge, B.A., Lubchenco, J., Chan, F., Bane, J.M., Kirincich, A.R., McManus, M.A., Nielsen, K.J., Pierce, S.D., Washburn, L.: Delayed upwelling alters nearshore coastal ocean ecosystems in the northern California current. Proc. Natl. Acad. Sci. U. S. A. **104**, 3719–3724 (2007). https://doi.org/10.1073/pnas.0700462104
6. Uiboupin, R., Laanemets, J.: Upwelling parameters from bias-corrected composite satellite SST maps in the Gulf of Finland (Baltic Sea). IEEE Geosci. Remote Sens. Lett. **12**, 592–596 (2015). https://doi.org/10.1109/LGRS.2014.2352397
7. Kozlov, I.E., Kudryavtsev, V.N., Johannessen, J.A., Chapron, B., Dailidien, I., Myasoedov, A.G.: ASAR imaging for coastal upwelling in the Baltic Sea. Adv. Sp. Res. **50**, 1125–1137 (2012). https://doi.org/10.1016/j.asr.2011.08.017
8. Ningsih, R.K., Syah, A.F.: Karakteristik Parameter Oseanografi Ikan Demersal Di Perairan Laut Arafura Menggunakan Data Penginderaan Jauh. Juv. Ilm. Kelaut. dan Perikan. **1**, 122–131 (2020). https://doi.org/10.21107/juvenil.v1i1.6884
9. Li, X., Zhang, X., Fu, D., Liao, S.: Strengthening effect of super typhoon Rammasun (2014) on upwelling and cold eddies in the South China Sea. J. Oceanol. Limnol. **39**, 403–419 (2021). https://doi.org/10.1007/s00343-020-9239-x
10. Anjas Swara, I.G.M., Astawa Karang, I.W.G., Indrawan, G.S.: Analisis Pola Sebaran Area Upwelling di Selatan Indonesia Menggunakan Citra Modis Level 2. J. Mar. Res. Technol. **4**, 56 (2021). https://doi.org/10.24843/jmrt.2021.v04.i01.p09
11. Pond, S., Pickard, G.L.: Introductory Dynamical Oceanography: Second Edition. Introd. Dyn. Oceanogr. Second Ed. 1–329 (2013). https://doi.org/10.1016/C2009-0-24288-7
12. Taufikurahman, Q., Hidayat, R.: Coastal upwelling in Southern Coast of Sumbawa Island, Indonesia. IOP Conf. Ser. Earth Environ. Sci. 54, (2017). https://doi.org/10.1088/1742-6596/755/1/011001
13. Tjasyono, H.B.: The Character of Rainfall in the Indonesian Monsoon. Int. Symp. Equatorial Monsoon Syst. 1–11 (2008)
14. Marpaung, S., Prayogo, T.: Analisis Arus Geostropik Permukaan Laut Berdasarkan Data Satelit Altimetri. Semin. Nas. Pengindraan Jauh. 561–567 (2014)
15. Umaroh, Anggoro, S., Muslim: The Dynamics of Sea Surface Height and Geostrophic Current in the Arafura Sea. In: IOP Conference Series: Earth and Environmental Science. Institute of Physics Publishing (2017). https://doi.org/10.1088/1755-1315/55/1/012046
16. Alawiyah E A, Sasmito B, B.N.: Analisis Pola Arus Geostropik Perairan Samudera Hindia Untuk Identifikasi Upwelling Menggunakan Data Satelit Altimetri, https://ejournal3.undip.ac.id/index.php/geodesi/article/view/19300, last accessed 2021/08/27
17. Atmadipoera, A.S., Khairunnisa, Z., Kusuma, D.W.: Upwelling characteristics during El Nino 2015 in Maluku Sea. In: IOP Conference Series: Earth and Environmental Science, p. 012018 (2018). https://doi.org/10.1088/1755-1315/176/1/012018
18. Herlina, I.R., Rahmat, H., Ahmad, B., Tania, J.: Upwelling di Laut Banda dan Pesisir Selatan Jawa serta Hubungannya dengan ENSO dan IOD. Omni-Akuatika. **12**, 1559–1602 (2016)
19. Wirasatriya, A., Susanto, R.D., Kunarso, K., Jalil, A.R., Ramdani, F., Puryajati, A.D.: Northwest monsoon upwelling within the Indonesian seas. Int. J. Remote Sens. **42**, 5437–5458 (2021). https://doi.org/10.1080/01431161.2021.1918790
20. Castelao, R.M., Barth, J.A.: Upwelling around Cabo Frio, Brazil: The importance of wind stress curl. Geophys. Res. Lett. 33, (2006). https://doi.org/10.1029/2005GL025182
21. Samad, W., Amran, M.A., Muhiddin, A.H., Tambaru, R.: Dinamika Spasial Temporal Sebaran Klorofil-A Perairan Selat Makassar Kaitannya Dengan Lokasi Penangkapan Ikan. Pros. Semin. Nas. Pengelolaan Perikan. Pelagis **2016**, 35–39 (2016)
22. Purba, N.P., Pranowo, W.S.: Dinamika Oseanografi. UNPAD Press, Deskripsi Karakteristik Massa Air dan Sirkulasi Laut (2015)
23. Ningsih, N.S., Supangat, A., Niño, E.: Karakteristik Upwelling di Sepanjang Perairan Selatan NTT Hingga Barat Sumatera. Ilmu Kelaut. Indones. J. Mar. Sci. **10**, 17–23–23 (2005). https://doi.org/10.14710/ik.ijms.10.1.17-23

Chapter 48
Analysis of the Variation of the Upper Citarum Discharge and Its Relationship with La Niña

Edi Rikardo Sinaga, Rusmawan Suwarman, Edi Riawan, Yogi S. M. Simanjuntak, and Yudha Setiawan Djamil

Abstract From 1952 to 2002, there were 21 periods of La Niña events, and 7 of them were in the strong category. One of a strong of the La Niña events occurred in 2010–2011, and the significant flood event was reported this year at the Upper Citarum watershed. Thus, the motivation of this study is to understand the impact of La Niña on discharge variations in the Upper Citarum. What is meant by the La Niña period in this study is when the ONI index is below -0.5 for at least three consecutive months. The composite La Niña period is an independent La Niña period or years that have been filtered from Indian Ocean Dipole events. The result reveals that the La Niña has the effects of increasing rainfall and the discharge over Upper Citarum in the period from August to November. At The La Niña Occurs the peak of rainfall (discharge) shifts from November to October (March to November), while the peak of the release shifted from November to October However, on the contrary the La Niña has an impact on the decreased in the period from February to July.

E. R. Sinaga (✉)
Department of Meteorology, Faculty of Earth Science and Technology, Bandung Institute of Technology (ITB), Bandung, Indonesia
e-mail: Rikardo98sinaga@gmail.com

R. Suwarman · E. Riawan
Atmospheric Science Research Group, Faculty of Earth Science and Technology, Bandung Institute of Technology (ITB), Bandung, Indonesia

Y. S. M. Simanjuntak
Department of Earth Science, Faculty of Earth Science and Technology, Bandung Institute of Technology (ITB), Bandung, Indonesia

Y. S. Djamil
Research Center for Climate and Atmosphere, National Research and Innovation Agency (BRIN), Bandung, Indonesia

E. Yulihastin et al. (eds.), *Proceedings of the International Conference on Radioscience, Equatorial Atmospheric Science and Environment and Humanosphere Science, 2021*,
Springer Proceedings in Physics 275, https://doi.org/10.1007/978-981-19-0308-3_48

48.1 Introduction

Indonesian Maritime Continent is an area where climate variability is modulated mainly by El Niño Southern Oscillation ENSO [1] and Indian Ocean Dipole (IOD) [2]. The effect of ENSO on rainfall varies spatially in Indonesia, due to their differences in responses. For example, in the east of Indonesia, there will be an increase in rainfall, however in the west of Indonesia (northern Sumatra and parts of Kalimantan) experience a decrease and increase in rain [3].

There are 21 La Niña events recorded from 1952 to 2020 and, seven of them are categorized as intense La Niña events. One of the most substantial La Niña periods occurred in 2010–2011. In 2010, the Bandung Basin or commonly known as the Upper Citarum was reported to experience the worst flooding [4] and was the highest flood frequency during the 2002–2019, period according to DIBI in 2010 [5] (Fig. 48.1). Based on seasonal data, the highest flood frequency in Bandung occurs in the March–May (MAM) season. However, according to previous studies, the increase in river discharge occurred in September–November (SON) during La Niña [6].

In the SON season, there was an increase in the discharge of the Upper Citarum river during La Niña [6]. In addition, the other studies reveal that the La Niña has an impact on the rainfall during the SON season over the western part of Java Island [2, 3, 7–9]. In addition, Aldrian et al. in [7] and Hidayat et al. in [10] La Niña has an impact on increasing rainfall for the December to February season (DJF) and the March to May season (MAM). However, this is contrary to the results of research conducted by Supari et al. in [3], which reveals that La Niña had an impact on decreasing rainfall in that season. Thus, the motivation of this study is to clarify how La Niña affects rainfall and discharge variability over the Upper Citarum River.

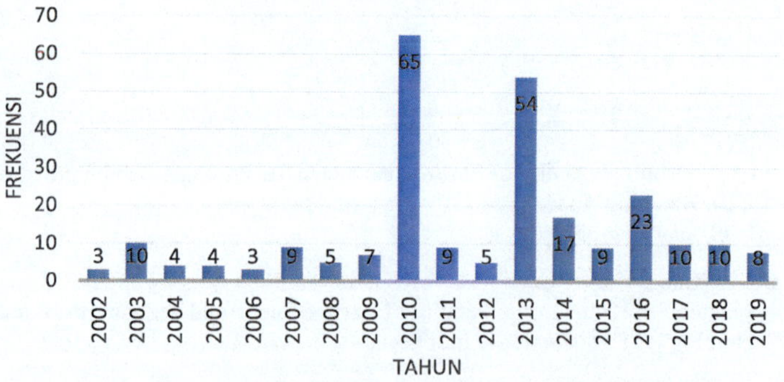

Fig. 48.1 Flood event over Citarum Hulu River (2002–2019) (*source* [5])

48.2 Data and Method

48.2.1 Data

The data used in this study include: daily in-situ observation of rainfall and river discharge data, monthly El Nino Southern Oscillation (ENSO) index, and monthly Indian Ocean Dipole (IOD) index. The detail of data used in this study can be seen in Table 48.1, while the location of the rain gauge stations and the Nanjung discharge outlet is shown in Fig. 48.2.

Table 48.1 Data used in this research

No	Data	Source	Period
1	In situ rainfall observation	BMKG	1980–2019
		PUSAIR	2000–2017
2	Discharge observation	PUSAIR	1980–2019
3	Index ENSO & IOD	National Oceanic and Atmospheric Administration (NOAA) can be accessed at [11]	1980–2019

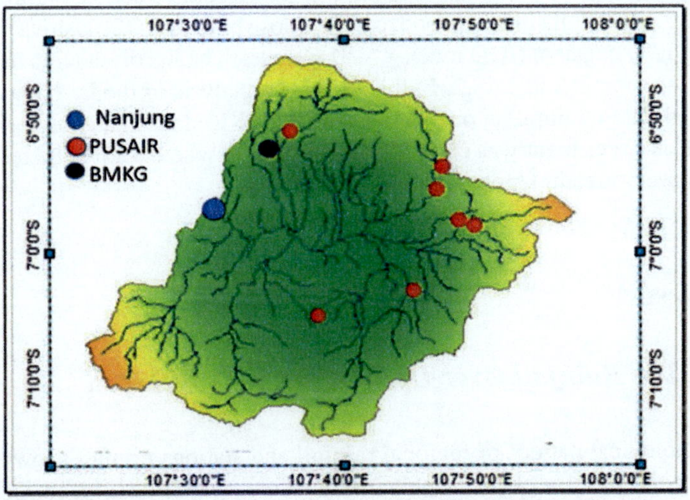

Fig. 48.2 Location of rain station and outlet Nanjung

Table 48.2 List of independent La Niña year periods

Year	Month
1983–1984	September–January
1988–1989	June–March
1995–1996	August–March
1999–2001	December–February
2005–2006	November–March
2007–2008	June–June
2008–2009	November–March
2010–2011	June–February
2011–2012	December–April
2016	August–December
2017–2018	December–April

48.2.2 Method

The period analyzed in this study is 40 years (1980–2019). An independent La Niña year is defined that there is no longer any influence from the IOD (Indian Ocean Dipole), both negative and positive IOD. [9]. The ENSO index obtained will be taken only in La Niña years with a minimum 3 consecutive active months (index of SOI $\leq -0{,}5$). The IOD index is used to decompose both the IOD positive (IOD index > 0.4) and IOD negative (IOD index ≤ -0.4) years. The list of La Niña independent years can be seen in Table 48.2. In this study, an analysis of the La Niña effect was also carried out on rainfall at one station (only BMKG station) and regional average rainfall. This experiment was conducted to answer of whether La Niña has the same impact in every area in Upper Citarum.

48.3 Result

48.3.1 The Rainfall Over Upper Citarum Watershed

The climatological pattern of regional rainfall and stations rainfall shows the same way. Peaks of rain occur twice, namely in November and March (Fig. 48.3). However, when La Niña occurs, the first peak shifts from November to October. However, the second peak of station rainfall in March continues until April, while regional rainfall experiences a downward trend in April, while regional rain experiences a downward trend in April. For the value of station rainfall and regional rainfall, Climatologically, the value of regional rainfall is higher than the value of station rainfall in the July to September period. On contrary when La Niña occurs, the station rainfall is more

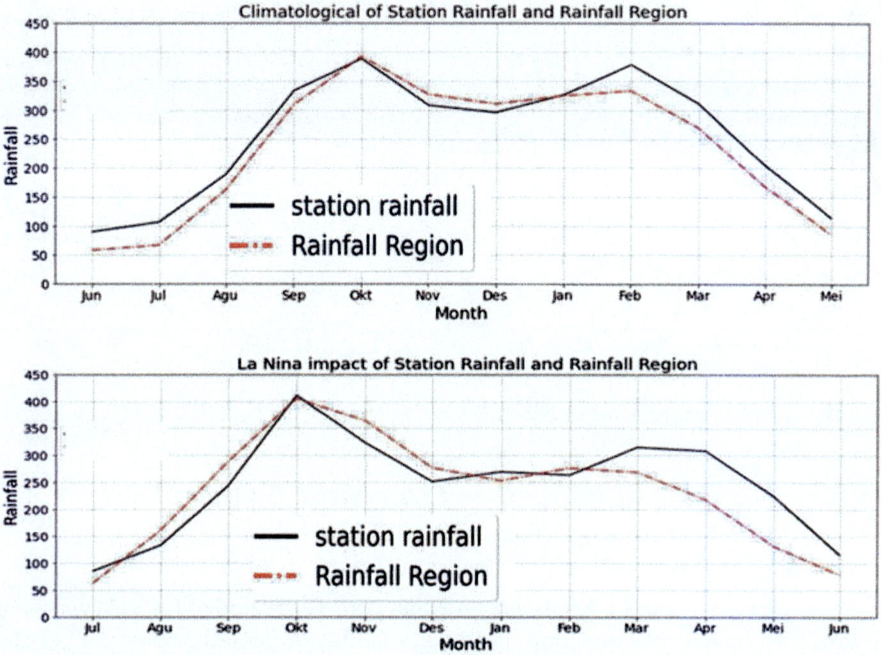

Fig. 48.3 Stational rainfall and regional rainfall in the La Niña years

significant than regional rainfall. But in the February to June period both, climatologically and when La Niña occurs, station rainfall tends to be lower than regional rainfall.

48.3.2 The Discharge of Upper Citarum in the La Niña years

The climatological pattern of the Upper Citarum river discharge is the same as the rainfall pattern (Fig. 48.4). From the climatological pattern, the discharge increases from September to March. After the peak of the discharge in March, the discharge tends to decrease. In the La Niña period, the peak occurs twice, namely November and March. However, in general, the discharge after November will tend to decrease until July. The difference between the discharge during La Niña and its climatology is when La Niña increases in August to December, and from January to July the discharge tends to decrease and will even be lower than the climatological discharge.

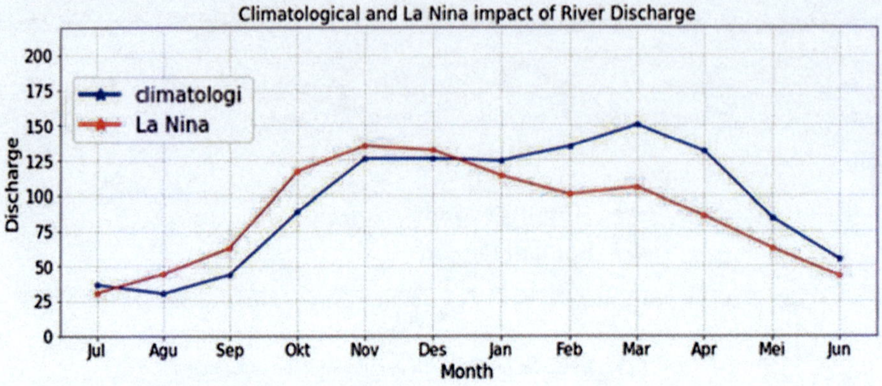

Fig. 48.4 River discharge in the La Niña years

48.3.3 The Ratio of Rainfall and Discharge Between La Niña and Climatology (Case 2000–2001)

Analysis for one year of La Niña cases in 2000–2001 is a period of La Niña years not followed by the IOD phenomenon, and the availability of rainfall data from PUSAIR from 2000–2017. So this year was taken to compare regional rainfall, station rainfall, and discharge.

In the period of La Niña from June 2000 to May 2001, an increase ratio of rainfall and discharge occurred in the period from July to October (Fig. 48.5). Both ratios, regional precipitation, and discharge, have the highest in August. However, this peak is not appeared at station rainfall. The increase in regional rain in August is nearly 240%, while for Station rainfall in September is about 120%. Figure 48.5 shows that station rainfall tends below the climatological rainfall, while regional precipitation increases almost throughout the year, except in, November and January.

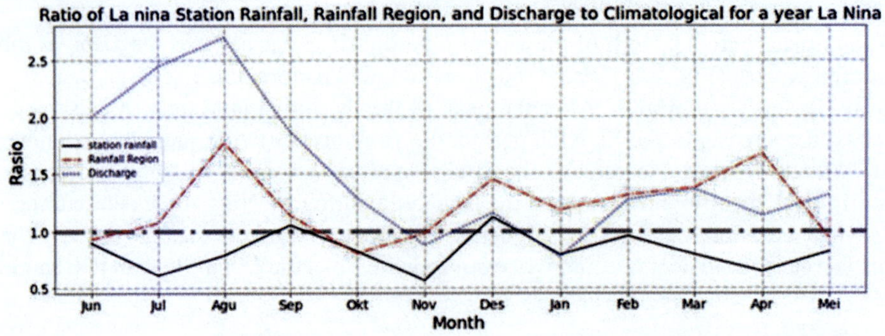

Fig. 48.5 Ratio station rainfall, region rainfall and river discharge in La Niña year (2000–2001)

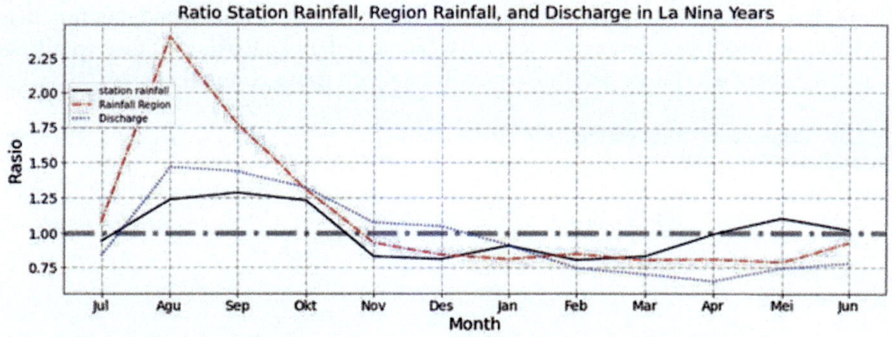

Fig. 48.6 Ratio station rainfall, region rainfall, and discharge in La Niña years

48.3.4 The Ratio of Rainfall and Discharge Between the Composite of La Niña years and Climatology

From one-year La Niña case and the composite, the station rainfall in the one-year La Niña case, the rainfall throughout the year is below the climatological level. In Contrast in the composite, there is an increase in the July to October period. At the composite rainfall station in the period July to October has increased. In the period from December to June, regional rainfall and discharge decreased in the La Niña composite. In contrast to one year in La Niña, regional rainfall and discharge increased.

In the composite analysis, both station rainfall, regional rainfall, and discharge have increased in the July to October period and will decrease in the February to June. The highest growth was in August for regional rainfall and discharge, while the highest increase in station rainfall was in September (Fig. 48.6). This is show that the station's rainfall cannot represent the discharge of the Upper Citarum River.

48.4 Conclusion

Based on the results obtained from the composite analysis of the Upper Citarum watershed rain and discharge, it shows that when La Niña occurs, the peak of rainfall (discharge) shifts from November to October (March to November). The increase in rainfall and discharge occurs in the period from August to November and will decrease in the period from February to June. From the composite analysis of La Niña ratios and one-year ratios, it is explained that there is a difference in the response of the rainfall from the Cemara station to the regional precipitation. In this case, the regional rainfall more closely follows the pattern of discharge.

In this study using observational rainfall data. We recommend adding data from high-resolution sources, such as weather radar data or satellite rainfall, so comparisons with station rainfall data will be much more comprehensive.

Acknowledgements This research is funded by the program of research, community services, and innovation (PPMI) ITB 2021.

References

1. Amirudin, A.A., Salim, E., Tangang, F., Juneng, L., Zuhairi, M.: Differential influences of teleconnections from the Indian and Pacific Oceans on rainfall variability in Southeast Asia. Atmosphere **11**(9), 886 (2020)
2. Hamada, J.I., Yamanaka, M.D., Matsumoto, J., Fukao, S., Winarso, P.A., Sribimawati, T.: Spatial and temporal variations of the rainy season over Indonesia and their link to ENSO. J. Meteorol. Soc. Jpn. Ser. II **80**(2), 285–310 (2002)
3. Supari, Tangang, F., Salimun, E., Aldrian, E., Sopaheluwakan, A., Juneng, L.: ENSO modulation of seasonal rainfall and extremes in Indonesia. Clim. Dyn. **51**(7), 2559–2580 (2018)
4. Deltares.: UCBFM Annex D,1D Modelling studies discharge extremes including 2009–2010. TA 7189-INO. Institutional Strengthening For Integrated Water Resources Management (IWRM), Bandung (2010)
5. Data Informasi Bencana Indonesia, https://dibi.bnpb.go.id/, 2021/08/20
6. Sahu, N., Behera, S.K., Yamashiki, Y., Takara, K., Yamagata, T.: IOD and ENSO impacts on the extreme stream-flows of Citarum river in Indonesia. Clim. Dyn. **39**, 1673–1680 (2012)
7. Aldrian, E.: Spatial patterns of ENSO impact on Indonesian rainfall. Jurnal Sains & Teknologi Modifikasi Cuaca **3**(1), 5–15 (2002)
8. Lestari, S., King, A., Vincent, C., Karoly, D., Protat, A.: Seasonal dependence of rainfall extremes in and around Jakarta, Indonesia. Weather Clim. Extremes **24**, 100202 (2019)
9. Kurniadi, A., Weller, E., Min, S.K., Seong, M.G.: Independent ENSO and IOD impacts on rainfall extremes over Indonesia. Int. J. Climatol. **41**(6), 3640–3656 (2021)
10. Hidayat, A.M., Efendi, U., Agustina, L., Winarso, P.A.: Korelasi Indeks Nino 3.4 dan Southern Oscillation Index (SOI) dengan Variasi Curah Hujan di Semarang. Jurnal Sains & Teknologi Modifikasi Cuaca **19**(2), 75–81 (2018)
11. Golden Gate Weather Services, https://ggweather.com/enso/oni.htm, 2021/09/15

Chapter 49
Evolution of Mesoscale Convective Complex and Its Atmospheric Conditions During Heavy Rain in Bandung, 22 February 2018

Fauziah Fangia Indra and Trismidianto

Abstract On February 22, 2018, there was a flood due to prolonged heavy rain in the Bandung Regency area (eastern part), which was identified as having a mesoscale convective complex. This paper describes the mesoscale convective complex that causes the flood and analyzes several weather parameters and conditions of global atmospheric phenomena when there are mesoscale convective clouds. The data used are rainfall data from GSMaP, cloud data from Himawari-8, and weather parameter data from ECMWF ERA-Interim. This study indicates that the rainfall at the time of the flood is included in the category of weighty rain caused by the mesoscale convective complex. The MCC began to develop from around 1600 LST from small clouds triggered by the convergent flow of inland winds that continued to grow and develop into a maximum-scale MCC. At 1900-2000 LST and decayed at midnight (2100-0100 LST). The formation of mesoscale convective clouds is supported by atmospheric conditions in divergence, vorticity, and vertical motion. The high specific humidity in the study area also allows the formation of convective clouds that produce heavy rain. There are convergent surface wind flows as the local convergence but not related with land breeze triggered the development of some of the clouds over Bandung Regency and its surrounding area. At the same time, the strong southerly wind on the surface triggered the development of this MCC. During MCC, reach the maximum stage, there is an upward with a high enough specific humidity that allows the formation of convective clouds that cause rain. At the same time, the vorticity in Bandung Regency, the level of 700 hPa, was negative. Several new convective systems were generated during the MCC decay stage at midnight due to the cold pool formed under the MCC system, which made the temperature under the MCC system very cold compared to the surrounding temperature.

F. F. Indra
Meteorogical Study Program, Bandung Institute of Technology, Bandung, Indonesia

Trismidianto (✉)
Research Center for Climate and Atmosphere, National Research and Innovation Agency (BRIN), Bandung, Indonesia
e-mail: trismidianto@brin.go.id

E. Yulihastin et al. (eds.), *Proceedings of the International Conference on Radioscience, Equatorial Atmospheric Science and Environment and Humanosphere Science, 2021*, Springer Proceedings in Physics 275, https://doi.org/10.1007/978-981-19-0308-3_49

49.1 Introduction

On February 22, 2018, several sub-districts in Bandung Regency were inundated by floods due to the overflowing Citarum River, which was triggered by persistent heavy rainfall from afternoon to morning the next day. The heavy rainfall is possibly caused by the Mesoscale convective complex (hereafter MCC) phenomenon, which is a particular case of mesoscale convective systems, which are the largest of the convective storms that form when clouds occur in response to convective instability amalgamate and organize upscale into a single cloud system with a huge upper cirriform cloud structure and rainfall covering large contiguous rain areas [1, 2]. Some of the papers examined showed that the passage of an MCC changes the surface conditions over a large region and produce heavy rainfall must be regarded as a cause of several of the hazards (flooding, localized flash floods, mudslides, and property and crop damage) produced by MCCs [e.g., 1, 3]. MCCs also play a significant role in the rainfall regimes of the tropics [4, 5]. The convection related to sea breeze convergence can aggregate into MCCs later in the days, which move off land to give the most significant precipitation during the local morning time over the oceans [6].

MCC can result in bad weather and continuous rain [1, 3, 7, 8]. Several previous studies have also concluded that MCC affects rainfall at the location where MCC occurs and affects rainfall in the area around the MCC occurrence [9]. MCC over the Indian Ocean affects rainfall along the western part of Sumatra Island. In another paper [8], while MCC above Kalimantan Island tends to migrate to Java Island when MCC is extinct, thus affecting the intensity of rainfall along Java Island [10]. MCC contributes in significant rainfall [7]. In Argentina and Northern Paraguay, MCC contributes 20% to the total precipitation [11], which is the same percentage as MCC contributes to Indonesian Maritime Continent areas [9]. During its life cycle, MCC tends to produce heavy rainfall [12].

The analyses of the environmental condition during MCC events have been described by several previous researchers in several regions in the world (e.g., Maddox 1983; Merritt 1985; Kane et al. 1987; Cotton et al. 1989; Augustine and Howard 1991; Laing 1996). In general, they found that the environmental conditions among some sites are very similar during MCC events, but several areas have different environmental conditions during MCC events. They show that specific thermodynamic patterns and dynamical features are usually present in large-scale environments.

However, from several studies that have been described above, no one has explained in detail the growth and movement of MCC and analysis of environmental conditions that are above Java Island. The study aims to explain in detail the development of MCC on the island of Java and the mechanism of its movement in causing heavy rainfall over Java Island on February 22, 2018, which caused flooding, in addition to explaining the analysis of the growth and development of MCC, this study will also explain the environmental condition of the atmospheric parameters during the MCC event.

49.2 Data and Method

In this study, we use black body temperature (hereafter T_{BB}) from the Himawari geostationary satellites, provided by the Japan Meteorological Agency (JMA), which supports weather forecasting, tropical cyclone tracking, and meteorology research. Most meteorological agencies in East Asia, Southeast Asia, Australia, and New Zealand use the satellites for their weather monitoring and forecasting operations. That satellite data has spatial and time resolutions $0.05° \times 0.05°$ and one hour, respectively. This data is used to identify MCC and analyze the development and movement of MCC. This study adapts the method from Trismidianto [8] to the identification of MCC with input temperature, latitude, and longitude values for each cloud shield pixel that obtained from IR to a modified version of a computerized MCC program for measuring the satellite-observed characteristics of MCCs [1], which stated that MCC has cloud shield with continuously $T_{BB} \leq -32$ °C (in this study converted to 241 K) must have an area $\geq$ 100,000 km^2 and the interior cold cloud $T_{BB} \leq -52$ °C (in this study converted to 221 K) must have an area $\geq$ 50,000 km^2. To distinguish MCC from the other MCS kind, thus, the MCC will have a elliptical with eccentricity ≥ 0.7 and a life cycle of MCC about ≥ 6 h. This program computes the area, eccentricity, and centroid positions of the − 32 °C and − 52 °C cold cloud shield associated with deep convection. The areal extend and eccentricity values were used to determine the duration and life cycle of the MCCs, while the centroid position indicated the path of propagation of the systems. From this method, it is found that the mesoscale convective event on February 22, 2018, is one type of MCC. During the mature stage, this MCC had a cloud shield with an area of around 214,755 km^2, and the interior cold cloud covered an area of around 70,137 km^2. The center of the MCC was around 7.21°S, 107.96°E with an eccentricity of around 0.73 and life cycle duration around 8 h, short-lived MCC.

The estimated rainfall data, corresponding to the MCCs, were obtained from the Global Satellite Mapping of Precipitation (GSMaP), provides a global hourly rain rate with a 0.1×0.1-degree resolution. GSMaP is a product of the Global Precipitation Measurement (GPM) mission, which offers global precipitation observations at three-hour intervals. Values are estimated using multi-band passive microwave and infrared radiometers from the GPM Core Observatory satellite and with the assistance of a constellation of other satellites. GPM's precipitation rate retrieval algorithm is based on a radiative transfer model. The gauge-adjusted rate is calculated based on optimizing the 24 h accumulation of GSMaP hourly rain rate to daily precipitation by NOAA/CPC gauge measurement. This data can be downloaded from https://sharaku.eorc.jaxa.jp/GSMaP/. As a validation of GSMaP data, this study uses observational data from the OGIMET synoptic data as a meteorological database 24. This data can be downloaded from http://www.ogimet.com/synopsc.phtml.en.

To analyses, the surface wind is utilized several data from Cross-Calibrated Multi-Platform (CCMP), which are available at 6-hourly intervals with 0.25° horizontal resolution and the environment's condition during MCC event is used several data from the European Centre for Medium-Range Weather Forecasts (ECMWF) ERA-Interim analysis fields, which are available at 6-hourly intervals with 1° × 1°° horizontal resolution [13].

49.3 Result and Discussion

49.3.1 T_{BB}, Rainfall and Surface Wind Data Analysis

49.3.1.1 MCC Evolution Analysis

Figure 49.1 shows that the indications of the occurrence of MCC in this case study began to appear over Bandung district at 1600 LST (Local Solar Time) refer to UTC + 7 h, on February 22, 2018, which was then referred to as the initial phase of the stage of the occurrence of MCC. These small-scale clouds began to grow and develop for about 2 h from 1700 to 1800 LST, and then this phase is referred to as the growth phase. MCC starts to reach its maximum size at midnight at 1900–2000 LST, starting now referred to as the MCC mature phase. From the time of the mature stage, it can conclude that the MCC activity that occurs, in this case,

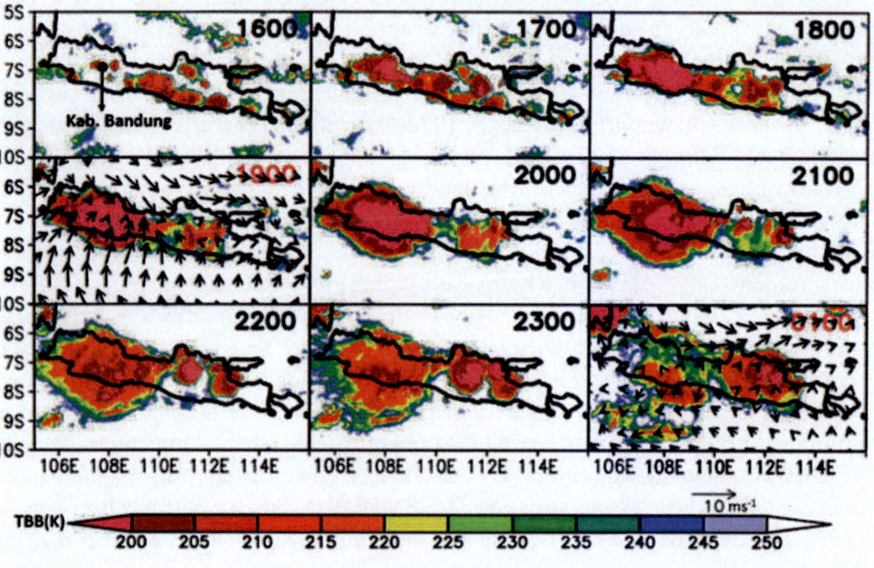

Fig. 49.1 Horizontal distribution of black body temperature (T_{BB}) for Mesoscale Convective Complex (MCC) criteria from infrared data obtained by Himawari 8 over Java Island on 1600–2300 LST, 22 February and 0100 LST, 23 February, 2018 (shaded) and wind vector (vector) from Cross-Calibrated Multi-Platform (CCMP) data during the occurrence of the Mesoscale Convective Complex (MCC) on 1900 LST 22 February, 2018 and 0100 LST, 23 February, 2018

is related to diurnal convection, which is also known as a diurnal cycle or diurnal variation. Most previous studies have shown that diurnal variation in the tropics peaks at night in land areas and the morning in sea areas [14–16]. Diurnal variations in the Indonesian Maritime Continent close relationship with regional rainfall distribution [17–19]. Diurnal variations in convective activity occur almost daily. In addition, the complex topography of the Indonesian Maritime Continent causes strong onshore wind circulation.

At 1900 LST, there were convergent surface wind flows as the local convergence but not related with land breeze triggered the development of some of the clouds over Bandung Regency and its surrounding area. At the same time, the strong southerly wind on the surface started the development of this MCC. Convergence means that more air is flowing into that region than flowing out. Since the air has to go somewhere, it rises, and rising air induces precipitation development. This local convergence is caused by a cold pool which, according to Engerer et al. [20], is an area of downdraft air cooled by evaporation that spreads out horizontally beneath a precipitating cloud. Also, the cold pools were associated with potential temperature decreases [21] and generated by these individual convective cells in an MCS typically spread out over the surface and combine to form a large mesoscale cold pool covering a contiguous area on the scale of the entire MCS [2]. MCC begins to decay at 2100-0100 LST, which is where the phase of MCC's extinction is also known as the decay phase. In this phase, MCC begins to break up into several new convective systems. The new convective system is believed to induce convective clouds around it so that it looks to develop into several reasonably large convective clouds towards the east.

Figure 49.2 indicates that the pattern of rainfall that occurs during the MCC event has the same pattern as the horizontal distribution of TBB, which has been shown in Fig. 49.1, where very dense rainfall or extreme rainfall occurs during the maximum phase of the MCC based on rainfall category of BMKG that shows on Table 49.1. This shows that MCC caused extreme rainfall over Bandung district on February 22, 2018, which resulted in flooding. This is also following the previous paper that MCC tends to cause extreme rainfall that can cause flooding [8, 9, 12]. The rainfall value also increases as the size of the MCC reaches its maximum, its intensity decreases as the MCC decays, and high rainfall reappears east of the MCC location due to the formation of new convection cells when the MCC decays. This indicates that MCC also contributes to diurnal rainfall on the mainland of Java Island, thus proving why more rainfall on the mainland occurs at night on Java Island. The movement of heavy rainfall also follows the movement pattern of the MCC.

49.3.1.2 MCC Propagation Analysis

Figures 49.3a, b show that the convective activity and rain clouds began to appear from 1600 LST over Bandung regency and became gradually more active and stronger during a mature stage in the morning over the MCC system area, which followed the increase of average rainfall. From this figure, it can see that the growth of MCC is

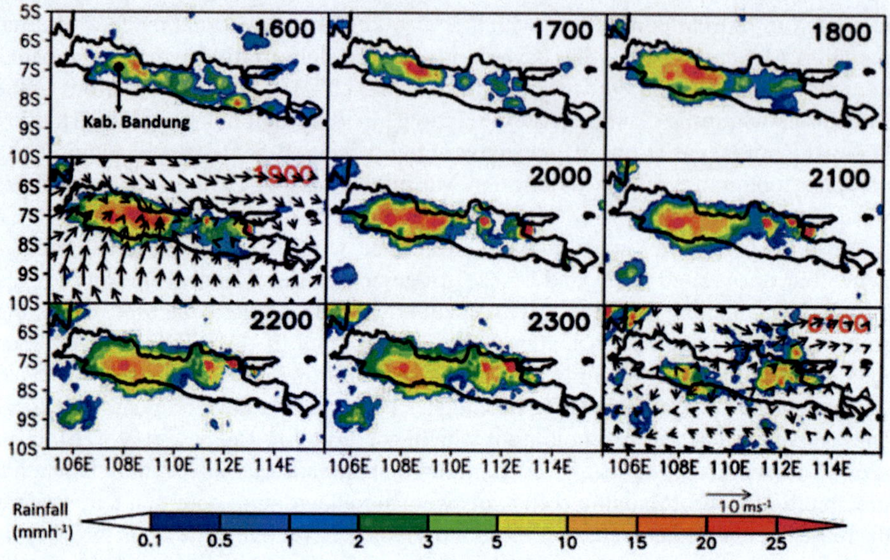

Fig. 49.2 Identical with Fig. 49.1 but for horizontal distribution of rainfall (shaded) from GSMaP data. Unit for rainfall is mm/h

Table 49.1 Rainfall threshold according to BMKG

Rainfall category	Intensity of rainfall	
	(mm/hour)	(mm/day)
Light	1–5	5–20
Medium	5–10	20–50
Heavy	10–20	50–100
Very dense	> 20	> 100

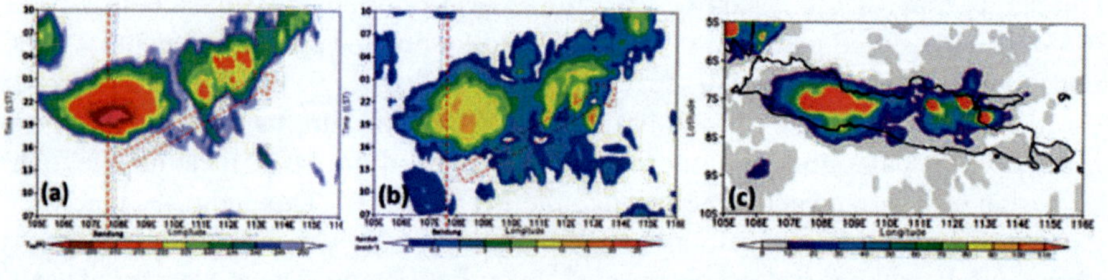

Fig. 49.3 Time-longitude cross-sections (averaged for latitude 7°S–5°S) during the MCC event, of; **a** T_{BB} from infrared data of Himawari-8 (unit Kelvin), **b** rainfall from GSMaP (unit mm/h), and **c** Horizontal distribution of rainfall accumulated during MCC event. Rainfall data obtained from GSMaP, unit for rainfall is mm /12 h

triggered by the growth of local convection without any support from the surrounding convective clouds because there is no convective activity around it. The growth of convective clouds is also followed by rain clouds with weighty or extreme intensity. During decay, new convective cells are formed, which cause movement and growth of other convective clouds towards the east. It can be said that there is eastward-propagating MCC in this case. The activity of the MCC is also followed by the growth and movement of rain clouds, which also causes high rainfall along with its movement so that there are also some weighty rains in several locations around the MCC after MCC decay. Figure 49.3c shows the rainfall accumulation during the MCC events from the Pre-MCC stage until 6 h after the post-MCC stage. The greatest rainfall accumulation is presented in the center of the MCC system area. It indicated that the significant influence of MCC on the rainfall is during MCC reached maximum extent in the mature stage. However, the high rainfall accumulation also appears in several regions near the MCC system area. It indicated that the effect of MCCs is not only over the MCC system area but also influencing the rainfall in the surrounding area of the MCC systems. In this research, the area in the MCC systems that get affected by the MCC is mentioned as affected area by the MCC.

49.3.2 The Environmental Condition During MCC Events

49.3.2.1 Analysis of Divergence, Geopotential Height and Streamline During MCC Event

Divergence is defined as the movement of air masses that spread over an area. Conversely, if the movement of air masses collects, then convergence occurs. A positive divergence value indicates a horizontal spreading movement, and conversely, a negative divergence value indicates a convergence or a meeting/compression of air masses. In the image below, the divergences are indicated by shaded, and streamlines are indicated by black lines. The streamline map shows shear lines around the Bandung area. A shear line is an area where there are sudden changes in wind direction and speed where the wind speed tends to decrease, causing a buildup of air masses that cause extreme weather [22].

Figure 49.4 shows the divergence at the low level (850-hPa). At 0700 LST in the Bandung Regency area, there was a negative divergence value. A negative value indicates that there has been a convergence in this region, which indicates the large number of air masses in this region. The convergent southerly wind flow supports the development of the MCC at 1900 LST, and there was a low-pressure center in the south of the island of Java, which could trigger the growth of rain clouds in the Bandung Regency area. At the medium level (500-hPa), negative divergences are seen at 0700 LST and 1300 LST. A negative value indicates that there has been a convergence, which indicates the large number of air masses in this region. In this layer, there is a low-pressure center at 07.00 LST, which can trigger the growth of rain

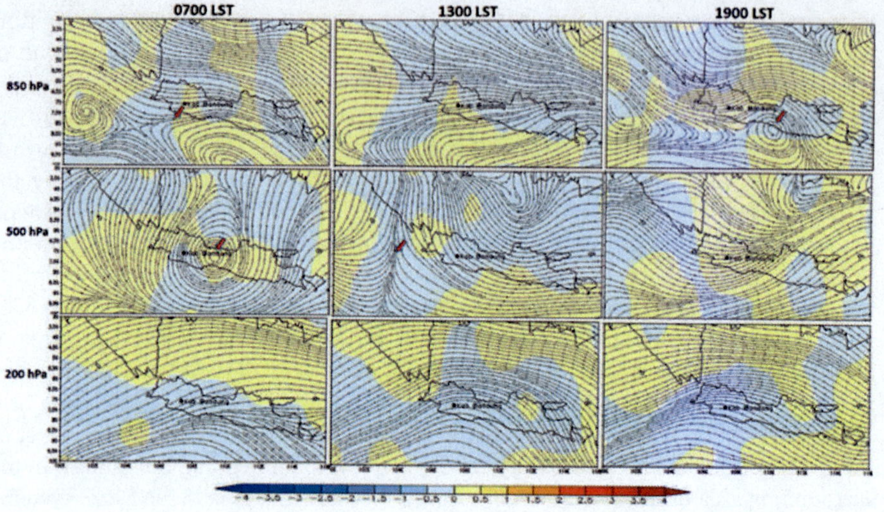

Fig. 49.4 Horizontal distribution of divergence (10^{-5} s^{-1}, shaded) and wind streamline (ms^{-1}, streamline) in pressure level 850, 500 and 700-hPa before MCC reaches maximum

clouds around Bandung Regency. At the high level (200 hPa), negative divergences are seen at 07.00 LST, 13.00 LST, and 19.00 LST.

Figure 49.5 shows that at 1900 LST in the Bandung Regency area at the high level of 200-hPa, the divergence is negative with relatively low geopotential from the surrounding area, the low geopotential height coincides with low pressure, as well as at the 700-hPa level. At this level, it is indicated that the occurrence of a convergence event is seen from a negative divergence. At the same time at the low level of 850-hPa, the divergence is positive, and this indicates that there is no condensation process and the formation of convective clouds. At 0100 LST at the high level of 200-hPa, the divergence was negative with relatively higher geopotential than the surrounding area. At the same time, at the 700 and 850-hPa levels, the divergence is positive. This shows that at the 700 and 850 hPa levels, there is no condensation process.

49.3.2.2 Analysis of Vertical Velocity and Vorticity During MCC Event

Figure 49.6 is a spatial plot of vertical velocity at the lower level of 850 hPa and medium level 700 hPa. At the 850 hPa level at 1900 LST, February 22, 2018 in Bandung Regency, the vertical velocity is negative with an absolute value of vertical velocity, which is greater than the surrounding area of 0.15 with a specific humidity of 13–13.5 (g/kg). At 0100 LST, February 23, 2018, the vertical velocity was also negative with relatively small absolute value of the vertical velocity of 0.06 Pa/s with a specific humidity of 12.5–13 (g/kg). At the 700 hPa level at 1900 LST, February 22, 2018 in Bandung Regency, the vertical speed is negative with relatively small absolute value of vertical speed, which is 0.09 with a specific humidity of 6.5–7

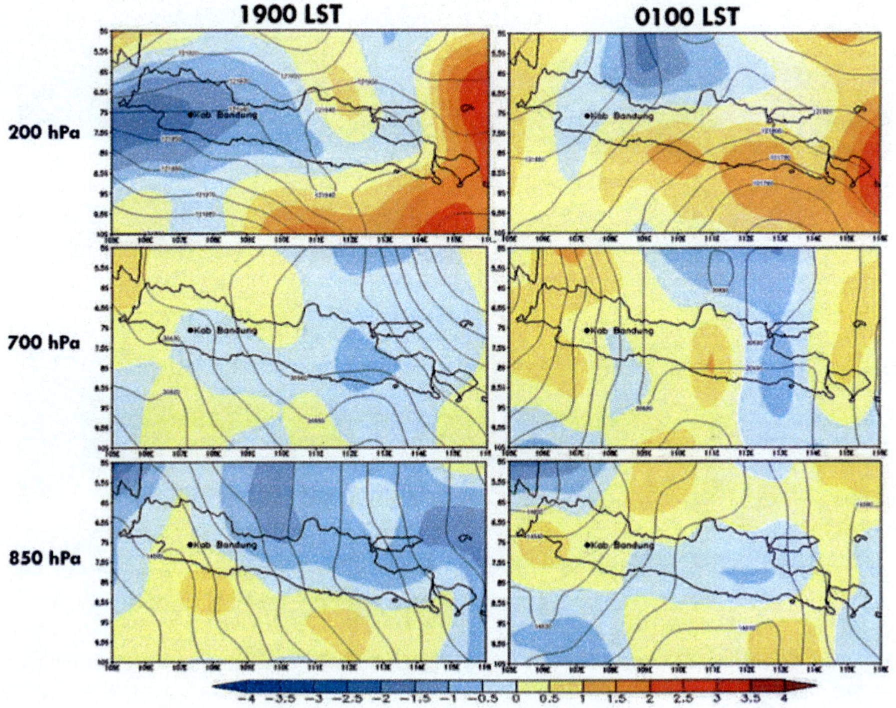

Fig. 49.5 Horizontal distribution of divergence (10^{-5} s^{-1}, shaded) and geopotential height (m, contour) in pressure level 850, 500, and 700-hPa during MCC reaches maximum and decay phase at 1900 LST and 0100 LST, respectively

(g/kg). At 0100 LST, February 23, 2018, the vertical speed was positive (0.03) with a specific humidity of 5.5–6 (g/kg). The vertical velocity with a negative value indicates an upward and a positive value indicates a downward. At the 850 hPa level at 1900, and 0100 LST and the 750 hPa level at 1900 LST, there is an upward with a high enough specific humidity that allows the formation of convective clouds that cause rain.

Figure 49.7 is a spatial plot of vorticity at the lower level of 850 hPa and medium level 700 hPa. Vorticity is closely related to the potential for the lifting of air masses in supporting the convection process that causes convective clouds. On February 22, 2018, in the 850 hPa level, the vorticity value in the Bandung Regency area showed a negative value between -1×10^{-5} and -1.5×10^{-5} Pa/s. Vorticity conditions are closely related to the potential for lifting air masses to support the convection process cause convective clouds. On February 22, 2018, at 19.00 LST, the vorticity in Bandung Regency, the level of 700 hPa was negative, ranging from -1.5×10^{-5} s^{-1} to -2×10^{-5} s^{-1} and the level of 850 hPa was negative, ranging from -0.5×10^{-5} s^{-1} to -1.5×10^{-5} s^{-1}. On February 23, 2018, at 01.00 LST, the vortices in Bandung Regency at 700 hPa, and 850 hPa levels were also negative, ranging from

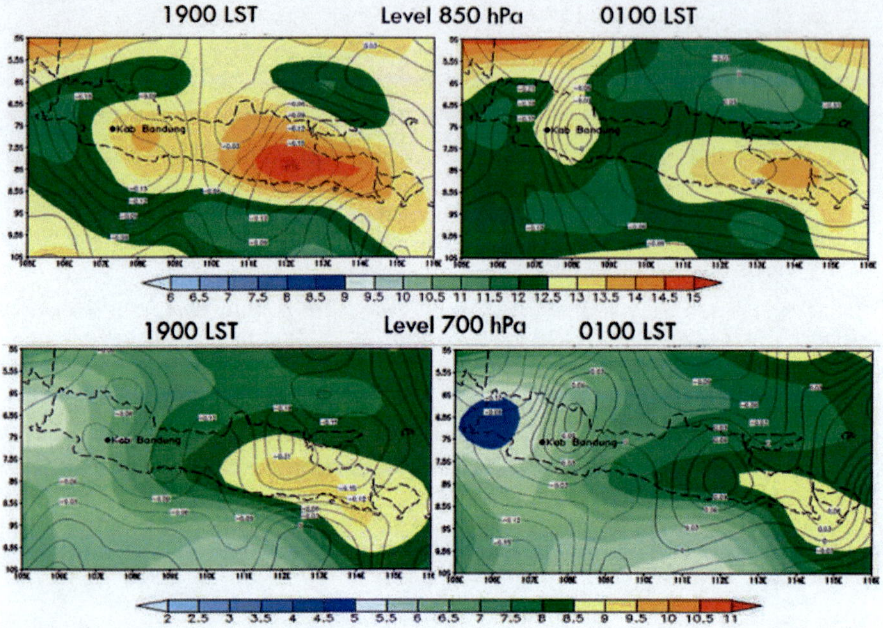

Fig. 49.6 Horizontal distribution of vertical velocity (10^{-2} Pa s^{-1}, shaded) and wind vector (ms-1, vector) in pressure level 850 and 700-hPa during MCC reaches maximum and decay phase at 1900 LST and 0100 LST, respectively

-0.5×10^{-5} s^{-1} to -1.5×10^{-5} s^{-1}. For southern latitudes, negative vortices indicate the presence of cyclonic, which results in convective clouds and vice versa for northern latitudes. Bandung area is located in the southern hemisphere, so that the negative value of vortices indicates an upward vertical motion, resulting in air masses being lifted and increasing the intensity of convective cloud growth in the 850 Mb and 500 Mb layers.

49.4 Conclusion

This study analyzes the occurrence of heavy rains in Bandung Regency caused by the MCC on February 22, 2018. The MCC began to develop from around 1600 LST from small clouds triggered by the convergent flow of inland winds that continued to grow and develop into a maximum-scale MCC. At 1900-2000 LST, which was triggered by the strengthening of the convergent wind flow around the Bandung Regency area towards the MCC system area. There are convergent surface wind flows as the local convergence but not related with land breeze triggered the development of some of the clouds over Bandung Regency and its surrounding area. At the same time, the strong southerly wind on the surface triggered the development of this MCC.

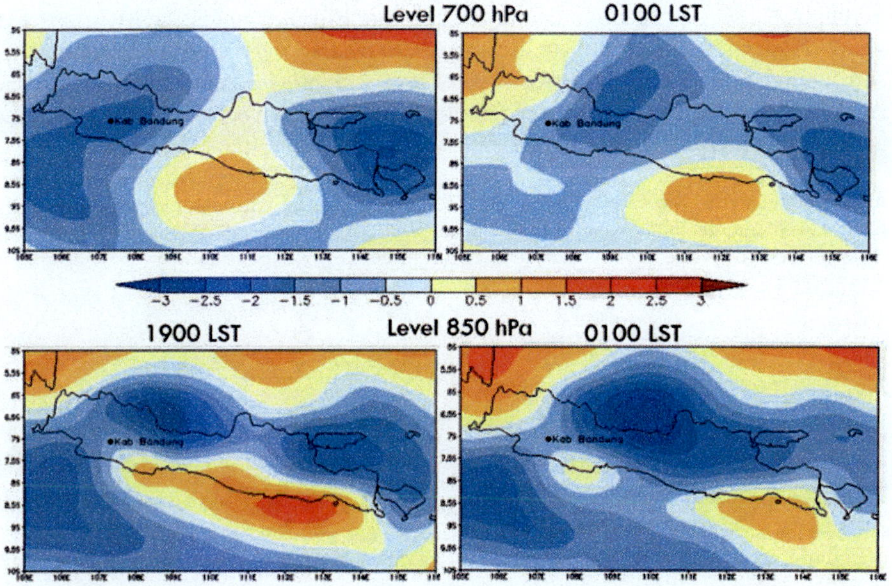

Fig. 49.7 Horizontal distribution of vorticity (10^{-5} s^{-1}, shaded) and wind vector (ms^{-1}, vector) in pressure level 850 and 700-hPa during MCC reaches maximum and decay phase at 1900 LST and 0100 LST, respectively

Convergence means that more air is flowing into that region than flowing out. Since the air has to go somewhere, it rises, and rising air induces precipitation development. The streamline map shows the presence of shear lines around the Bandung area. A shear line is an area where there are sudden changes in wind direction and speed where the wind speed tends to decrease, causing a buildup of air masses that cause extreme weather.

During MCC reach the maximum stage, there is an upward with a high enough specific humidity that allows the formation of convective clouds that cause rain. At the same time, the vorticity in Bandung Regency, the level of 700 hPa was negative, ranging from $-\ 1.5 \times 10^{-5}\ s^{-1}$ to $-\ 2 \times 10^{-5}\ s^{-1}$, and the level of 850 hPa was negative, ranging from $-\ 0.5 \times 10^{-5}\ s^{-1}$ to $-\ 1.5 \times 10^{-5}\ s^{-1}$. On February 23, 2018, at 01.00 LST, the vortices in Bandung Regency at 700 and 850 hPa levels were also negative, ranging from $-\ 0.5 \times 10^{-5}\ s^{-1}$ to $-\ 1.5 \times 10^{-5}\ s^{-1}$. For southern latitudes, negative vortices indicate the presence of cyclonic, which results in convective clouds and vice versa for northern latitudes. Bandung area is located in the southern hemisphere, so that the negative value of vortices indicates an upward vertical motion, resulting in air masses being lifted and increasing the intensity of convective cloud growth in the 850 and 500 hPa layers.

This MCC growth was also followed by the strengthening of convective activity, especially in the center area of the MCC system, which also caused heavy rains. Several new convective systems were generated during the MCC decay stage at midnight (2100 – 0100 LST) due to the cold pool formed under the MCC system, which made the temperature under the MCC system very cold compared to the surrounding temperature. The existence of a very significant temperature difference causes the wind flow out of the cold pool to form new convective clouds, which the approach is similar to the front process.

Acknowledgements We are grateful to ECMWF ERA-Interim (http://www.ecmwf.int/), Himawari-8 satellite (http://weather.is.kochiu.ac.jp/sat/GAME) and GSMaP satellite (http://sharaku.eorc.jaxa.jp/GSMaP) for available data to the public.

References

1. Maddox, R.A.: Mesoscale convective complexes. Bull. Am. Meteor. Soc. **61**, 1374–1387 (1980)
2. Houze, R.A. Jr.: Mesoscale convective systems Rev. Geophys. 42 (2004)
3. McAnelly, R.L., Cotton, W.R.: The precipitation life cycle of mesoscale convective complexes over the Central United States. Mon. Weather Rev. **117**, 784–808 (1989)
4. Houze, R.A., Jr.: Structure and dynamics of a tropical squall line system. Mon. Weather Rev. **105**, 1540–1567 (1977)
5. Chong, M.P., Amayenc, G.S., Testud, J.: A tropical squall line observed during the COPT 81 experiment in West Africa. Part I: kinematic structure inferred from dual-Doppler radar data. Mon. Weather Rev. **115**, 670–694 (1987)
6. Williams, M., Houze, R.A., Jr.: Satellite-observed characteristics of winter monsoon cloud clusters. Mon. Weather Rev. **115**, 505–519 (1987)
7. Fritsch, J.M., Kane, R.J., Chelius, C.R.: The contribution of mesoscale convective weather system to the warm season precipitation in the United States. J. Clim. Appl. Meteorol. **25**, 1333–1345 (1986)
8. Trismidianto, Tri, W.H., Ishida, S., Qoosaku, M., Atsuyoshi, M., Satoshi, I.: Development processes of oceanic convective systems inducing the heavy rainfall over the Western Coast of Sumatra on 28 October 2007. Sci. Online Lett. Atmos. (SOLA) **12**, 6−11 (2016). https://doi.org/10.2151/sola.2016-002
9. Trismidianto, E., Yulihastin, H., Satyawardhana, J., Nugroho, T., Ishida, S.: The Contribution of the Mesoscale Convective Complexes (MCCs) to total rainfall over Indonesian Maritime Continent. IOP Conf. Ser. Earth Environ. Sci. **54** (2017).https://doi.org/10.1088/1755-1315/54/1/012027
10. Trismidianto, Yulihastin, E., Satyawardhana, H., Ishida, S.: A composite analysis of the Mesoscale Convective Complexes (MCCs) development over the Central Kalimantan and its relation with the propagation of the rainfall systems. IOP Conf. Ser. Earth Environ. Sci. **54** (2017b).https://doi.org/10.1088/1755-1315/54/1/012036
11. Durkee, J.D., Mote, T.L., Shepherd, M.J.: The contribution of mesoscale convective complexes to rainfall across, subtropical South America. Int. J. Clim. **22** (2009)
12. Ashley, W.S.: A distribution of mesoscale convective complexe rainfall in the United State. Am. Meteorol. Soc. **131**, 3003–3301
13. Dee, D.P., Uppala, S.M., Simmons, A.J., Berrisford, P., Poli, S., Kobayashi, S., Andrae, U., Balmaseda, M.A., Balsamo, G., Bauer, D., Bechtold, P., Beljaars, A.C.M., Van de Berg, L., Bidlot, J., Bormann, N., Delsol, C., Dragani, R., Fuentes, M., Geer, A.J., Haimberger, L., Healy, S.B., Hersbach, H., Holm, E.V., Isaksen, L., Kallberg, K., Kohler, M., Matricardi, M., McNally, A.P., Monge-Sanz, B.M., Morcrette, J.J., Park, B.K., Peubey, C., De Rosnay, P., Tavolato, C., Thepaut, J.N., Vitart, F.: The ERA-interim reanalysis: configuration and performance of the data assimilation system Quart. J. Roy. Meteor Soc. **137**, 553–597 (2011). https://doi.org/10.1002/qj.828

14. Gray, W.M., dan Jacobson, R.W.: Diurnal variation of deep convective system in the tropics. Mon. Weather Rev. **105**, 1171–1188 (1977)
15. Murakami, M.: Analysis of the deep convective activity over the western Pacific and Southeast Asia. Part I: Diurnal variation. J. Meteorol. Soc. Jpn. **61**, 60–75 (1983)
16. Houze, R.A., Jr., Geotis, S.G., Marks, F.D., Jr., West, A.K.: Winter monsoon convection in the vicinity of north Borneo. Part I: structure and time variation of the clouds and precipitation. Mon. Weather Rev. **109**, 1595–1614 (1981)
17. Yang, G.-Y., Slingo, J.: The diurnal cycle in the tropics. Mon. Weather Rev. **129**, 784–801 (2001)
18. Collier, J.C., Bowman, K.P., North, G.R.: A comparison of tropical precipitation simulated by the community climate model with that measured by the tropical rainfall measuring mission satellite. J. Clim. **17**, 3319–3333 (2004)
19. Kikuchi, K., Wang, B.: Diurnal precipitation regimes in the global tropics. J. Clim. **21**, 2680–2696 (2008)
20. Engerer, N.A., Stensrud, D.J., Coniglio, M.C.: Surface characteristics of observed cold pools. Mon. Weather Rev. **136**, 4839–4849 (2008). https://doi.org/10.1175/2008MWR2528.1
21. Tompkins, A.M.: Organization of tropical convection in low vertical wind shears: the role of cold pools. J. Atmos. Sci. **58**, 1650–1672 (2001)
22. Hidayat, R., Ando, K., Masumoto, Y., Luo, J.J.: Interannual variability of rainfall over Indonesia: impacts of ENSO and IOD and their predictability. In: Workshop and International Seminar on Science of Complex Natural Systems, vol. 31 (2016). https://doi.org/10.1088/1755-1315/31/1/012043

Chapter 50
Flare Potentiality Associated to Different Sunspot Groups During Solar Cycle 24 Observed by BPAA Pasuruan

Amatul Firdausya Nur Cahyaningtyas, Siska Filawati, and Silvi Oktavia Hanum

Abstract Sunspot evolution is explained in the context of the development of active region undergoing morphological changes. The Zurich classification is a classification of sunspot groups based on their size, shape, and structure. Meanwhile, flare is an increase in the intensity of radiation emitted in a short time in the area around sunspots that can trigger Coronal Mass Ejection and cause radio signal disturbances in the Ionosphere. Therefore, it is essential to study sunspots and flares as a mitigation effort to anticipate the negative impact of sunspots and flares. This study uses flare data from NOAA and sunspot data from BPAA Pasuruan. The solar cycle 24, starting from 2008 to 2019, shows that the R-sunspot and flare number have the same trend, increasing at the beginning of the cycle. The maximum occurred in 2013–2014, marked by an increase in R-sunspot and flare, then decreased until 2019. We obtained that the pattern of the intensity of R-sunspot and flare occurrence has similarity. And the comparison of sunspot data grouped by Zurich classification and flares shows that flares occured during the solar cycle 24 are sequentially produced by sunspot classes of F, E, D, G, H, C, B, J, A. The early phase of the cycle (2008–2010) was dominated by flare B, while in the maximum phase was dominated by flare C, M, X which mostly came from sunspot class F, E, and D consecutively. The objective of this study is to analyze the flare potentiality for different sunspot groups.

50.1 Introduction

The solar cycle 24 began in December 2008 and ended in December 2019. It was observed as less activity compared to the 23rd solar cycle based on the sunspot number. Sunspot measures activity in a solar cycle. Therefore, the study of the solar cycle aims to predict the solar activity that affecting the Earth, such as the flare and

A. F. N. Cahyaningtyas (✉)
Astronomy Study Program, Institut Teknologi Bandung, Bandung 40116, Indonesia
e-mail: firdausyaam@gmail.com

S. Filawati · S. O. Hanum
BPAA Pasuruan, Pasuruan 67155, Indonesia

E. Yulihastin et al. (eds.), *Proceedings of the International Conference on Radioscience, Equatorial Atmospheric Science and Environment and Humanosphere Science, 2021*, Springer Proceedings in Physics 275, https://doi.org/10.1007/978-981-19-0308-3_50

coronal mass ejection (CME), in which both disturbances are the parameters of space weather.

Sunspot observation is held on several locations on Earth intending to investigate how sunspot evolution triggers disturbances on Sun that is negatively impacting Earth. One of them is held by Balai Pengamatan Antariksa dan Atmosfer (BPAA) Pasuruan since 1983, located at 7.567 °S, 112.675 °E. The sunspots were observed every day via a ground-based telescope, and the data was archived by calculating Sunspot relative numbers.

$$R = k(10g + f) \tag{50.1}$$

Equation (50.1) is referred to the Wolf Number equation, with k is variable scaling factor, g is the number of sunspot groups, and f is the number of sunspots.

Sunspot is a region on the solar photosphere with a high magnetic field concentration and two polarities (bipolar). It appears dark because of its low temperature caused by pressurized convection in the magnetized plasma region, heading to a lower energy flux towards the sunspots [1]. Sunspot appearance is closely related to active regions on the Solar photosphere, and its evolution is nothing but the evolution of energy from the active regions (ARs). Sunspot is a transient phenomenon that has a lifetime that varies from several days to several weeks or even lasts for months. Its diameter is in a range of 3500–60,000 km [2]. The Solar cycle represents the dynamic growth of the number and sunspot area. Research on sunspots within a pretty long period is utilized for observing the dynamical solar activity.

The sunspot evolution identifies by the changes in its shape, structure, and size. The birth and development undergone by an active region are related to a loop emerging from a deeply rooted toroidal flux strand. A single dark alignment in the photosphere and fibrils of filament in the chromosphere is produced by the thinnest loop. Inside the deeper region, exist many thin loops that emerge into a few bundles, and each bundle creates an emerging flux region (EFR) that produces an active region. A bipolar active region specified by a grand loop bundle is made up of various flux bundles that define an individual EFR [3].

Sunspot growth ends several days after all fluxes are combined. The flux loops that create the active region reach the atmosphere by rising convective granules. Probably, the upper side of the loop is rising because of the magnetohydrodynamic forces.

The decay evolution of sunspots starts when all fluxes reach the surface, and the active regions have achieved their maximum growth. The sunspots and the magnetic flux size will decrease; only one or two dominant spots remain and will likely grow by further merging in several days. The spectrum of magnetic elements becomes simpler: pores and micropores (magnetic knots) start to vanish, as well as most of the sunspots [3]. In many cases, a sunspot still exists for some time, slowly shrinking in several weeks or, in rare cases, can last for several months.

Sunspots are classified into some classes based on their appearance (including their sizes and structures). One method to classify sunspots into several classes is the Zurich classification described by Waldmeier in 1938. Zurich classification groups the sunspots into nine classes A, B, C, D, E, F, G, H, and J. This classification scheme is made based on the morphology and the evolution of sunspots. A sunspot belongs to class A and B are categorized as a small sunspot, the C class for medium sunspot, the big sunspots belong to D, E, and F classes, the G class for medium sunspots, and the H and J classes are for the small and single sunspot. The development of sunspots' increasing energy is related to the development of classes from A, B, C, D, E, and F consecutively. Meanwhile, the decrease in sunspot energy is associated with declining sunspot class from G, H, to J. A big and complex sunspot has a stronger magnetic field than the smaller one. Therefore, it can be concluded that sunspot groups included in D, E, and F class can be categorized as the strong and complex sunspot, whereas sunspot included to A, B, C, G, H, and J class is categorized as small and simple.

Flare is a sudden flash of increased intensity of radiation emitted in a short time in the surrounding sunspot area. The flare activity is related to sunspots. The big sunspot groups, including the classes D, E, and F, have the closest correlation with flare to the other classes of the sunspot group [1]. The energy stored inside the magnetic fields of sunspots suddenly released energetic particles and radiation across the entire spectrum of the Sun [5, 6]. In general, flares are produced by a magnetic reconnection, which causes a change in the topology of the magnetic field, and the energy stored in the magnetic fields is released, thereby accelerating the corona particles [7]. Magnetic energy storage in the active region is characterized by the degree of non-potentiality of sunspots' magnetic field [8].

Observations show that the emerging flux region and magnetic helicity are essential factors for the pre-flare state, with magnetic helicity being a measure of the magnetic topological complexity, i.e., the twists and vertices of magnetic field lines [9]. Observation of magnetic field in line of sight (abbreviated as LOS) shows that the field on the photosphere changes rapidly during the occurrence of flare [10]. Observing the magnetic field of a flare is very beneficial since the magnetic field is an excellent parameter to forecast a flare [11]. Since flare is produced by magnetic reconnection, its size and power depend on the energy of the sunspots' magnetic field; hence, it can be concluded that the big sunspot has a bigger potential to produce flare.

To investigate the further correlation between sunspots and their potentiality to produce flares, we analyze the data of flares obtained from NOAA's satellite observations, along with sunspot data observed by BPAA Pasuruan from 2008 to 2019 as a period of solar cycle 24.

50.2 Data and Methods

50.2.1 Data

The flare data was obtained from NOAA (*National Oceanic and Atmospheric Administration*) from 2008 to 2019 during the solar cycle 24. It is observed by the

GOES satellite at X-ray wavelength. The GOES satellite is located at an altitude of ~36,000 km or the Earth's geostationary orbit. GOES has an equal orbital period to the Earth's rotation period. NOAA flare data has a format of date and time in UT (start, end, and maximum) of the flare, the flare's location (degrees of latitude and longitude of the Sun), the flare type (based on X-ray energy class), the name of the observing satellite, the energy flux produced by the flare (J/m^2), and the NOAA sunspot region number. The flare data can be accessed through the National Centers for Environmental Information website: https://www.ngdc.noaa.gov/stp/space-weather/solar-data/solar-features/solar-flares/x-rays/goes/xrs/.

The sunspot data is obtained from observations using a sunspot telescope at BPAA Pasuruan from 2008 to 2019. The sunspot data recorded has a format of the date (Julian date), time (UT), weather conditions, observer initials, number of sunspot groups, number of individual spots, sunspot group location (degrees of latitude and longitude of the Sun), Zurich classification, and sunspot group number.

The Zurich classification is a well-known sunspot classification and has been widely used by most astronomers, advanced by Waldmeier in 1938 based on the sunspot group classification made by Cortie in 1901 [12].

This classification describes the evolutionary sequences of sunspot groups with nine types in this classification [1]. Based on Carrasco, et al. [12], the Zurich classification features can be detailed as follows:

1. Type A: One or two spots without penumbra and bipolar configuration.
2. Type B: the sunspots group without penumbra, with bipolar configuration.
3. Type C: sunspots group has a penumbra on their main spots, whereas the others do not.
4. Type D: bipolar groups with spots having penumbra, their size measured less than 10° long from the entire Sun's disk.
5. Type E: the bipolar groups with their main spots having penumbra, measured 10°–15° long.
6. Type F: large bipolar groups, measured more than 15° long.
7. Type G: the bipolar groups are containing penumbra and small spots among the main spots, with a size, measured more than 10° long.
8. Type H: the spots having single polarity, penumbra, and diameter more than 2.5°.
9. Type J: the spots having single polarity, penumbra, and diameter less than 2.5° (Fig. 50.1).

50.2.2 Methods

Sunspots are identified based on the sunspot number and Zurich classification. The sunspot number or R-sunspot can describe the phase characteristics in a solar cycle, while the Zurich classification divided the sunspot into 9 classes: A, B, C, D, E, F, G, I, and J. Data on R sunspot and Zurich classification were compared to obtain the Zurich classification that dominates the solar cycle.

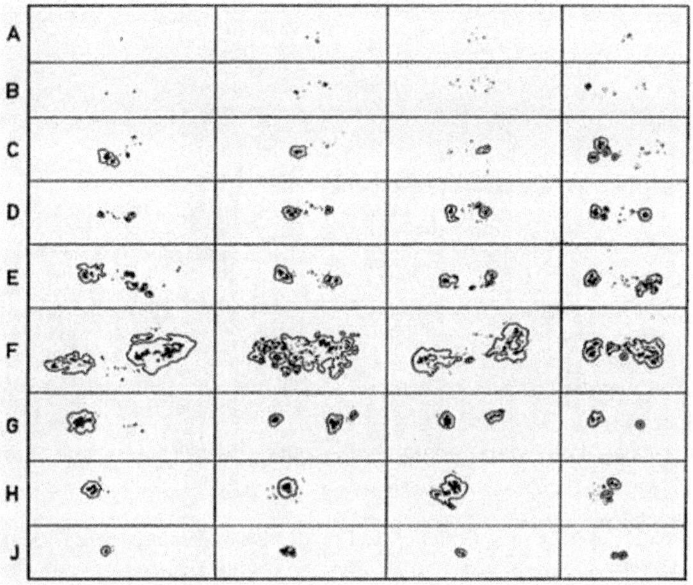

Fig. 50.1 Zurich classification for sunspot group [13]

Flares are identified based on x-ray classification, they are A, B, C, M, and X and compared with R sunspot to obtain the class of flare that dominates the solar cycle. Next, flare is identified by the source of sunspot by Zurich classification. It has the aim to understand which Zurich classification has dominance and significance to produce flares. Flares and sunspots are determined by their potentiality. The potentiality is calculated from the number of flares by the Zurich classification divided by the total number of flares.

50.3 Results and Discussion

50.3.1 Correlation of Flare Occurrence and R-Sunspot Number

The solar cycle 24 happened from 2008 until 2019 and its dynamics were investigated by observing its sunspot condition over a period. The sunspots observed by BPAA Pasuruan were categorized based on their Zurich class. Moreover, studying the solar cycle can also conduct by observing the flare. Sunspot and flare have the same pattern, low in the early phase, increase at the maximum stage, and decrease at the end of the cycle. From sunspot observation held at BPAA Pasuruan, we obtain that the total

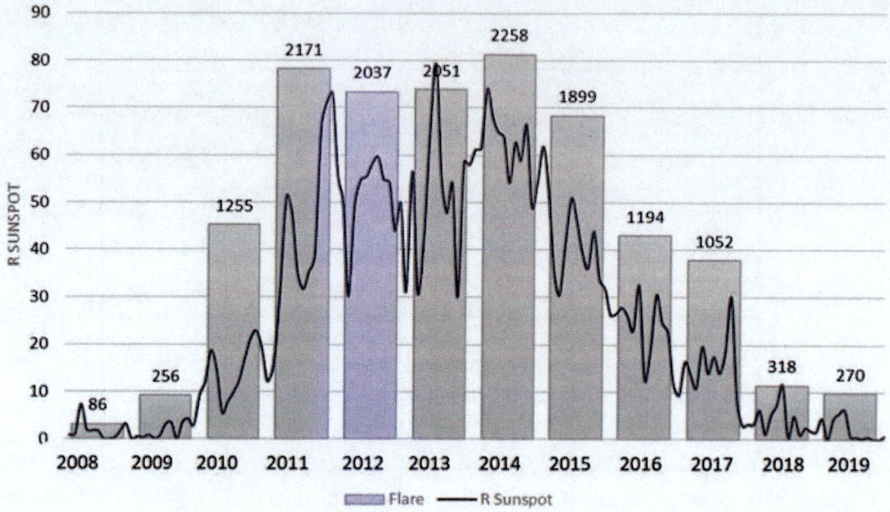

Fig. 50.2 Monthly average sunspot number and yearly total of flare occurrence during solar cycle 24

sunspot group that existed during cycle 24 is 9,882; and based on NOAA data, the total flares produced in cycle 24 is 11,747.

The solar cycle 24 had a period of ~11 years based on sunspot observations from BPAA Pasuruan. The solar cycle began in December 2008 until December 2019, with the maximum phase in May 2013. The maximum had a sunspot number of 79. Meanwhile, most of the flares occurred in 2014 with an accumulation of 2258 flares. Figure 50.2 shows the monthly averaged sunspot number obtained from sunspot observation of BPAA Pasuruan and flares data from the GOES satellite. Figure 50.2 shows that there is a similarity between the pattern of flare occurrence and the pattern of sunspot number.

From analyzing the R-sunspot value and flare occurrence, we can conclude that the flare occurrence is highly related to sunspot evolution. When sunspots reach their highest activity, flare occurrences also experience their maximum phase and vice versa. This conclusion highly corresponds with the theory that the active region, the region that has a strong magnetic concentration that later becomes sunspots, is most likely expected to be a place where the flare is growing (Fig. 50.3).

50.3.2 Flare Occurrence Correlation with Sunspot that Classified into Zurich

In this research, we match the flare data obtained from NOAA with the sunspot observation data that has been classified into Zurich Classification by BPAA Pasuruan. We divide the sunspot group into three main categories: the A, B, and C classes for small

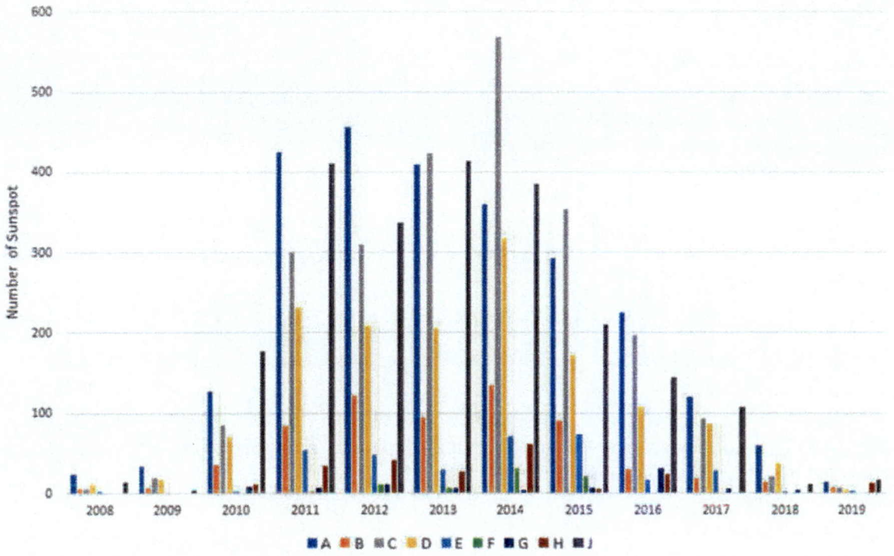

Fig. 50.3 The Zurich classification of sunspots during solar cycle 24

and simple sunspot undergoing their initial phase of growth; D, E, and F classes for big and complex sunspots which undergo their peak or maximum phase of growth; and the G, H, J for small and simple sunspots undergoing their end growth phase (the term complex and simple, refers to the morphology). From analysis, we found that 45.8% of total flares exist during cycle 24, and is produced by the D, E, and F sunspot class; 38.3% of total flare produced by the A, B, C, sunspot; and 15.9% flare produced by G, H, J sunspot.

Flares existed during cycle 24, produced mainly by the active region of D sunspot, except in 2014 and 2019. Cycle 24 ended in 2019, and it was indicated by the decline of the R-sunspot number; therefore, it also caused the decline of the active regions that produce flares.

Flares nearly have the same pattern as sunspots. In the early phase of the solar cycle, class B flares dominate compared to other classes in one year. Flare B increased from 2008 to 2010, then decreased from 2011 to 2014 with a total of 240 flares, increased again from 2015 to 2017, and decreased from 2018 to 2019. Flare C dominates the maximum of solar cycle 24, from 2011 to 2015. In 2014, flare C had the highest number, with a total of 1797 flares. Figures 50.4 and 50.5, flares in 2014 were dominated by flare C, with the dominating sunspots being C class sunspots.

Figure 50.6 shows flare potentiality by recognizing the Zurich class of sunspots. During solar cycle 24, the most potential sunspots that produce flare are D. Sunspot D has bipolar groups with penumbra, the size is less than 10° of Sun's disk.

From Figs. 50.4 and 50.5, we found that there is a difference between the sunspots' peak and flare's peak during cycle 24; that is, the sunspot peak happened in 2013; meanwhile, the flare's peak occurred in 2014. This can happen because, in 2014, the

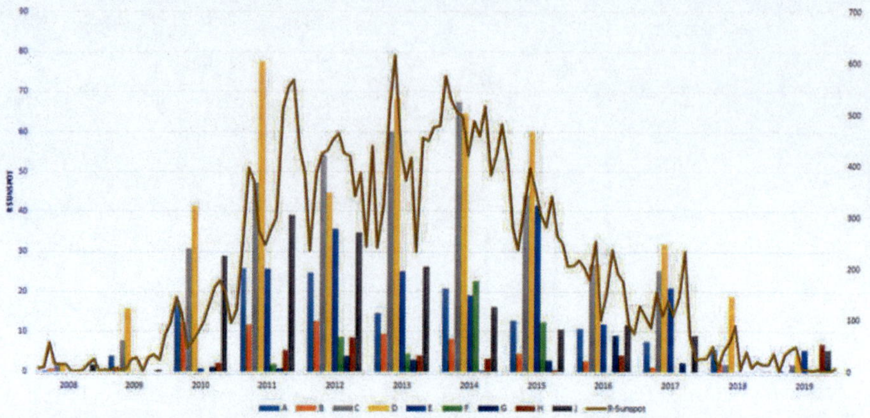

Fig. 50.4 The monthly average R-sunspot compared to flare occurrence produced by sunspot that classified by Zurich classification during solar cycle 24

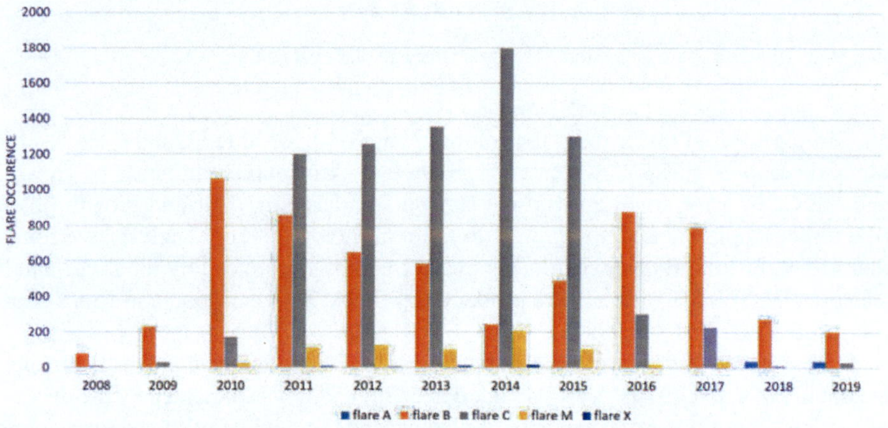

Fig. 50.5 The Zurich classification of sunspot and flare during solar cycle 24

number of large and complex sunspot groups, i.e., D, E, and F sunspots was more numerous than in 2013. The D, E, and F sunspot classes, have higher flare production potential than other sunspot classes. Therefore, we can infer that in 2014 when the peak of flare happened, the D, E, and F sunspot groups significantly contributed to producing flare, compared to other sunspot groups classes. These results are also supported by Eren et al. [1], which state that the D, E, and F sunspot classes have the highest flare potentiality production.

In this research, we define flare potentiality as the ability of a sunspot group to produce flare. Big potentiality means that the sunspot has a higher probability of producing a flare. We calculate the value of flare potentiality by dividing the number of flares produced by a certain sunspot class by the amount of sunspot in

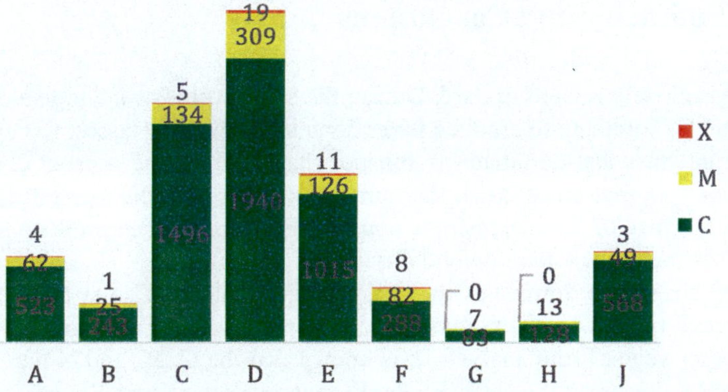

Fig. 50.6 Total of flare produced by different sunspots classes

that class itself. Figure 50.7 shows that the sunspots D, E, F have a higher potential to produce flare compared to other classes. This flare potentiality result is supported by the amount of flare produced by sunspots D, E, and F that has the highest value compared to others. In addition, Fig. 50.7 also shows that the highest flare potentiality was sunspot F. Sunspot F has the most complexity and the most expansive area than others. Still, sunspot F appeared at least as many as 68 during solar cycle 24. Thus, flare prediction in consequence of sunspot has the most possibility by sunspot F, then E and D.

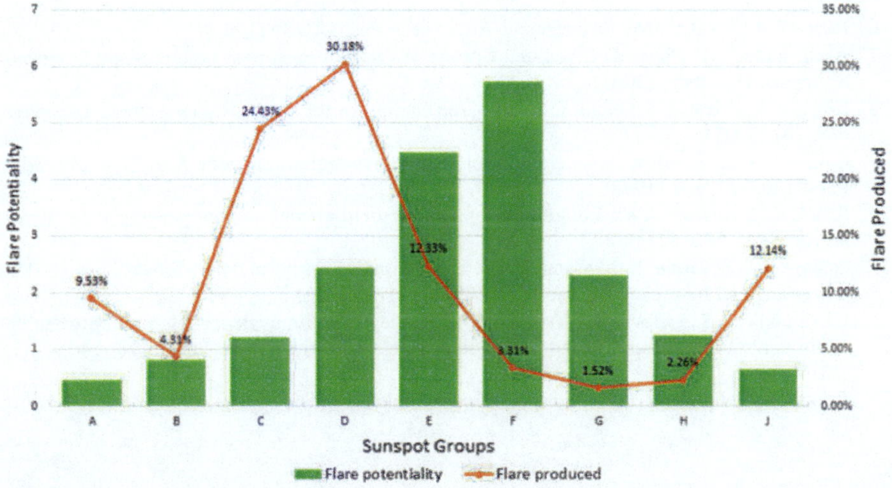

Fig. 50.7 Flare potentiality production for each sunspot class

50.4 Summary and Conclusions

Sunspot is closely related to flare. During the solar cycle 24, the big sunspot group (D, E, and F) dominate to produce flares compared to other classes. Except in 2014, the sunspot class that dominates is sunspot class C. Even the sunspot C dominated the sunspot population in 2014, the number of flares was the highest during solar cycle 24 because of the increasing amount of F sunspot which produce more flares since it has the highest flare potentiality.

The C-class flare dominates the entire flare produced in solar cycle 24. At the cycle's peak (in 2014), the C class flare is reaching its highest amount compared to the other years. From analysis, it is known that the C, M, and X flare is mainly produced by the sunspot class F, E, and D consecutively. From all explanations, we can conclude that sunspot F has the highest probability of producing flares.

References

1. Eren, S., Kilcik, A., et al.: Flare-production potential associated with different sunspot groups. Mon. Not. R. Astron. Soc. **456**, 65–75 (2017)
2. Solanki, S.K.: Sunspots: An overview. Astron. Astrophys. Rev. **11**, 153–286 (2003)
3. Zwaan, C.: The Evolution of Sunspot. Sunspot: Theory and Observations, pp. 75–100 (1992)
4. Setiahadi, B.: Solar Flare Occurrence by Considering the Zurich Classification, Area of Active Region, and Magnetic Topology: Warning to Solar Activity (2008)
5. Conlon, P.A., et al.: Multifractal Propcrtics of Evolving Active Regions. Solar Phys. **248**, 297 (2008)
6. Rust, D.M.: Solar flares: An overview. Adv. Space Res. **12**, 289 (1992)
7. Aschwanden, M.: Physics of the Solar Corona: An Introduction with Problems and Solutions. Springer, Heidelberg (2005)
8. Regnier, S., Priest, E.R.: Nonlinear force-free models for the solar Corona. Astron. Astrophys. **468**, 701 (2007)
9. Canfield, R.C., Pevstov, A.A.: Helicity of solar active-region magnetic field. Synoptic Solar Phys. **140**, 131–137 (1998)
10. Sudol, J.J., Harvey, J.W.: Longitudinal magnetic field changes accompanying solar flares. Astrophys. J. **635**, 647 (2005)
11. Gallagher, P.T., Moon, Y.-J., Wang, H.: Active-region monitoring and flare forecasting—I. Data processing and first results. Solar Phys. **209**, 171–183 (2002)
12. Carrasco, V.M.S., Lefevre, L., Vaquero, J.M., Gallego, C.: Equivalence relations between the cortie and zurich sunspot group morphological classifications. Sol. Phys. **290**, 1445–1455 (2015)
13. Bray, R.J., Loughhead, R.E.: Sunspots. Dover Publ. **228** (1964)

Chapter 51
Automatic True Color Composites Generation Based on NOAA JPSS Satellites Data

Karunika Diwyacitta, Budhi Gustiandi, and Andy Indradjad

Abstract True color composites is an image that shows natural color combinations as seen by human eyes. In satellite data context, true color composites is generated by combining the visible red, green and blue bands to associated red, green, and blue channels of a computer display. True colors products are mostly preferable for many people since they can quickly tell the information contained in it. Vegetation, waters, grounds, and lands are widely used information that can be obtained from true colors images. This paper proposes an automation system to generate true colors products from National Oceanic and Atmospheric Administration (NOAA) Joint Polar Satellite System (JPSS) series satellite data. The automation system uses an open-source module named Community Satellite Processing Package (CSPP) Polar2Grid Version 2.3 as its main processing system. Inputs needed for this system are VIIRS SDR band-I, band-I terrain corrected geolocation, and band-M in HDF5 format. The outputs generated from this system are reprojected true color images in GeoTIFF format. There are five types of modules inside the automated system. Those are searching, comparing, checking, processing, and storing modules. The system was installed in a server computer in the BRIN Remote Sensing Data Center. Monitoring activities had been done for a whole month, and no failure was found once the input data were available. During the monitoring activities, 109 data were carried out to analyze the correlation between processing time and the size of processed input data files. Bigger size input data files need a longer processing time to create a true color product. The proposed system has been proven to operate successfully and has been included in operational activities in BRIN remote sensing processing facilities. The products have been utilized as base maps in the current web-based low-resolution products dissemination system.

K. Diwyacitta (✉) · B. Gustiandi · A. Indradjad
Research Center for Remote Sensing, National Research and Innovation Agency, Jakarta, Indonesia
e-mail: karunika.diwyacitta@brin.go.id

E. Yulihastin et al. (eds.), *Proceedings of the International Conference on Radioscience, Equatorial Atmospheric Science and Environment and Humanosphere Science, 2021*, Springer Proceedings in Physics 275, https://doi.org/10.1007/978-981-19-0308-3_51

51.1 Introduction

A true color composite is an image that displays associated visible red, green, and blue bands from a remote sensing satellite sensor into red, green, and blue channels on a computer screen [1]. The resulting composite would look like what our eyes see naturally so that it can also be called a natural color composite. Water would be seen as dark blue features in the image. Vegetation would appear as green features in the image. Impervious ground and bare land would be seen as brown and light grey features in the image, respectively. Although slight differences in features are often difficult to distinguish because of their relatively low contrast in the image, natural or true color composites are still preferred by many people in some applications.

High-resolution true color images (0.2~0.34 m) from Google Earth imagery have been used to quantify water areas by extracting the features from the images to provide resources in monitoring lakes, reservoirs, and small ponds [2]. Medium resolution true color images from a commercial RGB camera attached to a small Unmanned Aerial Vehicle (UAV) were utilized to estimate forest canopy attributes (i.e., canopy cover, foliage clumping, and Leaf Area Index (LAI)) for forest stand monitoring [3]. Similar natural composites were also used to evaluate crop biomass by calculating the Normalized Green–Red Difference Index (NGRDI) and Leaf Area Index (LAI) from the images on agricultural areas [4]. Coarse-resolution natural images from Moderate Resolution Imaging Spectroradiometer (MODIS) instruments were used as an excellent metric for assessing a coupled biogeochemical model [5].

However, some scientists believe that coarse-resolution true or natural color composites imagery may have little utility for scientific inference [6]. Hence, these images are mostly used only as pretty pictures that support visualization of certain features on the Earth's surface or what we call a base map.

In this paper, a processing system was designed and implemented to generate true color composites from National Oceanic and Atmospheric Administration (NOAA) Joint Polar Satellite System (JPSS) satellite series (i.e., Suomi National Polar-orbiting Partnership/Suomi-NPP and NOAA-20) data to support coarse-resolution remote sensing products dissemination system. The satellites are the successor to the current National Aeronautics and Space Administration (NASA) Earth Observing System (EOS) and NOAA Polar-orbiting Operational Environ-mental Satellites (POES) satellites [7–9]. The composites act as near real-time base maps that will be updated when new data are available. Hence, latest products (such as active fires [10–14]) will be displayed or overlaid on top of the latest base map from the same satellite data.

The satellites data are received in direct broadcast mode four times daily by BRIN Remote Sensing Ground Station [15]. Two of the acquisition times are executed at midnight outside of office hours. The acquisitions also have to be done on week-ends, seven days a week. In this context, to support an optimal near real-time data generation, the processing system has to be designed to run automatically.

51.2 System Design and Implementation

The true color products from NOAA JPSS satellites are generated by an auto-mated processing system that uses an open-source module named Community Satellite Processing Package (CSPP) Polar2Grid Version 2.3 [16]. The general description of this automated system is shown in Fig. 51.1. Based on Fig. 51.1, inputs needed for the Polar2Grid module are VIIRS SDR Band-I, Band-M, and Terrain Corrected Geolocation files (GITCO*.h5 and GMTCO*.h5). VIIRS SDR Band-I consists of SVI01, SVI02, SVI03, SVI04, and SVI05. VIIRS SDR Band-M consists of SVM03, SVM04, SVM05, SVM07 and SVM11. All inputs come in Hierarchical Data Format 5 (HDF5) with.h5 file extension. All outputs come in Tag Image File Format (TIF) with.tif file extension.

This automated system is installed in a server computer which can easily be accessed by a desktop computer, as shown in Fig. 51.2. Server 192.168.7.40 is where the automated system is installed, while the others are where the SDR data from NOAA JPSS satellites are stored. NOAA-20 SDR data are stored in the 192.168.7.37 server and Suomi-NPP SDR data are stored in the 192.168.7.210 server. NOAA JPSS satellites data are acquired directly by BRIN Remote Sensing Ground Station in Sulawesi, Indonesia, and transferred to BRIN Remote Sensing Ground Station in Jawa, Indonesia. BRIN Remote Sensing Ground Station in Jawa then runs this automated system to generate true colors products from NOAA JPSS satellites data as one of its operational activities.

There are five modules and a configuration file used to build the automated system that generates true colors products of NOAA JPSS satellites. Those are searching, comparing, checking, processing, and storing modules. All modules and their connections are depicted in Fig. 51.3. The searching module searches all directories and their subdirectories defined in the configuration file. Found directories are then listed in a buffer file and sorted based on the most recent acquisition time. The comparing module compares all directories listed in the buffer file with a log file consisting of

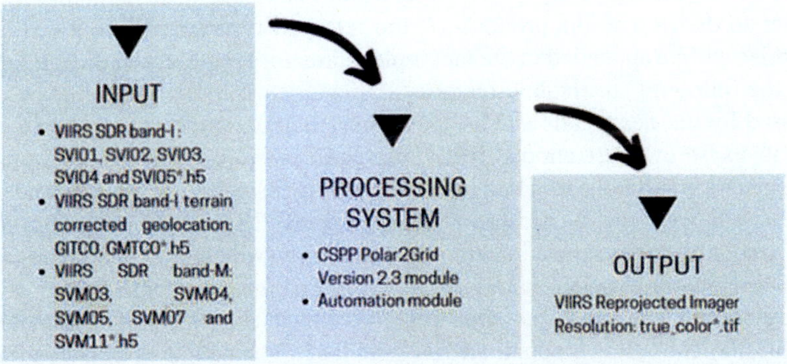

Fig. 51.1 General description of an automated system that generates true color products from NOAA JPSS satellites

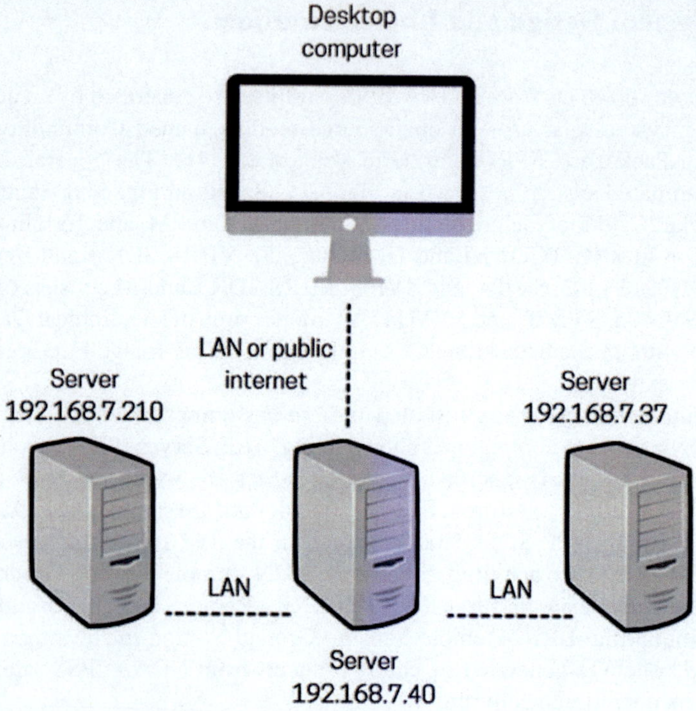

Fig. 51.2 Hardware configuration to run the automated system

a list of incomplete directories and another log file consisting of a list of processed directories. Incomplete directories are directories with incomplete inputs required by the processing module and processed directories are directories that have been processed previously by the automated system. If the directories are not included in both log files, they will be proceed further into the next module. The checking module has the main task to check whether the directories have all input files required to generate the true colors products. If the required files are incomplete, then the directories will be appended in the incomplete directory log file, and the module will check the following available directories. Directories with complete inputs will be processed by the next module. The processing module is the main module of this automated system which adopts CSPP Polar2grid version 2.3. True colors products will be generated after the required input data are processed by the module, and output products will be stored in corresponding directories defined by the storing module. Storing module extracts the name of output products (which have been standardized by the previous processing module) into several parameters consisting of acquisition year, month, date, and time. If a directory with the corresponding parameters are available, the true colors products will be stored in it. Then, if the associated directory is not available, a new directory will be created first, and true colors products will be stored in the new directory. The directory name will be appended to the processed

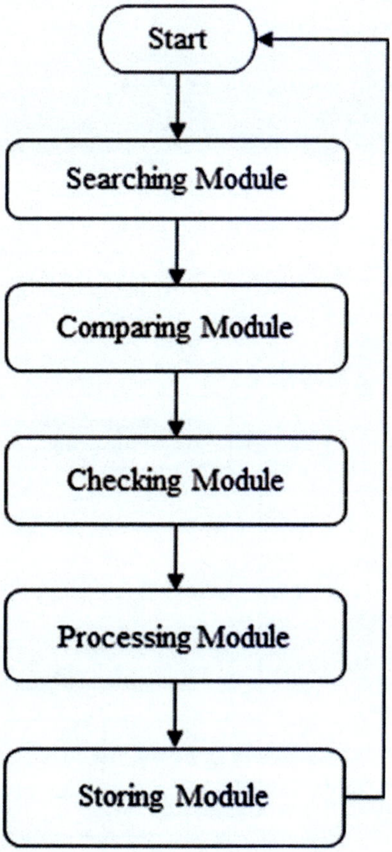

Fig. 51.3 The whole process of the automated system to generate true color products from NOAA JPSS satellites

log file. The process then will return to the first module again after all directories are processed.

All modules are combined into one entire automated system. The detailed flowchart of the whole system is illustrated in Fig. 51.4. The flowchart shows one cycle to generate true color products. The system will automatically repeat the entire process when there is a new directory available. The configuration file specifies the location of input and output directories. The directories and files configuration is shown in Fig. 51.5.

As shown in Fig. 51.5, a processing log is provided in the automated system. The log consists of records of all processing time for each processed directory. During a processing cycle, the time spent to complete generating associated products is recorded in this log. By associating the processing time and the size of the processed input data file, relations between two variables can be derived. In this paper, 109 processing cycles of NOAA JPSS satellites data during acquisition time from 1 to 31 May 2021 were analyzed.

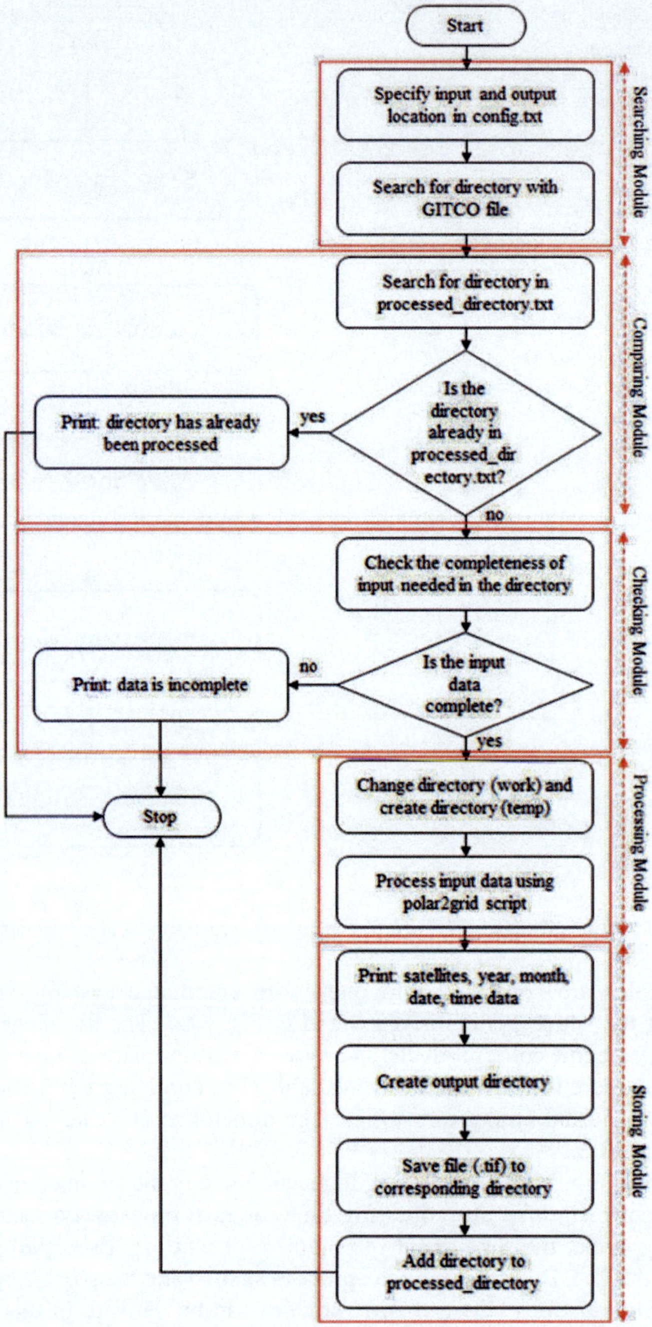

Fig. 51.4 Detailed flowchart of the automated system in generating true color products from JPSS NOAA satellites

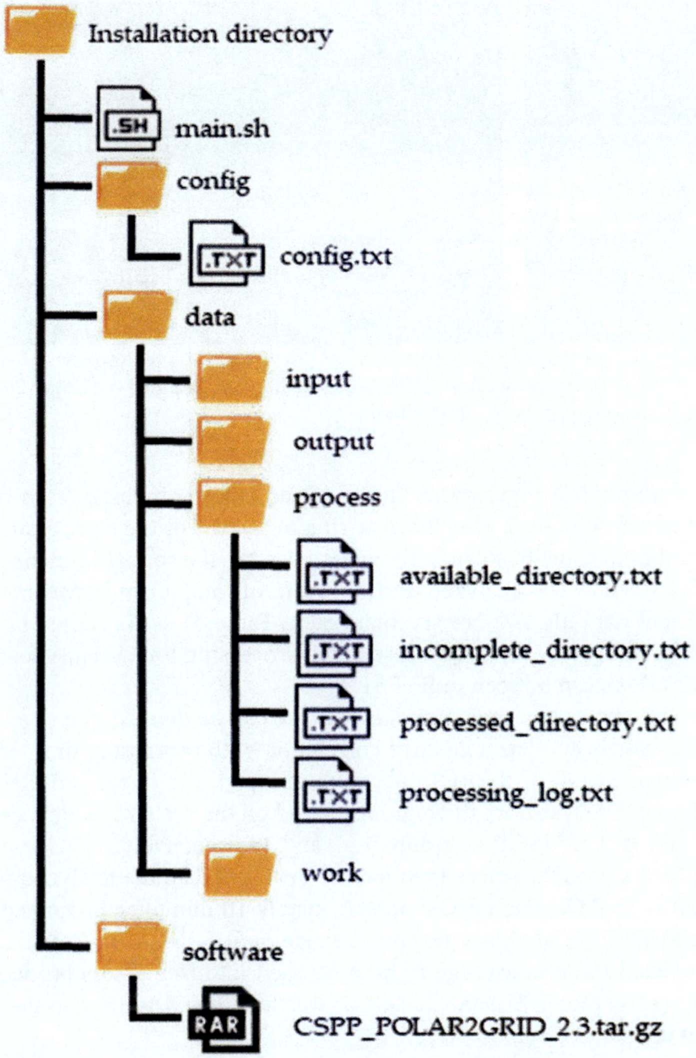

Fig. 51.5 Configuration of directories and files of the automated system used to generate true color products of JPSS NOAA satellites

51.3 Results

There are five files generated by the automated system for each completed directory (a directory with complete required input data). Those files are one true color file, three corrected reflectance files for band-M (M05, M03, M04), and one file for band-I (I03). All files are in GeoTIFF format with .tif file extension. Their spatial resolution is 375 m. One example of true color composites generated by the automated system

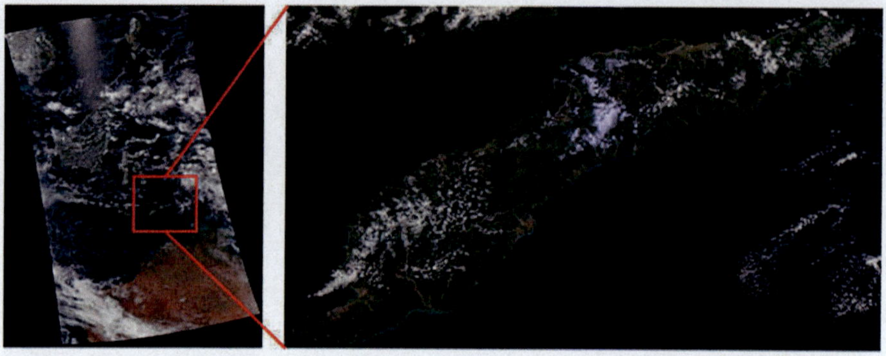

Fig. 51.6 True colors composite from NOAA-20 satellite data acquired on 2 May 2021 at 10.00 am and East Nusa Tenggara Island in Indonesia

is shown in Fig. 51.6. Figure 51.6 shows a true color composite from NOAA-20 satellite data acquired on 2 May 2021 at 10 a.m. A part of the compo-site image is zoomed further to show East Nusa Tenggara Island in the south of Indonesia region.

A total of 109 data was analyzed, and the results of comparison between processing time and input data file size are summarized in Table 51.1. As shown in the table, the size of input data files and their associated processing time vary. A scatterplot of their relationships can be seen in Fig. 51.7.

Figure 51.7 shows that the input data file size can be divided into several groups with similar values and has a positive correlation with processing time. In order to have a better view, all of the 109 data were grouped into three, and their average value was counted. There are three groups based on the file size, which are 1.1–5.0, 5.1–10.0, and 10.1–15.0 GB as groups 1, 2, and 3, respectively.

Figure 51.8 shows that more than half of the input data file analyzed was in the range of 10.1–15.0 GB and needed approximately 10 min to be processed into true colors products. The input data files in the range of 5.1–10.0 GB were 28 out of 109 data and needed 6.2 min average to be processed into true colors products. There were only eight out of 109 data whose size was less than 5.1 GB, and their average processing time was 1.8 min.

51.4 Discussion

The processing time taken to generate one true color product is affected by some factors. The main factor is the size of the input data file, as shown in Figs. 51.7 and 51.8. The bigger the input data file size, the longer it takes to generate true color products. The average processing time to create true colors products with input data files around 10.1–15.0 GB is approximately 10 min. The other factor that makes the processing time longer is the existence of different systems running on the same

Table 51.1 Comparison between input data size and processing time

NOAA JPSS satellites				
No.	Date	Time (UTC+7)	Input file size (GB)	Processing time (min)
1	1 May 2021	11.00	3.0	1.1
2		12.00	14.8	5.6
3	2 May 2021	12.00	14.8	5.2
4	3 May 2021	12.00	14.8	4.5
5		13.00	8.9	2.4
6	4 May 2021	11.00	11.8	3.6
7		13.00	10.3	3.4
8	5 May 2021	11.00	10.3	7.4
9		13.00	11.8	13.1
10	6 May 2021	11.00	7,4	3.1
	…	…	…	…
100	25 May 2021	12.00	15.0	13.3
101	26 May 2021	12.00	15.0	24.8
102	27 May 2021	11.00	15.0	15.8
103		13.00	8.9	7.0
104	28 May 2021	11.00	12.0	12.5
105		13.00	11.0	19.8
106	29 May 2021	11.00	11.0	7.5
107		13.00	15.0	10.7
108	30 May 2021	12.00	15.0	11.8
109	31 May 2021	12.00	14.0	14.3

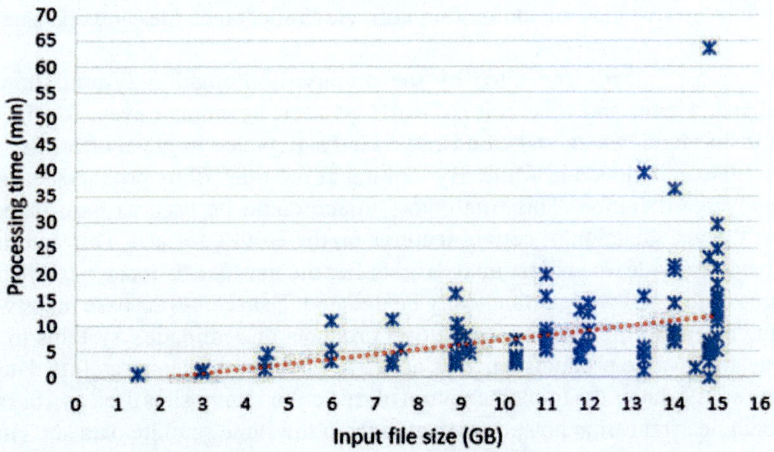

Fig. 51.7 A scatterplot for all the data that were analyzed

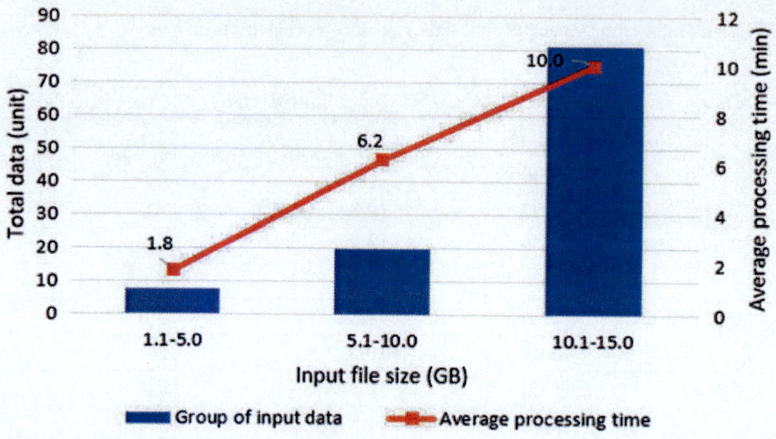

Fig. 51.8 A chart of average input data file size versus average processing time

server simultaneously. BRIN Remote Sensing Data Center has many operational activities done in a day so that multiple data processing systems can run at the same time. It makes the processing time longer even though the input data file size is relatively small. For some other cases, if the processing time is less than 5 min and the input data file size is less than 5 GB, the output of true colors products need to be checked. It is possible that the input data and output products are may be corrupt.

The developer of CSPP Polar2Grid claimed that the processing time should be less than 2 min to create nine corrected reflectance files output, including both true and false color images [16]. However, they do not specify the hardware used in their processing system. Most of the NOAA JPSS products that are avail-able in the NOAA Comprehensive Large Array-Data Stewardship System (CLASS) [17] can be accessed by the public after 3~6 h after the acquisition time. In our implementation, the outputs are five files, including four corrected reflectance files and one true color image.

True colors images generated by the proposed automation system shows the natural red, green, and blue colors clearly as seen by human eyes. It effectively distinguishes land, water, and clouds. The land is depicted in green or brown, water in dark blue, and clouds in white. By looking at the true colors products, each area can easily be identified. This true colors image could be used as base maps that support the visualization of certain features on the earth's surface. This automation system will provide true colors image right after the new data is received so that near real-time base maps will automatically be updated. Near real-time base maps will be used in current web-based low-resolution products dissemination systems to show other near-real-time products (such as active fires) from the same satellite data set.

Figure 51.9 shows the implementation of true color composites used as a base map to present near real-time hotspots detected from the same satellite data set. Hotspot detection was done by the VIIRS sensor onboard NOAA JPSS satellites. True color base maps generated by the automated system will support the visual interpretation

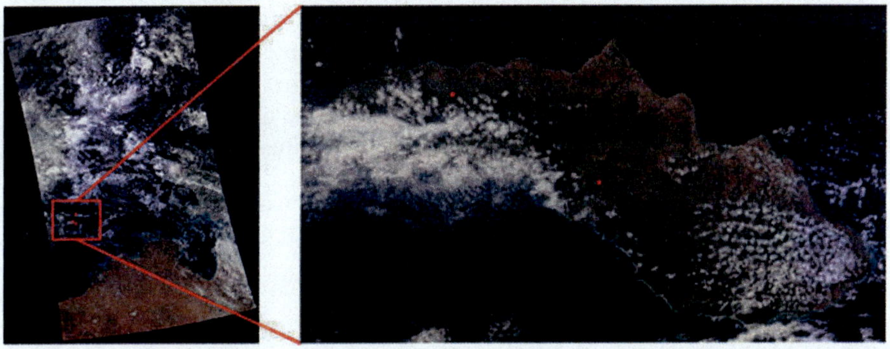

Fig. 51.9 Overlay of a hotspot feature (shown in a red dot) on a true color composite base map from the same data acquired on 13 August 2021 at 11.00 a.m. in Sumba Island, Indonesia

of the area where hotspots were detected. It will be easier for the users to view smoke around the hotspot area and raise their confidence that there was an active fire. In Fig. 51.9, the smoke is invisible in the hotspot area, so that further analysis needs to be carried out. Medium or high-resolution satellite data could be used to ensure the hotspot detected area.

The automated system has been run for several months. During the monitoring activities, the system runs successfully without any significant problem except for one condition. The condition was a technical problem during the data transfer process from BRIN Remote Sensing Ground Station in Sulawesi to BRIN Remote Sensing Data Center in Jawa. If data transfer process is troubled, the input data will not be available, and no data can be processed. This system does not have to be stopped when there is such a problem. It will normally run when the data transfer process returns to its normal condition. No failure was found once the input data were available.

Since all the processes are executed automatically, it can alleviate the work-load of the human resource. Repetitive preparation activities like logging into the processing system, opening terminal windows, entering parameters or variables required by the system, and executing the system every new satellite data available required in a cycle of the manual process can be reduced in the automated system [18]. Human errors can also be avoided, such as in typing parameters or variables required to execute the system.

51.5 Conclusions

An automated system to generate true color products from NOAA JPSS satellite series data has successfully been designed and implemented in the BRIN Research Center for Remote Sensing . Each area on the earth's surface can easily be distinguished by its different colors displayed as true color products. True color products could also be used as base maps to show the visualization of certain features on the earth's surface. Base maps will be available in near real-time through the automated system and will be used to support other near real time products from the same satellite data set.

The automated system has been implemented as one of the activities to support BRIN Remote Sensing Data Center operation. No failure has been found during the implementation of the system. All required input data have been processed successfully to generate near real-time true color composite products.

References

1. Humboldt State Geospatial Online, http://gsp.humboldt.edu/OLM/Courses/GSP_216_Online/lesson3-1/composites.html
2. Zhou, X., Zhou, Z., Miao, J., Apple, M.E., Spangler, L.: Quantification of water and exposed lined areas of coal-bed methane water ponds using regular true-color images by developing a novel uniformness based multi-component algorithm. J. Hydrol. **572**, 645–658 (2019). https://doi.org/10.1016/j.jhydrol.2019.03.015
3. Chianucci, F., Disperati, L., Guzzi, D., Bianchini, D., Nardino, V., Lastri, C., Rindinella, A., Corona, P.: Estimation of canopy attributes in beech forests using true colour digital images from a small fixed-wing UAV. Int. J. Appl. Earth Obs. Geoinf. **47**, 60–68 (2016). https://doi.org/10.1016/j.jag.2015.12.005
4. Jannoura, R., Brinkmann, K., Uteau, D., Bruns, C., Joergensen, R.G.: Monitoring of crop biomass using true colour aerial photographs taken from a remote con-trolled hexacopter. Biosyst. Eng. **129**, 341–351 (2015). https://doi.org/10.1016/j.biosystemseng.2014.11.007
5. Baird, M.E., Cherukuru, N., Jones, E., Margvelashvili, N., Mongin, M., Oubelkheir, K., Ralph, P.J., Rizwi, F., Robson, B.J., Schroeder, T., Skerratt, J., Steven, A.D.L., Wild-Allen, K.A.: Remote-sensing reflectance and true colour produced by a coupled hydrodynamic, optical, sediment, biogeochemical model of the Great Barrier Reef, Australia: Comparison with satellite data. Environ. Model. Softw. **78**, 79–96 (2016). https://doi.org/10.1016/j.envsoft.2015.11.025
6. McRoberts, R.E.: Satellite image-based maps: Scientific inference or pretty pictures? Remote Sens. Environ. **115**(2), 715–724 (2011). https://doi.org/10.1016/j.rse.2010.10.013
7. Goldberg, M.D., Kilcoyne, H., Cikanek, H., Mehta, A.: Joint polar satellite system: The United States next generation civilian polar-orbiting environmental satellite system. J. Geophys. Res. Atmos. **118**, 13463–13475 (2013). https://doi.org/10.1002/2013JD020389
8. Mitchell, G.: The joint polar satellite system overview. Int. Geosci. Remote Sens. Symp. 1581–1584 (2018). https://doi.org/10.1109/IGARSS.2018.8518787
9. Goldberg, M., Zhou, L.: The joint polar satellite system—Overview, instruments, proving ground and risk reduction activities. Int. Geosci. Remote Sens. Symp. 2776–2778 (2017). https://doi.org/10.1109/IGARSS.2017.8127573
10. Jones, S., Reinke, K., Mitchell, S., Mcconachie, F., Holland, C.: Advances in the remote sensing of active fires : A review. Bus. Coop. Res. Cent. Program. **40** (2017)
11. Csiszar, I., Schroeder, W., Giglio, L., Ellicott, E., Vadrevu, K.P., Justice, C.O., Wind, B.: Active fires from the suomi NPP visible infrared imaging radiometer suite: Product status and first evaluation results. J. Geophys. Res. **119**, 803–816 (2014). https://doi.org/10.1002/2013JD020453
12. Schroeder, W., Oliva, P., Giglio, L., Csiszar, I.A.: The new VIIRS 375m active fire detection data product: Algorithm description and initial assessment. Remote Sens. Environ. **143**, 85–96 (2014). https://doi.org/10.1016/j.rse.2013.12.008
13. Wooster, M.J., Roberts, G., Smith, A.M.S., Johnston, J., Freeborn, P., Amici, S., Hudak, A.T.: Thermal remote sensing of active vegetation fires and biomass burning events. Remote Sens. Digit. Image Process. **17**, 347–390 (2013). https://doi.org/10.1007/978-94-007-6639-6_18

14. Qu, J.J., Wang, W., Dasgupta, S., Hao, X.: Active fire monitoring and fire danger potential detection from space: A review. Front. Earth Sci. China **2**, 479–486 (2008). https://doi.org/10.1007/s11707-008-0044-7
15. Soleh, M., Suprijanto, A., Mahatmanto, B.P.A.: Preliminary design of remote sensing ground station system for the JPSS-1 (joint polar satellite system) data acquisition and processing. In: Proceedings of the 2nd International Conference of Indonesian Society for Remote Sensing. pp. 62–77. Indonesian Society for Remote Sensing (2016)
16. Webmaster, C.: Community Satellite Processing Package. Retrieved from Cooperative Institute for Meteorog-ical Satellite Studies, http://cimss.ssec.wisc.edu/cspp/download
17. NOAA: NOAA Comprehensive Large Array-data Stewardship System (CLASS)
18. Aguirre, S., Rodriguez, A.: Automation of a business process using robotic process automation (RPA): A case study. Commun. Comput. Inf. Sci. **742**, 65–71 (2017). https://doi.org/10.1007/978-3-319-66963-2_7

Chapter 52
Estimating the Impact of Spatio-temporal Land Cover Changes on Land Surface Temperature and Air Quality Using Satellite Data in Beas Valley, Himachal Pradesh, India

Sayanta Ghosh, Renu Lata, Isha Thakur, K. C. Gouda, and J. C. Kuniyal

Abstract Alteration in land use is one of the primary causes of global environmental change. Rapid variations in Land Use Land Cover (LULC) have also occurred in the already vulnerable Indian Himalayan Region (IHR) for the last few decades. These activities are highly influenced by anthropogenic activities and have more severe consequences on the air quality of the high-altitude area of IHR. In this study, the relationship between LULC alteration and changes in air quality (CO, SO2, O3, PM10, and PM2.5) have been assessed for Beas valley of Kullu district, IHR. Using LANDSAT imagery and MERRA-2 data, the impact of Spatio-temporal Land Cover Changes on air quality (CO, SO2, O3, PM10, and PM2.5) over 20 years for Beas valley of Kullu district, IHR has been monitored. Also, temporal changes in Land Surface Temperature (LST) over major LULC categories in the Beas Basin, where anthropogenic activities play a crucial role in altering a natural ecosystem, have been examined. Furthermore, the relationship of vegetation (NDVI), snow cover (NDSI), settlements (NDBI), and moisture content (NDMI) with LST have also been established using the coefficient of correlation for the study area. Observation data revealed that the average monthly surface concentration of CO showed a moderate increase of 15% in 2020 than that of 2000. Whereas MERRA-2 data revealed that the average concentration of particulate matters such as PM10, PM2.5 showed a huge

S. Ghosh · R. Lata (✉) · I. Thakur
G.B. Pant, National Institute of Himalayan Environment, Himachal Regional Centre, Mohal-Kullu, Himachal Pradesh, India
e-mail: renu15_negi@yahoo.co.in

K. C. Gouda
CSIR Fourth Paradigm Institute, NAL Belur Campus, Bangalore, India

J. C. Kuniyal
G.B. Pant, National Institute of Himalayan Environment, Kosi-Katarmat, Almora, Uttrakhand, India

E. Yulihastin et al. (eds.), *Proceedings of the International Conference on Radioscience, Equatorial Atmospheric Science and Environment and Humanosphere Science, 2021*, Springer Proceedings in Physics 275, https://doi.org/10.1007/978-981-19-0308-3_52

increase of 60.32% and 41.30% respectively in 2020 than that of 2000. The primary outcome of the paper, we believe, will be helpful to get the quantified information about the land-use alteration over the period and possible impact on the air quality in the Beas Valley.

52.1 Introduction

Variations in land-use have major role in global environmental changes. Alterations in the land-use pattern are also occurred rapidly in the already vulnerable Indian Himalayan Region (IHR) for the last few decades. These activities are highly influenced by anthropogenic activities and have more severe consequences on the air quality of the high-altitude areas of IHR. Land Surface Temperature (LST), a critical geophysical parameter which varies with the complexity of the land cover that changes naturally as well as due to anthropogenic activities, has a significant role in various studies such as hydrological cycle monitoring, climate change studies, land use/land cover dynamics, soil moisture and vegetation water stress monitoring, natural resources management, numerical weather modelling etc. [1–5]. On the other hand, land cover and land use changes play a major role in controlling the change of earth surface temperature [6]. Temporal change detection of LULC reveals the variation in composition of vegetation, waterbodies and built-up areas as well as anthropogenic activities [7]. These variations play an important role in temporal changes of LST as well as in climate change [8, 9]. Geographic Information System (GIS) and Remotely Sensed data are widely used tools to investigate LULC changes and LST derivation [10, 11]. Application of Thermal Remote Sensing technique provides opportunity to retrieve LST of different land cover at local scale as LST represents the land radiative skin temperature which is derived from solar radiation [12]. Few studies have carried out to analyse the impact of land cover changes on air quality in different parts of the world using satellite data. Xu et al. (2016) monitored the role of land use changes on the air quality of Wuhan city, China using GIS and satellite data [13] whereas Superczynski et al. (2011) explored the connection of LULC changes on air quality of Central Alabama, USA using satellite data and GIS [14]. But there's a lack of such type of study in high altitude areas of IHR. Henceforth in this present study, we have made an attempt to assess the impact of spatio-temporal variations of LULC on major air quality parameters (CO, SO2, O3, PM10, and PM2.5) and Land Surface Temperature (LST) for Beas valley of Kullu district, Indian Himalayan Region (IHR) for the better understanding of the land cover dynamics and monitoring the anthropogenic impact on the natural environment. For this purpose, using LANDSAT imagery and MERRA-2 reanalysis of data, the effect of spatiotemporal Land Cover Changes on various air quality parameters (CO, SO2, O3, PM10 and PM2.5) over the period of 20 years, i.e., from the year 2000 to 2020 for Beas valley of Kullu district, IHR have been monitored. Spatial variation in LST over different LULC classes in the catchment area of river Beas in Kullu District over the period of time where anthropogenic activities including rapid expansion agricultural land, tourism activities, hydro-electric power projects etc. alter the natural ecosystem have been examined. Moreover, the relationship of vegetation (NDVI), snow cover (NDSI), settlements (NDBI) and vegetation moisture content (NDMI) with LST have also established using coefficient of correlation for the study area. Database for various climatic parameters (Precipitation, Air Temperature, Relative Humidity) of the study area from 1981 to 2020 have been developed and respective representational charts have been prepared in MS Excel for a better understanding of the hydro-meteorological status of the study area. The outcome of the paper, we believe, will be helpful in monitoring the dynamics of land cover variations and its impact on the air quality parameters which is essential for sustainable environmental planning in the Beas Valley.

52.2 Methods and Materials

52.2.1 Study Area

The river Beas originated from the southern face of Rohtang Pass situated at an altitude of 4,361 m above mean sea-level in Kullu district and flows for some 470 km to the Sutlej River in Punjab, India. The drainage basin of the river covers an area of approximately 20,303 km^2 [15]. In Kullu district, the drainage basin of river Beas covers an area of 3274 km^2 which consist of the entire Kullu except the extreme south. The topographic elevation of the water basin varies from 959 m to over 6500 m (Fig. 52.1a). The basin is extended between latitudes 31° 43′ 30″ N to 32° 25′ 0″ N and longitudes between 76° 58′ 26.4″ E to 77° 51′ 44.399″ E. The location map and contours map (Fig. 52.1a, b) of the study area with altitude gradient have been developed in ArcGIS 10.8 using SRTM DEM satellite data of 30 m spatial resolution. SRTM-DEM data have been accessed from https://earthexplorer.usgs.gov/. Spatial distribution maps of annual Rainfall distribution, Relative Humidity (RH), average Air Temperature for the year 2020 have been developed to analyze the hydro-meteorological status using the Inverse distance weighting (IDW) interpolation method in ArcGIS 10.8 for Beas Basin (Fig. 52.1c–g). Total monthly rainfall, average monthly air temperature status, the status of average monthly relative humidity of the study area for the past 40 years have been presented in the form of a chart (Fig. 52.1h–j) to understand the hydro-meteorological scenario of the study area. As revealed from these maps and charts, the valley has climate variability with the elevation from Hot & sub-tropical in the lower elevation (959–1400 m) to warm temperate (1400–1900 m), cool temperate (1900–2500 m) and cold alpine to glacial (2500-6000 m) in the northern & eastern mountain ranges. The Beas valley is rich in plant diversity and dense forest because of high precipitation in the form of rainfall in July–September and heavy snowfall during the winter months of January–March [15]. Meteorological data have been collected from KVK Bajaura, Meteorological Centre, Shimla, District level survey document Kullu and NASA POWER Data

Access Viewer which is further based upon the Modern-Era Retrospective analysis for Research and Applications, version 2 (MERRA-2) data having spatial resolution of 0.5° × 0.625°.

52.2.2 Methodology and Data Used

Land Surface Temperature (LST) products have been calculated for Beas valley of Kullu district (Path-147 and Row-38) using single window/ single channel algorithm from Landsat-7 ETM+ and Landsat 8 TIRS/OLI satellite images of 30 m spatial resolution for 15th October, 2000 and 30th October, 2020 respectively (Fig. 52.2a, b) to analyze the spatio-temporal changes in LST.

LULC map of the Beas basin of Kullu district for October, 2000 (Fig. 52.3a) and October, 2020 (Fig. 52.3b) have been generated using Landsat 7 ETM+ at

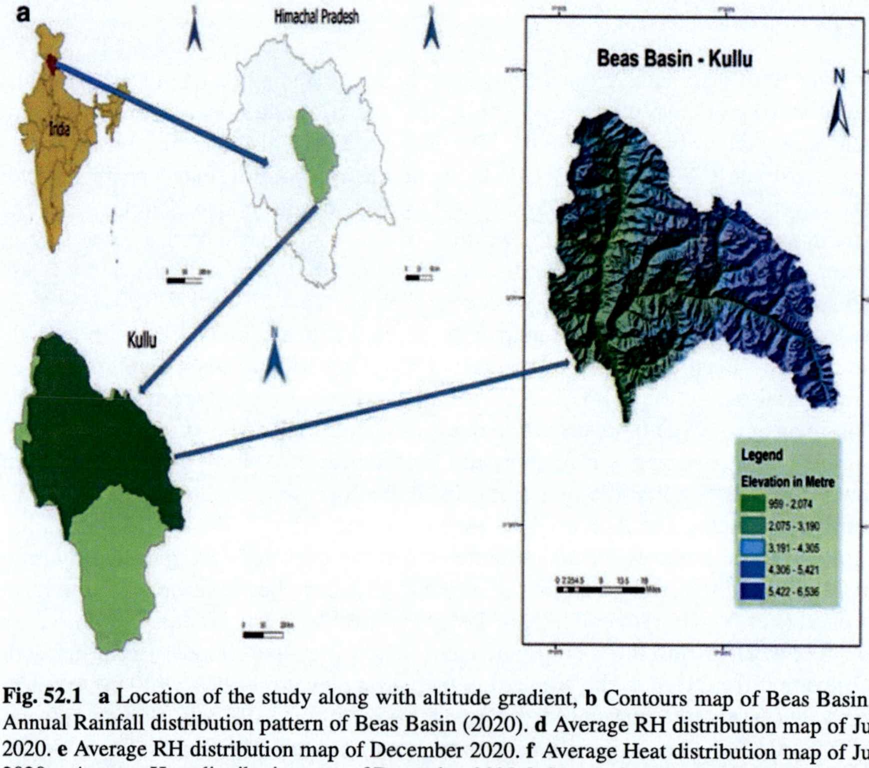

Fig. 52.1 **a** Location of the study along with altitude gradient, **b** Contours map of Beas Basin. **c** Annual Rainfall distribution pattern of Beas Basin (2020). **d** Average RH distribution map of July 2020. **e** Average RH distribution map of December 2020. **f** Average Heat distribution map of July 2020. **g** Average Heat distribution map of December 2020. **h** Status of total monthly rainfall at Beas Basin (January 1981 to December 2020). **i** Status of average air temperature at Beas Basin (January 1981 to December 2020). **j** Status of Relative Humidity at Beas Basin (January 1981 to December 2020)

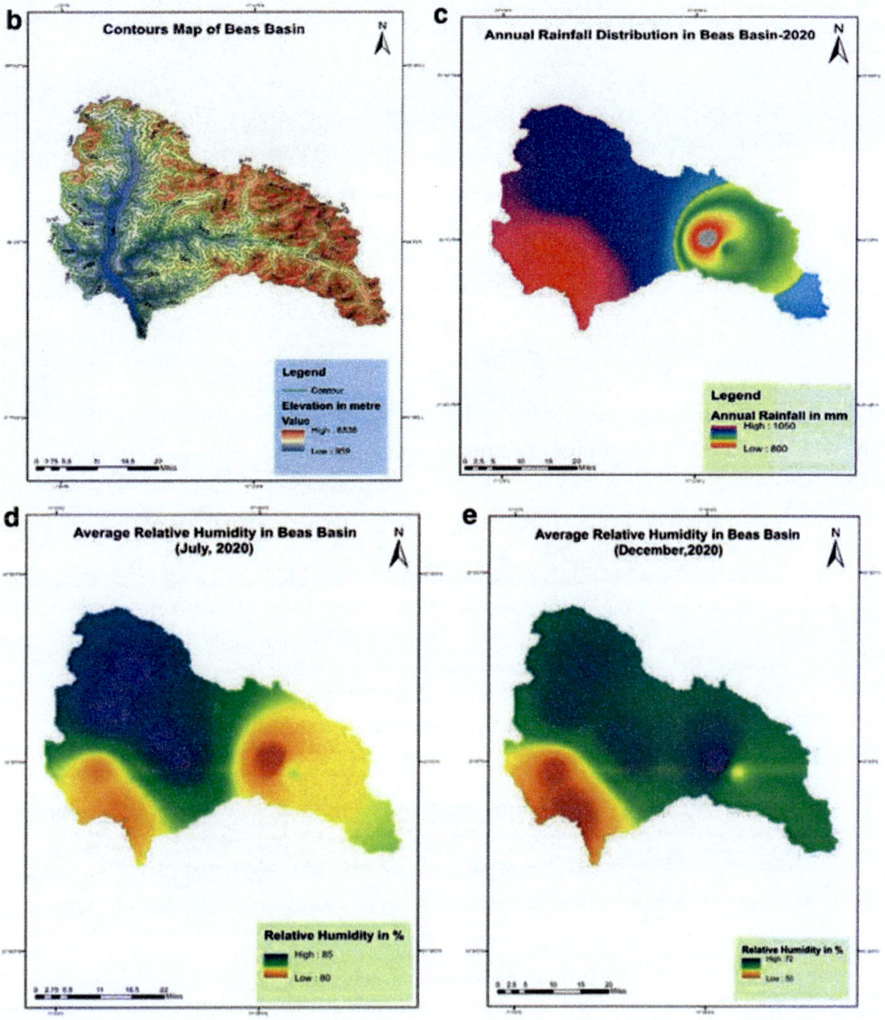

Fig. 52.1 (continued)

f
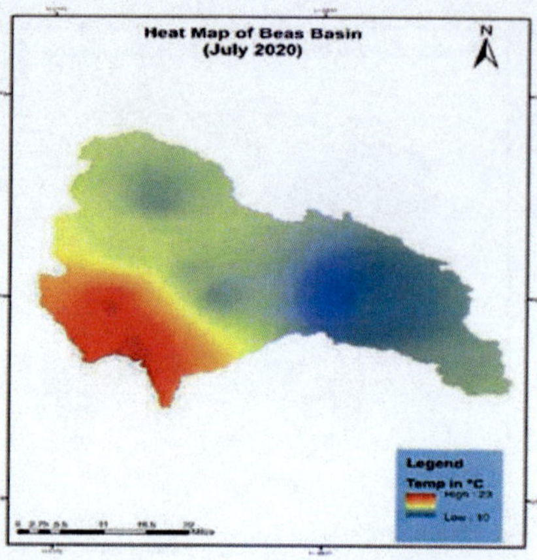

g
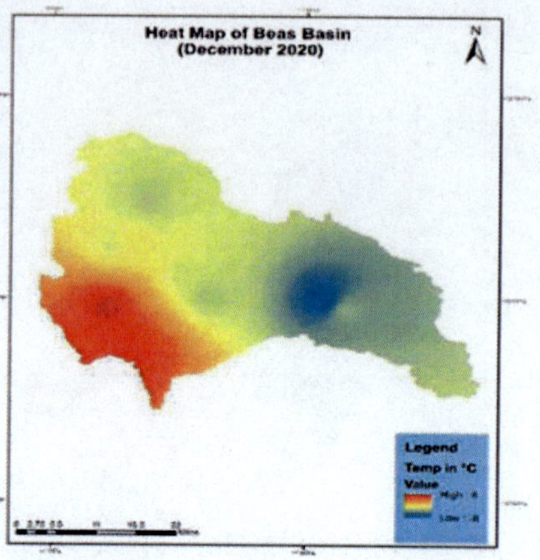

h
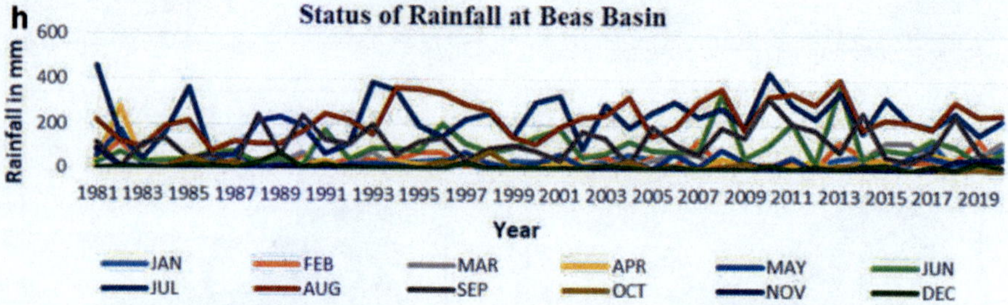

i
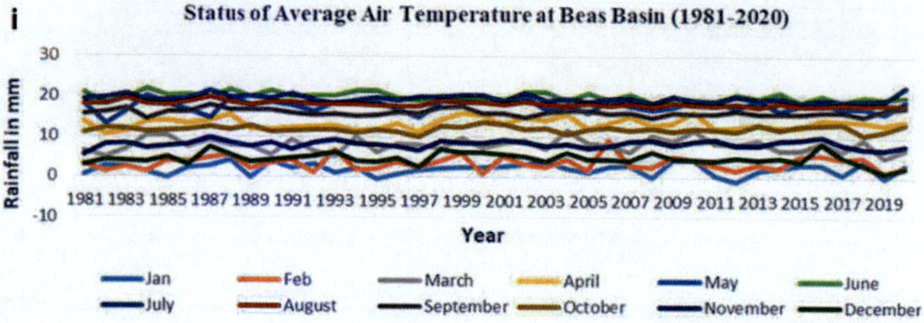

Fig. 52.1 (continued)

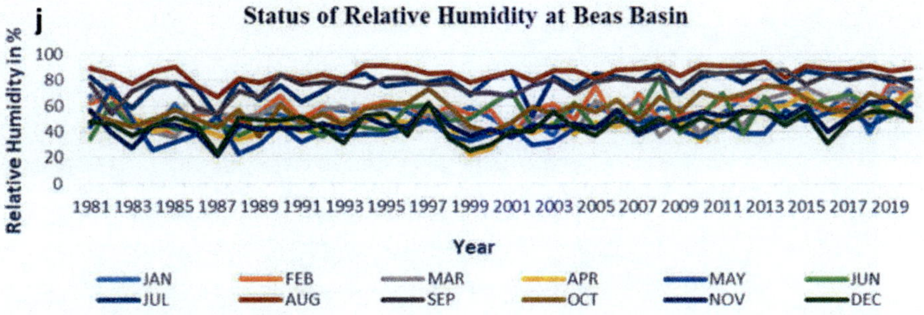

Fig. 52.1 (continued)

30 m resolution, ESA Sentinel-2 imagery at 10 m resolution satellite data. Eight major LULC classes have been pointed out and used here, namely (i) Glacier/Snow covered area (ii) Open/Rocky Land (iii) Water Bodies (iv) Forest Green area (v) Vegetation/Scrub/Grass land (vi) Barren land (vii) Built-up/Settlements (viii) Agriculture/Plantation Area. Landsat data has been accessed from https://earthexplorer.usgs.gov/.

Changes in air quality (CO, SO2,O3, PM10, and PM2.5) over 20 years have been monitored using Modern-Era Retrospective analysis for Research and Applications, version 2 (MERRA-2) data having spatial resolution of 0.5° × 0.625°. MERRA-2 data can be accessed from https://giovanni.gsfc.nasa.gov/giovanni/. MERRA-2 is a global atmospheric reanalysis data produced by the NASA Global Modeling and Assimilation Office (GMAO). The spatial distribution maps of various air quality

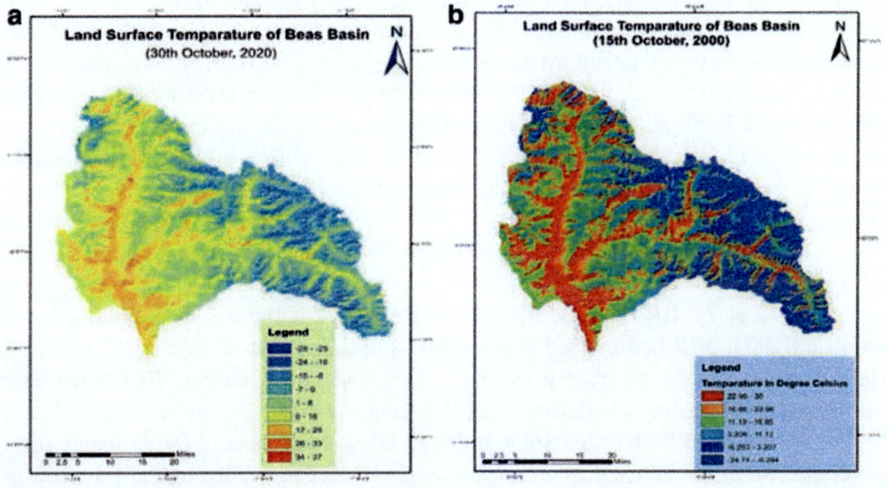

Fig. 52.2 **a** LST map of Beas Basin (15/10/00). **b** LST map of Beas Basin (30/10/20). **c** Flowchart for LST retrieval using Landsat 7 and Landsat 8 Satellite data

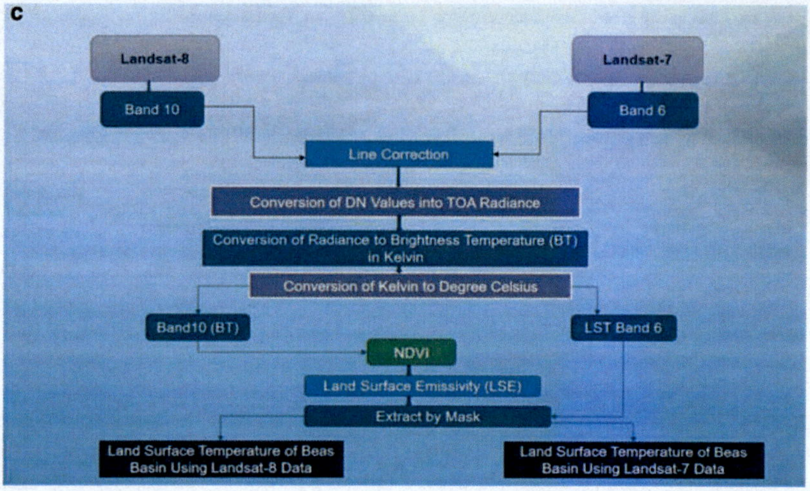

Fig. 52.2 (continued)

parameters over 20 years have been prepared in ArcGIS 10.8 using the Modern-Era Retrospective analysis for Research and Applications, version 2 (MERRA-2) data.

The Normalized Difference Vegetation Index (NDVI) is the most commonly used vegetation index to observe vegetation globally. Comparatively healthy vegetation strongly reflects in the near infrared region (NIR) of the electromagnetic spectrum. Whereas, vegetation has a low reflectance in red light of the electromagnetic spectrum. This index is calculated by subtracting Red band from NIR band, and dividing by the sum of Red band and NIR band (Eq. 52.1). NDVI values varies from + 1 to − 1, wherein negative values indicate water bodies and higher positive values indicate dense green–leafy vegetation. Moderate NDVI values (0.2 to 0.5) represent Shrubs and grasslands or senescing crops whereas lower NDVI values represent barren rocks, sand, or snow. Estimation of Normalized Difference Vegetation Index (NDVI) using Landsat 7 (ETM+) data for 15th October, 2000 and Landsat 8 (OLI/TIRS) data for 30th October, 2020 using Eq. 52.1 have been shown in Fig. 52.5a, b.

$$\text{NDVI} = (\text{NIRband} - \text{Red band})/(\text{NIRband} + \text{Red band}) \tag{52.1}$$

For Landsat 7, NIR band is band 4 and Red band is band 3 whereas in case of Landsat 8 data, NIR band is represented by band 5 and that of Red band is band 4. Here, the NDVI values for 15th October, 2000 and 30th October, 2020 have been calculated using raster calculator in ArcGIS 10.8.

The Normalized Difference Snow Index (NDSI) is very useful for delineate snow cover area and is often used as an important parameter for hydrological forecasting. This index is calculated by subtracting SWIR band from Green band, and dividing by the sum of Green band and SWIR band (Eq. 52.2). This is very essential for snow mapping as snow is not only very reflective in the visible parts specifically

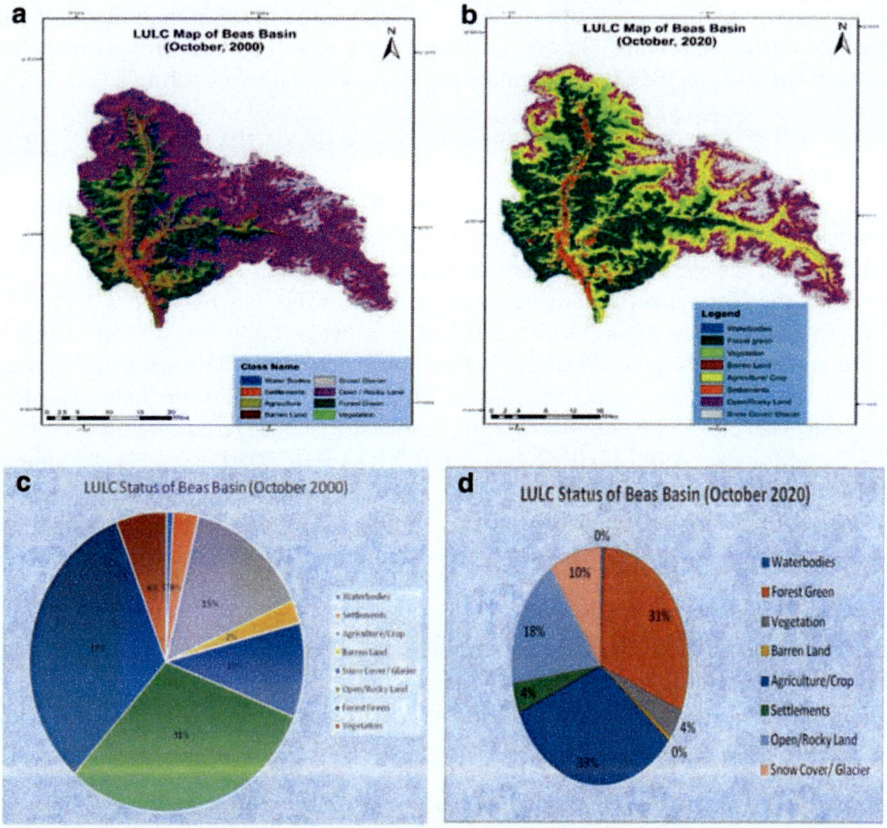

Fig. 52.3 **a** Status of LULC of Beas Basin (15/10/00). **b** Status of LULC of Beas Basin (30/10/20). **c** Area in % under different LULC categories (15/10/00). **d** Area in % under different LULC categories (30/10/20)

in Green band of the electromagnetic spectrum but also highly absorptive in the short-wave infrared (SWIR) part of the spectrum. Since snow absorbs most of the incident radiation in the SWIR while clouds do not, this allows NDSI to distinguish snow from clouds. NDSI values > 0.4 typically suggest the existence of snow in the area. Here, NDSI values of Beas Basin for 15th October, 2000 (Fig. 52.6a) and 30th October, 2020 (Fig. 52.6b) have been calculated using raster calculator in ArcGIS 10.8. NDSI is generally used in snow cover mapping application as well as in glacier monitoring.

$$\text{NDSI} = (\text{Green} - \text{SWIR})/(\text{Green} + \text{SWIR}) \tag{52.2}$$

For Landsat 7, Green band is Band 2 and SWIR band is represented by Band 5 whereas in case of Landsat 8 data Band 3 and Band 6 represent the Green band and SWIR band of the electromagnetic spectrum respectively.

Normalized Difference Built-up Index (NDBI) is the most common indexes for analysis the built-up areas. The build-up areas and bare soil reflects strongly in SWIR than that of NIR. NDBI can be calculated by using this following formula (Eq. 52.3):

$$NDBI = (SWIR - NIR)/(SWIR + NIR) \tag{52.3}$$

For Landsat 7, SWIR band is Band 5 and NIR band is represented by Band 4 whereas in case of Landsat 8 data Band 6 and Band 5 represent the SWIR band and NIR band of the electromagnetic spectrum respectively.

Also, the Normalize Difference Build-up Index value lies between −1 to + 1. Negative value of NDBI indicate water bodies as water body doesn't reflect in the Infrared spectrum region. Whereas higher value represents built-up areas. Lower NDBI values indicate vegetation. Normalized Difference Built-up Index (NDBI) maps of Beas basin of Kullu district using Landsat 7 (ETM+) data for 15th October, 2000 and Landsat 8 (OLI/TIRS) data for 30th October, 2020 using Eq. 52.3 have been shown in Fig. 52.7a, b.

Normalized Difference Moisture Index (NDMI) is used to determine vegetation water content. This formula uses the near infrared (NIR) and short wave infrared (SWIR) to capture the variations of moisture in vegetated areas. NDMI is applicable in applications looking at drought monitoring and subtle changes in vegetation moisture conditions. NDMI is also useful to determine fuel moistures which is further necessary for wildfire hazard assessments.

NDMI can be calculated by subtracting SWIR band from NIR band and then dividing by the sum of NIR band and SWIR band (Eq. 52.4).

$$NDMI = (NIR - SWIR)/(NIR + SWIR) \tag{52.4}$$

In case of Landsat 4 to 7, Band 4 and Band 5 represent the NIR and SWIR band respectively whereas in case of Landsat 8, NIR and SWIR bands are represented by Band 5 and Band 6 respectively.

Normalized Difference Moisture Index (NDMI) maps of Beas basin of Kullu district using Landsat 7 (ETM+) data for 15th October, 2000 and Landsat 8 (OLI/TIRS) data for 30th October, 2020 using Eq. 52.4 have been shown in Fig. 52.8a, b.

52.3 Results and Discussion

52.3.1 Assessment of Spatio-Temporal Changes of Land Surface Temperature in Beas Basin (Over 20 Years)

Here, the temporal spatio-variations of Land Surface Temperature (LST) have been monitored using Landsat 7 ETM+, Landsat 8 TIRS/OLI data for Beas valley of Kullu district, Indian Himalayan Region (IHR). LST products have been derived for Beas basin (Path-147 and Row-38) using single window/ single channel algorithm from Landsat 8 TIRS/OLI and Landsat-7 ETM+ satellite images for 30th October, 2020 and 15th October, 2000 respectively. The LST map of Beas basin for 15th October, 2000 indicates that the highest value was around 34.3 °C, which noticed in the lower Beas basin which consist of tourist place and settlements areas and the areas where Hydro-electric power projects and anthropogenic activities are continuing on; whereas, lowest LST value was observed to be − 24.7 °C in the glaciers cover areas of Beas basin (Fig. 52.2a). The LST map of Beas basin for 30th October 2020 shows that the highest value was around 36.23 °C, which is also observed in and around the lower Beas basin which comprise of barren land and settlements area, tourist place and the areas where development activities are going on; although, lowest LST value was observed to be − 27.26 °C in the glaciers cover areas of Beas basin (Fig. 52.2b). The Spatio-temporal Land Cover Changes over 20 years has also been analyzed in the later stage of study and the relationship of vegetation (NDVI), snow cover (NDSI), settlements (NDBI), and vegetation moisture conditions (NDMI) with LST were also established using the coefficient of correlation. The results reveal an increase in barren/open land, settlements areas, and other anthropogenic activities such as rapid growth of agricultural land over the years, which are the significant factors for the changes in land surface temperature.

52.3.2 *Temporal Change Detection for Eight Major Land Cover Types in Beas Basin (Over 20 Years)*

Land Use Land Cover (LULC) map of the Beas basin for October 2000 (Fig. 52.3a) and October 2020 (Fig. 52.3b) respectively has been generated using Landsat 7 ETM+ at 30 m resolution, ESA Sentinel-2 imagery at 10 m resolution satellite data. Eight Land use/ Land cover (LULC) categories have been identified and used in this study, namely (i) Glacier/Snow covered area (ii) Open/Rocky Land (iii) Water Bodies (iv) Forest Green area (v) Vegetation/Shrub/Grassland (vi) Barren land (vii) Built-up/Settlements (viii) Agriculture/Plantation Area. The results reveal a massive increase in Agricultural land, including orchard expansion, of 123% during the year 2020 than that of 2000. Also, there's a sharp increase of 40.63% in settlement areas which includes the tourism activities such as hotels, restaurants, etc. during the year 2020 than that of 2000. Moreover, it is also observed from LULC change detection that there's a sharp decline in the shrub vegetation and grassland about 45% during the year 2020 than that of 2000. These may be attributed to change in land use for the orchard development, encroachment in the forested area and disturbing project sites because of the developmental projects in study area. The change in the built-up area is attributed to human settlement and developmental projects such as hydropower projects and other schemes. Further, the forest green area and snow cover area are declined by 5.61% and 6.28%, respectively, during the year 2020. These are the

Table 52.1 Change in LULC categories (over the period of 20 years)

LULC categories	Area in square kilometer (October 2000)	Area in square kilometer (October 2020)
Glacier/snow covered area	**331.6**	**312**
Open/rocky land	1002.4	584.2
Water bodies	**26.3**	**15.6**
Forest green area	1063.2	1006.7
VEGETATION/SHRUB/GRASS LAND	**194.8**	**134.4**
Barren land	81	16.9
Built-up/settlements	**96**	**135**
Agriculture/plantation area	480	1070

major contributing factors towards the variation in Land Surface Temperature (LST) and Air Quality Parameters over 20 years (Table 52.1).

52.3.3 *Monitoring the Impact of Land Cover Changes on Air Quality Parameters (Over 20 years)*

It was revealed from the LULC change detection over 20 years (Fig. 52.4a) that there had been a vast increase in various anthropogenic activities during the present time than that of 2000, such as there was an increase in Agricultural land, settlement areas including hotels and other tourism activities, which are the significant factors for the variations in the mass concentration of various air quality parameters. The change in the built-up area is attributed to human settlement and developmental projects such as hydropower projects and other schemes. Further, there is a decrease in the forested area and a huge increase in agricultural activities and barren land. It may be due to the change in land use for the orchard development, encroachment in the forested area and disturbing project sites because of the developmental projects in the study area. The variation in the average monthly surface concentration of CO over 20 years as observed from MERRA-2 reanalysis of data with a spatial resolution of $0.5° \times 0.625°$ revealed that there was an increase of 15% (Fig. 52.4b, e, f). At the same time the spatial distribution maps of average monthly SO2 column mass density showed a rise of 15.87% in 2020 than 2000 (Fig. 52.4 g, h). The Change detection in the average monthly surface concentration of SO2 over 20 years is also presented in Fig. 52.4c. The spatial distribution maps of average monthly total column ozone (O3) for the years 2000 and 2020 show a slight increase of 2.25% in 2020 than that of 2000. The spatial distribution maps using the MERRA-2 reanalysis of data revealed that the average concentration of particulate matters such as PM2.5, PM10 showed a massive increase of 41.30% and 60.32% respectively in 2020 than that of 2000 (Fig. 52.4k–n).

52.3.4 Monitoring the Relationship of Vegetation (NDVI) with LST (Over 20 Years)

The relationship of vegetation (NDVI) with LST was established using the coefficient of correlation in ArcGIS 10.8 for 15th October 2000 (Fig. 52.5c) and 30th October 2020 (Fig. 52.5d) respectively. The result indicates a positive correlation between NDVI and LST with the value of correlation coefficient $R^2 = 0.2577$ in 2000, which increased to $R^2 = 0.5959$ in 2020. This positive relationship suggests that an increase in LST values provides favourable conditions for the growth of vegetation in the Beas basin during the pre-winter season 2000 and 2020 respectively.

52.3.5 Monitoring the Relationship of Snow Cover (NDSI) with LST (Over 20 Years)

The relationship of snow cover (NDSI) with LST was established using coefficient of correlation in ArcGIS 10.8 for 15th October 2000 (Fig. 52.6c) and 30th October 2020 (Fig. 52.6d) respectively. It shows a negative correlation with the strong correlation

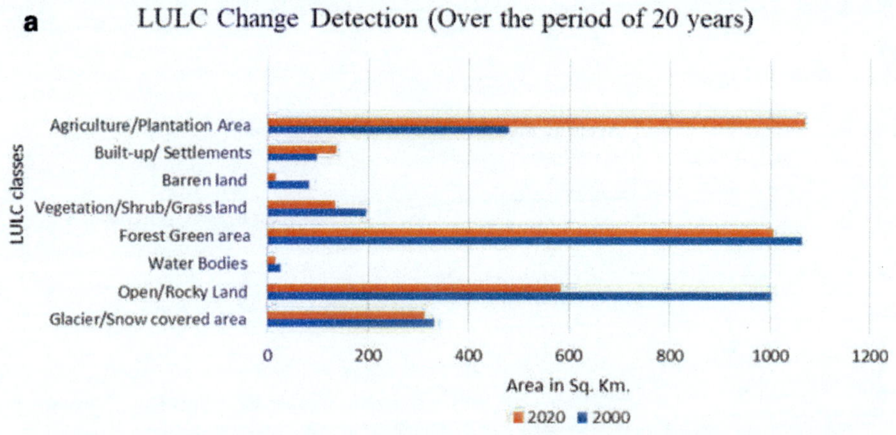

Fig. 52.4 **a** LULC change detection over 20 years. **4** Change detection in average monthly surface concentration of CO over 20 years. **c** Change detection in average monthly surface concentration of SO_2 over 20 years. **d** Change detection in average monthly total column ozone (O_3) over 20 years. **e** Average monthly surface concentration of CO during 2000. **f.** Average monthly surface concentration of CO during 2020. **g** Average monthly column mass density of SO2 during 2000; **h** Average monthly column mass density of SO_2 during 2020. **i** Average monthly total column ozone (O3) during 2000; **j.** Average monthly total column ozone (O_3) during 2020. **k** Average monthly PM 2.5 column mass density during 2000. **l** Average monthly PM 2.5 column mass density during 2020. **m** Average monthly dust column mass density during 2000. **n** Average monthly dust column mass density during 2020

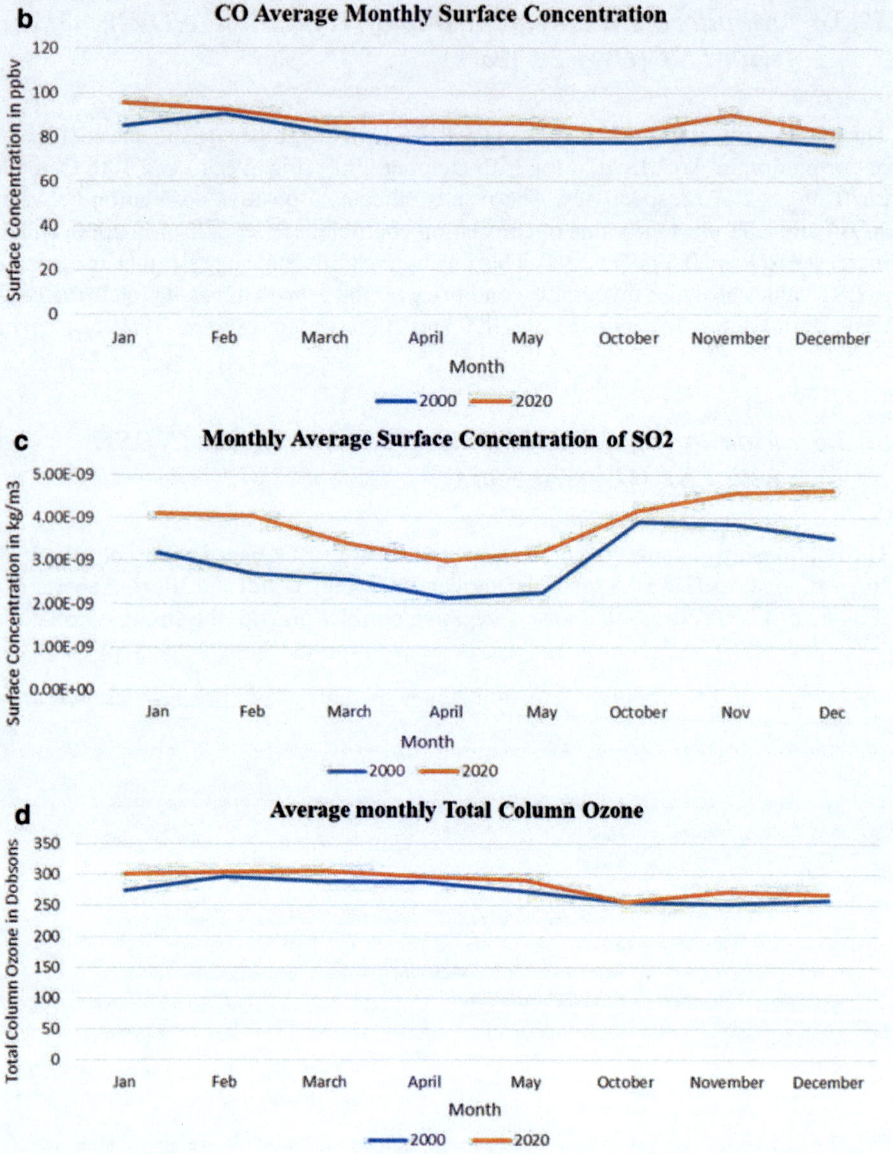

Fig. 52.4 (continued)

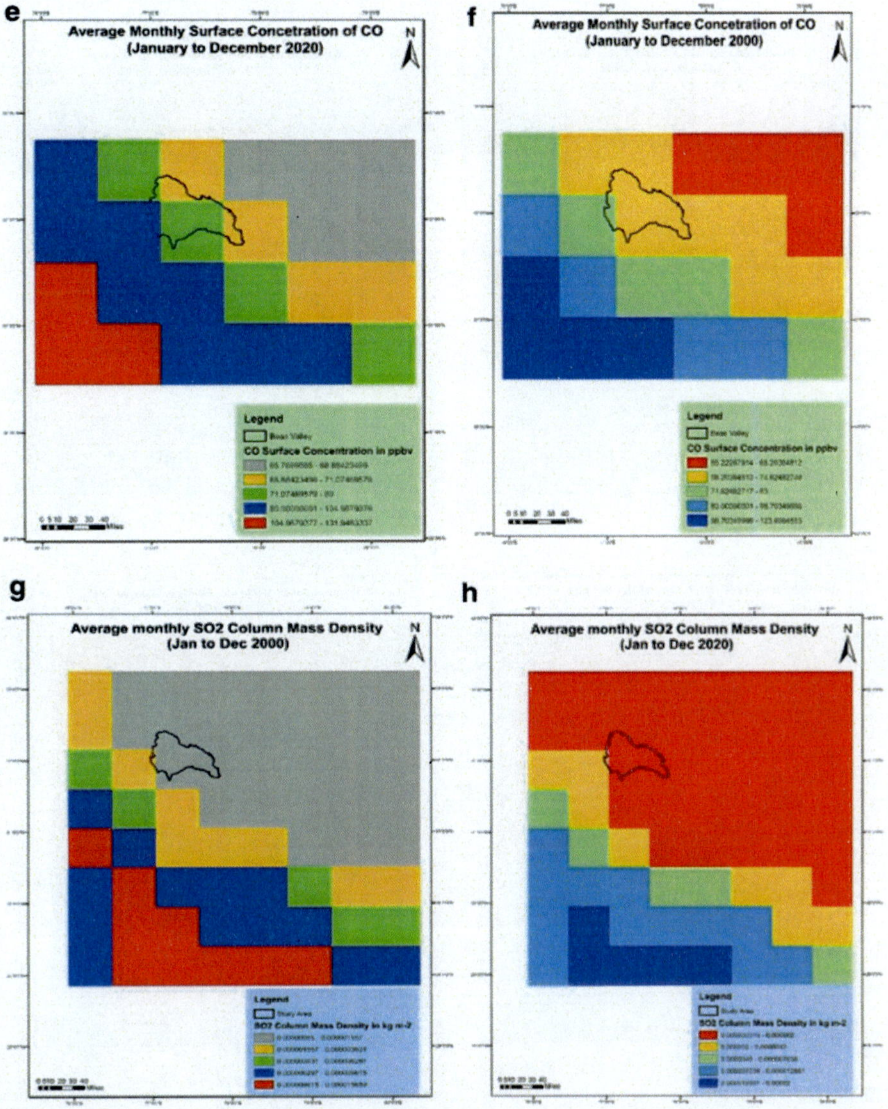

Fig. 52.4 (continued)

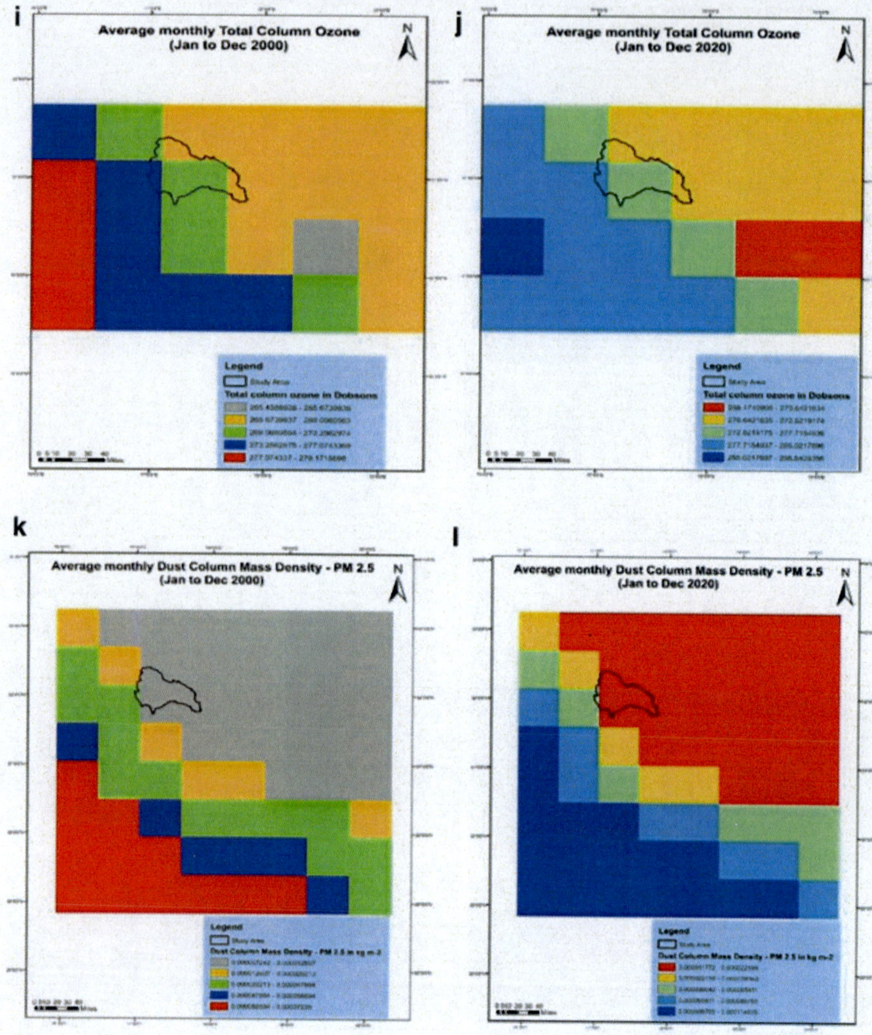

Fig. 52.4 (continued)

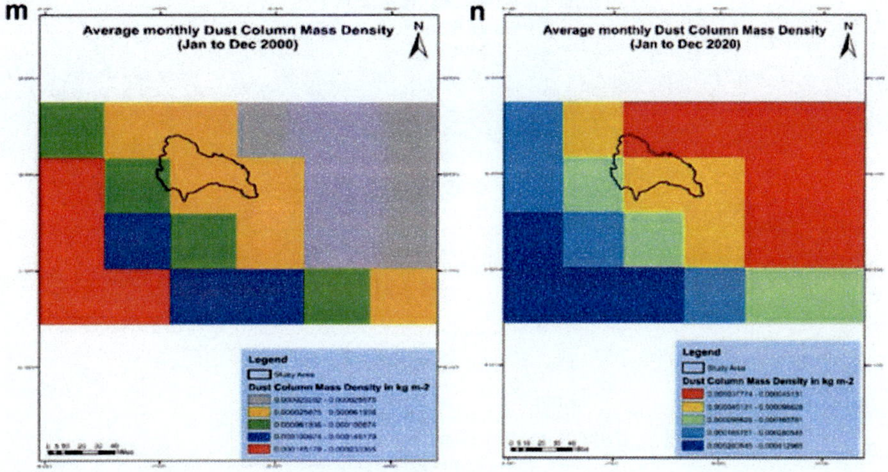

Fig. 52.4 (continued)

coefficients of $R^2 = 0.7072$ and $R^2 = 0.5642$ between NDSI and LST in the Pre-winter season of 2000 (15th October, 2000) and that of 2020 (30th October, 2020) respectively, indicating snow cover and LST have a negative correlation with regard to physical changes in the ground surface, surface emissivity and albedo of the region.

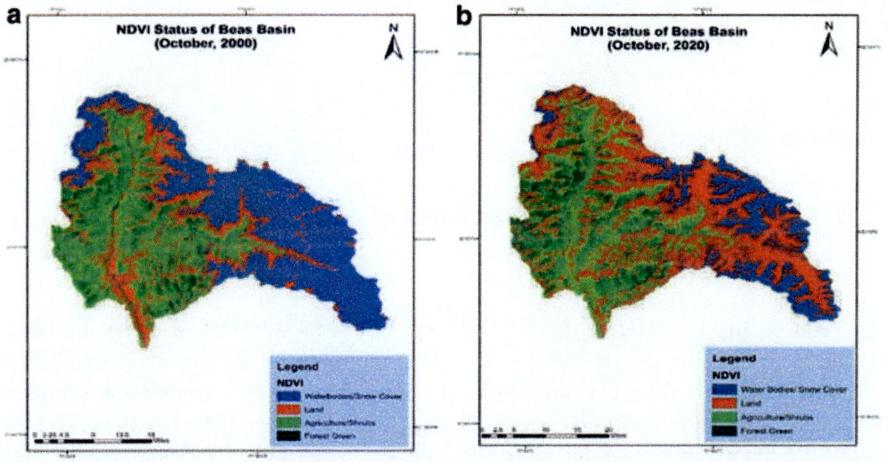

Fig. 52.5 **a** NDVI map of Beas Basin (15/10/00). **b** NDVI map of Beas Basin (30/10/20). **c** Relationship of vegetation (NDVI) with LST for 15th October 2000. **d** Relationship of vegetation (NDVI) with LST for 30th October 2020

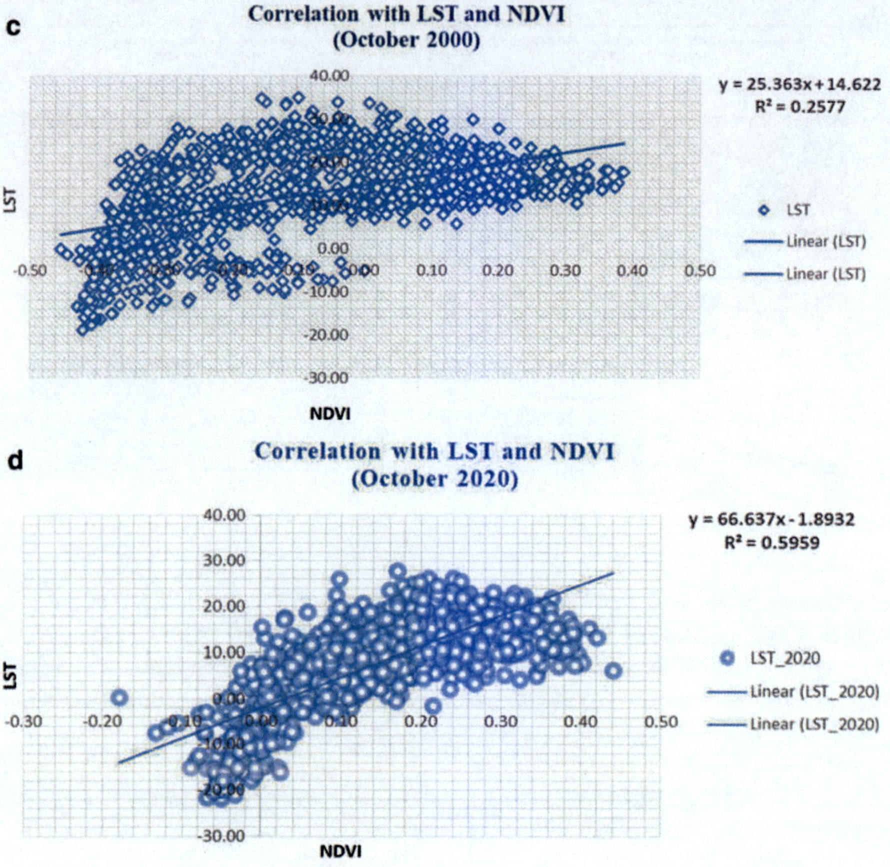

Fig. 52.5 (continued)

52.3.6 Monitoring the Relationship of Settlements (NDBI) with LST (Over 20 Years)

The relationship of settlements (NDBI) with LST was established using the coefficient of correlation in ArcGIS 10.8 for 15th October 2000 (Fig. 52.7c) and 30th October 2020 (Fig. 52.7d) respectively. It shows that positive correlation with the moderate correlation coefficients of $R^2 = 0.4382$ and $R^2 = 0.2081$ between NDBI and LST exist in the Pre-winter season of 2000 (15th October, 2000) and that of 2020 (30th October, 2020) respectively, indicating that, development activities, increasing built-up areas and LST have a positive correlation.

52.3.7 Monitoring the Relationship of Normalized Difference Moisture Index (NDMI) with LST (Over 20 Years)

The relationship of NDMI with LST was established using the coefficient of correlation in ArcGIS 10.8 for 15th October 2000 (Fig. 52.8c) and 30th October 2020 (Fig. 52.8d) respectively. The relationship of Vegetation Moisture Content (NDMI) with LST indicates a negative correlation between moisture content in the vegetation and LST with moderate correlation coefficients of $R^2 = 0.4382$ and $R^2 = 0.2266$ during the Pre-winter season of 2000 (15th October 2000) and that of 2020 (30th October 2020) respectively.

52.4 Conclusions

The present study was conducted in the Beas basin in Kullu District, Himachal Pradesh, India wherein an attempt has been made to monitor the variation in various air quality parameters with respect to alteration in LULC over 20 years as well as examined the change in LST over major LULC classes of the area over the time (20 years). Observation data revealed that the average monthly surface concentration of CO showed an increase of 15% in 2020 than that of 2000. At the same time, the average monthly Total Column Ozone showed a comparatively little rise of 2.25% in 2020 than that of 2000. The average monthly SO2 column mass density

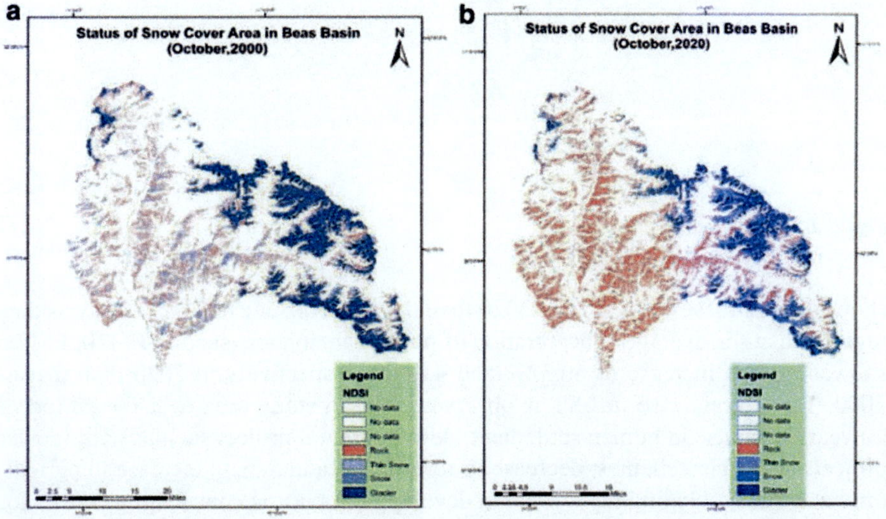

Fig. 52.6 **a** Snow Cover status of Beas Basin (15/10/00). **b** Snow Cover status map of Beas Basin (30/10/20). **c** Relationship of snow cover (NDSI) with LST (15/10/00). **d** Relationship of snow cover (NDSI) with LST (30/10/20)

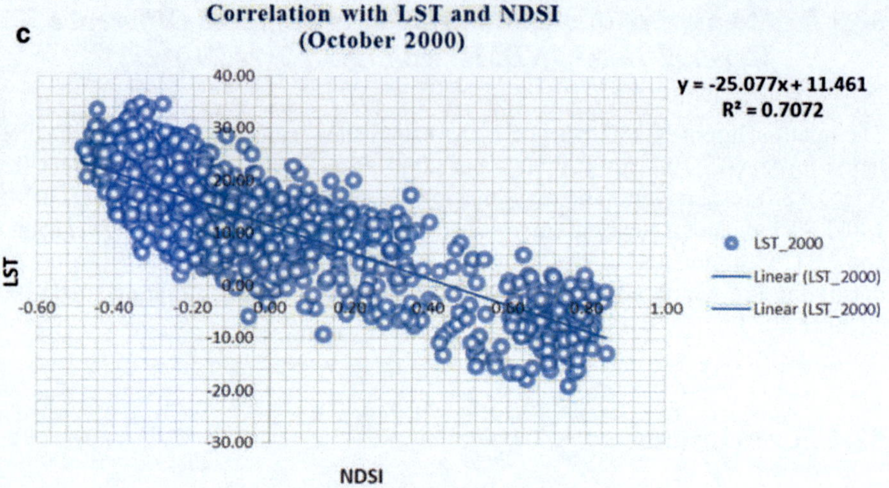

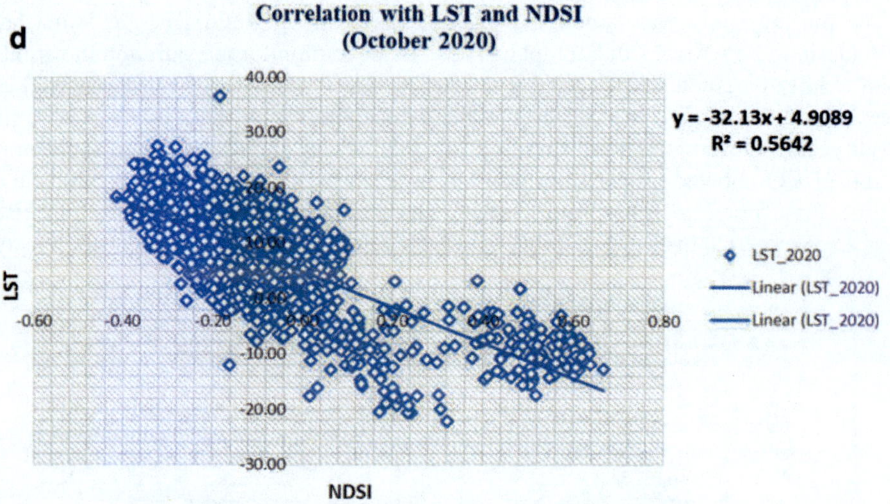

Fig. 52.6 (continued)

showed an increase of 15.87% in 2020 than that of 2000. MERRA-2 reanalysis data revealed that the average concentration of particulate matters such as PM10, PM2.5 showed a huge increase of 60.32% and 41.30% respectively in 2020 than that of 2000. The average rise in LST is observed 2 °C in study area over the period of 20 years. Increase in human settlement, developmental project such as hydropower projects and other schemes, decrease in forested area and huge increase in agricultural activities including rapid orchard development play as contributing factors for the rise in LST. The relationship of vegetation (NDVI) with LST indicate positive correlation between NDVI and LST with the value of correlation coefficient $R2 = 0.2577$ in 2000, which increased to $R2 = 0.5959$ in 2020. This positive relationship

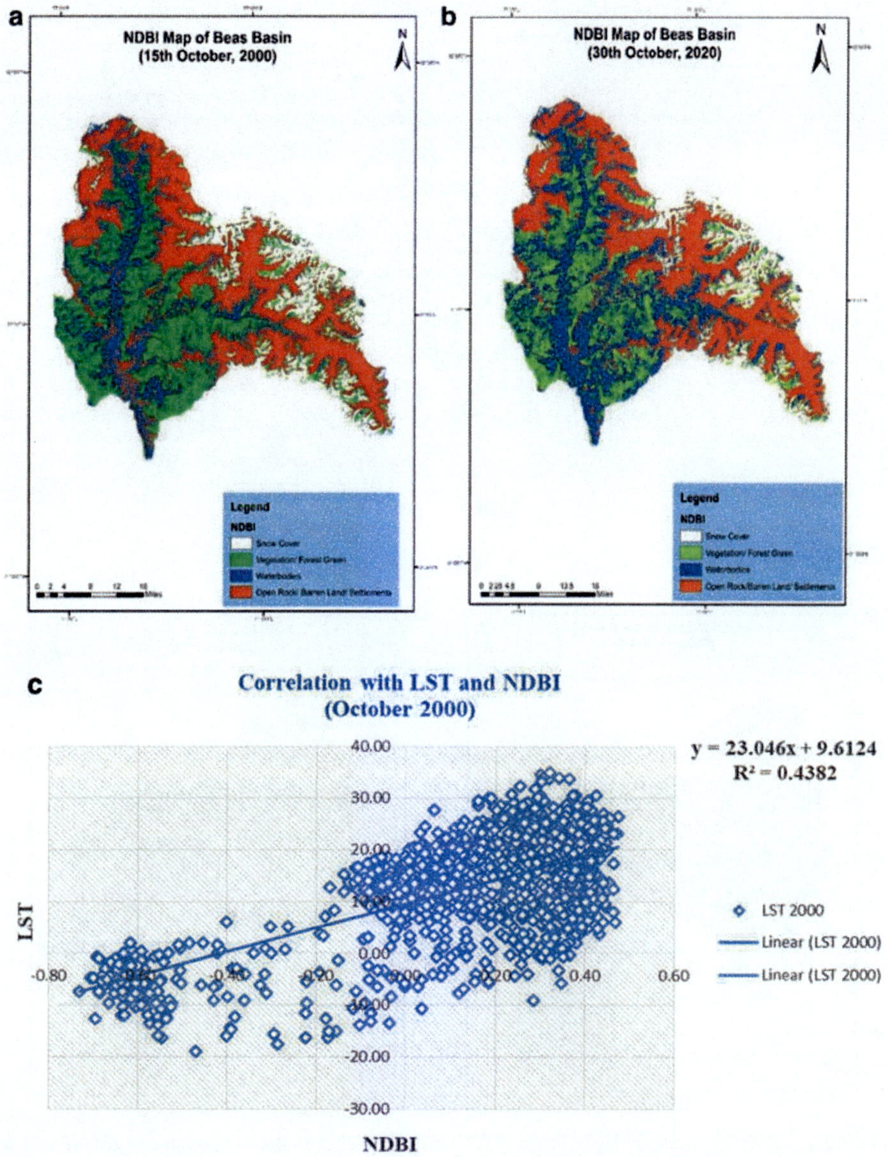

Fig. 52.7 **a** NDBI map of Beas basin(15/10/00). **b** NDBI map of Beas basin (30/10/20). **c** Relationship of settlements/built-up areas (NDBI) with LST (15/10/00). **d** Relationship of settlements/built-up areas (NDBI) with LST (30/10/2020)

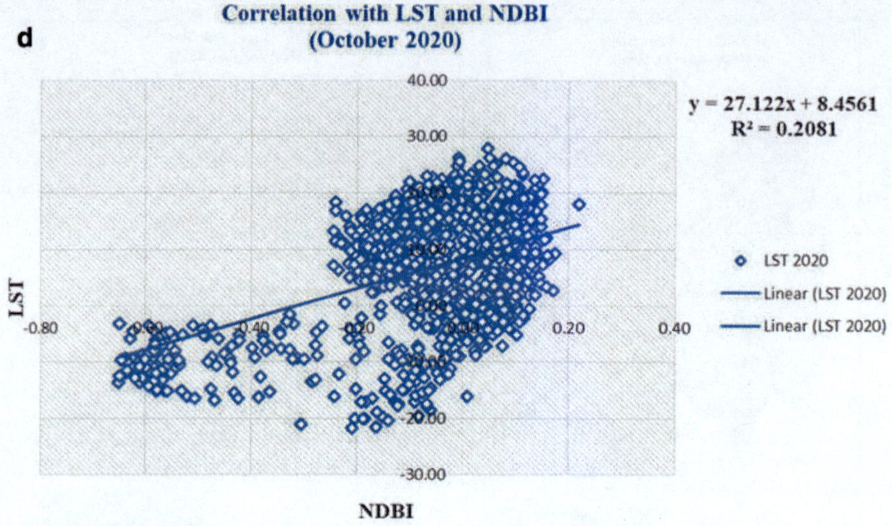

Fig. 52.7 (continued)

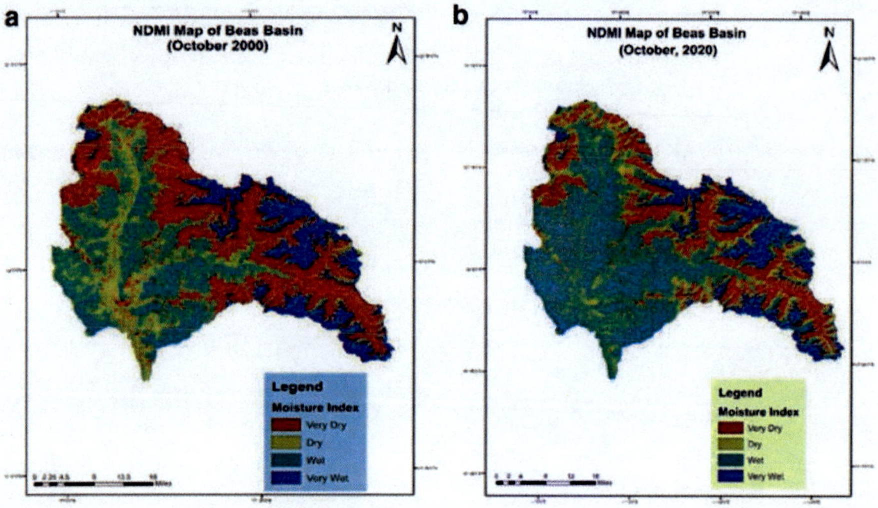

Fig. 52.8 **a** NDMI maps of Beas basin (15/10/00). **b** NDMI maps of Beas basin (30/10/20). **c** Relationship of vegetation moisture index (NDMI) with LST for 15th October 2000. **d** Relationship of vegetation moisture index (NDMI) with LST for 30th October 2020

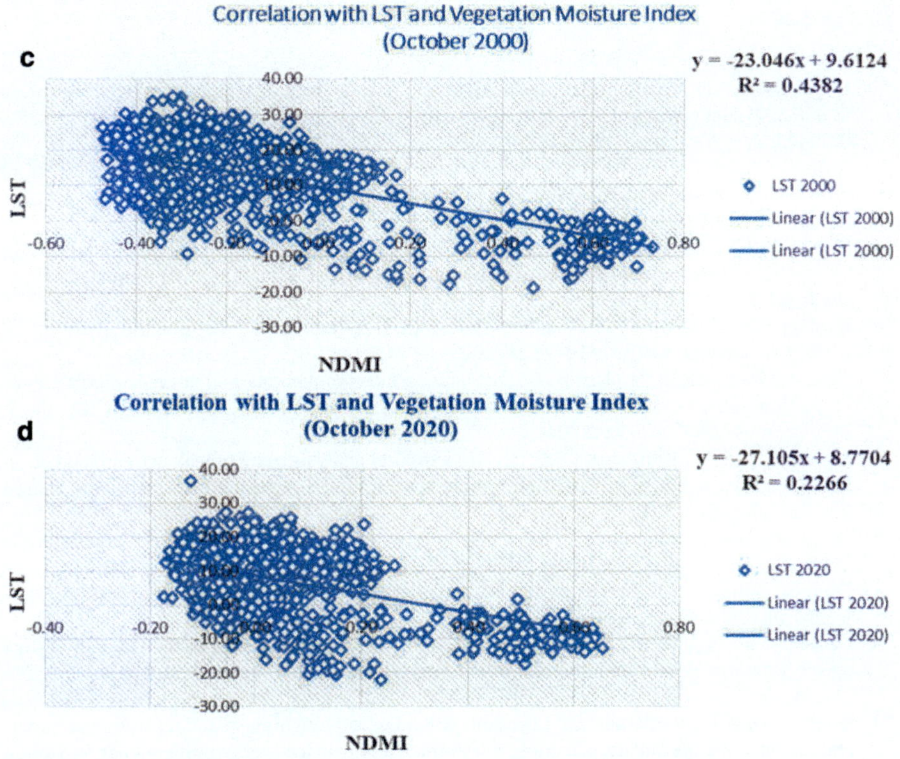

Fig. 52.8 (continued)

indicates that a rise in LST provides favourable conditions for the growth of vegetation in the valley during the pre-winter season of 2000 and 2020 respectively. The relationship of snow cover (NDSI) with LST show that negative correlation with the strong correlation coefficients of $R2 = 0.7072$ and $R2 = 0.5642$ between NDSI and LST in Pre-winter season of 2000 and that of 2020 respectively. The relationship of Normalized Difference Built-up Index (NDBI) with LST indicate positive correlation between Settlements and LST which implies that the increase in settlement areas contribute to increasing LST. The relationship of Vegetation Moisture Content (NDMI) with LST indicate moderately negative correlation between moisture content in the vegetation and LST.

References

1. Jeevalakshmi, D., Reddy S.N., Manikiam B.: Land surface temperature retrieval from LANDSAT data using emissivity estimation. Int. J. Appl. Eng. Res. **12**(20), 9679–9687. ISSN 0973-4562
2. Karnieli, A., Agam, N., Pinker, R.T., Anderson, M., Imhoff, M.L., Gutman, G.G.: Use of NDVI and land surface temperature for drought assessment: merits and limitations. J. Clim. **23**, 618–633 (2010)
3. Nikam, B.R., Ibragimov, F., Chouskey, A., Garg, V., Aggarwal, S.P.: Retrieval of land surface temperature from Landsat 8 TIRS for the command area of Mula irrigation project. Environ. Earth Sci. **75**, 1169 (2016)
4. Ramanathan, V., Crutzen, P., Kiehl, J., Rosenfeld, D.: Aerosols, climate and the hydrological cycle. Science **294**(5549), 2119–2124 (2001)
5. Wan, Z., Wang, P., Li, X.: Using MODIS land surface temperature and normalized difference vegetation index products for monitoring drought in the southern Great Plains, USA. Int. J. Remote Sens. **25**(1), 61–72 (2004)
6. Purwanto, U.D.H., Kurniawan, B.R.: Spatio temporal analysis trend of land use and land cover change against temperature based on remote sensing data in Malang city. Proc. Soc. Behav. Sci. **227**, 232–238 (2016)
7. Pal, S., Ziaul, S.: Detection of land use and land cover change and land surface temperature in English Bazar urban centre. Egypt J. Remote Sens. Space Sci. **20**(1), 125–145 (2017)
8. Igun, E., Williams, M.: Impact of urban land cover change on land surface temperature. Glob. J. Environ. Sci. Manag. **4**(1), 47–58 (2018)
9. Nzoiwu, C.P., Agulue, E.I., Mbah, S., Igboanugo, C.P.: Impact of land use/land cover change on surface temperature condition of Awka Town, Nigeria. J. Geogr. Inf. Syst. **09**(06), 763–776 (2017)
10. Bhuiyan, M.A.E., Witharana, C., Liljedahl, A.K.: Use of very high spatial resolution commercial satellite imagery and deep learning to automatically map ice-wedge polygons across Tundra vegetation types. J. Imaging **6**(12), 137 (2020)
11. Hegazy, I.R., Kaloop, M.R.: Monitoring urban growth and land use change detection with GIS and remote sensing techniques in Daqahlia governorate Egypt. Int. J. Sustain. Built. Environ. **4**(1), 117–124 (2015)
12. John, J., Bindu, G., Srimuruganandam, B., Wadhwa, A., Rajan, P.: Land use/land cover and land surface temperature analysis in Wayanad district, India, using satellite imagery. Ann. GIS **26**(4), 343–360 (2020). https://doi.org/10.1080/19475683.2020.1733662
13. Xu, G., Jiao, L., Zhao, S., Yuan, M., Li, X., Han, Y., Zhang, B., Dong, T.: Examining the impacts of land use on air quality from a spatio-temporal perspective in Wuhan. China. Atmos. **7**, 62 (2016)
14. Superczynski, S., Christopher, S.: Exploring land use and land cover effects on air quality in Central Alabama using GIS and remote sensing. Remote Sens. **3**, 2552–2567 (2011)
15. Survey Documents District Kullu, https://emerginghimachal.hp.gov.in/miningstone/survay_docs/kullu.pdf

Chapter 53
Assessing the Impacts of Anthropogenic Activities on Air Quality: A Study During the Lockdown in 2nd Wave of COVID-19 Pandemic in the Kullu Valley of North-Western Himalaya

Isha Thakur, Renu Lata, Jagdish Chandra Kuniyal, and Sayanta Ghosh

Abstract Criteria pollutants such as CO, SO_2, NO_2, PM_{10}, and $PM_{2.5}$ were monitored in the Kullu valley, a famous tourist destination in Himachal Pradesh, during the lockdown period of the 2nd wave of COVID-19 pandemic. The Pre-Lockdown Period (PLD) from 1 January to 6 May 2021 is taken here as a reference period, while the period from 7 May to 30 June 2021 is the Lockdown Period (LD). The present study was carried out to assess the anthropogenic impact on gaseous and particulate pollutants. Sulfur dioxide (1 h time average) and carbon monoxide (8 h average) were monitored using Thermo Fisher Scientific Gas Analyzers which uses pulse fluorescence technology and gas filter technology, respectively. While NO_2 (24 h average) was analysed by Jacob and Hochessier method, and PM_{10} (24 h average) and $PM_{2.5}$ (24 h average) were monitored using gravimetric method. The concentrations of pollutants were further analyzed incorporating mean $\pm$ standard error. HYSPLIT model was used to determine back trajectory of long-range transport of air pollutants. The results revealed average concentration of CO 0.59 ± 0.03 ppm in the PLD period, while it decreased to 0.35 ± 0.01 ppm in the LD period indicating how CO is largely influenced by vehicular emissions in the valley. Also, particulate matter such as PM_{10} and $PM_{2.5}$ showed a decrease in lockdown period by 65.31% and 52.5%, respectively. The overall air quality was improved in the Kullu valley during lockdown period of the second wave of the COVID-19 pandemic as tourist-related activity and anthropogenic activities were restricted. The MERRA-2 data during the pre-lockdown and the lockdown period was analyzed in ArcGIS 10.8, which indicates almost similar result.

I. Thakur · R. Lata (✉) · S. Ghosh
G.B. Pant National Institute of Himalayan Environment, Himachal Regional Centre, Mohal-Kullu, Himachal Pradesh, India
e-mail: renu15_negi@yahoo.co.in

J. C. Kuniyal (✉)
G.B. Pant National Institute of Himalayan Environment, Kosi-Katarmal, Almora, Uttrakhand, India
e-mail: jckuniyal@gmail.com

E. Yulihastin et al. (eds.), *Proceedings of the International Conference on Radioscience, Equatorial Atmospheric Science and Environment and Humanosphere Science, 2021*, Springer Proceedings in Physics 275, https://doi.org/10.1007/978-981-19-0308-3_53

53.1 Introduction

As we all know how badly COVID-19 pandemic affected our life socially, economically, physically, and mentally in the year 2020. In the year 2021, several countries witnessed 2nd wave of the COVID-19 pandemic, which again led to restrictions and lockdown. India saw 2nd wave of COVID-19 pandemic where the daily count on April 15, 2021, doubled the value during first peak [1]. In Himachal Pradesh, strict lockdown was imposed from 7 May 2021 to 30 June 2021.This time the lockdown was named as 'CORONA CURFEW'. In the year 2020, several studies reported improved air quality as anthropogenic activities were restricted. In 2021, when the lockdown was imposed in Kullu, Himachal Pradesh which is a famous tourist Valley, the tourists' vehicular movement was restricted during the lockdown. Criteria pollutants such as PM_{10}, $PM_{2.5}$, SO_2, CO, and NO_2 were monitored both in pre-lockdown when all anthropogenic activities were ongoing (1 January 2021 to 6 May 2021) and during lockdown period (7 May 2021 to 30 June 2021). This showed a decrease in the concentration of pollutants when all the vehicular movements were restricted, and tourists were not allowed in the valley.

53.2 Study Site

The present study was carried out at Govind Ballabh Pant National Institute of Himalayan Environment (GBP-NIHE), Himachal Regional Centre, Mohal-Kullu (1154 m, 31°54′ N, 77°07′ E). It is located near a famous tourist place Kullu hill spot in Himachal Pradesh (Fig. 53.1). It is one of the stations for monitoring ambient

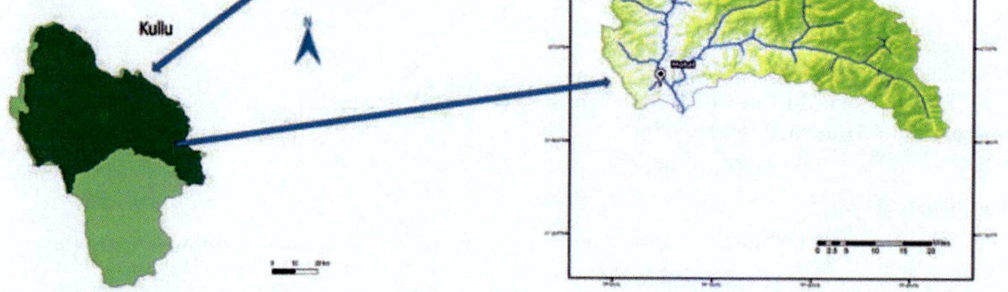

Fig. 53.1 Experimental site of the study at Mohal-Kullu, Himachal Pradesh

air quality using Thermofisher scientific gas analyzers funded by the Indian Space Research Organization (ISRO) under the ISRO-GBP-Atmospheric Chemistry Transport and Modeling (AT-CTM), and Aerosol Radiative Forcing over India (ARFI) Programs. The Kullu-Manali tourist region comprises of the Beas basin in the Kullu valley of Himachal Pradesh in the northwestern Himalaya. The Beas valley begins from Larji (957 m) in the lower Beas basin and stretches upto Rohtang crest (4038 m) within upper Beas basin [2]. The Kullu Valley is a rain shadow region as it is flanked by mountains from either side of the Kullu valley and is a major tourist destination [3]. It is also famous for its international Dusshera fair, which is celebrated every year. In the year 2008, the number of tourists increased significantly in the valley [4]. The Kullu valley, which earlier was an example of a pristine valley with clean air, is now on the way to some of the hill towns in terms of pollution due to significant traffic, hotel construction, mining, introducing lot of hydropower projects, and tourist-related activities.

53.3 Methodology

Sulfur dioxide (SO_2) was measured using an online gas analyzer manufactured by Thermo Scientific Model 43i. SO_2 was monitored for 1 h time average. This model operates on the principle that SO_2 molecule absorbs ultraviolet light and become excited at one wavelength then decays to a lower energy state emitting ultraviolet light at a different wavelength, specifically fluorescence [5]

$$SO_2 + hv_1 \rightarrow SO_2^* - SO_2 + hv_2$$

Model 43i measures concentration from 50 to 1000 ppb with an average time 100–300 s with a flow rate ranging between 0.5 and 1.0 L/min.

The concentration of carbon monoxide (CO, 8 h average) was measured using CO Analyser (Thermo scientific, model 48i). The model 48i is based on the principle that carbon monoxide absorbs infrared radiation at a wavelength of 4.6 micron because infrared absorption is a non-linear measurement technique. The basic analyzer needs to be transformed into a linear output. It has an internally stored calibration curve to accurately linearize the instrument output over any range up to a concentration of 10,000 ppm. Model 48i measures concentration from 0–1 ppm to 100 ppm with its averaging time from 10 to 3000 s and flow rate 0.5–2 L/min. The concentration of nitrogen dioxide (NO_2, 24 h average) was measured using the modified Jacob and Hochhesier method. Ambient NO_2 is collected by bubbling air through a solution of sodium hydroxide and sodium arsenite. The concentration of nitrite ion ($NO^-{}_2$) produced during sampling is determined calorimetrically by reacting the nitrite ion with phosphoric acid, sulfanilamide, and N-(1-naphthyl)-ethylene diamine dihydrochloride (NEDA) and measuring the absorbance of the highly colored azo-dye at 540 nm.

Particulate Matter (PM_{10}) (24 h time average) concentration in ambient air was determined by the gravimetric method using Respirable Dust Sampler (for 24-h average) and concentration is measured in $\mu g/m^3$ by using:

$$C\ (\mu g/m^3) = (w_f - w_i) * 10^6/V$$

where

- C concentration of PM_{10} in $\mu g/m^3$
- w_f final weight of filter in g
- w_i Initial weight of filter in g
- V Volume of air sampled in m^3.

Particulate Matter ($PM_{2.5}$) (24 h time average) concentration in ambient air is also determined by the gravimetric method. The formula used in the calculation of $PM_{2.5}$ is as follows.

$$PM_{2.5} = M_{2.5}/V$$

53.3.1 Air Quality Index (AQI)

The Air Quality Index (AQI) was calculated to determine the overall improvement in air quality during the lockdown. The AQI employs SO_2, O_3, NO_2, NH_3, $PM_{2.5}$, and PM_{10}, with a minimum of three pollutants available. On a scale of 0–500, the concentrations were expressed in number. The Air Quality Index (AQI) for each pollutant, such as, PM_{10}, $PM_{2.5}$ and SO_2 was calculated as follow:

$$AQI = IN_{HI} - In_{LO}/B_{HI} - B_{LO} * (Ci\ - B_{LO}) + IN_{LO}$$

where C_i is the concentration of pollutant and 'i' is the pollutant.

B_{HI} and B_{LO} are breakpoint concentrations greater and smaller than Ci and IN_{HI} and IN_{LO} are corresponding AQI values. The overall AQI is maximum AQIi, and the corresponding pollutant is the dominating pollutant. The AQI categorized six categories; good, satisfactory, moderate, poor, very poor, and severe depending on whether the AQI falls between 0–50, 51–100, 101–200, 201–300, 301–400, and 401–500, respectively [6].

53.4 Results and Discussion

53.4.1 Air Quality During COVID-19 Lockdown

Several previous studies have analyzed the indirect effect of COVID-19 during lockdown on air quality [7–9]. From a recent research on black carbon in the Kullu valley, it is evident that there is a decrease in its concentration during COVID-19 lockdown [10]. Also, studies carried out in the Indo-Gangetic Plain during COVID-19 lockdown shows a reduction in pollutants as well as in Aerosol Optical Depth (AOD) [11]. Most of the recent studies during the COVID-19 lockdown period shows a decrease in the concentration of pollutants such as NO_2, SO_2, CO, $PM_{2.5}$ and PM_{10} in most of the cities, the reason being its reduction took place in transportation, industry and commercial activities [12]. A recent study on the concentration of $PM_{2.5}$ shows a decrease in the concentration of $PM_{2.5}$ during COVID-19 lockdown [13]. Similar results were observed in the Kullu Valley during 2nd wave of the pandemic when the lockdown was imposed.

53.4.2 Sulfur Dioxide (SO_2)

In the year 2021, sulfur dioxide showed an average concentration 2.62 ± 0.10 ppb in the Pre-Lockdown (1 January 2021 to 6 May 2021) period, while it decreased to 2.16 ± 0.01 ppb during lockdown (7 May 2021 to 31 June 2021). Figure 53.2a shows a fall in diurnal variation of sulfur dioxide as vehicular movement was restricted. SO_2 showed a decrease by 17.17% in the average concentration of pollutants during lockdown period.

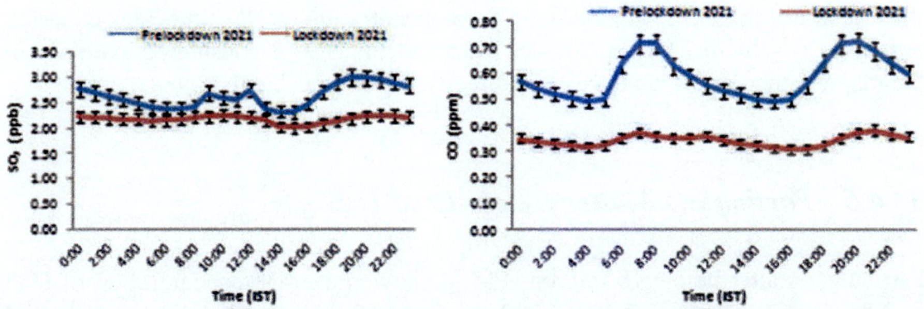

(a) (b)

Fig. 53.2 **(a)** Diurnal variation of sulphur dioxide (ppb), and **(b)** carbon monoxide (ppm) during pre-lockdown and lockdown in the year 2021

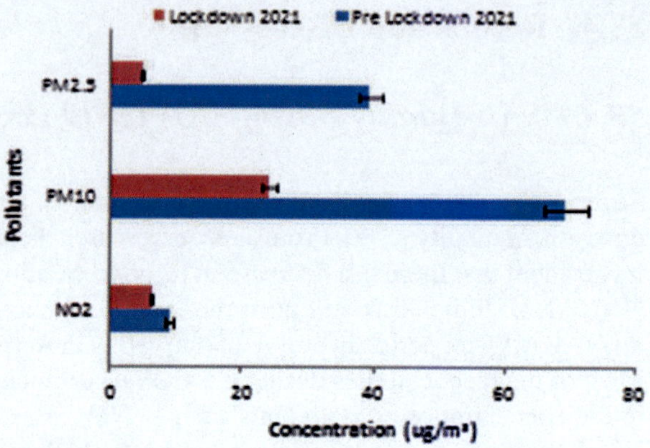

Fig. 53.3 Average concentration of NO_2, PM_{10} and $PM_{2.5}$ during pre-lockdown and lockdown in the year 2021

53.4.3 *Carbon Monoxide (CO)*

Carbon monoxide showed an average concentration 0.59 ± 0.03 ppm in the Pre-Lockdown (1 January 2021 to 6 May 2021) period, while it decreased to 0.35 ± 0.01 ppm during lockdown (7 May 2021 to 31 June 2021). Figure 53.2b shows a decrease in diurnal variation of carbon monoxide. An apparent decrease is visible when the anthropogenic activities were restricted. CO showed a decrease by 40.67% in the average concentration of pollutants during lockdown period.

53.4.4 *Nitrogen Dioxide (NO_2)*

Nitrogen Dioxide (24 h) showed an average concentration of 9.0 ± 0.73 $\mu g/m^3$ during pre-lockdown period while it decreased to 6.2 ± 0.4 $\mu g/m^3$ during the lockdown as illustrated in Fig. 53.3. A decrease in the concentration of NO_2 indicates emission reduction as vehicular movement was restricted and NO_2 is considered to be one of the major vehicular pollutants as it is directly linked to vehicular movement.

53.4.5 *Particulate Matter Below 10 M (PM_{10})*

Particulate matter below 10 micron (PM_{10}) showed average concentration of 69.3 $\pm$ 2.85 $\mu g/m^3$ in PLD (1 January 2021 to 6 May 2021) period, while it decreased up to 24.08 ± 0.64 $\mu g/m^3$ during lockdown (7 May 2021 to 31 June 2021). A clear decrease in PM_{10} is visible when the anthropogenic activities were restricted. PM_{10} showed a decrease by 65.31% during lockdown period as compared to pre-lockdown period (Fig. 53.3).

53.4.6 Particulate Matter Below 2.5 M ($PM_{2.5}$)

In the year 2021, $PM_{2.5}$ showed an average concentration of 39.4 ± 4.78 ug/m^3 in the PLD (1 January 2021 to 6 May 2021) period, while it decreased to 18.7 ± 2.64 μg/m^3 during lockdown (7 May 2021 to 31 June 2021). An apparent decrease is visible when all the anthropogenic activities were restricted. $PM_{2.5}$ showed a decrease by 52.5% in the lockdown period (Fig. 53.3).

53.4.7 Analysis of MERRA-2 Data Using ArcGIS10.8

MERRA-2 data were analyzed using ArcGIS 10.8, which indicated similar results. MERRA-2 data analysis was made for carbon monoxide indicating concentration from 69.75 to 95.77 ppb during pre-lockdown 2021. These values decreased from 66.15 to 80.17 ppb during lockdown when anthropogenic activities were restricted. Figure 53.4a, b show average concentration of CO during lockdown and pre-lockdown periods using MEERA-2 data.

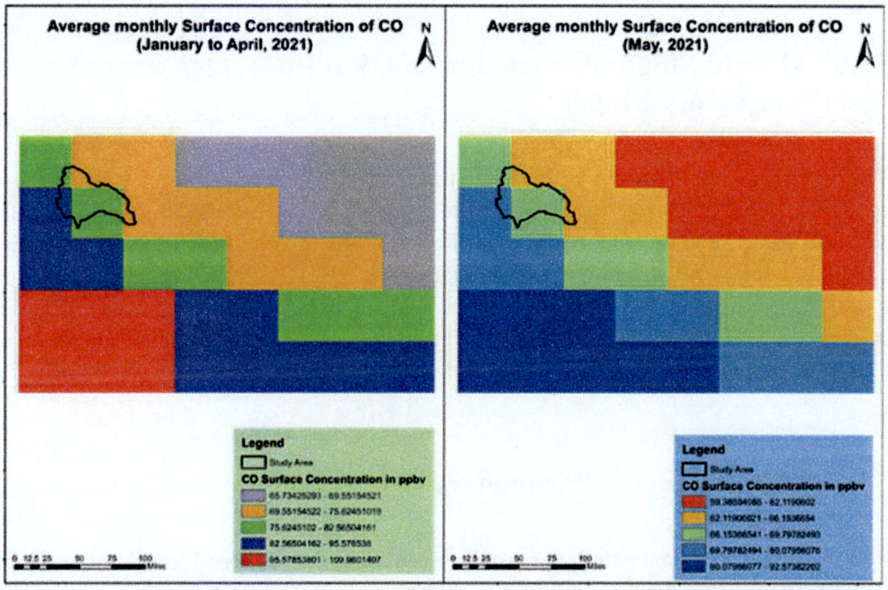

(a) **(b)**

Fig. 53.4 **(a)** Average concentration of carbon monoxide during pre-lockdown, and **(b)** lockdown as observed from MERRA-2 reanalysis of data

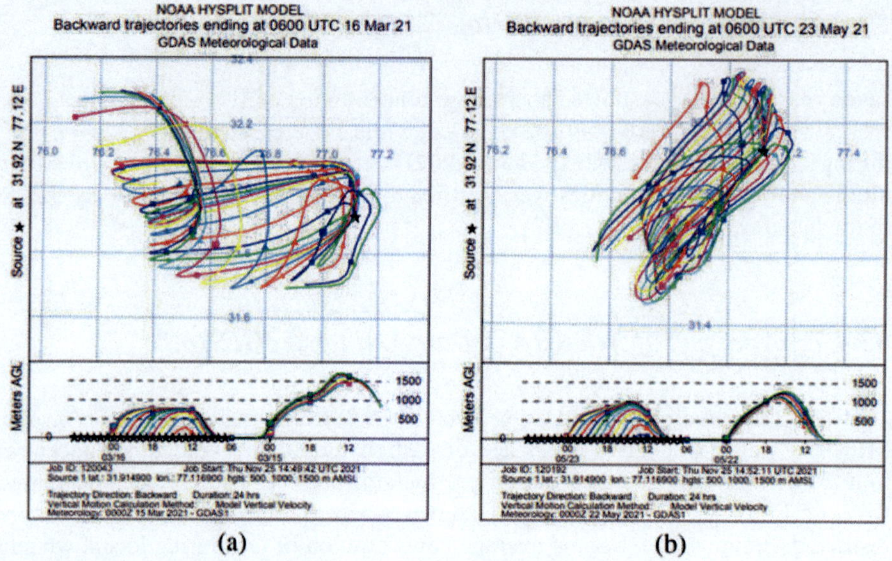

Fig. 53.5 HYSPLIT back trajectory during: **(a)** pre-lockdown (16 March 2021), and **(b)** lockdown (23 May 2021) on high pollution days

53.5 Hybrid Single-Particle Lagrangian Integrated Trajectory Model

The Hybrid Single-Particle Lagrangian Integrated Trajectory model was used to compute back trajectories of air parcels to determine how far and in what direction a parcel of air, and subsequently air pollutants will travel. Back trajectories were calculated for periods of high pollution days in 2021, both before and after the lockdown. During the pre-lockdown period, PM_{10} reached a maximum concentration of 127.4 $\mu g/m^3$ on March 16, 2021 (Fig. 53.5a). However, during the lockdown period, PM_{10} showed a maximum of 28.6 $\mu g /m^3$ on May 23, 2021 (Fig. 53.5b).

53.6 Meteorological Parameters

Specific meteorological parameters such as rainfall and wind speed sometimes lead to temporary reduction in pollutants. The pollutants decreased mainly due to restriction in anthropogenic activities rather than rainfall and wind speed. In the present study area, wind speed is very low in both-during pre-lockdown and lockdown periods. In the study area, wind speed was light, so the pollutants could not blown away by the wind. Hence, wind speed shows no significant correlation with the pollutants,

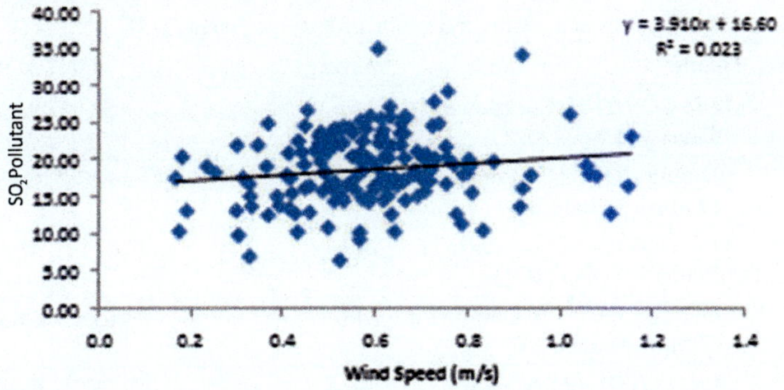

Fig. 53.6 Correlation between wind speed and SO_2 pollutant in the study area

Table 53.1 Average rainfall in the Kullu valley during pre-lockdown and lockdown, and concentration of pollutants

	Pre-lockdown 2021	Lockdown 2021
Average Rainfall (mm)	342.9	154
SO_2 (ppb)	2.62 ± 0.10	2.16 ± 0.01
CO (ppm)	0.59 ± 0.03	0.35 ± 0.01
NO_2 (μg/m^3)	9.0 ± 0.73	6.2 ± 0.4
PM_{10} (μg/m^3)	69.3 ± 2.85	24.08 ± 0.64
$PM_{2.5}$ (μg/m^3)	39.4 ± 4.78	18.7 ± 2.64

thus these pollutants are not going to be affected due to wind speed in the region. Figure 53.6 shows correlation between SO_2 pollutant and wind speed in the region.

Also, the average rainfall in the Kullu valley during pre-lockdown period was 342.9 mm (Meteorology Department, Shimla), while the average rainfall was 154 mm during lockdown period. Hence, decrease in pollutants was not influenced by rainfall during lockdown period which was directly related to anthropogenic activities. Table 53.1 shows average rainfall in the Kullu valley during pre-lockdown and lockdown, and the decrease in the concentration of pollutants was also observed.

53.7 Air Quality Index

The Air Quality Index (AQI) is a basic tool for effective communication of the status of air quality. The AQI converts complex air quality data from multiple pollutants into a single number (index value) nomenclature, as well as, indicates its health impacts. Good, satisfactory, moderately polluted, extremely bad, and severe are the six AQI categories. The value of ambient concentration of air pollutants and their negative health consequences determine these categories (health breakpoints). The air quality was measured during pre-lockdown and lockdown periods. During lockdown time, air quality improved a lot and its index shifted from moderate to good (Table 53.2).

Table 53.2 AQI categories and health breakpoints for the pollutants [14]

Pollutants	Time	Average	AQI	AQI Category
PM_{10}	Pre-COVID-19 Lockdown (1 January–6 May 2021)	24 h	58	Satisfactory
	During COVID-19 Lockdown (7 May–31 June 2021)	24 h	22	Good
$PM_{2.5}$	Pre-COVID-19 Lockdown (1 January–6 May 2021)	24 h	111	Moderate
	During COVID-19 Lockdown (7 May–31 June 2021)	24 h	65	Good
NO_2	Pre-COVID-19 Lockdown (1 January-6 May 2021)	24 h	13.5	Good
	During COVID-19 Lockdown (7 May–31 June 2021)	24 h	7.75	Good
Overall AQI	Pre-COVID-19 Lockdown (1 January–6 May 2021)	1 January–6 May 2021	111	Moderate
	During COVID-19 Lockdown (7 May–31 June 2021)	7 May–31 June 2021	65	Good

53.8 Conclusion

The study summaries as under:

1. PM_{10} showed a decrease by 65.31% during lockdown period as compared to pre-lockdown of the year 2021.
2. $PM_{2.5}$ showed a decrease by 52.5% during lockdown period.
3. CO showed a decrease by 40.67% concentration.
4. SO_2 showed a decrease by17.17% concentration.
5. MERRA-2 data was analyzed using ArcGIS 10.8, which showed a decrease in its concentration during lockdown period in 2021.
6. Overall, air quality was found to be improved during lockdown when anthropogenic activities were restricted. There were no tourists in the valley. Air Quality Index shifted from 'moderate' to 'good' during lockdown in 2021.
7. All the pollutants were found within prescribed limits of National Ambient Air Quality Standards.

Acknowledgements The authors sincerely thank to the Director, G.B. Pant National Institute of Himalayan Environment (NIHE), Kosi-Katarmal, Almora, Uttarakhand for providing the necessary facilities. Authors acknowledge with thanks to ISRO, Bengaluru for providing partial financial support to carry out the present work under two programmes of ISRO-GBP, i.e., Atmospheric Chemistry Transport and Modeling (AT-CTM) and Aerosol Radiative Forcing over India (ARFI). The authors are also thankful to the organizers of the International Conference on Radioscience, Equatorial Atmospheric Science and Environment (INCREASE) and the 6th Asia Research Node, LAPAN and Kyoto University for giving the opportunity to present this paper in the International Conference. We are thankful to National Oceanographic and Atmospheric Administration (NOAA) for using Hybrid Single-Particle Lagrangian Integrated Trajectory (HYSPLIT) model as well as MERRA-2 data for being used in the present publication.

References

1. Ranjan, R., Sharma, A., Verma, M.K.: Characterization of the second wave of covid-19 in India (2021). https://doi.org/10.1101/2021.04.17.21255665
2. Kuniyal, J.C., Jain, A., Shannigrahi, A.: Environmental impacts of tourism in Kullu-Manali complex in North Western Himalaya, India. Part 1: The adverse impacts. Int .J. Fieldwork Stud. **1**, 1–21 (2003)
3. Kuniyal, J.C., Vishvakarma, S.C.R., Badola, H.K., Jain, A.P.: Tourism in Kullu Valley. An environmental assessment (Bishen Singh & Mohindra Pal Singh, Dehradun, India), pp. 1–210 (2004)
4. Sharma, N., Kuniyal, J.C., Singh, M., Sharma, P., Chand, K., Negi, A., Sharma, M., Thakur, H.: Atmospheric ultrafine aerosol number concentration and its correlation with vehicular flow at two sites in the Western Himalayan region: Kullu-Manali, India. J. Earth Syst. Sci. **120**(2), 281–290 (2011). https://doi.org/10.1007/s12040-011-0046-9
5. Lv, H., Wang, Z., Gao, N.: A fluorescence method of detecting sulfur dioxide derivatives. IOP Conf. Ser. Earth Environ. Sci. **191**, 012107 (2018). https://doi.org/10.1088/1755-1315/191/1/012107
6. Sharma, S., Zhang, M., Anshika, Gao, J., Zhang, H., Kota, S.H.: Effect of restricted emissions during COVID-19 on air quality in India. Sci. Total Environ. **728**, 138878 (2020). https://doi.org/10.1016/j.scitotenv.2020.138878
7. Fattorini, D., Regoli, F.: Role of the chronic air pollution levels in the COVID-19 outbreak risk in Italy. Environ. Pollut. **264**, 114732 (2020). https://doi.org/10.1016/j.envpol.2020.114732
8. Bashir, M.F., Ma, B.J., Bilal, Komal, B., Bashir, M.A., Farooq, T.H., Iqbal, N., Bashir, M.: Correlation between environmental pollution indicators and COVID-19 pandemic: a brief study in Californian context. Environ. Res. **187**, 109652 (2020). https://doi.org/10.1016/j.envres.2020.109652
9. Collivignarelli, M.C., Abbà, A., Bertanza, G., Pedrazzani, R., Ricciardi, P., Carnevale Miino, M.: Lockdown for Covid-2019 in Milan: what are the effects on air quality? Sci. Total Environ. **732**, 139280 (2020). https://doi.org/10.1016/j.scitotenv.2020.139280
10. Gogoi, M.M., SureshBabu, S., Arun, B.S., Moorthy, K.K., Ajay, A., Ajay, P., Suryavanshi, A., Borgohain, A., Guha, A., Shaikh, A., Pathak, B., Gharai, B., Ramasamy, B., Balakrishnaiah, G., Menon, H. B., Kuniyal, J. C., Krishnan, J., RamaGopal, K., Maheswari, M., Naja, M., Kaur, P., Bhuyan, P.K., Gupta, P., Singh, P., Srivastava, P., Singh, R.S., Kumar, R., Rastogi, S., Kundu, S.S., Kompalli, S.K., Panda, S., Rao, T.C., Das, T., Kant, Y.: Response of ambient BC concentration across the Indian region to the nation-wide lockdown: results from the ARFINET measurements of ISRO-GBP. Curr. Sci. **120**(2), 341 (2021). https://doi.org/10.18520/cs/v120/i2/341-351
11. Thomas, A., Kanawade, V.P., Sarangi, C., Srivastava, A.K.: Effect of covid-19 shutdown on aerosol direct radiative forcing over the Indo-Gangetic plain outflow region of the Bay of Bengal. Sci. Total Environ. **782**, 146918 (2021). https://doi.org/10.1016/j.scitotenv.2021.146918
12. Fu, F., Purvis-Roberts, K.L., Williams, B.: Impact of the COVID-19 pandemic lockdown on air pollution in 20 major cities around the world. Atmosphere **11**(11), 1189 (2020). https://doi.org/10.3390/atmos11111189

13. Patra, S.S., Das, T., Ramasamy, B.: Estimating the inadvertent decrement in mortality due to reduction in ambient fine particulate concentrations during COVID-19 lockdown in India. Aerosol Sci. Eng. **5**(2), 247–252 (2021). https://doi.org/10.1007/s41810-021-00097-3
14. CPCB: National Air Quality Index report Central Pollution Control Board, New Delhi, India, pp. 1–58 (2014)

Chapter 54
Ozone Variability at 10 hPa Altitude in Indonesia Based on MERRA-2 Data

Prawira Yudha Kombara and Ninong Komala

Abstract The variability of ozone at 10 hPa altitude in the Indonesian region has been studied in this research. This study use MERRA-2 data from 2003 to 2020 for Indonesia region with coordinates 95°–145° E and 12° S–12° N. MERRA-2 is a reanalysis data which was generated from global models and NASA assimilation systems. MERRA-2 has the advantage of higher spatial resolution compared to satellite data. The method used in this research is Fast Fourier Transform (FFT) and Wavelet analysis. Based on the results of FFT and Wavelet analysis, we achieved the results that during the period 2003–2020, ozone in the territory of Indonesia at an altitude of 10 hPa has a seasonal pattern. It is also supported by averaging the area of the entire region of Indonesia for the same height and then did monthly composite, which results in two peaks in the one-year graph with six months interval. Then, we achieved the other results based on the spatial pattern. During Asian winter monsoon period, the most minimum of ozone concentration exists in the northern region. The another one is when Australian winter monsoon the most minimum of ozone concentration exists in the southern region.

54.1 Introduction

Ozone is one of the gases found in the earth atmosphere. Ozone is very abundant quantities is found in the stratosphere layer, where this layer has a role to filter ultraviolet-B rays from the sun [5]. Because the amount is very abundant, it is often referred to as the ozone layer [5]. Learning the pattern of the ozone layer is very important to provide a better understanding of the earth atmosphere. Learning the ozone layer can be done in several ways, such as by processing observational data. However, the observation of the ozone profile cannot be carried out in all areas such as in the tropics, especially in Indonesia. The ozone observation in Indonesia is still

P. Y. Kombara (✉) · N. Komala
Research Center for Climate and Atmosphere, National Research and Innovation Agency (BRIN), Bandung, Indonesia
e-mail: praw005@brin.go.id

E. Yulihastin et al. (eds.), *Proceedings of the International Conference on Radioscience, Equatorial Atmospheric Science and Environment and Humanosphere Science, 2021*, Springer Proceedings in Physics 275, https://doi.org/10.1007/978-981-19-0308-3_54

limited and rarely done. To anticipate the lack of observation data, using satellite observation data or reanalysis can be an alternative. Even so, satellite observation data still has a limitation, such as the spatial resolution is still rough. Generally, the satellite observation data produced by NASA has a spatial resolution of 1 degree or 111 km × 111 km. For studies in a wide area, data with a resolution of 1 degree is good enough to describe the climatological patterns. However, for the research needs the data with a higher spatial resolution, the reanalysis data can be used. One of the reanalysis data that can be used is The Modern-Era Retrospective Analysis for Research and Applications, version 2 (MERRA-2) data developed by NASA in the GMAO project. The MERRA-2 continues its predecessor and keep the GMAO's commitment to provide ongoing and real-time reanalysis data [4]. The MERRA-2 data has a spatial resolution of 0.5 × 0.625 degrees, and based on the evaluation conducted by Wargan et al. [8] and Sudeepkumar et al. [7], the MERRA-2 data can describe well the climatological pattern of ozone and can be used for relevant applications [1, 3, 7–9]. In addition, there is other study that use Merra-2 too for learn ozone profile variability from surface [6]. Therefore, the aim of this study is to examine the variability of ozone at 10 hPa altitude for the territory of Indonesia in 2003–2020 using MERRA-2 data.

54.2 Data and Method

In this study, we use MERRA-2 reanalysis data which can be downloaded on the web page: https://disc.gsfc.nasa.gov/ [4]. This data has a spatial resolution of 0.5° × 0.625 ° and temporal resolution of monthly. The range of data is from 2003 to 2020. To learn the spatial and temporal variability of ozone at an altitude of 10 hPa in Indonesia in the period of 2003–2020 the MERRA-2 data will be processed using the Fast Fourier Transform (FFT) wavelet analysis and Exploratory Data Analysis (EDA) methods.

The FFT method changes the time domain to the frequency domain of data. The change of domain has a purpose to find the dominant signal from data. Furthermore, the year when the most dominant signal occurs can be found with use wavelet analysis method. Wavelet analysis method has been used widely in atmospheric science in the last few decades. This method uses same principal with FFT method. The time domain of data is changed become frequency domain to find the dominant signal of data [2]. Next, the dominant signal is shown over time of data. Therefore, we can see from the 2003 until 2020 period, the most of dominant signal of data appears. In addition, the EDA method is used to find the pattern of data. For example, in one year what month the peak of data occurs and how the pattern of data in every season, etc. The season of this study is divided become four seasons that consists of December–January–February (DJF), March–April-May (MAM), June-July–August (JJA), and September–October–November (SON).

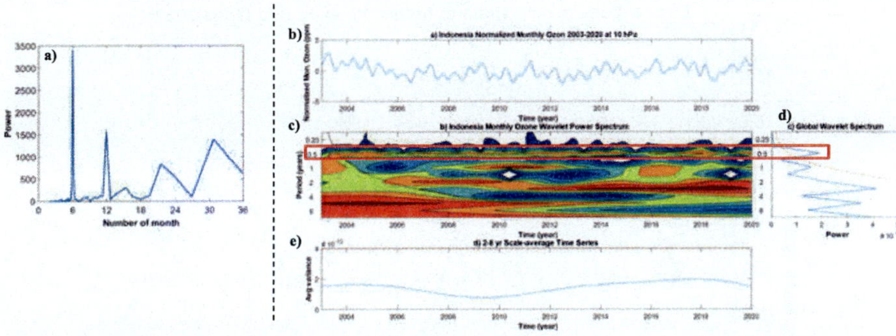

Fig. 54.1 FFT result (**a**), wavelet analysis result consists of normalized of Indonesia ozone from 2003 to 2020 (**b**), Indonesia monthly ozone wavelet power spectrum (**c**), global wavelet spectrum (**d**), and Scale average of time series (**e**)

54.3 Result and Analysis

54.3.1 FFT and Wavelet Analysis

The result of data processing uses FFT and Wavelet is shown in Fig. 54.1. As mention earlier, FFT converts the time domain of data to the frequency domain for finding the most of dominant signal from the data.

Based on the FFT result as shown in Fig. 54.1a, the Ozone from Merra-2 data has two dominant signals. The first signal is six months and the second signal is 12 months period. However, the most of dominant signal is six months periods. Therefore, the ozone from Merra-2 data has two peaks in one year. This result is confirmed by the wavelet analysis result that shown in Fig. 54.1c, d. The wavelet result shows that the six months period appears in every year consistently. Both of these results confirm a fact that the ozone in Indonesia at an altitude of 10 hPa has a dominant period of six months. The same thing was also found in previous research conducted by Wulandari and Dewi [10]. In previous studies using MSL/Aura (NASA) satellite data and the same method, ozone in Indonesia at an altitude of 10 hPa has a dominant period of six months [10].

54.3.2 Time Series and Composite Analysis

Ozone in Indonesia at an altitude of 10 hPa has a dominant period of six months which means that in a year, it has two peaks. To find out what month the peak occurs in a year, it can be done by monthly averaging data in a number of years. Figure 54.2 shows the monthly composite results of ozone data in Indonesia at an altitude of 10 hPa in the 2003–2020 range.

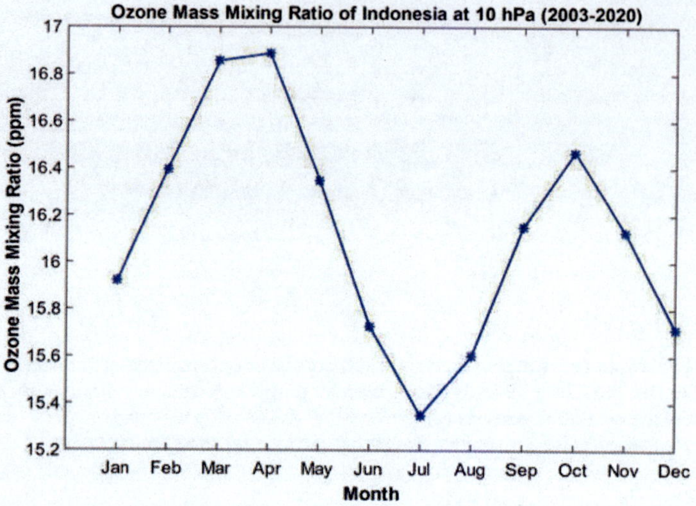

Fig. 54.2 The monthly composite of ozone (2003–2020)

Based on Fig. 54.2, the first peak of ozone occurs in April and the second peak in October. Even though, the reason why the first peak is higher than the second peak still unknown. There is a possibility that the first peak is higher than second peak caused by ultraviolet (UV) radiation from sun. However, that guess still need further investigation. Besides, this result is similar to previous research was conducted by Wulandari and Dewi [10]. Their result show two peaks in one year and occur in March and October. However, the peaks of ozone at 10 hPa in Indonesia occur in March and October relate to the pseudo motion of the sun, or it is known as the equinox. The equinox is the natural phenomenon where the sun position above the equator line exactly. In a year, the equinox occurs two times in March and September. With the sun position above the equator line exactly, the ozone's concentration increases quickly above the equator region. Therefore, it gives the result of the enhancement of Ozone mass of mixing ratio in March and October every year [10].

54.3.3 Spatial and Seasonal Pattern

In the previous subchapter, the time-series image shows two peaks in one year. If We see the results of the spatial visualization in Fig. 54.3 from January to December, it can be seen that there are two times ozone concentration increase as shown in Fig. 54.2. The two months are March and September, in which the equinox phenomenon occurs. This spatial image shows that the equinox phenomenon affects the mixing ratio of ozone at an altitude of 10 hPa in Indonesia. When the equinox in March and September occurs, the position of the sun directly above the equator makes the chemical reaction of ozone at an altitude of 10 hPa increase significantly [5]. If we

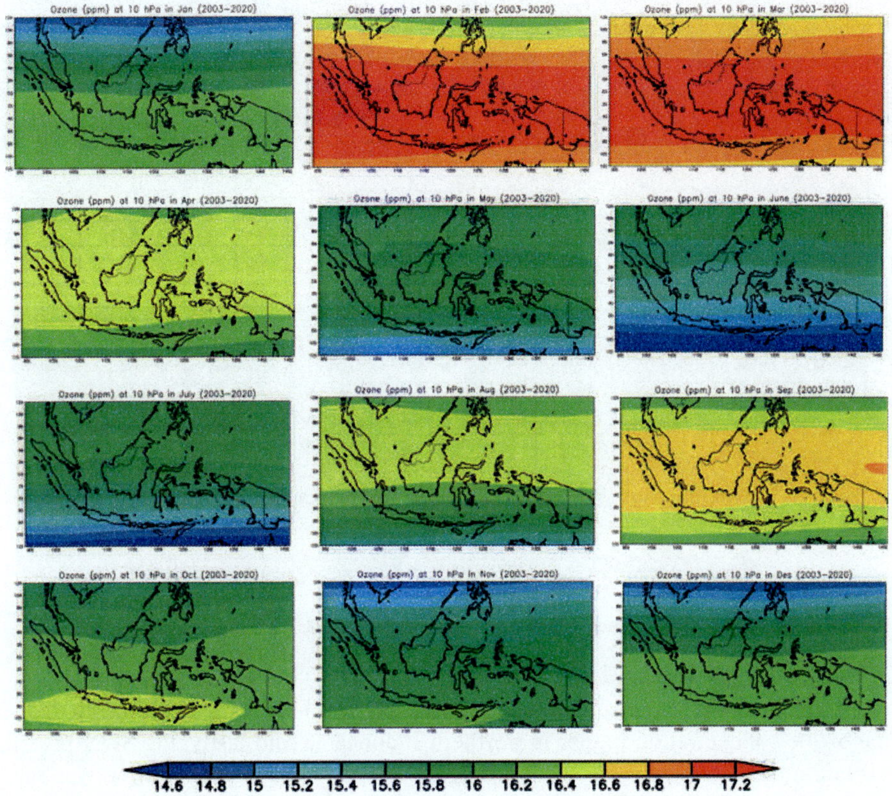

Fig. 54.3 Monthly composite of ozone at 10 hPa spatially from January to December for 2003–2020 period

look closely, the spatial pattern of ozone from January to December follows the apparent movement of the sun towards the earth. As in January, areas of lower ozone concentration are in northern Indonesia. In January, the Asian winter monsoon is occurring, and the sun is in the southern hemisphere, so that the ozone concentration will be higher in southern Indonesia. Likewise, the opposite happens in July, which is the month in which the Australian winter monsoon is occurring and the sun is in the northern hemisphere. Therefore, higher ozone concentrations occur in northern Indonesia.

Figure 54.4 shows the seasonal spatial visualization starting from DJF, MAM, JJA, and SON. If you look at the seasonal pattern, the effect of the apparent motion of the sun on the ozone concentration in Indonesia at altitude of 10 hPa is more obvious. During the DJF period, the Asian winter monsoon is occurring and the sun's position is in the southern hemisphere so that the ozone concentration is higher in the southern part of Indonesia. Then for the MAM period, where the equinox occurs, the sun's position is directly above the equator which causes the ozone concentration in Indonesia to be higher in the middle of Indonesia. Then in the JJA period, the Australian winter

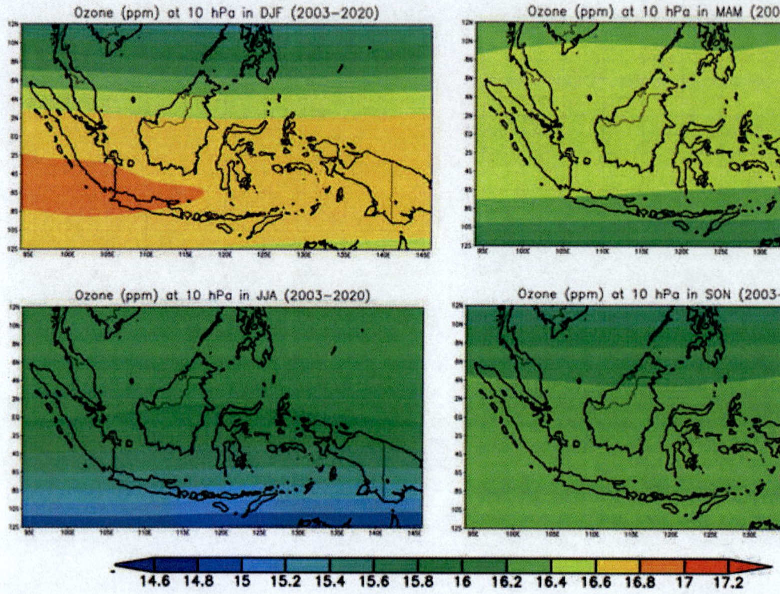

Fig. 54.4 Seasonally composite of ozone at 10 hPa from DJF, MAM, JJA, to SON

monsoon occurs and the sun's position is in the northern hemisphere. This makes the ozone concentration higher in northern Indonesia. And lastly, the SON period, the equinox occurred for the second time so that the ozone concentration was seen more evenly throughout Indonesia.

54.4 Conclusion

Using the FFT method the dominant period of ozone in Indonesia at an altitude of 10 hPa can be confirmed. The dominant periods that appear are the six-month and 12-month periods. However, the most dominant period is the six-month period. Then by using the wavelet analysis method, it can be seen that the six-month period appears consistently every year. In addition, the MERRA-2 reanalysis data is able to show the spatial pattern of ozone in each month for the territory of Indonesia at altitude of 10 hPa. From the results of monthly composite time series and spatial patterns, ozone in Indonesia at altitude of 10 hPa reaches its maximum value in March and September or when the equinox phenomenon occurs. From the results of the spatial plot, it can also be seen that the ozone pattern in Indonesia follows the apparent motion of the sun in one year.

References

1. Coldewey-Egbers, M., Loyola, D.G., Labow, G., Frith, S.M.: Comparison of GTO-ECV and adjusted Merra-2 total ozone columns from the last 2 decades and assessment of interannual variability. Atmos. Meas. Tech. **13** (2020)
2. Domingues, M.O., Mendes, Jr, O., da Costa, A.M.: On wavelet techniques in atmospheric science. Adv. Space Res. **35** (2005)
3. Gelaro, R., McCarty, W., Suarez, M.J., et al.: The modern-era retrospective analysis for research and applications, Version 2 (MERRA-2). J. Clim. **30** (2017)
4. Global Modeling and Assimilation Office (GMAO): MERRA-2 instM_3d_asm_Np: 3d, Monthly mean, Instantaneous, Pressure-Level, Assimilation, Assimilated Meteorological Fields V5.12.4, Greenbelt, MD, USA, Goddard Earth Sciences Data and Information Services Center (GES DISC) (2015). https://doi.org/10.5067/2E096JV59PK7
5. Komala, N.: Analysis of Java Island's ozone layer and ultra violet index variability based on satellite data. In: E3S Web of Conferences, vol. 76, 04001 (2019)
6. Stauffer, R.M., Thompson, A.M., Witte, J.C.: Characterizing global ozonesonde profile variability from surface to the UT/LS with a clustering technique and Merra-2 Reanalysis. J. Geophys. Res.: Atmos. (2018)
7. Sudeepkumar, B.L., Babu, C.A., Varikoden, H.: Atmospheric boundary layer height and surface parameters: trends and relationships over the west coast of India. Atmos. Res. **245**(2020), 105050 (2020)
8. Wargan, K., Labow, G., Frith, S., Pawson, S., Livesey, N., Partyka, G.: Evaluation of the ozone fields in NASA's MERRA-2 reanalysis. J. Clim. **30** (2017)
9. Wargan, K.: Southern polar ozone in Merra-2. Research Brief, GMAO (2016)
10. Wulandari, R.A., Dewi, K.: Analysis of stratospheric ozone and ozone depletion substances variability in quality of UV index over Indonesia related to Montreal protocol. Jurnal Teknik Lingkungan **24**(2) (2018)

Chapter 55
Variation of the Wind Profiles in the Tropical Tropopause Layer Associated with QBO-MJO Connection: An Equatorial Atmosphere Radar Observation

Arlif Nabilatur Rosyidah, Nurjanna Joko Trilaksono, and Noersomadi

Abstract The Quasi-Biennial Oscillation (QBO) is known as the alternating zonal wind in the stratosphere that is discovered to have a connection with tropospheric Madden–Julian Oscillation (MJO), particularly during the northern hemisphere winter. Discussion on the variation of wind profiles in the tropical tropopause layer (TTL), the transition layer between troposphere-stratosphere at 14–18.5 km, related to QBO-MJO connection is still limited due to the availability of high vertical and temporal resolution of observation data. We investigate 3-D wind profiles by Equatorial Atmosphere Radar (EAR) with 150 m vertical resolution during extended boreal winter (November to March) throughout 2003–2009 (16 years). We find five MJO active cases in the easterly QBO phase (QBOE), six during the westerly QBO phase (QBOW), and 20 in the transition phase or neutral QBO (QBON). The QBO index is defined as the change of zonal wind at 18.7 km obtained from EAR data. We observed radar tropopause (maximum echo power within TTL) to be lifted to 17.4 km during MJO active associated with QBOE compared to MJO active period during QBOW (16.7 km) and QBON (16.8 km). We also discover that vertical wind is stronger above the tropopause, representing more updraft activity (0.02–0.03 m/s) in QBOE than QBOW and QBON.

A. N. Rosyidah (✉)
Postgraduate Program in Earth Science, Institut Teknologi Bandung, Bandung, Indonesia
e-mail: arlifnabila05@gmail.com

N. J. Trilaksono
Atmospheric Science Research Group, Faculty of Earth Sciences and Technology, Institut Teknologi Bandung, Bandung, Indonesia

Noersomadi
Research Center for Climate and Atmosphere, National Research and Innovation Agency (BRIN), Bandung, Indonesia

E. Yulihastin et al. (eds.), *Proceedings of the International Conference on Radioscience, Equatorial Atmospheric Science and Environment and Humanosphere Science, 2021*, Springer Proceedings in Physics 275, https://doi.org/10.1007/978-981-19-0308-3_55

55.1 Introduction

QBO is an alternating zonal wind in the stratosphere with a 24–28-month period. The QBO is driven by the momentum fluxes of upward-propagating equatorial waves such as Rossby, Kelvin, Rossby-gravity [1]. Anomalous meridional circulation of the QBO possibly modulate the convective system by adjusting the tropopause height as well as change in static stability near tropopause due to adiabatic cooling and heating induced by the upward and downward motion of anomalous meridional circulations change [2, 3]. Thus, QBO is suspected to be in association with Madden Julian Oscillation (MJO) convective system. MJO is an eastward-moving large convective system in the equator with a periodicity of 30–100 days [4]. As major fluctuation in tropical weather on intraseasonal timescales, the MJO convective system possibly associate with extreme precipitation and lead to environmental hazards [5–7].

A previous study found around 40% of MJO activities are related to QBO in boreal winter [8]. MJO activities are significantly stronger during QBO easterly phase (QBOE) than the westerly phase (QBOW). During QBOE, the tropopause is higher and colder that can lead to lower static stability supported by weaker wind shear. Hence the convective activities are stronger than in QBOW and in the QBO transition/normal phase (QBON) [1, 9]. This phenomenon is an example of stratosphere-troposphere coupling. The alternating zonal wind in the stratosphere also gives impact on the troposphere through the connection in the transition layer known as tropical tropopause layer (TTL) [1, 10]. Therefore, understanding the TTL state could provide a better picture of how these two layers interact.

TTL is a transitional layer between the troposphere and stratosphere spanning from 14 to 18.5 km [11]. The TTL acts as a gate to the stratosphere and understanding all relevant processes in the QBO-MJO connection are essential for better weather prediction. The downward propagating zonal wind oscillation of QBO is found to affect wind profile in TTL that can both disturb or enhance tropical convection. However, discussion on the variation of wind profiles in the TTL related to QBO-MJO connection is still limited due to the availability of high vertical and temporal resolution of observation data. Hence, this study investigates 3-D wind profiles using high-resolution Equatorial Atmosphere Radar (EAR).

55.2 Data and Methods

This study is mainly based on observational data analyses from EAR located in Agam, West Sumatra, Indonesia (0.2° S, 100.32° E, 865 m) from November 2001 to December 2019. Additional analysis of the convective system was carried by using precipitation data from Tropical Rainfall Measuring Mission (TRMM) 3B42 product and Outgoing Longwave Radiation from European Centre for Medium-Range Weather Forecast (ECMWF).

55.2.1 Radar Tropopause

EAR is a circular antenna array radar approximately 110 m in diameter that incorporates 560 three-element Yagi antennas. Each EAR antenna is driven by a solid-state transmitter-receiver module with 180 W peak output power to produce a total peak output power of 100 kW. The active phased-array antenna system enables the EAR to steer radar beams on a pulse-to-pulse basis [12]. EAR is a high-resolution Doppler radar in spatial and time, up to 150 m and 10 min. The radar has five output variables: echo power, spectral width, and 3D wind calculated from Doppler velocity received in five beam directions (vertical, north, south, west, east). This study used a northward beam to minimize the shear broadening effect following Fujiwara et al. [13].

The data are presented relative to the height of the tropopause. The tropopause height estimated from echo power of EAR vertical beam following Yamamoto et al. [14]. In the hydrostatically stable stratosphere, VHF radars preferentially receive strong echoes in their vertical beams by reflection from stable layer of radio refractivity [14]. Echoes received in the vertical beam of VHF radars vertically increase near the tropopause due to the increase of static stability. Therefore, the tropopause can be determined with the height of maximum echo power within TTL.

55.2.2 Identifying MJO

The MJO phase and amplitude are determined by the Real-Time Multivariate MJO Index (RMMI) from the Bureau of Meteorology (BOM) [15]. MJO is considered in the active phase when it reaches phase 4, following EAR location, and the amplitude is more than 1. We defined Lag-0 as the first day MJO convective passing through EAR. To obtain larger sample sizes than those from the traditional definition of the winter season (December to February), This research focused on QBO-MJO connection during extended boreal winter (November to March). With this definition, we identified a total of 31 MJO events from 2003 to 2019 analysis period.

55.2.3 Identifying QBO

In this study, the QBO is defined by monthly averaged zonal wind at 18.6 km ($U_{18.6}$) obtained from the EAR dataset. When the monthly mean $U_{18.6}$ is smaller than 0.5 standard deviation, it is set to the QBOE phase. Likewise, the opposite is set to the QBOW [8]. We grouped all identified MJO events into QBOE and QBOW. A given MJO event is considered to be in QBOE (QBOW) phase if its mid-date, defined as the midpoint between its starting and ending dates, falls within QBOE (QBOW) month or when the zonal wind is less than −4.6 (greater than 3.06). By these definitions, there are 5 MJO events during QBOE, 6 events during QBOW, and 20 events during QBON.

55.2.4 OLR and Precipitation

MJO can be identified by eastward propagating negative anomalies of outgoing long-wave radiation (OLR). This study used OLR data from ERA5 ECMWF. Precipitation events are identified using the daily Tropical Rainfall Measuring Mission (TRMM) 3B42 version 7 (3B42v7) Multi-satellite Precipitation Analysis (https://trmm.gsfc.nasa.gov) [16]. Both datasets cover a period of 2003–2019 with 0.25° × 0.25° spatial resolution. The datasets are then rescaled into daily temporal resolution. After that, datasets classification was carried to identify QBO phases and averaged between 5° N–5° S and 97.5° E–102.5° E to represent the EAR region.

55.3 Results

Section 55.3.1 will discuss the representation of alternating zonal wind phases in the EAR. The TTL wind profiles through MJO active phase in different QBO cases are shown in Sect. 55.3.2. Section 55.3.3 shows the respond of tropopause height in different MJO cases also in connection of MJO-QBO.

55.3.1 Zonal Wind Oscillation Observed from EAR

We will first show the result of zonal wind oscillation observed with EAR. Figure 55.1 shows the time series of zonal wind in 18.7 km determined as the index. The red area

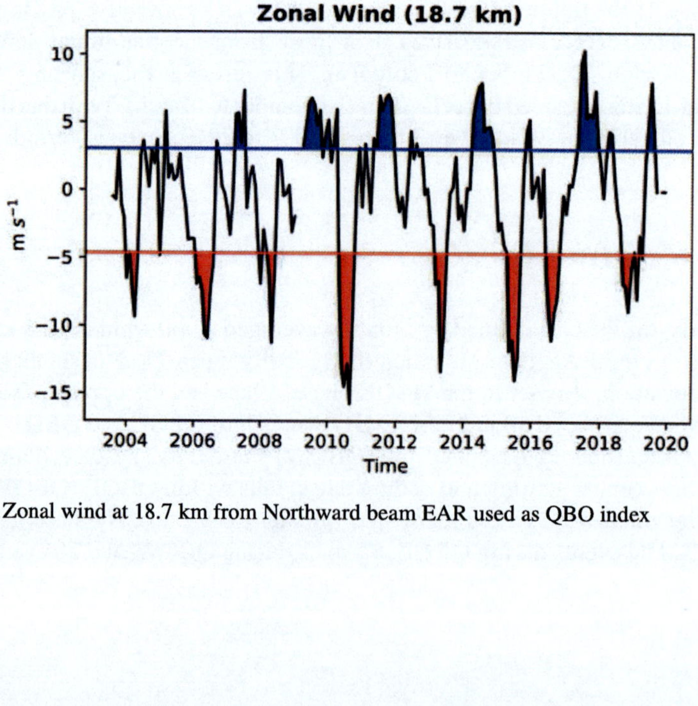

Fig. 55.1 Zonal wind at 18.7 km from Northward beam EAR used as QBO index

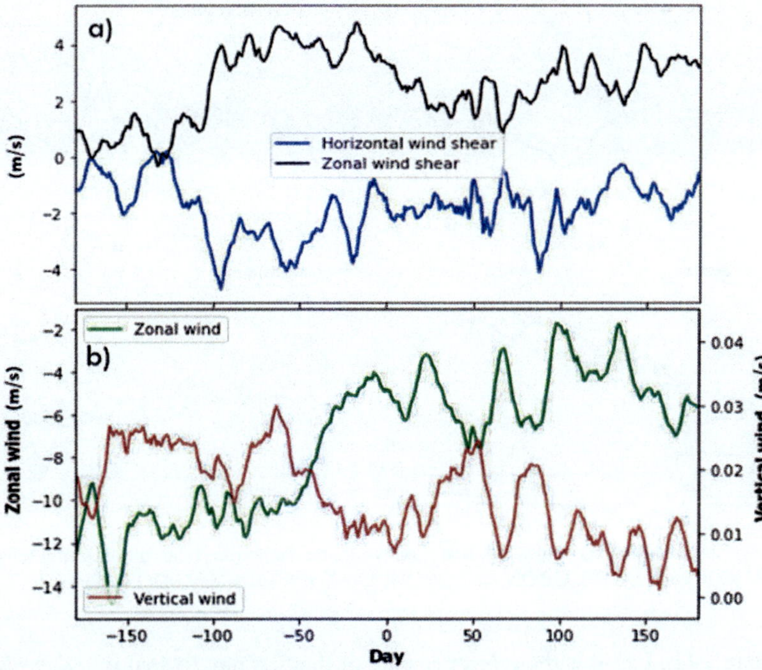

Fig. 55.2 Mean TTL: **a** wind shear and **b** zonal & vertical wind evolution from QBOE to QBOW. Lag-0 is the day when zonal wind is zero or nearly zero

is defined as QBOE, and the blue area is QBOW. From this point, we can clearly observe the oscillation of zonal wind. The 18.7 km index is selected for better TTL state representation. EAR observed seven cycles of QBO throughout 2003–2019 as well as the QBO disruption phenomenon in 2016–2017.

To evaluate the difference in wind variables between QBO phases, we show the composite of zonal wind, wind shear, and vertical wind in QBO transition period (Fig. 55.2). The wind shear is calculated using difference within altitude. These variables then presented as the average value in TTL (14.5–18 km). Lag-0 is the reference date when zonal wind is 0 (or the transition period). These results support the previous study [1, 9] because we observed higher vertical wind during negative phase of zonal wind than the positive phase. The zonal wind significantly increases near the transition period (Lag-0). It also gives agreement to the previous studies that horizontal wind shear is slightly higher during QBOW [9].

55.3.2 TTL Wind Profile During QBO-MJO Interaction

To evaluate QBO-MJO interaction, we separate MJO events into 3 groups of QBO phases. Figure 55.3 shows composite of zonal wind, vertical wind, and wind shear

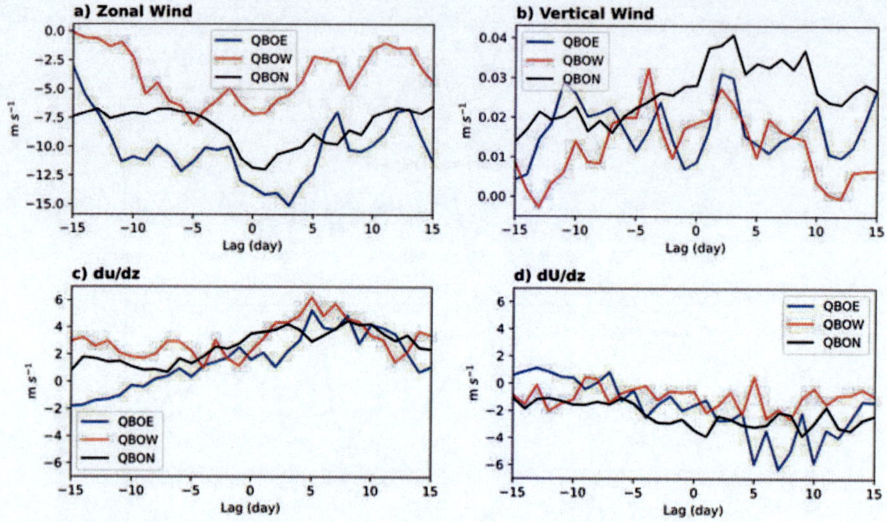

Fig. 55.3 Composite of **a** zonal wind, **b** vertical wind, **c** zonal wind shear, **d** horizontal wind shear during MJO in each QBOE, QBON, and QBOW. Lag-0 is the first day MJO in phase 4

profiles in TTL. Lag-0 is the reference date of the first day RMMI index in phase 4, then taken 15 days before and after to see the evolution of MJO. Lowest magnitude of Zonal wind is observed near Lag-0 in all QBO. Wind shear in all QBO phases remain similar but has a different peak of magnitude. The magnitude peak of zonal wind shear is seen after lag-5 in all QBO phases, while the horizontal wind shear peak in lag-15 days. The vertical wind is likely to be higher in QBON after MJO passing through EAR. There is no significant difference in wind shear variables. This study uses relatively short period (16 years) data therefore the significant difference is difficult to present.

55.3.3 Wind Profile Composite

Figure 55.4 shows wind profiles during active MJO (or when MJO in phase 4) and inactive MJO. The reference point is tropopause height calculated with echo power radar [14]. The zonal wind is stronger during MJO inactive phase (amplitude < 1), both more negative under tropopause and more positive during zonal wind. This condition leads to a greater wind shear value. The vertical wind magnitude stronger during MJO active associated with more convective activities.

Son et al. [8] demonstrated more unstable atmosphere associate with stronger deep convection during MJO active in QBOE rather than QBOW using reanalysis data, as also reported in Nishimoto and Yoden [11]. Based on these findings, we analyze wind profile using EAR observation. Figure 55.5 shows composite of zonal wind shear

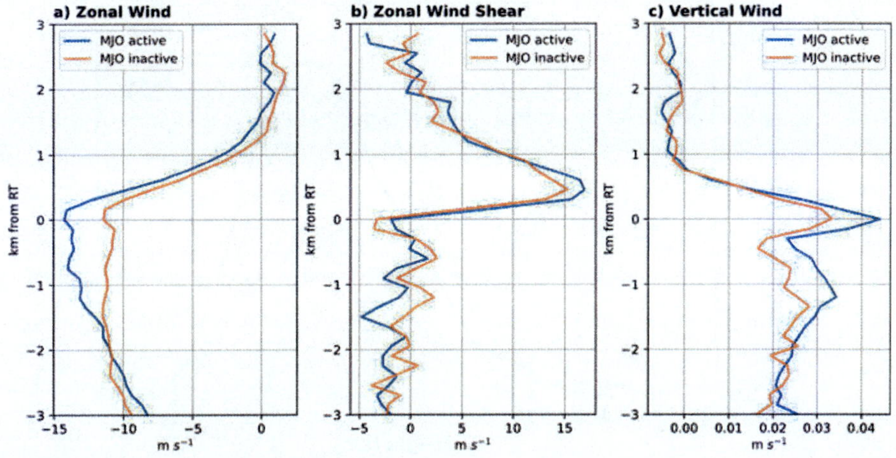

Fig. 55.4 Composite of **a** zonal wind, **b** vertical wind, **c** zonal wind shear profile near TTL during MJO active and MJO inactive

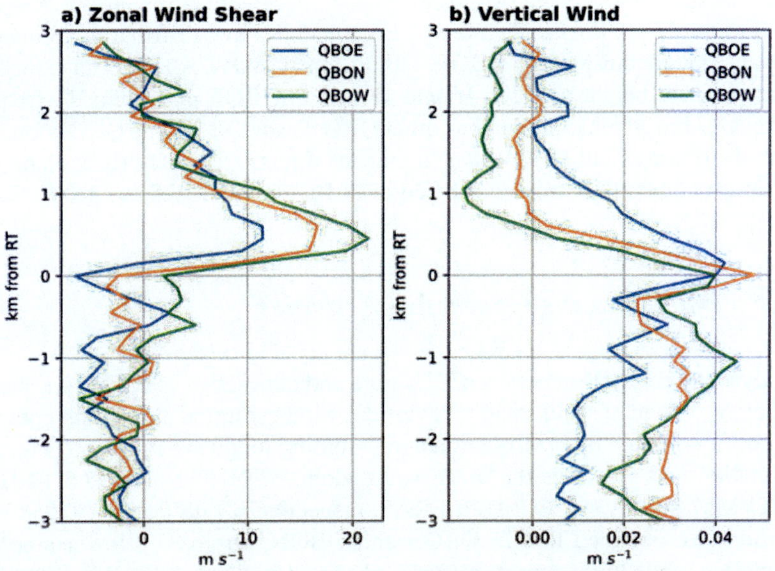

Fig. 55.5 Composite of zonal wind shear (**a**) and vertical wind (**b**) near TTL during MJO active in each QBOE, QBON, and QBOW

and vertical wind in 3 km near the tropopause during QBO-MJO interaction. Vertical wind tends to be higher (up to 0.03 m s^{-1}) during QBOE. At the same time, zonal wind shear weakens in QBOE that also favorable for deep convective formation [2].

Stronger vertical wind in the upper troposphere during QBOE also associate with tropopause height change. Therefore, we calculated tropopause height using radar

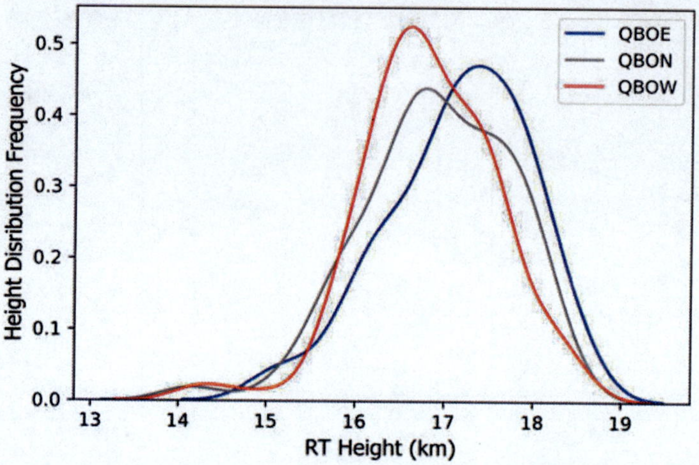

Fig. 55.6 Radar Tropopause (RT) height distribution frequency in MJO active for each QBOE, QBON, and QBOW

echo power at 13–19 km height interval. Figure 55.6 shows distribution frequency of tropopause height during each QBOE, QBON, and QBOW. All QBO have more than 20% tropopause height between 16 and 18 km. in QBOE maximum RT frequency befall in 17.4 km which is higher than other QBO phase. QBON and QBOW both have similar distribution, but QBOW has higher maximum frequency (0.5) than QBON (0.4) for each maximum tropopause height of 16.7 km and 16.8 km, respectively.

55.3.4 Impact on the Convective Activities

To analyze the connection between TTL state and convective activities, we show the time height section of wind shear and vertical wind evolution along with convective activities represented by precipitation and Outgoing longwave radiation (Fig. 55.7). Precipitation and OLR are the mean value along 5° N–5° S and 97.5°–102.5° E from TRMM 3B42 v7 & ECMWF-ERA5, respectively with 0.25° × 0.25° spatial resolution. We obtained that In QBOE and QBON, observed a lower amount of precipitation while higher precipitation is observed just before lag-0 in QBOW.

55.4 Conclusion

A recent study found QBO-MJO has a strong connection in boreal winter. The key concept on how these phenomena interact is in the TTL. Thus, this study highlights the variation of 3D wind in TTL using high vertical and temporal resolution observation

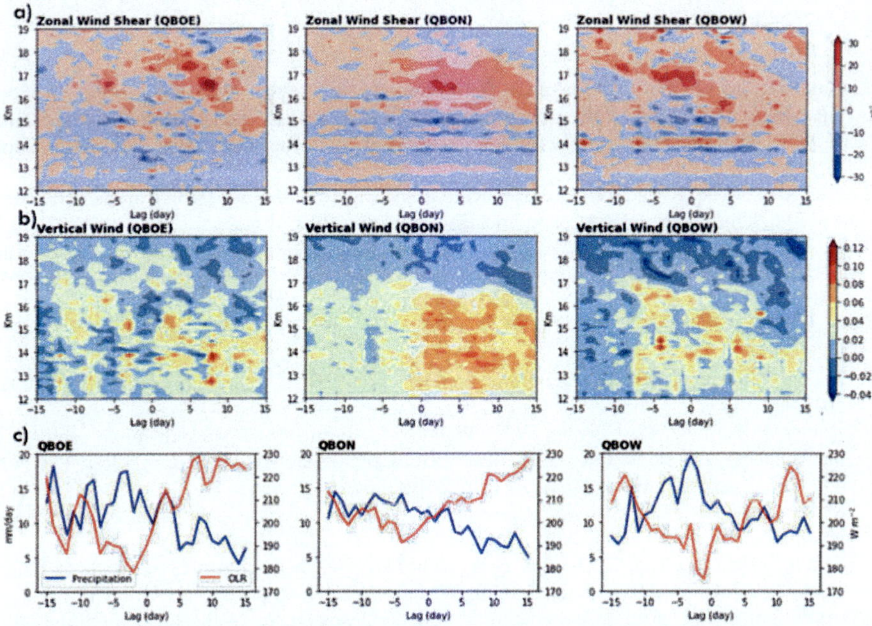

Fig. 55.7 Evolution of **a** zonal wind shear, **b** vertical wind, **c** OLR and precipitation during each QBOE, QBON, and QBOW. Lag-0 is the first day MJO in phase 4

data from EAR. The zonal wind oscillation is captured in EAR zonal wind data and determined as QBO index. The TTL state then analyze with convective activities from precipitation and OLR variable to further examine the connection within QBO-MJO. The results support the previous study that vertical wind and tropopause height tends to be higher in QBOE. Vertical wind tends to be higher during QBOE due to a more updraft and stronger deep convection caused by more unstable atmosphere supported by lower wind shear than QBOW.

Acknowledgements This research was supported by Research, Community Service, and Innovation Program (P3MI), Bandung Institute of Technology (ITB) FY2020, Center of Atmospheric Science and Technology, National Institute of Aeronautics and Space (LAPAN), No. 42 FY2020, Research Priority Program of Ministry Research and Technology/National Agency of Research and Innovation (RISTEK-BRIN) No. 253/E1/PRN/2020, and Collaborative Research based on the MU Radar and Equatorial Atmosphere Radar No. 2020-RISH-MU/EAR-General-00014.

References

1. Baldwin, M.P., Gray, L.J., Dunkerton, T.J., Hamilton, K., Haynes, P.H., Randel, W.J., Holton, J.R., Alexander, M.J., Hirota, I., Horinouchi, T., Jones, D.B.A., Kinnersley, J.S., Marquardt, C., Sato, K., Takahashi, M.: The quasi-biennial oscillation. Rev. Geophys. (2001). https://doi.org/10.1029/1999RG000073
2. Giorgetta, M.A., Bengtsson, L., Arpe, K.: An investigation of QBO signals in the east Asian and Indian monsoon in GCM experiments. Clim. Dyn. (1999)
3. Lu, H., Bracegirdle, T.J., Phillips, T., Bushell, A., Gray, L.: Mechanisms for the Holton-Tan relationship and its decadal variation. J. Geophys. Res. **119**(6) (2014). https://doi.org/10.1002/2013JD021352
4. Zhang, C.: Madden-Julian oscillation. Rev. Geophys. (2005). https://doi.org/10.1029/2004RG000158
5. Zhang, C., Zhang, B.: QBO-MJO connection. J. Geophys. Res. Atmos. https://doi.org/10.1002/2017JD028171
6. Xavier, P., Rahmat, R., Cheong, W.K., Wallace, E.: Influence of Madden-Julian oscillation on Southeast Asia rainfall extremes: observations and predictability. Geophys. Res. Lett. **41**, 4406–4412 (2014). https://doi.org/10.1002/2014GL060241
7. Muhammad, F.R., Lubis, S.W., Setiawan, S.: Impacts of the Madden–Julian oscillation on precipitation extremes in Indonesia. Int. J. Climatol. **41**, 1970–1984 (2021)
8. Son, S.W., Lim, Y., Yoo, C., Hendon, H.H., Kim, J.: Stratospheric control of the Madden-Julian oscillation. J. Clim. (2017). https://doi.org/10.1175/JCLI-D-16-0620.1
9. Collimore, C.C., Martin, D.W., Hitchman, M.H., Huesmann, A., Waliser, D.E.: On the relationship between the QBO and tropical deep convection. J. Clim. (2003). https://doi.org/10.1175/1520-0442(2003)016%3c2552:OTRBTQ%3e2.0.CO;2
10. Nishimoto, E., Yoden, S.: Influence of the stratospheric quasi-biennial oscillation on the Madden-Julian oscillation during austral summer. J. Atmos. Sci. (2017). https://doi.org/10.1175/JAS-D-16-0205.1
11. Fueglistaler, S., Dessler, A.E., Dunkerton, T.J., Folkins, I., Fu, Q., Mote, P.W.: Tropical tropopause layer. Rev. Geophys. (2009). https://doi.org/10.1029/2008RG000267
12. Fukao, S., Hashiguchi, H., Yamamoto, M., Tsuda, T., Nakamura, T., Yamamoto, M.K., Sato, T., Hagio, M., Yabugaki, Y.: Equatorial atmosphere radar (EAR): system description and first results. Radio Sci. (2003). https://doi.org/10.1029/2002RS002767
13. Fujiwara, M., Yamamoto, M.K., Hashiguchi, H., Horinouchi, T., Fukao, S.: Turbulence at the tropopause due to breaking Kelvin waves observed by the Equatorial Atmosphere Radar. Geophys. Res. Lett. (2003). https://doi.org/10.1029/2002GL016278
14. Yamamoto, M.K.: High time resolution determination of the tropical tropopause by the Equatorial Atmosphere Radar. Geophys. Res. Lett. **30**(21). https://doi.org/10.1029/2003gl018072
15. Wheeler, M., Kiladis, G.N.: Convectively coupled equatorial waves: analysis of clouds and temperature in the wavenumber-frequency domain. J. Atmos. Sci. (1999)
16. Huffman, G.J., Bolvin, D.T., Nelkin, E.J., Wolff, D.B., Adler, R.F., Gu, G., Hong, Y., Bowman, K.P., Stocker, E.F.: The TRMM multisatellite precipitation analysis (TMPA): quasi-global, multiyear, combined-sensor precipitation estimates at fine scales. J. Hydrometeorol. **8**(1), 38–55 (2007)

Chapter 56
Effect of Meteorology Parameters on Air Pollutant Standard Index in the Urban Area (Case Study in Jakarta)

Dessy Gusnita

Abstract This paper aims to analyze the significant correlation between air quality and meteorological variables (temperature, RH, radiation) in the DKI Jakarta areas time 2017–2020. Data analysis used secondary data on air pollutant standard index from Provincial Environment Service DKI Jakarta consisting of PM_{10}, NO_2, SO_2, CO, and O_3 parameters. Meteorological data uses data from BMKG, which consists of meteorological parameters, namely: solar radiation duration, temperature, and relative humidity. Air temperature, humidity, and radiation are part of the meteorological parameters that can affect the concentration of pollutants in the air. The results of the correlation of pollutants with meteorology parameters show that PM_{10} has the strongest correlation with the 3 meteorology components. Sun's duration and PM_{10} have a correlation of 0.69. Temperature and PM_{10} correlation have correlation of 0.55. And the correlation between PM_{10} and RH is inversely proportional, and the correlation value is 0.82.

56.1 Introduction

The city of Jakarta, as the capital city of Indonesia, has various kinds of progress in various sectors, both industry, tourism, education and the environment. In a higher population rate, which of course also increases the mobility and activities of people living in the city of Jakarta and surrounding cities such as Depok, Bekasi, Bogor. The rapid progress of the city of Jakarta has an impact on the environment, especially the air quality in DKI Jakarta. Increasing industrial and transportation activity was associated with air pollution, especially in the urban and industrial areas. Biomass burning and urban air pollution that occur, consecutively, in several wildfire-prone provinces and big cities in Indonesia are likely worsened in the future [1].

D. Gusnita (✉)
Research Center for Climate and Atmosphere, National Research and Innovation Agency (BRIN), Bandung, Indonesia
e-mail: nitagusnita@gmail.com

E. Yulihastin et al. (eds.), *Proceedings of the International Conference on Radioscience, Equatorial Atmospheric Science and Environment and Humanosphere Science, 2021*, Springer Proceedings in Physics 275, https://doi.org/10.1007/978-981-19-0308-3_56

Table 56.1 Air pollutant standard index range

Category	Range
Good	0-50
Moderate	51-100
Unhealthy	101-199
Very unhealthy	200-299
Dangerous	300–more

The Air Pollutant Standard Index is a number that does not have units that describe the condition of ambient air quality at specific location and time based on the impact on human health, aesthetic value, and other living things. The air quality standard index officially used in Indonesia is the Air Pollutant Standard Index (ISPU); this is following the Ministerial Decree State Environment: KEP 45/MENLH/1997 concerning Standard Index Air Pollutants [2] (Table 56.1).

Air pollution in an urban area is a dynamic mixture of pollutants emitted from numerous sources, including motor vehicles and other anthropogenic activities. The impact of air pollution can be one of the reasons for civilization diseases [3].

Besides emission factors, local meteorological conditions, especially wind, air temperature, rainfall, and radiation factors, also affect the concentration and distribution of pollutants in the air [4]. The concentration of pollutants in the atmosphere is not only influenced by the number of pollutant sources, but also by meteorological parameters, namely air temperature and wind speed [5]. In addition, the increase in temperature and rainfall also affects the increase in CO_2 from the surface [6]. From several locations, temperature and RH are also significant contributors to the variability of the concentration of CO_2. Mahesh et al., research the correlation of meteorological factors with CO_2, statistical analysis of the data shows that precipitation and relative humidity independently correlated 55% ($r = -0.55$) and 32% ($r = -0.32$), respectively [7]. Influences of prevailing meteorology (air temperature, wind speed, wind direction, and relative humidity) on GHGs have also been investigated. CO_2 and CH_4 show a strong positive correlation during winter, pre-monsoon, monsoon, and post-monsoon with correlation coefficients (Rs) equal to 0.80, 0.80, 0.61, and 0.72 respectively [8].

Radiation also affects the concentration of NO_x in the atmosphere. In summer, NO converted to NO_x increases with the increase in solar radiation. For example, in some large cities in Japan, due to the increase in urban population, the atmospheric temperature tends to be increased, which is strongly correlated with high NO_2 concentration It is also supported by research in Bahrain, which showed the highest NO_x concentration was in the urban areas with high traffic density [9].

Wind will affect the dispersion of pollutants (transport process) and determine which direction and how high the pollutant concentration is. SO_2, aerosols, nitrogen oxides, and hydrocarbons in the atmosphere will form a photochemical haze with the help of solar energy (radiation). SO_2 also, if it reacts with rainwater, will increase the acidity of rain water which can cause acidification water sources and reduction of soil nutrients; also cause corrosion metal and other building materials [10].

56.2 Data and Methods

56.2.1 Data

This study uses secondary data consisting of 2 primary data, namely meteorological data and air pollution data for time January 1, 2017; Dec 31, 2020 time period. Meteorological data comes from data released by the central BMKG consisting of relative humidity (RH), temperature, and duration of solar radiation, while air pollution data obtained from Provincial Environment Service DKI Jakarta in the form of air pollutant standard index (APSI) data of PM_{10}, O_3, CO, NO_2, and SO_2. These two types of meteorology and pollutant data use data for the 2017–2020 period due to the completeness of the available data.

56.2.2 Methods

To analyze the correlation between meteorological factors and air pollutant standard index, used Heil-Sen Siegel regression method [11].

Suppose the equation for the linear relationship of the independent variable (X) to the dependent variable (Y) is

$$Y = Q_o + Q_1 X + s$$

where β_o is the intercept, β_1 is the regression coefficient, and ε is residual; the value of β_1 is estimated by calculating the gradient of n(n − 1)/2 datum pairs, finding the median of the resulting gradient for each gradient between the datum and other n − 1 datums, then calculating the median of the n medians that have been obtained. The estimated value of a robust regression coefficient can be obtained by calculating the median of the following least-squares estimate

$$\tilde{\beta}_n = Median_i \left\{ \tilde{\beta}_1 = \frac{(Y_i - Y_j)}{(X_j - X_i)} : X_i \neq X_j,\ 1 \le i < j \le n \right\}$$

$$\tilde{\alpha}_n = Median_i \left\{ \tilde{\beta}_0 = \frac{(Y_j X_i - Y_i X_j)}{(X_j - X_i)} : X_i \neq X_j,\ 1 \le i < j \le n \right\}$$

Thus, the Theil-Sen Siegel regression equation is written as follows:

$$YTSS = \tilde{\alpha}n + \tilde{\beta}X$$

56.3 Result and Discussion

The meteorological parameters discussed in this study are sunshine duration, temperature, and relative humidity (RH). These three parameters will be associated with several pollutants, namely particulate matter (PM_{10}), ozone (O_3), nitrogen dioxide (NO_2), sulfur dioxide (SO_2), and carbon monoxide (CO), to find out how the three meteorological parameters relate to the five pollutants. Thus, there are eight research variables. Each variable was observed from January 1, 2017; December 31, 2020. To obtain a clear pattern, the daily data were averaged per month (Table 56.2).

Figure 56.1 presents average seasonal data for relative humidity, temperature, and sunshine duration for 2017–2020. For seasonal data, the length of the sun's duration shows a similar pattern during 2017–2020. This pattern repeatedly happened for four years of observation. It's just that in 2020 there will be a slight decrease in the length of the sun's duration. The peak duration of solar radiation occurs in September–October November (purple) in transition season. In contrast the minimum value of solar radiation occurs in the wet season, namely Dec–Jan–Feb (the red one).

Temperature meteorological analysis data shows a pattern/trend of temperature increases, especially in 2019 during the observation period. The maximum temperature generally occurs in the dry season of June–July–August, with the highest temperature 29.64 °C (in 2019). Meanwhile, minimum temperature occurred during December–February (rainy season) with the lowest temp at 27.2 °C.

The relative humidity pattern in Jakarta from 2017 to 2020 shows a relatively constant. The results of the RH data analysis showed that the exact peak of RH occurred in February from 2017 to 2020. However, in February 2019, the peak RH value fell to 80%, lower than in 2017–2018, 83%. And peak RH increased in February 2020 by 85%. The increase in RH is thought to be due to the decrease in the standard index of air pollutants in the city of Jakarta during the Covid 19 pandemic.

Data for Air Pollutant Standard index PM_{10}, O_3, CO, NO_2, and SO_2 are presented in Fig. 56.2. PM_{10} shows the highest concentration in the dry season of JJA, this is caused by the accumulation of particulate dust. The increasing the PM_{10} pollutant index during this season. O_3 shows the highest pollutant index in July. This is to be linear with the high sunshine duration in the transition season. Increased surface ozone concentration (O_3) in July 2018 in line with the decrease in CO gas concentration in July 2018 related to the photo chemical reaction of CO oxidation [12].

In NO_2 and SO_2, there was a significant increase in the pollutant index in October 2020; this related to the resumption of community activities after the Covid19 (New Normal policy) pandemic. Ozone concentrations at the ground level, depend on the formation and dispersion processes. Formation process, mainly depends on the precursor sources, whereas, the dispersion of ozone depends on meteorological

Table 56.2 Meteorological data and air pollutant standard index in Jakarta in 2017–2020

Month	Sun's duration	Temperature	RH	PM_{10}	O_3	NO_2	SO_2	CO
Jan-17	3.76	28.15	77.52	40.32	51.06	11.94	30.94	31.42
Feb-17	3.42	27.25	82.07	46.25	59.93	13.50	31.00	39.61
Mar-17	4.45	28.09	78.42	58.74	78.26	17.58	37.39	42.68
Apr-17	4.09	28.55	77.60	58.43	91.73	16.97	36.67	38.20
May-17	4.15	29.05	75.90	73.90	97.61	17.16	43.48	36.19
Jun-17	4.37	28.70	76.87	65.13	77.70	13.80	46.00	38.73
Jul-17	4.50	28.63	73.94	67.74	101.94	16.00	47.94	36.84
Aug-17	5.32	28.69	69.39	71.65	102.84	12.48	50.94	35.06
Sep-17	6.70	28.83	72.20	72.87	101.90	15.47	51.67	30.67
Oct-17	5.53	28.96	74.61	62.61	101.58	18.32	51.90	38.65
Nov-17	3.61	28.27	78.17	52.63	89.83	16.27	53.30	35.07
Dec-17	3.61	28.22	76.52	38.42	51.19	12.03	45.10	29.74
Jan-18	3.05	27.77	76.48	35.71	57.77	12.19	46.13	21.06
Feb-18	3.70	27.15	82.57	47.75	68.75	14.39	30.75	43.50
Mar-18	4.17	27.90	79.74	58.19	109.68	17.23	29.39	31.48
Apr-18	5.00	28.63	78.57	62.50	103.93	17.80	29.30	25.50
May-18	4.99	29.40	73.52	76.52	129.42	16.19	26.19	22.74
Jun-18	4.58	28.87	75.17	66.67	133.00	13.97	26.83	19.40
Jul-18	5.19	28.21	70.84	80.35	158.00	18.84	30.23	21.16
Aug-18	5.34	28.18	70.10	72.23	157.90	15.90	31.55	17.77
Sep-18	6.70	28.76	68.37	68.33	149.80	15.37	34.97	21.33
Oct-18	5.53	29.48	71.39	69.06	141.52	15.39	32.06	21.42
Nov-18	3.61	28.90	75.53	62.00	120.60	17.80	26.77	25.93
Dec-18	3.61	28.74	74.81	51.58	95.29	15.19	25.77	20.45
Jan-19	3.05	27.95	79.97	51.90	92.45	9.39	11.77	17.24
Feb-19	3.70	28.28	80.21	58.79	90.64	9.89	14.50	21.79
Mar-19	4.17	28.16	78.55	54.52	107.03	9.52	14.65	17.61
Apr-19	5.00	28.88	78.43	58.23	90.64	9.57	14.40	21.33
May-19	4.99	29.64	73.77	67.97	92.39	9.77	18.90	22.58
Jun-19	4.58	29.18	71.90	69.63	102.83	8.67	19.30	17.80
Jul-19	5.19	28.71	68.48	73.48	97.29	9.35	21.97	16.77
Aug-19	5.34	28.28	68.61	74.06	121.32	8.50	21.27	13.97
Sep-19	6.56	28.84	68.67	74.40	125.50	5.00	22.93	12.40
Oct-19	6.41	29.55	68.90	72.32	143.97	8.32	21.29	8.79
Nov-19	3.40	29.55	71.00	68.87	132.77	9.20	20.93	18.29

(continued)

Table 56.2 (continued)

Month	Sun's duration	Temperature	RH	PM_{10}	O_3	NO_2	SO_2	CO
Dec-19	3.50	28.48	79.10	63.55	104.84	8.52	18.32	24.68
Jan-20	2.92	27.88	83.06	56.42	71.33	11.97	34.97	32.06
Feb-20	2.91	27.67	84.25	56.52	92.72	14.14	25.72	35.62
Mar-20	4.14	28.57	79.55	60.16	113.03	15.55	22.19	37.55

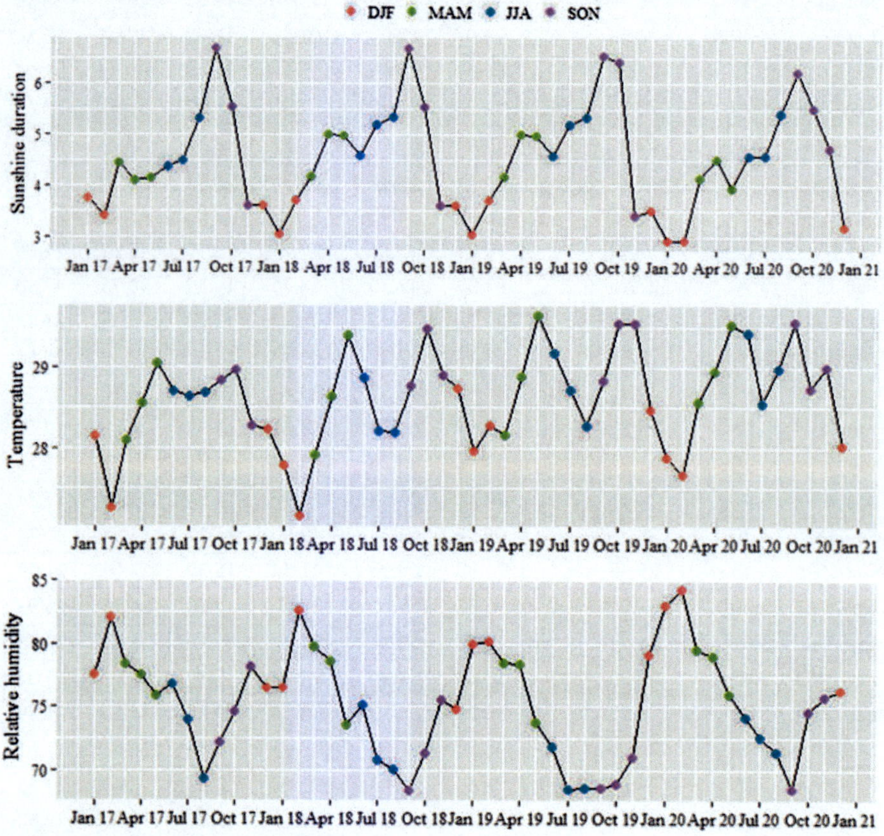

Fig. 56.1 Seasonal meteorological data

factors. In addition, the level of ozone concentration at the surface can be estimated by the result of source and sink mechanism, which predominately rely on the meteorological conditions of the environment [13]. The high pollutant index of NO_2 and SO_2 is thought to come from transportation activities.

The implementation impact of the Working from home policy regarding the Covid-19 pandemic on air quality conditions in Jakarta can be seen qualitatively

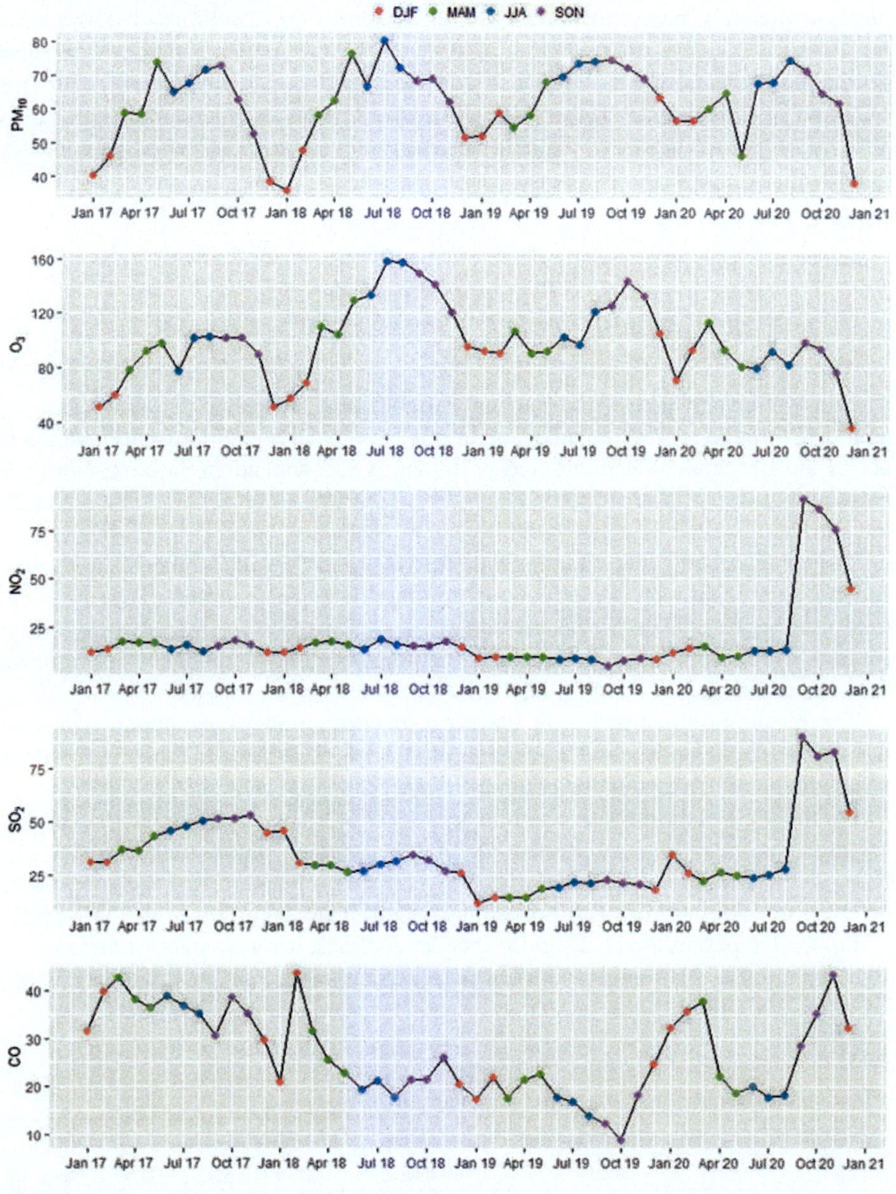

Fig. 56.2 Seasonal air pollutant standard index

and quantitatively, especially decrease in PM10 concentration levels [14]. Based on air pollutant standard index data, it is known after the implementation of the new normal policy, there was a significant increase in pollutant in Jakarta, especially the SO2 and NO2 parameters (Fig. 56.2). CO concentration was increased significantly with the implementation of the new normal policy during the period of covid-19 pandemic.

56.3.1 The Correlation Between the Meteorology with Pollutants

The correlation between the independent variables and the dependent variables can be analyzed through regression analysis. In this case, each meteorological parameters act as an independent variable and pollutant as the dependent variable, with the correlation pattern assumed to be linear. From the way of correlation that are built, the correlation value between variables can be received.

The air pollutant standard index data for NO_2 and SO_2 in October 2020 experienced a significant increase, in contrast to other data. The solution to this problem is to use the Thiel -Sen-Siegel method [11]. This method is also relatively better for data that is not normally distributed and data with non-homogeneous variance (heteroscedasticity).

The results of the analysis in the form of correlation values obtain based on the Theil-Sen-Siegel method regression model are presented in Table 56.3 with a visualization of the correlation pattern in Figs. 56.3, 56.4 and 56.5 (Table 56.4).

For the sun's duration with a significance level of 5%, it was found that there was a significant linear relationship between the length of sun's duration and PM_{10}, O_3, NO_2, CO, while the relationship with SO_2 is not significant. The correlation value between sun's duration and air pollution standard index is presented in Table 56.5.

For the temperature with a significance level of 5%, it was found that there was a significant linear relationship between the temperature on PM_{10}, O_3, and CO, while

Table 56.3 Summary statistics of research variables

Research variable	Min	Max	Mean	Std Dev
Suns's duration	2.91	6.70	4.53	1.02
Temp	27.15	29.64	28.62	0.60
RH	68.37	84.25	75.28	4.27
PM_{10}	35.71	80.35	62.02	10.95
O_3	35.29	158.00	98.79	27.36
NO_2	5.00	91.70	18.32	18.34
SO_2	11.77	90.33	34.12	17.39
CO	8.79	43.50	26.65	9.12

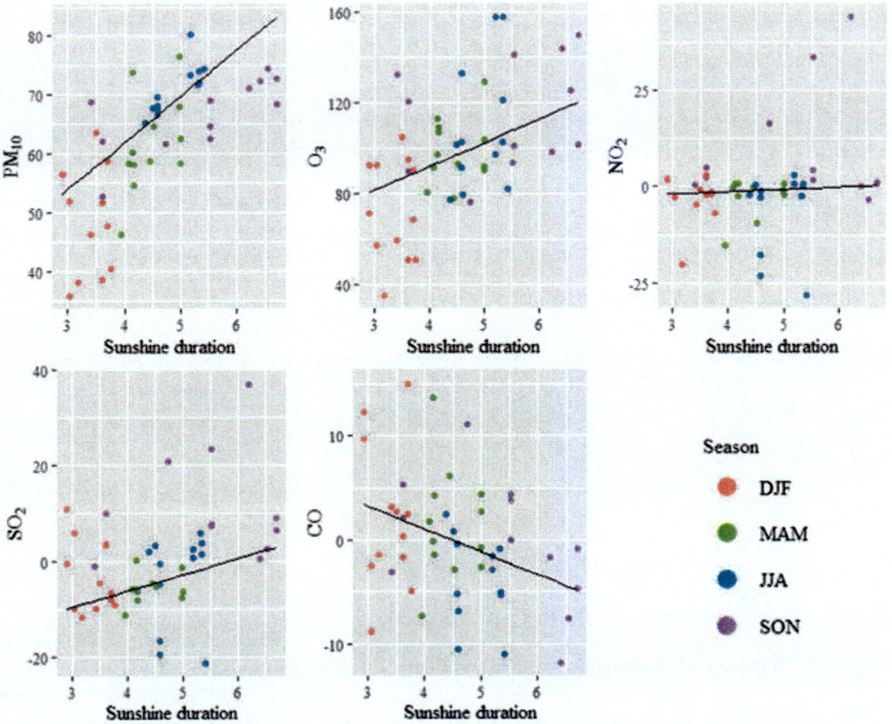

Fig. 56.3 Correlation between sunshine duration and air pollutant standard index PM_{10}, O_3, NO_2, SO_2, CO

the relationship with SO_2 and NO_2 is not significant. The correlation value between T and air pollution standard index is presented in Table 56.6.

Relative humidity (RH), with a significance level of 5%, found a significant linear correlation between temperature and PM_{10}, O_3, and CO, while correlation with NO_2 and SO_2 is not significant. The correlation value between RH and air pollution standard index is presented in Table 56.7.

56.4 Summary and Conclusion

The results of the correlation analysis between meteorological parameters data and air pollutant standard index parameters show the effect of pollutants are positively correlated with the sun's durations, except parameter pollutant CO that has negative correlation. The correlation between temperature and air pollutants is positive, meaning that the higher temperature, the higher the concentration. RH is negatively correlated with the air pollutant standard index for all parameters, except for CO.

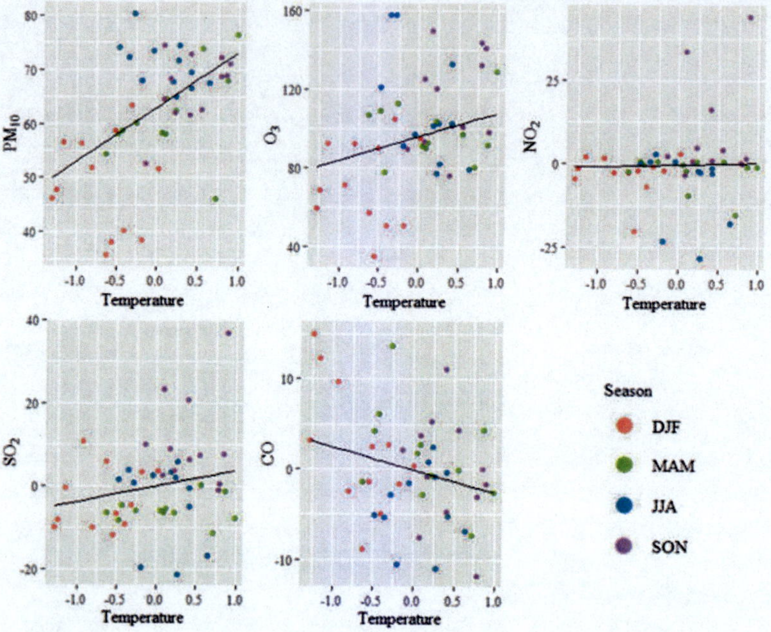

Fig. 56.4 Correlation between temperature and PM_{10}, O_3, NO_2, SO_2, CO

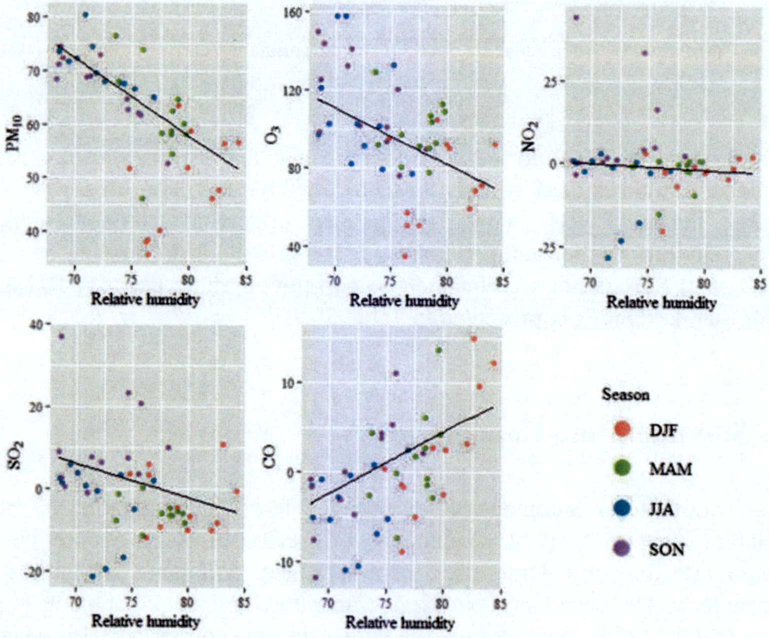

Fig. 56.5 Correlation between RH and PM_{10}, O_3, NO_2, SO_2, CO

Table 56.4 Correlation values based on Theil-Sen-Siegel method

	Sunshine duration	Temperature	RH
PM_{10}	0.696 (3.55 × 10–14)	0.552 (6.99 × 10–09)	−0.823 (1.68 × 10–09)
O_3	0.465 (4.75 × 10–09)	0.311 (0.0011)	−0.478 (5.00 × 10–08)
NO_2	0.103 (0.0198)	0.001 (0.988)	−0.013 (0.32036)
SO_2	0.045 (0.239)	−0.156 (0.375)	−0.065 (0.046053)
CO	−0.296 (4.90 × 10–05)	−0.300 (2.67 × 10–05)	0.539 (3.22 × 10–10)

Table 56.5 Correlation between sun's duration and air pollutant standard index

Meteorology	PM_{10}	O_3	CO	NO_2
Sun's duration	Moderate correlation (0.69) unidirectional More pro-longed sun-shine duration, the greater lPM_{10} level	Moderate correlation (0.46) unidirectional More pro-longed sun-shine duration, the greater the O_3 content	weak correlation (0.29) not in the same direction, more prolonged sun's duration, the lower the CO content	Very weak correlation (0.10) unidirectional more pro-longed sun's duration, the greater the NO_2 content

Table 56.6 Correlation between temperature and air pollutant standard index

Meteorology	PM_{10}	O_3	CO
Temperature	Moderate correlation (0.552) Uni directional The higher temp, the greater the level of PM_{10} Index	Weak correlation (0.31) Uni directional The higher the temperature, the greater the O_3 index	Weak correlation (−0.3) not in the same direction, The higher the temperature, the lower the CO level

Table 56.7 Correlation between RH and air pollutant standard index

Meteorology	PM_{10}	O_3	CO
Relative Humidity	Strong correlation (0.82) not in the same direction. The higher the relative humidity, the lower the PM_{10} level	Moderate correlation (−0.48) not in the same direction. The higher the relative humidity, the lower the O_3 content	Moderate correlation (0.54) unidirectional. The higher the relative humidity, the greater the CO level

Acknowledgements The authors acknowledge Provincial Environment Service DKI Jakarta and BMKG for data support.

References

1. Kusumaningtyas, S.D.A.: The recent state of ambient air quality in Jakarta. Aerosol Air Qual. Res. **18**, 2343–2354 (2018)
2. Kementerian Lingkungan Hidup: Kepmen LH No 45 tahun 1997 tentang Indeks Standar Pencemar Udara (ISPU) (1997)
3. Lim, S.S., et al.: A comparative risk assessment of burden of disease and injury attributable to 67 risk factors and risk factor clusters in 21 regions, 1990–2010: a systematic analysis for the Global Burden of Disease Study 2010. Lancet **380**, 2224–2260 (2012)
4. Wark, K., Warner, C.F.: Air pollution its origin and control, 2nd edn., pp. 1–136. Harper and Row Publisher, New York (1981)
5. Neiburger, M., Edinger, J.G., dan Bonner, W.D.: Memahami Lingkungan Atmofser Kita. diterjemahkan oleh: Ardina Purbo. Penerbit ITB, Bandung (1994)
6. Martin, D., Beringer, J., Hutley, L., McHugh, I.: Carbon cycling in mountain ach forest: analysis of below ground respiration. Agric. Forest Meteorol. **147**, 58–70 (2007)
7. Pathakoti, M., et al.: Influence of meteorological parameters on atmospheric CO_2 at Bharati, The Indian Antartic Research Station. Polar Res. **37**(1), 14442072. https://doi.org/10.1080/17518369.2018.1442072 (2018)
8. Sreenivas, G., et al.: Influence of meteorology and interrelationship with greenhouse gases (CO_2 and CH_4) at a suburban site of India. J. Atmos. Chem. Phys. **16**, 3953–3967 (2016)
9. Madany, I.M., Danish, S.: Spatial and temporal patterns in nitrogen dioxide concentrations in a hot desert region. J. Atmos. Environ. **27A**(15), 2385–2391 (1993)
10. Gow, T., Pidwirny, M.: Acid rain and deposition. Living landscape: Thompson-Okanagan (1996)
11. Farooqi, A.: A comparative study of Kendall-Theil Sen, Siegel vs quantil regression with outliers. Wayne State University, Detroit, Disertasi (2019)
12. Sulistiyono, A., dkk: Studi Profil Ozon Permukaan (O_3) Dan Gas Karbon Monoksida (CO) Antara Kota Bandung Dan Bukit KototabangTahun 2008. Jurnal ilmu lingkungan **17** (2019)
13. Radaideh, J.A.: Effect of meteorological variables on air pollutants variation in arid climates. J. Environ. Anal. Toxicol. **7**(4) (2017). https://doi.org/10.4172/2161-0525.100047
14. Umara Firman Rizi, D., et al.: Analisis dampak diterapkannya kebijakan working from home saat pandemi covid-19 terhadap kondisi kualitas udara di Jakarta. Jurnal Meteorologi Klimatologi dan Geofisika **6**(3) (2019)

Chapter 57
Atmospheric Conditions Associated with Northerly Surge, Borneo Vortex and Madden Julian Oscillation During the Extreme Rainfall Cases in Early 2021 Over the Western Part of the Maritime Continent

Anis Purwaningsih, Albert Klein Tank, and Jordi Vila

Abstract During early 2021 (January to February), high rainfall intensity resulted in flooding over several locations over the western part of the Maritime Continent (East Aceh, South Borneo, Aceh Tamiang, and Jakarta). The atmospheric conditions were specifically different, but the pattern of strong northeasterly wind over the South China Sea (SCS) appears during these all flooding events. During the flooding event over East Aceh and South Borneo, the high rainfall intensity is mainly associated with the vortex over the Borneo Island, denoted by high wind speed (more than 10 m/s), closed-counterclockwise circulation, and high vorticity over Borneo (over 25×10^{-6} s^{-1}). Moreover, during the Aceh Tamiang's flooding, there was a local convective activity and strong wind from the Indian Ocean. The wind from the Indian Ocean deflected southward along the west coast of Sumatera Island. This northwesterly wind over the Indian Ocean (west part of Sumatra Island) was stronger than the other flooding cases. Furthermore, the vorticity magnitude over this area (18×10^{-5} s^{-1}) was also considerably higher than the other cases. Moreover, during the Jakarta flooding event, the Borneo vortex did not appear, and the northeasterly wind over the South China Sea turned to southward wind and elongated to the southern part of Java Island. The meridional SST gradient along the 105° E resulted in high wind speed over the western part of the Java Island and played a role in low-level wind convergence at the beginning of the Jakarta flooding episode.

A. Purwaningsih (✉)
Research Center for Climate and Atmosphere, National Research and Innovation Agency (BRIN), Jakarta, Indonesia
e-mail: anis.purwaningsih@brin.go.id

A. K. Tank
Met Office Hadley Centre for Climate Science and Services, Exeter, UK

J. Vila
Wageningen University, Wageningen, The Netherlands

E. Yulihastin et al. (eds.), *Proceedings of the International Conference on Radioscience, Equatorial Atmospheric Science and Environment and Humanosphere Science, 2021*, Springer Proceedings in Physics 275, https://doi.org/10.1007/978-981-19-0308-3_57

57.1 Introduction

Indonesian Maritime Continent is highly vulnerable to hydrological disasters [1, 2]. Floods, drought, flash floods, and landslides are hydrometeorological disasters mainly triggered by extreme precipitation [1]. Such extreme events accounted for more than 70% of the total natural disasters in the Indonesia Maritime Continent [1]. In early 2021 (January–April), 89% of Indonesia's natural disasters were mainly categorized as hydrometeorological disasters, including floods, landslides, and tornadoes [2]. Over this period, the natural disasters caused approximately 479 casualties, with 60 people missing and 12.900 were wounded, not to mention that 5 million people became homeless [2]. In addition, those extreme events also exert economic losses owing to the damage caused by the disasters to public and private facilities [2]. Future prediction on precipitation (over the monsoonal area in a warming area) indicates a tendency of increased rainfall due to enhanced moisture convergence [3]. Therefore, according to this projection, the risk of hydrometeorological disasters in Indonesia will be higher in the future.

There were several simultaneous hydrometeorological disasters over the western part of the Maritime Continent during the period of the Asian Winter Monsoon early this year/2021 [2]. In January, several flooding events were found in some locations over the western part of Indonesia. For instance, floods triggered by heavy rainfall occurred several times in January 2021 over Aceh, Sumatera Island, as reported by Aceh's Agency for Disaster Management/BPBA [4]. The flooding events were on the 1st–3rd January over the East Aceh regency and surrounding areas [5], and on the 20th–23rd January over Aceh Tamiang [6, 7]. According to BPBA data, 23 villages, including hundreds of citizens of East Aceh Regency, were affected by this flood [8]. Based on Regional Agency for Disaster (BPBD), the flooding over the Aceh Tamiang reached 0.5–1 m. This flooding was because of the overflow of the Aceh Tamiang river [9]. Consequently, about 3.075 people were affected by this flood, according to the report from National Agency for Disaster Management (BNPB) [7]. Another flooding case, in early 2021 (January 12–15), was a big flood in Southern Borneo impacting more than a thousand inhabitants in ten cities [10–12]. This big flood was mainly because of the high precipitation intensity [10]. Moreover, the Meteorological, Climatological, and Geophysical Agency (BMKG) also released information that the flood was triggered by light to heavy rainfall since January 10th, reaching about 174 mm/day [13]. Furthermore, one of the other cases was flooding over Jakarta (February 19–20), mainly caused by heavy rainfall on 18–19 February from night to early morning [14]. The rainfall intensity reached 226 mm/day in Pasar Minggu, 176 mm/day in Halim, and 154 mm/day in Lebak Bulus observational stations [14]. BMKG explained several atmospheric conditions attributed to this disaster, including the surge from the northern hemisphere, Rossby wave, stability of the Western part of Java, low pressure over Australia, and active phase of La Nina [14]. Therefore, the leading cause of these flooding events is heavy precipitation, although it can also be amplified by anthropogenic activities such as poor land-use management, poor water management, etc.

Understanding the atmospheric condition is the primary key to understanding the precipitation mechanism resulting in hydrometeorological disaster over the western part of The Maritime Continent. Based on the previous study, during the period of Asian winter monsoon (December–January–February), the monsoon engine results in low-level anticyclone over the southern Siberian with high pressure (greater than 1035 hPa), which generates Cold Surge (CS) over the northern hemisphere of Asian area [15]. This low-level anticyclone moves southward; consequently, the surface temperature drops over 6 °C over southern China within 24–48 h [15]. Based on climatology, the occurrence of this surge is quite intensive. During the East Asian winter monsoon (from October to April 1979–1995), there is at least 13 Cold Surge (CS) in each winter season with 5–14 days period on average [15]. Over Northern Sumatra and Peninsular Malaysia, this surge generates low level-convergence because of the orographic condition over these locations [16]. The cold surge extending to the equatorial area forms Cross-Equatorial Northerly Surge (CENS) and influences precipitation over the Maritime Continent [17–20]. Climatology of CENS showed 62 events during October–March 1999–2009 [17].

Furthermore, synoptic-scale disturbances such as Borneo Vortexes and planetary waves such as Madden Julian Oscillation (MJO) influence the precipitation over the western part of the Maritime Continent. The modulation of Borneo vortexes (BVs) enhances precipitation intensity by 50% of the total rainfall over Borneo's western coast [21]. Over the northwest coast of the Malay Peninsula and the Iran Mountains region of Indonesia across the northwest of Borneo have the highest precipitation rate during BVs event (more than 9 mm/day on October–March 1979–2014) [21]. BVs also influence about 60–70% of extreme precipitation over these locations. Moreover, the active-MJO increases the convective activity over its propagation locations, thus increasing the rain over there. For example, MJO can enhance precipitation and cloud formation over the Maritime Continent [22, 23]. Based on The Tropical Rainfall Measuring Mission (TRMM) precipitation data filtered by Wu and Hsu [23], the western part of the Maritime Continent is affected by the MJO phases 2, 3, and 4. Based on the MJO phase/propagation, the MJO phases 3, 4, and 5 covers the Eastern part of The Indian Ocean and the Maritime Continent. A positive anomaly of precipitation is supported by a negative OLR anomaly representing the increasing cloud existence over the western part of the Maritime Continent. More specifically, the study shows that phases 3 and 4 of MJO enhance precipitation over the northern part of Sumatra Island, the Southern part of Borneo, and the northern part of Java Island during December–January–February [24]. Over Borneo and Java, the enhancement of rainfall intensity is about 1–3 mm/day (10–30%) [22]. However, the effect of precipitation enhancement is more profound over the seas than over the lands in the MC [22]. Regarding extreme precipitation, active MJO increases the 70% probability of excessive rainfall over the western and central parts of the MC [25]. The frequency of extreme precipitation events is spatially higher over the Northern Sumatera, East Java, and Southern Borneo during phase 3 MJO [24].

The BV, CS, CENS, and MJO interact and impact each other. For example, CS or MJO phase 3–5 broaden the track density of BVs, but with a lower magnitude during MJO than during CS [21]. On the other hand, the combination of CS and MJO

Table 57.1 The four flooding episodes over the western part of the Maritime Continent in early 2021

Flooding events	Location	Coordinates	Periods
Event-1	East Aceh	4.5° N, 97.6° E	1–3 January 2021
Event-2	South Borneo	3.6° S, 115° E	12–15 January 2021
Event-3	Aceh Tamiang	4.2° N, 97.6° E	21–23 January 2021
Event-4	Jakarta	6.1° S, 106.2° E	February 2021

doubled the daily mean track density of BVs during October–March, intensified the magnitude (greater maximum wind speed at 10 m/s and 925 hPa), and prolonged the lifetime of BV [21]. Furthermore, the interaction between BVs, CS, and MJO impacts precipitation over several locations. For example, BVs and CS result in 5.6 mm/day higher than all BVs over the Southern China Sea and more than 12 mm/day over the east coast of the Malay Peninsula [21]. In comparison, BVs and MJO increase precipitation over North Borneo by more than 4.2 mm/day [21].

The surge with other climate phenomena such as Borneo Vortex and Madden–Julian Oscillation (MJO) modulated high precipitation over Northern Sumatra. As a result, it resulted in flooding over southern Peninsular Malaysia from the end of 2006 until early 2007 [16, 26]. Regarding the distinctive mechanisms of each extreme rainfall and the complicated interaction between atmospheric phenomena mentioned previously, therefore, in this research, we tried to understand the atmospheric condition during each flooding event in Table 57.1 on several locations (Fig. 57.1). More specifically, this research aims to study whether atmospheric phenomena (such as CS, CENS, BVs, and MJO) affect rainfall and extreme intensity over the western part of the Maritime Continent, particularly during flooding events (Table 57.1). Thus, the main research question is, How are precipitation, wind, and sea surface temperature conditions related to CS, CENS, BVs, and MJO over the western part of the Maritime Continent during several flooding events (Table 57.1)?

57.2 Data and Method

The daily precipitation data were obtained from Global Satellite Mapping Precipitation (GSMaP) MVK v7 [27] and were used to examine the precipitation rate during the four flooding episodes. We then used Wind data (925 hPa) obtained from National Centers for Environmental Prediction (NCEP) Reanalysis data [28]. For assessing MJO, we used the interpolated Outgoing Longwave Radiation (OLR) data (daily, $0.25° \times 0.25°$) from National Center for Atmospheric Research (NCAR) [29] (provided by the NOAA/OAR/ESRL PSL, Boulder, Colorado, USA, from their Web site). Finally, the Sea Surface Temperature data was obtained from the

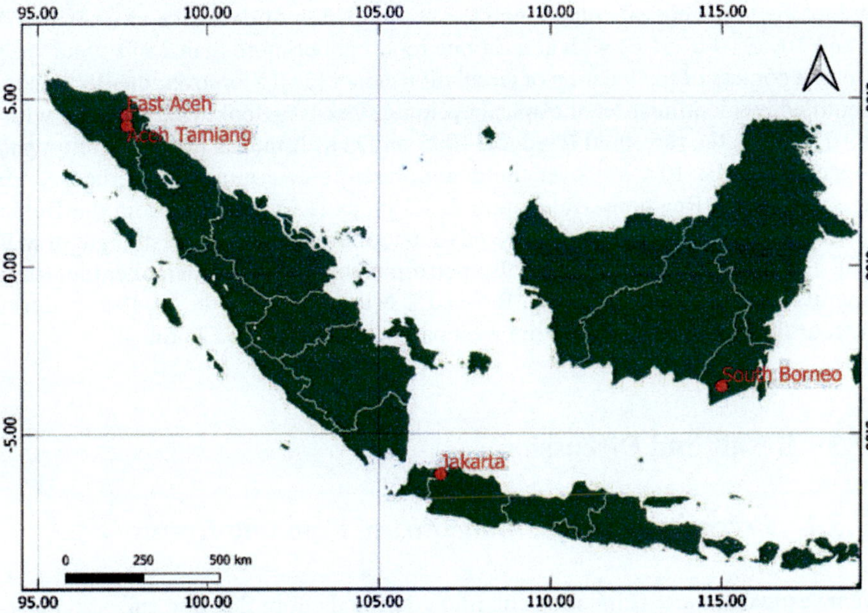

Fig. 57.1 Maps of Western Region of Maritime Continent and location of floods (red dots)

National Oceanic and Atmospheric Administration (NOAA), the OISST (Optimum Interpolation Sea Surface Temperature) v2.1 [20].

We identified the atmospheric conditions resulting in flooding by looking at the precipitation, wind, and sea surface temperature over the western part of the Maritime Continent and relating it to CS, CENS, BVs, and MJO. The precipitation condition during the flooding events was assessed by calculating the 5-days rainfall accumulation in each flooding event and the daily time series of precipitation over the point of locations (East Aceh: 4.5° N, 97.6° E, South Borneo: 3.6° S, 115° E, Aceh Tamiang: 4.2° N, 97.6° E, and Jakarta: 6.1° S, 106.2° E).

Cold surge is associated with high pressure, low surface temperature, and enhanced northern wind over Southern China [15]. A study shows the cooling of SST along the tropical South China Sea is associated with the early morning precipitation and potentially results in flooding events over Jakarta and its surroundings [19]. Moreover, the cold surge over the South China Sea also can be identified by observing the meridional wind over this location, where the maximum northerly wind is higher than 7 m/s in the region of 10°–20° N and 110°–120° E [15]. Therefore sea surface temperature (SST) and wind (speed, direction, and divergence) conditions were assessed to determine cold surges along the South China Sea. We used the surface wind (995 hPa) and examined the speed, direction, and divergence of meridional and zonal wind for each flood case.

Moreover, the wind vorticity was calculated to identify the Borneo vortex. The vortex pattern over Borneo Island can be identified using surface wind data (925 hPa) by looking at the closed counter-clockwise circulation of domain area (2.5° S–7.5° N and 107.5°–117.5° E) with at least one report about more than 2 m/s wind speed over the corners of the location of circulation center [26]. Moreover, the BVs also be identified by identification of tropical cyclonic system by looking at vorticity with 5×10^{-6} s^{-1} as the threshold (Hodges, 1995 on [21]). It means that a vorticity value greater than 5×10^{-6} s^{-1} over the domain area is determined as Borneo vortex. And finally, the Realtime Multivariate MJO index (RMM index) from the Bureau of Meteorology Australia was analyzed to determine the status and strength of MJO [30]. The MJO's propagation was also performed by making a hovmoller plot (time-longitude diagram) of anomaly OLR over 15° N to 15° S along the 50–190° E, during the four flooding episodes over the west part of the Maritime Continent.

57.3 Result and Discussion

57.3.1 Precipitation Condition During Flooding Events

We investigated the precipitation condition during the four flooding episodes during early 2021 (see Table 57.1). The five days of precipitation over the west part of the Maritime Continent were accumulated during the episode of each case (Fig. 57.2). The first episode was during the flooding in Aceh (1–3 January 2021) (event-1), with the 5-days accumulation of precipitation over the East Aceh being around 120 mm/5 days. In the second episode (event-2), the flood was over South Borneo

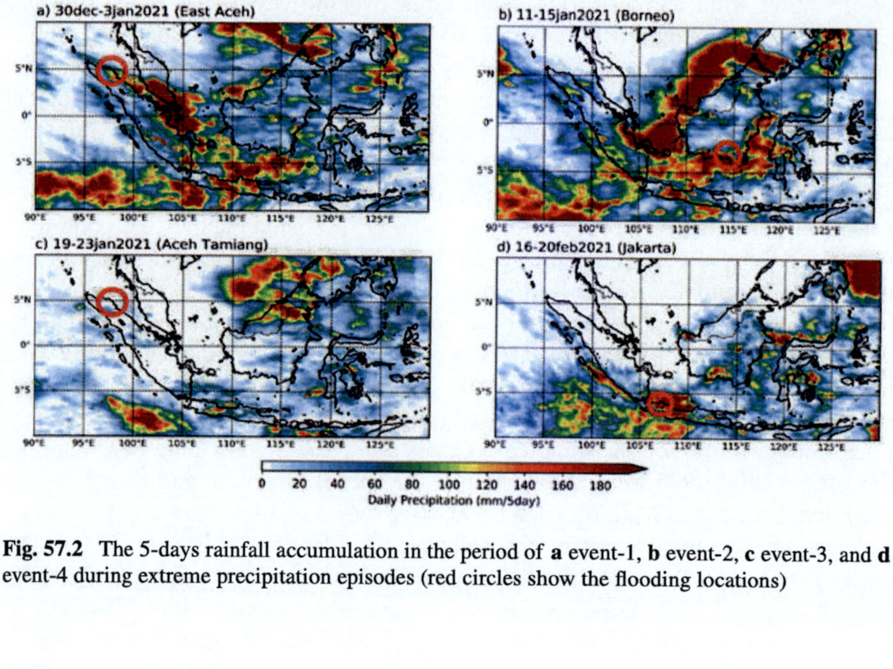

Fig. 57.2 The 5-days rainfall accumulation in the period of **a** event-1, **b** event-2, **c** event-3, and **d** event-4 during extreme precipitation episodes (red circles show the flooding locations)

on 12–15 January 2021, and the accumulation of precipitation was about 80–140 mm/5 days. In the third episode (event-3), precipitation over the Aceh Tamiang was 40–60 mm/5 days, and the flood occurred on 21–23 January 2021. Finally, the fourth episode was the flood in Jakarta (18–20 February 2021) (event-4), with more than 120 mm/5 days. Therefore we can conclude that the rainfall accumulated during event-1, event-2, and event-4 were considerably high. However, during event-3, it was low.

Figure 57.3 shows the time series of observed precipitation amounts over the flooding locations. The daily precipitation rates from several days before the extreme

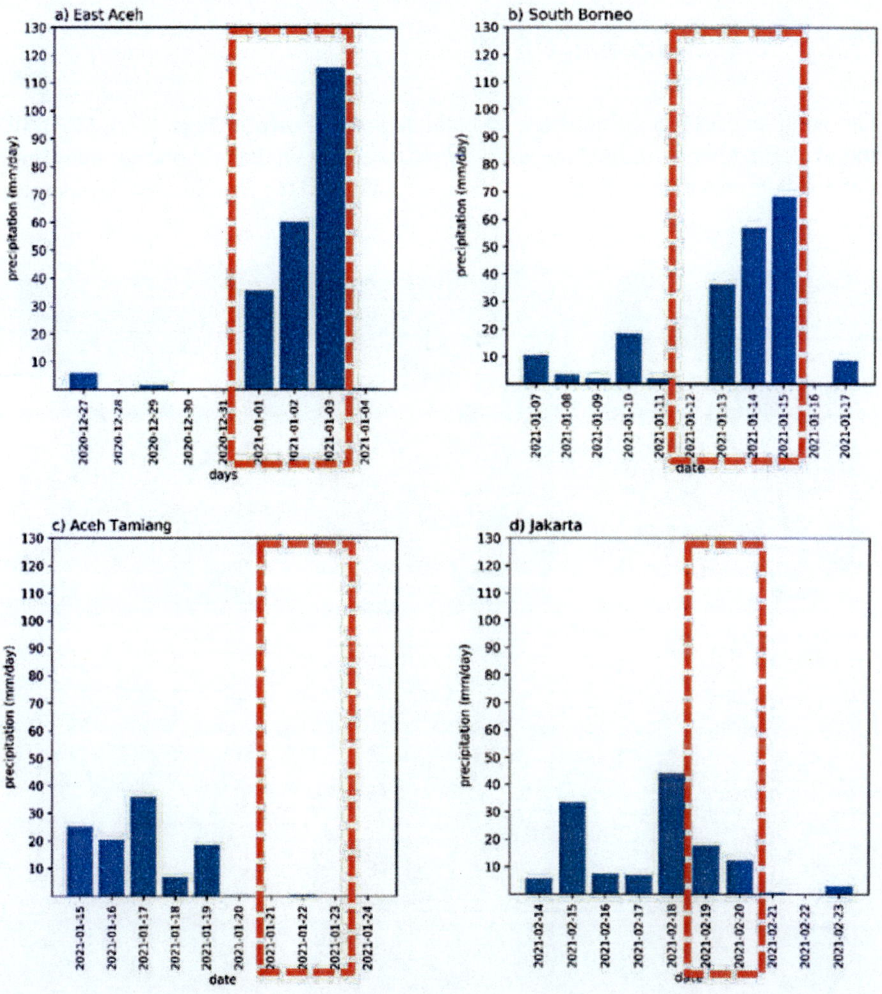

Fig. 57.3 The daily precipitation from several days before the extreme event to the day of the extreme event during **a** event-1, **b** event-2, **c** event-3, and **d** event-4. The red boxes show the flooding event dates

event to the day of the extreme event are shown. For example, daily precipitation over flooding location is high during event-1 (reaching 110 mm/day on January 3rd, 2021) and event-2 (nearly 70 mm/day on January 15th, 2021). However, during event-3 and event-4, precipitation over the flooding locations was considerably low, but the rainfall during several days before the events was high. Thus, water is accumulated from rain several days before extreme events.

The atmospheric parameters depicting Cold Surge, CENS, Borneo Vortex, and MJO are explained in the upcoming subsections to understand the atmospheric condition during the flooding events.

57.3.2 Cold Surge and CENS

The wind and SST conditions are investigated and shown in Figs. 57.4, 57.5, 57.6 and 57.7. First, we examined the wind speed, direction, and divergence of meridional and zonal wind for each flood case. Next, we analyzed the Sea Surface Temperature

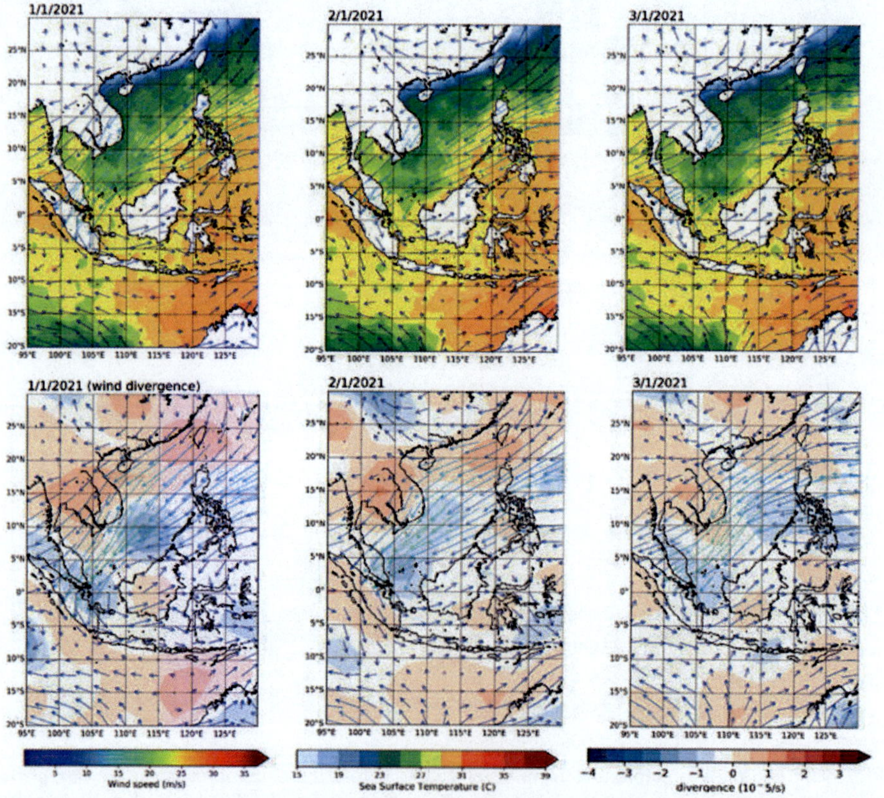

Fig. 57.4 The 925 hPa wind speed (arrow's shade) and direction (arrow), and SST (contour) conditions during event-1 (East Aceh's flood 1–3 January 2021)

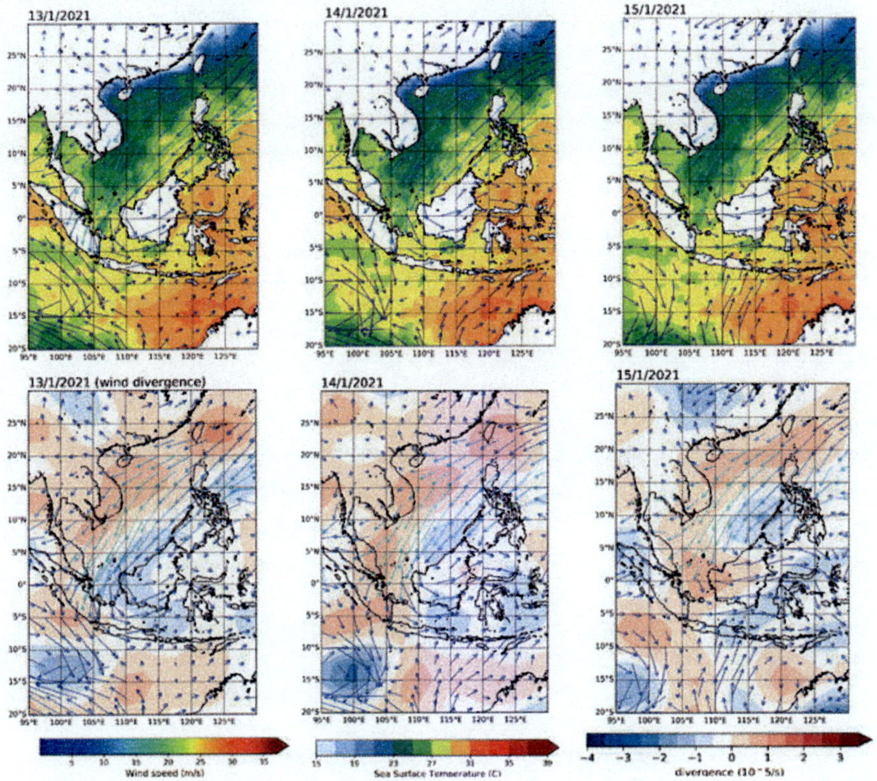

Fig. 57.5 The 925 hPa wind speed (arrow's shade) and direction (arrow), and SST (contour) conditions during event-2 (South Borneo's flood 13–15 January 2021)

(SST) by looking at the spread of warmer and colder SST spatially and relating it with the air pressure gradient, thus affecting wind direction. The divergence (convergence) represents the downward (upward) air movement, which correlates with precipitation (convection) activity. For example, during the flood over East Aceh's flood (event-1), the warmer SST (around 30 °C) was around the northern part of Sumatera Island and colder SST (about 27 °C) around the South China Sea (Fig. 57.4). Consequently, the surface pressure gradient results in a stronger northeasterly wind speed (around 20 m/s) over the South China Sea than the other surrounding area. Over East Aceh, the wind speed was high, about 10 mm/s. On January 1st 2021, there was a low-level convergence system over the northern part of Sumatera with a magnitude of around $1 \times 10^{-5}\ s^{-1}$. Moreover, there was a confluence wind from the Indian Ocean with northeasterly wind from the South China Sea (more than 10 m/s) (Fig. 57.4 bottom). This convergence system was expected to trigger cloud formation resulting in heavy precipitation over the Northern part of Sumatra Island, with moisture sources from the Indian Ocean and the South China Sea, concluding from the wind direction. However, on January 2nd, 2021, there was a low-level divergence over this location,

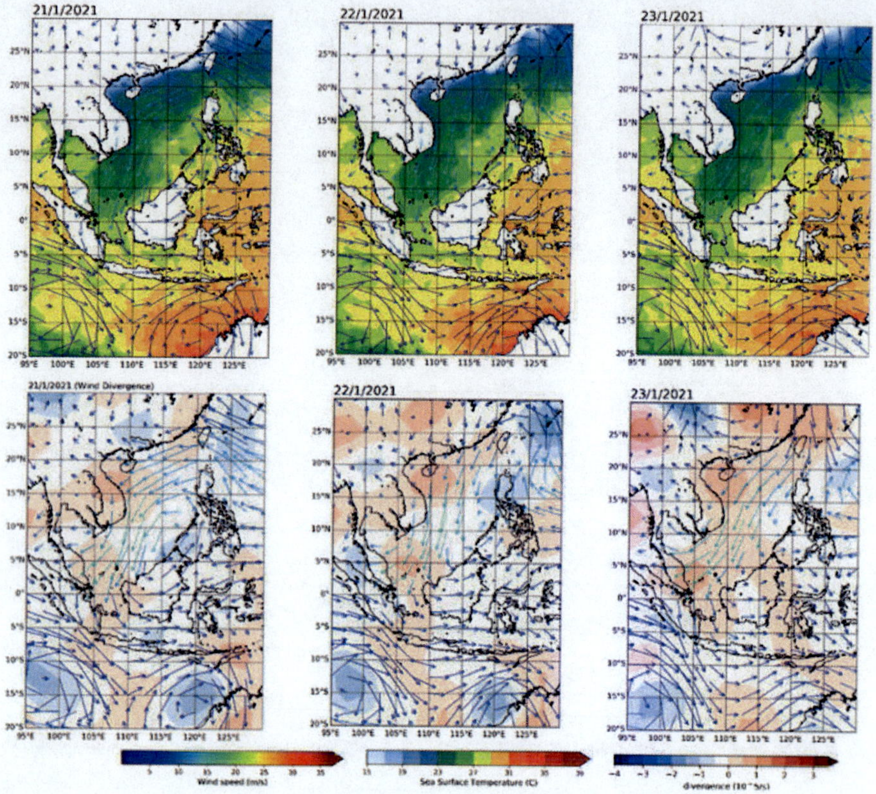

Fig. 57.6 Wind speed (arrow's shade) and direction (arrow), and SST (contour) conditions during event-3 (Aceh Tamiang's flood 21–23 January 2021)

representing the downdraft movement of air masses correlating with precipitation phenomena.

Figure 57.5 represents the wind and SST conditions during flooding over South Borneo (event-2). The warmer SST (around 30 °C) in the eastern part of Borneo covered a wider area than during the other flood events. The colder SST was also formed over the South China Sea around the equator, resulting in zonal SST contrast (between west and east Borneo). Thus the wind moved eastward over the equatorial Borneo Island (around 10 m/s). Over the flood area in South Borneo, the wind convergence ($1 \times 10^{-5}\ s^{-1}$) with wind speed around 10 m/s. Moreover, there was a closed-wind-system above Borneo island with a counter-clockwise wind direction during January 13th and 14th, 2021. This vortex system also appeared during East Aceh's flood (January 1st and 2nd, 2021 in Fig. 57.5). We further analyzed this vortex system by looking at the other wind parameters, such as vorticity, to determine the curl of wind speed or rotation (within Sect. 57.3.3).

Figure 57.6 represents the wind and SST conditions during event-3. Although Aceh Tamiang (event-3) is quite near East Aceh (area of event-1), the atmospheric

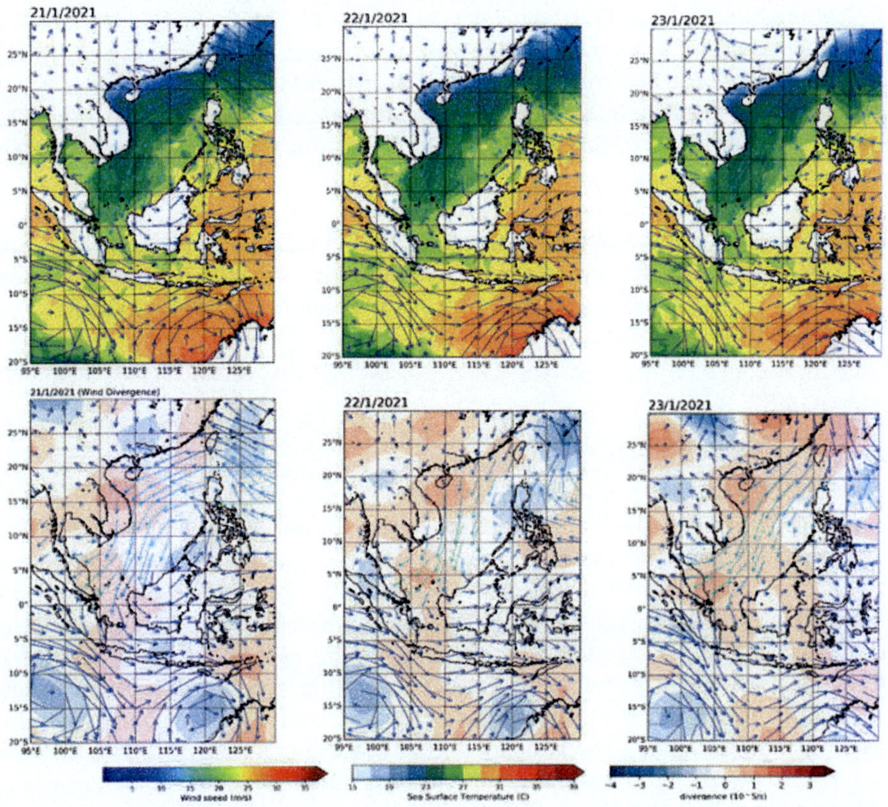

Fig. 57.7 Wind speed (arrow's shade) and direction (arrow), and SST (contour) conditions during event-4 (Jakarta's flood 18–20 February 2021)

conditions during both events are different. Compared to event-1, there was stronger westerly wind from the Indian Ocean and weaker northeasterly wind from the South China Sea toward the Northern part of Sumatera Island during event-3 (Fig. 57.6). Moreover, a convergence system over Aceh Tamiang and the Indian Ocean was shown by blueish colour in Fig. 57.6. Therefore, the precipitation source was from the convergence system related to convective activity forming clouds in these areas. The stronger westerly wind can explain that the moisture sources are from the Indian Ocean. Meanwhile, during event-1, the moisture was from the Indian Ocean and the South China Sea (Fig. 57.4). Furthermore, among the three consecutive days (Fig. 57.6), the strongest divergence is on January 22nd, more than 1×10^{-5} s^{-1}. This downward motion (divergence magnitude) during Aceh Tamiang's flood is similar to the previous study about extreme precipitation over Sumatra and Peninsular Malaysia. During Sumatera and southern Peninsular Malaysia's flood at the end of 2006 and early 2007, the divergence over flood areas reached more than 1×10^{-5} s^{-1} [16].

During event-4 (Fig. 57.7), the colder SST (23–27 °C) on the South China Sea reached the equator and around 105°E, while the warmer SST (approximately 31 °C) on the Indian Ocean near Australia extended westward to about 105°E. This colder SST expanded southward than during the other flood cases (events 1–3 on Figs. 57.4 until 57.6). On January 20th, the area coverage of this colder SST expanded over the Java Sea in the southern part of Java Island. This condition resulted in a stronger Southward wind direction toward Jakarta (western and south Java Island). Consequently, the low-level convergence system ($1 \times 10^{-5}\ s^{-1}$) appeared on January 18th, influencing the cloud formation over this location. It became a divergences system ($1 \times 10^{-5}\ s^{-1}$) on January 19th and 20th with the stronger northerly wind toward Jakarta from the South China Sea.

57.3.3 Borneo Vortexes (BVs)

We plotted the wind speed, direction, and divergence during flooding events 1–4 to identify the Borneo vortexes (BVs). The result indicates the closed counter-clockwise circulation over the Borneo during event-1 (East Aceh's flood) and event-2 (Borneo's flood) (Fig. 57.8). The wind speed and vorticity value in Fig. 57.8 explained the strength of this vortex. The vortex was stronger during event-2, with the wind speed reaching 10–20 mm/s over the center of the vortex (Fig. 57.8b), while during event-1, the wind speed was 10–15 mm/s. Moreover, the vorticity over Borneo was also stronger during event-2. The stronger vorticity was around $39 \times 10^{-6}\ s^{-1}$ during event-2, while during event-1, the vorticity was about $27 \times 10^{-6}\ s^{-1}$. Therefore, although it can be concluded as BVs during those two events, the vorticity magnitude was considerably lower than in the previous study. The strong off-shore Borneo vortex was determined based on the previous research by setting the threshold at the magnitude of vorticity around $50 \times 10^{-6}\ s^{-1}$ [31, 32]. Therefore, the vorticity magnitude and wind speed imply active Borneo vortexes during event-1 and event-2 but are not categorized as strong. The modulation of BVs results in higher precipitation over south Borneo, where the event-2 is located [21].

During the other flood cases (event-3/Aceh Tamiang and event-4/Jakarta), there was no BV. However, the vorticity magnitude was high to form vortexes over Borneo Island ($25 \times 10^{-6}\ s^{-1}$) (Fig. 57.8c). The vortexes appeared within a larger area (15° N–2° S and 108–120° E) than the area defined as Borneo Vortex by Chang et al. [26]. Thus the vortexes during event-3 (Fig. 57.8c) cannot be categorized as BVs. In the other flooding case over Jakarta (event-4/Fig. 57.8d), the vorticity magnitude over Borneo island was higher than the other locations ($20 \times 10^{-6}\ s^{-1}$). Still, there was no closed counter-clockwise circulation above the Borneo Island.

Moreover, over the flooding locations, the vorticity magnitude was around $10 \times 10^{-6}\ s^{-1}$ (over Aceh Tamiang) and $20 \times 10^{-6}\ s^{-1}$ (over Jakarta), which are categorized as higher than other surrounding locations. Therefore the tendency of wind to rotate in the flooding location is also higher, defined by the vorticity magnitude. This wind rotation triggered a local convergence system (based on Figs. 57.6 and 57.7) over the

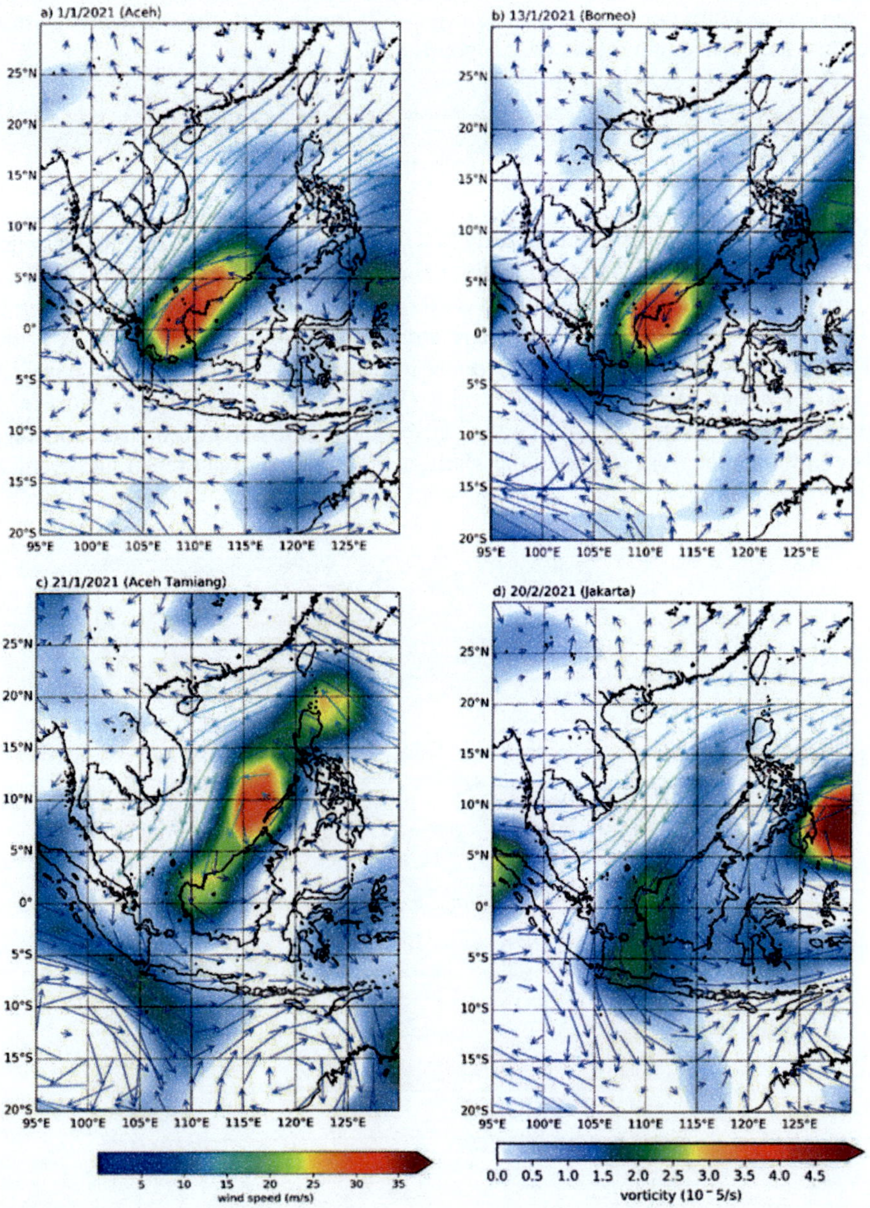

Fig. 57.8 Wind speed, direction and potential vorticity during **a** event-1, **b** event-2, **c** event-3, and **d** event-4 floodings

flooding location during event-3 and event-4. It implies that the convective system is on a local scale rather than on a synoptic scale.

57.3.4 *Madden–Julian Oscillation*

MJO phases and strength during early 2021 are shown in the diagram RMM Index (Fig. 57.9). From Fig. 57.9 and additional details on BOM's website, we summarized the Madden–Julian Oscillation (MJO) phase and state during flood shown in Table 57.2. The result shows no active and strong MJO over the west part of the Maritime Continent during flood events. Instead, MJO was active over phases 3 (East of Indian Ocean), phases 7, and 8 (Western Pacific). However, because of the eastward movement of MJO, the active phase 3 of MJO may affect the deep convection in the western part of the Maritime Continent during event-1 and event-2.

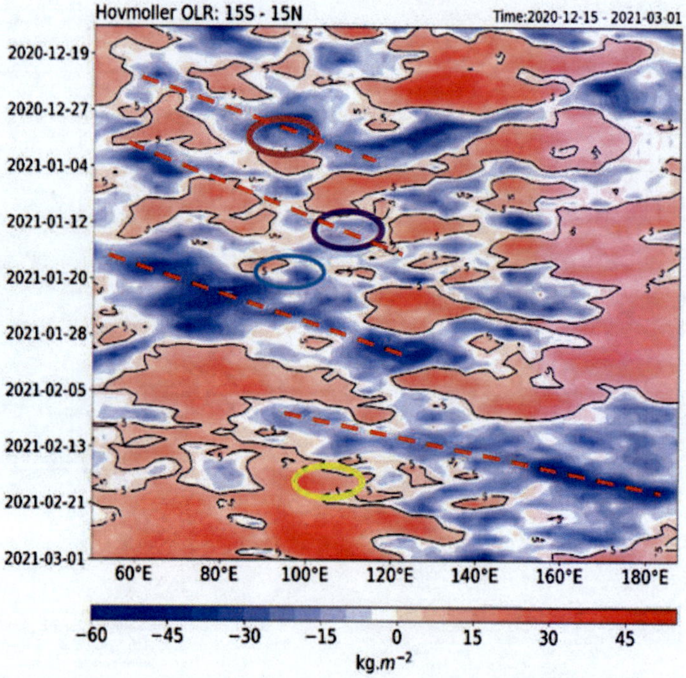

Fig. 57.9 Hovmoller diagram of outgoing longwave radiation (OLR) anomalies (the OLR average over 15° N to 15° S along the 50–190° E) during the four flood episodes over the west of the Maritime Continent. Ellipsis marks show the location of flooding events (the orange mark: event-1/East Aceh at 97.6°E, January 1st–3rd 2021, the purple mark: event-2/South Borneo at 115° E, January 12th–15th 2021, blue mark: event-3/Aceh Tamiang at 97.6° E, January 21st–23rd 2021, and yellow mark: event-4/Jakarta at 106.25° E, February 18th–20th 2021)

Table 57.2 MJO phase and state during flooding events

Flood events	Dates	MJO phase	MJO State
1 (East Aceh)	1–2 January 2021	Phase 3 (East of the Indian Ocean)	Weak
2 (South Borneo)	11–12 January 2021	Phase 3 (East of the Indian Ocean)	Strong
3 (Aceh Tamiang)	21–22 January 2021	Phase 6 (Western Pacific)	Strong
4 (Jakarta)	19–20 February 2021	Phase 7 (Western Pacific)	Strong

Particularly during event-2 (South Borneo), the MJO was in phase 3 with a strong state.

To show the movement of cloud clusters indicating the MJO's phases, we plotted the hovmoller diagram of the Outgoing Longwave Radiation (OLR) anomaly from December 15th, 2020, until March 1st, 2021 (in Fig. 57.9). The lower OLR anomaly magnitude (more blueish) shows more cloud clusters than the average during this season (December–January–February). The OLR anomaly generally shows more clouds over the flood locations (blueish color in Fig. 57.9, showing by the orange, purple and blue ellipses), except over Jakarta (reddish color displayed by the yellow ellipse). The eastward movement of the cloud cluster appears during this period (indicated by the red dashed line in Fig. 57.9), particularly during the middle of January 2021 (in coincidence with event-1 and event-2 (East Aceh and South Borneo flood events). Further explanation about MJO condition during flooding events is elaborated with relating it to the active Borneo vortexes (in the discussion part).

57.3.5 Discussion

Generally, during all flooding events, the northeasterly wind was associated with the cold surge noticed by the wind speed exceeding 10 m/s and colder SST over the vicinity of the South China Sea. Moreover, the Borneo Vortex pattern appeared during the East Aceh (event-1) and South Borneo (event-2) floods in early 2021. The wind speed (more than 10 mm/s) and vorticity magnitude (39×10^{-6} s^{-1} during event-2 and 27×10^{-6} s^{-1} during event-1) were high over Borneo Island. In addition, the strong northeasterly wind from the South China Sea reached Northern Sumatra. This northeasterly wind deflected over the equator along 105–110° E. The wind deflection resulted in the westerly and southwesterly wind toward Borneo Island. This deflection is strengthened by the active MJO phase 3 (the eastward movement of the wave) over the eastern part of the Indian Ocean toward the western part of the Maritime Continent. Therefore, the active Borneo vortexes were intensified by the active phase MJO (phase 3), was confirmed by the previous study [21]. Generally, MJO propagates in several ways when entering the Maritime Continent region, which is in consequence of the islands' orography [23, 33, 35]. Firstly, there is a blocking mechanism of MJO due to the Maritime Continent's topography; thus, the eastward movement prevails to the south of the MC or does not cross the MC, and it generates distinctive vorticity and convergence distribution [23]. Secondly, sometimes it breaks off around the MC [33, 34].

Furthermore, these active BVs intensify convective activity over the trajectory, thus enhancing precipitation [21]. During event-1, the convergence system which induces the convective activity appeared over the flooding location and its surrounding area on the 1st and 3rd January 2021 (Fig. 57.4). Meanwhile, during event-2, the convergence system also appeared above the flooding location and vicinity, from 13th until January 15th 2021 (Fig. 57.5).

Moreover, there is a relation between wind direction and Sea Surface Temperature (SST) gradient over a particular area. During event-1, the northeasterly wind over the South China Sea was stronger than during event-2 due to the higher sea surface temperature spread over, the greater area over the surrounding of Northern Sumatera than during event-2. Therefore, low-level wind convergence was more intensive at the beginning of event-1 ($1–2 \times 10^{-5}\ s^{-1}$). During event-2, there was a warmer sea surface temperature (compared to the East Aceh flood) over the eastern part of Borneo island (around the equator, east of 118° E). Moreover, there was colder sea surface temperature over the southern part of the flood location (South Borneo), which did not exist during the event-1. Moreover, low-level wind convergence was associated with the high rainfall accumulation over this location and its vicinity. Therefore, there is an indication of local convective activity during event 2.

The interaction of MJO and BVs did not appear during Aceh Tamiang's flood (event-3). During event-3, there was a vortex over Borneo and the surrounding area but with broader coverage than the BVs definition by Chang et al. [26]. Therefore it is not categorized as BVs. Moreover, the MJO was inactive over phase 3; thus, the vortex was less strong than other cases (event-1 and event-2). Instead, the wind from the Indian Ocean deflected southward along the west coast of Sumatera Island. In addition, this northwesterly wind over the Indian Ocean was stronger than the other flood cases. Moreover, the vorticity magnitude of this area ($18 \times 10^{-5}\ s^{-1}$) was also considerably higher than the other cases. A convergence system related to convective activity forming clouds was over the East Indian Ocean and North Sumatra island. Therefore, based on the wind direction, convergence, and vorticity, the precipitation source was expected from local convective activity with the moisture source from the Indian Ocean.

Furthermore, based on the 5-days precipitation accumulation from January 19th to 23rd, 2021, the precipitation rate was considerably low over the flood location and its surrounding (Fig. 57.6). However, the time series of daily precipitation at the flood's site shows the high accumulation of daily rainfall from January 15th to 19th, 2021, reaching 100 mm during 5 days (Fig. 57.7). Therefore the excessive water was not from precipitation during the flooding event, but the water was accumulated from the previous days before the flood.

Compared to the other flooding cases, there was a different wind pattern over the South China Sea (SCS) during Jakarta's flood (event-4). The northeasterly wind over the SCS deflected into the northerly wind and reached Java Island, where Jakarta is located. This strong northerly wind did not appear in the other flood cases. This wind pattern is associated with SST conditions over the SCS and the Indian Ocean. The

SST during event-4 showed a different pattern from the other flood cases. Warm sea surface temperature over the Indian Ocean in southern Java extended westward, reaching 105° E around 10–15° S, and the colder sea surface temperature over the South China Sea elongated to the Java Sea. Thus there was a southerly gradient of SST along the 105E, which results in the northerly wind along this longitude. The wind condition during this flooding case represents the Cross Equatorial Northerly Surge. This finding is consistent with the previous study that CENS attributes to a higher precipitation rate during DJF over several locations [17] and also causes some extreme precipitation events over the Southern part of Java Island [17–19].

57.4 Conclusion

We explained the rainfall, wind, and sea surface temperature conditions (which relate to CS, BVs, and MJO conditions) over the western part of the MC during flooding events over several locations: East Aceh (event-1), South Borneo (event-2), Aceh Tamiang (event-3) and Jakarta (event-4) in the early of 2021. The atmospheric conditions are different during the flooding events over the locations (events 1–4). Generally, the strong northeasterly wind over the South China Sea (SCS) appears during these all flooding events. Moreover, during the flooding event (event 1 and 2), there were active BVs that is denoted by high wind speed (more than 10 m/s), closed-counterclockwise circulation, and high vorticity over Borneo ($39 \times 10^{-6}\ s^{-1}$ during event-2 and $27 \times 10^{-6}\ s^{-1}$ during event-1). These BVs were suspected to be intensified by active MJO phase 3. Thus the intensified BVs result in higher precipitation over several locations over the western part of the MC (such as the west coast of Borneo, northwest coast of the Malay Peninsula, and the southeast Malay Peninsula). Therefore, high precipitation was detected during event-1 and event-2.

While during the event-3, there was a local convective activity and strong wind from the Indian Ocean. There was a distinct wind pattern during this flooding event compared to the other cases. The stronger wind can be found over the Indian Ocean near Sumatera with high vorticity magnitude ($18 \times 10^{-5}\ s^{-1}$). Therefore, the convective activity is more local and triggered by the strong wind from the Indian Ocean. The precipitation on the flooding event was considerably low, but the 5-day rainfall accumulation from January 15th to 19th, 2021, was high, reaching 100 mm/5 days.

Moreover, during event-4, increasing precipitation over Jakarta is due to the role of CENS. The BVs do not appear; however, the northeasterly wind over SCS turned to the southward wind. This wind deflected and had a stronger wind speed, which elongated to the southern part of Java Island. The meridional SST gradient along the 105° E attributed to this high wind speed toward the western part of the Java island. Therefore, it played a role in low-level wind convergence at the beginning of the Jakarta flooding episode. This strong-northerly wind and the meridional SST gradient do not appear in the other flooding cases.

References

1. BNPB: Risiko bencana Indonesia (2016), [Online]. Available: http://inarisk.bnpb.go.id/pdf/BukuRBI_Final_low.pdf
2. Jati, R.: Bencana Hidrometeorologi Dominan Sepanjang Awal Januari Hingga Akhir April 2021 - BNPB (2021). https://bnpb.go.id/berita/bencana-hidrometeorologi-dominan-sepanjang-awal-januari-hingga-akhir-april-2021. Accessed 22 July 2021
3. Christensen, J.H., et al.: (17) (PDF) Regional climate projections. Climate change 2007: the physical science basis (2007). https://www.researchgate.net/publication/284652431_Regional_climate_projections_Climate_change_2007_The_physical_science_basis. Accessed 2 July 2021
4. BPBA: BPBA | Banjir Menjadi Bencana Yang Paling Banyak Terjadi Diawal Tahun 2021 (2021). https://bpba.acehprov.go.id/index.php/news/read/2021/02/01/949/banjir-menjadi-bencana-yang-paling-banyak-terjadi-diawal-tahun-2021.html. Accessed 2 March 2021
5. RRI: Awal Tahun 2021, Banjir Kepung Aceh Timur - Tanggap Bencana | (2021). https://rri.co.id/tanggap-bencana/954700/awal-tahun-2021-banjir-kepung-aceh-timur. Accessed 22 July 2021
6. Lhokseumawe, M.: 9 Kecamatan di Aceh Tamiang Terendam Banjir (2021). https://regional.kompas.com/read/2021/01/21/15183201/9-kecamatan-di-aceh-tamiang-terendam-banjir. Accessed 22 July 2021
7. Mufarida, B.: BNPB: 3.075 Jiwa Terdampak Banjir di Kabupaten Aceh Tamiang : Okezone Nasional (2021). https://nasional.okezone.com/read/2021/01/21/337/2348305/bnpb-3-075-jiwa-terdampak-banjir-di-kabupaten-aceh-tamiang. Accessed 2 March 2021
8. CNN: 23 Desa dan Ribuan Warga Terdampak Banjir Aceh Timur. CNN Indonesia (2021). https://www.cnnindonesia.com/nasional/20210103204913-20-589000/23-desa-dan-ribuan-warga-terdampak-banjir-aceh-timur. Accessed 10 May 2021
9. Lhokseumawe, M.: Banjir di Aceh Tamiang Meluas, 4.000 Orang Mengungsi (2021). https://regional.kompas.com/read/2021/01/22/18215791/banjir-di-aceh-tamiang-meluas-4000-orang-mengungsi. Accessed 2 March 2021
10. BNPB: [Update] - 10 Kabupaten/Kota Terdampak Banjir di Kalimantan Selatan - BNPB (2021). https://bnpb.go.id/berita/-update-10-kabupaten-kota-terdampak-banjir-di-kalimantan-selatan. Accessed 1 March 2021
11. CNN: Banjir Kalimantan Selatan: 5 Warga Tewas, 112 Ribu Mengungsi (2021). https://www.cnnindonesia.com/nasional/20210117071801-20-594710/banjir-kalimantan-selatan-5-warga-tewas-112-ribu-mengungsi. Accessed 1 March 2021
12. Luxiana, K.M.: Cek Penyebab Banjir di Kalsel, Ini Hasil Penelusuran Bareskrim (2021). https://news.detik.com/berita/d-5344263/cek-penyebab-banjir-di-kalsel-ini-hasil-penelusuran-bareskrim. Accessed 1 March 2021
13. Adilah, R.Y.: Penjelasan Ahli Soal Penyebab Banjir di Kalimantan Selatan Halaman 2 | merdeka.com (2021). https://www.merdeka.com/peristiwa/penjelasan-ahli-soal-penyebab-banjir-di-kalimantan-selatan.html?page=2. Accessed 1 March 2021
14. Dewi, R.K.: Banjir Jakarta, Waspada Hujan Lebat hingga 25 Februari 2021 Halaman all - Kompas.com (2021). https://www.kompas.com/tren/read/2021/02/20/185200265/banjir-jakarta-waspada-hujan-lebat-hingga-25-februari-2021?page=all. Accessed 2 March 2021
15. Zhang, Y., Sperber, K.R., Boyle, J.S.: Climatology and interannual variation of the East Asian winter monsoon: results from the 1979–95 NCEP/NCAR reanalysis. Mon. Weather Rev. **125**(10), 2605–2619 (1997). https://doi.org/10.1175/1520-0493(1997)125%3c2605:CAIVOT%3e2.0.CO;2
16. Tangang, F.T., et al.: On the roles of the northeast cold surge, the Borneo vortex, the Madden-Julian Oscillation, and the Indian Ocean Dipole during the extreme 2006/2007 flood in southern Peninsular Malaysia. Geophys. Res. Lett. **35**(14), 1–6 (2008). https://doi.org/10.1029/2008GL033429

17. Hattori, M., Mori, S., Matsumoto, J.: The cross-equatorial northerly surge over the maritime continent and its relationship to precipitation patterns. J. Meteorol. Soc. Japan **89**(A), 27–47 (2011). https://doi.org/10.2151/jmsj.2011-A02
18. Wu, P., et al.: The impact of trans-equatorial monsoon flow on the formation of repeated torrential rains over java Island. Sci. Online Lett. Atmos. **3**(February), 93–96 (2007). https://doi.org/10.2151/sola.2007-024
19. Yulihastin, E., Wahyu Hadi, T., Sari Ningsih, N., Ridho Syahputra, M.: Early morning Peaksin the diurnal cycle of precipitation over the northern coast of West Java and possible influencing factors. Ann. Geophys. Discuss., 1–13 (2019). https://doi.org/10.5194/angeo-2019-107
20. Huang, B., et al.: Improvements of the daily optimum interpolation sea surface temperature (DOISST) Version 2.1. J. Clim. **34**(8), 2923–2939 (2021). https://doi.org/10.1175/JCLI-D-20-0166.1
21. Ju, L.: Climatology of Borneo vortices in the HadGEM3-GC31 general circulation model. J. Clim. (2021). https://doi.org/10.1175/JCLI-D-20-0604.1
22. Hidayat, R., Kizu, S.: Influence of the Madden-Julian Oscillation on Indonesian rainfall variability in austral summer. Int. J. Climatol. **30**(12), 1816–1825 (2010). https://doi.org/10.1002/joc.2005
23. Wu, C.H., Hsu, H.H.: Topographic influence on the MJO in the maritime continent. J. Clim. **22**(20), 5433–5448 (2009). https://doi.org/10.1175/2009JCLI2825.1
24. Purwaningsih, A., Harjana, T., Hermawan, E., Andarini, D.F.: KONDISI CURAH HUJAN DAN CURAH HUJAN EKSTREM SAAT MJO KUAT DAN LEMAH: DISTRIBUSI SPASIAL DAN MUSIMAN DI INDONESIA. J. Sains Teknol. Modif. Cuaca **21**(2), 85–94 (2020). https://doi.org/10.29122/jstmc.v21i2.4153
25. Muhammad, F.R., Lubis, S.W., Setiawan, S.: Impacts of the Madden–Julian oscillation on precipitation extremes in Indonesia. Int. J. Climatol. **41**(3), 1970–1984 (2021). https://doi.org/10.1002/joc.6941
26. Chang, C.P., Harr, P.A., Chen, H.J.: Synoptic disturbances over the equatorial South China Sea and western maritime continent during boreal winter. Mon. Weather Rev. **133**(3), 489–503 (2005). https://doi.org/10.1175/MWR-2868.1
27. Kubota, T., et al.: Satellite Precipitation Measurement, vol. 1. Springer (2020)
28. Kalnay, E., et al.: The NCEP/NCAR 40-Year Reanalysis Project. - NASA/ADS. Bull. Am. Meteor. Soc. (1996). https://ui.adsabs.harvard.edu/abs/1996BAMS...77..437K/abstract. Accessed 20 May 2021
29. Liebmann, B., Smith, C.: Description of a complete (interpolated) outgoing Longwave radiation dataset. Bull. Am. Meteor. Soc. **77**, 1275–1277 (1996)
30. Wheeler, M.C., Hendon, H.H.: An all-season real-time multivariate MJO index: development of an index for monitoring and prediction (2004)
31. Mower, R.N., Chu, J.-H., Martin, D.W., Auvine, B.: Mean state of the troposphere over southeast Asia and the East Indies, December 1978. Q. J. R. Meteorol. Soc. **110**(466), 1023–1033 (1984). https://doi.org/10.1002/qj.49711046613
32. Braesicke, P., See Hai, O., Abu Samah, A.: Properties of strong off-shore Borneo vortices: a composite analysis of flow pattern and composition as captured by ERA-Interim. Atmos. Sci. Lett. **13**(2), 128–132 (2012). https://doi.org/10.1002/asl.372
33. Ahn, M.S., Kim, D., Ham, Y.G., Park, S.: Role of maritime continent land convection on the mean state and MJO propagation. J. Clim. **33**(5), 1659–1675 (2020). https://doi.org/10.1175/JCLI-D-19-0342.1
34. Kerns, B.W., Chen, S.S.: Large-scale precipitation tracking and the MJO over the Maritime Continent and Indo-Pacific warm pool. J. Geophys. Res. **121**(15), 8755–8776 (2016). https://doi.org/10.1002/2015JD024661
35. Inness, P.M., Slingo, J.M.: The interaction of the Madden-Julian Oscillation with the Maritime Continent in a GCM. Q. J. R. Meteorol. Soc. **132**(618 A), 1645–1667 (2006). https://doi.org/10.1256/qj.05.102

Chapter 58
Analysis of Multi-scale Meteorological Parameter Triggering Seroja Tropical Cyclone and Its Effect on Extreme Rainfall Over Nusa Tenggara Timur

Luthfiyah Jannatunnisa and Trismidianto

Abstract From April 1–6, 2021, a tropical cyclone caused extreme rainfall resulting in flooding in several areas in Nusa Tenggara Timur. This paper explained the analysis of multi-scale meteorological parameters during tropical cyclone Seroja events and their effects on rainfall on the island of NTT. Convective activity and rainfall during tropical cyclones were analyzed using Himawari-8 IR1 data and GSMaP data, respectively. Analysis of atmospheric parameters used ECMWF ERA-5 data, and several indices were used to see their interactions with global atmospheric phenomena. There are four stages of development of the Seroja tropical cyclone, namely disturbance, tropical low, mature, and decay phases, with the peak of cyclone occurrence on April 5, 2021. High convective activity began to be seen in the south of Nusa Tenggara Timur during the disturbance and low tropical phases and spread throughout Nusa Tenggara Timur during the mature stage and began to move further south during the decay phase. Tropical cyclones occur during strong Madden Julian Oscillation transitions between phases 4 and 5, which are triggers for intense convective activity during hurricanes, besides being influenced by low-level solid divergence, contributed by winds from the north and south. This tropical cyclone also causes the weak effect of the Borneo vortex around Kalimantan due to the strong attraction of the mass of water vapor by the tropical cyclone. Tropical cyclones also cause changes in the movement of monsoon winds across the island of Sumatra. Compared to rainfall data, the intensity of rainfall over several areas in Nusa Tenggara Timur began to increase ever since the tropical cyclone peak and continued to rise although the cyclone was slowly decaying. With this result, we hope future, we can anticipate the impact of any tropical storms that arise near Indonesia.

L. Jannatunnisa
Meteorological Study Program, Bandung Institute of Technology, Bandung, Indonesia

Trismidianto (✉)
Research Center for Climate and Atmosphere, National Research and Innovation Agency (BRIN), Bandung, Indonesia
e-mail: trismidianto@brin.go.id

E. Yulihastin et al. (eds.), *Proceedings of the International Conference on Radioscience, Equatorial Atmospheric Science and Environment and Humanosphere Science, 2021*, Springer Proceedings in Physics 275, https://doi.org/10.1007/978-981-19-0308-3_58

58.1 Introduction

A tropical cyclone (TC) is a synoptic-scale meteorological phenomenon in the form of a large storm and grows on warm sea levels (> 26.5 °C) with convective cloud areas. The wind speed around the center of the TC itself is more than 63 km/h (34 knots). The average TC lives between 3 and 18 days, which weakens when entering cold water areas or land. The average lifespan of a TC from growth to extinction is about seven days but can vary from 1 to 30 days. The terms used to describe TC's also vary globally; for example, in the North Atlantic Ocean, this event is called Hurricane, etc. For Indonesia itself, the naming of TC's varies throughout the world; in Indonesia, alone TC's are named after the names of flowers and fruits so that they are easy to remember by the public.

TC occurs in tropics or subtropics areas in the form of storm winds. It begins with the presence of tropical or low-pressure centers that are intensive over the oceans, thus triggering an intensive process of convection and cloud formation. The study of TC has long been introduced by Fasiq (1913) in a book that contains traces, the beginning and destructive action of TC. Many studies of TC are also carried out separately in various cyclone basins of this Earth. However, based on historical events, generally, TC is formed effectively above the 10° N and 10° S region. Therefore, Indonesia's territory should be free from TC formation, but still, there is some part of Indonesia are adjacent to the formation and trajectory of TC [1]. TC's are among the most deadly and destructive natural disasters in terms of loss of human life and economic destruction [2]. Thus, interest in research on tropical cyclones has increased substantially, especially true regarding the study of intraseasonal and interannual variations in the intensity and occurrence of tropical storms, as associated with such phenomena as the Madden Julian Oscillation (MJO—e.g., Maloney and Hartmann [3]), the Quasi-biennial Oscillation (QBO—e.g., Gray et al. [4]), the El Niño and Southern Oscillation (ENSO), and the Pacific Decadal Oscillation (PDO)/North Atlantic Oscillation (NAO) (e.g., Gray et al. [4]; Chan [5]; O'Brien et al. [6]; Houghton et al. [7]).

On April 1, 2021, TC Seroja began to grow in the waters in Luwu Sea near Nusa Tenggara Timur (NTT). In general, TC occurs at latitudes that are more than 10°, but this TC Seroja is a rare cyclonic phenomenon because it occurs at low latitudes. The previous TC Seroja seed of TC 99S that appeared on April 2, 2021, which began to form in the south of NTT, Indonesia, on April 3, 2021. This TC caused floods and hurricanes in several NTT, Indonesia, and Timor Leste, which brought a powerful rainstorm. Rain and wind hit almost the entire scope of NTT, even to Timor Leste, with moderate, heavy, to extreme intensity. The Savu Sea is turbulent. There are waves as high as 4–6 m in the middle of the sea that produce tidal waves with waves as high as 1–1.5 m on the beach. On land, the TC Seroja wreaks havoc on rain-wind for hours, accompanied by lightning, and wreaks havoc. River overflows, tidal waves, flash floods, even cold lava floods, and landslides on Lembata Island. Fallen trees, houses were washed away, bridges fell by the swift river water, mud floods, and some other impacts occurred almost simultaneously in the early hours of the morning until the morning of Sunday April 4, 2021. Based on data from the National Disaster Management Agency (BNPB), on April 28, 2021, there were 20 districts and one city hit by the TC Seroja, in which 182 people died. Forty-seven people were missing, 132 people were injured, 84,876 people were displaced, and 3516 units of public and social facilities were damaged.

TC dramatically affects the surrounding weather conditions. Some studies have examined the contribution of TCs to precipitation at the regional scale [8, 9]. Villarini and Denniston [10] showed a higher probability of TC-induced annual maximum rainfall events in Australia during La Niña. Yin et al. [11] pointed out that the number of TC-heavy rain days is higher during El Niño years in eastern China; besides that, the phenomenon of TC's can also be related to local, meso, or other global phenomena [12]. TC's affect not only the trajectory area directly but also affects/affected the surrounding weather conditions. Therefore, TCs affect the weather conditions in the Indonesian territory and can occur outside the appropriate period; the storm monitoring / tropical cyclone around Indonesia is essential to do, so that information anticipates a possible impact on the form of bad weather such as floods and landslides [1]. The TC Seroja is also expected to affect the surrounding weather patterns, but the trigger has not been clearly explained. For this reason, an in-depth study of the occurrence of this TC is needed. Based on this, this paper explained the analysis of multi-scale meteorological parameters during TC Seroja events and their effects on rainfall on the island of NTT.

58.2 Data and Method

There are four stages of the TC life cycle which are also seen in the formation of TC Seroja. The division of the stages of the TC cycle is based on information from BMKG at the link http://tcwc.bmkg.go.id/siklon/learn/03/id, while for the time event for the stage, it is categorized based on information on TC Seroja from the link https://zoom.earth/storms/seroja-2021/#layers=daily so that it can be concluded in Fig. 58.1a. The four stages are the formation stage, immature stage, mature stage, and weakening stage. During the formation stage, an atmospheric disturbance was observed. If observed by satellite, it can be seen that there is a convective area with Cumulonimbus clouds. In the immature stage, the cyclone eye begins to appear, and the strong convective region forms a circular or almost circular cloudy belt. When the TC reaches the mature stage, the cyclonic circulation and gale-force wind area expand. If viewed through satellite imagery, it can be seen that the cloud conditions are more regular and symmetrical and if the intensity is stronger, cyclone eyes can be observed. During the weakening stage, the warm center of the cyclone begins to dissipate, air pressure increases, and the region with maximum wind speed expands and widens away from the center of the cyclone. This stage can occur quickly if a tropical cyclone passes in an area that is not conducive to growth, such as entering high latitude water areas with cold sea surface temperatures or entering the land. These four stages can be seen in the development of TC Seroja; the formation stage

Category	Stages	Time (LST)
Disturbance	Formation stage	April 1 at 19 - April 3 at 13
Tropical Low	Immature stage	April 3 at 19 WIB - April 4 at 19 WIB
Category 1 TC	Mature stage	April 5 at 01 - April 6 at 01
Category 2 TC	Weakening stage (Far away to the southwest)	April 6th at 07.00
Category 1 TC		April 6 at 13 - April 8 at 01

(a)

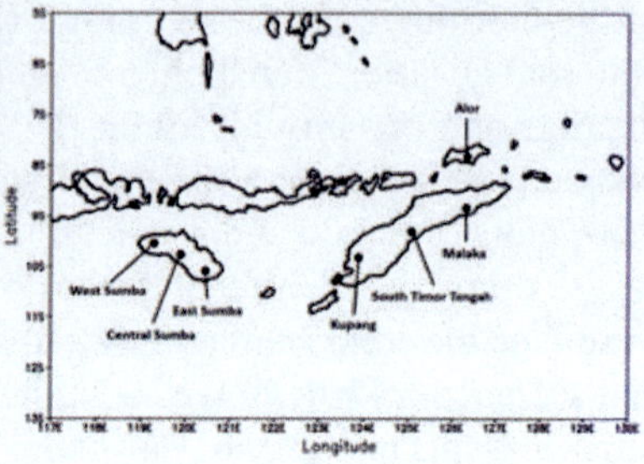

(b)

Fig. 58.1 **a** TC Seroja life cycle (source: zoom.earth), **b** NTT and the surrounding area in the study area

occurs in the period of April 1, at 1900 LST (Local Solar Time)–April 3, at 1300 LST with wind speeds of 45 km/h, then on April 3, at 1300 LST–April 4, at 1900 LST there was an increase in rate to 55–65 km/h at immature stage. The peak or ripe stage occurs on April 5, at 0100 WIB–April 6, at 0100 WIB with wind speeds of 85–95 km/h. After that, the impact weakened in the region as the typhoon moved further southwest, despite higher wind speeds. This has an impact in particular on the NTT region. This study examines the effect of tropical cyclones on rainfall in NTT and the surrounding area, as shown in Fig. 58.1b.

In this study, three primary data were used, namely brightness temperature (T_{BB}) from the Himawari-8 satellite, rainfall estimation data from GSMaP, and meteorological parameter data from ECMWF ERA 5. All data is processed using GrADS. The T_{BB} data used in this study was taken from the Himawari-8 satellite operated by the Japan Meteorological Agency (JMA) with a horizontal resolution of 0.05° × 0.05° and temporal resolution of 1 h for infrared (IR) imagery. The Himawari-8 satellite data can be downloaded for free at http://weather.is.kochiu.ac.jp/sat/GAME. Analysis of the development of TC Seroja was carried out by visual observation method on the results of plotting Himawari-8 data on IR1 channels by looking at the TBB value. Rainfall estimation data used in this study comes from GSMaP (Global Satellite Mapping of Precipitation) satellite data Near Real-Time (NRT). GSMaP Near Real-Time (GSMaP_NRT) has been implemented since October 2008. Data sets are provided in real-time (about four to five hours after observation) by file transfer protocol (FTP). This GSMaP NRT data from JAXA has been validated and calibrated with both the intelligent and radar data in Japan, with validation results likely to be quite good. GSMaP data can be accessed at http://sharaku.eorc.jaxa.jp/GSMaP/ (JAXA, 2014). The TC Seroja on the surrounding rainfall is seen by visual observations of rainfall estimates from GSMaP satellite data during the TC Seroja. The movement of the TC Seroja can be analyzed using a hovmoller diagram or cross-section during the occurrence of the TC Seroja.

The meteorological parameter data is obtained from ECMWF ERA5. ERA5 is the fifth generation ECMWF atmospheric reanalysis of the global climate covering January 1950 until present. ERA5 is produced by the Copernicus Climate Change Service (C3S) at ECMWF. ERA5 provides hourly estimates of a large number of atmospheric, land, and oceanic climate variables. The data cover the Earth on a 30 km grid and resolve the atmosphere using 137 levels from the surface up to a height of 80 km. ERA5 includes information about uncertainties for all variables at reduced spatial and temporal resolutions. This data has a time resolution of 6-hourly, a horizontal resolution of $1° \times 1°$ in latitude and longitude in 27 pressure level that include 1000, 975, 950, 925, 900, 875, 850, 825, 800, 775, 750, 700, 650, 600, 550, 500, 450, 400, 350, 300, 250, 225, 200, 175, 150, 125 and 100 hPa. The environmental parameters used in this study are the wind, relative Humidity, temperature, specific Humidity, divergence, vertical velocity, vorticity, and geopotential height. These data can be downloaded at https://cds.climate.copernicus.eu/#!/search?text=ERA5&type=dataset. Meteorological conditions analysis is also carried out with the spatial analysis of ECMWF ERA5 data combined with other data.

To analysis the condition of ENSO and IOD, we use the ENSO index and IOD index. The ENSO index is measured from the SST anomaly, Nino 3.4, and the Tahiti-Darwin IOS, which is defined as the difference in Sea Level Pressure (SLP) anomaly in the East Pacific, Tahiti (17.6° S, 149.6° W) with SLP in the West Pacific, Darwin (12.4° S, 130° E). Nino 3.4 covers part of the Nino 3 and Nino 4 areas, which are located at 120° W–170° W and 5° S–5° N. Monthly ENSO index data are obtained from the website http://www.cpc.ncep.noaa.gov. The IOD index is determined from the difference in the western equatorial Indian Ocean SST anomaly (50° E–70° E/10° S–10° N) with the Indian Ocean SST anomaly off the coast of Sumatra (90° E–110° E/10°S–Equator). Monthly IOD Index data are obtained from the BOM (Australian Bureau of Meteorology) website http://www.bom.gov.au/climate/iod/. To analyze of MJO phase, we use the MJO index from the Real-time Multivariate MJO series 1 (RMM1) and 2 (RMM2). The RMM1 and RMM2 together provide information on the state of the MJO in all parts of the tropics, and have been demonstrated by Wheller and Hendon [13] to be applicable in all seasons. As long as MJO activity was identified, RMM1 led RMM2 around the fourth cycle, indicating MJO propagation eastward along the equator. MJO index data was obtained from the CPC (Climate Prediction Center) website, NOAA (https://www.cpc.ncep.noaa.gov/products/precip/CWlink/MJO/). Analysis of monsoon condition in this study based on the data from this link http://apdrc.soest.hawaii.edu/projects/monsoon/seasonal-monidx.html.

58.3 Result and Discussion

58.3.1 T_{BB} and Precipitation Analysis

The evolution of the TC Seroja can be analyzed using T_{BB} from the Himawari-8 satellite data shown in Fig. 58.2. TC Seroja seeds began to appear at night at 1900 LST on April 1, 2021, which is the beginning of the formation stage of TC Seroja. This stage lasts long enough until April 3, 2021, at 1300 LST. At this stage, an atmospheric disturbance has been seen. From the night of April 1, 2021, until the early morning of April 2, 2021, it has been seen that there are convective areas with small convective clouds in the land area above the city of Kupang and around the Timor Sea. These small convective clouds continued to grow large scale until noon on April 3, 2021. At this stage, there was no visible center of circulation, but sometimes a spiral-shaped end of the cloud belt can be seen, as shown in figures f1–h1, from morning to noon on April 3, 2021.

On the evening of 1900 LST on April 3, 2021, the immature stage began to occur. At this stage, it is clear that convective solid areas form a protective belt or are almost round and spread in several regions around the TC area. The eye of the cyclone at the circulation center has also begun to be seen and formed since the morning at 0700 LST on April 4, 2021, as shown in figures d2–h2. The mature stage of TC began to occur in the early hours of April 5, 2021, at 0100 LST. At this stage, the cyclonic circulation was seen expanding, and the cloud conditions were more regular, and the intensity was more vigorous. The cyclone eye began to look strong on April 5, 2021,

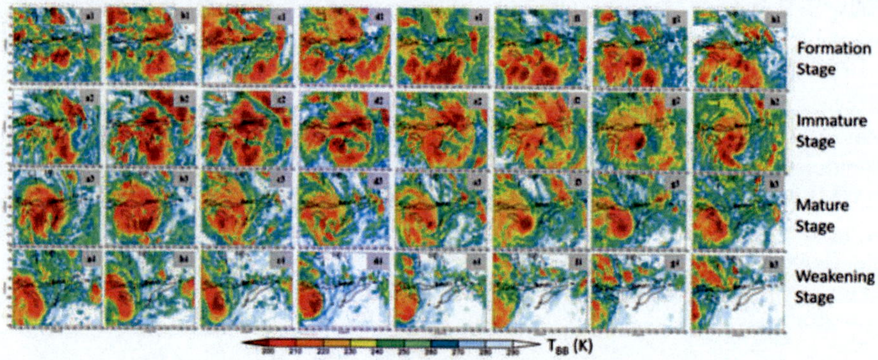

Fig. 58.2 Horizontal distribution of brightness (T_{BB}) temperature on 1–8 April, 2021 during TC Seroja. Figure a1–h1 shows the formation stage where a1–b1 on April 1 at 1900 LST and April 2 at 0100 LST, c1–e1 on April 2 at 0700, 1300, 1900 LST, and figure f1–h1 on April 3 at 0100, 0700, 1300 LST. Figure a2–h2 shows the immature stage where a2 on April 3 at 1900 LST, b2–h2 on April 4 at 0100, 0400, 0700, 1000, 1300, 1600, 1900 LST. Figure a3–h3 shows the mature stage, a3–h3 April 5 at 0100, 0400, 0700, 1000, 1300, 1600, 1900, 2200 LST. Figure a4–h4 shows the weakening stage where the cyclone moving southwest, a4-h4 April 6 at 0100, 0400, 0700, 1000, 1300, 1600, 1900, 2200 LST. T_{BB} unit is Kelvin (K)

over the island of Rote Ndao. TC Seroja started to weaken in the morning at 0700 LST on April 6, 2021. The cyclone's eyes began to shrink, and convective clouds started to decrease in intensity, and the spiral cloud belt began to disappear. At this stage of attenuation, the movement of cyclones to the surrounding area, the south coast of the island of NTT, and its surroundings are also visible.

Figure 58.3 shows an influence of TC Seroja on rainfall, and the pattern follows the TBB pattern in Fig. 58.2. From Fig. 58.3, we can see that at the formation stage, namely from 1 to 3 April 2021, there has been moderate-intensity rain over Kupang and several areas around it. East Sea. Then at the immature stage since the night of April 3, 2021, until the evening of April 4, 2021, there was rain with moderate to a heavy intensity which had spread over NTT and the surrounding area, especially in the area around the eye of the storm, then at the ripe stage since the early hours of the morning. April 5, 2021, it was already seen that the intensity of hefty rainfall occurred across almost all areas in NTT and the surrounding area. During the weakening stage, the power of precipitation began to decrease significantly, and TC Seroja began to leave NTT and the surrounding area towards Australia.

Figure 58.4a shows time series rainfall in several affected areas from TC Seroja. From the graph, we can see that in Rote Ndao Regency, it rained with very heavy intensity from the formation stage to the immature stage, besides that at the immature stage there was also hefty rain in East Sumba, West Sumba, Sabu Raijua. During the mature stage of TC Seroja, it was seen that there was very heavy rainfall intensity in Central Sumba, East Sumba, and West Sumba. Rainfall intensity decreased significantly in these areas when the TC started to weaken. Reinforced by daily rainfall observations from several BMKG observation stations in NTT as shown in Fig. 58.4b, we can see that in Kupang, Rote Ndao, and East Flores, rainfall with extreme intensity

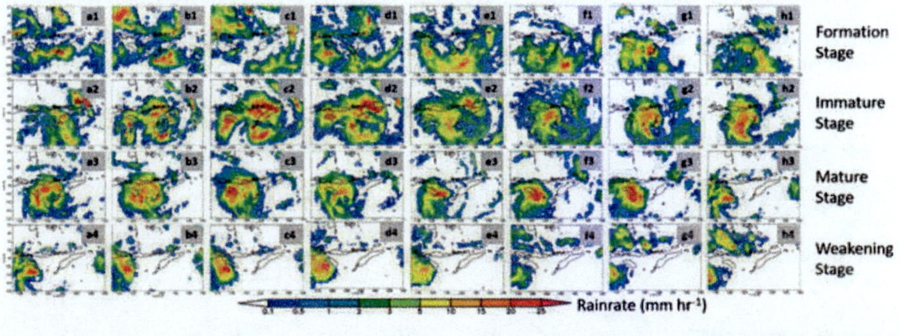

Fig. 58.3 Horizontal distribution of rainfall estimation from GSMaP data on 1–8 April, 2021 during TC Seroja. Figure a1–h1 shows the formation stage where a1–b1 on April 1, at 1900 LST and April 2, at 0100 LST, c1–e1 on April 2, at 0700, 1300, 1900 LST, and figure f1–h1 on April 3, at 0100, 0700, 1300 LST. Figure a2–h2 shows the immature stage where a2 on April 3 at 1900 LST, b2–h2 on April 4 at 0100, 0400, 0700, 1000, 1300, 1600, 1900 LST. Figure a3–h3 shows the mature stage, a3–h3 April 5 at 0100, 0400, 0700, 1000, 1300, 1600, 1900, 2200 LST. Figure a4–h4 shows the weakening stage where the cyclone was moving southwest, a4–h4 April 6, at 0100, 0400, 0700, 1000, 1300, 1600, 1900, 2200 LST. Rainrate unit is mm h^{-1}

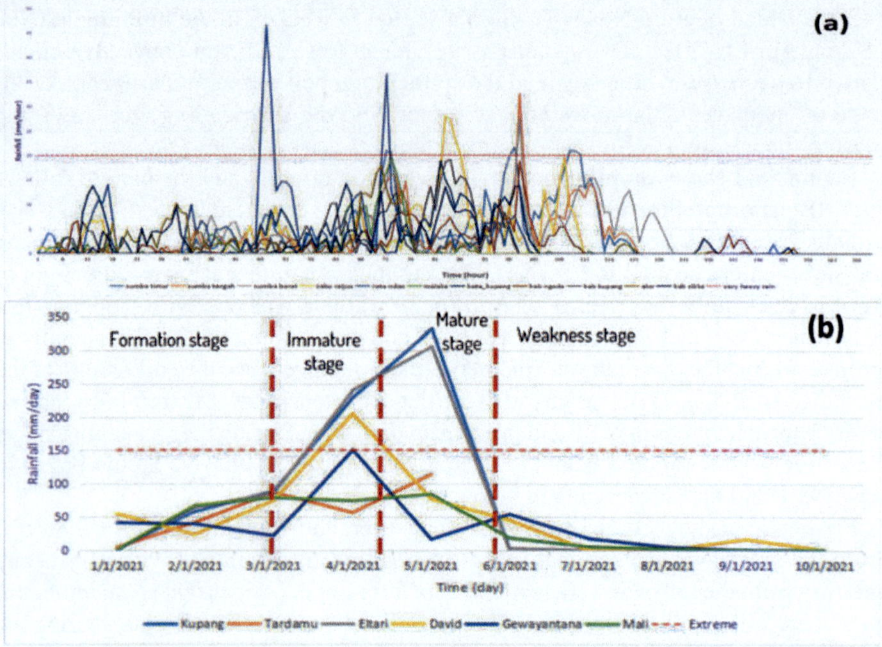

Fig. 58.4 **a** Hourly rainfall time series from GSMaP data, and **b** daily rainfall observation from BMKG data, during TC Seroja occurred

is even more during the period of the Seroja tropical cyclone, namely 4–5 April 2021. And after the TC weakened, the intensity of rainfall began to decrease drastically.

One way to show the movement of a rainfall cloud or TC is to create and analyze a hovmoller diagram, as shown in Fig. 58.5a. In this study, a hovmoller diagram was carried out by averaging the longitude of the area of the TC Seroja from 117° to 130°

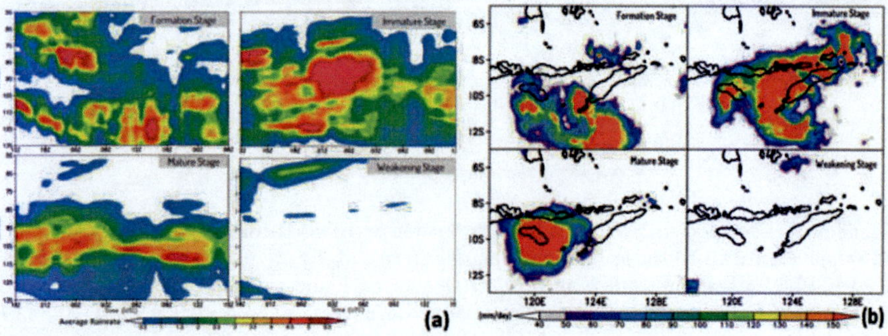

Fig. 58.5 **a** Time–latitude cross-section in longitude averaged between 117° and 130° for rainfall and **b** rainfall accumulation in every stage during TC Seroja occurred

E. The hovmoller diagramming is carried out evenly because the cyclone movement is estimated from top to bottom with time changes. The formation stage of the TC Seroja was triggered by the incorporating of several rainfall clouds from the land over NTT at a latitude of around 7°–8° S with the presence of rainfall clouds from the waters of the Timor Sea around latitude 11°–13° S with high rainfall intensity. The rainfall cloud combination triggers the occurrence of the TC Seroja with extreme rainfall intensity at its center over NTT during the immature and mature stages. This high rainfall occurs after the peak of the TC Seroja. During the Weakening stage of the TC, the rainfall intensity decreases significantly. Rainfall accumulation for eight days (April 1–8, 2021) using GSMaP satellite data was conducted to see which areas had the most influence during the occurrence of the TC Seroja. Figure 58.5b shows that high rainfall intensities during the formation stage were seen over the Rote Ndao islands and the Timor Sea. And the power of hefty rainfall extends in several areas in NTT; during the mature stage, the power of precipitation is very high. The power of precipitation over NTT has decreased drastically, and there is even no more rainfall over NTT.

58.3.2 Atmospheric Condition Analysis

Based on the streamline plot in Fig. 58.6, we can see that the direction of the wind flows at the time of the TC Seroja incident. Seroja tropical cyclone is seen causing the gathering of air masses originating from the South China Sea, Indian Ocean, and Southern Pacific Ocean. This condition shows that the energy of tropical cyclone Seroja is quite large, even after moving away from Indonesia, the Fujiwhara effect occurs, where tropical cyclones Seroja and Odette rotate cyclonically with each other, according to the definition of the National Weather Service is a binary interaction in which tropical cyclones are within a certain distance (300–750 nautical miles depending on the size of the tropical cyclone) each other begin to rotate around the exact midpoint. Finally, tropical cyclone Seroja was able to survive and continue to move to mainland Australia. During the occurrence of TC Seroja, there was also a Borneo vortex, but it did not last long, and there was no significant effect of the Borneo vortex on the surrounding rainfall; this was due to the weakening of the Borneo vortex from TC Seroja. From the streamline, we also seen that there is the power of TC Seroja to attract the wind in the Indian Ocean so that it moves across the oceanic islands towards the center of the occurrence of TC.

The divergence parameter at the 850 hPa level indicates the level of diffused air, positive values indicate the presence of diffuse/divergent perspective, and harmful for concentrated/convergent air. At this level, we can see from the tropical disturbance stage (Figures a and b) that the air is concentrated and increasingly dark blue / increasingly hostile until the Seroja cyclone moves away to the southwest. This convergent condition affects the occurrence of an updraft. Figure 58.6 also shows the atmospheric parameter of divergence at an altitude of 200 hPa, where figures a and b are dated April 3, at 0700 and 1300 LST. Figures c and d are dated April 4, at

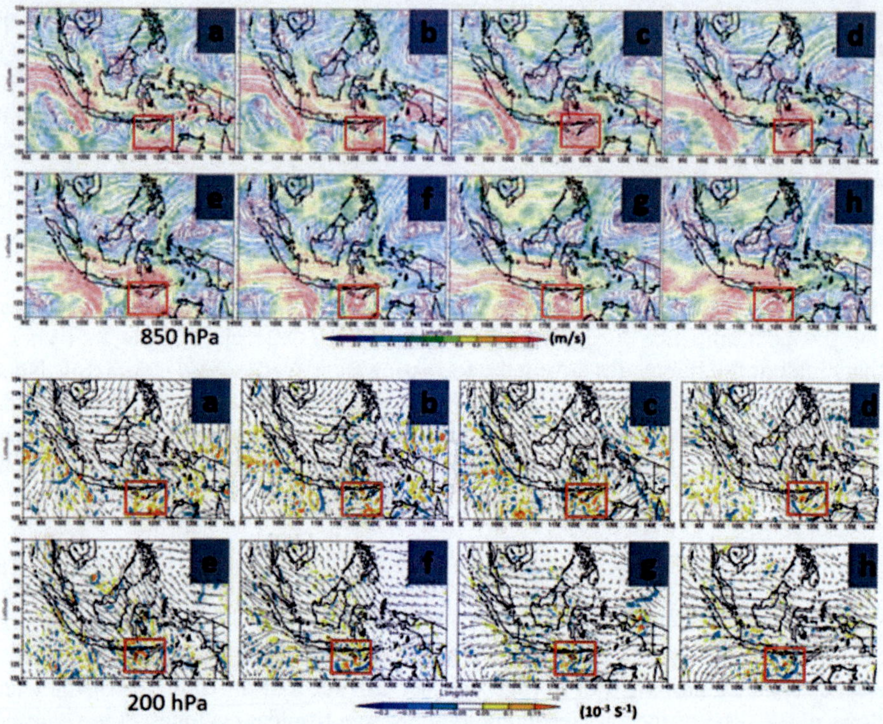

Fig. 58.6 The streamline colored with wind speed and divergence (shaded) at an altitude of 850 hPa and 200 hPa from the reanalyzed ERA5. Figure **a** and **b** on April 3, at 0700 and 1300 LST. Figure **c** and **d** April 4, at 0700 and 1900 LST. Figure **e** and **f** on April 5, at 0700 and 1900 LST. Figure **g** and **h** April 6, at 0700 and 1900 LST

0700 and 1900 LST. Images e and f on April 5, at 0700 and 1900 LST. Images g and h April 6, at 0700 and 1900 LST. Then at an altitude of 200 hPa, the air distribution in all directions gets stronger until TC Seroja moves away to the southwest, which means that the intensity of updraft events decreases.

Figure 58.7 shows the composite atmospheric parameters between temperature (shaded) and geopotential (contour) at an altitude of 500 hPa, where Figures a and b show conditions on April 3, at 0700 and 1300 LST. Figures c and d show conditions on April 4, at 0700 and 1900 LST. Figures e and f show conditions on April 5, at 0700 and 1900 LST. Figures g and h show conditions on April 6, at 0700 and 1900 LST. Low geopotential areas are named with the notation LOW or L, which is indicated by the arrangement of closed geopotential contours getting smaller and smaller. The area with a low geopotential height coincides with an area of low pressure which indicates the presence of a tropical cyclone, for temperature, we can see that the conditions in which tropical cyclones have a higher temperature than other environments. Figure 58.7 also shows the relative humidity conditions during the TC Seroja incident. Relative humidity shows the ratio between water vapor pressure (e)

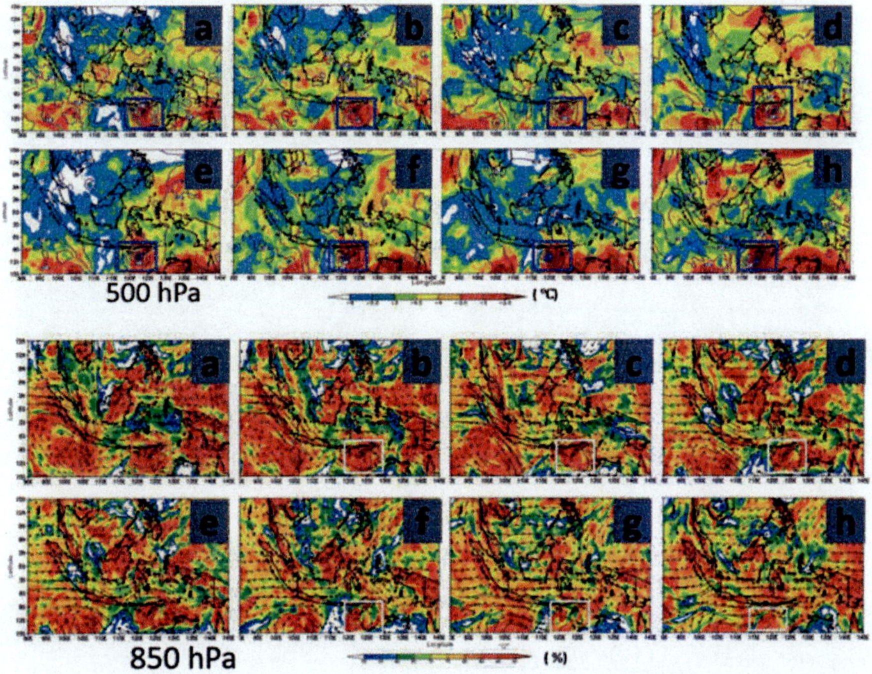

Fig. 58.7 The atmospheric composite parameter between temperature (shaded) and geopotential (contour) at an altitude of 500 hPa (Figure above), and relative humidity at a height of 850 hPa (Figure below) where **a** and **b** on April 3, at 0700 and 1300 LST. Figures **c** and **d** show condition on April 4, at 0700 and 1900 LST. Images **e** and **f** on April 5, at 0700 and 1900 LST. Images **g** and **h** April 6, at 0700 and 1900 LST

and saturated water vapor pressure (ice) at the same temperature. We can see that the RH during the TC Seroja shows more than 95%, which means the air is saturated or humid.

Figure 58.8 shows that at the stage of cyclone formation, especially on April 2, at 0100 and 0700 LST, it is seen that there is a strong convergence at an altitude of around 950–800 hPa and a substantial divergence at the 300–100 hPa layer. Then when the immature stage was seen on April 3, at 1900 LST–April 4, at 0700 LST, there was an expansion of the convergence strengthening area and on April 4, at 1100–1600 LST, there was also a strong convergence in the 500–350 hPa layer. And in the mature stage, strong convergence occurs at the level of 1000–950 hPa, and there is an expansion of the divergence area at the level of 250–100 hPa. By the time the TC has started to weaken where the cyclone is beginning to move away, it seems that there are no events, either convergent or divergent. For the vertical velocity contour, we can see that from the cyclone formation stage until it matures, the value is negative, which means an updraft occurs, where the moist air mass will rise, and when the cyclone moves away to the southwest, the area that is not affected by the

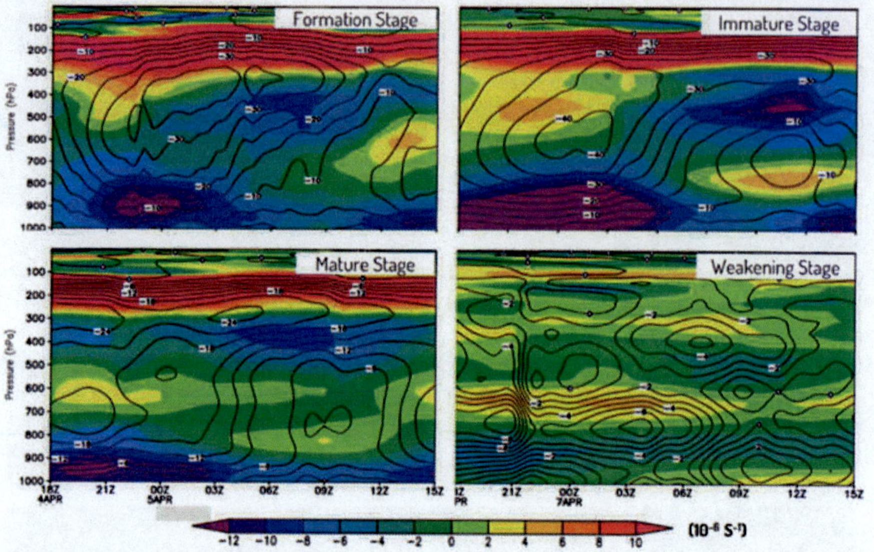

Fig. 58.8 Pressure–time cross-section for composite between the divergence (shaded, 10^{-6} S^{-1}) and the vertical velocity (contour, 10^{-2} Pa s^{-1}) in every stage during TC Seroja occurred. By averaging the values of divergence and vertical velocity for latitude and longitude in the area of TC Seroja (15° and 5° S) and 117°–130° E)

tropical cyclone shows an increasing value towards positive where updraft activity decreases.

58.3.3 Interannual and Interseasonal Variability Analysis

In general, the causes of rainfall in Indonesia are influenced by several phenomena, including ENSO and IOD as the coupled ocean–atmosphere modes in the tropical Indo-Pacific region, which have a significant impact on the rain over the maritime continent [14, 15]. ENSO is a leading mode of tropical climate variability at interannual timescales and is characterized by sea surface temperature (SST) and surface pressure anomalies across the Pacific Ocean. Its positive "El Ni˜no" phase occurs when SSTs are warm in the eastern tropical Pacific Ocean, contrasted by its negative "La Ni˜na" phase when they are cool. The IOD is another coupled ocean–atmosphere mode [16], with a positive (negative) phase characterized by warm (cool) SSTs over the western Indian Ocean and cool (warm) SSTs in the eastern Indian Ocean. The two phenomena activities often co-occur, resulting in a decrease in the amount of rainfall (drought) and increased rainfall (floods) in some parts of Indonesia. Another phenomenon that also dramatically affects rainfall in Indonesia is the Madden–Julian Oscillation (MJO) as the significant intraseasonal (30–60 days) variability in the tropics, characterized by the eastward propagation of cloud clusters and precipitation from the Indian Ocean to the western Pacific Ocean [17]. The influence of the MJO on Indonesian precipitation variability has been generally discussed by several authors [18–20].

Figure 58.9a shows that the IOD Index was positive in April when the Seroja TC occurred; this indicates that the west coast of Sumatra has high pressure, while the east coast of the African continent has low pressure so that airflow from the west of Sumatra will go east of Africa which results in cloud formation. Convective clouds over Africa and produce above-normal rainfall. On the other hand, in the western region of Sumatra, there is a drought after the mass of water vapor fails to reduce it as rain. The existence of the high-pressure area of the west coast of Sumatra also supports the movement of wind or water vapor masses to regions of low pressure, including towards TC Seroja. *This* strengthens Fig. 58.6, where there is wind movement from the west coast of Sumatra across the island of Sumatra to the TC Seroja area. Figure 58.9b shows that the ENSO Index in April 2021 was in a weak La Nina condition where the easterly winds along the equator became stronger than usual and generally impacted the form of increased rainfall in the western equatorial Pacific, including Indonesia. La Nina conditions also trigger the growth of convective clouds in Indonesia, especially in central Indonesia; this condition is possible one of the triggers for the development of convective clouds around the area where TC Seroja occurs during the formation stage and also makes the life cycle of TC also long.

Figure 58.10 shows the sea surface temperature (SST) on March 29–April 4, 2021 around the NTT area. The SST value is in the range of 29–30 °C, where this condition makes the NTT region have lower pressure than other regions and this high SST is a contributor to high rainfall because it supports the growth of a lot of convective clouds that trigger the occurrence of TC Seroja.

According to Khromov [21], monsoon areas are areas where the dominant wind direction reverses at least 120° between January and July. January is the winter maximum in the northern hemisphere with the lowest average temperature in the

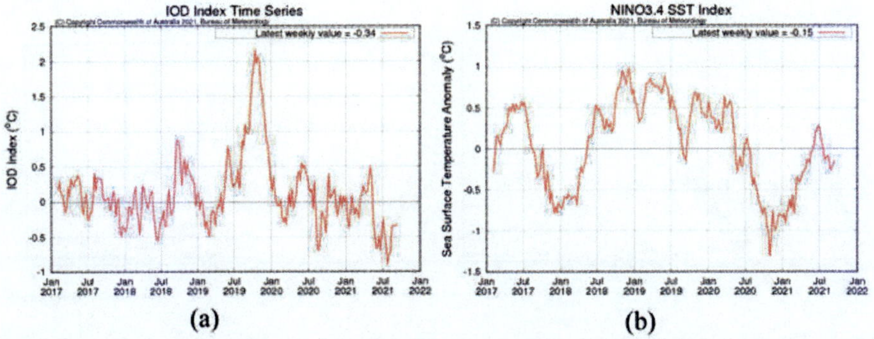

Fig. 58.9 **a** IOD index and **b** NINO 3.4 Index from the Bureau of Meteorology (BOM)

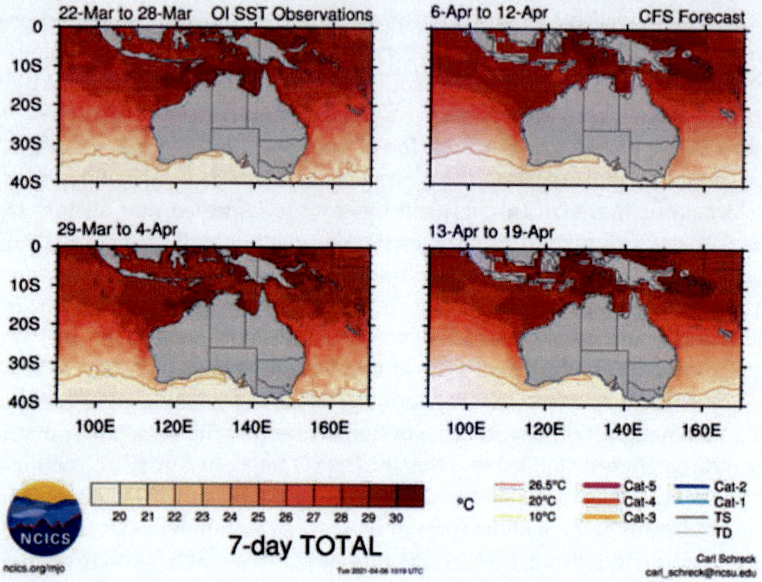

Fig. 58.10 Sea surface temperature during TC Seroja occurred

northern hemisphere, and July is the summer maximum with the highest average temperature in the northern hemisphere. Ramage [22] defines of monsoon by adding wind strength criteria and identifying areas including Africa, Asia, and Australia as areas that meet both the reverse wind criteria and the monsoon rain criteria. Figure 58.11 shows the WPMI value (Asian monsoon index) during April 2021 getting stronger until May 2021, and for the Australian monsoon, on the contrary, it is weakening. *This* shows the impact that when the Asian monsoon winds flow, Indonesia will experience the rainy season in April, which is also the trigger for many convective clouds before the Seroja TC.

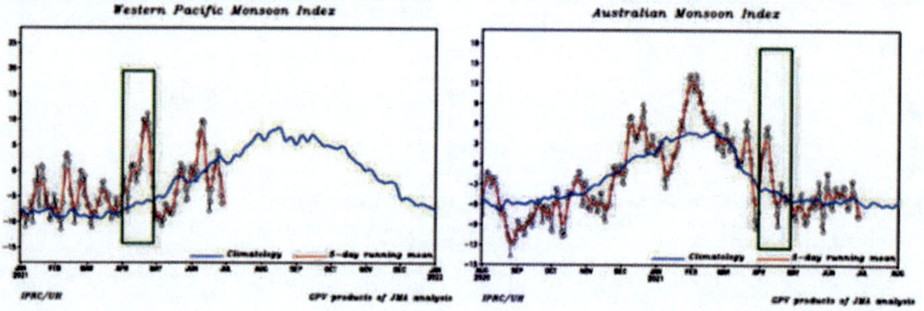

Fig. 58.11 Western Pacific Monsoon index (left) and Australian Monsoon index (right) from Japan Meteorology Agency (JMA)

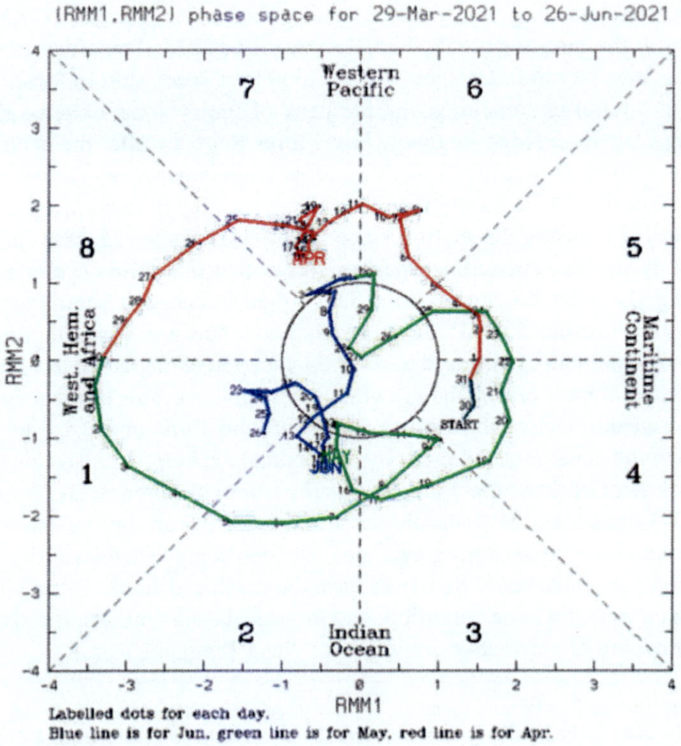

Fig. 58.12 MJO index during TC Seroja occurred

The MJO index on April 1–5, 2021, shows that the condition of the MJO is entering phases 4 and 5 as shown in Fig. 58.12. The index values for Phase 4 and 5, also show values above 1, which means that there is a strong MJO in the Indonesian maritime continent. So with the active MJO, it increases the possibility of many convective clouds in Indonesia, which is believed also to trigger the development of convective clouds into TC Seroja.

58.4 Conclusions

TC Seroja seeds began to appear at night at 1900 LST on April 1, 2021, which is the beginning of the formation stage of TC Seroja. This stage lasts long enough until April 3, 2021, at 1300 LST. On the evening of 1900 LST on April 3, 2021, the immature stage began to occur. At this stage, it is clear that strong convective areas form a protective belt or are almost round and spread in several regions around the TC area. The mature stage of TC began to occur in the early hours of April 5, 2021, at 0100 LST. At this stage, the cyclonic circulation was seen expanding, and the cloud conditions were more regular, and the intensity was more vigorous. TC Seroja began to weaken in the morning at 0700 LST on April 6, 2021. The effect of the Seroja tropical cyclone on rainfall is seen when the cyclone seeds start to form, there is an increase in rainfall in some areas, at the peak of the cyclone, extreme rain occurs in Kupang City, Rote Ndao Regency, East Flores Regency after the cyclone moved away, rain fell still drastically in the NTT area. High convective activity began to be seen in the south of NTT during the disturbance and low tropical phases and spread throughout NTT during the mature stage, and began to move further south during the decay phase. The streamlined pattern shows that the Seroja cyclone causes air masses to gather from the South China Sea, Indian Ocean, and South Pacific Ocean. The divergence at the 200 hPa layer strengthens, and a diffuse air mass is seen. Geopotential contours can be used to show the location of the sacred tropical cyclone, and the temperature is higher than in other environments. This tropical cyclone also causes the weak effect of the Borneo vortex around Kalimantan due to the strong attraction of the mass of water vapor by the tropical cyclone. TC also causes changes in the movement of monsoon winds across the island of Sumatra. TC occurs during strong MJO transitions between phases 4 and 5, which are one of the triggers for strong convective activity during cyclones, besides being influenced by strong low-level divergence, contributed by winds from the north and south. During TC Seroja, ENSO is in a weak La Nina condition, and increased Asian monsoon activity causes the strengthening of convective activity over Nusa Tenggara Timur.

Acknowledgements Thank you so much for the three sources of data, which we can get for free, and the data is beneficial to this research.This research uses data from ECMWF ERA-Interim (http://www.ecmwf.int/), Himawari-8 satellite (http://weather.is.kochiu.ac.jp/sat/GAME) and GSMaP satellite (http://sharaku.eorc.jaxa.jp/GSMaP/).

References

1. Haryani, N.S., Any, Z.: Dinamika Siklon Tropis di Asia Tenggara Menggunakan Data Pengindraan Jauh. Pusbangja LAPAN, 324 (2012)
2. Bengtsson, L., Boztet, M., Esch, M.: Will green-house gas-induced warming over the next 50 years lead to higher frequency and greater intensity of hurricanes? Tellus **48A**, 57–73 (1996)
3. Maloney, E.D., Hartmann, D.L.: The Madden-Julian Oscillation, barotropic dynamics, and North Pacific tropical cyclone formation. Part I: Observations. J. Atmos. Sci. **58**, 2545–2557 (2001)
4. Gray, W.M.: Atlantic season hurricane frequency. Part 1: El Nino and 30 mb Quasi Biennial Oscillation influences. Mon. Wea. Rev. **112**, 1649–1668 (1984)
5. Chan, J.C.L.: Tropical cyclone activity in the Northwest Pacific in relation to the El Nino/Southern Oscillation phenomenon. Mon. Wea. Rev. **113**, 599–606 (1985)
6. O'Brien, J.J., Richards, T.S., Davis, A.C.: The effect of El Nino on U.S. landfalling hurricanes. Bull. Amer. Meteor. Soc. **77**, 773–774 (1996)
7. Houghton, J.T., et al. (eds.): Climate Change 2001: The Scientific Basis, 857pp. Cambridge University Press, Cambridge, UK (2001)

8. Chen, F., Fu, Y.: Contribution of tropical cyclone rainfall at categories to total precipitation over the western North Pacific from 1998 to 2007. Sci. China Earth Sci. **58**, 2015–2025 (2015). https://doi.org/10.1007/s11430-015-5103-9
9. Ng, B., Walsh, K., Lavender, S.: The contribution of tropical cyclones to rainfall in northwest Australia Int. J. Climatol **35**, 2689–2697 (2015). https://doi.org/10.1002/joc.4148
10. Villarini, G., Denniston, R.F.: Contribution of tropical cyclones to extreme rainfall in Australia. Int. J. Climatol. **36**, 1019–1025 (2016). https://doi.org/10.1002/joc.4393
11. Yin, Y., Gemmer, M., Luo, Y., Wang, Y.: Tropical cyclones and heavy rainfall in Fujian Province, China. Quat. Int. **226**, 122–128 (2010). https://doi.org/10.1016/j.quaint.2010.03.015
12. Curtis, S., Adler, R.F.: Evolution of El Nino precipation relationships from satellite and gauges. J. Geoph. Res. **108** (2003)
13. Wheeler, M.C., Hendon, H.H.: An all-season real-time multivariate MJO index: development of an index for monitoring and prediction. Mon. Weather Rev. **132**, 1917–1932 (2004). https://doi.org/10.1175/1520-0493(2004)132%3c1917:AARMMI%3e2.0.CO;2
14. Aldrian, E., Susanto, D.: Identification of three dominant rainfall regions within Indonesia and their relationship to sea surface temperature. Int. J. Climatol. **23**, 1435–1452 (2003)
15. Nur'utami, M.N., Hidayat, R.: Influences of IOD and ENSO to Indonesian rainfall variability: role of atmosphere-ocean interaction in the Indo-Pacific sector. Procedia Environ. Sci. **33**, 196–203 (2016). https://doi.org/10.1016/j.proenv.2016.03.070
16. Saji, N.H., Goswami, B.N., Vinayachandran, P.N., Yamagata, T.: A dipole mode in the tropical Indian Ocean. Nature **401**(6751), 360–363 (1999)
17. Madden, R.A., Julian, P.R.: Detection of a 40–50 day oscillation in the zonal wind in the tropical Pacific. J. Atmos. Sci. **28**, 702–708 (1971). https://doi.org/10.1175/1520-0469(1971)028%3c0702:DOADOI%3e2.0.CO;2
18. Jones, C., Waliser, D.E., Lau, K.M., Stern, W.: Global occurrences of extreme precipitation and the Madden-Julian oscillation: observations and predictability. J. Clim. **17**, 4575–4589 (2004). https://doi.org/10.1175/3238.1
19. Hidayat, R., Kizu, S.: Influence of the Madden-Julian oscillation on Indonesian rainfall variability in austral summer. Int. J. Climatol. **30**, 1816–1825 (2010). https://doi.org/10.1002/joc.2005
20. Xavier, P., Rahmat, R., Cheong, W.K., Wallace, E.: Influence of Madden-Julian oscillation on Southeast Asia rainfall extremes: observations and predictability. Geophys. Res. Lett. **41**, 4406–4412 (2014). https://doi.org/10.1002/2014GL060241
21. Khromov, S.P.: Die geographische Verbreitung der Monsune. Petermanns Geographische Mitteilung **101**, 234–237 (1957)
22. Ramage, C.S.: Role of a tropical "Maritime Continent" in the atmospheric circulation. Mon. Weather Rev. **96**, 365–369 (1968)

Chapter 59
Estimating Rainfall Data Using Tropical Rainfall Measuring Mission (TRMM) Data: A Study Case in Pesawaran Meteorology and Geophysics Agency

Ali Rahmat and Fajar Setiawan

Abstract The availability of climate information is essential to support various sectors such as disaster mitigation, the agricultural sector, plantations, mining, and other strategic sectors. One of the important climate elements is rainfall. Because the information about rainfall is important, measurement at many points and continuously becomes a necessity. However, direct monitoring cannot be applied because of the limitation of funds and the equipment. One way to monitor rainfall occurrence is using satellite data; one uses Tropical Rainfall Measuring Mission (TRMM) data. This research aims to estimate the rainfall from TRMM data and evaluate the accuracy. The data was used in this research is monthly rainfall data from Pesawaran Meteorology and Geophysics Agency (BMKG Pesawaran) and monthly data from TRMM for 2000–2010. The results show that the R^2 of the original TRMM and observation data is 0.6561 with RSME 123.19 mm. After applying Machine Learning with random forest fitting, the R^2 was increased to 0.8876, and RSME decreased to 41.2 mm. The estimation data can be used as general information from these results, and the model estimation can be improved through machine learning.

59.1 Introduction

Nowadays, climate information has become vital information to support human activities. Furthermore, the availability of climate information is essential to support various sectors such as agriculture, plantation, mining, and other strategic industries. For example, agriculture and plantations require rainfall data to determine the right planting time and irrigation planning [1]. In addition, rainfall data is also crucial as information in determining disaster mitigation [2, 3]. Therefore, providing climate information that is accurate, precise, and easily accessible to users in various sectors

A. Rahmat (✉) · F. Setiawan
Reseach Center for Limnology and Water Resources, National Research and Innovation Agency, Jakarta, Indonesia
e-mail: alirahmat911@gmail.com

E. Yulihastin et al. (eds.), *Proceedings of the International Conference on Radioscience, Equatorial Atmospheric Science and Environment and Humanosphere Science, 2021*, Springer Proceedings in Physics 275, https://doi.org/10.1007/978-981-19-0308-3_59

is a challenge in itself. Climate information is used by relevant sectors to make policies, strategies, and action plans.

Rainfall is one of the most critical climate elements, but spatially and temporally is still challenging to predict. In addition to its dynamic nature, the physical processes involved are also very complex. Because information about rainfall is necessary, multiple and continuous measurements are necessary. However, continuous, direct monitoring cannot be applied because of funds and equipment limitations. For example, in Lampung province, there are only four stations that give free access data, but start 2013, many data were missing. As an alternative, estimated rainfall data from the satellite can be used. The Tropical Rainfall Measuring Mission (TRMM) satellite is one satellite data designed to assess rainfall data.

TRMM is a space collaboration project between NASA and the Japan Aerospace Exploration Agency (JAXA) intended to monitor and study rainfall in the tropics within the framework of NASA's long-term effort to observe the earth as a global system [4]. The TRMM program is for long-term research designed to study land, sea, air, ice, and the total system of life on earth [5].

Since its launch in 1997, TRMM data has been widely used in various studies of weather and climate problems in Indonesia. For instance, TRMM satellite data is used to analyze extreme weather conditions [6, 7]. Gunawan [8] has compared monthly rainfall observations with the TRMM satellite and the NOAH surface model and proved that TRMM data has the potential to fill in temporally empty surface data for a location and spatial voids for areas that lack surface rainfall observations. Therefore, this research aims to estimate the rainfall from TRMM data and evaluate the accuracy in Pesawaran Meteorology and Geophysics Agency.

59.2 Methods

This research was carried out in Hydroinformatic Laboratory, Research Center for Limnology, Badan Riset dan Inovasi Nasional (BRIN), Cibinong, Bogor, Indonesia. This research was conducted from July–September 2021. We collected data from the ground observation station and remotely sensed data. We obtained the ground measurement data from Pesawaran Meteorology and Geophysics Agency (BMKG Pesawaran) by downloading the official website link (https://dataonline.bmkg.go.id/home).

We collected free access remotely sensed TRMM data from Giovanni's website (https://giovanni.gsfc.nasa.gov/giovanni/). We analyzed monthly rainfall data from June 2002 to July 2013. We used 148 monthly rainfall data for model calibration and 13 data for model validation. We used Microsoft Excel and R software (https://www.r-project.org/) to process the data. A linear regression model and two machine learning fitting models (k-NN, Random Forest) were used to select the best model to estimate rainfall from TRMM data. We calculated the coefficient of determination (R^2) and Root Square Mean Error (RSME) values to check the model performance.

We conducted a visual analysis to observe the pattern between estimated and ground rainfall observation temporal data.

59.3 Results and Discussions

Figure 59.1 provides the scatterplots of the observed monthly rainfall data and the corresponding estimated rainfall from the TRMM satellite in the model calibration. Figure 59.1a. shows the observed data and the original TRMM data, the coefficient of determination (R^2) values is 0.651, and the RMSE is 123.19 mm. This low R^2 and large RMSE indicate that TRMM data can not be directly used as a source of rainfall data for further research application. As an alternative, we should improve the TRMM

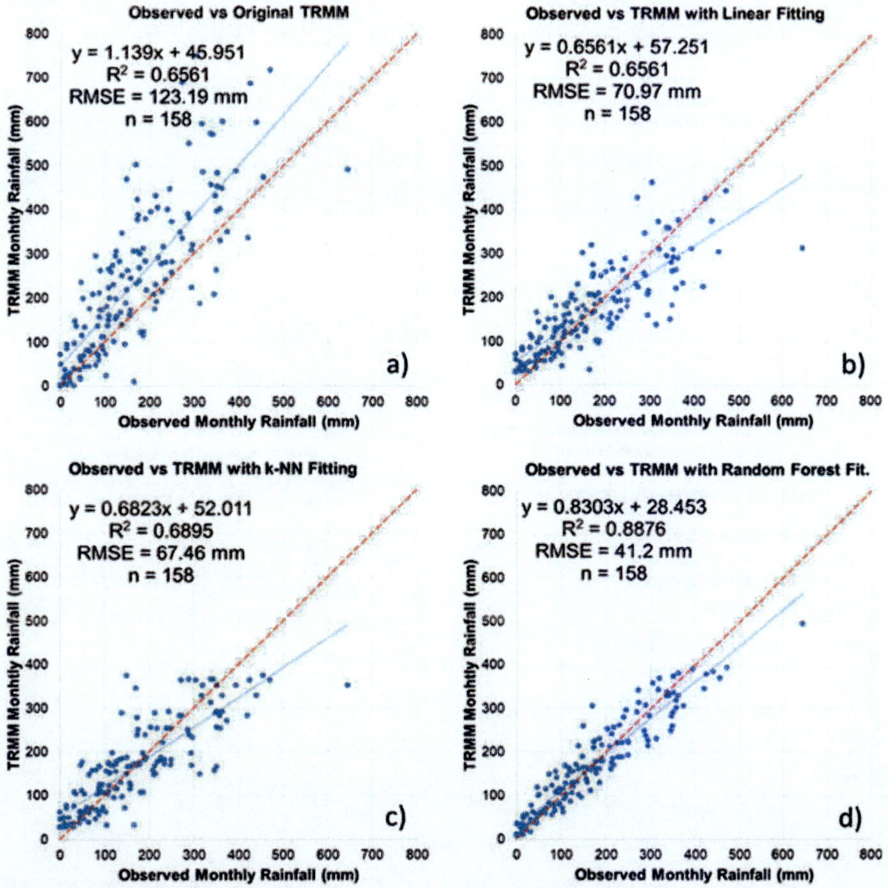

Fig. 59.1 Relationship between observed (BMKG data) and estimated data (TRMM) in the model calibration

data quality by fitting the data using several popular fitting models, for example, linear regression, machine learning k-NN, and Random forest fitting. Figure 59.1b shows a simple linear regression model fitting result. We obtained the same R^2; however, the RMSE values decreased from 123.19 mm to 70.97 mm. Figure 59.1b revealed that a simple linear fitting was insufficient to improve the rainfall estimation.

Figure 59.1c shows the improvement of estimated rainfall from TRMM data using Machine Learning with K-NN fitting. We observed that using Machine learning with k-NN fitting slightly increased the rainfall estimation; however, the under-over estimate was still considerable ($R^2 = 0.689$ and RMSE = 67.46 mm). Figure 59.1d shows the estimated rainfall from TRMM data using Machine Learning with Random forest fitting. We pointed out that the best R^2 and RMSE from the Random Forest fitting (0.887 and 41.2 mm, respectively).

Figure 59.2 illustrates the validation result of our developed model to estimate rainfall using TRMM data. In line with the calibration result, we observed that the

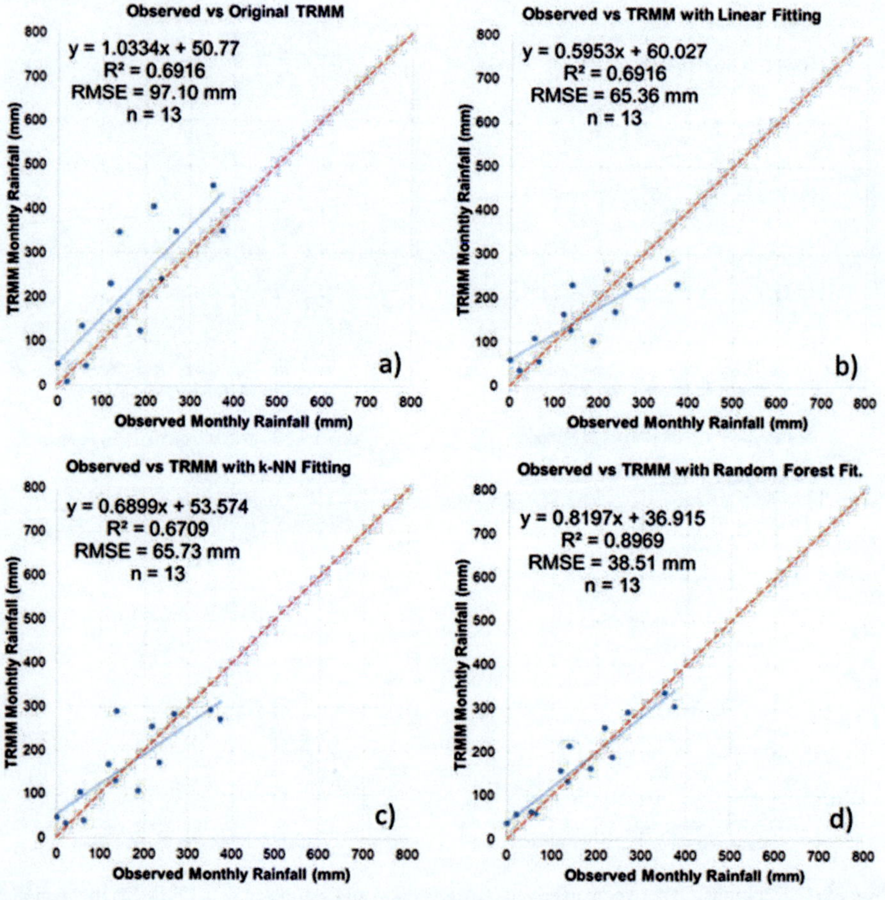

Fig. 59.2 Relationship between observed (BMKG data) and estimated data (TRMM) in the model validation

Fig. 59.3 The pattern of rainfall from observed data (BMKG data) and TRMM original (estimation)

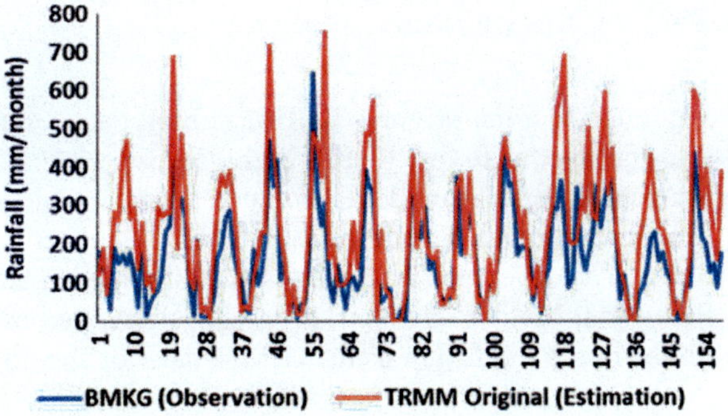

Fig. 59.4 The pattern of rainfall from observed data (BMKG data) and TRMM random forest (estimation)

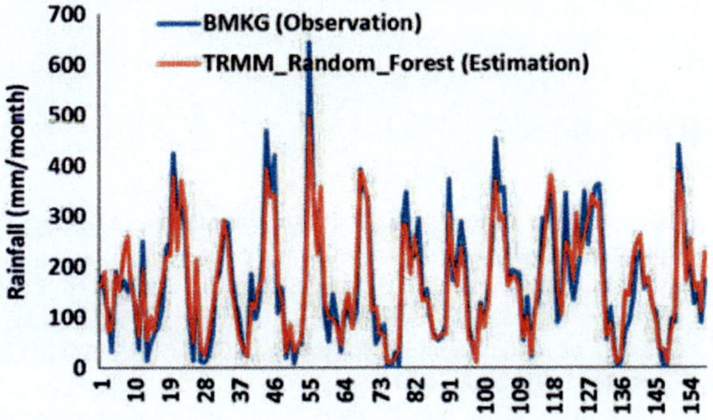

best rainfall estimation is from TRMM with Random Forest fitting (Fig. 59.2d). The best R^2 and RMSE in the validation process is 0.896 and 41.2 mm, respectively (Figs. 59.3 and 59.4).

Based on Fig. 59.3, in general, the estimated monthly rainfall pattern or data from the original TRMM has the same temporal pattern as the observation data at BMKG. However, the gap between the two data is still relatively high, as reflected from the RMSE value of 123.19 mm. This large gap is probably related to the TRMM satellite coverage for one grid area is 25 km $\times$ 25 km. As a consequence, there are many biases or differences in values. To improve the TRMM data quality, we apply linear regression and two machine learning fitting models (k-NN, Random Forest) in the R program. Finally, we found that the Random forest fitting can accurately estimate rainfall using TRMM data. Random forest fitting has a better performance than the original TRMM data, TRMM with linear model fitting, TRMM with k-NN model fitting both in calibration and validation procedure (Figs. 59.1 and 59.2).

59.4 Conclusions

We found that the original TRMM rainfall data can not be directly used as general information due to low R^2 and high RSME.

We developed a model to produce an acceptable accuracy of rainfall estimation from TRMM data. The developed model used machine learning with Random forest fitting. The developed model has successfully estimated more accurate rainfall in the Pesawaran Meteorological station. Our developed model has immense potential to fill the rainfall data gap using TRMM data for specific purposes.

Authorship
Ali Rahmat and Fajar Setiawan are main contributor/author in this article.

References

1. Cahyono, P., Astuti, N.K., Purwito, Rahmat, A.: Mapping the rainfall distribution for irrigation planning in dry season at pineapple plantation, Lampung Province, Indonesia (Study case at Great Giant Pineapple Co. Ltd.). IOP Conf. Ser.: Earth Environ. Sci. **129**, 012017 (2018)
2. Rahmat, A., Zaki, M.K., Effendi, I., Mutolib, A., Yanfika, H., Listiana, I.: Effect of global climate change on air temperature and precipitation in six cities in Gifu Prefecture, Japan. J. Phys.: Conf. Ser. **1155**, 012070 (2019)
3. Dung, B.X., Kim, T.D.T:. Runoff and soil erosion response to clear cutting period of acacia plantation in a headwater mountain of Vietnam. Appl. Res. Sci. Technol. **1**(1), 12–25 (2021)
4. NASDA: TRMM data users handbook. Earth Observation Center, National Space Development Agency of Japan (2001)
5. Xie, P., Yatagai, A., Chen, M., Hayasaka, T., Fukushima, Y., Liu, C., Yang, S.: A gauge-based analysis of daily precipitation over East Asia. J. Hydrometeorol. **8**, 607–626 (2007)
6. Renggono, F., Syaifullah, M.D.: Kajian Meteorologis Bencana Banjir Bandang Di Waisor Papua Barat. Jurnal Meteorologi dan Geofisika **12**, 33–41 (2011)
7. Marpaung, S., Satiadi, D., Harjana, T.: Analisis Kejadian Curah Hujan Ekstrim di Pulau Sumatera Berbasis data Satelit TRMM dan Observasi Permukaan. Jurnal Sains Dirgantara **9**, 127–138 (2012)
8. Gunawan, D.: Perbandingan Curah Hujan Bulanan Dari Data Pengamatan Permukaan, Satelit TRMM dan Model permukaan NOAH. Jurnal Meteorologi dan Geofisika **9**, 65–77 (2008)

Chapter 60
GSMaP Seasonal Rainfall Verification Over Western Java

Arifatus Hikmah Rusmanansari, Rusmawan Suwarman, Yudha Setiawan Djamil, and Yan Firdaus Permadhi

Abstract West Java, the most developed region in Indonesia, faced a great challenge in handling hydrological issues such as water supply (crops production, municipal water system, etc.) and hydrometeorological hazards (flood, erosion, etc.). Continuous large-scale rainfall estimation is crucial to handle hydrological issues, which is provided by a high-resolution rainfall satellite product such as GSMaP. However, the quality of GSMaP remains unclear since it lacks ground-based verification. Widely distributed rain gauge network in West Java with almost 38 stations from PUSAIR and 12 stations from Meteorology and Geophysics is used to verify the quality of GSMaP. Based on ground observation, in the southern of West Java, the first peak of rainfall pattern occurs in November while in the northern coastal area of West Java. Spatially from GSMaP data also shows the same thing where in November there is a fairly high intensity of rain in the southern region of West Java. Therefore, the conditions are detected both from ground observation and GSMaP thus rainfall seasonal variability of West Java is also captured. Although GMaPs magnitude underestimated rainfall mainly in the southern region of West Java, the GSMaP data is adequate to understand the pattern and variability of rainfall over West Java.

A. H. Rusmanansari (✉)
Meteorology Department, Bandung Institute of Technology (ITB), Bandung, Indonesia
e-mail: arifatusrusmanansari@gmail.com

R. Suwarman
Atmospheric Science Research Group, Faculty of Earth Science and Technology, Bandung Institute of Technology (ITB), Bandung, Indonesia

Y. S. Djamil
Research Center for Climate and Atmosphere, National Research and Innovation Agency (BRIN), Bandung, Indonesia

Y. Firdaus Permadhi
Meteorological, Climatology, and Geophysical Agency (BMKG), Jakarta, Indonesia

E. Yulihastin et al. (eds.), *Proceedings of the International Conference on Radioscience, Equatorial Atmospheric Science and Environment and Humanosphere Science, 2021*, Springer Proceedings in Physics 275, https://doi.org/10.1007/978-981-19-0308-3_60

60.1 Introduction

West Java is one of the provinces in Indonesia prone to various disasters. Landslides, floods, and droughts are hydrometeorological disasters that often occur in West Java. Based on West Java Open Data, landslides were the most frequent disasters in 2020, with 3.232 incidents. These hydrometeorological disasters often occur at the peak of the rainy season and the transitional season, which is a study of temporal variability. The spatial and temporal variability of rainfall is important for the development of disaster adaptation and mitigation strategies. The key to understanding rainfall variability across various spatial and temporal scales relies on high a quality and widely covered rainfall dataset [1].

In addition to disaster mitigation purposes, monitoring and measuring rainfall is also crucial for managing hydrological and water sources [2]. In general, there are two sources to collect rainfall data i.e., conventional rain gauge networks and remote sensing systems, such as ground-based weather radar and satellites [3]. Basically, observation data for monitoring is requisite, however availability of observation data is limited and does not cover a wide area. Spatial limitation of the ground-based observation is the main motivation behind satellite-derived rainfall data developed in the last two decades [4].

Satellite-based rainfall products have been developed and fill numbers of data gaps by providing homogeneous spatio-temporal coverage of both ocean and land areas [5]. One of the satellite data that produces rainfall products and is often used in several studies is Global Satellite Mapping of Precipitation (GSMaP). The GSMaP and its global precipitation maps are therefore appealing for a wide range of hydrological applications, such as flood monitoring and forecasting [6]. The results of the comparison of GSMaP monthly rainfall in the Jakarta-Bogor area with surface rainfall show a correlation value of more than 0.60 in coastal and plain areas where the GSMaP rainfall is always lower than the intensity of surface rainfall [7]. GSMaP rainfall has a better correlation to ground-based observation than CHIRPS and TRMM in Bandung, West Java [8].

However, the accuracy of the satellite rainfall product is limited, hence it is necessary to evaluate and correct satellite rainfall forecasts in different climatic and geographical areas [9]. A way to verify the satellite rainfall is by verifying the seasonal pattern of satellite rainfall data with observational data. The objective of this study is verifying the GSMaP seasonal rainfall data, especially in areas Western Java not covered by ground-based observation and to investigate the rainfall characteristic.

60.2 Data and Method

This study using some data such as:

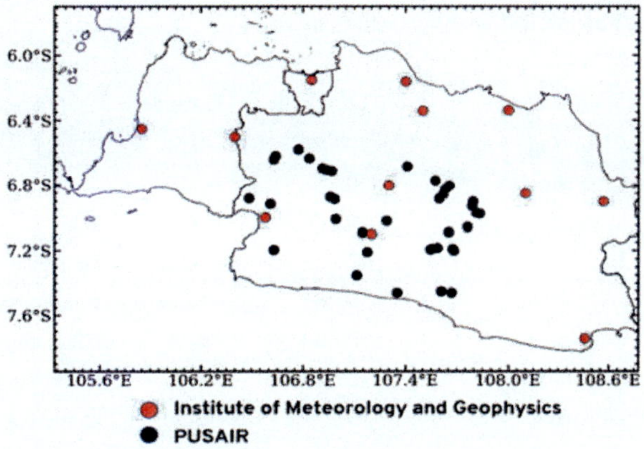

Fig. 60.1 Rain gauge distribution in West Java Province, Indonesia. Black dots are gauge networks maintained by the PUSAIR, while red dots are by the Institute of Meteorology and Geophysics

1. Global Satellite Mapping of Precipitation (GSMaP) rainfall data with a spatial resolution of 0.10° × 0.10° or equivalent to 11.06 × 11.06 km, which is sufficient to resolve the terrestrial complexity of the region [10]. This study used GSMaP_NRT (GSMaP Near Real Time) monthly rainfall data for 18 years (2000–2017), which is openly accessible in ftp://hokusai.eorc.jaxa.jp.
2. Ground-based observation or rain gauge data comes from two sources. The first is daily in-situ rainfall data (rain gauge) from 38 locations in the West Java region for 18 years (2000–2017) (see Fig. 60.1) and the detail coordinates are shown at Table 60.1. This data was compiled by the Research and Development Center for Water Resources (PUSAIR) Bandung [8, 11]. Then the second is monthly in-situ rainfall data from 12 locations widely spread on the West Java Province for around 23 years (see Fig. 60.1). This data was compiled by the Institute of Meteorology and Geophysics [12].

Seasonal verification of the GSMaP monthly product is conducted by simply comparing their precipitation climatology data and ascertaining the correlation of it with in-situ rainfall data.

60.3 Results

60.3.1 Spatio-Seasonal Rainfall Over Western Java

Monthly rainfall over West Java is known to be dominated by annual mode [13]. GSMaP and rain gauges data show three months of time difference between the annual peak of rainfall over the northern and southern coast of West Java (Fig. 60.2). The first rainfall peak is in November and it is located in the southern part of West Java (Fig. 60.2a).

Table 60.1 The list of station location

PUSAIR			Institute of Meteorology and Geophysics		
Station name	Longitude	Latitude	Station name	Longitude	Latitude
Cisomang	107.41	−6.68	Menes	105.85	−6.45
Cipeusing	107.57	−6.77	Jasinga	106.4	−6.5
Margahayu	107.66	−6.80	Jakarta	106.85	−6.15
Kayu Ambon	107.63	−6.82	Jatisari	107.5	−6.25
Dago Pakar	107.62	−6.86	Pondokbales	107.4	−6.16
Cemara	107.60	−6.88	Cikencreng	108.4	−7.8
Tanjungsari	107.80	−6.90	Plabuhan Ratu	106.58	−7
Jatiroke	107.79	−6.93	Cempaka	107.2	−7.1
Rancaekek	107.75	−6.96	Ciranjang	107.3	−6.8
Cicalengka	107.83	−6.98	Darmaradja	108.1	−6.85
Cipaku-Paseh	107.76	−7.06	Anjatan	108	−6.34
Cisondari	107.65	−7.09	Ciawigebang	108.57	−6.9
Cibeureum	107.68	−7.19			
Kertamanah	107.68	−7.20			
Cipanas Pangalengan	107.59	−7.18			
Cileunca	107.54	−7.19			
Cidadap Montoya	107.29	−7.02			
Sukanegara	107.15	−7.09			
Geger Bitung	106.99	−7.01			
Batu Karut	106.99	−6.89			
Cisalada	106.96	−6.87			
Gunung mas	106.97	−6.71			
Panjang	106.94	−6.70			
Pasir muncang	106.91	−6.69			
Katulampa	106.84	−6.63			
Cibanteng Hilir	106.77	−6.58			
Bungbulang	107.61	−7.45			
Cianten	106.63	−6.64			
Cibarusah	106.63	−7.20			
Cidamar	107.35	−7.46			
Cikelat	106.49	−6.88			
Cikiray	106.61	−6.92			
Cikoronjo	107.18	−7.21			
Cileunca	107.55	−7.19			
Citoe	106.49	−6.88			

(continued)

Table 60.1 (continued)

PUSAIR			Institute of Meteorology and Geophysics		
Station name	Longitude	Latitude	Station name	Longitude	Latitude
Kampung Cijulang	107.12	−7.35			
Pakenjeng	107.67	−7.46			
Sentral Kracak	106.64	−6.62			

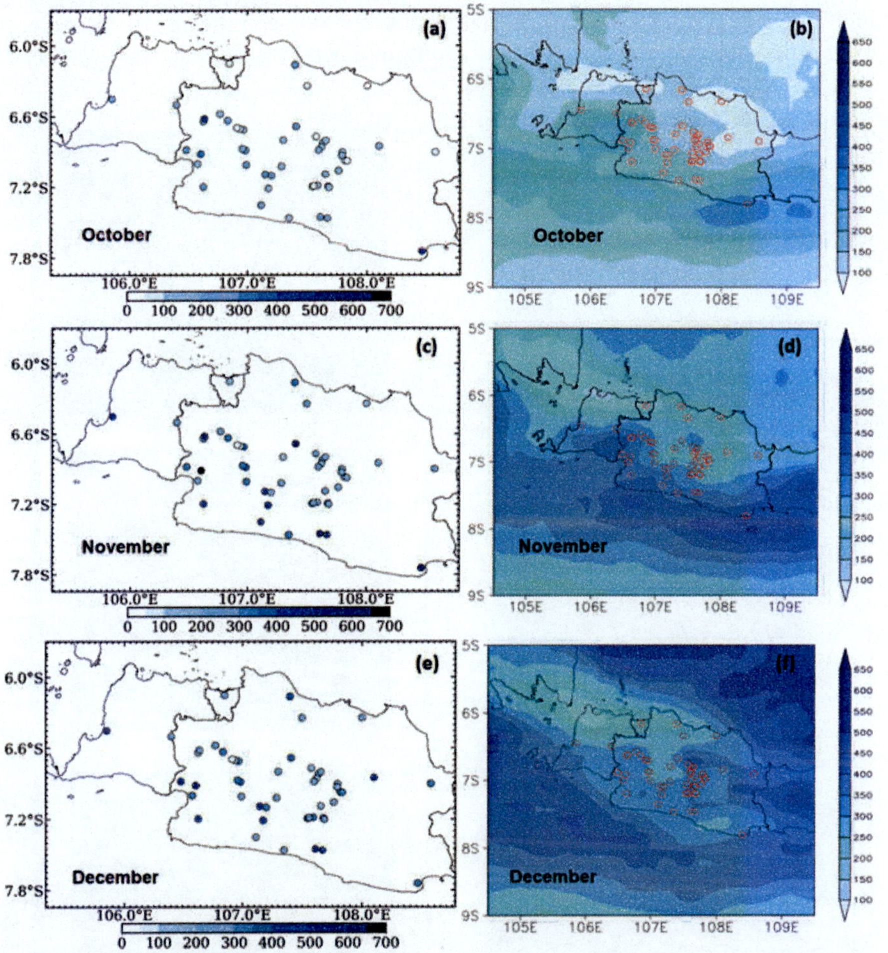

Fig. 60.2 Rain gauge (left) and GSMaP (right) rainfall in **a** and **b** October, **c** and **d** November, **e** and **f** December

Rainfall over the southern region starts to develop from October and rises towards November. The intensity of rainfall rose from about 100–200 mm to over 500 mm. Meanwhile, the northern coastal area did not experience a significant change in the value of rainfall intensity from October to November. This condition persisted until December where the southern coastal area had a higher rainfall intensity than the northern coastal area.

In January the intensity of rainfall in the southern coast decreased, while the northern coastal increased (see Figs. 60.3a, b and 60.4). However, the increase in rainfall on the north coast from GSMaP data is not very significant when compared to ground-based observation data. In addition, during the period from October to

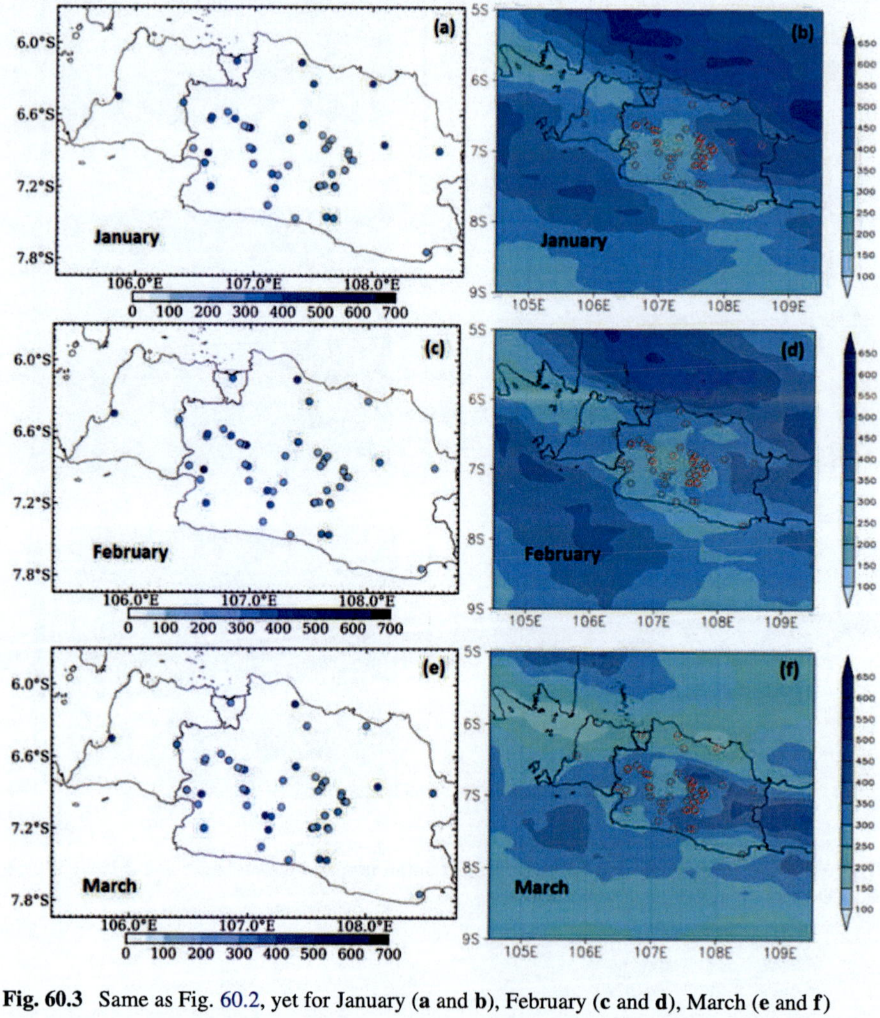

Fig. 60.3 Same as Fig. 60.2, yet for January (**a** and **b**), February (**c** and **d**), March (**e** and **f**)

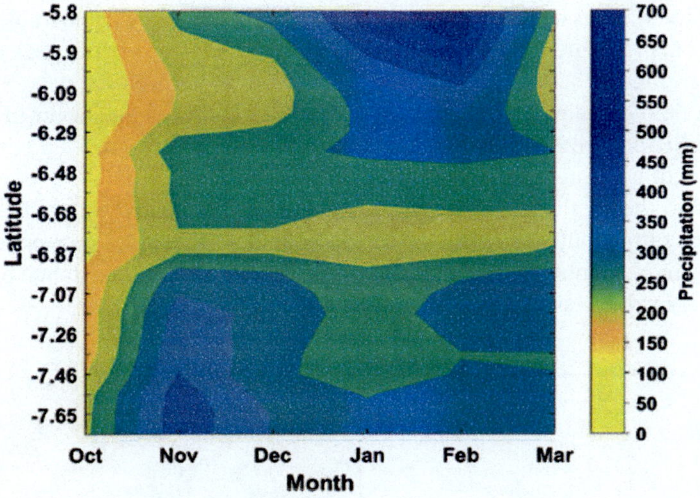

Fig. 60.4 Hovmoller plot that indicate propagation of precipitation using GSMaP data

March there is a propagation of high rainfall intensity from the north coast to the south coast observed in both datasets.

Based on different time appearances, the first peak over western Java region can be divided into three clusters. The division of the three clusters is due to the different rainfall patterns from north to south of western Java. The northern and southern cluster are dominated by observation stations in the coastal areas, while the middle or center clusters is dominated by observation stations in the hilly areas (see Fig. 60.5). Even though the center cluster and southern cluster tend to have the same rainfall pattern,

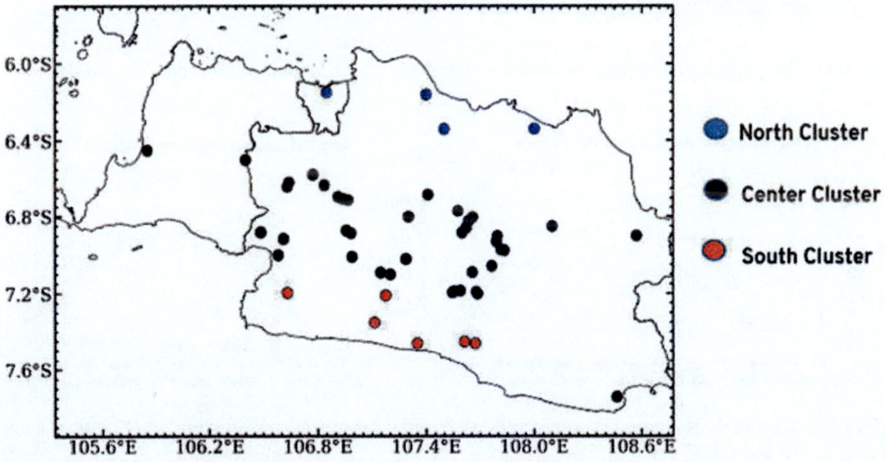

Fig. 60.5 Three regional clusters are based on rainfall patterns

the intensity still has differences (see Fig. 60.7). This result is comparable with the research that in the Greater Jakarta area (northern cluster), the highest peak of rainfall occurred in January [14].

Figure 60.6 explains more detail about rainfall anomalies that occur in the north and south. It can be seen that in November the northern region tends to have a negative anomaly value (orange) which means the rainfall intensity is low when compared to the southern region which tends to have a positive value (blue) which means the rainfall value increases. Whereas in January the opposite happened to the conditions in November, although the northern region did not show much blue color, the anomaly value was clearly different from the southern region.

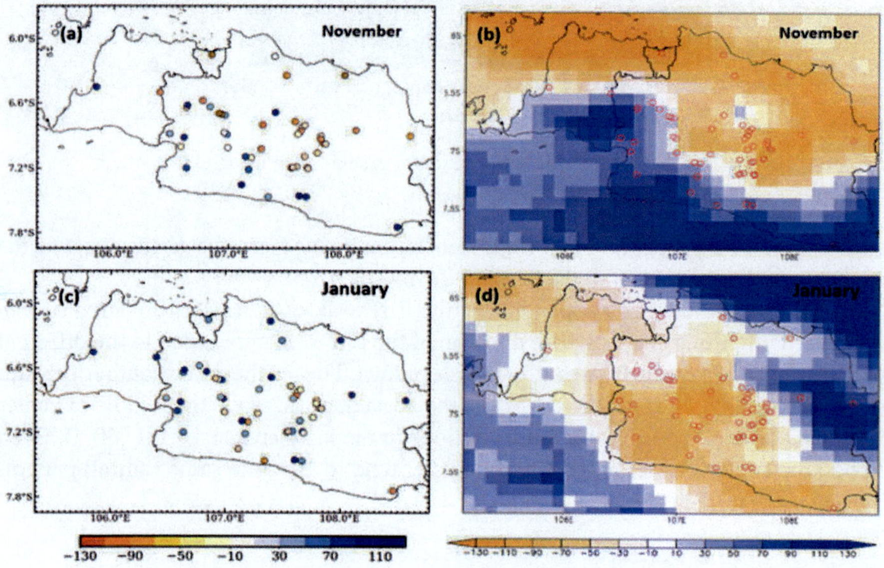

Fig. 60.6 Precipitation anomaly from rain gauge data (**a** and **c**) and GSMaP data (**b** and **d**)

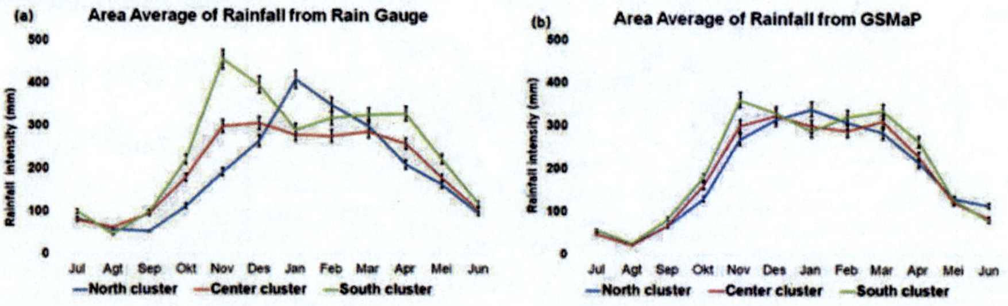

Fig. 60.7 Climatology of area averaged rainfall based on three clusters as shown in Fig. 60.5 from **a** rain gauge and **b** GSMaP

60.3.2 Correlation of Rainfall Intensity Between GSMaP and Rain Gauge

The correlations between the GSMaP data and in-situ data in the north and south coastal area are 0.93 and 0.96, respectively. It suggested that the patterns are similar. However, the magnitude of both datasets should be evaluated based on their climatology. Figure 60.7 shows rainfall climatology from GSMaP and rain gauges over West Java clustered in three regions. In general, GSMaP rainfall over West Java tends to underestimate ground-based observation except for the center cluster, it is adhered to with the statement that GSMaP rainfall is always lower than the intensity of surface rainfall [7]. This can be seen when in November from rain gauge data, rainfall in the southern cluster was above 400 mm. Meanwhile, from GSMaP data, the rainfall is below 400 mm. Similar conditions occurs for the northern cluster, especially in January, where rainfall from rain gauges reached 400 mm while from GSMaP data only reached 300 mm.

60.4 Conclusion

Based on this research, we can conclude a few things:

- The rain gauge (in-situ data) and GSMaP data show that the rainfall pattern between the north coast and the south coast have the same pattern, which means the patterns show the appearance of their monthly rainfall peaks. First peak in the north cluster occurs in January, while in the south cluster the first peak occurs in November.
- Rainfall climatology of GSMaP product underestimates in-situ data over north cluster and south cluster, while in the center cluster GSMaP product tends to overestimate compared by in-situ data. However, both data show a similar pattern of rainfall climatology.
- GSMaP shows the movement and rainfall peaks changing in each cluster. In the southern cluster, the rainfall peak appears in November, then in the center cluster rainfall peak appears in December, while in the northern cluster rainfall peak occurs in January. This indicates the movement of rainfall peaks from south to north.

Acknowledgements This research was funded by the ITB Research Program in 2021.

References

1. Kikuchi, K., Wang, B.: Diurnal precipitation regimes in the global tropics. J. Clim. **21**, 2680–2696 (2008). https://doi.org/10.1175/2007JCLI2051.1
2. Kidd, C., Huffman, G.: Global precipitation measurement. Meteorol. Appl. **18**(3), 334–353 (2011)
3. Sugiartha, N., Ogawara, K., Tanaka, Tc., Mahendra, M.S.: Article application of GSMaP product and rain gauge data for monitoring rainfall condition of flood events in Indonesia. Int. J. Environ. Geosci. **1**(1), 36–47 (2017)
4. Sorooshian, S., Hsu, K.-L., Gao, X., Gupta, H., Imam, B., Braithwaite, D.: Evaluation of PERSIANN system satellite-based estimates of tropical rainfall. Bull. Amer. Meteor. Soc. **81**, 2035–2046 (2000)
5. Kucera, P.A., Ebert, E.E., Turk, F., Levizzani, V., Kirschbaum, D., Tapiador, F.J., Loew, A., Borsche, M.: Precipitation from space: advancing earth system science. Bull. Amer. Meteor. Soc. **94**, 365–375 (2013). https://doi.org/10.1175/BAMS-D-11-00171.1
6. Tian, Y., Peters-Lidard, C.D., Adler, R.F., Kubota, T., Ushio, T.: Evaluation of GSMaP precipitation estimates over the contiguous United States. J. Hydrometeorol. **11**(2), 566–574 (2010)
7. Wibowo, Y.A.: Evaluasi Curah Hujan GSMaP dan TRMM TMPA Dengan Curah Hujan Permukaan Wilayah Jakarta – Bogor (2010)
8. Rinaldi, Y., Suwarman, R., Abdillah, M.R.: Koreksi Curah Hujan Menggunakan Data Satelit Terhadap Observasi Permukaan (Studi Kasus: Bandung Jawa Barat). Digilib Institut Teknologi Bandung (2019)
9. You, Q.L., Min, J., Zhang, W.: Comparison of multiple datasets with gridded precipitation observations over the Tibetan Plateau. Clim. Dyn. **45**, 791–806 (2015)
10. Kachi, M.: Overview of global satellite mapping of precipitation (GSMaP) side event at the Japan Pavillion on the 6th World Water Forum (2012)
11. Ferdiansyah, A., Yuningsih, S.M.: The local potential flow Saguling reservoir using rainfall runoff model. Jurnal Sumber Daya Air **16**(1) (2020)
12. Yasunari, T.: Temporal and spatial variations of monthly rainfall in Java, Indonesia. Southeast Asian Stud. **19**(2) (1981)
13. Aldrian, E., Susanto, R.D.: Identification of three dominant rainfall regions within Indonesia and their relationship to sea surface temperature. Int. J. Climatol. **23**, 1435–1452 (2003)
14. Al Habib, A.H., Cahyo, K.N., Mahron, H.N., Nugraheni, I.R.: Identifikasi Pola Curah Hujan Diurnal Menggunakan Citra Satelit GSMaP (Global Satellite Mapping of Precipitation) di Wilayah Jabodetabek. Prosiding Seminar Nasional Geografi III, UGM (2020)

Chapter 61
Prediction of Extreme Rainfall of September 9, 2020 in Padang City Based on Clouds Brightness Temperature Difference from Himawari-8 Satellite Data

Nining Jumianti, Marzuki Marzuki, Wendi Harjupa, Risyanto, and Muhammad Fadhlan Putranto

Abstract Padang city has an equatorial rain pattern with high convective cloud activity. Extreme rainfall often occurs in this region, which impacts human life in material losses and loss of life. Himarawi-8 satellite observation allows detecting cloud growth causing extreme rain faster than using radar or surface rain observations. This study observed the extreme rain phenomenon in Padang city on September 9, 2020, using the Himawari-8 satellite data. This extreme rain had rainfall intensity above 100 mm/day, which caused floods and landslides in some areas in the city. Convective clouds that cause extreme rain are detected using brightness temperature data from the IR1 band. Then brightness temperature difference method is used to provide complete information about cloud particle types. Himawari-8 satellite data can show convective clouds with cloud top temperatures experiencing significant changes, with the lowest cloud top temperatures reaching 200 K. In addition, clouds with strong updrafts and ice particles in these clouds were also obtained. The brightness temperature difference value can provide a clear picture of convective clouds so that observations of clouds that produce extreme rain on September 9, 2020, are well observed. The results were also compared with Integrated Multi-satellitE Retrievals for GPM (IMERG) data.

N. Jumianti · M. Marzuki (✉)
Department of Physics, Faculty of Mathematics and Natural Science, Universitas Andalas, Padang, Indonesia
e-mail: marzuki@sci.unand.ac.id

W. Harjupa · Risyanto · M. F. Putranto
Research Center for Climate and Atmosphere, National Research and Innovation Agency (BRIN), Bandung, Indonesia

E. Yulihastin et al. (eds.), *Proceedings of the International Conference on Radioscience, Equatorial Atmospheric Science and Environment and Humanosphere Science, 2021*, Springer Proceedings in Physics 275, https://doi.org/10.1007/978-981-19-0308-3_61

61.1 Introduction

Indonesia is a region with unique weather and climate characteristics. As a tropical region, Indonesia allows convective clouds that cause extreme weather, such as heavy rainfall. Extreme rain is rain with a very heavy intensity that can impact human life [1, 2] in material losses and loss of life. Heavy rain events with high rainfall can impact flood events in several regions in Indonesia, one of which is Padang City. Extreme rain events can be predicted by analyzing the characteristics of clouds that can cause rain. Cloud observation can be done using satellites, including the Himawari-8 satellite by the Japan Meteorology Agency (JMA).

Himawari-8 was successfully launched from Japan's Tanegashima Space Center using an H-IIA rocket on October 7, 2014, and settled in geostationary orbit on October 16, 2014 and located at 140° E. This satellite has an Advance Himawari Imager (AHI) instrument with several bands and temporal and spatial resolution better than the previous generation. AHIs have 16 observation bands, with 3 for visible (VIS) band, 3 for near-infrared (NIR) band, and 10 for infrared (IR) band, with a spatial resolution for Indonesia territory 1 km and 2 km for the VIS band, 4 km for the NIR and IR band, while the resolution Himawari-8 satellite temporal 10 min [3]. Therefore, the Himawari-8 satellite can observe cloud activity near-real-time to see clouds that can cause extreme rain.

Several methods can detect extreme rain events using Himawari-8 satellite data, including Brightness Temperature Difference (BTD). One of the data outputs of the Himawari-8 satellite is IR1 product data from band 13 (B13). This band shows the cloud top temperature from observations at a wavelength of 10.4 μm. In addition, the BTD value is used on bands 11, 13, and 15 (B11, B13, and B15). BTD1, B11 subtract by B13, in the occurrence of cloud, the higher absorption by cloud particle of B13 then B13 will result in the positive value which means the detection of ice particle. The detection of water particles in the cloud will result in a negative BTD1 [4]. BTD between band 13 and band 15 provides information on whether the cloud phase is water or ice; band 15 observes ice particles when cumulus clouds are in the mature stage, while band 13 may give the information about water particles in infant cumulus. The BTD2 contain ice particle it will result as a positive value. BTD3 uses three-band observations from the Himawari-8 satellite, where when BTD2 is higher than BTD1, then cloud contains water particles; conversely, if the value of BTD2 is smaller than BTD1, then the cloud has particles of ice [4].

This study examines the application of the BTD method to predict the extreme rain event on September 9, 2020, in Padang city. National News (Liputan6, 2020) reported that this long duration rain event with very high intensity occurred in Padang City on September 9, 2020, night until September 10, 2020, morning. The BMKG noted that this heavy rain has a rainfall intensity of about 100 mm/day in several regions in Padang City. This extreme rain also caused a disaster in Padang City. On September 10, 2020, morning, several points were flooded, and 3 points were hit by landslides [5]. This study looked at the cloud that is expected to provide complete information using Brightness Temperature value data to see the distribution of convective clouds that can cause extreme rain, knowing the value of the cloud top temperature, and knowing the type of hydrometeor contained in the cloud.

61.2 Data and Method

61.2.1 Data and Software

This study used the Himawari-8 satellite data from Japan Meteorology Agency (JMA), which has started operations from 2015 to 2022. Himawari-8 satellite data, including 16 channels, geometry, latitude, longitude, and time. Himawari-8 has 16 observation bands, including 3 visible bands, 3 near-infrared bands, and 10 infrared bands. For the territory of Indonesia, the resolution is 1 km and 2 km for VIS band, 4 km for the NIR and IR bands, while the resolution Himawari-8 satellite temporal 10 min [3]. Data were obtained from the Japan Meteorological Satellite Center Website ftp://ftp.ptree.jaxa.jp. Himawari-8 satellite data were processed using software python, which is assisted by using the jupyter notebook tool.

In addition, we also used precipitation data from Integrated Multi-satellitE Retrievals for GPM (IMERG). GPM data is processed using Matlab software. In this study IMERG data were used to observe rain data for the same date throughout of Padang City.

61.2.2 Case Selection

On Wednesday, September 9, 2020, there has been raining with a very high rainfall intensity of about 100 mm/day, accompanied by strong winds since Wednesday night which has caused floods and landslides at several points in Padang city. With the availability of more continuous weather data and information, we can use this rain event on September 9, 2020 to study how weather phenomena occur during rain events so that we can detect extreme rain events early and reduce the adverse effect of these rain events using brightness temperature (BT) value.

61.2.3 Hydrometeor Type

This study used rightness temperature in bands 11, 13, and 15 (B11, B13, B15) from the Himawari-8 satellite. This method can observe the growth of cumulus clouds that cause rain [6]. We used the different data from this band to determine the condition of clouds top and whether the clouds detected by the satellite are in the ice phase or the water phase. Table 61.1 shows the three BTD parameters from Himawari-8.

Table 61.1 Parameter for cloud phase detection

No	Parameter	Value for water	Value for ice
BTD1	B11-B13	Negative	Positive
BTD2	B13-B15	Negative	Positive
BTD3	BTD1-BTD2	Negative	Positive

Furthermore, Fig. 61.1 shows a comparison of cloud top conditions using three different bands.

BTD1 is a subtraction of B11 and B13. B11 uses a wavelength of 8.6 μm, and B13 uses a wavelength of 10.4 μm. The absorption of cloud particles at B13, which is higher than B11, will produce a positive value indicating ice particles in the cloud. If the cloud contains water particles, the BTD1 value is negative. BTD2 is a subtraction of B13 and B15. B15 uses a wavelength of 12.4 μm to detect ice particles in adult cumulus clouds and B13 to detect water particles in baby cumulus. A positive BTD2 value indicates that the cloud contains ice particles, while a negative value contains water particles [7].

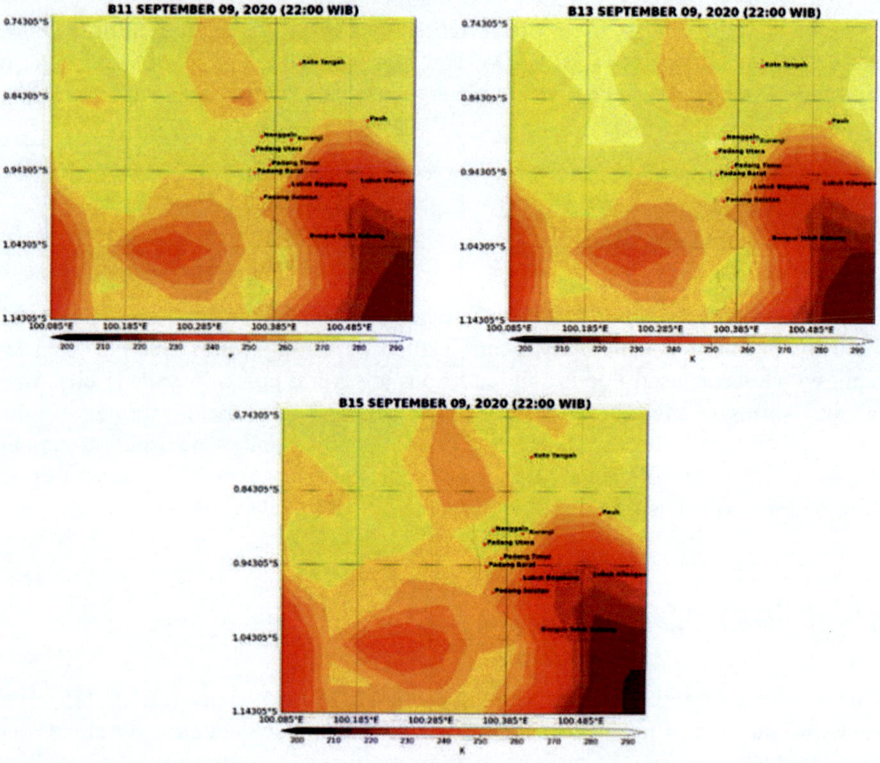

Fig. 61.1 Brightness temperature B11, B13, and B15

The BTD3 is commonly referred to as the 'tripectral technique' [4]. BTD3 uses three observation bands to detect cloud particles. When BTD2 is higher than BTD1, the cloud contains water particles, and if the BTD2 value is less than BTD1, the cloud contains ice particles. In addition to BTD1, BTD2, and BTD3, we also used BTD31 which is a subtraction of B08 and B13, to detect the distribution of convective clouds during a rain event on September 9, 2020 [8].

61.2.4 Evaluation Between BTD and IMERG

The BTD obtained during the September 9, 2020 rain event will be evaluated with the IMERG data. The cloud top temperature before the heavy rain event was obtained from the Himawari-8 satellite data. We use several assessment indicators: the probability of detection (POD), False Alarm Ratio (FAR), and Critical Success Index (CSI). These parameters use a confusion matrix, as shown in Table 61.2. Because the resolution of Himawari-8 data and IMERG data is different, we conducted up-scaling process. Initial temporal and spatial resolutions of Himawari-8 data are 0.05 degrees and 10 min. Moreover, IMERG data has a spatial resolution of 0.1 degrees and 30 min temporal resolution.

True Positive (TP) represents cloud convective detected by Himawari-8 satellite and precipitation detected by IMERG. False Negative (FN) represents convective cloud detected, False Positive (FP) represents precipitation detected only by IMERG, and True False (TF) represents no cloud and precipitation detected. As for statistical analysis, BTD TP represents BTD value positive from Himawari-8 data and precipitation detected by IMERG. FN represents BTD value positive, TF represents BTD value negative, and precipitation detected. Finally, TF represents BTD value negative, and no precipitation detected.

Table 61.2 Statistical metrics

Metrics	Equation	Perfect value
Probability of detection (POD)	TP/(TP + FN)	1
False alarm ratio (FAR)	FP/(TP + FN)	0
Critical success index (CSI)	TP/(TP + (FP + FP)	1

61.3 Result and Discussion

61.3.1 *Distribution of Rainfall from IMERG Data on September 9–10, 2020*

Figure 61.2 shows the spatial distribution of rainfall data from IMERG on September 9, 2020. The rain event had occurred in Padang City at 18.00 local western Indonesia Time (from now on referred to as WIB) with a rainfall intensity > 100 mm/h. This rain event occurred over in the administrative area of Padang city, namely in the southwest of Padang city. It continued to expand until it reached all sub-district in Padang City. The rain incident began to approach the Bungus Teluk Kabung sub-districts. After a few hours, at 23.00 WIB and one hour later at 00 WIB on September 10, 2020, the rain had evenly occurred in all sub-district in Padang city. Rain events continued to happen until the next day, on September 10, 2020, morning.

61.3.2 *Cloud Top Temperature Analysis*

One of the Himawari-8 satellite bands is band 13 or IR1. The IR1 shows the cloud top temperature from radiation observation at a wavelength of 10.4 μm. When the cloud top temperature is colder, the higher the cloud growth and the potential for Cumulonimbus clouds that cause rain. In Addition, IR1 can also use to see patterns of cloud top temperature in the study of weather warning system innovation [9]. This study shows a plot of cloud top temperature before and during the rain to see the value and pattern of cloud top temperature from the Himawari-8 satellite.

Figure 61.3 shows the cloud top temperature plot. At 21.00 WIB, the cloud top temperature decreased to 240 K in the southwest and south of Padang city. The decrease in cloud top temperature also began to occur in the Bungus Teluk Kabung dan Pauh sub-district. Then 10 min later, at 21.10 WIB, the cloud top temperature dropped further in the Bungus Teluk Kabung, Lubuk Begalung, Lubuk Kilangan dan Pauh sub-district area with temperatures below 230 K. At 21.20 WIB, the south-eastern of Padang city also experienced a reduction in cloud top temperature to 230 K and a reduction in Kuranji, Pauh, and Lubuk Kilangan sub-district; the cloud top temperature was getting colder. At 21.50 WIB, the cloud top temperature was getting colder in southeastern reach 220 K. The cloud top temperature in Padang city is decreasing every 10 min. Until 22.00 WIB, cloud top temperature in the southeast reaches 200 K. It extends to all sub-district in Padang city at 22.40 WIB the cloud top temperature is very cold, covering the whole sub-district in Padang city, and the hardest is in Bungus Teluk Kabung sub-district.

Cloud top temperature decreases continuously until 22.40 WIB, indicating that the clouds are overgrowing in a short time. IMERG plot shows that rain occurred in all sub-districts in Padang city at 23.00 WIB. Thus, the clouds experienced dissipation

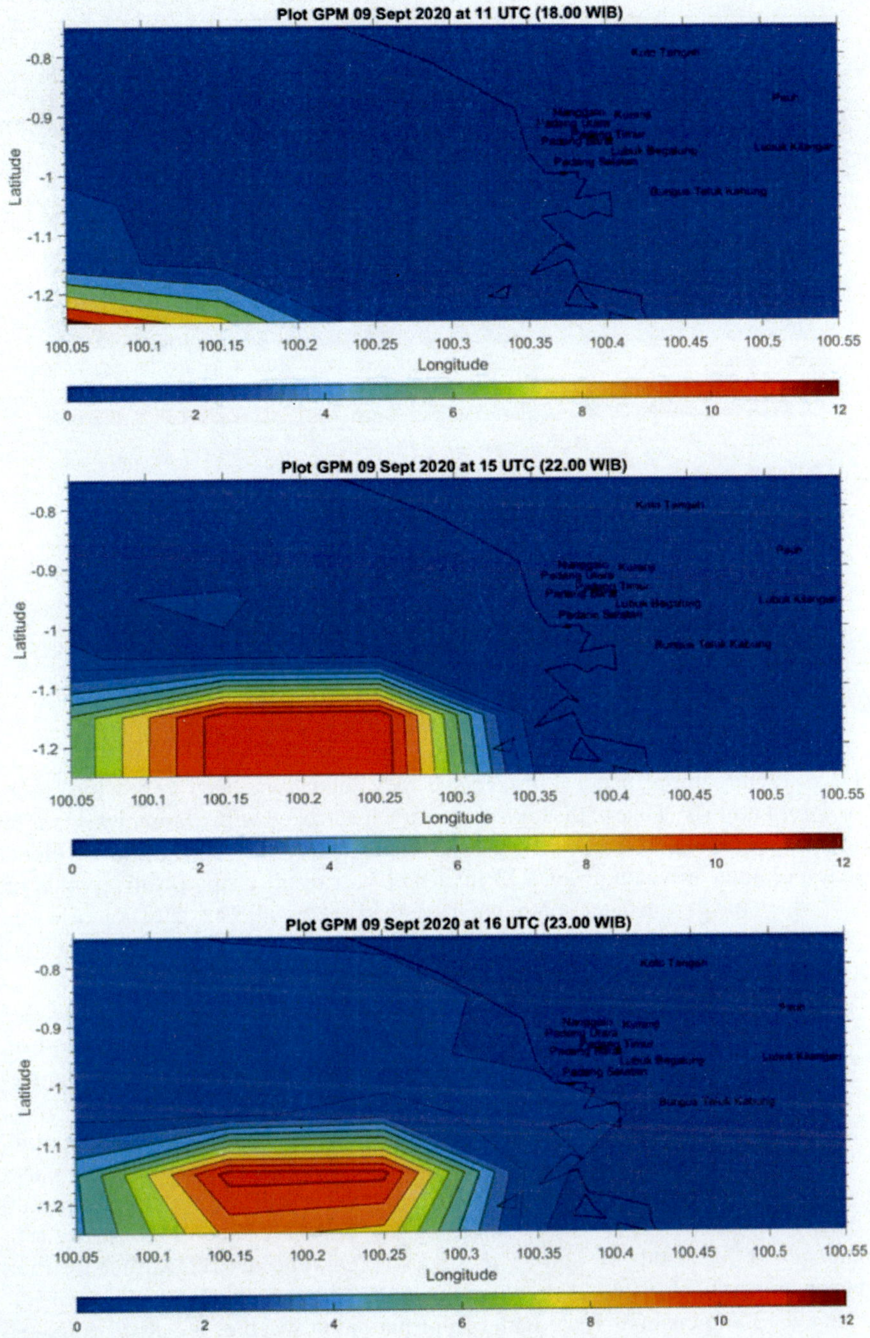

Fig. 61.2 Spatial distribution of rainfall from IMERG data on September 9, 2020 and September 10, 2020

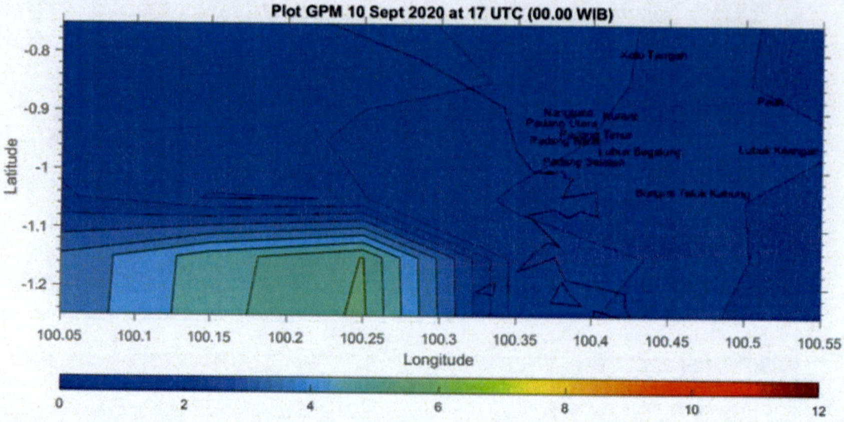

Fig. 61.2 (continued)

20 min later. This shows that the cloud will overgrow with very cold cloud top temperatures before the heavy rain.

61.3.3 BTD1, BTD2 and BTD3 Analysis

BTD1 is a reduction of bands 11 and 13. This value can be used to provide cloud phase information because of the high absorption in the band with a larger wavelength during the ice phase. In the occurrence of cloud, the higher absorption by cloud particle in longer wavelength of B13 than B11 will result in the positive value of BTD1, which means the detection of ice particle in the cloud. The detection of water particles in the cloud will result in a negative BTD1. Figure 61.4 shows the spatial distribution of BTD1 for rain events on September 9, 2020.

Before 21.00 WIB, the BTD1 value below zero indicates that at 21.00, the sky is clear/no clouds. Until 21:50, the BTD1 value in the southeast and southwest of the Padang city area had a positive value. Ten minutes later, the BTD1 value in the Bungus Teluk Kabung, Lubuk Kilangan, and Lubuk Begalung areas were positive. At 22.00 WIB, 22.10 WIB, 22.30 WIB, 22.40 WIB, when the cloud top temperature gets colder, as seen from the IR1 plot, the BTD1 value is positive in Padang City, indicating that the lower the cloud top temperature value will contain clouds with an ice phase.

The subtracting bands 13 and 15 produces BTD2, and the BTD2 plot for rain events on September 9, 2020, is shown in Fig. 61.5. The reduction in the value of BT on channel 13 and channel 15 provides information on whether the cloud contains water or ice; this is because channel 15 observes ice particles when cumulus clouds are in a mature state. In contrast, channel 13 keeps cloud particles in cumulus clouds in the early stages of formation.

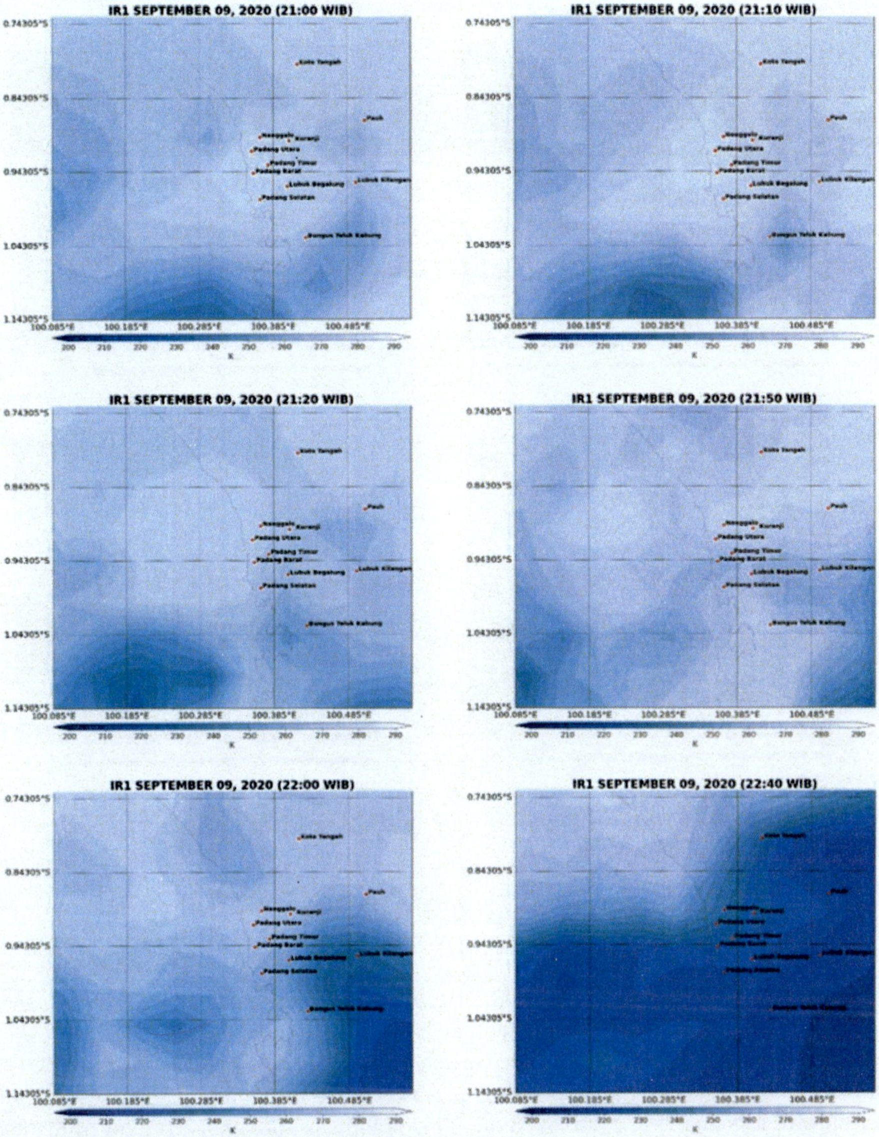

Fig. 61.3 Cloud top temperature plot from Himawari-8 satellite IR1 on September 9, 2020

Based on the Himawari-8 data plot, the BTD2 values in Padang City before and during the rain event were all positive but with different values (Fig. 61.5). At 21.00 WIB, before the rain incident, the BTD 2 value in all sub-districts in the city of Padang was with a significant positive value and lasted for several minutes. When the cloud top temperature is getting colder, according to the IR1 plot, the BTD2 value is positive but not significant (towards zero). The plots of BTD2 values indicate that

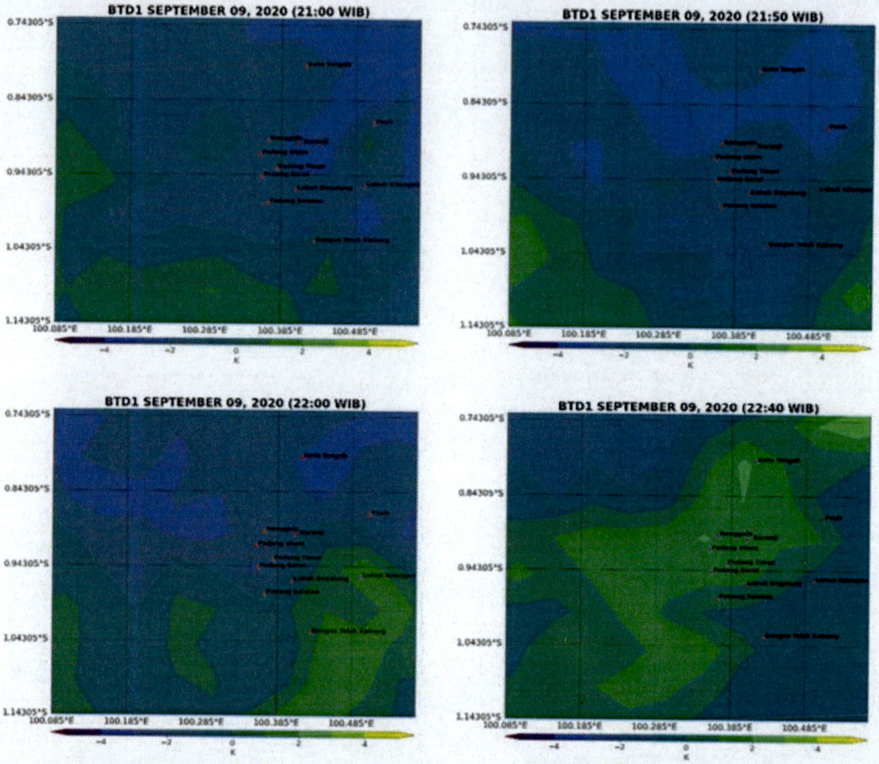

Fig. 61.4 BTD1 plot on September 9, 2020

clouds contain an ice phase, but this is different from the BTD1 plot, which shows that the BTD1 value before the rain event also shows no clouds when there is no rain at 21.00 WIB.

Figure 61.6 shows the plot of BTD3. The BTD3 provides information on hydrometeor parameters. When BTD2 is higher than BTD1, the cloud contains water particles and vice versa; the cloud contains ice particles if the BTD2 value is less than BTD1. Before the rain incident at 21.00 WIB, the BTD3 value throughout Padang City was very low to less than −4, 21.10 WIB the BTD3 plot was not much different from the previous 10 min. When the cloud top temperature value is very cold, the BTD3 plot obtained towards midnight begins to rise towards 0, but no plot depicts a positive BTD3 value or exceeds zero for all plots obtained. It can be said that the use of BTD3 shows that the BTD3 values are all lower than zero, but when the cloud top temperature is freezing, the BDT3 value will go to zero.

All values of BTD2 are larger than zero, indicating that the BTD2 cannot detect the clouds with the water phase. A similar pattern was observed for the BTD3. All BTD3 value before, during, and after the rain event was negative. Thus, BTD3 could not detect clouds with ice phase, but it only shows that when the cloud top temperatures are very cold, the BTD3 value obtained reaches zero. This result is

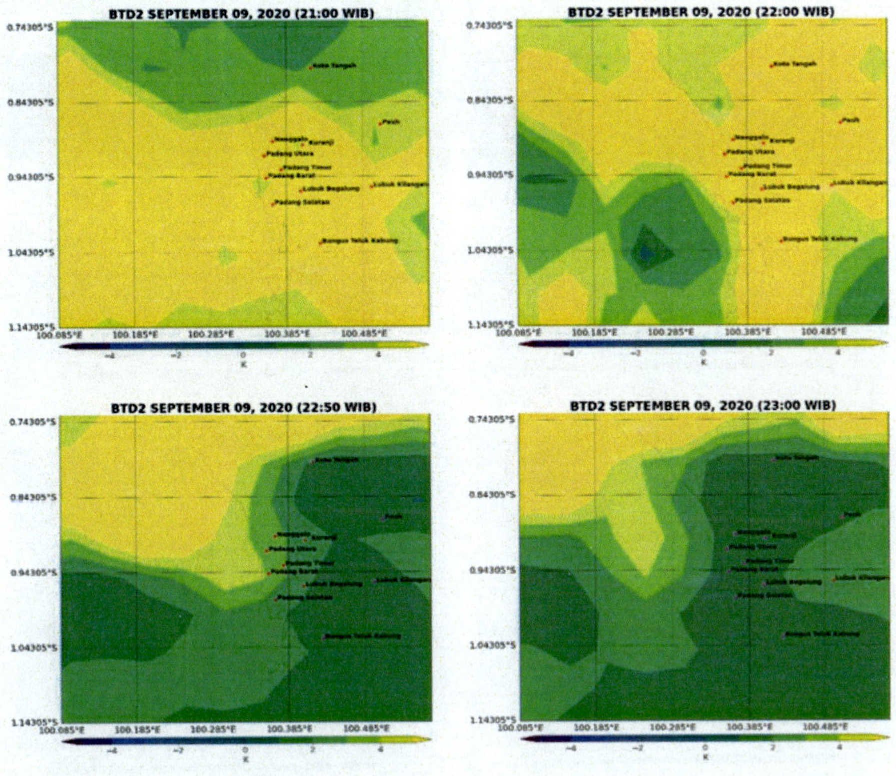

Fig. 61.5 BTD2 plot on September 9, 2020

different from some previous studies [4, 6, 7], which may be due to differences in the data resolution.

61.3.4 Convective Cloud Analysis

The Himawari-8 satellite data processing results using BTD31 show an image of cloud cover, and we can see the convective distribution over the Padang City area (Fig. 61.7). When the rain had not yet reached the sub-districts in Padang City (21.00 WIB), the BTD2 value was very large as observed in many sub-districts such as Nanggalo, Kuranji, Padang Utara, Padang Timur, Padang Barat, Padang Selatan, and Lubuk Begalung. Every 10 min, the BTD31 shows a very significant change until it rained in the middle of the night until the following day, on September 10, 2020. This BTD31 plot is very helpful in seeing how changes in convective clouds cause rain events. By midnight and the next day, the value of BTD2 was no longer more significant than the value of BTD1, so almost the same value in every sub-district

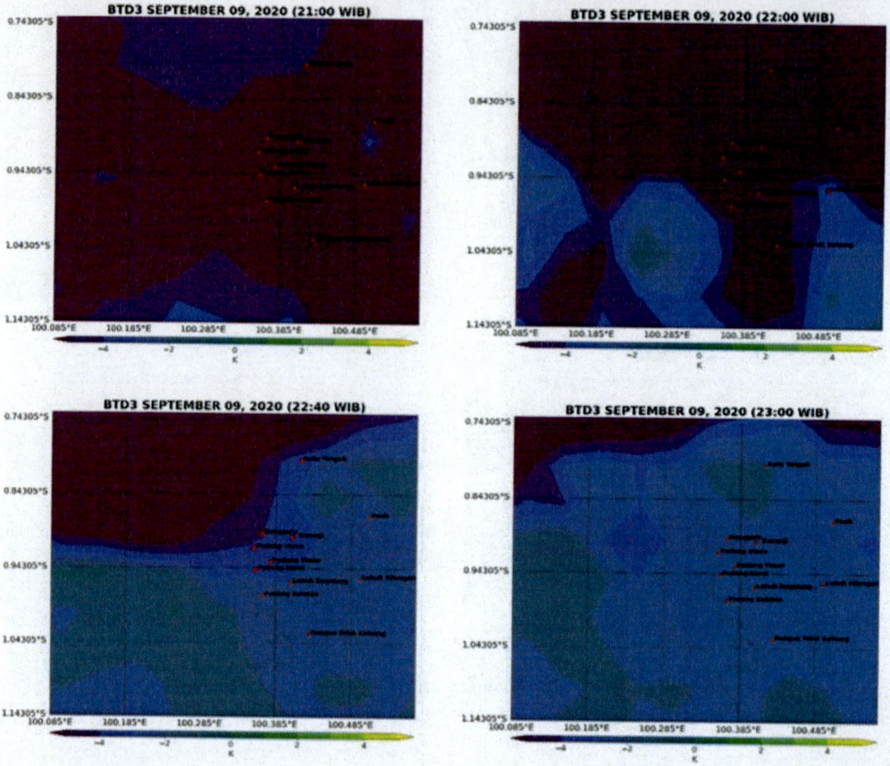

Fig. 61.6 BTD3 plot on September 9, 2020

in Padang City was observed. Thus, the BTD31 plot can help us to see how the distribution of convective clouds that can cause rain with 10 min of data captured by the Himawari-8 satellite.

61.3.5 Statistical Evaluation

POD, FAR, and CSI between cloud top temperature IR1 and IMERG data are shown in Fig. 61.8. The figure is presented with several lag-time to find the best parameter and the possible duration before the cloud experiencing the dissipation stage. The POD between IR1 and IMERG (Fig. 61.8a) at lag-0 is close to 1 in all areas of Padang city. This is also in accordance with the respective observation of cloud top temperature and spatial distribution plot from IMERG data. Lag-1 shows the lowest POD value. The FAR value for the Padang city area is low at lag-0 and increases at lag-3. The lower the FAR value, the better. Likewise, for the critical success index CSI (Fig. 61.8c), the best values are also obtained at lag-0.

Statistical evaluation between BTD1 and IMERG value is shown in Fig. 61.9.

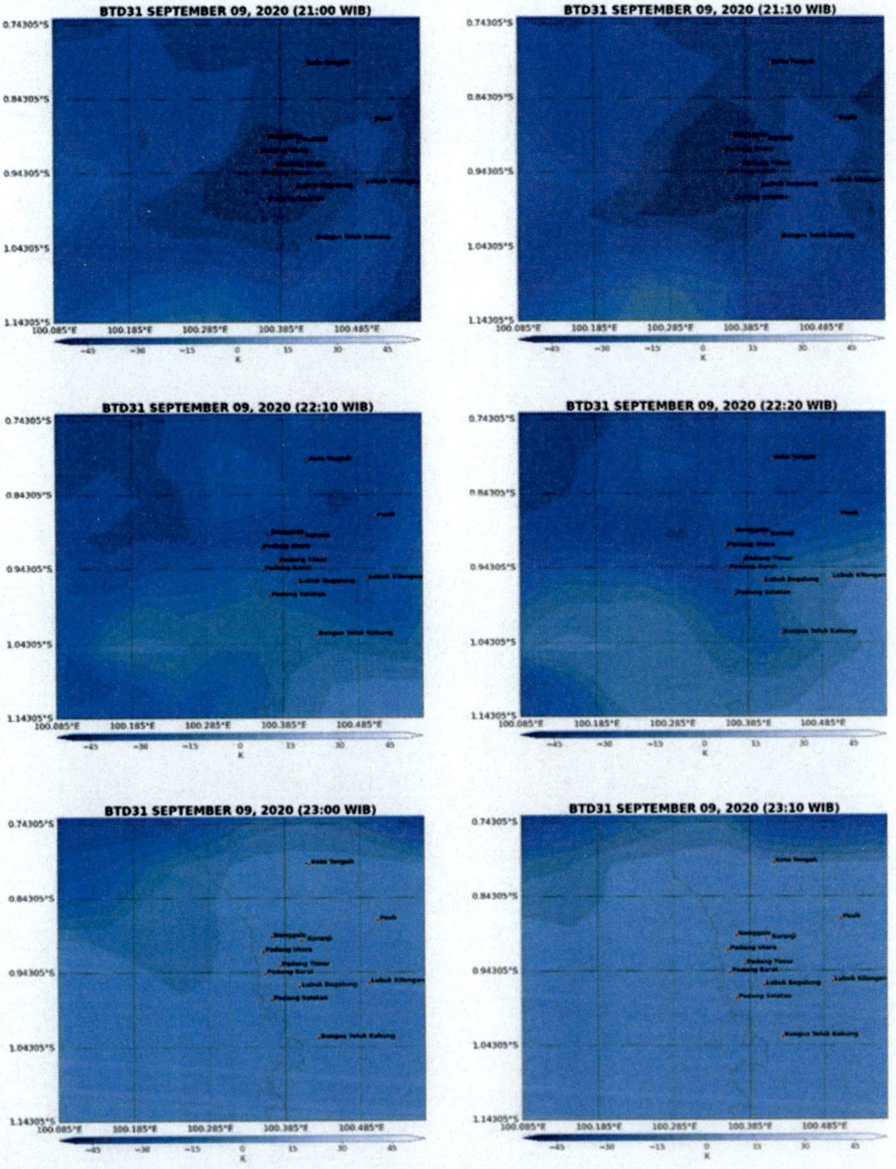

Fig. 61.7 BTD31 plot on September 9, 2020

The POD parameter of BTD1 and IMERG data are shown in Fig. 61.9a. Plot with 0 lag time has the best value for all areas in Padang city. For the FAR parameter in all sub-districts in Padang city, the value of lag-0 ranges from 0.4 to 0.6. This value is almost similar to lag-1, lag-2, and lag-3. The best value of CSI is also at lag-0 for the mainland area in Padang city.

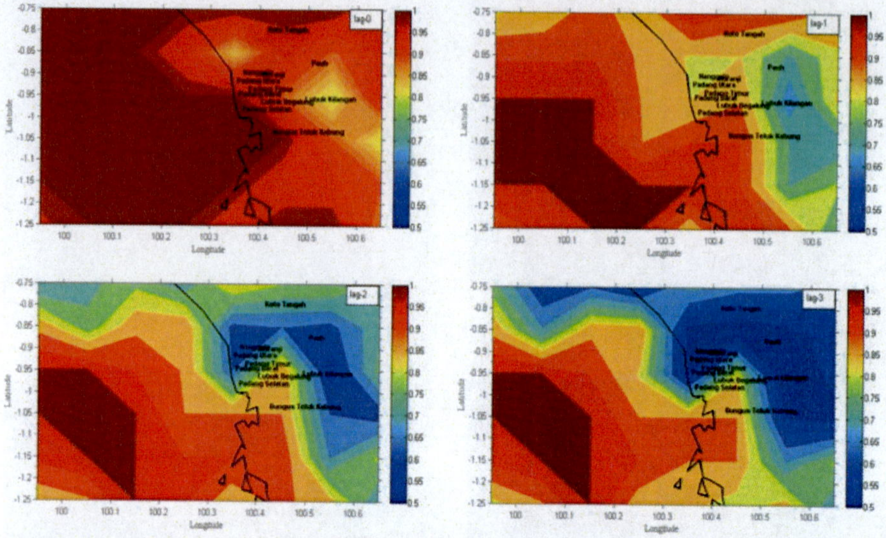

(a)

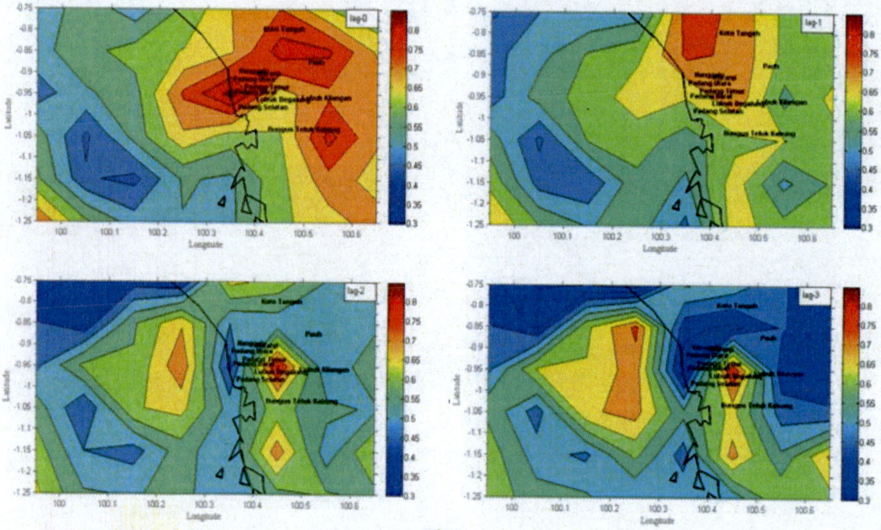

(b)

Fig. 61.8 **a** POD and **b** FAR between IR1 Himawari-8 data and IMERG data. **c** CIS between IR1 Himawari-8 data and IMERG data

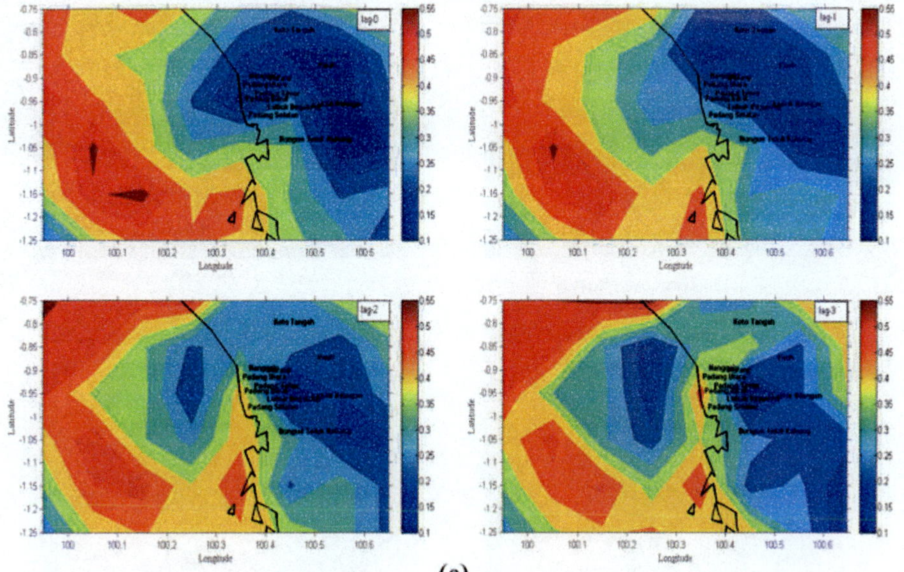

(c)

Fig. 61.8 (continued)

However, because all BTD2 data are positive and all BTD3 are negative, we cannot plot the data to analyze particles from ice contained in clouds. This is due to the inability to visualize the algorithm in the confusion matrix.

61.4 Conclusion

The result shows that the BTD method successfully observed extreme rain events in Padang city. Cloud monitoring using Himawari-8 satellite data during rain events on September 9, 2020, shows that extreme rain events occur when the cloud top temperature is lower than −60 °C. When the cloud's top temperature is low, the cloud contains ice particles. BTD value can also provide a clear picture of how the clouds spread from before the rain incident to when it rained in Padang. Statistical evaluation using IMERG data shows that the best POD, FAR, and CIS were found with 0 lag time. BTD1 also found the best POD and CIS values at 0 lag time, but different in FAR parameter.

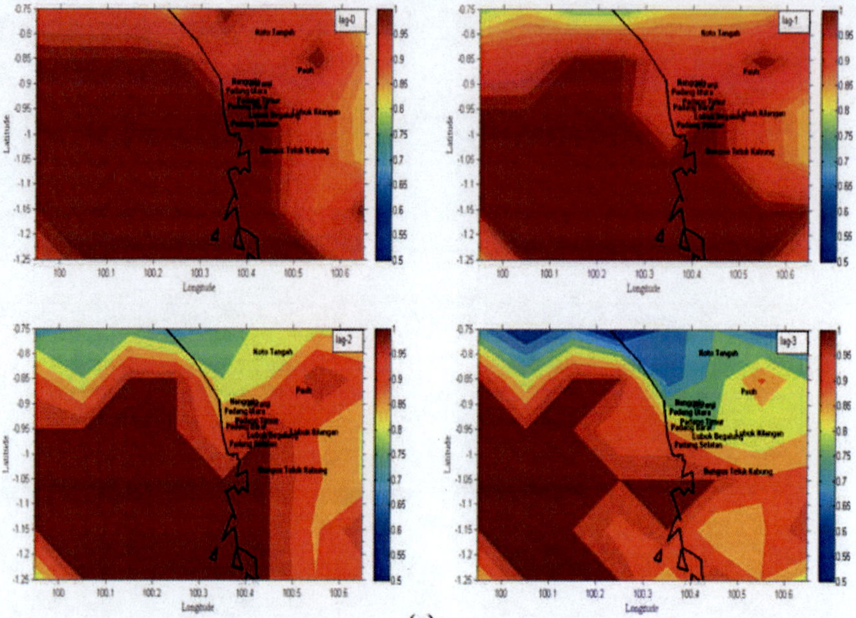

(a)

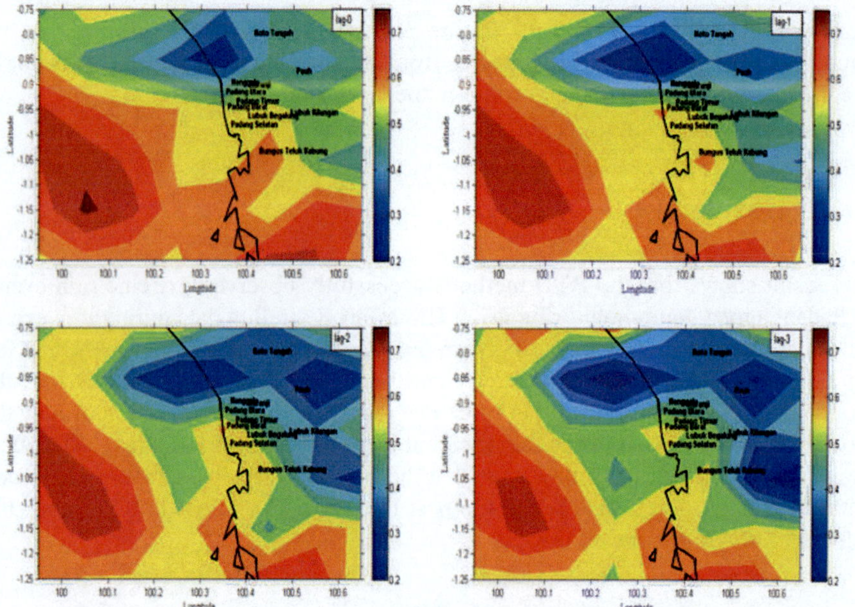

(b)

Fig. 61.9 **a** POD and **b** FAR between BTD1 and IMERG data. **c** CIS between BTD1 and IMERG data

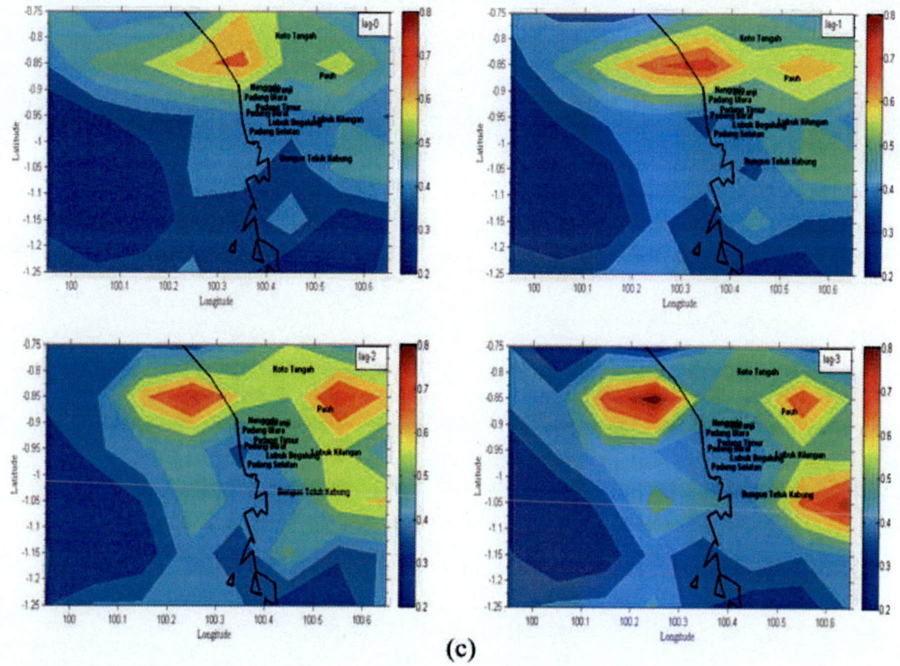

(c)

Fig. 61.9 (continued)

References

1. Ahrens, C.D., Samson, P.J.: Extreme weather and climate. Books Cole, USA (2010)
2. Baranowski, D.B., Flatau, M.K., Flatau, P.J., et al.: Social-media and newspaper reports reveal large-scale meteorological drivers of floods on Sumatra. Nat. Commun. **11**, 2503 (2020). https://doi.org/10.1038/s41467-020-16171-2
3. Bessho, K., Date, K., Hayashi, M., Ikeda, A., Imai, T., Inoue, H., Kumagai, Y., Miyakawa, T., Murata, H., Ohno, T., Okuyama, A., Oyama, R., Sasaki, Y., Shimazu, Y., Shimoji, K., Sumida, Y.: An introduction to Himawari-8/9. J. Meteorol. Soc. Jpn **2**(94), 151–183 (2016)
4. Strabala, K.I., Ackerman, S.A., Menzel, W.P.: Cloud properties inferred from 8–12 μm data. J. Appl. Meteorol. **33**, 212–229 (1994)
5. https://www.liputan6.com/regional/read/4352846/bmkg-minangkabau-hujan-di-padang-masuk-kategori-ekstrem. Last accessed 21 Feb 2021
6. Harjupa, W., Nakakita, E., Sumida, Y., Masuka, A.: Trial utilization of rapid scan observation of Himawari-8 for obtaining information on cumulus life stage. J. Jpn. Soc. Civil Eng. Ser. B1 (Hydraul. Eng.) **5**(74), I_283–I_288 (2018)
7. Mecikalski, J.R., Bedka, K.M.: Forecast. Convective initiation by monitoring the evolution of moving cumulus in daytime GEOS imagery. Mon. Weather Rev. **134**, 49–78 (2016)
8. Yasuhiko, S., Hiroshi, S., Takahito, I., Akira, S.: Convective cloud information derived from Himawari-8 data. Meteorol. Satellite Center Tech. Note **62**, 19–37 (2017)
9. Efendi, A.N., Fachrurozi, M.: Pemanfaatan Data Satelit Ir 1 Himawari-8 untuk Mengurangi Dampak Kerugian Materil dan Jiwa Akibat Bencana Alam di Wilayah Melawi Kalimantan Barat. Seminar nasional Penginderaan Jauh ke-4 Tahun 2017 (2017)

Chapter 62
Simulation of Solar Flare Mechanism Based on Ideal Magnetohydrodynamics State by Disrupting the Stability of Magnetic Field Due to the Plasma Momentum Injection

Ni'matus Sholikhah and Bambang Setiahadi

Abstract The flare is one of the forms of solar activity that has a direct influence on the Earth. Electromagnetic energy released by flare could induce the disturbance of the Earth's atmosphere. The flare mechanism is slightly tricky to perceive because of the complexity of the flare process, which has non-linear characteristics and also requires some parameters. The condition of magnetohydrodynamic could be used to describe the phenomenon of flare formation and other Solar activities. The observation result does not show a sufficient illustration of the flare mechanism formation. In order to obtain further understanding of the mechanisms of flare formation, the numerical approximation simulation could be applied by using some approximation such as statistical empirical and theoretical. In this research, we are using numerical approaches such as the finite difference method to solve the ideal magnetohydrodynamic equations. The result of the numerical solution is used to simulate the process of flare formation. Many physical processes of the Solar are suspected as the cause of flare formation. This paper outlines the process of flare formation as a result of the disturbance of the magnetic field generated by the plasma injection momentum.

62.1 Introduction

The Solar is the largest source of energy for the survival of life on the Earth. The energy produced by the Solar comes from the result of the thermonuclear reaction in the Solar's core. Because of the large amount of energy produced, it will encourage releasing energy which leads emergence of various kinds of solar surface activity.

N. Sholikhah (✉)
Department of Physics, Faculty of Mathematics and Natural Sciences, Brawijaya University, Malang 65145, Indonesia
e-mail: sholikhahnimatus25@gmail.com

B. Setiahadi
Indonesian National Institute of Aeronautics and Space (LAPAN), Pasuruan 67155, Indonesia

E. Yulihastin et al. (eds.), *Proceedings of the International Conference on Radioscience, Equatorial Atmospheric Science and Environment and Humanosphere Science, 2021*, Springer Proceedings in Physics 275, https://doi.org/10.1007/978-981-19-0308-3_62

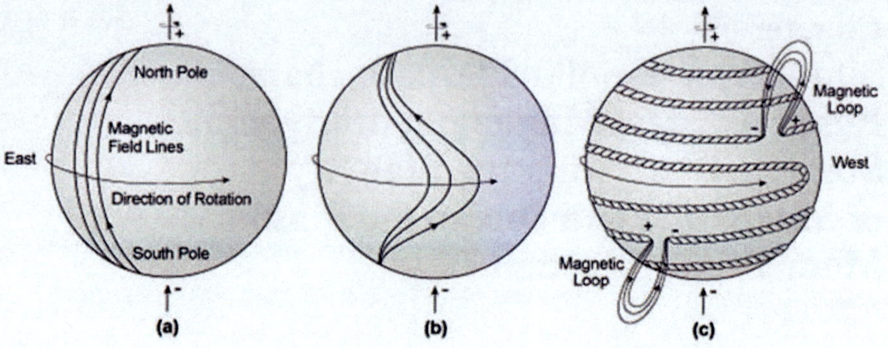

Fig. 62.1 Process of sunspot formation triggered by differential rotation of the solar [2]

Flare is one of the forms of Solar surface activity that has a direct influence on the Earth. During the flare process, it can release magnetic energy up to 10^{30} erg s^{-1}. This energy is released in various forms, not only thermal energy but also kinetic energy, which can be observed as hard X-ray (HXR), soft X-ray (SXR), etc. [1].

Sunspot can be suspected as an early indicator of the emergence of solar surface activity. In the beginning, the Solar has a poloidal magnetic field, as shown in Fig. 62.1a. The differential rotation of the Solar in the convective zone can be suspected to cause the emergence of a sunspot. Due to the differential rotation of the Solar, where the rotational speed at the equatorial region is faster than at the poles, it causes a change in the topology of the magnetic field from a poloidal magnetic field to a toroidal magnetic field (see Fig. 62.1b). After several rotations, the magnetic field becomes twisted, such as a rope. Furthermore, the magnetic field is distorted and stressed until it becomes unstable. Then through the buoyancy force magnetic flux tube rises and penetrates the surface of the convective zone, and it is observed as a sunspot in the photosphere layer (see Fig. 62.1c) [2]. Sunspot with large size also stores enormous magnetic energy inside.

Sunspot magnetic loop has a stable magnetic field topology with the footpoints resting on the dense photosphere. However, as time went on, the sunspot magnetic loop evolved and became unstable because of photospheric motion controlled by differential rotation. Photospheric slow-motion gives a perturbation to the sunspot magnetic loop stability through the shear perturbation, the twist perturbation, and the plasma pumping perturbation. As a result, the magnetic field becomes massive in the active region and causes a decrease in temperature from 6000 to 4000 K in the photosphere layer. Therefore, the sunspot looks darker than the surrounding area. A darker and extensive sunspot has a strong magnetic field and potentially stores enormous magnetic energy. Because of the distribution of energy concentration, the magnetic flux tube can break at any moment, and the energy will release abruptly. This mechanism is known as Flare [3].

Various kinds of research have been done to explain the mechanism of solar flare both observation and simulation. Both are useful for the development research in this field. In 1997, Hori [4] created a 2D model of solar flare loops from 1D hydrodynamics calculation. This model uses semicircular loops isolated by thermal conditions. This model simulates the density, temperature, and pressure of flare loops. The distribution of Soft X-Ray (SXR) emissions is also counted in the computation process. Figure 62.2 below shows the multiple loop configuration in the 2D model with nine magnetic loops. The simulation program is used to study the response of the hydrodynamics condition through the assumption of flare energy release both along the lengths of the loops and radially over the system of loops. In this model, to create a magnetic reconnection process, heat deposition is assumed as the trigger of releasing the energy of flare. The heat deposition in the innermost loop spread to the outermost loop with bringing the energy and the energy will release as flare energy.

Magnetic reconnection is considered one of the promising mechanism of a solar flare. There are various kinds of magnetic reconnection models that try to explain solar the flare mechanism. The first model was revealed by Sweet and Parker in 1958 which is known as the Sweet-Parker reconnection model. Then, in 1964 Petschek published his model which is known as the Petschek model that bring the idea of slow MHD standing shock waves. Another model of magnetic reconnection is called CSHKP model which is a standard model of magnetic reconnection (see Fig. 62.3) [5].

The CSHKP model explains that an open coronal magnetic field line reconnects to form a closed loop when magnetic reconnection is in progress. This causes the solar wind along the reconnected field lines to collide with each other and generate a shock inside the closed loop. The X point is the point where the reconnection occurs then the reconnection reforms at the higher altitude. This causes increasing in the height of the closed loops [5].

In this research, a simulation of the flare mechanism is created by providing perturbation to the stable sunspot magnetic loop through the injection of plasma momentum at a point in the photosphere layer. The model of this simulation is in the 2D model with semicircular loop configuration. A Flare is a form of evolution from a statical state to a dynamical state. Therefore, we need to develop the Magnetohydrostatics (MHS) state to the Magnetohydrodynamics (MHD) state. The differences between the MHS state and MHD state are the MHS only focused on space solutions. Meanwhile, the MHD focused on space and time solutions. In order to create a physical state of the solar in the simulation, we are applying a numerical approach to solving the MHD equation. The statical state interprets the initial condition of a sunspot.

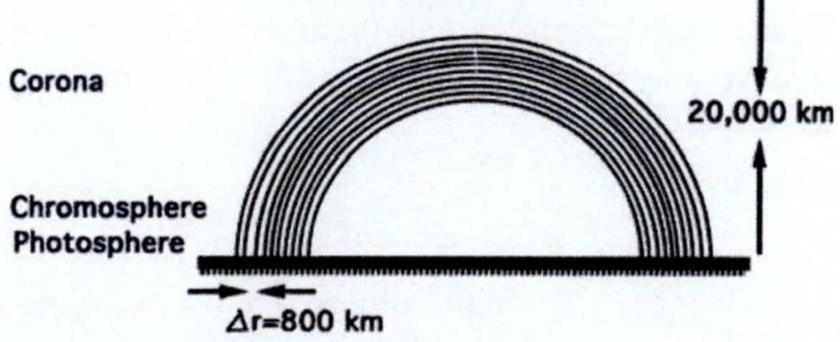

Fig. 62.2 Multiple loop configuration of 2D model [4]

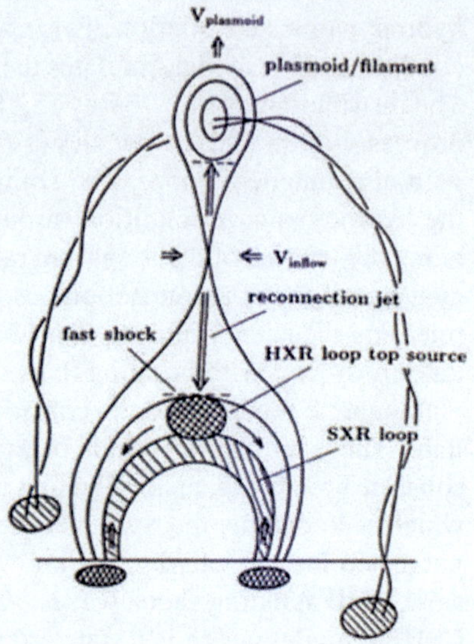

Fig. 62.3 Schematic picture of CSHKP reconnection model [5]

In other words, the statical state is a condition where the plasma flow and magnetic field are in a steady-state. However, before the dynamic process takes place, the static state needs to be applied first.

62.1.1 Magnetohydrostatics

According to Setiahadi [6], MHS is a branch of plasma physics in which there is a relation between electron-plasma and magnetic field in the plasma steady-state. Both magnetic field and charged particles in the plasma are assumed as fluid in the static state. Based on this assumption, there are fourth influential parameters. The parameters are plasma viscosity, plasma pressure, magnetic force, and gravitational forces. Equation 62.1 below represents the state of MHS equilibrium.

$$\nabla^2 \rho v + \nabla p + (\nabla \times B) \times B + \rho g = 0 \tag{62.1}$$

Because plasma speed is lower than light speed, the velocity vector is assumed to equal zero. Furthermore, with the assumption that the magnetic field is very low in the solar corona, the gravitational stratification is fulfilled. Then Eq. (62.1) is re-written as below,

$$\nabla p + (\nabla \times B) \times B = 0 \tag{62.2}$$

Equation (62.2) can be outlined into the following equation,

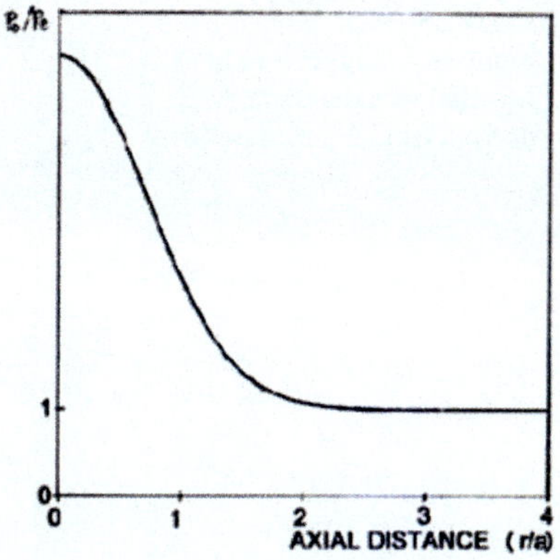

Fig. 62.4 A graph represents pressure as a function of axial distance derived from MHS state [6]

$$\frac{r}{dr}\left[p(r) + \frac{1}{2}B(r)^2\right] + \frac{1}{r}B(r)^2 = 0 \tag{62.3}$$

where r is a radius of the magnetic loop. Furthermore, from Eq. (62.3), an integral form can be written as follows,

$$B^2r^2 = -2\int r^2\frac{dp}{dr} \tag{62.4}$$

Finally, we obtain the equation for the pressure as described below,

$$p = p_e + (p_0 - p_e)\exp\left(-r^2/a^2\right) \tag{62.5}$$

Pressure p has a maximum value while $r = 0$. Then, in otherwise pressure p has a minimum value while r has a large value. It can be seen as the Fig. 62.4,

Meanwhile, for magnetic field solution can be written as Eq. (62.6) below,

$$B^2 = 2(p_0 - p_e)\left(\frac{a}{r}\right)^2 - 2(p_e - p_0)\left[1 + \left(\frac{a}{r}\right)^2\right]\exp\left[-(a/r)^2\right] \tag{62.6}$$

Figure 62.5 represents the meaning of Eq. (62.6)

62.1.2 Magnetohydrodynamics

The process that occurs in the flare is a form of evolution from a statical state to a dynamical state which is arranged by losing MHS equilibrium then evolving dynamical state, which is represented by MHD states. The basics of the MHD equation are the magnetic induction equation and continuity equation. The magnetic induction

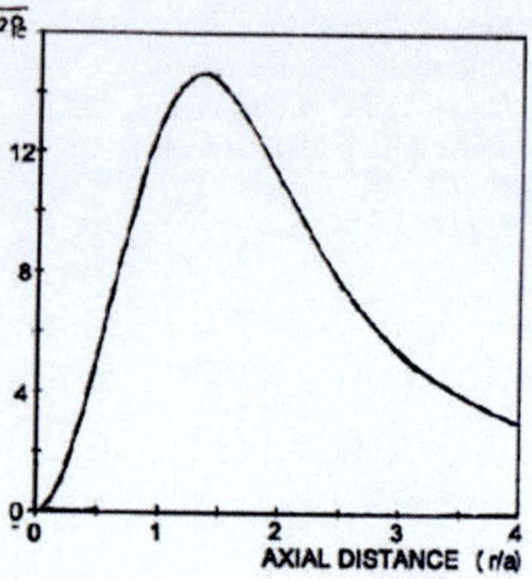

Fig. 62.5 A graph represents magnetic field as a function of axial distance derived from MHS state [6]

equation is used to describe the magnetic field of the Solar. Meanwhile, the continuity equation is used to describe the plasma flow of the Solar. The combination of magnetic induction equation and continuity equation will produce MHD equations as written below [6],

$$\frac{\partial \rho}{\partial t} + \nabla.\rho V = 0 \tag{62.7}$$

$$\frac{\partial \rho V}{\partial t} + \nabla.\rho V = -\nabla p + (\nabla \times B) \times B \tag{62.8}$$

$$\frac{\partial B}{\partial t} + \nabla.BV = (B.\nabla)V - B(\nabla.V) \tag{62.9}$$

$$\frac{\partial p}{\partial t} + \nabla.pV = -(\gamma - 1)p(\nabla.V) \tag{62.10}$$

Based on the interactions that occur in the MHD state, at least four basic plasma physics parameters are needed to explain the frozen-in concept, which explains the interaction between the magnetic field and the electron cloud. The four parameters are plasma density (ρ), momentum density (ρV), magnetic field density (B), and plasma pressure density (p) [6].

In the MHD state, we are applying fast mode speed to make propagation waves where the propagation is the combination between hydrodynamic component and magnetodynamic component, which is expressed by the equation below [6],

$$V_{FM}{}^2 = C_S{}^2 + V_A{}^2 \tag{62.11}$$

With C_S is hydrodynamic speed following equation below,

$$C_S{}^2 = \gamma \frac{p_0}{\rho_0} \tag{62.12}$$

And V_A is Alfen speed which represents magnetodynamic component and follows the expression below,

$$V_A{}^2 = \frac{B_0{}^2}{\rho_0} \tag{62.13}$$

The fast mode speed V_{FM} will determine the computation stability each time. Therefore, it is necessary to monitor the fast mode speed that evolves in the computation process.

62.1.3 *Finite Difference Method*

A Differential equation is not easy to solve analytically. Therefore, in this research, we are applying the finite difference method to solve the differential equation of the MHD equations. In the Finite difference method, the numerical solution is obtained by replacing the derivatives in the equation with discrete approximation and evaluating each grid to develop a computation system [7].

In general, partial differential equations are not only containing one domain but also containing several domains. For example, if we have a function with two domains (spatial and temporal), we can apply the finite difference method to obtain the solution for both the spatial domain and time domain [8]. By using a forward difference type, we obtain the numerical solution in the spatial domain as Eq. 62.14 below,

$$\frac{\partial}{\partial x}u(x_i, t_n) = \lim_{\Delta x \to 0} \frac{\partial u}{\partial x} = \frac{u_{i+1}^n - u_i^n}{\Delta x} \tag{62.14}$$

With $i = 0, \ldots, N_x$ and $n = 0, \ldots, N_t$ represent points of the mesh as two-dimensional in the x, t plane [8].

62.2 Data and Methods

The MHD equation is a form of the partial differential equation which is not easy to solve analytically. Therefore, a numerical method is needed to be done to obtain its solution. The numerical solution is obtained by discretizing the MHD equation then it is solved by finite difference method.

Furthermore, by adopting the concept of Finite Difference method, we obtain the numerical solution which can be used to solve the MHD equation as following,

$$\nabla_i \Psi_i \to \frac{\partial \Psi_i}{\partial x_i} \to \frac{(\Psi_{i+1} - \Psi_i)}{\left[x_{i+1} + v_{i+1}^{1/2}\left(\frac{1}{2}\delta t\right)\right] - \left[x_i + v_i^{1/2}\left(\frac{1}{2}\delta t\right)\right]} \tag{62.15}$$

$$\Gamma_i(\nabla_i \otimes \Psi_i) \to \frac{\partial \Psi_i}{\partial x_i} \to \frac{\left[\frac{1}{2}(\Psi_{i+1} - \Psi_i)\right]\left[\Gamma_{i+1} - \Gamma_i\right]}{\left[x_{i+1} + v_{i+1}^{1/2}\left(\frac{1}{2}\delta t\right)\right] - \left[x_i + v_i^{1/2}\left(\frac{1}{2}\delta t\right)\right]} \tag{62.16}$$

$$\Omega_i \to \frac{\left[\frac{1}{2}(\Omega_{i+1} - \Omega_i)\right]\left[x_{i+1} - x_i\right]}{\left[x_{i+1} + v_{i+1}^{1/2}\left(\frac{1}{2}\delta t\right)\right] - \left[x_i + v_i^{1/2}\left(\frac{1}{2}\delta t\right)\right]} \tag{62.17}$$

For reasons of ease of operation, the basic physical parameters $p, \rho, \rho V, B$ are symbolized to a single quantity as Q. The quantity will be executed numerically by a specific method to obtain the numerical representation, which is not far from the original differential forms. Then the various terms on the right side of Eqs. (62.7–62.10) are represented by S_i. Furthermore, in the general form, the Eqs. (62.7–62.10) is re-written as bellow,

$$\frac{\partial Q}{\partial t} + \nabla.QV = S_i \tag{62.18}$$

In the finite difference method to obtain the final and definitive solution, we develop a global point-to-point solution both in space and in time. Q as any quantity, which will be found the solution derivatively, constantly subjected to floating number round-off errors and the error may propagate from point to point in the computational domain and from time to time during a running of computer code [3]. It will cause numerical instability. Therefore, we need to recheck the numerical stability of the numerical solutions by some automatic procedure.

First of all, we need to ensure that Q is not a negative quantity everywhere in the computational domain at any moment. Because depending on the physical state, the parameters $p, \rho, \rho V, B$ should not be negatives. Then we need to control numerical flux coded by f that entering and leaving numerical cell boundaries during a computation process. Firstly, the numerical flux built from Q define as bellow,

$$f^0_{i+\frac{1}{2}} = 2\eta\left(Q^{t-1}_{i+1} - Q^t_i\right) \tag{62.19}$$

$$f^0_{i-\frac{1}{2}} = 2\eta\left(Q^t_i - Q^{t-1}_{i+1}\right) \tag{62.20}$$

Then advance Q using finite difference describe as expression below,

$$Q^{t+1}_{i+1} = Q^{t-1}_{i+1} - \frac{\Delta t}{\Delta x}\left(Q^t_{i+2} V^t_{i+2} - Q^t_i V^t_i\right) \tag{62.21}$$

$$Q^{t+2}_i = Q^t_i - \frac{\Delta t}{\Delta x}\left(Q^{t+1}_{i+1} V^{t+1}_{i+1} - Q^{t+1}_{i-1} V^{t+1}_{i-1}\right) \tag{62.22}$$

In addition, we need to compute and define the anti-diffusive numerical fluxes as the expression below,

$$f^{1}_{i+\frac{1}{2}} = 2\eta\left(Q^{t+1}_{i+1} - Q^{t+2}_{i}\right) \tag{62.23}$$

$$f^{1}_{i-\frac{1}{2}} = 2\eta\left(Q^{t+2}_{i} - Q^{t+1}_{i-1}\right) \tag{62.24}$$

Furthermore, we can apply the numerical diffusion as below,

$$Q^{t+1}_{i+1} = Q^{t+1}_{i+1} + f^{0}_{i+\frac{1}{2}} - f^{0}_{i-\frac{1}{2}} \tag{62.25}$$

$$Q^{t+2}_{i} = Q^{t+2}_{i} + f^{0}_{i+\frac{1}{2}} - f^{0}_{i-\frac{1}{2}} \tag{62.26}$$

To obtain second-order precision, we need to take the first difference of the numerical diffused transported Q, which is depicted below,

$$\Delta_{1+\frac{1}{2}} = Q^{t+1}_{i+1} - Q^{t+2}_{i} \tag{62.27}$$

$$\Delta_{1-\frac{1}{2}} = Q^{t+2}_{i} - Q^{t12}_{i-1} \tag{62.28}$$

Then we need to limits and corrects the fluxes by seeking the unphysical negative fluxes following the expression below,

$$S = sign\left(f^{1}_{i+\frac{1}{2}}\right) \tag{62.29}$$

$$f^{c}_{i+\frac{1}{2}} = S.max\left[0, min\left(S.\Delta_{i-\frac{1}{2}}, \left|f^{1}_{i+\frac{1}{2}}\right|, S.\Delta_{i+\frac{1}{2}}\right)\right] \tag{62.30}$$

In conclusion, we can apply the numerical anti-diffusion by using the corrected fluxes f^{C} which are written as follow:

$$Q^{t+1}_{i+1} = Q^{t+1}_{i+1} - f^{C}_{i+\frac{1}{2}} + f^{C}_{i-\frac{1}{2}} \tag{62.31}$$

$$Q^{t+2}_{i} = Q^{t+2}_{i} - f^{C}_{i+\frac{1}{2}} + f^{C}_{i-\frac{1}{2}} \tag{62.32}$$

In this research, to obtain a stable and convergent result, Courant-Friederichs-Lewy (CFL) condition needs to be applied. The CFL condition will control the fast mode MHD through algorithm fast-mode time-step control in every time new computational cycle. Therefore, the fast mode propagation speeds have to be limited to no more than 0.5 Δx to avoiding computational instability that emerged since the perturbation entering the next computational grids in one time step. This condition is following expression below,

$$\Delta t \leq \frac{1}{2}\frac{\Delta x}{V_{FM}} \tag{62.33}$$

The MHD interactions need at least four basic physical parameters. They are the pressure density (p), the plasma density (ρ), the momentum density (ρV), and the magnetic field density (B). In this research, we consider works where the boundary and initial conditions on low solar coronal magnetic arcade. Therefore, other forces such as solar differential momentum force and gravitational force are neglected. The solar differential momentum is a long time scale process. Meanwhile, in this research, we will create a short-duration flare model.

Takasao et al. [9] had published their analysis about flare which occur on 2010 August 18. This flare is a kind of short-duration flare with extreme-ultraviolet radiation. Some plasma blobs were observed in the sheet structure above the hot loops during the rising phase of the flare (see Fig. 62.6). They also found that the plasma blobs collide which each other, it was released from the sheet structure. They have an assumption that the plasma blobs are plasmoid or magnetic islands which are transported toward interplanetary space.

Computation code in this research is made by Fortran programming language to generate numerical data from the numerical solution of MHS and MHD equation. Furthermore, numerical data will be plotted as a contour to produce a 2D visual image using Python programming language. This visual image can be used to obtain further understanding of the mechanisms of flare formation.

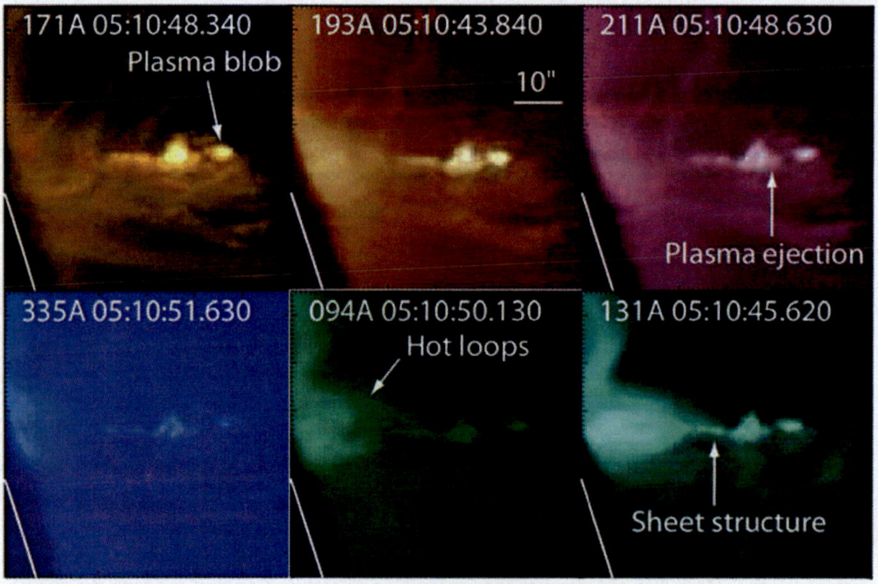

Fig. 62.6 Close-up images of the reconnection site in six different wavelengths (171, 193, 211, 335, 94, and 131 Å) of AIA [9]

62.3 Result and Analysis

The MHS state represented by Eqs. (62.5–62.6) is applied as the initial state of the stable sunspot magnetic loop. The graph shown in Fig. 62.4 gives a result that the pressure value is maximum while the radius is zero. In other words, pressure in the center is very large where the center is meaning photosphere layer. It is in accordance with the fact that the density of particles in the photosphere layer is denser than the layer above such as the chromosphere and corona. Nevertheless, the magnetic field has a different form. Figure 62.5 shows a result that the magnetic field is very low in the center or the photosphere layer, then the value increases and reaches the maximum value before decay to zero at the far from center radius. The result of the 2D model is shown in the Fig. 62.7,

After the MHS state is reached, the dynamical process can be applied by running MHD computation. In order to create the simulation of the flare mechanism, we are disturbing the stable sunspot magnetic loop by giving a perturbation. Furthermore, the perturbation will be executed dynamically until the MHD state is achieved. In this research, the perturbation is given in the form of momentum injection at a point in the sunspot magnetic loop. Based on the dynamic state, the four basic parameters in the MHD state are connected each other to arrange the condition of the Sun. As a result, the perturbation given to one of the basic parameters which in this case is given to momentum will affect other parameters. The perturbation is causing a change in the sunspot magnetic field topology, inspect Fig. 62.8 when $t = 0.0095$. It means that the perturbation affects the stability of the magnetic field.

Furthermore, the numerical simulation process continues to obtain the illustration of the flare mechanism. The instability of the magnetic field will trigger magnetic reconnection in the center of perturbation ($t = 0.1168$ of Fig. 62.8). The reconnection magnetic process will change configuration of magnetic field line topology in the

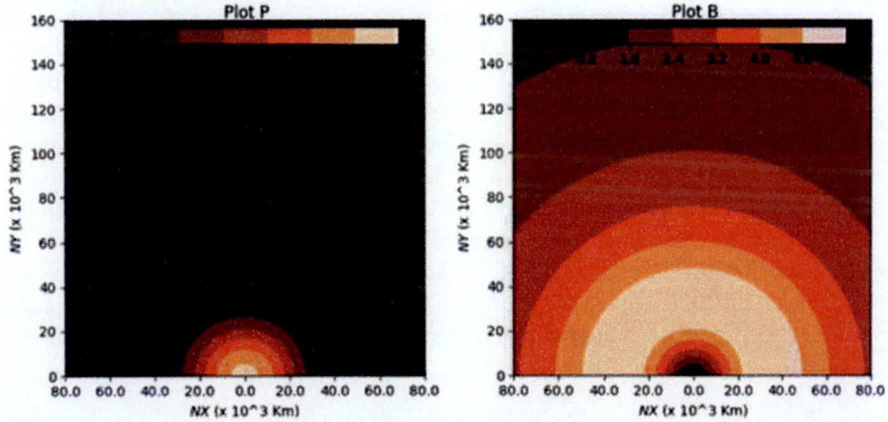

Fig. 62.7 2D model of pressure p which is derived from MHS state (left). 2D model of sunspot magnetic loop B which is derived from MHS state (right)

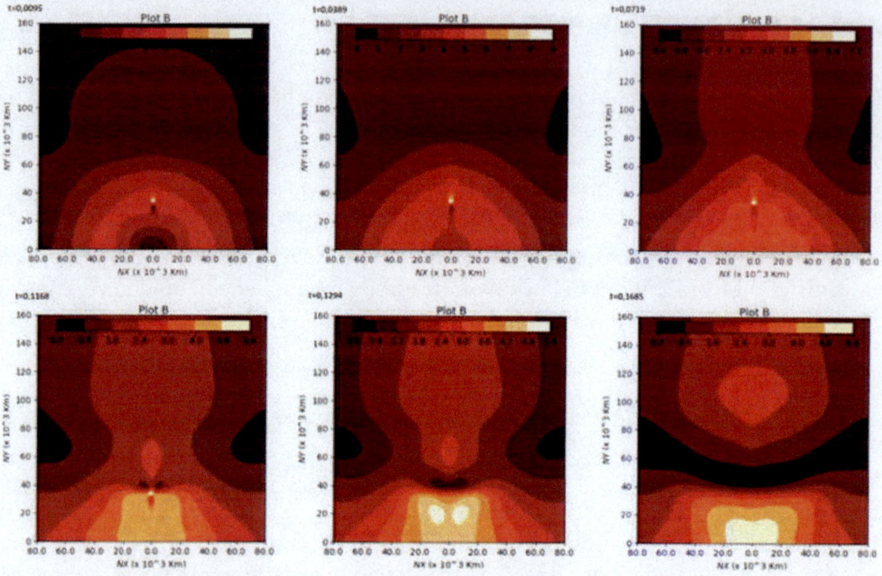

Fig. 62.8 Time evolution of the sunspot magnetic loop from the 2D MHD model

magnetic structure. The perturbation will encourage comprehensive evolution of the basic structure of the magnetic field.

The magnetic reconnection give an impact to the transformation of magnetic energy. It will transform the magnetic energy into kinetic energy, thermal energy and fast particle energy [10]. Furthermore, the energy released by magnetic reconnection process will be distributed in the forms such as radiation, thermal conduction, mass ejection, wave propagation and the generation of high-energy particles [5].

According to Shibata [5], in the reconnection magnetic process a part of magnetic energy is transported downward along the magnetic field line through thermal conduction, high-energy particles, and heating chromospheric plasma. Meanwhile, another part of the energy magnetic is transported upward as ejecting plasma blobs called plasmoid. It can be observed in the Fig. 62.8 when t = 0.1294, where the magnetic field looks brighter at the bottom. We assume that the magnetic energy subjected to collision with the dense particles in the photosphere layer. In the observation result, it can be observed as hard X-Ray (HXR) and soft X-Ray (SXR). In the Fig. 62.8 when t = 0.1685 also can be observed that another part of the energy will be released towards the interplanetary space. This result looks similar as the observation of flare occur on 2010 August 18 (see Fig. 62.7). However, further research is needed to find out the forms of energy which is released towards the interplanetary space.

62.4 Conclusion

The flare mechanism can be explained through the simulation result by creating the development of the MHS state to the MHD state. The MHS state will be given a slight perturbation by injection momentum to trigger the instability of the magnetic field until the dynamic state is achieved and executed to find out the process that occurs during flare. This perturbation will affect the four basic parameters that arrange the MHD state.

In conclusion, we can see that the perturbation will cause a change of the sunspot magnetic topology. It will trigger a magnetic reconnection in the perturbation center which is converting the magnetic energy to not only thermal energy but also kinetic energy. The result of the simulation shows that the magnetic energy is transported downward as high-energy particles which is subjected to collision with the dense particle in the photosphere layer. Meanwhile, another magnetic energy is transported toward interplanetary space. This energy give an influence for the Earth's atmosphere especially the magnetosphere.

References

1. Dennis, B.R., Schwartz, R.A.: Solar flares: the impulsive phase. In: Proceedings of the 104th Colloquium of the International Astronomical Union, pp. 75–94. Kluwer Academic Publisher, Dordrecht, The Netherlands (1989)
2. Lang, K.R.: Sun, Earth and Sky, 2nd edn. Springer, New York (2006)
3. Setiahadi, B.: Advances and frontiers in solar-terrestrial magnetohydrodynamics computer simulations and space early warnings at Lapan Watukosek. In: Semi-local Proceedings of Simulating, Computing Technology and Applications, pp. 32–58 (2005)
4. Hori, K., Yokoyama, T., Kosugi, T., Shibata, K.: Pseudo-two-dimensional hydrodynamic modelling of solar flare loops. Astrophys. J. **489**, 426–441 (1997)
5. Shibata, K., Magara T.: Solar flares : magnetohydrodynamics processes. Living Rev. Solar Phys. **8** (2011)
6. Setiahadi, B.: Magnetohydrodynamic computer simulation of erupting low solar coronal magnetic fields. In: Proceeding of National Seminar on Biology and science Education, pp. 424–429. Muhammadiyah University of Surakarta, Surakarta (2017).
7. Sauer, T.: Numerical Analysis, 3rd edn. Pearson, Hoboken (2018)
8. Langtangen, H.P., Linge, S.: Finite Difference Computing with PDEs. Springer, Cham (2017)
9. Takasao, S., Asai, A., Isobe H., Shibata K.: Simultaneous observation of reconnection inflow and outflow associated with the 2010 August 18 solar flare. Astrophys. J. Lett. **745** (2012)
10. Priest, E.R., Forbes, T.G.: The magnetic nature of solar flares. Astron. Astrophys. Rev. **10**, 313–377 (2001)

Chapter 63
Diurnal Variation of Fine Particulate Matter in Indonesia Based on Reanalysis Data

Nani Cholianawati

Abstract Fine Particulate Matter (PM2.5) has great effects on human health. Globally, PM2.5 cycles have a morning peak, an early night time peak, and an afternoon minimum (bimodal pattern). Knowing diurnal variations will improve our understanding of PM2.5 exposure. This research used hourly PM2.5 surface concentrations during 2019–2020 from Ground Observation in Central Jakarta and Reanalysis Data MASINGAR JMA with spatial resolution 0.375° × (0.37147°–0.37461°). We validated reanalysis data by comparing it with ground observation in Central Jakarta. This binned data set is the average of hourly data for January (rainy season) and July (dry season). Descriptive analysis for diurnal variations of PM2.5 was done both measured from ground observations and reanalysis. The result was hourly reanalysis PM2.5 fitted enough to ground observation data ($r = 0.67$). PM2.5 reanalysis was underestimated comparing to ground observation data. Both reanalysis and in situ PM2.5 showed a unimodal pattern in the dry season, with a maximum peak occurred at midnight and an afternoon minimum. PM2.5 cycles in the rainy season showed a low bimodal pattern with a night peak, a morning peak, and an afternoon minimum. PM2.5 concentration from morning to afternoon in July 2020 is lower than in 2019. This reduction is plausible related to the lockdown policy (COVID19 pandemic).

63.1 Introduction

Fine particulate matter (PM2.5) has great impacts on public health. Epidemiological and toxicological studies [1] have shown that PM2.5 induces cardiopulmonary disorders, diabetes mellitus, and adverse birth out-comes. PM2.5 may still pose a hazard to public health even at very low levels (far below national standards) of exposure. PM2.5 exposure has caused 4.2 million premature deaths worldwide per year in 2016 [2].

N. Cholianawati (✉)
Research Center for Climate and Atmosphere, National Research and Innovation Agency (BRIN), Bandung, Indonesia
e-mail: nani.cholianawati@brin.go.id

E. Yulihastin et al. (eds.), *Proceedings of the International Conference on Radioscience, Equatorial Atmospheric Science and Environment and Humanosphere Science, 2021*, Springer Proceedings in Physics 275, https://doi.org/10.1007/978-981-19-0308-3_63

World Health Organizations applied standards for 24 h and annual PM2.5 concentrations to keep the health of society. As no threshold for PM2.5 has been identified below which no harm to health is observed, the recommended values should be regarded as representing acceptable and achievable objectives to minimize health effects in the context of local constraints, capabilities and public health priorities [3]. In 2021, WHO has updated global guidelines to offer health-based air quality guideline levels for six key air pollutants: PM2.5, PM10, O_3, NO_2, SO_2, and CO [4]. The recommendation is a short-term (24-h) PM2.5 air quality guidelines level of 15 $\mu g/m^3$ and a long-term (annual) PM2.5 air quality guidelines level of 5 $\mu g/m^3$ [4].

Diurnal variation of surface PM2.5 concentration (diurnal PM2.5) could significantly affect aerosol radiative and health impacts [5]. It could also reflect the physical and chemical mechanisms of air pollution formation and evolution. Knowing diurnal variations will improve our understanding of PM2.5 exposure. We can do some steps for mitigating poor air quality, such as wearing a PM2.5 mask and limiting outdoor activities for certain time.

PM2.5 concentrations vary over the course of a day on a global scale based on ground observations data with a morning peak between 7:00 and 10:00 local solar time (LST) and an early night time peak between 21:00 and 23:00 LST, with an afternoon minimum occurring between 15:00 and 17:00 LST [6]. These diurnal cycles are related to dynamics of the atmospheric mixed layer, diurnal emission patterns, and topography. From modeling results, the diurnal variation of anthropogenic emissions and injection height of power plant emissions can affect the diurnal cycle of surface PM2.5 concentrations to some extent. Still, the impact is much smaller than that of PBL mixing [5]. On regional average in China, minimum surface PM2.5 concentration occurred in the afternoon [5].

For the Indonesia region, previous studies focused on Jakarta using Air Quality Monitoring Station data collection. Some studies using PM2.5 data monitored by US Embassy in Central and South Jakarta showed the average value of PM2.5 concentration each year is always greater than 10 $\mu g/m^3$ or above WHO air quality guidelines [7, 7]. PM2.5 in Jakarta was influenced by temperature, NDVI, humidity, and residential area [9]. The distribution of PM2.5 and PM10 during the night tends to be higher than during the day though the traffic road is slower [10]. This condition is due to the reduction of temperature in night then particle diffusion decreases. Based on Ministry of Environment and Forestry report, annual PM2.5 concentrations on several cities also exceeds WHO recommendations [11].

The reanalysis datasets are used to understand the evolution of the state of the atmosphere during the modern observing period, based on available surface and upper-air observations from a variety of satellite instruments and in situ observing systems [12]. By using reanalysis product, we can consider characteristics of a species not only in one site (like ground observations) but also in broader regions with particular spatial and temporal resolution. This research was aimed to investigate diurnal variations of fine particulate matter over Indonesia based on reanalysis data.

63.2 Methodology

63.2.1 Data

We used hourly PM2.5 surface concentrations ($\mu g/m^3$) during 2019–2020 from ground observation in Central Jakarta owned by US Embassy. This station was located at 6°11′S and 106°50′E. This data can be downloaded directly at the *airnow* website. PM2.5 concentrations have been recorded hourly since 21 September 2015. Data was saved in Comma Separated Values files and appended every time there is new update data.

We also used hourly PM2.5 concentrations from global reanalysis data provided by Japan Meteorological Agency (JMA) with spatial resolution 0.375° × (0.37147°–0.37461°) [13]. This product is the forecast (every one hour) of aerosol properties by the MRI/JMA global aerosol model called Model of Aerosol Species IN the Global AtmospheRe (MASINGAR) and it can be downloaded at Japan Aerospace Exploration (JAXA) P-Tree System. MASINGAR is an online chemical transport model for tropospheric aerosol species and is coupled with an atmospheric general circulation model (MRI/JMA GCM 98) and a four-dimensional data assimilation system for atmospheric part [14]. MRI/JMA98 (MJ98) GCM was developed by the Meteorological Research Institute (MRI) and the Japan Meteorological Agency (JMA) based on the JMA global weather forecasting model [14]. Emission inventories used by MASINGAR were derived from some research about emission processes of anthropogenic sulfur compounds and carbonaceous aerosols (see [14] for details).

Another data used in this study was rainfall data which was taken from Statistic Indonesia (BPS). All data were collected during 2019–2020.

63.2.2 Methods

First, we identify the maximum and minimum monthly rainfall accumulation for both years to represent rainy (wet) and dry seasons. Data validation is calculated by binning collocated reanalysis data with ground observation in Central Jakarta using Pearson's Correlation Coefficient. Then we do average hourly data for the month which represent the rainy season and the dry season for both years. This process is useful to evaluate the performance of reanalysis data compared to in situ measurement before doing analysis for broader region (Indonesia).

For reanalysis data, we calculate when PM2.5 reaches maximum (at hour 00, 01, etc.) over the Indonesia region for both seasons. Descriptive analysis was done for diurnal variations of PM2.5, both measured from ground observations and reanalysis. Further analysis is done by investigating space–time variability of ambient PM2.5 diurnal pattern seen with Hovmöller diagram.

63.3 Result and Discussion

63.3.1 Analysis Data from Ground Observation

Rainfall accumulation (mm/month) in Jakarta reached a high value in January (represent rainy season) and minimum value in July (represent dry season) (shown in Fig. 63.1 left). The station which recorded this data is located as shown in the map (Fig. 63.1 right). Though in February 2020, rainfall reaches maxima but we prefer to choose January for the next analysis because on another side, February 2019 has lower rainfall than January 2019.

Hourly PM2.5 in Central Jakarta during 2019–2020 (see Fig. 63.2) has a range concentration between 0 and 228 μg/m^3. PM2.5 in January 2019 have range of 0–92 μg/m^3, January 2020 has wider range, namely 0–142 μg/m^3. Others, PM2.5 range concentrations in July 2019 is 17–146 μg/m^3 and in July 2020 is 9–115 μg/m^3.

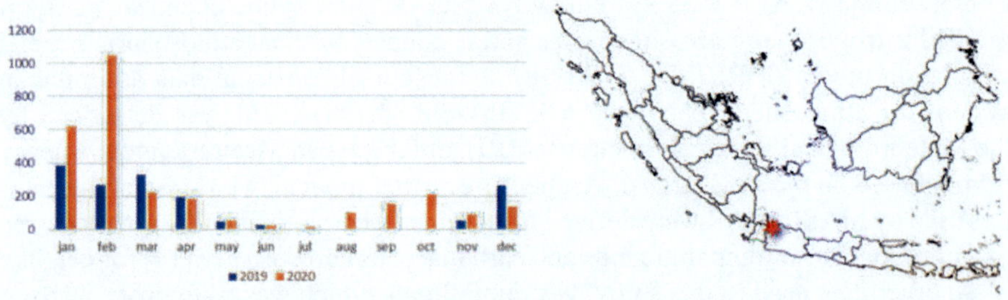

Fig. 63.1 Rainfall accumulation (mm/month) in Jakarta Province (left); PM2.5 instrument location at 6° 11′S and 106° 50′ E (right)

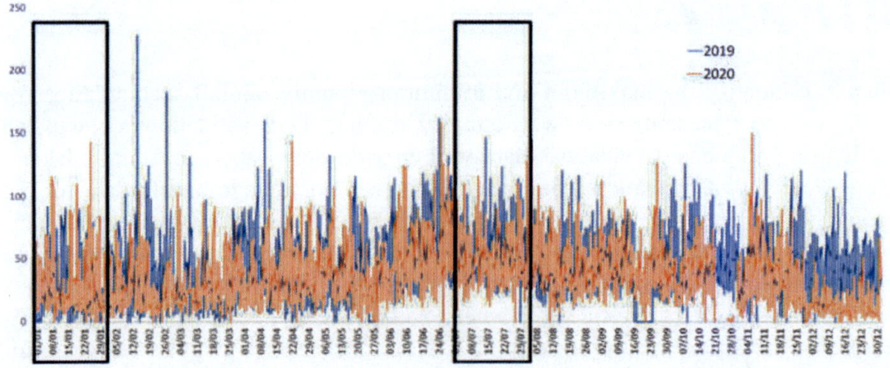

Fig. 63.2 Hourly PM2.5 concentrations Central Jakarta during 2019–2020 (μg/m^3)

63.3.2 Diurnal Variations of PM2.5 Over Central Jakarta

Diurnal variations of PM2.5 over Central Jakarta based on ground observation data are shown in Fig. 63.3. For all months, PM2.5 has a minimum value from midday until early night. This result is following previous findings [8]. These conditions are also shown in Fig. 63.4, which is marked by blue color for every hourly PM2.5 average. The red color indicates higher PM2.5 concentrations compared to other times.

In January 2019, maximum PM2.5 occurred late at night (2 a.m.), then low decreasing for a moment and reached maximum again in the early morning (7 a.m.). It also occurred in January 2020, with the highest PM2.5 is in midnight and early morning (6 a.m.). In July 2019 and 2020, PM2.5 reaches high values at midnight until early morning. From this analysis, we can conclude that in the rainy season, PM2.5 has two maximum or low bimodal pattern. We use the 'low' word due to

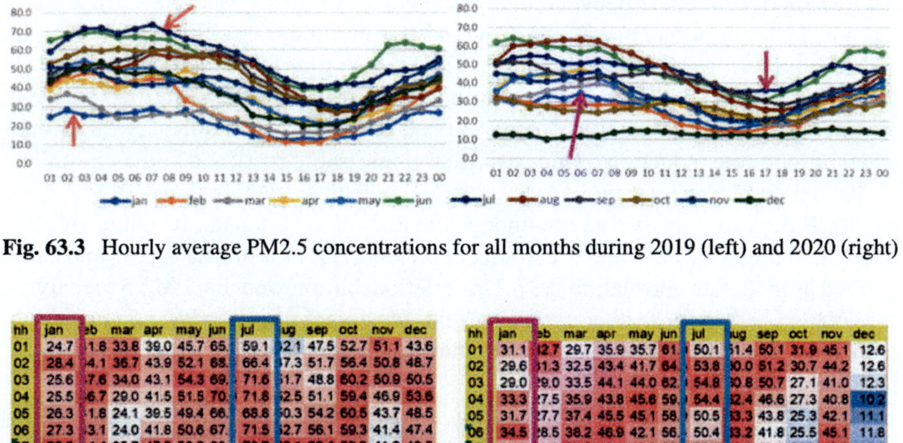

Fig. 63.3 Hourly average PM2.5 concentrations for all months during 2019 (left) and 2020 (right)

hh	jan	eb	mar	apr	may	jun	jul	ug	sep	oct	nov	dec
01	24.7	1.8	33.8	39.0	45.7	65.	59.1	2.1	47.5	52.7	51.1	43.6
02	28.4	4.1	36.7	43.9	52.1	68.	66.4	7.3	51.7	56.4	50.8	48.7
03	25.6	7.6	34.0	43.1	54.3	69.	71.6	1.7	48.8	60.2	50.9	50.5
04	25.5	6.7	29.0	41.5	51.5	70.	71.8	2.5	51.1	59.4	46.9	53.6
05	26.3	1.8	24.1	39.5	49.4	66.	68.8	0.3	54.2	60.5	43.7	48.5
06	27.3	3.1	24.0	41.8	50.6	67.	71.5	2.7	56.1	59.3	41.4	47.4
07	28.8	4.4	25.7	47.5	50.2	66.	73.7	8.1	60.4	59.0	41.6	48.5
08	25.7	4.1	25.9	45.0	44.8	65.	69.2	6.3	59.7	60.5	42.1	47.4
09	28.0	3.1	26.2	49.1	41.8	61.	65.9	6.1	56.9	57.0	43.4	43.7
10	22.2	9.0	28.5	45.3	44.7	56.	63.7	6.2	57.4	57.5	45.8	41.6
11	20.0	4.8	23.5	37.4	38.5	57.	61.5	6.9	59.3	53.8	45.1	36.8
12	16.9	3.1	22.5	34.4	35.1	53.	58.5	4.7	55.3	51.0	42.5	33.9
13	14.6	7.8	20.5	33.8	32.3	52.	54.3	2.2	45.9	43.1	39.2	25.6
14	13.4	2.7	17.8	27.9	28.1	47.	49.1	3.9	41.7	39.0	36.5	24.2
15	12.1	1.1	15.7	22.9	24.1	41.	44.1	9.6	34.9	35.6	32.5	23.1
16	13.0	0.5	16.4	21.3	23.3	39.	40.8	4.1	32.8	33.7	31.6	19.8
17	12.4	0.9	16.2	22.0	20.9	39.	40.3	8.7	32.4	29.5	30.7	20.1
18	13.6	4.8	17.4	20.9	21.9	39.	41.6	6.9	29.0	27.8	29.8	20.2
19	14.0	7.2	18.0	22.8	23.9	46.	41.7	7.6	30.7	28.8	29.8	23.7
20	16.5	1.8	21.4	25.9	29.4	52.	43.0	9.0	31.2	36.5	33.9	32.2
21	19.4	6.0	24.8	28.6	33.4	62.	48.3	2.0	35.6	38.5	39.0	33.6
22	24.6	8.3	28.1	38.3	41.4	64.	49.2	4.7	38.4	40.5	44.4	37.8
23	27.5	5.0	28.8	40.0	42.5	62.	49.9	4.4	39.9	41.7	48.0	39.6
00	26.7	9.8	33.0	40.4	47.0	60.	55.4	9.1	45.8	42.4	53.0	44.7

hh	jan	eb	mar	apr	may	jun	jul	ug	sep	oct	nov	dec
01	31.1	2.7	29.7	35.9	35.7	61.	50.1	1.4	50.1	31.9	45.1	12.6
02	29.6	1.3	32.5	43.4	41.7	64.	53.8	0.0	51.2	30.7	44.2	12.6
03	29.0	9.0	33.5	44.1	44.0	62.	54.8	0.8	50.7	27.1	41.0	12.3
04	33.3	7.5	35.9	43.8	45.6	59.	54.4	2.4	46.6	27.3	40.8	10.2
05	31.7	7.7	37.4	45.5	45.1	58.	50.5	3.3	43.8	25.3	42.1	11.1
06	34.5	8.5	38.2	46.9	42.1	56.	50.4	3.2	41.8	25.5	45.1	11.8
07	31.6	8.3	39.5	45.4	42.6	58.	52.1	2.6	43.1	24.4	46.3	11.6
08	30.0	8.6	38.1	40.6	39.0	54.	50.6	8.8	44.0	26.0	41.6	13.7
09	29.2	7.0	31.7	33.2	33.5	48.	49.0	6.2	44.2	26.4	36.7	14.9
10	28.7	6.0	28.6	30.3	30.6	46.	49.1	1.6	45.3	27.7	32.4	14.6
11	24.5	2.8	29.4	29.8	32.7	43.	49.5	8.6	44.8	29.4	31.2	14.4
12	21.5	8.1	28.5	28.3	30.6	40.	45.7	2.4	42.4	30.3	30.8	13.2
13	20.2	6.7	24.6	27.1	26.2	40.	43.7	6.2	41.5	27.9	26.3	12.6
14	16.7	4.3	23.6	23.9	21.2	37.	37.8	4.9	39.6	26.9	22.0	12.1
15	15.3	4.5	22.5	23.2	20.4	34.	35.5	0.6	35.4	24.3	20.6	12.5
16	17.4	4.2	20.2	21.2	18.9	31.	35.5	8.2	33.4	22.8	20.2	12.5
17	18.3	6.1	18.5	21.5	19.1	30.	36.3	5.9	30.6	19.3	21.1	12.4
18	19.6	7.8	17.7	22.4	22.5	34.	38.1	5.6	28.2	21.6	22.0	12.4
19	19.8	7.2	19.3	23.3	25.3	37.	40.0	7.3	30.3	19.8	27.2	13.2
20	24.9	2.8	22.8	26.8	29.0	41.	45.1	9.7	33.0	24.0	31.8	15.0
21	26.2	4.7	24.5	30.3	32.5	48.	49.8	1.5	36.1	25.4	34.1	15.7
22	31.0	6.0	26.5	31.9	33.4	56.	49.3	5.2	37.9	28.9	37.0	14.9
23	33.0	8.4	29.1	34.2	35.8	57.	45.9	8.1	42.3	26.7	36.2	14.5
00	33.1	2.8	30.3	37.1	38.3	56.	47.5	4.8	47.1	28.7	41.4	13.6

Fig. 63.4 Hourly average PM2.5 concentrations in μg/m^3 on 2019 (left) and 2020 (right); the red highlight color indicates a higher PM2.5 concentration throughout the day, otherwise the blue highlight color indicates a lower PM2.5 concentration throughout the day

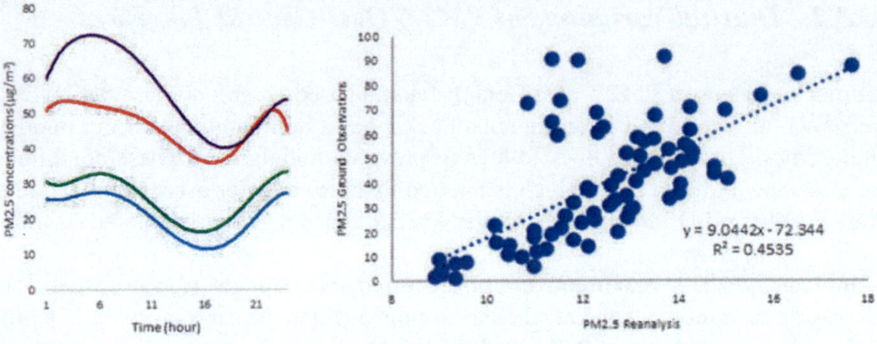

Fig. 63.5 Diurnal variations of PM2.5 concentrations ($\mu g/m^3$) in Central Jakarta in January 2019 (blue line), January 2020 (green line), July 2019 (purple line), and July 2020 (red line) (left); Scatter plot PM2.5 concentrations ($\mu g/m^3$) reanalysis data collocated with ground observation (right)

the less sharp difference between the two maxima. In the dry season, PM2.5 has a unimodal pattern or one maximum peak and one minimum (see Fig. 63.5 left).

In January (Fig. 63.5 left), both years had the similar low bimodal pattern with PM2.5 on 2020 is slightly greater than on 2019. In July, PM2.5 in 2020 is much lower than 2019. This may be due to COVID-19 Large-Scale Social Restrictions, which implies the reduction of industrial activities, traffic road, and human mobility.

Validation between ground observation and reanalysis data is shown in Fig. 63.5 (right). PM2.5 reanalysis data are underestimate with in situ data. Its range value is nearly five times lower. The correlation coefficient (r) for these data is 0.67 and categorized as moderate correlation [15]. This relationship showed that PM2.5 reanalysis data could be used for further analysis. Next, we do not consider PM2.5 concentrations quantitatively since reanalysis products has a range value much smaller than in situ observation.

For reanalysis data, diurnal variations of PM2.5 in Central Jakarta on January have the low bimodal pattern with night and morning peaks and minimum at noon (see Fig. 63.6). Otherwise, in July, based on reanalysis data, showed a unimodal pattern with night maximum and noon minimum. This result is in accordance with the result based on ground observation data. Higher PM2.5 at night is related to minimum temperature [16], which can reduce particle diffusion over the surface area. But we should consider sensitivity of atmospheric model output due to low mixing height.

63.3.3 Diurnal Variations Over Indonesia

Figure 63.7 shows at what hour is PM2.5 concentrations reach maximum (00–23 UTC). The further analysis, we use Local Time (LT) or UTC+7. In January 2019, generally the maximum peak of PM2.5 occurred at early night 7–9 p.m. LT (green color). In January 2020, there are more variations; for, example in Central and East

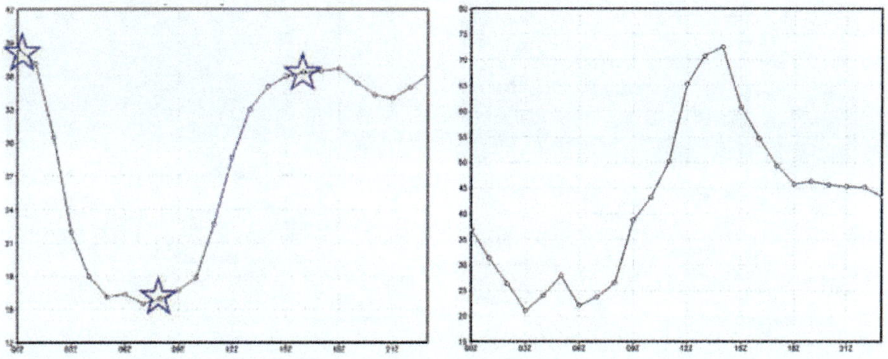

Fig. 63.6 Diurnal variations of PM2.5 in Central Jakarta ($\mu g/m^3$) derived from reanalysis data on January (left) and July (right). The X-axis is time in 24 h UTC start from 00 UTC—23 UTC

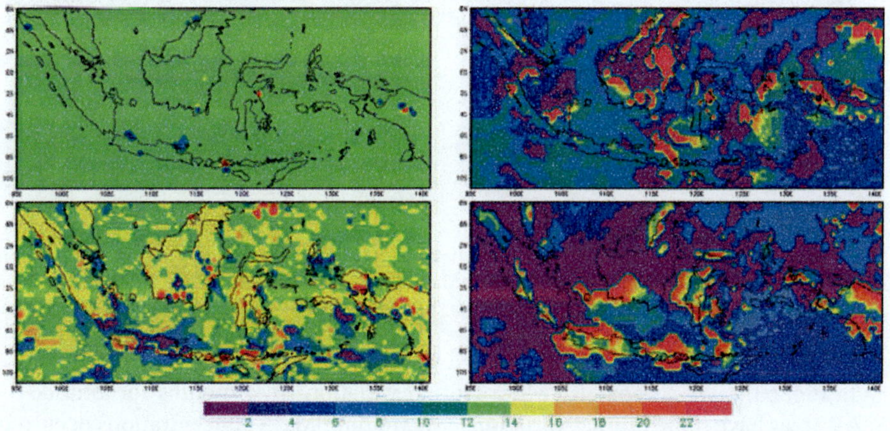

Fig. 63.7 Maximum time (UTC) for hourly PM2.5 average concentrations January 2019 (upper left), July 2019 (upper right), January 2020 (lower left), and July 2020 (lower right)

Java (purple color) have the highest PM2.5 in the morning (7–9 a.m. LT), but other areas reach maximum at night (yellow color). In July, both years have different times show the highest PM2.5 concentrations.

To investigate diurnal variations of PM2.5, we used the Hovmoller Diagram time–longitude (constant latitude) for two areas: megapolitan (Jakarta) and fire prone area (Ogan Komering Ilir). Y-axis is time in format 00–23 UTC. In Fig. 63.8, we compared PM2.5 diurnal variations in 4 months for latitude 6°12′ S. In the noon, PM2.5 has a minimum value and reaches a maximum value at night. The highest PM2.5 concentrations are found at Jakarta (Longitude 106° E) and its surrounding areas. As the capital of Indonesia, Jakarta has many anthropogenic activities, mainly industry and transportation, resulting in high emissions and poor air quality than other cities.

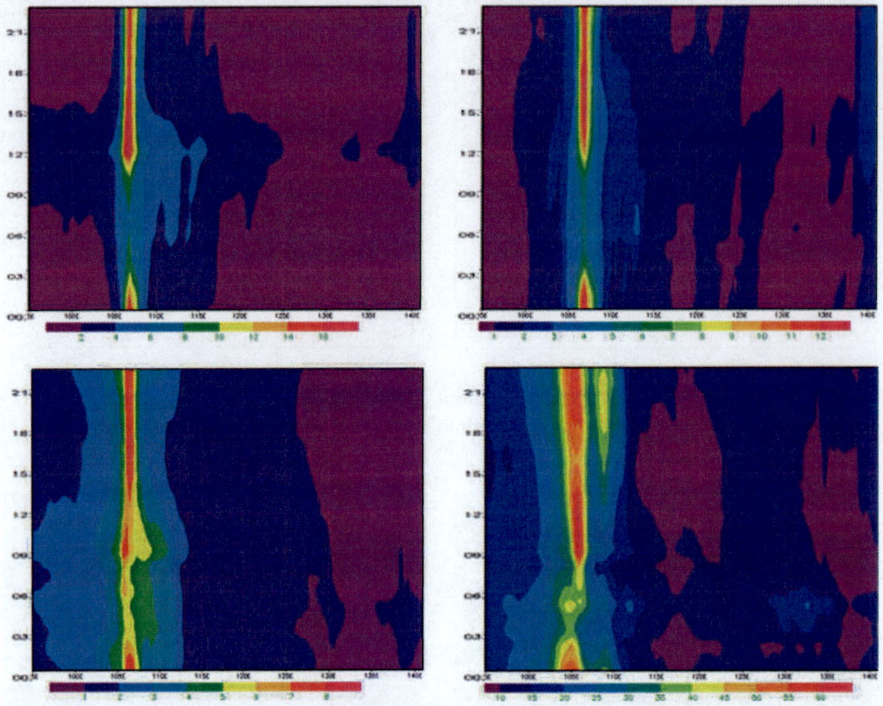

Fig. 63.8 Hovmoller diagram of diurnal variations of PM2.5 concentrations at latitude 6.2° S in January 2019 and January 2020 (upper figures), July 2019 and July 2020 (lower figures). Y-axis is time (hour in UTC) and X-axis is longitude

In Fig. 63.9, we compared PM2.5 diurnal variations in 4 months for latitude 3° S. All areas have more diurnal variations. Maximum PM2.5 concentrations occurred at South Sumatera (104° E) and Central—South Kalimantan (115° E). Smoke from biomass burning is the main cause of poor air quality for those areas. The forest and land fires have been recurring almost every year in Indonesia particularly in peat land.

63.4 Conclusion

Hourly reanalysis PM2.5 fitted enough to ground observation data ($r = 0.67$). PM2.5 reanalysis is underestimated comparing to ground observation data. Both reanalysis and ground PM2.5 data showed a unimodal pattern in the dry season, with a maximum peak occurred at midnight and an afternoon minimum. PM2.5 cycles in the rainy season showed a low bimodal pattern with a night peak, a morning peak, and an afternoon minimum. The PM2.5 concentration from morning to afternoon in July

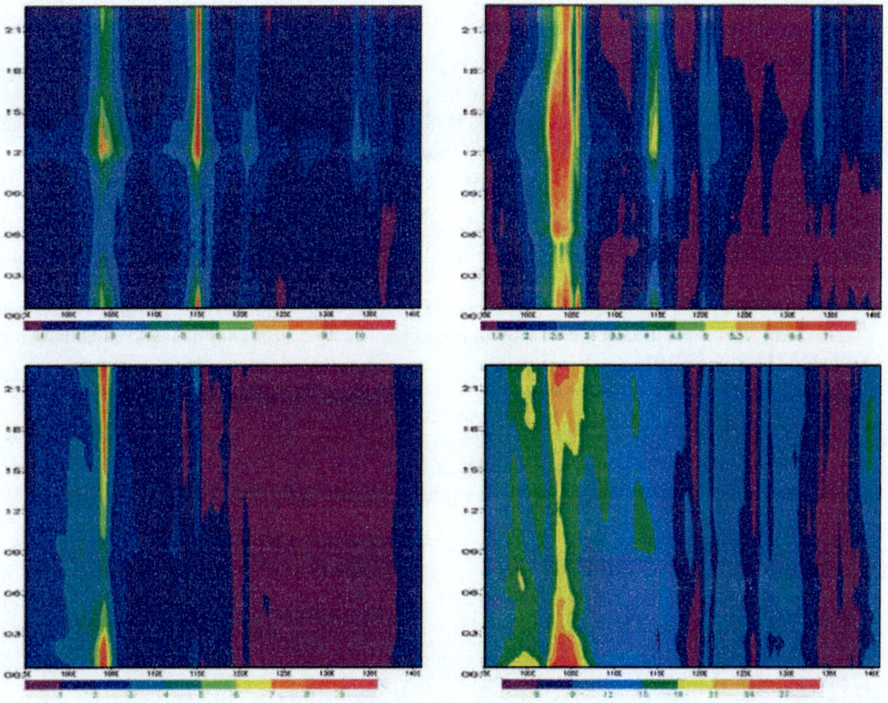

Fig. 63.9 Hovmoller diagram of diurnal variations of PM2.5 concentrations at latitude 3° S in January 2019 and January 2020 (upper figures), July 2019 and July 2020 (lower figures). Y-axis is time (hour in UTC) and X-axis is longitude

2020 related to the lockdown policy (COVID19 pandemic) is lower compared to the same month in 2019.

References

1. Feng, S., Gao, D., Liao, F., Zhou, F., Wang, X.: The health effects of ambient PM2.5 and potential mechanisms. Ecotoxicol. Environ. Saf. **128**, 67–74 (2016)
2. WHO Homepage, https://www.who.int/news-room/fact-sheets/detail/ambient-(outdoor)-air-quality-and-health. Last accessed 2021/10/06
3. World Health Organization: health effects of particulate matter, policy implications for countries in Eastern Europe, Caucasus and Central Asia. WHO Regional Office for Europe, Denmark (2013)
4. World Health Organization. WHO global air quality guidelines: particulate matter (PM2.5 and PM10), ozone, nitrogen dioxide, sulfur dioxide and carbon monoxide. World Health Organization (2021). https://apps.who.int/iris/handle/10665/345329
5. Du, Q., Zhao, C., Zhang, M., Dong, X., Chen, Y., Liu, Z., Hu, Z., Zhang, Q., Li, Y., Yuan, R., Miao, S.: Modeling diurnal variation of surface PM2.5 concentrations over East China with

WRF-Chem: impacts from boundary-layer mixing and anthropogenic emission. Atmos. Chem. Phys. **20**, 2839–2863 (2020)
6. Manning, M.I., Martin, R.V., Hasenkopf, C., Flasher, J., Li, C.: Diurnal patterns in global fine particulate matter concentration. Environ. Sci. Technol. Lett. **5**(11), 687–691 (2018)
7. Alfiandy, S., Nurjaman, A.: Analisis Pola Konsentrasi PM 2.5 di Wilayah Jakarta Selatan dan Jakarta Pusat Periode Data Januari 2016–Desember 2017. 4 2019. 16–20 (2019)
8. Hutauruk, R.C.H., Rahmanto, E., Pancawati, M.C.: Variasi Musiman dan Harian PM2.5 di Jakarta Periode 2016–2019. Bull. GAW Bariri **1**(1), 20–28 (2020)
9. Kusuma, W.L., Wu, C.D., Zeng, Y.T., Handayani, H.H., Jaelani, L.M.: PM2.5 pollutant in Asia—a comparison of metropolis cities in In-donesia and Taiwan. Int. J. Environ. Res. Public Health **16**. MDPI AG: 4924 (2019)
10. Sari, I.R.J., Fatkhurrahman, J.A., Andriani, Y.: Pola Sebaran Polutan PM2.5 dan PM10 Harian Terhadap Faktor Suhu Dan Kelembaban. Prosiding SNST ke-10. Fakultas Teknik Universitas Wahid Hasyim (2019)
11. KLHK Official Website, https://ditppu.menlhk.go.id/portal/read/kondisi-kualitas-udara-di-beberapa-kota-besar-tahun-2019. Last accessed 2021/11/29
12. Poli, P., Hersbach, H., Dee, D.P., Berrisford, P., Simmons, A.J., Vitart, F., Laloyaux, P., Tan, D.G.H., Peubey, C., Thépaut, J., Trémolet, Y., Hólm, E.V., Bonavita, M., Isaksen, L., Fisher, M.: ERA-20C: an atmospheric reanalysis of the Twentieth Century. J. Clim. **29**(11), 4083–4097 (2016)
13. JAXA Homepage, https://www.eorc.jaxa.jp/ptree/userguide.html. Last accessed 2021/10/06
14. Tanaka, T.Y., Sekiyama, T.T., Kiyotaka, S., Chiba, M.: MASINGAR, a global tropospheric aerosol chemical transport model coupled with MRI/JMA98 GCM: model description. Meteorol. Geophys. **53**(4), 119–138 (2003)
15. Schober, P., Boer, C., Schwarte, L.A.: Correlation coefficients: appropriate use and interpretation. Anesth. Analg. **126**(5), 1763–1768 (2018)
16. Siswanto, Van Oldenborgh, G.J., Schrier, G., Jilderda, R., Hurk, B.: Temperature, extreme precipitation, and diurnal rainfall changes in the urbanized Jakarta city during the past 130 years. Int. J. Climatol. **36**, 10 (2015)

Chapter 64
Weather Prediction System Using Thomas-Fiering Model to Determine Initial Planting Recommendations in Bandung City

Candra Nur Ihsan and Nova Agustina

Abstract Bandung City is the largest metropolitan city in West Java and is the capital city of the province. The phenomenon that has occurred in the last few years is that the city of Bandung has failed to harvest, resulting in the price of agricultural products increasing. The major impact of changing seasons needs to be minimized, one of which is to make weather predictions to recommend carly planting. In this study, researchers made weather predictions consisting of temperature, humidity and rainfall using the Thomas-Fiering model to determine the start of planting in 2022. The evaluation was conducted using the homogeneity test (T-test) to compare two different population samples, namely the training data population and the prediction data. As a result, both data have the same variance (homogeneous). After the weather prediction data is obtained, the initial planting recommendations can be applied based on the criteria for temperature, humidity, rainfall, and length of time for planting. The results of the initial planting recommendations were successfully visualized in the form of a Planting Calendar.

64.1 Introduction

Various studies observe the impact of climate change affecting different sectors, for example, the vegetation [1], property [2], to crops [3]. Uncertain climate change is a phenomenon that is currently being discussed, thus attracting the attention of researchers to examine various techniques to minimize the negative impacts of climate change. The Massachusetts General Hospital (MGH) Institute of Health Professions revealed that the combined effects of higher temperatures, increased

C. N. Ihsan (✉)
Research Center for Climate and Atmosphere, National Research and Innovation Agency (BRIN), Bandung, Indonesia
e-mail: candra.nur.ihsan@brin.go.id

N. Agustina
Department of Informatics, Sekolah Tinggi Teknologi Bandung, Bandung, Indonesia

E. Yulihastin et al. (eds.), *Proceedings of the International Conference on Radioscience, Equatorial Atmospheric Science and Environment and Humanosphere Science, 2021*, Springer Proceedings in Physics 275, https://doi.org/10.1007/978-981-19-0308-3_64

rainfall volatility, and extreme weather adversely affect crops [4]. In the extreme, climate change can result in total crop failure.

The climate in Indonesia is affected by three types, monsoon climate, marine climate and tropical climate. Tropical climate combined by fertile soil makes the wealth of natural resources in Indonesia will be abundant if it can be managed properly. Natural wealth that is used properly will have a good impact on Indonesian economic growth. The agricultural sector is one of growth factor for regional economy [5]. Almost all land in various areas can be used for farming. This argument is backed by data taken from the Indonesian Central Statistics Agency (BPS), the area for rice fields in Indonesia on 2019 reached 7,463,948 ha. [6]. According to Prihatin in his research showed that currently the city of Bandung has very small agricultural land remaining, just a few hectares located on east Bandung region [7].

The diminish in agricultural land has made the Mayor of Bandung, Oded M. Danial, take steps to take full advantage of the remaining land to plant in the remaining land, even in the yards of the residents of Bandung City. The Mayor of Bandung formed a new lifestyle for the community, especially when the situation was difficult in dealing with Covid-19 by forming 190 Farmer Groups (Poktan) spread over 151 urban villages [8]. However, agricultural production management still needs to be done more seriously, considering that many factors will affect the success of a farming process. One of the factors that affect the success of farming is the conditions of temperature, humidity and rainfall.

The problem is the phenomenon that occurs when the seasons in Indonesia experience erratic changes. The impact of seasonal changes affects agricultural production and national food security [9]. For example, in January 2021, the price of cayenne pepper in Bandung increased by 100% due to crop failure [10]. Rapid price increases can be detrimental to consumers and entrepreneurs who always use chili in their sales. To minimize adverse impacts on harvest caused by uncertain seasonal change, early planting predictions can be made based on future weather predictions[11, 11].

Weather data has been provided by the Meteorology, Climatology and Geophysics Agency (BMKG). The city of Bandung has the main agricultural production, such as: Spinach, Cayenne Pepper, Red Beans, Water Spinach, Cauliflower, Cucumber [13]. Considering the water content required for planting seeds varies, some can be gradual with a little water or need a lot of water to survive, it is necessary to do planting predictions combined with weather predictions. Planting prediction can be done by knowing future weather conditions, but before making planting prediction, first we must predicting the supporting commodities for planting predictions, namely rainfall, humidity and temperature [14]. The Thomas-Fiering model can be used to make these predictions [15].

In summary, our contributions to this research are: (1) We will predict rainfall, humidity and temperature in 2022 using the Thomas-Fiering model; (2) Predictions are taken from historical data on the amount of rainfall, average humidity, and average temperature per month obtained from the Bandung BMKG Geophysics Station for 40 years (from 1980 to 2020) used as training data; (3) A prediction system is built on a web-based application; (4) Predicted farm commodities are: Spinach, Cayenne Pepper, Red Beans, Wet Spinach, Cauliflower, Cucumber, and Shallots; (5)

Predictions will be displayed in the form of a planting calendar based on the amount of rainfall, average humidity, and average temperature.

64.2 Data and Method

In this study, we took data on temperature, humidity, and rainfall from the BMKG (https://dataonline.bmkg.go.id/) "UPT-96783-Stasiun Geofisika Bandung" located in Bandung. Training data are taken from January 1, 1980, to December 31, 2020 (40 years). We reprocess the data taken from the BMKG to make data on average temperature, average humidity, and the amount of monthly rainfall. An example of the data we took from the BMKG website can be seen in Table 64.1.

In addition to input data (temperature, humidity, and rainfall) we also collect standard data on temperature, humidity, and rainfall for food crops growing in the city of Bandung (Spinach, Cayenne pepper, Red Beans, Water Spinach, Broccoli, Cucumber) as supporting data. The data is taken from the Center for Research and Development of Agricultural Land Resources (BBSDLP). The data can be seen in Table 64.2.

From the data on average temperature, average humidity and total monthly rainfall in the previous year, we will make predictions for 5 years, then prediction data in 2022 will be used as a reference to predict the start of planting in that year based on temperature, humidity standards and Monthly Rainfall of each plant. The method

Table 64.1 Data from BMKG on average temperature, average humidity, and monthly rainfall in Bandung

	Temperature (°C)	Humidity (%)	Rainfall (mm)
Jan 1980	22.6	81.6	119
Feb 1980	23.1	76.5	58
Mar 1980	22.6	80.9	393
Apr 1980	23.1	81.1	274
...	...	...	...
Dec 2020	23.7	77.5	261.8

Table 64.2 Standard of average temperature, average humidity, and total monthly precipitation [16]

	Temperature (°C)	Humidity (%)	Rainfall (mm)	Harvest time (month)
Spinach	20–32	24–90	100–200	1
Cayenne pepper	18–30	60–85	100–250	3
Red Beans	16–25	70–90	50–250	2
Water Spinach	15–30	>60	100–250	1
Broccoli	12–27	80–90	100–200	2
Cucumber	12–27	42–90	200–400	2

used to build the system is Extreme Programming (XP) [17] method which consists of planning, design, coding, testing stages, the last is software increment.

64.2.1 *Planning*

To find out the pattern of temperature, humidity and monthly rainfall in 2020 to 2025, we use the Thomas-Fiering model to be implemented in the system. The first step is to determine the average temperature, humidity, and monthly rainfall in the number of years (N) that will be predicted. The steps for predictions using Thomas-Fiering are:

a. Determine the Average of the data for each month (j) of the year (i)

$$\overline{R} = \frac{\sum i R_{ji}}{N} \tag{64.1}$$

b. Determine the Standard Deviation (S) grouped each month (j).

$$Sj = \sqrt{\frac{\sum i(R_{ji} - R_j)^2}{(N-1)}} \tag{64.2}$$

c. Determining the Correlation Coefficient (r_j)

$$r_j = \frac{\sum i(R_{ji} - \overline{R}_j)(R_{j+1,i} - \overline{R}_{j+1})}{\sqrt{\{\sum i(R_{ji} - \overline{R}_j)^2 . \sum i(R_{j+1,i} - \overline{R}_{j+1})^2\}}} \tag{64.3}$$

d. Determining the Regression Coefficient (b_j)

$$b_j = \frac{r_j S_{j+1}}{S_j} \tag{64.4}$$

e. Generating Normal Random Numbers (Zi)

$$\mathrm{f(x)} = \frac{1}{\sigma\sqrt{2\pi}} e^{\frac{-(x-\mu)^2}{2\sigma^2}} \mathrm{f(x)} = \frac{1}{\sigma\sqrt{2\pi}} e^{\frac{-(x-\mu)^2}{2\sigma^2}} \tag{64.5}$$

where π is mean of x, σ is standard deviation of x, $\pi = 3.14$, $e = 2.71$.

f. Calculate Thomas-Fiering

$$R_{i+1} = \overline{R}_{j+1} + b_j(R_i - \overline{R}_j) + Z_i S_{j+1}\sqrt{1 - r_j^2} \cdots . \tag{64.6}$$

Table 64.3 Oldeman classification [18]

Rainfall (mm)	Classification
>200	Wet
100–200	Humid
<100	Dry

Table 64.4 Calendar of early planting recommendations in Bandung City in 2022

Commodity	Month											
	Jan	Feb	Mar	Apr	Mei	Jun	Jul	Aug	Sept	Oct	Nov	Dec
Spinach				■	■	■			■			■
Cayenne Pepper				■								
Red Beans				■	■							■
Water Spinach				■	■	■			■			■
Brocolli				■	■							
Cucumber	■	■									■	
■	Initial Planting Recomendation											

In addition to specifying a model for making predictions, we use the Oldeman Climate Classification to define monthly criteria. Monthly criteria in the Oldeman classification are determined based on the amount of rainfall. In Table 64.3 we can see the Oldeman classification rules.

64.2.2 Design

The system we built generates initial planting recommendations based on the predicted temperature, humidity, and rainfall. First, the system will predict air temperature, humidity, and rainfall. After that, the system will recommend the start of planting based on the predictions of air temperature, humidity, and rainfall. The data we get from BMKG, which has been converted into a monthly average, we store in the database server. In Fig. 64.1, we can see that the system will retrieve data from the database. Then, the system will calculate the standard deviation, correlation coefficient, and regression as the parameters needed for forecasting. In addition to calculating these parameters, the system can also generate random values. If the required parameters and random values have been developed successfully, the calculation of the Thomas-Fiering model can start. The forecast result will be tested using a T-test, and if it passes, then a planting calendar prediction can be made. If not, the system must regenerate other random values.

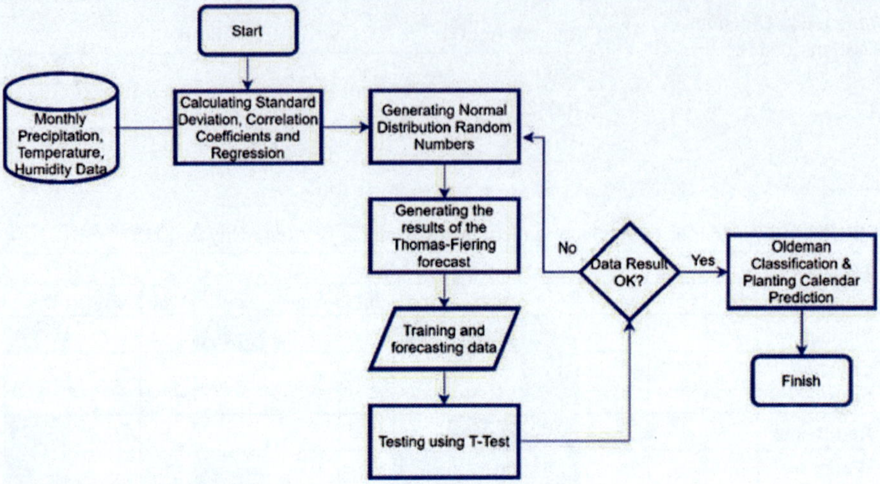

Fig. 64.1 Design system

64.2.3 Coding

The programming language we use is Vue JS, and MySQL as a DBMS (Database Management System) to run SQL commands to access base data.

64.2.4 Testing

In addition to testing the system using the black-box test, model evaluation was also carried out to measure the significance level of an analysis of temperature, humidity, and rainfall predictions. The evaluation technique is to test the Thomas-Fiering prediction results using a T-test for the same variance because we have to prove that if the prediction results come from the same population as the training data, this is done using the Polled Variance formula:

$$t = \frac{\overline{X_1} - X_2}{\sqrt{\frac{(n_1-1)S_1^2+(n_2-1)S_2^2}{n_1+n_2-2}\left(\frac{1}{n1}+\frac{1}{n2}\right)}} \quad (64.7)$$

$\overline{X_1}$ = average in the first variant,
X_2 = average in the second variant,
n_1 = number of samples in the first variant,
n_2 = number of samples in the second variant,
S_1^2 = standard deviation of the first variant, and
S_2^2 = standard deviation of the second variant.

In the case of this study, the first variant is the training data (temperature, humidity, and rainfall for 40 years), and the second variant is the prediction data for five years.

64.3 Result and Discussion

We have successfully implemented functions to predict temperature, humidity, and rainfall in the system. We tested the system using a black box test with the results that all functionalities on the system functioned well. In addition, we also compare the results of calculations on the system with our manual calculations to test the accuracy of the model calculations. After the model is applied, a T-Test is performed to measure the significance level of analysis. As a result, calculations using the Thomas-Fiering model on the system are also in accordance with the expected results. Figure 64.2 visualizes the results of the initial planting recommendation and monthly season in Bandung city after calculating the Thomas-Fiering model.

After the predictions of temperature, humidity and rainfall were obtained, we made predictions of initial planting recommendations based on these three categories for spinach, cayenne pepper, kidney beans, kale, broccoli, and cucumber. In Table 64.2, we visualize a planting calendar that illustrates the initial recommendations for planting information. Overall, almost all months of 2022 can be planted, except for July and August which are classified as dry months.

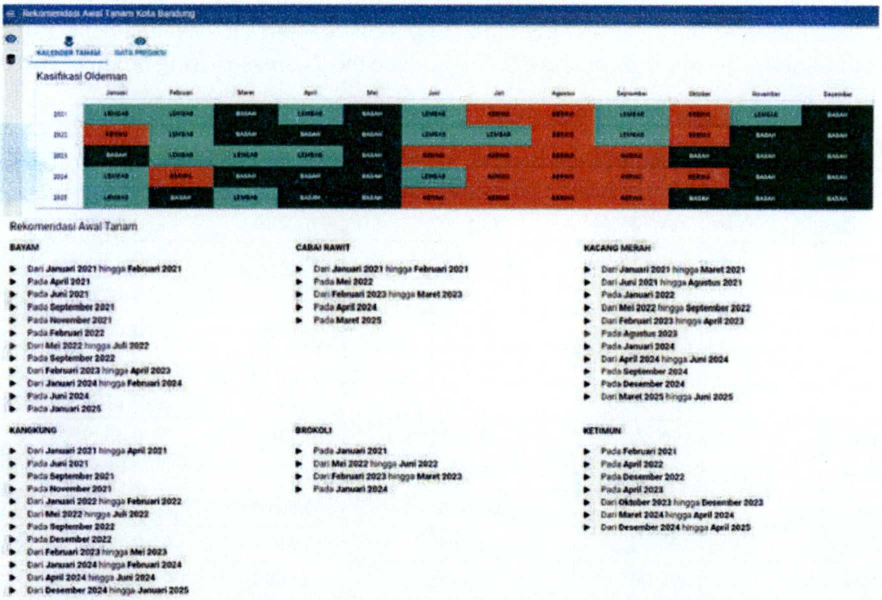

Fig. 64.2 System implementation, showing planting recomendation

To maintain good productivity of agricultural products, farmers need to periodically evaluate the availability of water to maintain the conditions needed by plants. In fact, if farmers want to plant in a month that is not recommended, farmers can still grow crops but need to pay attention to substituting the necessary needs with additional support systems, such as irrigation to manage water availability or greenhouses to maintain temperature and humidity.

64.3.1 Thomas-Fiering Predictions and Oldeman Classification

Thomas-Fiering is used to predict future temperature, humidity, and rainfall. The training data that we use are taken from data on temperature, humidity, and rainfall from 1980 to 2020. In this subsection, we take an example of calculating rainfall predictions. Table 64.5 shows the calculated average (R), Standard Deviation (S_j), Correlation Coefficient (r_j) and Regression Coefficient (b_j).

After the mean, standard deviation, relation coefficient and regression are obtained, the next process is to generate Normal Random Numbers (Zi) as another input data to Thomas-Fiering model. We use the same Normal Random Number for temperature, humidity and precipitation predictions. Generated Random Normal Numbers can be seen in Table 64.6.

The next step is to generate monthly temperature, humidity and precipitation predictions using Thomas-Fiering. This prediction will be the basis for the initial determination of the planting of certain commodities (spinach, cayenne pepper, red beans, kale, broccoli, and cucumber). To produce the Thomas-Fiering prediction, the

Table 64.5 Calculation results

Month	$\overline{R}$	S_j	r_j	b_j
January	214.725	89.8097	−0.0299	−0.0371
February	204.115	111.6016	0.2256	0.225
March	288.485	111.3003	0.136	0.1055
April	256.13	86.2965	−0.152	−0.1485
May	185.625	84.2634	0.006	0.0041
June	89.0475	57.693	0.1114	0.1168
July	70.43	60.4919	0.5769	0.4539
August	52.765	47.6015	0.4512	0.8592
September	88.6175	90.6531	0.4818	0.6018
October	175.7375	113.2335	0.4127	0.5108
November	290.095	140.1443	0.3698	0.3341
December	259.5775	126.6201	−0.111	−0.0787

Table 64.6 Normal random numbers (Z_i) rainfall in 2022

Month	Z
January	−1.3027
February	0.09155
March	1.21552
April	0.52860
May	0.214366
June	0.576589
July	0.300070
August	−0.632942
September	0.333125
October	−1.04106
November	1.535738
December	−0.285393

formula $R_{i+1} = \overline{R}_{j+1} + b_j\left(R_i - \overline{R}_j\right) + Z_i S_{j+1}\sqrt{1 - r_j^2} \cdots$.. is used, An example of applying the Thomas-Fiering formula for January 2022 rainfall predictions:

$$R_{jan2022} = 214.725 + (-0.0371 * (558.5 - 259.5775)) + \left(-1.3027 * 89.8097\sqrt{1 - (-0.0299)^2}\right) = 86.6$$

The value of 558.5 is obtained from the prediction results for December 2021, which is 1 month before January 2022 (R_i) using Thomas-Fiering. Predicting temperature and humidity also performs the same steps as predicting rainfall. The results can be seen in Fig. 64.3, the X-Axis representing Temperature, Humidity, and Rainfall. The Y axis represents the predicted value generated using Thomas-Fiering in 2022. In general, temperatures have increased from January to December with a predicted value range of 22–23 °C. Uniquely, humidity and rainfall have the same graphic pattern, for example when humidity increases, rainfall also increases. The value generated on humidity using Thomas-Fiering in 2022 is in the range of 70–85, rainfall is in the range of 50–450.

After the Thomas-Fiering prediction value is obtained, the last step is to predict the type of monthly based on rainfall using the Oldeman Classification. Complete data on predictions of temperature, humidity, rainfall, and Oldeman's classification can be seen in Table 64.7 which predicts that in 2022 Bandung will be dominant in wet and humid months.

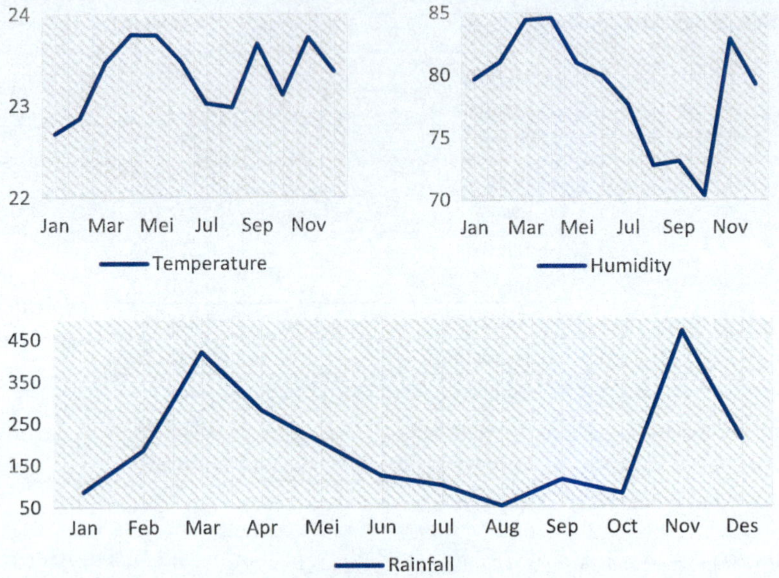

Fig. 64.3 Monthly temperature, humidity, and rainfall forecast in 2022

Table 64.7 Thomas-Fiering prediction table and Oldeman classification for 2022

	Temperature (°C)	Humidity (%)	Rainfall (mm)	Oldeman classification
January	22.69	79.66	86.68	Dry
February	22.86	81.04	185.26	Humid
March	23.46	84.40	420.52	Wet
April	23.77	84.56	281.61	Wet
May	23.77	80.97	203.79	Wet
June	23.48	79.96	124.22	Humid
July	23.03	77.65	101.22	Humid
August	22.98	72.75	52.339	Dry
September	23.67	73.1	114.82	Humid
October	23.12	70.35	81.74	Dry
November	23.74	82.9	458.65	Wet
December	23.37	79.2	210.39	Wet

64.3.2 Thomas-Fiering Evaluation

The evaluation of the Thomas-Fiering model was carried out using a 5% T-test, with a total of 45 years of data (40 years of training data, 5 predictive data). Thus, the value of T-table (a) is 2.01063. Next, we calculate t-count to determine whether the predicted results of temperature, humidity, and rainfall generated using the Thomas-Fiering model are acceptable. The results of the tests that have been carried out can be seen in Table 64.8, with the benchmark t-count must be smaller than the t-table value (a).

Judging from the test results obtained, we conclude that predictions using the Thomas-Fiering model for temperature, rainfall, and humidity get acceptable results according to the T-test.

To test the model's suitability with the actual data, we compared the predicted results of Thomas-Fiering with the observational data taken from the BMKG data. We divided the data into two groups from the 40 years of data that we took, namely training data and testing data. The scenario is that 35 years are used as training data (1980–2015), and five years are used as test data (2016–2020). Overall, Thomas-Fiering's predictions produce a pattern similar to the observed data. For example, we visualize the comparison pattern of predictions and observations of rainfall data in 2020, which can be seen in Fig. 64.4.

Table 64.8 T-test result

Month	Temperature	Humidity	Rainfall	Result
January	0.08	0.60	0.91	Accepted
February	0.07	0.22	0.25	Accepted
March	0.17	0.58	0.71	Accepted
April	0.28	1.01	0.82	Accepted
May	0.02	1.02	1.51	Accepted
June	0.03	0.23	0.06	Accepted
July	0.1	0.17	0.46	Accepted
August	0.12	0.33	0.7	Accepted
September	0.05	0.09	0.04	Accepted
October	0.24	0.57	0.68	Accepted
November	0.31	0.92	1.73	Accepted
December	0.39	0.83	1.01	Accepted

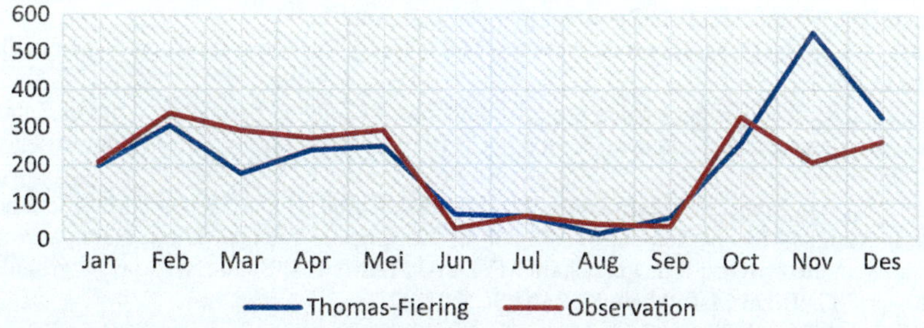

Fig. 64.4 Comparison between Thomas-Fiering prediction and observation in 2020

64.4 Conclusion

Based on the results of the research presented, predictions of temperature, humidity, and rainfall using the Thomas-Fiering model proved to have acceptable predictive results, as evidenced by the acceptance of all prediction results on the T-test which showed that the predicted data were in the same population as the training data. In addition, the application of Thomas-Fiering and the planting calendar to the system produces an analysis that is in accordance with the results of the analysis on manual calculations, so that the system has met the needs of future analysis by using the Thomas-Fiering model for prediction of planting recommendations.

For future research, we suggest predicting temperature, humidity and rainfall in other areas to expand the supporting area that can be given initial planting recommendations. Another suggestion is to compare the Thomas-Fiering Model with other predictive models, e.g., ARIMA, ARIMAX, etc.

References

1. Wang, X., Guo, N., Wang, J., Yang, J.: Impacts of the climate change on the vegetation in Maqu County in the upper reaches of Yellow River. In: International Geoscience Remote Sensing Symposium, pp. 3412–3416 (2007). https://doi.org/10.1109/IGARSS.2007.4423578
2. Akpodiogaga-a, P., Odjugo, O.: General overview of climate change impacts in Nigeria. J. Hum. Ecol. **29**(1), 47–55 (2010). https://doi.org/10.1080/09709274.2010.11906248
3. Tebaldi, C., Lobell, D.B.: Towards probabilistic projections of climate change impacts on global crop yields. Geophys. Res. Lett. **35**(8), 2–7 (2008). https://doi.org/10.1029/2008GL033423
4. Surmaini, E., Runtunuwu, E.: "Upaya sektor Pertanian dalam Menghadapi Perubahan Iklim", Upaya Sekt. Pertan. dalam Menghadapi Perubahan Iklim **30**(1), 1–7 (2015). https://doi.org/10.21082/jp3.v30n1.2011.p1-7
5. Nadziroh, M.N.: Peran Sektor Pertanian Dalam Pertumbuhan Ekonomi Di Kabupaten Magelang. J. AGRISTAN **2**(1), 52–60 (2020)
6. BPS, Executive summary of paddy harvested area and production in Indonesia 2019 (2020)
7. Prihatin, R.B.: Alih Fungsi Lahan Di Perkotaan (Studi Kasus Di Kota Bandung Dan Yogyakarta). J. Aspir. **6**(2), 105–118 (2016). https://doi.org/10.22212/aspirasi.v6i2.507
8. Samuel, L.: Tidak Ada Lahan Pertanian Tersisa, Kota Bandung Kini Punya 190 Kelompok Tani, Ini Rahasianya DeskJabar (2021)
9. Rizatus Shofiyati, D.S.P.: Effendi Passandaran, "Teknologi Penginderaan Jauh untuk Pemantauan Pola pertanaman Padi Sawah akibat perubahan Ikllim"
10. Putra, W.: Harga Cabai Rawit di Bandung Melambung, Pemkot: Ada Gagal Panen detiknews (2021)
11. Ruminta, R.: Analisis penurunan produksi tanaman padi akibat perubahan iklim di Kabupaten Bandung Jawa Barat. Kultivasi **15**(1), 37–45 (2016). https://doi.org/10.24198/kultivasi.v15i1.12006
12. Dwi, B., Nugroho, A., Maftukhah, R.: Prediksi Curah Hujan Dengan Metode Thomas-Fiering untuk Menentukan Jadwal Tanam di Kecamatan Tabukan Utara Kabupaten Kepulauan Sangihe Provinsi Sulawesi Utara MYDSAL TAMUSA, Bayu Dwi Apri Nugroho, STP., M.Agr.,Ph.D;Rizki Maftukhah, STP., M.Sc (2018)
13. A. S. for H. S.-S. BPS-Statistics Indonesia, "Produksi Tanaman Sayuran Menurut Kecamatan dan Jenis Tanaman di Kota Bandung (kuintal), 2019 dan 2020," BPS Kota Bandung (2021)

14. Schillerberg, T.A., Tian, D.: Changes of crop failure risks in the United States associated with large-scale climate oscillations in the Atlantic and Pacific Oceans. Environ. Res. Lett. **15**(6) (2020). https://doi.org/10.1088/1748-9326/ab82cd
15. S. C. Noviadi, "Rainfall modeling based on early predicted and season zone characteristic in the BMKG season zone over Lombok river basin," IOP Conf. Ser. Earth Environ. Sci., vol. 622, no. 1, 2021, doi: https://doi.org/10.1088/1755-1315/622/1/012043.
16. Djaenudin, D.M.H., Suhardjo, H., Hidayat, A.: Petunjuk Teknis Evaluasi Lahan untuk Komoditas Pertanian (2011)
17. Lebens, M., Finnegan, R.: Agile Processes in Software Engineering and Extreme Programming, vol. 419. Springer International Publishing, Cham (2021)
18. Sasminto, R.A., Tunggul, A., Rahadi, J.B.: Spatial analysis for climate determination of Schmidt-Ferguson and Oldeman classifications in Ponorogo City. J. Sumberd. Alam dan Lingkung **1**(1), 51–56 (2014)

Chapter 65
Correlation Investigation Between Aerosols Properties, Meteorological, and Environmental Variables During Summer Monsoon in Riau Province, Indonesia

Wilin Julian Sari, Waluyo Eko Cahyono, Prawira Yudha Kombara, Emmanuel Adetya, and Rahaden Bagas Hatmaja

Abstract Aerosols are affected by local meteorological and environmental phenomena. To better understand the contribution of local aerosol properties in Riau Province, Indonesia toward global climate change, this study investigated the relationship between the aerosol optical depth (AOD), enhanced vegetation index (EVI), wind speed (WS), Angstrom exponent (AE) and relative humidity (RH) during the summer monsoon period of 2011 to 2020. The data used in this study are monthly data of Modis-Terra Deep Blue (DB) land only AOD at 550 nm (MOD08_M3 v6.1), Modis-Terra EVI product (MOD13C2 v006), WS MERRA 2 reanalysis data (M2IMNXLFO v5.12.4), DB AE for land (MOD08_M3 v6.1) and RH AIRS3STM v006, which were all retrieved from NASA's Giovanni website. The results of this study show that the highest AOD value of 3.022 was observed in September 2015 and during the studied period, EVI coincided with AOD with a small value of correlation. Meanwhile, a highly negative correlation coefficient (R) is observed between AOD and RH, revealing that RH helped decreasing the amount of atmospheric aerosols. Therefore, all of the meteorological and environmental variables in this study, during the summer monsoon period, affect Riau's local aerosol amount/properties.

W. J. Sari (✉)
Research Center for Quantum Physics, National Research and Innovation Agency (BRIN), South Tangerang, Indonesia
e-mail: willin1231@gmail.com

W. E. Cahyono · P. Y. Kombara · E. Adetya · R. B. Hatmaja
Research Center for Climate and Atmosphere, National Research and Innovation Agency (BRIN), Bandung, Indonesia

E. Yulihastin et al. (eds.), *Proceedings of the International Conference on Radioscience, Equatorial Atmospheric Science and Environment and Humanosphere Science, 2021*, Springer Proceedings in Physics 275, https://doi.org/10.1007/978-981-19-0308-3_65

65.1 Introduction

Aerosols are atmospheric particles of pollutants formed by multi-phase system suspension that affects the energy budget balance of our earth by absorbing and scattering the sunlight, and it also affects the environment, such as visibility, as well as human health [1, 2]. Further, aerosols affect global climate change via their cooling and warming effect on the atmosphere [3]. And to better understand the global effect of aerosols towards climate change, it is imperative to first study its local contribution due to te fact that the amount of aerosols in the atmosphere over a particular area is depent on that area's sources of aerosols and most importantly is depent on the area's meteorological, such as conditions like a monsoon (times of the year) [1, 4] and environmental factors, where one exact same factor can give different contribution towards its local aerosols characteristics and amount. Aerosols are generated both naturally, such as forest fires, and anthropogenically, such as crop residue burning. In Indonesia, the amounts are normally higher during the dry season (summer monsoon) due to land clearing, crop residue burning, agricultural fires, and, more importantly, forest fires [5].

Aerosol optical depth (AOD) data are regularly used to express the aerosol amount in the atmosphere [6]. There are two types of methods generally used in detecting aerosol, namely ground-based observations and satellite remote sensing, where each of them has its limitation. Ground-based observations can produce accurate aerosol optical depth (AOD) for an exact location, but it will be challenging to gain AOD for more expansive areas because of the instrument limitations and the Spatio-temporal of aerosol distribution. Meanwhile, satellite remote sensing can provide AOD for wider areas, but the resulted AOD might be topography and surface covered. The technique o Moderate Resolution Imaging Spectroradiometer (MODIS) on the Terra and Aqua satellites has been developed and commonly used by researchers to obtain AOD data of land and sea [1, 7]. AOD products from MODIS have been extensively verified using ground measurements, such as Aerosol Robotic Network (AERONET), and numerous studies have proven MODIS products are in agreement with AERONET. Hence, MODIS data can be utilized to analyze aerosol properties [1, 8].

Although a means of studying local aerosol has existed, to the best of the authors' knowledge, there has not been a study that discusses how Indonesia's meteorological and environmental factors affect its local aerosol properties. Therefore, the relation and/or the impact of meteorological and environmental variables on the Indonesia's local aerosol properties should be investigated to provide information of the effect of Indonesia's local aerosols on global climate change. Riau Province was chosen as the study area because of the forest fires that frequently occurred during the summer monsoon period in this location. As such, aerosol amounts would be higher, and its relation with meteorological and environmental variables can be clearly seen.

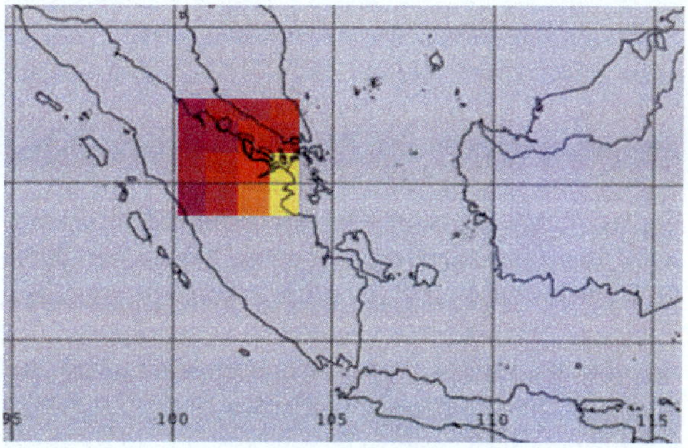

Fig. 65.1 Area of this study expressed in AOD time-averaged map

65.2 Data and Methods

65.2.1 Data

Monthly mean values of MODIS-Aqua Deep Blue (DB) land only AOD at 550 nm (MYD08_M3 v6.1), DB Angstrom exponent (AE) for land (MOD08_M3 v6.1), relative humidity (RH) (AIRS3STM V006), enhanced vegetation index (EVI) (MYD13C2 V005), wind speed (WS)(M2IMNXLFO 5.12.4) with a spatial resolution of one degree from June 2011 to September 2020, accessed from NASA Giovanni website, were used in this study. Figure 65.1 shows the area of this study (expressed in AOD time-averaged map), namely Riau Province, Indonesia (100.1258, −1.1985, 103.9128, 2.7215).

65.2.2 Methods

Correlation investigation in this study focuses on the summer monsoon period, namely June–July–August–September (JJAS) between 2011 and 2020. AOD at 550 nm is one of the most reliable aerosol properties used across many studies. AE is Angstrom wavelength that construes the aerosol size where AE less than 1 reflects coarse mode aerosol originated from dust or sea spray and AE more than 1 reflects fine mode aerosols originated from fossil fuel and biomass burning [1, 3, 9].

The mean value method was utilized to calculate the multi-year of all variables values. Further, Pearson product-moment analysis was used to analyze the relationship between AOD, and meteorological variables where the correlation coefficient is symbolized as R. AOD is deemed to have a high correlation, either positive or negative, to the meteorological variables when R is large [10].

65.3 Results and Discussion

AOD and AE values during the summer monsoon period from 2011 to 2020 in Riau Province are presented in Table 65.1. The table shows that the AOD values are always high during the studied period (all values are above 0.2) and are specifically higher during forest fire events in 2011, 2014, 2015, and 2019 (highlighted in yellow), with the highest value of 3.022 in September 2015. An AOD value of less than 0.05 suggested a clear sky with good visibility and a very low amount of aerosols, whilst a value of 1 represented hazy conditions. And a value of more than 2 or 3 revealed an intensely bad air condition with a very high amount of aerosols [3]. Rain rarely occurred during the summer monsoon period that made every burning and fire incident in Riau Province worse [11, 12]. Owing to the El Nino events during the summer monsoon (dry season) in 2015 and 2019 [13], the forest fires in Riau were getting more fierce hence generating more aerosols which then manifested in exceptionally high AOD values [1].

Times series of area-averaged of AOD and AE in Fig. 65.2a, greater than 1 value of AE (Table 65.1) and a high positive R-value of 0.69 show that the value of AE

Table 65.1 AOD and AE values during the summer monsoon period from 2011 to 2020 in Riau Province

Month/Year	AOD	AE	Month/Year	AOD	AE
June/2011	0.329	1.697	June/2016	0.230	1.67
July/2011	0.452	1.705	July/2016	0.304	1.702
August/2011	0.648	1.706	August/2016	0.306	1.632
September/2011	1.109	1.685	September /2016	0.280	1.589
June/2012	0.425	1.705	June/2017	0.267	1.636
July/2012	0.334	1.693	July/2017	0.235	1.641
August/2012	0.719	1.667	August/2017	0.349	1.673
September/2012	0.782	1.621	September/2017	0.259	1.662
June/2013	0.627	1.716	June/2018	0.244	1.652
July/2013	0.353	1.656	July/2018	0.363	1.691
August/2013	0.477	1.636	August/2018	0.310	1.597
September/2013	0.417	1.639	September/2018	0.435	1.593
June/2014	0.297	1.717	June/2019	0.258	1.662
July/2014	0.318	1.681	July/2019	0.459	1.644
August/2014	0.280	1.629	August/2019	0.780	1.68
September/2014	1.196	1.686	September/2019	2.101	1.706
June/2015	0.354	1.687	June/2020	0.254	1.565
July/2015	0.517	1.691	July/2020	0.297	1.652
August/2015	1.382	1.667	August/2020	0.272	1.569
September/2015	3.022	1.776	September/2020	0.236	1.503

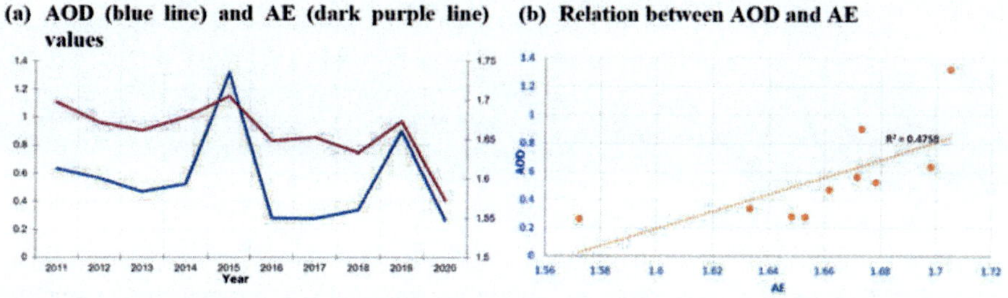

Fig. 65.2 **a** Time series of area-averaged of AOD (blue line) and AE (dark purple line) values (both are unitless) in Riau Province, Indonesia, for ten years from June 2011 to September 2020 during the summer monsoon period, **b** linear regression of AOD and AE

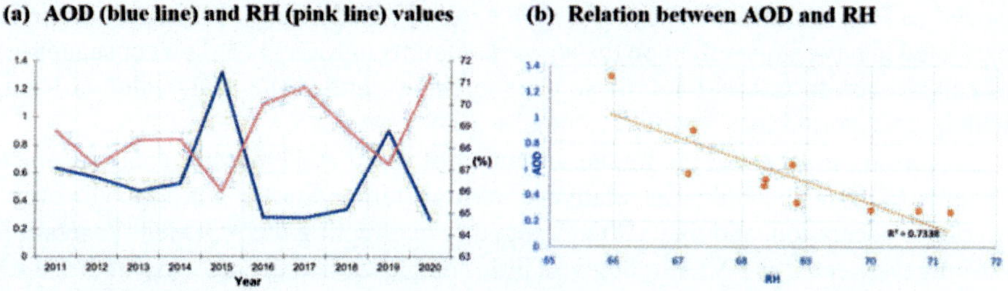

Fig. 65.3 **a** Time series of area-averaged of AOD (blue line) and RH (pink line) values in Riau Province, Indonesia, for ten years from June 2011 to September 2020 during the summer monsoon period, **b** linear regression of AOD and RH

was increasing as the value of AOD enhanced, which means the aerosols type in this study was fine mode aerosols originated from biomass burning.

With the R-value of −0.86, RH and AOD displayed a very significant negative correlation (see Fig. 65.3). The decrease of AOD as the RH increased is due to wet removal of aerosols through atmospheric aerosol scavenging processes, which removed aerosol particles by precipitation [14]. This result is similar to previous studies in several areas in China conducted by Shen et al. and Li et al. where they also found a very high negative R-value (between −0.70 and −1.00) [1, 6]. Yet, this type of correlation will not always be the case, in which other studies have shown a high positive R-value for the relation between AOD and RH. This positive R-value is because of the hygroscopic growth of aerosol particles, which then enhanced the aerosol particles' ability to scatter light, thus reducing visibility (higher AOD value) [3, 15].

During the period of the summer monsoon, aerosols can be carried from the south-eastern part of Indonesia to the north-western part by the dispersal function generated from the monsoon circulation, and the wind speed can impact the dispersion velocity of the aerosol particles produced from the biomass burning [1, 16]. Hence, the high value of AOD in this period is also influenced by the transportation of aerosols particles (regulated by the wind direction and wind speed) from Riau's neighboring

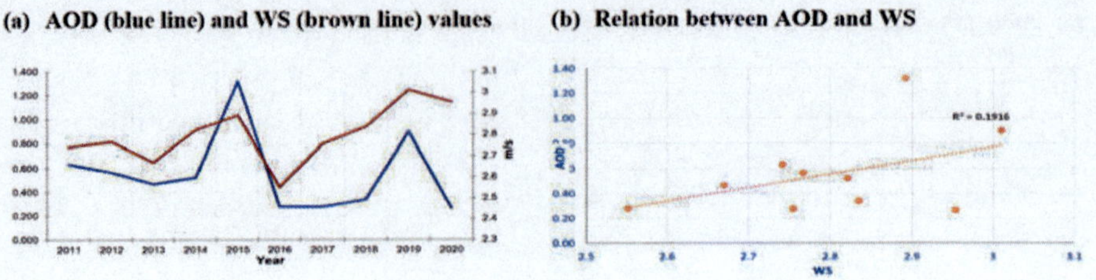

Fig. 65.4 **a** Time series of area-averaged of AOD (blue line) and WS (brown line) values in Riau Province, Indonesia, for ten years from June 2011 to September 2020 during summer monsoon period, **b** linear regression of AOD and WS

provinces. For example, South Sumatra also experienced forest fires during a similar period as Riau in 2019, and therefore the easterly wind of the summer monsoon is predicted to have some effect on the aerosol amounts in Riau [17]. As a consequence, as can be seen from Fig. 65.4, these phenomena expanded the AOD value of Riau, which is supported by a positive value of R (0.44) between AOD and WS.

According to Li et al., the impact of meteorological and environmental variables towards AOD can be ranked as relative humidity > temperature > wind speed > evaporation > vegetation, and they all work together in affecting the AOD [6]. Vegetation coverage, reflected as EVI, might give a little contribution to the decrease of the AOD because it provides friction and resistance that will reduce the wind speed, therefore stops the transportation of the aerosol particles [1, 18]. This is why the relation between EVI and AOD is negatively correlated ($R = -0.28$), which indicates that although it is not highly significant, vegetation index plays some role in regulating the aerosol properties. However, vegetation coverage's role is not significant enough because aerosol particles are affected by other plantation features, such as leaves characteristic of the vegetation, rainfall, and soil characteristics. For instance, leaf surface area and leaf size will determine leaf ability to capture aerosols. Another example, trees that have a concentrated network of roots closed by soil layers will anchor the soil strongly, thus lower the soil/dust chance to become aerosol particles due to its better cohesion (lowering the amount of aerosols in the atmosphere) [3] (Fig. 65.5).

On the other hand, other studies resulted in a positive R of the relation between AOD and EVI. This is because the aerosol particles presented in their study areas might be nutrient-rich particles which would better the plant growth or intensify the radiation that enhances the efficiency of the plant's photosynthesis [19, 20].

65.4 Conclusion

Both meteorological and environmental variables impact the atmospheric aerosols, which are expressed as AOD. Relation between AOD and AE suggested that aerosol particles in Riau mostly came from fossil fuel and/or biomass burning, such as crop

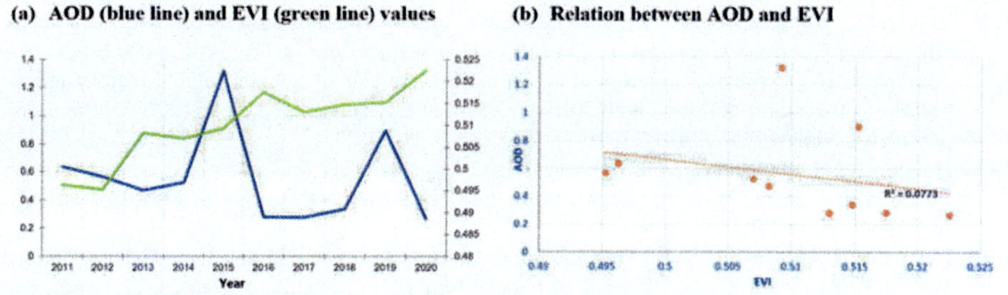

Fig. 65.5 **a** Time series of area-averaged of AOD (blue line) and EVI (green line) values in Riau Province, Indonesia, for ten years from June 2011 to September 2020 during the summer monsoon period, **b** linear regression of AOD and EVI

residue burning and forest fires. In addition, RH gave the highest impact towards AOD reflected with a high negative R-value where this is caused by the atmospheric scavenging that washes out the aerosols in the atmosphere via precipitation. On the contrary, due to the characteristic of the summer monsoon, wind speed would help increasing AOD values through its dispersion velocity. Lastly, although in a small amount, EVI still gave a little negative contribution towards AOD where this phenomenon helped decreasing the amount of atmospheric aerosols.

References

1. Shen, Y., Zhang, L., Fang, X., Zhao, Z., Li, X., Wang, J., Chai, Q.: Long-term analysis of aerosol optical depth over the Huaihai economic region (HER): possible causes and implications. Atmosphere (Basel) **9**, 1–20 (2018). https://doi.org/10.3390/atmos9030093
2. Sitnov, S.A., Mokhov, I.I., Likhosherstova, A.A.: Exploring large-scale black-carbon air pollution over Northern Eurasia in summer 2016 using MERRA-2 reanalysis data. Atmos. Res. **235**, 104763 (2020). https://doi.org/10.1016/j.atmosres.2019.104763
3. Tariq, S., Nawaz, H., Ul-Haq, Z., Mehmood, U.: Investigating the relationship of aerosols with enhanced vegetation index and meteorological parameters over Pakistan. Atmos. Pollut. Res. **12**, 101080 (2021). https://doi.org/10.1016/j.apr.2021.101080
4. Yang, S.H., Jeong, J.I., Park, R.J., Kim, M.J.: Impact of meteorological changes on particulate matter and aerosol optical depth in seoul during the months of june over recent decades. Atmosphere (Basel) **11**, 1–19 (2020). https://doi.org/10.3390/atmos11121282
5. Arisman: Analisis Tren Kebakaran Hutan dan Lahan di Indonesia Periode Tahun 2015–2019 (Trend analysis of forest and land fires in Indonesia periods 2015–2019). J. Sains Teknol. Lingkung. **6**, 1–9 (2020)
6. Li, J., Ge, X., He, Q., Abbas, A.: Aerosol optical depth (AOD): spatial and temporal variations and association with meteorological covariates in Taklimakan desert. China. PeerJ. **9**, 1–22 (2021). https://doi.org/10.7717/peerj.10542
7. Kaufman, Y.J., Tanré, D., Boucher, O.: A satellite view of aerosols in the climate system. Nature **419**, 215–223 (2002). https://doi.org/10.1038/nature01091
8. Boiyo, R., Kumar, K.R., Zhao, T.: Statistical intercomparison and validation of multisensory aerosol optical depth retrievals over three AERONET sites in Kenya. East Afr. Atmos. Res. **197**, 277–288 (2017). https://doi.org/10.1016/j.atmosres.2017.07.012

9. Zhang, M., Wang, L., Gong, W., Ma, Y., Liu, B.: Aerosol optical properties and direct radiative effects over Central China. Remote Sens. **9** (2017). https://doi.org/10.3390/rs9100997
10. Akpinar, E.K., Akpinar, S., Öztop, H.F.: Statistical analysis of meteorological factors and air pollution at winter months in elaziğ Turkey. J. Urban Environ. Eng. **3**, 7–16 (2009). https://doi.org/10.4090/juee.2009.v3n1.007016
11. Hendon, H.H.: Indonesian rainfall variability: impacts of ENSO and local air-sea interaction. J. Clim. **16**, 1775–1790 (2003). https://doi.org/10.1175/1520-0442(2003)016%3c1775:IRVIOE%3e2.0.CO;2
12. Heriyanto, E., Nuryanto, D.E.: Prediksi Sebaran Asap Kebakaran Hutan/Lahan Menggunakan WRF/CHEM (Studi Kasus: Tanggal 14 dan 20 Juni 2012, Pekanbaru-Riau). J. Meteorol. dan Geofis. **15**, 51–58 (2014). https://doi.org/10.31172/jmg.v15i1.172
13. Stiegler, C., Meijide, A., Fan, Y., Ashween Ali, A., June, T., Knohl, A.: El Niño–Southern oscillation (*ENSO*) event reduces CO_2 uptake of an Indonesian oil palm plantation. Biogeosci. Discuss. **2015**, 1–27 (2019). https://doi.org/10.5194/bg-2019-49
14. Santachiara, G., Prodi, F., Belosi, F.: Atmospheric aerosol scavenging processes and the role of thermo- and diffusio-phoretic forces. Atmos. Res. **128**, 46–56 (2013). https://doi.org/10.1016/j.atmosres.2013.03.004
15. Yoon, S.C., Kim, J.: Influences of relative humidity on aerosol optical properties and aerosol radiative forcing during ACE-Asia. Atmos. Environ. **40**, 4328–4338 (2006). https://doi.org/10.1016/j.atmosenv.2006.03.036
16. Liu, X., Yan, L., Yang, P., Yin, Z.Y., North, G.R.: Influence of Indian summer monsoon on aerosol loading in East Asia. J. Appl. Meteorol. Climatol. **50**, 523–533 (2011). https://doi.org/10.1175/2010JAMC2414.1
17. Tjasyono, H.B.: The character of rainfall in the Indonesian Monsoon. In: International Symposium Equatorial Monsoon System, pp. 1–11 (2008)
18. Chi, Y., Zuo, S., Ren, Y., Chen, K., Tariq, S.: The spatiotemporal pattern of the aerosol optical depth (AOD) on the canopies of various forest types in the Exurban National Park: a case in Ningbo City. East. China. Adv. Meteorol. **2019** (2019). https://doi.org/10.1155/2019/4942827
19. Kanniah, K.D., Beringer, J., North, P., Hutley, L.: Control of atmospheric particles on diffuse radiation and terrestrial plant productivity: a review. Prog. Phys. Geogr. **36**, 209–237 (2012). https://doi.org/10.1177/0309133311434244
20. Artaxo, P., Butt, N.: Fires increase Amazon forest productivity. Geophys. Res. Lett. 4654–4662 (2015). https://doi.org/10.1002/2015GL063719 Received

Chapter 66
Simulation and Evaluation of Particulate Matter (PM): A Case Study of Haze During September 2019 in Palangkaraya City, Indonesia

Waluyo Eko Cahyono, Prawira Yudha Kombara, Wilin Julian Sari, and Emmanuel Adetya

Abstract Atmospheric remote sensing offers a unique opportunity to estimate air quality indirectly, which is very important for managing and monitoring air quality in major cities of developing countries, particularly in Indonesia. Meteorology and chemical transport online calculation of Weather Research and Forecasting with Chemistry (WRF-Chem Model) was conducted in Palangkaraya and validated against satellite observations for aerosol meteorology concentrations (PM2.5). This study has an objective to assess PM2.5 pollution at the city level for an extreme case of continuous smog that occurred in September 2019 in Palangkaraya. The results show that there is a good relationship between the AOT from the satellite and the PM model and measurements on the surface.

66.1 Introduction

Air pollution has become one of the most important concerns of local authorities in different cities in Indonesia because poor air quality can have a negative impact on human health including asthma, impaired lung function, cardiorespiratory disease and increased mortality [1], as well as on ecosystems and climate [2]. Therefore, further research is needed to estimate the impact of anthropogenic emissions on air pollutant concentrations in urban areas, because the largest contribution of anthropogenic emissions comes from urban sources that emit various gases and emissions of various particulate species [3]. Therefore, urban impacts require detailed modeling studies. Fine and surface aerosol particles called PM2.5, which are 2.5 m in diameter or less, are a well-known parameter for understanding air pollution levels [4]. Therefore, characterization of urban pollutant impacts requires detailed modeling studies. The main pollutants emitted into the atmosphere in urban areas are sulfur oxides

W. E. Cahyono (✉) · P. Y. Kombara · W. J. Sari · E. Adetya
Research Center for Climate and Atmosphere, National Research and Innovation Agency (BRIN), Bandung, Indonesia
e-mail: walu002@brin.go.id

E. Yulihastin et al. (eds.), *Proceedings of the International Conference on Radioscience, Equatorial Atmospheric Science and Environment and Humanosphere Science, 2021*, Springer Proceedings in Physics 275, https://doi.org/10.1007/978-981-19-0308-3_66

(SO_x), nitrogen oxides (NO_x), carbon monoxide (CO), volatile organic compounds (VOCs), metal oxides and particulate matter [5]. PM/aerosols are mainly composed of black carbon, sulfate, nitrate and organic matter. These chemical pollutants include: primary PM, which is directly emitted into the atmosphere from various sources (e.g., road traffic, construction sites, traffic dust, fires), and secondary PM, which is formed in the atmosphere through chemical reactions. In the gas and water phases, these compounds will turn the oxidation of precursors such as sulfur dioxide (SO_2), nitrogen oxides (NO_x), and volatile organic compounds (VOC) into nonvolatile and semi-volatile species. Economic development in Indonesia has recently accelerated due to the rapid growth of heavy industrial operations [6]. Increasing urbanization in this country is likely to have far-reaching environmental consequences. These areas have suffered from severe air pollution problems over the last decade, and as industrial activity and car numbers increase, PM, SO_2, VOC and NO_x emissions will also increase significantly. PM10, PM2.5, and UFP, respectively, are areas of great interest to scientists in recent years due to issues related to human health, climate change, and atmospheric visibility [6]. Therefore, PM simulation is needed to develop an effective PM control strategy. The PM simulation can be conducted by using dynamic model such as WRF-Chem. WRF-Chem is a dynamic model that fuse weather and chemistry components. This model can be used to simulate the dispersion of air pollution, predict the haze and smoke from forest fire, predict the dispersion of volcanic ash, etc. The WRF-Chem has been used by many institutions and researchers to simulate or predict how the dispersion of air pollutant and learn the complex interaction between chemistry and physic in the atmosphere [7, 8]. This study reveals the condition of air pollution, especially aerosols and PM as pollutants including aerosol on September 16, 2019, in Palangkaraya.

66.2 Data and Methods

In this study, we will use WRF-Chem version 4.2.2 model with the study area in the new capital of Indonesia, namely Palangkaraya, Central Kalimantan, Indonesia (Fig. 66.1). The meteorological data used is the Final Reanalysis (FNL) issued by NCEP-NCAR and can be downloaded at http://rda.ucar.edu/datasets/ds083.2. The FNL data has a spatial resolution of about 25 km and a time resolution of 6 h. The FNL data will be the input data for the WRF-Chem model. Results from WRF-Chem will have a resolution of 5 km with hourly time interval.

This model uses several atmospheric and chemical parameters such as convection, microphysics, boundary layers, radiation, and so on [9]. The emission data that will be used in this study consist of biogenic data using Megan data and anthropogenic data using EDGAR-HTAP. The configuration of WRF-Chem model used in this study is shown in Table 66.1. Simulations were carried out to project the amount of emissions PM2.5 in 16 September 2019 in the study area. Results from WRF-Chem were compared with NASA's MERRA model, MoEF monitoring, and

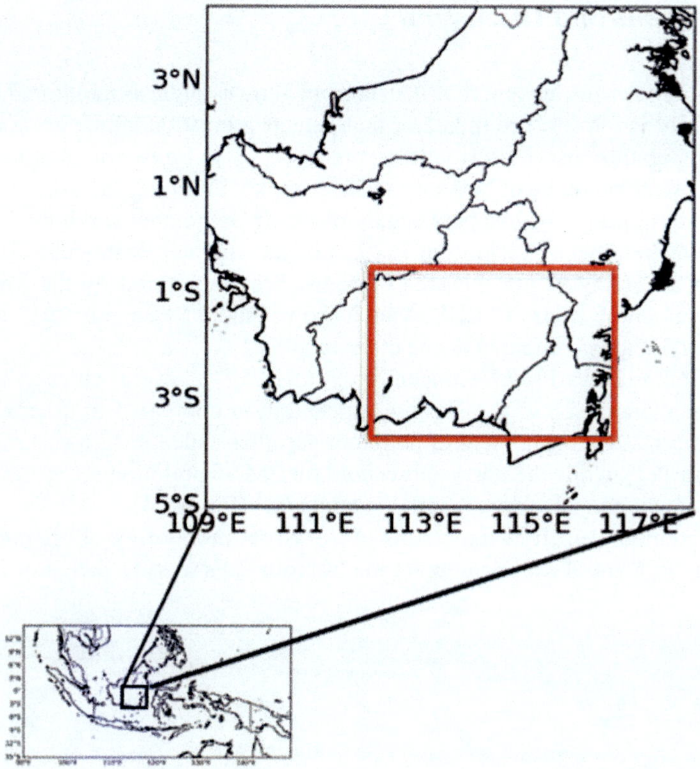

Fig. 66.1 The study area is Palangkaraya City, Central Kalimantan, Indonesia and the domain for WRF-Chem model (box with red line)

Table 66.1 The configuration of WRF-Chem in this study

Parameterization	Domain 1
Cumulus	Grell 3
Microphysics	Purdue Lin
Boundary layer	YSU
Radiation (shortwave and longwave)	RRTMG
Surface	NOAH MP
Chemical	GOCART couple with RACM-KPP

AERONET observations (https://aeronet.gsfc.nasa.gov). The AERONET (AErosol RObotic NETwork) is a federation of ground-based remote sensing aerosol networks established by NASA.

66.3 Results and Discussion

Figure 66.2 shows the hotspot matrix data from Terra satellite as monitored by Lapan [10]. Hotspot is a forest fire indicator that detects a location which has a relatively higher temperature than its surrounding areas. As such, forest and land fires can be indicated by the presence of hotspots and fire smoke. From the data matrix, it can be seen that starting in July, hotspots began to rise, in September reaching the highest peak with the number of hotspots at 1620, and gradually decreasing until December. This is in accordance with the area of forest fires monitored by the Ministry of Environment and Forestry (KLHK) which shows that in September 2019 it reached the highest fire area compared to the other months.

Figure 66.3 shows PM2.5 Concentration on 2019 in several cities in Indonesia [11], Palangkaraya shows the highest concentration compared to other cities. On 16 September 2019, the content of pollutant particles called PM2.5 was recorded at 1413 $\mu g/m^3$[11], while the normal threshold for PM2.5 pollution set by the Ministry of Environment and Forestry (KLHK) is 65 $\mu g/m^3$ (Fig. 66.4).

On September 16, from the results of Airvisual monitoring of air pollution in Palangkaraya, Central Kalimantan, for the last four days was at a dangerous level, 20

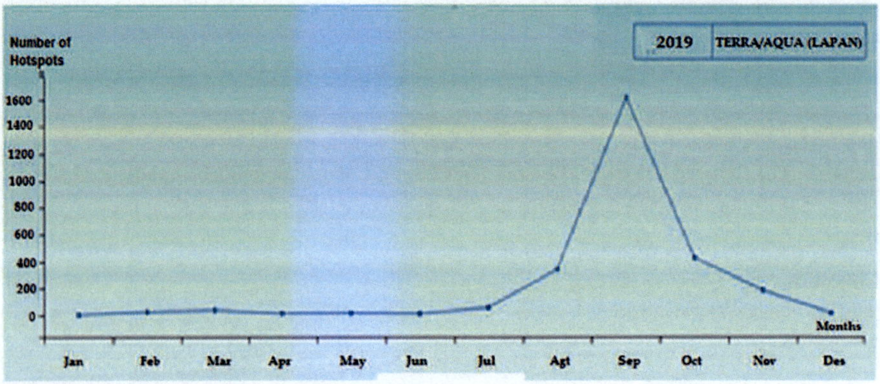

Fig. 66.2 Hotspot matrix data from Terra satellite as monitored by Lapan [10]

Fig. 66.3 PM2.5 concentration on 2019 in several cities in Indonesia

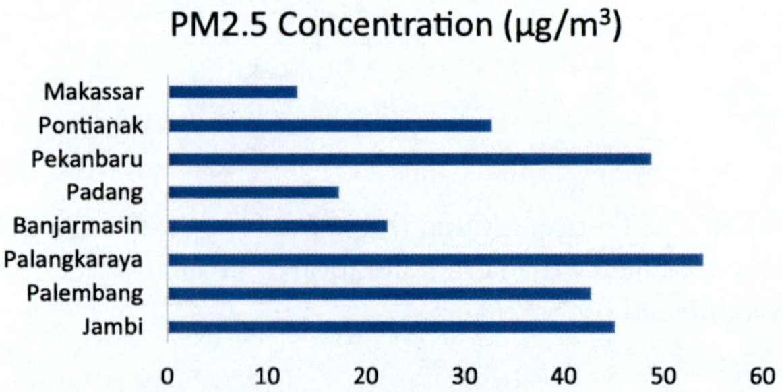

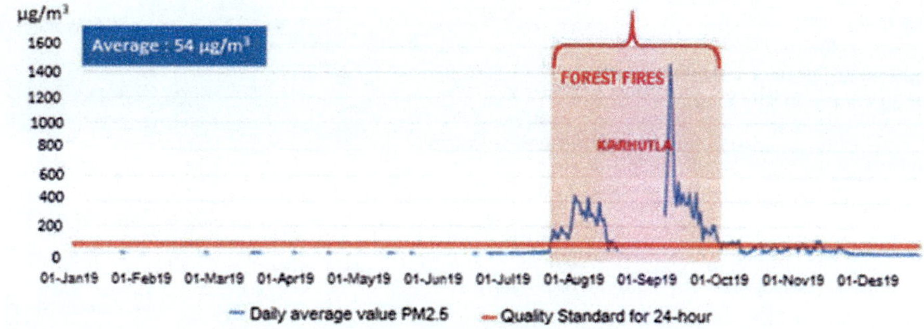

Fig. 66.4 PM2.5 daily average value PM2.5 in Palangkaraya. [11] *Source* KLHK

Fig. 66.5 Aerosol monthly average value on 2019 in Palangkaraya (AERONET)

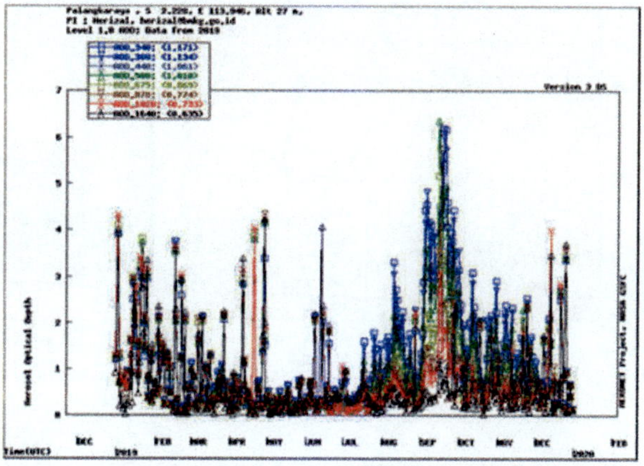

times above the normal threshold, due to smoke from forest and land fires. Similarly, the results from Aeronet in September 2019 showed high aerosol content (Fig. 66.5).

Since aerosol optical thickness (AOT) is a key parameter of aerosols and description of the Earth's radiation budget, it is widely measured from ground sun photometer network NASA/AERONET and from satellite. Fine and surface level aerosol particle called PM2.5, whose diameter is 2.5 μm or less, is a well-known parameter for understanding polluted level of air [9].

Based on the results of the Merra model, in mid-September 2019 showed a high AOT, as well as the results of the Modis satellite monitoring which showed the same result, namely a high AOT value (Fig. 66.6). The results of the WRF-Chem model are also seen in September indicating a high PM2.5 content of 4.3 mg/m^3 (Fig. 66.7).

Comparing MERRAero to PM2.5 data collected by the Chemical Speciation Network (CSN) offers greater insight on the species MERRAero predicts well and those for which there are biases relative to the EPA observations. To further understand discrepancies between the reanalysis and observations, we use complimentary data to assess two important aspects of MER- RAero that are of relevance to the diagnosis of PM2.5, in particular AOD and vertical structure [12].

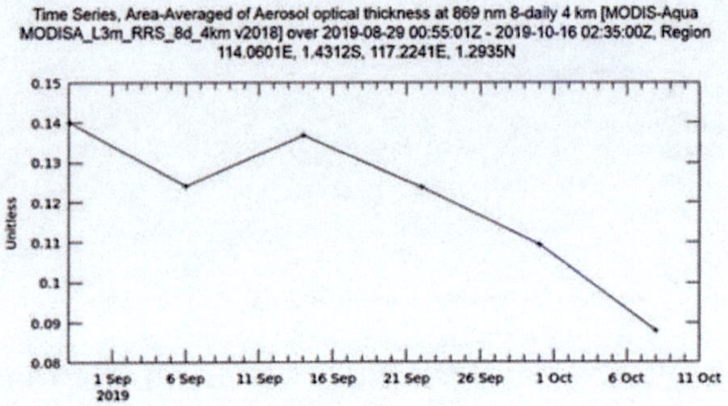

Fig. 66.6 Aerosol daily average value on September-Oktober 2019 in Palangkaraya Modis satellite

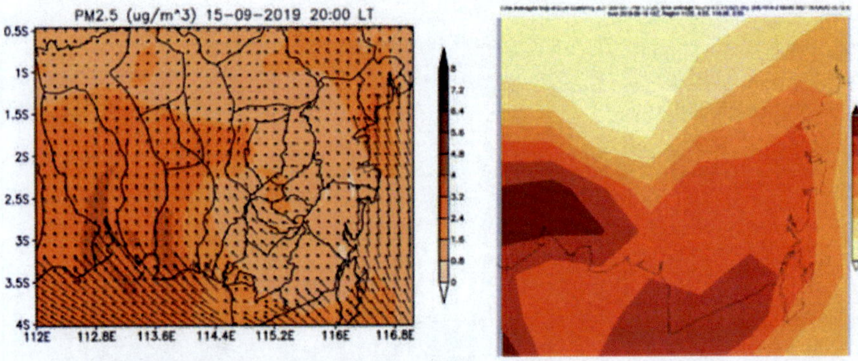

Fig. 66.7 WRF-Chem model (left) and Merra Model

Figure 66.7 shows the PM2.5 dispersion between WRF-Chem (left) and Merra-2 (right) model. Both of figures show the similarity pattern, particularly in the western region. Those show PM2.5 is more concentrated in the western region of Palangkaraya although there is differences in PM2.5 values. We found that PM2.5 pollutant exceeded normal threshold levels within days over the studied time intervals. Our evaluation reveals that MERRA-2's raw estimates of PM2.5 on a monthly time scale or longer are helpful in long-term air quality studies [13]. Finally, we verified that the atmospheric model with WRF-Chem chemical coupling, configured for the Palangkaraya study area with a 4 km $\times$ 4 km domain, simulated acceptable and reasonable accuracy [14].

66.4 Conclusion

We applied WRF-Chem to simulate air quality in September 2019, when intensive biomass burning occurred in Palangkaraya. The TERRA/AQUA satellite dataset supports hotspot observation data. PM10 and PM2.5 are higher at times and areas with a more significant number of fires found. Based on the Merra model, mid-September 2019 showed a high AOT, and the Modis satellite monitoring results showed the same result, namely a high AOT value. The results of the WRF-Chem model were also seen in September, which showed a high PM2.5 content of 4.3 mg/m^3. Our calculation used WRF-Chem and remote sensing satellite data in air quality modeling and captured pollution events. We advocate the integration of diverse data sources, such as satellite remote sensing data and ground-based observation data, in conjunction with an assimilation approach in an online model system for building innovative operational air quality forecasts in the region to address haze impacts and health concerns.

Acknowledgements First of all, we would like to thank the Director of the Research and Technology Center for Atmospheric for all the support and enthusiasm during our research activities in Bandung. We are grateful for the funding received from the International Network for Government Science Advice (INGSA) and the International Science Council.

References

1. Mindell, J., Joffe, M.: Predicted health impacts of urban air quality management. J. Epidemiol. Commun. Health **58**, 103–113 (2004)
2. Menon, S., Unger, N., Koch, D., Francis, J., Garrett, T. Sednev, I., Shindell, D., Streets, D.: Aerosol climate effects and air quality impacts from 1980 to 2030. Environ. Res. Lett. **3**, 024004 (2008). https://doi.org/10.1088/1748-9326/3/2/024004
3. Steinke, Seinfeld, J.H., Pandis, S.N.: Atmospheric Chemistry and Physics: From Air Pollution to Climate Change (1326 p). Wiley, USA (1998)
4. Susetyo, K.E., Kusin, K., Nina, Y., Jagau, Y., Kawasaki, M., Naito, D.: 2019 Peatland and Forest Fires in Central Kalimantan, Indonesia, S.: 08 2019. 2019, 1–12 (2020)
5. Baklanov, A., Schlünzen, K., Suppan, P., Baldasano, J., Brunner, D., Aksoyoglu, S., et al.: Online coupled regional meteorology chemistry models in Europe: current status and prospects. Atmos. Chem. Phys. **14**(1), 317–398 (2014). https://doi.org/10.5194/acp-14-317-2014
6. Heriyanto, E., Syaufina, L., Sobri, M.: Forecasting simulation of smoke dispersion from forest and land fires in Indonesia. Procedia Environ. Sci. **24**, 111–119. https://doi.org/10.1016/j.proenv.2015.03.015
7. Kumar, A., Jiménez, R., Belalcázar, L.C., Rojas, N.Y.: Application of WRF-Chem model to simulate PM10 concentration over Bogota. Aerosol Air Qual. Res. **16**, 1206–1221. https://doi.org/10.4209/aaqr.2015.05.0318
8. Lecoeur, E., Seigneur, C.: Dynamic evaluation of a multi-year model simulation of particulate matter concentrations over Europe. Atmos. Chem. Phys. **13**, 4319–4337 (2013)
9. Mukai, M.: Suspended particulate matter sampling at an urban AERONET Site in Japan, Part 2: relationship between column aerosol optical thickness and $PM_{2.5}$ concentration. J. Appl. Remote Sens. **4**(1): 043504. https://doi.org/10.1117/1.3327930
10. Direktorat PKHL Kementerian Lingkungan Hidup dan Kehutanan RI. Data Matrik Titik Panas TERRA/AQUA (LAPAN). Accessed 26 Apr. 2020, from sipon-gi.menlhk.go.id website (2020) http://sipongi.menlhk.go.id/hotspot/matrik_tahunan
11. Direktorat PPU Kementerian Lingkungan Hidup dan Kehutanan RI. Accessed 26 Jul. 2020 (2020). https://ditppu.menlhk.go.id/portal/read/kondisi-kualitas-udara-di-beberapa-kota-besar-tahun-2019

12. Cristóvão, R., Cidália, B., Ramiro, M., Rui, B.: Pollution prevention and wastewater treatment in fish canning industries of Northern Portugal. In: International Proceedings of Chemical, Biological and Environmental Engineering, vol. 32(1), pp. 12–16 (2012). https://doi.org/10.7763/IPCBEE
13. Navinya, C.D., Vinoj, V., Pandey, S.K.: Evaluation of Pm2.5 surface concentrations simulated by Nasa's Merra version 2 aerosol reanalysis over India and its relation to the air quality index. Aerosol Air Qual. Res. **20**(6), 1329–1339 (2020). https://doi.org/10.4209/aaqr.2019.12
14. Buchard, V., da Silva, A.M., Randles, C.A., Colarco, P., Ferrare, R., Hair, J., Hostetler, C., Tackett, J., Winker, D.: Evaluation of the surface PM2.5 in version 1 of the NASA MERRA aerosol reanalysis over the United States. Atmos. Environ. **125**, 100–111 (2016). https://doi.org/10.1016/j.atmosenv.2015.11.004

Chapter 67
Determination of Priority Locus of Handling Stunting and Regional-Based Stunting Management Strategies in Serang City

Erti Nurfindarti and Nugrahana Fitria Ruhyana

Abstract One of the main problems of development in Serang City is that the quality of human resources is not yet optimal, which is marked by the low quality of public health, including 2794 stunting cas1es in 2021. The incidence of stunting is related to various impacts of climate change, such as food insecurity, availability of clean water, and sanitation. This study aims to determine thc priority locus of stunting management based on the concept of adaptation to climate change and regional-based stunting management strategies. The research method is a mixed sequential explanatory method by combining the results of quantitative and qualitative analysis in determining the priority locus of stunting handling and regional-based stunting management strategies. The results of the quantitative analysis showed that there were 4 (four) urban villages that had a high risk of stunting, and based on a qualitative analysis of the condition of the four urban villages, the stunting management strategies were accelerate the information and education about stunting and climate change adaptation, prioritize government activities for handling stunting (infrastructures, preventing programs, data resilience, budget support, and improve healthy living behaviour).

67.1 Introduction

Stunting is a condition that occurs in children under five years because of failure to grow because of chronic malnutrition so that the children's height was under height standard for their age [1]. This condition is indicated from the measurement data of children showing height according to the age range below the WHO standard [2]. The incidence of stunting is influenced by many factors, including infectious

E. Nurfindarti (✉)
Regional Development Planning Agency of Serang City, Serang, Indonesia
e-mail: erti.nurfindarti@gmail.com

N. F. Ruhyana
Regional Research and Development Planning Agency of Sumedang Regency, Sumedang, Indonesia

E. Yulihastin et al. (eds.), *Proceedings of the International Conference on Radioscience, Equatorial Atmospheric Science and Environment and Humanosphere Science, 2021*, Springer Proceedings in Physics 275, https://doi.org/10.1007/978-981-19-0308-3_67

diseases and poor hygiene practices [3]. Another influential factor is the parenting pattern of feeding toddlers [4]. Good psychosocial parenting (emotional reactions, positive encouragement, affection, and a comfortable atmosphere) can also reduce the risk of stunting which is related to family harmony and resilience [5].

Stunting is an indicator of the second goal of Sustainable Development Goals: Eliminating Hunger, Achieving Food Security and Good Nutrition, and Improving Sustainable Agriculture [6], because of the relation between stunting and climate change [6]. Climate change will affect food security, give impact food accessibility, especially for the poor, so that it will lead to malnutrition in children from these groups. Besides that, the availability of clean water, healthy living behavior, and sanitation infrastructure are problems related to climate change that can affect the occurrence of stunting [7].

Stunting will affect the cognitive abilities and intelligence of children [8]. Children's learning capacity is not optimal and decreased performance during school will affect children's productivity in adulthood [9]. This decline in productivity will cause economic losses for the country [10] and makes stunting something that must be done and handled seriously by the state.

Based on the Global Nutrition Report in 2018, Indonesia's Stunting Prevalence of 132 countries was ranked 108th, while in the Southeast Asia region, Indonesia's stunting prevalence was second highest after Cambodia. The results of the 2018 Basic Health Research in Indonesia showed a decline in the prevalence of stunting at the national level by 6.4% over a five-year period, from 37.2% (2013) to 30.8% (2018) [11]. However, the COVID-19 pandemic in 2020 is predicted to increase the prevalence of stunting worldwide, including Indonesia. This condition can be exacerbated by the existence of climatic phenomena such as la nina which causes environmental conditions to deteriorate such as floods. The government needs to anticipate the impact of climate change on the possibility of increasing stunting. Due to the limited government budget, data on priority locations is needed for more effective stunting management.

This research will study the priority locus of stunting handling and regional-based stunting management strategies in Serang City. Determination of priority locus for stunting handling has been carried out in previous studies in Sumedang Regency through a climate change adaptation approach [12] and stunting management strategies in Bandung [13]. The difference with previous research is the use of more diverse variables since the availability of data sources is more supported by the Serang City government, and the scoring method for determining the risk index. Whereas in previous studies using the t-score scoring method without variable weighting, this study uses a scoring method with an index of 1–5, and each variable is given a weighting based on qualitative expert judgement. In addition, this study determines stunting management strategies based on the conditions of each region. Determining the priority locus of stunting handling based on the climate change adaptation approach is very important because the issue of climate change is a strategic issue that must continue to be seriously addressed by the region and even the country has given the magnitude of the impact of climate change on various aspects of life. If the priority locus of stunting has been found, the government and the community can take steps and policies to deal with stunting that is more targeted, effective, and efficient. These policies need to be supported by a regional-based stunting management strategy because the condition of the area (urban village) has different social, economic, and cultural characteristics. Therefore, a specific strategy for dealing with stunting is needed based on the characteristics of the area concerned. In addition, the weight gain on each variable is also a consideration in determining stunting management strategies.

67.2 Data and Method

The data used in this study consisted of primary data and secondary data. Secondary data contains information on all urban villages in Serang City, amounting to 66 urban villages as the unit of analysis in this study. The data was obtained from several agencies, namely data on short and very short toddlers in 2021 (Health Department), data on village potential in 2020 (BPS), data on Serang City slums in 2021 (Bappeda), and data on poor households in decile one and decile two in 2021 (Social Service). The primary data was obtained from the results of interviews and observations in the urban village which, according to the results of quantitative data processing, were categorized in areas with a high risk of stunting.

The research method used is a mixed-method with a sequential explanatory design approach where the quantitative approach is carried out first and then followed by a qualitative approach [14]. In this study, the quantitative approach becomes a priority because the results of data processing with a scoring technique will determine the locations most at risk of stunting from all urban villages in Serang City. Furthermore, to obtain more in-depth information on conditions in priority locations, a qualitative approach was used through interviews with stakeholders related to areas most at risk or those with a "high" category of stunting risk index, namely regional heads and related agencies, in this case Bappeda as agency in charge of regional planning. The results of the interviews were then combined with various suggestions and inputs obtained during FGDs and deliberation. The use of mixed methods has the advantage that the data obtained becomes more complete when the priority location information obtained from secondary data is then confirmed with primary data from stakeholders in priority locations. The results of this quantitative and qualitative analysis are very helpful in providing recommendations and developing more relevant strategies in regional-based stunting management in Serang City.

The selection of variables to measure potential hazards and vulnerabilities along with their constituent components is based on the results of previous studies [11, 12]. However, there are adjustments and additions to new variables, including the number of floods in the last five years and the density of slum settlements as a determinant of sensitivity, the ratio of early childhood education (*Pendidikan Anak Usia Dini/PAUD*) and toddlers, waste management, the proportion of the population of college graduates, waste management, and the ratio of health cadres as a determinant of adaptive capacity. Areas with a high density of slums will be more sensitive to stunting because children with malnutrition are more likely to be found in slum areas [15–18]. The high frequency of flooding in an area also potentially causes many children to be malnourished [19–24], which is the main cause of stunting.

In adaptive capacity, the existence of early childhood education and integrated health services (*Pos Pelayanan Terpadu/Posyandu*) cadres in an area has an important role in anticipating and reducing stunting in children under five years through socialization and education about nutrition and healthy lifestyles, especially for mothers with low education [15, 25, 26]. This also confirms that the high level of public education in an area will increase the adaptive capacity of the community in reducing the risk of stunting. Waste management is also closely related to a higher potential for stunting [27, 28]. The conceptual framework of environmental factors used in this study is described as follows (Fig. 67.1).

The concept of climate change adaptation was adopted from the Intergovernmental Panel on Climate Change, the fifth report or known as IPCC AR-5 [29]. The data is processed by determining the scale of each hazard and vulnerability component which includes exposure, sensitivity, and adaptive capacity of each urban village. Vulnerability increases when exposure and sensitivity are high while adaptive capacity is low. When hazards and vulnerabilities are at a very high level then the risk is also very high. For standardization of scores for each variable that has different unit values, a scale of 1 (very low) to 5 (very high) is used on the hazard and vulnerability index and its constituent components. From the results of this risk

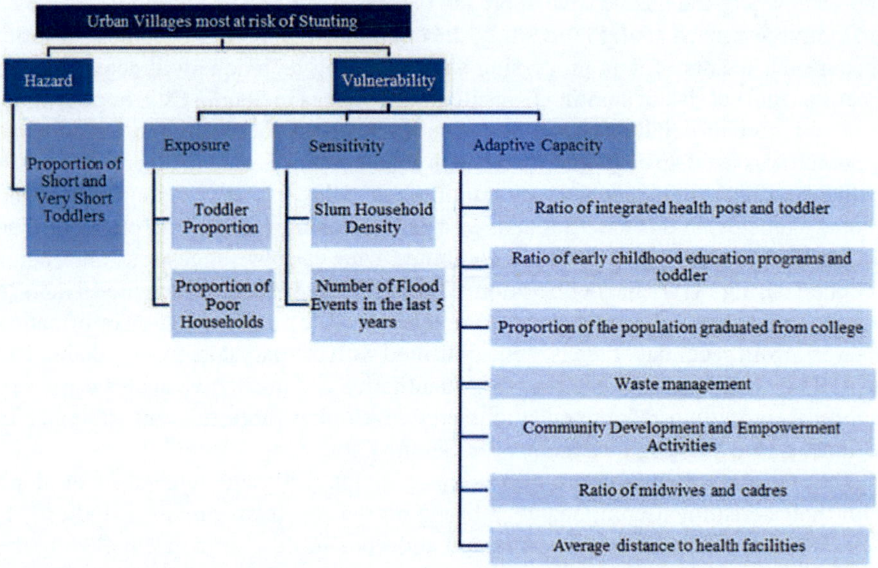

Fig. 67.1 Conceptual framework of environmental factors that potentially cause stunting in Serang City based on available literature and data

Table 67.1 Risk scoring via hazard and vulnerability score overlay

Vulnerability \ Hazard	Very high (5)	High (4)	Medium (3)	Low (2)	Very low (1)
Very vulnerable (5)	Very high (10)	Very high (9)	High (8)	High (7)	Medium (6)
Vulnerable (4)	Very high (9)	High (8)	High (7)	Medium (6)	Low (5)
Quite vulnerable (3)	High (8)	High (7)	Medium (6)	Low (5)	Low (4)
Less vulnerable (2)	High (7)	Medium (6)	Low (5)	Low (4)	Very low (3)
Not vulnerable (1)	Medium (6)	Low (5)	Low (4)	Very low (3)	Very low (2)

Source Developed from National Disaster Management Agency (BNPB) [30]

analysis, a matrix of the relationship between hazard and vulnerability analysis is generated (Table 67.1).

67.3 Result and Discussions

The process of determining the priority locus for stunting handling in Serang City using a climate change adaptation approach was conducted using quantitative analysis to map the risk of stunting which further elaborated using the qualitative analysis such as interviews, FGDs, and deliberation. The results of the exploration of various variables that determine the risk of stunting, especially in the aspect of vulnerability, can be a concern for stakeholders to formulate strategies to improve conditions so that each kelurahan becomes more resilient. Some data related to vulnerability that are not yet available in this study due to data limitations, both those that can reduce exposure and sensitivity or that can increase adaptive capacity, can be added to the next study so that the information provided becomes more comprehensive.

67.3.1 Quantitative Analysis

The analytical process begins with hazard mapping by scoring the proportion of short and very short toddlers in each urban village which is an indication of the number of stunting toddlers . This proportion is the percentage of short and very short toddlers

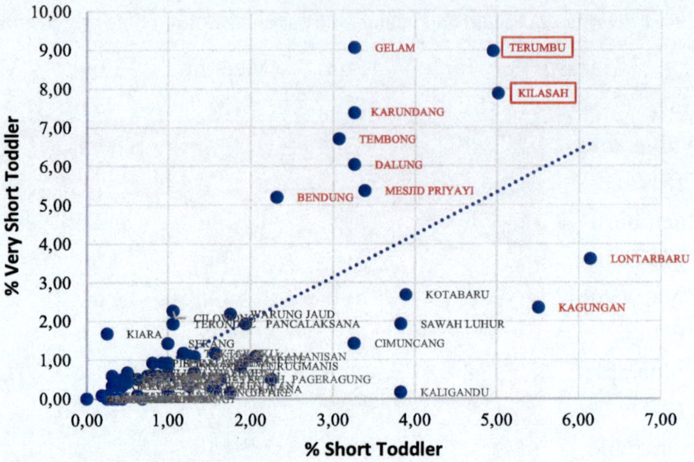

Fig. 67.2 The proportion of toddlers indicated for stunting in Serang City

in an urban village compared to the total toddlers in Serang City in 2021 based on data from the Health Office as shown in Fig. 67.2. The proportion of very short toddlers is given a higher score, which is 0.6, compared to the proportion of short toddlers given a weight of 0.4 considering the higher chance of stunting in very short toddlers.

From the scatter plot (Fig. 67.2), it is identified that the urban village in the upper right corner is the urban village with the highest proportion of short and very short toddlers (red). The highest proportion of very short toddlers is in Kasemen District (33.89%), while the highest proportion of short toddlers is in Serang District (27.48%). There are two urban villages in Kasemen District that have a higher stunting potential than other areas, Terumbu and Kilasah urban village.

After obtaining a mapping of the urban village from the hazard aspect, in this case, an indication of stunting in children under five in each urban village, a vulnerability analysis was carried out which included exposure, sensitivity, and adaptive capacity. Exposure is determined by the proportion of poor households that is assigned a weight of 0.75 and the proportion of children under five who are assigned a weight of 0.25. This weighting is based on the fact that children under five are exposed to stunting more in poor households due to a lack of chronic nutritional intake so that the number of poor households in an urban village will increase the potential for stunting exposure and will be even higher if in an urban village. It also has a high proportion of toddlers. Urban villages indicated by high stunting exposure are in the upper right position, namely urban villages Banten, Warung Jaud, Mesjid Priyayi, and Kilasah, all of which are in Kasemen District (Fig. 67.3).

The sensitivity of the stunting urban village in Serang City is determined by two variables, namely the higher the density of slum settlements and the more frequent occurrences of flood disasters in the last five years, the higher the sensitivity of the

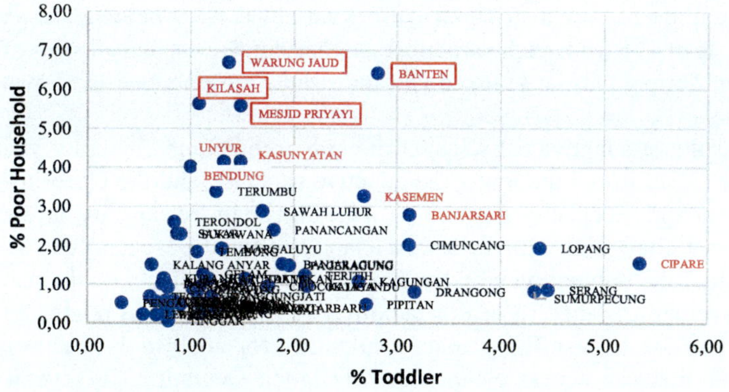

Fig. 67.3 Village with stunting exposure

urban village to stunting events. The density of slums is given a higher weight (0.7) because it is considered to increase the sensitivity of stunting more than flood events which are given a lower weight (0.3). The urban village with the highest sensitivity level is dominated by urban villages located in Kasemen District, namely urban villages Bendung, Kasunyatan, Margaluyu, and Warung Jaud (Fig. 67.4).

Meanwhile, adaptive capacity is determined by more variables due to the availability of data that can describe the resilience of the region in minimizing stunting. There are seven variables used with the largest weight being on the ratio of integrated health services and toddlers and the ratio of early childhood education and toddlers

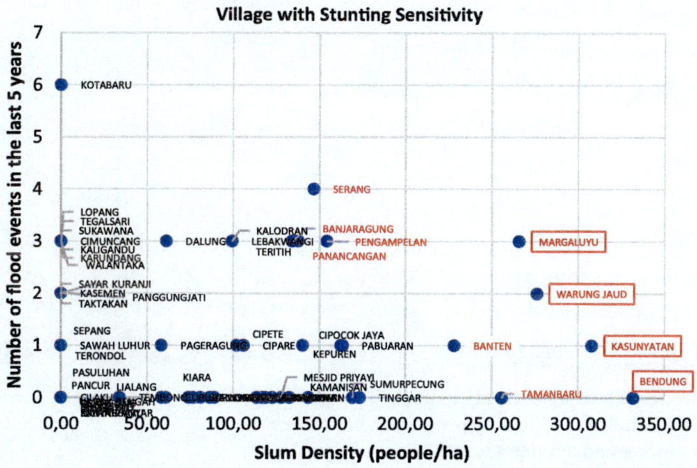

Fig. 67.4 Village with stunting sensitivity

each 0.2, and the rest are distributed to other variables. After being given a weighting from a scale of 1–5 from each variable, then the adaptive capacity scores of 66 urban villages in Serang City are ranked and the results are obtained as shown in Table 67.2.

The results of adaptive capacity scoring are accompanied by color visualization where the closer to red the lower the adaptive capacity, and the closer to green the higher the adaptive capacity. Regions with the best adaptive capacity have significant differences with regions with the lowest adaptive capacity, especially in terms of the proportion of the population who have graduated from college, waste management, and the average distance to health facilities. Another variable related to adaptive capacity but due to insufficient data availability for analysis is the proportion of households that can access clean water and good sanitation. In several previous

Table 67.2 Best 5 and lowest 10 in adaptive capacity score

Rank	Sub District	Urban Village	a	b	c	d	e	f	g	Adaptive Capacity Score
1	Serang	Lontarbaru	3.6	4.0	5.0	4.1	2.20	3.2	4.90	3.9
2	Serang	Unyur	4.8	4.6	4.0	5.0	1.00	1.1	4.37	3.9
3	Cipocok Jaya	Cipocok Jaya	4.2	4.7	4.1	1.9	3.80	3.0	4.21	3.8
4	Serang	Cimuncang	3.6	3.3	3.7	5.0	3.40	2.1	4.91	3.7
5	Serang	Serang	3.7	3.2	4.7	5.0	1.00	2.5	4.95	3.7
57	Kasemen	Kasunyatan	2.7	3.9	1.8	1.0	2.20	1.8	3.93	2.6
58	Kasemen	Mesjid Priyayi	3.6	2.5	1.2	1.0	3.80	3.2	2.73	2.5
59	Kasemen	Kilasah	3.9	3.2	1.0	1.0	1.00	2.9	3.55	2.5
60	Curug	Kamanisan	3.1	4.2	1.7	1.0	1.00	1.4	3.11	2.5
61	Curug	Tinggar	2.9	4.8	1.7	1.0	1.00	1.5	2.29	2.4
62	Kasemen	Margaluyu	3.0	2.8	1.3	1.0	2.20	2.3	3.57	2.4
63	Curug	Cipete	2.9	3.8	1.5	1.0	1.00	2.1	3.05	2.4
64	Kasemen	Banten	2.8	1.8	1.9	2.2	2.20	2.1	3.27	2.3
65	Kasemen	Bendung	4.1	1.6	1.1	1.0	1.80	3.3	1.78	2.1
66	Curug	Curugmanis	2.5	3.6	1.3	1.5	1.00	1.9	2.09	2.1

[a]Ratio of integrated health post and toddler
[b]Ratio of early childhood education programs and toddler
[c]Proportion of the population graduated from college
[d]Waste management
[e]Community development and empowerment activities
[f]Ratio of midwives and cadres to population
[g]Average distance to health facilities

studies, low access to clean water and poor sanitation can increase the risk of stunting [31, 32].

Data from 66 urban villages overlayed the hazard and vulnerability index, then obtained four urban villages that have a high risk of stunting and all of them are in Kasemen District. The two urban villages with the highest risk are urban village Kilasah which has a very high level of danger (scale 5), while urban village Bendung has a very high vulnerability (scale 5).

Each urban village has different problems related to the potential for stunting, so specific interventions are needed. However, areas with a high risk of stunting have high poverty rates, many slums, low public education, no waste management, and on average have relatively long distances to health facilities.

These results are in line with several studies that have been conducted. Socio-economic differences, especially in poor households and low education levels, have been shown to have a higher risk of stunting [33–35]. Another aspect that is needed so that the risk of stunting in children under five can be prevented is by providing access to drinking water sources and better management of household solid waste to minimize the occurrence of repeated infections [26, 27]. The results of another study using data from the Indonesian Family Life Survey revealed that maternal education, water and sanitation conditions, household poverty, and access to health facilities strongly influence chronic malnutrition among children in Indonesia [36].

67.3.2 Qualitative Analysis (Interviews, FGDs, Deliberation)

Interviews were conducted with three leaders of the urban village with a "high" category of stunting risk index, namely urban villages Bendung, Kilasah, Mesjid Priyayi, and Warung Jaud. In addition, interviews were also conducted with officials at the Regional Development Planning Agency as an organization that has a planning task, in this case, related to stunting management planning in Serang City.

The interview has been recorded, and keywords were obtained, which became the core of the information. The results obtained from these interviews are:

1. Delivery of information regarding stunting cases in urban village areas

Information about stunting cases in an area must be conveyed to regional leaders immediately so that action would be taken directly. Faster information will delivered with the help of information technology. Research in Tasikmalaya, West Java, Indonesia showed a significant change in the attitude of cadres from before and after the use of the Si Centing (Stunting Detection and Prevention) application, which means that this application has a significant positive effect.

2. Handling vs. preventive measures

Based on information obtained from regional leaders, if there are cases of stunting, the urban village will coordinate with the community health center (*pusat kesehatan masyarakat/puskesmas*) to determine the steps for handling it. Stunting is a case of chronic malnutrition that must be prevented from pregnancy, even from a young age.

3. Understanding of climate change adaptation

All respondents stated that they had never received information about determining priority locus for stunting handling based on the concept of climate change adaptation. It is necessary to share knowledge with stakeholders so they can understand the concept. This method needs to be known because, through the climate change adaptation approach, stunting prevention and management can be carried out simultaneously. Prevention is carried out on locations categorized as medium and low risk, while handling is carried out on locations with high and very high categories.

The Focus Group Discussion (FGD) has been held by the Serang City Health Office by involving all regional apparatus organizations (*organisasi perangkat daerah/OPD*) in the Serang City Government. The FGD discussed the situation analysis of stunting data and other supporting data owned by all regional apparatus organizations of Serang City. This activity obtained some information, suggestions, and input related to stunting.

1. The potential for assistance is quite enormous, but often cannot meet the data requirements.

Assistance from the central government or provincial government in managing residential areas (sanitation, clean water supply, and handling slum areas) is quite a lot, but all of them require regional data before assistance is given. The obstacle that occurs is the inability of the region to provide this data. Of course, this is very unfortunate. Therefore each region needs to have robust data related to the condition of its territory, so that it is easy to get assistance.

2. Stunting is a case of chronic malnutrition so that preventive measures are prioritized.

Handling stunting is not the same as handling malnutrition. Handling stunting have to start from adolescence period, pregnancy, until the first 1000 days of the baby's birth, related to diet and parenting. In addition, it must pay attention to clean and healthy living behavior, environmental hygiene, and adequate sanitation. Therefore, the Health Office runs a program of giving blood boost tablets for young women, and pregnant women, as a stunting prevention measure.

3. The importance of education, campaigns, or socialization regarding early detection, prevention, and handling of stunting.

Dissemination of information about stunting needs to be done through formal education, health campaigns or promotions, and socialization. Formal education needs to be included in the curriculum regarding stunting, early detection, prevention, and treatment. Health promotion and socialization need to be carried out continuously to the community.

4. Improvement of clean and healthy life behaviour as the culture of the community.

Sanitation for the community has often been carried out, in the form of public bathrooms, but in fact, it is not used by the community. This condition becomes an evaluation that the community still needed an education about importance of sanitation, and also clean and healthy life behavior (not open defecation, washing hands, throwing garbage in its place). Any massive development will not succeed without people's understanding of clean and healthy living.

5. Local government budget support is needed in handling stunting

Prevention and handling of stunting is not only the task of one regional apparatus but is the collaboration of all parties. All regional apparatus organizations that play a role in preventing and handling stunting must prepare a program of activities so that a budget must be prepared to implement it.

The Stunting Deliberation (*Rembug/musyawarah*) is held to expose the program of activities and commitments with all stakeholders involved in handling stunting. The difference with the FGD is that the participants in the deliberation consist of elements of local government and non-local government (philanthropy, academics, NGOs, community). The discussion obtained in this deliberation are:

1. Potential involvement of Islamic Philanthropic Institutions in stunting prevention and management

One of the Islamic philanthropic institutions that attended this meeting expressed its willingness to contribute to stunting prevention and treatment efforts. In 2019 this institution also collaborated with the Serang City Health Office in handling malnutrition. This collaboration shows a positive opportunity for the Serang City government to get assistance in handling stunting with non-government institutions such as Islamic philanthropic institutions.

2. The willingness of academics to take part in stunting handling and prevention

The academic stated that his party is ready to cooperate in the prevention and handling of stunting. The Serang City Government can take collaboration with the academics for studies or research on stunting prevention and management. Students undergoing Real Work Lecture (*kuliah kerja nyata/KKN*) assignments can be volunteer to carry out health promotions regarding stunting to the community. The willingness of academics to contribute to the handling of stunting is an opportunity for the Serang City Government.

3. Urban village leaders expect a regulation about utilization of Corporate Social Responsibility (CSR) funds in the handling of stunting in their area so that they can become the other fund than the legitimate Regional Government Expenditure Budget (*Anggaran Pendapatan Belanja Pemerintah Daerah*/APBD).

67.3.3 Regional-Based Management Strategy

The regional-based stunting management strategy considers the results of quantitative and qualitative research. The results of quantitative research showed considered variables in stunting handling based on the concept of climate change adaptation. The risk index of stunting is obtained from the sum of the hazard and vulnerability indices. Meanwhile, susceptibility is obtained from the product of exposure and sensitivity, then divided by adaptive capacity.

The four urban villages with a high risk of stunting, as shown in Table 67.3, the strategy refers to the results of the exposure, sensitivity, and adaptive capacity scores of each urban village. In terms of exposure, the variable poor households (decile 1), urban village Warung Jaud has the highest score (5.0), followed by Mesjid Priyayi (4.3), Kilasah (4.2), and Bendung (3.0) so that the handling of poverty for stunting is prioritized according to this order. Poverty handling for stunting management means that priority assistance is given to households that have children with stunting cases or who have teenage girls, and pregnant or breastfeeding mothers.

Table 67.3 The urban village with the highest risk index score

Sub District	Urban Village	Hazard		Exposure		Sensitivity		Adaptive Capacity							Risk (Score)
		a	b	c	d	e	f	g	h	i	j	k	l	m	
Kasemen	Kilasah	4.3	4.5	1.6	4.4	2.4	1.0	3.9	3.2	1.0	1.0	1.00	2.9	3.55	High (8)
Kasemen	Bendung	2.5	3.3	1.5	3.4	5.0	1.0	4.1	1.6	1.1	1.0	1.80	3.3	1.78	High (8)
Kasemen	Mesjid Priyayi	3.2	3.4	1.9	4.3	2.5	1.0	3.6	2.5	1.2	1.0	3.80	3.2	2.73	High (7)
Kasemen	Warung Jaud	2.1	2.0	1.8	5.0	4.3	2.3	3.7	2.7	1.4	4.3	3.80	2.1	2.58	High (7)

[a]Proportion of short toddler
[b]Proportion of very short toddler
[c]Proportion of toddler
[d]Proportion of poor households
[e]Slum density (person/hectare)
[f]Num of flood events last 5 years
[g]Ratio of integrated health post and toddler
[h]Ratio of early childhood education programs and toddler
[i]Proportion of population graduated from college
[j]Waste management
[k]Community development and empowerment activities
[l]Ratio of mid-wives and cadres to population
[m]Average distance to health facilities

In terms of sensitivity, the variable density of slum areas, urban village Bendung has the highest score (5.0), followed by Warung Jaud (4.3), Mesjid Priyayi (2.5), and Kilasah (2.4), so that handling priority slum areas for handling stunting is given in this sequence. Handling slum areas for stunting management means that slum areas to be handled must prioritize areas with households that have children with stunting cases, or have teenage girls, or have pregnant or breastfeeding mothers. In the variable number of flood events, the urban village of Warung Jaud has a high score (2.3), compared to three other urban villages (1.0), so that flood management and prevention have to be prioritized in the urban village of Warung Jaud.

Adaptive Capacity shows region's resilience to the occurrence of stunting, and there are several variables. The lowest ratio of early childhood education programs and toddlers is in the urban village Bendung, so this area needs to increase access to early childhood education so that education about clean and healthy living behavior, prevention of stunting, and provision of additional food can be given.

The variable population that graduated from college in the four urban villages are equally small, so the community needs to be encouraged to achieve the highest education, and give access or information on scholarships, especially for the poor. Variable of waste management in the three urban villages; Kilasah, Bendung, and Mesjid Priyayi are low, due to the shortage of temporary waste disposal sites, lack of waste banks, and lack of community involvement in waste management. Thus, it is need intervention through the provision of waste disposal sites, initiation of waste banks, and increasing community involvement in waste management in the area.

Community empowerment activities are still low in the urban village Kilasah (the score is 1.0), and the urban village Bendung (1.8). This variable consists of community empowerment activities in the provision of sanitation and cleans water facilities, services for educational, cultural, and health infrastructure, and management of the urban village housing environment. Thus, the interventions that need to be improved in these two urban villages are community empowerment in the provision of sanitation and clean water infrastructure, service of educational, health, and cultural facilities, and the management of the housing environment.

The variable ratio of health workers (midwives and cadres) to the population is lowest in the urban village of Warung Jaud (the score is 2.1), signs that this urban village still lacks midwives and cadres. So it is necessary to recruit new midwives and cadres in the urban village of Warung Jaud. Meanwhile, on the variable distance to health facilities, the urban village Bendung has the lowest score (1.78), so the urban village Bendung requires new health facilities whose location is prioritized in an location where there are stunting cases (Table 67.4).

Table 67.4 Regional-based stunting management strategy

Regional-based stunting management strategy	Priority Locus			
	1st priority	2nd priority	3rd priority	4th priority
Poverty alleviation	Warung Jaud	Mesjid Priyayi	Kilasah	Bendung
Handling slum areas	Bendung	Warung Jaud	Mesjid Priyayi	Kilasah
Flood control	Warung Jaud			
Increasing the number of early childhood education	Bendung			
Access to higher education scholarships (S1)	Kilasah	Bendung	Mesjid Priyayi	Warung Jaud
Waste management	Bendung, Kilasah, Mesjid Priyayi			
Community empowerment activities	Kilasah	Bendung		
Increasing the number of health workers (midwives and cadres)	Warung Jaud	Kilasah		
Increasing the number of health facilities	Bendung	Warung Jaud	Mesjid Priyayi	Kilasah

Determination of regional-based stunting management strategies in addition to the results of quantitative analysis, stunting management strategies are also obtained from the results of qualitative analysis. Strategies based on the results of qualitative analysis are more general which are applied to each priority location for handling stunting, namely:

1. Fast information

Accelerate the delivery of information related to stunting cases to urban village leader.

2. Education

Carry out campaigns/counseling/education on the meaning of stunting, the dangers of stunting, the urgency of preventing and handling stunting through the school curriculum, counseling at an integrated health service.

3. Priority program in the priority location

Prioritizing government activity programs in priority locations for stunting treatment based on the level of urgency per location (urban village), which can be seen from the results of quantitative calculations.

4. Data toughness

Prepare priority areas to have the necessary data so that they can qualify for assistance.

5. Improve culture

Counseling about clean and healthy living behaviour for improving community culture.

6. Improve stunting prevention programs

Improving stunting prevention programs starting from adolescence and pregnancy (giving blood boost tablets, nutritious food, clean and healthy living behavior).

7. Government budget support

Provide local government budget support in stunting management activities.

8. Sharing knowledge about handling stunting with adaptation to climate change

Providing knowledge about stunting handling with a climate change adaptation approach to support sustainable development.

9. Multi-stakeholder cooperation

The Serang City Government, should take advantage of the potential role of non-government stakeholders to make collaboration for handling stunting.

67.4 Conclusion

There are four urban villages with a high category of stunting risk index, namely the urban village Bendung, Kilasah, Mesjid Priyayi, and Warung Jaud, all located in Kasemen District. Regional-based strategies pay attention to the location's condition on the elements of exposure, sensitivity, and adaptive capacity (quantitative analysis result). The elements with a low score becomes a strategic step through interventions, and then accompanied by general strategic steps obtained from the results of qualitative analysis. For example, the urban village of Warung Jaud has priority in the flood control program, so it must be accompanied by budgetary support from the government to implement the program.

References

1. Purwandari, E.S., Estiningtyas Sakilah Adnani, Q., Yuli Astutik, R.: Analysis of maternal age at married, number of children, history of breastfeeding, mother's education and high risk of pregnancy with incidence of stunting in children under five-years. Women, Midwives, Midwifery **1**(1), 21–30 (2021)
2. WHO Child Growth Standards: Dev. Med. Child Neurol. **51**(12), 1002–1002 (2009)
3. Audiena, N.P., Siagian, M.L.: Association between infectious disease and hygiene practice on stunting toddler aged 24–59 months. 149–157 (2021)

4. Permatasari, T.A.E.: The effect of feeding parenting on the incidence of stunting in toddlers. Jurnal Kesehatan Masyarakat Andalas. **14**(2), 3 (2021)
5. Helmizar, R., Putra, D.A.: Risk factors for stunting in children aged 3–5 years in Tanah Datar Regency in 2018 (a follow-up study). Jurnal Kesehatan. **13**(2), 10–16 (2019)
6. National Development Planning Agency (Bappenas): Metadata Indikator. Kementerian PPN/BAPPENAS, Jakarta (2017)
7. Idris, I.: Areas and population groups in Pakistan most exposed to combined effects of climate change, food insecurity and COVID-19 (2021)
8. Ahmad, A.S., et al. Analysis of risk factors for the incidence of stunting in toddlers. J. Health Sci. Prevent. **5**(1), 10–14 (2021)
9. Rafika, M.: The impact of stunting on children's psychological conditions. Bul Jagaddhita. **1**(1), 1–4 (2019)
10. Renyoet, B.S., Martianto, D., Sukandar, D.: Potential economic losses due to stunting in toddlers in Indonesia in 2013. J Gizi dan Pangan. **11**(3), 247–254 (2016)
11. Aditianti, S., Raswanti, I., Izwardy, D., Irianto, S.E.: Prevalence and stunting risk factors in children 24–59 months in Indonesia: analysis of basic health research data 2018. Nutr Food Res [Internet]. **43**(1), 29–40 (2020)
12. Ruhyana, N.F., Nurfindarti, E., Essa, W.Y.: Study on priority locus for handling stunting in Sumedang Regency with climate change adaptation risk assessment approach. J Borneo Adm. **17**(1), 65–88 (2021)
13. Essa, W.Y,, Nurfindarti, E., Ruhyana, N.F.: Strategies for handling stunting in Bandung City. **13**, 15–28 (2021)
14. Soeracmad, Y., Ikhtiar, M., Bintara, A.: The relationship of household environmental sanitation with stunting incidents in toddlers at the Wonomulyo Health Center, Polewali Mandar Regency in 2019. J-KESMAS **5**(2), 138 (2019). Available from: https://journal.lppm-unasman.ac.id/index.php/jikm/article/view/519
15. Julia, M., van Weissenbruch, M.M., Delemarre-van de Waal, H.A., Surjono, A.: Influence of socioeconomic status on the prevalence of stunted growth and obesity in Prepubertal Indonesian children. Food Nutr. Bull. [Internet]. **25**(4), 354–60 (2004). Available from: http://journals.sagepub.com/doi/https://doi.org/10.1177/156482650402500405
16. Woldehanna, T., Behrman, J.R., Araya, M.W.: The effect of early childhood stunting on children's cognitive achievements: Evidence from young lives Ethiopia. Ethiop. J. Heal. Dev. **31**(2), 75–84 (2017)
17. Chirande, L., Charwe, D., Mbwana, H., Victor, R, Kimboka, S., Issaka, A.I., et al.: Determinants of stunting and severe stunting among under-fives in Tanzania: evidence from the 2010 cross-sectional household survey. BMC Pediatr. [Internet]. **15**(1), 165 (2015). Available from: http://bmcpediatr.biomedcentral.com/articles/https://doi.org/10.1186/s12887-015-0482-9
18. Mittal, A., Singh, J., Ahluwalia, S.: Effect of maternal factors on nutritional status of 1–5-year-old children in urban slum population. Indian J. Community Med. [Internet]. **32**(4), 264 (2007). Available from: http://www.ijcm.org.in/text.asp?2007/32/4/264/37691
19. Jones, A.D., Yosef, S.: The implications of a changing climate on global nutrition security. In: The Fight Against Hunger and Malnutrition [Internet], pp. 432–466. Oxford University Press (2015). Available from: http://www.oxfordscholarship.com/view/https://doi.org/10.1093/acprof:oso/9780198733201.001.0001/acprof-9780198733201-chapter-20
20. Tirado, M.C., Hunnes, D., Cohen, M.J., Lartey, A.: Climate change and nutrition in Africa. J Hunger Environ. Nutr. [Internet]. **10**(1), 22–46 (2015). Available from: https://doi.org/10.1080/19320248.2014.908447
21. Myers, S.S., Smith, M.R., Guth, S., Golden, C.D., Vaitla, B., Mueller, N.D., et al.: Climate change and global food systems: potential impacts on food security and undernutrition. Annu. Rev. Public Health [Internet]. **38**(1), 259–277 (2017). Available from: https://www.annualreviews.org/doi/https://doi.org/10.1146/annurev-publhealth-031816-044356
22. Phalkey, R.K., Aranda-Jan, C., Marx, S., Höfle, B., Sauerborn, R.: Systematic review of current efforts to quantify the impacts of climate change on undernutrition. Proc. Natl. Acad. Sci. [Internet]. **112**(33), E4522–E4529 (2015). Available from: http://www.pnas.org/lookup/doi/https://doi.org/10.1073/pnas.1409769112

23. Rodriguez-Llanes, J., Ranjan-Dash, S., Mukhopadhyay, A., Guha-Sapir, D.: Flood-exposure is associated with higher prevalence of child undernutrition in Rural Eastern India. Int. J. Environ. Res. Public Health [Internet]. **13**(2), 210 (2016). Available from: http://www.mdpi.com/1660-4601/13/2/210
24. Cooper, M.W., Brown, M.E., Hochrainer-Stigler, S., Pflug, G., McCallum, I., Fritz, S., et al.: Mapping the effects of drought on child stunting. Proc. Natl. Acad. Sci. [Internet]. **116**(35), 17219–17224 (2019). Available from: http://www.pnas.org/lookup/doi/https://doi.org/10.1073/pnas.1905228116
25. Chang, S.M., Walker, S.P., Grantham-McGregor, S., Powell, C.A.: Early childhood stunting and later behaviour and school achievement. J. Child Psychol. Psychiatry Allied Discip. **43**(6), 775–783 (2002)
26. Sugianto, M.A.: Policy analysis of stunting prevention and control in Indonesia: by approach what is the problem represented to be? Jurnal Ekonomi, Manajemen, Bisnis, dan Sosial (Embiss) **1**(3), 197–209 (2021)
27. Irianti S, Prasetyoputra P, Dharmayanti I, Azhar K, Hidayangsih PS. The role of drinking water source, sanitation, and solid waste management in reducing childhood stunting in Indonesia. IOP Conf Ser Earth Environ Sci. 2019;344(1):0–9.
28. Ivankova, N.V., Creswell, J.W., Stick, S.L.: Using mixed-methods sequential explanatory design: from theory to practice. Field Methods [Internet]. **18**(1), 3–20 (2006). Available from: http://journals.sagepub.com/doi/https://doi.org/10.1177/1525822X05282260
29. IPCC: Summary for policy makers. In: Field, C.B., Barros, V.R., Dokken, D.J., K.J. A, (eds.) LLW, editors. Climate Change 2014: Impacts, Adaptation, and Vulnerability Part A: Global and Sectoral Aspects Contribution of Working Group II to the Fifth Assessment Report of the Intergovernmental Panel on Climate Change [Internet], pp. 1–32. Cambridge University Press, Cambridge, United Kingdom and New York, NY, USA; 2014. Available from: https://www.cambridge.org/core/product/identifier/CBO9780511976988A014/type/book_part
30. National Disaster Management Agency of the Republic of Indonesia (BNPB): Indonesia Disaster Risk Index Tahun 2018 [Internet]. 2018; 115 p. Available from: https://www.bnpb.go.id//uploads/renas/1/BUKURENASPB.pdf
31. Mbuya, M.N.N., Humphrey, J.H.: Preventing environmental enteric dysfunction through improved water, sanitation and hygiene: an opportunity for stunting reduction in developing countries. Matern. Child Nutr. [Internet]. **12**, 106–120 (2016). Available from: https://onlinelibrary.wiley.com/doi/https://doi.org/10.1111/mcn.12220
32. Rahayu, R.M., Pamungkasari, E.P., Wekadigunawan, C.: The biopsychosocial determinants of stunting and wasting in children aged 12–48 Months. J Matern. Child Heal [Internet]. **03**(02), 105–118 (2018). Available from: http://thejmch.com/index.php?journal=thejmch&page=article&op=view&path%5B%5D=85&path%5B%5D=95
33. Ruth Sharon Apio, B., Mawa, R., Lawoko, S., Nand Sharma, K.: Socio-economic Inequality in stunting among children aged 6–59 months in a Ugandan population based cross-sectional study. Am. J. Pediatr. **5**(3), 125 (2019)
34. Chowdhury, T.R., Chakrabarty, S., Rakib, M., Afrin, S., Saltmarsh, S., Winn, S.: Factors associated with stunting and wasting in children under 2 years in Bangladesh. Heliyon [Internet]. **6**(9), e04849 (2020). Available from: https://linkinghub.elsevier.com/retrieve/pii/S2405844020316923
35. Sunguya, B.F., Zhu, S., Mpembeni, R., Huang, J.: Trends in prevalence and determinants of stunting in Tanzania: an analysis of Tanzania demographic health surveys (1991–2016). Nutr J [Internet]. **18**(1), 85 (2019). Available from: https://nutritionj.biomedcentral.com/articles/https://doi.org/10.1186/s12937-019-0505-8
36. De Silva, I., Sumarto, S.: Child malnutrition in Indonesia: can education, sanitation and healthcare augment the role of income? J Int Dev. **30**(5), 837–864 (2018)

Chapter 68
Palu Koro Fault Zone Microzonation and Its Relation to Potential Hazard

Muhammad Hanif, Adrin Tohari, and Dadan Dani Wardhana

Abstract The Mw 7.4 earthquake on September 28, 2018 in Palu-Donggala has resulted in significant fatalities and damages due to liquefaction and ground oscillation. Small-scale subsurface geological features on the zone are presumed to be a controlling factor for liquefaction. Therefore, we ran 130 microtremor measurements around the southern Palu region to study the microzonation related to hazard-prone areas, especially for liquefaction and collapsed construction. A series of single-station surveys were conducted to delineate the region's predominant period, amplification factor, and seismic vulnerability distribution. The resulting microzonation maps were cross-validated with geological maps, liquefaction zone, and collapsed building distribution. The study suggested several hazard-prone areas that should be paid attention to, especially for construction planning.

68.1 Introduction

The Palu Koro fault is one of the most active geological structures in Indonesia. The slip rate reaches up to 42 mm/year [9] and caused numerous big earthquakes in the region. The 7.4 Mw earthquake on September 28th 2018 ruptured 67 km inland fault segments [3, 6] and resulted in 1407 fatalities and 65.733 damaged construction in the whole Palu, Sigi, and Donggala regency. Due to the 2018 earthquake, the southern Palu area was strike by some major derivate damages such as tsunami and building collapsed from liquefaction and ground amplification.

Flow type liquefaction generally occurs in areas with a thin layer of sand beneath the dense soil layer. This type of liquefaction stroked some areas in the Palu and Sigi region (Fig. 68.1) such as in Balaroa and Petobo. The less damaging type of liquefaction, sandboiling also occurred on several points during the 2018 event. The controlling factors of liquefaction are influenced by geological, hydrological,

M. Hanif (✉) · A. Tohari · D. D. Wardhana
Research Center for Geological Disaster, Research Organization for Earth Sciences and Maritime, National Research and Innovation Agency (BRIN), Bandung 40153, Indonesia
e-mail: muha170@brin.go.id

E. Yulihastin et al. (eds.), *Proceedings of the International Conference on Radioscience, Equatorial Atmospheric Science and Environment and Humanosphere Science, 2021*, Springer Proceedings in Physics 275, https://doi.org/10.1007/978-981-19-0308-3_68

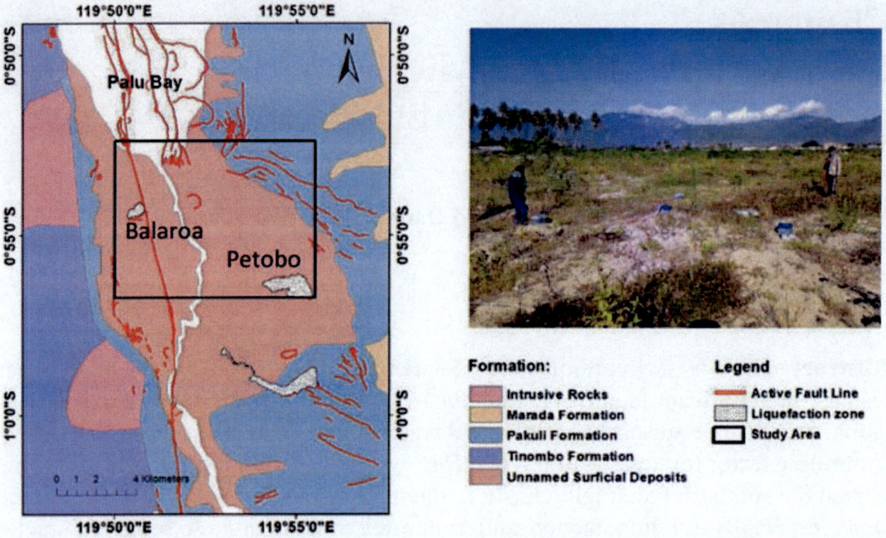

Fig. 68.1 Left: Geological map of Palu Quadrangle (after [10] showing the surface rock distribution, liquefaction area (black polygon), and active fault line (red line) [6]. The study area (black square) indicates Balaroa in NW and Petobo in SE direction. Right: Photo of microtremor data acquisition in the Balaroa area and geological map legend

and subsurface soil structure conditions. On the other hand, the building collapsed also caused many casualties due to its ground oscillation. Thus, it is essential to understand the factors of geological conditions and subsurface structures that caused liquefaction and ground shaking in Palu and the surrounding region. Such knowledge is essential to the regional development and hazard mitigation plan from the earthquake-collateral hazard of the region.

The southern Palu area is composed of alluvial deposits in the valley, granite and granodiorite rocks in the west to north area, schist-phyllitic and molasses in the eastern part [10]. The distribution of surface geological rocks can be seen in Fig. 68.1.

The microtremor method utilizes traffic, industry, and other human activities to illuminate subsurface geological structures [4]. The method is used to study the characteristics of dynamic subsoil surface, such as resonance frequency and seismic vulnerability index, using its horizontal to vertical spectral ratio (HVSR) analysis [5]. The damage associated with an earthquake depends not only on the magnitude of the earthquake and the distance of the epicenter but also on its local geological conditions [1]. The phenomenon is known as the local site effect, caused by the contrast impedance of fine sedimentary material layers on the bedrock.

The liquefaction potential is accurately gained from geotechnical techniques such as standard penetration tests and cone penetration tests. These geotechnical methods provide detailed subsurface features, such as soil types, thickness, strength, and groundwater level. However, these methods are expensive and time-consuming compared to other geophysical surveys such as microtremor. Many previous studies have attempted the microtremor method to elucidate subsurface conditions in Palu liquefaction-prone areas [2, 7, 8, 11]. Using the microtremor method, [11] analyzed the HVSR result using year 2013 data and concluded that the high predominant period is associated with alluvial fan deposited zone. Rusydi et al. [8] utilized the average shear wave velocity on 30-m soil (Vs30) to delineate the site class distribution in Palu and concluded that the whole Palu city lies on thick, soft soil. [7] concluded from the seismic vulnerability scale (Kg) that no area in Palu is safe from earthquake-triggered hazards. Moreover, [2] using pre-2018 data, delineate the ridge-like local subsurface structure, which probably blocked the groundwater flow and triggered liquefaction. However, the studies primarily utilized pre-2018 earthquake microtremor data while the dynamic of soil layers and their physical properties might differ after some earthquakes. Thus, no previous studies have been attempted to illuminate microzonation factors controlling the liquefication and ground oscillation using post-2018 Palu earthquake microtremor data.

This paper presents the results of a microzonation study to evaluate subsurface conditions controlling liquefaction and building collapsed prone areas using the microtremor method. The objectives of this study are to (1) obtain the predominant period, amplification factor, and seismic vulnerability map and (2) evaluate the areas prone to those hazards.

68.2 Data and Method

The microtremor data acquisitions were conducted on 130 locations around southern Palu area, particularly in the regency of Ulujati, East Palu, South Palu, Tatanga, and small part of West Palu and northern Sigi Regency (Fig. 68.2). The measurement took 15–20 min at each location using the McSEIS MT Neo accelerometer and Geodaqs velocitymeter. The SeisImager/SW and Geonet software were utilized to process the 3-component data to obtain the horizontal to vertical ratio, dominant amplitude, and period of each location.

The time-domain data were segmented into 30–40 parts with 20 s each. The segments with big noises were unselected to the spectral calculation processing. The least noise segments on every dataset were chosen and subjected to Fast Fourier Transform to analyze its spectral variation.

The horizontal to vertical (H/V) spectral ratio r(f) is derived from its NS, EW, and vertical component amplitude (F_{NS}, F_{EW}, F_{UD}, respectively) using the following Eq. 68.1:

$$r(f) = \frac{\sqrt{F_{NS}(f) + F_{EW}(f)}}{F_{UD}(f)} \tag{68.1}$$

The H/V spectra on the location were calculated over those best segments to obtain the predominant amplitude and period on each microtremor station. The resulting map was cross-matched between the distribution of geological map, damaged

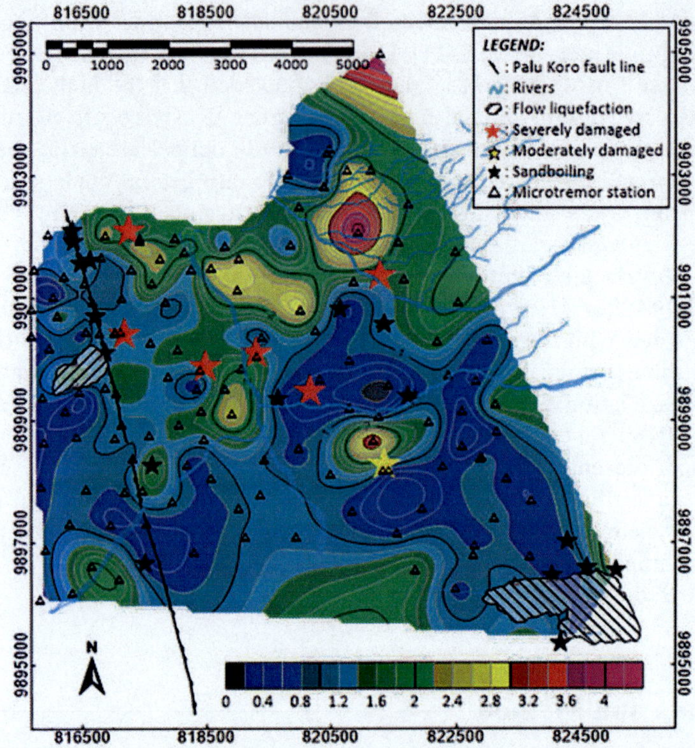

Fig. 68.2 The predominant period map on southern Palu. Black dashed lines indicate the Palu Koro fault. Black stars indicate the location of liquefaction, red and yellow stars indicate severely and moderately damaged buildings, respectively. Coordinate in WGS84 UTM Zone 50S

building, and liquefaction location to better understand hazard distribution in the area. Field observation was done to obtain the damaged buildings and liquefaction location data.

The seismic vulnerability index was calculated from Eq. 68.2, below:

$$Kg = \frac{Ao^2}{fo} \tag{68.2}$$

where Ao and fo are predominant amplitude and frequency, respectively, which is extracted from H/V spectra on each station.

68.3 Results and Discussions

The spectral transformation allows the predominant frequency and amplitude to be analyzed. Those parameters are essential to elucidate soil amplification features on the area, especially on the high probability hazard zone such as Palu. Basically, the west and eastern part of study area are dominated by the Quartnerary old alluvium fan deposit. The Tertiary granitoid and schist-phylitics formation are on the west- and eastward, respectively. The predominant frequency, amplification factor, and seismic vulnerability map are shown on Figs. 68.2, 68.3, and 68.4, respectively.

The predominant period value in southern Palu city has a range of 0.2–4.4 s. The Palu Koro fault zone is mainly situated on a 0.8–1.2 s predominant period area (Fig. 68.2). These relatively low dominant period zones were slightly isolated 100–2000 m from the fault line and associated with liquefaction area. The liquefaction phenomena close to the Palu Koro fault line (black star symbol near the black dashed line on Fig. 68.2) were sand boiling type. On the other hand, the flow type liquefaction on the Balaroa zone (western side of fault line) has a larger damage area than the

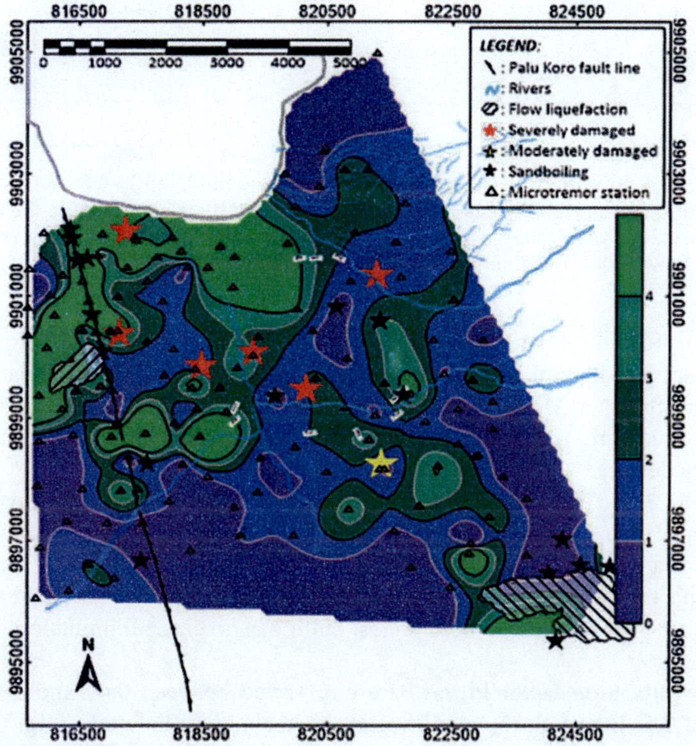

Fig. 68.3 The amplification factor map in southern Palu. Black stars indicate the location of liquefaction, red and yellow stars indicate severely and moderately damaged buildings, respectively. Coordinate in WGS84 UTM Zone 50S

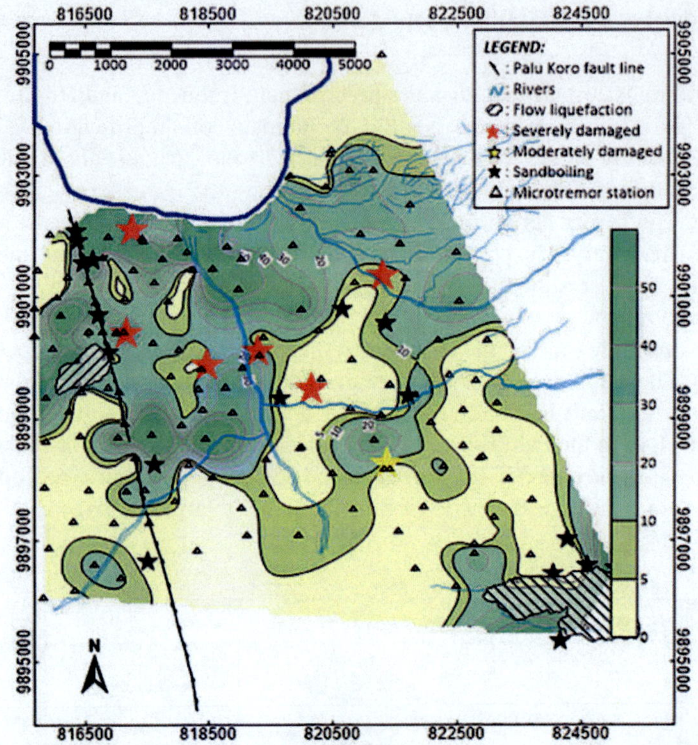

Fig. 68.4 Seismic vulnerability index map in southern Palu. Black stars indicate the location of liquefaction, red and yellow stars indicate severely and moderately damaged buildings, respectively. Coordinate in WGS84 UTM Zone 50S

latter, also situated on the low predominant period. However, the post-liquefaction predominant period map could not be clearly indicated and might indicate a more complex mechanism. Moreover, the flow type liquefaction in the Petobo area (south-east of the study area) also has a low dominant period value but no clear delineation due to data density.

Figure 68.2 also indicates the presence of a subsoil zone with a predominant period value greater than 1.4 s from the coast to the center of Palu City. This indicates the presence of a thick-soft soil layer so that there might be a subsurface topography that controls the thickness of the soft soil layer at this location. Most of the damaged buildings are also situated in these zones, but it has no clear illumination in several locations.

The amplification factor in the Palu city varied between 0.49 and 10 with an average of 2.5. Based on the amplification factor map in Palu city (Fig. 68.3), the location of damaged buildings was associated with the high amplification factor area (>3 scale). Additionally, the liquefaction zone on Balaroa and Petobo is also located in a high category amplification factor zone, which corresponds to high ground shaking when triggered by relatively strong earthquake. However, half of the sand boiling locations are on the low amplification factor map (<2 scale). Consequently, the sand boiling liquefaction was better illuminated in the predominant period map than the amplification factor results. In contrast, the damaged building zone was better delineated on the amplification map.

According to the seismic vulnerability map interpretation in Fig. 68.4, the high vulnerability zones were associated with a seismic vulnerability index of higher than 10 scale, medium zones were in 5–10 scale, and low zones were below 5 scale. Primarily in the low seismic vulnerability areas, there was no report for severely and moderately damaged buildings. However, the low zone bordered with the high zone also has hazard potential due to its isolation from the high seismic vulnerable zone. From the above microzonation results, the ridge-like structure areas might be indicated by a boundary zone between low, medium, and high categories of the predominant period, amplification factor, or seismic vulnerability zone. This subsurface ridge-like structure might block and accumulate groundwater in the sediment, which caused it to be highly water-saturated [2]. Those ridge-like structure areas need to be well studied for liquefaction and ground amplification potential for future building and development.

68.4 Conclusions

The low-scale predominant period map (0.8–2.1 s) was associated with liquefaction zones. On the other hand, a high category of amplification factor (>2 scale) was related to damaged building-prone areas. Additionally, the high seismic vulnerability zone (>5 scale) indicated more conclusive areas prone to building collapse and liquefaction triggered by an earthquake. Looking at the Palu basin geometry from the microzonation map, we may locate the possible vulnerable zone for the liquefaction and ground amplification. However, more dense stations coupled with various geological and geophysical methods (e.g., well-log, resistivity, penetration test, and ground penetrating radar) should be undertaken to better understand the potential hazard and controlling mechanism of each hazard in the Palu city.

Acknowledgements The authors wish to thank Prof. Danny Hilman Natawidjaja and Dr. Mudrik R. Daryono for providing excellent Palu-Koro active fault data. The research is supported by INSINAS Kemenristekdikti (070/RPL-LIPI/INSINAS-1/II/2019) research funding. All the authors are equally contributed to this manuscript.

References

1. Bard, P.Y.: Microtremor measurements: a tool for site effect estimation. In: The Effects of Surface Geology on Seismic Motion, vol. 3, pp. 1251–1279 (1999)

2. Cipta, A., Rudyanto, A., Afif, H., Robiana, R., Solikhin, A., Omang, A., Supartoyo, Hidayati, S., 2020. Unearthing the buried Palu–Koro Fault and the pattern of damage caused by the 2018 Sulawesi Earthquake using HVSR inversion. Geol. Soc. London, Spec. Publ. SP501-2019-70. https://doi.org/10.1144/sp501-2019-70
3. Gunawan, E., Widiyantoro, S., Supendi, P., Nishimura, T.: Identifying the most explainable fault ruptured of the 2018 Palu-Donggala earthquake in Indonesia using coulomb failure stress and geological field report. Geod. Geodyn. (2020) https://doi.org/10.1016/j.geog.2020.04.004
4. Nakamura, Y.: Seismic vulnerability indices for ground and structures using microtremor. World Congr. Railw. Res. (1997)
5. Nakamura, Y.: Method for dynamic characteristics estimation of subsurface using microtremor on the ground surface. Q. Rep. RTRI (Railw. Tech. Res. Inst.) (1989)
6. Natawidjaja, D.H., Daryono, M.R., Prasetya, G., Udrekh, Liu, P.L.F., Hananto, N.D., Kongko, W., Triyoso, W., Puji, A.R., Meilano, I., Gunawan, E., Supendi, P., Pamumpuni, A., Irsyam, M., Faizal, L., Hidayati, S., Sapiie, B., Kusuma, M.A., Tawil, S.: The 2018 Mw7.5 Palu 'supershear' earthquake ruptures geological fault's multi-segment separated by large bends: results from integrating field measurements, LiDAR, swath bathymetry, and seismic-reflection data. Geophys. J. Int. (2020) https://doi.org/10.1093/gji/ggaa498
7. Nugraha, J., Jaya, A., Alimuddin, I.: Identification of Palu-Koro fault mechanism and earthquake hazard vulnerability of. Int. J. Eng. Sci. Appl. **6** (2019)
8. Rusydi, M., Efendi, R., Sandra, Rahmawati: Earthquake hazard analysis use Vs30 data in Palu. J. Phys. Conf. Ser. **979** (2018). https://doi.org/10.1088/1742-6596/979/1/012054
9. Socquet, A., Simons, W., Vigny, C., McCaffrey, R., Subarya, C., Sarsito, D., Ambrosius, B., Spakman, W.: Microblock rotations and fault coupling in SE Asia triple junction (Sulawesi, Indonesia) from GPS and earthquake slip vector data. J. Geophys. Res. Solid Earth **111**, 1–15 (2006). https://doi.org/10.1029/2005JB003963
10. Sukamto, R.: Reconnaissance geologic map of Palu Area, Sulawesi–scale 1: 250,000. Geol. Surv. Indones. Dir. Miner. Resour. Geol. Res. Dev. Centre, Bandung, Open File (1973)
11. Thein, P.S., Pramumijoyo, S., Brotopuspito, K.S., Wilopo, W., Kiyono, J., Setianto, A.: Site response characteristics of H/V spectrum by microtremor single station observations at Palu City. Indonesia. J. Appl. Geol. **5**, 1–9 (2015). https://doi.org/10.22146/jag.7202

Chapter 69
Utilizing the Global Positioning System (GPS) Data from Sumatran GPS Array (SuGAr) to Measure Seismic Deformation Related to the Mw7.6 Padang Earthquake 2009

Galang P. Refindo, Nabilla F. Syafitri, Tsany N. A. Yenuar, Muhammad Hamidi, Marzuki Marzuki, Elistia L. Namigo, and Deasy Arisa

Abstract The Sumatran GPS Array (SuGAr), widely distributed in Sumatra, has provided the possibility to measure various mechanisms of plate deformations. This study analyzes deformation in the pre-seismic, co-seismic, and post-seismic phases related to the Mw7.6 Padang earthquake, September 30, 2009. GPS data were used to understand the deformation at several areas with various distances from the epicenter. Ten SuGAr data in RINEX format is processed using GAMIT/GLOBK software to obtain three-dimensional daily position data. The value of postfit nrms <0.25 for 96.98% data, Wide Lane >90%, and Narrow Lane >85% confirmed the reasonable processing procedures and the excellent ability to correct and eliminate errors due to ambiguity. This study demonstrated a stable rate of plate movement before the earthquake, followed by a sudden change at the co-seismic phase, where the GPS stations ABGS, BTHL, KTET, JMBI, LAIS, LNNG, MNNA, MSAI, NGNG, and PKRT showed significant deformations. GPS stations in the Mentawai Backthrust zone (BTHL, MSAI, NGNG, KTET, PKRT) showed the 2.62–47.13 mm southwestward movement at the co-seismic phase, the GPS stations in the west coast of Sumatra (ABGS, LNNG, LAIS, and MNNA) have been moved 2.26–25.13 mm to southeastward (ABGS), northwestward (LNNG), northeastward (LAIS), and southwestward (MNNA), and the GPS station in the north coast of Sumatra (JMBI) have been moved 3.26 mm to northwestward. The results also show that the further away from the epicenter, the smaller the deformation. The result also indicates that the plate movement returns to its pre-seismic stable rate immediately after the earthquake.

G. P. Refindo · N. F. Syafitri · T. N. A. Yenuar · M. Hamidi · M. Marzuki (✉) · E. L. Namigo
Department of Physics, Andalas University, Padang 25163, Indonesia
e-mail: marzuki@sci.unand.ac.id

D. Arisa
Research Center for Geotechnology, Indonesian Institute of Sciences (LIPI), Bandung 40135, Indonesia

E. Yulihastin et al. (eds.), *Proceedings of the International Conference on Radioscience, Equatorial Atmospheric Science and Environment and Humanosphere Science, 2021*, Springer Proceedings in Physics 275, https://doi.org/10.1007/978-981-19-0308-3_69

69.1 Introduction

The tectonic setting of West Sumatra is quite complex where the Mentawai Megathrust zone traverses this area in the southwest of the Mentawai Islands, the Mentawai Backthrust zone from Siberut to South Pagai, and the Sumatran Fault zone, which passes through several regions of West Sumatra. The Mentawai Megathrust Zone is a subduction area of the Australian plate to Sumatra, which is above the Eurasian Plate. This zone has caused several major earthquakes, such as the 1797 earthquake (Mw = 8.4) and the 1833 earthquake (Mw = 8.9). The Mentawai Backthrust Zone is an active fault zone covering South Pagai Island, North Pagai Island, Sipora Island, and Siberut Island in the Mentawai Islands [1]. The Sumatra Fault Zone is divided into small segments that pass through several areas in West Sumatra, namely the Angkola segment, the Barumun segment, the Suliti segment, and the Siulak segment [2]. These small segments are active faults that have the potential to produce earthquakes with a magnitude <7.7, which are close to residential areas, such as the Padang earthquake in 2009 (Mw = 7.6) [3].

The Padang earthquake occurred on September 30, 2009, at 10:16:09 UTC with a magnitude of 7.6 and the epicenter position at 0.720° south latitude and 99.867° east longitude. The epicenter of this earthquake was 30 km southwest of Pariaman, 60 km from Padang City, with a depth of 81 km from the earthquake's hypocenter that could be felt up to ± 600 km from the earthquake epicenter [4]. The 2009 Padang earthquake (Mw = 7.6) resulted in losses of 3550 people, 13,154 buildings heavily damaged, 3877 buildings moderately damaged, and 6787 buildings lightly damaged, landslides and power cuts in the Padang—Pariaman area [5]. This earthquake was caused by the strike-slip movement of the Sumatran Faults and the influence of oblique movements from the Australian to the Eurasian plate, and not the release of the 200-year energy accumulation from the Mentawai in the Mentawai Segment [6].

Earthquakes cause the surrounding earth's crust to deform in both vertical and horizontal directions. Deformation caused by earthquakes can be estimated using a geodetic approach using the Global Positioning System (GPS). Monitoring with GPS can be done periodically or continuously. The periodic method is carried out by observing the movement of the earth's plates through changes in the coordinates of several points on the plate from time to time or a specific time interval. In contrast, in the continuous method, observations are carried out per day. GPS can be used to observe ground deformation due to earthquakes by measuring changes in the position of GPS stations around the earthquake area during pre-seismic, co-seismic, and post-seismic, which are processed in graphical format [7].

Several studies on GPS observations related to deformation caused by the Padang earthquake, 2009 (Mw = 7.6) have been carried out by Maiyudi et al. [8], Iqbal et al. [9], and Arisa et al. [10]. They have analyzed the deformation of the pre-seismic, co-seismic, and post-seismic phases caused by the Padang 2009 (Mw = 7.6) earthquake using GAMIT/GLOBK. This study also analyzes the deformation due to the Padang 2009 (Mw = 7.6) earthquake using the same technique as the previous study. However, this study adds a correlation analysis between the distance of the earthquake source and the velocity of deformation. The results of this study are expected to complement the results of previous research on the observation of deformation due to the Padang 2009 (Mw = 7.6) earthquake by geodetic approach.

69.2 Data and Methods

This study uses a geodetic approach based on the relative changes of GPS receiver points to satellites in the pre-seismic, co-seismic, and post-seismic phases. The 2009 Padang earthquake (Mw = 7.6) data and the parameters were obtained from the USGS (United States Geological Survey) catalog. Furthermore, we also used GPS data from Sumatran GPS Array (SuGAr) observation in RINEX (Receiver Independent Exchange) format. Other data were used for data correction. The GNNS (GPS) data has a precision scale of deformation changes up to millimeters.

69.2.1 Data

Earthquake Data

The 2009 Padang (Mw = 7.6) earthquake occurred on September 30, with the epicenter at −0.788° N, 99.961° E, and a hypocenter of 80 km [4]. The location of the earthquake can be seen in Fig. 69.1.

Rinex Data

We used rinex data from SuGAr (Sumatran GPS Array) and IGS (International GNSS Service) stations for pre-seismic, co-seismic, and post-seismic phases. The pre-seismic phase was at DoY 001–269 before the earthquake, i.e., January 01, 2009–September 26, 2009. On the day of the main-shock, the co-seismic phase was at DoY 270–274 (September 30, 2009). The post-seismic phase was at DoY 275–365, namely October 2, 2009–December 31, 2009. The rinex data that can be downloaded via FTP (File Transfer Protocol) provided by CDDIS (Crustal Dynamics Data Information System), SOPAC (Scrips Orbit and Permanent Array), and UNAVCO (University NAVSTAR Consortium) for 10 (ten) SuGAr stations are ABGS, BTLH, JMBI, KTET, LAIS, LNNG, MNNA, MSAI, NGNG, and PKRT. The station location can be seen in Fig. 69.2. In addition, we also used the rinex data from 12 IGS stations as references point from SuGAR stations to satellites. The IGS stations are ALIC, BJFS, CCJM, COCO, DGAR, IISC, IRKM, NNOR, POHN, PNGM, POL2, XMIS, which can be seen in Fig. 69.3. A reference point was used to estimate the absolute position of the SuGAR station and IGS station to reduce bias in determining the position of each station to the position of the satellite.

The rinex data consists of the SuGAr station's, IGS station's rinex raw, and IGS satellite orbit data. Observation data has the extension *.YYo (YY: 2 digit year, o: observation file) and navigation data has the extension *.YYn (YY: 2 digit year, n:

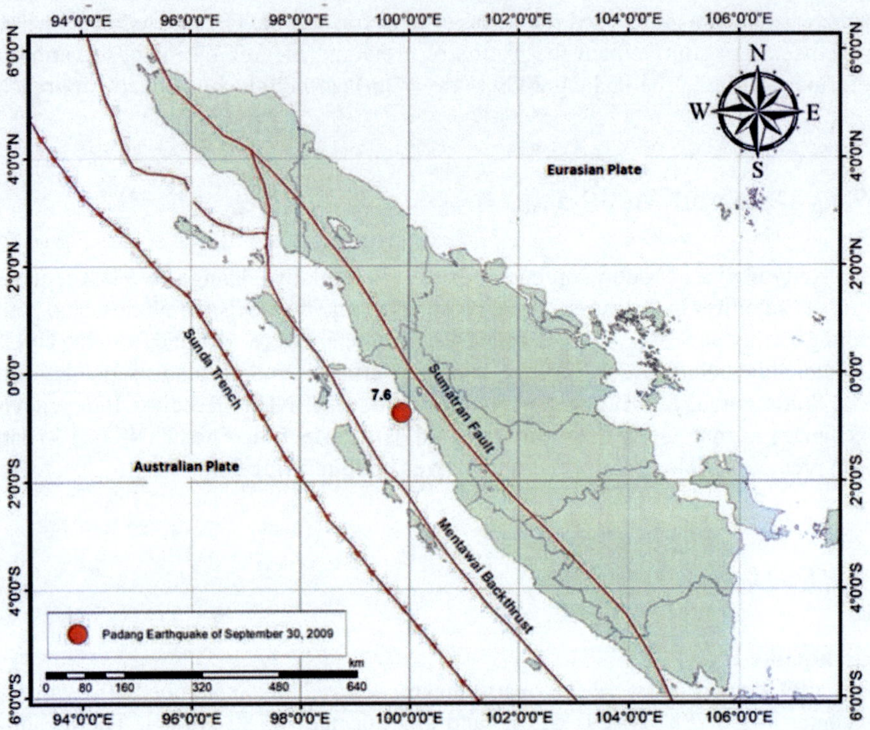

Fig. 69.1 Location of the epicenter of the 2009 Padang earthquake (Mw = 7.6) (red circle)

navigation file), while orbital data has the extension *Y.DDD.fit (Y: 1 digit year, DDD: 3 digit day, fit: file orbit), extension *Y.DDD.rms (Y: 1 digit year, DDD: 3 digit day, rms: file orbit), and extension *Y.DDD (Y: 1 digit year, DDD: 3 digit day). We also use the model data of atmospheric, tidal, and weather as correction data. Correction data are helpful for reducing the bias generated by orbital errors, clock errors, and ambiguity cycles of atmospheric and ionosphere effects to obtain results close to the actual value.

Instruments and Software

The Generic Mapping Tools (GMT) software were used to plot the position of the earthquake, SuGAR station, IGS station, and plotting the time-series graph from the results of GAMIT/GLOBK processing. GAMIT/GLOBK software can be downloaded at http://geoweb.mit.edu/gg/pre.php. This study used GAMIT/GLOBK software in processing SuGAR and IGS rinex data to obtain three-dimensional position calculation data.

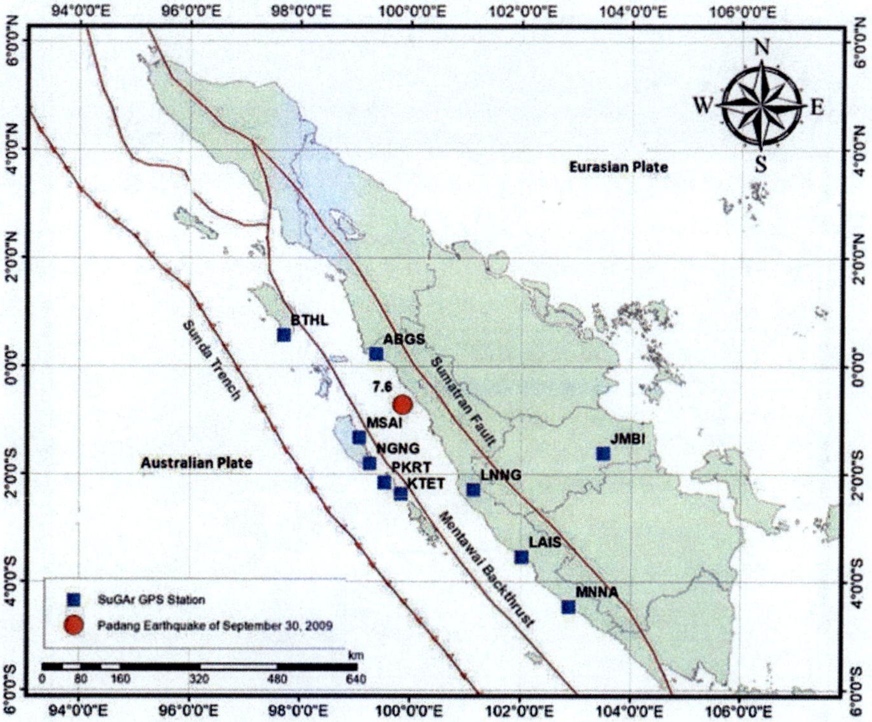

Fig. 69.2 SuGAR observation station used in the study (blue square) and 2009 Padang earthquake (Mw = 7.6) epicenter (red circle)

69.2.2 Procedure

Data Preparation

The first step is collecting earthquake parameters data from the USGS catalog https://earthquake.usgs.gov/earthquakes/eventpage/usp000h237/executive. Earthquake parameters were used are origin time, epicenter position (latitude and longitude), depth (hypocenter), focus mechanism (strike, dip, rake). The second step is collecting SuGAr data via FTP provided by the CDDIS (Crustal Dynamics Data Information System), SOPAC (Scrips Orbit and Permanent Array), and UNAVCO (University NAVSTAR Consortium). The last step is collecting data correction via ftp://everest.mit.edu/pub/GRIDS.

Data Processing

Data processing consists of two stages, namely (1) data processing using GAMIT/GLOBK software and (2) data plotting using a generic mapping tool (GMT). Before data processing, make sure the 8 (eight) required input data are complete. The eight input data are rinex GPS data, l-file, station.info, session.info, navigation, sestbl,

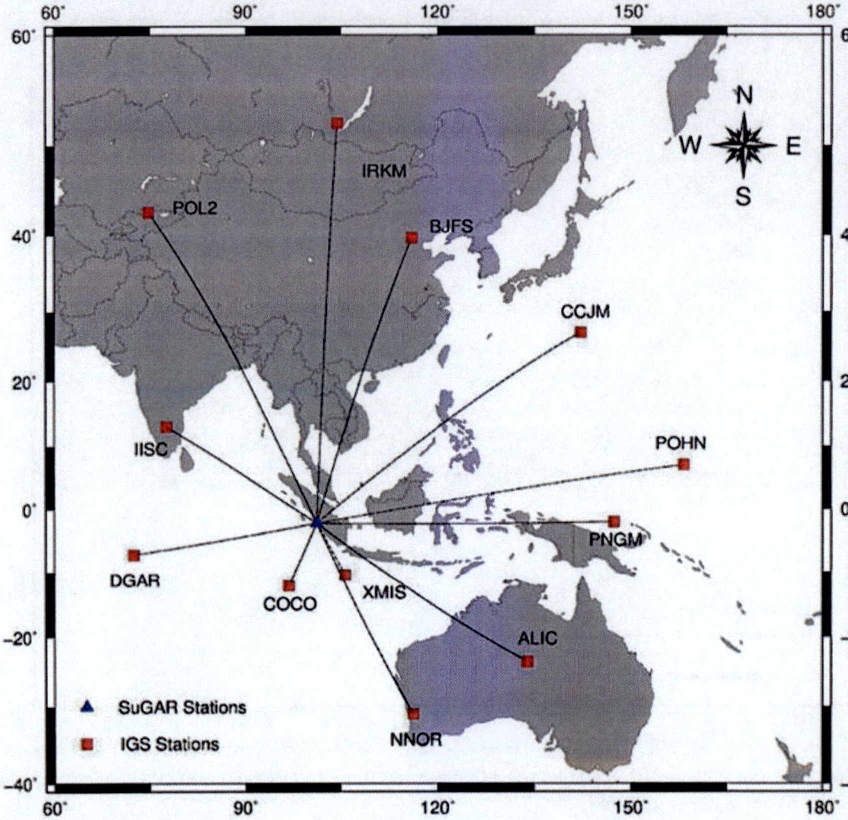

Fig. 69.3 Location of the International GNSS Service station (red square) as references points to the SuGAR station (blue triangle)

sittbl, and ephemeris data. Data processing using GAMIT/GLOBK consists of several stages, which can be seen in Fig. 69.4. After getting the GAMIT/GLOBK processing results, we plot the results using GMT to get the time series to see deformation caused by the 2009 Padang earthquake (Mw = 7.6).

69.3 Results

69.3.1 Time Series of Direction Shift of Receiver's Location

The GAMIT process was produced two results; automatic batch processing and repeatability processes. The automated batch processing produces a summary file (sh_gamit < DoY > .summary) with a reasonably good quality value in the postfit

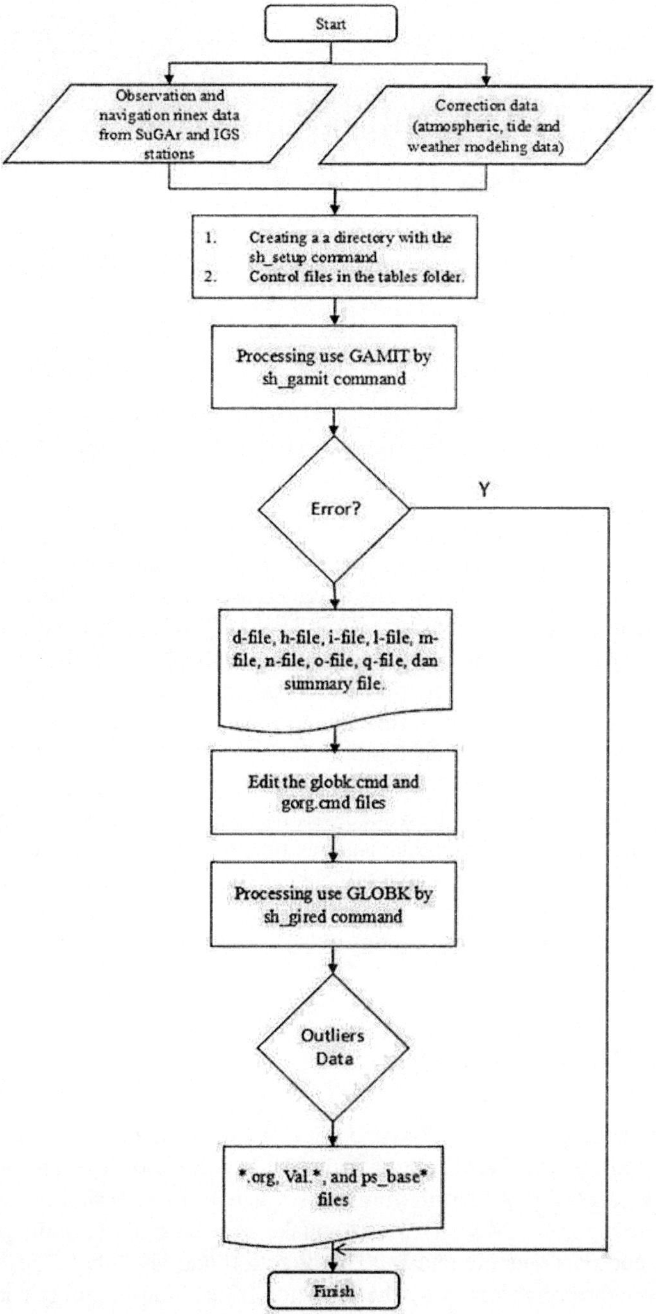

Fig. 69.4 Data processing by GAMIT/GLOBK

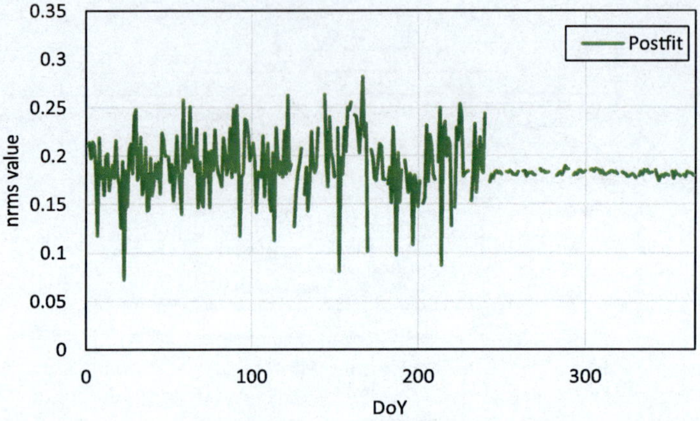

Fig. 69.5 Postfit nrms value (green line) per-Days of Year (DoY) observations for 2009

nrms (normalized root mean square) average DoY <0.25. Although there are 10 (ten) postfit nrms with a value >0.25, the DoY data with a postfit nrms value <0.25 has 96.98%, indicating that the results obtained are reasonable. The highest postfit nrms value from the observations was 0.28181 at 167 DOY, while the smallest value was 0.07245 at 23 DoY. The postfit nrms value is smaller than the prefit nrms for each DoY, indicating that automatic batch processing has successfully eliminated the cycle slips effect. The postfit nrms value for each DoY can be seen in Fig. 69.5.

GAMIT processing in the repeatabilities process obtained values of the wide lane (WL) >90% and narrow lane (NL) >85%, which represents the percentage of error removal due to ambiguity has been successfully removed. The percentage values of WL and NL can be seen in Fig. 69.6. The postfit, WL, and NL values are pretty good even though the observed data is incomplete caused by data voids in several DoY at several stations. Thus, this limitation does not really affect data processing results because WL and NL obtained are pretty good.

69.3.2 GLOBK Processing

GLOBK processing generates *.pos output in the form of geocentric and local topocentric coordinates. This result provides the position movement for each SuGAr station on the east (mm) and North (mm) axes. Changes in the position of each station due to the Padang 2009 (Mw = 7.6) earthquake were carried out in the pre-seismic, co-seismic, and post-seismic phases. The direction and velocity values of changes or deformation are represented by the direction of each observation station toward east and north. The values for the direction and velocity movement of each station can be seen in Fig. 69.6.

Figure 69.7 shows the time series for each station with the Kalman Filtering Method. The error values and NRMS as an indication of the error level of the results were also given. Several stations have WRMS values >5 mm, i.e., BTHL, ABGS,

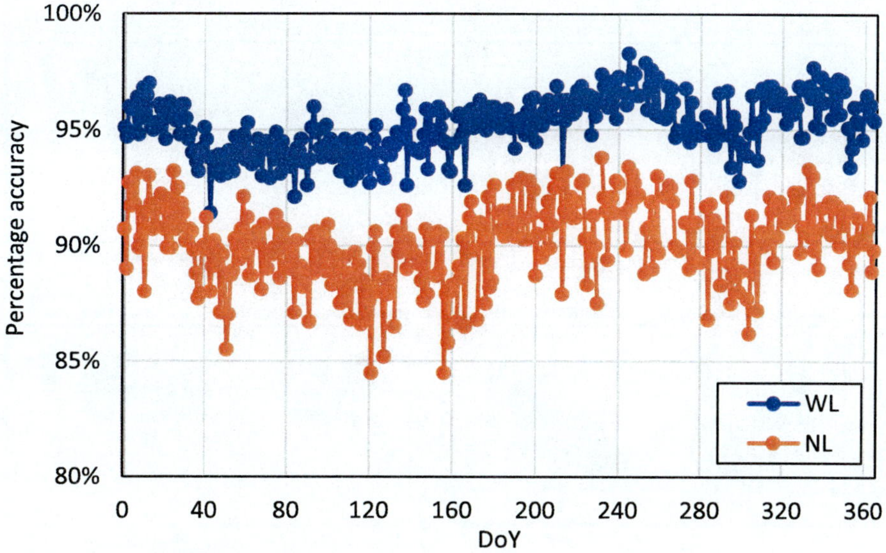

Fig. 69.6 The Percentage value of WL (green line) and NL (orange line) for each day

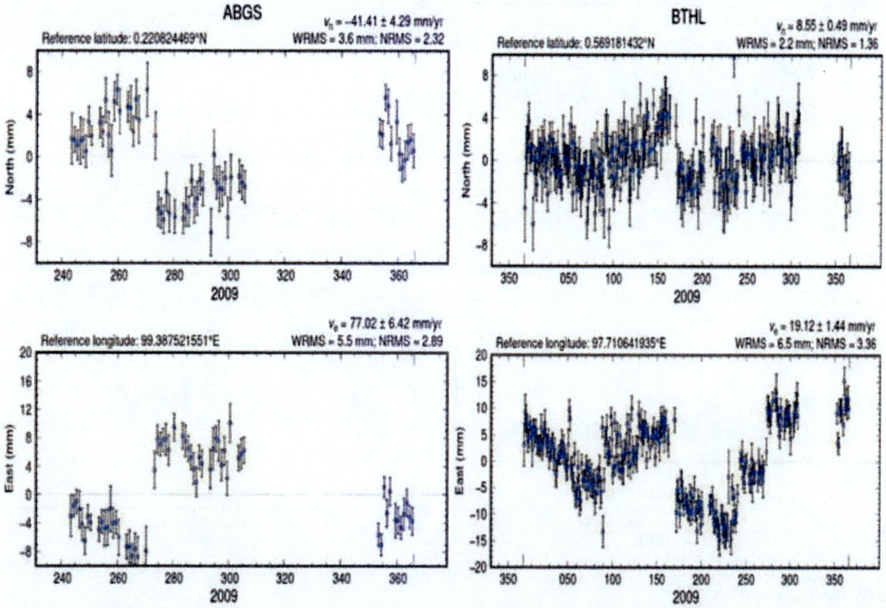

Fig. 69.7 Time series of the direction shift of the receiver's location at ABGS, BTHL, JMBI, KTET, LAIS, LNNG, MNNA MSAI, NGNG, and PKRT stations during 2009 from GLOBK. The x-axis is the time of observation, and the y-axis is the direction of shift of the receiver's location in the North direction and the East direction

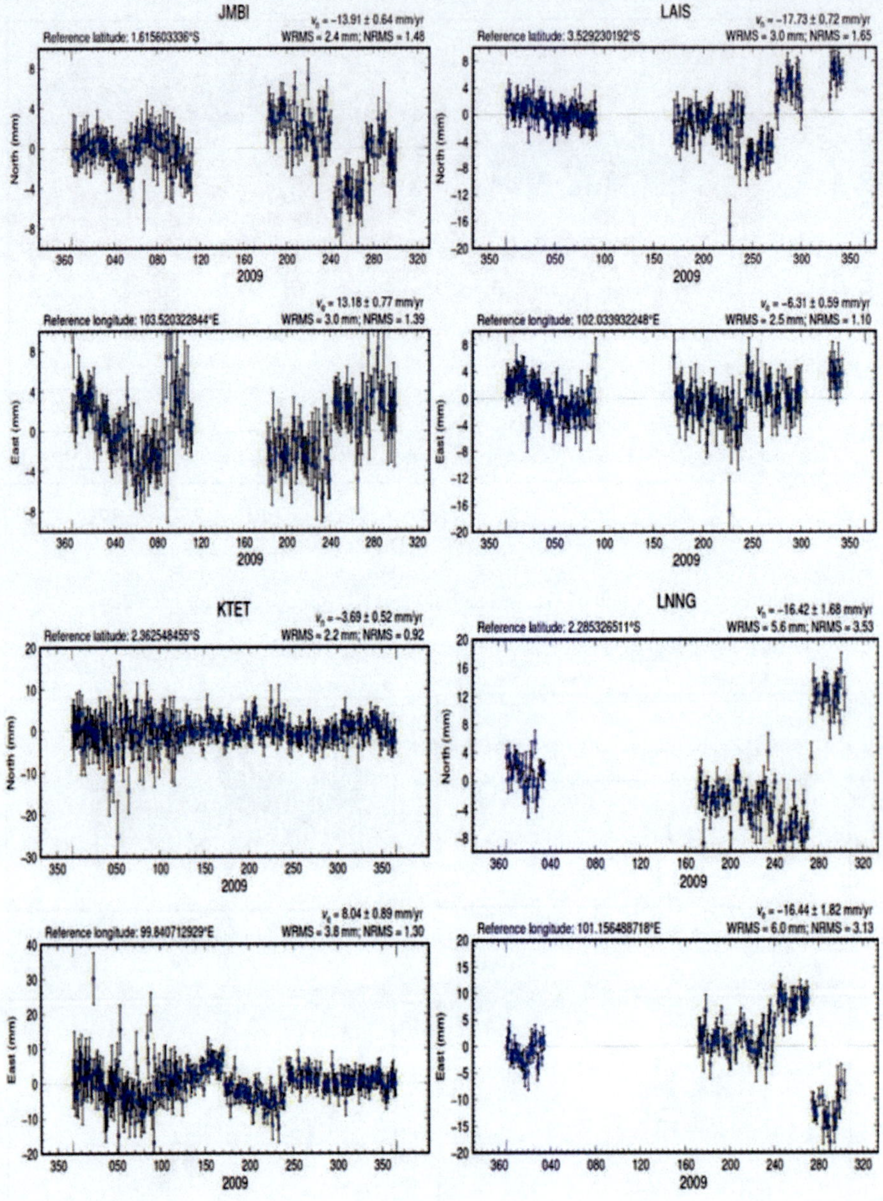

Fig. 69.7 (continued)

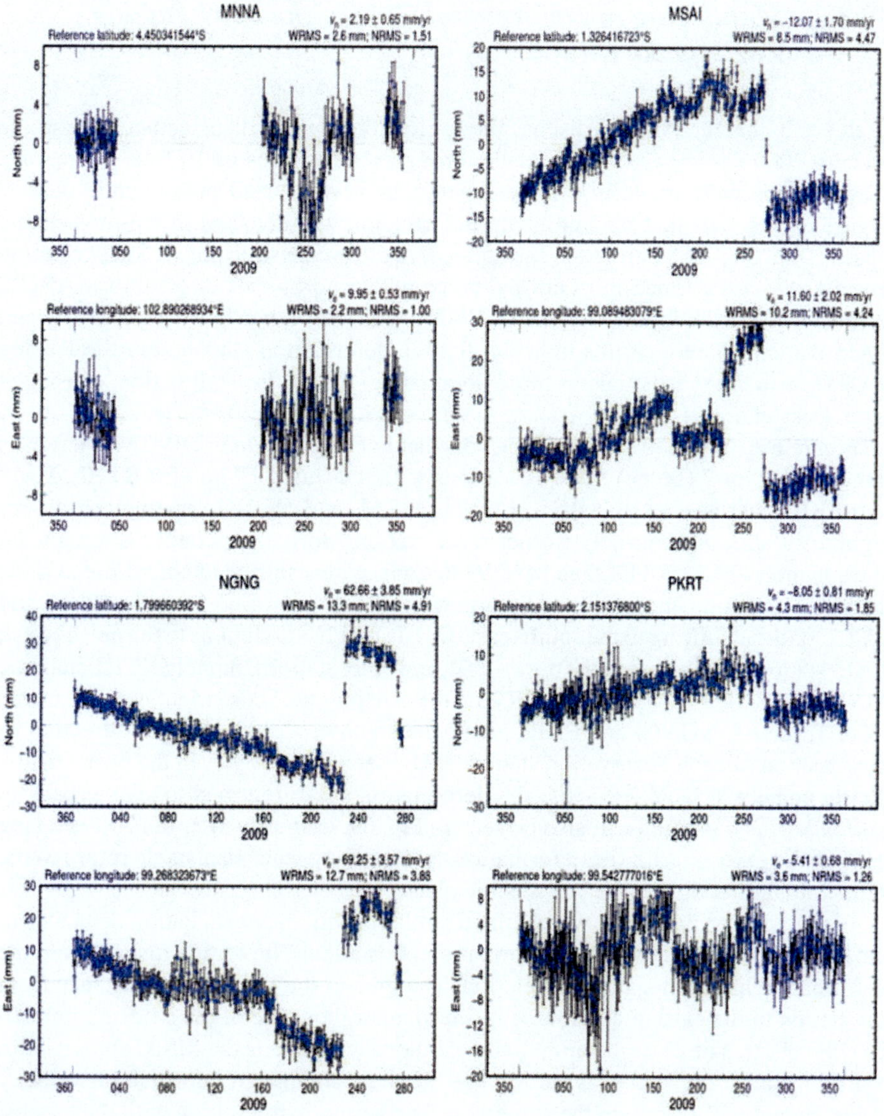

Fig. 69.7 (continued)

and LNNG stations which may be due to some empty DoY data. MSAI and NGNG stations also have very large WRMS, which may be due to another earthquake before the 2009 earthquake. The source of such earthquakes is close to these two stations [9]. In addition, NRMS values <3 mm were also obtained at JMBI, KTET, LAIS, and PKRT stations.

69.3.3 *Monitoring of Deformation Due to the 2009 Padang Earthquake (Mw = 7.6)*

In this sub-chapter, we will discuss the estimation of crust displacement due to the Padang 2009 (Mw = 7.6) earthquake through observations of 10 SuGAr stations. Observations were made in the pre-seismic, co-seismic, and post-seismic phases. Figure 69.7 shows that no significant deformation was detected in the pre-seismic phase (DoY 001–269 before the earthquake) in the pre-seismic phase. SuGAr station experienced a deformation in normal value with a north–east direction for ABGS, BTHL, JMBI, KTET, MNNA, MSAI, NGNG, and PKRT, north–west for LNNG and LAIS stations. These results indicate that the deformation shift before the Padang 2009 (Mw = 7.6) earthquake is more dominant in the northeast direction, consistent with Arisa et al. [9].

Figure 69.7 also shows that in the co-seismic phase, at DoY 270–274, there was a movement for 7 (seven) SuGAR stations with the shift >10 mm for BTHL, LAIS and PKRT stations, even ABGS, LNNG, MSAI, and NGNG experienced a very significant shift >20 mm. The other three stations do not experience a significant shift, namely JMBI, KTET, and MNNA stations, with a shift value of <5 mm. Of the 10 SuGAr stations observed, the deformation directions varied, namely ABGS and BTHL stations shifting to the south–east, JMBI and LNNG stations to the north–west, LAIS stations deforming to the north–east, and other stations, namely KTET stations, MNNA, MSAI, NGNG, and, PKRT to the southwest. Thus, Mentawai segments (KTET, MSAI, NGNG, and PKRT) have the southwest direction. The direction of deformation of each observation station can be seen in Fig. 69.8. In the post-seismic phase, namely at DoY 275–365, the deformation speed returned to stable (back to normal). This condition can also be seen in Fig. 69.7. Shortly after the 2009 Padang earthquake (Mw = 7.6), the deformation speed slowly leads to a stable return to the same initial direction as the pre-seismic phase.

In the co-seismic phase, we obtained a relationship between station distances to the earthquake source with the deformation per station. The station distance and the deformation have a negative correlation. The greater the value of the distance from the station to the earthquake source is, the smaller the value of the deformation will be (Fig. 69.9). The closest station to the earthquake source is the MSAI station, with a distance of 109.5 km. This station has the most significant deformation velocity value, namely 47.13 mm to the southwest. Furthermore, the station with the farthest distance is the MNNA station, with a distance of 532.1 km, and this station has a displacement of 2.26 mm. Thus, the farther the distance from the earthquake source, the smaller the perceived deformation velocity.

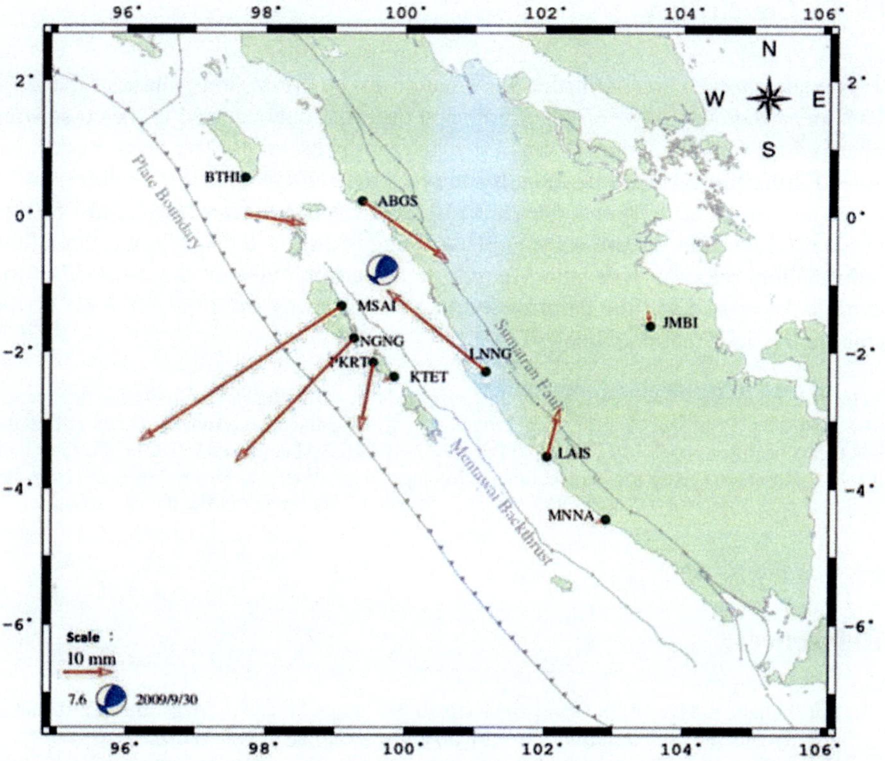

Fig. 69.8 The direction of deformation of each station caused by the 2009 Padang earthquake

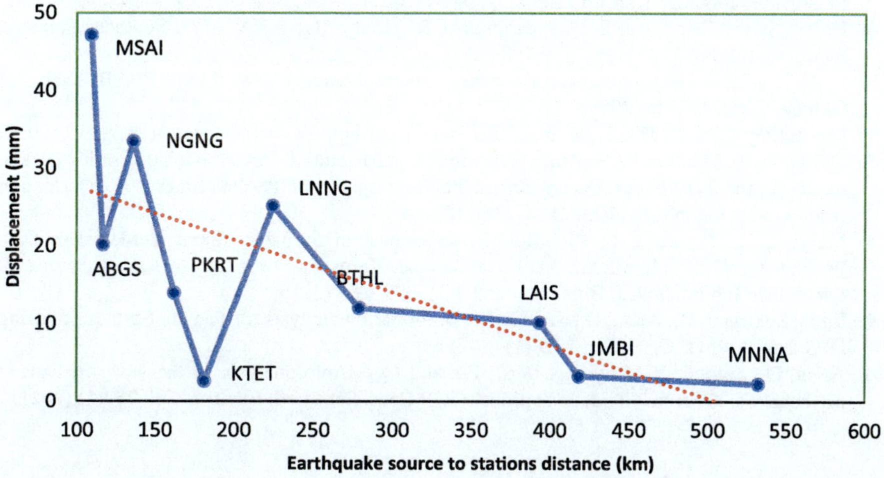

Fig. 69.9 The correlation between the distance of the station to the source of the earthquake with the displacement caused by the 2009 Padang earthquake (Mw = 7.6)

69.4 Conclusion

This study showed no significant deformation in the pre-seismic phase of the 2009 Padang earthquake (Mw = 7.6). Significant deformation occurred in the co-seismic phase with 2.62–47.13 mm deformation to the southwest for the Mentawai Backthrust zone. Furthermore, the Sumatra fault zone experienced 3.26–25.13 mm deformation to the northwest, 20.20 mm to the southeast, 10.14 mm to the northeast, and 2.56 mm to the southwest. In the pre-seismic phase, there is a shift to the original state of the deformation velocity. This study found a relationship between the distance of the earthquake source and the deformation; the farther away from the epicenter is, the smaller the deformation value will be.

Acknowledgements The present study was supported by 2021 Research Grants for Publication to Support the Professor Acceleration Programme from Universitas Andalas (PDU-KRP2GB-UNAND)(contract no: T/11/UN.16.17/PP.IS-PDU-KRP2GB-Unand/LPPM/2021). Furthermore, thanks to Research Center for Geotechnology, Indonesian Institute of Sciences (LIPI) and Nanyang Technological University's Earth Observatory of Singapore for operating the SuGAr network.

References

1. Natawidjaja, D.H. et al.: Source parameters of The Great Sumatran Megathrust Earthquakes of 1797 and 1833 Inferred from Coral Microatolls. J. Geophys. Res. **111**(6), 1–37 (2006)
2. Natawidjaja, D.H.: Updating active fault maps and sliprates along the Sumatran Fault Zone, Indonesia. IOP Conf. Ser. Earth Environ. Sci. **118**(1) (2018)
3. Natawidjaja, D.H.: Tectonic setting Indonesia dan Pemodelan Sumber Gempa dan Tsunami, Pelatihan Pemodelan Run-up Tsunami, 20–24 (2007)
4. United States Geological Survey, earthquake Summary Map XXX: M7.6 Southern Sumatra, Indonesia (2009)
5. Badan Nasional Penanggulangan Bencana, Laporan Harian PUSDALOPS BNPB: Sabtu, 24 Oktober 2009, Jakarta (2009)
6. McCloskey J., et al.: The September 2009 Padang earthquake, Nat. Geosci. **3**(2), 70–71 (2010)
7. Gunawan, E., Maulida, P., Meilano, I., Irsyam, M. and Efendi, J.: Analysis of coseismic fault slip models of the 2012 Indian Ocean earthquake: importance of GPS data for crustal deformation studies. Acta Geophys. **64**(6), 2136–2150 (2016)
8. Maiyudi, R., Permana, I., Ridholfi M.: Deformasi di Sumatera Barat Berdasarkan Data Pengamatan GPS Tahun 2002-2009 dan Dampak Kerusakan Lingkungan Akibat Gempa 30 September Tahun 2009. J. Bina Tambang **4**(3), 379–284 (2019)
9. Iqbal, Syafriani, R., Arisa, D., Setiadi, B.: Deformation analysis for Padang Earthquake using GPS data. J. Phys. Conf. Ser. **1876**(1) (2021)
10. Arisa, D., Setiadi, B., Priyanto, W.S.: Pre-and Coseismic analysis of the Mw 7.6 Padang earthquake 2009 from Geodetic approach. IOP Conf. Ser. Earth Environ. Sci. **789**(1) (2021)

Chapter 70
The Characteristic of Solar Flare and CMEs that Caused SPE During 24th Solar Cycle

Neflia

Abstract SPE is one of the most severe hazards in the space environment. Such events tend to occur during periods of intense solar activity and can lead to high radiation doses in short time intervals. The proton enhancements produced by these solar events may last several days and are very hard to predict in advance, and they also can cause harm both satellites and humans in space. The most significant proton sources in the interplanetary medium are both solar flares and interplanetary shocks driven by coronal mass ejections (CMEs). In this study, I try to find the characteristic of flare and CME that can cause the proton events in the interplanetary medium. For my preliminary study, I will search flare characteristics such as class and position as SPE cause. I also researched with CME characteristics such as Angular Width (AW) and linear speed. During solar cycle 24, the solar activity remains very low with several large flares and halo CME. This low activity also occurs on solar proton events in interplanetary medium. From January 2009 to December 2019, there were 46 SPEs with flux range from 11—6530 pfu (10 meV). The solar flares during these events varies from C to X- class flare. From 45 X-class flares that occurs during 2009—2019, only 11 flares cause the SPE. Most of the active region locations are at the solar western hemisphere (35/46). There are 40 from 323 halo CME (AW = 360°), 37 from 203 with linear speed >1000 km/s, and 34 from 109 Halo CME with linear speed >1000 km/s cause SPE. Although the probability of SPE from all flares and CMEs during this range of time is small, but they have three common characteristics, i.e., most of the SPE has active region position at the solar western hemisphere, the CME have AW = 360°, and have a high linear speed.

70.1 Introduction

Energetic Particles in the interplanetary medium are observed with energies range from supra-thermal up to 10E+20 GeV. The main contituents are protons, α-particles

Neflia (✉)
Space Research Center, National Research and Innovation Agency, Bandung, Indonesia
e-mail: neflia@brin.go.id

E. Yulihastin et al. (eds.), *Proceedings of the International Conference on Radioscience, Equatorial Atmospheric Science and Environment and Humanosphere Science, 2021*, Springer Proceedings in Physics 275, https://doi.org/10.1007/978-981-19-0308-3_70

and electrons. These particle populations originates in a different sources, such as Galactic Cosmic Rays (GCR), Anomalous Cosmic Rays (ACR), Solar Energetic Particles (SEPs), Energetic Storm Particles (ESP), and Corotating Interaction Region (CIR) [1]. SPE is one of the most severe hazards in the space environment. Such events tend to occur during periods of intense solar activity and can lead to high radiation doses in short time intervals. The proton enhancements produced by these solar events may last several days and are very hard to predict in advance, and they also can cause harm both satellites and humans in space. The most significant proton sources in the interplanetary medium are both solar flares and interplanetary shocks driven by coronal mass ejections (CMEs) [1].

Many researches about solar energetic particles have been done before, and it is still continuing until today. Several research only uses CME parameters as the prompt of solar proton events, e.g. [2–5]. Some other research using flares as the source of SPE, e.g. [6–9]. There is also research that combined flares and CME parameters as the source of SPE, e.g. [10–13].

In this paper, I study the characteristics of flare and CME that can cause SPE from their parameters such as flare class and location and CME angular width and linear speed using statistical methods. Flare class and location, and CME angular width and linear speed are believed to have a lot of contribution to the proton event in the interplanetary medium.

70.2 Data

The Solar Proton Events (SPE) during the 24th solar cycle from January 2009 to December 2019 from Coordinated Data Analysis Workshops (CDAW) is used in this study. This SPE data consists of a list of Major SPE events with particle intensity in the $E > 10$ meV proton channel exceeds ten particles/(cm^2 s sr) and can be obtained from https://cdaw.gsfc.nasa.gov/CME_list/sepe/. Other than SPE, Coronal Mass Ejections (CME) from telescope C2 and C3 that carried by the Large Angle and Spectrometric Coronagraph (LASCO) onboard the Solar and Heliospheric Observatory (SOHO) mission is used to characterize the CME that can cause SPE. To characterize the CME that can cause SPE, CME parameters such as, Angular Width (AW), and Linear Speed (v), will be analyzed. The CME that has been cataloged, generated, and maintained at Coordinated Data Analysis Workshops (CDAW) can be obtained from https://cdaw.gsfc.nasa.gov/CME_list/. The last data used in this study is solar X-ray flare in the 1–8 Å wavelength from National Centers for Environmental Information (NCEI), formerly named National Geophysical Data Center (NDGC). The flare parameters such as, class and position will be analyzed to characterize the flare parameter that can cause SPE. This flare event data can be obtained at https://ftp.swpc.noaa.gov/pub/warehouse/. Table 70.1. Show the SPE and flare and CME associated with SPE during the 24th solar cycle.

To find the characteristics of flare and CME that caused SPE, CME angular width is grouped into three groups, i.e. $AW > 120°$; $120 < AW < 360°$, and $AW = 360°$.

Table 70.1 SPE, CME, and Flare during 24th solar cycle

SPE					CME					Flare				
Start time		Max time		Flux	Time		Speed	Width	Loc	Start time		Peak time		Class
14/08/2010	11:05	14/08/2010	12:45	14	14/08/2010	10:12	1205	H	N17W52	14-Aug	09:38	14-Aug	10:05	C4.4/–
07/03/2011	21:45	08/03/2011	08:00	50	07/03/2011	20:00	2125	H	N31W53	07-Mar	19:43	07-Mar	20:12	M3.7/–
21/03/2011	04:10	22/03/2011	01:35	14	21/03/2011	02:24	1341	H	>W90	– /–	– :–	– /–	– :–	– –/–
07/06/2011	07:20	07/06/2011	18:20	72	07/06/2011	06:49	1255	H	S21W54	07-Jun	06:16	07-Jun	06:41	M2.5/2 N
04/08/2011	04:30	05/08/2011	21:50	96	04/08/2011	04:12	1315	H	N19W36	04-Aug	03:49	04-Aug	03:57	M9.3/2B
09/08/2011	08:20	09/08/2011	12:10	26	09/08/2011	08:12	1610	H	N17W69	09-Aug	07:59	09-Aug	08:05	X6.9/2B
22/09/2011	17:55	26/09/2011	11:55	35	22/09/2011	10:48	1905	H	N09E89	22-Sep	10:29	22-Sep	11:01	X1.4/–
26/11/2011	08:15	27/11/2011	01:25	80	26/11/2011	07:12	933	H	N17W49	26-Nov	06:09	26-Nov	07:10	C1.2/–
23/01/2012	04:45	24/01/2012	15:30	6310	23/01/2012	04:00	2175	H	N28W21	23-Jan	03:38	23-Jan	03:59	M8.7/–
27/01/2012	18:55	28/01/2012	02:05	795	27/01/2012	18:27	2508	H	N27W71	27-Jan	18:03	27-Jan	18:37	X1.7/1F
07/03/2012	02:50	08/03/2012	11:15	6530	07/03/2012	00:24	2684	H	N17E27	07-Mar	00:13	07-Mar	00:24	X5.4/–
13/03/2012	18:05	13/03/2012	20:45	469	13/03/2012	17:36	1884	H	N17W66	13-Mar	17:12	13-Mar	17:41	M7.9/–
17/05/2012	01:55	17/05/2012	04:30	255	17/05/2012	01:48	1582	H	N11W76	17-May	01:25	17-May	01:47	M5.1/1F
26/05/2012	23:25	27/05/2012	10:45	14	26/05/2012	20:57	1966	H	>W90	– /–	– :–	– /–	– :–	– –/–
14/06/2012	23:25	16/06/2012	22:30	15	14/06/2012	14:12	987	H	S17E06	14-Jun	12:52	14-Jun	14:35	M1.9/1 N
07/07/2012	00:05	07/07/2012	07:45	25	06/07/2012	23:24	1828	H	S13W59	06-Jul	23:01	06-Jul	23:08	X1.1/–
08/07/2012	18:10	09/07/2012	04:30	19	08/07/2012	16:54	1497	157	S17W74	08-Jul	16:23	08-Jul	16:32	M6.9/1 N
12/07/2012	17:25	12/07/2012	22:25	96	12/07/2012	16:48	885	H	S15W01	12-Jul	16:16	12-Jul	16:49	X1.4/–
17/07/2012	15:30	18/07/2012	06:00	136	17/07/2012	13:48	958	176	S28W65	17-Jul	13:19	17-Jul	15:59	C9.9/1F
19/07/2012	06:40	19/07/2012	14:25	80	19/07/2012	05:24	1631	H	S13W88	19-Jul	04:17	19-Jul	05:58	M7.7/–

(continued)

Table 70.1 (continued)

SPE					CME					Flare				
Start time		Max time		Flux	Time		Speed	Width	Loc	Start time		Peak time		Class
23/07/2012	08:00	23/07/2012	21:45	12	23/07/2012	02:36	2003	H	>W90	– /–	– :–	– /–	– :–	– –/–
01/09/2012	01:25	02/09/2012	08:50	60	31/08/2012	20:00	1442	H	S25E59	31-Aug	19:45	31-Aug	20:43	C8.1/2F
28/09/2012	01:20	28/09/2012	04:45	28	28/09/2012	00:12	947	H	N06W34	27-Sep	23:36	27-Sep	23:57	C3.7/1F
15/03/2013	19:40	17/03/2013	07:00	16	15/03/2013	07:12	1063	H	N11E12	15-Mar	05:46	15-Mar	06:58	M1.1/1F
11/04/2013	08:25	11/04/2013	16:45	114	11/04/2013	07:24	861	H	N09E12	11-Apr	06:56	11-Apr	07:16	M6.5/3B
15/05/2013	06:35	17/05/2013	17:20	42	15/05/2013	01:48	1366	H	N12E64	15-May	01:25	15-May	01:48	X1.2/2 N
22/05/2013	14:20	23/05/2013	06:50	1660	22/05/2013	13:25	1466	H	N15W70	22-May	13:08	22-May	13:32	M5.0/–
23/06/2013	08:30	24/06/2013	05:20	14	21/06/2013	03:12	1900	>207	S16E73	21-Jun	02:30	21-Jun	03:14	M2.9/–
30/09/2013	00:25	30/09/2013	20:05	182	29/09/2013	22:12	1179	H	N17W29	29-Sep	20:42	29-Sep	23:32	C1.3/–
28/12/2013	19:00	28/12/2013	23:15	29	28/12/2013	17:36	1118	H	>W90	– /–	– :–	– /–	– :–	– –/–
06/01/2014	08:15	06/01/2014	16:00	42	06/01/2014	08:00	1402	H	>W90	06-Jan	07:30	06-Jan	07:45	C2.1/–
07/01/2014	19:55	09/01/2014	03:40	1026	07/01/2014	18:24	1830	H	S15W11	07-Jan	18:04	07-Jan	18:32	X1.2/–
20/02/2014	08:15	20/02/2014	09:25	22	20/02/2014	08:00	948	H	S15W73	20-Feb	07:26	20-Feb	07:56	M3.0/–
25/02/2014	03:50	26/02/2014	00:10	24	25/02/2014	01:25	2147	H	S12E82	25-Feb	00:39	25-Feb	00:49	X4.9/–
18/04/2014	13:40	19/04/2014	01:50	58	18/04/2014	13:25	1203	H	S20W34	18-Apr	12:31	18-Apr	13:03	M7.3/–
10/09/2014	21:35	12/09/2014	15:55	126	10/09/2014	18:00	1267	H	N14E02	10-Sep	17:21	10-Sep	17:45	X1.6/–
01/11/2014	13:55	02/11/2014	21:10	11	01/11/2014	06:00	740	>160	N27W79	– /–	– :–	– /–	– :–	– –/–
18/06/2015	04:35	18/06/2015	14:45	17	18/06/2015	01:25	1714	195	S16W81	18-Jun	00:33	18-Jun	01:27	M1.2/–
21/06/2015	04:05	22/06/2015	19:00	1066	21/06/2015	02:36	1366	H	N12E16	21-Jun	02:06	21-Jun	02:36	M2.6/–

(continued)

Table 70.1 (continued)

SPE					CME					Flare				
Start time		Max time		Flux	Time		Speed	Width	Loc	Start time		Peak time		Class
25/06/2015	10:05	27/06/2015	00:30	22	25/06/2015	08:36	1627	H	N09W42	25-Jun	08:02	25-Jun	08:16	M7.9/–
29/10/2015	03:05	29/10/2015	10:00	24	29/10/2015	02:36	530	202	>W90b	– /–	– :–	– /–	– :–	– –/–
02/01/2016	00:15	02/01/2016	04:50	22	01/01/2016	23:24	1730	H	S25W82	01-Jan	23:10	02-Jan	00:11	M2.3/–
14/07/2017	04:40	14/07/2017	23:20	22	14/07/2017	01:25	1200	H	S06W29	14-Jul	01:07	14-Jul	02:09	M2.4/–
04/09/2017	22:30	05/09/2017	19:20	210	04/09/2017	20:12	1418	H	S10W12	04-Sep	20:28	04-Sep	20:33	M5.5/–
06/09/2017	12:35	07/09/2017	23:25	844	06/09/2017	12:24	1571	H	S08W33	06-Sep	11:53	06-Sep	12:02	X9.3/–
10/09/2017	16:25	11/09/2017	11:40	1490	10/09/2017	16:00	3163	H	>W90b	10-Sep	15:35	10-Sep	16:06	X8.2/–

CME Linear Speed also is grouped into three groups, v < 500 km/s; 500 < v < 1000 km/s, and v > 1000 km/s. Solar flares will be grouped according to their class A, B, C, M, X class and based on their location, i.e. North-west (NW), South-West (SW), North-East (NE), and South-East (SE).

70.3 Results and Discussion

Solar activity is quite low during 24th solar cycle, with only several large flares and halo CME. From January 2009 to December 2019, there were 14,277 solar flares and 16,093 CMEs. This low activity also occurs on solar proton events in the interplanetary medium. There were only 46 SPEs with fluxes ranging from 11—6530 pfu (10 meV). This highest proton flux is much lower than the highest flux on the 23rd cycle that reached 30,000 pfu. Flare and CME are believed as the main factors that cause SPE. However, the correlation of flare and CME to the SPE is small if seen from the total of flares and CMEs events to the total SPEs. It is only 3.2% and 2.8% for flare and CME, respectively. But it is still necessary to know the common characteristics of flare and CME that have the potential to cause SPE. In the next session, I will discuss flare and CME that have the potential to cause SPE based on CME angular width (AW), CME liner speed (v), flare class, and position.

70.3.1 CME

The CME that caused SPE is relatively small if comparing to the total of CME events during the 24th solar cycle, as shown in Table 70.2. No CME that causes SPE for the angular width smaller than 120°. For Angular width ranging from 120 to 360°, there are only 6 SPEs that occur from the total of 1011 CMEs with the same category (0.5%). The same result also showed by the halo CME with an angular width of 360°. There are only 40 SPEs or 12% of the total of CME with this category.

The low correlation also occurs to CME linear speed as shown in Table 70.3. There is no CME that causes SPE for CME linear speed smaller than 500 km/s. For linear speed ranging from 500 to 1000 km/s, there are 9 CMEs or 4% from the total 2265 CMEs with the same category that causes SPE. At the same time, there are 37 CME or 18% from the total 203 CME that caused SPE with the same category.

Table 70.2 Number of events of CME angular width with AW > 120°, 120 < AW < 360° and AW = 360°

AW	Number of event	Number of event (SPE)
AW < 120	14,713	0
120 < AW < 360	1011	6
AW = 360	323	40

Table 70.3 Number of events of CME linear speed with v > 500 km/s, 500 < v < 1000 km/s and v > 1000 km/s

Speed	Number of events	Number of events (SPE)
v < 500	13,623	0
500 < v < 1000	2265	9
v > 1000	203	37

Table 70.4 Number of events of CME linear speed with v > 500 km/s, 500 < v < 1000 km/s and v > 1000 km/s for AW = 360° (halo CME)

Speed	Number of events	Number of events (SEP)
v < 500	55	0
500 < v < 1000	159	6
v > 1000	109	34

CME and SPE correlation are quite different with the category of linear speed with angular width 360° (halo CME) as shown in Table 70.4. There are 6 halo CMEs with linear speed ranging from 500 to 1000 km/s that causes SPE from the total of 159 CMEs with the same category (3.8%). Whereas for linear speed, v > 1000 km/s, there are 34 CMEs from the total of 109 CMEs that causes SPE or it's about 31%. Although correlation of CME linear speed with angular width 360° to SPE is not significant and the overall CME that causes SPE in this category is still small, i.e. only 12%, but this result showed that halo CME with high linear speed has a more significant potential to cause SPE in the interplanetary medium.

70.3.2 Flare

The correlation between flare and SPE show almost identical to the correlation between CME and SPE, the overall correlation is very low, as seen in Table 70.5. From 14,277 flares recorded by GOES X-ray, there are only 40 flares that cause SPE or about 0.2%, with 7 SPEs or 0.04% are caused by the C-class flares, 22 SPEs or 0.14% are caused by the M-class flares and 11 SPEs or 0.08% are caused by X-class flares. When viewed from the total incidence of flares by class, X-class flare caused the most occurrence of SPE, which was about 26% with a total number of 45 X-class flare events. Meanwhile, the percentage of SPE caused by M-class and C-class flare

Table 70.5 Number of events of flare based on the flare class classification

Class	Number of event	Number of events (SPE)
A	74	0
B	5866	0
C	7588	7
M	701	22
X	45	11

is about 0.1% and 3%, respectively, with the total incidence in each class being 7588 for C-class flares and 701 events for M-class flares. From Table 70.5 we are also able to see that during the 24th solar cycle, most SPE is caused by the M-class flare. It's about 52% of the total SPE caused by M-class flare, whereas X-class flare causes SPE only 30% of total SPE. However, because the incidence of X-class flare is less often than M-class flare, so the percentage of X-class flare is bigger than M-class flare from the total of Flare in the same category.

There are 14,277 flare events recorded by GOES during the 24th solar cycle, but there are only 4769 flare events that can be identified as their location, as can be seen in Table 70.6. All of the identified flares are located almost in all directions in the solar hemisphere as shown in Fig. 70.1. The total number of flare events is also quite the same in all directions. However, from 46 SPE that occur during the 24th solar cycle, there are 35 SPEs or 76% caused by flares that located in the western solar hemisphere, with 14 SPEs is caused by the flares located in the northwest, 14 SPEs caused by flares located in the southwest, and 7 SPEs caused by flares located in the western side of the solar hemisphere. Whereas there are only 11 SPEs or 24% caused by flares that are located in the east solar hemisphere. The distribution of flares based on their location can be seen in Fig. 70.2.

The overall result showed that both Flare and CME are not the only factor that causes SPE in the interplanetary medium. The low correlation of flare and CME to

Table 70.6 Number of events of flare based on flare location (NW, SW, NE, SE)

Location	Number of events	Number of events (SPE)
NW	1016	14
SW	1212	14
NE	1120	7
SE	1421	4
W > 90		7

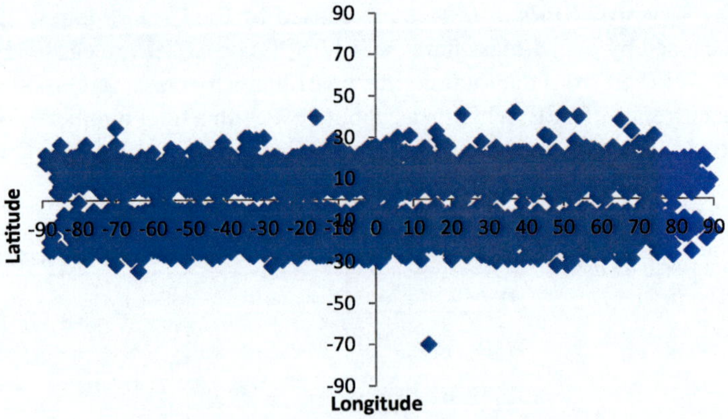

Fig. 70.1 The identified flare location during the 24th solar cycle

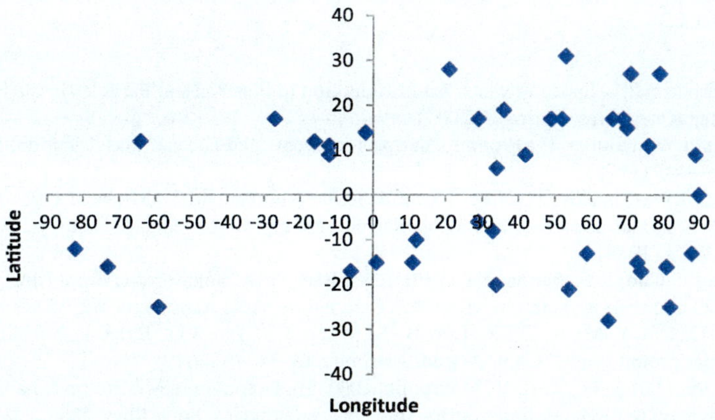

Fig. 70.2 Location of flare that caused SPE during the 24th solar cycle

SPE might be related to the mechanism of propagation and acceleration of flares and CME from their sources in the sun to the interplanetary medium [14, 15]. It also might be related to their sources in the sun [16, 17]. From several papers also stated that the CME and flare that caused SPE also accompany by Type-II, type-III, and type IV radio burst [18–21], Although their statistical correlation from the total of SPE, is low, but there are still show the characteristic of flare and CME that has the possibility to cause SPE. The common characteristics of CME and flare that cause the SPE is the flare located in the western solar hemisphere and CME has a high linear speed with angular width 360° or halo CME.

70.4 Conclusion

Although the correlation between total flares and CME events to SPE is low, there are still three common characteristics of flares and CME that have the possibility to cause SPE, i.e., most of the flares that cause SPE located in the western solar hemisphere, whereas CME have angular width 360° and have a high linear speed. The comprehensive result, however, will be achieved if all parameters of interaction between solar and space environment are concluded.

Acknowledgements CME and SPE used in this study are generated and maintained at the CDAW Data Center by NASA and the Catholic University of America in cooperation with the Naval Research Laboratory. The X-ray flare data from National Centers for Environmental Information (NCEI), formerly named National Geophysical Data Center (NDGC), and Space Weather Prediction Center (SWPC), National Oceanic and Atmospheric Administration (NOAA), is utilized. The author would like to thank Santi Sulistiani for the discussion.

References

1. Kallenrode, M.B., Space Physics. An Introduction to Plasmas and Particles in the Heliosphere and Magnetospheres. Springer (2004)
2. Kahler. S.W., Hildner, E.: Prompt solar proton events and coronal mass ejection. Solar Phys. **87**, 429–443 (1978)
3. Kahler. S.W., Sheeley, Jr. N.R., Howard, R.A., Koomen, M.J., Michels, D.J.: Association between coronal mass ejections and solar energetic proton events. J. Geophys. Res. **87**(11), 9683–9693 (1984)
4. Hwang, J., Cho, K.S., Moon, Y.J., Kim, R.K., Park, Y.D.: Solar proton events during the solar cycle 23 and their association with CME Parameters. Acta Astronautic **67**, 353–361 (2010)
5. Wang, C., Cui, Y.-M., Ao, X.-Z., Luo, B.-X., Wang, J.-J., Liu, S.Q.: Relationship of Halo CMEs and solar proton events. Chin. Astron. Astrophy **43**, 34–46 (2019)
6. Belov. A., Garcia, H., Kurt, V., Mavromichalaki, H., Gerontidou, M.: Proton enhancement and their relation to the x-ray flares during three last solar cycles. Solar Phys. **229**, 135–159 (2005)
7. Smart, D.F., Shea, M.A.: The Helio-longitudinal distribution of solar flares associated with solar proton events. Adv. Space. Res. **17**(2), (2) 113–(2) 116 (1996)
8. Park, J., Moon, Y.J., Lee, D.H., Youn, S.: Dependence of Solar proton events on their associated actvities: flare parameters. J. Geophys. Res. **115**, A10105 (2010)
9. Park, J., Moon, Y.J., Gopalswamy, N.: Dependence of solar proton events on their associated activities: flare parameters. J. Geophys. Res. **1175**, A08108 (2012)
10. Cane, H.V., Erricson, W.C., Prestage, N.P.: Solar flares, type III radio burst, coranal mass ejections, and energetic particles. J. Geophys. Res. **107**(A10), 1315 (2002)
11. Papaioannou, A., Sandberg, I., Anastasiadis, A., Kouloumvakos, A., Georgoulis, M.K., Tziotziou, K., Tsiropoula, G., Jiggens, P., Hrlgers, A.: Solar flares, coronal mass ejections and solar energetic particle event characteristic. J. Space Weather Space Clim. **6**, A42 (2016)
12. Klein, K.L., Dalla, S.: Acceleration and propagation of solar energetic particles. Space Sci. Rev. (2017)
13. Reames, D.V.: Solar Energetic Particles: A Modern Primer on Understanding Sources, Acceleration and Propagation, Springer (2017)
14. Reames, D.V.: The Two Sources of Solar Energetic Particles, Space Science Rev, Springer (2013)
15. Vlahos, L., Anastasiadis, A., Papaioannou, A., Kouloumvakos, A., Iliker, H.: Sources of solar energetic particles. Phil. Trans. R. Soc. A **377**, 20180095 (2019)
16. Winter, L.M., Ledbetter, K.: Type II and Type III radio burst and their correlation with solar energetic proton events. The Astrophys. J. 809–105 (2015)
17. Ameri, D., Valtonen, E., Pohlojainen, S.: Properties of high-energy solar particle events associated with solar radio emissions. Solar Phys. **294**, 122 (2019)
18. Trottet, G., Samwel, S., Klein, K.-L., Dudok de Wit, T., Miteva, R.: Statistical evidence for contibutions of flares and coronal mass ejections to major solar energetic particle events. Solar Phys. **290**, 819–839 (2014)
19. Dierckxsens, M., Tziotsiou, K., Dalla, S., Patsou, L., Marsh, M.S., Crosby, N.B., Malandraki, O., Tsiropoula, G.: Sol. Phys. **290**, 841–874 (2015)
20. Ramli, N., Hamidi, Z.S., Abidin, Z.Z., Shahar, S.N.: The relation between solar radio burst type II, III and IV due to solar activity, IconSpace, 10–12 (2015)
21. Miteva, R., Samwel, S.M., Krupar, V.: Solar energetic particles and radio burst emission. J. Space Weather Space Clim. **7**, A37 (2017)

Chapter 71
Continuum Study on Uranus at the Millimeter/Submillimeter Wavelength with ALMA Data

Farahhati Mumtahana and Taufiq Hidayat

Abstract Uranus is a unique solar system body with its extreme rotational axis, low luminosity, asymmetrical internal magnetic field and magnetosphere, and questionable internal structure. Yet unlike the other Solar System planets, Uranus is the least understood and least visited planet that was only be attained once by Voyager 2 flyby mission in 1986. However, with the development of observation techniques and instruments both in space and ground, some advanced telescopes presently observe Uranus routinely. Moreover, most of the exoplanets detected so far have similarities with the ice giant type. Atacama Large Millimeter/submillimeter Array (ALMA) is one high-end radio interferometer that frequently observes Uranus since it is often chosen as a calibrator. With the abundance of ALMA data, this study is aimed to reconstruct the Uranus images and build the Uranus continuum model from the distribution of flux density over several frequencies on ALMA wide-range bands. Common Astronomy Software Application (CASA) data processing tool with a Python interface is used to achieve these goals. The ALMA data used in this research was Uranus observational data from 2012 to 2018 at bands 3, 4, 6, 7, and 8. After careful data selection and calibration, we obtained 110 observational data points. Due to the high angular resolution that can be achieved by ALMA, some well-resolved radio continuum images of Uranus were also obtained. Furthermore, the deconvolution result of the combined 78-point continuum data is sufficiently representative to build the continuum model, showing the higher the frequency the higher the flux density.

71.1 Introduction

Uranus is a unique solar system body, but it is also the least understood, even when compared to the twin, Neptune [5]. Uranus has an extreme rotational axis as it tilts

F. Mumtahana (✉) · T. Hidayat
Department of Astronomy, InstitutTeknologiBandung, FMIPA, Bandung, Indonesia
e-mail: fara006@brin.go.id

F. Mumtahana
Space Research Center, National Research and Innovation Agency (BRIN), Bandung, Indonesia

E. Yulihastin et al. (eds.), *Proceedings of the International Conference on Radioscience, Equatorial Atmospheric Science and Environment and Humanosphere Science, 2021*, Springer Proceedings in Physics 275, https://doi.org/10.1007/978-981-19-0308-3_71

around 90‘°, almost on the orbit plane. It has very low luminosity, unclear interior structure from the core until the outermost part, and asymmetric internal magnetic field and magnetosphere [1]. This seventh planet is located at 19 AU from the Sun and receives solar flux around 3.4 W/m^2 [14]. From radio emission detected on its magnetosphere, Uranus rotates 14.24 h [10].

The internal heat source of Uranus is also the lowest, probably almost zero [4]. It is estimated due to slower internal circulation in the ice mantle [6]. This condition may explain the difference regarding atmospheric dynamics and hydrocarbon specimens of the planet's atmosphere [10]. It can also answer the fewer features on the Uranus atmosphere since Neptune has internal heat flux and high air content to trigger turbulence and storms. This very weak internal heat source also indicates that most of the dynamics are influenced by the Sun [4, 19].

Above all, the ice giant type turned out to be the most typical in exoplanet detection. Presently, the exoplanet database stated the most finding are categorized as sub-Neptunian size, similar to Uranus and Neptune. It makes Uranus more essential to study, in order to have better knowledge about the kind of ice giant in our Solar System. As a local laboratory, Solar System holds a vital role in understanding the nature of general planetary systems [1]. Within the Solar System itself, Uranus also holds a crucial role in studying its origin and evolution. [5, 7].

The only spacecraft to visit Uranus remains Voyager 2 on the flyby mission in January 1986 [10]. This spacecraft encountered the Uranian system on the polar side during the solstice [12]. Consequently, it generated images of the pale blue featureless object in the visible light, without any spectacular cloud bands or storms. This observation still leaves many questions unanswered about ice giant planets. Nevertheless, more striking phenomena were expected to appear on the equator during the equinox [1].

The rapid development of instrumentation technology, both on space and ground, made it possible to continue observing Uranus after Voyager 2. Some advanced large telescopes accompanied by Adaptive Optics (AO) technology will be able to reveal the dynamic features of the Uranus atmosphere [8]. For example, some phenomena such as storms were observed in Infrared with Keck Telescope [4], and some bright spots were also seen with Hubble Space Telescope [9, 16]. Meanwhile, the high contrast image observed from Keck and Gemini Telescopes has also revealed so many features on its equator during equinox [17].

Under the radio wavelength, Uranus has also been frequently observed with Atacama Large Millimeter/submillimeter Array (ALMA). This radio interferometer has a wide band of observation windows and a long baseline to achieve small angular resolution. ALMA observes Uranus from time to time since it is often chosen as a calibrator object due to its brightness and relatively close distance. With the abundance of Uranus observation from ALMA, this study is aimed to reconstruct the Uranus images containing flux contour features at several bands. Surprisingly we also found Uranus as a scientific target in the ALMA project, to study its ring and Troposphere observed by Molter et al. [13]. They observed Uranus three times at different bands of ALMA, while 5 data were observed with VLA [4]. They also used some previous observation data at very wide frequency range between 1 and hundreds of GHz to obtain the brightness temperature profile of Uranus. Within the frequency range of ALMA, we also aim to build such Uranus continuum model from the distribution of flux density as well as Brightness Temperature using more ALMA data.

71.2 ALMA Radio Interferometer

The Atacama Large Millimeter/submillimeter Array (ALMA) is a highly sensitive and complex radio interferometer on Earth, located in Chajnantour Plateau, Chile. ALMA uses 66 high-precision dish antennas consisting of 54 antennas of 12 m diameter and 12 antennas of 7 m diameter. With the extreme dryness at the latitude above 5000 m, it can cover an extensive frequency range over some receiver bands as in Table 71.1. The ALMA array configuration will reach the most extended baseline, around 16 km. It is possible to attain the angular resolution within arcsecond to milliarcsecond, with the pointing accuracy less than 0.6 arcsecond [2].

ALMA works in several cycles, starting with *Cycle 0* in September 2011 at the bands 3, 6, 7, and 9. After further development and additional bands, ALMA is currently working in *Cycle 8* with more advanced capability from band 3–10. Apparently, band 1 and 2 will be added in the future. ALMA uses a technique called radio interferometry. The following section (Sect. 2.1) will explain brief basic interferometry, following by the description of ALMA data selection in Sect. 2.2.

71.2.1 *Radio Interferometry and Aperture Synthesis*

For a single dish telescope with diameter, D, and wavelength, λ, the angular resolution in radians can be calculated by the diffraction limit, $\theta \sim 1.22\ \lambda/D$ [3]. A radio

Table 71.1 ALMA interferometer specification

Parameters	Specifications
Total collecting area	6 600^{2}m
Frequency range	31–950 GHz
Angular resolution	0.2–0.004 arc seconds
Array configurations	150–16 km
ALMA bands	3 [84–116 GHz] 4 [125–163 GHz] 6 [211–275 GHz] 7 [275–373 GHz] 8 [385–500 GHz] 9 [602–720 GHz]

telescope has longer wavelength, therefore to achieve the exact resolution as in other electromagnetic wavelengths, a radio telescope needs a larger diameter. However, the bigger telescope is pricier and more challenging. The interferometer technique can be the solution to this issue as the longest baseline (*b*) can replace the diameter of the dish in the limit diffraction equation. With the longer distance between two or more elements, resolution can be smaller. In interferometer, each antenna observes the same source, S_0,with angle θ from the meridian. There will be a signal delay from the first and second or more antennas. The geometrical delay is calculated by $\tau_g = b.S_0/c$. The signals from each antenna arrive at the correlator with the same phase [15]. The technique used in interferometry, called Aperture Synthesis, will be summarize from Remijan et al., in the following paragraph.

Different from the single dish, the interferometer does not measure direct intensity but coherence. The correlator calculates a quantity, known as the complex visibility. In this case, what is measured is the complex visibility function on Earth, *V(u,v)*, which is the 2D Fourier transform of the sky emission or the intensity distribution of sky brightness, *I(l,m)*. The components of *u* and *v* are the projected lengths of the antenna separations measured at the time of observation in units of the wavelength. They are identified as specific spatial frequency components on the sinusoid, while *l* and *m* are direction cosines, relative to a reference position in the East–West for *u* and *l* and North–South direction for *v* and *m*, respectively. Generally, the on-axis position the value of *l* and *m* are zero, as the phase center. The relationship between them is explained by the *van Cittert-Zernike theorem* as the basis of aperture synthesis, which is mathematically written by:

$$V(u, v) = \int\int I(l, m)e^{2\pi i(ul+vm)}dldm$$
$$I(l, m) = \int\int V(u, v)e^{-2\pi i(ul+vm)}dudv \tag{71.1}$$

Thus, *V(u,v)* is the double integral of *I(l,m)* times an exponential factor of series of sinus and cosines, integrated over some angles *(l,m)* on the sky. Fourier transform works in another way, so the true sky brightness will be obtained by measuring all of the complex visibilities. However, it practically depends on the number of antennas and baseline combination which samples the UV-plane. Typically, there is incompleteness on spatial frequency sampled from observation. The observer needs to choose the array configuration that can recover features or details of the object. The longer array baseline can recover small features, while the more compact configuration can recover extensive target features [15].

Table 71.2 Data selection criteria

Parameters	Criteria
Source name	Uranus
Spectral resolution	>1000 kHz
Integration times	>300 s
Option	Public data only
Selected bands	3, 4, 6, 7, 8, 9

71.2.2 ALMA Data Selection

The ALMA data can be downloaded at its archive, which can be accessed through the ALMA website (https://almascience.nrao.edu/aq/?resultview=observation). Since Uranus were often observed on several projects in ALMA, particularly as calibrators, we need to shortlist them. Thus, it is important to select more criteria in the query form as listed in Table 71.2. As seen on the table, in this data mining we chose spectral resolution of more than 1000 kHz due to the continuum criteria and integration time of more than 300 s to filter the abundant data. The ALMA data that available to be acquired are the public data which usually publish some year after observation period. We selected several bands especially from the early cycle, as there were definitely more available.

Due to the immense size of each data, we used the download script which was run under Ubuntu. The script provided by ALMA archive are able to automatically re-download three times, if the connection drops sometimes. However, it is necessary to make sure that the data package was fully complete. It should contain all of the ALMA data structure, including several directories of product, calibration, and raw data. More about ALMA Data can be seen at ALMA Data Archive Manual [18]. For this study, we obtained ALMA data mostly from bands 3 and 6, following by bands 4, 7, and 8, which total sized about 5 TB.

71.3 Data Processing

71.3.1 Data Reduction with CASA

All of the data were processed with Common Astronomy Software Application (CASA) by Mc Mullin et al. 2007 [11]. In this research, various data processing was carried out with CASA version 4.7.2, 5.1.0, and 5.4.0, which can be downloaded at the CASA website (https://casa.nrao.edu/casaobtaining.shtml). The various version was necessary due to the highly diverse of observation date, in which particular version usually mentioned in *README* file on each directory.

It can be seen in the flowchart on Fig. 71.1, ALMA data as input will undergo a flagging process first to eliminate unnecessary signals and generate a flagging table. ALMA provides raw data, named Alma Science Data Models (ASDM), in which after being successfully calibrated, will have output extension as measurement set (*.ms*). This file is a dataset consisting of several tables of visibility data from observations and their supporting files.

The procedure of calibration includes the flux and phase calibration using quasars and scientific targets which varies on each observation project. It was carried out by applying calibration task in CASA for older data and CASA pipeline for the latest data. In our study, the calibration itself was the most time-consuming, though it depends on the amount and size of the raw data. Some calibration failure has been recalibrate using different CASA version. Of all the projects downloaded, nearly half were successfully calibrated, resulting in 110 observation data within various bands and observation periods. Some data with similar observation time merged with the task called concat, resulting in the final 78 Uranus data points. This number is

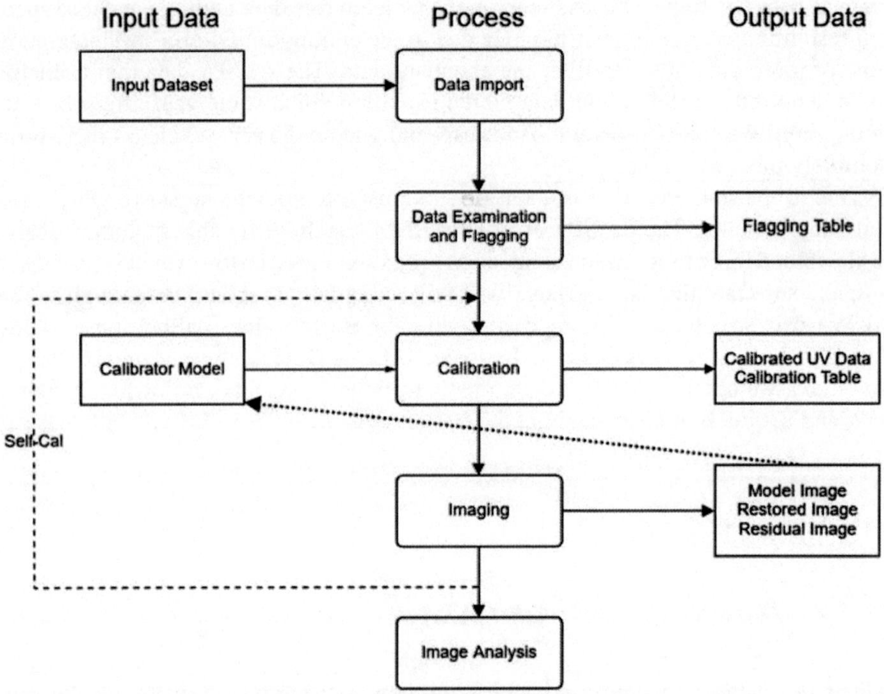

Fig. 71.1 Flowchart ALMA data processing with CASA (casa.nrao.edu). The program infrastructure is built based on the CASA core with C++ tools combined with an interactive Python interface for data reduction

appropriate for our research, particularly to build the continuum model. The following process is continuum imaging deconvolution that will be explained more in Sect. 3.1.

71.3.2 Deconvolution Cleaning and TB Conversion

As it is described on the Sect. 2.1, deconvolution refers to the process of reconstructing a model of the sky brightness distribution, given a dirty image called residual and the Point-Spread-Function (PSF) of the instrument. This process is called deconvolution, because under certain conditions, the residual image can be written as the result of a convolution of the true sky brightness and the PSF of the instrument. Deconvolution forms the minor cycle of iterative image reconstruction in CASA.

Some input parameters to run the CASA clean task in this research consist of Hogbom deconvolver algorithm, multi-frequency synthesis (mfs) mode, Briggs weighting, and standard gridder. Once we started the deconvolution command, the dirty image would appear on the CASA Graphic User Interface (GUI). The next task was masking, a process to tell the algorithm the position and strength of the source and estimate the side lobe. A further step was subtracting iteratively until it appears uniform between residual noise and masked image, as it is seen on the left panels of Fig. 71.2. After that, we could end the process, and finally obtained the clean image containing the model and residual image. This cleaned image contains statistics for further analysis. The result of flux value (I) in Jansky/beam can also be converted to the brightness temperature (Tb) in Kelvin with the following Eq. 71.2.

$$Tb = 1.22 \times 10^3 \frac{I}{v^2 \theta_{\max} \theta_{\min}} \tag{71.2}$$

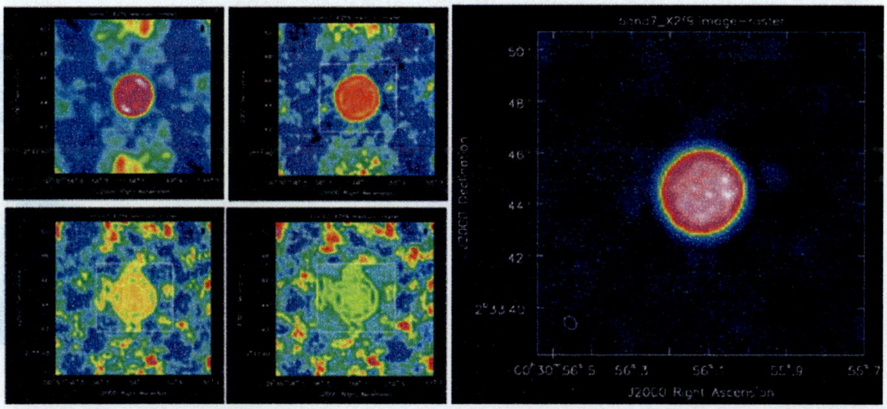

Fig. 71.2 CLEAN process with CASA from dirty image to subtracted image (right panels) and the clean image result (right panel)

The value we obtain in the flux density field is the flux in Jansky of the entire region, therefore in this equation, theta (θ) is identified as the angular size of the entire Uranus diameter. Each θ value depends on the Uranus distance which varies with the observation times. Using Planetary Data System (PDS) NASA (https://pds-rings.setikemud.org/tools/viewer2ura.html), we can see the Uranus ephemeris in order to calculate them. Different diameters of Uranus for each observational data were calculated by converting the observed distance value seen from PDS NASA. We can also obtain the geometric shapes or orientations as well as some related information on certain ephemeris for analyzing the images as well. Therefore, if the previous flux density result has no correction to the ephemeris, this *Tb* conversion has been considered them.

71.4 Results and Analysis

Of all the projects downloaded, nearly half were successfully calibrated, resulting in 110 observation data within various bands and observation periods. Some data with similar observation time merged with the task called concat, resulting in the final 78 Uranus data points. This number is appropriate for our research, particularly to build the continuum model. In addition, we also had one project targeting Uranus as their scientific object by Molter et al. [13].

After all of the cleaning process, we had a diameter of Uranus range between 3.33 to 3.67″. Each image has various beam width depends on the frequency and array configuration explained in Sect. 71.2. With the Uranus diameter of 3.5″, the results were divided into three categories: well resolved (<1″), marginally resolved (″1<3″), and unresolved (>3.5″). From the distribution plot of beam result in Fig. 71.3, it appears that the resolved data is less than unresolved, showing by more data points above Uranus diameter presented in a red dashed line. However, the entire data can be used to create the Uranus continuum plot, presented in Sect. 4.1.

Table 71.3 shows the most well-resolved observational data represent from each band and observation time. There are 6 data of Uranus observation from each band (3, 4, 6, 7, and 8), observed from 2012 to 2018. On the project ID, some data at the bands 3, 4, and 6 were observed from the project 2017.1.00855 [13], which was the only project targeting Uranus as the scientific object in this research.

Other parameters such as diameter array in m, and longest baseline (B) in km, integration time in second, center frequency in GHz, and beam in arcsecond (″) were listed in Table 71.3. The longest *baseline* of the interferometer along with frequency will determine those beam values. Although most recent continuum data use a 7 m array, the resolved data mainly were provided from the 12 m array observation. Thus, the following selected images were mostly from the 12-m diameter array.

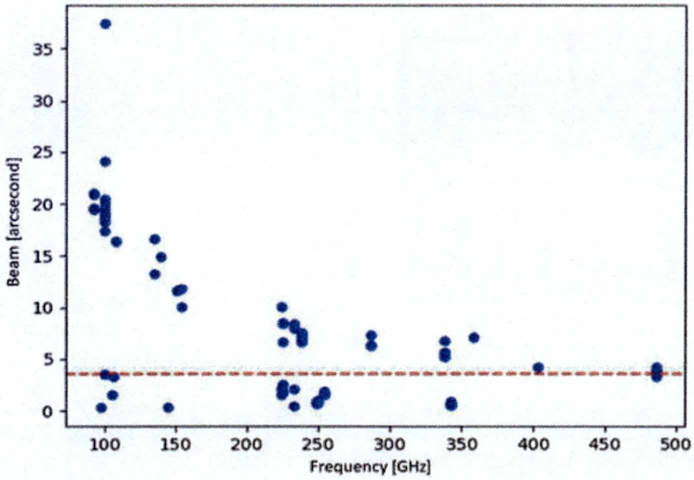

Fig. 71.3 Observation data points of Uranus with ALMA Data. It appears that most data have beam width larger than Uranus, showing by more dots above Uranus diameter, presented in red dashed line of 3.5"

Table 71.3 Selected Uranus continuum image at various frequency bands from 2012 to 2018

Band	ALMA project ID	Array (m)	B (km)	Observation date	Integration time (s)	Frequency (GHz)	Beam (″)
3	2017.1.00855	12	3	12/03/17	262,061	97.49	0.301
4	2017.1.00855	12	2.2	12/27/17	1506.82	144.99	0.313
6	2017.1.00855	12	1	09/13/18	2182.85	233.00	0.466
7	2011.0.00539	12	0.4	07/02/12	367.488	342.99	0.539
8	2015.1.00384	7	0.4	08/09/16	6674.4	486.57	3.306

71.4.1 Continuum Imaging Results

The six images in Table 71.3 are shown in Figs. 71.4, 71.5, 71.6, 71.7 and 71.8 on the left panel. The result images are the combination of raster and contour, with RMS noise sigma depend on the respective unit contour level or background RMS values. Those raster images were processed with the *Rainbow2* colorize on CASA, in which color difference indicates the flux value in Jansky/beam (Jy/beam). Each image is equipped with color wedges on the right axis to give information of the flux density. The contour lines indicate sigma values, starting from 3 sigmas on the outermost line. There is also a tiny white circle on the lower-left corner of each image that represents the beam width in arcseconds, where the values can also be seen in Table 71.3. Meanwhile, the orientation position from PDS NASA is displayed on the right panel of each image. These geometry illustrations describe Uranus's position in particular observation time, which can be related to the contour images on the left.

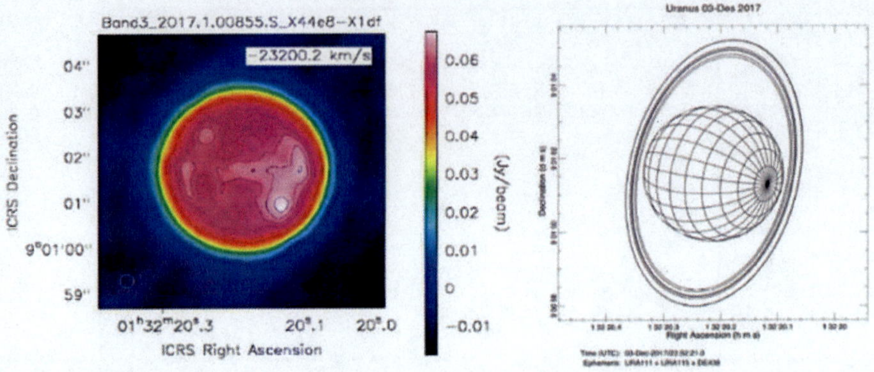

Fig. 71.4 Uranus continuum with beam 0.301" from project 2017.1.00855.S/X44e8 at the Band 3. The image was processed with unit contour level of 0.003 Jy/beam in the CASA program (left). The Uranus orientation from PDS NASA on Dec, 3 2017 (right)

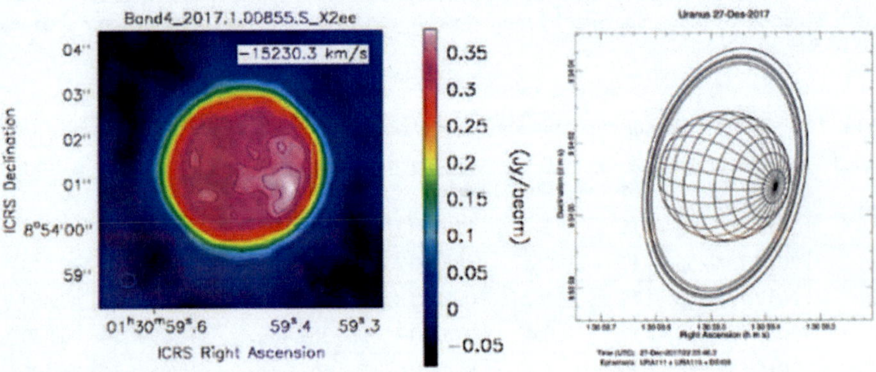

Fig. 71.5 Uranus continuum with beam 0.313″ from project 2017.1.00855.S/X2ee at the Band 4. The image was processed with a contour level of 0.02 Jy/beam in the CASA program (left). The Uranus orientation from PDS NASA on December 2, 2017 (right)

Band 3

At band 3, the best resolved image as seen in Fig. 71.4 was from the *project* 2017.1.00855.S with filename X44e8. With the unit contour level 0.003 Jy/beam, Fig. 71.4 has contour line of 0.009, 0.027, 0.045, 0.051, 0.057, 0.06, 0.063, 0.066 Jy/beam, from the outer to inner line.

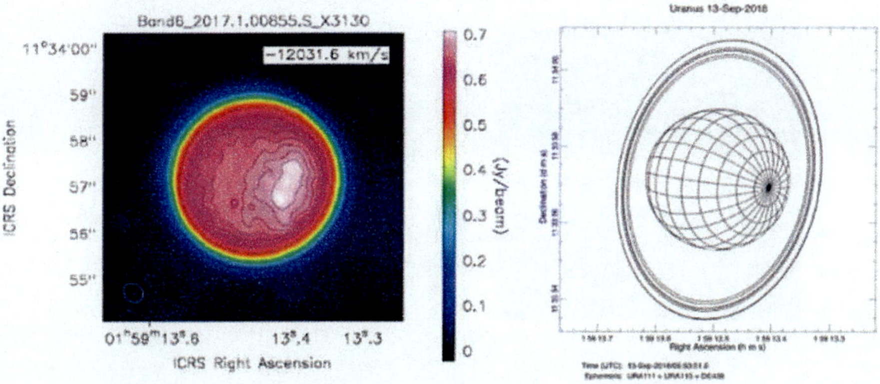

Fig. 71.6 Uranus continuum with beam 0.466″ from project 2017.1.00855.S/X3130 at the Band 6. The image was processed with a unit contour level of 0.003 Jy/beam in the CASA program (left). The Uranus orientation from PDS NASA on September 13, 2018 (right)

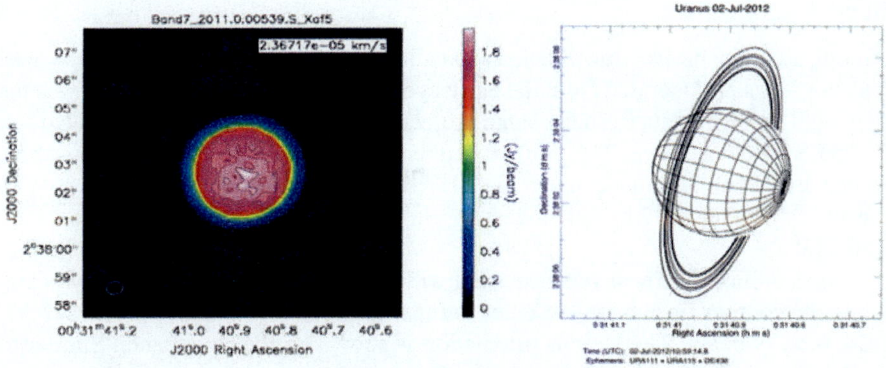

Fig. 71.7 Uranus continuum with beam 0.536″ from project 2011.0.00539.S/Xaf5 at the Band 7. The image was processed with a unit contour level of 0.003 Jy/beam in the CASA program (left). The Uranus orientation from PDS NASA on July 2, 2012 (right)

The Uranus image from this project was observed on December 3, 2017, obtaining the well-resolved image with a beam of 0.301". The array diameter used was 12-m, with the longest baseline around 3 km. The raster image shows bright color with a higher flux value in Jansky per beam, indicating the polar region. This interpretation is confirmed by orientation images of Uranus at the observation time which shows the polar side on the right.

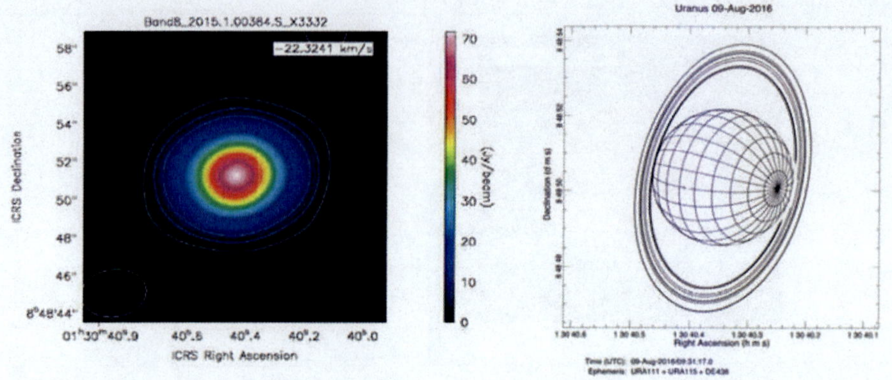

Fig. 71.8 Uranus continuum with beam 0.536" from project 2015.1.00384.S/X3332 at the Band 8. The image was processed with a unit contour level of 0.003 Jy/beam in the CASA program (left). The Uranus orientation from PDS NASA on August, 9 2016 (right)

Band 4

Among all of the project downloaded and calibrated, the data from band 4 is the least because it has not operated from the early cycle. The image on the Fig. 71.5 was the only well-resolved data from the same project with previous band 3 images, 2017.1. 00,855.S filename X2ee. This higher frequency observation was taken about three weeks after band 3, on December 27, 2017, with integration times of 1500 s. With the array diameter of 12 m and longest baseline around 2.2 km, resulting in the beam of 0.313".

Image on Fig. 71.5 was reconstructed with a unit contour level of 0.02 Jy/beam, while the contour lines from the outer to inner are 0.06, 0.14, 0.24, 0.26, 0.3, 0.32, 0.34, 0.36 Jy/beam. The Uranus orientation is shown on the right panel. Similar to the previous image, it describes the polar feature. Therefore, the area on the contour image is brighter with a higher flux density.

Band 6

Our data are mostly generated from band 6, as it has started since *Cycle 0* in 2012. We obtained several well-resolved images from band 6, yet the best image was also from project 2017.1.00855.S with filename X3130. The image on Fig. 71.6 was observed on September 13, 2018. The ellipse on the left bottom corner displays the observation beam of 0.466", obtained from the observation with 12-m diameter and baseline around 1 km. The contour lines represent the value of 3, 5, 7, 10, 12, 12.5, 13, 13.5 Jy/beam, with unit contour level of 0.003 Jy/beam.

Even though it has a similar feature as band 3 and 4, the polar side on this band 6 image slightly looks shifted to the left. It is confirmed by the Uranus geometry-image on the right panel. It can be seen that, in the following year, the Uranus pole became more Earth-side than in Fig. 71.4 and 71.5.

Band 7

This band 7 image was taken 367.488 s, the fastest integration time among other bands in this section. It was obtained from project 2011.0.00539.S with filename Xaf5. The observation was conducted in the early cycle on July 2, 2013, with a 12-m diameter array and the longest baseline only 400 m. Nevertheless, it was able to reach the well-resolved feature with a small beam width of 0.536″.

The displayed image has a value of unit contour level 0.003 Jy/beam with each contour line value of 0.06, 0.3, 0.6, 1.4, 1.6, 1.8, 1.84 and 1.9 Jy/beam. Different from the previous images, it appears that observations were made at the equinox or pointing towards the equator, as is seen in the orientation image. The higher flux value in the center of the image is probably represents the equator of Uranus which exposes more sunlight during the season.

Band 8

Band 8 is one of the extended observation frequency ranges that has been applied to ALMA in a follow-up cycle. Accordingly, the available data on the ALMA archive was not as much as the other bands. Of all the band 8 data obtained, none of them was categorized as well resolved. Therefore, Fig. 71.8 will display the highest resolution image band 8, from project 2015.1.00384.S with the filename X3332.

With the unit contour level of 0.01 Jy/beam, the contour image of Uranus is obtained with each line from outside to inside are 2.7, 4.5, 9, 13.5, 18, 27, 54 and 63 Jy/beam. There are no special features shown in the image because it looks homogeneous. However, it makes the reconstruction of the Uranus continuum image has completed in various bands and observation periods. Even though band 9 was also the early cycle, no image of Uranus was obtained with our criteria specified in this work. The complete information regarding all the results of continuum data processing is shown at the Sect. 4.2.

71.4.2 Uranus Continuum Models

Both resolved and unresolved results, the flux values of 78 calibrated data points were plotted to reconstruct a continuum model of Uranus. Some parameters result from deconvolution process are listed on Table 71.4 at the Appendix section. Figure 71.9 describes the distribution of flux density over various ALMA wavelength ranges, divided into several red dashed vertical lines to separate each band (3, 4, 6, 7, and 8). It appears that the flux value in Jansky tends to be higher as the frequency or band number increases. This tendency of rising is accompanied by a higher uncertainty. It can be seen in the observation data below 200 GHz have a smaller error range,

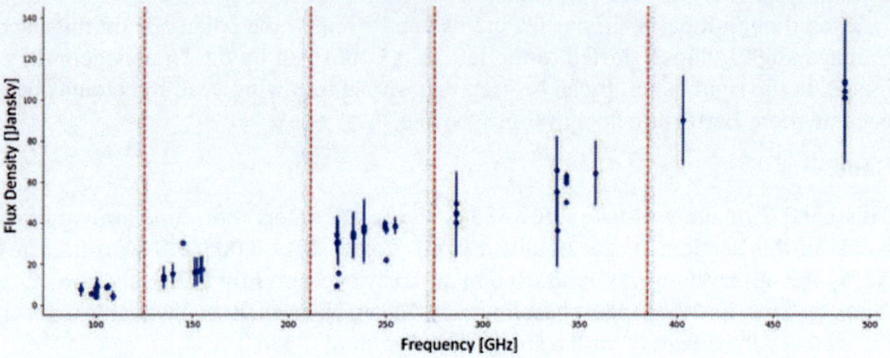

Fig. 71.9 Distribution plot of flux density over various ALMA bands from 78 Uranus observation data points

even some data points have very small *uncertainty*. Yet, this error range gets bigger in the higher frequency band, around 300 GHz and above. This is because the higher the frequency of observations, it will encounter a greater instrumentation error along with atmospheric effect, resulting in a higher range of uncertainty values. The flux density value can be converted to a brightness temperature, as it is mentioned in earlier Sect. 3.2 with Eq. 71.2.

Figure 71.9 is the result of the Uranus continuum that has been corrected for the ephemeris of each observation data. The brightness temperature decreases with frequency as would be expected for a planet with a dense atmosphere like Uranus. It has the characteristic of higher temperatures at lower altitudes. On the curve, we get an expected decrease profile for higher frequencies (Fig. 71.10).

This plot result is entirely appropriate because the *Tb* values span between the Uranus equilibrium temperature about 50 K. The appearance of slightly spread-out points on the curves is due to the seasonal fluctuations of Uranus which cause changes

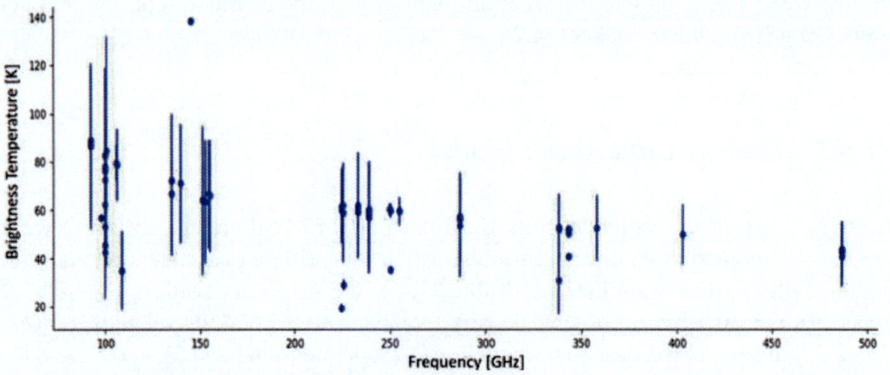

Fig. 71.10 Distribution plot of brightness temperature over various ALMA bands from 78 Uranus observation data points

in the direction of view of the bright poles. And the high error value in the lower band, is due to the larger telescopic beam width. Though the error value is high enough, they were roughly estimated from the calibration error which generally can reach 10%. This result, in a certain range of ALMA, is in line with Uranus continuum model plotted from various Uranus observation data [13].

71.5 Discussion and Summary

- Uranus observation data on ALMA archive, which is very abundant, provides an excellent opportunity to conduct various studies, such as Uranus continuum imaging. The wide frequency range and antenna configuration of ALMA can resolve Uranus in a small angular resolution.
- We obtained the data from 2012 to 2018 at band 3, 4, 6, 7, and 8. Among all the projects that fulfill our criteria, only one project observed Uranus as a science target. Meanwhile, the rest projects observed Uranus as the bandpass or flux calibrators. Half of the downloaded Uranus data were successfully calibrated with CASA. Apparently, the calibration failures in some observational data were caused by incomplete downloaded files, differences in CASA versions, and other uncertain reasons.
- The results of image processing with cleaning algorithm on CASA are categorized as well-resolved, marginally resolved and unresolved. Well-resolved data can be used to see the contours of the difference in flux values, with the highest flux being a bright polar representation. Meanwhile, all data including marginally resolved and unresolved images were used to plot the continuum of Uranus over the ALMA wavelength range.
- On the continuum curve of flux density distribution with frequency, it is obtained that the higher the ALMA band, the higher the flux density value in Jansky, but accompanied by a high uncertainty value as well. On the other hand, when converted to brightness temperature in Kelvin, the plot tends to decrease from low to high frequency. These results agree with the equation of flux—*Tb* conversion which involves the frequency and diameter of Uranus on each ephemeris time and distance. Our curve also seems reasonable when compared to the Uranus brightness temperature profile from Molter et al [13], within the range of ALMA and with more observation data.
- The distribution of data points both the flux density and brightness temperature plots looks scattered. This is due to seasonal fluctuations that cause changes in the direction of view of the bright pole of Uranus. Nevertheless, the acquisition of these two plots which mostly from calibration observation data is quite successful in establishing an outline model of the Uranus continuum over ALMA bands.
- On the cleaned flux density contour images, Uranus looks very bright with sharp edges. To make further analysis of those results, it is significant to have additional step before cleaning, to subtract a uniform limb-darkened disk from the data in UV-plane. Due to the Fourier transform in CASA from the UV-plane into the

image plane, ringing artefacts will arise, moreover if the cleaning process is too deep. This additional process will also affect the results on the continuum plot which supposed to be slightly more accurate.

Acknowledgements This paper makes use of the following ALMA data: ADS/JAO.ALMA2012.1.00720.S, ADS/JAO.ALMA2015.1.01025.S, ADS/JAO.ALMA2013.1.01329.S, ADS/JAO.ALMA2016.1.00911.S, ADS/JAO.ALMA2016.1.01123.S, ADS/JAO.ALMA2015.1.01187.S, ADS/JAO.ALMA2017.1.00855.S, ADS/JAO.ALMA2016.2.00105.S, ADS/JAO.ALMA2016.2.00172.S, ADS/JAO.ALMA2011.0.00243.S, ADS/JAO.ALMA2012.1.00482.S, ADS/JAO.ALMA2012.1.00142.S, ADS/JAO.ALMA2013.1.00586.S, ADS/JAO.ALMA2013.1.00588.S, ADS/JAO.ALMA2013.1.01358.S, ADS/JAO.ALMA2016.1.00803.S, ADS/JAO.ALMA2016.2.00200.S, ADS/JAO.ALMA2017.1.00855.S, ADS/JAO.ALMA2011.0.00539.S, ADS/JAO.ALMA2012.1.00097.S, ADS/JAO.ALMA2013.1.00099.S, ADS/JAO.ALMA2015.1.00997.S, ADS/JAO.ALMA2016.1.01305.S, ADS/JAO.ALMA2016.1.01285.S, ADS/JAO.ALMA2015.1.00384.S.

ALMA is a partnership of ESO (representing its member states), NSF (USA) and NINS (Japan), together with NRC (Canada), MOST and ASIAA (Taiwan), and KASI (Republic of Korea), in cooperation with the Republic of Chile. The Joint ALMA Observatory is operated by ESO, AUI/NRAO and NAOJ. This study was also supported by LAPAN Scholarship as a part master thesis project in Astronomy magister program at ITB.

Appendix

See Table 71.4.

Table 71.4 Results on the 78 Uranus continuum imaging from ALMA data, processed with CASA

No	Raw filename	Array (m)	Observation date	Int time(s)	Frequency (GHz)	Beam (″)	Flux density (Jy)	TB (K)
BAND 3								
1	X1b1e-X1c87	12	07/08/13	7303,2	105,51	1,54	8,89E + 00	79,497
2	X2d4	12	12/13/13	157,344	106,11	3,36	9,11E + 00	78,754
3	X2369	7	09/07/16	302,4	100,59	18,27	9,33E + 00	84,616
4	8446X	7	12/02/16	337,44	92,07	21,03	8,02E + 00	88,072
5	X6903	7	12/09/16	337,68	92,07	19,57	8,02E + 00	88,951
6	X129e-Xe6f	7	11/14/16	87,537,9	108,58	16,36	4,50E + 00	34,807
7	X8446	7	12/02/16	337,44	92,07	20,92	7,94E + 00	87,150
8	X6903	7	12/09/16	5243,23	92,07	19,48	7,79E + 00	86,399
9	X2574 -X49f- X454	12	12/25/15	92,803,2	99,99	3,55	7,81E + 00	76,092
10	X49e6-X4e7d	7	12/27/15	8180,16	99,99	20,02	8,41E + 00	82,205
11	X625-X561	7	01/11/16	91,228,9	99,99	24,14	4,56E + 00	45,628
12	X57b8-X3e6	7	01/17/16	82,431,7	99,99	19,10	4,22E + 00	42,726
13	X4460-X4894	7	01/19/16	6489,79	99,99	18,55	7,14E + 00	72,583
14	X6f0	7	01/18/16	4174,7	99,99	19,43	8,15E + 00	82,708
15	X4fde	7	04/28/16	4350,53	99,99	37,44	7,07E + 00	77,380
16	X61a8	7	05/03/16	302,4	99,99	17,40	7,15E + 00	78,069
17	X30e5-X3380 -X53b	7	07/07/16	426,108	99,99	20,49	6,16E + 00	62,141
18	X71b0–X77ec-X2e2c	7	09/23/16	85,652	99,99	18,89	5,88E + 00	53,241
19	X44e8- X1df	12	12/03/17	262,061	97,49	301	5,84E + 00	56,785
20	X529	12	01/06/17	184.56	97,49	2,35	1,32E + 00	13,650
21	X1651	7	11/05/16	9240	107,85	14,23	9,72E + 00	75,605

(continued)

Table 71.4 (continued)

No	Raw filename	Array (m)	Observation date	Int time(s)	Frequency (GHz)	Beam ($''$)	Flux density (Jy)	TB (K)
BAND4								
22	X2ee	12	12/27/17	1506,82	144,99	329	3,03E + 01	138,061
23	X43a1	7	09/22/17	338,496	152,44	11,69	1,63E + 01	63,396
24	X24d2	7	09/12/17	337,2	154,19	10,12	1,73E + 01	66,628
25	X139b	7	09/21/17	337,152	155,01	11,79	1,75E + 01	65,884
26	X4638	7	09/22/17	337,344	139,65	14,93	1,53E + 01	70,991
27	Xffc	7	09/21/17	336,912	151,13	11,65	1,62E + 01	64,078
28	X4cc1-X766	7	08/03/17	81,657,3	134,99	16,57	1,25E + 01	66,585
29	X2803-X2f9e	7	08/24/17	11,302,7	134,99	13,33	1,40E + 01	72,123
BAND 6								
30	X201	12	07/04/12	5106,82	249,50	0,72	3,77E + 01	60,336
31	X78b	12	07/13/12	235,2	249,50	0,91	3,81E + 01	60,066
32	X674	12	08/10/12	237,12	249,50	0,84	3,98E + 01	60,110
33	Xc78	12	06/16/12	5161,39	250,15	1,04	3,65E + 01	59,878
34	X85c-X4c	12	07/03/12	90,691,6	250,15	0,84	2,21E + 01	35,299
35	X1484	7	10/06/13	302,832	224,49	10,06	3,43E + 01	61,656
36	X41ca	12	01/03/15	4129,39	224,49	2,12	3,01E + 01	59,587
37	X271- X192- X4c2	12	01/13/15	258,265	224,50	1,52	9,55E + 00	19,261
38	X4d3-X548	12	12/09/14	512,829	225,45	1,73	1,55E + 01	29,179
39	Xa89	12	12/15/14	3684,14	233,00	2,13	3,34E + 01	59,443
40	X509f	7	07/07/14	2273,28	233,00	8,01	3,30E + 01	60,977
41	X95	7	07/16/14	4626,62	233,00	8,38	3,41E + 01	61,945
42	X522a	12	01/04/15	4105,97	254,58	1,92	3,86E + 01	59,504

(continued)

Table 71.4 (continued)

No	Raw filename	Array (m)	Observation date	Int time(s)	Frequency (GHz)	Beam (″)	Flux density (Jy)	TB (K)
43	X562	12	01/13/15	157,344	254,59	1,65	3,82E + 01	59,911
44	X1734-X193f-X150a	7	10/25/16	11,430	238,78	7,49	3,82E + 01	60,290
45	X7e6	7	11/07/16	5032,66	238,78	6,70	3,64E + 01	57,805
46	X2990	7	11/08/16	337,344	238,78	6,76	3,72E + 01	59,201
47	Xb3a3	7	11/09/16	336,336	238,78	7,44	3,79E + 01	60,298
48	X1627	7	11/15/16	337,344	238,77	7,12	3,56E + 01	56,970
49	X2c47	7	07/29/17	4757,66	225,25	8,49	3,06E + 01	58,899
50	X2f06	7	07/23/17	336,336	225,26	6,62	3,03E + 01	58,799
51	X1e40	7	07/01/17	5101,54	225,27	8,40	3,07E + 01	61,841
52	X3130	12	09/13/18	2182,85	233,00	453	3,50E + 01	58,931
51	X264 -Xb6	12	12/23/14	1,049,600	226,83	1,74	3,24E + 01	61,735
52	X5ee5-Xe1f	12	01/17/15	122,500	224,47	2,16	2,95E + 01	59,812
53	X18d-X6db	12	01/01/15	4926.67	258,91	1,40	4,03E + 01	59,852
54	X123b-Xa05	12	12/31/14	3,121,320	255,16	1,42	3,87E + 01	59,154
55	X43e6	7	08/30/17	25,350	229,73	8,32	3,01E + 01	52,989
56	X5473	7	07/21/17	5280	238,69	6,09	3,28E + 01	56,982
BAND 7								
57	X98b	12	07/30/12	371,568	342,99	0,57	5,01E + 01	40,702
58	X28f	12	08/11/12	372,384	342,99	0,88	6,29E + 01	50,235
59	X7ab	12	06/17/12	367,488	342,99	0,60	5,94E + 01	51,777
60	X2f9	12	06/18/12	367,488	342,99	0,57	5,99E + 01	52,153

(continued)

Table 71.4 (continued)

No	Raw filename	Array (m)	Observation date	Int time(s)	Frequency (GHz)	Beam (″)	Flux density (Jy)	TB (K)
61	Xaf5	12	07/02/12	367,488	342,99	0,54	6,15E + 01	52,246
62	X8d7	7	10/08/13	302,832	338,07	5,61	6,55E + 01	51,994
63	X1bb0	7	04/26/14	6921,6	338,09	6,80	5,51E + 01	52,874
64	X271c	7	06/08/14	302,832	358,02	7,04	6,41E + 01	52,439
65	X3f5c	7	04/14/16	345,024	286,36	6,30	4,00E + 01	53,552
66	X4a7a	7	06/04/16	5505,46	286,38	6,33	4,44E + 01	57,335
67	X9566	7	06/15/16	348,288	286,38	7,27	4,50E + 01	57,328
68	X1339	7	08/25/16	344,496	286,38	6,26	4,94E + 01	56,116
69	X65fb	7	07/26/17	5950,9	338,21	5,16	3,62E + 01	31,043
70	X675c	7	07/09/17	7800	336,55	4,19	6,11E + 01	54,450
71	X2b4a	7	07/09/17	82,786.4	337,44	4,15	6,09E + 01	53,996
BAND 8								
72	X451-X1da1	7	10/11/16	11,792,6	402,78	4,26	9,00E + 01	49,812
73	Xa728	7	08/10/16	6626,35	486,57	3,72	1,09E + 02	43,775
74	X3332	7	08/09/16	6674,4	486,57	3,31	1,01E + 02	40,860
75	X1019	7	07/29/16	6630,77	486,57	3,39	1,04E + 02	42,536
76	X293	7	07/16/16	6704,64	486,58	3,76	1,01E + 02	42,558
77	X277f	7	08/11/16	6641,28	486,57	4,18	1,08E + 02	43,464
78	X8e84	7	08/12/16	6618,14	486,57	3,94	1,09E + 02	43,689

References

1. Arridges, C.S., et al.: The science case for an orbital mission to Uranus: exploring the origins and evolution of ice giant planets. Planet. Space Sci. **104**, 122–140 (2014)
2. Baudry, A.: The ALMA radio telescope. In: Proceeding of Science (2009)
3. Condon, J.J., Ransom, S.M.: Essential Radio Astronomy, vol. C7. Princeton University Press, Princeton (2016)
4. De Pater, I., et al.: Record-breaking storm activity on Uranus in 2014. Icarus **252**, 121–128 (2015)
5. Guillot, T.: Uranus and Neptune are key to understand planets with hydrogen atmospheres. Space Sci. Rev. **1**, 159–196 (2019)
6. Helled, H.B: Long-term atmospheric variability on Uranus and Neptune, Icarus, **186**, 291–301 (2020)
7. Hofstadter, M., et al.: Uranus and Neptune missions: a study in advance of the next planetary science decadal survey. Planet. Space Sci. **177**, 104680 (2019)
8. Irwin, P.G.J., et al.: Spectral analysis of Uranus' 2014 bright storm with VLT/SINFONI. Icarus **264**, 72–89 (2014)
9. Karkoschka, E.: Uranus apparent seasonal variability in 25 HST filters. Icarus **151**, 84–92 (2001)
10. Lissauer, J.J., de Pater, I.: Fundamental Planetary Sciences: Physics, Chemistry, and Habitability, pp. 219–222. Cambridge University Press, Cambridge (2013)
11. McMullin, J.P., et al.: CASA architecture and applications. Astronom. Data Anal. Softw. Syst. XVI ASP Conf. Series **376**, 127–130 (2007)
12. Meeus, J.: Equinoxes and soltices on Uranus and Neptune. J. Br. Astron. Assoc. **107**, 6 (1997)
13. Molter, E.M., et al.: Tropospheric composition and circulation of Uranus with ALMA and the VLA. Planetary Sci. J. **2**(1), 3 (2021)
14. Mousis, O., et al.: Scientific rationale for Uranus and Neptune in situ explorations. Planetary Space Sci. **155**, 12–40 (2017)
15. Remijan, A. et al.: ALMA Cycle 8 technical handbook. ALMA Technical Handbook, ALMA Doc. 8.3, ver. 1.0. (2020)
16. Sromovsky, L.A., et al.: Post-equinox dynamics and polar cloud structure on Uranus. Icarus **220**, 694–712 (2012)
17. Sromovsky, L.A., et al.: High S/N Keck and Gemini AO imaging of Uranus during 2012–2014: new cloud patterns, increasing activity, and improved wind measurements. Icarus **258**, 192–223 (2015)
18. Stoehr, F.: ALMA Science Archive Manual, ALMA Doc. 8.15 v1.0 (2020)
19. Tice, D.S.: Remote Sounding of the Atmospheres of Uranus and Neptune. Icarus **165**, 168–180 (2010)

Chapter 72
Spatial Modeling of Tidal Flood Due to the Fishpond to Tidal Waves in Bengkalis, Indonesia

Giant Amor, Muhammad Hanif, Arie Yulfa, Rizki Athorik Hidayat, and Poppy Indrayani

Abstract Almost 99% of the reported disasters are hydrometeorological types [11]. tidal flooding is one of the potentials that has recently become a frightening specter in the coastal area of Bengkalis. tidal flooding is an annual disaster that occurs in the coastal region of the Bengkalis Regency. (BPBD Bengkalis, 2020). it often happens and it has a serious problem for people living around coastal and tidal flats. This condition aggravated In 2020, most of the coastal areas in Bengkalis Island by the conversion of mangrove land into Fishponds, which is a classic problem in the research area for support the economy of the Bengkalis community. Moreover, The Sea level rise reported by IPCC (2007) increases by 15–90 cm with a definite increase of 48 cm in 2100. The method in this research is the spatial analysis model of coastal flooding use scenario optimum high tidal and estimation impact of conversion mangrove to become fishpond. All parameter data from the geospatial information agency (BIG), as land cover, coastal line, digital elevation model, and tides is analyzed using Cost Distance analysis with a geographic information system (GIS) in Arcgis. The aim of this research is to model the coastal flooding and the impact of the fishpond on tidal floods. For the final decision, we use the extreme scenario using the highest tidal discharge, which is 3.4 m to result in the Spatial Modeling Of Tidal Flood In Bengkalis. The results of this spatial modeling succeeded in showing the coastal flooding area, with the highest to lowest inundation range This model is suitable for representing the worst scenario of high tides in the coastal area of Bengkalis.

G. Amor (✉) · A. Yulfa · R. A. Hidayat
Department of Geography, Universitas Negeri Padang, Padang 25163, Indonesia
e-mail: giantamor84@gmail.com

P. Indrayani
Department Architecture, University of Fajar, Makassar, Indonesia

M. Hanif
Department of Geo-Informatics, Khon Kaen University, Khon Kaen 40002, Thailand

E. Yulihastin et al. (eds.), *Proceedings of the International Conference on Radioscience, Equatorial Atmospheric Science and Environment and Humanosphere Science, 2021*, Springer Proceedings in Physics 275, https://doi.org/10.1007/978-981-19-0308-3_72

72.1 Introduction

Sumatra is the second largest area of mangrove land. Where: "Indonesia has more than 4 million hectares": 15.46% in Sumatra, 2.35% in Sulawesi, 2.35% in Maluku, 9.02% in Kalimantan, 1.03% in Java, 0.18 Bali and Nusa Tenggara, and 69.43% in Irian Jaya. The conversion of mangrove forests into fish ponds strongly supports the community's economy is being empowered by the business sector in Bengkalis. From an economic perspective, the fishery business is profitable and it could increase people's income. However, this activity might lead to environmental degradation. On the other hand, Bengkalis is an area that is vulnerable to tidal flooding. Based on the latest news, flood conditions have occurred again since December 2019 until early 2020. The height of the sea water that rises to the land varies from 15 cm to 30 cm. High tide floods like this are subscribed every year. (Riaupos.com. 2021).

Based on information from the local community, this tidal flood has begun to be felt in the last 10 years, this tidal event mainly occurs in coastal areas that are prone to flooding, including Teluk Latak Village, Pedekik, Damon, Senggoro, Wonosari, Sebauk, Simpang Ayam, Meskom, Deluk, Jangkang, Bantan Sari, Bantan Air, Ulu Pulau, Pambang Baru, and Batu Panjang Village on Rupat Island. (Coverage 6. 2018). The tidal flood incident also occurred on October 30, 2019, the activities of residents in Bandar Jaya Village, Siak Kecil District, Bengkalis Regency were paralyzed because the main road in the area was flooded and water reached a height of 20 cm inundating a number of local residents' houses (Antarariau.com. 2019).

Geospatial technology such as geographic information systems is very effective for representing and analyzing spatial information, and very effective for calculating spatial data, and also for modeling as a basis for analysis and policy making, especially in disaster analysis and estimation of disaster slopes in settlements. Seeing the problem of the disaster which is very clearly a strategic issue in Bengkalis Regency, the author is interested in conducting research with the title "Spatial Modeling Of Tidal Flood Due To The Fishpond To Tidal Waves In Bengkalis, Indonesia".

Thus, carrying out The mapping of coastal flood hazards in Bengkalis is a tool used to determine flood area limits inland and in other areas exposed to coastal floods due to different risks such as storms, surge waves, sea-level rise caused by climate change. So far there has been no research that discusses flooding in Bengkalis. In fact, flood modeling in Bengkalis is not only useful for disaster mitigation, but also in the context of saving environmental degradation in coastal areas.

Therefore, this study aims to identify the spatial distribution of the Rob flood inundation on the Bengkalis Island coast due to high tides using spatial modeling by using the method Cost Distance.

72.2 Materials and Methods

72.2.1 *Land Surface Roughness Design and Cost Distance Analysis*

Modeling potential vulnerability of coastal areas assisted by GIS technology in interpreting constituent variables such as digital elevation model, slope, coastal line, land cover, and tides. The data as a roughness coefficient map is obtained from the land use map resulting from the extraction of land cover Digital Elevation Model. It is a product of the Geospatial Information Agency to serve the availability of elevation data in Indonesia. This data is needed in the analysis process that will design a spatial model that is the topic of this research, such as data on land cover, shoreline, the slope of the research area, and others. Type data: 32 bits Float. Spatial Resolution: 10 m (bands 1 Panchromatic). This Research Aims To Simulate Or Predict The Inundation Of Rob Flood That Runs On The Land When Tide Occurs As A Mathematically Measurable Energy Release.

Cost distance mapping and spatial analysis often involve more than calculating distances and elevations. Distance costs can also be a variety of variables, ranging from economic, social, cultural, environmental, or other barriers that can create this type of 'cost' in travel or movement [1]. Based on the suitability for analyzing between variables, this study is suitable for calculating the reduced surface reduction power due to being a pond that can inhibit the movement of tides.

72.2.2 *Damage of Mangrove Forests and Fish Pond Status in Bengkalis Regency*

No	District	Land Area	Mangroves	Large of fish ponds	
		(Ha)	Damaged	Total	Activities
1	Rupat Utara	12,784	25.3	3.32	0.06
2	Rupat	5806	1.5	3.38	0.45
3	Bantan	5804	20.7	62.63	61.29
4	Bengkalis	4254	0.5	23.33	15.34
5	Pinggir	2125	5.3	3.32	0.06
6	Siak Kecil	3452	7.8	0.05	0.05
7	Bukit Batu	898	7.8	3.23	1.37

(continued)

(continued)

No	District	Land Area	Mangroves	Large of fish ponds	
		(Ha)	Damaged	Total	Activities
Total		**41,718**	**64.1**		
2018				92.94	78.56
2017				61.28	44.59
2016				64.12	33.08
2015				64.49	32.65
2014				50.25	31.90

Source Department of Forestry Service Reg Riau. Source Fishery Service of Bengkalis Regency
Notes Condition December 2018

Mangroves are found at the forefront in terms of their position relative to many coastal hazards. The fact that they thrive in many coastal settings provides some indication of their ability to overcome the danger, or at least to recover from big impact [2]. Mangroves may contribute to reducing loss of life and damage to property from storms and cyclones as they reduce the impacts of waves, storm surges, and high winds.

Figure 72.1 reveals the fact that changes in field existence of the mangrove ecosystem is highly threatened by the conversion of mangroves become fishponds carried out by the community for business opportunities. Waves are most rapidly reduced when they pass through a greater density of obstacles [2]. This Means The structure of the mangrove forest is important in determining the potentialrisk reduction by mangrove.

Fig. 72.1 Damage of mangrove forests to become fishpond.on Coastal Bengkalis, 2021

Table 72.1 Coefficient land surface roughness

Roughness	Land cover	Coefficient value
R1	Water body	0.007
R2	Shrub	0.04
R3	Forest	0.07
R4	Plantation	0.035
R5	Bare land	0.015
R6	Agricultural area	0.025
R7	Built-Up	0.045
R8	**Mangroves**	0.025
R9	**Fish pond**	0.01

Source Barryman (2006)

72.2.3 Analytical Methods

At this stage, Geographic Information Systems (GIS) can be developed in urban hydrology applications, especially making a model [3]. In tidal flood modeling, the data used include surface feature data (Digital Elevation Model), topographic maps, land use maps, flow discharge models, runoff coefficients. The modeling technique uses cost distance modeling [4]. Data in the form of contours showing the elevation lines of the area are processed to become DEM, and modeling to look for runoff, accumulation of flow, and concentration of water flow (Table 72.1).

The following is a mathematical process for making data on the coefficient of land surface roughness [5].

$$\boldsymbol{H_{loss}} = ((167n2/H01/3) + 5\sin S)$$

Description:

H_{loss}: Reduction of water level and energy in the interval of shoreline
n: Coefficient of land surface roughness
H0: Maximum scenario of high water level of the tidal wave
S: Slope of the land surface.

72.2.3.1 Coastal Flooding Innundation Design

The basic principle of calculating the cost distance is to calculate the cost of each cell that is passed, the fundamental equation of the cost distance behind the ArcGIS software [1] (Fig. 72.2).

$$\boldsymbol{Cost\ Distance} = cds + \frac{crs + cra}{2}$$

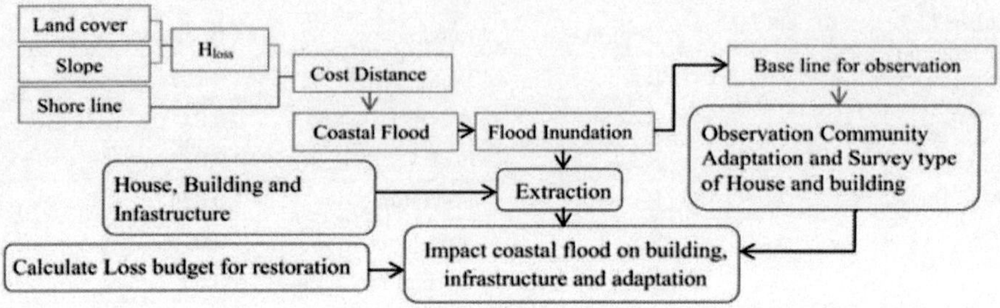

Fig. 72.2 Research procedure

where:

Cds = is cost distance sources
Crs = raster cost

$$CoastalFlood = cd - Maxcd$$

where: cd = cost distance and Max cd = Maximum value of cost distance. After the results of the coastal flood analysis are obtained. These results must be subtracted from the flood height data that occurred at the research location [11]. As in the location of this study, coastal flooding has a height of up to 100 cm above sea level, so data extraction is carried out at the height of 1 m. In this study, the land elevation was extracted from DEM images with a resolution of 10 m.

72.3 Results and Discussions

72.3.1 Spatial Modeling Tidal Flood Map in Coastal Bengkalis

We found the topography of the Bengkalis area is generally a flat and low area, as is the characteristic of the sloping landscape of the east coast of Sumatra. And it is dominated by peatland, making this research area vulnerable to tidal floods that hit coastal areas. The distribution of tidal inundation on the Bengkalis coast by an inundation map that has been modeled with mathematical calculations in Arcgis 10.4 software using Cost Distance Analysis.

From this modeling, it can be seen that the zone of exposure of the area to tidal inundation is modeled with a maximum tide scenario of 3 m. This is achieved by considering the variation in the distribution of tides along the coastline which is generally gentle and not more than 3 m high from land to sea (Fig. 72.3).

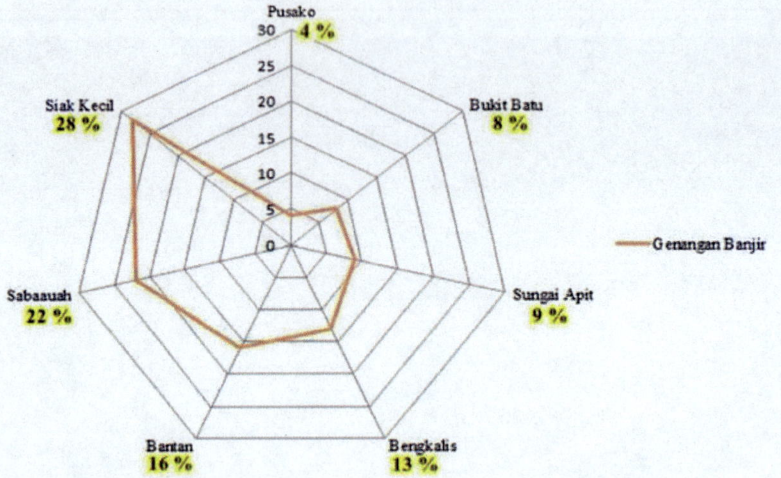

Fig. 72.3 Percentage areas impact of tidal flood inundation in coastal Bengkalis

From the picture above, it is clear that the administrative areas are affected by the tidal flood disaster. So for the prevention of choice of other places in the event of a greater tidal flood disaster. The flood vulnerability is relatively high for the major coastal cities. The values of all three assessment components together are more significant. Directions for city development, especially residential areas, also need to be reviewed and accommodate disaster zones as a disaster mitigation strategy in the form of developing development policies. The limitation of this research is which not use high- resolution data. For better coastal flood modeling, we should use the multiscale, multi- physics aspects of this study are considerable and highly informationally provocative. Coastal storm surges are on the order of a meter or more in the vertical and extend alongshore for coastal distances of several kilometers, with persistence of over several hours to several days

72.4 Conclusions

Tidal flood inundation mapping model with an approach to the condition of the basin/surface elevation and ground surface roughness, so that it can provide information on the area of the tidal flood inundation from the estimated tidal overflow, as well as calculate the settlements affected by the flood. puddle model. This method is appropriate in spatial modeling of flood inundation because with a very dynamic land cover, it calculates the coefficient of surface roughness based on the land cover can reduce the energy of the incoming waves. Currently, in Bengkalis, the Mangrove Land Cover has turned into a fish Pond which will have an impact on the amount of energy carried by the land waves.

Acknowledgements The authors would like to thank Geospatial Information Agency (BIG) for a accessibility Data, and INCREASE who has given the author opportunity to present the results of the research carried out

The author also would like to thank Dr. Erma Yulihastin with coaching in writing for supervising the who has given the opportunity to present the study that has been carried out by the author.

Authors Contribution All authors contributed equally as the main contributors to this work.

References

1. Johnston, K.M.: Spatial Analyst—Cost Distance Analysis. Esri. (2015)
2. Spalding, M., McIvor, A., Tonneijck, F., Tol, S., Eijk, P. van: Mangroves for coastal defence. Wetl. Int. Nat. Conserv. **42** (2014)
3. Marfai, M.A., Mardiatno, D., Cahyad, A., Nucifera, F.: Pemodelan Spasial Bahaya Banjir Rob Berdasarkan Skenario. Bumi Lestari. **13**, 244–256 (2017)
4. Hanif, M., Putra, B. G., Hidayat, R. A., Ramadhan, R., Shafrina, W., Moh, W., Hermon, D., Suhana, E., Makassar, U. F.: Impact of Coastal Flood on Building, Infrastructure, and Community Adaptation. Jurnal Geografi Gea, 21(2). https://ejournal.upi.edu/index.php/gea
5. Berryman, K.R.: Review of Tsunami hazard and risk in New Zealand. Igns. **139** (2006)
6. Online News; https://riaupos.jawapos.com/bengkalis/05/12/2021/263234/pasang-keling-rendam-ratusan-rumah-dan-ruas-jalan-di-bengkalis.html
7. Moh Holli, (2012), Pemodelan Genangan Banjir Pasang Air Laut Di Kabupaten Sampang Menggunakan Citra Alos Dan Sistem Informasi Geografis. Semnas Kedaulatan Pangan Dan Energi 2012universitas Trunojoyo, Madura, Juni 2012
8. Lee, S.J., Choi, H.I.: Development and Implementation of an Assessment Framework for Coastal Flood Vulnerability. Establishment and application of flood vulnerability evaluation structure. J. Korean Soc. Hazard Mitig. **18**(6), 291–299 (2018). https://doi.org/10.9798/KOSHAM.2018.18.6.291
9. Haigh, I.D., Nicholls, R.J., Penning-Roswell, E., Sayers, P.: Impacts of climate change on coastal flooding, relevant to the coastal and marine environment around the UK. MCCIP Sci. Rev. **2020**, 546–565 (2020). https://doi.org/10.14465/2020.arc23.cfl
10. Badan Informasi Geospasial. (2021). Kabupaten Bengkalis, https://tanahair.indonesia.go.id/portal-web
11. Indonesia Disaster Management (BNPB), https://bnpb.go.id/berita/bencana-hidrometeorologi-dominan-sepanjang-awal-januari-hingga-akhir-april-2021-
12. 12utra, a. (2021). literature reviews : hydrometeorological disasters and climate change adaptation efforts literature reviews : hydrometeorological. July, 1–https://doi.org/10.24036/sjdgge.v5i1.363

Chapter 73
Indo-Australian Plate Velocity Measurement During Interseismic Phase in 2010–2014 Using Sumatran GPS Array (SuGAr) Data

Vira Friska, Deasy Arisa, Marzuki Marzuki, and Fadilla Monica

Abstract The movement of the Indo-Australian Oceanic Plate towards the Eurasian Continental Plate in the western part of Sumatra Island causes this area has a high potential for earthquakes. The magnitude of an earthquake depends on the energy accumulation in the interseismic period. This study aims to determine the plate velocity and direction of movement at the stable/interseismic phase during 2010–2014, using Sumatran GPS Array (SuGAr) data. RINEX data from GPS was processed using GAMIT/GLOBK software. There are eight SuGAr observation stations which uses across the west coast of Sumatra Island. The result shows that annual velocity during an interseismic phase of some SuGAr station which are BTHL, MLKN, NGNG, and PTLO stations moved northeastward. The direction of this movement is consistent with the direction of subduction of the Indo-Australian Plate and based on the time series graph, all of SuGAr stations during the annual interseismic phase (2012–2014) indicated stable movement.

73.1 Introduction

Sumatran Island is part of the edge Sunda arc of the southern part of the Eurasian plate. The circuit includes an island arc system with a subduction mechanism between the Indo-Australian plate against the Eurasian plate in the north [1] with the rate of 50–70 mm/yr [2]. The energy of the Indo-Australian plate subduction zone is often released, and it is related to the relative velocity of the plate movement. This subduction mechanism causes the Sumatran Island to have the potential for earthquakes. The 9.1 Mw Aceh earthquake that occurred on December 26th, 2004, was one of the largest earthquakes due to the Indo-Australian subduction activity, which caused

---

D. Arisa
Research Center of Geotechnology, Indonesian Institute of Sciences (LIPI), Jakarta, Indonesia

V. Friska · M. Marzuki (✉) · F. Monica
Department of Physics, Faculty of Mathematics and Natural Sciences, Universitas Andalas, Padang, Indonesia
e-mail: marzuki@sci.unand.ac.id

E. Yulihastin et al. (eds.), *Proceedings of the International Conference on Radioscience, Equatorial Atmospheric Science and Environment and Humanosphere Science, 2021*, Springer Proceedings in Physics 275, https://doi.org/10.1007/978-981-19-0308-3_73

a Tsunami [3]. Sumatra's seismic activities are also caused by the Sumatran Fault (Semangko Fault) and the Mentawai Fault's activity [4]. The earthquake can cause the earth's crust to deform, namely a change of position or an object's movement in absolute or relative, such as the earth's plates movement [5]. In a cycle of earthquakes, the deformation process can be divided into fourth phases: interseismic, preseismic, cseismic, and postseismic [6]. There are several methods to observe the surface deformation due to the earthquake, and one of the methods is taking advantage of GPS (Global Positioning System) data. GPS data can record movements with a submillimeter's precision per year and is proven as an indispensable tool for crust deformation analysis. In Sumatran Island, we can use the Sumatran GPS Array (SuGAr), GPS station observation, for deformation analysis, which records GPS data every day with high accuracy. Figure 73.1 shows the station map of the study area.

The level of seismicity in the subduction zone is linear with the relative velocity of plate movement, indicating that the energy of the plate motion is being converted into

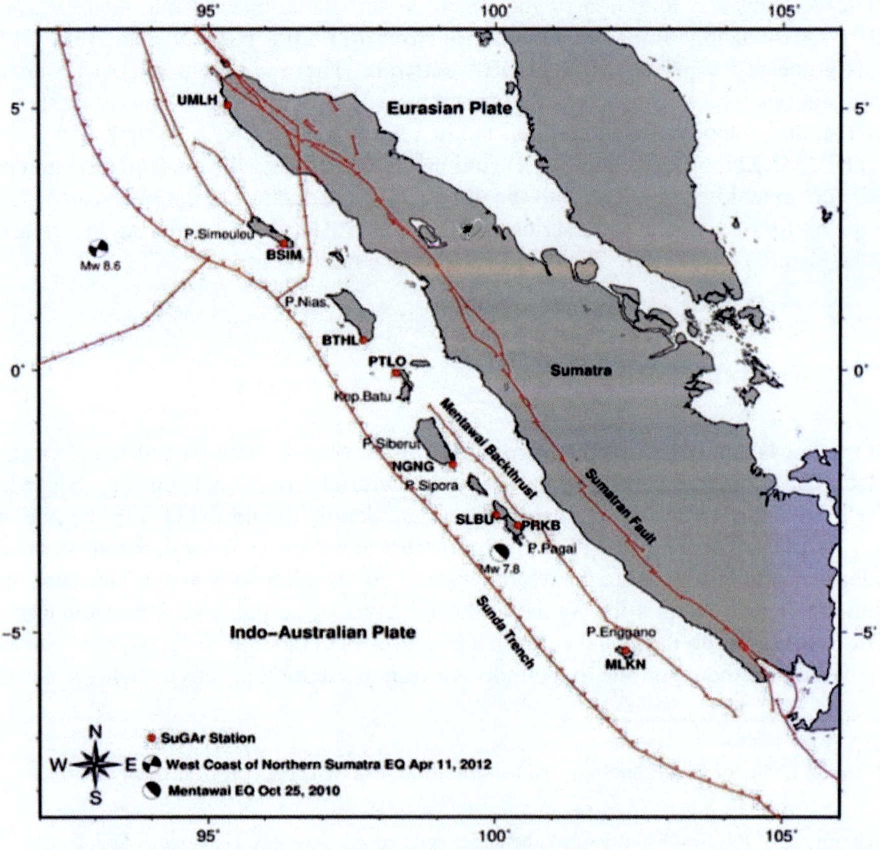

Fig. 73.1 Station map of the study area. The red circle shows the location of the SuGAr station used in this study

accumulated energy, which could be released in future earthquakes [7]. Sajagat [8] also explained that the earthquake's magnitude depends on the interseismic process accumulation of its energy. Therefore, determining the plate velocity and direction of movement is essential. This research investigated Indo-Australian plate velocity during the interseismic phase in 2010–2014 using the observation of the SuGAr station.

73.2 Data and Processing

73.2.1 Observation Data

Sumatran GPS Array (SuGAr). SuGAr is a GPS station spread over 1300 km along the west coast of Sumatran Island. SuGAr was initiated by Profesor Kerry Sieh et al. from the California Institute of Technology (CalTech) Tectonic Observatory (TO) in collaboration with the Indonesian Institute of Sciences (LIPI) in 2002. An open source rinex data of the SuGAr is provided by SOPAC, CDDIS, and UNAVCO. Table 73.1 shows eight SuGAr stations used in this study.

International GNSS Service (IGS). IGS (International GNSS Service) is an international organization that collects permanent data sources from GNSS (Global Navigation Satelite System) station and maintains GNSS station. The primary data provided by IGS are GPS phase codes and measurements in rinex files format for each station. IGS stations are used as binding points that bound the observation points.

Table 73.1 List of SuGAr stations in this study

SuGAr station name	Location	Lon (deg)	Lat (deg)	First Epoch	Last Epoch
BSIM	Bandara Simeulue	96.3262	2.4092	2010 01 01	2014 12 31
BTHL	Botohilithano, Nias	97.7107	0.5692	2010 01 01	2014 12 31
MLKN	Malakoni, Pulau Enggano	102.2770	-5.3527	2010 01 01	2014 12 31
NGNG	Nyang Nyang, Siberut	99.2683	-1.7996	2010 01 01	2014 12 31
PRKB	Parak Batu, Pagai Selatan	100.4000	-2.9666	2010 01 01	2014 12 31
PTLO	Pulau Telo	98.2800	-0.0546	2010 01 01	2014 12 31
SLBU	Silabu, Pagai Utara	100.0000	-2.7663	2010 01 01	2014 12 31
UMLH	Udjung Muloh, Aceh	95.3391	5.0531	2010 01 01	2014 12 31

Thus, the observation point moves relative to the reference point. IGS information could be found on the IGS network website.

73.2.2 Data Processing

The differential GPS Positioning (DGPS) method is used in this study, an accurate positioning method in GPS observation [9]. The DGPS method uses more than one receiver to determine the station's position based on its distance from other stations that be used as a corrected station.

This study used GAMIT (GPS Analysis of Massachusett Institute of Technology) software to processing GPS data in RINEX (Receiver Independent Exchange) format to obtain the daily position data. The GAMIT process results are the daily coordinates of all RINEX data from GPS stations in a loosely constrained solution file (h-file) containing the covariance matrix and the average parameter. Moreover, h-file as input data is processed based on the Kalman Filtering method using GLOBK (Global Kalman Filter VLBI and GPS analysis program). The result of GLOBK processing was displayed in a time series of deformation in horizontal (Eastward and Northward) and vertical (Upward) positions from the SuGAr station. In addition, the horizontal and vertical velocity of the SuGAr station from GLOBK processing and the direction of velocity mapping using Generic Mapping Tools (GMT).

73.3 Result

73.3.1 Seismic Cycle

Figure 73.2 shows the time series of seismic cycles from SuGAr station during 2010–2014. The red curve is the deformation in northward, the green curve for eastward deformation, and the blue curve for upward deformation. The coseismic jump was observed in some SuGAr stations (Fig. 73.2). Light blue dashed line indicates the 8.6 Mw earthquake that occurred on April 11th, 2012, off the west coast of northern Sumatra. Furthermore, the black dashed line on some stations such as SLBU, PRKB, and NGNG was 7.8 Mw Mentawai earthquake on October 25th, 2010. In this research, the velocity of the SuGAr station during the interseismic phase has been calculated. The first calculation is measurement of SuGAr station's movement velocity along earthquake of October 25th, 2010 to earthquake of April 11th, 2012. The second one is calculated interseismic phase is the annual interseismic of SuGAr station with the first epoch Mei 11th, 2012, and the last epoch December 31th, 2014. In this period, no significant earthquake occurred.

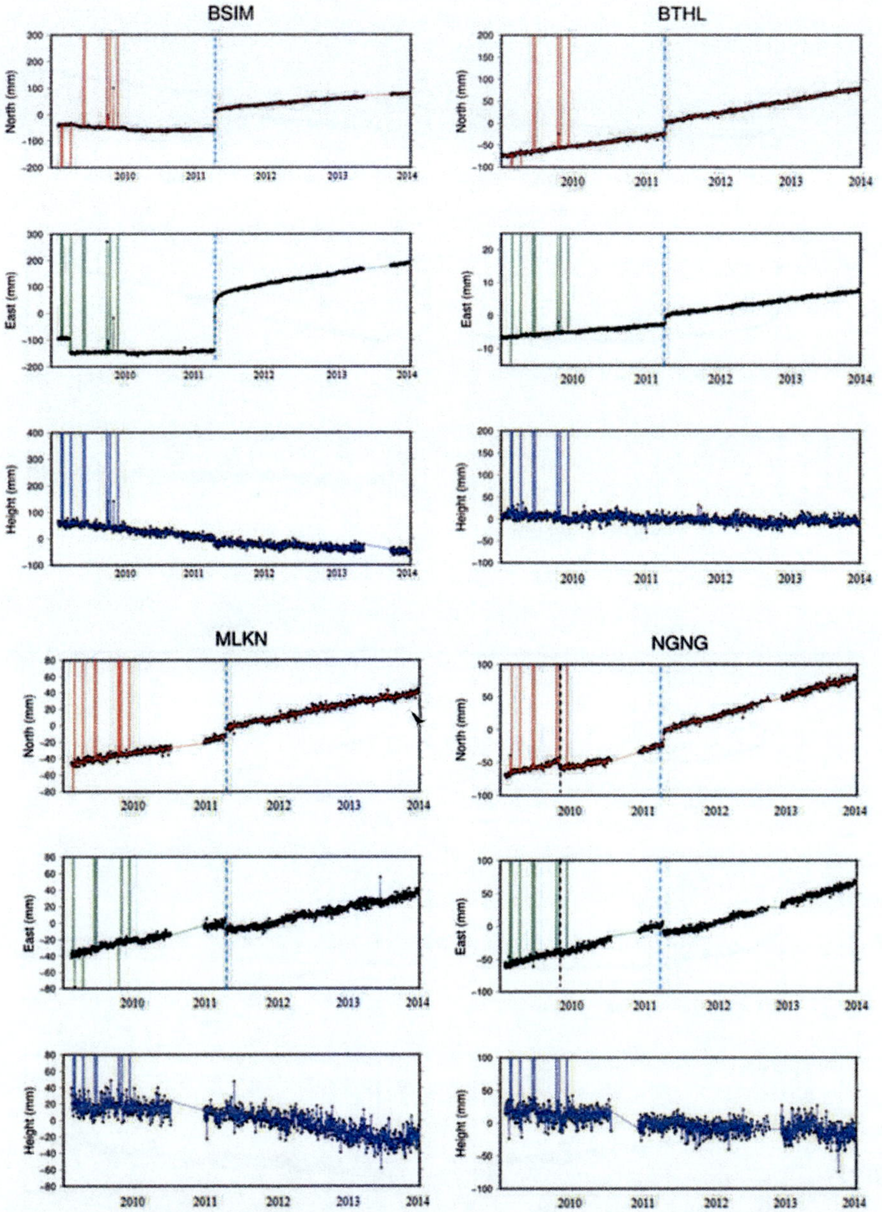

Fig. 73.2 Timeseries of a seismic cycle (inter, pra, co-, and post- seismic) recorded by SuGAr, light blue dashed line was off West Coast of Sumatra Earthquake on April 11th, 2012 and the black dashed line is Mentawai Earthquake on October 25th, 2010

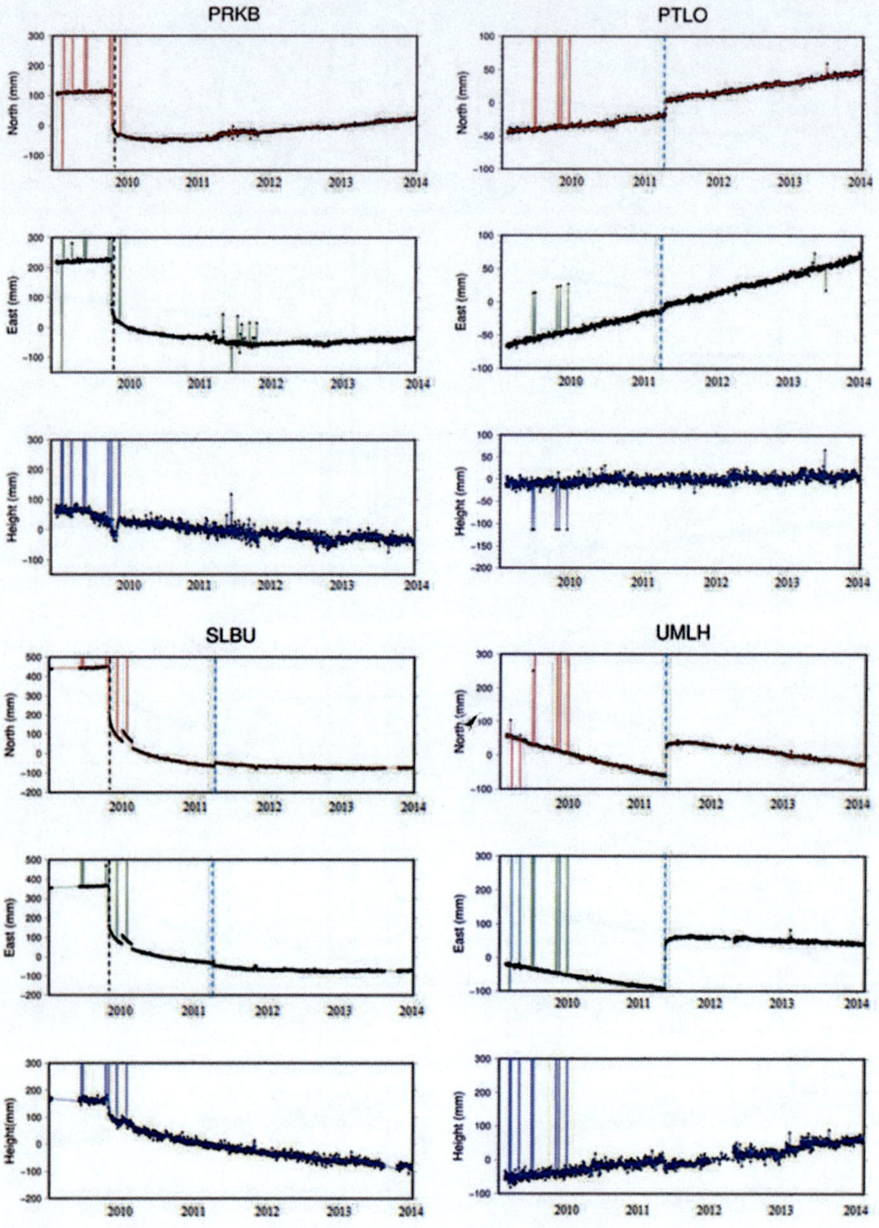

Fig. 73.2 (continued)

Table 73.2 The Velocity of SuGAr station during interseismic phase

SuGAr stations	The movement velocity of SuGAr station from 2010 to 2012 (mm/yr)	Interseismic of annual condition (mm/yr)
BSIM	6.64	76.58
BTHL	26.91	43.74
MLKN	20.69	26.17
NGNG	39.43	42.21
PRKB	42.30	57.65
PTLO	26.89	34.81
SLBU	143.35	143.11
UMLH	58.87	24.46

Table 73.2 shows the velocity of the SuGAr station during an interseismic process. SuGAr stations have different horizontal velocities during the interseismic phase for each condition, but all indicated stable movement before the earthquake (Fig. 73.2).

73.3.2 *Velocity of the SuGAr Station During an Interseismic Phase in 2010–2014*

Velocity of the SuGAr station during 2010–2012. The direction of the SuGAr station movement can be seen in Fig. 73.3 (represented by a red arrow). The direction of BTHL, NGNG, MLKN, and PTLO stations in this period leads toward the northeast according to the slope of the Indo-Australian plate. The average velocity in this interseismic process is relatively stable, with a 45.60 mm/yr value.

Velocity of the SuGAr station for the annual interseismic phase. The direction of the SuGAr station movement can be seen in Fig. 73.4 (represented by a red arrow). Some SuGAr stations' (BTHL, PTLO, NGNG, MLKN) direction toward the northeast is following the direction of the oblique subduction of the Indo-Australian plate, with an average velocity of 36.73 mm/year for these four stations. The UMLH station, which is located in Udjung Muloh, Aceh, has a direction of movement to the northwest at a velocity of 24.46 mm/year. This direction tends to be parallel to the Sumatran Fault line. It was concluded that the UMLH station was not only influenced by the Indo-Australian subduction but also by the Sumatran fault. Based on [10], also explains that the contact area in the southern segment is relatively stronger than the northern segment. The velocity of all SuGAr stations obtained varies with the largest velocity of 143.11 mm/yr was observed by the SLBU station (Pagai Utara Island), followed by the BSIM station (Simeuleu) station (76.58 mm/yr) and the smallest velocity was observed by UMLH stations (24.46 mm/yr). The velocity of SuGAr station is still influenced by the movement of the regional plate or the Sunda Block, which has a shift speed of $\pm$ 20 mm/yr relative to the east.

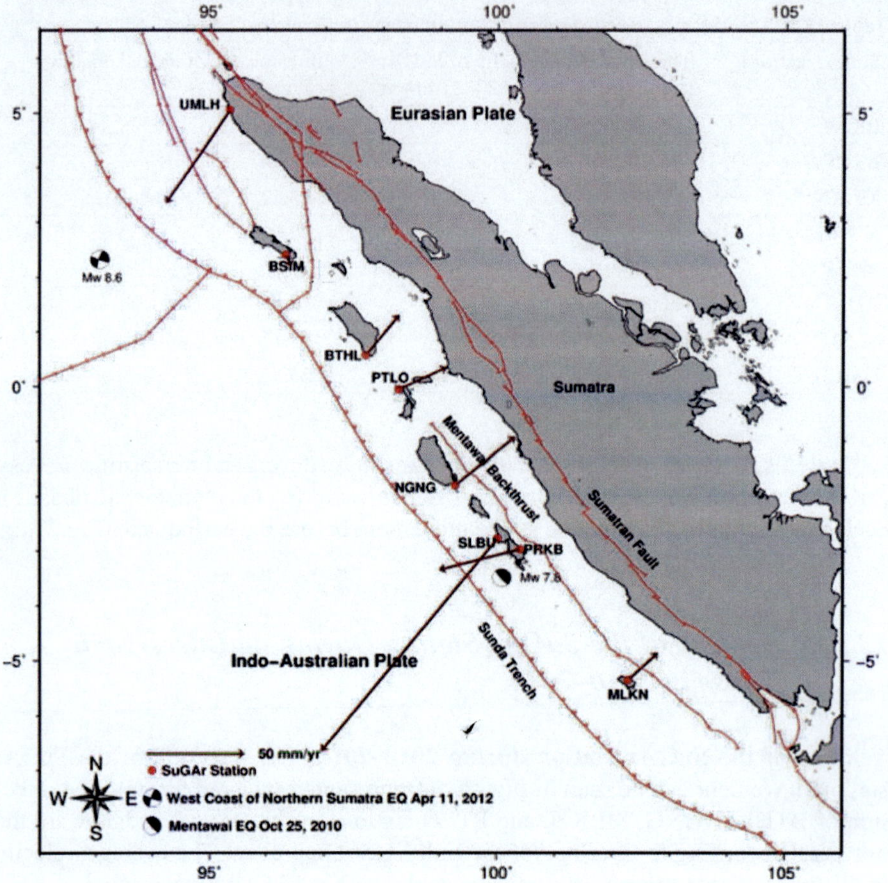

Fig. 73.3 Map of the direction of SuGAr station movement during 2010–2012

73.4 Conclusion

This study shows that most SuGAr stations are stable and moving toward the north-east during the annual interseismic phase. The velocity of SuGAr station in the northern such a UMLH is smaller than the station in southern (NGNG), because the UMLH station was not only influenced by the Indo-Australian subduction but also by the Sumatran fault. The average annual velocity during 2010–2012 is smaller than the velocity during the annual interseismic phase, because the plate's movement is slowing down when approaching the release of energy such as the 2012 earthquake.

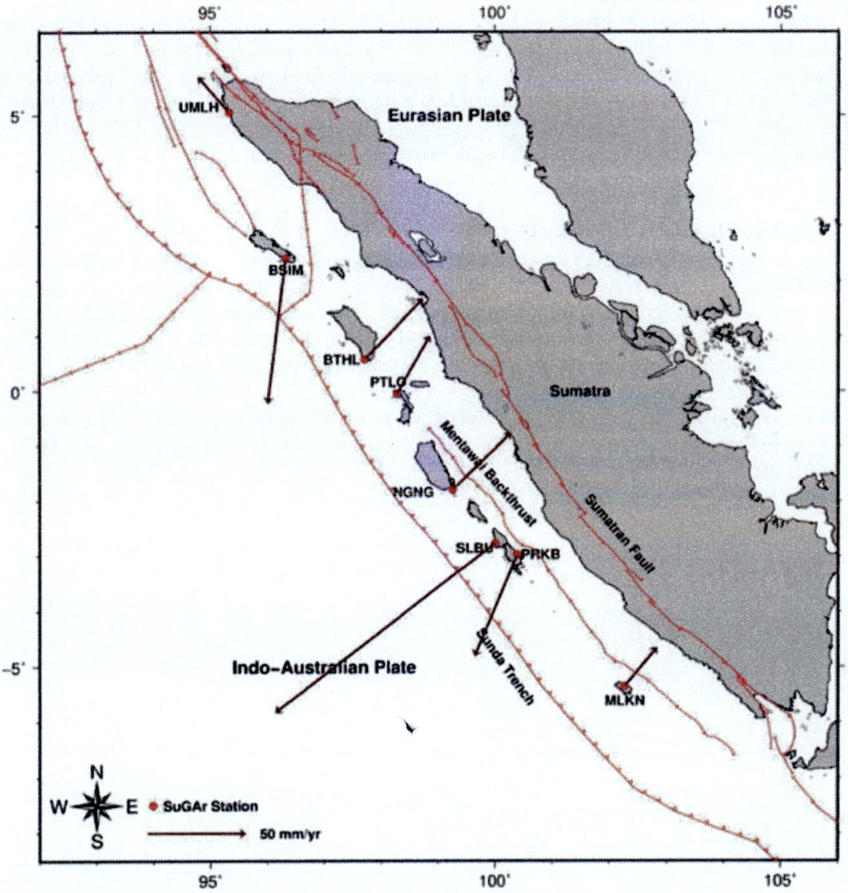

Fig. 73.4 Map of the direction of SuGAr station movement during annual interseismic phase (May 11th, 2012–December 31st, 2014)

Acknowledgements The present study was supported by 2021 Research Grants for Publication to Support the Professor Acceleration Programme from Universitas Andalas (PDU-KRP2GB-UNAND) (contract no: T/11/UN.16.17/PP.IS-PDU-KRP2GB-Unand/LPPM/2021). Furthermore, thanks to Research Center for Geotechnology, Indonesian Institute of Sciences (LIPI), and Nanyang Technological University's Earth Observatory of Singapore for operating the SuGAr network.

References

1. Mulyana, B.: Extension tektonik selat sunda. Bull. Sci. Contribution **4**(2), 137–145 (2006)
2. Prawirodirdjo, L., McCaffrey, R., Chadwell, C.D., Bock, Y., Subarya, C.: Geodetic observations of an earthquake cycle at the Sumatra subduction zone: role of interseismic strain segmentation. J. Geophys **115**(B03414) (2010)

3. United States Geological Survey (USGS) Earthquake Catalogue of 9.1Mw Aceh, https://earthquake.usgs.gov/earthquakes/eventpage/usp000hnj4/executive. Last accessed 2021/09/01.
4. Satria, L.A., Yogaswara, S., Ubaya, T., dan Anggraini, F.: Aktifitas Gempabumi Sumatera Barat Berdasarkan Sumber dari Januari Hingga Juni 2018, Stasiun Geofisika Kelas I Silaing Bawah (2018)
5. Kuang, S.: Geodetic Network Analysis dan Optimal Design. Ann Arbor Press Inc., Michigan (1996)
6. Mori, J.,: Earthquake Prediction. Lecture notes on KAGI 21 Summer School, Institute of Technology Bandung, Indonesia (2004)
7. Muzli.: Hubungan Kecepatan Relatif Pergerakan Lempeng dengan Tingkat Seismisitas di Zona Subduksi. Jurnal Meteorologi dan Geofisika **15**(3), 187–192 (2014)
8. Sajagat, M.J., Awaluddin, M., Yuwono, B.D.: Hitungan Kecepatan Pergerakan Stasiun SuGAr Akibat Proses Interseismik Gempa Mentawai 2007. Jurnal Geodesi Undip **5**(4), 196–206 (2016)
9. Misra, P., Enge, P.: Global Positioning System Signals, Measurements, and Performance. Ganga Jamuna Press, MA (2001)
10. Prawirodirdjo, L., McCaffrey, R., Chadwell, C.D., Bock, Y., Subarya, C.: Geodetic observations of interseismic strain segmentation at the Sumatra subduction zone. Geophys. Res. Lett. **24**(21), 2601–2604 (1997)

Chapter 74
Deformation Analysis During the Pre-, Co- and Post-Seismis Phases Associated with the 2019 Mw6.0 Mentawai Earthquake Using Satellite Geodetic Technology from Sumatran GPS Array (SuGAr) Data

Fadilla Monica, Deasy Arisa, Marzuki Marzuki, and Vira Friska

Abstract Subduction activity near Sumatra Island has frequently generated earthquakes, making this region an active tectonic area in the world. As one of the disaster mitigation actions in earthquake zone societies, geodynamic analysis is needed to model an earthquake probability. The geodynamic study is easier to complete with a new Global Positioning System (GPS) technology. By using the continued GPS method, it can analyze a deformation at the earthquake phases, the silent/stable period with no earthquake occurrence (preseismic), the period after the earthquake (postseismic), and the period during the earthquake rupture (coseismic). This work conducted deformation analyzes of pre-, co-, and postseismic associated with the Mw 6.0 Mentawai earthquake using Sumatran Global Positioning System Array (SuGAr) data. The earthquake occurred on February 2, 2019, through a thrust fault mechanism. SuGAr data were processed using GAMIT/GOLBK 10.7 software. Several stations around the earthquake's epicenter were use, i.e., KTET, LAIS, LNNG, MKMK, MLKN, PPNJ, SIOB, SLBU, SMGY, TARA, TIKU, and TRTK. During the preseismic phase, the deformation vector is about 2.37–7.20 mm northeastward, the same direction as that of subduction of the Indo-Australian plate towards Eurasia. The largest deformation vector in the coseismic phase is 25.3 mm southwestward was observed at the SMGY station. After the earthquake (postseismic), the vector deformation was 1.13–16.33 mm to the southwest, indicating that there is still a release of earthquake energy in this postseismic phase, which still lasts for 30 days.

F. Monica · M. Marzuki (✉) · V. Friska
Department of Physics, Universitas Andalas, Padang 25163, Indonesia
e-mail: marzuki@sci.unand.ac.id

D. Arisa
Research Center of Geotechnology, Indonesia Institute of Sciences (LIPI), Jakarta 40135, Indonesia

E. Yulihastin et al. (eds.), *Proceedings of the International Conference on Radioscience, Equatorial Atmospheric Science and Environment and Humanosphere Science, 2021*, Springer Proceedings in Physics 275, https://doi.org/10.1007/978-981-19-0308-3_74

74.1 Introduction

West Sumatra Province is one of the regions in Indonesia with a high earthquake potential due to the presence of three active earthquake generator sources in this region, which are part of the Eurasian Plate and the Indian-Australian Plate. The three primary sources of these earthquakes are the Subduction zone (Megathrust), the Mentawai Backthrust, and the great Sumatran Fault [1]. Subduction zone form when the Indian-Australian Ocean plate descends under the Eurasian plate. In the subduction zone, the two plates cause the formation of a trough commonly known as the Sunda trench (Fig. 74.1a). The continuous subduction of the Indian-Australian Plate against the Eurasian Plate causes an accumulation of energy which causes the island of Sumatra, especially West Sumatra, to experience frequent earthquakes.

Earthquakes will usually cause the surrounding earth's crust to deform vertically and horizontally [2]. Deformation reflects the distribution of earthquake energy and the resulting damage. The method widely used to observe deformation due to earthquakes is monitoring changes in GPS (Global Positioning System) observation points installed around potential earthquake areas [3]. Sumatra and the Mentawai islands already have a GPS observation station known as SuGAr (Sumatran GPS Array). GPS can provide deformation information in horizontal and vertical directions with precision to the order of millimeters and good consistency [4]. This method can also describe the condition of the observation point during a tectonic earthquake cycle (earthquake cycle). There are several phases of earthquakes in an

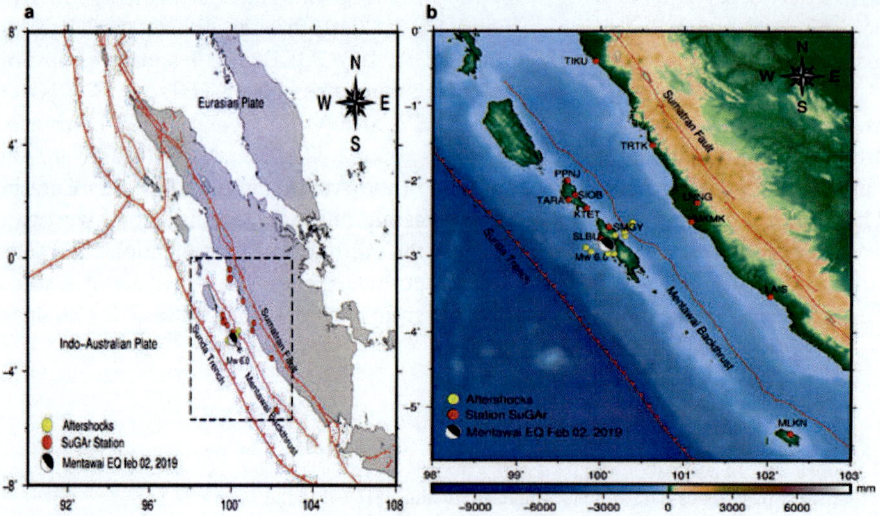

Fig. 74.1 **a** The position of Sumatra Island to the tectonic plates. The black dotted line shows the part of the deformation observation from the Mentawai Earthquake 2019 (main shock) **b** Topographic map and the location of the SuGAr station used in the deformation observation. SuGAr GPS station locations are marked with red dots and aftershock are marked with yellow dots

earthquake cycle, namely inter-seismic, preseismic, coseismic, and postseismic. The inter-seismic phase is a phase of energy accumulation at the boundary between two plates. The preseismic phase is the phase just before the earthquake occurs. The time of the earthquake is called the coseismic phase, and the postseismic phase releases earthquake energy and leads to initial equilibrium, respectively [5].

Research on the deformation phase of the earthquake has been carried out by several previous researchers, such as the 2008 Mentawai earthquake with a magnitude of 7.3 Mw. This study shows that during the earthquake (coseismic), deformation occurs in the area around the Mentawai segment of about 214 mm, and during the postseismic phase, this area takes a long time to return to a stable state [6]. Research on the 2010 Mentawai earthquake with a magnitude of 7.3 Mw showed 22 $\pm$ 0.5 cm deformation to the southwestward [7]. Not only earthquakes with enormous strengths, deformation studies for small earthquakes that have just occurred are also needed to see the cycle of the earthquake, whether the earthquake is a postseismic phase of a large earthquake that has happened before.

This study observes deformation due to the Mentawai earthquake 2 February 2019 (mainshock) using SuGAr station observation data. The SuGAr stations are spread along the west coast of Sumatra Island, so it is hoped that any changes that occur due to the movement of the Indo-Australian and Eurasian plate boundaries can be observed. The results of the BMKG monitoring show that on February 2, 2019, 52 aftershocks had occurred. Among them, there were 5 earthquake activities whose shocks were felt by the community, namely magnitude 5.3, 5.8, 53, 5.9 and 5.0 (Fig. 74.1b). In this study, data processing was carried out using the scientific software GAMIT 10.74. The purpose of this study was to observe seismic deformation in the Mentawai area using the GPS method. Through deformation observations, it will be able to understand how long the earthquake cycle occurred in the area caused by an earthquake with a magnitude of 6.0Mw followed by 52 aftershocks. This information will provide input in earthquake cycle modeling and earthquake modeling in the Sumatran subduction zone which is useful for calculating stress changes due to tectonic activity in the Sumatran subduction zone. Then this information can be used to map the accumulation of stress due to shifting of tectonic plates along the Sumatran subduction zone.

74.2 Data and Methods

The data used to observe the deformation of the Mentawai earthquake on February 2, 2019, during the preseismic, coseismic, and postseismic phases are SuGAr station observation data. The Mentawai earthquake occurred at DoY (Day of Year) 033; preseismic observations started from DoY 001 to DoY 032 (January 1–February 1). postseismic observations were carried out for 92 days beginning from DOY 034 to DOY 125 (February 3–May 5). The data used for the study were observation data from 12 SuGAr stations, namely KTET (Katiet), LAIS (Lais), LNNG (Lunang), MKMK (Muko Muko), MLKN (Malakoni), PPNJ (Pulau Panjang), SIOB (Sioban), SLBU

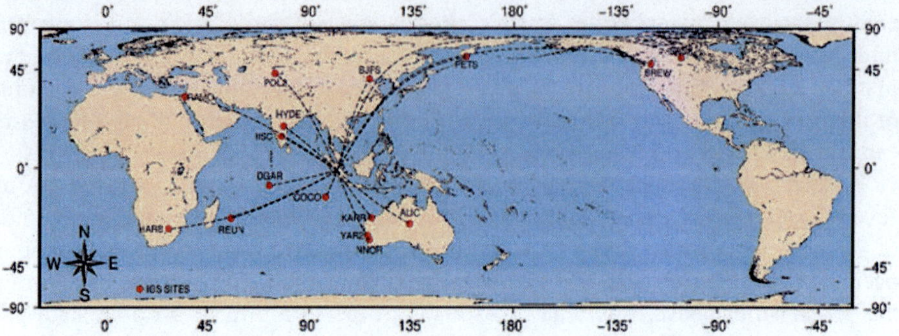

Fig. 74.2 Locations of International GNSS Service (IGS) stations around the world

(Silabu), SMGY (Saumanganya), TARA (Taraet), TIKU (Tiku), and TRTK (Taratak). The position of the SuGAr station used in this study can be seen in Fig. 74.1b. SuGAr data is available and can be downloaded from the Nanyang Technology (NTU) web at ftp://ftp.earthobservatory.sg.

This study also uses observation data from the International GNSS Service (IGS) station as a tie point, orbital IGS data (Precise Ephemeris), and navigation data (Broadcast Ephemeris). The data can be download from the data provider, the SOPAC archive on RINEX format via the GAMIT software. 16 IGS stations have been select as correction stations with distribution in all directions from the observation point. The IGS stations used include ALIC, BJFS, BREW, COCO, DGAR, HARB, HYDE, IISC, KARR, NNOR, PETS, POL2, RAMO, REUN, SASK, and YAR2. The distribution of these stations can be seen in Fig. 74.2.

The determination of the point of the SuGAr station, which is used periodically, must be corrected for a position with other stations that are relatively stationary from the observation station. The more correct stations are used, the more accurate station position data we get. In addition to the number, the direction of the fixing station that corrects in all cardinal directions will produce a good correction. This research also uses signal correction data such as atmospheric modeling, tidal, and weather modeling, which can be download at ftp://everest.mit.edu/pub/GRIDS. This signal correction data is used to minimize noise and clutter from signals captured by satellites from those emitted by GPS stations.

RINEX data processing from GPS is carried out with a series of commands using GAMIT/GLOBK 10.74 software run on the Linux Ubuntu 16.04 operating system. The result of the GAMIT process is the daily coordinates of all RINEX data for GPS stations in the form of a loosely constrained solution (h-file) containing the covariance matrix and the average parameter. Results of GAMIT data processing in the format of an h-file are used as input for GLOBK data processing. In the processing by GLOBK, a combination of DoY and plotting of coordinate parameters will be conduct out with a Kalman filter. The data processing results are displayed in a time series format from the horizontal and vertical position parameters of the GPS station. Deformation vector mapping is done using GMT (Generic Mapping Tools).

74.3 Result

The results of SuGAr data processing using GAMIT/GLOBK software are the data on changes in the daily coordinate position of the SuGAr station (latitude and longitude) and time-series graphs. Figure 74.3 shows time-series graphs of observations of SuGAr stations (SMGY and SLBU) closest to the earthquake epicenter from DoY 001 to DoY 125. There are three graphs showing movement in the east (marked green) and north (red) directions and up (yellow). Coseismic jumps were seen at both stations during the 033 DoY, which means an earthquake occurred in the 033 DOY. The quake caused a change in the direction of movement of the station before and after the quake. Based on the time series, it was initiate that the trend of deformation before the earthquake was to the northeast, following the direction of the subduction of the plate (9). After the earthquake, several stations experienced movements in different directions from the tip of subduction of the plate.

The amount of deformation in the pre-, co-, and postseismic phases can be seen in Table 74.1. During the preseismic phase (DoY 001 to DoY 032), the magnitude of the shift that occurs is 2.37–7.20 mm. In the preseismic phase shown in Fig. 74.4, several SuGAr stations' deformation direction is the same as the deformation rate

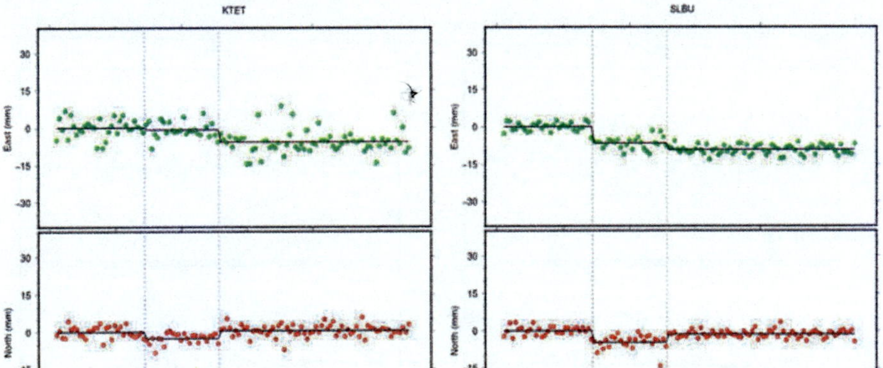

Fig. 74.3 Time series graph of KTET and SLBU SuGAr station during the co-, pre-, and postseismic phases. The X-axis is the DOY observation, and Y-axis is a horizontal and vertical shift for SMGY, MKMK, SIOB and PPNJ SuGAr Stations for LNNG, TRTK, TIKU and TARA SuGAr Stations for MLKN and LAIS SuGAr Stations

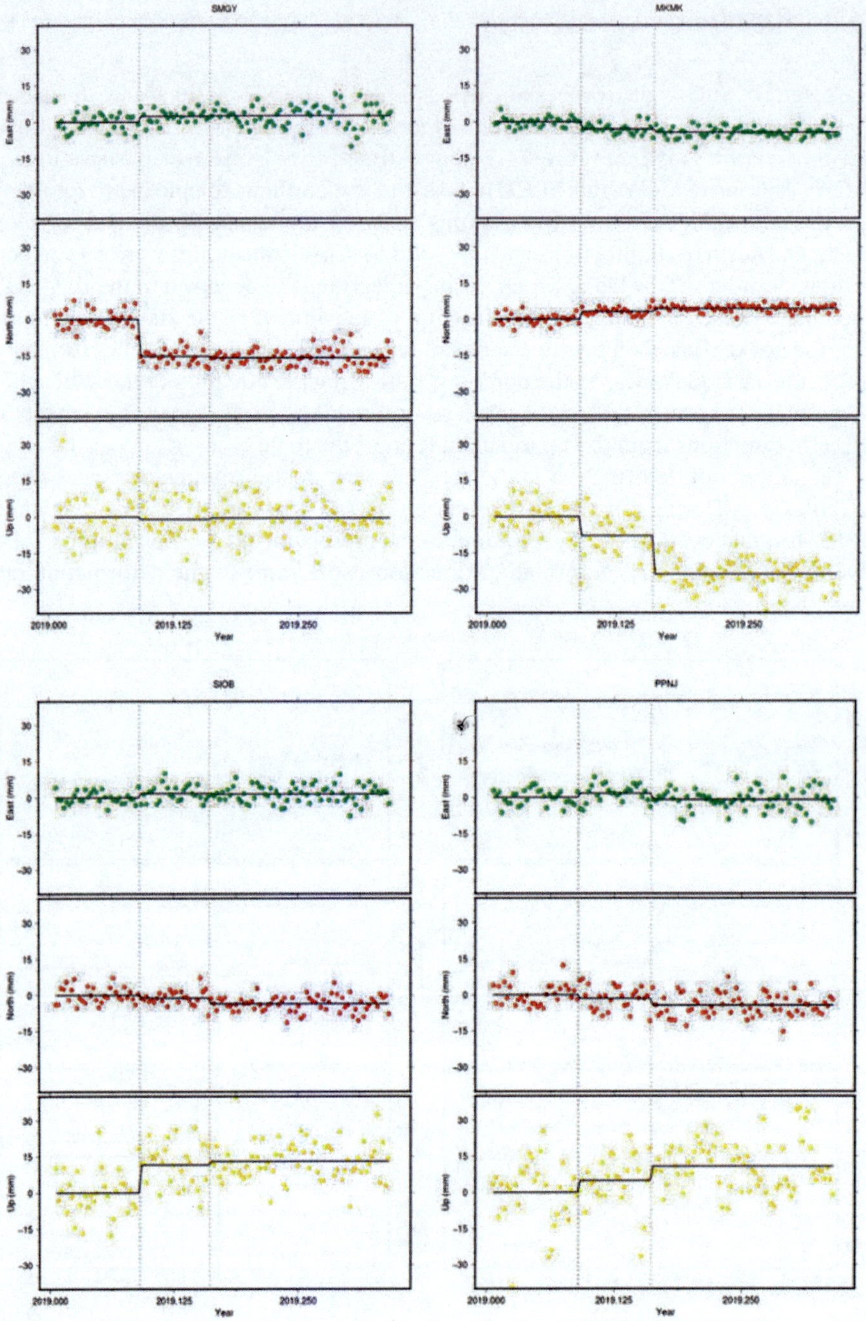

Fig. 74.3 (continued)

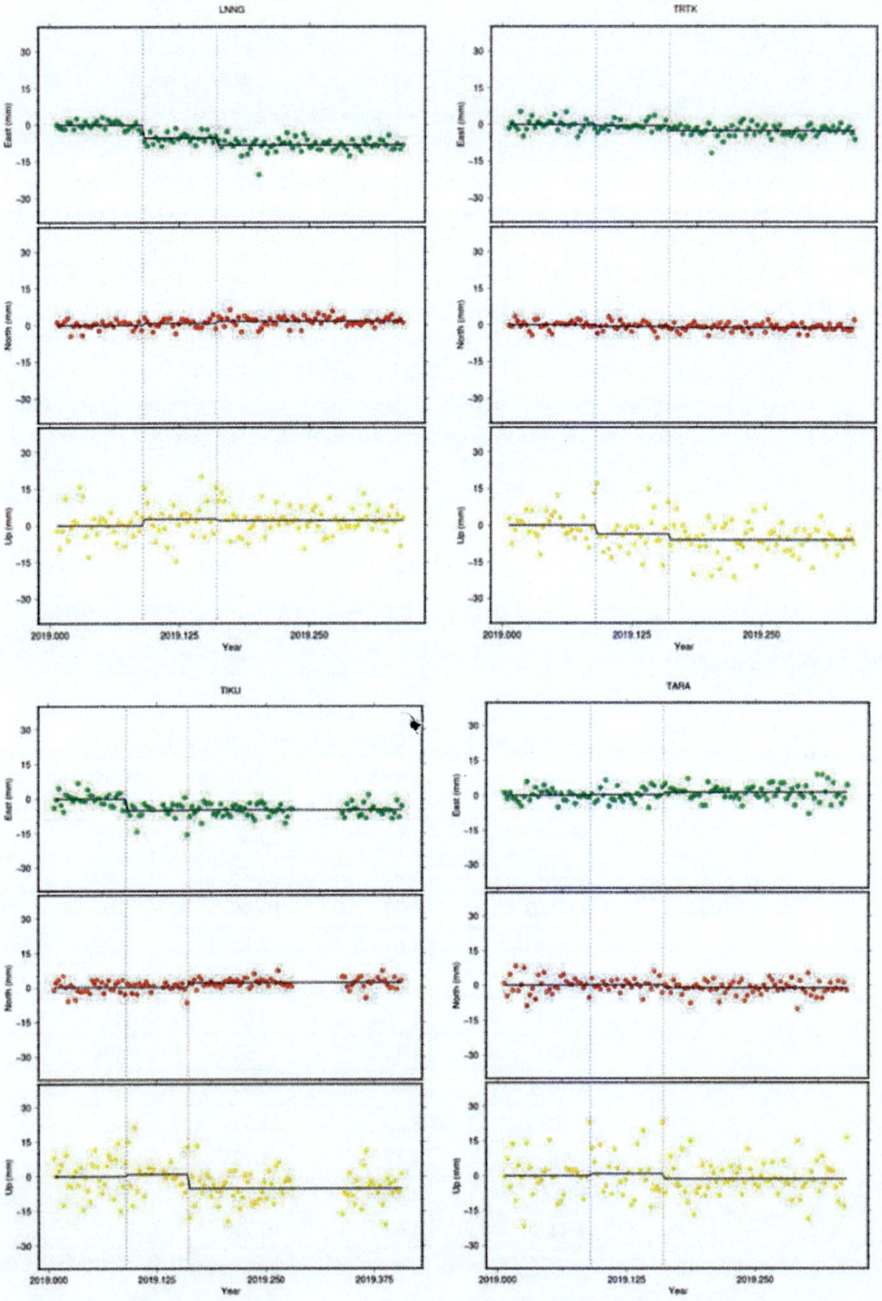

Fig. 74.3 (continued)

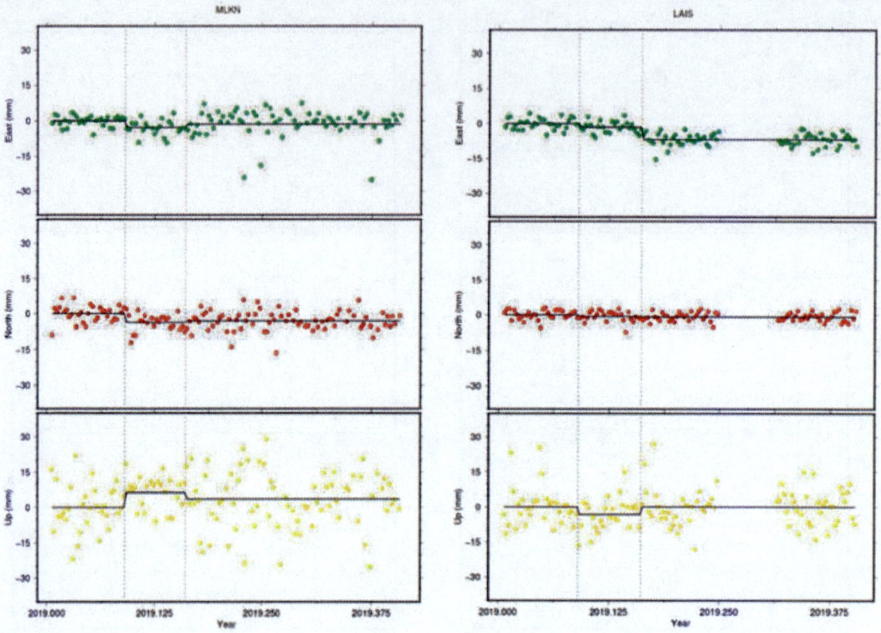

Fig. 74.3 (continued)

Table 74.1 Displacements value during pre-, co- and postseismic

Station name	Horizontal movement (mm)		
	Preseismi c(31 days)	Coseismic	Postseismic (30 days)
KTET	4.25	10.06	5.02
LAIS	6.45	7.73	3.54
LNNG	2.37	10.06	5.02
MKMK	4.61	7.14	3.22
MLKN	3.86	6.79	1.33
PPNJ	4.16	10.69	6.43
SIOB	4.47	9.81	6.94
SLBU	7.20	10.69	6.43
SMGY	3.53	25.34	16.33
TARA	4.17	8.72	7.11
TIKU	2.84	6.86	1.13
TRTK	4.62	10.77	3.89

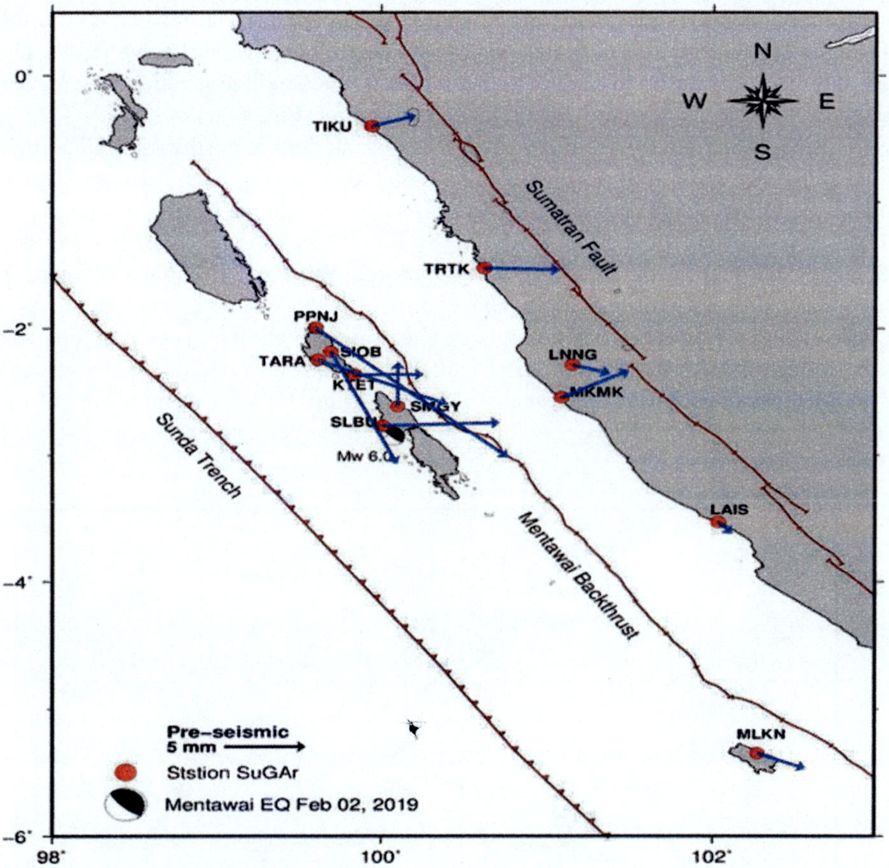

Fig. 74.4 Vector deformation of the 2019 Mentawai earthquake in the preseismic phase

due to the subduction of the Indian-Australian Plate towards Eurasia, namely towards the northeast. The magnitude of the shift from DoY 031 and DoY 032 gets smaller as it approaches the coseismic earthquake, which indicates that the accumulated energy is getting bigger. At this stage, the rock or material can no longer withstand the forces caused by relative motion. Then the accumulated energy will be release in the form of earthquake waves that propagate in all directions. Thus, DoY 032 is in the locked phase and releases energy (earthquake) at DoY 033, called the coseismic phase.

One of the stations with the most enormous shift when an earthquake occurs (coseismic) is the SMGY station, with a displacement of 25.34 mm. This station is 26 km from the epicenter of the quake. The station with the minor shift during an earthquake (coseismic) is the MLKN with a change of 6.79 mm. This station is the farthest station from the earthquake epicenter. In the coseismic phase, the magnitude of the deformation is influence by the distance of the station to the earthquake epicenter, the closer the station position from the earthquake epicenter, the more significant the shift that occurs at the station (Table 74.1). The farther the station from the epicenter of the earthquake is the smaller the change will be generate. The magnitude of the deformation of the SuGAr station also illustrates the deformation that occurs in the area where the station was installed (Fig. 74.5).

The direction of deformation of several SuGAr stations is northwest, indicating a change in the direction of plate deformation due to the earthquake. As a result of its elastic nature, the rock causes movement directions to change along the subduction zone from the trough to a depth of 40 km [8]. Based on earthquake data from United States Geological Survey (USGS), on February 2, 2019, an earthquake occurred at a depth of 26 km, which indicates that the rock is still elastic at that depth. As a consequence, if it gets tremendous energy, elastic deformation will occur in the rock.

The postseismic phase of the 2019 Mentawai earthquake is calculated several days after the earthquake occurred, namely at DOY 034 to DOY 125. The results of the large deformation can also be seen in the table, and the deformation vector mapping

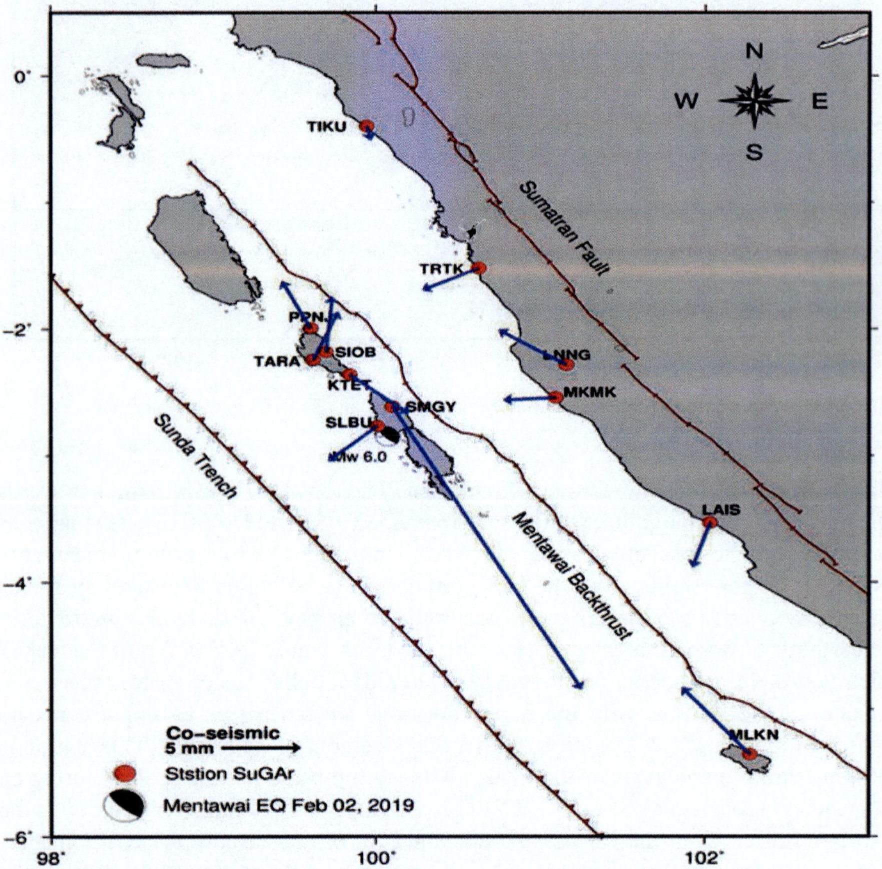

Fig. 74.5 Vector deformation of the 2019 Mentawai earthquake in the coseismic phase

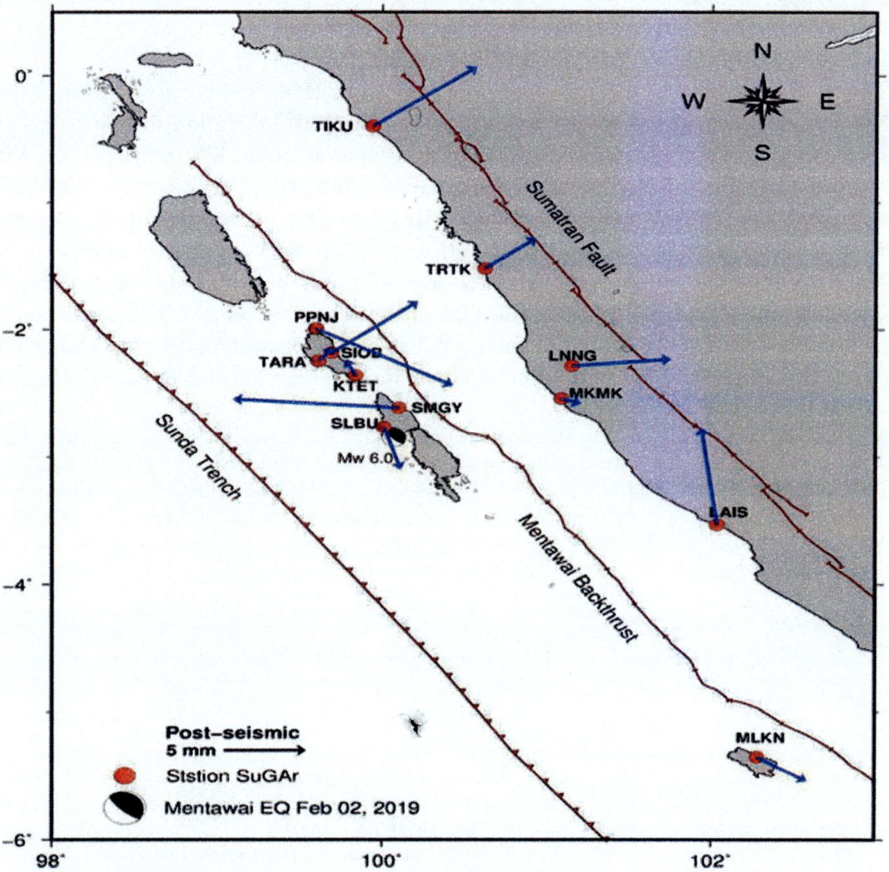

Fig. 74.6 Vector deformation of the 2019 Mentawai earthquake in the postseismic phase

can be seen in Fig. 74.6. The postseismic phase of the SuGAr station shifts back to the northeast following the direction of the plate subduction. The magnitude of the shift becomes smaller because enormous energy has been released in the coseismic phase, and the postseismic phase is the release of the remaining energy. In the postseismic phase, the most significant deformation occurred at the SMGY station, 16.33 mm, and the smallest one was at the TIKU station with a displacement of 1.13 mm. During the 92 days of observation, the postseismmic phase of this earthquake only lasted 30 days. Ths indicate that the postseismic phase for smal magnitude earthquake only takes a short time to realease residual energy and acumulation energy.

74.4 Conclusion

This study shows that the most significant deformation associated with the 2019 Mw6.0 Mentawai earthquake was observed at the SMGY station in North Pagai, Mentawai, as a 25.34 mm southwestward coseismic jump. The shift of SuGAR stations is affected by the distance between the station and the epicenter. If it is close to the earthquake epicenter, the change will be even more significant. During the 30 days after the earthquake, plate relaxation at the postseismic phase was observed. The magnitude of the change of the postseismic phase becomes smaller because a large amount of energy has been released in the coseismic phase. The postseismic phase is the release of the remaining energy. The shifted postseismic trend is indicated by a change in the direction of plate movement compared to the preseismic trend.

Acknowledgements The present study was supported by 2021 Research Grants for Publication to Support the Professor Acceleration Programme from Universitas Andalas (PDU-KRP2GB-UNAND) (contract no: T/11/UN.16.17/PP.IS-PDU-KRP2GB-Unand/LPPM/2021). Furthermore, thanks to Research Center for Geotechnology, the Indonesian Institute of Sciences (LIPI), and Nanyang Technological University's Earth Observatory of Singapore for operating the SuGAr network.

References

1. Prawirodirdjo, L., Bock, L., Genrich, J.F., Puntodewo, S.S.O., Rais, J., Subarya, C., Sutisna, C.: One century of tectonic deformation along the Sumatran fault from triangulation and global positioning system surveys. J. Geophys. Res., **105**(B12) (2000)
2. Bock, L.Y., Prawirodirjo, J.F.: Crustal motion in Indonesia from Global positioning system measurements. J. Geophys. Res., **108**(8), (2003)
3. McCaughet, J.: Sumatran earthquake with GPS. Earth observatory of Singapore. Singapore (2012)
4. Segal, Davis: Applications for Geodynamics and Earthquake Studies. Standford.edu (1997)
5. Krasnoperov, R., Earth crust motion and deformation analysis based on space geodesy methods. In: Proceedings of the International Conferences Electronic Geophysical Year: State of The Art and Result, Peereslavl-Zaessky, Rusia (2009).
6. Pujiastuti, D., Arisa, D., Anggriani, R.R., Hilma, A.Z.: Analysis of preseismic deformation of GPS station due to the earthquake in mentawai segment February 2008. J. Phys. Sci. **12**(1), 35–43 (2020)
7. Ardika, M., Meilano, I., Gunawan, E., postseismic Deformation Parameter of The 2010 M7.8 Mntaawai, Indonesia, Earthquake Inferred from Continuous GPS Observations. Asian J. Earth Sci., **8**(4), 127–133 (2015)
8. Natawidjaja, D.H.: The Sumatran fault zone from source to Hazard. J. Earthquake Tsunami **1**(1), 21–47 (2007)

Chapter 75
Assessing Potential Human and Economic Loss of Tsunami Impact in Cilacap City, Indonesia a Geospatial Approach

Ranie Dwi Anugrah and Stevani Anggina

Abstract Cilacap is located in Southern Java, Indonesia that notably prone to tsunami hazards. Since Cilacap is a densely populated area that vulnerable to tsunami hence, it is necessary to assess the risk of tsunami in its coastal zone. This study aims to estimate the risk of tsunami in the Cilacap coastal area by integrating the function of hazard, which is based on tsunami probability occurrence and tsunami inundation, with exposure data by using InaSAFE tools. Furthermore, it provides estimated displaced population and minimum needs information. The data used in this research consists of statistical data, maps, and images that are collected through literature studies and secondary data. The result shows with a scenario of 7 m height of the tsunami; there are four villages at the high and very high risk of loss throughout tsunami events with 176.73 ha area. The estimated loss from its area affects 21,690 people and the economic loss from physical damage by $345,071.

75.1 Introduction

Indonesia is located within three major tectonic plates, such as Eurasian, Pacific, and Australian-Indian Plate, and with some plates like the Philippine Sea plate. In the Southeast Asia region, all of them move towards each other, which creates a complex network of plate boundaries [1]. These phenomena led Indonesia to have a high level of seismicity and prone to a tsunami in coastal areas. Tsunamis in Indonesia are caused mainly by a tectonic earthquake along subduction zones and

These authors equally contributed to this work.

R. D. Anugrah (✉)
Detailed Spatial Planning of Environmental Carrying Capacity Area - Region 1, Ministry of Agrarian Affairs and Spatial Planning/National Land Agency, Jakarta, Indonesia
e-mail: raniedwianugrah@gmail.com

S. Anggina
Center for Aerospace Policy Studies, LAPAN, BRIN, Jakarta 10330, Indonesia
e-mail: stevani.anggina@lapan.go.id

E. Yulihastin et al. (eds.), *Proceedings of the International Conference on Radioscience, Equatorial Atmospheric Science and Environment and Humanosphere Science, 2021*, Springer Proceedings in Physics 275, https://doi.org/10.1007/978-981-19-0308-3_75

other active seismic areas. Some coastal areas of Indonesia that have a high potential to be affected are the islands' coastal areas that are facing the Pacific Ocean and the Indian Ocean.

One of the tsunamis incidents occurred on the south coast of Java, Indonesia, on July 17th, 2006, caused by an earthquake (Mw = 7.7) brought massive damages to many provinces' coastal areas, i.e. West Java, Central Java, and Yogyakarta Special Province. The subsequent tsunami killed about 664 people and resulted in the number of destruction to coastal villages and towns such as Pangandaran, Batukaras, Ciamis, and Cilacap [2]. Cilacap city is one of the affected areas, and there had been many damages and casualties due to the South Java tsunami occurrence in 2006.

Since Cilacap is one of area that vulnerable to tsunami, there were some studies assess tsunami incidents in Cilacap. There was a study that developed modeling to create risk assessment to reduce the impact of a tsunami based on hazard assessment and population modeling, which resulted in evacuation planning for hazard zone. Also the study determined if surface roughness and population distribution are essential input for inundation modeling [3]; a study that creates modeling for tsunami run-up was done which resulted in estimation of inundation tsunami distance and height [4]; study to analyze the potential of tsunami in Cilacap coastal area and construct evacuation route plan [5]; however, it is little to know the study about the estimation of human and economical physical damaged that could exist if tsunami disaster happened. There was a study to measure human loss and financial loss for building in Cilacap [6], yet, there were no estimated need for a displaced population that could be a benefit in the preparedness for tsunami disaster.

For that reason, this research paper has a specific question to analyze the tsunami risk; what is the human and economic loss that could happen when tsunami disaster strikes coastal area in Cilacap City? The purpose of this paper is to measure the potential human and economic loss of tsunami disaster in Cilacap City. Moreover, this paper provides information about the displaced population and the minimum needs for displaced people with a geospatial approach based on result from InaSAFE tool.

Study using a geospatial approach to measure tsunami risk and impact has been done by some scholars. A mapping of tsunami risk in the Special Region of Yogyakarta was obtained with two approach, first, with integration of hazard and vulnerability, second, with additional of exposure factor [7]. The study showed the classification of risk zone of tsunami from very low to very high risk area. InaSAFE application as one of geospatial tool has been used to estimate number of people affected from some disaster scenario like earthquake in Padang and tsunami in Maumere [8]. The paper stated if InaSAFE can be used for estimating loss with the scenario based on historical data which is might contribute to prepare mitigation if in the future similar disaster scenario happen. It also gave example about estimated number of inundated buildings and people if Maumere affected by tsunami caused by earthquake with magnitude of 8.1.

Therefore, our study is expected to be used as preliminary assessment that able to give an overview of evaluation of requirements to increase awareness and readiness from stakeholders for tsunami disaster mitigation in Cilacap City.

Table 75.1 Data list

Data	Format	Source
Population density	Raster	Worldpop
Digital elevation model (DEM)—TerraSAR (9 m)	Raster	Geospatial information agency (BIG)
Land cover	Vector	Bing imagery and geospatial information agency (BIG)
Historical tsunami event	Tabular	USGS, ITDB, NOAA

75.2 Data and Methodology

75.2.1 Data

The data used in this research are secondary data obtained from government institutions and publicly available data. The data consists of maps, statistical data, and images. The following is a list of data used in this research (Table 75.1).

75.2.2 Methodology

The methodology used in this paper combines the calculation of tsunami hazard analysis and exposure data through InaSAFE tools. Tsunami hazard measurement is based on tsunami inundation that was added up with three parameters, namely slope, tsunami wave height, and surface roughness, calculate of run-up height and inundation distances. Historical data from USGS, ITDB, and NOAA is used to decide the worst cases tsunami wave height scenario for tsunami inundation. The worst cases scenario is based on the wave height that occurred in the South Coast of Java on July 17th, 2006, which is 7 m.

The maximum horizontal distance inland that is reached by tsunami; generally measured perpendicularly to the coastline, an inundation distance, and run-up height is the maximum elevation of tsunami waves above sea level. The method used to determine the tsunami inundation using the two equations. The equation for the inundation distance developed by Pignatelli [10]:

$$X_{\max} = (H_t)^{1.33} n^{-2} 0.06 \cos S \tag{75.1}$$

Equation for run-up height developed by McSavaney and Rattenburry [10]:

$$Y_{loss} = \left(\frac{167 n^2}{H_t^{1/3}} \right) + 5 \sin S \tag{75.2}$$

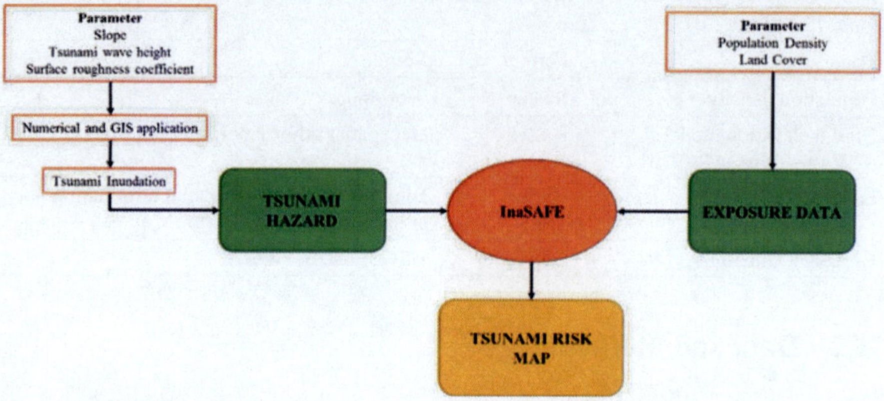

Fig. 75.1 Research framework

where $X_{\max}$ is the maximum distance that the wave height can penetrate inland (meter), H_t is the wave height at the coast (meter), n is the surface roughness coefficient, and S is the slope of the land surface (degree). Y_{loss} is the loss in wave height per meter of inundation distance (meter). The slope is the ratio between the height differences (vertical distance) of the land with the horizontal distance. The surface roughness is influenced by morphological features and vegetation textures. The data is obtained by extracting land cover and classified by the value of the surface roughness coefficient. The surface roughness coefficient shows an effective decrease for large waves with the size of the obstacles in their paths (Fig. 75.1).

Following process to assess the risk, this research uses InaSAFE tools by inputting exposure data with parameters of population density, landcover, and the hazard data that prior already been calculated. InaSAFE is the software to produce a real impact scenario that can be used for disaster preparedness. According to National Disaster Agency (BNPB), Australia-Indonesia Facility for Disaster Reduction (AIFDR), and Humanitarian OpenStreetMap Team (HOT) [11], InaSAFE performance is affected by data input and output in the form of hazard, exposure, and impact (the result of InaSAFE process). InaSAFE can be used to predict the effect of hazard scenario to exposure data by estimating the number of affected people, buildings, and roads; also able to provides list of minimum needs for the displaced population [12]. Another example of a study which utilizing InaSAFE tools to measure tsunami impact is the study in Manado city coastal area that can analyze tsunami prone zone, building exposure, and road exposure [13].

75.3 Result and Discussion

75.3.1 *Tsunami Hazard Analysis*

Based on the result of inundation analysis, it is found, with the scenario of 7 m height of tsunami wave, the coastal area could be inundated by 459.32 ha (see Table 75.2). The percentage of area in the low category, medium, high, and very high is 69%, 10%, 10%, and 11%, respectively. Thus, the most inundated area would be in the lowest depth category with 316.56 Ha area (Table 75.3).

Determination of classification at the level of tsunami waves using the equal interval data classification method from ArcGIS software to divide the range of attribute values into equal-sized sub-ranges. The classification of the low category is with less than 2 m inundation depth, the medium category is 2–4 m, 2–6 m in the high category, and more than 6 m for the very high category. Hazard map in Fig. 75.2 shows there are 7 villages inundated by tsunami, namely, Mertasinga, Gumlir, Kebonmanis, Tegalmulyan, Gunung Sumping, Sidanegara, and Sidakaya. The area that is marked with red color is the most inundated area with more than 6 m depth, while the majority area with dark green color sign shows 69% area affected by inundation. The maximum inundation distance from the coastline is 778.26 m. Thus, the more near the village distance to the coastline, the more risk area more prone to be exposed by deepest inundation.

Table 75.2 List of tsunami events in cilacap and surrounding

Year	Earthquake (Mw)	Wave height (m)	Location of epicentrum	Distance of epicentrum to cilacap (km)	Category	Data source
1963	6.5	2.00	Java	433.876	Sea	Rynn (2002) [9]
1994	7.8	3.70	East of Java	525.142	Sea	Rynn (2002) [9]
2004	9.1	0.51	Off west coast of Sumatera	1904.931	Sea	NOAA
2006	7.7	7.00	Pangandaran	251.861	Sea	NOAA
2007	8.5	5.00	Bengkulu	975.188	Sea	NOAA

Table 75.3 Tsunami inundation area

Level	Category	Area (Ha)	Percentage
1	Low	316.56	69
2	Medium	45.58	10
3	High	46.91	10
4	Very high	50.27	11
Total		459.32	100

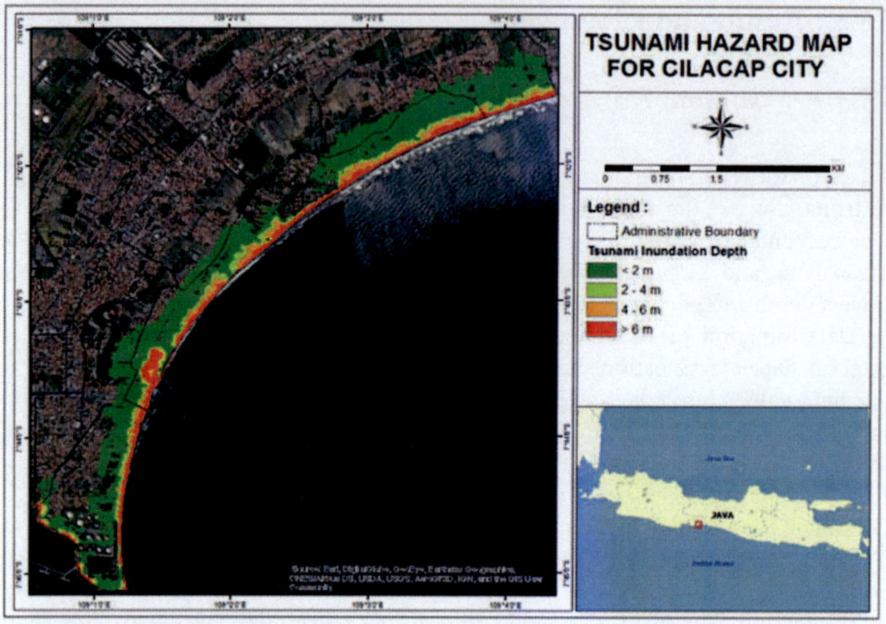

Fig. 75.2 Tsunami Hazard Map

75.3.2 Tsunami Risk Analysis

Risk Analysis

After adding exposure data, the following process is to determine the risk area of tsunami. The result of tsunami risk analysis shows the area in the very high category is 51.95 Ha or 29 percentage among total risk area. The area in the high class is 34.3 Ha with 19%. In the medium category, the area is 26% of total affected area, and in the low category is 44.84 Ha with 25% of total affected area (Table 75.4).

Generally, it can be seen if the majority of the affected areas would be in a very high-risk category. In Fig. 75.3, the widest area belong to Tegalkamulyan village is located in the very high-risk category.

Table 75.4 Risk area of tsunami

Level	Category	Area (Ha)	Percentage
1	Very high	51.95	29.40%
2	High	34.3	19.41%
3	Medium	45.64	26.82%
4	Low	44.84	25.37%
Total		176.73	100.00%

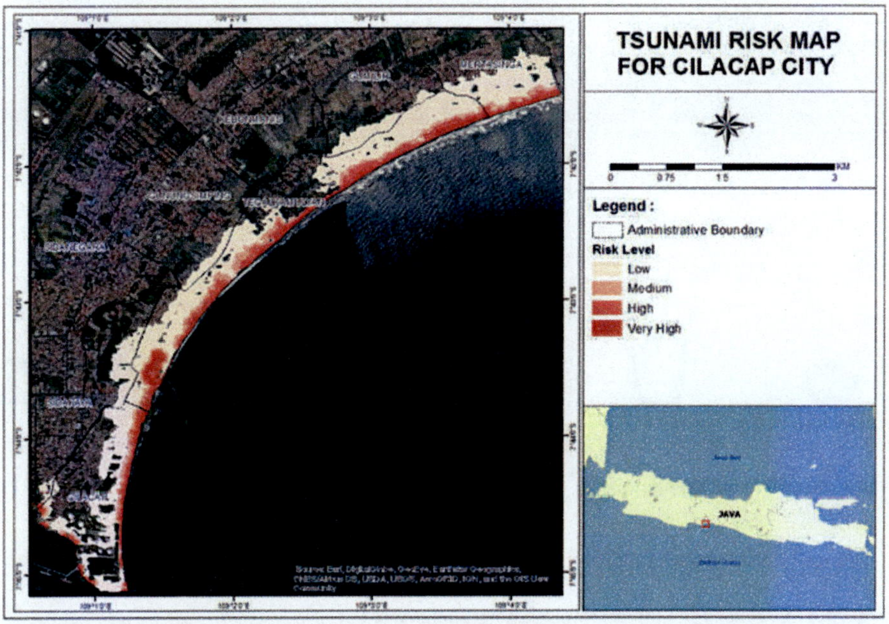

Fig. 75.3 Tsunami risk map

Human Impact

The estimation of human risk is measured by a number of people affected in the Cilacap coastal area, assessment of displaced population, and minimum needs. Affected people are people who get direct and indirect impact, who experience consequences in terms of economic, physical, and social loss to their lives. In InaSAFE, people who are killed during the tsunami events are considered as the affected people. The calculation of people who affected by tsunami majority would be in low category with 19,000 people. The very high class consists of estimated 620 people influenced.

The number of displaced people estimation can be seen in Table 75.5. Displace people are people who have to migrate from their place of residence because of risk or disaster. Displaced people in this paper means people who need to be evacuated. The demographic and minimum needs reports are obtained from displaced/evacuated

Table 75.5 Number of people affected

Category	People	Percentage
Low	19,000	88
Medium	1100	5
High	970	4
Very high	620	3
Total	21,690	100

Table 75.6 Number of people displaced

No	Villages	People
1	Cilacap	1800
2	Sidakaya	40
3	Tegalkamulyan	720
4	Mertasinga	140
	Total	2700

Fig. 75.4 Estimated number of minimum needs per week

people calculation. Totally, in this report, it is predicted 2700 displaced people in Cilacap coastal area (Table 75.6).

Figure 75.4 shows the estimated number of minimum needs per week according to PERKA BNPB 7/2008. They would need 7400 kg of rice, 45,900-L of drinking water, 176,000-L of clean water, 530 units of family kits, and 140 units of emergency toilets each week. The purpose of the minimum needs measurement is to provide a quick method to calculate the support requirements for displaced persons.

Economic Impact

The economic impact in this paper is measured from physical damage loss, such as loss in each for land cover. Four types of landcover that are affected by the tsunami, namely, commercial, farmland, industrial, and residential. The calculation of cost from damage using the method from BAPPENAS [14] is summarized in Fig. 75.5.

The estimated land cover affected by the tsunami in Cilacap coastal area is 380.420 Ha. The type of land use most affected by the tsunami was farmland area, covering an area of 171.981 ha. After obtaining the extent area of the affected land cover type, we calculated the damage value caused by the tsunami with the monetary value. The estimated total cost of damage is $ 345,071.36, with the highest loss cost be owned by the commercial area with $ 304,532.86. Although the commercial land cover has less affected area compared to farmland and residential, however, most commercial area is located in the high affected category area (Table 75.7).

Land cover type	Low Damage	Moderate Damage	High Damage
Residential	Area (ha) x Damage to house building, furniture and home appliances is estimated at $360 x 20%	Area (ha) x Damage to houses, furniture and home appliances is estimated at $720 x 40%	Area (ha) x Damage to house building, furniture and home appliances is estimated at $1440 x 60%
Agricultural	Area (ha) x Production cost of $504 per hectare and yield of 5 tons per hectare, assuming the price of the harvest is $ 0.1728 per kg x 20%	Area (ha) x Production cost of $504 per hectare and yield of 5 tons per hectare, assuming the price of the harvest is $ 0.1728 per kg x 40%	Area (ha) x Production cost of $504 per hectare and yield of 5 tons per hectare, assuming the price of the harvest is $ 0.1728 per kg x 60%
Commercial	Area (ha) x Cleaning and re-painting fee of $360 x 20%	Area (ha) x Repair, cleaning and re-painting costs of $3600 x 40%	Area (ha) x The cost of dismantling, rebuilding, and cleaning costs is $21,600 x 60%
Industrial	Area (ha) x Cleaning and re-painting fee of $360 x 20%	Area (ha) x Repair, cleaning and re-painting costs of $3600 x 40%	Area (ha) x The cost of dismantling, rebuilding, and cleaning costs is $21,600 x 60%

Fig. 75.5 References of economic impact calculation

Table 75.7 Number of affected area and loss

Land cover	Affected area (Ha)	Economic loss
Commercial	55.963	$ 304,532.86
Farmland	171.981	$ 28,064.68
Industrial	0.849	$ 61.14
Residential	151.627	$ 12,412.69
Total	380.420	$ 345,071.36

75.4 Conclusion

The results of a combination between a tsunami hazard and a tsunami vulnerability can estimate the areas affected by tsunami risk, the affected population and the losses incurred in the event of a tsunami disaster. According to the result of tsunami risk analysis, there would be four villages at the risk area by tsunamis, such as Cilacap Village, Tegalkamulyan Village, Sidakaya Village, and Mertasinga Village, with the scenario of 7 m heights of tsunami. In general, the economic impact caused in the event of a tsunami would be $ 345,071.36 in Cilacap City; signficant economic loss would be come from the commercial area. For human impact, in general, it would affect 21,690 people in the risk area. Stakeholders should probably prepare the minimum needs for displaced people in the risk area as tsunami disaster mitigation.

References

1. Bock, Y., Prawirodirdjo, L., Genrich, JF., Stevens, CW., McCaffrey, R., Subarya, C., Puntodewo SSO., Calais, E.: Crustal motion in Indonesia from global positioning system measurements. J. Geophys. Res.: Solid Earth, **108**(B8) (2003)
2. Lavigne, F., Gomez, C., Giffo, M., Wassmer, P., Hoebreck, C., Mardiatno, D., Paris, R., et al.: Field observations of the 17 July 2006 Tsunami in Java. Natl Hazards Earth Syst. Sci. **7**(1), 177–183 (2007)
3. Kongko, W., Schlurmann, T., Khomarudin, R.: Tsunami model of cilacap-Indonesia: inundation and its mitigation. In AGU Fall Meeting Abstracts, Vol. 2009, pp. NH31B-1114 (2009)
4. Fadhil, E.: Pemodelan Run Up Tsunami: Studi Kasus Kota Cilacap. Undergraduate thesis, Institut Teknologi Sepuluh Nopember (2018)
5. Hilmi, E., Hendarto, E., Riyanti, Sahri, A.: Analisis potensi bencana abrasi dan Tsunami di pesisir cilacap. Jurnal Penanggulangan Bencana **3**(1), 34–42 (2012)
6. Sihombing, R. G. P.: Generating reliable database for loss estimation and mitigation planning of tsunami effects: case study Cilacap City Indonesia, Doctoral dissertation, Universitas Gadjah Mada (2010)
7. Steinritz, V., Pena-Castellnou, S., Marliyani, G. I., Reicherter, K.: GIS-based study of tsunami risk in the Special Region of Yogyakarta (Central Java, Indonesia): In IOP Conference Series: Earth and Environmental Science Vol. 851, No. 1, p. 012007. IOP Publishing (2021)
8. Pranantyo, I. R., Fadmastuti, M., & Chandra, F.: InaSAFE applications in disaster preparedness. In: AIP Conference Proceedings, Vol. 1658, No. 1, p. 060001, AIP Publishing LLC (2015)
9. Rynn, J.: A preliminary assessment of tsunami hazard and risk in the Indonesian region. Sci. Tsunami Haz. **20**, 193–215 (2002)
10. Pignatelli, C., Sansò, P., Mastronuzzi, G.: Evaluation of tsunami flooding using geomorphologic evidence. Mar. Geol. **260**(1–4), 6–18 (2009)
11. McSaveney, M., Rattenbury, M.: Tsunami impact in Hawke's Bay, Institute of Geological and Nuclear Sciences. In client Report 2000/77 (2000)
12. AIFDR Document, https://www.dfat.gov.au/sites/default/files/aus-indonesia-facility-drr-design-doc.pdf. Last accessed 16 Nov 2021
13. Kumaat, J.C., Kandoli, S. T. B., Laeloma, F.: Spatial modeling of Tsunami impact in manado city using geographic information system. In: IOP Conference Series: Materials Science and Engineering, Vol. 306, No. 1, p. 012069, IOP Publishing (2018)
14. BAPPENAS: Master plan for the rehabilitation and reconstruction of the regions and communities of the province of Nanggroe Aceh Darussalam and the island of Nias, Province of North Sumatera. Jakarta (2005)

Chapter 76
Effects of El-Nino and La-Nina on the Velocity Potential at 200 hPa over Maritime Continent

Eddy Hermawan, Tyo Maulana, Rahmat Gernowo, and Dita Fatria Andarini

Abstract Rainfall is one of the climatological parameters that play a role in climate variability in Indonesia. Climate dynamics in Indonesia require a comprehensive analysis to determine the factors causing extreme events in Indonesia. The velocity potential of 200 hPa can describe the convective level of clouds in the troposphere. To analyze the dynamics of the Indonesian atmosphere, a temporal, spatial, and spectrum analysis was carried out at a potential velocity of 200 hPa. The results showed that the velocity potential of 200 hPa had oscillations for six months with peak oscillations in MAM and SON. In the case of extreme climates in Indonesia, especially in El Niño in 2015 and La Niña in 2020, the velocity potential of 200 hPa can capture changes in climate conditions both in Indonesia and in the Pacific Ocean as seen from the results of spatial analysis and the Hovmöller diagram shown. The velocity potential of 200 hPa has a negative correlation with rain. This explains that if the rainfall increases, the velocity potential of 200 hPa will decrease and vice versa. Based on the coefficient of determination test results, the potential velocity of 200 hPa contributes around 10–20%. In general, a potential velocity of 200 hPa can to rate the dynamics of the atmosphere that occurs in Indonesia, especially in the case of El Niño in 2015 and La Niña in 2020.

E. Hermawan (✉) · R. Gernowo
Research Center for Climate and Atmosphere, National Research and Innovation Agency (BRIN), Bandung, Indonesia
e-mail: eddy001@brin.go.id

T. Maulana · D. F. Andarini
Departement of Physics, Diponegoro University, Semarang, Indonesia

E. Yulihastin et al. (eds.), *Proceedings of the International Conference on Radioscience, Equatorial Atmospheric Science and Environment and Humanosphere Science, 2021*, Springer Proceedings in Physics 275, https://doi.org/10.1007/978-981-19-0308-3_76

76.1 Introduction

76.1.1 Background

Indonesia is a maritime country located between two continents and two oceans and is traversed by the equator which causes Indonesia to have complex atmospheric dynamics. One of the impacts is the determination of the dry season and the rainy season which are increasingly difficult to predict. One of the factors that influence this is the intensity of rainfall. The intensity of precipitation is affected by global climate variability that occurs in Indonesia, including the El Nio Southern Oscillation (ENSO), Indian Ocean Dipole (IOD), Madden-Jullian Oscillation (MJO), Kelvin Wave, and so on [1].

ENSO is the interaction of the ocean with the atmosphere in the equatorial Pacific Ocean. This interaction makes ENSO has three phases, namely El Nino, La Nia, and neutral. When El Nino occurs, the Indonesian region experiences cooling of sea surface temperatures which causes drought. Meanwhile, during La Nina, the sea surface temperature in the Indonesian region warms which causes an increase in rainfall [2]. While IOD is the interaction between the atmosphere and the oceans in the area around the Indian Ocean. IOD has two phases, namely positive IOD (+) and negative IOD (−). A positive IOD causes a decrease in rainfall in Indonesia. On the other hand, negative IOD causes an increase in rainfall in Indonesia [3].

Indonesia is divided into three rainfall patterns. The three patterns are Monsunal, Equatorial, and Local (Anti-Monsunal) [2]. Rainfall is associated with several atmospheric parameters, including air humidity, Dew Point (Dew temperature), air pressure, and Precipitable Water Vapor (PWV) [4]. In addition to the previously mentioned parameters, there are other parameters that can indicate an increase in rainfall, one of which is Velocity Potential [5].

Velocity Potential is an atmospheric parameter that describes the convergence of the zonal wind [6]. In several previous studies, Velocity Potential was used as a parameter to indicate the presence of climate phenomena, including MJO, ENSO, and Heavy Rainfall Extreme (HRE) [6]. Velocity Potential describes the convective level of the cloud. The convective level of clouds affects the formation of cumulonimbus clouds which are the cause of rain [7, 8].

Research on the characteristics of the velocity potential, especially in the 200 hPa layer, has not been studied much. In this study, an analysis of the characteristics of the velocity potential at the 200 hPa layer, especially over the Indonesian Maritime was carried out and then related to its impact on three rainfall patterns in Indonesia and associated with extreme climate events in Indonesia.

76.2 Data and Method

76.2.1 Data

The research area domain covers the territory of Indonesia with geographical coordinates 80°–150° east longitude and 20° north latitude—20° south latitude. In addition, three cities were taken from each type of rainfall pattern. The monsoon region is represented by the cities of Semarang, Yogyakarta and Surabaya. The equatorial region is represented by the cities of Pontianak, Ternate, and Padang. Meanwhile, the local area (anti monsoon) was represented by the cities of Sorong, Ambon, and Fak-Fak. The time domain of the study is monthly data from January 1981 to September 2020 and daily data for years with extreme climatic conditions. El Nino extreme climatic conditions are represented by daily data from January 2015 to December 2015. Meanwhile, La Nina is represented by daily data from January 2020 to October 2020. Rainfall data comes from CHIRPS, Velocity Potential from NCEP NCAR, and ENSO and IOD from NOAA.

76.2.2 Procedure

Data Preparation

The downloaded data is then selected based on the determined research area and time or downscaled. The research area covers the Indonesian Maritime Continent, Semarang City, Yogyakarta, Surabaya, Pontianak, Ternate, Padang, Sorong, Ambon, and Fak-Fak as shown on the map in Fig. 3.1. Meanwhile, the ENSO and DMI indices use a global scale.

Data Processing

Rainfall, velocity potential, ENSO index, and DMI data are plotted in time series to see variations and anomalous values of the tested parameters with a time span from January 1981 to September 2020. The data is then selected to obtain the period of year of occurrence of extreme climatic conditions in Indonesia. Furthermore, the daily data is plotted to determine the monthly variation pattern of rainfall and velocity potential. Then the rainfall data and velocity potential are plotted spatially to determine the distribution of the spatial variation values.

Spectral analysis using the Power Spectral Density (PSD) technique. This PSD technique uses monthly data from rainfall and velocity potential from 1981 to 2020. The PSD technique aims to determine the periodization of rainfall and velocity potential parameters using the Fast Fourier Transform (FFT) principle. This analysis is divided into three regions according to the type of rainfall pattern to determine the differences in the oscillations of each rainfall pattern.

Velocity potential data is plotted with a Hovmöller diagram to determine the evolution of the observed parameters over time and spatially at the same time. This diagram allows to find out the pattern of oscillation or propagation of waves in the atmosphere. Thus, it can be analyzed the factors that influence the occurrence of extreme climate events in Indonesia. The temporal span of the plot focuses daily data on the 2015 El Niño and 2020 La Nina events.

In this correlation analysis used the coefficient of determination test technique. The coefficient of determination test is a technique to measure the strength of the relationship between two variables with a certain scale. If the coefficient of determination is close to zero, it means that the influence of the independent variable (velocity potential 200 hPa) on the dependent variable (rainfall) is weak. Meanwhile, if the coefficient of determination is close to one, it means that the influence of the independent variable (velocity potential 200 hPa) on the dependent variable (rainfall) is strong. This analysis aims to determine the magnitude of the influence of the velocity potential on the occurrence of rain in the territory of Indonesia.

76.3 Result and Discussion

Characteristics of velocity potential 200 hPa

Figure 76.1 shows the monthly time series of the 200 hPa velocity potential in nine regions in Indonesia. It can be seen that there are similar patterns (oscillations) in the nine areas, but the periodization of the patterns changes is not clear.

Figure 76.2 shows the distribution of the monthly mean velocity potential of 200 hPa in one year. Based on the distribution pattern formed, it shows that there is a similar distribution pattern in all regions of Indonesia. In Fig. 76.3, the peak velocity potential anomaly of 200 hPa is seen in March, April, May, September, October, and November (MAM, SON). This shows that the velocity potential of 200 hPa has the same oscillation period throughout Indonesia.

Spectrum analysis aims to determine the periodicity of the velocity potential of 200 hPa. Figure 76.4 shows the results of the spectrum analysis of the velocity potential of 200 hPa in nine regions. Figure 76.3a shows that the monsoon region has a dominant oscillation of about 12 months. In Fig. 76.3b, the equatorial region has a dominant 12-month oscillation as well, but the six-month oscillation is also quite high. Figure 76.3c shows the dominant oscillation for the local area (anti-monsoon) for about 12 months as well. This shows that the average oscillation of the velocity potential of 200 hPa is around 12 months.

Analysis of 2015 El Niño Events

The time series for the interaction of ENSO and IOD during the 2015 El Nio event is shown in Fig. 76.5. Based on Fig. 76.4, it can be seen that the ENSO index is above the normal limit of 2 °C. While the index of the IOD is at a position of around 1.1 °C. This shows that 2015 was an active El Niño year.

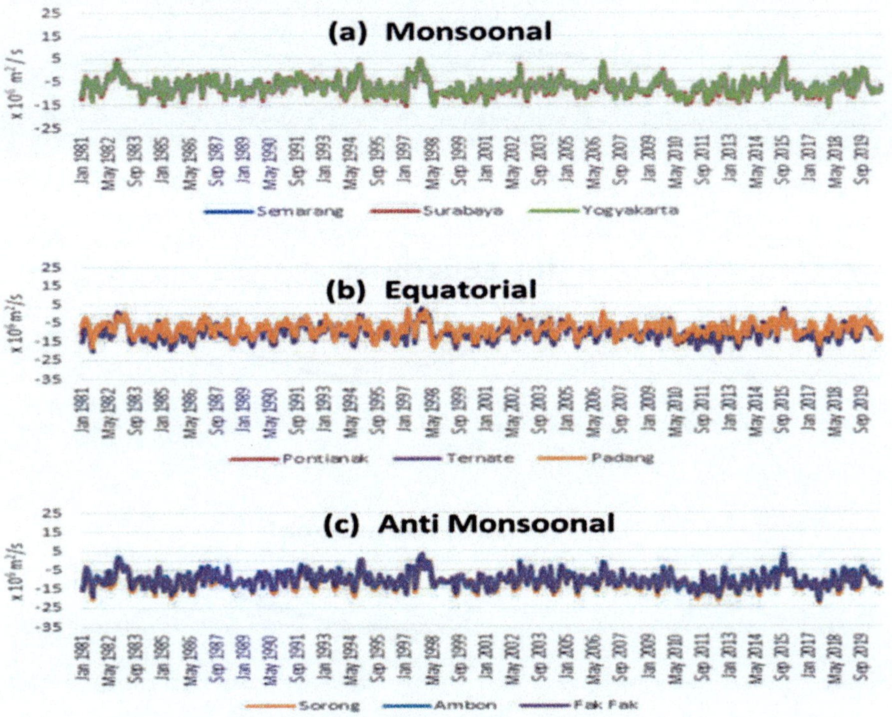

Fig. 76.1 Velocity potential time series 200 hPa period January 1981 to September 2020

The active El Niño in the Pacific Ocean Region, causes convection clouds to move towards the Pacific Ocean region. This causes a decrease in rainfall (long drought) in Indonesia. Figure 76.5 shows the daily time series of rainfall and velocity potential of 200 hPa in each area of rainfall type in 2015. Based on Fig. 76.4, from June 2015 to October 2015, there is a negative anomaly of rainfall which indicates a decrease in rainfall. rainfall. Meanwhile, the velocity potential anomaly value of 200 hPa is positive which indicates a decrease in cloud convection in the 200 hPa layer. This indicates that the territory of Indonesia is experiencing drought.

Spatial analysis is used to determine the distribution of the parameters in two dimensions. Figure 76.6 illustrates the distribution of rainfall and velocity potential of 200 hPa in 2015. As seen in Fig. 76.6a, rainfall experienced a significant decrease in October 2015 marked by a red spatial anomaly. While in Fig. 76.6b, it shows the distribution of the velocity potential of 200 hPa. Seen in October 2015, Indonesia has a low level of cloud convection marked by a red anomaly.

The Hovmöller diagram is used to determine the evolution of the observed parameters over time and spatially at the same time. Figure 76.7 shows the evolution of the velocity potential of 200 hPa in 2015. Based on Fig. 76.7, the positive anomaly (red) has been visible since July and started to strengthen until October. This indicates that from July to October, the level of cloud convection in Indonesia has decreased,

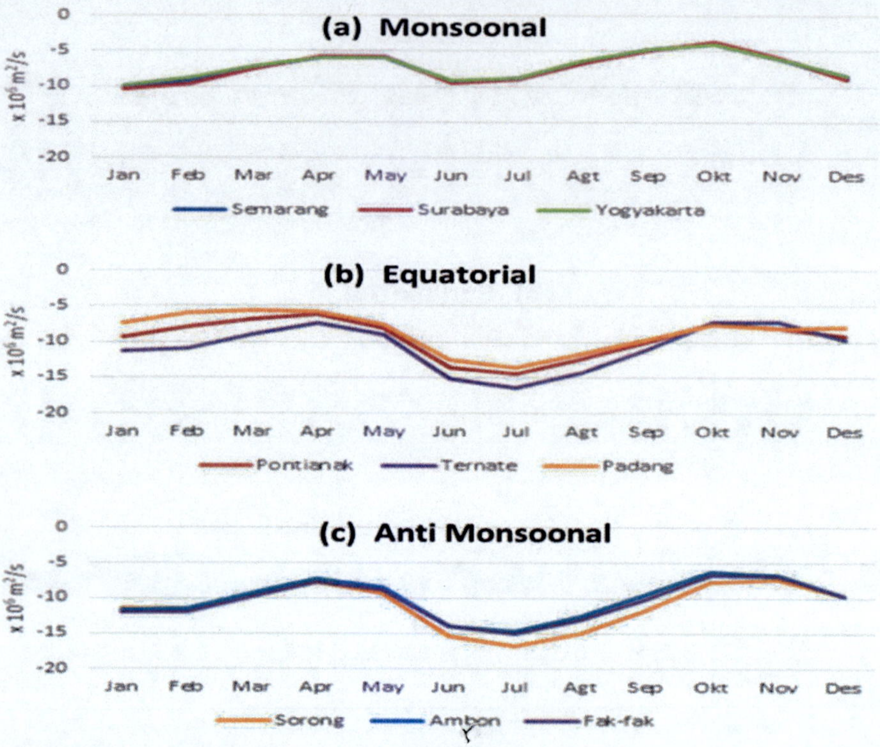

Fig. 76.2 Annual distribution of velocity potential 200 hPa

although not evenly. In addition, there is a negative anomaly (blue) that appears in the east of Indonesia, precisely around the Pacific Ocean. This negative anomaly indicates an increase in convective clouds around the Pacific Ocean.

Analysis of La Niña Events in 2020

The time series for the interaction of ENSO and IOD at the time of the La Nia event in 2020 is shown in Fig. 76.9. Based on Fig. 76.8, it can be seen that the ENSO index is below the normal limit at −1.2 °C. Meanwhile, the IOD index tends to be in a neutral position (between 0.8 °C to −0.8 °C). This indicates that La Niña will become active in 2020.

Figure 76.9 shows the daily time series of rainfall and velocity potential of 200 hPa in each area of rainfall type in 2020. Based on Fig. 76.7, from September 2020 to October 2020, there is a positive anomaly of rainfall which indicates an increase in rainfall. rainfall. Meanwhile, the velocity potential anomaly value of 200 hPa is negative which indicates an increase in cloud convection in the 200 hPa layer. This indicates that Indonesia has experienced a significant increase in rainfall intensity.

Based on the spatial analysis of rainfall distribution and velocity potential of 200 hPa in 2020 in Fig. 76.10, it can be seen that in Fig. 76.10a, rainfall experienced

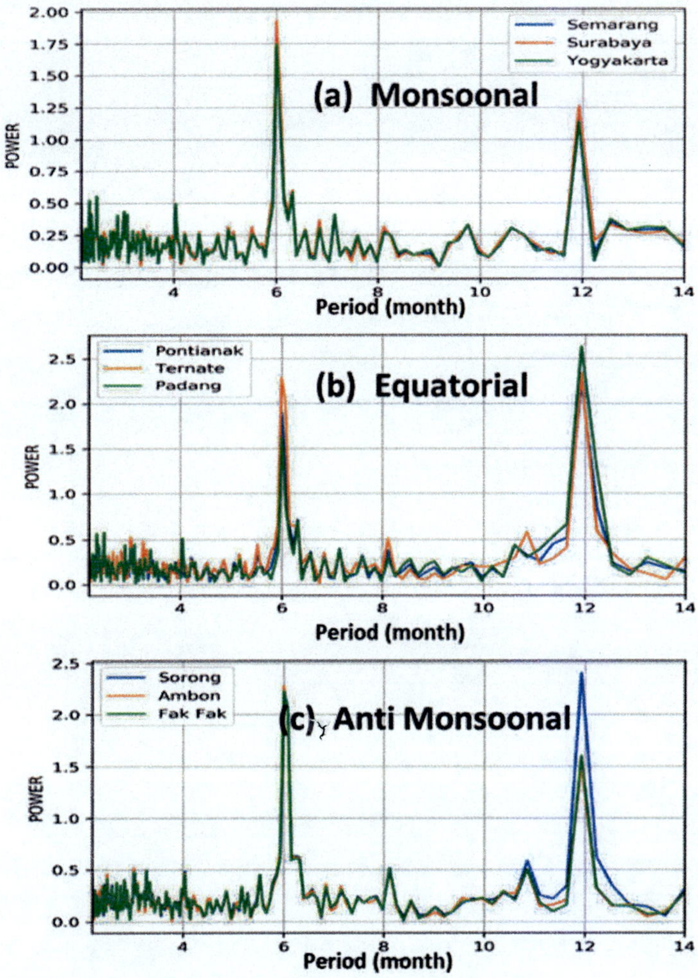

Fig. 76.3 Power spectral density (PSD) velocity potential 200 hPa

a significant increase from June to October 2020 marked by a blue spatial anomaly. While in Fig. 76.10b, shows the distribution of the velocity potential of 200 hPa. Seen from September to October 2020, the Indonesian region has a high level of cloud convection marked by a blue anomaly.

Based on the Hovmöller diagram in Fig. 76.11, it shows the evolution of the velocity potential of 200 hPa in 2020. Based on Fig. 76.11, the negative anomaly (blue) has been visible since August and began to strengthen until October. This indicates that from August to October, the level of cloud convection in Indonesia has increased although not evenly. In addition, there is a positive anomaly (red) that

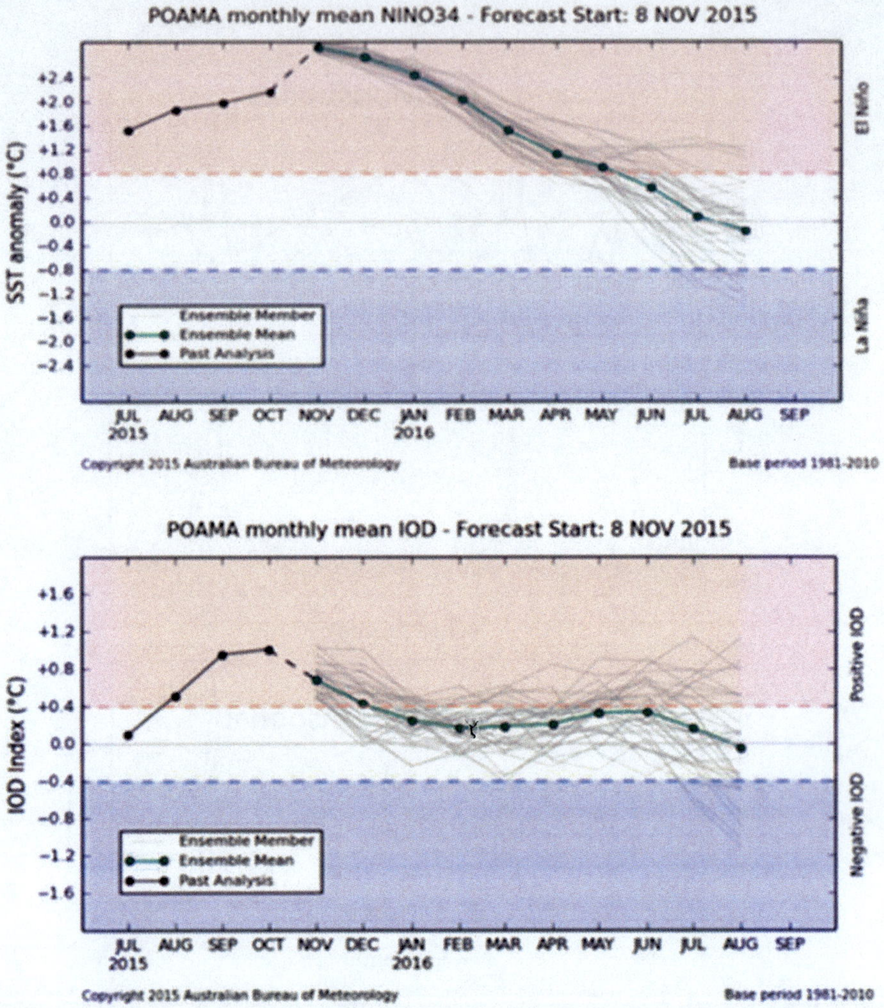

Fig. 76.4 ENSO and IOD interaction time series during the 2015 El Niño

appears in the east of Indonesia, precisely around the Pacific Ocean. The positive anomaly indicates a decrease in convective clouds around the Pacific Ocean.

Correlation Analysis

Correlation analysis was conducted to find out how much influence the velocity potential of 200 hPa has on rainfall in Indonesia. The correlation analysis used is based on the coefficient of determination. By using different regional classifications, different coefficients of determination will be obtained. In the monsoon region based on Table 76.1, the coefficient of determination obtained is 0.15 or 15%. This means that in the monsoon area, the occurrence of rainfall is influenced by the velocity

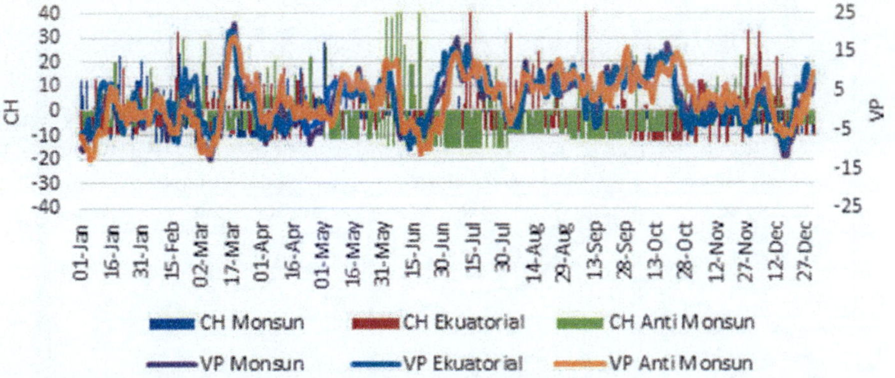

Fig. 76.5 Daily time series of rainfall anomalies and velocity potential of 200 hPa in 2015

potential of 200 hPa by 15%. Then in the equatorial region, based on Table 76.2, the coefficient of determination value is 0.12 or 12%. This means that in the equatorial region, the occurrence of rainfall is influenced by the velocity potential of 200 hPa by 15%. While in the local area (anti-monsoon) based on Table 76.3, the coefficient of determination obtained is 0.201 or 20.1%. This means that in the local area (anti-monsoon) the incidence of rainfall that occurs is influenced by the velocity potential of 200 hPa by 20.1%. Based on the above analysis, it can be concluded that the velocity potential of 200 hPa has an influence on the intensity of rainfall in Indonesia, ranging from 10 to 20% which indicates that there are other factors that are more dominant in influencing rainfall in Indonesia which were not examined in this study.

76.4 Conclusion

1. Velocity potential is a global atmospheric parameter that describes the convergence of zonal wind flows. Velocity potential has an oscillation period of six months. The positive velocity potential anomaly explains the convergent flow that causes the low level of cloud convection around the area. In contrast, the negative anomaly describes the divergent flow that causes the high level of cloud convection around the room.
2. Velocity potential and rainfall are negatively correlated. When rainfall increases, the velocity potential value decreases, and vice versa. This illustrates that when the level of cloud convection increases, the intensity of rainfall will increase. Velocity potential has a coefficient of determination ranging from 10 to 20%. This shows that the velocity potential of 200 hPa statistically affects 10–20% of the rainfall events, while the other 80–90% are other factors not examined in this study.

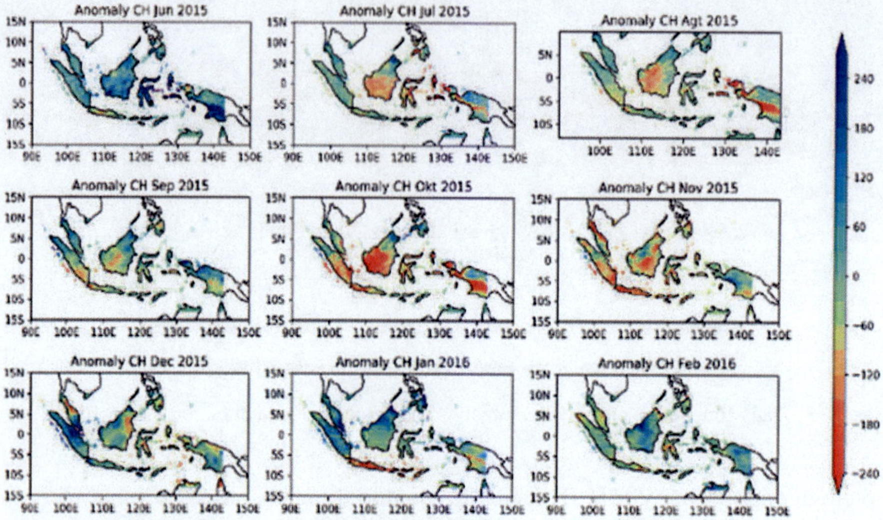

(a) Rainfall

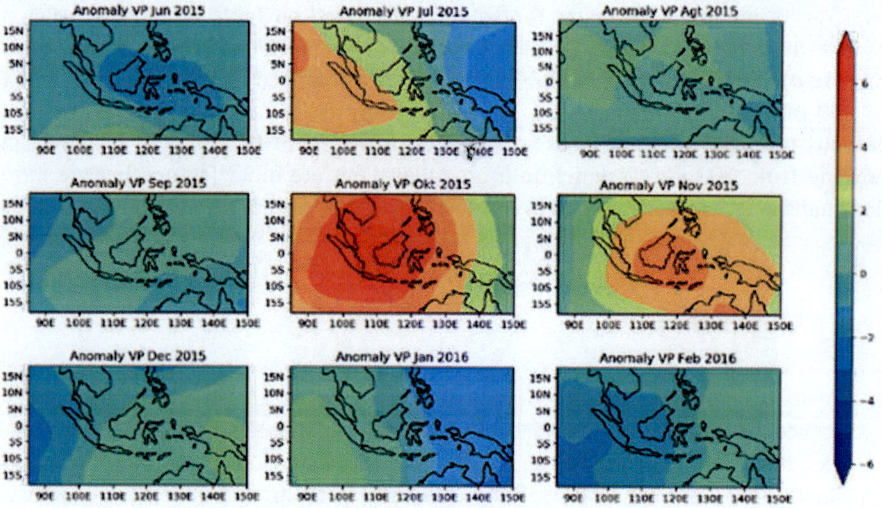

(b) Velocity potential 200 hPa

Fig. 76.6 2015 spatial distribution map

3. The velocity potential of 200 hPa can describe the dynamics of the atmosphere in Indonesia, especially when experiencing El Nio 2015 and La Nia 2020 in both temporally, spatially, and Hovmöller.

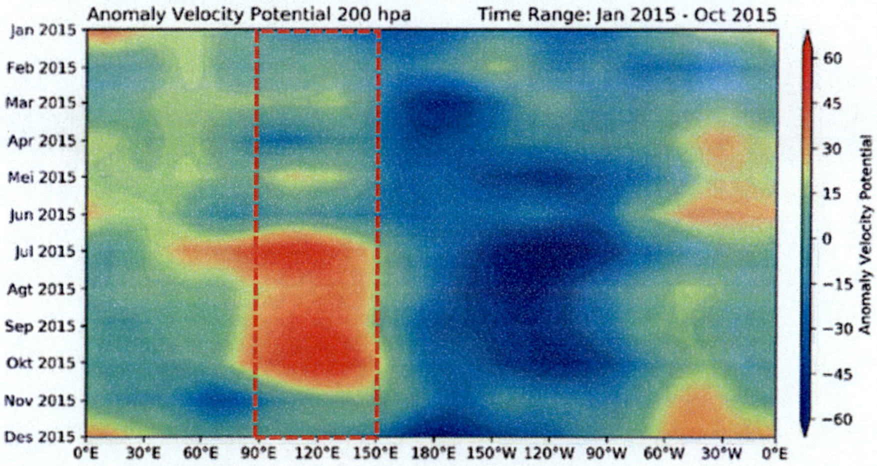

Fig. 76.7 2015 Hovmöller velocity potential diagram of 200 hPa

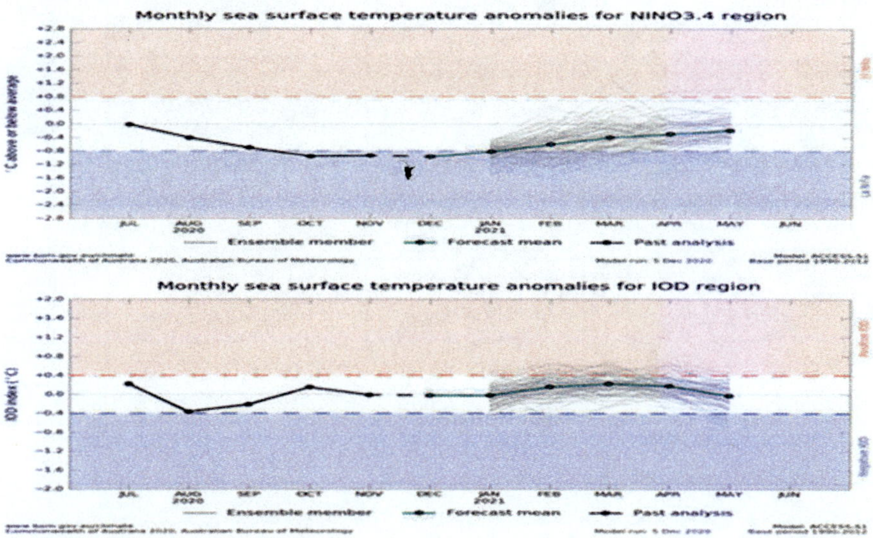

Fig. 76.8 ENSO and IOD interaction time series during La Nina in 2020

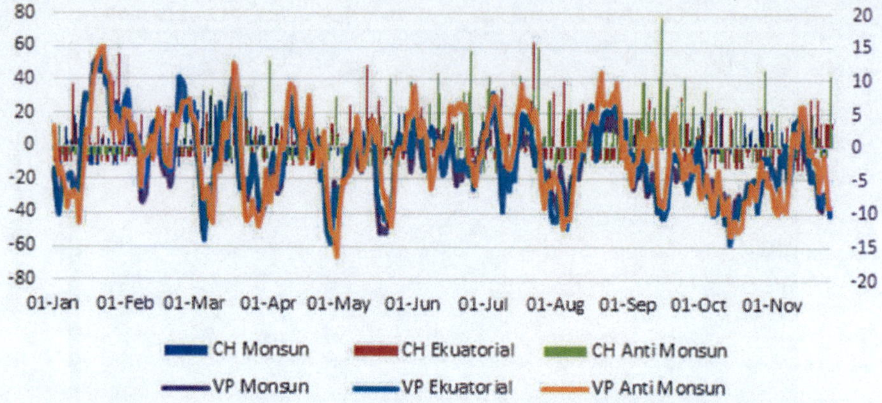

Fig. 76.9 Daily time series of rainfall anomalies and velocity potential of 200 hPa in 2020

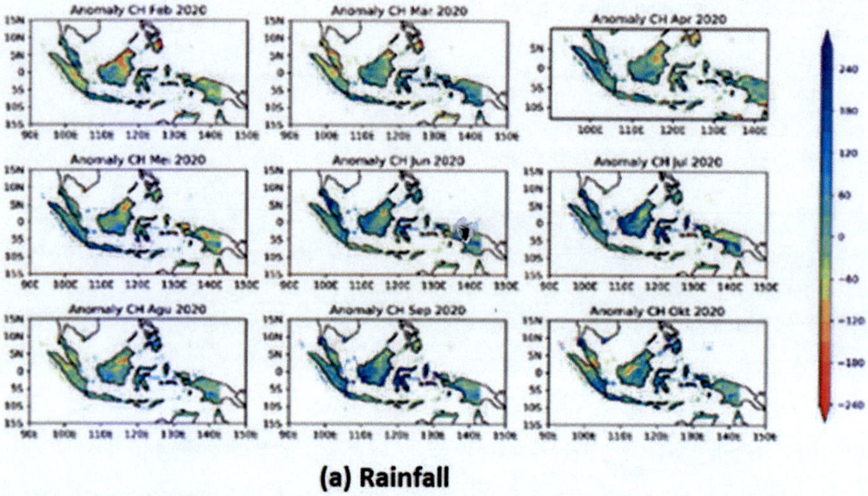

Fig. 76.10 2020 spatial distribution map

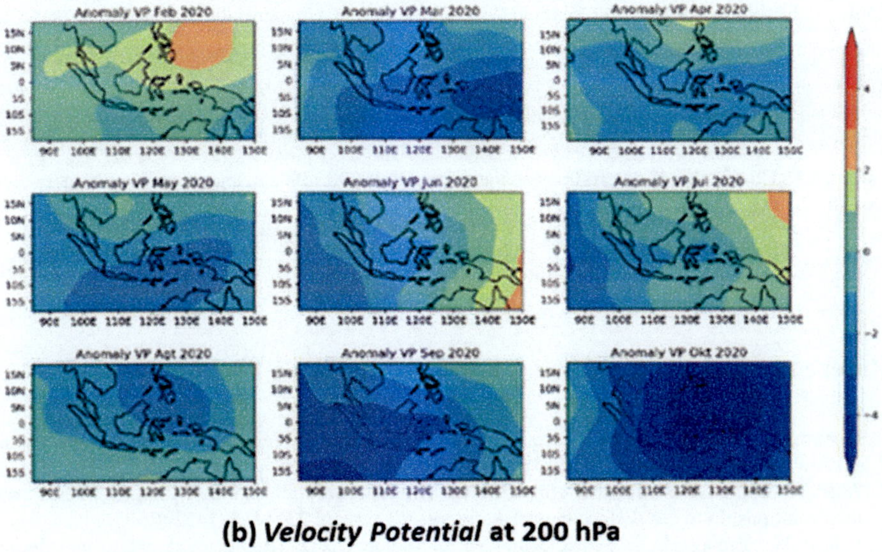

Fig. 76.10 (continued)

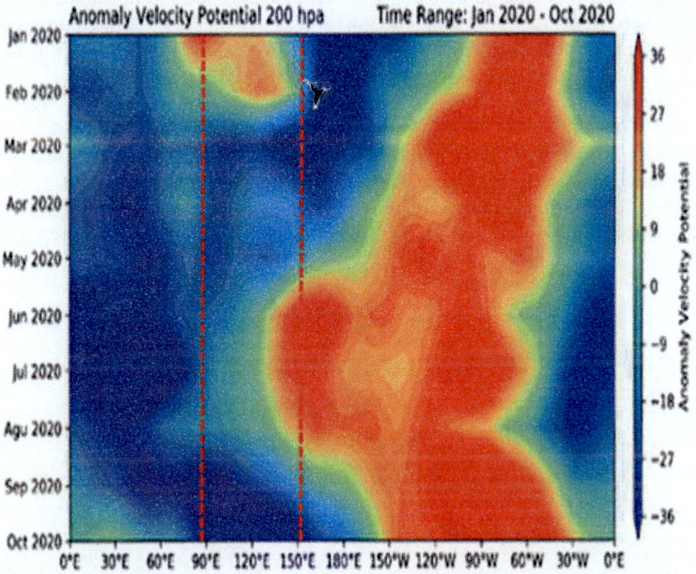

Fig. 76.11 Hovmöller velocity potential 200 hPa chart in 2020

Table 76.1 Velocity potential determinant coefficient of 200 hPa monsoon to monsoon rainfall

R	R square	Adjusted R square
0.388	0.15	0.148

Table 76.2 Velocity potential determinant coefficient 200 hPa equatorial to equatorial rainfall

R	R square	Adjusted R square
0.388	0.12	0.010

Table 76.3 Coefficient of determinant velocity potential 200 hPa anti-monsoon to anti-monsoon rainfall

R	R Square	Adjusted R square
0.449	0.201	0.200

References

1. Lestari, S., King, A., Vincent, C., et al.: Seasonal dependence of rainfall extremes in and around Jakarta Indonesia. Weather Clim. Extrem. **24**, 100202 (2019)
2. Aldrian, E., Dwi, S.R.: Identification of three dominant rainfall regions within Indonesia and their relationship to sea surface temperature. Int. J. Climatol **23**, 1435–1452 (2003)
3. Saji, N.H., Yamagata, T.: Possible impacts of Indian Ocean Dipole mode events on global climate. Clim. Res. **25**, 151–169 (2003)
4. Hashim, F.R., Nik Daud, N.G., Ahmad, K.A., et al.: Prediction of rainfall based on weather parameter using artificial neural network. J. Fundam. Appl. Sci. **9**, 493 (2018)
5. Sohn, B.J., Yeh, S.W., Lee, A., et al.: Regulation of atmospheric circulation controlling the tropical Pacific precipitation change in response to CO_2 increases. Nat Commun **10**
6. Mafuru, K.B., Guirong, T.: Assessing prone areas to heavy rainfall and the impaction of the upper warm temperature anomaly during March–May rainfall season in Tanzania. Adv. Meteorol., (2018)
7. Ventrice, M.J., Wheeler, M.C., Hendon, H.H., et al.: A modified multivariate Madden-Julian oscillation index using velocity potential. Mon. Weather Rev. **141**, 4197–4210 (2013)
8. Fan, F., Dong, X., Fang, X., et al.: Revisiting the relationship between the South Asian summer monsoon drought and El Niño warming pattern. Atmos. Sci. Lett. **18**, 175–182 (2017)

Chapter 77
Soil Tillage System and Water Management Effect on Soil Physical Properties and Yield in Rainfed Paddy Field

Nourma Al Viandari, Anicetus Wihardjaka, Heru Bagus Pulunggono, and Suwardi

Abstract The dry season rice yield production is generally lower than the wet season crop on the rainfed fields. The low yield of paddy is due to drought stress during the reproductive growth phase. This study aimed to determine the effect of water management and tillage system on rice yields at the rainfed fields. The experiment was conducted at the Jaken Experimental Station of the Indonesian Agricultural Environment Research Institute, Pati, Indonesia, from February to June 2019 with a split-plot design, repeated four times. The treatment consisted of water management as the main plot and soil tillage as a sub-plot. The treatment of water management was continuous flooding and depended on rainfall flooding. The treatment of the tillage system was normal tillage (0–10 cm) and deep tillage (25–30 cm). The tillage system and water management treatment did not significantly affect the yield of IR64 rice grain, with an average grain yield of 4.1 tons ha^{-1}. Tillage system such deep and normal tillage significantly reduces bulk density during the maximum tillering phase. Even it has not yet given significant results, deep tillage and water management have a great potential to improve soil physical properties. Further research in the next crop season is needed to complete the data and obtain more in-depth information about the study.

77.1 Introduction

Indonesia has two seasons as a tropical area. The climates are dry season from April to October and rainy season during November to April. Agricultural systems

N. Al Viandari (✉) · H. B. Pulunggono · Suwardi
Department of Soil Science and Land Resources, Faculty of Agriculture, IPB University, Dramaga, Bogor, Indonesia
e-mail: nourmaviandari@apps.ipb.ac.id

N. Al Viandari · A. Wihardjaka
Indonesian Agricultural Environment Research Institute, Pati, Indonesia

E. Yulihastin et al. (eds.), *Proceedings of the International Conference on Radioscience, Equatorial Atmospheric Science and Environment and Humanosphere Science, 2021*, Springer Proceedings in Physics 275, https://doi.org/10.1007/978-981-19-0308-3_77

in Indonesia, especially the rainfed paddy system, highly depend on climate conditions. Crop system in the rainfed area mostly has three cropping patterns annually. The first crop season are a *Gogo Rancah*, a direct-seeded rice crop system cultivated from November to February, the second crop season called a Walik Jerami, a transplanted rice during the dry season cultivated from March to June, and the last crop season are depended on the farmer. They usually plant maize and legume [1, 2]. The productivity of rice in the rainfed fields around 3–4 t ha^{-1}, which is relatively lower when compared to irrigated rice fields, especially at Walik Jerami crop season [3]. Even though have a lower rice yield on the second crop season, the rainfed paddy fields has a great prospect, especially in Pati, Indonesia, since the area of the rainfed rice fields is around 21,025 ha [4].

Several factors caused the low productivity of the rainfed fields, including: (a) a low soil fertility [5], especially micronutrient deficiencies [6], (b) relatively high soil density with a bulk density between 1.35 and 1.75 g cm^{-3} [7], lack of water [8] and reduce soil moisture content [9]. The farmers do minimum rice cultivation processing such as minimum tillage, especially on the Walik Jerami crop season, to accelerating planting time and prevent drought in the planting.

On the other side, minimum tillage has several weaknesses, including: (a) the higher bulk density and it can cause soil compaction, especially in layers more than 20 cm, which becomes an obstacle in the nutrient transformation process [10], (b) weed increase [11], and the production of biomass and tillers from each clump is relatively reduced [12]. Soil tillage management has several impacts on soil properties, especially on physical properties, such as soil permeability, soil infiltration, soil structure, and create the optimum condition for plant growth [13].

Multiple solutions are needed to keep plant growth well. One of the increasing tillage systems by deep tillage. The deep tillage system is an essential strategy in the rainfed rice fields to prevent drought due to the unpredictable rain pattern, especially rainfed areas so dependent on rainwater availability. By applying deep tillage system, it is expected to stimulate the root system as well as the distribution of soil porosity, soil water holding capacity [14]. Deep tillage also can help the plant roots grow and expand the root area to access water and soil nutrients [15].

Several technologies had applied to improving yield in the rainfed paddy field system especially during drought stress and also pest infestation. One of them is applying a variety of IR64. Hu et al. [16] reported that gene of IR64 resistance to Bacterial Blight (BB) diseases. Some studies also reported that IR64 has drought tolerance. Uga et al. [17] reported that any gene who has a deeper rooting and drought tolerance was transferred to IR64. Moreover, this rice has been dominantly planted in various areas, especially rainfed rice fields. IR64 has extensively grown in rainfed condition [18].

Applying deep tillage and water management are expected to improving soil physical properties of rainfed fields so that could improve yield of paddy. This study aimed to determine the effect of deep tillage and water management in improving the productivity of paddy at Walik Jerami, and to determine physical properties changing of the soil after the paddy was harvested.

77.2 Materials and Methods

77.2.1 Experimental Site Condition

This study was run at the Jaken Experimental Station in Pati, Central Java, Indonesia, from February to June 2019 on the first dry season. The classification of soil is Aeric Tropaquepts, with the characteristics of topsoil are silt loam with 26% of sand, 62% of silt, and 12% of clay), pH-H_2O around 5.23, low organic C and total N around 3.31 mg g-1, a low cation exchange capacity 4.08 cmol-kg^{-1}, and exchangeable cations such as Ca^{2+}, Mg^{2+}, K^+, Na^+ are low (3.07; 0.33; 0.10; 0.06 cmol-kg^{-1}). Volumetric water content at field capacity (pF 2.540 is 2.54, and permanent witting point (PWP pF 4.2) is 5.74%, and bulk density at 0–10 cm of depth around 1,45 g cm3, 10–20 cm of depth are 1.60 g cm3.

77.2.2 Experimental Design, Treatment, and Cultivation Details

The experimental design used a split-plot design, repeated four times. The main plots were water treatment and consisted of continuous flooding and the other one is depending on rainfall (as rainfed conditions). The subplot consisted of normal tillage and the deep tillage. The normal tillage treatment was commonly used by the farmer which 0–15 of depth. The other treatment is deep tillage treatment which has a depth around 15–30 cm. For the irrigation treatment, irrigation system started from a week before a rimordial phase and kept until a week before harvesting.

Paddy was planted on a plot which has 4 × 8 m^2 size. Plot preparation was done by a hoe to till and overturn the soil before planting. The ground also leveled a week before planting. 21 Day After Planting (DAP) of IR64 seedlings were transplanted at 20 × 20 cm. All of the plots were fertilized with a dose of 90 kg ha^{-1} N, 22.5 ha^{-1} kg P_2O_5, 60 kg ha^{-1} of K2O, and 20 kg ha^{-1} S. Applied Nitrogen (N) fertilizer three times which the first ¼ N before planting, ½ N was given when active tillers phase around 20 to 25 DAP, and the last ¼ N fertilizer was given a week before primordial. All the phosphorus (P) fertilizers were applied simultaneously with the first application of N fertilizer. Kalium (K) fertilizer was given twice, the first ½ K was given before planting, and the rest of K was given at the active tillers stage. Sulfur (S) fertilizer was given at the same time as the first N and P fertilizer. The pest was controlled by considering field conditions. Combination of post-growing herbicides and manual control was done as preventive action. Paddy was harvested at 90 DAP.

As a soil sample observation, composite soil samples at a depth of 0–15 and 15–30 cm were put at the beginning. Soil parameters were pH, C-organic, total N, CEC, P Bray 2, exchangeable cations (K, Na, Ca, Mg). Agronomical parameters measured included plant height and number of tillers at the primordial and harvested stage, soil water content, bulk density in each tillage depth treatment, both depth on 0–15 and 15–30 cm at maximum tillers and harvesting. Harvesting components were 14% MC of yield, a weight of paddy straw, the number of grains per panicle, filled grain percentage per panicle, and 1000 grain weight.

77.2.3 Statistical Analysis

32 samples and data has been collected. The effect of the treatment on the soil physical properties and yield components were analyzed using the GenStat 64-bit Release 20.1 (PC/Windows 8–10) with the tillage and water management as a factor and plot as repeated measurement to measure the variance on each treatment and Tukey test at <0.05 level for the post hoc test.

77.3 Result and Discussion

Water is one of the most affected things for the availability of soil nutrients. [19] reported that low water availability reduces the rate of nutrient cycling and decomposition process on the soil. The water content of the soil both at rainfed paddy system and irrigation system are presented in Fig. 77.1. Figure 77.1 showed that soil water content decreases until harvest and approaches the permanent wilting point. The x-axis is Day After Planting (DAP) of paddy and for the y-axis is amount of the soil water content. However, the deep tillage, both at the rainfed paddy system treatment and irrigation system treatment gave a higher water content level compared than normal tillage at 57 DAP until 90 DAP (harvesting time). This condition also occurs in the irrigation system. Deep tillage can increase water content when rice is 57 DAP to 90 DAP. The increase in water content is thought to increase soil infiltration to stimulate paddy growth and yield. Busscher and Bauer [20] reported that deep tillage increasing soil infiltration and stimulating water and soil nutrient for the plants. Soil water content also stimulate the moisture to soil aeration status and nutrient availability to plant growth [21].

Water management does not significantly affect the height and number of tillers (Table 77.1). Nevertheless, on the flip side, the tillage system significantly affected the number of tillers, both maximum and productive, but not significantly on plant height, however the normal tillage system treatment had relatively lower growth than the deep tillage system. Further research is needed to observe the water content level on the next crop season. The further research also are needed to study more deeply the effect of deep tillage on the availability of plant growth. Martinez et al. [22] described there was a significant effect between tillage systems and depth in three years of observations. Deep tillage treatment stimulates the plant root grow deeper to reach soil nutrient easily [23].

Table 77.2 showed that water management and tillage systems has insignificantly effect on crop yields and yield components. The average of grain yield reaches

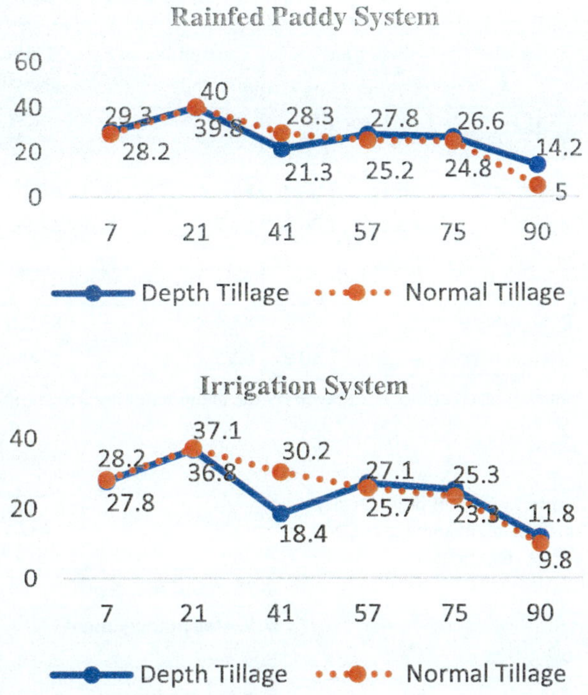

Fig. 77.1 Water content of tillage system treatment and water management treatment at the first dry season

Table 77.1 Plant height and number of rice tillers IR64 on tillage and water management treatments

Treatment	Height		Tillers number	
	35 DAP	Harvesting	Maximum phase	Generative phase
Water management				
Flooding	57.5 a	87.6 a	16.3 a	12.8a
Rainfed condition	56.2 a	86.2 a	15.5 a	12.3a
Soil tillage				
Deep tillage	57.8 a	87.8 a	17.0 a	13.4 a
Normal tillage	55.9 a	85.9 a	14.8 b	11.6 b

values in each column followed by the same letter are not significantly different according to LSD 5%

4 t ha^{-1}. Rainfed and normal tillage treatment have lower grain yields and yield components compared than irrigated systems or flooding and deep tillage.

Soil bulk density is one of the factors that affect root growth. The tillage treatments had a significant effect on the bulk density on a depth of 0–15 cm at 40 DAP or the maximum tillering phase. However, at a depth of 16–30 cm, the normal soil tillage had a significant effect on lowering the bulk density compared to deep tillage (Table 77.3). With a decrease in bulk density, it is expected to reduce soil density which

Table 77.2 Yield and yield components of IR64 on water management and soil tillage treatment

Treatment	Yield t ha^{-1}		The number of grains per panicle	% filled grain per panicle	1000 grain weight (g)
	Grain	Straw			
Water management					
Flooding	4.31a	7.92 a	85.3 a	83.2 a	28.0 a
Rainfed condition	4.15a	7.70 a	81.7 a	82.9 a	27.9 a
Soil tillage					
Deep tillage	4.33a	7.82 a	84.4 a	83.2 a	28.1 a
Normal tillage	4.13a	7.80 a	82.5 a	82.9 a	27.8 a

values in each column followed by the same letter are not significantly different according to LSD 5%

Table 77.3 Soil bulk density in water management and soil tillage treatment

Treatment	Soil bulk density (g ml^{-1})		
	5 DAP	40 DAP	harvesting
0–15 cm of soil depth			
a. Water management			
Flooding	1.49 a	1.52 a	1.49 a
Rainfed condition	1.46 a	1.52 a	1.46 a
b. Soil tillage			
Deep tillage	1.48 a	1.58 a	1.48 a
Normal tillage	1.47 a	1.45 b	1.47 a
15–30 cm of soil depth			
a. Water management			
Flooding	1.60 a	1.72 a	1.65 a
Rainfed condition	1.58 a	1.62 b	1.64 a
b. Soil tillage			
Deep tillage	1.62 a	1.69 a	1.76 a
Normal tillage	1.57 b	1.64 b	1.53 a

values in each column followed by the same letter are not significantly different according to LSD 5%

can affect the ability of plant roots to penetrate the soil. Mouazen and Ramon [24] reported that increasing bulk density will increase soil compactness and decrease soil permeability. Bulk density rises in line with soil depth as the reducing of organic matter and aggregation.

77.4 Conclusion

Tillage system and water management did not significantly affect the grain yield of rice plants IR64, with an average grain yield of 4.10 t ha^{-1}. Tillage system such deep and normal tillage significantly reduces bulk density during the maximum tillering phase. Even it has not yet given significant results, deep tillage and water management have a great potential to improve soil physical properties. Further research in the next crop season is needed to complete the data and obtain more in-depth information about the study.

References

1. Wihardjaka, A., Idris, K., Rachim, A., Partohardjono, S.: Pengelolaan Jerami dan Pupuk Kalium pada Tanaman Padi di Lahan Sawah Tadah Hujan. Pengelolaan Jerami dan Pupuk Kalium pada Tanam. Padi di Lahan Sawah Tadah Hujan Kahat K **4**(1), 26–32 (2002)
2. Boling, A., Tuong, T.P., Jatmiko, S.Y., Burac, M.A.: Yield constraints of rainfed lowland rice in Central Java, Indonesia. F. Crop. Res. **90**(2–3), 351–360 (2004). https://doi.org/10.1016/j.fcr.2004.04.005
3. USDA Homepage. https://ipad.fas.usda.gov/highlights/2016/03/Indonesia/In-dex.htm. Accessed 2021/09/10
4. Pati, B.P.S.: Kecamatan Pucakwangi dalam angka 2019. Pati: Badan Pusat Statistik Pati (2019)
5. Mutert, E., Fairhurst, T.: 12–15 pp. 99–110 (2010)
6. Bustami, B., Sufardi, S., Bakhtiar, B.: Serapan hara dan Efisiensi Pemupukan Phosfat serta pertumbuhan padi varietas local. J. Manaj. Sumberd. Lahan **1**(2), 159–170 (2012)
7. Unger, P.W.: Unger rhob vs P.pdf. Soil Tillage Res. **37**, 67–75 (1996)
8. Farobyfalatehan, A., Raswatie, F.D., Sari, D.A.P.: Planting and consumption patterns of upland rice farmer in Indonesia **1**(1) (2017)
9. Rossato, L., et al.: Impact of soil moisture on crop yields over Brazilian semiarid. Front. Environ. Sci. **5**(Nov), 1–16, (2017). https://doi.org/10.3389/fenvs.2017.00073
10. FAO & ITPS, Intergovernmental Technical Panel on Soils. Status of the World's Soil Resources (2015)
11. Sebayang, H.T., Rifai, A.P.: Journal of degraded and mining lands management Azotobacter population, soil nitrogen and groundnut growth in mercury-contaminated tailing inoculated with Azotobacter. J. Degrad. Min. L. Manag. **5**(53), 2502–2458 (2018). https://doi.org/10.15243/jdmlm
12. Moukoumbi, Y.D., Sie, M., Vodouhe, R., Bonou, W., Toulou, B., Ahanchede, A.: Screening of rice varieties for their weed competitiveness. Afr. J. Agric. Res. **6**(24), 5446–5456 (2011). https://doi.org/10.5897/AJAR11.1162
13. Badalíková, B.: Influence of soil tillage on soil compaction (Jan 2010), 19–30, (2010). https://doi.org/10.1007/978-3-642-03681-1_2
14. Baumhardt, R.L., Jones, O.R., Busscher, W., Frederick, J.: Long-term benefits of deep tillage on soil physical properties and crop yield. In: Proceedings 27th Southern Conservation tillage Systems Conference Florence, South Carolina, USA, 27–29 June, 2005, no. Jan 2005, 95–101 (2005)
15. Barison, J., Uphoff, N.: Rice yield and its relation to root growth and nutrient-use efficiency under SRI and conventional cultivation: An evaluation in Madagascar. Paddy Water Environ. **9**(1), 65–78 (2011). https://doi.org/10.1007/s10333-010-0229-z
16. Hu, K., Cao, J., Zhang, J., Xia, F., Ke, Y., Zhang, H., Xie, W., Liu, H., Cui, Y., Cao, Y.,Sun, X., Xiao, J., Li, X., Zhang, Q., Wang, S.: Improvement of multiple agronomic traits by a disease

resistance gene via cell wall reinforcement. Nat. Plants 3 (2017). https://doi.org/10.1038/nplants.2017.1039

17. Uga, Y., Sugimoto, K., Ogawa, S., Rane, J., Ishitani, M., Hara, N., Kitomi, Y., Inukai, Y., Ono, K., Kanno, N., Inoue, H., Takehisa, H., Motoyama, R., Nagamura, Y., Wu, J., Matsumoto, T., Takai, T., Okuno, K., Yano, M.: Control of root system architecture by DEEPER ROOTING 1 increases rice yield under drought conditions. Nat. Genet. **45**, 1097–1102 (2013)
18. Mackill, D.J., Khush, G.S.: IR64: a high-quality and high-yielding mega variety. Rice **11**(1) 2018. https://doi.org/10.1186/s12284-018-0208-3
19. Guntiñas, M.E., Gil-Sotres, F., Leirós, M.C., Trasar-Cepeda, C.: Sensitivity of soil respiration to moisture and temperature. J. Soil Sci. Plant Nutr. **13**(2), 445–461 (2013). https://doi.org/10.4067/S0718-95162013005000035
20. Busscher, W.J., Bauer, P.J.: Soil strength, cotton root growth and lint yield in a southeastern USA coastal loamy sand. Soil Tillage Res. **74**(2), 151–159 (2003). https://doi.org/10.1016/j.still.2003.06.002
21. Voroney, P.: Soils for horse pasture management. Horse Pasture Manag 65–79 (2019). https://doi.org/10.1016/b978-0-12-812919-7.00004-4
22. Ingrid Martínez, G., et al.: Influence of conservation tillage and soil water content on crop yield in dryland compacted Alfisol of Central Chile. Chil. J. Agric. Res. **71**(4), 615–622 (2011). https://doi.org/10.4067/s0718-58392011000400018
23. Schneider, F., Don, A., Hennings, I., Schmittmann, O., Seidel, S.J.: The effect of deep tillage on crop yield—What do we really know? Soil Tillage Res. **174**, 193–204 (2017)
24. Mouazen, A.M., Ramon, H.: A numerical-statistical hybrid modelling scheme for evaluation of draught requirements of a subsoiler cutting a sandy loam soil, as affected by moisture content, bulk density and depth. Soil Tillage Res. **63**(3–4), 155–165 (2002). https://doi.org/10.1016/S0167-1987(01)00243-4

Chapter 78
Geochemical Weathering Indices of Volcanic Soil After the Eruption from Mount Sinabung in 2020

Retno Leodita Lubis, Juniarti Juniarti, Saftia Laila Rajmi, Aldi Nanda Armer, Novika Yulanda, Fakhrijal Rizki Hidayat, Hazi Zulhakim, Ichsan Faishal Syukri, Frisa Irawan Ginting, and Dian Fiantis

Abstract The weathering process of volcanic material release plant nutrients and then fertilize the soil. Volcanic materials weather to form volcanic soils. The weathering of volcanic soils can be estimated by calculating the weathering indices and using them to evaluate the stage of soil development, nutrient mobility and indicate the intensity of chemical weathering in the soil. In this paper, we analyze the total elemental oxides and the degree of weathering of the soil affected by the eruption of Mount Sinabung for ten years (2010–2020). The samples were taken using the grid sampling method with an interval of 1 km, and 34 samples were collected. In this study, the calculation of the chemical weathering indices includes the Ruxton Ratio, Bases Loss, and Desilication Indices (DI). The results of this study indicate that the average total soil oxide is in decreasing order $SiO_2 > Al_2O_3 > Fe_2O_3 > CaO > K_2O > MgO > P_2O_5 > TiO_2$. The average soil weathering indices calculated from the Ruxton ratio are between 1.90 and 2.61, Base losses are around 3.32–6.99, and desilication indices are 1.45–1.92. The volcanic soils of Mount Sinabung have a moderate level of weathering. The level of soil weathering in the vicinity of Mt. Sinabung from highest to lowest is Southeast > South > East > Northeast.

78.1 Introduction

Mount Sinabung is one of the most active volcanoes in Indonesia. Mount Sinabung is in Karo Regency North Sumatra Province. After dormancy for more than four hundred years, Mount Sinabung first erupted in August 2010 [1] and is still erupting (December 2021). Of the 129 volcanoes (69 active and 60 inactive volcanoes)

R. L. Lubis · J. Juniarti · S. L. Rajmi · A. N. Armer · N. Yulanda · F. R. Hidayat · H. Zulhakim · I. F. Syukri · F. I. Ginting · D. Fiantis (✉)
Department of Soil Science, Faculty of Agriculture, Andalas University, Limau Manis, Padang, Indonesia
e-mail: dianfiantis@agr.unand.ac.id

E. Yulihastin et al. (eds.), *Proceedings of the International Conference on Radioscience, Equatorial Atmospheric Science and Environment and Humanosphere Science, 2021*, Springer Proceedings in Physics 275, https://doi.org/10.1007/978-981-19-0308-3_78

in Indonesia, Mount Sinabung is the volcano with the most prolonged eruption period [2]. Continuous eruptions change the morphology and topography of volcanoes, deposition of volcanic materials [3]. The volcanic activities influence soil geochemical properties in the volcano's vicinity.

The types of rocks and minerals control geochemical properties. Generally, volcanic glass dominates the mineral content in volcanic soils, and they are considered easily weather minerals with high levels of porosity and permeability [4]. In addition, volcanic materials contain silica glass, an amorphous solid of irregularly formed silicate oxide [5]. The parent material, which has high silica-based glass, is at the core of the development of volcanic soils [5]. Silica content in Mount Sinabung ash is relatively high. Anda [2] reported that Sinabung ash deposits have silica content ranging from 57.78 to 60.38%. Rajmi et al. [6] noted that the volcanic ash of Mount Sinabung contains 51.51–67.51% SiO_2, which is classified as basalt to dacite. Rock classification is determined based on silica content [7]. Volcanic ash also contains primary minerals, and they have the potential to enrich plant nutrients into the soil as weathering process prevails.

Assessment of elemental oxides can be analyzed using XRF (X-Ray Fluorescence). The results of XRF analysis are widely used to measure the total concentration of elements in the soil that are important for soil fertility [8]. The primary composition of volcanic ash is SiO_2, Al_2O_3, FeO_3, and minor amount of MgO, CaO, Na_2O, and S which will weather and increase plant nutrients into the soil [9]. Estimating the level of weathering can be done by calculation the weathering indices. The weathering indices can provide information about soil development and the mobility of rock constituent elements during the weathering process. Information about the degree of weathering is very beneficial in the environmental research process [10]. Calculating the weathering indices is one way to characterize the level of weathering based on the value of elemental oxide [11].

This article provides information on the geochemical properties of the soil after being affected by the eruption of Mount Sinabung for ten years. The objective of this study was to assess the weathering degree of the soils after ten years affected by the explosion.

78.2 Materials and Methods

78.2.1 Soil Sampling and Analysis

This research was conducted at Mount Sinabung with an area of 4.517.25 ha. A total of 34 samples were collected in an area within a radius of 3–7 km (Fig. 78.1). Soil samples were taken at a 0–20 cm depth using an auger after cleaning up the volcanic material on the soil surface. Soil samples were taken from the East, Northeast, Southeast, and South sides. Each area has a different thickness of volcanic ash. In 5–7 km,

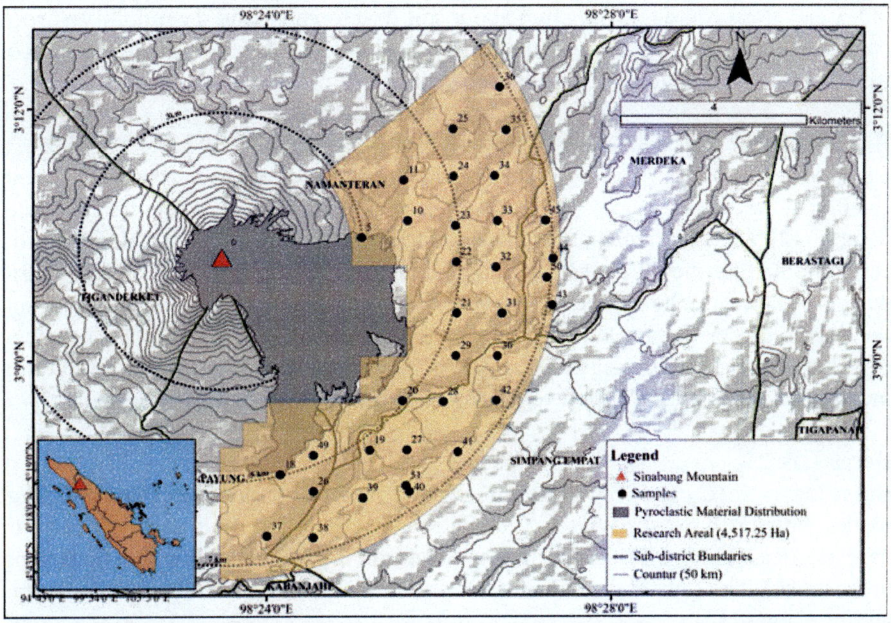

Fig. 78.1 Soil sampling points collected in an area affected by Mount Sinabung eruptions

the samples are generally located in horticultural planting areas and have undergone various agriculture practices.

The type of rock material determines soil geochemical properties. According to Latif et al. [12], rock formation affects soil properties assessed on their mineral and chemical content. Pure volcanic ash with a high SiO_2 content is used to group rock types. A previous analysis reported that the volcanic ash of Mount Sinabung has SiO_2 (51.51–67.51%) considered as mafic to felsic materials [13]. Fiantis et al. [14], reported pure volcanic ash also shows that the SiO_2 content is relatively high, namely 52–61%, classified as basaltic-andesite type.

The concentration of total elemental oxide was analyzed using the X-Ray Fluorescence (XRF) spectrometer P Analytical Epsilon 3 to analyze geochemical properties. The soil sample was first dried and then sieved using a 200 μm sieve. The results of XRF analysis are widely used to measure the total concentration of elements in the soil that are important for soil fertility, including macro and micronutrients [8].

78.2.2 Chemical Weathering Indices

Weathering indices are a way to measure the condition of volcanic ash at the weathering stage and assess the soil fertility level so that the levels of soil nutrient sources can be predicted [10]. In addition, according to Stockmann et al. [8], the weathering index can also provide information about soil development and the mobility of rock constituent elements during the weathering process. In this study, chemical indices were evaluated to characterize weathering degrees:

Ruxton Ratio
Using formula of $R = SiO_2/Al_2O_3$, this method was introduced by Ruxton (1986) who calculated the rate of loss of Silica (SiO_2) with the loss of resistant elemental oxide, namely Al_2O_3 in the weathering process [15]. The reliability of the Ruxton ratio is determined by the uniformity of the parent material, the resistance of aluminum during weathering, the degree of difference between moles of silica to alumina, and the degree of correlation between silica loss and elemental oxide [16]. The range of values in this weathering index is 0–10. The lower the value of the weathering indices, the more optimal the level of weathering and otherwise [10].

Bases Loss
Bases loss $= Al_2O_3/(CaO + MgO + K_2O)$, this method connects resistant elemental oxides with easily washed bases. Calculations using this formula show the level of base loss (Ca, Mg, and K) and the increase in aluminum content during the weathering process [14]. The greater the value of base loss indicates that the Alumina content is getting bigger [15].

Desilication Indices
$DI = SiO_2/(Fe_2O_3 + Al_2O_3 + TiO_2)$, is the molar ratio of movable Si to immobile oxides (Fe, Al and Ti) [8]. Al, Fe, and Ti as immobile elements will increase during the weathering process of silicate minerals and base cations [8, 17]. The lower the DI value, the more optimal the level of weathering [15].

78.2.3 Geochemical Data Analysis

Geochemical statistical analysis of the soil samples in this study used the JMP Pro 14 software. The total elemental oxides of the soil samples used were analyzed by discriminant analysis. The weathering level map was obtained by interpolating method with GIS Program.

78.3 Result and Discussion

78.3.1 Total Elemental Oxide

The elemental oxide composition showed a significant negative correlation between Al_2O_3 and K_2O, SiO_2 and CaO (Table 78.1); when the Al_2O_3 content is higher, the

Table 78.1 Correlation between total elemental oxide

	MgO	Al_2O_3	SiO_2	P_2O_5	K_2O	CaO	Fe_2O_3
MgO	1	−0.237	−0.188	−0.089	0.094	0.05	0.296
Al_2O_3	−0.237	1	−0.781	−0.26	−0.735	−0.791	0.381
SiO_2	−0.188	−0.781	1	0.043	0.56	0.729	−0.8
P_2O_5	−0.089	−0.26	0.043	1	0.289	0.164	−0.027
K_2O	0.094	−0.735	0.56	0.289	1	0.784	−0.246
CaO	0.05	−0.791	0.729	0.164	0.784	1	−0.469
Fe_2O_3	0.296	0.381	−0.8	−0.027	−0.246	−0.469	1

K_2O, SiO_2, and CaO content tend to be lower. An increase in Al_2O_3 concentration causes a decrease in SiO_2 concentration [18]. The process of losing mobile elements (Ca, Mg and Na) and increasing the concentration of immobile elements (Si, Al, and Fe) is believed to bc thc initial stage of the geochemical weathering process of the soil [19]. Furthermore, there is a significant positive correlation between CaO and K_2O.

Linear discriminant analysis was used to compare the geochemical elements of the soil affected by the eruption in the East, Northeast, Southeast, and South sectors. Seven chemical composition data were analyzed using discriminant analysis: MgO, Al_2O_3, SiO_2, P_2O_5, K_2O, CaO, and Fe_2O_3. Discriminant analysis differentiates South soil from other soils (Fig. 78.2). At the same time, East soils exhibit similar characteristics to the Northeast soils but differ from the Southeast and South soils (Fig. 78.2). The Northeast soils have a higher K_2O content than the South soils.

Canonical correlation analysis is a statistical tool to quantify and identify the relationship between two types of variable groups. The canonical correlation test was carried out between elemental oxide and the position of soils in this study. The results of canonical analysis varied with the data, respectively: canonical 1: 51.348%, II: 35.878%, and III: 12.775%, with the eigenvalues, are 1.9, 1.3, and 0.5, respectively. The higher the eigen value, the more variation in the dependent variable which the discriminant function could explain. Hannon et al. [20], reported linear discriminant analysis can categorize the geological properties of several variables with 95% accuracy. Significant canonical correlations $P \ll 0.0001^*$, 0.0007* and 0.0588 indicate that there are significant differences between sectors (Table 78.2). According to Fiantis et al. [15], the canonical correlation of $P < 0.0001$ and 0.083 between the chemical characteristics of the soil has a significant difference.

78.3.2 Weathering Indices

Based on the calculation of weathering levels using the Ruxton ® method soils in the Northeast and East with Southeast sectors are different (Fig. 78.3). The average Ruxton index from highest to lowest is Northeast, East, South, and Southeast with

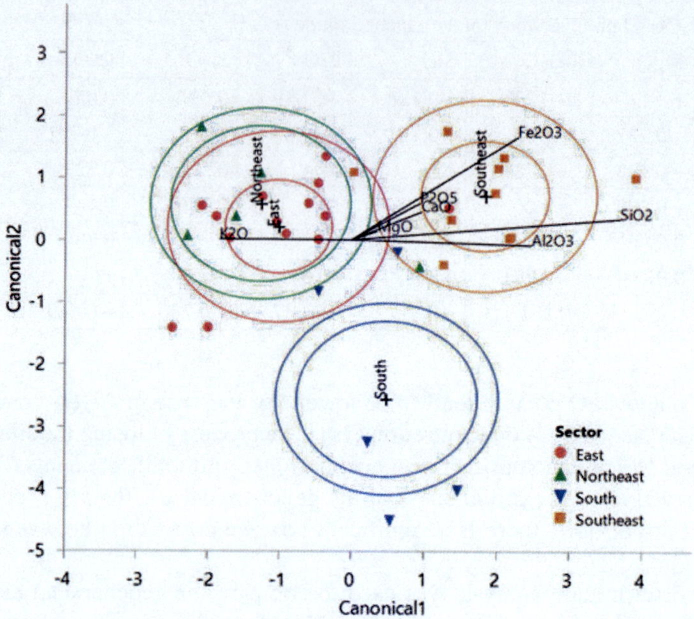

Fig. 78.2 Differences in total elemental oxides in various sectors

Table 78.2 Cannonical detail

	Cannonical variate		
	I	II	III
Eigenvalue	1.9	1.3	0.5
Percent	51.348	35.878	12.775
Cum percent	51.348	87.226	100.000
Canonical corr	0.810	0.756	0.567
Likelihood ratio	0.100	0.291	0.678
Approx. F	4.069	3.562	2.468
NumDF	21	12	5
DenDF	69.465	50.000	26.000
Prob > F	<0.0001*	0.0007*	0.0588

values of 2611, 2366, 2.166, and 1.902%, respectively. The highest level of weathering was found in the Southeast soils, and the lowest was in the Northeast soils. The level of weathering in the research area is still classified as moderate. Following [10], the weathering level of soils in the research area is average. Weathering indices value close to 0 means it has a weathering level close to optimal [11]. Figure 78.4 shows the weathering index's distribution map based on the Ruxton method. The distribution map shows that the soil on the Southeast side has a higher Silica loss level than other sides.

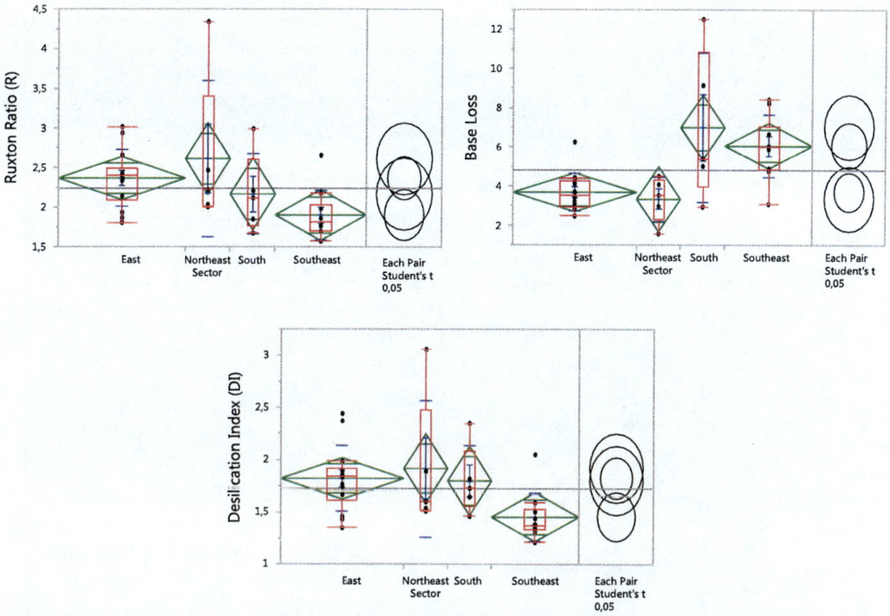

Fig. 78.3 Oneway analysis of weathering indices by volcanic soil

The volcanic ash on Mount Sinabung is classified as mafic (basalt) to felsic (dacite) material with SiO_2 content (51.51–67.51%) with a total elemental composition of Al_2O_3 (15.54–23.41%), Fe_2O_3 (2.84–10.02%), and CaO (2.94–6.46%) [6]. Ilham et al. [21] reported that the volcanic ash of Mount Sinabung had an oxide content of SiO_2 49.33%; Al_2O_3 15.93%; Fe_2O_3 6.48%; CaO 5.87%. Along with further weathering, the silica content will decrease rapidly because it has more mobile properties than Al, Fe, and Ti. Anda [2] reported that the volcanic ash of Mount Sinabung collected from different years (2010–2014) had relatively the same SiO_2 content ranging from 57 to 60%, indicating that the volcanic ash of Mount Sinabung is relatively resistant to weathering.

During the chemical weathering process, there is a loss or leach of some alkaline minerals from the parent material, which causes a decrease in basic elements in the soil. This study found that the level of loss in the South to Southeast soil was considered equal, but they are different from soils in the East and Northeast. Optimal weathering levels are found in the South and Southeast sectors (6991 and 6051); this explains that the base loss in the East and Northeast sectors is lower. Furthermore, soils with high levels of bases loss have dominantly prevailed in the Southeast to the South sides (Fig. 78.5). According to report Lubis et al. [13], the East and Northeast sides have higher exchangeable Ca and Mg than the South and Southeast sectors. Non-alkaline elements have the property of being easy to move; as such, the soil leads to advance weathering with low non-alkaline elements content. Ca, Mg, and K will decrease in line with the progress of weathering [22, 23].

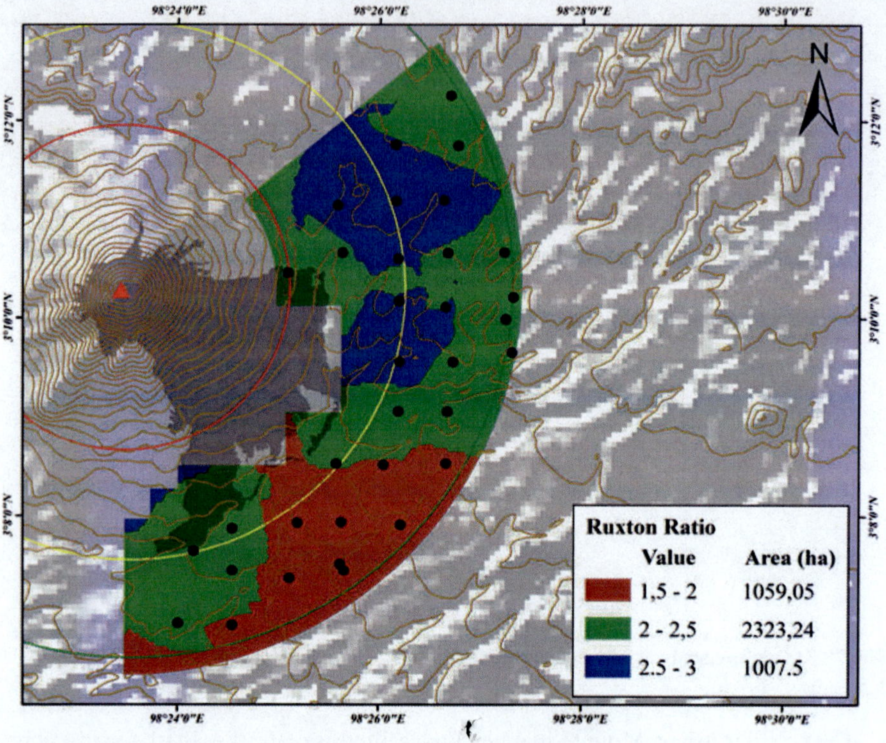

Fig. 78.4 Weathering indices distribution map based on Ruxton ratio method

The Desilication Index Ratio (DI) is used to calculate the weathering index by comparing mobile silica with resistant oxides, namely Al, Fe, and Ti. The DI ratio value in the Northeast and East soils is significantly different from the Southeast, with the lowest average in the Southeast soils (Fig. 78.3). The weathering of silica minerals is lower in the Northeast and East sides than in the Southeast soils. The DI ratios with the highest to the lowest scores were Northeast > South East > Southeast with 1.916, 1.821, 1.789, and 1.452%, respectively. Furthermore, soils with high levels of silica loss are dominant in the Southeast to a South sector (Fig. 78.6).

This study found that soils with the lowest total SiO_2 oxide are on the Southeast side, which means that Silica loss in the Southeast was higher than in other sides. According to Stockmann et al. [8], the higher the Silica ratio, the lower the weathering rate so that the presence of immobile elements (Al, Fe, and Ti) are low.

This study used well-distributed soil data to study the volcanic material deposit from Mount Sinabung eruptions started in 2010. The data enable the mapping of volcanic material weathering indices in an area from the Northeast to the South of the volcano and about 8 km. The parameter evaluated includes the Ruxton ratio, bases loss, and desilication indices, using the data's geochemical content. The study concluded that the Mount Sinabung volcanic material has a moderate level of weathering.

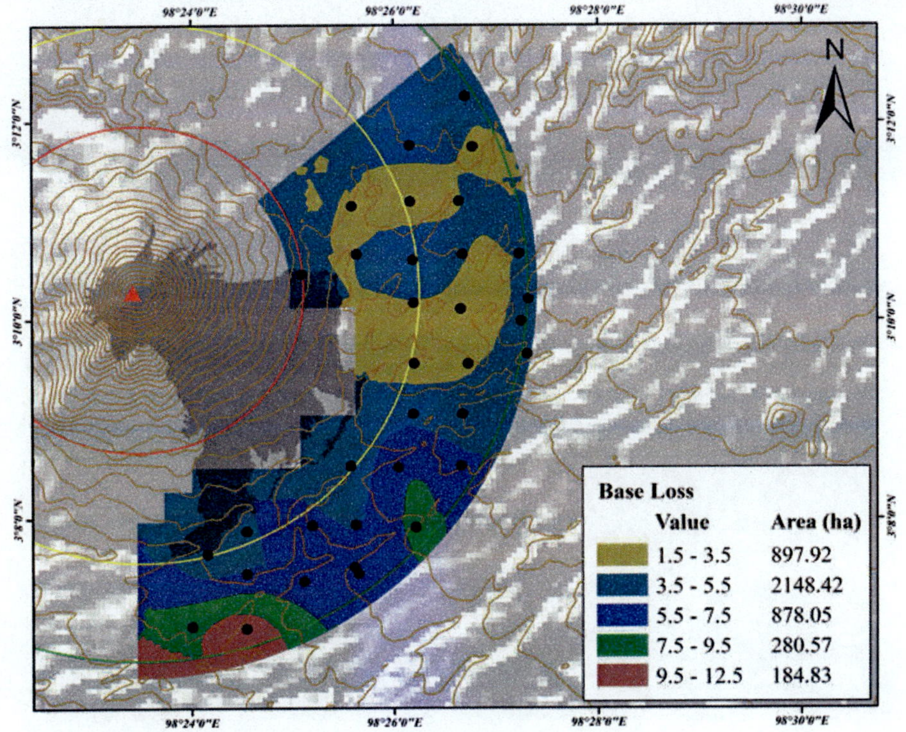

Fig. 78.5 Map of the distribution of the level of base loss in the soil

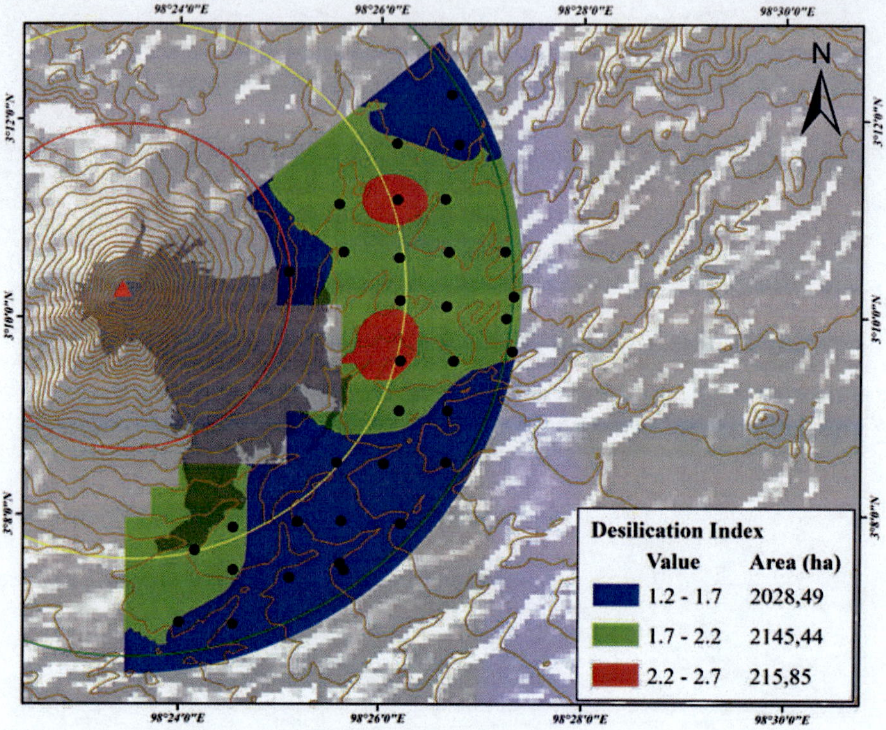

Fig. 78.6 Weathering indices distribution map based on desilication indices method

78.4 Conclusion

The weathering index of the soil affected by the inter-sectoral eruption has almost the same value. The weathering level is moderate in the three weathering indices calculations used (Ruxton, base loss, and desilication indices). The level of weathering from highest to lowest is Southeast > South > East > Northeast. The potential plant nutrient content in the soil is Ca > Mg > K > S > P. The paper is original and gives significance for further understating the volcanic deposit weathering process around Mount Sinabung. However, further research is needed to provide more detailed data.

Acknowledgements This work was financed by the Directorate for Research and Community Service, Deputy for Field Affairs Strengthening Research and Development, Ministry of Research and Technology/National Research and Innovation Agency based on the SKIM Research Implementation Contract Master Thesis No.34/SP2H/LT/DRPM/2020 to Dian Fiantis.

References

1. Sutawidjaja, I.S., Prambada, O., Siregar, D.A.: The August 2010 phreatic eruption of Mount Sinabung. North Sumatra. Indones. J. Geosci. **8**, 55–61 (2013)
2. Anda, M.: Characteristics of pristine volcanic materials: beneficial and harmful effects and their management for restoration of agro ecosystem. Sci. Total Environ. **543**, 480–492 (2016)
3. Ginting, F.I., Nelson, M., Minasny, B., Fiantis, D.: Changes in Anak Krakatau landscape after December 2018 eruption. In: IOP Conference Series: Earth and Environmental Science, p. 12088. IOP Publishing (2021)
4. Nanzyo, M., Dahlgren, R., Shoji, S.: Chemical characteristics of volcanic ash soils. In: Developments in Soil Science, pp. 145–187. Elsevier (1993)
5. Delmelle, P., Opfergelt, S., Cornelis, J.-T., Ping, C.-L.: Volcanic soils. In: The Encyclopedia of Volcanoes, pp. 1253–1264. Elsevier (2015)
6. Rajmi, S.L., Gusnidar, G., Lubis, R.L., Ginting, F.I., Hidayat, F.R., Zulhakim, H., Armer, A.N., Yulanda, N., Syukri, I.F., Fiantis, D.: Improving volcanic soil chemistry after the eruption of Mt. Sinabung, North Sumatera in 2020. In: IOP Conference Series: Earth and Environmental Science. p. 12042. IOP Publishing (2021)
7. Shoji, S., Kodayashi, S., Yamada, I., Masui, J.: Chemical and mineralogical studies on volcanic ashes I. Chemical composition of volcanic ashes and their classification. Soil Sci. Plant Nutr. **21**, 311–318 (1975)
8. Stockmann, U., Cattle, S.R., Minasny, B., McBratney, A.B.: Utilizing portable X-ray fluorescence spectrometry for in-field investigation of pedogenesis. CATENA **139**, 220–231 (2016)
9. Di Figlia, M.G., Bellanca, A., Neri, R., Stefansson, A.: Chemical weathering of volcanic rocks at the island of Pantelleria, Italy: Information from soil profile and soil solution investigations. Chem. Geol. **246**, 1–18 (2007)
10. Fiantis, D., Nelson, M., Shamshuddin, J., Goh, T.B., Van Ranst, E.: Determination of the geochemical weathering indices and trace elements content of new volcanic ash deposits from Mt. Talang (West Sumatra) Indonesia. Eurasian Soil Sci. 43, 1477–1485 (2010)
11. Price, J.R., Velbel, M.A.: Chemical weathering indices applied to weathering profiles developed on heterogeneous felsic metamorphic parent rocks. Chem. Geol. **202**, 397–416 (2003)
12. Latif, D.O., Rifa'i, A., Suryolelono, K.B.: Chemical characteristics of volcanic ash in Indonesia for soil stabilization: morphology and mineral content. Int. J. Geomate. 11, 2606–2610 (2016)

13. Lubis, R.L., Rajmi, S.L., Armer, A.N., Hidayat, F.R., Zulhakim, H., Yulanda, N., Syukri, I.F., Fiantis, D.: Chemical Properties of Volcanic Soil After 10 Years of the Eruption of Mt. Sinabung (North Sumatera, Indonesia). In: IOP Conference Series: Earth and Environmental Science. p. 12043. IOP Publishing (2021)
14. Fiantis, D., Ginting, F.I., Nelson, M., Van Ranst, E., Minasny, B.: Geochemical characterization and evolution of soils from Krakatau Islands. Eurasian Soil Sci. 1–15 (2021)
15. Fiantis, D., Malone, B., Pallasser, R., Van Ranst, E., Minasny, B.: Geochemical fingerprinting of volcanic soils used for wetland rice in West Sumatra. Indonesia. Geoderma Reg. **10**, 48–63 (2017)
16. Ruxton, B.P.: Measures of the degree of chemical weathering of rocks. J. Geol. **76**, 518–527 (1968)
17. Nesbitt, H.W., Markovics, G.: Chemical processes affecting alkalis and alkaline earths during continental weathering. Geochim. Cosmochim. Acta. **44**, 1659–1666 (1980)
18. Jayawardena, U. de S., Izawa, E.: A new chemical index of weathering for metamorphic silicate rocks in tropical regions: a study from Sri Lanka. Eng. Geol. **36**, 303–310 (1994)
19. Chorover, J., Amistadi, M.K., Chadwick, O.A.: Surface charge evolution of mineral-organic complexes during pedogenesis in Hawaiian basalt. Geochim. Cosmochim. Acta. **68**, 4859–4876 (2004)
20. Hannon, J.S., Huff, W.D., Sturmer, D.M.: Geochemical relationships in Cretaceous bentonites as inferred from linear discriminant analysis. Sediment. Geol. **390**, 1–14 (2019)
21. Ilham, D.J., Kautsar, F.R., Januarti, J., Anggarini, U., Fiantis, D.: The potential use of volcanic deposits for geopolymer materials. In: IOP Conference Series: Earth and Environmental Science. p. 12035. IOP Publishing (2020)
22. Shoji, S., Nanzyo, M., Dahlgren, R.A.: Volcanic ash soils: genesis, properties and utilization. Elsevier (1994)
23. Fedo, C.M., Wayne Nesbitt, H., Young, G.M.: Unraveling the effects of potassium metasomatism in sedimentary rocks and paleosols, with implications for paleoweathering conditions and provenance. Geol. **23**, 921–924 (1995)

Chapter 79
2D and 3D Subsurface Geological Model of Baribis Fault Zone Using the Gravity Method

Muhammad Hanif and Lina Handayani

Abstract The Baribis Fault is one of the active faults in West Java province. Research has been conducted on the identification of the fault using the gravity method. The survey aimed to illuminate geological structures such as fault, basin, horst. In this study, hundreds point around the Bandung to Pamanukan region covered the probable Baribis active fault line in the Subang-Purwakarta segment. Subsurface modeling is carried out to find out the buried faults geometry in the region. The horizontal gradient attributes, 2D, and 3D modeling were subjected to map the subsurface structures based on variation of rock density values. The resulting 2D and 3D model of Baribis active fault zone evaluated by several a priori published data revealed subsurface rock density distribution and probable area prone to surface rupture. This study suggested the inferred buried Baribis Fault lineament in the segment which might be useful for future infrastructure development and hazard mitigation.

79.1 Introduction

The Baribis Thrust Fault Zone (BFZ) is one of active fault in West Java province, Indonesia. It is based upon the seismic activities [17], comprehensive geological-geophysical study [3], historical earthquake simulations [13], geodetic observations [1, 7, 10], and integrated active fault study [12]. The studied region located in the central part of West Java and extend from Bandung to Subang (107.43283, −6.88465 to 108.03434, −6.1938) (Fig. 79.1). The fault is located in the south of Pamanukan High, northward border of Bogor basin, and the Quarternary volcanic area [11]. Based on remote sensing and geological mapping studies, the geomorphology of the Baribis fault zone (BFZ) in the Subang to Purwakarta region was diminished which probably due to intensive erosion and tectonic activities [4].

M. Hanif (✉) · L. Handayani
Research Center for Geological Disaster, Research Organization for Earth and Marine Sciences, National Research and Innovation Agency (BRIN), Bandung 40153, Indonesia
e-mail: muha170@brin.go.id

E. Yulihastin et al. (eds.), *Proceedings of the International Conference on Radioscience, Equatorial Atmospheric Science and Environment and Humanosphere Science, 2021*, Springer Proceedings in Physics 275, https://doi.org/10.1007/978-981-19-0308-3_79

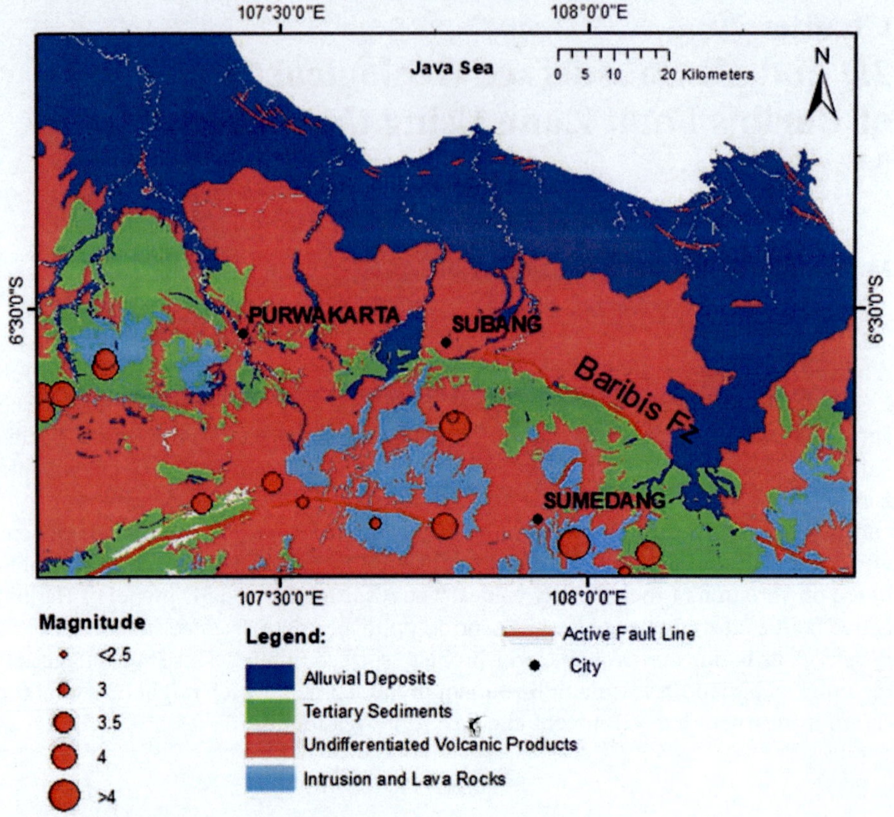

Fig. 79.1 Simplified surface geological map on West Java area [11], active fault line [12] and seismicity around the Baribis fault zone [17]

The complete Bouguer anomaly (CBA) of gravity analysis can illuminate the subsurface density from shallow geological structures to the deep earth crust. It depends on the objective and survey scale. Particularly in the active fault study, some researchers had delineate some geological structures target [2, 9, 14]. It is one of versatile method to delineate groundwater formation, basement depth, and some cavity exploration. Investigating the active fault structure is critical for infrastructures building and safety. However, spatial and subsurface geometry of the Baribis fault in West Java has not been extensively studied. Two published 2D model in the region has limitations on the unclear processing step (i.e., anomaly decomposition method) and resulting in different interpretation of Bogor Basin basement depth of ~4 km [15] and ~10 km [16]. Thus, the processing steps, 2D, and 3D subsurface model from gravity anomaly needs to be clearly analyzed to better understand the basin depth and fault geometry.

79.2 Method

The primary gravity reduction techniques such as drift, tide, free-air, terrain and Bouguer correction were preprocessed to obtain the complete Bouguer anomaly (CBA). The CBA separation method utilized the upward continuation filter to generate the regional deeper geological structure and the residual shallower feature map. In this process, upward continuation elevation was 5000 m to obtain the regional anomaly maps (Fig. 79.2). Additionally, residual anomalies are obtained from the reduction between complete Bouguer anomalies with its regional map.

Line section and rectangle location on residual maps (Fig. 79.2) were chosen crossing the Baribis fault zone to get better understanding of the fault on 2D and 3D section. Both modeling was built using GM-SYS, Oasis Montaj software module. To investigate fault zone spatial distribution, filtering was done by using first horizontal and total horizontal derivative (THD) techniques. The maximum value of THD indicate the steepest gradient of Bouguer anomaly which is a sign of source boundaries. On the other hand, the minimum one indicates the flat zone.

$$THD = \sqrt{\left(\frac{dP}{dx}\right)^2 + \left(\frac{dP}{dy}\right)^2} \tag{79.1}$$

where the total horizontal derivative is a function of dP/dx or the first horizontal derivative in x direction and dP/dy on the y directions [5].

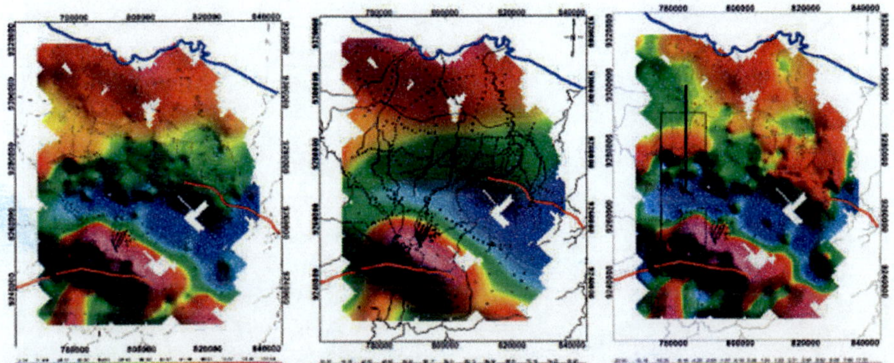

Fig. 79.2 Complete Bouguer anomaly map (left), regional anomaly (middle), and residual anomaly (right) with the black line and box indicating the 2D and 3D model subset. Coordinates in WGS84 UTM Zone 48S

79.3 Results and Discussion

According to the regional geological structures, the regional anomaly from the upward continuation 5000 m was clearly matched the Pamanukan high, Bogor basin, and Tangkuban Parahu complex (Fig. 79.2, middle). The resulting residual anomaly (Fig. 79.2, right) was used to generate the 2D and 3D subsurface geological model, particularly on black line and box section. The results of 2D and 3D profile are depicted on Figs. 79.4 and 79.5. The radially averaged spectral method reveals the average regional depth was up to 9 km [8] and slightly different with previous result [16]. Therefore, the 2D forward model of Bogor basin and the surrounding features was interpreted for up to 9 km depth.

The first and total horizontal derivatives techniques (Fig. 79.3) revealed an inferred Baribis fault zone interpretation. The first horizontal derivative attribute resulting in more clear delineation of inferred fault than the total horizontal derivative. It indicates that the fault is buried in quite a deeper depth. This attribute also has a good structure delineation as tilt derivative and analytic signal from previous study [8]. Although the surface manifestation of the fault in this segment has been disappeared due to intensive geological factors (e.g., tectonic activities, volcanic deposit, and erosion), the resulting inferred buried fault (Fig. 79.3) might be an areas probable to surface rupture.

The 2D forward modeling resulting in 5 layers. The first layer is indicated as an alluvial and tuff deposit with a density of 2.12 g/cm^3. In the second layer is indicated as tuffaceous sand with a density of 2.52 g/cm^3. The third layer indicated

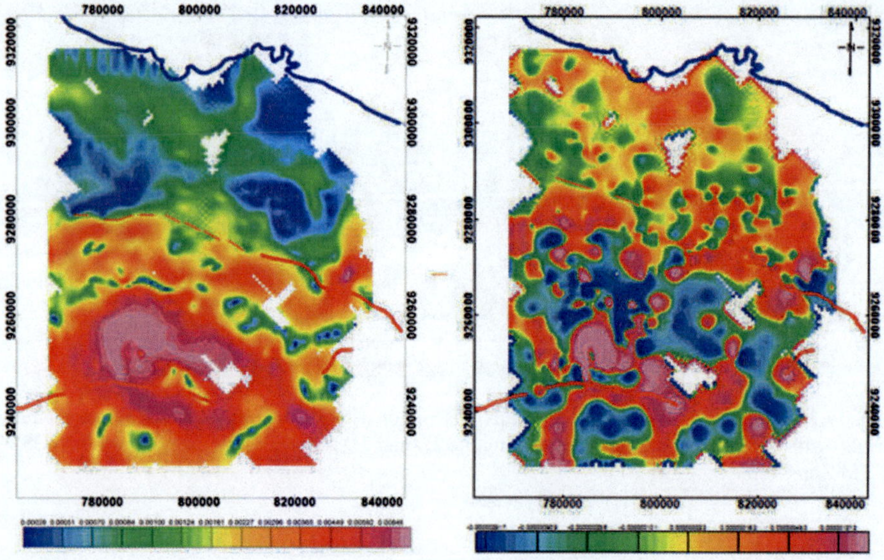

Fig. 79.3 First horizontal derivative (left) and total horizontal derivative (right) attribute. Red dashed lines depict the inferred Baribis continuation. Coordinates in WGS84 UTM Zone 48S

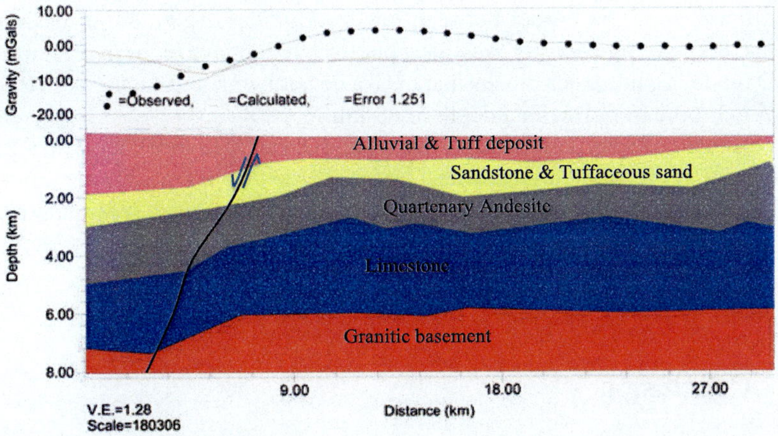

Fig. 79.4 2D forward model showing the alluvial and tuff deposit (pink), sandstone and tuffaceous sand (light brown), Quartenary andesite (grey), limestone (blue), and basement (red). The interpreted subsurface Baribis fault delineation (black line) depicts a subtle thrusting structure

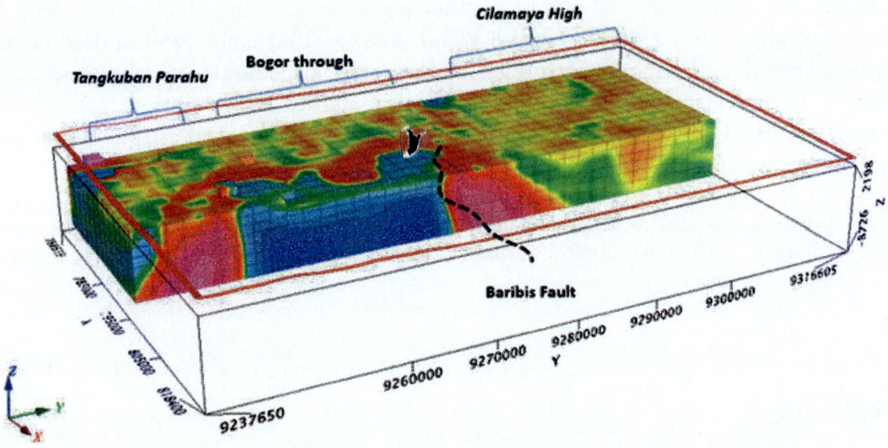

Fig. 79.5 3D model result. The inferred Baribis fault zone depicted by black dashed line

as quartenary andesite with a density of 2.66 g/cm^3. In the fourth layer, interpreted as limestone with a density of 2.8 g/cm^3. The fifth layer indicated as basal basement with a density of 2.99 g/cm^3. In the 2D modeling also indicated the existence of fault structure although there is no clear rock discontinuity on the layers (Fig. 79.4). This result is in agreement with the spatial fault continuity interpretation using horizontal gradient attribute (Fig. 79.3). The fault is assumed to be the lineament of Baribis fault zone (Fig. 79.4). Additionally, the 3D model has clearly defined some regional geological features such as Cilamaya high, Bogor through, and Tangkuban Parahu volcanic complex (Fig. 79.5). The interpreted Baribis fault zone depicted with black dashed line lies on the border between low average density of Bogor basin in the south and high average density West Java platform (Cilamaya high) in the northward. The 3D modeling produces blocks that have a density with a value between 2.00 and 3.00 g/cm3 (blue to red) with a depth of ~9 km.

According to the horizontal derivative map, 2D, and 3D model, the active Baribis fault is exist in the studied area. Thus, it is important to pay more attention on the building or infrastructure planning. However, more studies need to be undertaken to better understand the maximum probable magnitude on this segment for example by using fault length scaling relationship [6].

79.4 Conclusion

The horizontal derivative method revealed the probable area of buried Baribis fault in the Subang-Purwakarta segment. Additionally, the 2D forward model confirmed the fault vicinity and the 9 km depth of basement. Based on the density modeling, the layers consist of alluvial and tuff deposit with a density of 2.12 g/cm^3 tuffaceous sand with a density of 2.52 g/cm^3, Quartenary andesite with a density of 2.66 g / cm^3, limestone with a density of 2.8 g/cm^3 and basal basement with a density of 2.99 g/cm^3. The 3D model also revealed the regional geological background and Baribis fault lineament in the subsurface. More studies need to be undertaken to investigate the maximum probable earthquake magnitude as well as the mitigation and building planning strategy.

Acknowledgements The research is funded by DIPA Research Origanization for Earth Sciences and Maritime SP DIPA-124.01.1.690501/2022 (17 November 2021), BRIN. All authors are equally contributed to this manuscript.

References

1. Abidin, H.Z., Andreas, H., Kao, T., Ito, T., Meilano, I., Kimata, F., Natawidjaya, D.H., Harjono, H.: Crustal deformation studies in java (Indonesia) using GPS. J. Earthq. Tsunami **3**, 77–88 (2009). https://doi.org/10.1142/S1793431109000445
2. Aboud, E., El-Masry, N., Ahmed, S.: MicroGravity mapping for King Abdulaziz University campus: imaging subsurface. Arab. J. Geosci. 11 (2018). https://doi.org/10.1007/s12517-018-4029-x
3. Aribowo, S., Husson, L., Natawidjaja, D.H., Authemayou, C., Daryono, M.R., Puji, A.R., Valla, P.G., Pamumpuni, A., Wardhana, D.D., de Gelder, G.: Active back-arc thrust in North West Java, in: EGU General Assembly 2021. pp. EGU21–10469 (2021). doi.org/https://doi.org/10.5194/egusphere-egu21-10469
4. Armandita, C., Raharjo, B., Satyana, A.H., Nugraha, W., Graha, S., Racmat, S.: Perkiraan inversi Sesar Baribis serta perannya terhadap proses sedimentasi dan kemungkinan adanya "reworked source" pada endapan turbidit lowstand setara Talang Akar (Studi pendahuluan di daerah Sumedang dan sekitarnya). Bul. Geol. Inst. Teknol. Bandung 34 (2002)

5. Blakely, R.J.: Potential theory in gravity and magnetic applications. Cambridge University Press (1995)
6. Gunawan, E.: An assessment of earthquake scaling relationships for crustal earthquakes in Indonesia. Seismol. Res. Lett. (2021). https://doi.org/10.1785/0220200267
7. Gunawan, E., Widiyantoro, S.: Active tectonic deformation in Java, Indonesia inferred from a GPS-derived strain rate. J. Geodyn. **123**, 49–54 (2019). https://doi.org/10.1016/j.jog.2019.01.004
8. Hanif, M., Handayani, L., Gaol, K.L., Arisbaya, I., Aulia, A.N., Wardhana, D.D.: Tracing Baribis active fault continuity in the Subang region using gravity edge detection analysis. In: International Conference on the Ocean and Earth Sciences. IOP Conf. Series: Earth and Environ. Sci. **789**, 1–6. doi.org/https://doi.org/10.1088/1755-1315/789/1/012062
9. Kebede, S., Travi, Y., Asrat, A.: Groundwater origin and fl ow along selected transects in Ethiopian rift volcanic aquifers 55–73 (2008). doi.org/https://doi.org/10.1007/s10040-007-0210-0
10. Koulali, A., Mcclusky, S., Susilo, S., Leonard, Y., Cummins, P., Tregoning, P., Meilano, I., Efendi, J., Wijanarto, A.B.: The kinematics of crustal deformation in Java from GPS observations : implications for fault slip partitioning. Earth Planet. Sci. Lett. **1**, 1–11 (2016). https://doi.org/10.1016/j.epsl.2016.10.039
11. Martodjojo, S.: Evolusi Cekungan Bogor Jawa Barat. Penerbit ITB Bandung (2003)
12. Nasional, P.S.G.: Peta Sumber dan Bahaya Gempa Indonesia Tahun 2017. Pusat Litbang Perumahan dan Pemukiman, Kementerian Pekerjaan Umum dan Perumahan Rakyat (National Center for Earthquake Studies. Indonesian Seismic Sources and Seismic Hazard Maps 2017. Center for Research. ISBN 978–602–5489–01–3
13. Nguyen, N., Griffin, J., Cipta, A., Cummins, P.R.: Indonesia's historical earthquakes: modelled examples for improving the national hazard map, Record 2015/23 (2015)
14. Nugraha, G.U., Handayani, L., Lubis, R.F., Wardhana, D.D., Gaol, K.L.: Basement characteristics of Jakarta groundwater basin based on satellite gravimetry data. Indones. J. Geogr. **52**, 42–52 (2020). https://doi.org/10.22146/ijg.46672
15. Patmosukismo, S., Yahya, I.: The basement configuration of the North West Java area, In: Proceedings Indonesian Petroleum Association. pp. 129–152 (1974). doi.org/https://doi.org/10.29118/ipa.2164.129.152
16. Smyth, H.R., Hall, R., Nichols, G.J.: Cenozoic volcanic arc history of East Java, Indonesia: the stratigraphic record of eruptions on an active continental margin. Spec. Pap. Geol. Soc. Am. **436**, 199–222 (2008). https://doi.org/10.1130/2008.2436(10)
17. Supendi, P., Nugraha, A.D., Puspito, N.T., Widiyantoro, S., Daryono, D.: Identification of active faults in West Java, Indonesia, based on earthquake hypocenter determination, relocation, and focal mechanism analysis. Geosci. Lett. **5**, 31 (2018). https://doi.org/10.1186/s40562-018-0130-y

Chapter 80
Risk Management of Spaceport in Indonesia: Vulnerability Level of Biak Spaceport

Intan Perwitasari, Stevani Anggina, and Arif Nur Hakim

Abstract Indonesia plans to build and operate a spaceport as mentioned in Indonesia Space Master Plan 2017, and Biak, Papua, is one of the candidates of Spaceport location. Safety factors become the standard that must be met in the construction and operation of the spaceport. To get a spaceport construction license or an operator license, the applicant must inform the vulnerability of the location. Vulnerability of spaceport as a part of risk management information is used to safety measures. The operation of the spaceport will depend on several factors including vehicle technology, location (distance to settlement, density), and others. The purposes of this study are to analyze the level of vulnerability and the need for safety standards in the construction of a spaceport in Biak. Data used are primary and secondary data from previous studies and questionnaires. The methods used are quantitative and qualitative by using spatial data, overlay, and the AHP method. The conclusion of this study is North Biak District has low vulnerability as a location of Spaceport. Strategies to risk management are required in the construction and operation of the spaceport to reduce vulnerability. Several strategies such as increasing adaptive capacity with safety training, building some infrastructure (health facilities, transportation infrastructure), and partnership between Biak Numfor Regency, LAPAN and other stakeholder, are needed as a part of the spaceport management system.

I. Perwitasari (✉) · S. Anggina
National Research and Innovation Agency (BRIN), Jakarta 10340, Indonesia
e-mail: intanperwita@gmail.com; intan004@brin.go.id

S. Anggina
e-mail: stevani.anggina@yahoo.com

A. N. Hakim
Rocket Technology Center, National Research and Innovation Agency (BRIN), Bogor, Indonesia
e-mail: arif.nur.hakim@brin.go.id

E. Yulihastin et al. (eds.), *Proceedings of the International Conference on Radioscience, Equatorial Atmospheric Science and Environment and Humanosphere Science, 2021*, Springer Proceedings in Physics 275, https://doi.org/10.1007/978-981-19-0308-3_80

80.1 Introduction

80.1.1 Background

Regarding Indonesia space law, Indonnesia has a national spaceport program in medium and long-term. Previous studies mention that Biak has the potential to spaceport location in Indonesia [1]. This choice is based on result of several studies that chose Biak as one of the location candidates of the spaceport. The selection is base multi criteria, technical and non-technical factors (safety, spatial, infrastructure, demographic, meteorology) [2, 3]. In the licensing of the spaceport, it will have to consider the safety aspects. One of the regulations of spaceport related to safety is zonation, which is divided into 3 zones, namely zone 1, zone 2, and zone 3 and under the regulations as a prohibited area. In this area, everyone is prohibited from constructing buildings or perform other activities that may cause launch failure [4]. The determination of this zoning is determined by technical factors and safety factor in the operation of the spaceport in Biak in the future. In practice, each spaceport has a different distance to the settlement or zoning because they have different level technology launcher and have different fascilities in the location. The importance of the zonation is related to the need for range control license in spaceport location such as in the UK [5]. Figure 80.1, shows the zoning arrangement by considering the existence of facilities such as fuel storage and the location of the launch pad.

The Safety Management System (SMS) according to ENAC (2011) is a fundamental part of the spaceport management system [7]. Risk Management is a part

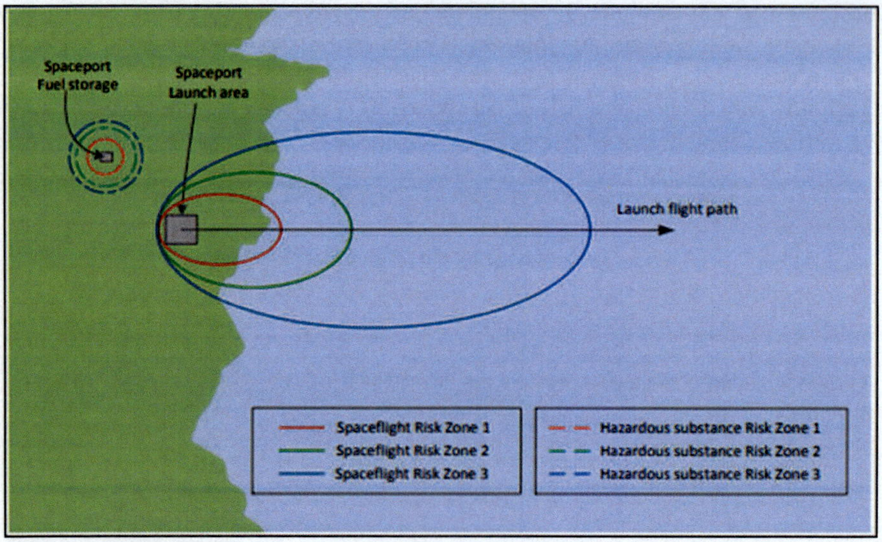

Fig. 80.1 Spaceport Risk Zones Concept [6]

of SMS dan Vulnerability is a risk-forming parameter. Various types of disasters according to Griffith, et. al (2018) in spaceport mostly attributed to rocket storage and launches, which are strongly related to the propellant influence, so that the charging, launch and re-entry processes are crucial stages in operation [8]. This underlies the need for zones and control within it.

The National Disaster Management Agency (BNPB) in national risk management has mapped the types of vulnerabilities including high technology [9]. Spaceport is a high-technology area and the measurement of vulnerabilities a round the location area is important. International practice, the applicant in the UK must provide information regarding this vulnerability as part of potential disasters and risks in the spaceport and demonstrate its management efforts. The environment around a spaceport (nature and build environment, population) should be drawn in relation to its vulnerability, and how many [5]. This Vulnerability is important to know as one of the factors that affect disaster risk. The new disasters will occur if the danger occurs in vulnerable conditions [10]. The level of vulnerability determines the level of risk. The applicant must submit the proposal and approval of the environmental analysis document (AMDAL) which includes information on location, area, risk and public acceptance. The purpose of this study is to analyze the level of vulnerability and the need for safety standards in the construction of a spaceport in Biak.

80.2 Theory

80.2.1 Vulnerability of Spaceport

BNBP [9] states that vulnerability is influenced by social, economic, physical, and ecological parameters. But the vulnerability of spaceport has not been regulated yet. Spaceport is a high-tech transportation infrastructure where elements of potential disaster and risk must be anticipated and measured. Vulnerability according to the IPCC is shaped by exposure levels, sensitivity, and adaptive capacity. This measurement is formulated as follows [11].

$$\text{Vulnerability (V)} = \frac{Exposure(E) + Sensitivity(S)}{CapacityAdaptive(CA)} \tag{80.1}$$

V is defined as a condition (environment, community) that leads to the inability to deal with the threat of disaster [9]. E is the probability that a system interacts with a disturbance. S level is defined as an internal condition in a system. CA is defined as the ability of the system to adapt according to changes. The importance of vulnerability research in disaster mitigation is to provide an illustration of how vulnerable people are affected to minimize the risk of disasters due to technology failure [12].

80.2.2 Safety Management System

The Safety Management System is a fundamental part of the spaceport management system. It is proportionated to the dimension of the spaceport operator organization and its related activities [7]. The Safety Management System includes:

- determination of the safety policies of the spaceport operator;
- allocation of responsibilities and tasks as well as the issuance of the provisions for the staff, sufficient for the implementation of the company policies and safety standards;
- analysis of the risks related to the activities;
- continuous monitoring of safety standards; and
- recording and analysis of the deviations from the applicable standards;

80.2.3 AHP and GIS

Analytical Hierarchy Process (AHP) is used as multi criteria decision-making tools, to analysis related the most important selection attributes. AHP is utilized to solve problems by creating hierarchies which able to be used for assessing contribution of parameter with certain weights [13] This paper used AHP tools to organize the criteria and sub-criteria to determine weights of each parameter, then this paper calculated the result using scoring framework, which after that the result was mapped by using GIS tools. The GIS framework in this paper utilized overlay strategy. The goal of overlay analysis is to display different features in one map. In general, overlay process uses spatial indexed data from different types of information and then presented to one map which based on question [14].

80.3 Research Methodes

80.3.1 Location

Locus analysis is Districts in Biak Numfor Regency, Papua. Biak Spaceport is location in North Biak Districts. According BPS (2017), Saukobye population is 117. The nears resident from launch pad (based on asumtion/scenario) is has range more than 2 km or in zone 3.

Table 80.1 Source data

Data	Type of data		Source
District Administration Biak Numfor (village)	Secondary data	Vector	Dukcapil 2019
Biak Numfor Regency Administrative Boundary	Secondary data	Vector	BIG 2021
District capital	Secondary data	Vector (plot)	BIG 2021
Settlement	Secondary data	Vector	BIG 2021
Road network	Secondary data	Vector	BIG 2022
Population density	Secondary data	Tabular	BPS 2021
Number of health facilities	Secondary data	Tabular	BPS 2021
Level of education	Secondary data	Tabular	BPS 2022
Range (Distance to launcpad, zona1, zona 2, zona3)	Primer	Lokasi vector	Measurement with QGIS from the sub-district capital point to the launcher point
Location koordinat (Launch pad)	Primer	Vector (plot)	LAPAN
Weight	Primer	Quesionary	Expert/key person

80.3.2 Data

The data used in this paper are secondary and primary data from several source: literature reviews, statistical data (population density, number of health facilities, level of education, road network), spatial data (range, district administration, settlement) and qualitative survey from expert or key person (quesionary). Data obtained will be processed through weighted overlay as the integration of Analytical Hierarchy Process (AHP) and geographical information system (GIS). The scoring index for weighted overlay will be obtained from expert review (Table 80.1).

80.3.3 Methodes

This paper uses qualitative and quantitative data by employing geospatial approach. The research framework combined from Macbeth [6] which assess individual risk of spaceport with scenario of zonation and weighted overlay method. The criteria of vulnerability is show in Fig. Exposure is measured by the criteria range or distance to the launch point. Sensitivity is measured by the criteria for population and adaptive capacity is measured by the approach of the existence of health facilities, human resources conditions and the existence of transportation facilities with the approach of road conditions (Figs. 80.2 and 80.3).

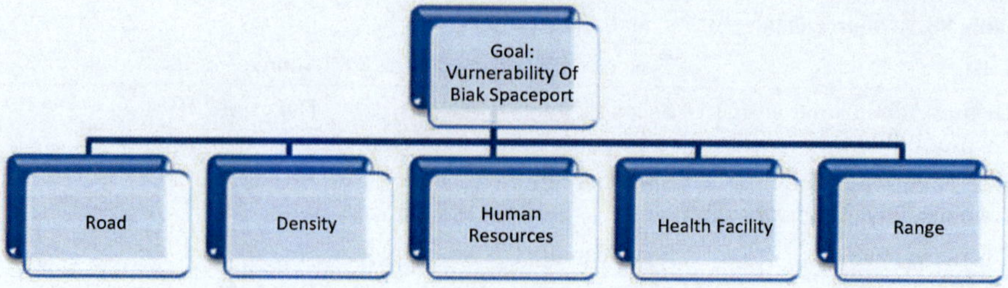

Fig. 80.2 Hierarchy of vulnerability

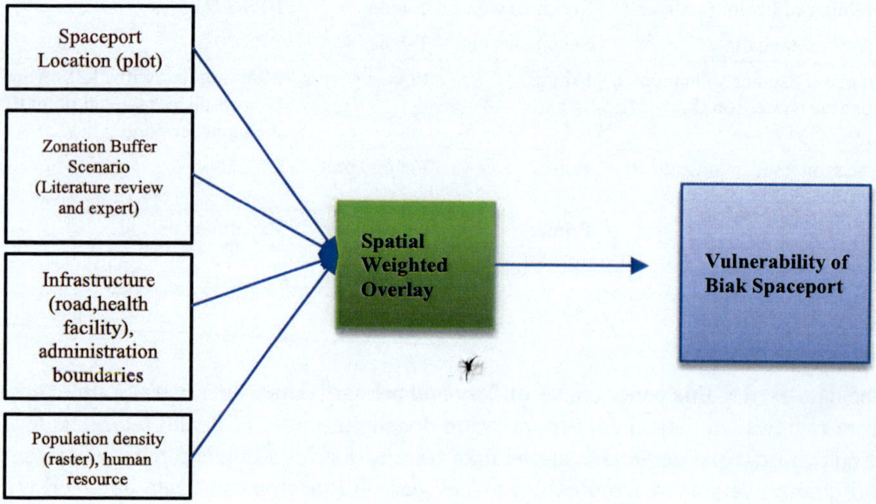

Fig. 80.3 Research framerwork Macbeth [6], modified by author

In this study, the author modifies the model developed by Macbeth, taking into account the vulnerability determination variables according to the IPCC and other selected variable.

80.4 Result and Analysis

80.4.1 International Practices for Spacepor Management

Based on survei, Biak Spaceport has the closest distance to the settlement as far as 2.5 km. When compared to some spaceports in some countries such as Japan (see Table 80.2), the distance with nearest resident/ settlement considered quite safe until now and still operating. Uchinoura Space Centre has the nearest distance to the

Table 80.2 Benchmarking of distance spaceport to resident and SMS

Spaceport indicator	Naro	Uchinoura	Kouru	Alcantara	Biak
Range/distance to resident (closed)	4 km	1.5 km	12 km	7 km	2 km
Rocket performance		Epsilon	Soyuz, Vega		Sonda, RPS
SMS	Yes	Yes	Yes	Yes	To be proposed

Table 80.3 Demographics of Saukobye village in 2016 [15]

Indicators	Number
Population	117
Density	62
Land area	8.5 km^2
Village/RT	2/0
Bamuskan/Ormas	1/2

settlement as far as 1.5 km. Naro, South Korea has a distance of 4 km, and others has more. In the operation of spaceport, Alcantara, Uchinoura and Naro, Kouru Guyana use Safety Management System. This justification is based on the type of vehicle that will be launched in Biak. LAPAN in the short and medium term will launch sonda rockets and satellite launchers which are still under or same with the Epsilon type or same. The launch direction is to the east and the drop zone is a free sea area. Based on the survey and technic team, zone 1 arrangement scenario was determined, which is a forbidden area for anyone to enter during the launch, which is 1 km away. Zone 2, which is a prohibited zone and is only intended for launch safety officers as far as 2 km, and zone 3, a forbidden zone that is obtained from entering the area only with permits (Table 80.3).

80.4.2 Criteria Weights

Based on the view of the keypersons, through AHP obtained the result (see Figure) that the distance factor with the launch becomes the first factor with a weight of 0.39 in a spaceport. This is based on the needs of range control set related to zoning rules. Zone 1 which is 1 km is a closed area for launch activities. Zone 2 which is 2 km is a limited area and not for public activities. And the zoning needs of 2 km in Biak is based on the results of LAPAN research [11].

Figure 80.4 illustrates human resource as second rank parameter of vulnerability and with a weight of 0.23. It has to do with the need for public support and acceptance, which are considerations in terms of vulnerability in Biak. Biak has customs and cultures that deserve attention. The population density criteria carries a weight of 0.18, followed by health facilities with a weight of 0.09, and road criteria with a

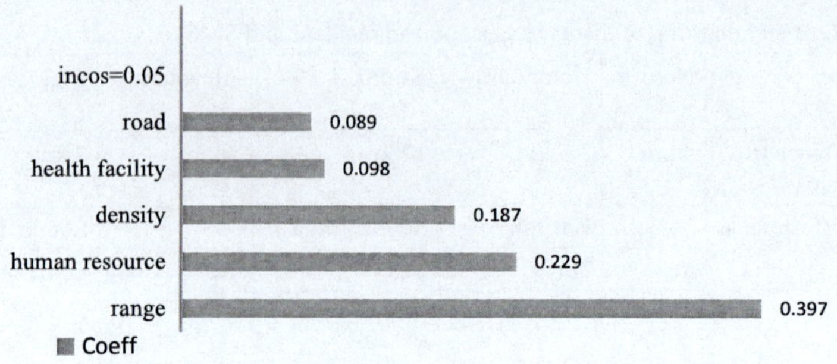

Fig. 80.4 Figure criteria weight from AHP

weight of 0.08. Its spaceport generally consider location far from settlements and low populations. Spaceports are ranked by their vulnerability based on the need for a drop zone of launch activities. This weight determines the priority in determining the vulnerability.

80.4.3 Vulnerability

The level of exposure of the Biak spaceport is low. The highest density populated settlement is in zone 3. International practice, the time of operation or launch, the usual strategy is to evacuate residents or stop all residents' activities around the location moments before the launch activity. The determination of this zoning needs to be supported by the arrangement of supporting regional spaces. Villages that fall into zone 3 in the Northern Biak District region are Korem and Saukobye. In accordance with the Conservation Act, zone 3 is still allowed for public activities but is limited. Zone 3 is 3 km from the launch pad point further than the launch point distance with the settlement at Uchinoura Spaceport. However, here it should be noted is the probability and reability factor of rocket technology launched there.

The results of the sensitivity level mapping show that location around the spaceport in general had a low level of sensitivity (see Fig. 80.5). Mapping using zone buffers 1, 2, and 3 shows that resident housing is found in zone 3. The results of field surveys in 2019 and 2020, found that there are any house/ buildings within zone 1, and 2 areas ranging 1–2 km from launchpad. The building is only to transit while they farming or sailing. However, according to interviews the building is a temporary halfway house. Saukobye village has a population of 117 people with a population density of 62 people/km^2 or under the low classification (Figs. 80.6 and 80.7).

Assessment of Biak Numfor adaptive capacity level is generally categorized as high. This is indicated by the existence of road carrying capacity, health facilities that are considered fulfilling. In addition, the capacity of humans reources through the level of education is relatively good where almost sub-districts have a percentage above 40% who graduate from junior high school or have reached compulsory 9 years

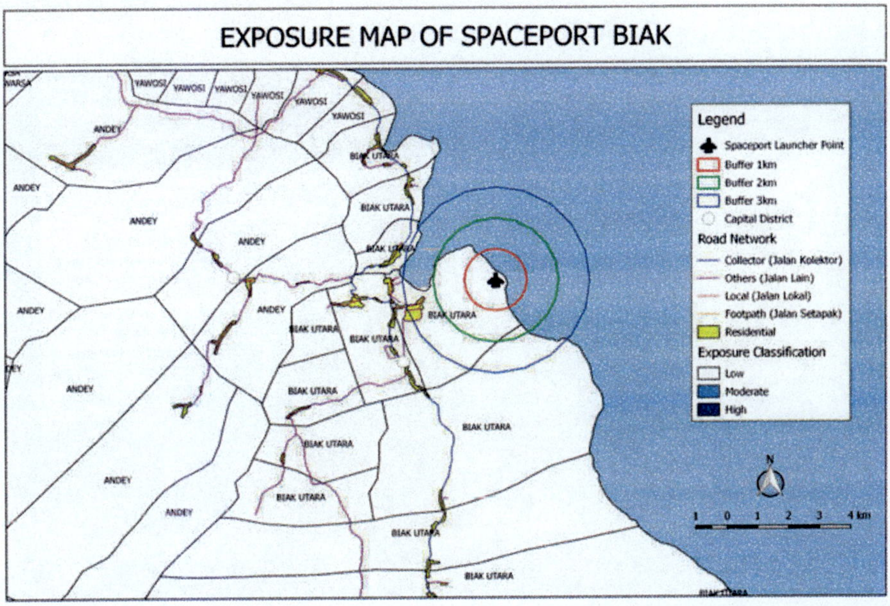

Fig. 80.5 Exposure map

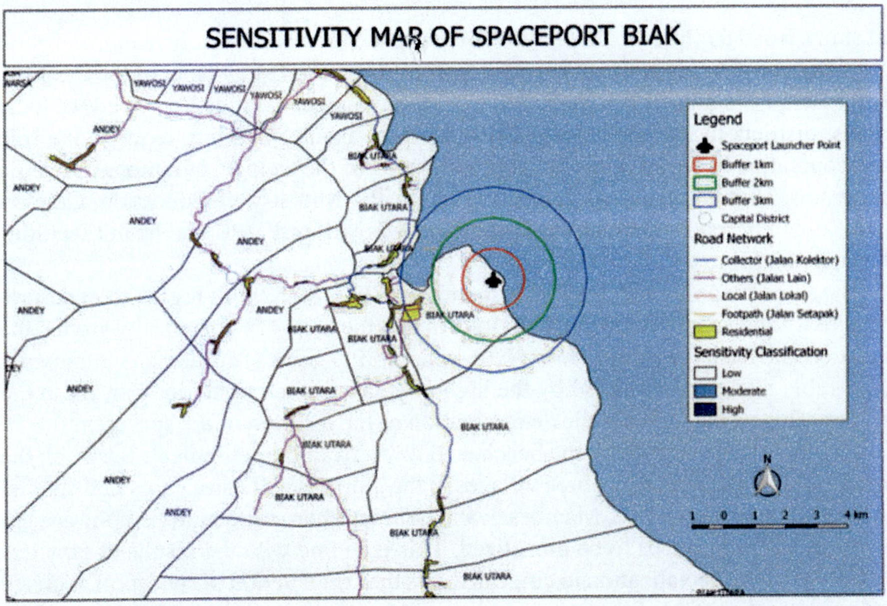

Fig. 80.6 Sensitivity map

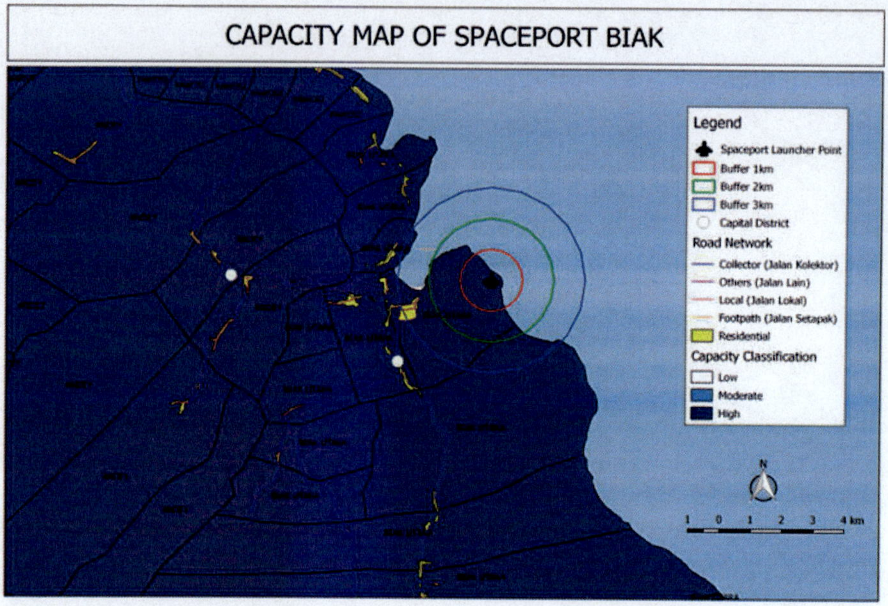

Fig. 80.7 Capacity adaptive map

of study. Road facilities to the location of the spaceport are relatively ready, and good from the entrance from the port and airport in the capital Biak Kota to the location. Increasing the status of land becomes a solution to meet the logistical needs of local roads, districts to national roads. This adaptive capacity building requires the role of cooperation between relevant agencies as part of the support of national strategic programs. Human resources capacity through the Ministry of Education, Culture, and Research, road construction through the Ministry of PUPR and health facilities through the Ministry of Health.

Based on mapping the level of vulnerability (Fig. 80.8) with regard to exposure, sensitivity and adaptive capacity, it is known that the level of vulnerability around the location and Biak Regency is low. High vulnerability values is Biak Kota Subdistrict (see Table 80.4). It is affected by the high population and population density in the region. This result supports the determination of the location of the spaceport one of its factors is to be near or by the beach, and away from the settlement. Based on the results of this vulnerability, three villages in the North Biak District area that entered zone 3, namely: Saukobye, Mambesak and Korem. Thus, community empowerment in these 3 regions needs to be prioritized. This is in line with the results of previous research, where socialization to citizens wants the construction of spaceport to create jobs [11] in the affected areas.

Zones needs related to zone 1, and 2 are absolutely done with the procurement of additional land outside the current LAPAN land. The zone is an area that must be free through land procurement for the needs of spaceport. This is to ensure the sustainability of the program and safety and security factors in the spaceport.

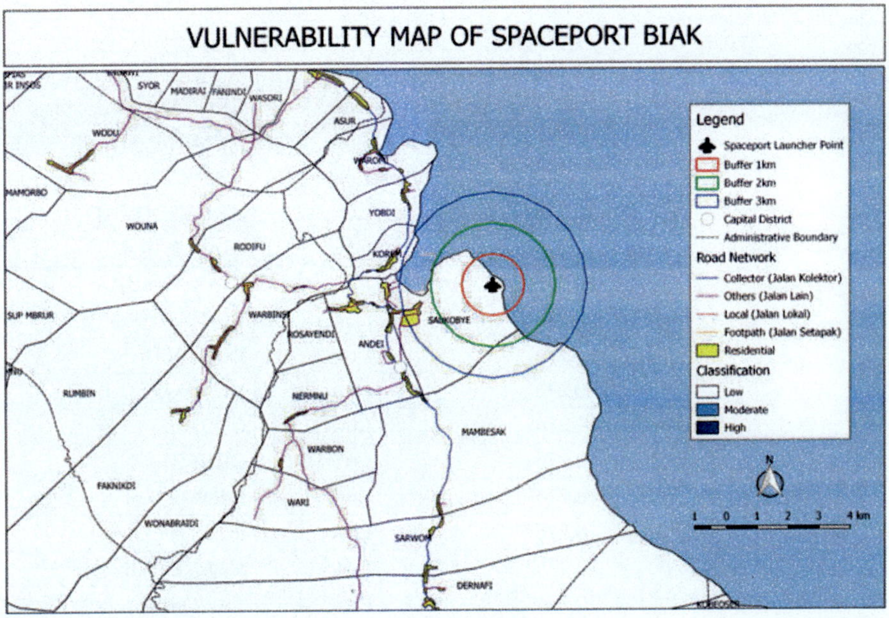

Fig. 80.8 Vulnerability Map

80.4.4 Strategy

Vulnerability is part of risk management measurement. In building a Spaceport Management System, based on the SMS approach, several strategies are needed. Based on the results of the measurement of the proximity, the strategy in reducing the level of vulnerability is to build adaptive capacity and zoning arrangements and control over spaceport. Some support facilities absolutely must be provided by the state. This is in accordance with the provisions of the Keantariksaan Law regarding basic facilities and supporting facilities [5]. The construction of spaceport locations requires space arranged in zoning. Road as a mode of transportation to facilitate logistics lines in development and operation. Health facilities are available as needed by safety standards. The National Coordination Meeting on 2019 has mapped the role of each stakeholder both biak numfor local government, ministry of PUPR, LAPAN, including industry [16]. Thus, such efforts have seen the importance of governance over risk management through WBS mapping and paying attention to zoning. The PUPR Ministry itself has taken a support step by increasing the status of the road into a national road through state financing. This is also done by the New Zealand government [4, 16] as a form of support and government role in the spaceport.

Another strategy in SMS related to policy management parameters, is to build a spaceport business scheme and regulatory tools related to spaceport (see Figure). Related to safety aspects, space booker registries both conducted by LAPAN and other parties must meet environmental regulations related to environmental impact

Table 80.4 Vulnerability Score

Vulnerability score	Category	Sub district
0.222	Low	ANDEY
0.222	Low	BRUYADORI
0.222	Low	NUMFOR TIMUR
0.222	Low	WARSA
0.276	Low	AIMANDO PADAIDO
0.276	Low	BIAK BARAT
0.276	Low	BIAK TIMUR
0.276	Low	BIAK UTARA
0.276	Low	BONDIFUAR
0.276	Low	NUMFOR BARAT
0.276	Low	ORIDEK
0.276	Low	ORKERI
0.276	Low	PADAIDO
0.276	Low	POIRU
0.276	Low	SWANDIWE
0.276	Low	YAWOSI
0.276	Low	YENDIDORI
0.324	Low	BONDIFUAR
0.324	Low	SAMOFA
0.591	High	BIAK KOTA

analysis assessments and investment-related provisions. Currently the Copyright Act, has cut the bureaucratic path of investment registration including in the spaceport. Spaceportin Indonesia itself enter the classification of economic activities regulated in the Standard Classification of Industrial Fields (KBLI) [17]. Some countries that have spaceport make guidelines and regulations related to licensing such as in the United States (FAA, 2000; Space Frorida, 2021) UK [5] and more. The purpose of fulfilling the agreement is to obtain a license (spaceport licency, operator spaceport), and the main purpose is safety and security. Currently LAPAN under the coordination of BRIN develops policy tools related to the development and operation of spaceport in accordance with national legal norms with due regard to international regulations (Fig. 80.9).

The strategy in marketing is related to the potential for commercialization. Spaceport are high-cost and high-risk projects, so efforts are needed to invite investors to get non-state financing opportunities. Strategic dialogue is necessary because the launch technology is dual use and strategic. Public dialogue is needed related to community empowerment efforts and public acceptance of the existence of the spaceport.

The strategy in managing the operation of spaceport is to conduct safety training to implementers and communities and establish cooperation. This cooperation is absolutely necessary because the suplay chain of this launch technology is global

Fig. 80.9 Strategy of SMS. *Source* : analysis data

and the launch technology is a sensitive technology. Safety is important for safety in this activity. The international practice of some spaceport is developed through several types of international cooperation such as cooperation between two countries, and public private partnership (PPP) cooperation. In developing space tourism, cooperation and discussions are needed related to spatial planning and licensing. This is related to provisions related to zoning 3 and beyond.

80.5 Conclusion

Indonesia has potential location spaceport in Biak. Safety is one of factors that determined to set safety range or zonation in Biak and the problem information how level vulnerability is limited. The result of this study Biak Numfor Regency has low level vulnerability from spaceport development and operational. This according to several criteria: density, health facilities, human resource/ education, distance to resident, zonation, and transportation infrastructure (road). The vulnerable regions where in the spaceport zonas are North Biak district (Korem, Saukobye village). This regions are located in zone 3 or the distance from launchpad is estimated 2–3 km. From SMS perspective, the strategy to spaceport management system are to build supporting policies and regulations, conduct risk management, by approaching international and national cooperation in terms of marketing spaceport for commercialization prospects and increasing capacity through human resource management and training.

Acknowledgements The authors would like to thank the key persons and inform an to this research.

Author Contribution Introduction, Theory, Methods, Collecting, Analysis, Review: INT Collecting data, Methods, Analysis: ST, AN.

References

1. LAPAN.: Studi Kelayakan Bandar Antariksa Equator Biak (1990)
2. Perwitasari, Intan.: Indonesia spaceport selection based on multicriteria analysis: a study on relative importance and priority regarding spaceport selection location attributes utilizing AHP. In: Proceedings of the 3rd International Conference on Indonesian Social & Political Enquiries (ICISPE 2018). Atlantis Press (2019). https://www.atlantis-press.com/proceedings/icispe-18/125922564
3. Dachyar, M, Purnomo, H.: Spaceport site selection with analytical hierarchy process decision making. Indian J. Sci. Technol (2018). https://doi.org/10.17485/ijst/2018/v11i10/96506
4. LAPAN.: Indonesia Space Act Number 21 Year 2013 (2013)
5. Civil Aviation Authority: Guidance for Spaceport Licence Applicants and Spaceport Licences (2021)
6. Macbeth, R.: Spaceport: keeping people safe. Health and Safety Laboratory (2018)
7. ENAC.: Regulation on Construction and Operation of Spaceport (2021)
8. Griffith, E., Alicea, F., Seth, L., Joseph, D., Michael, J.G.: Literature review on spaceport fire safety. Final Report. US (2018)
9. BNPB.: Peraturan Kepala Badan Nasional Penanggulanan Bencana Nomor 02 Tahun 2012 Tentang Pedoman Umum Pengkajian Risiko Bencana. Jakarta. (2012)
10. Nurjanah, S., Kuswanda, D., Siswanto. B.P.: Analysis of social vulnerability and climate change adaptation in Gemblakan Atas Village, Yogyakarta City. Proc. Biol. Educ. Conf. **16**, Nomor 1 (2013)
11. Perwitasari, Intan and Dini Susanti.: Vulnerability of Spaceport Contruction in Biak Island. Int. J. Innov. Sci. Res. Technol (2020). https://ijisrt.com/assets/upload/files/IJISRT19DEC563.pdf
12. IPCC.: Climate Change 2001 : Synthesis Report Climate Change 2001 : Synthesis Report. (2001)
13. Duc, T.T.: Using GIS and AHP technique for land-use suitability analysis. In International symposium on geoinformatics for spatial infrastructure development in earth and allied sciences, pp. 1–6 (2006)
14. Ahlqvist, O.: Overlay (in GIS). In: Kitchin, R., Thrift, N. (eds.) International Encyclopedia of Human Geography, vol. 8, pp. 48–55. Elsevier, Oxford (2006)
15. BPS Kab Biak Numfor: Kabupaten Biak Dalam Angka (2017)
16. LAPAN: Naskah Urgensi Perencanaan Pembangunan Bandar Antariksa di Kabupaten Pulau Biak. Report.Unpublished (2020)
17. BPS: Klasifikasi Baku Lapangan Usaha Industri (2020)